The Physiology
and
Biochemistry *of*
Prokaryotes

The PHYSIOLOGY *and* BIOCHEMISTRY *of* PROKARYOTES

THIRD EDITION

David White
Indiana University

New York   Oxford
OXFORD UNIVERSITY PRESS
2007

Oxford University Press, Inc., publishes works that further
Oxford University's objective of excellence
in research, scholarship, and education.

Oxford   New York
Auckland   Cape Town   Dar es Salaam   Hong Kong   Karachi
Kuala Lumpur   Madrid   Melbourne   Mexico City   Nairobi
New Delhi   Shanghai   Taipei   Toronto

With offices in
Argentina   Austria   Brazil   Chile   Czech Republic   France   Greece
Guatemala   Hungary   Italy   Japan   Poland   Portugal   Singapore
South Korea   Switzerland   Thailand   Turkey   Ukraine   Vietnam

Published by Oxford University Press, Inc.
198 Madison Avenue, New York, New York 10016
www.oup.com

Oxford is a registered trademark of Oxford University Press

Library of Congress Cataloging-in-Publication Data

White, David, 1936–
    The physiology and biochemistry of prokaryotes / David White.—
3rd ed.
       p.;cm.
    Includes bibliographical references and index.
    ISBN-13: 978-0-19-530168-7 (cloth : alk. paper)
    ISBN-10: 0-19-530168-4 (cloth : alk. paper)
    1. Prokaryotes—Physiology.   2. Microbial metabolism.
    [DNLM: 1. Bacteria—metabolism.   2. Archaea—physiology.
    3. Prokaryotic Cells—physiology. QW 52 W583p 2006]   I. Title.
QR88.W48 2006
571.2'93—dc22
                                                    2005013746

1   3   5   7   9   8   6   4   2
Printed in the United States of America on acid-free paper

CONTENTS

BOXED MATERIAL

PREFACE

The prokaryotes are a diverse assemblage of organisms that consists of the bacteria (also called eubacteria) and the archaea (also called archaebacteria). The text provides an updated account of the major aspects of the prokaryotes, such as cell structure, metabolism, developmental biology, adaptation to environmental changes, and biofilm formation. The text also covers cellular signaling pathways that allow individual bacterial cells to sense and respond to the environment, and also to signal each other so that they can respond as a cooperating population of organisms. It is written primarily for advanced undergraduate and beginning graduate students, and those who teach them. Topics include the structure and function of the prokaryotic cell (Chapter 1), growth and cell division (Chapter 2), membrane and cytosol bioenergetics (Chapters 3 and 7), electron transport and photosynthetic electron transport (Chapters 4 and 5), metabolic regulation (Chapter 6), intermediary metabolism (Chapters 8, 9, and 12–14), the metabolism of DNA, RNA, and protein, as well as how chromosomes are segregated during the cell cycle (Chapter 10), cell wall, capsule, and polysaccharide biosynthesis (Chapter 11), homeostasis (Chapter 15), solute transport (Chapter 16), protein export and secretion (Chapter 17), adaptive and developmental changes, multicellular behavior, and the signaling that underlies these events (Chapter 18), and how bacteria respond to environmental stress (Chapter 19). The text emphasizes the underlying principles of chemistry and physics in explaining the various physiological and metabolic features of prokaryotic cells, and thus provides the background for further advanced studies in this area.

To help students learn the material and prepare for examinations, each chapter ends with a summary and study questions, as well as a section of references and notes. The literature cited can be consulted for more data supporting the conclusions stated in the text and the notes expand on topics mentioned in the text. The notes are particularly important. Not only do they contain references to review articles and research papers, but they also have vital background information as well as detailed explanations of text material, including descriptions of experimental approaches. Students are therefore urged to examine the notes, particularly when the text directs them to do so.

This third edition of the text also contains boxes, which offer interesting historical perspectives on earlier discoveries that established the central tenets of biochemistry from the end of the nineteenth century into the 1960s, and the people who made these discoveries. From this material, the reader should gain an appreciation of the life-long effort of the many people who have contributed to our understanding of the biochemical bases of how cells function. Other boxes explain in more detail topics discussed in the text. See the listing of boxed material following the Table of Contents.

Most of the metabolic pathways that exist in the prokaryotes are the same as those in all living organisms, reflecting a unity in biochemistry dictated by principles of chemistry and (with respect to bioenergetics) physics. On the other hand, there is also a great deal of physiological diversity. The diversity of the prokaryotes is in part due to adaptations to the different habitats in which they grow. The habitats range in pH from approximately 1 to 12, temperatures from below 2 °C to over 100 °C, pressures as great as several thousand atmospheres (at the bottom of the oceans), habitats without oxygen, and habitats with salt concentrations that can reach saturating levels.

The physiological types among the prokaryotes mirror the diversity in the habitats and the sources of nutrient. Thus there are aerobes, anaerobes, and facultative anaerobes; heterotrophs, autotrophs, phototrophs, and chemolithotrophs; and alkaliphiles (alkalophiles), acidophiles, and halophiles. Some of these physiological types are found only among the prokaryotes, with rare exceptions. For example, among microorganisms that have been cultured in the laboratory, the ability to live indefinitely in the absence of air (anaerobic growth) is confined almost entirely to prokaryotes. Also, growth using the energy extracted from inorganic compounds (lithotrophy) seems not to have evolved in the eukaryotes and is restricted to certain groups of prokaryotes. The same can be said about the ability to reduce nitrogen gas to ammonia (nitrogen fixation), although this characteristic is more widespread among the prokaryotes than is lithotrophy. Another typically prokaryotic capability is the use of inorganic compounds such as nitrate and sulfate as electron acceptors during respiration (anaerobic respiration).

The organization of the text is according to topics rather than organisms, although the physiology of specific groups of prokaryotes is emphasized. This pattern of organization lends itself to the elucidation of general principles of physiology, metabolism, responses to environmental challenges, and cellular/multicellular development.

Most of the carboxyl groups are drawn as nonionized and the primary amino groups as nonprotonated. However, at physiological pH these groups are ionized and protonated, respectively. The names of the organic acids indicate that they are ionized (e.g., acetate rather than acetic acid).

There are two distinct evolutionary lines of prokaryotes commonly referred to as eubacteria and archaebacteria. However, this terminology implies a specific relationship between eubacteria and archaebacteria, whereas at the molecular level, the archaebacteria are not any more related to the eubacteria than they are to the eukaryotes. In recognition of this fact, Woese et al. suggested that the eubacteria be referred to simply as bacteria and that the archaebacteria be called archaea.[1] That terminology is followed in this text.

Acknowledgments

I would like to express gratitude to the following individuals who kindly and carefully reviewed portions of this edition: Carl Bauer, Yves Brun, Jim Drummond, Martin Dworkin, Pat Foster, Clay Fuqua, Heidi Kaplan, Larry Shimkets, and Ashley Williams, as well as the anonymous reviewers. I also want to thank the many individuals who helped me by reviewing the first two editions. I would also like to thank the team at Oxford University Press for helping me with this edition, including the senior editor for life sciences, Peter Prescott, Kaity Cheng, the editorial assistant for Life Sciences, Barbara Mathieu, the production editor, and Brenda Griffing, the copy editor. This edition, like the first two editions, was illustrated by Eric J. White.

REFERENCE

1. Woese, C. R., Kandler, O., and M. L. Wheelis. 1990. Towards a natural system of organisms: proposal for the domains Archaea, Bacteria and Eucarya. *Proc. Natl. Acad. Sci. USA* 87:4576–4579.

SYMBOLS

c	speed of light (3.0×10^8 m/s)
C	coulomb
cal	calorie (4.184 J)
E_0	standard redox potential (reduction potential) at pH 0
E_0'	standard redox potential at pH 7
E_m	midpoint potential at a specified pH (e.g., $E_{m',7}$); for pH 7, also written as E_m', which is numerically equal to E_0'
E_h	actual redox potential at a specified pH; for pH 7, E_h' or $E_{h,7}$
eV	electron volt. The work required to raise one electron through a potential difference of one volt. It is also the work required to raise a monovalent ion (e.g., a proton) through an electrochemical potential difference of one volt. One electron volt = 1.6×10^{-19} J.
F	Faraday constant (approximately 96,500 Cs). The charge carried by one mole of electrons or monovalent ion. It is the product of the charge carried by a single electron and Avogadro's constant.
g	generation time (time/generation)
ΔG_0	standard Gibbs free energy change at pH 0 (J/mol or cal/mol)
$\Delta G_0'$	Gibbs free energy at pH 7
ΔG_p	phosphorylation "potential" (energy required to synthesize ATP via physiological concentrations of ADP, inorganic phosphate, and ATP)
h	Planck's constant (6.626×10^{-34} J·s)
J	joule: one coulomb-volt ($C \times V$). The work required to raise one coulomb through a potential difference of one volt.
k	instantaneous growth rate constant (time^{-1})
kJ	kilojoule
K	absolute equilibrium constant

K'	apparent equilibrium constant at pH 7
K	degrees kelvin (273.16 + °C)
$\Delta\mu_{ion}$	electrochemical potential difference, expressed in joules, between two solutions of an ion separated by a membrane
$\Delta\mu_{ion}/F$	electrochemical potential difference expressed as volts or millivolts
$\Delta\mu_{H^+}/F$	the proton motive force. Electrochemical potential difference in volts or millivolts of protons between two solutions separated by a membrane. Also written as Δp.
N	Avogadro's number (6.023×10^{23} particles/mol)
Δp	See $\Delta\mu_{H^+}/F$
ΔpH	difference in pH between the inside and outside of the cell; usually, $pH_{in} - pH_{out}$
R	the ideal gas constant (8.3144 J K^{-1} mol^{-1} or 1.9872 cal K^{-1} mol^{-1})
V	volt. The potential difference across an electric field
$\Delta\Psi$	membrane potential; usually $\Psi_{in} - \Psi_{out}$

CONVERSION FACTORS, EQUATIONS, AND UNITS OF ENERGY

Electrode potential at pH 7

$E'_h = E'_0 + [RT/nF]\ln([ox]/[red])$ volts, where n refers to the number of electrons, and [ox] and [red] refer to the concentrations of oxidized and reduced forms, respectively. When [ox] = [red], then $E'_h = E'_0 = E'_m$:

$E'_h = E'_0 + (60/n)\log_{10}([ox]/[red])$ mV at 30 °C

Electron, charge

1.6023×10^{-19} C

Gibbs energy

For the reaction $aA + bB \leftrightarrow cC + dD$,

$\Delta G = \Delta G_0 + RT\ln[C]^c[D]^d/[A]^a[B]^b$

Gibbs energy and equilibrium constant

$\Delta G'_0 = -RT\ln K'_{eq}$ or $-2.303RT\log_{10}K'_{eq}$

Gibbs energy for solute uptake and concentration gradient

$\Delta G = RT\ln[C]_{in}/[C]_{out}$ at 30 °C

$\Delta G = 5.8\log_{10}[C]_{in}/[C]_{out}$ kJ/mol

or

$\Delta G/F = 60\log_{10}[C]_{in}/[C]_{out}$ mV

Growth

$g(k) = 0.693$

$x = x_0 2^Y$

the equation for exponential growth, where Y is the number of generations and x is mass or any parameter that changes linearly with mass

$Y = t/g$

Light, energy in a quantum

$E = \nu h = hc/\lambda$

where ν = frequency (c/λ), h = Planck's constant, λ = wavelength

$E\ (kJ) = 1.986 \times 10^{-19}/\lambda$

where λ is in nanometers

$E \text{ (eV)} = 1.24 \times 10^3 / \lambda$

where λ is in nanometers

Light, energy in an einstein

$E = Nh\nu = Nhc/\lambda = 1.197 \times 10^5 \text{ kJ}/\lambda$

where λ is the wavelength in nanometers, N is Avogadro's number, and c is the speed of light

$E \text{ (kJ)} = 1.196 \times 10^5 / \lambda$

where λ is in nanometers

Nernst equation

$\Delta\Psi = -(RT/nF) \ln[S]_{in}/[S]_{out} \text{ V}$ or $\Delta\Psi = -60/n \log [S]_{in}/[S]_{out} \text{ mV at 30 °C}$

where [S] is the concentration of a diffusible cation ($S_{in} > S_{out}$) of valence n, and $\Delta\Psi$ is the membrane potential, inside negative. The equation states that at equilibrium the chemical driving force due to the outward diffusion of S along its concentration gradient is equal to the electrical driving force drawing S into the cell. According to the equation, each 10-fold concentration difference of a permeant monovalent cation corresponds to a potential difference of 60 mV.

Phosphorylation potential

$\Delta G_p = \Delta G_0' + RT \ln[\text{ATP}]/[\text{ADP}][P_i]$

Proton potential

$\Delta p = \Delta\mu_{H^+}/F = \Delta\Psi - 60 \, \Delta\text{pH mV at 30 °C}$

Proton potential and ΔE_h at equilibrium

$-n\Delta E_h = y\Delta p$

where n is the number of electrons transferred over a redox potential difference of ΔE_h volts, and y is the number of protons translocated over a proton potential difference of Δp volts.

2.303RT

5.8 kJ/mol or 1.39 kcal/mol at 30 °C

2.303RT/F

0.06 V or 60 mV at 30 °C

DEFINITIONS

Acetogenic bacteria Anaerobic bacteria that synthesize acetic acid from CO_2 and secrete the acetic acid into the medium.

Acidophile Organism that grows between pH 1 and 4 and not at neutral pH.

Aerobe Organism that uses oxygen as an electron acceptor during respiration.

Aerotolerant anaerobe Organism that cannot use oxygen as an electron acceptor during respiration but can grow in its presence.

Anaerobe Organism that does not use oxygen as an electron acceptor during respiration.

Alkaliphile (alkalophile) Organism that grows at pH above 9, often with an optimum between 10 and 12.

Antibiotic Antimicrobial compound produced by a microorganism.

Antimicrobial agents Antibiotics and chemically synthesized antimicrobial compounds, among others.

Autotroph Organism that uses CO_2 as sole or major source of carbon.

Chaperone proteins Proteins that transiently bind to other proteins and assist in proper folding of the target protein and/or transport to a correct cellular site.

Chaperonins Multisubunit complexes of chaperone proteins.

Chemolithotroph See **Lithotroph**.

Cytoplasm The fluid material enclosed by the cell membrane.

Cytosol The liquid portion of the cytoplasm.

Dalton (Da) All atomic and molecular weights refer to the carbon isotope, ^{12}C, which is 12 Da or 1.661×10^{-24} g. Daltons are numerically equal to molecular weights and can be used as units when molecular weight units of grams per mole are not appropriate (e.g., when one is referring to ribosomes).

Dehydrogenase Dehydrogenases are enzymes that catalyze oxidation–reduction reactions in which hydrogens as well as electrons are transferred. They are named after one of the substrates (e.g., pyruvate dehydrogenase).

Drug Any chemical that affects the physiology of an organism. This includes drugs such as alcohol and caffeine, as well as therapeutic drugs to treat disease, and antimicrobial drugs to treat infections. Antimicrobial drugs include antibiotics, which are made by microorganisms and kill or prevent the growth of other microorganisms, and synthetic drugs.

Einstein One "mole" of light (6.023×10^{23} quanta).

Electrode potential The tendency of a molecule to accept an electron from another molecule is given by its electrode potential, E, also called the reduction potential, the redox potential, or the oxidation–reduction potential.

Facultative anaerobes Organisms that can grow anaerobically in the absence of oxygen or will grow by respiration if oxygen is available.

Facultative autotroph Organisms that can grow on CO_2 as sole or major source of carbon or on organic carbon.

Gel electrophoresis Procedure in which macromolecules such as proteins or nucleic acids are applied to a polyacrylamide or agarose gel and subjected to an electrical field. The macromolecules having a net electrical charge migrate in the electrical field and can be separated according to their size.

Growth yield constant, Y The amount of dry weight of cells produced per weight of nutrient used.

Halophile Organism that requires high salt concentrations for growth.

Heat-shock proteins Proteins that transiently increase in amount relative to most cell proteins when the temperature is elevated. Several are chaperone proteins.

Heterotroph Organism that uses organic carbon as a major source of carbon.

Holliday junction An intermediate in homologous recombination in which a cruciform (cross-shaped) structure is formed.

Hyperthermophile Organism whose growth temperature optimum is 80 °C or greater.

Isomerase Enzymes that catalyze the transfer of chemical groups (e.g., hydrogen or phosphate) within molecules to produce isomeric forms of a molecule with the same chemical formula. For example, glucose ($C_6H_{12}O_6$) and fructose ($C_6H_{12}O_6$) are isomers of each other because they have the same chemical formula, although their chemical structures are different.

Leader region The region of messenger RNA for an operon that is 5′ of the coding region for the first gene.

Lithotroph Organism that oxidizes inorganic compounds as a source of energy for growth.

Mesophile	Organism whose growth temperature optimum is between 25 and 40 °C.
Midpoint potential (E_m)	The electrode potential when the molecule is 50% oxidized.
Mutases	A subclass of isomerases; enzymes that transfer a functional group (e.g., a phosphate group) from one part of a molecule to another.
Neutrophile	Organism that grows at a pH optimum near neutrality.
Obligate anaerobes	Organisms that will grow only in the absence of oxygen but are not necessarily killed by oxygen
Oligopeptide	Refers to a molecule of a few amino acids joined together by peptide bonds. Also referred to as a peptide. A pentapeptide is an example of an oligopeptide. When many amino acids are joined together, the molecule is referred to as a polypeptide (MW < 10,000) or a protein (MW > 10,000).
Phosphorylation potential	The energy required to phosphorylate one mole of ADP by using physiological concentrations of ADP, P_i, and ATP.
Photon	Quantum; a particle of light.
Photosynthesis	The use of light as a source of energy for growth.
Photoautotroph	Organism that uses light as a source of energy for growth and CO_2 as the source of carbon.
Photoheterotroph	Organism that uses light as a source of energy for growth and organic carbon as the source of cell carbon.
Phototroph	Organism that uses light as the source of energy for growth.
Psychrophile	Organism that grows best at temperatures of 15 °C or lower, and does not grow above 20 °C.
Quantum	A particle of light (photon).
Regulon	A set of noncontiguous genes controlled by the same transcription regulator.
Standard conditions	All reactants and products are in their "standard states." This means that solutes are at a concentration of one molar (1 M = one mole of solute per liter) and gases are at one atmosphere. Biochemists usually take the standard state of H^+ as 10^{-7} M (pH 7). By convention, if water is a reactant or product, its concentration is set at 1.0 M, even though it is 55.5 M in dilute solutions.
Strict anaerobes	Organisms that will grow only in the absence of oxygen and are killed by traces of oxygen.
Tautomerism	Refers to an isomerization in which the isomeric forms are easily interconvertible and an equilibrium exists between the isomers. An example is the enol–keto tautomerization in which enol pyruvic acid readily converts to pyruvic acid.
Thermophile	Organism that can grow at temperatures greater than 55 °C.
Y_m	Molar growth yield constant; grams of dry weight of cells produced per mole of nutrient used.

The Physiology
and
Biochemistry *of*
Prokaryotes

1

Structure and Function

The prokaryote (procaryote) domains are the Bacteria and the Archaea. Prokaryotes are defined as organisms that have no membrane-bound nucleus. (The prefix *pro*, borrowed from Greek *pro*, means earlier than or before; *karyote*, borrowed from Greek *káryon*, means kernel or nut.) Besides lacking a nucleus, prokaryotes are devoid of organelles such as mitochondria, chloroplasts, and Golgi vesicles. However, as you will learn from reading this chapter, their cell structure is far from simple and reflects the evolution of prokaryotes into quite sophisticated organisms that are the most successful in inhabiting diverse ecological niches.

Despite the absence of organelles comparable to those found in eukaryotic cells, prokaryotes' metabolic activities, cell division activities, and intracellular signaling activities that regulate metabolism and cell development are compartmentalized and/or localized to specific sites within the cell. For example, as this chapter will point out, compartmentalization occurs within multienzyme granules that house enzymes for specific metabolic pathways, in intracellular membranes within the cytosol, within the cell membrane, within a special compartment called the periplasm in gram-negative bacteria, within the cell wall itself, and within various inclusion bodies that house specific enzymes, storage products, or photosynthetic pigments.

As this chapter points out, the external structure of prokaryotic cells also has specialized structures, called appendages. The appendages include pili and flagella. Pili (sing., pilus), which are of different types, serve different functions. Depending upon the type, pili are used for adhesion to other cells when that becomes necessary for colonization, for movement on solid surfaces via a form of gliding motility called twitching, and for mating. Flagella (sing. flagellum) are used by single cells to swim in liquid, as well as a form of group swimming on moist solid surfaces called swarming.

In addition, it turns out that the poles of non-spherical bacteria cells are physiologically and structurally different from the rest of the cell, and sometimes the cell poles in the same cell differ from each other. This is obvious, for example, when flagella or pili protrude from one or both cell poles, or when a cell pole is distinguished by having a stalk, as is the case for *Caulobacter*, discussed in Chapter 18. In other cases the differences between poles are less obvious, as shown, for example, in the discussion of the polar localization of the chemotaxis proteins in Chapter 18. As also discussed in Chapter 18, in *Myxococcus* the two gliding motility engines, the A and S engines, are not only in opposite cell poles, but actually switch poles on a regular basis. This is an example of the dynamic interplay that can take place between the various parts of a prokaryotic cell.

It has become recently clear that prokaryotes even have an internal protein cytoskeleton, a property previously thought to be restricted to

Table 1.1 Major subdivisions of Bacteria

Bacteria and their subdivisions

Purple bacteria (now referred to as the division or phylum Proteobacteria)

α subdivision
Purple nonsulfur bacteria (*Rhodobacter, Rhodopseudomonas*) rhizobacteria, agrobacteria, rickettsiae, *Nitrobacter, Thiobacillus* (some), *Azospirillum, Caulobacter*

β subdivision
Rhodocyclus (some), *Thiobacillus* (some), *Alcaligenes, Bordetella, Spirillum, Nitrosovibrio, Neisseria*

γ subdivision
Enterics (*Acinetobacter, Erwinia, Escherichia, Klebsiella, Salmonella, Serratia, Shigella, Yersinia*), vibrios, fluorescent pseudomonads, purple sulfur bacteria, *Legionella* (some), *Azotobacter, Beggiatoa, Thiobacillus* (some), *Photobacterium, Xanthomonas*

δ subdivision
Sulfur and sulfate reducers (*Desulfovibrio*), myxobacteria, bdellovibrios

Gram-positive eubacteria

A. High (G + C) species
Actinomyces, Streptomyces, Actinoplanes, Arthrobacter, Micrococcus, Bifidobacterium, Frankia, Mycobacterium, Corynebacterium

B. Low (G + C) species
Clostridium, Bacillus, Staphylococcus, Streptococcus, mycoplasmas, lactic acid bacteria

C. Photosynthetic species
Heliobacterium

D. Species with gram-negative walls
Megasphaera, Sporomusa

Cyanobacteria and chloroplasts
Oscillatoria, Nostoc, Synechococcus, Prochloron, Anabaena, Anacystis, Calothrix

Spirochaetes and relatives

A. Spirochaetes
Spirochaeta, Treponema, Borrelia

B. Leptospiras
Leptospira, Leptonema

Green sulfur bacteria
Chlorobium, Chloroherpeton

Bacteroides; flavobacteria and relatives

A. Bacteroides group
Bacteroides, Fusobacterium

B. Flavobacterium group
Flavobacterium, Cytophaga, Saprospira, Flexibacter

Planctomyces and relatives

A. Planctomyces group
Planctomyces, Pasteuria

B. Thermophiles
Isocystis pallida

Chlamydiae
Chlamydia psittaci, C. trachomatis

Radio-resistant micrococci and relatives

A. Deinococcus group
Deinococcus radiodurans

B. Thermophiles
Thermus aquaticus

Green nonsulfur bacteria and relatives

A. Chloroflexus group
Chloroflexus, Herpetosiphon

B. Thermomicrobium group
Thermomicrobium roseum

Archaea subdivisions

Extreme halophiles
Halobacterium, Halococcus morrhuae

Methanobacter group
Methanobacterium, Methanobrevibacter, Methanosphaera stadtmaniae, Methanothermus fervidus

Methanococcus group
Methanococcus

"Methanosarcina" group
Methanosarcina barkeri, Methanococcoides methylutens, Methanotrhix soehngenii

Methospirillum group
Methanospirillum hungatei, Methanomicrobium, Methanogenium, Methanoplanus limicola

Thermoplasma group
Thermolasma acidophilum

Thermococcus group
Thermococcus celer

Extreme thermophiles
Sulfolobus, Thermoproteus tenax, Desulfurococcus mobilis, Pyrodictium occultum

Source: Hodgson, D. A. 1989. Bacterial diversity: the range of interesting things that bacteria do, pp. 4–22. In: *Genetics of Bacterial Diversity*. D. A. Hopwood and K. F. Chater (Eds.), Academic Press, London.

eukaryotic cells. As described in this chapter and in Chapter 2, the cytoskeleton is critical for maintaining cell shape as well as for cell division.

Thus, as has been pointed out many times in recent years, the prokaryotic cell is not simply a "bag of enzymes," in accordance with earlier descriptions, but rather, a cell that is sophis-

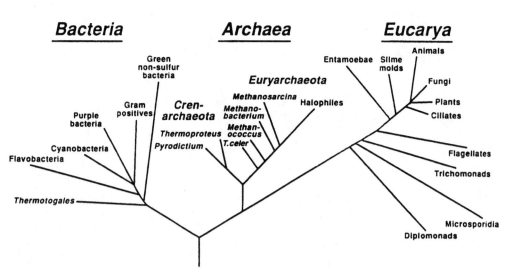

Fig. 1.1 Phylogenetic relationships among life forms based upon rRNA sequences. The line lengths are proportional to the evolutionary differences. The position of the root in the tree is approximate. The "Purple bacteria" are now referred to as the phylum Proteobacteria, and as summarized in Table 1.1, comprise a wide variety of gram-negative organisms including phototrophic, chemoheterotrophic, and chemolithotrophic bacteria. *Source*: Woese, C. R., and N. R. Pace. 1993. Probing RNA structure, function, and history by comparative analysis. *The RNA World*. Cold Spring Harbor Press, Cold Spring Harbor, NY.

ticated and dynamic both structurally and physiologically. Furthermore, as Chapter 18 explains, depending upon the growth conditions, prokaryotes are capable of living either as single cells, when suspended in liquid, or as interacting cells in multicellular populations called biofilms on solid surfaces. The study of prokaryotes presents a challenge to the investigator to experimentally learn how the organisms are temporally and spatially organized so that myriad activities can take place within the cells and between the cells to enable survival in the face of environmental challenges.

1.1 Phylogeny

Figure 1.1 shows a current phylogeny of life-forms based up a comparison of ribosomal RNA (rRNA) nucleotide sequences. For a more complete explanation of Fig. 1.1 and how it was derived, see Box 1.1. Notice that three lines (domains) of evolutionary descent—Bacteria (eubacteria), Eucarya (eukaryotes), and Archaea (archaebacteria)—that diverged in the distant past from a common ancestor.[1–3] (The term *archaean* may be used to describe particular

archaea.) Archaea differ from bacteria in ribosomal RNA nucleotide sequences, in cell chemistry, and in certain physiological aspects, described in Sections 1.1.1 and 1.2.

Table 1.1 lists examples of prokaryotes in the different subdivisions within the domains Bacteria and Archaea. Notice that the gram-positive bacteria are a tight grouping. Although there is no single grouping of gram-negative bacteria, most of the well known gram-negative bacteria are in the Proteobacteria division.

1.1.1 Archaea

Phenotypes

From a morphological point of view (size and shape), archaea (archaebacteria) look like typical bacteria (eubacteria). However, they form a group of organisms phylogenetically distinct from both bacteria and eukaryotes. The archaea commonly manifest one of three phenotypes: *methanogenic, extremely halophilic,* and *extremely thermophilic,* which also correspond to phylogenetic groups. These will now be described.

The *methanogenic* archaea (kingdom Euryarchaeota), also referred to as methanogens,

BOX 1.1 PHYLOGENY

The evolutionary relationships among all living organisms have been deduced by comparing the ribosomal RNAs of modern organisms. The structures of the ribosomal RNA molecules from different living organisms are sufficiently conserved in certain regions, conserved sequences and secondary structures can be aligned to permit comparison of the differences in base sequences between RNA molecules in homologous regions. Researchers analyzed the number of positions that differ between pairs of sequences, as well as other features (e.g., which positions vary, the number of changes that have been made in going from one sequence to another). The number of nucleotide differences between homologous sequences is used to calculate the evolutionary distance between the organisms and to construct a phylogenetic tree. Most recently published phylogenetic trees are based upon 16S rRNA sequences.

There are two major ways in which phylogenetic trees are constructed from the nucleotide differences. In the *evolutionary distance* method, the number of nucleotide differences is used as a measure of the evolutionary distance between the organisms. The second method, the *maximum parsimony* method, is more complicated. It takes into account not only the nucleotide differences but also the positions at which the differences occur and the nature of the differences. Parsimony means "stinginess," going to extreme, for the sake of economy, and the method attempts to find the simplest evolutionary tree that can explain the differences in nucleotide sequences. The simplest tree is the one that requires the smallest number of nucleotide changes to evolve the collection of extant sequences from a postulated ancestral sequence.

It has been pointed out that that it is difficult to assess the validity of a phylogenetic tree. There are several reasons for this. For example, none of the methods accounts for the possibility that multiple changes may have taken place at any single position. This is likely to result in an underestimate of the distances.

Underestimation of the distances, in turn, might bring two distantly related lineages much closer, and in fact might make them appear to be specifically related to each other when they are not. Other ambiguities in phylogenetic trees could arise if different positions in the sequence alignments had changed at very different rates. Furthermore, the branching pattern itself can differ if a different algorithm is used, even with the same data-base. The outcome is that there exist a variety of phylogenetic trees, and no single published tree has been accepted as perfectly representing the 4 billion years or so of bacterial evolution. Despite these problems, there is consistency in the trees that are based upon 16S rRNA sequences, and it is believed that most of the phylogenetic trees derived from sequencing bacterial 16S rRNA are at least plausible. However, the assumption remains unproven that the relationships between the 16S rRNAs represent the phylogenies of the organisms rather than simply the phylogeny of a given 16S rRNA. Because of this, it is imperative that the relationships between the different organisms be tested by means of other characters (e.g., other appropriate molecules besides the 16S rRNA, and various phenotypic characteristics of the organisms). For a further discussion of the various methods of constructing phylogenetic trees and the associated problems, see refs. 1 through 3. For a discussion of lateral gene transfer between groups of prokaryotes, and how it affects important physiological characteristics, see ref. 4.

REFERENCES

1. Woese, C. R. 1991. Prokaryotic systematics: The evolution of a science, pp. 3–11. In: *The*

Prokaryotes, Vol. I. A, Balows, H. G. Trüper, M. Dworkin, W. Harder, and K.-H. Schleifer (Eds.). Springer-Verlag, Berlin.

2. Stackebrandt, E. 1991. Unifying phylogeny and phenotypic diversity, pp. 19–47. In: *The Prokaryotes*, Vol. I. A, Balows, H. G. Trüper, M. Dworkin, W. Harder, and K.-H. Schleifer (Eds.). Springer-Verlag, Berlin.

3. Felsenstein, J. 1982. Numerical methods for inferring evolutionary trees. *Q. Rev. Biol.* 57:379–404.

4. Boucher, Y., C. J. Douady, R. T. Papke, D. A. Walsh, M. E. R. Boudreau, C. I. Nesbo, R. J. Case, and W. F. Doolittle. 2003. Lateral gene transfer and the origins of prokaryotic groups. *Annu. Rev. Gen.* 37:283–328.

produce methane. This is important for their survival, because they derive energy from the process. The methanogens produce methane by reducing carbon dioxide to methane, or by converting acetate to carbon dioxide and methane. Their metabolism is explained in Chapter 13. Methanogens are obligate anaerobes that grow in environments such as anaerobic groundwaters, swamps, and sewage.

The *extremely halophilic archaea* (also in the kingdom Euryarchaeota) require very high sodium chloride concentrations (at least 3–5 *M*) for growth. They grow in salt lakes and solar evaporation ponds. The halophilic archaea have unique light-driven proton and chloride pumps called *bacteriorhodopsin* and *halorhodopsin*, respectively, Bacteriorhodospsin creates an electrochemical proton gradient used to drive ATP synthesis, and halorhodopsin is used to accumulate chloride intracellularly to maintain osmotic stability. These pumps are described in more detail in Chapter 3 (Sections 3.8.4 and 3.9). Although most extreme halophiles are archaea, there are exceptions, such as the bacterium *Ectothiorhodospira* and the alga *Dunaliella*.

The *extremely thermophilic* archaea (kingdom Crenarchaeota) grow in thermophilic environments (generally 55–100 °C).[4] Some of these have an optimal growth temperature near the boiling point of water! They use inorganic sulfur either as an electron donor or as an electron acceptor in energy yielding redox reactions. (The pathways for sulfate reduction and sulfur oxidation are described in Sections 12.2.2 and 12.4.1, respectively.) Some of these archaea, which are also called *sulfur dependent*, oxidize inorganic sulfur compounds such as elemental sulfur and sulfide by using oxygen as the electron acceptor and derive ATP from the process.

Examples include *Sulfolobus* and *Acidianus*. Other extreme thermophiles are anaerobes that use elemental sulfur or thiosulfate as the electron acceptor to oxidize hydrogen gas. These include *Thermoproteus*, *Pyrobaculum*, *Pyrodictium*, and *Archaeoglobus*. (*Pyrobaculum* and *Pyrodictium* use S° as an electron acceptor during autotrophic growth, i.e., growth on CO_2 as the carbon source, whereas *Archaeglobus* uses $S_2O_3^{2-}$.) Archaea belonging to the genus *Pyrodictium* have the highest growth temperature known, being able to grow at 110 °C. A few of the sulfur-oxidizing archaea are *acidophiles*, growing in hot sulfuric acid at pH values as low as 1.0. They are called *thermoacidophiles*, indicating that they grow optimally in hot acid. For example, *Sulfolobus* grows at pH values of 1 to 5 and at temperatures up to 90 °C in hot sulfur springs, where it oxidizes H_2S (hydrogen sulfide) or S° to H_2SO_4 (sulfuric acid). Although most of the extreme thermophiles are obligately sulfur dependent, some are facultative. For example, *Sulfolobus* can be grown heterotrophically on organic carbon and O_2 as well as autotrophically on H_2S or S°, O_2, and CO_2. Interestingly, some of the sulfur-dependent archaea have metabolic pathways for sugar degradation not found among the bacteria. These are mentioned in Section 8.4. It should be pointed out that although most extremely thermophilic prokaryotes are archaea, some, (e.g., *Thermotoga* and *Aquifex*) are bacteria.

Recently, a new genus of thermoacidophilic archaea was discovered.[5] *Picrophilus*, a member of the kingdom Euryarchaeota, order Thermoplasmales, is an obligately aerobic, heterotrophic archaeon that grows at temperatures between 45 and 65 °C at a pH of 0 with an optimum pH of 0.7. It was isolated from hot geothermally heated acidic solfataras[6] in Japan. Despite the sulfur content of the habitat, *Picrophilus* is not sulfur dependent.

Comparison of domains Archaea, Bacteria, and Eucarya

Bacteria can be distinguished from archaea on the basis of the following structural and physiological differences.

1. Whereas the lipids in the membranes of bacteria and eukaryotes are *fatty acids ester-linked to glycerol* (see later: Figs. 1.16 and 9.5), the archaeal lipids are methyl-branched, isopranoid *alcohols ether-linked to glycerol* (Fig. 1.18).[7] Archaeal membranes are discussed in Section 1.2.5, and the biosynthesis of archaeal lipids is discussed in Section 9.1.3.

2. Archaea lack peptidoglycan, *a universal component of bacterial cell walls*. The cell walls of some archaea contain pseudomurein, a component absent from bacterial cell walls (Section 1.2.3).

3. Archaea contain histones that resemble eukaryal histones and bind archaeal DNA into compact structures resembling eukaryal nucleosomes.[8,9]

4. The archaeal RNA polymerase differs from bacterial RNA polymerase by having 8 to 10 subunits, rather than 4 subunits, *and* by not being sensitive to the antibiotic rifampicin. The difference in sensitivity to rifampicin reflects differences in the proteins of the RNA polymerase. It is interesting that the archaeal RNA polymerase resembles the eukaryotic RNA polymerase, which also has many subunits (10–12) and is not sensitive to rifampicin.

5. Some protein components of the archaeal protein synthesis machinery differ from those found in the bacteria. Archaeal ribosomes are not sensitive to certain inhibitors of bacterial ribosomes (i.e., erythromycin, streptomycin, chloramphenicol, and tetracycline). These differences in sensitivity to antibiotics reflect differences in the ribosomal proteins. In this respect, archaeal ribosomes resemble cytosolic ribosomes from eukaryotic cells. Other resemblances to eukaryotic ribosomes are the use of

methionine rather than formylmethionine to initiate protein synthesis, and the requirement for an elongation factor, EF-2, that can be ADP-ribosylated by diphtheria toxin. In contrast, bacteria use formylmethionine to initiate protein synthesis, and their EF-2 is not sensitive to diphtheria toxin. However, the archaeal ribosomes are similar to bacterial ribosomes in being 70S; that is, they sediment in a centrifugal field of a velocity of 70 svedberg units.

6. The halophilic archaea have light-driven ion pumps not found among the bacteria. The pumps are bacteriorhodopsin and halorhodopsin, which pump protons and chloride ions, respectively, across the membrane. The proton pump serves to create a proton potential that can be used to drive ATP synthesis.

7. The methanogenic archaea have several coenzymes that are unique to archaea. The coenzymes are used in the pathway for the reduction of carbon dioxide to methane and in the synthesis of acetyl–CoA from H_2 and CO_2. These coenzymes and their biochemical roles are described in Section 13.1.5.

In addition to the differences just listed, there are some similarities between archaeal and eukaryotic RNA polymerase, and archaeal and eukaryotic ribosomes (Table 1.2).

1.2 Cell Structure

Much of the discussion of cell structure will refer to the more well-studied bacteria because structural studies of archaea are fewer; however, there are well-known differences between archaea and bacteria with respect to cell walls, cell membranes, and flagella, and these will be pointed out. The structure and function of the major cell components will be described, beginning with cellular appendages (pili, and flagella) and working our way into the interior of the cell.

1.2.1 Appendages

Numerous appendages, each designed for a specific task, extend from bacteria surfaces. We

Table 1.2 Comparison between Bacteria, Archae, and Eucarya

Characteristic	Bacteria	Archaea	Eucarya
Peptidoglycan	Yes	No	No
Lipids	Ester linked	Ether linked	Ester linked
Ribosomes	70S	70S	80S
Initiator tRNA	Formylmethionine	Methionine	Methionine
Introns in tRNA	No	Yes	Yes
Ribosomes sensitive to diphtheria toxin	No	Yes	Yes
RNA polymerase	One (4 subunits)	Several (8–12 subunits each)	Three (12–14 subunits each)
Ribosomes sensitive to chloramphenicol, streptomycin, kanamycin	Yes	No	No

shall describe two classes of appendages: flagella and pili. The flagella are used by single cells for swimming in liquid, and a type of group swimming, called swarming, on moist solid surfaces. Depending upon the type, the pili (sometimes called fimbriae) are used for adhesion to surfaces, including adhesion to the surfaces of animal cells; in certain cases they serve for a form of gliding motility called twitching (type IV pili), and for mating (sex pili).

Flagella

For a review of all known prokaryotic motility structures, see ref. 10. For a review of bacterial flagella, see ref. 11. Swimming bacteria have one or more flagella, which are organelles of locomotion that protrude from the cell surface. Flagella allow single cells to swim in liquid and can also be used for swarming on a solid surface. Swarming, a means of group swimming by which bacterial colonies can spread, is discussed later in this section.

1. An overview

The bacterial flagellum is a stiff, helical filament (either a left-handed helix or a right-handed helix, depending upon the species), approximately 20 nm in diameter, that rotates like a propeller. The bacterial flagellum is unrelated to the eukaryotic flagellum in composition, structure, and mechanism of action. (For a description of eukaryotic flagella and cilia, see note 12 in the section References and Notes, at the end of the chapter.) The word "flagellum" was first used to describe the bacterial filament in 1852; in Latin, *flagellum* means whip. The bacterial organelle, however, it is more like a stiff propeller than like a whip, which is a flexible rod. When the bacterial flagellum

rotates, a helical wave travels from the proximal to the distal end (outward from the cell), and as a consequence the cell is pushed forward as illustrated later. (See Fig. 18.19A.) The flagellum is a very complex machine driven by a tiny rotating motor embedded in the membrane. Both its structure and the mechanism of its motility will be described. The flagella that are studied in most detail are those of *Escherichia coli* and *Salmonella typhimurium*, and we will begin with a discussion of these flagella. The flagella of other bacteria, except for the spirochaetes, are similar in general structure to those of *E. coli* and *S. typhimurium*. Archaeal flagella are somewhat different, as will be described. For a description of structural cell components required for nonflagellar motility, see Box 1.2.

2. General structure

The proteins are named after genes found in *Escherichia coli* and *Salmonella typhimurium*.[13] Mutations in the *mot* (motility) genes result in paralyzed flagella, and as we shall see later, the Mot proteins provide the torque that causes the flagellum to rotate. The flagellum consists of a *basal body, a hook, a filament, a motor, a switch, an export apparatus, capping proteins*, and *junction proteins*. We discuss some of these ports now.

The basal body. Examine Fig. 1.2. At the base of the flagellum there is a basal body embedded in the membrane. In gram-negative bacteria, the basal body consists of three stacked rings (C, M, and S rings) and a central rod. The M and S rings are actually one ring called the MS ring, made from different domains of the FliF protein. This scheme is supported by electron microscopic evidence. The term "MS"

BOX 1.2   NONFLAGELLAR MOTILITY

Many bacteria do not have flagella, and yet are capable of motility. Depending upon the bacterium, the motility may take place on a solid surface (twitching and gliding) or in liquid (swimming). For a review, consult refs. 1 and 2. The mechanistic bases for these movements differ and are related to certain cell structure features present in the particular cell.

Type IV pili, twitching motility, and gliding motility

For a review, see ref. 3. See ref. 4 for an article about type IV pili and gliding motility in the unicellular cyanobacterium *Synechocystis*. For a discussion of the assembly and retraction of type IV pili, read note 5, and also Section 17.5.6. Also, read the discussion of type IV pili and social motility in myxobacteria in Section 18.16.2. Type IV pili are fibrillar protein appendages located at either pole of certain gram-negative pathogenic bacteria that infect animals, plants, and fungi, such as *Pseudomonas aeruginosa*, *Bacteroides ureolyticus*, *Legionella pneumophila*, *Neisseria meningitidis*, *Ralstonia solanacearum*, and *Vibrio cholerae*. They are also found in many nonpathogenic gram-negative bacteria such as *Myxococcus xanthus*, which is a social bacterium that constructs multicellular fruiting bodies (discussed in Section 18.16), and in the unicellular cyanobacterium *Synechocystis*. Type IV pili comprise the mechanosystem for twitching motility, which is a form of gliding that takes place on moist surfaces. Twitching motility in *M. xanthus* is called social motility (S-motility), described in Section 18.16.2. Surfaces upon which twitching motility can occur include agar, epithelial tissue, plant tissue, and various other surfaces such as glass, plastics, and metal. Motility on such surfaces is important for rapid colonization, for the formation of biofilms (discussed in Section 18.22.5),

and for the building of fruiting bodies by myxobacteria (Section 18.16.2).

Twitching motility in *P. aeruginosa* and several other bacteria is characterized by short, intermittent jerks (hence the name "twitching") of the cells. Gliding motility in myxobacteria is different in that it is a smooth movement of cells either forward or backward on a solid surface. Gliding motility in *M. xanthus* actually consists of two systems that contribute toward the movement of the cells, adventurous (A-motility), and a system called social (S-motility) which is similar to the twitching system of *P. aeruginosa*. These are described in Section 18.16.2. In both twitching motility and S-motility, the cells move during the retraction of the pili, as if they were being pulled forward. This has been demonstrated by using cells whose type IV pili were labeled with an amino-specific fluorescent dye and visualized using fluorescence microscopy.[6] In one model, the tip of the pilus adheres to another cell (or an inanimate object), and when the pilus retracts the cell is pulled forward. In agreement with this, cells rarely move via twitching motility unless they are close together, approximately the length of the type IV pilus.[7] An explanation of the twitching movements is given in note 8.

Pore complexes, slime extrusion, and gliding motility

Gliding motility by filamentous cyanobacteria, and myxobacteria adventurous motility (A-motility), are apparently due to the secretion of slime through pores, called junction pores, near the septa that separate the cells in the filament, or through nozzle like pore complexes at the cell poles of myxobacteria cells. One model for A-motility in myxobacteria is that the cell is propelled ("pushed") as a result of to hydration of the polyelectrolyte slime that is secreted from pores at the rear pole of the cell. However, it must be emphasized that at this time the mechanism of A-motility in myxobacteria is not understood. (See Section 18.16.2.)

Gliding in the green fluorescent bacteria (GFP) group

Bacteria in the GFP group glide rapidly on solid surfaces.[2,9] Some species rotate as they glide. The cells suspended in liquid can propel adsorbed latex beads in multiple paths around the cell, indicating that cell surface molecules, perhaps polymers or fibers, to which the beads attach, move. Perhaps the same molecules are part of the machinery that propels the cells when they are on a solid surface.

REFERENCES AND NOTES

1. Trachtenberg, S. 1998. Mollicutes—Wall-less bacteria with internal cytoskeletons. *J. Struct. Biol.* **124**:244–256.

2. McBride, M. J. 2001. Bacterial gliding motility: Multiple mechanisms for cell movement over surfaces. *Annu. Rev. Microbiol.* **55**:49–75.

3. Mattick, J. S. 2002. Type IV and twitching. *Annu. Rev. Microbiol.* **56**:289–314.

4. Bhaya, D. 2004. Light matters: phototaxis and signal transduction in unicellular cyanobacteria. *Mol. Microbiol.* **53**:745–754.

5. The type IV pilus is constructed of subunits, which in *Pseudomonas aeruginosa* and *Myxococcus xanthus* are called PilA (or pilin). One can think of the production of type IV pili as occurring in three post-transcriptional stages. Stage one involves the removal of the leader peptide from prepilin, the precursor to the pilin subunit. Stage two is the assembly of PilA into the pilus in the periplasmic space. And stage three, which is coupled to stage two, is extrusion of the pilus through the outer envelope. As described in Chapter 17 (Section 17.5), several proteins required for pilus assembly are homologous to proteins involved with type II protein secretion. When the pilus retracts, it disassembles at its base and the PilA subunits remain in the inner membrane (cell membrane) and are recycled. *M. xanthus* and *P. aeruginosa* have a putative two-component system that regulates the expression of the *pilA* gene. In both organisms, PilR is suggested to be the positive transcription regulator that is part of the two-component signaling system. Wu and Kaiser reported that the putative sensor histidine kinase, PilS, is a negative regulator of *pilA* expression in *M. xanthus* (Wu, S. S., and D. Kaiser, 1997. Regulation of expression of the *pilA* gene in *Myxococcus xanthus*. *J. Bacteriol.* **179**:7748–7758). Disassembly and retraction of the pilus require energy, thought to be provided by ATP. One of the proteins required for retraction is PilT, which is an inner membrane protein that is thought to mediate the disassembly of the pili. PilT is a nucleotide-binding protein suggested to be an ATPase. Mutations in the *pilT* gene result in hyperpiliated cells because the pili do not retract. Another protein, PilQ, forms the export pore in the outer membrane and is necessary for the extrusion of pili. Homologous proteins involved in the biogenesis and function of type IV pili in different bacteria may have different names.

6. Skerker, J. M., and H. C. Berg. 2001. Direct observation of extension and retraction of type IV pili. *Proc. Natl. Acad. Sci. USA* **98**:6901–6904.

7. Semmler, A. B., Whitchurch, C. B., and J. S. Mattick. 1999. A re-examination of twitching in *Pseudomonas aeruginosa*. *Microbiology* **145**:2863–2873.

8. Various bacterial species, including *P. aeruginosa*, use twitching to move as groups of cells. Sometimes these groups are called "rafts" of cells. The rafts are approximately 10 to 50 cells wide and by using a microscope are readily seen projecting outward from the edge of the spreading colony. Similar rafts of cells are seen at the edge of expanding colonies (called "swarms") of *M. xanthus* (and other myxobacteria). The rafts are often narrower at the outer tips that are moving forward and have been called "spearhead-like." Within the rafts, the cells are in close contact, aligned along their lengths. Deeper within the colonies, the cells move as smaller groups. If a cell moves from one group or raft toward a different group of cells at an angle, and touches another cell at its pole, it "jerks" into an aligned position with the group of cells. Hence, the characteristic "twitching" movements seen with *P. aeruginosa*. As mentioned earlier, *M. xanthus* is characterized by smooth movements of cells, rather than twitching, even though one of this microorganism's two motility systems (S-motility) is similar to the *P. aeruginosa* system. Thus, other factors contribute toward the smooth gliding of *M. xanthus*.

9. McBride, M. J., T. F. Braun, and J. L. Brust. 2003. *Flavobacterium johnsoniae* GldH is a lipoprotein that is required for gliding and chitin utilization. *J. Bacteriol.* **185**:6648–6657.

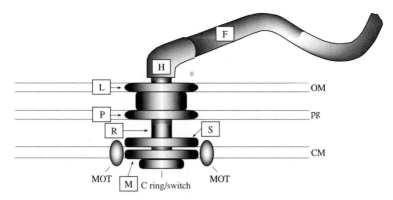

Fig. 1.2 Bacterial flagellum in a gram-negative envelope. The basal body itself, to which the hook–flagellum assembly is attached, is 22.5 nm × 24 nm in size and is composed of four rings, L, P, S, and M, connected by a central rod. The M ring is embedded in the cell membrane and the S ring appears to lie on the surface of the membrane. The S and M rings are actually one ring, the MS ring. The P ring may be in the peptidoglycan layer, and the L ring seems to be in the outer membrane. The P and L rings may act as bushings that allow the central rod to turn. Gram-positive bacteria have similar flagella but lack the L and P rings. The MotA and MotB proteins form complexes (Mot) that couple the influx of protons to the rotation of the rotor. The rotor consists of the MS ring with the FliG protein attached to its cytoplasmic surface, and the C ring attached to the cytoplasmic surface of the MS ring. The switch complex consists of three peripheral membrane proteins, FliG (also part of the rotor), FliM, and FliN, which probably are closely apposed to the cytoplasmic face of the M ring. Not shown are hook accessory or adaptor proteins (HAP1 and HAP3) between the hook and filament, and a protein cap (HAP2) on the end of the filament. The flagellum is assembled from the proximal to the distal end, with the filament being assembled last. It appears that the HAP1 and HAP3 proteins are required for the proper assembly of the filament onto the hook. Abbreviations: OM, outer membrane; pg, peptidoglycan; CM, cell membrane; R, central rod; MOT, MotA and MotB; H, hook; F, filament.

indicates the ring's location: membrane and supramembranous. The P ring (FlgI protein) and the L ring (FlgH protein) are also named according to their location: peptidoglycan and lipopolysaccharide. The L ring in *S. typhimurium* has been shown to be a lipoprotein.[14] Presumably the lipid portion helps to anchor the protein in the lipid regions of the outer envelope. A central rod, made from the FlgB, FligC, and FlgF proteins, passes through the rings and leads to the hook portion (H) of the flagellum on the outside of the cell. The outermost two rings (P and L) may act as bushings, allowing the central rod to rotate in the peptidoglycan and outer membrane. Gram-positive bacteria do not have P and L rings. The basal body transmits torque to the hook and filament, causing them to rotate.[11]

The motor. The fascinating part of the flagellar apparatus is the tiny motor, approximately 50 nm in diameter, that lies at the base of the flagellum and causes it to rotate. For a review of the flagellar motor, see ref. 15. The motor consists of two parts: a nonrotating part called the stator (made of the Mot proteins) and a rotating part called the rotor, which includes the FliG proteins that transmit torque to the MS ring.

The stator consists of two different proteins, MotA and MotB, indicated as Mot in Fig. 1.2. These exist as particles of multiple complexes, $(MotA)_4(MotB)_2$, that span the cell membrane and surround the MS ring. The number of such particles in various bacteria has been reported to be 12 to 16. Such rings of particles surrounding the MS ring were seen in electron micrographs of freeze-fractured cell membranes, but not when either MotA or MotB was missing.[16] It has been suggested that the large periplasmic domain of MotB is attached noncovalently to the peptidoglycan, explaining why the MotA/MotB complex does not rotate when the motor turns.

In agreement with the conclusion that MotA and MotB are part of the motor, mutations in the *motA* and *motB* genes result in paralyzed flagella (Mot⁻ phenotype). The MotA/MotB complexes are believed to conduct protons across the membrane, and to use this proton movement to provide the torque to rotate the rotor. Extreme alkaliphiles and some marine bacteria use a sodium ion current. This is reviewed in ref. 17. (As discussed in Chapter 3, bacteria use a proton current across the cell membrane to do other kinds of work, e.g., ATP synthesis, in addition to rotating flagella.) For an explanation of the conclusion that MotA and MotB form a complex, see note 18. Membrane vesicles prepared from strains synthesizing wild-type MotA were more permeable to protons than were vesicles prepared from strains synthesizing mutant MotA, indicating that MotA is likely to be part of a proton channel.[19] Some of the evidence in support of the conclusion that the Mot proteins form a complex that is a transmembrane proton channel is in ref. 20.

How does passage of protons through the Mot complex generate torque? One model proposes that as protons pass through the MotA/MotB complexes, conformational changes occur in the complexes that drive the attached rotor through rotational steps. For a model of how this might occur, consult ref. 21.

What is the rotor? Sometime the C ring (with the FliG proteins) is referred to as the rotor, and sometimes both the MS and C rings are meant. (See note 22 for references.) An essential rotor component is FliG. FliG interacts with the Mot proteins and transmits torque generated by the Mot complex to rotate the rotor. The FliG proteins are part of the C ring and attach the C ring to the MS ring. As explained next, the C ring functions as a switch that reverses the direction of rotation of the rotor.

The switch. In certain bacteria, such as E. coli and S. typhimurium, the motor spontaneously changes its direction of rotation periodically. The frequency of switching is influenced by attractants and repellents that the bacterium might encounter, and is important for chemotaxis, described in Section 18.12. Mutants that fail to change flagellar rotation map in three genes, fliG, fliM, and fliN, which code for the complex of switch proteins (FliG, FliM, and FliN).[23] It appears that the three switch proteins form a complex of peripheral membrane proteins closely associated with the cytoplasmic side of the MS ring. In particular, FliG seems to be bound to the MS ring itself, and is also bound to FliM and FliN. The latter two form a cup-shaped cytoplasmic ring, called the C ring (cytoplasmic), located directly beneath the basal body attached to the MS ring. The protein CheY-P, which is important in switching flagellar rotation during chemotaxis (Section 18.12), binds to FliM. (For more information about these proteins, including their function, see note 24.)

The hook and the HAP proteins. The central rod is attached to an external curved flexible hook made of multiple copies of a special protein called the hook protein, FlgE, which is the product of the flgE gene. There are also two hook-associated proteins: HAP1, also called FlgK (product of the flgK gene) and HAP3, also called FlgL (product of the flgL gene). These proteins are necessary to form the junction between the hook and the filament, and a third HAP, HAP2, also called FliD (product of the fliD gene), which caps the flagellar filament. Mutants that lack the HAPs secrete flagellin into the medium. The length of the hook is regulated by the cytoplasmic C ring, mentioned earlier as housing the switch proteins. According to the model, the hook subunits, FlgE, fill the C ring and then are transferred en masse to the growing hook through the export apparatus, described later, which is in the middle of the C ring. (See later subsection entitled *The export apparatus for flagellar components*.)

The filament and the capping proteins. Attached to the hook is a rigid, hollow, helical filament, that along with the hook, protrudes from the cell. When it rotates, it acts as a propeller and pushes the cell forward. The protein in the filament is called *flagellin* (which in *Escherichia* and *Salmonella* is known as FliC) and is present in thousands of copies. Flagellin is not identical in all bacteria. For example, the protein can vary in size from 20 to 65 kD a depending upon the species of bacterium. Furthermore, although there is homology between the C-terminal and N-terminal ends of most flagellins, the central part can vary considerably and is distinguished immunologically in different bacteria. In some cases, there is no homology at all. For example, nucleotide-derived amino acid sequences for the flagellins from *Rhizobium meliloti* show almost no relationship to flagellins from *E. coli*,

S. typhimurium, or *Bacillus subtilis*, but are 60% similar to the N and C termini of flagellin from *Caulobacter crescentus*.[25]

3. Brief summary of assembly of the flagellum

The flagellum and its associated components are assembled in a precise order, beginning with the components closest to the cell membrane. The first components assembled, probably in a coordinated fashion, are the MS and C rings. This is followed by construction of the transport apparatus for export of the remainder of the flagellar components through a channel in the center of the MS ring. Then the rod is assembled, followed by the hook. The hook is not completed until the P and L rings have been assembled. When the hook is complete, the flagellin monomers are exported and the filament is assembled. MotA and MotB are assembled late in the assembly process, to coincide with the appearance of the filament. As described next, the timing of assembly of the components is related to the timing of the transcription of the genes encoding these components.

4. Brief summary of timing of transcription of genes required for assembly of flagellum and its associated components

As we have seen, the flagellum and its components are assembled in a precise order. Interestingly, the genes for the flagellum and its associated components are expressed in the order that the gene products are used. How is this done? The genes are in three operons that are sequentially expressed. The operons are denoted as belonging to class 1, 2, or 3. The genes in class 1 (*flhDC*) comprise an operon that encodes transcriptional activators of the class 2 operon. Class 2 genes are the genes for the basal body and hook, as well as a sigma factor (FliA) required for transcription of class 3 genes. Class 3 genes are genes required for the synthesis of the flagellin monomers (FliC) and the torque generating unit (MotA and MotB).

5. Growth of the flagellum

Although one might expect new flagellin subunits to be added to the growing filament at the base next to the cell surface, this is not the case. The flagellin subunits actually travel through a hollow core in the basal body, hook, and filament and are added at the distal tip. The capping protein (HAP2) is important in this regard because it prevents the flagellin monomers from leaking out into the medium. Somehow, the capping protein must be able to move out from the growing filament to allow extension of the filament, neither becoming detached nor allowing flagellin subunits to leak out into the medium. Models by which this may be done have been suggested.[26,27]

How can growth at the tip of the filament be demonstrated? Growth at the tip has been demonstrated by the use of fluorophenylanine, a phenylalanine analogue, or radioisotopes. Incorporation of fluorophenylalanine by *Salmonella* resulted in curly flagella that had only half the normal wavelength. When the analogue was introduced to bacteria that had partially synthesized flagella, the completed flagella were normal at the proximal ends and curly at the distal ends, implying that the fluorophenylalanine was incorporated at the tips during flagellar growth.[28] When *Bacillus* flagella were sheared off the cells and allowed to regenerate for 40 min before the addition of radioactive amino acid ([³H]leucine), radioautography showed that all the radioactivity was at the distal region of the completed flagella.[29] The flagellin monomers are possibly transported through the central hole in the filament to the tip, where they are assembled. For more information about the biosynthesis of flagella, consult the review by Macnab.[11]

The export apparatus for flagellar components.

In the center of the C ring on the cytoplasmic side is a knob composed of several Flh and Fli proteins (FlhA, FlhB, FliH, FliI, FliO, FliP, FliQ, FliR). The ATPase FliI is negatively regulated by FliH, and its activity is required for flagellar assembly. (See note 30.) FliI and FliH, as well as a chaperone protein, FliJ, are soluble proteins, and the others are integral membrane proteins. The knob is the transport apparatus used for the transport through a central channel in the MS ring of all the flagellar components (basal body, rod, hook, filament) that are assembled beyond the cytoplasmic membrane. One of the exported proteins is a muramidase, called FlgJ, and it presumably functions to allow the nascent rod to penetrate the peptidoglycan. The transport system for the bulk of flagellar materials is a flagellum-specific type III export system. (See the description of type III export systems in Section 17.5.2.) The P and L rings are assembled using the Sec system of transport, described in Section 17.1.

6. Differences in flagellar structure

Although the basic structure of flagella is similar in all bacteria thus far studied, there are important species-dependent differences. For example, some bacteria have *sheathed flagella*, whereas others do not. In some species (e.g., *Vibrio cholerae*) the sheath contains lipopolysaccharide and appears to be an extension of the outer membrane.[31] In spirochaetes the sheath is made of protein. (See later.)

Bacteria also vary with respect to the number of different flagellins in the filaments. Depending upon the species, there may be only one type of flagellin, or two or more different flagellins in the same filament. For example, *E. coli* has one flagellin; *R. meliloti* and *Bacillus pumilis* each have two different flagellin proteins, and *Caulobacter crescentus* has three.[32-34] The data for *B. pumilis* and *C. crescentus* support the conclusion that the different flagellins reside in the same filament. It has been proposed that the filaments of *R. meliloti* are composed of heterodimers of the two different flagellins. The flagella of most bacteria that have been studied (e.g., *E. coli*) show a smooth surface under electron microscopy and are called *plain filaments*. However certain bacteria (e.g., *Rhizobium lupini*, *R. meliloti*) have "complex" filaments with obvious helical patterns of ridges and grooves on the surface. Flagella with plain filaments rotate either clockwise or counterclockwise, whereas flagella with complex filaments rotate only clockwise, with intermittent stops. It is thought that complex filaments are more rigid (because they are brittle) than plain filaments, and better suited for propelling bacteria in viscous media such as the gelatinous matrix through which *R. lupini* and *R. meliloti* must swim to infect root hairs of leguminous plants.[35] A discussion of the role of bacterial flagella in pathogenicity can be found in the review by Moens and Vanderleyden.[35]

Although *E. coli* and *S. typhimurium* have three rings (MS, P, and L) through which the central rod passes, other bacteria may have fewer, or more. For example, gram-positive bacteria lack the outer two rings (P and L). Additional structural elements of unknown function (e.g., additional rings or arrays of particles surrounding the basal body) have been observed in certain bacteria.

Spirochaete flagella. A major difference between spirochaete flagella and those found in other bacteria is that in spirochaetes, the flagella do not protrude from the cell; rather, they are in the periplasm wrapped around the length of the protoplasmic cylinder next to the cell membrane and are closely surrounded by an outer lipid bilayer membrane called an outer membrane sheath. The spirochaete flagella are called *axial filaments*. For a review, see ref. 36. For a list of diseases caused by spirochaetes, see note 37.

Most spirochaete cells are helically coiled and have two or more flagella (some have 30 or 40 or more) inserted subterminally near each cell pole. Some species have a flat meandering waveform, rather than being helically coiled. The number of flagella inserted at opposite poles is the same. The flagella are usually more than half the length of the cell and overlap in the middle. The spirochaete flagellum is often surrounded by a *proteinaceous sheath*. *Borrelia burgdorferi*, the spirochaete that causes Lyme disease, is somewhat different. Its axial filaments are not surrounded by proteinaceous sheaths. Furthermore, this organism has a planar waveform shape rather than the corkscrew type.[38] The rotation of the rigid periplasmic flagella between the outer membrane sheath and the cell cylinder is thought to move the cell by propagating a helical wave backward down the length of the highly flexible cell cylinder, propelling the cell forward, which allows the cells to "corkscrew" through viscous media, such as mud, sediments, and connective tissue in animals. (For a more detailed model, see note 39.) There are five antigenically related flagellins in the axial filaments, but it is not known whether individual filaments contain more than one type of flagellin.[40] Despite these differences, the spirochaete flagella and those found in other bacteria are structurally similar insofar as they have a basal body composed of a series of rings surrounding a central rod, plus a hook and a filament.

7. Site of insertion of flagella and the number of flagella

The site of insertion of the flagellum and the number of flagella vary with the bacterium. Some rod-shaped or curved cells have flagella that protrude from one or both of the cell poles. A bacterium with a single, polar flagellum is said to be *monopolar*. Bacteria with a bundle of flagella at a single pole are *lophotrichous*, from the Greek word *lophos* (crest or tuft) and *trichos* (hair). Bacteria with flagella at both poles are said to be *bipolar*. They may have either single or bundles of flagella at the poles. *Amphitrichous* refers to bundles of flagella at both poles. (*Amphi* in Greek means on both sides.)

Some bacteria (e.g., spirochaetes) have *subpolar* flagella, which are inserted near but not exactly at the cell poles, and some curved bacteria (e.g., *Vibrio*) have a single, *medial* flagellum. If the flagella are arranged laterally all around the cell, (e.g., as in *Escherichia* and *Salmonella*) they are said to be *peritrichous*. (*Peri* in Greek means around.) Peritrichous flagella coalesce into a trailing bundle during swimming.

8. Role of flagella in tactic responses and in virulence

Swimming bacteria are capable of tactic responses: that is, they swim toward environments more favorable with respect to nutrient, light, and electron acceptors, and away from toxic or less favorable environments. These swimming responses occur because bacteria can sense environmental signals, transfer these signals to the flagellum motor, and modify the rotation of their flagella to swim in a particular direction. How this occurs for bacteria of different types is discussed in detail in Sections 18.12, 18.13, and 18.14. Flagella can also make important contributions to the virulence of pathogenic bacteria.[41] For example, it has been suggested that the ability of spirochaetes (spiral-shaped bacteria) such as *Treponema pallidum*, the causative agent of syphilis, to swim in a corkscrew fashion though viscous liquid aids in their dissemination (e.g., through connective tissue or the junctions between endothelial cells).

9. Flagella and swarming

Swarming with flagella is a type of social swimming in which cells move on solid surfaces in groups (called rafts) of physically interacting cells. (For a review of the different motility systems that allow bacteria to move on solid surfaces, read ref. 42, for a review of swarming behavior, and how swarmer cells differ from cells grown in liquid, consult ref. 43.) Swarming allows bacterial populations to

rapidly spread as multicellular populations, rather than as single cells, over a wet, solid surface, for example in biofilms. (As discussed in Sections 18.19.2, and 18.22.7, gliding represents another means for bacteria to spread on a solid surface.) Swarming was first described in *Proteus* species more than 100 years ago (1885). Now it is recognized to be a widespread phenomenon among many bacteria genera, both gram-positive and gram-negative.

The cells in swarming populations are called swarmer cells, and they are morphologically different from swimmer cells grown in liquid. The morphological changes that occur when swimming cells from a liquid culture are inoculated onto an agar plate and convert to swarmer cells can be more or less pronounced depending upon the bacterium and the concentration of the swarming agar. In general, cells that convert from swimming cells to swarming cells become nonseptate filaments, multinucleoid, and hyperflagellated with lateral flagella. (*Bacillus subtilis* swarmer cells differ less dramatically from the nonswarmer cells in being only slightly larger and having only two nuclei.) The lateral flagella are critical, because to migrate as populations of cells, the cells physically interact with each via the lateral flagella.

Swarming is facilitated by the production of a surfactant that reduces surface tension at the aqueous periphery of the colony on hydrophilic surfaces such as agar, allowing the colony to expand. The surfactant is one of the components in the extracellular slime produced by the bacteria. For example, *Serratia marcescens* produces a cyclic lipopeptide (3-hydroxydecanoic acid attached to five amino acids), and *B. subtilis* produces a lipopeptide surfactant called surfactin.[44,45] Isolated swarmer cells rarely move. One of the differences between isolated cells and cells in a group is that the cells in the group are encased in much more slime than is found around single cells. It could be that wetting agents in the slime hydrate the external medium sufficiently for the flagella to rotate, allowing swarming to take place on the agar.

To demonstrate swarming on agar, one inoculates petri plates containing agar at a concentration that is wet enough to allow swarming to take place (e.g., 0.5–1%, or sometimes 2%), but not so dilute that cells can swim as individuals. For example, if the agar concentration is between 0.2 and 0.3%, individual bacteria can swim through water-filled pores in the agar. Such very dilute agar plates are referred to as "swim plates." Swarming is easily recognized because a population of swarming cells spreads over the agar plate, whereas nonswarmers grow as discrete colonies when inoculated in the center of the plate. As reported for *B. subtilis*, some bacteria may lose the ability to swarm when they are maintained as laboratory cultures.[45]

A different flagellar system is sometimes produced for swarming. Although many bacteria, such as *Escherichia*, *Proteus*, *Salmonella*, and *Serratia*, use the same lateral flagella system for both swimming in liquid and for swarming on a solid surface (albeit they make many more flagella for swarming), other bacteria, such as species of *Aeromonas*, and *Azospirillum*, *Rhodobacter centenum*, *Vibrio alginolyticus*, and *V. parahaemolyticus*, make a special lateral flagella system for swarming and growth in viscous environments. (Reviewed in ref. 46.) These bacteria are polarly flagellated when grown in liquid but become both polar and peritrichously flagellated when grown on surfaces. For example, *V. alginolyticus* and *V. parahaemolyticus* produce many unsheathed lateral flagella that aid in swarming when these micro organisms are grown on a solid surface. For some bacteria, such as *Vibrio* species, the polar and lateral flagella are determined by different genes. The increase in production of lateral flagella may be necessary to overcome surface friction and/or the viscosity of the slime produced by a swarming population. Perhaps some sort of mechanosensory system, activated by surface friction and/or slime viscosity, signals the expression of the genes for lateral flagella. It has been suggested that the inhibition of flagellar rotation on the solid surface might be part of the mechanosensory signaling system in *Vibrio*.

10. Archaeal flagella
Archaeal flagella are different from bacterial flagella.[47] The primary sequences of the flagellins from several archaea show no homology at all to bacterial flagellins, although they do show sequence homology to each other at the N-

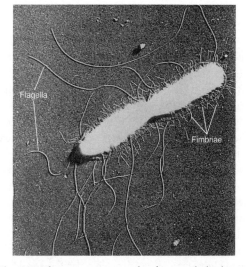

Fig. 1.3 Electron micrograph of a metal-shadowed preparation of *Salmonella typhi murium* showing flagella and fimbriae. The cell is about 0.9 μm in diameter. (Reprinted with permission of J. P. Duguid.)

terminal end.[48] Also, the basal body structure in *Methanospirillum hungatei* appears to be simpler than for bacterial flagella in that a simple knob is present rather than rings.[49]

Fimbriae, pili, filaments, and fibrils

Protein fibrils extending from the cell surface have been called various names, including fimbriae, pili, filaments, and simply fibrils. They extend from the surface of most gram-negative bacteria (Fig. 1.3).[50-53] Similar fibrils are rarely seen in gram-positive bacteria, but are present in the gram-positive *Corynebacterium renale* and *Actinomyces viscosus*.[54,55] (*C. renale* is the causative agent of bovine pyelonephritis and cystitis, and *A. viscosus* is part of the normal flora of the human mouth, where it adheres to other bacteria in plaque and to teeth.) All such appendages are not alike, nor are their functions known in all instances. The size varies considerably. They can be quite short (0.2 μm) or very long (20 μm), and differ in width from 3 to 14 nm or greater. Some originate from basal bodies in the cell membrane. However, most seem not to originate in the cell membrane at all, and how they are attached to the cell is not known. Furthermore, they are not always present. Although freshly isolated strains frequently have fibrils, they are often lost during subculturing in the laboratory. They are apparently useful in the natural habitat but dispensable in laboratory cultures. What do they do?

1. Pili (fimbriae)

Many of the fibrils can be observed to mediate attachment of the bacteria to other cells (e.g., other bacteria, or animal, plant, or fungal cells). Although adhesive fibrils are sometimes referred to as fimbriae to distinguish them from the sex pili that are used in bacterial conjugation, in this discussion, all such fibrils will be referred to as pili. Pili are important for colonization because they help the bacteria to stick to surfaces. In nature most bacteria grow while attached to surfaces where the concentration of nutrients is frequently highest. Attachment can be via adhesive pili and/or nonfibrillar material that may be part of the cell wall or glycocalyx (Section 1.2.2). Adhesive pili have *adhesins*, that is, molecules that cause bacteria to stick to surfaces. The adhesins are proteins in the pili, often minor proteins at the tip, that recognize and bind to specific receptors on the surfaces of cells. Adhesion is studied in the laboratory under experimental situations of two types: (1) attachment of bacteria to erythrocytes, causing them to clump (*hemagglutination*), and (2) attachment of bacteria in vitro to host cells to which they normally attach in vivo. The pili presumably attach to the erythrocytes because the erythrocytes carry cell surface molecules resembling the adhesin receptors found on the natural host cells. Often researchers who study such attachments have found that when specific monosaccharides and oligosaccharides are added to the suspension, they inhibit attachment or hemagglutination. The implications are that at least some, if not all, pili bind to oligosaccharides in cell surface receptors, and that the added monosaccharides and oligosaccharides are inhibitory because they compete with the receptor for the adhesin. Receptors on animal cell surfaces include *glycolipids* and *glycoproteins*, which are embedded in animal cell membranes via the lipid and protein portions, and present their oligosaccharide moieties to the outer surface. It has been reported that many strains of *E. coli*, *Salmonella*, and *Shigella* carry pili whose hemagglutinin activity is prevented by D-mannose and methyl-α-mannoside. These pili, therefore, are believed to attach to mannose glycoside residues on the cell surface receptors.

They are called *common pili*, or *mannose-sensitive pili*, or (more frequently) *type 1 pili*. Other pili recognize different cell surface receptors and are sometimes called mannose-resistant pili.

Medical significance. Pilus proteins (and other adhesins) are of great medical significance, as the following discussion will illustrate. An example of attachment to host cells via bacterial pili is the attachment of the causative agent of gonorrhea, *Neisseria gonorrhoeae*, to epithelial cells of the urogenital tract. Another example is *E. coli*. Most clinical isolates of *E. coli* from the urinary and gastrointestinal tracts possess type 1 pili. Many also (or instead) have galactose-sensitive pili, called *P-type pili*.[56,57] P-type pili bind the α-D-galactopyranosyl-(1–4)-β-D-galactopyranoside in the glycolipids on cells lining the upper urinary tract.[58]

Other known pili include the *type 4 pili (type IV pili)* produced by *N. gonorrhoeae*, *Pseudomonas aeruginosa*, *Bacteroides nodosus*, and *Moraxella bovis* (see note 59) and the *Tcp (toxin coregulated pili) pili* produced by *Vibrio cholerae*. (see note 60 for the diseases caused by these bacteria.) The Tcp pili are necessary for *V. cholerae* to colonize the intestinal mucosa, and mutants that do not make these pili do not cause disease. Thus it is evident that pili of different types are specialized for attachment to specific receptors and can account for specificity of bacterial attachment to hosts and tissues. Pili can be distinguished by inhibition of binding by mono- and oligosaccharides, as well as by a variety of other methods, including morphology, antigenicity, molecular weight of the protein subunit, isoelectric point of the protein subunit, and amino acid composition and sequence. Since gram-positive bacteria generally do not have pili, their adhesins are part of other cell surface components (e.g., the glycocalyx, to be described in Section 1.2.2).

2. Sex pili

Bacteria are capable of attaching to each other for the purpose of transmitting DNA from a donor cell to a recipient cell. This is called mating. Some bacteria (not all) use pili for mating attachments. The pili that mediate attachment between mating cells are different from the other pili and are called sex pili. The requirement of sex pili for mating is found for enteric bacteria

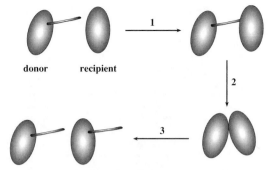

Fig. 1.4 F-pilus-mediated conjugation. Transfer of a sex plasmid. The donor cell has a plasmid and an F pilus encoded by plasmid genes. 1, The F pilus binds to the recipient cell. 2, A depolymerization of the pilus subunits. Causes the pilus to retract, bringing the two cells together. 3, The plasmid is transferred as it replicates so that when the cells separate, each has a copy of the plasmid and is a potential donor.

such as *E. coli*, and pseudomonads, but it is not universal among the bacteria. For example, gram-positive bacteria do not have sex pili. The sex pilus grows on "male" strains that donate DNA to recipient ("female") strains. In *E. coli*, it is coded for by a conjugative transmissible plasmid, the F plasmid, that resides in the donor strains. (see note 61 for an explanation of plasmids.)

Figure 1.4 illustrates how the F pilus in *E. coli* works. The tip of the F pilus adheres to receptor molecules on the recipient cell surface. This is followed by a retraction of the pilus (depolymerization of the F-pilin subunits into the cell membrane), bringing the two cells closer together until their surfaces are in contact. DNA transfer takes place at the site of contact (not through the sex pilus). Sex pili seem to be designed for mating in cell suspension, where the bacteria are not in intimate contact. The sex pilus presumably helps to stabilize the mating pairs of bacteria. These pili are not needed for many bacteria, perhaps because they mate in colonies or aggregates on solid surfaces where the cells are in close contact. However, the gram-positive bacterium *Streptococcus faecalis* (now called *Enterococcus faecalis*) forms efficient mating aggregates in liquid suspension and does not have sex pili. In other words, the adhesins are located on the cell surface rather than on pili. Mating in *E. faecalis* is of additional interest because it is induced by a sex pheromone secreted into the medium by recipient cells. The sex pheromone signals the donor cells to synthesize cell surface adhesins that promote cell aggregate formation (clumping) and subsequent DNA transfer.[62] Mating interactions mediated by surface adhesins is widespread in the microbial world. Other well-studied examples include *Chlamydomonas* mating and yeast mating.[63]

1.2.2 The glycocalyx

The term "glycocalyx" is often used to describe all extracellular material that is *external to the cell wall*.[64–67] (The word *calyx* is from the Greek *kályx*, meaning husk or outer covering.) Such material includes the extracellular matrix (ECM) in biofilms, described in Section 18.22. The polymers in glycocalyces are predominantly polysaccharides and/or proteins. All bacteria are probably surrounded by glycocalyces as they grow in their natural habitat, although they often lose these external layers when cultivated in the laboratory. The extracellular polymers may be in the form of S layers, capsules, slime, or a loose network of fibrils.

The S layers, which are found on the cell wall surfaces of a wide range of gram-positive and gram-negative eubacteria, are arrays of protein or glycoprotein subunits on the cell wall. They are also present in the archaea, where the S layer sometimes covers the cell membrane and serves as the cell wall itself. If the S layer is the wall itself, it is not called a glycocalyx.

Capsules are composed of fibrous material at the cell surface (Figs. 1.5 and 1.6). They may be rigid, flexible, integral (i.e., very closely associated with the cell surface), or peripheral (i.e., loose) material that is sometimes shed into the medium.

Material that loosely adheres to the cell wall is sometimes called *slime or slime capsule*. If the loosely adhering or shed material is polysaccharide, it is often referred to as extracellular polysacharide (EPS), such as exists in biofilm matrices. (In discussing biofilms, "EPS" also serves as a general term to refer to e̲xtracellular p̲olymeric s̲ubstance, not necessarily polysaccharide.) The capsular polysaccharide is usually covalently bound to phospholipid or phospholipid A, which is embedded in the surface of the cell. This is not the case for extracellular

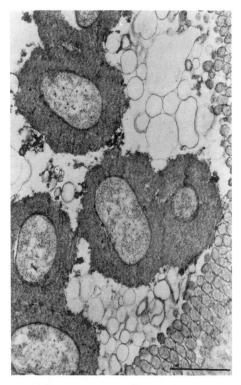

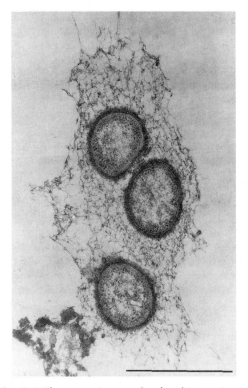

Fig. 1.5 Electron micrograph of a thin section of *E. coli* stained with a ruthenium red dye. The cells are adhering to neonatal calf ileum. *Source*: Costerton, J. W., T. J. Marrie, and K.-J. Cheng. 1985. Phenomena of bacterial adhesion, pp. 3–43. In: *Bacterial Adhesion*. D. C. Savage and M. Fletcher (Eds.). Plenum Press, New York and London.

Fig. 1.6 Electron micrograph of a thin section of bacteria adhering to a rock surface in a subalpine stream. The sample, showing fimbriated glycocalyx, was stained with ruthenium red dye. *Source*: Costerton, J. W., T. J. Marrie, and K.-J. Cheng. 1985. Phenomena of bacterial adhesion, pp. 3–43. In: *Bacterial Adhesion*. D. C. Savage and M. Fletcher (Eds.). Plenum Press, New York and London.

polysaccharide. The synthesis of extracellular polysaccharides is discussed in Section 11.3. For a review of bacterial capsules and their medical significance, see ref. 68.

Chemical composition

The glycocalyces from several bacteria have been isolated and characterized. Although many are polysaccharides, some are proteins. For example, some *Bacillus* species form a glycocalyx that is a polypeptide capsule. Also, pathogenic *Streptococcus* species have a fibrillar (hairlike) protein layer, the M protein, on the external face of the cell wall. (See note 69 for a description of the role of M protein in *S. pyogenes* pathogenesis.)

The capsular polysaccharides are extremely diverse in their chemical composition and structure.[70] Some (homopolymers) consist of one type of monosaccharide whereas heteropolymers are composed of more than one

type of monosaccharide. The monosaccharides are linked together via glycosidic linkages to form straight-chain or branched molecules. They can be substituted with organic or inorganic molecules, further increasing their diversity. For example, *E. coli* strains make more than 80 different capsular polysaccharides, called K antigens. Sometimes the same capsular polysaccharide is made by different bacterial species. For example, the K1 polysaccharide of *E. coli* is identical to the group B capsular polysaccharide of *Neisseria meningitidis*.

Role

An important role for the glycocalyx is adhesion to the surfaces of other cells or to inanimate objects to form a biofilm. Such adhesion is necessary for colonization of solid surfaces or growth in biofilms (Figs. 1.5 and 1.6). The bacteria can adhere to a nonbacterial surface or to each other via these attachments. An example is the complex ecosystem of several different bacteria maintained by intercellular adherence in human oral plaques.[71] Advantages to such adherence include the higher concentrations of nutrients that may be found on the surfaces.

Another role of the glycocalyx is protection from phagocytosis. For example, mutants of pathogenic strains of bacteria that no longer synthesize a capsule, such as unencapsulated strains of *Streptococcus pneumoniae*, are more easily phagocytized by white blood cells, making the pathogens less virulent.

The glycocalyx can also prevent dehydration of the bacterial cell, an important role in the soil. This is because polyanionic polysaccharides, including polysaccharide capsules, are heavily hydrated.

1.2.3 Cell walls

Most bacteria are surrounded by a cell wall that lies over the external face of the cell membrane and protects the cell from bursting due to the cell's internal turgor pressure.[72,73] The turgor pressure exists because bacteria generally live in environments that are more dilute than the cytoplasm. As a consequence, there is a net influx of water that results in a large hydrostatic pressure (turgor) of several atmospheres directed out against the cell membrane. Two different types of wall exist among the bacteria. One type of wall can be stained by using the Gram stain procedure. Bacteria having walls of this type are called gram-positive. Bacteria possessing the second wall type do not stain and are called gram-negative.

We introduce the Gram stain next, following up with a description of peptidoglycan, a cell wall polymer found in both gram-positive and gram-negative walls, that is responsible for the strength of bacterial cell walls. Then we consider, in turn, the gram-positive and gram-negative cell walls.

The Gram stain

The Gram stain was invented in 1884 by a Danish physician, Christian Gram, to allow the use of ordinary bright-field microscopy in the visualization of bacteria in tissues. For more about Christian Gram, see Box 1.3.

When appropriately stained, bacteria with cell walls can usually be divided into two groups, depending upon whether they retain a crystal violet–iodine stain complex (gram-positive) or do not (gram-negative) (Fig. 1.7). Gram-negative bacteria are visualized with a pink counterstain called safranin. During the staining procedure the complex of crystal violet and iodine can be removed with alcohol or acetone from gram-negative cells, but not from gram-positive cells. This result is thought to be related to the thickness and composition of the gram-positive wall in comparison to the much thinner gram-negative wall.

BOX 1.3 HISTORICAL PERSPECTIVE: CHRISTIAN GRAM

Christian Gram was a Danish physician who worked at the morgue of the City Hospital of Berlin. In 1894 he published a procedure he had devised for staining bacteria. The method allowed the staining of all bacteria in tissue preparations to differentiate them from nuclei. Some bacteria did not retain the stain, however, making the procedure less effective than Gram had desired.

It is not clear who realized that the Gram stain could be used to distinguish between different bacteria, which later became known as gram-positive and gram-negative, but in 1886 Carl Flügge published a textbook in which he pointed out that Gram's staining procedure is useful for the differential staining of bacteria in tissues and for diagnostic purposes. Christian Gram died in 1935.

Source: Lechevalier, H. A., and M. Solotorovsky. 1965. *Three Centuries of Microbiology*. McGraw-Hill Book Company, New York.

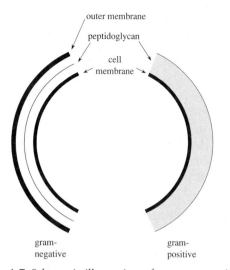

Fig. 1.7 Schematic illustration of a gram-negative and a gram-positive bacterial cell wall. Note the presence of an outer membrane (also called outer envelope) in the gram-negative wall and the much thicker peptidoglycan layer in the gram-positive wall.

The difference in Gram staining is also related to the outer wall layer of gram-negative bacteria (also called the outer envelope or outer membrane), which is rich in phospholipids and thus made leaky by the lipid solvents, alcohol, and acetone. The archaea can stain either gram-positive or gram-negative, but their cell wall composition is different from that of the bacteria, as will be described later.

Peptidoglycan

The strength and rigidity of bacterial cell walls is due to molecules called peptidoglycan or murein, which consist of glycan chains cross-linked by peptides (see later: Figs. 1.18, 11.3).

Although Fig. 1.8 depicts the glycan chains running parallel with the cell membrane, this is a controversial point. Another model, the scaffold model, proposes that the glycan chains run perpendicular to the cell membrane.[74] The glycan consists of alternating residues of N-acetylglucosamine (GlcNAc or G) and N-acetylmuramic acid (MurNAc or M) attached to each other via β-1,4 linkages. Attached to the MurNAc is a tetrapeptide that cross-links the glycan chains via peptide bonds. The tetrapeptide usually consists of the following amino acids in the order that they occur from MurNac: L-alanine, D-glutamate, L-R$_3$ (an amino acid that varies with the species: see note 75), and D-alanine. The structural diversity of the peptidoglycan, as well its synthesis, is discussed in more detail in Section 11.1. The peptidoglycan forms a three-dimensional network surrounding the cell membrane, covalently bonded throughout by the glycosidic and peptide linkages. It is the covalent bonding that gives the peptidoglycan its strength. The shape of the peptidoglycan maintains the shape of the cell as well as shape changes that occur during cellular morphogenesis (e.g., during the formation of round cysts from rod-shaped vegetative cells). In addition to peptidoglycan, certain cytoskeletal components such as MreB and crescentin are involved in shape determination. This is discussed later in this chapter in Section 1.2.7. Destruction of the peptidoglycan by the enzyme lysozyme (which hydrolyzes the glycosidic linkages) or interference in its synthesis by antibiotics (such as penicillin, vancomycin, or bacitracin) results in the inability of the cell wall to restrain the turgor pressures. Under these circumstances, the influx of water in

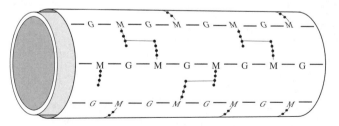
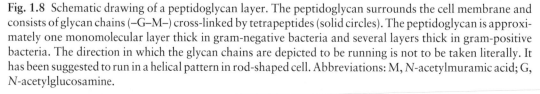

Fig. 1.8 Schematic drawing of a peptidoglycan layer. The peptidoglycan surrounds the cell membrane and consists of glycan chains (–G–M–) cross-linked by tetrapeptides (solid circles). The peptidoglycan is approximately one monomolecular layer thick in gram-negative bacteria and several layers thick in gram-positive bacteria. The direction in which the glycan chains are depicted to be running is not to be taken literally. It has been suggested to run in a helical pattern in rod-shaped cell. Abbreviations: M, N-acetylmuramic acid; G, N-acetylglucosamine.

dilute media causes the cell to swell and burst. Since the strength of the peptidoglycan is due to its covalent bonding throughout, changes in its shape, or expansion during growth of the cell, must be accompanied by hydrolysis of some of the covalent bonds and synthesis of new bonds. However, because of the high intracellular turgor pressure, this must be done very carefully; other wise a lethal weakness in the peptidoglycan structure will be produced when its bonds are cleaved. This problem is discussed by Koch.[76]

There are some important differences between the peptidoglycans in gram-positive and gram-negative bacteria. The peptidoglycan of gram-negative bacteria can be isolated as a sac of pure peptidoglycan that surrounds the cell membrane in the living cell. This receptacle, called the *murein sacculus*, is elastic and is believed to be under stress in vivo because of the expansion due to turgor pressure against the cell membrane.[77] In contrast, the peptidoglycan from gram-positive bacteria is covalently bonded to various polysaccharides and teichoic acids and cannot be isolated as a pure murein sacculus (See the discussion of the gram-positive wall in the next subsection.)

The peptidoglycan from gram-negative bacteria differs from the peptidoglycan from gram-positive bacteria in two other ways: (1) diaminopimelic acid is generally the diamino acid in gram-negative bacteria, whereas it is the diamino acid in only some of the gram-positive bacteria; and (2) the cross-linking is generally direct in gram-negative bacteria, whereas there is usually a peptide bridge in gram-positive bacteria (see later: Fig. 11.3).

As described in Chapter 11, the peptidoglycan in gram-negative bacteria is attached noncovalently to the outer envelope via lipoprotein. Whether it is also attached to the cell membrane is a matter of controversy. The evidence for attachment to the cell membrane is that when *E. coli* is plasmolyzed (placed in hypertonic solutions so that water exits the cells), and prepared for electron microscopy by chemical fixation followed by dehydration with organic solvents, the cell membrane adhers to the outer envelope in numerous places, while generally shrinking away in other locations. It is reasonable to suggest that the zones of adhesion may be areas in which the peptidoglycan

bonds the cell membrane to the outer membrane because the peptidoglycan lies between the two membranes. It has also been argued, however, that the zones of adhesion seen in electron micrographs of plasmolyzed, chemically fixed, and dehydrated cells are artifacts, since they were not visible in cells prepared for electron microscopy by rapid cryofixation (also called freeze substitution) after plasmolysis.[78] In cryofixation the water in and around the cells is rapidly frozen by liquid helium and osmium tetroxide in acetone is substituted for the frozen water.

Gram-positive walls

1. Chemical composition

The gram-positive cell wall in bacteria is a thick structure approximately 15 to 30 nm wide and consists of several polymers, the major one being peptidoglycan (Figs. 1.8 and 1.9). The kinds and amounts of other polymers in the wall vary according to the genus of bacterium. The nonpeptidoglycan polymers can comprise up to 60% of the dry weight of the wall. The ones most commonly found are usually covalently bound to the glycan chain of the peptidoglycan. These polymers include teichoic acids, teichuronic acids, neutral polysaccharides, lipoteichoic acids, and glycolipids of different types (Fig. 1.10).

Teichoic acids. Teichoic acids are polyanionic polymers of either ribitol phosphate or glycerol phosphate joined by anionic phosphodiester bonds. They can comprise 30 to 60% of the dry weight of the wall and may have multiple roles, as described in ref. 79. Teichoic acids vary structurally with respect to the extent and type of molecules covalently bonded to the hydroxyl groups of the glycerol phosphate or ribitol phosphate in the backbone. Frequently, the amino acid D-alanine is attached, as well as the sugar glucose or N-acetylglucosamine. The teichoic acids are attached to the peptidoglycan via covalent bonds between the phosphate of glycerol phosphate or ribitol phosphate to the C6 hydroxyl of N-acetylmuramic acid (MurNac).

Teichuronic acids. Teichuronic acids are polyanionic acidic polysaccharides containing uronic acids (e.g., some have N-acetylgalactosamine

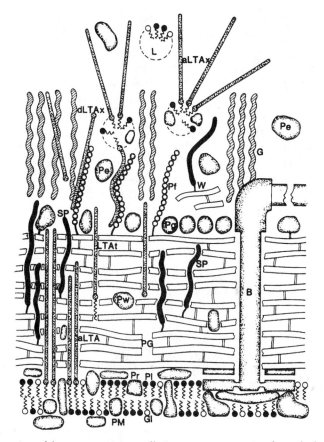

Fig. 1.9 Schematic drawing of the gram-positive wall. Components, starting from the bottom, are as follows: PM, plasma cell membrane consisting of protein (Pr), phospholipid (Pl), and glycolipid (Gl). Overlying the cell membrane is highly cross-linked peptidoglycan (PG) to which are covalently bound teichoic acids, teichuronic acids, and other polysaccharides (SP). Acylated lipoteichoic acid (aLTA) is bound to the cell membrane and extends into the peptidoglycan. Some of the LTA is in the process of being secreted (LTAt). LTA that already has been secreted into the glycocalyx is symbolized as aLTAx. Some of the LTA in the glycocalyx is symbolized as dLTAx because it has been deacylated (i.e., the fatty acids have been removed from the lipid moiety). Within the glycocalyx can also be found lipids (L), proteins (Pe), pieces of cell wall (W), and polymers that are part of the glycocalyx proper (G). Also shown is an inserted flagellum (B). *Source:* Wicken, A. 1985. Bacterial cell walls and surfaces, pp. 45–70. In: *Bacterial Adhesion.* D. C. Savage and M. Fletcher (Eds.). Plenum Press, New York and London.

and D-glucuronic acid). When *Bacillus subtilis* is grown in phosphate-limited media, the teichoic acid is replaced by teichuronic acid (reviewed in ref. 79).

Neutral polysaccharides. Neutral polysaccharides are particularly important for the classification of streptococci and lactobacilli, where they are used to divide the bacteria into serological groups (e.g., groups A, B, and C streptococci).

Lipoteichoic acids. A polyanionic teichoic acid found in most gram-positive walls is lipoteichoic acid (LTA), which is a linear polymer of phosphodiester-linked glycerol phosphate covalently bound to lipid. The C2 position of the glycerol phosphate is usually glycosylated and/or D-alanylated. Because of the negatively charged backbone of glycerol phosphate and the hydrophobic lipid, the molecule is amphipathic (i.e., it has both a polar and a nonpolar end). The lipid portion is bound hydrophobically to the cell membrane, whereas the polyglycerol phosphate portion extends into the cell wall. Unlike the other polymers thus far discussed, LTA is not covalently bound to the peptidoglycan. Its biological role at this location is not understood, but in some bacteria it is

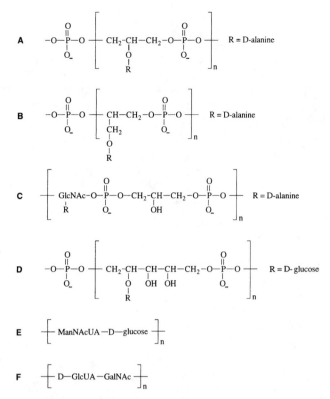

Fig. 1.10 Some teichoic and teichuronic acids found in different gram-positive bacteria. (A) Glycerol phosphate teichoic acid with D-alanine esterified to the C2 of glycerol. (B) Glycerol phosphate teichoic acid with D-alanine esterified to the C3 of glycerol. (C) Glycerol phosphate teichoic acid with glucose and N-acetylglucosamine in the backbone subunit. D-alanine is esterified to the C6 of N-acetylglucosamine. (D) Ribitol phosphate teichoic acid, with D-glucose attached in a glycosidic linkage to the C4 of ribitol. (E) Teichuronic acid with N-acetylmannuronic acid and D-glucose. (F) Teichuronic acid with glucuronic acid and N-acetylgalactosamine. It is believed that teichoic and teichuronic acids are covalently bound to the peptidoglycan through a phosphodiester bond between the polymer and a C6 hydroxyl of muramic acid in the peptidoglycan.

secreted and can be found at the cell surface, where it is thought to act as an adhesin. For example, LTA is secreted by *S. pyogenes*, where it binds with the M protein and acts as a bridge to receptors on host tissues. (See note 80 for a further explanation.) Under these circumstances one should consider the secreted LTA to be part of the glycocalyx along with the M protein.

Other glycolipids. There is a growing list of gram-positive bacteria that do not contain LTA but have instead other amphiphilic glycolipids that might substitute for some of the LTA functions, whatever they might be.[81] Bacteria having these cell surface glycolipids (also called macroamphiphiles or lipoglycans) belong to various genera, including *Micrococcus*, *Streptococcus*, *Mycobacterium*, *Coryne-*

bacterium, *Propionibacterium*, *Actinomyces*, and *Bifidobacterium*.

Bacteria belonging to the genus *Mycobacterium* include the causative agents of tuberculosis (*M. tuberculosis*) and leprosy (*M. leprae*). *M. tuberculosis* infects approximately one-third of the world's population and causes about 3 million deaths annually. Between 12 million and 13 million people worldwide are infected with *M. leprae*. For more complete descriptions of tuberculosis and leprosy, see Boxes 1.4 and 1.5, respectively. The cells walls of mycobacteria consist of waxy lipids that can comprise up to 40% of the dry weight of the cell wall. The waxy lipids are responsible for resistance to dehydration, acids, and alkalies. Indeed, treatment of sputum with dilute sulfuric acid or dilute sodium hydroxide enriches for

BOX 1.4 TUBERCULOSIS

Tuberculosis is a lung disease caused by *Mycobacterium tuberculosis*, but the bacteria can spread to other tissues. The lung tissue is slowly destroyed by a complex process involving macrophages activated by the immune system. The symptoms are the coughing up of mucus, which may be bloody, low-grade fever, night sweats, and weight loss (wasting). Tuberculosis used to be called *consumption* and also the *white plague*.

Pathogenesis

The bacteria enter the respiratory tract via inhalation of nasopharyngeal secretions from individuals with active disease. Once bacteria have entered the lungs, small lesions, frequently called *primary lesions*, are produced, where the bacteria reproduce. At this stage of the infection, the bacteria are engulfed by nonactivated alveolar macrophages. A certain fraction of the bacteria survive and grow inside the macrophages at the site of the lesions. When they grow inside the macrophages, the macrophages are killed and release the bacteria. Since, however, the cellular immune system reacts against the bacteria, the alveolar macrophages are activated, becoming better killers of the mycobacteria. This stops the spread of the infection. The lesions heal and form granulomas, also called tubercles, containing large activated macrophages called epithelioid cells because they resemble epithelial cells, and multinucleated giant cells derived from the fusion of epitheloid cells (Langhans giant cells). The tubercles also contain lymphocytes and fibroblasts, the latter forming fibrous tissue in the tubercles. There are relatively few bacteria present in the tubercles, and these are mostly outside the macrophages. However, the activated macrophages eventually destroy the lung tissue in the center of the tubercles, causing it to become semisolid or "cheesy." The "cheesy" tissue is said to be *caseated*. (The verb "to caseate"

comes from the Latin *caseus*, which means cheese.) The caseated lesions may heal, becoming infiltrated with fibrous tissue and, often, acquiring deposits of calcium, which form nodules (large tubercles) that can be seen in X-rays images. The walled-off tubercles remain in the lung for the rest of the individual's life and contain viable bacteria. Most active cases of tuberculosis are due to the reactivation of tubercles. This may happen in individuals with weakened immune systems, such as people with AIDS, or in the elderly.

During an early stage in the progress of the disease—that is, during the primary lesion stage—the bacteria may spread into regional lymph nodes, and from there into the blood-stream. The bacteria may also spread into the blood-stream from caseated lung tissue. The circulatory system can bring the bacteria to various organs and tissues, including the bone marrow, spleen, liver, meninges, and kidneys. The bacteria can even return from these organs, to the lungs via the blood circulatory system, and this can contribute to the reactivation of the disease in the lungs years after the initial infection. An asymptomatic period of years frequently precedes the appearance of symptoms in various body organs. Symptoms, when they do occur, are caused by malfunction of the organs or tissues due to necrotizing lesions, or to damage of blood vessels that feed the tissues.

The role of the cellular immune response in damaging infected tissue

The macrophages are activated by lymphocytes that are activated during the immune response to the bacteria. These are $CD4^+$ (T-helper cells) and $CD8^+$ (T-cytolytic) lymphocyte. The T-helper (T_H) cells activate the other lymphocytes, namely, B lymphocytes, which mature into antibody-secreting cells, T_C cells, and macrophages. The B lymphocytes are part of the humoral response, and they mature to make antibodies. The T-cytolytic lymphocytes (T_C cells) are part of the cellular immune response, and their major role is to attach to and kill virus-infected cells and foreign tissues. Both the T_H and T_C cells produce interferon gamma (IFN-γ), which is a major activator of macrophages. The macrophages are said to be active when they become capable of new activities, such as the engulfment and killing of bacteria. Hence, activated macrophages are also part of the cellular immune response and limit the spread of the infection. Tissue necrosis (destruction) is not due to toxins produced by the mycobacteria. Rather, it results when hydrolytic enzymes and toxic forms of oxygen are released by the activated macrophages in the granulomas. These factors are made by macrophages to kill the pathogens they have engulfed; but when leaked into the interstitial fluid, they kill surrounding tissue. Activated macrophages also release factors that cause blood clotting (procoagulant factors), and when the clots form in local blood vessels, blood supply to the tissues is depressed, resulting in death of the tissues.

Laboratory diagnosis

The identification of acid-fast rods in smears of sputum is a tentative diagnosis of tuberculosis. To identify acid-fast rods, the Ziehl–Neelson or Kinyoun method is used to stain the smears. The cells in the smears can also be stained with a fluorescent rhodamine–auramine dye and viewed with a fluorescence microscope. In part because there may be only very few organisms present in the smears, culturing the specimens is more definitive. The cultures can be grown on medium containing egg yolk or oleic acid and albumin. However, the bacteria grow very slowly (the cells divide appproximately once a day), and it may be 3 to 6 weeks before growth is visible. A more rapid test is to use the polymerase chain reaction (PCR) to amplify the bacterial DNA. The PCR assay can be applied to sputum material, and specific DNA probes can be used to detect *M. tuberculosis* DNA sequences. The reagents can be purchased in a kit.

Who gets sick

Tuberculosis is a major worldwide killer, certainly one of the most important of the lethal infectious diseases. The World Health Organization (WHO) estimates that about 10 million new *active* cases of tuberculosis occur each year, and 3 million people die. In the twentieth century alone there were about 1 billion deaths due to tuberculosis. It has been estimated that about half the world's population of approximately 6 billion people is infected, but that only 30 million of infected people have active cases. This means that about 99% of people who become infected with *Mycobacterium tuberculosis* do not become ill and are healthy carriers, although as noted, the disease can become activated later in life. Estimates are that approximately 15 million people in the United States are infected. A skin test called the *tuberculin test* can detect whether a person has been infected with *M. tuberculosis*.

How they get sick

The infection is transmitted from someone with an active case of tuberculosis who coughs or sneezes, producing respiratory droplets that the next victim inhales. Some people who have been infected are free of symptoms for many years, and may die without developing active tuberculosis. Activation of the disease is associated with malnutrition, old age, and a suppressed immune system, as in people with AIDS. The number of new active cases in the United States each year exceeds 20,000. In 1995 the U.S. Centers for Disease Control and Prevention (CDC) reported 22,900 new active cases.

Treatment

Antituberculosis drugs must be taken for at least 6 months. The drugs that can be taken are isoniazid (Nydrazid), rifampin (Rifadin and Rimactane), ethambutol (Myambutol), streptomycin, and pyrazinamide. Usually more than one medication is taken.

Tuberculin skin test

Tuberculin (called *old tuberculin*) refers to material isolated from *M. tuberculosis* by boiling the cells. A purified protein derivative (PPD) of old tuberculin is used for the skin test. It is a mixture of cell wall proteins and polysaccharides prepared by fractionation with trichloroacetic acid (TCA), ammonium sulfate, and alcohol. The PPD is injected into the skin. If the person being tested has not been exposed to *M. tuberculosis*, there is no immediate reaction, and a positive skin reaction occurs 3–5 weeks after inoculation. If the individual has been infected, a delayed-type hypersensitivity reaction occurs at the site of injection within a couple of days. This is manifested as an indurated (raised, hard) area that is often red (erythema tons). The diameter of the induration, measured within 2 to 3 days, is interpreted according to the following list of ranges:

> 0–4 mm = negative
> >10 mm = positive
> 5–10 mm = uncertain; must be retested

For an immunocompromised person (e.g., an AIDS patients), a 5 mm diameter is considered to be a positive test.

BCG vaccine

There exists a vaccine for tuberculosis: the *bacille Calmette–Guérin* (BCG) vaccine, named after the French microbiologists who developed the drug. The vaccine is made from a live, attenuated bovine strain of tubercle bacillus (*M. bovis*). "Attenuated" means that the strain has been grown in culture for a long time and has lost so much virulence that it cannot cause tuberculosis. BCG is widely used world wide; in developing countries it is routinely given to children at birth. The WHO has estimated that 85% of children born in 1990 received the BCG vaccination during the first year of their life. However, there is a great deal of uncertainty about BCG's efficiency, which has reportedly ranged from 0% (one study in India) to 90%. The results studies conducted in the United Kingdom indicate protection against infection of approximately 70% of vaccinated individuals. All tuberculin-negative children in the United Kingdom receive the BCG vaccination at 10 to 13 years of age.

The BCG vaccine is not routinely used in the United States. The reasons for this in the past have been the relatively low incidence of the disease (<1% of children and young adults give a positive tuberculin skin test), and the fact that any person who has been vaccinated will test positive in the tuberculin test. Under the latter circumstances, it would not be possible to use the tuberculin test to ascertain the spread of the infection. Nor would it be possible to ascertain whether an individual has become recently infected, and therefore should receive preventive therapy with isoniazid.

History

The people who have died from tuberculosis between 1805 and 1967 include many who have been very important in music, writing, and the theater: the English poet John Keats (died in 1821), the English author D. H. Lawrence (1930), the English actress Vivien Leigh (1967), the English novelist and essayist George Orwell (1950), the American playwright Eugene O'Neill (1953), the Polish pianist and composer Frederic Chopin (1849), the Scottish novelist and poet Walter Scott (1832), the Italian violinist and composer Nicolò Paganini (1840), the German poet and playwright Johann Wolfgang von Goethe (1832), the Russian playwright and novelist Anton Pavlovich Chekhov (1904), and the German poet and playwright Johann Christoph Freidrich von Schiller (1805).

BOX 1.5 LEPROSY

Leprosy, a chronic disease caused by *Mycobacterium leprae*, can manifest itself in cutaneous lesions, nerve impairment resulting in sensory loss, and tissue destruction. The bacteria infect the skin, peripheral nerves, eyes, and mucous membranes, and cartilage, muscle, and bone may be destroyed. Humans are the only known source of *M. leprae*. The bacteria are obligate intracellular parasites and grow well in the peripheral nerves, specifically in the Schwann cells, which make up the myelin sheaths.

Symptoms

There are two forms of leprosy: *lepromatous* and *tuberculoid*, but these are two extremes, and infected people can display symptoms intermediate between them.

Tuberculoid leprosy. This is a self-limiting disease that may regress. Skin lesions are present but contain very few organisms. The lesions, which can occur anywhere on the body, are usually one or two anesthetic (without sensation), hypopigmented macules (flat spots) with raised reddish or purple edges. The lesions vary in size from a few millimeters to larger ones that that may cover the entire trunk. There is palpable thickening of peripheral nerves due to the growth of bacteria in nerve sheaths. Skin samples subjected to biopsy show tuberculoid granulomas consisting of activated macrophages and lymphocytes. These resemble the granulomas formed in the lungs in response to *M. tuberculosis* infection and are partly attributable to the ability of the bacteria to grow inside the macrophages. The cell-mediated immune response against the bacteria results in nerve damage due to inflammation. As a consequence, there is loss of sensation in the area of the lesions.

Lepromatous leprosy. The skin lesions are anesthetic, extensive, large, nodular, and disfiguring. The face becomes disfigured (lion's face) owing to thickening, nodules, and plaques. Eyebrows and eyelashes are lost. Lepromatous leprosy results when the immune system fails to mount an effective cell-mediated defense. It is a progressive disease, that is fatal if not treated. The bacteria disseminate throughout the body and can be found in all the body organs. The bacteria also grow in the mucous membranes of the nose. The cartilaginous septum is destroyed, resulting in severe nasal deformities, including collapse of the nose. There is also damage to underlying cartilage and bone. Fingers or toes may be lost.

Laboratory diagnosis

Diagnosis depends upon detecting the bacteria in skin smears and skin biopsies. Generally such tests are positive only for lepromatous leprosy. To make a skin smear, the dermis is cut and tissue fluid is scraped from the slit and smeared on a microscope slide. The smears are stained for acid-fast bacilli by using a modification of the Ziehl–Neelson stain. For the skin biopsy, material is taken from the lesion and fixed in 10% formalin and stained for acid-fast bacilli. The bacteria cannot be cultured in cell-free cultures (i.e., *in vitro*). They can be grown in mouse footpads, however, by injecting a suspension of bacteria derived from a skin biopsy. It takes about 6 months for maximum growth to occur. Growing the bacteria in mouse footpads allows diagnosticians to test the sensitivity of the infecting organism to therapeutic drugs such as dapsone and rifampin. *M. leprae* can also be cultivated in the armadillo.

Who gets sick

Between 12 million and 13 million people are infected worldwide, mostly in developing countries, especially those in tropical or semi tropical areas. In the past, however, leprosy has not been confined to the warmer climates. It is most prevalent in India, where in 1994 the WHO put the number of estimated cases at 1,167,900. In the United States about 100 to 200 new cases a year are reported each year to the CDC (mostly among immigrants).

How they get sick

Leprosy is thought to be transmitted via nasal secretions and contact with skin lesions, although the disease is not nearly as contagious as once thought. Spread by skin contact would probably require a broken skin on the recipient for the bacteria to penetrate.

Treatment

The drug Dapsone, which is related to the sulfonamides, will successfully treat leprosy. For strains that show some resistance to Dapsone, a combination of drugs, including Dapsone, rifampin, and clofazimine may be used.

History

Leprosy is an ancient and infamous disease often referred to as *Hansen's disease*, after the Norwegian scientist Gerhard Hansen, who discovered the causative agent in 1874. The disease apparently started in the Far East prior to 600 BC and spread to the Near East, Africa, and finally to Europe, where incidence peaked during the Middle Ages. It was probably introduced into the Americas with the early Spanish explorations and the slave trade. The first reference to leprosy in the United States was in the Floridas in 1758. A hospital for the treatment of leprosy was established in New Orleans in 1785. The Louisiana Leper Home was established in 1894, near the village of Carville, and in 1921 this facility became the National Leprosarium, a federal facility. It now exists as the Gillis W. Long Hansen's Disease Center at Carville.

mycobacteria, as opposed to other bacteria in the respiratory flora.

For the following discussion, refer to Fig. 1.11. The main waxy lipids are branched-chain hydroxy fatty acids called mycolic acids, also found in *Nocardia*, *Corynebacterium*, and *Rhodococcus*. The mycolic acids are a major class of lipid, and they form on the external face of the cell wall a hydrophobic layer that can be esterified to a polysaccharide called *arabinogalactan*, which is a copolymer of arabinose and galactose, or to trehalose, which is a disaccharide of D-glucose. If the layer is bound to trehalose, the compound is called *cord factor*, which is described next. The arabinogalactan itself is covalently bonded to acetyl or glycolyl groups in the peptidoglycan via phosphodiester bonds. There are variations in the size of individual mycolic acids among the different species of *Mycobacterium*. In *M. tuberculosis* the mycolic acid is a 3-hydroxy fatty acid with two alkyl branches. The first branch (R_1) is at C2 and is a hydrocarbon side chain of 24 carbons ($-C_{24}H_{49}$). The second branch (R_2) is at C3 and is a hydroxylated hydrocarbon with 60 carbons in the chain ($-C_{60}H_{120}-OH$).

In addition to mycolic acid–arabinogalactan, the *M. tuberculosis* cell wall has a mycolic acid containing cell surface glycolipid called *cord factor* (Fig. 1.11). Cord factor consists of the glucose disaccharide, trehalose, to which is covalently bonded two molecules of mycolic acid. The characteristic *M. tuberculosis* colony appearance of long, intertwining cords, due to the side-by-side interactions of long chains of cells forming serpentine ropelike rods, is generally attributed to cord factor. Hence the name. As reviewed in ref. 82, cord factor is also found

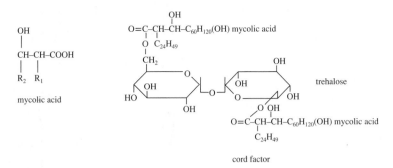

Fig. 1.11 Chemical structures of mycolic acid and cord factor. Mycolic acid is a β-hydroxy fatty acid with alkyl groups R_1 attached to the α carbon and R_2 attached to the β carbon. In *Mycobacterium tuberculosis* R_1 is $C_{24}H_{49}$ and R_2 is $C_{60}H_{120}(OH)$. Cord factor is a glycolipid in the cell walls of *M. tuberculosis*. It is a disaccharide of D-glucose [Glc(α_1–α_1)Glc], called trehalose, to which mycolic acid is attached via ester linkage to the C6 hydroxyl of the sugar.

in noncording mycobacteria. It has been suggested that cord factor is responsible for the wasting, fever, and lung damage symptomatic of tuberculosis. Cord factor is required for virulence, and it is lacking in avirulent mycobacteria. When injected intravenously into rabbits, cord factor causes weight loss and granulomas in the liver and lungs.[83]

When mycobacteria are appropriately stained, the dye (basic fuschin) is not removed by dilute hydrochloric acid in ethanol (acid alcohol) because of the presence of the waxy lipids. Therefore these microorganisms are called *acid-fast* bacteria. Acid-fast staining is an important diagnostic feature for the identification of *Mycobacterium* in clinical specimens, such as sputum. Acid-fast bacteria also do not take the Gram stain unless the wall lipids are removed with alkaline ethanol.

In summary, gram-positive cell walls have diverse types of neutral and acidic polysaccharides, glycolipids, lipids, and other compounds either free in the wall or covalently bound to the peptidoglycan. The functions of most of these polymers are largely unknown, although some may act as adhesins and/or presumably affect the permeability characteristics of the cell wall, whereas others, such as mycolic acid derivatives in *M. tuberculosis*, contribute to resistance properties and virulence.

Gram-negative wall

The gram-negative cell wall is structurally and chemically complex. It consists of an outer membrane composed of lipopolysaccharide, phospholipid, and protein, and an underlying peptidoglycan layer (Fig. 1.12). Between the outer and inner membrane (the cell membrane) is a compartment called the *periplasm*, wherein the peptidoglycan lies.

1. Lipopolysaccharide structure and function

Lipopolysaccharide (LPS) consists of three regions: *lipid A*, *core*, and a *repeating oligosaccharide*, sometimes called *o-antigen* or *somatic antigen*; its chemical structure and synthesis are described in Section 11.2. In the Enterobacteriaceae (e.g., *E. coli*), the lipopolysaccharide is confined to the outer leaflet of the outer membrane and is arranged so that the lipid A portion is embedded in the membrane as part of the lipid layer, and the core and oligosaccharide extend into the medium (Fig. 1.12).

Mutants that lack the oligosaccharide experience loss of virulence, and it is believed that the LPS can increase pathogenicity. The lipid A portion of the LPS is an endotoxin as explained in note 84. Loss of most of the core and oligosaccharide in *E. coli* and related bacteria is associated with increased sensitivity to hydrophobic compounds (e.g., antibiotics, bile salts, and hydrophobic dyes such as eosin and methylene blue). This is because the LPS provides a permeability barrier to hydrophobic compounds. (A model for how this might occur is described later.) It is advantageous for the enteric bacteria to have such a permeability barrier because they live in the presence of bile salts in the intestine. In fact, the basis for selective media for gram-negative bacteria is

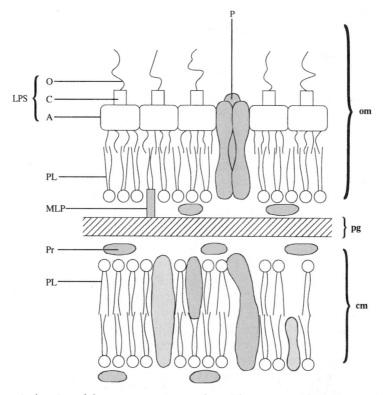

Fig. 1.12 Schematic drawing of the gram-negative envelope. The outer membrane consists of lipopolysaccharide, phospholipid, and proteins, most of which are porins. Underneath the outer membrane is the peptidoglycan layer, which is noncovalently bonded to the outer membrane via murein lipoproteins, themselves covalently attached to the peptidoglycan. The cell membrane is composed of phospholipid and protein. The area between the outer membrane and the cell membrane is called the periplasm. The wavy lines are fatty acid residues, which anchor the phospholipids and lipid A into the membrane. Abbreviations: LPS, lipopolysaccharide; O, oligosaccharide; C, core; A, lipid A; P, porin; PL, phospholipid; MLP, murein lipoprotein; pg, peptidoglycan; Pr, protein; om, outer membrane; cm, cell membrane.

the resistance of these bacteria to bile salts and/or hydrophobic dyes, which are included in the media. The bile salts and dyes inhibit the growth of gram-positive bacteria, but not gram-negative bacteria because of the LPS. An example of such a selective medium is eosin–methylene blue (EMB) agar, which is used for the isolation of gram-negative bacteria because it inhibits the growth of gram-positive bacteria. As explained in note 85 and summarized in Section 16.5, drug efflux pumps are another important reason that bacteria are resistant to many antibiotics.

Apparently the outer membrane to hydrophobic compounds has low permeability because the phospholipids are confined primarily to the inner leaflet of the outer envelope, whereas the LPS, which is a permeability barrier to hydrophobic substances, is in the outer leaflet. Since lipid A contains only saturated fatty acids, the LPS presents a somewhat rigid matrix, and it has been suggested that this property, plus the tendency of the large LPS molecules to engage in lateral noncovalent interactions, makes it difficult for hydrophobic molecules to penetrate between the LPS molecules to the phospholipid layer. Mutants that lack a major region of the oligosaccharide and core are more permeable to hydrophobic compounds because there is more phospholipid in the outer leaflet of these mutants; it is likely, as well that there is less lateral interaction among the LPS molecules. The asymmetric distribution of phospholipid to the inner leaflet of the outer envelope seems to be an adaptive evolutionary response of the enteric bacteria to hydrophobic toxic substances in the intestine of animals. Accordingly, not all gram-negative

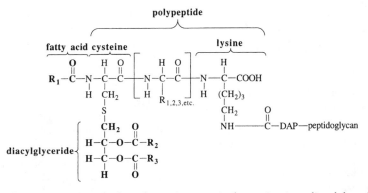

Fig. 1.13 Murein lipoprotein. Attached to the amino-terminal cysteine is a diacylglyceride in thioether linkage and a fatty acid in amide linkage. The lipid portion extends into the outer envelope and binds hydrophobically with the fatty acids in the phospholipids and lipopolysaccharide. The carboxy-terminal amino acid is lysine, which can be attached via an amide bond to the carboxyl group of diaminopimelic acid (DAP) in the peptidoglycan. The murein lipoprotein therefore holds the outer envelope to the peptidoglycan.

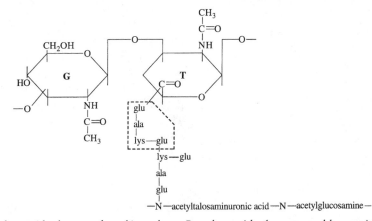

Fig. 1.14 Pseudopeptidoglycan as found in archaea. Pseudopeptidoglycan resembles peptidoglycan in being a cross-linked glycopeptide. It differs from peptidoglycan in the following ways: (1) N-acetyltalosaminuronic acid replaces N-acetylmuramic acid. The glycosidic linkage is β1-3 instead of β1-4. There are no D-amino acids. Abbreviations: G, N-acetylglucosamine; T, N-acetyltalosaminuronic acid. The peptide subunit is enclosed in dashed lines.

bacteria have an outer envelope with an asymmetric distribution of lipopolysaccharide and phospholipid.

2. Lipoproteins

In addition to lipopolysaccharide and phospholipid, a major component of the outer membrane is protein, of which there are several different kinds. One of the proteins is called the *murein lipoprotein*. This is a small protein with lipid attached to the amino-terminal end (Figs. 1.13 and 1.14). The lipid end of the molecule extends into and binds hydrophobically with the lipids in the outer envelope. The protein end of some of the molecules is covalently bound to the peptidoglycan, thus anchoring the outer envelope to the peptidoglycan. In *E. coli*, about one-third of the murein lipoprotein is bound to the peptidoglycan. Mutants unable to synthesize the murein lipoprotein have unstable outer envelopes that bleb off into the medium at the cell poles and septation sites. Therefore, the murein lipoprotein may play a structural role in keeping the outer membrane attached to the cell surface.

There are a small number of other outer membrane or cell membrane lipoproteins.[86] These were discovered by chemically cross-linking the peptidoglycan in whole cells to closely associated proteins with a bifunctional cross-linking reagent. (Note 87 explains how cross-linking reagents work.) There is no evidence that these additional lipoproteins are covalently bonded to the peptidoglycan, and their functions remain to be elucidated.

4. Porins and other proteins

The major proteins in the outer envelope are called porins. The porins form small non-specific hydrophilic channels through the outer envelope, allowing the diffusion of low molecular weight (<600 Da) neutral and charged solutes, such as sugars and ions. The channels are necessary to allow passage of small molecules into and out of the cell. E. coli has three major porins: OmpF, OmpC, and PhoE. Each porin makes a separate channel. Thus, there are OmpF, OmpC, and PhoE channels.

The OmpC channel is approximately 7% smaller than the OmpF channel and is expected to make the outer envelope less permeable to larger molecules. Both OmpF and OmpC are present under all growth conditions, although the ratio of the smaller OmpC to the larger OmpF increases in high osmolarity media and at high temperature. The increased amounts of OmpC relative to OmpF presumably also occur in the intestine, where osmolarity and temperature are higher than in lakes and streams that also harbor E. coli. This may confer an advantage to the enterics because the smaller OmpC channel should present a diffusion barrier to toxic substances in the intestine, whereas the larger OmpF channel should be advantageous in more dilute environments outside the body.

The protein PhoE is produced only under conditions of inorganic phosphate limitation. This is because PhoE is a channel for phosphate (and other anions) whose synthesis under limiting phosphate conditions reflects the need to bring more phosphate into the cell. Porins appear to be widespread among gram-negative bacteria, although they are not all identical. As mentioned earlier, E. coli regulates the amounts of the various porins according to growth conditions. This is discussed in Chapter 18.

Since the porins exclude molecules with molecular weights larger than 600 Da, one would expect to find other proteins in the outer membrane that facilitate the translocation of larger solutes across the outer membrane. This is the case. E. coli has an outer membrane protein called the LamB protein that forms channels for maltose and maltodextrins. Other proteins in the outer membrane of E. coli facilitate the transport of vitamin B_{12} (BtuB), nucleosides (Tsx), and several other solutes.

Archaeal cell walls

Archaeal cell walls are not all alike and are very different from bacterial cell walls. For example, *no archaeal cell wall contains peptidoglycan.* Archaeal cell walls may be either pseudopeptidoglycan, polysaccharide, or protein (the S layer). Pseudopeptidoglycan (also called pseudomurein) resembles peptidoglycan in consisting of glycan chains cross-linked by peptides (Fig. 1.14). However, the resemblance stops here. In pseudopeptidoglycan, although one of the sugars is N-acetylglucosamine as in peptidoglycan, the other is N-acetyltalosaminuronic acid instead of N-acetylmuramic acid. Furthermore, the sugars are linked by a β-1,3 glycosidic linkage rather than a β-1,4, and the amino acids in the peptides are L-amino acids rather than D-amino acids. (The latter are present in peptidoglycan; compare Figs. 1.14 and 11.2.) Thus, pseudomurein is very different from murein.

1.2.4 Periplasm

Lying between the cell membrane and the outer membrane of gram-negative bacteria is a separate compartment called the *periplasm*[88] (Fig. 1.15). It appears as a space in electron micrographs of thin sections of cells but should be considered to be an aqueous compartment containing protein, oligosaccharide, salts, and the peptidoglycan. It seems that the peptidoglycan and oligosaccharides may exist in a hydrated state, forming a periplasmic gel.[89]

The literature discusses whether the periplasm has zones of adhesion (called Bayer's patches) between the inner and outer membranes.[89,90] Whether or not zones of adhesion are seen depends upon how the cells are prepared for electron microscopy. One school of thought holds that the zones are artifacts of fixation; the counterargument is that the zones are seen only if the proper techniques are employed to preserve the rather fragile adhesion sites. This is an important issue because the zones of adhesion have been postulated to be sites of the translocation apparatuses that export

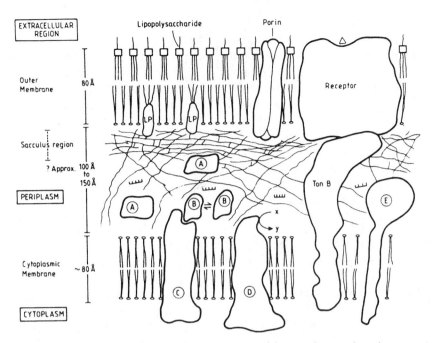

Fig. 1.15 A model of the periplasm in *E. coli*. The outer region of the periplasm is thought to consist of cross-linked peptidoglycan attached to the outer envelope via lipoprotein (LP) covalently bound to the peptidoglycan. The inner region of the periplasm (approaching the cell membrane) is believed to consist of less cross-linked peptidoglycan chains and oligosaccharides that are hydrated and form a gel. The gel phase is thought to contain periplasmic proteins (e.g., A and B). Thus A might be a periplasmic enzyme and B a solute-binding protein that interacts with a membrane transporter (C); D and E are integral membrane proteins, perhaps enzymes. The outer envelope is depicted as consisting of lipopolysaccharide, porins, and specific solute transporters (receptors), which require a second protein (TonB) for uptake. *Source*: Ferguson, S. J. 1991. The periplasm, pp. 311–339. In: *Prokaryotic Structure and Function, A New Perspective*. S. Mohan, C. Dow, and J. A. Coles (Eds.). Cambridge University Press, Cambridge.

lipopolysacharide, polysaccharide (capsule), and protein through the inner and outer membranes.

The periplasm should be to be considered a cellular compartment with specialized activities. These activities include oxidation–reduction reactions (Chapters 4 and 12), osmotic regulation (Chapter 15), solute transport (Chapter 16), protein secretion (Chapter 17), and hydrolytic activities such as those mediated by phosphatases and nucleases. The phosphatases and nucleases degrade organophosphates and nucleic acids that might enter the periplasm from the medium and transport the hydrolytic products into the cell.

Periplasmic components

The periplasm is chemically complex and carries out diverse functions. The following list of components and their functions, which reflects the importance of the periplasm, is simply a partial inventory emphasizing the periplasmic functions about which most is known.

1. Oligosaccharides

The oligosaccharides in the periplasm are thought by some to be involved in osmotic regulation of the periplasm because their amounts decrease when the cells are grown in media of high osmolarity. This complex subject is discussed more fully in Section 15.2.

2. Solute-binding proteins

Solute-binding proteins in the periplasm assist in solute transport by binding to solutes (e.g., sugars and amino acids that have entered the periplasm through the outer envelope) and delivering the solutes to specific transporters (carriers) in the cell membrane. This important means of bringing nutrients into the cell is discussed in Section 16.3.3.

3. Cytochromes c

Some of the enzymes in the periplasm are cytochromes c that oxidize carbon compounds or inorganic compounds and deliver the electrons to the electron transport chain in the cell membrane. These oxidations are called periplasmic oxidations. There are other oxidoreductases in the periplasm as well, but the various cytochromes c are very common. Periplasmic oxidations are important for energy metabolism in many different gram-negative bacteria and are discussed in Chapters 4 and 12.

4. Hydrolytic enzymes

Hydrolytic enzymes in the periplasm degrade nutrients to smaller molecules that can be transported across the cell membrane by specific transporters. For example, the enzyme amylase is a periplasmic enzyme that degrades oligosaccharides to simple sugars. Another example is alkaline phosphatase, which removes phosphate from simple organic phosphate monoesters. The inorganic phosphate is then carried into the cell via specific inorganic phosphate transporters.

5. Detoxifying agents

Some periplasmic enzymes are detoxifying agents. For example, the enzyme to degrade penicillin (β-lactamase) is a periplasmic protein.

6. TonB protein

An interesting periplasmic protein anchored to the cell membrane in E. coli is the TonB protein; its mechanism of action is not understood. (See Fig. 1.15.) It is known that the protein is required for the uptake of several solutes that do not diffuse through the porins; rather, they require specific transport systems (also called receptors) in the outer envelope. All these solutes have molecular weights larger than 600 Da, which is the upper limit for molecules that enter via the porins. Examples of solutes with specific outer membrane receptors that require TonB for uptake are iron siderophores and vitamin B_{12} (cobalamin). (For an explanation of iron siderophores, see note 91.) Interestingly, these solutes are brought into the periplasm against a large concentration gradient, sometimes 10^3 times higher than the concentration outside the cell. In a way that is not understood, the TonB protein couples the electrochemical energy (the proton motive force, Δp) in the cell membrane to the uptake of certain solutes through the outer envelope and into the periplasm.[92–95] TonB is thought to be an energy transducer. One suggestion is that TonB is energized by the electrochemical potential that exists across bacterial cell membranes and that in the energized state, TonB causes a conformational change in the outer membrane receptor protein that results in translocation of the solute (e.g., vitamin B_{12}) or ligand-bound solute (e.g., iron–siderophore complexes) through the receptor channel into the periplasm.[95] Accessory proteins in the cell membrane, which in E. coli are called ExbB and ExbD, interact with TonB and may use the proton motive force Δp (i.e., uptake of H^+ through ExbB/D) to convert TonB into an energized conformation that somehow energizes the uptake of material through transporters in the outer membrane. This conception was reviewed in 2003 ref. 95, which also discusses two experimentally based models designed to explain the mechanism by which TonB acts as an energy transducer. One model proposes that TonB in an energized conformation leaves the ExbB/D complex in the cell membrane and moves to the transporter in the outer membrane. It then returns to the ExbB/D complex to be reenergized. This has been called the "shuttle" model. An alternative model proposes that one part of TonB remains associated with the ExB/D complex in the cell membrane and one part reaches across to the transporter in the outer membrane.

Is there a periplasm in gram-positive bacteria?

In the past it has been assumed that gram-positive bacteria lack a space between the cell membrane and the cell wall equivalent to the periplasm of gram-negative bacteria. This is because thin sections of gram-positive cells do not indicate that such a space exists, and of course gram-positive bacteria do not have an outer envelope. Evidence suggests, however, that perhaps gram-positive bacteria do have a compartment analogous to the periplasm found in gram-negative bacteria.[96,97]

The evidence in favor of the existence of a periplasm in Bacillus subtilis includes the release of putative periplasmic proteins after protoplasts of the cells have been made with lysozyme. (For a definition of protoplasts and how they are stabilized, read note 98.) Proteins solubilized by removing the cell wall during protoplast formation may include proteins resident in the area between the cell membrane and the cell wall. The proteins released by protoplast formation include nucleases, which are distinct from cytoplasmic nucleases.

Additional evidence has been obtained by using ultrarapid freezing and electron microscopy (cryo–transmission electron microscopy, or cryo-TEM) of frozen–hydrated thin sections of *B. subtilis*. When this was done, the area outside the cell membrane was seen to be bipartite, consisting of a low-density 22 nm region surrounded by a 33 mm high-density outer wall zone, which is thought to consist of peptidoglycan, teichoic acid, and cell wall proteins.[97] Further studies must be undertaken to elucidate the exact nature of this putative periplasm, including its contents and the relationships of the proteins found therein with the cell membrane and cell wall, as well as the similarities to the gram-negative periplasm.

1.2.5 Cell membrane

We now come to what is certainly the most functionally complex of the cell structures, the cell membrane. The cell membrane is responsible for a broad range of physiological activities including solute transport, electron transport, photosynthetic electron transport, the establishment of electrochemical gradients, ATP synthesis, biosynthesis of lipids, biosynthesis of cell wall polymers, secretion of proteins, the secretion and uptake of intercellular signals, and responses to environmental signals. To refer to the cell membrane simply as a lipoprotein bilayer does not do justice to the machinery embedded in the lipid matrix, a complex mosaic of parts whose structure and interactions at the molecular level are not well understood.

As expected, the protein composition of cell membranes is complex. There can be more than 100 different proteins. Many of the proteins are clustered in functional aggregates (e.g., the proton translocating ATPase, the flagella motor, electron transport complexes, certain of the solute transporters). At the molecular level, the membrane is certainly a complex and busy

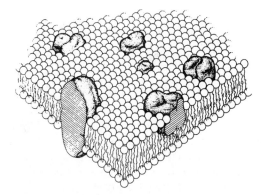

Fig. 1.16 Model of the cell membrane showing bimolecular lipid leaflets and embedded proteins; the phospholipid molecules are interacting with one another via their hydrophobic (apolar) "tails." The hydrophilic (polar) "heads" of the phospholipids face the outside of the membrane, where they interact with proteins and ions. Proteins can span the membrane or be partially embedded. *Source:* Singer, S. J., and G. L. Nicolson. 1972. The fluid mosaic model of the structure of cell membranes. *Science* **175**:720–731. Copyright 1972 by the Association of Academies of Science.

place. What follows is a general description of the membrane, without reference to its microheterogeneity.

Bacterial cell membranes

Bacterial cell membranes consist primarily of phospholipids and protein in a fluid mosaic structure in which the phosphlipids form a bilayer (Fig. 1.16). The structure is said to be fluid because there is extensive lateral mobility of bulk proteins and phospholipids. Nevertheless, certain protein aggregates (e.g., complex solute transporters and electron transport aggregates) remain as aggregates within which the proteins interact to catalyze sequential reactions.

1. The lipids

The phospholipids are fatty acids esterified to two of the hydroxyl groups of phosphoglycerides (Fig. 1.17). The structure and synthesis of phospholipids is described in detail in Section 9.1.2. The third hydroxyl group in the glycerol backbone of the phospholipid is covalently bound to a substituted phosphate group, which makes one end of the molecule very polar owing to a negative charge on the ionized phosphate group. Because the

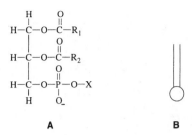

Fig. 1.17 Phospholipids have both a polar and a nonpolar end. (A) Phospholipid with two fatty acids (R) esterified to glycerol. The phosphate is replaced by X, which determines the type of phospholipid. In bacteria, X is usually serine, ethanolamine, a derivative of glycerol, or a carbohydrate derivative. See Section 9.1.3 for a more complete description of bacterial phospholipids. (B) Schematic drawing of a phospholipid showing the polar (circle) and nonpolar (straight lines) regions.

phospholipids are polar at one end and nonpolar at the other end (the end with the fatty acids), they are said to be *amphipathic*, able to spontaneously aggregate with their nonpolar fatty acid regions interacting with each other by hydrophobic bonding, while their polar phosphorylated regions face the aqueous phase, where ionic interactions occur with cations, water, and polar groups on proteins. Phospholipids accomplish all this by spontaneously forming lipid bilayers in water solutions or in cell membranes.

2. The proteins

There are two classes of proteins in membranes, *integral* and *peripheral*. Integral proteins are embedded in the membrane and bound to the fatty acids of the phospholipids via hydrophobic bonding. They can be removed only with detergents or solvents. Peripheral proteins, attached at membrane surfaces to the phospholipids by ionic interactions, can be removed by washing the membrane with salt solutions. The insertion of the proteins into the membrane during membrane synthesis is discussed in Section 17.2.

3. Permeability

The phospholipid bilayer acts as a permeability barrier to virtually all water-soluble molecules. Thus most solutes diffuse or are carried across the membrane through or on special protein transporters that bridge the phospholipid bilayer. These modes are discussed in the context of solute transport in Chapter 16.

(However, the lipid bilayer is permeable to water molecules, gases, and small hydrophobic molecules.) An important consequence of the lipid matrix is that ions do not freely diffuse across the membrane unless they are carried on or through protein transporters. Because of this, the membrane is capable of holding a charge that is due to the unequal transmembrane distribution of ions. This is discussed in Chapter 3.

4. Aquaporins (water channels)

Although the lipid bilayer allows rapid equilibration of water, there do exist in *E. coli* and other bacteria water channels, called aquaporins, that are similar to the aquaporins found in eukaryotes and enhance the rapid equilibration of water across the cell membrane. (Reviewed in ref. 99.) The gene coding for the water channel protein in *E. coli* is *aqpZ*. The expression of the *aqpZ* gene under different extracellular osmolarity conditions and its requirement for viability have been investigated.[100] Null mutants of *aqpZ* are viable, although the colonies are smaller than the wild-type strain. Interestingly, when *E. coli* is grown in media of high osmolarity, the synthesis of the aquaporin channels is repressed. This may help to protect the cell from hypo-osmotic stress in the event of a sharp decrease in the osmolarity of the external medium.

5. Mechanosensitive channels

For reviews of mechanosensitive channels, read refs. 99 and 101. As will be explained in Section 15.2, bacteria adjust the internal osmotic pressure so that it is always higher than the external osmotic pressure. This keeps water flowing into the cell via osmosis and maintains the high internal turgor pressure that is important for growth. The internal osmotic pressure is kept high by the accumulation of certain solutes such as K^+, glutamate, glutamine, proline, trehalose, and betaine.

One consequence of maintaining a high internal osmotic pressure is that a sudden decrease in the external osmotic pressure can endanger the cell by promoting a sudden increase in the influx of water, leading to overexpansion of the cell wall and subsequent lysis of the cell. This is sometimes referred to as *hypo-osmotic stress* or *hypo-osmotic shock*. The cells are protected from this form of destruction by mechanosensitive (MS) channels that open under conditions of hypo-osmotic stress and provide a means for internal solutes to rapidly exit the cell, thus lowering the internal osmotic pressure. (See note 102 for an explanation.) Mutants that do not have MS channels lyse when subjected to hypo-osmotic stress.

Mechanosensitive channels are present in most bacteria as well as archaea. *E. coli* has three such channels: a large channel called MscL, a small channel called MscS, and a "mini" channel called MscM (L refers to large, S to small, and M to mini).

Archaeal cell membranes

1. The lipids

Archaeal membrane lipids differ from those found in bacterial membranes.[103–105] The archaeal lipids consist of *isopranoid alcohols* (either 20 or 40 carbons long), *ether-linked* to one glycerol to form monoglycerol diethers or to two glycerols to form diglycerol tetraethers. These are illustrated in Fig. 1.18. Their synthesis is described later, in Section 9.1.3. (Recall that bacterial glycerides are fatty acids esterified to glycerol. Refer to Fig. 1.16 for a comparison.) The C_{20} alcohol is a fully saturated hydrocarbon called *phytanol*. The C_{40} molecules are two phytanols linked together head to head in the diglycerol tetraether lipids. Thus, the lipids are either phytanyl glycerol diethers or diphytanyl diglycerol tetraethers. The diethers and tetraethers occur in varying ratios depending upon the bacterium. For example, there may be from 5 to 25 different lipids in any one cell. This is really quite a diverse mix and can be contrasted to the lipid complement of a typical bacterium, which has only four or five different phospholipids. The diversity of the archaeal lipids is due to the different polar head groups that exist, as well as to the mix of core lipid to which the head groups are attached (Fig. 1.18). Although the polar head group is responsible for the polarity of most phospholipids, there is some polarity at one end of the archaeal lipids without a polar head group because of the free hydroxyl group on the glycerol. (Recall that hydroxyl groups are capable of forming hydrogen bonds with water and proteins.) It is usually stated that the ether linkages, which

are more stable to hydrolytic cleavage, are an advantage over ester linkages in the acidic and thermophilic environments in which archaea live. It is clear, however, that ether-linked lipids are not necessary for growth at high temperatures. A bacterium called *Thermotoga* does not have ether-linked lipids, yet grows in geothermally heated marine sediments alongside the sulfur-dependent thermophilic archaea. This finding emphasizes that the correlation between ether-linked lipids and habitat is not precise.

2. The proteins

There is little information regarding archaeal membrane proteins. It is known, however, that in bacteriorhodopsin and halorhodopsin in *Halobacterium*, the conformational array in the cell membrane is dependent upon interaction with polar membrane lipids.[104] The functions of these two proteins are discussed in Sections 3.8.4 and 3.9.

3. The membrane

The thermoacidophilic archaea and some methanogens have tetraether glycerolipids in the cell membrane. These lipids have a polar head group at both ends and span the membrane, forming a *lipid monolayer* (Fig. 1.19). This is the only known example of a membrane having no midplane region. Since there is no midplane region, the lipid monolayer is more resistant to levels of heat that would disrupt the hydrophobic bonds holding the two lipids in the lipid bilayer together. The increased resistance to heat of the lipid monolayer may confer an advantage to organisms living at high temperatures. However, it cannot be claimed that diether lipids or tetraether lipids are a *specific* adaptation to high temperatures, although they may be advantageous in these environments. This is because some mesophilic methanogens have tetraether lipids, whereas two extremely thermophilic archaea, *Methanopyrus kandleri* and *Thermococcus celer*, do not have tetraether lipids.

1.2.6 Cytoplasm

The cytoplasm is defined as everything enclosed by the cell membrane. Cytoplasm is a viscous material containing a heavy concentration of protein (100–300 mg/mL),[106] salts, and metabolites. In addition, there are large

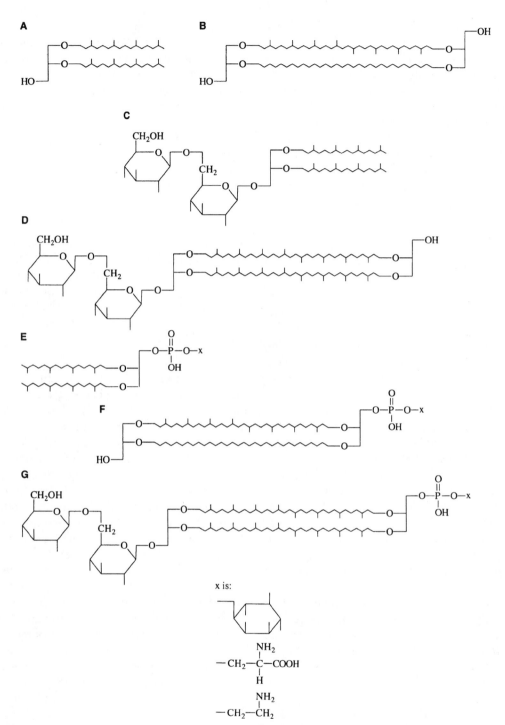

Fig. 1.18 Major lipids of *Methanobacterium thermoautotrophicum*: (A) glycerol diether (archaeol), (B) diglycerol tetraether (caldarchaeol), (C) a glycolipid (gentiobiosyl archaeol), (E) a phospholipid (archaetidyl–X), where X can be inositol, serine, or ethanoamine, (F) a phospholipid (caldarchaetidyl–X), (G) a phosphoglycolipid (gentiobiosyl caldarchatidyl–X). *Source*: Nishihara, M., H. Morii, and Y. Koga. 1989. Heptads of polar ether lipids of an archaebacterium, *Methanobacterium thermoautotrophicum*: structure and biosynthetic relationship. *Biochemistry* **28**:95–102.

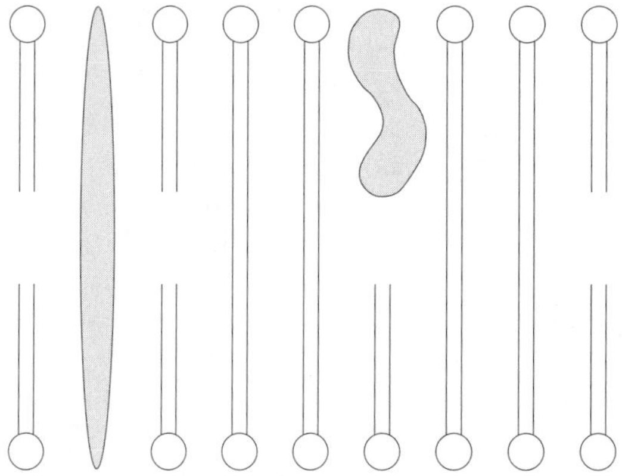

Fig. 1.19 Lipid layer with membrane proteins (shaded areas) in archaebacteria membranes. The glycerol diethers form a lipid bilayer, and the tetraethers form a monolayer. Some archaebacteria (e.g., the extreme halophiles) contain only the diethers. Most of the sulfur-dependent thermophiles have primarily the tetraethers, with only trace amounts of the diethers. Many methanogens have significant amounts of both the di- and tetraethers.

aggregates of protein complexes designed for specific metabolic functions, various inclusions, and highly condensed DNA. Intracytoplasmic membranes are also present in many prokaryotes. The soluble part of the cytoplasm is called the *cytosol*. We will begin with the intracytoplasmic membranes.

Intracytoplasmic membranes

Many prokaryotes have intracytoplasmic membranes that have specialized physiological functions.[107] Intracytoplasmic membranes are often connected to the cell membrane and are generally believed to be derived from invaginations of chemically modified areas of the cell membrane. Connections to the cell membrane are not always seen, however, and it is unknown whether the intracytoplasmic membranes are derived from an invagination of the cell membrane or are synthesized independently of the cell membrane (e.g., the thylakoids of cyanobacteria). A few prokaryotes with intracytoplasmic membranes and their physiological roles are listed.

1. Methanotrophs
Bacteria that grow on methane as their sole source of carbon (methanotrophs) possess intracytoplasmic membranes that are suggested to function in methane oxidation. Methane oxidation is discussed later, in Section 13.2.1.

2. Nitrogen fixers
Bacteria for which nitrogen gas serves as a source of nitrogen use an oxygen-sensitive enzyme called *nitrogenase* to reduce the nitrogen to ammonia, which is subsequently incorporated into cell material. Many of these organisms have extensive intracytoplasmic membranes. One such nitrogen-fixing bacterium is *Azotobacter vinelandii*, whose intracytoplasmic membranes increase with the degree of aeration of the culture. Since respiratory activity is localized in the membranes, it is probable that an important role for *Azotobacter* intracytoplasmic membranes is to increase the cellular respiratory activity, to provide more ATP for nitrogen fixation and to remove oxygen from the vicinity of the nitrogenase. Nitrogen fixation is discussed in Section 12.3.

3. Nitrifiers
Intracellular membranes are also found in nitrifying bacteria (i.e., bacteria that oxidize ammonia and nitrite as the sole source of electrons: *Nitrosomonas*, *Nitrobacter*, *Nitrococcus*). Several of the enzymes that catalyze ammonia and nitrite oxidation are in the membranes. This is discussed in Section 12.4.

4. Phototrophs
In bacteria that use light as a source of energy for growth (phototrophs), the intracytoplasmic membranes are the sites of the photosynthetic apparatus. The membrane structure varies: flat membranes, vesicles, flat sacs (thylakoids in cyanobacteria), and tubular invaginations of the cell membrane (photosynthetic bacteria). See the discussion of photosynthesis and photosynthetic membranes in Chapter 5, especially Fig. 5.16.

Inclusion bodies, multienzyme aggregates, and granules

Certain bacteria contain specialized organelles in the cytoplasm. Some researchers refer to these entities as inclusion bodies rather than organelles. These differ from eukaryotic organelles in not being surrounded by a lipid bilayer–protein membrane, although they do have a membrane or coat. In addition, there are numerous large aggregates and multienzyme complexes in all bacteria.

1. Gas vesicles

Aquatic bacteria such as cyanobacteria, certain photosynthetic bacteria, some nonphotosynthetic bacteria, and certain archaea have *gas vesicles* surrounded by a simple protein coat consisting primarily of ga<u>s</u> <u>v</u>esicle protein A (GsvA), a small hydrophobic protein that is highly conserved among the diverse groups of organisms. Gas vesicles are hollow, spindle-shaped structures about 100 nm long, filled with gas in equilibrium with the gases dissolved in the cytoplasm. The gas vesicles allow the organisms to float in lakes and ponds at depths that support growth because of favorable light, temperature, or nutrients. For example, the green sulfur bacterium *Pelodictyon phaeoclathratiforme* forms gas vesicles only at low light intensities.[108] Perhaps this allows the bacteria to float at depths where the light is optimal for photosynthesis. Many bacteria and cyanobacteria with gas vesicles are plentiful in stratified freshwater lakes, but they are not as abundant in isothermally mixed waters. Other prokaryotes containing gas vesicles (e.g., the halophilic archaeon *Halobacterium*) live in hypersaline waters, and a few marine species of cyanobacteria belonging to the genus *Trichodesmium* have gas vesicles. When gas vesicles are collapsed by experimentally subjecting cells to high hydrostatic pressure or turgor pressure, the cells are no longer buoyant and sink. Collapsed vesicles do not recover, and the cells acquire gas-filled vesicles only by *de novo* synthesis of new vesicles. Thus, during synthesis of the vesicles, water is somehow excluded, presumably because of the hydrophobic nature of the inner protein surface.

2. Carboxysomes

Bacteria that obligately grow on CO_2 as their sole or major source of carbon (strict autotrophs) sometimes have large (100 nm) polyhedral inclusions called *carboxysomes*.[109] These inclusions have been observed in nitrifying bacteria, sulfur oxidizers, and cyanobacteria. The distribution appears to be species specific (e.g., not all sulfur oxidizers have carboxysomes). Ribulose-1,5-bisphosphate carboxylase (RuBP carboxylase), the enzyme in the Calvin cycle that incorporates CO_2 into organic carbon, is stored in carboxysomes. The enzyme is discussed in Section 13.1.1. The physiological role

for carboxysomes is not clear, since many autotrophs do not have them.

3. Chlorosomes

Green sulfur photosynthetic bacteria (e.g., *Chlorobium*) have ellipsoid inclusions called *chlorosomes* (formerly called chlorobium vesicles) that lie immediately underneath the cytoplasmic membrane. The chlorosomes are surrounded by a nonunit membrane of galactolipid, with perhaps some protein. At one time it was believed that such vesicles were found only in the green sulfur photosynthetic bacteria. However, similar vesicles have been found in the green photosynthetic bacterium, *Chloroflexus*. The major light-harvesting photopigments are located in the chlorosomes, whereas the photosynthetic reaction centers are in the cell membrane. This means that during photosynthesis in these organisms light is absorbed by pigments in the chlorosomes and energy is transmitted to the reaction centers in the cell membrane, where photosynthesis takes place. In photosynthetic bacteria that do not have chlorosomes, the light-harvesting pigments surround the reaction centers in the cell membrane. The structure and function of chlorosomes are discussed in Section 5.6.

4. Magnetosomes

See ref. 110 for a review of magnetosomes in bacteria. Certain marine and freshwater bacteria have magnetic particles called magnetosomes within their cytoplasm. Most of these bacteria have not yet been cultured in the laboratory. The magnetic particles are a string of crystals of magnetite (Fe_3O_4), each one of which is surrounded by a specialized, complex membrane containing phospholipids and proteins. Magnetosomes should be thought of as a navigational device that orient the bacteria with the earth's magnetic field so that they swim in a particular direction, a behavior described as magnetotactic. It is clear that swimming is important for the occurrence of magnetotacticity, because dead cells are not pulled by a magnetic field. The earth's magnetic field is such that when the bacteria are swimming, the magnetosomes allow orientation in the magnetic field, guiding the microorganisms to swim downward in their natural aqueous habitat. All the magnetotactic bacteria are microaerophilic or anaerobic, and it is thought that magnetotaxis to lower levels is beneficial because there is less oxygen at greater depths. Magnetosomes have been been best studied in *Magnetospirillum* spp., which belong to the α-Proteobacteria group.

5. Granules and globules

Bacteria often contain cytoplasmic granules whose content varies with the bacterium. Many bacteria store a lipoidal substance called *poly-β-hydroxybutyric acid* (PHB) in the granules as a carbon and energy reserve. Other bacteria may store *glycogen* for the same purpose. Other granules found in bacterial cells can include *polyphosphate* and, in some sulfur-oxidizing bacteria, *elemental sulfur globules*.

6. Ribosomes

Ribosomes are the sites of protein synthesis. They are small ribonucleoprotein particles, approximately 22 nm by about 30 nm, or about the size of the smallest viruses. Ribosomes consist of over 50 different proteins and three different types of RNA (23S, 16S, and 5S). Bacterial ribosomes sediment in a centrifugal field at a characteristic velocity of 70S, as opposed to eukaryotic cytosolic ribosomes, which are 80S. Bacterial ribosomes are very similar regardless of the bacterium. However, there are some differences from archaeal ribosomes, which are also 70S. These differences were described in Section 1.1.1.

7. The nucleoid

The site of DNA and RNA synthesis is the nucleoid, an amorphous mass of DNA unbounded by a membrane, lying approximately in the center of the cell. Faster-growing bacteria may contain more than one nucleoid, but each nucleoid has but one chromosome, and all the chromosomes are identical. The DNA is very tightly coiled. Indeed, if the DNA from *E. coli* were stretched out, it would be 500 times longer than the cell!

DNA-binding proteins influence nucleoid structure and gene regulation. The DNA in the nucleoids is bound to several proteins as well as to nascent chains of RNA. Of course, among the proteins bound to the DNA are RNA polymerase molecules engaged in transcription, but several other proteins can be present as well.

When nucleoids are isolated from *E. coli* under low salt conditions (which preserves the attachment of proteins to the DNA), membrane proteins are present as well as the DNA-binding proteins HU, IHF, H-NS, and Fis. HU, often referred to as a histonelike protein because it has physical properties and an amino acid composition similar to eukaryotic histones, is the major DNA-binding protein present in *E. coli*.[111] It binds to DNA without any apparent sequence specificity, wrapping the DNA and causing it to bend. Integration host factor (IHF) binds to specific sequences and also bends DNA. The DNA-binding proteins, HU, H-NS, and Fis, are called histonelike because they resemble histone in their electrostatic charge, binding to DNA, and because they have a low molecular weight and a high copy number. However, unlike HU itself, the other histonelike proteins are not similar to histone in their amino acid composition or structure. For example, the histonelike protein H-NS binds to curved or bent DNA. The structure of H-NS, its role in DNA superstructure, and its role in gene regulation is reviewed in ref. 112.

It is believed that all the DNA-binding proteins facilitate bending of DNA and/or have a high affinity for bent DNA, or restraining of DNA supercoils in the nucleoid, and for the compact structure of the nucleoid.[113] The DNA-binding proteins are important not only for nucleoid structure but also for the regulation of expression of certain genes. See the discussion of the H-NS and Fis proteins in Section 2.2.2 (ribosomal RNA genes), the discussion of H-NS and IHF in Section 18.11.5 (virulence genes), and the discussion of IHF in Section 18.3 (nitrate regulation), Section 18.4 (Ntr regulon), and Section 18.7 (porin biosynthesis).

Another protein important for DNA structure is DNA gyrase (topoisomerase II), which is responsible for negative supercoiling, that is, the twisting of the double helix about its axis in the direction opposite to the right-handed double helix (Section 10.1). Cellular DNA is mostly in a negative supercoil and otherwise would be much more difficult to unwind to obtain single strands for replication and transcription (Section 10.1) There is also an enzyme, topoisomerase I, that removes negative supercoils. Supercoiling of DNA and topoisomerases are discussed in Chapter 10.

8. Multienzyme complexes

It should not be thought that the enzymes in the cytoplasm are a random mixture of proteins. There are many examples of enzymes in the same pathway forming stable multienzyme complexes, reflecting strong intermolecular bonding.[114] For example, *pyruvate dehydrogenase* from E. coli is a complex of three different enzymes, each present in multiple copies (50 proteins total), that oxidizes pyruvic acid to acetyl–CoA and CO_2 The size of the pyruvate dehydrogenase complex is $(4.6–4.8) \times 10^6$ Da. Contrast this with the size of a 70S ribosome (another multienzyme complex), which is about 2.7×10^6 Da. Other enzyme complexes that catalyze a consecutive series of biochemical reactions include the α-*ketoglutarate dehydrogenase* complex, which consists of three different enzymes present in multiple copies (i.e., 48 proteins), 2.5×10^6 Da (in E. coli), and oxidizes α-ketoglutarate to succinyl–coalnzyme A (acetyl–SCoA) and CO_2. Another example is *fatty acid synthase* in yeast. It consists of seven different enzymes (2.4×10^6 Da) and synthesizes fatty acids from acetyl–SCoA. (For comparison to bacteria, see note 115.) One of the advantages of a multienzyme complex is that it facilitates the channeling of metabolites, thus increasing the efficiency of catalysis. For example, in these stable enzyme complexes the biochemical intermediates in the pathway are transferred directly from one enzyme to the next without entering the bulk phase. Thus, there is no dilution of intermediates, nor is there reliance on random diffusion to reach a second enzyme.

Cytosol

The liquid portion of the cytoplasm, the cytosol, can be isolated in diluted form as the supernatant fraction obtained after broken cell extracts have been centrifuged at $105,000 \times g$ for 1 to 2 h, which should result in sedimentation of the membranes, the DNA, the ribosomes, very large protein aggregates, and other intracellular inclusions. In the cytosol are found the enzymes that catalyze a major portion of the biochemical reactions in the cell, such as the enzymes of the central pathways for the metabolism of carbohydrates (glycolysis, the pentose phosphate pathway, and the Entner–Doudoroff pathway) as well as the central pathways for organic acid metabolism (citric acid cycle and glyoxylate pathways), and enzymes for other pathways such as the biosynthesis and degradation of amino acids, lipids, and nucleotides.

The concentration of proteins in the cytosol is very high, making the material viscous, and it is expected that extensive protein–protein interactions occur among the enzymes, even those that do not exist in tight complexes. If such interactions exist, however, they must be weak, since most of the enzymes that catalyze metabolic pathways in the cytosol (e.g., the glycolytic enzymes that catalyze the degradation of glucose to pyruvic acid or lactic acid) cannot be isolated as complexes. It has generally been assumed that these enzymes exist either as independent proteins or in loose associations that are easily disrupted during cell breakage and accompanying dilution of the proteins. (Bear in mind that the concentration of proteins in broken cell extracts is orders of magnitude lower than in the aqueous portion of the unbroken cell because of the addition of buffers during the washing and suspension of the cells prior to breakage.)

1.2.7 Cytoskeleton

Eukaryotic cytoskeleton

To place the discussion of the prokaryotic cytoskeleton in context, we will begin with a brief review of the eukaryotic cytoskeleton. Eukaryotic cells have a network of protein fibers, collectively called a cytoskeleton, extending throughout the cytoplasm. Three types of protein filament are present in the eukaryotic cytoskeleton, and they are distinguished in part by their size. In increasing diameter, they are actin filaments (7 nm diameter), intermediate filaments (10–12 nm diameter), and microtubules (25 nm diameter). The filaments are made from different protein subunits: actin (actin filaments), tubulin (microtubules), vimentin, lamin (intermediate filaments), and so on. The eukaryotic cytoskeleton, along with motor proteins such as myosin, dynein, and kinesin, has many functions, briefly reviewed in note 116. It is responsible for cell shape, the movements of cilia and flagella, muscle contraction, endocytosis, and the movements of components such as vesicles from one part of the cell to another.

Prokaryotic cytoskeletal components similar to eukaryotic cytoskeletal components

Prokaryotes possess proteins that show some similarity to the eukaryotic cytoskeletal components actin, tubulin, and intermediate filament components, suggesting a prokaryotic origin to these cytoskeletal components. For reviews see refs. 117, 118, and 120.

Proteins

FtsZ. FtsZ is a cell division protein related to tubulin. It assembles as a ring at the site of cell division and recruits other proteins to form a contractile septal ring that constricts the cell during division.

MreB. A protein called MreB has sequence patterns similar to that of actin, is a member of the actin superfamily, and can be found in many rod-shaped, filamentous, and helical bacteria. It has been reported to form a cytoskeleton in some bacteria, encircling the cell as spirals under the cell membrane along the longitudinal axis, and it contributes toward determining the shapes of nonspherical bacteria.[119] Mutations in the gene encoding MreB causes *E. coli* to lose its rod shape and become spherical. Mutants of rod-shaped *B. subtilis* cause the cells to round up and lyse. Examination of genome sequences indicates that *mreB* is not present in coccoid bacteria.

Crescentin. In addition to MreB, the loss of which leads to a loss of the rod shape, resulting in lemon-shaped cells, *Caulobacter crescentus* has a cytoskeleton protein called crescentin, which is responsible for its vibrioid shape.[120–122] Crescentin has certain characteristics similar to intermediate filaments. The screening of transposon–insertion libraries revealed mutants that grow as rod-shaped bacteria rather than vibrios. This led to the discovery of crescentin. Immunofluorescence microscopy revealed that crescentin exists as a helical filament along the concave side of the cell membrane.

Cytoskeleton and cell shape

All the preceding statements beg the question, what is the relationship between the recently discovered cytoskeleton components in prokaryotes, the peptidoglycan, and cell shape? It is clear that both the peptidoglycan and the cytoskeleton components are important for the particular shape of the bacterium. (For a discussion of peptidoglycan, see Section 1.2.3 and Fig. 1.8.) For example, if the peptidoglycan of rod-shaped bacteria such as *E. coli* is experimentally destroyed by lysozyme, the cell rounds up and assumes the shape of a sphere. (The experiment is performed in an isotonic medium to prevent cell lysis.) These and other experiments leave no doubt that the peptidoglycan is responsible for maintaining cell shape.

As described earlier, however, mutations that lead to a loss in MreB or crescentin can also alter cell shape. One possibility is that when peptidoglycan is synthesized, its conformational arrangement (perhaps a helical pattern running lengthwise along the cell), and therefore the shape of the cell, is dictated by cytoskeleton proteins lying against the inner surface of the cell membrane.[122]

Cytoskeletal components unrelated to eukaryotic cytoskeletal components are present in Mollicutes

Mollicutes

The Mollicutes (*Spiroplasma*, *Mycoplasma*, *Acholeplasma*) are a class of very small, wall-less bacteria. They may be free-living, but many are parasitic on animals (including humans), and plants, and can cause disease. Mollicutes can be found as part of the natural flora in the mouth, throat, and genitorurinary tract of mammals and birds. (See note 123 for more information about diseases caused by Mollicutes.) Several strains of *Mycoplasma* can glide, and *Spiroplasma* can swim. These bacteria are also chemotactic. However, they do not have analogues to genes responsible for chemotaxis or in other bacteria. All Mollicutes stain gram-negative; however, 16S rRNA analysis indicates that they are closely related to gram-positive bacteria, most notably *Clostridium*, from which they are believed to have evolved.

Cytoskeleton

Mollicutes have an internal protein cytoskeleton that determines the shape of the cell and in some cases (e.g., *Spiroplasma*) can function in motility. They do not have homologues to any of the eukaryotic genes for cytoskeletal proteins or motor proteins, such as actin, microtubule proteins, myosin, or dynein. Thus, their cytoskeleton is made from protein unrelated to cytoskeletal proteins found in eukaryotes.

Spiroplasma

Spiroplasma cells are very thin, round, helical, tubelike cells. The cell diameter is approximately that of a bundle of 10 to 12 flagella. Swimming is by means of helical contractions and extensions of the cell, and rotation about its helical axis. For a discussion of this type of motility, including the forces and energetics, the student is referred to refs. 124, 125, and 132. Movement is more rapid in viscous media.

What causes movements in *Spiroplasma*? Inside each cell is a single flat, cytoskeletal ribbon made of seven paired parallel protein fibers, attached to the inner surface of the cell membrane along its innermost (and shortest) helical side. The ribbon is continuous and spans the entire length of the cell. Nonmotile mutants do not have the cytoskeletal ribbon. The cytoskeletal ribbon with its associated proteins is a linear motor, as opposed to flagella, which have rotary motors. What this means is that the cytoskeletal ribbon undergoes linear contractions and extensions, and because it is attached to the cell membrane, the cell moves forward in the fluid. It has been proposed that conformational changes in the monomeric cytoskeletal proteins are what produces the length changes in the ribbon.

Mycoplasma

Mycoplasma cells are very small, somewhat spherical cells. One end is extended as a narrow neck, giving the cell a flasklike appearance. Many are nonmotile. However, some are capable of movement on a solid surface. *Mycoplasma* has an internal cytoskeleton composed of protein fibers that originate at the tip of the "neck" region. These fibers are present in both motile and nonmotile *Mycoplasma*. *Mycoplasma* adheres to host tissues at the tip of the "neck" region, and it has been suggested that the cytoskeletal fibers play a role in adhesion. How the fibers might be related to motility is not clear.

1.3 Summary

There are two evolutionary lines of prokaryotes, the Bacteria and the Archaea. Archaea are similar to bacteria, but they differ in certain fundamental aspects of structure and biochemistry. These include differences in ribosomes, cell wall chemistry, membrane lipids, and coenzymes. In addition, certain archaea have metabolic pathways (e.g., methanogenesis) not found in bacteria.

Surrounding most bacteria are pili and extracellular material called a glycocalyx. The pili are protein fibrils. The glycocalyx encompasses all extracellular polymers beyond the cell wall, polysaccharide or protein, including capsules and slime layers. The pili and glycocalyx anchor the cell to specific animal and plant cell surfaces, as well as to inanimate objects, and in several cases to each other.

Some bacteria possess a fibril called a sex pilus that attaches the cell to a mating partner and retracts to draw the cells into intimate contact. When the cells are in contact, DNA is transferred unilaterally from the donor cell (i.e., the one with the sex pilus) to the recipient.

Protruding from many bacteria are flagella filaments that aid the bacterium in swimming. At the base of the flagellum is a basal body that is embedded in the cell membrane. At the base of the flagellum is a rotary motor that in most bacteria runs on a current of protons. Extreme halophiles and some marine bacteria use a sodium ion current instead of a proton current. Some bacteria have flagella covered by a membranous sheath (e.g., *Vibrio cholerae*, *V. parahaemolyticus*, *Helicobacter*, *Bdellovibrio bacteriovorus*). There is a proteinaceous sheath covering the flagella (axial filaments) of most spirochaetes (an exception is *Borrelia burgdorferi*, the causative agent of Lyme disease) which is actually located in the periplasm rather than outside of the cell. Spirochaetes are shaped like a coil, and when the flagellum turns, the cell moves in a corkscrewlike fashion, which apparently is suited for movement through high-viscosity media. (Again, *B. burgdorferi* is an exception and does not move in a corkscrewlike fashion.[126]) Bacterial flagella are important for the pathogenicity of certain bacteria and this is discussed in the review by Moens and Vanderleyden.[35] Many genera of bacteria contain species that swim as interacting populations of cells on moist solid surfaces such as dilute agar, and other surfaces where biofilms form. The movement is called swarming and is due to lateral flagella.

There exist several other mechanisms of besides flagella-driven motility. These include "twitching," a form of gliding motility powered by type IV pili, gliding motility driven by extracellular slime secretion through pores in the cell surface, and swimming in certain Mollicutes driven by a cytoskeletal linear motor.

There are two types of bacterial cell wall. One kind has a thick peptidoglycan layer to which there are covalently bonded polysaccharides, teichoic acids, and teichuronic acids. Bacteria with such walls stain gram-positive. A second type of wall has a thin peptidoglycan layer and an outer envelope consisting of lipopolysaccharide, phospholipid, and protein. Cells with such walls do not stain gram-positive and are called gram-negative. There is also a separate compartment in gram-negative bacteria called a periplasm, between the outer envelope and the cell membrane. The periplasm is the location of numerous proteins and enzymes, including proteins required for solute transport and enzymes that function in the oxidation and degradation of nutrients in the periplasm. Archaea have very different cell walls. No archaeon has peptidoglycan, although some have a similar polymer called pseudopeptidoglycan or pseudomurein.

The cell membranes of the bacteria are all similar. They consist primarily of phosphoglycerides in a bilayer and protein. The phosphoglycerides are generally fatty acids esterified to glycerol phosphate. Archaeal cell membranes have lipids that are long-chain alcohols, etherlinked to glycerol. Archaeal membranes can be a bilayer or a monolayer, or perhaps a mixture of the two.

The cytoplasm of prokaryotes is a viscous solution of protein. Salts, sugars, amino acids, and other metabolites are dissolved in the proteinaceous cytoplasm. In many bacteria intracytoplasmic membranes are present. In many cases these membranes are invaginations of the cell membrane. They are sites for electron transport and specialized biochemical activities. Numerous inclusion bodies are also present. These include ribosomes, which are the sites of protein synthesis, large aggregates of enzymes that catalyze short metabolic pathways, gas vesicles, and carbon and energy reserves (e.g., glycogen particles). Some bacteria have magnetic particles called magnetosomes in the cytoplasm. Also present in the cytoplasm is DNA, very tightly packed, and referred to as a nucleoid. There is evidence that the cytoplasm has a protein cytoskeleton that plays a role in maintaining cell shape, as well as serving other functions. It is clear that although the cytoplasm of prokaryotes is not compartmentalized into membrane-bound organelles like the eukaryotic cytoplasm, it nonetheless represents a very complex pattern.

The central metabolic pathways that are described in the ensuing chapters (viz., glycolysis, the Entner–Doudoroff pathway, the pentose phosphate pathway, and the citric acid cycle) all take place in the liquid part of the cytoplasm called the cytosol. Also found in the cytosol are most of the other pathways, including the enzymatic reactions for the synthesis and degradation of amino acids, fatty acids, purines, and pyrimidines.

The cell membrane is the site of numerous other metabolic pathways, including phospholipid biosynthesis, protein secretion, solute transport, electron transport, cell wall biosynthesis, the generation of electrochemical ion gradients, and ATP synthesis. The prokaryotic cell can therefore be considered to have three major metabolic domains: cytosol, particulate (ribosomes, enzyme aggregates, etc.), and membrane.

Study Questions

1. What chemical and structural differences distinguish the Archaea from the Bacteria?

2. What are the differences between the gram-positive and gram-negative cell walls?

3. Which cell wall polymers use D-alanine (as opposed to L-alanine) as part of their structure? (*Hint*: Polymers of one type are in both gram-positive and gram-negative walls; those of the other are present only in gram-positive walls.)

4. What structures enable bacteria to adhere to surfaces? What is known about chemistry,

location, and receptors of the molecules that mediate the adhesion?

5. Contrast the functions and cellular location of peptidoglycan, phospholipid, and lipopolysaccharide. What is it about the chemical structure of these three classes of compounds that is suitable for their functions and/or cellular location?

6. What are porins, and what do they do? Are they necessary for gram-positive bacteria? Explain.

7. What are the physiological and enzymatic functions associated with the periplasm?

8. The protein TonB is an energy transducer in gram-negative bacteria that transfers energy from the cell membrane to specific transporters in the outer membrane. Speculate about how this might occur. The subject is discussed in ref. 95.

9. Summarize the cell compartmentalization of metabolic activities in prokaryotes.

10. What are some multienzyme complexes? What advantage do they confer?

11. Which metabolic pathways or activities are found in the cytosol and which in the membranes?

12. What distinguishes twitching from swimming?

13. What is the evidence that bacteria possess a cytoskeleton? Is there more than one type of cytoskeleton? Explain.

14. What is the experimental baseis for the conclusion that MotA and MotB form a complex? What is the experimental basis for the conclusion that MotA conducts protons? Why has it been concluded that FliG is the rotor portion of the flagellar motor? Why are the MS and C rings sometimes included as part of the rotor?

REFERENCES AND NOTES

1. Woese, C. R., O. Kandler, and M. L. Wheelis. 1990. Towards a natural system of organisms: proposal for the domains Archaea, Bacteria, and Eucarya. *Proc. Natl. Acad. Sci. USA* 87:4576–4579.

2. Woese, C. R. 1987. Bacterial evolution. *Microbiol. Rev.* **51**:221–271.

3. Pace, N. R., D. A. Stahl, D. J. Lane, and G. J. Olsen. 1985. Analyzing natural microbial populations by rRNA sequences. *ASM News* **51**:4–12.

4. Stetter, K. O. 1989. Extremely thermophilic chemolithoautotrophic archaebacteria, pp. 167–176. In: *Autotrophic Bacteria*. H. G. Schlegel and B. Bowien (Eds.). Springer-Verlag, Berlin.

5. Schleper, C., G. Puehler, I. Holtz, A. Gambacorta, D. Janekovic, U. Santarius, H.-P., Klenk, and W. Zillig. 1995. *Picrophilus* gen. nov., fam. nov.: a novel aerobic, heterotrophic, thermoacidophilic genus and family compromising archaea capable of growth around pH 0. *J. Bacteriol.* **177**:7050–7059.

6. Solfataras, which are acidic, geothermally heated habitats containing inorganic sulfur compounds, including sulfuric acid, may be volcanic fissures, springs, basins, or dried soil that previously contained solfataric water. They have an acidic pH in the aerobic portions because sulfuric acid arises from the oxidation of hydrogen sulfide, which occurs spontaneously and also as a result of the growth of sulfur-oxidizing bacteria.

7. Koga, Y., M. Nishihara, H. Morii, and M. Akagawa-Matsushita. 1993. Ether polar lipids of methanogenic bacteria: structures, comparative aspects, and biosynthesis. *Microbiol. Rev.* **57**:164–182.

8. Grayling, R. A., K. Sandman, and J. N. Reeve. 1996. Histones and chromatin structure in hyperthermophilic archaea. *FEMS Microbiol. Rev.* **18**:203–213.

9. Pereira, S. L., R. A. Grayling, R. Lurz, and J. N. Reeve. 1997. Archaeal nucleosomes. *Proc. Natl. Acad. Sci. USA* **94**:12633–12637.

10. Bardy, S. L., S. Y. M. Ng, and K. F. Jarrell. 2003. Prokaryotic motility structures. *Microbiology* **149**:295–304.

11. Macnab, R. M. 2003. How bacteria assemble flagella. *Annu. Rev. Microbiol.* **57**:77–100.

12. Eukaryotic flagella differ from bacterial flagella in many ways. Rather than being stiff, rotating filaments, they are flexible and have an undulating wave motion from base to tip that pushes the cell forward. The movement is driven by ATP rather than by a proton current; the structure is very different from that of prokaryotic flagella. The core of the filament is called an axoneme. The axoneme, which has an internal system of rods called microtubules, composed of tubulin subunits and associated proteins, is covered by a sheath that is an extension of the cell membrane. Each axoneme has nine fused pairs of microtubules that are attached to each other as a bundle, surrounding two central microtubules, all of which extend the length of the filament. This is called the 9 + 2 arrangement. Numerous proteins

project along the length of the microtubule doublets. Some of these are cross-links that attach the doublets to each other. Others, called dynein, cause the bending. Dynein hydrolyzes ATP as a source of energy and moves along the length of one microtubule doublet, pulling the attached doublet so that it bends. Other attached proteins constitute a relay system that controls the bending so that the appropriate waveform is produced.

Cilia are constructed in a similar fashion, and the molecular basis for their movement is the same, although their movement is more whiplike than the movement of flagella.

13. Iino, T., Y. Komeda, K. Kutsukake, R. M. Macnab, P. Matsumura, J. S. Parkinson, M. I. Simon, and S. Yamaguchi. 1998. New unified nomenclature for the flagellar genes of *Escherichia coli* and *Salmonella typhimurium*. *Microbiol. Rev.* 52:533–535.

14. Schoenhals, G. J., and R. M. Macnab. 1996. Physiological and biochemical analyses of FlgH, a lipoprotein forming the outer membrane L ring of the flagellar basal body of *Salmonella typhimurium*. *J. Bacteriol.* 178:4200–4207.

15. Berg, H. C. 2003. The rotary motor of the bacterial flagella. *Annu. Rev. Biochem.* 72:19–54.

16. Khan, S., M. Dapice, and T. S. Reese. 1988. Effects of *mot* gene expression on the structure of the flagellar motor. *J. Mol. Biol.* 202:575–584.

17. Ivey, D. M., M. Ito, R. Gilmour, J. Zemsky, A. A. Guffanti, M. G. Sturr, D. B. Hicks, and T. A. Krulwich. 1998. Alkaliphile bioenergetics, pp. 181–210. In: *Extremophiles: Microbial Life in Extreme Environments*. K. Horikoshi and W. D. Grant (Eds.). John Wiley & Sons, New York.

18. Suppressor mutations can indicate whether proteins interact in a complex. With respect to MotA and MotB, suppressor mutations in *E. coli* were isolated that suppressed *motB* missense mutations, that would have resulted in resulting paralyzed or partially paralyzed flagella. DNA sequencing of the suppressor mutants showed that they caused single amino acid changes in MotA. The conclusion is that MotA and MotB interact as a complex. Electron micrographic evidence also indicates that MotA and MotB form a complex. Electron micrographs show 10 to 12 particles surrounding the M ring in the cytoplasmic membrane. The particles are not present when either MotA or MotB is absent in mutants. For more information, read Garza, A. G., L. W. Harris-Haller, R. A. Stoebner, and M. D. Manson. 1995. Protein interactions in the bacterial flagellar motor. *Proc. Natl. Acad. Sci. USA* 92:1970–1974.

19. Blair, D. F., and H. C. Berg. 1990. The MotA protein of *E. coli* is a proton-conducting component of the flagellar motor. *Cell* 60:439–449.

20. Garza, A. G., L. W. Harris-Haller, R. A. Stoebner, and M. D. Manson. 1995. Motility protein interactions in the bacterial flagellar motor. *Proc. Natl. Acad. Sci. USA* 92:1970–1974.

21. Kojima, S., and D. F. Blair. 2001. Conformational change in the stator of the bacterial flagellar motor. *Biochemistry* 40:13041–13050.

22. Some researchers refer to the FliG proteins, which are bound to the peripheral inner surface of the MS ring, as well as to the C ring, as the rotor. Some researchers refer to the C ring as the rotor, and some researchers refer to both the MS and C rings as the rotor. See Bardy, S. L., S. Y. M. Ng, and K. F. Jarrell. 2003. Prokaryotic motility structures. *Microbiology* 149:295–304. Berg, H. C. 2003. The rotary motor of the bacterial flagella. *Annu. Rev. Biochem.* 72:19–54. Grünenfelder, B., Gehrig, S., and U. Jenal. 2003. Role of the cytoplasmic C terminus of the FliF motor protein in flagellar assembly and rotation. *J. Bacteriol.* 185:1624–1633.

23. Francis, N. R., G. E. Sosinsky, D. Thomas, and D. J. DeRosier. 1994. Isolation, characterization and structure of bacterial flagellar motors containing the switch complex. *J. Mol. Biol.* 235:1261–1270.

24. FliG, FliM, and FliN are thought to be involved in the turning of the flagellar motor (generating torque), because certain mutations result in paralysis. However, the evidence suggests that only FliG is directly involved in generating torque. It has been concluded that FliG, FliM, and FliN interact with one another in a complex because certain mutations in *fliG, fliM, and fliN* can be suppressed by other mutations in any of the *fli* genes. CheY-P, the molecule that causes the flagellar motor to reverse its direction of rotation, binds to FliM (Section 18.12). FliG, FliM, and FliN are probably also involved in flagellar assembly, since null mutants do not have flagella. This may be because the switch complex must be made prior to more distal portions of the flagellum.

25. Pleier, E., and R. Schmitt. 1989. Identification and sequence analysis of two related flagellin genes in *Rhizobium meliloti*. *J. Bacteriol.* 171:1467–1475.

26. Yonekura, K., S. Maki, D. G. Morgan, D. J. DeRosier, F. Vonderviszt, K. Imada, and K. Namba. 2000. The bacterial flagellar cap as a rotary promoter of flagellin self-assembly. *Science* 290: 2148–2152.

27. Hughes, K. T., and P. D. Aldridge. 2001. Putting a lid on it. *Nat. Struct. Biol.* 8:96–97.

28. Iino, T. 1969. Polarity of flagellar growth in *Salmonella*. *J. Gen. Microbiol.* 56:227–239.

29. Emerson, S. U., K. Tokuyasu, and M. I. Simon. 1970. Bacterial flagella: polarity of elongation. *Science* 169:190–192.

30. It has been proposed that ATP hydrolysis by FliI may drive the export of proteins through the translocation apparatus, as well as being necessary in the process of bringing proteins to the translocation site.

FliH binds to FliI and is a negative regulator of its activity so that the ATPase activity functions only when necessary. See Minamino, T., and R. M. Macnab. 2000. Interactions among components of the *Salmonella* flagellar export apparatus and its substrates. *Mol. Microbiol.* 35:1052–1064. Minamino, T., and R. M. Macnab. 2000. FliH, a soluble component of the type III flagellar export apparatus of *Salmonella*, forms a complex with FliI and inhibits its ATPase activity. *Mol. Microbiol.* 37:1494–1503.

31. Fuerst, J. A., and J. W. Perry. 1988. Demonstration of lipopolysaccharide on sheathed flagella of *Vibrio cholerae* 0:1 by protein A–gold immunoelectron microscopy. *J. Bacteriol.* 170:1488–1494.

32. Weissborn, A., H. M. Steinman, and L. Shapiro. 1982. Characterization of the proteins of the *Caulobacter crescentus* flagellar filament. *J. Biol. Chem.* 257:2066–2074.

33. Pleier, E., and R. Schmitt. 1989. Identification and sequence analysis of two related flagellin genes in *Rhizobium meliloti*. *J. Bacteriol.* 171:1467–1475.

34. Driks, A., R. Bryan, L. Shapiro, and D. J. DeRosier. 1989. The organization of the *Caulobacter crescentus* flagellar filament. *J. Mol. Biol.* 206:627–636.

35. Moens, S., and J. Vanderleyden. 1996. Functions of bacterial flagella. *Crit. Rev. Microbiol.* 22:67–100.

36. Charon, N. W., and S. F. Goldstein. 2002. Genetics of motility and chemotaxis of a fascinating group of bacteria: the spirochaetes. *Annu. Rev. Gen.* 36:47–73.

37. Some spirochaetes cause disease. Diseases caused by spirochaetes include syphilis, caused by *Treponema pallidum*; Lyme disease, caused by *Borrelia burgdorferi*; relapsing fever, caused by *Borrelia* spp.; leptospirosis, caused by *Leptospira* spp., and periodontal disease, caused by *Treponema denticola*. In addition to living in animals, spirochaetes live in soil, fresh water, and salt water, and attached to protozoa in the termite gut. The spirochaetes attached to the protozoa in the termite gut enable the protozoa to swim.

38. Goldstein, S. F., N. W. Charon, and J. A. Kreiling. 1994. *Borrelia burgdorferi* swims with a planar waveform similar to that of eukaryotic flagella. *Proc. Natl. Acad. Sci. USA* 91:3433–3437.

39. One model for spirochaete motility is based upon the finding that the periplasmic flagella influence the shape of the cell. In other words, the periplasmic flagella act as a periplasmic skeleton. As an example, consider *Borrelia burgdorferi*. *B. burgdorferi* has 7 to 11 flagella attached to each end of the cell, and the flagella overlap at the cell center to form a continuous bundle. The cells themselves have a flat-wave shape, rather than being "corkscrew" helical coils. Mutant cells that lack flagella owing to an insertion in *flaB*, which is the gene that encodes the major flagellar filament protein, no longer resemble flat wave but are rod shaped. Thus, the flagella influence the shape of the cell. Bearing this in mind, recall that the flagellum rotates between the outer membrane sheath and the flexible protoplasmic cell cylinder. One model is based on the configuration of the flagella: side by side with the flexible protoplasmic cylinder and held there owing to containment by the outer membrane sheath. The model proposes the following: (1) the flagella have a left-handed helical shape (going away from the observer) wrapped around the protoplasmic cylinder; (2) the flagella rotate counterclockwise as viewed from the back of the cell; (3) as a result of flagella rotation, backward-moving waves are propagated down the length of the flexible protoplasmic cylinder, and the cell is pushed forward in the viscous medium. For a more detailed discussion of the model, see ref. 36.

40. Parales, J., Jr., and E. P. Greenberg. 1991. N-Terminal amino acid sequences and amino acid compositions of the *Spirochaeta aurantia* flagellar filament polypeptides. *J. Bacteriol.* 173:1357–1359.

41. Moens, S., and J. Vanderleyden. 1996. Functions of bacterial flagella. *Crit. Rev. Microbiol.* 22:67–100.

42. Harshey, R. M. 2003. Bacterial motility on a surface: many ways to a common goal. *Annu. Rev. Microbiol.* 57:249–273.

43. Belas, R. 1997. *Proteus mirabilis* and other swarming bacteria, pp. 183–219. In: *Bacteria as Multicellular Organisms*. J. A. Shapiro, and M. Dworkin (Eds.). Oxford University Press, New York.

44. Matsuyama, T., K. Kaneda, Y. Nakagawa, K. Isa, H. Hara-Hotta, and I. Yano. 1992. A novel extracellular cyclic lipopeptide which promotes flagellum-dependent and -independent spreading growth of *Serratia marcescens*. *J. Bacteriol.* 174:1769–1776.

45. Kearns, D. B., and R. Losick. 2003. Swarming in undomesticated *Bacillus subtilis*. *Mol. Microbiol.* 49:581–590.

46. McCarter, L. M. 2005. Multiple modes of motility: a second flagellar system in *Escherichia coli*. *J. Bacteriol.* 187:1207–1209.

47. Jarrell, K. F., D. P. Bayley, and A. S. Kostyukova. 1996. The archaeal flagellum: a unique structure. *J. Bacteriol.* 178:5057–5064.

48. Faguy, D. M., K. F. Jarrell, J. Kuzio, and M. L. Kalmokoff. 1994. Molecular analysis of archaeal flagellins: similarity to the type IV *n*-transport superfamily widespread in bacteria. *Can. J. Microbiol.* 40:67–71.

49. Faguy, D. M., S. F. Koval, and K. F. Jarrell. 1994. Physical characterization of the flagella and flagellins from *Methanospirillum hungatei*. *J. Bacteriol.* 176:7491–7498.

50. Pearce, W. A., and T. M. Buchanan. 1980. Structure and cell membrane-binding properties of bacterial fimbriae, pp. 289–344. In: *Bacterial Adherence*. E. H. Beachey (Ed.). Chapman & Hall, London.

51. Ottow, J. C. G. 1975. Ecology, physiology, and genetics of fimbriae and pili. *Ann. Rev. Microbiol.* 29:79–108.

52. Jones, G. W., and R. E. Isaacson. 1983. Proteinaceous bacterial adhesins and their receptors. *Crit. Rev. Microbiol.* 10:229–260.

53. Hultgren, S. J., S. Abraham, M. Caparon, P. Falk, J. W. St. Geme III, and S. Normark. 1993. Pilus and nonpilus bacterial adhesins: assembly and function in cell recognition. *Cell* 73:887–901.

54. Yanagawa, R., and K. Otsuki. 1970. Some properties of *Corynebacterium renale*. *J. Bacteriol.* 101:1063–1069.

55. Cisar, J. O., and A. E. Vatter. 1979. Surface fibrils (fimbriae) of *Actinomyces viscosus* T14V. *Infect. Immun.* 24:523–531.

56. Eisenstein, B. I. 1987. Fimbriae, pp. 84–90. In: *Escherichia coli and Salmonella typhimurium: Cellular and Molecular Biology*. Vol. 1. Neidhardt, F. C., et al. (Eds.). ASM Press, Washington, DC.

57. Hultgren, S. J., S. Normark, and S. N. Abraham. 1991. Chaperone-assisted assembly and molecular architecture of adhesive. *Annu. Rev. Microbiol.* 45:383–415.

58. Hultgren, S. J., S. Abraham, M. Caparon, P. Falk, J. W. St. Geme III, and S. Normark. 1993. Pilus and nonpilus bacterial adhesins: assembly and function in cell recognition. *Cell* 73:887–901.

59. Type IV pili are also used by certain bacteria, such as *Pseudomonas*, to form biofilms, and by myxobacteria for gliding motility. This is discussed in Chapter 18.

60. Pathogenic strains of *Escherichia coli* cause gastrointestinal and urinary tract infections, *Neisseria gonorrhoeae* causes gonorrhoeae, *Bacteroides nodosus* causes bovine foot rot, *Moraxella bovis* causes bovine keratoconjunctivitis, *Vibrio cholerae* causes cholera, and *Pseudomonas aeruginosa* causes infections of the urinary tract and wounds. The adherence of these bacteria to their host tissue is the first stage in pathogenesis.

61. In addition to the bacterial chromosome, most bacteria carry smaller extrachromosomal DNA molecules called *plasmids*. There are many types of plasmid. They all carry genes for self-replication. Conjugative plasmids also carry genes for DNA transfer into recipient cells (transmissible plasmids). Other genes that may be carried by plasmids include genes for antibiotic resistance (carried by resistance transfer factors or R-factors) and genes for the catabolism of certain nutrients. For example, *Pseudomonas* carries plasmids for the degradation of camphor, octane, salicylate, and napththalene. Some plasmids confer virulence because the plasmids carry genes for toxins or other virulence factors, such as pili. The F plasmid in *E. coli* carries genes for self-transmission, including genes for the F pilus, and is also capable of integrating into the chromosome and promoting the transfer of chromosomal DNA.

62. Dunny, G. M. 1991. Mating interactions in gram-positive bacteria. pp. 9–33. In: *Microbial Cell Interactions*. M. Dworkin (Ed.). ASM Press, Washington, DC.

63. An excellent reference for microbial cell–cell interactions is the book edited by Dworkin: *Microbial Cell–Cell Interactions*, 1991. M. Dworkin (Ed.). American Society for Microbiology, Washington, DC.

64. Costerton, J. W., T. J. Marrie, and K.-J. Cheng. 1985. Phenomena of bacterial adhesion, pp. 3–43. In: *Bacterial Adhesion: Mechanisms and Physiological Significance*. D. C. Savage, and M. Fletcher (Eds.). Plenum Press, New York.

65. Sleytr, U. B., and P. Messner. 1983. Crystalline surface layers on bacteria. *Annu. Rev. Microbiol.* 37:311–339.

66. Costerton, J. W., R. T. Irvin, and K.-J. Cheng. 1981. The bacterial glycocalyx in nature and disease. *Annu. Rev. Microbiol.* 35:299–324.

67. Roberts, I. S. 1995. Bacterial polysaccharides in sickness and in health. *Microbiology* 141:2023–2031.

68. Bacterial capsules. 1990. *Current Topics in Microbiology and Immunology*. K. Jann and B. Jann (Eds.). Springer-Verlag, Berlin.

69. *Streptococcus pyogenes* (also called group A streptococcus) is a human pathogen that causes a wide range of diseases such as pharyngitis, impetigo, erysipelas (acute infection of the skin), cellulitis (infection of subcutaneous tissues), and necrotizing fasciitis (the strain that causes necrotizing fasciitis is called "flesh-eating bacteria", it also causes subcutaneous infection resulting in muscle destruction and organ failure) and toxic shock syndrome (a drastic decrease in blood pressure, i.e., shock, followed by organ failure). *S. pyogenes* produces several virulence factors, among which is M protein. M protein is a cell surface protein that is released from the bacteria via a proteolytic enzyme secreted by *S. pyogenes*. M protein binds a plasma protein called fibrinogen, and the complex activates neutrophils (polymorphonuclear leukocytes, or PMNs). The activated PMNs release heparin-binding protein, which causes vascular leakage resulting in the release of plasma and blood cells from the blood vessels, resulting in shock that contributes to subsequent organ failure. For more information, read: Herwald, H., H. Cramer, M. Mörgelin, W. Russel, U. Sollenberg, A. Norrby-Teglund, H. Flodgaard, L. Lindbom, and L. Björck. 2004. M protein, a classical bacterial virulence determinant, forms

complexes with fibrinogen that induce vascular leakage. *Cell* **116**:367–379.

70. Roberts, I. S. 1996. The biochemistry and genetics of capsular polysaccharide production in bacteria. *Annu. Rev. Microbiol.* **50**:285–315.

71. Kolenbrander, P. E., N. Ganeshkumar, F. J. Cassels, and C. V. Hughes. 1993. Coaggregation: specific adherence among human oral plaque bacteria. *FASEB J.* **7**:406–409.

72. Nikaido, H., and M. Vaara. 1987. Outer membrane, pp. 7–22. In: *Escherichia coli and Salmonella typhimurium: Cellular and Molecular Biology*, Vol. 1. F. C. Neidhardt et al. (Eds.). ASM Press, Washington, DC.

73. Weidel, W., and H. Pelzer. 1964. Bagshaped macromolecules—A new outlook on bacterial cell walls. *Adv. Enzymol.* **26**:193–232.

74. Dmitriev, B. A., F. V. Toukach, K.-J. Schaper, O. Holst, E. T. Rietschel, and S. Ehlers. 2003. Tertiary structure of bacterial murein: the scaffold model. *J. Bacteriol.* **185**:3458–3468.

75. The amino acid in position three in the peptidoglycan peptide subunit is referred to as L-R$_3$. It varies with the species. Gram-negative bacteria usually have *meso*-diaminopimelic acid (*meso*–DAP), or in the case of some spirochaetes, L-ornithine. There is much more variability in L-positive bacteria, and L-R$_3$ can be L-alanine, L-homoserine, L-diaminobutyric acid, L-glutamic acid, L-ornithine, L-lysine, or DAP.

76. Koch, A. L. 1983. The surface stress theory of microbial morphogenesis. *Adv. Microb. Physiol.* **24**:301–366.

77. Koch, A. L., and S. Woeste. 1992. Elasticity of the sacculus of *Escherichia coli*. *J. Bacteriol.* **174**:4811–4819.

78. Kellenberger, E. 1990. The "Bayer bridges" confronted with results from improved electron microscopy methods. *Mol. Microbiol.* **4**:497–705.

79. Neuhaus, F. C., and J. Baddiley. 2003. A continuum of anionic charge: structures and functions of D-alanyl-teichoic acids in gram-positive bacteria. *Microbiol. Mol. Biol. Rev.* **67**:686–723.

80. *S. pyogenes* is an important pathogen that causes most streptococcal infections in humans, including "strep throat." The M protein is a cell surface adhesin that is thought to bind to the LTA, which in turn binds to host cell surfaces.

81. Sutcliffe, I. C., and N. Shaw. 1991. Atypical lipoteichoic acids of gram-positive bacteria. *J. Bacteriol.* **173**:7065–7069.

82. Karakousis, P. C., W. R. Bishai, and S. E. Dorman. 2004. *Mycobacterium tuberculosis* cell envelope lipids and the host immune response. *Cell. Micobiol.* **6**:105–116.

83. Hamasaki, N., K. Isowa, K. Kamada, Y. Terano, T. Matsumoto, T. Arakawa, K. Kobayashi, and I. Yano. 2000. In vivo administration of mycobacterial cord factor (trehalose 6,6′-dimycolate) can induce lung and liver granulomas and thymic atrophy in rabbits. *Infect Immun.* **68**:3704–3709.

84. Virulence is defined as a quantitative measure of a pathogen's ability to cause disease. Bacterial pathogens produce virulence factors of several types. These include antiphagocytic capsules, enzymes that degrade host tissues, and toxins. The lipopolysaccharide (LPS) is a toxin that can also contribute toward virulence. LPS is referred to as an endotoxin because it is not secreted by the bacteria into the extracellular medium. It is released from dying bacteria and can be shed as blebs from living bacteria. The endotoxic activity of the LPS resides in the lipid A portion. Endotoxin causes macrophages to produce tumor necrosis factor (TNF-α), which accumulates in the bloodstream. The increased TNF-α in the bloodstream causes septic shock, which is a drastic drop in blood pressure due to leakage of blood vessels. This is a very dangerous condition because the drop in blood pressure results in organ failure.

85. Although the outer membrane is a permeability barrier to antibiotics and other toxic substances, it is not the only reason for the resistance of gram-negative bacteria to these substances. A second mechanism is the presence of periplasmic β-lactamases that hydrolyze the earlier β-lactam antibiotics that were used to treat infections. A third mechanism, and one that makes very important contributions to the resistance of pathogenic bacteria to antimicrobial agents, is the presence of drug efflux pumps. Gram-positive bacteria also have drug efflux pumps. A wide range of toxic substances can be removed from the bacterial cytoplasm of both gram-positive and gram-negative bacteria by using these pumps. Some pumps transport a single drug, whereas others transport several unrelated drugs. Antimicrobial agents that are pumped out of cells include basic dyes, quarternary ammonium compounds, and a large number of different antibiotics including puromycin, chloramphenicol, tetracycline, and erythromycin. For a review of this subject, read Section 16.5.

86. Leduc, M., K. Ishidate, N. Shakibai, and L. Rothfield. 1992. Interactions of *Escherichia coli* membrane lipoproteins with the murein sacculus. *J. Bacteriol.* **174**:7982–7988.

87. Cross-linking reagents can be useful for determining which molecules are physically associated. These are bifunctional reagents (i.e., they have a chemical group at both ends of the molecule that can form a covalent bond to functional groups on proteins or other molecules such as peptidoglycan), thus able to cross-link two molecules that are within the length of the cross-linking reagent. For example, the reagent dithio-bis-succinimidylpropionate (DSP) forms covalent cross-links between amino groups that are less than 12 angstrom units (1.2 nm) apart.

Therefore, DSP can cross-link proteins that are physically associated with the peptidoglycan. After treatment with DSP, the peptidoglycan and its cross-linked proteins can be isolated. DSP has two arms held together by a disulfide bond that can be cleaved with a cleavable cross-linking reagent called mercaptoethanol. The cross-linked proteins that were held to the peptidoglycan by the DSP are released after mercaptoethanol treatment and can be analyzed by gel electrophoresis. Thus, the proteins cross-linked to the peptidoglycan can be identified. In addition to various lipoproteins, including the murein lipoprotein, an outer membrane protein, OmpA, can be cross-linked to the peptidoglycan, indicating close association. The cross-linking of OmpA to the peptidoglycan reflects its extension through the outer envelope.

88. Ferguson, S. J. 1992. The periplasm, pp. 311–339. In: *Procaryotic Structure and Function: A New Perspective*. S. Mohan, C. Dow, and J. A. Cole (Eds.). Society for General Microbiology Symposium 47, Cambridge University Press, Cambridge.

89. Hobot, J. A., E. Carleman, W. Villiger, and E. Kellenberger. 1984. Periplasmic gel: new concept resulting from the reinvestigation of bacterial cell envelope ultrastructure by new methods. *J. Bacteriol.* 160:143–152.

90. Bayer, M. E. 1991. Zones of membrane adhesion in the cryofixed envelope of *Escherichia coli*. *J. Struct. Biol.* 107:268–280.

91. Bacteria secrete specific iron chelators called siderophores for scavenging iron from the environment. Special transport systems in the outer membrane bind the iron siderophores and bring the ligand-complex iron into the periplasm. These are activated by TonB.

92. Postle, K. 1990. TonB and the gram-negative dilemma. *Mol. Microbiol.* 4:2019–2025.

93. Braun, V. 1995. Energy-coupled transport and signal transduction through the gram-negative outer membrane via TonB-ExbB-ExbD-dependent receptor proteins. *FEMS Microbiol. Rev.* 16:295–307.

94. Bradbeer, C. 1993. The proton motive force drives the outer membrane transport of cobalamin in *Escherichia coli*. *J. Bacteriol.* 175:3146–3150.

95. Postle, K., and R. J. Kadner. 2003. Touch and go: tying TonB to transport. *Mol. Microbiol.* 49:869–882.

96. Merchante, R., H. M. Pooley, and D. Karamata. 1995. A periplasm in *Bacillus subtilis*. *J. Bacteriol.* 177:6176–6183.

97. Matias, V. R. F., and T. J. Beveridge. 2005. Cryo–electron microscopy reveals native polymeric cell wall structure in *Bacillus subtilis* 168 and the existence of a periplasmic space. *Mol. Microbiol.* 56:240–251.

98. Protoplasts are cells from which the cell wall has been removed. They can be stabilized by suspension in isotonic buffer (i.e., buffer that is iso-osmolar with the cell contents, hence preventing water from rushing in and lysing the protoplasts).

99. Booth, I. R., and P. Louis. 1999. Managing hypoosmotic stress: aquaporins and mechanosensitive channels in *Escherichia coli*. *Curr. Opin. Microbiol.* 2:166–169.

100. Calamita, G., B. Kempf, M. Bonhivers, W. R. Bishai, E. Bremer, and P. Agre. 1998. Regulation of the *Escherichia coli* water channel gene *aqpZ*. *Proc. Natl. Acad. Sci. USA* 95:3627–3631.

101. Edwards, M. D., I. R. Booth, and S. Miller. 2004. Gating the bacterial mechanosensitive channels: MscS a new paradigm? *Curr. Opin. Microbiol.* 7:163–167.

102. When a cell is subjected to sudden hypo-osmotic stress, the rapid flow of incoming water exerts additional outward pressure against the cell membrane, causing distortion of the membrane, and this results in the immediate opening of the mechanosensitive channels. (Mechanosensitive channels are sometimes referred to as "gated" channels.) This results in outward efflux of internal solutes along their diffusion gradients, thus lowering the internal osmotic pressure and the influx of water.

103. De Rosa, M., A. Gambacorta, and A. Gliozzi. 1986. Structure, biosynthesis, and physiochemical properties of archaebacterial lipids. *Microbiol. Rev.* 50:70–80.

104. Sternberg, B., C. L'Hostis, C. A. Whiteway, and A. Watts. 1992. The essential role of specific *Halobacterium halobium* polar lipids in 2D-array formation of bacteriorhodopsin. *Biochim. Biophys. Acta* 1108:21–30.

105. Gambacorta, A., A. Gliozzi, and M. De Rosa. 1995. Archaeal lipids and their biotechnological applications. *World J. Microbiol.* 11:115–131.

106. Westerhoff, H. V., and G. R. Welch. 1992. Enzyme organization and the direction of metabolic flow: Physiochemical considerations, pp. 361–390. In: *Current Topics in Cellular Regulation*, Vol. 33. E. R. Stadtman and P. B. Chock (Eds.). Academic Press, New York.

107. Reviewed by Drews, G. 1991. Intracytoplasmic membranes in bacterial cells, pp. 249–274. In: *Prokaryotic Structure and Function, A New Perspective*. S. Mohan, D. Dow, and J. A. Coles (Eds.). Cambridge University Press, Cambridge.

108. Overmann, J., S. Lehmann, and N. Pfennig. 1991. Gas vesicle formation and buoyancy regulation in *Pelodictyon phaeoclathratiforme* (green sulfur bacteria). *Arch. Microbiol.* 157:29–37.

109. Codd, G. A. 1988. Carboxysomes and ribulose bisphosphate carboxylase/oxygenase, pp. 115–164. In: *Advances in Microbial Physiology*, Vol. 29. A. H. Rose and D. W. Tempest (Eds.). Academic Press, New York.

110. Schüler, D. 2003. Molecular analysis of a subcellular compartment: the magnetosome membrane in *Magnetospirillum gryphiswaldense*. *Arch. Microbiol.* **181**:1–7.

111. Rouviere-Yaniv, J., and F. Gros. 1975. Characterization of a novel, low-molecular-weight DNA-binding protein from *Escherichia coli*. *Proc. Natl. Acad. Sci. USA* **72**:3428–3432.

112. Rimsky, S. 2004. Structure of the histone-like protein H-NS and its role in regulation and genome superstructure. *Curr. Opin. Microbiol.* **7**:109–114.

113. Pettijohn, D. E. 1996. The nucleoid, pp. 158–166. In: *Escherichia coli and Salmonella: Cellular and Molecular Biology*. F. C. Neidhardt et al. (Eds.). ASM Press, Washington, DC.

114. Molecular interactions among enzymes is reviewed by Srivastava, D. K., and S. A. Bernhard. 1986. Enzyme–enzyme interactions and the regulation of metabolic reaction pathways, pp. 1–68. In: *Current Topics in Cellular Regulation*, Vol. 28. B. L. Horecker and E. R. Stadtman (Eds.). Academic Press, New York.

115. In bacteria, fatty acid synthesis is catalyzed by similar enzymatic reactions as found in yeast and mammals, but the enzymes cannot be isolated as a complex.

116. Motor proteins hydrolyze ATP and use the released energy for internal conformational changes that allow the proteins to move ("walk") in one direction along linear macromolecules such as protein filaments. Motor proteins in eukaryotic cells include myosin, dynein, and kinesin. Myosin moves along protein microfilaments (actin filaments) and plays several roles in doing so. These roles include (a) the contraction of skeletal muscle cells, (b) the contraction of one cell into two during cytokinesis, (c) ameboid movement, and (d) the movement of organelles and vesicles along actin microfilaments in the cytoplasm during cytoplasmic streaming. Dynein is a motor protein that moves along microtubules in cilia and flagella, and as a consequence the cilia and flagella bend. Kinesin, like cytoplasmic dynein, is a motor protein that binds to organelles and vesicles

and moves along cytoplasmic microtubules, pulling organelles and vesicles through cytoplasm.

117. Norris, V., G. Turnock, and D. Sigee. 1996. The *Escherichia coli* enzoskeleton. *Mol. Microbiol.* **19**:197–204.

118. Lewis, P. J. 2004. Bacterial subcellular architecture: Recent advances and future prospects. *Mol. Microbiol.* **54**:1135–1150.

119. van den Ent, F., L. A. Amos, and J. Löwe. 2001. Prokaryotic origin of the actin cytoskeleton. *Nature* **413**:39–44.

120. Lutkenhaus, J. 2003. Another cytoskeleton in the closet. *Cell* **115**:648–650.

121. Ausmees, N., J. R. Kuhn, and C. Jacobs-Wagner. 2003. The bacterial cytoskeleton: An intermediate filament-like function in cell shape. *Cell* **115**:705–713.

122. Figge, R. M., A. V. Divakaruni, and J. W. Gobel. 2004. MreB, the cell shape-determining bacterial actin homologue, co-ordinates cell wall morphogenesis in *Caulobacter crescentus*. *Mol. Microbiol.* **51**:1321–1332.

123. Mollicutes are part of the normal flora in healthy people. However, they can cause disease. Diseases in humans caused by Mollicutes include primary atypical pneumonia (*Mycoplasma pneumoniae*), pelvic inflammatory disease (*M. hominis*), and nongonococcal urethritis (*Ureaplasma urealyticum*). *Spiroplasma* causes diseases in insects and plants.

124. Gilad, R., Porat, A., and S. Trachtenberg. 2003. Modes of *Spiroplasma melliferum* BC3: a helical, wall-less bacterium driven by a linear motor. *Mol. Microbiol.* **47**:657–669.

125. Trachtenberg, S., R. Gilad, and N. Geffen. 2003. The bacterial linear motor of *Spiroplasma melliferum* BC3: From single molecules to swimming cells. *Mol. Microbiol.* **47**:671–697.

126. Golstein, S. F., N. W. Charon, and J. A. Kreiling. 1994. *Borrelia burgdorferi* swims with a planar waveform similar to that of eukaryotic flagella. *Proc. Natl. Acad. Sci. USA* **91**:3433–3437.

2

Growth and Cell Division

The study of the growth of microbial populations is at the heart of microbial cell physiology because population growth characteristics reflect underlying physiological events in the individual cells. In fact, these cellular events are frequently manifested to the investigator only by changes in the growth of populations. It is therefore important to understand the changes in growth that microbial populations undergo and to be able to measure population growth accurately. Additionally, to investigate certain aspects of cell physiology (e.g., the interdependencies between the rates of synthesis of the individual classes of macromolecules such as DNA, RNA, and protein), the investigator must be able to manipulate population growth (e.g., by placing the culture in balanced growth or continuous growth). This chapter describes methods to measure growth, presents an analysis of exponential growth kinetics, discusses growth in batch culture and in continuous culture, introduces some important aspects of growth physiology under certain nutritional conditions, and covers cell division.

2.1 Measurement of Growth

Growth is defined as an increase in mass, and one can measure any growth parameter provided it increases proportionally to the mass of the culture. The most commonly used measurements of growth are turbidity, total and viable cell counts, dry weight, and protein.

2.1.1 Turbidity

Turbidity measurements are routinely, used to ascertain bacterial growth because of the simplicity of the procedure. Within limits, the amount of light scattered by a bacterial cell is proportional to its mass. Therefore, light scattering is proportional to the total mass. (The relationship between total mass and light scattering deviates from linearity at very high cell densities.) If the average mass per cell remains a constant, one can also use light scattering to measure changes in cell number. What is actually measured is the *fraction of incident light transmitted* through the culture (i.e., not the scattered light). The more dense the culture, the less light is transmitted. This relationship is given by the *Beer–Lambert law* (eq. 2.1). Thus if I_0 is the incident light and I is the transmitted light, then the Beer–Lambert law states that in a population whose cell density is x, the fraction of light that is transmitted (I/I_0) will decrease as the logarithm of x to the base 10.

$$I/I_0 = 10^{-xl} \qquad (2.1)$$

where I is the light path in centimeters.

That is, if one takes the log to the base 10 of both sides of eq. 2.1, then $\log (I/I_0) = -xl$, where I/I_0 is the fraction of incident light that is transmitted. The situation can be drawn schematically as shown in Fig. 2.1. Note that the fraction of light that is transmitted decreases as a logarithmic function (base 10); the fraction decreases exponentially, that is,

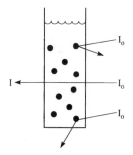

Fig. 2.1 Illustration of light scattering. The incident light is I_0 and the transmitted light is I. The spectrophotometer usually provides the logarithm of the reciprocal of the fraction of transmitted light, that is, $\log(I_0/I)$, and this is called the optical density or turbidity.

with the density of the culture. Bacteriologists, however, prefer to measure something that increases with cell density. They therefore use the reciprocal of $\log(I/I_0)$ which is $\log(I_0/I)$. The $\log(I_0/I)$ is called *turbidity* or *absorbancy* or *optical density* and has the symbol OD (for optical density) or A (for absorbancy). Thus turbidity or OD is directly proportional to cell mass in the culture (or cell density, if the ratio of mass to cell remains constant) and is written as follows:

$$OD = A = xl \qquad (2.2)$$

Turbidity is measured with a colorimeter or spectrophotometer. In practice, one constructs a standard curve by measuring the turbidity of several different cell suspensions, counting the cell number independently and using it as a measure of cell mass. A straight line is generated, as shown in Fig. 2.2. However, the line deviates from linearity at high cell densities. This is because when the cell density is too high, some of the light is rescattered and directed toward

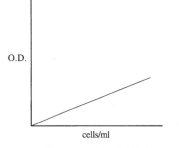

Fig. 2.2 Standard curve of optical density versus cells per milliliter. The line deviates from linearity at very high cell densities.

the phototube, thus lowering the turbidity reading.

Whenever turbidity measurements are made on an unknown sample, a standard curve must be constructed to determine the cell density. Turbidity measurements are the simplest way to measure growth. Methods to determine the cell density by direct cell counts are described next.

2.1.2 Total cell counts

If the mass per cell is constant, one can measure growth by sampling the culture over a period of time and counting the cells. There are two ways to do this: total cell counts and viable cell counts. For *total cell counts* one can use a counting chamber. This is simply a glass slide divided into tiny square wells of known area and depth. A drop of culture is placed on a slide, and a cover slip is applied. Each square well in the grid then holds a known volume of liquid. The number of cells is counted microscopically and the number of cells per milliliter is obtained by using a conversion factor based on the total number of square wells counted and their volume. Although this is a simple procedure, and one used widely, it has two limitations:

1. It does not distinguish between live and dead cells.

2. It cannot be performed on populations whose cell density is too low to count microscopically ($<10^6$ cells/mL).

Electronic cell counting is a more sophisticated method. The bacteria, suspended in a saline solution, are placed in a chamber with an electrode separated by a second chamber filled with the same saline solution and also provided with an electrode. A microscopic pore separates the two chambers. The bacterial suspension is pumped through the pore into the second chamber. Whenever a bacterium passes through the pore, the electrical conductivity of the circuit decreases because the electrical conductivity of a bacterial cell is less than that of the saline solution. This change in conductivity results in a voltage pulse that is counted electronically. The electronic counter has the advantage of allowing one also to measure the *size* of the bacterium. This is because the size of the pulse is proportional to the size of the bacterium. Thus, one can obtain both a total cell count and a size distribution.

2.1.3 Viable cell counts

A *viable cell count* is one in which the cells are serially diluted and then deposited on a solid growth medium. Each viable cell grows into a colony and the colonies are counted. Viable cell counts are routinely performed in microbiology laboratories. However, it is important to recognize that they may underestimate the number of viable cells for the following reasons:

1. A viable cell count will underestimate the number of live cells if the cells are clumped, since each clump of cells will give rise to a single colony.

2. Some bacteria plate with poor efficiency; that is, single cells give rise to colonies with low frequency.

2.1.4 Dry weight and protein

The most direct way to measure growth is to quantify the *dry weight* of cells in a culture. The cells are harvested by either centrifugation or filtration, dried to a constant weight, and carefully weighed. Since protein increases parallel with growth, one can also measure growth by doing *protein measurements* of the cells. The cells are either harvested by centrifugation or filtration or (more usually) precipitated first with acid or alcohol, then recovered by centrifugation or filtration. A simple colorimetric test for protein is then performed.

2.2 Growth Physiology

The vast topic of growth physiology includes the regulation of rates of synthesis of macromolecules, the regulation of the timing of DNA synthesis and cell division, and such adaptive physiological responses to nutrient availability as changes in gene expression, homeostasis adaptations to the external environment, and the coupling of the rates of biosynthetic pathways to the rates of utilization of products for growth (metabolic regulation). From this point of view, most of this text is concerned with the physiology of growth, and we therefore return to the topic in subsequent chapters. What follows here is an introduction to growth physiology as it pertains to the synthesis of macromolecules during steady state growth, and very general adaptations of cells to nutrient depletion. We begin with a discussion of the sequence of growth phases through which a population of bacteria progresses when inoculated into a flask of fresh media.

2.2.1 Phases of population growth

When one measures the growth of populations of bacteria grown in batch culture, a progression through a series of phases can be observed (Fig. 2.3). The first phase is frequently a *lag phase*, in which no net growth occurs (i.e., no increase in cell mass). This is followed by a phase of *exponential growth*, in which cell mass increases exponentially with time. Following exponential growth, the culture enters the *stationary phase*, a phase of no net growth. After a stationary period, cell death occurs in

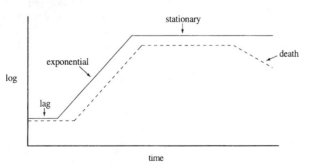

Fig. 2.3 Growth kinetics in batch culture: solid line, mass; dashed line, viable cells. Note that if we define growth as an increase in mass, then only the solid line accurately reflects the growth of the culture. The dashed line reflects growth only when it is parallel to the solid line.

a final stage called the *death phase*. Notice in Fig. 2.3 that, prior to the exponential phase of growth, when cell division occurs, the cells increase in mass (solid line); that is, they grow larger. At the end of exponential growth, the cells continue to divide after growth has ceased (dashed line), but they become smaller. It would seem to be an advantage to a population of bacteria to continue to divide after growth has ceased in the population, since in this way more cells can be produced for distribution to new sites, where conditions for continued growth may be better. One consequence of the uncoupling of growth from cell division during the lag and late log phases is that the size of the cell varies during growth in batch culture. For the investigator, the noncoincidence of growth and cell division during stages of batch growth has a practical consequence: namely, cell counts are not always a valid measurement of growth.

Lag phase

When cells in the stationary phase of growth are transferred to fresh media, a lag phase often occurs. In this situation, the lag phase is due basically to the time required for the physiological adaptation of stationary phase cells in preparation for growth. Usually the longer the cells are kept in the stationary phase, the longer is the lag phase when they are transferred to fresh media. Some of the physiological changes that occur in the stationary phase are described in Section 2.2.2.

There are several possible reasons for the lag phase. The lag phase can be due to the time required for recovery of cells from toxic products of metabolism, such as acids, bases, alcohols, or solvents, that may accumulate in the external medium. Sometimes, new enzymes or coenzymes must be synthesized before growth resumes. This will be necessary if, for example, the fresh medium is different from the inoculum medium and requires a change in the enzyme composition of the cells. If significant cell death occurs in the stationary phase, an apparent lag phase will be measured because the inoculum includes dead cells that contribute to the turbidity. The lag phase can be avoided if the inoculum is taken from the exponential phase of growth and transferred to fresh medium of the same composition.

Stationary phase

Cells stop growing and enter the stationary phase for various reasons. Among these are exhaustion of nutrients, limitation of oxygen, and accumulation of toxic products (e.g., alcohols, solvents, bases, acids). The accumulation of toxic products is frequently a problem for fermenting cells because instead of being converted to cell material, most of the nutrient is excreted as waste products. The excretion of fermentation end products is discussed in Chapter 14. For a review of the physiology of stationary phase cells, read ref. 1.

Death

Death can result from several factors. Common causes of death include the depletion of cellular energy and the activity of autolytic (self-destructive) enzymes. Some bacteria begin to die within hours of entering the stationary phase. However, many bacteria remain viable for longer periods. For example, some bacteria sporulate or form cysts when exponential growth ceases. The spores and cysts are resting cells that remain viable and germinate in fresh media. As discussed next, even nonsporulating bacteria can adapt to nutrient depletion and remain viable for long periods in stationary phase.

2.2.2 Adaptive responses to nutrient limitation

In the natural environment bacteria are frequently faced with starvation conditions and enter intermittent periods of no growth or very slow growth. In some environments the generation time may be many days or even months because the nutrient levels are so dilute. When a bacterial culture faces the depletion of an essential nutrient, the culture may keep growing by inducing specific uptake systems to scavenge the environment for the nutrient or a source thereof, and/or it may induce the synthesis of enzymes that can use an alternative source of the nutrient. For example, bacteria starved for inorganic phosphate may induce the synthesis of a high-affinity inorganic phosphate uptake system as well as the synthesis of enzymes capable of degrading organic phosphates to release inorganic phosphate (phosphatases). (See the discussion of the Ntr and Pho regulons in Sections 18.4 and 18.5.) If the bacteria cannot bring into the cells the essential nutrient, then the population will stop growing and dividing. Under these circumstances, the population enters stationary phase or, in the case of certain bacteria, the cells may sporulate or encyst. (See Section 18.16 for a description of sporulation.)

Lately, increased attention is being given to physiological changes that occur in bacteria that enter the stationary phase when they are experimentally subjected to starvation.[2-5] These bacteria undergo physiological changes that result in metabolically less active cells that are more resistant to environmental hazards. This property has been associated for a long time with bacteria that sporulate or form desiccation-resistant cysts upon nutrient deprivation. The spores and cysts represent metabolically inactive or less active stages of the life cycle of the organism. When nutrients become available once more, the spores and cysts germinate into vegetative cells that grow and divide. More recently, it has been discovered that even some bacteria that do not form spores (e.g., E. coli, Salmonella, Vibrio, Pseudomonas) undergo adaptive changes when faced with nutrient deprivation and thereupon enter stationary phase. Some of these changes also result in resistant, metabolically less active cells, and such responses may be common in most if not all bacteria. However, not all the effects can be rationalized in terms of survivability, and their physiological role is not yet known. Some of the changes that occur in cells that are starved are described next. For a review of starvation in bacteria, see ref. 6.

Changes in cell size

As discussed earlier, cells that enter stationary phase upon carbon exhaustion generally become smaller because they undergo reductive division; that is, they keep dividing for 1 to 2 h after growth has ceased. Reductive division results in the production of more cells, which may be advantageous for dispersing the population. In some cases there may be several cell divisions in the absence of growth so that the cells size differs radically from the growing cell. Some bacteria decrease in length from several micrometers (e.g. 5–10 μm), to approximately 1 to 2 μm or even less. (see note 7.) As discussed in ref. 1, in addition to size reduction due to continued division in the absence of growth, bacteria can become smaller during starvation after reductive division has been completed. The additional decrease in size results from self-digestion of cell material, including the cell envelope. Some gram-negative bacteria (e.g., Pseudomonas) bleb off outer membrane vesicles as they become smaller. Frequently, the decrease in size can be accompanied by a change from rod-shaped cells to coccoid-shaped cells, as discussed next.

Morphological changes

Along with a reduction in size, some bacteria (e.g., Arthrobacter) change from rod-shaped cells to coccoid cells in stationary phase. Similar morphological changes from rods to small coccoid shapes can occur upon starvation in several other bacteria, including Klebsiella, Escherichia, Vibrio, and Pseudomonas.

Changes in surface properties

When certain marine bacteria are starved, the cell surface becomes hydrophobic and the cells are more adhesive.[8] Vibrio synthesizes surface fibrils and forms cell aggregates when starved for a long time. These changes in the surfaces of starved cells make them more adhesive. Presumably, it is advantageous for nutrient-limited cells to adhere to particles that have adsorbed nutrient on the surfaces of the particles.

Changes in membrane phospholipids

In E. coli all the unsaturated fatty acids in the membrane phospholipids become converted to the cyclopropyl derivatives by means of the methylation of the double bonds. The advantage to the cyclopropane fatty acids is not known, since mutants unable to synthesize cyclopropane fatty acids appear not to be at a survival disadvantage when faced with environmental stresses such as starvation and high or low oxygen tension.

Changes in metabolic activity

When bacteria enter stationary phase, their overall metabolic rate slows. In addition, during starvation many bacteria experience a significant increase in the turnover (metabolic breakdown and resynthesis) of protein and RNA. Presumably, in starved cells the protein and ribosomal RNA can serve as an energy source to maintain viability and crucial cell functions. The latter would include solute transport systems, an energized membrane, and ATP pool levels.

Changes in protein composition

Bacteria may synthesize 50 to 70 or more new proteins under conditions of carbon, nitrogen, or phosphate starvation. Many of the proteins serve specific functions related to the nutrient that is in low concentration. For example, phosphate starvation induces the synthesis of PhoE porin, which is an outer membrane channel for anions, including phosphate (Sections 1.2.3 and 18.5). This helps the bacterium to bring in more phosphate. Another example is the nitrogen fixation genes that are induced when the cells are starved for nitrogen (Sections 12.3 and 18.4). In addition to the proteins of known function, there are also many proteins made under all conditions of starvation for which there is presently no known function, although many of these are presumed to be involved in resistance to stress.

Changes in resistance to environmental stress

Cells entering stationary phase also become more resistant to environmental stresses such as high temperatures, osmotic stress, and certain chemicals. For example, *E. coli* reportedly becomes more resistant in stationary phase to high temperature, hydrogen peroxide, high salt, ethanol, solvents such as acetone and toluene, and acidic or basic pH. These resistant properties are due to the synthesis of a starvation sigma factor that has four names: RpoS, σ^s, σ^{38}, and KatF.

1. The rpoS gene in E. coli encodes a sigma factor required for transcription of genes expressed during stationary phase and starvation

See Section 10.2.4 for a discussion of sigma factors. The reasons for concluding that *rpoS* encodes a sigma factor are as follows: (1) the predicted amino acid sequence of the *rpoS* gene product, based upon nucleotide sequence data, suggests that it is a sigma factor and (2) studies with purified protein have confirmed that the protein binds to RNA polymerase and directs transcription of *rpoS*-dependent genes. RpoS is required for the transcription of a regulon that includes at least 50 genes that encode proteins induced by carbon starvation when cells enter stationary phase; it is also necessary for the transcription of several genes whose activities increase during phosphate and nitrogen starvation.[9-12] Accordingly, the protein is also called σ^s, for starvation sigma factor. It is also known as σ^{38} because it has a molecular weight of 38 kDa. Sometimes called the "master regulator of the stationary phase response," RpoS has been found in other gram-negative bacteria in addition to *E. coli*, but not in gram-positive bacteria.[12] Gram-positive bacteria use a different sigma factor (σ^B) to regulate the transcription of genes homologous to some of the σ^s-dependent genes in *E. coli*.[13]

Proteins whose synthesis depends upon σ^s include a catalase made during stationary phase (HPII), which presumably accounts for resistance to H_2O_2, an exonuclease III that can repair DNA damage due to H_2O_2 and near-UV radiation, and an acid phosphatase. Because wild-type *rpoS* mutants do not have stationary phase resistance properties (measured as percentage survival) to heat, high salt (osmotic shock), near-UV, H_2O_2, or prolonged starvation (e.g., several days starvation for carbon or nitrogen), it has been concluded that the products of *rpoS*-dependent genes are necessary for survival under stressful conditions during this phase.[8] The transcription of the *rpoS* gene (measured using a *rpoS–lacZ* fusion plasmid) increases dramatically in a nutrient-rich medium as cells approach and enter stationary phase and during starvation for specific nutrients such as phosphate.[14,15] (See note 16 for a description of *lacZ* gene fusions.) Homologues to *rpoS* exist in other gram-negative bacteria, including pathogens. Mutations in *rpoS* in the intestinal pathogen *Salmonella* result in the attenuation of virulence in mice.[17,18] One might suppose that RpoS aids pathogens to survive stress such as oxidative stress, which they might encounter inside host cells, or pH stress (e.g., in the stomach).

2. RpoS is a global regulator that is important for stress reponses in exponentially growing cells as well as in stationary phase

The RpoS subunit of RNA polymerase should be viewed as a *global regulator* that also functions during responses to stress during exponential growth.[19] Increased RpoS levels during exponential growth can be caused by slow growth, temperature upshift from 30 °C to 42 °C, and high osmolarity. For example, during osmotic upshifts in exponentially growing cultures RpoS levels increase. Such cultures become not only osmotolerant but also thermotolerant and resistant to H_2O_2.

3. Regulation of RpoS levels can be at the transcriptional, post-transcriptional, and post-translational levels

Interestingly, the reason for the increased levels of RpoS during stress responses is not simply increased transcription of the *rpoS* gene. The levels of RpoS are regulated not only at the transcriptional level but also at the post-transcriptional and post-translational levels. Thus, the regulation of RpoS levels is very complex. For example, the rate of degradation of RpoS decreases when cells enter stationary phase. The half-life of RpoS in exponential cultures of *E. coli* growing at 37 °C is less than 2 min. However, the half-life is greater than 30 min when cells enter stationary phase. See Box 2.1 and the references therein for a discussion of how the levels of RpoS are adjusted, including transcriptional regulation of the *rpoS* gene, regulation of translation of the RpoS messenger RNA, and the role of proteases in the degradation of RpoS.

Other regulators for stationary phase gene expression

Although σ^s is certainly a major transcription factor for gene expression during stationary phase, additional regulatory factors exist (reviewed in ref. 5). This conclusion is suggested partly because the variation in the timing of expression during stationary phase of σ^s-dependent genes, depending upon the gene in question, clearly points to regulatory factors in addition to σ^s. In fact, there are several σ^s-dependent genes that are also controlled by other global regulators, both positive and negative. One of these, cyclic AMP (cAMP)–C-AMP

receptor protein (CRP) complex, stimulates the expression of approximately two-thirds of the genes expressed during carbon starvation. Furthermore, not all genes whose expression increases during stationary phase require σ^s. It is clear from the available data that gene expression during stationary phase is complex and poorly understood at the present time. Nevertheless, expression during stationary phase is extremely important because bacteria in their natural habitat are frequently in this phase or growing extremely slowly owing to limitations of carbon, phosphate, or nitrogen sources. How bacteria survive such nutritional stress is fundamental to an understanding of their physiology in natural ecosystems.

The control of ribosomal RNA synthesis

There is much controversy surrounding the question of what regulates the rates of synthesis of ribosomes, although it is clear that ribosome synthesis is coupled to growth rates (Section 2.2.3).[20] As shown in Fig. 2.4, faster growing cells have more ribosomes (reflected in cellular RNA) per unit cell mass. For example, the number of ribosomes in an *E. coli* cell can vary from less than 20,000 to about 70,000 depending upon the growth rate. In addition to growth rate control, amino acid starvation results in the inhibition of ribosome synthesis. Probably several factors are involved in regulating rates of ribosome synthesis. For example, Fis, a DNA-binding protein whose synthesis is increased in rich media, positively regulates transcription of rRNA genes. Another DNA-binding protein, H-NS, antagonizes Fis stimulation (see later). A very important regulator is guanosine tetraphosphate (ppGpp), which increases when cells are subjected to amino acid starvation or carbon and energy limitation, and slows the synthesis of rRNA (and tRNA). The global regulator ppGpp is best known as the effector that inhibits rRNA and tRNA synthesis during the *stringent response*, an effect originally discovered as the inhibition of rRNA and tRNA synthesis due to amino acid starvation.[21] See note 22 for other physiological effects of ppGpp.

1. The stringent response

The stringent response in *E. coli* is a temporary inhibition in the synthesis of ribosomal RNA

BOX 2.1 TRANSCRIPTIONAL, TRANSLATIONAL, AND POST-TRANSLATIONAL REGULATION OF LEVELS OF RPOS

The regulation of transcription of *rpoS* (*katF*) appears to be complicated and is not well understood. There are probably several signaling pathways that influence the transcription of the *rpoS* gene, allowing the cells to respond to a variety of environmental challenges. The transcription of *rpoS* does not always increase markedly when cells enter stationary phase because expression during growth may be high. What happens during transcription depends upon how the cells are grown. Anaerobiosis has been reported to stimulate *rpoS* expression in exponential cells grown in a rich medium. Several other factors appear to influence the levels of *rpoS* expression. For example, depending upon the strain of *E. coli*, *rpoS* expression may be high during exponential growth in a minimal medium and increase only slightly during stationary phase. Additionally, starvation induced by transferring exponential phase cells to a medium lacking a specific nutrient might stimulate *rpoS* expression significantly or only marginally, depending on which component is missing from the medium.

The amounts of RpoS also increase in response to an upshift in osmolarity, even in exponentially growing cells. The increase is due to an increase in translation of the RpoS mRNA as well as to a decrease in the turnover of RpoS due to ClpXP. This, then, is an example of translational and post-translational control over the levels of a protein.

The stability of the RpoS protein increases during stationary phase. Apparently RpoS is sensitive to proteolytic digestion by the ClpXP protease during exponential growth but less so during stationary phase. The evidence for this is that RpoS is stabilized during exponential growth in *clpP* and *clpX* mutants. For a discussion of the role of RpoS as well as how the levels of RpoS can be regulated, see refs. 1 through 5.

REFERENCES

1. Loewen, P. C., and R. Wenge-Aronis. 1994. The role of the sigma factor σ^s (KatF) in bacterial global regulation. *Annu. Rev. Microbiol.* **48**:53–83.

2. Wilmes–Riesenberg, M. R., J. W. Foster, and R. Curtiss III. 1997. An altered *rpoS* allele contributes to the avirulence of *Salmonella typhimurium* LT2. *Infect. Immun.* **65**:203–210.

3. Schweder, T., K.-H. Lee, O. Lomovskaya, and A. Matin. 1996. Regulation of *Escherichia coli* starvation sigma factor (σ^s) by ClpXP protease. *J. Bacteriol.* **178**:470–476.

4. Zhou, Y., and S. Gottesman. 1998. Regulation of proteolysis of the stationary-phase sigma factor RpoS. *J. Bacteriol.* **180**:1154–1158.

5. Venturi, V. 2003. Control of *rpoS* transcription in *Escherichia coli* and *Pseudomonas*: why so different? *Mol. Microbiol.* **49**:1365–2958.

and transfer RNA when the cells are shifted to a medium in which they are starved for an amino acid. It works in the following way. When cells are shifted from a medium rich in amino acids to a minimal medium, "uncharged" tRNA accumulates, causing an insufficiency of aminoacylated tRNA, which in turn results in "idling" or "stalled" ribsomes. An enzyme on the "stalled" ribosomes called RelA [(p)ppGpp synthetase I, or PSI], which is the product of the *relA* gene, becomes activated by uncharged tRNA that binds to the A site. Activated RelA synthesizes the signaling molecule ppGpp and to a lesser extent the pentaphosphate (i.e., pppGpp). Together they are referred to as (p)ppGpp. The ppGpp somehow slows the transcription of ribosomal RNA (and transfer RNA). This leads to an inhibition of ribosomal protein synthesis by an interesting mechanism called translational coupling, as explained in note 23.

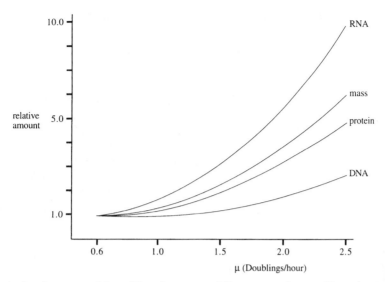

Fig. 2.4 Macromolecular composition of *E. coli* grown at different growth rates. The values are amounts per cell and have been normalized to values at a doubling time of 0.6 doubling/h. *Source*: Neidhardt, F. C., J. L. Ingraham, and M. Schaechter. 1990. *Physiology of the Bacterial Cell*. Sinauer Associates, Sunderland, MA.

2. Carbon and energy limitation cause an increase in the levels of ppGpp in a pathway independent of RelA

The regulator (p)ppGpp also accumulates during carbon and energy limitation on a pathway that occurs in *relA* mutants, hence is independent of RelA, and requires the product of the *spoT* gene [(p)ppGpp synthetase II, or PSII]. The SpoT protein is most probably a bifunctional enzyme that both synthesizes and degrades (p)ppGpp. The ratio of biosynthetic to degradative activity is somehow regulated by carbon and energy limitation. This may account at least in part for the lower number of ribosomes in cells growing slowly in poor medium. It is therefore clear that the inhibition of rRNA and tRNA synthesis coincides with essentially any nutritional condition (not simply amino acid starvation) that slows the growth rate. See note 24 for a further discussion of the synthesis of (p)ppGpp and the contributions of RelA and SpoT.

3. Other factors that control rRNA synthesis: Fis, H-NS

It must be emphasized that the control of ribosomal RNA synthesis is complex and that ppGpp is only one factor in the regulation. Other proteins involved in transcriptional regulation of the ribosomal RNA genes include the Fis and H-NS proteins.[25] (See note 26 for a

further discussion of Fis.) Fis binds to DNA sequences upstream of the rRNA promoter and activates the transcription of the operon. (Fis regulates the expression, both positively and negatively, of many genes besides rRNA genes.) Importantly, Fis synthesis is under growth rate control. Synthesis increases when exponentially growing cells are shifted to a nutritionally richer medium and when stationary phase cells, which have very low Fis levels, are subcultured into a nutrient-rich medium.[27] One would expect an increase in the rate of rRNA synthesis under these conditions, and Fis accounts in part for the increased synthesis.

Fis-dependent activation of ribosomal RNA genes is antagonized by another protein called H-NS (also called H1), which can therefore act as a repressor of rRNA transcription.[28] (H-NS, discussed in Section 1.2.6 as part of the nucleoid, does not bind to specific nucleotide sequences but does bind to curved or bent DNA.) The cellular levels of H-NS are highest under conditions of slow rRNA synthesis (e.g., stationary phase). H-NS proteins belong to a remarkable group of DNA-binding proteins that have other cellular functions besides the regulation of rRNA gene transcription. (See Section 1.2.6 and note 29.) So, although ppGpp somehow slows ribosomal RNA synthesis, it appears to be only one component of one or more complex regulatory systems that must control ribosomal RNA synthesis under a variety of growth conditions.

2.2.3 Macromolecular composition as a function of growth rate

Much has been learned about growth physiology by observing changes in macromolecular composition of cells whose growth rates have been altered by changing the nutrient composition of the medium, or by manipulation of the dilution rates of cells growing in continuous culture (Section 2.4).

Figure 2.4 illustrates the changes due to differences in the nutrient composition of the growth media. In the ratios of cell mass, RNA, protein, and DNA of *E. coli* grown at increasingly faster growth rates. The ordinate gives relative amounts of macromolecules *per cell*, normalized to cells undergoing 0.6 doubling per hour. The abscissa gives the growth rate μ, in units of doublings per hour. Notice that the mass of the cell as well as the relative concentrations of the macromolecules increase as exponential functions of the growth rate. As shown in Fig. 2.4, faster growing cells have a higher ratio of RNA to protein, a larger mass, and more DNA.

Why faster growing cells have a larger RNA-to-protein ratio

Faster growing cells are enriched for RNA with respect to the other cell components. This reflects a higher proportion of ribosomes in faster growing cells, as well as the composition of ribosomes: approximately 65% RNA and 35% protein by weight. The reason for the increase in ribosomes is that, over a wide range of rapid growth rates, a ribosome polymerizes amino acids at an approximately constant rate. When a cell increases or decreases its growth rate, and therefore the number of proteins it must make per unit time, it adjusts the number of ribosomes rather than making each ribosome work substantially slower or faster.

Why faster growing cells have more DNA

Figure 2.4 also shows that faster growing cells have more DNA per cell. The increased DNA per cell is rationalized in the following way. In *E. coli*, the minimum time required between the initiation of DNA replication and completion of cell division (cell separation) for rapidly growing cells (i.e., those with doubling times between 20 and 60 min) is about 60 min.[30] (See note 31 for an explanation of the C and D periods.) If the generation time is shorter than 60 min, then DNA replication must begin in a nearlier cell cycle and cells are born with partially replicated DNA, which increases the average amount of DNA per cell. This is illustrated in Fig. 2.4, which shows that cells with a generation time of less than 60 min have more DNA.

Why faster growing cells have more mass

As seen in Fig. 2.4, cells that are grown at a faster growth rate have a greater mass (i.e., they are larger). For example, *E. coli* when growing at a doubling rate of 2.5 doublings per hour is approximately six times larger in mass than when it is growing at 0.6 doubling per hour. Why is that? It is because, first, the initiation of DNA replication requires a minimum cell mass per replication origin, called the critical cell mass (M_i); in addition, faster growing cells have several replication origins, and for the shorter doubling times (<60 min), replication must begin in an earlier cell cycle. (See note 32 for a more complete explanation.) (The significance of M_i was reported by Donachie in 1968.[33] See note 34 for how this was done. Whether it is an absolute constant or whether it decreases slightly with increasing growth rate is not known with certainty.[35,36] For this discussion, we will assume that it is constant.)

2.2.4 Diauxic growth

Many bacteria, including *E. coli*, will grow preferentially on glucose when presented with mixtures of glucose and other carbon sources. The result is a biphasic growth curve as shown in Fig. 2.5 for glucose and lactose. Preferential growth on one carbon source before growth on a second carbon source is called *diauxic growth*. The bacteria first grow exponentially on glucose, then enter a lag period, and finally grow exponentially on the second carbon source.

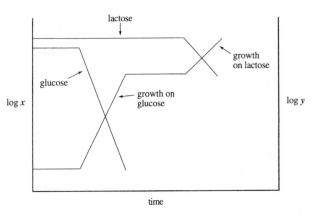

Fig. 2.5 Diauxic growth on glucose and lactose: x, bacterial mass, y, glucose or lactose concentration in the medium. There are two phases of growth. Initially the cells grow on the glucose. When the glucose is sufficiently depleted, a lag phase occurs followed by growth on lactose. During the lag phase the cells induce the synthesis of enzymes necessary for growth on lactose. Glucose prevents the induction of the genes necessary for growth on lactose. The inhibition by glucose is called catabolite repression and is discussed in Section 16.3.4.

2.2.5 *Catabolite repression by glucose*

It is now known that in many bacteria glucose represses the synthesis of enzymes required to grow on certain alternative carbon sources such as lactose, and this can account for the diauxic growth that discussed.

Glucose also inhibits the uptake of certain other sugars into the cell. The result is that growth on the second carbon source does not proceed until the glucose has been exhausted from the medium. During the lag period following the exponential growth on glucose, the bacteria synthesize the enzymes required to grow on the second carbon source. The repression of genes by glucose is called *catabolite repression*, or *glucose repression*, and is indeed widespread among bacteria. The inhibition of uptake of alternative sugars is called *inducer exclusion*. Catabolite repression in gram-negative and gram-positive bacteria is discussed Sections 18.10.

A rationale for glucose repression points out that glucose is one of the most common carbon sources in the environment and that bacteria frequently grow more rapidly on glucose than on other carbon sources, especially other sugars. Therefore, bacteria in certain ecological habitats that preferentially utilize glucose may be at a competitive advantage with respect to the other cells that depend upon glucose for carbon and energy. Additionally, the enzymes required to metabolize glucose are generally constitutive

(i.e., present under all growth condition), and the cell is prepared to grow on glucose at any time.

However, not all bacteria preferentially grow on glucose, nor are the glucose catabolic enzymes necessarily constitutive. Several obligately aerobic bacteria, when given a mixture of glucose and an organic acid, will grow first on the organic acid. This is also called catabolite repression except that it is the organic acid that is the repressor of glucose utilization. One example is presented by the nitrogen-fixing bacteria belonging to the genus *Rhizobium*. These bacteria, which live symbiotically in root nodules of leguminous plants, grow most rapidly in laboratory cultures on C_4 carboxylic acids that are part of the citric acid cycle such as succinate, malate, and fumarate. When presented with a mixture of succinate and glucose, *Rhizobium* shows diauxic growth by growing on the succinate first and the glucose second. This is because in these bacteria, succinate represses key glucose-degrading enzymes of the Entner–Doudoroff and Embden–Meyerhof–Parnas pathways.[37] *Pseudomonas aeruginosa* offers another example of a species in which diauxic growth occurs first on organic acids and then on carbohydrates (e.g., glucose). If presented with glucose or other carbohydrates and any one of several organic acids such as acetate, or one of the intermediates of the citric acid cycle, *P. aeruginosa* will grow on the organic acid first and will repress the enzymes required to degrade the carbohydrate.[38]

2.3 Cell Division

Two major events that occur in the cell cycle are DNA replication and cell division (also referred to as cytokinesis). DNA replication is discussed in Section 10.1. Cell division is the splitting of a mother cell into two daughter cells separated by a septum. For bacteria that grow as single cells, cell division is accompanied by, or followed by, cell separation. There are a large number of proteins involved in the division of a mother cell into two daughter cells and their subsequent separation. This is clearly a very complex business, which must be carefully coordinated. As will become evident from the following discussion, cell division is imperfectly understood. Most research attention has been focused on the gram-negative bacteria *E. coli* (and the closely related *Salmonella typhimurium*), and *Caulobacter crescentus*, and the gram-positive bacterium *Bacillus subtilis*. We will begin with *E. coli*, and what happens up to the point at which the septum is formed.

2.3.1 The period before septum formation

When *E. coli* reaches a critical cell mass during growth, DNA synthesis is initiated (Section 2.2.3). Initiation requires the DnaA protein, which binds to the replication origin (*oriC*) and opens up the duplex, allowing other proteins required for DNA replication to enter (Section 10.1.3). During replication the sister chromosomes move toward opposite halves of the cell, so that when cell division takes place, each daughter cell receives a chromosome. The movement of the sister chromosomes to opposite cell poles, called *chromosome partitioning*, is discussed in Section 10.1.6. As discussed earlier (note 31), rapidly growing *E. coli* requires approximately 40 min (the C period) to replicate its chromosome, followed by a period (the D period) of about 20 min before cell division is complete.

2.3.2 Cell division (cytokinesis)

For a review of the proposed steps involved in cell division (cytokinesis), see ref. 39.

Septum formation

The first morphological sign of cell division is the centripetal synthesis of a septum at midcell, which forms by inward growth of the cell membrane and peptidoglycan layers. (The factors that determine the site of septum formation will be discussed later.) In *E. coli*, this begins soon after chromosome replication is complete (i.e., at the end of the C period). In many gram-negative bacteria, including *E. coli*, the outer envelope also invaginates at this time (i.e., cell separation and septation occur at approximately the same time) so that septa usually are not seen in thin sections of cells prepared for electron microscopy. Invagination is manifested as a constriction, which becomes narrower as septation proceeds (Fig. 2.6A). Initially there is a double layer of peptidoglycan in the septum, but this splits in half as constriction proceeds and the cells separate. Septation and peptidoglycan synthesis are coupled; that is, inhibitors of septal peptidoglycan synthesis prevent septation. (Later we shall discuss PBP3, also called FtsI, the enzyme in *E. coli* required for septal peptidoglycan synthesis. See the subsection entitled *FtsI and FtsW* and note 65, cited therein, which give a list of the inhibitors of PBP3.)

In *E. coli*, septation occupies during approximately the first 10 to 13 min of the D period, depending upon the growth rate (slower growing cells have a longer period of septation). Daughter cells separate within 7 min of completion of the septum.

Cell division in *B. subtilis* differs from that in *E. coli* in that septation precedes cell separation, although cell separation may begin before septation is complete (Fig. 2.6B).[40] The result of all of this is that thin sections of dividing *B. subtilis* cells are seen under electron microscopy to have partially completed septa, and there is no visible constriction at the beginning of septation. As in gram-negative bacteria, cell separation in *B. subtilis* (and other gram-positive bacteria) is effected by the splitting of the septum. Under certain conditions of rapid growth, septum formation in *B. subitilis* can actually be completed before cell separation begins, and as a consequence, chains of cells grow attached to each other via their septa.

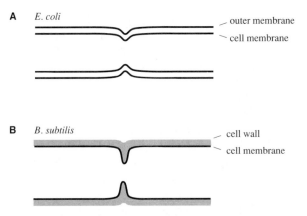

Fig. 2.6 Cell division in gram-negative and gram-positive bacteria (A) Many gram-negative bacteria divide by constriction as the outer membrane invaginates with the cell membrane. Thus, septation and cell separation occur together. (B) In gram-positive bacteria, a septum consisting of cell membrane and peptidoglycan forms, but there is generally little constriction in the early stages. Thus, septation precedes cell separation.

2.3.3 Proteins required for septum formation and cell division

It must be emphasized that the physiological events underlying cell division in bacteria are incompletely understood. However, there has been important progress in identifying the genes and proteins involved.[41-43] Most research in the area has focused on *E. coli*, where several temperature-sensitive division mutants have been isolated, as well as *Bacillus* and *Caulobacter*. (See note 44 for a description of temperature-sensitive mutants.) At the restrictive temperature, temperature-sensitive mutations in genes required for cell division result in the growth of filamentous cells (i.e., cells that grow longer but do not divide). Since mutations in genes not directly involved in septation or cell separation can also result in filamentous growth at restrictive temperatures (e.g., mutations that cause a defect in DNA synthesis), one must be careful in assigning a role for a particular gene in cell division, per se. (See note 45 for additional comments.) Several of the cell division genes are called *fts* genes, for <u>f</u>ilamentation <u>t</u>emperature-<u>s</u>ensitive phenotype. At the restrictive temperature, the cells do not divide but grow into long filaments that eventually lyse. There are at least 10 proteins in *E. coli* known to be associated with the septal ring and essential for cell division. They are FtsZ, FtsA, ZipA, FtsK, FtsQ, FtsL, FtsB (also called YgbQ), FtsW, FtsI (also called PBP3), and FtsN. (See note 46 for a discussion of two nonessential septal ring proteins involved with cell division.)

FtsZ and FtsA are cytoplasmic, and the others are inner membrane proteins associated at the constriction site or the septum. Importantly, FtsZ forms a ring in the cell center. The FtsZ ring is made before invagination and is believed to contract, leading to the constriction of the cell in the center. As we shall see shortly, other cell division proteins join the FtsZ ring and are also important for septum formation. The FtsZ ring with the other proteins is called a *septal ring*. As will soon become clear, we do not yet know the precise roles that most of the cell division proteins play in cell division. Figure 2.7 is a model of how the cell division proteins might be arranged at the site of septation.

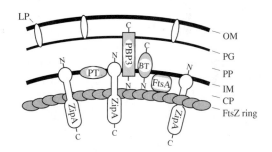

Fig. 2.7 Septal ring model in which a ring of FtsZ proteins is connected to the cell membrane by ZipA. Other abbreviations as follows: PT, FtsK and FtsW division proteins; BT, FtsL, FtsN, and FtsQ division proteins; PP, periplasm; CP, cytoplasm; LP, lipoprotein; OM, outer membrane; PG, peptidoglycan; IM, inner (cell) membrane. *Source*: Hale, C. A., and P. A. J. de Boer. 1997. Direct binding of FtsZ to ZipA, an essential component of the septal ring structure that mediates cell division in *E. coli*. *Cell* 88:175–185.

The proteins required for septum formation and cell division in *E. coli* will be described next. However, it must be pointed out that although all bacteria and archaea thus far examined have the *ftsZ* gene, and those with peptidoglycan have the *ftsI* gene (involved in peptidoglycan synthesis), this is not true of the other cell division genes. Genome sequences reveal that some bacteria and archaea lack some or all of the other cell division genes identified in *E. coli*. It is not known whether prokaryotes missing some of own 10 cell division proteins have evolved different proteins to take their place or whether these proteins simply are not always necessary. Erickson discusses the occurrence of the *E. coli* cell division genes in eight other bacteria and one archaean, as well as the implications of their absence.[47]

FtsZ

The FtsZ protein is responsible for initiating the formation of what will become the septal ring. Refer again to Fig. 2.7 for a model of how FtsZ and the other cell division proteins might interact to form the septal ring. FtsZ is a cytoplasmic protein that polymerizes on the inner surface of the cell membrane at the division site to form a circumferential ring early in the cell cycle, and it remains at the leading edge of the invaginating membrane during septation (Fig. 2.8).[48,49] Polymerized FtsZ recruits FtsA and ZipA to form the FtsZ ring. After the other proteins have been added, the ring is referred to as the septal ring. The septal ring is not present in newborn cells or at the ends of cells, indicating that it is disassembled after septation and is assembled again in the daughter cells prior to the next cell division. The cellular location of FtsZ can be visualized by means of immunoelectron microscopy, immunofluorescence, and GFP fusions. (See the subsequent discussion of ZipA for an explanation of GFP fusions, and FtsI for immunofluorescence microscopy.) That mutations in *ftsZ* lead to filaments without septa or constrictions is a reflection of the importance of the FtsZ protein. It is now clear that between the daughter chromosomes, FtsZ forms a contractile ring that constricts the cell during division. (See note 50 for a discussion of the proposed models for the mechanism of constriction of the FtsZ ring.) FtsZ is related to tubulin, which forms the eukaryotic cytoskeletal microtubules, and it has been suggested that tubulin and FtsZ may have evolved from the same protein. (See note 51 for a discussion of tubulin and FtsZ.)

FtsZ is present in all bacteria thus far examined, including mycoplasma, as well as in the plant *Arabidopsis*, where it was suggested to be a choroplast protein and is presumably involved in fission of the chloroplast, and in members of the domain Archaea.[52,53] However, FtsZ alone is not sufficient for division, since mutations in the other *fts* genes will also lead to filamentous growth even though such mutants still form a ring of FtsZ in the center of the cell. Indeed, there exist immunofluorescence data suggesting that FtsA and FtsN are also located in the FtsZ ring after it forms. According to one model, the FtsZ ring forms first, perhaps with ZipA, and then the other proteins are recruited to the septal ring in sequence by proteins that are in the ring (see Section 2.3.4). There is no invagination or septation until all the cell division proteins have been localized at the site of cell division. This has been demonstrated by means of null mutants.

FtsA

Another protein in the septal ring is FtsA. It appears that the recruitment of FtsA to the septal ring requires the prior localization of FtsZ. FtsA possesses an amino acid sequence domain found in ATPases, and FtsA isolated from *B. subitilis* has ATPase activity.[54] How the protein functions is not known, but it appears that FtsA interacts with FtsZ. This suggestion is prompted by the observation that for cell division to continue, the ratio of FtsZ to FtsA must be within a narrow range, and that FtsA,

Fig. 2.8 Schematic drawing showing recruitment of FtsZ to the leading edge of the invaginating cell membrane during septation.

along with FtsN, becomes localized at the site of septum formation in a ring with FtsZ.[55,56] Research with the yeast two-hybrid system has shown that FtsZ and FtsA interact, and it has been demonstrated that the C-terminal portion of FtsZ is important for its interaction with FtsA.[57,58] (See note 59 for a description of the yeast two-hybrid system.)

ZipA

ZipA (Fts**Z** **i**nteracting **p**rotein **A**) is a cell membrane protein in *E. coli* that binds via its C-terminal end to the FtsZ ring, perhaps as it is forming.[60] ZipA, which may be an anchor in the membrane to which FtsZ attaches (Fig. 2.7), was identified by its ability to bind to radioactively labeled FtsZ after separation of the membrane proteins by gel electrophoresis. It was also shown to be located in the septal ring using a fluorescent ZipA–GFP fusion protein.[61] (See note 62 for a further description of these experiments.) Fluorescent ZipA–GFP rings are found in young cells before invagination is seen, and they appear to coincide with the FtsZ rings. Mutants lacking ZipA form filaments without constrictions. Thus, it appears that the recruitment of ZipA to the septal ring is an early event that may occur at around the same time that FtsZ is recruited, or shortly after ward. The primary sequence of ZipA has not generally been conserved among the bacteria. One exception is *Haemophilus influenzae*, which does possess a ZipA homologue. Other bacteria probably possess functional homologues of ZipA that play a role in anchoring FtsZ to the membrane. Perhaps one of the transmembrane proteins at the septal site plays such a role.

FtsI and FtsW

FtsI and FtsW are located at the site of septum formation, along with FtsZ, FtsA, FtsN, and ZipA.[63,64] As explained in note 65, their localization was determined using immunofluorescence microscopy. The *ftsI* gene, also called *phpB*, codes for the FtsI protein, which is also called PBP3, for **p**enicillin-**b**inding **p**rotein. This is an enzyme required for the formation of peptide cross-links during peptidoglycan synthesis specifically in the septum. (See Section 11.1.2 for a discussion of peptidoglycan synthesis.) The equivalent protein in *B. subtilis* is

called PBP2B and is encoded by the *pbpB* gene. (See note 66 for a further discussion of the roles played by the penicillin-binding proteins during growth and cell division.) Inactivation of FtsI, either with antibiotics or via conditional mutants, results in the cessation of cell division, but the cells maintain their rod shape and can elongate, indicating that the enzyme is not involved in forming peptidoglycan cross-links elsewhere in the cell wall. FtsI (PBP3) may act together with FtsW to cross-link the peptidoglycan strands in the septum. One reason for suggesting that FtsW acts with FtsI is that FtsW is homologous to RodA, an integral membrane protein that augments the cross-linking activity of PBP2, which is a penicillin-binding protein involved in peptidoglycan cross-link synthesis during cell elongation. (Inactivation of PBP2 and mutations in *rodA* result in spherical *E. coli* cells.) It has also been demonstrated that FtsW is required to recruit FtsI to the septal ring.[67]

FtsB, FtsK, FtsL, FtsN, and FtsQ

The exact functions of the proteins encoded by *FtsB, FtsK, FtsL, FtsN*, and *FtsQ*, the five genes required for cell division are not as well understood as are the functions of the other cell division proteins. It is known that their protein products are all membrane associated at the division site. Each has a cytoplasmic domain at the N-terminal end, a single transmembrane domain, and a C-terminal domain in the periplasm. At least two of them, FtsK and FtsN, are associated with FtsZ as well as FtsA in the septal ring.[68] They were recognized either because mutations resulted in an inhibition of cell division, causing filamentation, or because they were able to suppress other cell division mutants. One of the cell division functions of FtsK in *E. coli* is to activate the separation of chromosome dimers that can form If an unequal number of recombinations have occurred between daughter circular chromosomes during DNA replication. FtsK does this by activating the *dif*–XerCD recombinase system. (See the discussion of chromosome separation in Section 10.1.6.)

EnvC

In 2004 another protein, EnvC (YibP), was identified as a septal ring factor required for septal murein cleavage, an activity that is important for outer membrane constriction and separation of daughter cells.[69]

EnvA

The product of the *envA* gene is necessary for cell separation. (Mutants in *envA* form chains of cells.) As discussed in Section 11.2.2, the gene actually codes for an enzyme required for an early step in lipopolysaccharide biosynthesis (lipid A synthesis), suggesting that synthesis of the outer envelope lipopolysaccharide is necessary, perhaps in an indirect way, for cleavage of the peptidoglycan in the septum and consequent cell separation. Thus, *envA* should be considered to be a gene that is involved in lipopolysaccharide biosynthesis and has pleiotropic effects.

2.3.4 A model for the sequence of events in the assembly of the proteins at the division site

A model for the sequence of events in the assembly of the cell division proteins at the division site for *E. coli* postulates an ordered sequence of events during which the prior localization of the earlier added proteins is necessary for the recruitment of the later added proteins. First FtsZ forms the ring, perhaps with the help of ZipA. Then FtsA is recruited by FtsZ and joins the ring, followed by FtsK. Next to join is FtsQ, followed by FtsL and FtsB; then FtsW is added, followed by FtsI, and last FtsN. The assembly of FtsL and FtsB is codependent. FtsQ is found only in bacteria with a cell wall, and it may be involved with cell wall synthesis. Shortly after or around the time of the addition of FtsN, constriction begins at the septal site. See note 70 for a summary of evidence for the model. In *B. subtilis* the situation is different, in as much as the proteins are recruited to the septal ring in a concerted manner, rather than sequentially. See note 71 for experimental evidence in support of this conclusion.

2.3.5 Determining the site of septum formation; the nucleoid occlusion system and the Min system

A major goal of cell division research has been to discover the factors that determine the site of septum formation. The site of septum formation is determined by two separate systems, called *nucleoid occlusion* and *Min*. These two systems exert a negative control. That is, they *prevent* septum formation in parts of the cell other than the cell center, thus ensuring that septum formation takes place only at the cell center.[72]

Nucleoid occlusion

"Nucleoid occlusion" refers to cytological evidence indicating that the FtsZ ring does not form in the parts of the cell containing a dense nucleoid mass. Somehow, the nucleoid mass interferes with the polymerization of the FtsZ ring on the inner surface of the cell membrane.[73] (In 2004 a DNA-binding protein was identified in *B. subtilis* that prevents FtsZ ring formation in the vicinity of the nucleoid.[74]) Because of nucleoid occlusion, the septum can form only at polar regions of the cell or in the space between segregated nucleoids. This has been demonstrated in temperature-sensitive mutants that are defective in DNA segregation at the nonpermissive temperature (*parC* mutants). In these mutants, which are filaments at the nonpermissive temperature and contain a large, central unsegregated nucleoid, the FtsZ rings were positioned primarily to either side of the nucleoid.[75]

Min system

The Min system, which has been best studied in *E. coli*, consists of cell-membrane-associated proteins that prevent the formation of the FtsZ ring in locations other than the cell center, such as at the cell poles. *E. coli* has an operon, called the *minB* operon. The *minB* operon consists of three genes, *minC*, *minD*, and *minE*, and it is the protein products of these genes that determine the site of septation.[41,42,76,77] MinC blocks the formation of the FtsZ ring in vivo and interferes with the polymerization of the FtsZ subunits in vitro. Because MinC interferes with polymerization of FtsZ without affecting GTP hydrolysis (required for FtsZ polymerization), it has been suggested that MinC stimulates the disassembly of the FtsZ ring. Thus, it is MinC that determines whether an FtsZ ring will form at any particular site. Its presence says "no" and its absence says "yes." However, MinC does not work alone. It requires MinD in vitro to be effective at physiological concentrations; and in vivo it prevents FtsZ ring formation only after binding to membrane-associated MinD, which forms a MinCD complex that is bound to the inner surface of the cell membrane. Membrane-bound MinD recruits MinC to form the membrane-bound MinCD complex. At any single moment, there is less MinCD complex at the cell center, and therefore this is where the FtsZ ring forms. In *E. coli*, but not necessarily in other bacteria (see later), the amount and location of the MinCD complex is regulated by MinE. MinE accumulates as a ring near midcell and keeps the midcell region free of MinCD. According to a current model, MinE stimulates the release of MinCD from the membrane into the cytosol. The evidence that MinCD prevents FtsZ ring formation, and that MinE relieves the inhibition of FtsZ ring formation, is as follows.

1. Deletion of the *min* locus results in cells that divide at the poles as well as in the center to produce anucleate minicells (hence the name *min* genes), as well as large multinucleate cells. Thus it can be concluded that the Min proteins are not required for septation per se, but rather for the localization of the septum to the center of the cell.

2. When MinE is made in excess, inhibition by MinCD is relieved both in the center and at the poles and minicells are produced. This is the same phenotype that is obtained with *minC* and/or *minD* mutants.

3. When MinE is absent, MinCD coats the entire inner surface of the cell membrane, and long filamentous cells without FtsZ rings form.

4. When MinCD is in excess, cell division is prevented. Excess FtsZ can overcome the inhibition by MinCD.

The Min proteins oscillate between the two polar regions of the cell

With the aid of GFP fusions to each of the Min proteins, and fluorescence microscopy, the intracellular location of the MinC, MinD, and MinE proteins in live cells has been monitored. [See the earlier discussion of ZipA (note 62) for an explanation of the use of the green fluorescent protein from the jellyfish *Aequorea victoria* to localize proteins in situ.] What has been learned is that in *E. coli*, the Min proteins are not static, but move in a defined manner. (See Fig. 2.9, and note 78 for a discussion of the mechanism by which this occurs.)

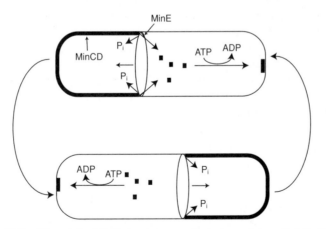

Fig. 2.9 Oscillation of MinCD and the MinE ring between the halves of the cell. MinCD (heavy line) forms a cylindrical coat on the inner surface of the cell membrane in one half of the cell, and a ring of MinE has formed at one end of the MnCD cylinder, close to the center of the cell. The arrow indicates that the MinE ring moves sideways toward the cell pole, causing the removal of MnCD from the membrane in that half of the cell. Membrane-associated MinD has bound ATP, and MinE promotes the removal of MinCD from the membrane by stimulating MinD ATPase activity. The removal of MinCD from one half of the cell is followed by its reassembly at the other half, beginning as a polar cap. This requires the exchange of MinD-bound ADP with cytosolic ATP. The result is that the entire membrane is covered by MinCD most of the time, for the cell center, where MinE prevents the formation of MinCD. *Source*: adapted from Errington, J., R. A. Daniel, and D.-J. Scheffers. 2003. Cytokinesis in bacteria. *Microbiol. Mol. Biol. Rev.* 67:52–65.

Membrane-bound MinD rapidly oscillates from pole to pole with a periodicity that can be as short as 20 s, so that at any one time MinD has coated the inner membrane surface of half the cell like liquid in a test tube.[79,80] Because MinD recruits MinC to the membrane, MinC (which inhibits FtsZ ring formation) also oscillates back and forth from one end of the cell to the other. At the start of a cycle, the MinCD complex, which covers the inner membrane from near the midcell region to one of the cell poles, progressively disassembles from the midcell region toward the cell pole. When the disassembly process reaches the cell pole, a new complex of MinCD begins to assemble at the opposite pole and rapidly spreads toward the midcell region, covering the other half of the cell. The oscillation is such that the amounts of MinCD are lowest at the cell center, allowing a ring of FtsZ to form there. The experiments, and a model explaining the results is reviewed in ref. 81. The oscillation of MinCD between both halves of the cell is due to the oscillation of a ring of MinE that forms at midcell and moves along the membrane at the rim of the "test tube" of MinCD toward a cell pole, causing the release of MinCD from the membrane (Fig. 2.9). This is followed by the reassembly of a MinCD cap at the opposite pole. (The oscillation of MinCD and MinE is described in ref. 38.) It has been suggested that MinE does this by stimulating the release of MinD from the membrane, which requires prior stimulation of MinD ATPase activity (see note 78). In this model, the oscillating MinE ring keeps the membrane in the cell center clear of MinCD, thus allowing the FtsZ ring to form there.

2.3.6 Some differences with respect to the regulation of the site of septum formation in bacteria

A similar, and at the same time somewhat different, situation holds for *B. subtilis*. The *minC* and *minD* genes are found in *B. subtilis*, whose genome has been completely sequenced. As in *E. coli*, MinCD in *B. subtilis* functions to ensure that the FtsZ ring assembles only at the cell center during division of growing cells. However, in contrast to *E. coli*, *B. subtilis* does not have *minE*, and there is no oscillation of MinCD. *B. subtilis* does have a protein called

DivIVA that retains MinCD at the cell poles. For a more complete discussion of the Min system in *B. subtilis*, and differences from *E. coli*, see note 82.

There are several other bacteria whose genome sequences indicate that they are missing one or more of the *min* genes. For example, *minC*, *minD*, and *minE* are missing from *Mycoplasma genitalium*, *Mycoplasma pneumoniae*, and *Haemophilus influenziae*, whose genomes have been 100% sequenced, as well as from *Streptococcus pyogenes*, over 95% of whose genome has been sequenced. There are no homologues of *minC* or *minE* in *Streptomyces coelicolor*.

2.4 Growth Yields

When a single nutrient (e.g., glucose) is the sole source of carbon and energy, and when its quantity limits the production of bacteria, it is possible to define a *growth yield constant*, Y. The growth yield constant is the amount of dry weight of cells produced per weight of nutrient used (i.e., Y is the weight of cells made divided by the weight of nutrient used). It has no units, since the weights cancel out. [The molar growth yield constant is Y_m, which is the dry weight of cells produced (in grams) per *mole* of substrate used.] For example, the $Y_{glucose}$ for aerobically growing cells is about 0.5, which means that about 50% of the sugar is converted to cell material, and 50% is oxidized to CO_2. For certain sugars and bacteria, the efficiency of conversion to cell material can be much lower (e.g., 20%). These differences are thought to be related to the amount of ATP generated from unit weight of the carbon source.

The more ATP made, the greater the growth yields. This makes sense if you consider that growth is, after all, an increase in dry weight, and a certain number of ATP molecules is required to synthesize each cell component. A value of 10.5 g of cells per mole of ATP (called the Y_{ATP} or molar growth yield) has been determined for fermenting bacterial cultures growing on glucose as the energy source. The experiments are performed in media in which all the precursors to the macromolecules (e.g., amino acids, purines, pyrimidines) are supplied, so that the glucose serves only as the energy source, and essentially all the glucose carbon is accounted for as fermentation end products.

Knowing the amount of ATP produced in the fermentation pathways, it is possible to calculate the Y_{ATP} from the $Y_{glucose}$. For example, 22 g of *Streptococcus faecalis* cells is produced per mole of glucose. This organism ferments a mole of glucose to fermentation end products using the Embden–Meyerhof–Parnas pathway that produces two moles of ATP (Section 8.1). Thus, the Y_{ATP} is 22/2 or 11. On the other hand, *Zymomonas mobilis* produces only 8.6 g of cells per mole of glucose fermented. These organisms use a different pathway for glucose fermentation, the Entner–Doudoroff pathway, which yields only 1 ATP per mole of glucose fermented (Section 8.6). Thus, the Y_{ATP} is 8.6. As mentioned, a comparison of several different fermentations by bacteria and yeast has produced an average Y_{ATP} of 10.5. Although this value can vary, knowledge of the growth yields and the Y_{ATP} has allowed some deductions regarding which fermentation pathway might be operating. Also, an unexpectedly high growth yield in fermenting bacteria can point to unrecognized sources of metabolic energy (Section 3.8.3).

2.5 Growth Kinetics

2.5.1 The equation for exponential growth

During exponential growth, the mass in the culture doubles in each generation. The equation for exponential growth is

$$x = x_0 2^Y \qquad (2.3)$$

where x is anything that doubles each generation, x_0 is the starting value, and Y is the number of generations; x can be cell number, mass, or some cell component (e.g., protein, DNA). Taking $\log_{10}$ of both sides,[83] we write

$$\log x = \log x_0 + 0.301Y \qquad (2.4)$$

Let us define g as the generation time (i.e., the time per generation). Therefore, $g = t/Y$, and thus $Y = t/g$, where t is time elapsed.

Therefore, eq. 2.4 becomes

$$\log x = \log x_0 + (0.301/g)(t) \qquad (2.5)$$

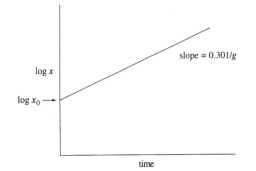

Fig. 2.10 Exponential growth plotted on a semilog scale: g, generation time (time per generation).

You will recognize this as an equation for a straight line. When x is plotted against t on semilog paper (which is to base 10), the slope is $0.301/g$ and the intercept is x_0. See Fig. 2.10.

The generation time, g

The generation time, the time needed for the population of cells to double, is an important parameter of growth and is probably the one most widely used by microbiologists. For this reason, the student should learn how to find the generation time. The simplest way is graphically. In practice, one determines the generation time from inspection of a plot of x versus t on semilog paper (Fig. 2.3 or 2.9) and simply reads off the time it takes for x to double. Or, if x, x_0, and t are known accurately for the exponential phase of growth, one can use eq. 2.3.

The growth rate constant, k

The growth rate constant, k, is a measure of the *instantaneous* growth rate and has the units of reciprocal time. Another widely used parameter of growth, k is not to be confused with the generation time, g, which is the *average* time needed for the population of cells to double. To clarify the distinction, recall that since each bacterial cell gives rise to two cells, the rate at which the population is growing at any instant, that is, the instantaneous rate of growth (dx/dt), must be equal to the number of cells at that time (x) times a growth rate constant (k). Thus:

$$dx/dt = kx, \quad \text{or} \quad dx/x = k\,dt \qquad (2.6)$$

Upon integration, we have

$$\ln x = kt + \ln x_0$$

or $$x = x_0 e^{kt} \qquad (2.7)$$

Taking $\log_{10}$ of both sides of eq. 2.7 converts the expression from the natural log ($\log_e$) to log base 10, so that x can be plotted against t on semilog paper:

$$\log x = 0.4342(kt) + \log x_0 \qquad \text{or}$$

$$\log x = kt/2.303 + \log x_0 \qquad (2.8)$$

Note that this gives the same line as in Fig. 2.10. However, the slope is equal to $k/2.303$. Since the slope in Fig. 2.10 is equal to $0.301/g$, it follows that $k/2.303$ must be equal to $0.301/g$. Thus $k = 0.693/g$ (or $kg = 0.693$). Therefore, k and g vary as the reciprocal of each other with the proportionality constant of 0.693. The slope of the curve in Fig. 2.10 can give you either k (the instantaneous growth rate) or g (the doubling time for the population).[84]

Using the growth equations

The growth equations can be used for finding g and k. In another, more routine, circumstance, one is growing a culture to be used at a later time and must choose the proper size inoculum. Convenient approaches to remember for this calculation have been given (eq. 2.3 or 2.5). Suppose you have an exponentially growing culture at a density of 10^8 cells/mL and you would like to subculture it so that 16 hr later the density of the new culture will also be 10^8 cells/mL. If $g = 2$ h, what should x_0 be?

One way to do this problem is to estimate the number of generations (Y) and then use eq. 2.3 or 2.5. For example, since $Y = t/g$, the number of generations is 16/2, or 8. By using eq. 2.3, we find $10^8 = x_0 2^8$ or $x_0 = 10^8/2^8 = 3.9 \times 10^5$ cells/mL. This means that the initial cell density in the growth flask should be 3.9×10^5 cells/mL. Since, however, the cell density of the inoculum is 10^8/mL, the inoculum must be diluted $10^8/3.9 \times 10^5$ or 2.6×10^2 times. To grow 1 liter of cells, the inoculum size would have to be 3.8 mL (see note 85). Most routine growth problems can be solved with the following equations:

$$x = x_0 2^Y$$

$$Y = t/g$$

$$g(k) = 0.693$$

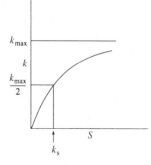

Fig. 2.11 Variation of growth rate as a function of substrate concentration: k_{max}, maximal growth rate constant; k, specific growth rate constant; k_s, nutrient concentration that gives $\frac{1}{2}k_{max}$; S, nutrient concentration. The concentrations at which growth rate is proportional to nutrient concentration are very low, in the micromolar range.

2.5.2 The relationship between the growth rate (k) and the nutrient concentration (S)

In the natural environment the concentrations of nutrients is so low that the growth rates are limited by the rates of nutrient uptake or the rate at which a stored nutrient is used. At very low nutrient concentrations, the growth rate can be shown to be a function of the nutrient (substrate) concentration (Fig. 2.11). The curve approximates saturation kinetics and is probably due to the saturation, in the membrane, of a transporter that brings the nutrient into the cell. The curve can be compared with the kinetics of solute uptake on a transporter, as described in Section 16.2.

One can rationalize the kinetics by assuming that at very low concentrations of nutrient, the growth rate is proportional to the percentage of transporter that has bound the nutrient. At high nutrient concentrations, the rate of nutrient entry approaches its maximal value because all of the transporter has bound the nutrient. Therefore the growth rate becomes independent of the nutrient concentration and is how limited by some other factor. The curve in Fig. 2.11 can be described by the following equation, where k is the specific growth rate constant, k_{max} is the maximal growth rate constant, S is the nutrient concentration, and K_s is the nutrient concentration that gives $0.5\,k_{max}$:

$$k = \frac{k_{max}S}{K_s + S} \qquad (2.9)$$

These are the same kinetics used to describe the saturation of an enzyme by its substrate, i.e. Michaelis–Menten kinetics (Section 6.2.1). Equation 2.9 can be used to calculate growth yields during continuous growth (Section 2.4).

2.6 Steady State Growth and Continuous Growth

A culture is said to be in steady state growth (balanced growth) when all its components double at each division and maintain a constant ratio with respect to one another. Steady state growth is usually achieved when a culture is maintained in exponential growth by subculturing (i.e., when it is not allowed to enter stationary phase) or during continuous growth.

2.6.1 The chemostat

Continuous growth takes place in a device called a chemostat (Fig. 2.12). A reservoir continuously feeds fresh medium into a growth chamber at a flow rate (F) set by the operator. In the chemostat the concentration of the limiting nutrient in the reservoir is kept very low so

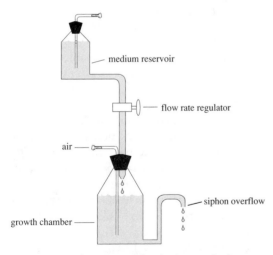

Fig. 2.12 A chemostat. The device works because the growth of the cells in the growth chamber is limited by the rate at which a particular limiting nutrient (e.g., glucose) is supplied. That is, the concentration of the limiting nutrient in the reservoir is low enough to ensure that there is never an excess in the growth chamber. At each moment when fresh medium comes into the growth chamber, the incoming limiting nutrient is rapidly utilized and growth cannot continue until fresh nutrient enters.

that growth is limited by the availability of the nutrient. When a drop of fresh medium enters the growth chamber, the growth-limiting nutrient is immediately used up and the cells cannot continue to grow until the next drop of medium enters. This set of conditions allows one to manipulate the growth rate by adjusting the rate at which fresh medium enters the growth chamber.

The dilution rate, D

The dilution rate is equal to F/V, where F is the flow rate and V is the volume of medium in the growth chamber. For example, if the flow rate (F) is 10 mL/h and the volume (V) in the growth chamber is 1 liter, then the dilution rate (D) is 10 mL h^{-1}/1000 mL, or 0.01 h^{-1}. Notice that the units of D are reciprocal time. If one multiplies the dilution rate D by x, the number of cells in the growth chamber, the product Dx represents the rate of loss of cells from the outflow and has the units of cells/time.

Relationship of dilution rate (D) to growth rate constant (k)

In the steady state, $k = D$. To show this, consider that the rate of growth is $dx/dt = kx$. When a steady state is reached in the growth chamber, the rate of formation of new cells (kx) equals the rate of loss of cells from the outflow (Dx). That is, $kx = Dx$, and therefore, $k = D$. Thus in a chemostat, the growth rate of a culture can be changed merely by changing the flow rate, F, which determines the dilution rate, D.

Dependence of cell yield, x, on concentration of limiting nutrient in reservoir, S_r

The actual concentration of cells in the growth chamber is manipulated by changing the concentration of limiting nutrient in the reservoir. Let S_r be the concentration of limiting nutrient in the reservoir and S be the concentration of the nutrient in the growth chamber. The difference, $S_r - S$, must be equal to the amount of nutrient used up by the growing bacteria. If we define the growth yield constant Y as being equal to the mass of cells (x) produced per amount of nutrient used up ($S_r - S$), then $Y = x/(S_r - S)$ or, rearranging,

$$x = Y(S_r - S) \tag{2.10}$$

Where S can be calculated from Eq. 2.9 by substituting D for k. Since S is much smaller than S_r, it can be conveniently ignored in most calculations.

Varying the cell density (x) and growth rate constant (k) independently

Equation 2.10 predicts that if the concentration of rate-limiting nutrient in the reservoir (S_r) increases, the cell density in the growth chamber will increase. When S_r increases, more nutrient enters the growth chamber per unit time. Initially, S must increase. Since the growth rate is limited by the concentration of S (Fig. 2.11), the cells will respond initially by growing faster and the cell density will increase. A new steady state will be reached in which the increased number of cells can and does use up all the available nutrient as it enters. Thus, even though S_r is increased, the dilution rate still controls the growth rate. What has been accomplished, therefore, when S_r is increased, is a higher steady state value of x, according to eq. 2.10, but no change in the growth rate constant (k). There are two conclusions to be drawn:

1. The only way to change the steady state growth rate constant is to change the dilution rate (D), because this changes the rate of supply of S.

2. The only way to change the steady state growth yield is to change the concentration of limiting nutrient in the reservoir, S_r.

2.7 Summary

Growth is defined as an increase in mass and can be conveniently measured turbidometrically. Other methods can be used, provided they measure something that parallels mass increase, such as the rate of increase in cell number (viable or total) or specific macromolecules (e.g., protein, DNA, RNA). Steady state (continuous) growth is defined as the growth of a population of cells during which all the components of the cell double at each division.

Growth in batch culture can progress through a lag and log to stationary phase, where net growth of the population (measured as mass or its equivalent) has ceased. Eventually, the viable cells may decrease in number, and this is referred to as the death phase. The availability of nutrients, the need to synthesize specific enzymes to metabolize newly encountered nutrients, and the accumulation of toxic end products in the medium can explain the different stages of the growth curve. In addition, when cells enter the stationary phase, nutrient depletion causes specific adaptive physiological changes to occur.

Growth during the log phase can be described by a simple exponential equation depicting a first-order autocatalytic process. That is, the mass doubles at each generation. From this equation one can derive a generation time and an instantaneous growth rate constant. These constants are used to characterize growth under different physiological situations and to predict growth yields at specific times for experimental purposes.

A variety of physiological and morphological changes take place in bacteria when they are subjected to nutrient limitation and enter stationary phase. These may include the induction of specific uptake systems to scavenge the environment for limiting nutrient or ions, sporulation, or encystment. Even bacteria that do not sporulate or encyst undergo significant physiological changes when they are starved. Cells may become metabolically less active and more resistant to environmental stresses such as heat, osmotic stress, and certain chemicals (e.g., hydrogen peroxide). Bacteria starved for a required amino acid, or for a carbon and energy source (stringent response), may also undergo inhibition of ribosomal RNA synthesis, which is correlated with increased synthesis of guanosine tetra- and pentaphosphates. The response to starvation is mediated in part by an increase in the transcription of certain genes. The transcription of many of these genes requires σ^s also called σ^{38}. Cells also become smaller and may change their morphology when they are starved. There may also be changes in the chemistry of the cell surface.

When cells are grown at different growth rates because of nutritional alterations or chemostat growth, the macromolecular composition changes. Faster growing cells are larger, have proportionally more ribosomes, and have more DNA per cell. This observation can be rationalized in terms of an approximate constancy of ribosome efficiency in protein synthesis, and an almost constant period between the initiation of DNA replication and cell division.

Diauxic growth is characterized by two phases of population growth separated by a stationary phase during which the cells are incubated with certain pairs of carbon and energy sources. In diauxic growth with glucose, for example, the cells grow on the glucose first because it represses the expression of the genes necessary to grow on the second carbon source and because it prevents the uptake of other sugars. Repression under these conditions is called catabolite repression or glucose repression. Prevention of sugar uptake is called inducer exclusion. There are at least two possible rationales for this. Glucose is the most widely used carbon and energy source, and cells are better able to outgrow their neighbors if they use the glucose first. In doing so, moreover, they lower the supply of glucose to other cells. Furthermore, many bacteria always express the genes to metabolize glucose (i.e., they are constitutive); the genes to metabolize other carbon sources, however, are often not expressed unless the carbon source is present in the medium. This lowers the energy burden of carrying genetic information. However, glucose is certainly not a universal catabolite repressor among the bacteria. For example, several obligately aerobic bacteria (e.g., *Rhizobium*) are known to grow preferentially on organic acids such as succinate, malate, and fumarate, when given a mixture of glucose and one of these acids. This makes physiological sense for these organisms because growth on the C_4 carboxylic acids is faster than growth on glucose.

E. coli begins DNA replication when it reaches a critical mass in the cell cycle. Following DNA replication, the sister chromosomes are partitioned to the ends of the cells and cell division takes place. Cell division is a complex process requiring at least 10 genes in *E. coli* to form a septum in the center of the cell. Most of the gene products, including FtsZ, are part of a septal ring that constricts the cell, forming two cells.

Cells can be grown in continuous culture by using a chemostat. Two advantages to growing cells this way are as follows: (1) the cells can be maintained in balanced growth and (2) the growth rate constant can be easil changed simply by altering the flow rate. The growth yields can be separately manipulated by changing the concentration of limiting nutrient in the reservoir. Growth of continuous cultures is possible when the growth rate of the culture is limited by the supply of a nutrient that is continuously fed into the growth chamber.

Study Questions

1. What is the generation time (g) of a culture with a growth rate constant (k) of 0.01 min^{-1}?

   ans. 69.3 min

2. Assume you wanted to grow a culture to 10^8 cells/mL in 3 h. The generation time is 30 min. What should be the starting cell density (x_0)?

   ans. 1.6×10^6

3. Assume an inoculum whose cell density is 10^8/mL. The generation time is 30 min. If you started with a 10^{-2} dilution, how many hours would you grow the culture to reach 10^8/mL?

   ans. 3.3 h

4. Assume you have a stock culture at 5×10^9 cells per milliliter and you wish to inoculate 1 liter of fresh medium so that in 15 h the cell density will be 2×10^8/mL. Assume a generation time of 3.5 h. What should be the dilution? What size inoculum should be used?

   ans. 1/500 or 2×10^{-3}; 2 mL

5. Assume the yield coefficient for glucose ($Y_{glucose}$) is 0.5 g of cells per gram of glucose consumed. In a glucose-limited chemostat, what should be the concentration of glucose in the reservoir to produce a mass of cells in the growth chamber (x) of 0.1 mg/mL? For this problem, ignore S in eq. 2.9 because it is much smaller than S_r.

   ans. 0.2 mg/mL

6. In a 500 mL chemostat, what should be the flow rate (F) in minutes for a generation time (g) of 6 h?

   ans. 0.95 mL/min

7. Suppose you were operating a chemostat with S as the limiting nutrient. Assume that D is 0.2 h^{-1}, K_s is 1×10^{-6} M, and k_{max} is 0.4 h^{-1}.

    a. What is the concentration of S in the growth chamber?

    b. If the cell density were 0.25 mg/mL, what would be the concentration of S in the reservoir for $Y_S = 0.5$?

    c. What is the concentration of S in the growth chamber when the cells are growing only half as fast (i.e. $D = 0.1$ h^{-1})?

    ans. a. 1×10^{-6} M; b. 0.5 mg/mL;

       c. 3.3×10^{-7} M

8. During which phases of population growth would you not use cell number as an indicator of growth?

9. What is the physiological role of RpoS? When is it made?

10. What are three models that have been proposed for the mechanism of contraction of the FtsZ ring?

11. Compare the ways used by *E. coli* and *B. subtilis* to ensure that septation takes place at midcell.

REFERENCES AND NOTES

1. Nyström, T. 2004. Stationary-phase physiology. *Annu. Rev. Microbiol.* **58**:161–181.

2. Siegele, D. A., and R. Kolter. 1992. Life after log. *J. Bacteriol.* **174**:345–348.

3. Matin, A. 1991. The molecular basis of carbon-starvation-induced general resistance in *Escherichia coli. Mol. Microbiol.* **5**:3–10.

4. Kolter, R., D. A. Siegele, and A. Tormo. 1993. The stationary phase of the bacterial life cycle. *Annu. Rev. Microbiol.* **47**:855–874.

5. Hengge-Aronis, R. 1996. Regulation of gene expression during entry into stationary phase, pp. 1497–1512. In: *Escherichia coli and Salmonella: Cellular and Molecular Biology.* F. C. Neidhardt et al. (Eds.). ASM Press, Washington, DC.

6. Kjelleberg, S. 1993. *Starvation in Bacteria.* Plenum Press, New York and London.

7. This is especially obvious among the cytophages and flexibacteria, the latter sometimes decreasing in length from over 100 µm to approximately 10 to 30 µm.

8. Kjelleberg, S., M. Hermansson, and P. Marden. 1987. The transient phase between growth and nongrowth of heterotrophic bacteria, with emphasis on the marine environment. *Annu. Rev. Microbiol.* **41**:25–49.

9. McCann, M. P., J. P. Kidwell, and A. Matin. 1991. The putative σ factor KatF has a central role in development of starvation-mediated general resistance in *Escherichia coli. J. Bacteriol.* **173**:4188–4194.

10. Loewen, P. C., and R. Hengge-Aronis. 1994. The role of the sigma factor σ^s (KatF) in bacterial global regulation. *Annu. Rev. Microbiol.* **48**:53–80.

11. O'Neal, C. R., W. M. Gabriel, A. K. Turk, S. J. Libby, F. C. Fang, and M. P. Spector. 1994. RpoS is necessary for both the positive and negative regulation of starvation survival genes during phosphate, carbon, and nitrogen starvation in *Salmonella typhimurium. J. Bacteriol.* **176**:4610–4616.

12. Hengge-Aronis, R. 2000. The general stress response in *Escherichia coli*, pp. 161–178. In: *Bacterial Stress Responses.* G. Storz and R. Hengge-Aronis (Eds.). ASM Press, Washington, DC.

13. Price, C. W. 2000. Protective function and regulation of the general stress response in *Bacillus subtilis* and related gram-positive bacteria, pp. 179–197. In: *Bacterial Stress Responses.* G. Storz and R. Hengge-Aronis (Eds.). ASM Press, Washington, DC.

14. Mulvey, M. R., J. Switala, A. Borys, and P. C. Loewen. 1990. Regulation of transcription of *katE* and *katF* in *Escherichia coli. J. Bacteriol.* **172**:6713–6720.

15. Gentry, D. R., V. J. Hernandez, L. H. Nguyen, D. B. Jensen, and M. Cashel. 1993. Synthesis of the stationary-phase sigma factor σ^s is positively regulated by ppGpp. *J. Bacteriol.* **175**:7982–7989.

16. Gene fusions are valuable probes to monitor the expression of genes of interest. Consider a *lacZ* fusion. The fused gene has the promoter region of the target gene but not the promoter for the *lacZ* gene. Expression of the fused gene is therefore under control of the promoter region of the target gene. The fusions produce a hybrid protein, its amino-terminal end is derived from the target gene and the carboxy-terminal end from β-galactosidase. The hybrid protein has β-galactosidase activity. Therefore, an assay for β-galactosidase is a measure of the expression of the target gene. Thus, one can measure the expression of virtually any gene simply by constructing the proper gene fusion and performing an assay for β-galactosidase. One can construct gene fusions in vitro or in vivo. In vitro construction involves using restriction endonucleases to cut out a portion of the gene with its promoter region from a plasmid containing the cloned DNA. The excised portion is ligated to a *lacZ* gene, without its promoter, or ribosome-binding site, in a second plasmid. The plasmid containing the fused gene is then introduced into the bacterium, and transformants are selected on the basis of resistance to an antibiotic-resistant marker on the plasmid and the production of β-galactosidase.

17. Fang, F. C., S. J. Libby, N. A. Buchmeier, P. G. Loewen, J. Switala, J. Harwood, and D. G. Guiney. 1992. The alternative sigma factor KatF (RpoS) regulates *Salmonella* virulence. *Proc. Natl. Acad. Sci. USA* **89**:11978–11982.

18. Wilmes-Riesenberg, M. R., J. W. Foster, and R. Curtiss III. 1997. An altered *rpoS* allele contributes to the avirulence of *Salmonella typhimurium* LT2. *Infect. Immun.* **65**:203–210.

19. Hengge-Aronis, R. 1996. Back to log phase: σ^s as a global regulator in the osmotic control of gene expression in *Escherichia coli. Mol. Microbiol.* **21**:887–893.

20. Gourse, R. L., T. Gaal, M. S. Bartlett, J. A. Appleman, and W. Ross. 1996. rRNA transcription and growth rate-dependent regulation of ribosome synthesis in *Escherichia coli. Annu. Rev. Microbiol.* **50**:645–677.

21. Cashel, M., D. R. Gentry, V. J. Hernandex, and D. Vinella. 1996. The stringent response, pp. 1458–1496. In: *Escherichia coli and Salmonella: Cellular and Molecular Biology*, Vol. 1. F. C. Neidhardt et al. (Eds.). ASM Press, Washington, DC.

22. In addition to an inhibition of ribosomal RNA synthesis, ppGpp seems to be responsible for the following:

 a. A large decrease in the rate of synthesis of protein.
 b. A temporary cessation in the initiation of new rounds of DNA replication.
 c. An increase in the biosynthesis of amino acids.
 d. A decrease in the rates of synthesis of phospholipids, nucleotides, peptidoglycan, and carbohydrates.
 e. A stimulation of development in the myxobacterium *Myxococcus xanthus*. This is discussed in Section 18.16.3.
 f. A stimulation of expression virulence genes in *Salmonella*. (See Pizzaro-Cerdá, J., and K. Tedin. 2004. The bacterial signal molecule, ppGpp, regulates *Salmonella* virulence gene expression. *Mol. Microbiol.* **52**:1827–1844.)
 g. Stimulation of the synthesis of sigma factor σ^s, which is induced during starvation. (See Mulvey, M. R., J. Switala, A. Borys, and P. C. Loewen. 1990. Regulation of transcription of *katE* and *katF* in *Escherichia coli. J. Bacteriol.* **172**:6713–6720.)

One must therefore conclude that the increase in levels of ppGpp and the responses due to those increases can be viewed as a general response to conditions that limit growth, not simply a result of amino acid starvation. This is because ppGpp levels rise as a consequence of any nutrient or energy limitation (e.g., nitrogen limitation) or a shift to a poorer carbon or energy source. Despite the obvious importance of ppGpp or one or more metabolically related compounds in mediating the stringent response, little is known concerning how ppGpp is responsible for the myriad effects that appear to be correlated with increased of this molecule. One might think of ppGpp as a second messenger that receives a signal from the environment (i.e., nutrient or energy depletion) and transfers the signal to the genome, either activating or inhibiting transcription of relevant genes, or affecting some other cellular process such as the activity of an enzyme. Of course, ppGpp need not act directly on the target, and perhaps it is one component in a longer signal transduction sequence.

23. Ribosomal RNA and ribosomal proteins are synthesized independently of one another and then assembled into ribosomes. However, the synthesis of ribosomal proteins depends on continued synthesis of rRNA. Hence, a decrease in the synthesis of rRNA leads to an inhibition of ribosomal protein synthesis. This is because in the absence of rRNA to which they normally bind, certain ribosomal proteins will bind to ribosomal mRNA and inhibit translation. To undertand this, it is important to recognize that ribosomal proteins are encoded in operons in which the translation of the genes is coupled. *E. coli* has at least 20 such operons that encode the ribosomal proteins, and in certain instances also DNA primase, RNA polymerase, and elongation factors.

In translational coupling, the operon produces a polycistronic mRNA, and the translation of the gene immediately upstream is required for translation of the downstream gene. A model of translational coupling for the downstream gene postulates that the translational start region, including the start codon, is inside a hairpin loop in the mRNA and thus cannot bind to a ribosome. However, when the ribosome translating the upstream gene comes to the stop codon of the upstream gene then the ribosome disrupts the secondary structure of the mRNA and "opens it up," making the translational start region for the downstream gene accessible to a second ribosome. One of the ribosomal proteins encoded by the operon is believe to be able to prevent translation of the entire operon when the protein is present in excess. It is therefore a translational repressor.

The repressor ribosomal protein can bind not only to rRNA but also to the initiating region of the mRNA of one of the genes (usually the first one) in the operon. It appears that in at least some instances the sequence in the mRNA to which the repressor protein binds is similar to the sequence in the rRNA, thus accounting for the ability of the repressor ribosomal protein to bind to both the ribosomal RNA and the mRNA. As an example of such translational autoregulation, consider translation of the *rplK–rplA* operon which encodes the 50S ribosomal subunit proteins L11 (*rplK*) and L1 (*rplA*). The *rplA* gene is downstream of the *rplK* gene, and its translation requires prior translation of the *rplK* gene. The L1 protein regulates the translation of the entire operon. It can bind to either rRNA or the mRNA in the translational initiation region of the *rplK* gene, and it can block translation not only of the *rplK* gene but also of the *rplA* gene that follows it. All the ribosomal protein operons are regulated in a similar manner by one of the protein products.

24. The ppGpp can be synthesized on ribosomes that are "stalled" because of a restriction in the supply of aminoacylated tRNA. This occurs during amino acid starvation (e.g., by restricting amino acids to auxotrophs) and requires the product of the *relA* gene. The *relA* gene was discovered in a search for mutants that failed to respond to amino acid starvation with the stringent response. These mutants were termed "relaxed," hence the name of the gene. The RelA protein [also called (p)ppGpp synthetase I or PSI] is a ribosome-associated protein that synthesizes either ppGpp or pppGpp by displacing AMP and transferring a pyrophosphoryl group from ATP to the 3'-OH of either GDP or GTP:

$$GDP + ATP \rightarrow ppGpp + AMP$$

Because the pool size for GDP in *E. coli* is small relative to GTP, it is believed that the product is pppGpp, which is then dephosphorylated by a 5'-phosphohydrolyase (gpp) to ppGpp. This suggestion is in agreement with the kinetics of appearance of pppGpp and ppGpp. The synthetase is activated in starved (stalled) ribosomes when aminoacylated tRNA becomes limiting during amino acid starvation. The levels of (p)ppGpp may also rise upon carbon and energy starvation (e.g., glucose starvation) even in null *relA* mutants, indicating alternative ways to synthesize (p)ppGpp. Thus, even in a relaxed strain, RNA accumulation stops when the cells are shifted from a good carbon and energy source to a poorer medium (but not when starved for an amino acid). This constitutes the evidence for a relA-independent (p)ppGpp synthetase, called PSII. PSII is thought to be encoded by the *spoT* gene and is believed to be a 3'-pyrophosphotransferase that uses GTP as the acceptor and synthesizes pppGpp. SpoT is also a pyrophosphohydrolase that degrades (p)ppGpp to GDP and GTP. In other words, it has been proposed that SpoT is a bifunctional enzyme that takes part in either the synthesis or the degradation of ppGpp, and the ratio of biosynthetic to degradative activity is somehow controlled by energy starvation. The result is an increase during energy starvation of ppGpp. See the following review article for a discussion of these points: Cashel, M., D. R. Gentry, V. J. Hernandex, and D. Vinella. 1996. The stringent response, pp. 1458–1496. In: *Escherichia coli and Salmonella: Cellular and Molecular Biology*, Vol. 1. F. C. Neidhardt et al. (Eds.). ASM Press, Washington, DC.

25. Xu, J., and R. C. Johnson. 1995. Identification of genes negatively regulated by Fis: Fis and RpoS comodulate growth-dependent gene expression in *Escherichia coli. J. Bacteriol.* 177:938–947.

26. Fis negatively regulates the expression of the *fis* gene and positively or negatively regulates the expression of several other genes. Two-dimensional gel electrophoresis of *fis* mutants has revealed over 20 proteins whose synthesis is decreased (indicating Fis-dependent transcription) and a similar number of proteins whose synthesis has increased (indicating Fis repression). The evidence suggests that Fis binds to both the rRNA DNA and the RNA polymerase and stimulates transcription by interacting with the polymerase rather than by causing an alteration in DNA structure such as bringing upstream DNA closer to the polymerase. See ref. 25; and Gourse, R. L., T. Gaal, M. S. Bartlett, J. A. Appleman, and W. Ross. 1996. rRNA transcription and growth rate–dependent regulation of ribosome synthesis in *Escherichia coli. Annu. Rev. Microbiol.* 50:645–677.

27. Ball, C. A., R. Osuna, K. C. Ferguson, and R. C. Johnson. 1992. Dramatic changes in Fis levels upon nutrient upshift in *Escherichia coli. J. Bacteriol.* 174:8043–8056.

28. Reviewed in: Wagner, R. 1994. The regulation of ribosomal RNA synthesis and bacterial cell growth. *Arch. Microbiol.* 161:100–109.

29. In addition to being a transcriptional activator for rRNA genes, Fis stimulates *Hin*-, *Gin*- and *Cin*-mediated inversion and site-specific recombination of phage λ (both excision and integration) with the bacterial chromosome. Fis also binds to the *E. coli* origin of replication (*oriC*) and possibly plays a role in the initiation of DNA replication. H-NS is involved in transcriptional regulation of several genes whose activity is modulated by cellular stress (e.g., gene regulation in stationary phase, osmotic shock). See the discussion of the nucleoid in Section 1.2.6.

30. Cooper, S., and C. R. Helmstetter. 1968. Chromosome replication and the division cycle of *Escherichia coli* B/r. *J. Mol. Biol.* 31:519–539.

31. The length of time between the onset of DNA replication and the completion of cell division is the sum of two time periods [i.e., the time it takes for the chromosome to replicate (the C period) and the interval between the end of a round of DNA replication and the completion of cell division (the D period)]. The shortest C and D periods are approximately 40 and 20 min, respectively, in rapidly growing *E. coli*. Hence 60 min is the minimum time required between the onset of a round of DNA replication and the completion of cell division. These times do change with growth rate. The C period decreases from approximately 67 min at 0.6 doubling per hour to about 42 min at 2.5 doublings per hour. The D period also decreases (but less so) as the growth rate increases: from 30 min at 0.6 doubling per hour to 23 min at 2.5 doublings per hour. See: Bipatnath, M., P. P. Dennis, and H. Bremer. 1998. Initiation and velocity of chromosome replication in *Escherichia coli* B/r and K-12. *J. Bacteriol.* 180:265–273, and Bremer, H., and P. P. Dennis. 1996. Modulation of chemical composition and other parameters of the cell by growth rate. In: *Escherichia coli and Salmonella typhimurium: Cellular and Molecular Biology*, 2nd ed., Vol. 2, pp. 1553–1569. F. C. Neidhardt et al. (Eds.). ASM Press, Washington, DC.

32. For illustrative purposes, let us assume that the C and D periods are constants at 40 and 20 min, respectively, so that a cell divides 60 min after it has initiated DNA replication. We will also assume that each cell grows exponentially and that $x = x_0 2^Y$, where x_0 is the mass at birth and x is the mass at a point in the cell cycle where Y is the fraction of cell cycle time. For example, if the cell is half-way through the cell cycle, $x = x_0 2^{0.5}$. Let us call a cell dividing every 60 min a 60 min cell, a cell dividing every 40 min a 40 min cell, and a cell that divides every 30 min a 30 min cell. The 60 min cell must be born with one chromosome (and one replication origin) and will begin replicating that chromosome at the beginning of the cell cycle so that 60 min later it divides. Let us say that the mass of the 60 min cell at the beginning of the cell cycle when it begins replicating the chromosome is x_0. This means that the mass (x) 20 min later (when, as we shall see, the 40 min cell begins DNA replication) is about $1.15x_0$ ($x = x_0 2^{0.2}$). The 40 min cell also needs 60 min between the initiation of a round of DNA replication and cell division. This means that the 40 min mother cell must initiate DNA synthesis at two replication origins (one for each of the daughter cell chromosomes) 20 min into the cell cycle. Now if you compare the 40 min cell with the 60 min cell, you will realize that the 40 min cell must on average be larger. This is because by 20 min, the 40 min cell must grow to a mass that will initiate replication at two origins of replication; that is, it must be $2x_0$, where x_0 is the size of the 60 min cell at birth. Since that size is reached by the 40 min cell half-way into the cell cycle, the 40 min cell must be 1.41 times larger than its mass at birth ($x = x_0 2^{0.5} = x_0 1.41$). That is, $2x_0$ is 1.41 times larger than the mass at birth of the 40 min cell. The mass at birth must therefore be $2x_0/1.41$, or about $1.41x_0$. This can be contrasted to the 60 min cell that was born with a mass of x_0. Now let's look at the 30 min cell. To ensure that the daughter cells can divide 60 min later, the 30 min cell must begin replicating the daughter chromosomes at the beginning of the cell cycle. Thus, replication must begin at two origins (one for each of the daughter chromosomes), and the mass at the beginning of the cell cycle must be $2x_0$ as opposed to $1.42x_0$ for the 40 min cell and x_0 for the 60 min cell.

33. Donachie, W. D. 1968. Relationship between cell size and time of initiation of DNA replication. *Nature* 219:1077–1079.

34. Donachie assumed that each cell grew exponentially and doubled in mass during the cell cycle. Knowing the size of the cell at the time of cell division, he was able to calculate the size at any time during the cell cycle. With this information and the knowledge that DNA initiation occurred 60 min before cell division, he was able to compute the mass of the cell at the time of initiation of DNA initiation. This mass, divided by the number of replication origins, was a constant.

35. Wold, S., K. Skarstad, H. B. Steen, T. Stokke, and E. Boye. 1994. The initiation mass for DNA repli-

cation in *Escherichia coli* K-12 is dependent on growth rate. *EMBO J* 13:2097–2102.

36. Cooper, S. 1997. Does the initiation mass for DNA replication in *Escherichia coli* vary with the growth rate? *Mol. Microbiol.* 26:1138–1143.

37. Chandra, N. M., and P. K. Chakrabartty. 1993. Succinate-mediated catabolite repression of enzymes of glucose metabolism in root-nodule bacteria. *Curr. Microbiol.* 26:247–251.

38. Collier, D. N., P. W. Hager, and P. V. Phibbs Jr. 1996. Catabolite repression control in the Pseudomonads. *Res. Microbiol.* 147:551–561.

39. Errington, J., R. A. Daniel, and D.-J. Scheffers. 2003. Cytokinesis in bacteria. *Microbiol. Mol. Biol. Rev.* 67:52–65.

40. Nanninga, N., L. J. H. Koppes, and F. C. deVries-Tijssen. 1979. The cell cycle of *Bacillus subtilis* as studied by electron microscopy. *Arch. Microbiol.* 123:173–181.

41. Donachie, W. D. 1993. The cell cycle of *Escherichia coli. Annu. Rev. Microbiol.* 47:199–230.

42. Vicente, M., and J. Errington. 1996. Structure, function and controls in microbial division. *Mol. Microbiol.* 20:1–7.

43. Lutkenhaus, J., and A. Mukherhjee. 1996. Cell division, pp. 1615–1626. In: *Escherichia coli and Salmonella: Cellular and Molecular Biology*, Vol. 1. F. C. Neidhardt et al. (Eds.). ASM Press, Washington, DC.

44. Temperature-sensitive mutants display the mutant phenotype at an elevated temperature (e.g., 42 °C) but have a wild-type phenotype at a lower temperature (e.g., 30 °C). They are especially useful for studying genes that are essential for growth, since the culture can be maintained at the lower growth temperature but the mutant phenotype can be studied by raising the temperature.

45. Many filamentous mutants are known, some of which are called *fts* mutants. However, because they are defective in other cell events that can affect cell division (e.g., DNA synthesis, nucleoid segregation, protein secretion, heat-shock response), *fts* mutants are not considered to be specifically involved in cell division.

46. In addition to the essential proteins that join the septal ring, two nonessential proteins that also join the septal ring have been identified. These are ZapA (YgfE) and AmiC. ZapA is a cytoplasmic protein that binds to FtsZ. AmiC is a periplasmic N-acetyl-muramoyl-L-alanine amidase. Its role is to cleave the peptide cross-links in the septal murein during daughter cell separation. AmiC cleaves the amide bond between the carboxyl group in the terminal D-alanine and an amino group from a neighbor peptide in the septal murein peptide. AmiC is recruited by FtsN to the septal ring. Various lytic transglycosylases and endopeptidases, as well as other amidases, can also participate in splitting the septal murein. For data on this subject, see: Heidrich, C., A. Ursinus,

J. Berger, H. Schwarz, and J. V. Höltje. 2002. Effects of multiple deletions of murein hydrolases on viability, septum cleavage, and sensitivity to large toxic molecules in *Escherichia coli*. *J. Bacteriol.* **184**:6093–6099.

47. Erickson, H. P. 1997. FtsZ, a tubulin homologue in prokaryote cell division. *Trends Cell. Biol.* **7**:362–367.

48. Bie, E., and J. Lutkenhaus. 1991. FtsZ ring structure associated with division in *Escherichia coli*. *Nature* **354**:161–164.

49. Lutkenhaus, J., and S. G. Addinall. 1997. Bacterial cell division and the Z ring. *Annu. Rev. Biochem.* **66**:93–116.

50. The mechanism of constriction of the FtsZ ring is not understood. One model for constriction proposes that FtsZ is anchored to the membrane and, in the presence of an unidentified motor protein, short FtsZ filaments slide past one another to reduce the circumference of the ring. A second model proposes that FtsZ is anchored to the membrane and that the ring reduces its circumference when the FtsZ filaments depolymerize at the anchor site and lose subunits into the cytoplasm, while the filament remains anchored to the membrane. A third model proposes that the ring constricts because the hydrolysis of GTP by FtsZ results in the bending of the FtsZ filaments.

51. Microtubules are found only in eukaryotic cells. They are hollow cylindrical tubes, about 25 nm in diameter, made from the protein tubulin. The tubulin is a dimer consisting of the subunits α-tubulin and β-tubulin. The dimers exist as 13 rows (called protofilaments) surrounding the central hollow core of the tube. Microtubules are found in the cytoplasm next to plant cell walls and in the cytoplasm of animal cells, where they participate, with kinesin, in moving organelles from one cellular location to another. Microtubules are also important for shape determination of cells, and they make up the spindle apparatus that is necessary for the separation of daughter chromosomes during nuclear division. Tubulin undergoes a GTP-dependent polymerization to extend microtubules (e.g., during the construction of the spindle apparatus). Depolymerization is associated with GTPase activity. Another location of microtubules in eukaryotic cells is in the cilia and flagella. The microtubules, arranged as nine outer microtubule doublets surrounding two singlets in the center of the flagellum or cilium, are responsible for the ATP-dependent bending of these structures during their whiplike beating. FtsZ is not tubulin. However, it resembles tubulin in that it has GTPase activity and is polymerized in vitro in the presence of GTP to form protofilaments that are straight, curved, or shaped in "mini-rings." A third resemblance to tubulin is that FtsZ has a glycine-rich sequence of seven amino acids (GGGTGTG) that is important for GTP binding and is similar to a sequence of seven amino acids in tubulins called the "tubulin signature" (G/AGGTG(S/A)G).

52. Osteryoung, K. W., and E. Vierling. 1995. Conserved cell and organelle division. *Nature* **376**:473–474.

53. Wang, X., J. Lutkenhaus. 1996. FtsZ ring: The eubacterial division apparatus conserved in archaebacteria. *Mol. Microbiol.* **21**:313–319.

54. Feucht, A., I. Lucet, M. D. Yudkin, and J. Errington. 2001. Cytological and biochemical characterization of the FtsA cell division protein of *Bacillus subtilis*. *Mol. Microbiol.* **40**:115–125.

55. Addinall, S. G., and J. Lutkenhaus. 1996. FtsA is localized to the septum in an FtsZ-dependent manner. *J. Bacteriol.* **178**:7167–7172.

56. Addinall, S. G., C. Cao, and J. Lutkenhaus. 1997. FtsN, a late recruit to the septum in *Escherichia coli*. *Mol. Microbiol.* **25**:303–309.

57. Din, N., E. M. Quardokus, M. J. Sackett, and Y. V. Brun. 1998. Dominant C-terminal deletions of FtsZ that affect its ability to localize in *Caulobacter* and its interaction with FtsA. *Mol. Microbiol.* **27**:1051–1063.

58. Wang, X., H. Jian, A. Mukherjee, C. Cao, and J. Lutkenhaus. 1997. Analysis of the interaction of FTsZ with itself, GTP, and FtsA. *J. Bacteriol.* **179**:5551–5559.

59. The yeast two-hybrid system is based upon two properties of the yeast GAL4 protein, which is a transcriptional activator of genes encoding enzymes for galactose metabolism: it has an N-terminal end that binds to the target DNA and a C-terminal end that activates transcription. The N- and C-terminal ends have been cloned separately, and they can be genetically fused to two different proteins (e.g., FtsA and FtsZ) to form hybrid proteins encoded on separate plasmids. If FtsA and FtsZ form a complex such that the N- and C-terminal ends of the GAL4 protein are in close proximity, gene transcription is activated when a yeast cell contains both hybrids (both plasmids). Transcription can be monitored in yeast deleted for GAL4 but containing the fusion reporter gene *GAL1-lacZ* and assaying for β-galactosidase.

60. Hale, C. A., and P. A. J. de Boer. 1997. Direct binding of FtsZ to ZipA, an essential component of the septal ring structure that mediates cell division in *E. coli*. *Cell* **88**:175–185.

61. Helm, R., A. B. Cubitt, and R. Y. Tslen. 1995. Improved green fluorescence. *Nature* **373**:663–664.

62. By using affinity blotting, Hale and de Boer were able to demonstrate that ZipA is capable of binding to FtsZ. The probe was radioactive FtsZ, which was a fusion protein in which a 54 amino acid tag sequence containing a stretch of histidine residues (HFKT) was fused to the N terminus of FtsZ. The tag sequence also had the substrate site for the catalytic portion of heart muscle kinase. The FtsZ fusion protein was purified by nickel chelation chromatography and labeled with ^{32}P via enzymatic phosphorylation. The membrane proteins were separated by means of sodium dodecyl sulfate polyacrylamide gel

electrophoresis (SDS-PAGE) and electrophoretically transferred to a nitrocellulose filter. The filters were then incubated with the radioactive FtsZ fusion protein, and the radioactivity was detected on X-ray film. The radioactive FtsZ bound to ZipA. To localize the ZipA protein in the cells, investigators created a fusion gene called *zipA*-GfpS65t, which encodes the green fluorescent protein GfpS65T, which in turn is fused to the C terminus of ZipA and has a promoter stimulated isopropyl-β-thiogalactoside (IPTG). The GFP is found naturally in the jellyfish *Aequorea victoria* and, when the gene for this protein is expressed in other organisms, such as *E. coli*, a prominent green fluorescence results. When the gene is correctly fused to a gene specifying a protein of interest such as a ZipA, a fused protein is made which can be visualized by its green fluorescence. The fused gene was introduced into *E. coli* on a phage and induced with IPTG. Cells were then observed with a fluorescence microscope, and a bright ring in the septal area indicated the presence of ZipA.

63. Weiss, D. S., K. Pogliano, M. Carson, L.-M. Gusman, C. Fraipont, M. Nguyen-Distèche, R. Losick, and J. Beckwith. 1997. Localization of the *Escherichia coli* cell division protein FtsI (PBP3) to the division site and cell pole. *Mol. Microbiol.* 25:671–681.

64. Wang, L., M. K. Khattar, W. D. Donachie, and J. Lutkenhaus. 1998. FtsI and FtsW are localized to the septum in *Escherichia coli*. *J. Bacteriol.* 180:2810–2816.

65. Immunofluorescence microscopy is more sensitive than immunoelectron microscopy for detecting proteins within the cell. Antibodies are raised against the protein (e.g., against FtsI). The cells are permeabilized (e.g., with lysozyme) and incubated with the specific antibody. The antibody–antigen complex is then visualized by using fluorescein-conjugated anti-IgG. For example, the IgG might be from rabbit, whereas the anti-rabbit IgG would be from another animal.

66. *E. coli* has three other transpeptidases in addition to PBP3 that are used for peptidoglycan synthesis and bind penicillin (viz., PBP1a, PBP1b, and PBP2). However, PBP3 preferentially binds certain β-lactams (viz., furazlocillin, cephalexin, or benzyl penicillin at low concentration). The result is the inhibition of septum formation but not of net peptidoglycan synthesis. This indicates a specific role for PBP3 in peptidoglycan synthesis in the septum. This conclusion is supported by the failure of mutants defective in the synthesis of PBP3 to form septa, although they do elongate. In contrast, mutants defective in the synthesis of PBP2 have round cells because they are defective in synthesizing peptidoglycan cross-links along the length of the cell. With respect to enzymatic activity, there are actually two classes of PBPs, class A and class B. The higher molecular weight PBPs (class A)—for example, PBP1a of *E. coli* and PBP1 of *B. subtilis*—have both transglycosylase and transpeptidase activity. Class B

PBPs (e.g., PBP2 and PBP3) have transpeptidase activity only.

67. Mercer, K. L. N., and D. S. Weiss. 2002. The *Escherichia coli* cell division protein FtsW is required to recruit its cognate transpeptidase, FtsI (PBP3), to the division site. *J. Bacteriol.* 184:904–912.

68. Yu, X.-C., A. H. Tran, Q. Sun, and W. Margolin. 1998. Localization of cell division protein FtsK to the *Escherichia coli* septum and identification of a potential N-terminal targeting domain. *J. Bacteriol.* 180:1296–1304.

69. Bernhardt, T. G., and P. A. J. DeBoer. 2004. Screening for synthetic lethal mutants in *Escherichia coli* and identification of EnvC (YibP) as a periplasmic septal ring factor with murein hydrolase activity. *Mol. Microbiol.* 52:1255.

70. Immunofluorescence microscopy can be used to visualize the assembled cell division proteins. For example, consider the FtsZ ring. Antiserum to purified FtsZ is added to cells previously fixed (e.g., with paraformaldehyde and gluteraldehyde), permeabilized with lysozyme, and fixed to slides. Then a fluorescein-labeled antibody to the anti-FtsZ antibody is added to locate the anti-FtsZ antibody in the cells. Fluorescence microscopy then reveals a bright band corresponding to the FtsZ ring in the center of the cell. A similar approach has been used to locate the FtsA and FtsN proteins to the FtsZ ring. Another cytological approach to view cell division proteins uses the green fluorescent protein to tag the cell division protein. Gene fusions between the GFP gene and the cell division gene are created in vitro, and the recombinant genes are introduced into test cells on a plasmid. The cell division protein can then be localized in whole cells because it fluoresces. By using immunofluorescence or GFP tagging, the various cell division proteins have been localized to the division site at midcell.

The evidence that FtsZ forms a ring independently of the other Fts proteins is that the ring forms even in mutants that do not make the other Fts proteins. The evidence that FtsZ recruits FtsA and FtsN is that if the formation of the FtsZ ring is blocked, then FtsA and FtsN do not localize to the septum. This can be shown using temperature-sensitive mutants that do not form a Z ring at the nonpermissive temperature or by the overproduction of SulA in wild-type cells. When the *sulA* gene carried on a plasmid is induced, the cells stop dividing, form filaments, and do not have Z rings. (The SulA protein interacts with FtsZ.) It is believed that FtsK arrives early at the ring because it does not require FtsQ, FtsL, FtsB, FtsW, FtsI, or FtsN to assemble there. However, FtsN requires all the proteins and is therefore believed to assemble last.

See: Addinall, S. G., and J. Lutkenhaus. 1996. FtsA is localized to the septum in an FtsZ-dependent manner. *J. Bacteriol.* 178:7167–7172, and Addinall, S. G., C. Cao, and J. Lutkenhaus. 1997. FtsN, a late recruit to the septum in *Escherichia coli. Mol. Microbiol.* 25:303–309.

71. *In B. subtilis* the cell division proteins are recruited in a concerted manner, rather than sequentially as appears to be the case for *E. coli.* Thus, in *B. subtilis* mutations or deletions of any of the cell division proteins (DivIB, DivIC, FtsL, PBP2B, and probably FtsW) prevent the recruitment of all of them. Also, in *B. subtilis* there is no homologue to FtsN, and the FtsK homologue (SpoIIIE) is not required for cell division.

72. Xuan-Chuan, Y., and W. Margolin. 1999. FtsZ ring clusters in *min* and partition mutants: role of both the Min system and the nucleoid in regulating FtsZ ring localization. *Mol. Microbiol.* 32:315–326.

73. Norris, V., C. Woldringh, and E. Mileykovskaya. 2004. A hypothesis to explain division site selection in *Escherichia coli* by combining nucleoid occlusion and Min. *FEBS Lett.* 561:3–10.

74. Wu, L. J., and J. Errington. 2004. Coordination of cell division and chromosome segregation by a nucleoid occlusion protein in *Bacillus subtilis. Cell* 117:915–925.

75. Sun, Q., Y. Xuan-Chuan, and W. Margolin. 1998. Assembly of the FtsZ ring at the central division site in the absence of the chromosome. *Mol. Microbiol.* 29:491–504.

76. Raskin, D. M., and P. A. J. De Boer. 1999. MinDE-dependent pole-to-pole oscillation of division inhibitor MinC in *Escherichia coli. J. Bacteriol.* 181:6419–6424.

77. Zhou, H., and J. Lutkenhaus. 2004. The switch I and II regions of MinD are required for binding and activating MinC. *J. Bacteriol.* 186:1546–1555.

78. Refer to Fig. 2.8B. ATP plays an important role in the oscillation of MinD from pole to pole. It is known that MinD has ATPase activity that is stimulated severalfold by MinE. The stimulation in vitro requires phospholipid vesicles, reflecting the fact that stimulation in vivo occurs when MinD is bound to the cell membrane. MinE mutants that are unable to stimulate the MinD ATPase activity do not oscillate the MinD. According to the current model, MinD binds ATP and dimerizes as a consequence, whereupon the ATP–MinD dimer polymerizes on the membrane and recruits MinC, which prevents FtsZ ring formation. The ring of MinE at the receding edge of the cylindrical coat of MinCD displaces MinC that is bound to MinD, and stimulates the hydrolysis of the MinD-bound ATP. As a consequence of the hydrolysis of ATP, the MinD dimers are disrupted, and depolymerized MinD is released from the membrane and enters the cytosol. In the cytoplasm, the ADP is replaced by ATP, MinD dimerizes, and a new layer of polymerized MinCD forms at the opposite pole. For a more complete discussion, read: Lutkenhaus, J., and M. Sundaramoorthy. 2003. MinD and role of the

deviant Walker A motif, dimerization and membrane binding oscillation. *Mol. Microbiol.* 48:295–303.

79. Hale, C. A., H. Meinhardt, and P. A. J. de Boer. 2001. Dynamic localization cycle of the cell division regulator MinE in *Escherichia coli. EMBO J.* 20:1563–1572.

80. Fu, X., Y.-L., Shih, Y. Zhang, and L. I. Rothfield. 2001. The MinE ring required for proper placement of the division site is a mobile structure that changes its cellular location during the *Escherichia coli* division cycle. *Proc. Natl. Acad. Sci. USA* 98:980–985.

81. Lutkenhaus, J. 2002. Dynamic proteins in bacteria. *Curr. Opin. Microbiol.* 5:548–552.

82. *B. subtilis* has MinD and MinC, which as in *E. coli* together prevent FtsZ ring formation. In contrast to *E. coli*, there is no MinE, and the MinCD proteins do not oscillate in *B. subtilis.* Instead of MinE, *B. subtilis* has a protein called DivIVA. The role of DivIVA is retain MinD, and therefore MinC, at the cell poles. It has been proposed that a gradient of MinCD occurs along the long axis of the cell such that the highest concentrations are at the cell poles and the lowest concentrations are at the cell center, enabling the FtsZ ring to assemble there. After the division apparatus has been assembled in the cell center, DivIVA and MinCD localize there at around the time of cell division and remain at the new cell poles when cell division is complete. The placement of the Min proteins at the division site requires the prior localization of FtsZ and PBP2B division proteins. PBP2B is required for peptidoglycan synthesis. MinCD and DivIVA are also at the old cell poles of the daughter cells owing to the previous cell division. For research on this subject, see: Harry, E. J., and P. J. Lewis. 2003. Early targeting of Min proteins to the cell poles in germinated spores of *Bacillus subtilis*: evidence for division apparatus-independent recruitment of Min proteins to the division site. *Mol. Microbiol.* 47:37–48, and Wu, L. J., and J. Errington. 2004. Coordination of cell division and chromosome segregation by a nucleoid occlusion protein in *Bacillus subtilis. Cell* 117:915–925. (DivIVA is also required for the attachment of the chromosome to the cell poles during sporulation. This is discussed in Section 10.1.7.)

83. Suppose you want to convert $\log_2$ to $\log_{10}$. First you would write the exponential equation (e.g., $x = x_0 2^Y$). Now take $\log_{10}$ of both sides of the equation: $\log_{10} x = \log_{10} x_0 + \log_{10} 2(Y)$. Since $\log_{10} 2 = 0.301$, the equation becomes $\log x = \log x_0 + 0.301 Y$. Note that $\log_{10}$ is usually written simply as log.

84. Some investigators use the symbol μ for the instantaneous growth rate constant and k for the reciprocal of the generation time: $k = 1/g$.

85. The dilution is 2.6×10^2. Assume the volume of inoculum is x. Therefore, $2.6 \times 10^2(x)$ must equal the final volume, which is 1000 mL $+ x$. Solving for x gives 3.8 mL.

3

Membrane Bioenergetics: The Proton Potential

A major revolution in our conception of membrane bioenergetics has taken place in the last 40 years as a result of the theoretical ideas of Peter Mitchell referred to as the chemiosmotic theory.[1-6] Briefly, the chemiosmotic theory states that energy-transducing membranes (i.e., bacterial cell membranes; mitochondrial and chloroplast membranes) pump protons across the membrane, thereby generating an electrochemical gradient of protons across the membrane (the proton potential) that can be used to do useful work when the protons return across the membrane to the lower potential. In other words, bacterial, chloroplast, and mitochondrial membranes are energized by proton currents.

Of course, the return of the protons across the membrane must be through special proton conductors that couple the translocation of protons to do useful cellular work. These proton conductors are transmembrane proteins. *Some membrane proton conductors are solute transporters, others synthesize ATP, and others are motors that drive flagellar rotation.* The proton potential provides the energy for other membrane activities besides ATP synthesis, solute transport, and flagellar motility (e.g., reversed electron transport and gliding motility).

Because the chemiosmotic theory is central to energy metabolism, it lies at the foundation of all bacterial physiology. As explained in this chapter, the chemiosmotic theory brings together principles of physics and thermodynamics in explaining membrane bioenergetics. A study

of the principles of the chemiosmotic theory will give the reader a deeper understanding of how the bacterial cell uses ion gradients to couple energy-yielding (exergonic) reactions to energy-requiring (endergonic) reactions. This chapter explains the principles of the theory. For a brief historical account, see Box 3.1.

3.1 The Chemiosmotic Theory

According to the chemiosmotic theory, protons are translocated out of the cell by exergonic (energy-producing) driving reactions, which are usually biochemical reactions (e.g., respiration, photosynthesis, ATP hydrolysis). Some of the translocated protons leave behind negative counterions (e.g., hydroxyl ions), thus establishing a membrane potential, outside positive. Protons may also accumulate electroneutrally in the extracellular bulk phase, establishing a proton concentration gradient that is high on the outside (outside acid). When the protons return to the inside, moving down the concentration gradient and toward the negative pole of the membrane potential, work can be done.

Figure 3.1 illustrates the proton circuit in a bacterial cell membrane. The cell membrane is similar to a battery in that it maintains a potential difference between the inside and outside, except that the current that flows is one of protons rather than electrons. In Fig. 3.1, the potential difference is maintained by reactions 1, which translocate protons to the outside.

BOX 3.1 HISTORICAL PERSPECTIVE: OXIDATIVE PHOSPHORYLATION

The first mechanism suggested to explain how electron transport was coupled to ATP synthesis was published in 1953. It was postulated that mitochondria make a high-energy phosphorylated derivative that donates a phosphoryl group to ADP. The model for this mechanism was based upon the mechanism of ATP synthesis in the cytosol catalyzed by triosephosphate dehydrogenase and phosphoglycerate kinase (see later: Fig. 8.2, reactions 6 and 7). However, no investigators were able to find the postulated phosphorylated intermediate despite many attempts.

A second postulated mechanism was published in 1961 by the English scientist Peter Mitchell (1920–1992). Mitchell hypothesized that electron transport is coupled to the generation of an electrochemical proton gradient, which in turn drives ATP synthesis (the chemiosmotic theory). In 1973 a third mechanism was suggested, namely, that the energy released during electron transport is trapped in a conformational change in an electron carrier protein, and this energy is then used to drive the synthesis of ATP from ADP and inorganic phosphate.

Even as late as 1977 there were some reservations. Mitchell's hypothesis was termed "an attractive possibility" that had not been proven.[1]

It was, however, accepted as being correct and of far-reaching importance, and in 1978 he won the Nobel Prize in Chemistry. For more information about the prevalent ideas in 1977 regarding electron transport and oxidative phosphorylation, the student is referred to a series of review articles.[2]

REFERENCES

1. Boyer, P. D. 1977. Coupling mechanisms in capture, transmission, and use of energy. *Annu. Rev. Biochem.* **46**:957–966.

2. Boyer, P. D., B. Chance, L. Ernster, P. Mitchell, E. Racker, and E. C. Slater. 1977. Oxidative phosphorylation and photophosphorylation. *Annu. Rev. Biochem.* **46**:955–1026.

Reactions 1 include redox reactions that occur during electron transport (Fig. 3.2, 1) and an ATP-driven proton pump (the ATP synthase) (Fig. 3.2, 7). These will be discussed later (Sections 3.7.1 and 3.7.2). The cell membrane does work via reactions 2. The work that is done by the protons that enter the cell includes the extrusion of sodium ions (Fig. 3.2, 3), solute transport (Fig. 3.2, 4), flagellar rotation (Fig. 3.2, 6), and the synthesis of ATP via the ATP synthase (Fig. 3.2, 7, and Section 3.6.2). As Mitchell emphasized (see Box 3.1), the membrane's low permeability to protons is important because the major route of proton reentry is via the energy-transducing proton transporters rather than by general leakage. This, of course, would be expected of a lipid bilayer that is relatively nonpermeable to protons. Some bacteria couple respiration or the decarboxylation of carboxylic acids to the extrusion of sodium ions (Fig. 3.2, 2). (See Sections 3.7.1 and 3.8.1.) The reentry of sodium ions can also be coupled to the performance of work (e.g., solute uptake Fig. 3.2, 5). Once established, the membrane potential can energize the secondary flow of other ions. For example, the influx of potassium ions can be in response to a membrane potential, inside negative, created by proton extrusion. Mitochondrial and chloroplast membranes are also energized by proton gradients. Therefore, this is a widespread phenomenon, not restricted to prokaryotes.

3.2 Electrochemical Energy

When bacteria translocate protons across the membrane to the outside surface, energy is conserved in the proton gradient that is established.

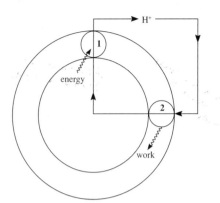

Fig. 3.1 The proton current. There is a proton circuit traversing the bacterial cell membrane. Protons are translocated to the cell surface, driven there by either chemical or light energy through a proton pump (**1**) and returned through special proton transporters (**2**) that do work. The accumulation of protons on the outside surface of the membrane establishes a membrane potential, outside positive. A pH gradient can also be established, outside acid. In several gram-negative bacteria oxidizing certain inorganic compounds (lithotrophs), or single-carbon compounds such as methanol, protons that are released into the periplasm via periplasmic oxidations contribute to the proton current (Chapter 12). In some cases, periplasmic oxidations are the sole provider of protons for inward flux.

The energy in the proton gradient is both electrical and chemical. The *electrical* energy exists because a positive charge (i.e., the proton) has been moved to one side of the membrane, creating a charge separation, and therefore a membrane potential. When the proton moves back into the cell toward the negatively charged surface of the membrane, the membrane potential is dissipated (i.e., energy has been given up and work can be done). The energy dissipated when the proton moves to the inside of the cell is equal to the energy required to translocate the proton to the outside.

Stated more precisely, energy is required to move a charge *against* the electric field (i.e., to the side of the same charge). This energy is stored in the electric field. The energy that is stored in the electric field is called *electrical energy*. Conversely, the electric field gives up energy when a charge moves *with* the electric field (i.e., to the opposite pole) and work can be done. The amount of energy is the same, but opposite in sign.

The same description applies to chemical energy. Energy is required to move the proton against its concentration gradient. This energy is stored in the concentration gradient. The energy that is stored in a concentration gradient is called *chemical energy*. When the proton returns to the lower concentration, the energy in the concentration gradient is dissipated and work can be done.

The sum of the changes in electrical and chemical energies is called *electrochemical energy*. The symbol for electrochemical energy is $\Delta\mu$, which is equal to $\mu_{in} - \mu_{out}$. For the proton, it would be $\Delta\mu_{H^+}$. Electrochemical energy is now expressed are joules per mole. The electrochemical energy is discussed in more detail in the following sections.

3.2.1 The electrochemical energy of protons

The proton motive force

The electrochemical work that is performed when an ion crosses a membrane is a function of both the membrane potential, $\Delta\Psi$, and the difference in concentration between the solutions separated by the membrane. For example, for one mole of protons:

$$\Delta\mu_{H^+} = F\Delta\Psi + RT \ln[H^+]_{in}/[H^+]_{out} \qquad J$$

$$(3.1)$$

In eq. 3.1, $F\Delta\Psi$ represents the electrical energy when one mole of protons moves across a potential difference of $\Delta\Psi$ volts, and $RT \ln[H^+]_{in}/[H^+]_{out}$ represents the chemical energy when one mole of protons moves across a concentration gradient of $[H^+]_{in}/[H^+]_{out}$.

To express eq. 3.1 in millivolts (mV), we simply divide by the Faraday constant $F(\approx 96,500$ C; see later). Since $RT/F \ln[H^+]_{in}/[H^+]_{out} = -60 \Delta pH$ at 30 °C, where $\Delta pH = pH_{in} - pH_{out}$, eq. 3.1 is expressed in millivolts as follows:

$$\Delta p = \Delta\mu_{H^+}/F = \Delta\Psi - 60 \Delta pH \qquad mV \text{ (at 30 °C)}$$

$$(3.2)$$

Usually, $\Delta\mu_{H^+}/F$ is called the proton motive force and is denoted as Δp. The Δp is the potential energy in the electrochemical proton gradient. When protons move toward the lower electrochemical potential, the Δp gives

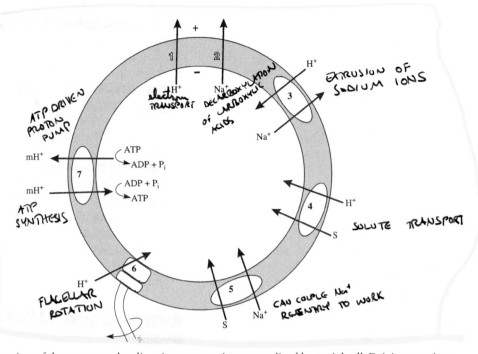

Fig. 3.2 An overview of the proton and sodium ion currents in a generalized bacterial cell. Driving reactions (metabolic reactions) deliver energy to create proton (**1**) and sodium ion (**2**) electrochemical gradients, high on the outside. The major driving reactions encompassed by reaction **1** are the redox reactions that occur during electron transport. The establishment of sodium ion potentials coupled to metabolic reactions such as respiration is not widespread. When the ions return to the lower electrochemical potentials on the inside, work can be done. Built into the membrane are various transporters (porters) that translocate protons and sodium ions back into the cell, completing the circuit, and in the process doing work. There are three classes of porters: (a) *antiporters*, which carry two solutes in opposite directions, (b) *symporters*, which carry two solutes (S) in the same direction, and (c) *uniporters*, which carry only a single solute. The Na^+/H^+ antiporter (**3**) is the major mechanism for extruding Na^+ in bacteria and also functions to bring protons into the cell for pH homeostasis in alkaliphilic bacteria. In most bacteria the Na^+/H^+ antiporter creates the sodium potential necessary for the Na^+/solute symporter because a primary Na^+ pump is not present. The antiporter uses the proton electrochemical potential as a source of energy. The H^+/solute symporter (**4**) uses the proton potential to accumulate solutes, and a Na^+/solute symporter (**5**) uses the sodium electrochemical potential to accumulate solutes. Also shown are a flagella motor that turns at the expense of the proton electrochemical potential (**6**) and an ATP synthase that synthesizes ATP at the expense of the proton electrochemical potential (**7**). The ATP synthase is reversible and can create a proton electrochemical potential during ATP hydrolysis.

up energy (is dissipated) and work can be done (e.g., flagellar rotation, ATP synthesis, solute transport). Cells must continuously replenish the Δp as it is used for doing work. One should view the Δp as a force pulling protons across the membrane into the cell toward their lower electrochemical potential. To replenish the Δp, an equal but opposite force must be exerted to push protons out of the cell toward the higher electrochemical potential (against the Δp). As will be discussed later, the force that generates the Δp (translocates protons out of the cell) can result from several exergonic reactions (reactions that give up energy), the most widely used being oxidation–reduction reactions in

the membrane that occur during respiration, and ATP hydrolysis. Bacteria maintain an average Δp of approximately -140 to $-200\,mV$. The values for respiring bacteria tend to be a little higher than those for fermenting bacteria.[7] Equation 3.2 is a fundamental equation in cell biology. It is derived in more detail later. (See eqs. 3.3–3.10.)

Units: what is meant by volts, electron volts, and joules?

It is important to distinguish between volts (V), electron volts (eV), and joules (J). Potential energy differences (e.g., ΔE, $\Delta \Psi$, Δp) are expressed as volts or millivolts. As long as charges are not moving, these remain as potential energy and work is not done. When charges are moving, work is being done, either on the charges or by the charges, depending upon whether the charges are moving toward a higher potential or a lower potential. The quantity of work that is done is proportional to the product of the amount of charge that moves and the potential difference over which the charge moves.

The units of work are either joules or electron volts. One joule (J) is defined as the energy required to raise a charge of one coulomb (C) through a potential difference of one volt (i.e., $J = C \times V$). (The older literature used calorie units instead of joules: 1 calorie = 4.184 J.) To calculate the change in joules when a *mole* of monovalent ions or electrons travels over a voltage gradient, one multiplies volts by F, the Faraday constant, since this is the number of coulombs of charge *per mole* of electrons or monovalent ions. (See note 8.) One electron volt is (rather than a coulomb) the increase in energy of a single electron or monovalent ion when raised through a potential difference of one volt. The charge on the electron or monovalent ion is approximately 1.6×10^{-19} C. Therefore, one electron volt is equal to 1.6×10^{-19} J. If the electrons are moving toward a lower energy level (i.e., toward a higher electrode potential), then work (e.g., the generation of a Δp) can be obtained from the system. The energy available from the electron flow is $n\Delta E$ eV or $nF\Delta E$ joules (if n refers to moles of electrons). An equivalent amount of work must be done to move the electrons to a higher energy level.

Similarly, if y protons move over a potential of Δp volts, then $y\Delta p$ electron volts of work is done. If y moles of protons move, then $yF\Delta p$ joules of work is done. Often work units are converted to volts or millivolts when one wishes to express the potential energy in a concentration gradient. For example, 17,400 J is required to move one mole of solute against a concentration gradient of 1,000. This can be expressed as the potential energy in the concentration gradient, 0.180 V (17,400/F). We can also think about volts or millivolts as a *force* that *pushes* a molecule down its electrical, electrochemical, or chemical gradient. The Δp, in millivolts, is the force that pushes protons, thus it is called the *proton motive force*.

A more detailed explanation of and derivation of eq. 3.2

1. The electrical component of the $\Delta\mu_{H^+}$

A membrane potential, $\Delta\Psi$, exists across the cell membrane where $\Delta\Psi = \Psi_{in} - \Psi_{out}$. By convention, the $\Delta\Psi$ is negative when the inner membrane surface is negative. The volt is the unit of $\Delta\Psi$. The work done on or by the electric field when charges traverse the membrane potential is equal to the total charges carried by the ions or electrons multiplied by the $\Delta\Psi$. If a single electron or monovalent ion such as the proton moves across the membrane, then the work done is $\Delta\Psi$ electron volts. (For a divalent ion, the work done would be $2\Delta\Psi$ electron volts.) The amount of work that is done *per mole* of protons that traverses the $\Delta\Psi$ is

$$\Delta G = F\Delta\Psi \qquad J \qquad (3.3)$$

Bearing in mind that coulombs × volts = joules (i.e., V = J/F), eq. 3.3 is often expressed as electrical potential energy (or force) in volts by dividing both sides of the equation by the Faraday constant:

$$\Delta G/F = \Delta\Psi \qquad V \qquad (3.4)$$

2. The chemical component of the $\Delta\mu_{H^+}$

Of course if a concentration gradient of protons exists, we must add the chemical energy to the electrical energy in eq. 3.3 to obtain the expression for the electrochemical energy, $\Delta\mu_{H^+}$. The chemical energy of the proton (or any solute) as a function of its concentration is

$$G = G_0 + RT\ln[H^+] \qquad J \qquad (3.5)$$

(See note 9 for a more complete discussion.)

The free energy change accompanying the transfer of one mole of protons between a solution of protons outside $[H^+]_{out}$ the cell and inside $[H^+]_{in}$ the cell is the difference between the free energies of the two solutions,

$$\Delta G = RT\ln[H^+]_{in} - RT\ln[H^+]_{out} \qquad J \quad (3.6)$$

or

$$\Delta G = RT\ln[H^+]_{in}/[H^+]_{out} \qquad J$$

Equation 3.6 refers to the free energy change when one mole of protons moves from one concentration to another, in which the concentration gradient does not change (i.e., as applies to a steady state). It does *not* refer to the total energy released when the concentration of protons comes to equilibrium. *Equation 3.6 can be used for the movement of any solute over a concentration gradient, not simply protons, and we will see this equation again when we discuss solute transport.* (In describing the movement of solutes other than protons, the symbol S may be substituted for H^+. Otherwise, the equation used is identical to eq. 3.6.)

Usually, eq. 3.6 is expressed in electrical units of potential (volts). To do this, one substitutes $8.3144 \, J \, K^{-1} \, mol^{-1}$ for R, 303 K (30 °C) for T, converts ln to $\log_{10}$ by multiplying by 2.303, and divides by F to convert joules to volts, thus deriving eq. 3.7:

$$\Delta G/F = 0.06 \log[H^+]_{in}/[H^+]_{out} \quad V$$

$$= 60 \log[H^+]_{in}/[H^+]_{out} \quad mV \quad (3.7)$$

Since $\log[H^+]_{in}/[H^+]_{out} = pH_{out} - pH_{in}$, eq. 3.7 can be written as

$$\Delta G/F = 60(pH_{out} - pH_{in})$$

$$= -60(pH_{in} - pH_{out})$$

$$= -60 \, \Delta pH \quad mV \quad (3.8)$$

3 Proton electrochemical energy, $\Delta \mu_{H^+}$
The sum of the electrical (**1**) and chemical energies (**2**) of the proton is the proton electrochemical energy. We are now ready to derive an expression for the proton motive force. The total energy change accompanying the movement of one mole of protons through the membrane is the sum of the energy due to the membrane potential (eq. 3.3) and the energy due to the concentration gradient (eq. 3.6). This sum is called the electrochemical energy, $\Delta \mu_{H^+}$:

$$\Delta \mu_{H^+} = F\Delta \Psi + RT \ln[H^+]_{in}/[H^+]_{out} \quad J \quad (3.9)$$

One can also express the electrochemical energy as a potential in volts or millivolts (proton motive force, or Δp) by dividing by the Faraday constant (or by adding eqs. 3.4 and 3.8):

$$\Delta p = \Delta \Psi - 60 \, \Delta pH \quad mV \text{ at } 30 \, °C \quad (3.10)$$

The same equation is used to express the electrochemical potential for any ion (e.g., Na^+),

$$\Delta \mu_{Na^+}/F = \Delta \Psi - 60 \, \Delta pNa \quad mV \quad (3.11)$$

where pNa is $-\log(Na^+)$.

By convention, the values of the potentials are always negative when the cell membrane is energized for that particular ion.

3.2.2 Generating a $\Delta \Psi$ and a ΔpH

We now consider some biophysical aspects of generating the proton motive force. First we will examine the formation of the $\Delta \Psi$, and then we will see how the capacitance of the membrane affects the $\Delta \Psi$ that develops. Finally, we will consider the establishment of a ΔpH.

Electrogenic flow creates a $\Delta \Psi$

The movement of an uncompensated charge creates the membrane potential. When this happens, the charge movement is said to be *electrogenic*. The moving charge can be either a proton or an electron. For example, a membrane potential is generated when a proton is translocated through the membrane to the outer surface, leaving behind a negative charge on the inner surface:

$$H^+_{in} \rightarrow H^+_{out}$$

A membrane potential can also develop if a molecule that has been reduced on the inner membrane surface, picking up cytoplasmic protons in the process, then diffuses across the membrane. This molecule is oxidized on the outer surface, releasing the protons to the exterior, while the electrons return electrogenically across the membrane to the inner surface:

$$A + 2e^- + 2H^+_{in} \rightarrow AH_2$$

$$AH_2 \rightarrow A + 2e^- + 2H^+_{out}$$

This is also electrogenic flow, but in this case it is the electron that is the moving charge rather than the proton. The work done is the same because the electron carries the same charge as the proton (i.e., 1.6×10^{-19} C). It makes no difference from the point of view of calculating the Δp whether one thinks in terms of protons moving as positive charges across the membrane from the inside to the outside, or simply being carried as hydrogen atoms in a reduced organic compound (e.g., AH_2). There is a net translocation of protons in either case. In both instances, the total energy necessary to move y protons against the Δp is $y\Delta p$ electron volts. We will see that bacteria use both electrogenic electron flow and electrogenic proton flow to create a membrane potential.

The size of the membrane potential depends upon the capacitance of the membrane

The membrane potential that develops when even a small number of protons move across the membrane can be more than 100 mV. This can be understood by considering the membrane to be a capacitor. The membrane is a capacitor because the membrane lipids prevent the protons from rapidly leaking back into the cell, and so the membrane stores positively charged protons on one surface, just as a capacitor stores charges on one surface.

The relationship between the charge that accumulates on one face of a capacitor and the voltage across the capacitor is

$$\Delta V = \Delta Q/C \qquad (3.12)$$

where ΔQ is the charge (in coulombs), ΔV is the voltage, and C is a proportionality constant called the capacitance. The value of C for biological membranes is low, about 1 microfarad per square centimeter ($\mu F/cm^2$) of membrane. The equation can be rewritten with different symbols and used to predict the theoretical membrane potential:

$$\Delta \Psi \text{ (volts)} = en/C \qquad (3.13)$$

where e is the charge per proton (1.6×10^{-19} C), n is the number of protons, and C is the capacitance. Assuming a membrane area of about 3×10^{-8} cm^2 (for a spherical cell the size of a typical bacterium), then $C \approx 3 \times 10^{-14}$ F. Therefore, only 40,000 protons translocated to the cell surface is sufficient to generate a membrane potential of -200 mV.[10,11] (By convention, the $\Delta \Psi$ is said to be negative when the inside potential is negative.)

The membrane potential that actually develops varies in magnitude from approximately -60 mV to about -200 mV, depending upon the bacterium and the growth conditions.[6] The membrane potential that develops when a relatively small number of protons are translocated limits electrogenic proton pumping and ultimately the size of the membrane potential itself. This is because the membrane potential, which is negative on the inside, pulls protons back into the cell.

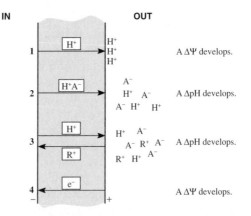

Fig. 3.3 A ΔpH in the bulk phase and a $\Delta \Psi$ cannot develop simultaneously by the movement of a proton. (1) Electrogenic movement of a proton. (2) Establishment of a ΔpH by the extrusion of both protons and counterions. (3) Establishment of a ΔpH by the exchange of protons for cations in the medium. Theoretically, a ΔpH could develop if the proton on the outer surface of the membrane exchanged with a cation from the bulk phase. But even under these circumstances, the membrane potential (positive outside) would limit further efflux of protons. (4) A $\Delta \Psi$ can also develop via electrogenic flow of an electron.

Generating a ΔpH

When a proton is translocated across the membrane, a $\Delta \Psi$ and a ΔpH cannot be created simultaneously. This statement is summarized graphically in Fig. 3.3. Let us first discuss creating a ΔpH (i.e., accumulating protons in the external bulk phase). Remember that the bulk external medium can become acidified during proton translocation only if electrical neutrality is conserved (i.e., each proton in the bulk phase must have a negative counterion). This can happen if the proton is pumped out with an anion (i.e., H^+/R^-) or if a cation enters the cell from the external bulk phase in exchange for the proton (i.e., H^+ in exchange for R^+) (Fig. 3.3). However, these would be electroneutral events and a charge separation, hence a $\Delta \Psi$, would not develop. Therefore, *in the absence of compensating ion flow*, the protons that are pumped out of the cell remain on or very close to the membrane, and a $\Delta \Psi$ rather than a ΔpH (measurable with a pH electrode) is created. (Theoretically, however, a ΔpH and a $\Delta \Psi$ could develop if a cation in the external medium exchanged for the proton on the membrane.)

Of course, some of the protons could be electroneutrally released into the bulk phase to create a ΔpH and some might remain at the outer membrane surface to create a $\Delta\Psi$; but even under these circumstances, a large ΔpH cannot be generated in the face of a large $\Delta\Psi$. This is because the positive charge on the outside surface inhibits further efflux of protons. In fact, to demonstrate the formation of a ΔpH experimentally as a result of proton pumping, a large $\Delta\Psi$ must not be allowed to develop. These conditions are achieved experimentally by making the membrane permeable either to a cation, so that the incoming cations can compensate electrically for the outgoing protons, or to an anion that moves in the same direction as the proton. For example, in many experiments the K^+ ionophore valinomycin is added to make the membrane permeable to K^+ (Section 3.4). When this is done, K^+ exchanges for H^+, and a ΔpH can develop.

Although bacteria cannot make both a $\Delta\Psi$ and a ΔpH with the same proton, they can create a $\Delta\Psi$ during proton translocation and then convert it to a ΔpH. Suppose a $\Delta\Psi$ is created because a few protons are translocated to the cell membrane outer surface. Proton translocation cannot proceed for very long because a membrane potential develops quickly which limits further efflux of protons. However, a cation such as K^+ might enter the cell electrogenically. This would result in a lowering of the membrane potential because of the positive charge moving in. Now more protons can be translocated out of the cell. The protons can leave the outer membrane surface and accumulate in the external phase because they are paired with the anion that was formerly paired with the K^+. Thus, a membrane potential can form during proton translocation and be converted into a ΔpH by the influx of K^+. This is an important way for bacteria to maintain a ΔpH, as described later in Section 15.1.3.

3.3 The Contributions of the $\Delta\Psi$ and the ΔpH to the Overall Δp in Neutrophiles, Acidophiles, and Alkaliphiles

Partly for the reasons stated in Section 3.2.2, the contributions of the $\Delta\Psi$ and the ΔpH to the Δp are never equal. Additionally, the relative contributions of the ΔpH and the $\Delta\Psi$ to the Δp

vary, depending upon the pH of the environment in which the bacteria naturally grow. Sections 3.3.1, 3.3.2, and 3.3.3 summarize, respectively, the relative contributions of the $\Delta\Psi$ and the ΔpH to the Δp in neutrophiles, acidophiles, and alkaliphiles,[12] notice that in acidophiles and alkaliphiles, the $\Delta\Psi$ (acidophiles) or the ΔpH (alkaliphiles) has the wrong sign and actually detracts from the Δp.

3.3.1 Neutrophilic bacteria

For neutrophilic bacteria (i.e., those that grow with a pH optimum near neutrality), the $\Delta\Psi$ contributes approximately 70 to 80% to the Δp, with the ΔpH contributing only 20 to 30%. This is reasonable when one considers that the intracellular pH is near neutrality and therefore the ΔpH cannot be very large.

3.3.2 Acidophilic bacteria

For acidophilic bacteria (i.e., those that grow between pH 1 and pH 4, and not at neutral pH), the $\Delta\Psi$ is positive rather than negative (below an external pH of 3, which is where they are usually found growing in nature) and thus *lowers* the Δp. Under these conditions the force in the Δp is due entirely to the ΔpH. Let us examine an example of this situation. *Thiobacillus ferrooxidans* is an aerobic acidophile that can grow at pH 2.0 (see Section 12.4.1). Because it has an intracellular pH of 6.5, the ΔpH is 4.5, which is very large (remember, $\Delta pH = pH_{in} - pH_{out}$). When the aerobic acidophiles grow at low pH, however, they have a positive $\Delta\Psi$, and for *T. ferrooxidans* the $\Delta\Psi$ is +10 mV.[13] (See note 14.) The contribution of the ΔpH to the Δp is $-60 \Delta pH$ or -270 mV. Since $\Delta p = \Delta\Psi - 60 \Delta pH$, the actual Δp would be -260 mV. In this case, the $\Delta\Psi$ lowered the Δp by 10 mV. As discussed in Section 15.1.3, the inverted membrane potential is necessary for maintaining the large ΔpH in the acidophiles.

3.3.3 Alkaliphilic bacteria

An opposite situation holds for the aerobic alkaliphilic bacteria (i.e., those that grow above pH 9, often with optima between pH 10 and pH 12). For these organisms the ΔpH is one to two units *negative* (because the cytoplasmic pH is less than 9.6) and consequently *lowers* the Δp, by 60 to 120 mV.[15] Therefore, in the alkaliphiles, the potential of the Δp may come entirely from the $\Delta \Psi$. In fact, as explained in Section 3.10, because of the large negative ΔpH, the calculated Δp in these organisms is so low that it raises conceptual problems regarding whether there is sufficient energy to synthesize ATP.

3.4 Ionophores

Before we continue with the discussion of the proton motive force, we must explain ionophores and their use. Ionophores are important research tools for investigating membrane bioenergetics, and their use has contributed to an understanding of the role of electrochemical ion gradients in membrane energetics. As mentioned earlier, membranes are poorly permeable to ions, and this is why the membrane can maintain ion gradients. Ionophores perturb these ion gradients. Most ionophores are organic compounds that form lipid-soluble complexes with cations (e.g., K$^+$, Na$^+$, H$^+$) and rapidly equilibrate these across the cell membrane (Fig. 3.4). The incorporation of an ionophore into the membrane is equivalent to short-circuiting an electrical device with

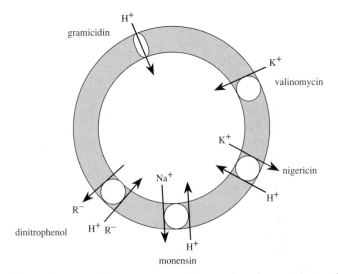

Fig. 3.4 Examples of ionophores and the transport processes that they catalyze. All the reactions are reversible. *Valinomycin*: transports K$^+$ and Rb$^+$. Valinomycin transports only one cation at a time. Since K$^+$ is positively charged, valinomycin carries out electrogenic transport (i.e., creates a membrane potential). If valinomycin is added to cells with high intracellular concentrations of K$^+$, then the K$^+$ will rush out of the cell ahead of counterions and create a transient membrane potential, outside positive, as predicted by the Nernst equation. In the presence of excess extracellular K$^+$ and valinomycin, the K$^+$ will rapidly diffuse into the cell, collapsing the existing potential. *Nigericin*: carries out an electroneutral exchange of K$^+$ for H$^+$. When nigericin is added to cells, one can expect a collapse of the pH gradient as the internal K$^+$ exchanges for the external H$^+$, but the membrane potential should not decrease. The combination of nigericin and valinomycin will collapse both the ΔpH and the membrane potential. *Monensin*: carries out an electroneutral exchange of Na$^+$ or K$^+$ for H$^+$. There is a slight preference for Na$^+$. *Dinitrophenol*: an anion (R$^-$) that carries out electroneutral influx of H$^+$ and R$^-$ into the cell and returns to the outside without H$^+$ (i.e., R$^-$). Therefore this anion will collapse both the ΔpH and the $\Delta \Psi$. This is the classic uncoupler of oxidative phosphorylation. *Gramicidin*: carries out electrogenic transport of H$^+$ > Rb$^+$, K$^+$, Na$^+$. Gramicidin differs from the other ionophores in that it forms polypeptide channels in the membrane. That is, it is not a diffusible carrier. Since the addition of gramicidin results in the equilibration of protons across the cell membrane, it will collapse the $\Delta \Psi$ and the ΔpH (i.e., the Δp). *Carbonyl cyanide-p-trifluoromethylhydrazone (FCCP)* (not shown): a lipophilic weak acid that exists as the nonprotonated anion (FCCP$^-$) and as the protonated form (FCCPH), both of which can travel through the membrane. Protons are carried into the cell in the form of FCCPH. Inside the cell the FCCPH ionizes to FCCP$^-$, which exits in response to the membrane potential, outside positive. The result is a collapse of both the ΔpH and the $\Delta \Psi$.

a copper wire. Another way of saying this is that the ionophore causes the electrochemical potential difference of the ion to approach zero. Since some ionophores are specific for certain ions, it is sometimes possible, by means of the appropriate use of ionophores, to identify the ion current that is performing the work. For example, if ATP synthesis is prevented by a proton ionophore, this implies that a current of protons carries the energy for ATP synthesis.

One can also preferentially collapse the $\Delta\Psi$ or ΔpH with judicious use of ionophores, and perhaps gain information regarding the driving force for the ion current. For example, nigericin, which catalyzes an electroneutral exchange of K^+ for H^+, will dissipate the pH gradient but not the $\Delta\Psi$.[16] Valinomycin plus K^+, on the other hand, will initially dissipate the $\Delta\Psi$ because it electrogenically carries K^+ into the cell, thus setting up a potential opposite to the membrane potential.

It is also possible to *create* a $\Delta\Psi$ by using valinomycin. The addition of valinomycin to starved cells or vesicles loaded with K^+ will induce a temporary $\Delta\Psi$ predicted by the Nernst equation, $\Delta\Psi = -60 \log[S_{in}]/[S]_{out}$ mV. (This is the concentration gradient expressed as millivolts at 30 °C.) What happens is that the K^+ moves from the high internal concentration to the lower external concentration in the presence of valinomycin (Fig. 3.5). Because the K^+ moves faster than its counterion, a temporary diffusion potential, outside positive, is created. The diffusion potential is temporary because of the movement of the counterions. The use of ionophores has helped researchers to determinine which ions are carrying the primary current that is doing the work, and to investigate the relative importance of the membrane potential and ion concentration gradients in providing the energy for specific membrane functions.

3.4.1 The effect of uncouplers on respiration

Uncouplers are ionophores that have the following effects:

1. They collapse the Δp and thereby inhibit ATP synthesis coupled to electron transport.

2. They stimulate respiration.

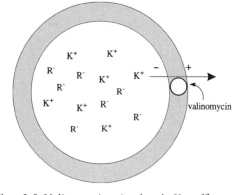

Fig. 3.5 Valinomycin-stimulated K^+ efflux can impose a temporary membrane potential. When valinomycin is added to cells or membrane vesicles loaded with K^+, the valinomycin dissolves in the membrane and carries the K^+ out of the cell. The efflux of K^+ creates a diffusion potential, since the negative counterions lag behind the K^+. The membrane potential is predicted by the Nernst equation, $\Delta\Psi = -60 \log(K^+)_{in}/(K^+)_{out}$. The membrane potential is transient because it is neutralized by the movement of counterions.

Why should uncouplers stimulate respiration? The flow of electrons through electron carriers in the membrane is *obligatorily* coupled to a flow of protons in a closed circuit (Section 4.5). This occurs at the coupling sites discussed in Section 3.7.1. Protons are translocated to the outer surface of the membrane and then reenter via the ATP synthase. If reentry is blocked by inhibitors of the ATP synthase [e.g., dicyclohexylcarbodiimide (DCCD)] or slowed by depletion of ADP, then respiration is slowed.

One possible explanation for the slowing of respiration is that, as the protons are translocated to the outside surface, the Δp rises and approaches the ΔG of the oxidation–reduction reactions. This might slow respiration, since the oxidation–reduction reactions are reversible. One can view this as the Δp producing a "back-pressure." The reentry of the protons via the ATP synthase can be viewed, in this context, as placing a limit on the rise of the Δp and thereby promoting respiration. Uncouplers might also stimulate respiration because they collapse the Δp. In the presence of uncouplers such as dinitrophenol, H^+ rapidly enters the cell on the uncoupler rather than through the ATPase.

3.5 Measurement of the Δp

Measurements of the size of the Δp are a necessary part of analyzing the role that the Δp plays in the overall physiology of the cell. The two components of the Δp, the ΔΨ and the ΔpH are measured separately.[17,18]

3.5.1 Measurement of ΔΨ

The membrane potential is measured indirectly because bacteria are too small for the insertion of electrodes. Suppose a membrane potential exists and an ion is allowed to come to its electrochemical equilibrium in response to the potential difference. Then, at equilibrium, the electrochemical energy of the ion is zero, and we can use the equation for electrochemical energy (for a monovalent ion):

$$\Delta\mu_s/F = 0$$
$$= \Delta\Psi + 60 \log_{10}[S]_{in}/[S]_{out} \qquad mV \quad (3.14)$$

Solving for ΔΨ, we write

$$\Delta\Psi = -60 \log_{10}[S]_{in}/[S]_{out} \qquad mV \text{ at } 30\ ^\circ C$$
$$(3.15)$$

Equation 3.15 is one form of the Nernst equation. It states that the measurement of the intracellular and extracellular concentrations of a permeant ion at equilibrium allows one to calculate the membrane potential. The bacterial cell membrane is relatively nonpermeable to ions; therefore, to measure the membrane potential, one must use either an ion plus an appropriate ionophore (e.g., K^+ and valinomycin) or a lipophilic ion (i.e., one that can dissolve in the lipid membrane and pass freely into the cell). When the inside is negative with respect to the outside, a cation (R^+) is chosen because it accumulates inside the cells or vesicles (Fig. 3.6). When the inside is positive, an anion (R^-) is used. It is important to use a small concentration of the ion to prevent the collapse of the membrane potential by the influx of the ion.

Cationic or anionic fluorescent dyes have also been used to measure the ΔΨ. The dyes partition between the cells and the medium in response to the membrane potential, and the fluorescence is quenched. The fluorescent dye will monitor relative changes in membrane potential (Fig. 3.7). When a fluorescent dye

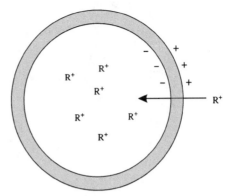

Fig. 3.6 Measurement of membrane potential by the accumulation of a lipophilic cation. The cation accumulates in response to the membrane potential until the internal concentration reaches equilibrium, that is, the point of which efflux equals influx. At this time the electrochemical energy of the cation is zero and $\Delta\Psi = -60 \log [(R^+)_{in}/(R^+)_{out}]$ mV at 30 °C. Any permeant ion can be be used as long as it diffuses passively, is used in small amounts, is not metabolized, and accumulates freely in the bulk phase on either side of the membrane. Permeant ions that have been used include lipophilic cations such as tetraphenylphosphonium (TPP+).

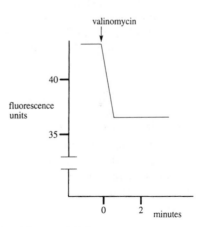

Fig. 3.7 The use of fluorescence to measure the membrane potential. An example of what one might expect if valinomycin were added to bacteria in buffer containing low concentrations of potassium ion in the presence of a fluorescent cationic lipophilic probe. The potassium inside would rush out, creating a diffusion potential as predicted by the Nernst equation. The cationic probe would enter in response to the membrane potential, and the fluorescence would be quenched. It has been suggested that the quenching of fluorescence is due to the formation of dye aggregates with reduced fluorescence inside the cell.

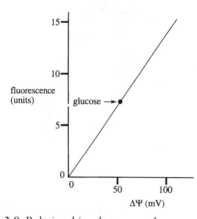

Fig. 3.8 Relationship between the membrane potential and fluorescence change of 1,1′-dihexyl-2,2′-oxacarbocyanine (CC₆). The changes in fluorescence, (ordinate) are plotted as a continuous line corresponding to different membrane potentials imposed by the addition of valinomycin to *Streptococcus lactis* cells. The membrane potentials were calculated by using the Nernst equation from potassium concentration ratios (in/out) in parallel experiments, where the intracellular K^+ concentrations were about 400 mM and the extracellular concentrations were varied. Also shown is the membrane potential caused by the addition of glucose (rather than valinomycin) to the cells. *Source*: Adapted from Maloney, P. C., E. R. Kashket, and T. H. Wilson. 1975. Methods for studying transport in bacteria, pp. 1–49. In: *Methods in Membrane Biology*, Vol. 5. E. D. Korn (Ed.). Plenum Press, New York.

is used to measure the absolute membrane potential, it is necessary to produce a standard curve. This involves measuring the fluorescence quenching in a sample for which the membrane potential is known—that is, has been measured by independent means, such as the distribution of a permeant ion (Fig 3.8).

3.5.2 Measurement of ΔpH

A common way to measure ΔpH is to measure the distribution of a weak acid or weak base at equilibrium between the inside and the outside of the cell. The assumption is that the uncharged molecule freely diffuses across the membrane but the ionized molecule cannot. Inside the cell the acid becomes deprotonated or the base becomes protonated, the extent of which depends upon the intracellular pH. For example,

$$AH \rightarrow A^- + H^+ \quad \text{or} \quad B + H^+ \rightarrow BH^+$$

On addition of a weak acid (or base) to a cell suspension, the charged molecule accumulates in the cell. One uses a weak acid (e.g., acetic acid) when pH_{in} exceeds pH_{out} because the acid will ionize more extensively at the higher pH, and a weak base (e.g., methylamine) when pH_{in} is less that pH_{out} because such a base will become more protonated at the lower pH. At equilibrium, for a weak acid,

$$K_a = [H^+]_{in}[A^-]_{in}/[AH]_{in}$$
$$= [H^+]_{out}[A^-]_{out}/[AH]_{out}$$

where K_a is the dissociation constant of the acid, HA.

If pH_{in} and pH_{out} are at least 2 units higher than the pK, most of the acid is ionized on both sides of the membrane and there is no need to take into account the un-ionized acid. Therefore, the ΔpH can be calculated by using eq. 3.16:

$$pH_{in} - pH_{out} = \Delta pH$$
$$= \log_{10}[A^-]_{in}/[A^-]_{out} \quad (3.16)$$

Thus, the $\log_{10}$ of the ratio of concentrations inside to outside the weak acid is equal to the ΔpH. In practice one uses radioactive acids or bases as probes and measures the amount of radioactivity taken up by the cells. A more complex equation must be used when the concentration of the un-ionized acid (AH) cannot be ignored.[19] Cytoplasmic pH is sometimes also measured by ^{31}P nuclear magnetic resonance (^{31}P NMR) of phosphate whose spectrum is pH dependent. (See note 20.)

3.6 Use of the Δp to Do Work

The Δp provides the energy for several membrane functions, including solute transport and ATP synthesis discussed in this section. (See Chapter 16 for a more complete discussion of solute transport.)

3.6.1 Use of the Δp to drive solute uptake

As an example of how the Δp can be used for doing work, consider symport of an uncharged solute, S, with protons (Fig. 3.2, 4). The total driving force on S at 30 °C is

$$y\Delta p + 60 \log[S]_{in}/[S]_{out} \qquad mV \qquad (3.17)$$

where y is the ratio of H^+ to S and $60 \log[S]_{in}/[S]_{out}$, the force involved in moving S from one concentration to another, is the same as eq. 3.7 but written for S, the solute, rather than for the proton.

At equilibrium, the sum of the forces is zero, and therefore,

$$y\Delta p = -60 \log[S]_{in}/[S]_{out} \qquad (3.18)$$

For $y = 1$ and $\Delta p = -0.180$ V, we write

$$3 = \log[S]_{in}/[S]_{out} \quad and \quad [S]_{in}/[S]_{out} = 10^3$$

Therefore, a Δp of -180 mV could maintain a 10^3 concentration gradient of S_{in}/S_{out} if the ratio of H^+/S were 1. If the ratio of H^+/S were 2, then a concentration gradient of 10^6 could be maintained. It can be seen that very large concentation gradients can be maintained by using the Δp.

3.6.2 The ATP synthase

Built into the cell membranes of prokaryotes and mitochondrial and chloroplast membranes is an enzyme complex that couples the translocation of protons down a proton potential gradient to the phosphorylation of ADP to make ATP. It is called the proton-translocating ATP synthase, or simply ATP synthase. As discussed later, in Section 3.7.2, the ATP synthase is reversible and can pump protons out of the cell, generating a Δp.

Description of the ATP synthase

For an overview of the ATP synthase, see ref. 21. The structure is briefly described here, but more details are provided shortly, when the mechanism of its function is discussed. (See later subsection entitled *Model of the mechanism of ATP synthase*.) Mitochondria, chloroplasts, and bacteria have a similar ATPase with homologous proteins. The ATP synthase consists of two regions, the F_0 and F_1 (Fig. 3.9).[22]

The F_0 region spans the membrane and serves as a proton channel through which the protons are pumped. The F_1 region is located on the inner surface of the membrane and is the catalytic subunit responsible for the synthesis and hydrolysis of ATP. The F_1 unit is also called the coupling factor. The F_1 subunit from *E. coli*

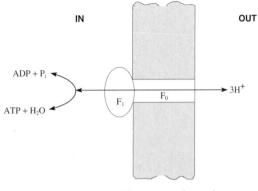

Fig. 3.9 ATP synthase. The proton channel F_0 spans the membrane. It consists of polypeptides of types a, b, and c. The catalytic subunit F_1, on the inner membrane surface, catalyzes the reversible hydrolysis of ATP. It consists of polypeptides of types α, β, γ, δ, and ε. Under physiological conditions, the ATP synthase reaction is poised to proceed in either direction. When Δp levels decrease relative to ATP, they can be restored by ATP hydrolysis. When the ATP levels decrease relative to Δp, more ATP can be made.

is made from five different polypeptides with the following stoichiometry: α_3, β_3, γ_1, δ_1, and ε_1. Each of the β-polypeptides contains a single catalytic site for ATP synthesis and hydrolysis, although the sites are not equivalent at any one time. The F_0 portion has three different polypeptides, which in *E. coli* are stoichiometrically a_1, b_2, and c_{10}. All in all, *E. coli* uses polypeptides of 8 different types to construct a 22-polypeptide machine that acts as a *reversible* pump driven by the proton potential or by ATP hydrolysis.

As discussed more completely in Section 3.7.2, the amount of energy to synthesize one mole of ATP is given by ΔG_p. Assume that ΔG_p is equal to approximately 50,000 J or (dividing by the Faraday Constant) 518 mV. Thus, $y\Delta p$ must be ≥ -518 mV, where y is the number of entering protons. If $y = 3$, then to synthesize one mole of ATP, the Δp must be at least $-518/3$ or -173 mV. (By convention the Δp is negative when energy is available to do work.) (See note 23.)

Evidence that either a $\Delta\Psi$ or a ΔpH can drive the synthesis of ATP via proton influx through the ATP synthase

It is possible to impose a $\Delta\Psi$ or a ΔpH on cells or membrane vesicles and demonstrate that the influx of protons through the ATP synthase

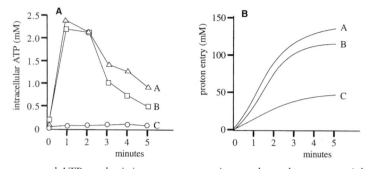

Fig. 3.10 Proton entry and ATP synthesis in response to an imposed membrane potential. A dense suspension of *Streptococcus lactis* cells was incubated in each of three buffers: A, B, and C. Average internal potassium was about 360 mM. For buffer B, the external K$^+$ was 0.4 mM. For buffers A and C, KCl was added to a final concentration of 3 mM. External pH was set at pH 5 (sample A) or pH 6 (samples B and C), and proton entry was measured as the amount of acid added to maintain the external pH at the initial value, using a pH stat. Valinomycin was added at time zero, and intracellular ATP levels (A) and proton entry (B) were measured. Upon addition of valinomycin, the cells immediately made ATP during proton influx (curves A and B). For sample A, the Δp was 200 mV, of which 125 mV was due to the measured membrane potential and 75 mV due to the measured ΔpH. For sample B, the Δp was also about 200 mV despite the smaller contribution from the ΔpH (15 mV). This was because the external K$^+$ concentration in sample B was 0.4 mM instead of 3 mM, thus raising the membrane potential. Notice that in samples A and B the amounts of ATP made and proton influx were approximately the same, despite the differences in the membrane potential and the ΔpH. This suggests that proton influx and ATP synthesis depend upon the Δp rather than on the individual values of the $\Delta\Psi$ or the ΔpH. Sample C had a Δp of only 140 mV (membrane potential, 125 mV; ΔpH, 15 mV) and made no ATP. This suggests a threshold value for Δp for ATP synthesis. Presumably, below the threshold value of Δp, the ATP synthase pumps protons out of the cell at the expense of ATP. *Source*: Maloney, P. C. 1977, Obligatory coupling between proton entry and the synthesis of adenosine 5'-triphosphate in *Streptococcus lactis*. *J. Bacteriol.* **132**:564–575.

results in the synthesis of ATP. A $\Delta\Psi$ can be imposed by using valinomycin and creating a potassium ion diffusion potential. (The use of valinomycin to impose a membrane potential is described in Section 3.4.) A ΔpH can be created by adding acid to the medium in which cells or membrane vesicles are suspended. These critical experiments were done with the lactic acid bacterium *Streptococcus lactis*.[24,25] Washed cells of *Streptococcus lactis* containing a high intracellular concentration of K$^+$ (300–400 mM) were suspended in a medium containing a low concentration of K$^+$ (0.3–0.4 mM). The initial Δp was close to zero. Valinomycin was then added. The valinomycin caused the rapid efflux of K$^+$, which as predicted by the Nernst equation resulted in a temporary potassium diffusion potential, outside positive. The cells responded to the valinomycin by making ATP (Fig. 3.10). This indicates that a $\Delta\Psi$ can drive ATP synthesis.

A similar experiment was done, this time in the presence of N,N′-dicyclohexylcarbodiimide

(DCCD), which is an inhibitor of the ATP synthase. The DCCD inhibited ATP synthesis *and* proton influx (Fig. 3.11). *This indicates that a major route of proton influx is through the ATP synthase and that the ATP synthase is responsible for ATP synthesis.* Notice that proton entry continued even in the presence of DCCD. It was assumed that proton entry into DCCD-treated cells resulted from passive inflow or leakage by routes other than the ATP synthase.

To demonstrate that a ΔpH can drive ATP synthesis, washed cells of *S. lactis* were suspended in media to which a small volume of sulfuric acid was added to lower the external pH to about 3.5. The internal pH was initially about 7.6, giving a ΔpH of 4.1 and a driving force (-60ΔpH) of about -246 mV.[26] The cells responded to the imposed ΔpH by making ATP (Fig. 3.12B) during proton influx (Fig. 3.12A). In these experiments it was necessary to add valinomycin to the samples to stimulate proton

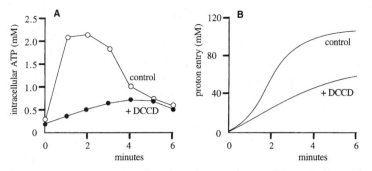

Fig. 3.11 Effect of DCCD on proton entry and ATP synthesis under conditions similar to those for Fig. 3.10. Cells were suspended in buffer A, the ratio of K_{in}^+ to K_{out}^+ was about 120, and valinomycin added at time zero. Some samples were incubated with DCCD, which is an inhibitor of the ATP synthase. DCCD blocked both ATP synthesis and proton entry, suggesting that a major route of influx of protons is through the ATP synthase and that this flow results in ATP synthesis. *Source*: Maloney, P. C. 1977. Obligatory coupling between proton entry and the synthesis of adenosine 5′-triphosphate in *Streptococcus lactis*. 1977. *J. Bacteriol.* **132**:564–575.

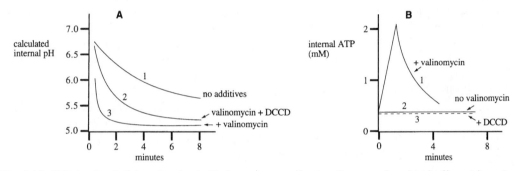

Fig. 3.12 ATP synthesis driven by the ΔpH. *Streptococcus lactis* cells were placed in buffer with various additions. At time zero the external pH was lowered from pH 8 to pH 3.5 with sulfuric acid. (A) The internal pH dropped very slowly (curve 1) unless valinomycin was added (curve 3). This difference can be explained by assuming that in the absence of valinomycin, the entering H⁺ imposed a membrane potential, inside positive, that limited the uptake of protons. Valinomycin allowed the exit of internal K⁺, thus diminishing the buildup of an internally positive potential. The addition of DCCD markedly inhibited proton uptake in the first minute, suggesting that most of the protons that enter rapidly are entering via the ATP synthase (curve 2). (B) The addition of the sulfuric acid at time zero resulted in an immediate synthesis of ATP (curve 1). No such increase in ATP was observed if valinomycin was omitted (curve 2), presumably because net proton influx was slow in the absence of valinomycin. The ATP synthase inhibitor DCCD also prevented ATP synthesis (curve 3). *Source*: Adapted from Maloney, P. C., and F. C. Hansen III. 1982. Stoichiometry of proton movements coupled to ATP synthesis driven by a pH gradient in *Streptococcus lactis*. *J. Membs. Biol.* **66**:63–75.

influx and consequent ATP synthesis. This is because, in the absence of valinomycin, the entering protons established a membrane potential, inside positive, that impeded further net influx of protons. Valinomycin allowed K⁺ to exit the cell in exchange for the proton, thus electrically compensating for the influx of the protons. In this case the exiting K⁺ did not produce a membrane potential because excess K⁺ (0.1 M) was added to the medium to lower

the K_{in}^+/K_{out}^+. Both ATP synthesis and the influx of protons were inhibited by the ATP synthase inhibitor DCCD, indicating that proton entry was primarily through the ATPase.

Model of the mechanism of ATP synthase

The mechanism of ATP synthesis/hydrolysis by the ATP synthase is complex. One model, the binding change mechanism, is based upon the finding that purified F_1 will synthesize tightly bound ATP in the absence of a Δp.[27] (The K_d for the high-affinity site is 10^{-12} M.) The equilibrium constant between enzyme-bound ATP and bound hydrolysis products (ADP and P_i) is approximately one, with the use of soluble F_1.[28] The tight binding of ATP and the equilibrium constant of approximately one suggest that the energy requirement for net ATP synthesis is for the release of ATP from the enzyme rather than for its synthesis. The model proposes that when protons move down the electrochemical gradient through the F_0F_1 complex, a conformational change occurs in F_1 that results in the release of newly synthesized ATP from the high-affinity site. This is illustrated in Fig. 3.13A.

To understand the proposed mechanism of how ATP synthesis is energized, it is necessary to add some more details about the structure of the ATP synthase. (See Fig. 3.13B.) In the F_1 region the three α and three β subunits alternate to form a hollow cylinder that surrounds the γ subunit, which interacts with the three α and three β subunits. The γ subunit forms a central stalk with the ϵ subunit and protrudes from the bottom of the cylinder, where it attaches to the c subunits, which are in the form of a cylindrical ring in the F_0 part embedded in the membrane. The a and b subunits are connected laterally (peripherally) to the c ring. The b subunits form a peripheral stalk connecting the a subunit to a complex of subunits α and δ.

As Fig. 3.13A points out, the "binding-change mechanism" postulates that there are three catalytic sites for synthesizing ATP, one on each of the three β subunits, each of which is paired with an α subunit. Each catalytic site on a β subunit must cycle through three conformational changes for ADP and P_i to be bound (L, or loose, conformation), ATP synthesized (T, or tight, conformation), and ATP released from the enzyme (O, or open, conformation).

What drives the conformational changes? A remarkable mechanism has come to light. Evidence suggests that the ATP synthase contains a centrally located rotary motor driven by proton translocation and that the rotation of the motor causes the conformational changes at the catalytic sites in the β subunits. The motor consists of the c_n subunits comprising a ring in F_0, linked to the γ subunit and the ϵ subunit in F_1. As mentioned, the γ subunit along with the ϵ subunit forms a central stalk in F_1. The idea is that proton flux through F_0 causes the c_n ring to rotate, and this in turn rotates $\gamma\epsilon$ in F_1. When the $\gamma\epsilon$ rotor turns as a consequence of the influx of protons, the γ subunit sequentially makes contact with the three different $\alpha\beta$ subunits, which are prevented from rotating with $\gamma\epsilon$, and force is exerted sequentially on each of the three β subunits, causing each one to undergo three sequential conformational changes (L, T, or O). As mentioned, each site binds ADP and P_i while in the L conformation, synthesizes ATP while in the T conformation, and releases ATP while in the O conformation. It takes approximately three H^+ ions traversing the F_0 portion of the ATP synthase to make one ATP. The molecular structure of the ATP synthase, as well as evidence that it contains a rotating motor, can be found in refs. 29, 30, 31, and 32.

As discussed later in Section 3.7.1, certain bacteria use a sodium motive force, rather than a proton motive force, to provide energy to the ATP synthase. The rotating motor model applies to the sodium motive force as well as the proton motive force. The student may recall that bacteria have another membrane-bound rotating motor driven by proton influx. It is the flagellum motor described in Section 1.2.1.

3.7 Exergonic Reactions that Generate a Δp

Section 3.6 described the influx of protons down a proton potential gradient coupled to the performance of work (e.g., solute transport or ATP synthesis). These activities dissipate the Δp. Let us now consider driving forces that generate a Δp, that is, forces that move protons out of the cell toward a higher potential. The major driving reactions in most prokaryotes are the redox reactions in the cell membranes of respiring organisms (electron transport), ATP hydrolysis in fermenting organisms, and several other less frequently used driving reactions considered in Section 3.8. At this time, let us consider some general thermodyamic features common to all the driving reactions that generate the proton gradient. The translocation of protons out of cells is an energy-requiring process that can be written as

$$y\text{H}^+_{\text{in}} \rightarrow y\text{H}^+_{\text{out}}$$

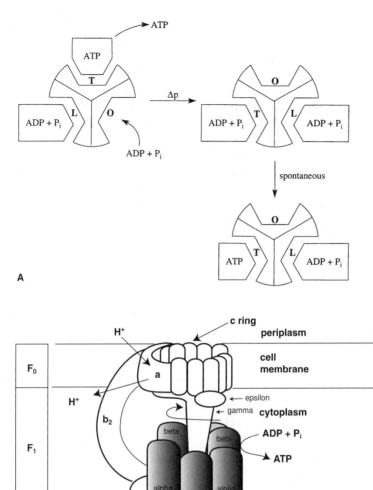

Fig. 3.13 (A) Binding-change mechanism for ATP synthase. Because there are three β subunits encoded by a single gene, and because there is only one catalytic site per β subunit, the maximum possible number of catalytic sites is three. However, the three catalytic sites are not functionally equivalent and are thought to cycle through conformational changes driven by the electrochemical proton gradient. As a result of the conformational changes, newly synthesized ATP is released from the catalytic site. The three conformational states are O (open, very low affinity for substrates), L (loose binding,), and T (tight binding, the active catalytic site). ATP is made spontaneously at site T, which converts to site O in the presence of a Δp and releases the ATP. The Δp also drives the conversion of the preexisting site O to site L, which binds ADP and P_i, as well as the conversion of the preexisting site L to site T. In this model, the only two catalytic sites that bind substrates are L and T. *Source*: Adapted from Cross, L. R., D. Cunningham, and J. K. Tamura. 1984. *Curr. Top. Cell Regul.* 24:335–344. (B). Rotary model for *E. coli* ATP synthase. The 10 c subunits in the F_0 portion form a cylindrical ring in the lipid bilayer. The ε and γ subunits interact with each other and form a central stalk in the F_1 portion that is connected to the c ring. Surrounding the εγ stalk are the three pets of alternating $\alpha_3\beta_3$ subunits. Subunits a and b are at the periphery of the c ring. The a subunit is thought to be confined to the membrane, and the two b subunits extend as a second stalk peripherally located between F_0 and F_1. The b stalk connects to a complex of subunits δ and α in the F_1 portion. The rotor is considered to be the c ring and the εγ stalk, and the stator against which the rotor moves is proposed to be $\alpha_3\beta_3\delta ab_2$. It is thought that protons move through the a and c subunits across the membrane. The flux of H^+ (or Na^+ in the case of certain bacteria that use a sodium motive force) through the F_0 portion is believed to cause the rotor to move with respect to the stator. This is believed to induce conformational changes in the catalytic subunits, α_3, resulting in ATP synthesis. *Source*: Adapted from Nicholls, D. G., and S. J. Ferguson. 2000. *Bioenergetics 3*. Academic Press, Son Diego, CA.

Whese ΔG, the amount of energy required to translocate y moles of protons out of the cell, is $yF\Delta p$ joules, or for y protons, $y\Delta p$ electron volts. (This is the same energy, but of opposite sign, that is released when the protons enter the cell.) Therefore, the ΔG of the driving reaction must be equal to or greater than $yF\Delta p$ joules or $y\Delta p$ electron volts. If the reaction is near equilibrium, which is the case for the major driving reactions, the energy available from the driving reaction is approximately equal to the energy available from the proton gradient and $\Delta G_{\text{driving reaction}} = yF\Delta p$. This relationship allows one to calculate the Δp if y and $\Delta G_{\text{driving reaction}}$ are known.

The most common classes of driving reactions, which we shall discuss first, are oxidation–reduction reactions that occur during respiration and photosynthetic electron transport, and ATP hydrolysis. In all cases we will write the ΔG for the driving reaction. Then we will equate the $\Delta G_{\text{driving reaction}}$ with $yF\Delta p$. Finally, we will solve for Δp.

3.7.1 Oxidation–reduction reactions as driving reactions

Electrode potentials and energy changes during oxidation–reduction reactions

The tendency of a molecule (A) to accept an electron from another molecule is given by its electrode potential E, also called the reduction potential, the redox potential, or the oxidation–reduction potential. Electrons spontaneously flow toward molecules with a higher electrode potential. Under standard conditions (1 M for all solutes and gases at 1 atm), the electrode potential has the symbol E_0 and is related to the actual electrode potential E_h as follows:

$$E_h = E_0 + [RT/nF] \ln[\text{ox}]/[\text{red}]$$

or

$$E_h = E_0 + 2.3[RT/nF] \log_{10}[\text{ox}]/[\text{red}]$$

where n is the number of electrons transferred per molecule, R is the gas constant, and F is the Faraday constant. Usually standard potentials at pH 7 are quoted and the symbol is E_0'. (See note 33 for more discussion of the relationship of E_0 to pH.)

You can see that it is important to know the concentrations of the oxidized and the reduced forms to be able to find the actual electrode potential. Molecules with more negative E_h values are better donors (i.e., electrons spontaneously flow to molecules with more positive E_h values). It is important to understand that when an electron moves over a ΔE_h (which is $E_{h,\text{acceptor}} - E_{h,\text{donor}}$) to a higher electrode potential, the electron is actually moving toward a lower potential energy and as a consequence, energy is released. The energy released is proportional to the potential difference (ΔE_h volts) over which the electron travels. When n electrons move over a potential difference of ΔE_h volts, the energy released is equal to $n\Delta E_h$ electron volts. (Recall that potential differences are given in volts but work units are in electron volts.) Since the total charge carried by one mole of electrons is F coulombs, the *work done per n moles* of electrons is $\Delta G = -nF\Delta E_h$ joules, where ΔG is the Gibbs free energy. The negative sign in the equation shows that energy is released (work is done) when the ΔE_h is positive. As described next, the movement of electrons during an oxidation–reduction reaction toward a lower potential energy level (higher electrode potential) can be coupled to the extrusion of protons toward a higher potential energy level (higher Δp), that is, to the generation of a Δp.

Oxidation–reduction reactions can generate a Δp

An important method of generating a Δp is to couple proton translocation to oxidation–reduction reactions that occur during electron transport. The details of electron transport are discussed in Chapter 4, but here we will simply compute the Δp that can be created. Energy released from oxidation–reductions can generate a Δp because the oxidation–reduction reactions are coupled to proton translocation. We postpone a description of the mechanism of coupling until Section 4.5, concentrating here on the energetic relationship between the ΔE_h and the Δp.

Let us do a simple calculation to illustrate how to estimate the size of the Δp that can be generated by an oxidation–reduction reaction that is coupled to proton translocation. Consider what happens in mitochondria and in many bacteria. A common oxidation–reduction reaction in the respiratory chain is the oxidation of reduced ubiquinone, UQH_2, by cytochrome c_1 (cyt c_1) in an enzyme complex called the *bc_1 complex*. The oxidation–reduction is coupled to the translocation of protons across the membrane, hence the creation of a Δp. As stated earlier, the total energy change during an oxidation–reduction reaction is $\Delta G = -nF\Delta E_h$ joules, where n is the number of moles of electrons. The total energy change during proton translocation is $yF\Delta p$ joules, where y is the number of moles of translocated protons. [One can convert to electrical potential (volts) and describe these reactions in terms of forces by dividing by the Faraday Constant.] This is summarized as follows:

1. $UQH_2 + 2$ cyt $c_1(ox)$

    $\rightarrow UQ + 2$ cyt $c_1(red) + 2H^+$

2. $yH_{in}^+ \rightarrow yH_{out}^+$

Reaction **1** gives the Gibbs free energy, in joules, and in reaction **2**, for $yF\Delta p$, the units are joules, as will.

These reactions (**1** and **2**) are coupled; that is, one cannot proceed without the other. What is the size of the Δp that can be generated?

Reactions **1** and **2** are close to equilibrium and can proceed in either direction. Therefore one can write that the total force available from the redox reaction is equal to the total force of the proton potential:

$$\Delta G/F = -n\Delta E_h = y\Delta p \qquad (3.19)$$

Equation 3.19 summarizes an important relationship between the ΔE_h of an oxidation–reduction reaction during respiration and the Δp that can be generated at a coupling site. We will return to this point, and to this equation, in Chapter 4 when coupling sites are discussed in more detail.

Now we solve for Δp. Four protons ($y=4$) are translocated for every two electrons ($n=2$) that travel from reduced quinone to cytochrome c_1. The ΔE_h between quinone and cytochrome c_1 is approximately $+0.2$ V (200 mV). Substituting this value for ΔE_h, and by using $n=2$ and $y=4$, we obtain a Δp of -0.1 V (-100 mV). That is, when two electrons travel down a ΔE_h of 200 mV and four protons are translocated, -100 mV is stored in the Δp. For another way of looking at this, assume that when two electrons

travel down a redox gradient of 0.2 V, there is 0.4 eV available for doing work (2×0.2 V). The 0.4 eV is used to raise each of four protons to a Δp of 0.1 V. Oxidation–reduction reactions in the respiratory chain that are coupled to proton translocation are called coupling sites, of which the bc_1 complex is an example. Coupling sites are discussed in more detail in Section 4.5.

Coupling of redox reactions during electron transport to proton translocation is the main process by which a Δp is created in respiring bacteria, in phototrophic prokaryotes, in chloroplasts, and in mitochondria.

Reversed electron transport

The fact that some of the redox reactions in the respiratory pathway are in equilibrium with the Δp has important physiological consequences. For example, this fact supports the expectation that the Δp can drive electron transport in reverse. That is, protons driven into the cell by the Δp at coupling sites can drive electrons to the more negative electrode potential (e.g., the reversal of reactions **2** and **1**, shown earlier for the oxidation of UQH_2 by cyt c_1 coupled to the extrusion of protons). In fact, one test for the functioning of reversed electron transport is its inhibition by ionophores that collapse the Δp (Section 3.4). Reversed electron transport commonly occurs in bacteria that use inorganic compounds (e.g., ammonia, nitrite, sulfur) as a source of electrons to reduce NAD^+ for biosynthesis of cell material (chemolithotrophs), since these electron donors are at a potential higher than NAD^+. (A list of electrode potentials of biological molecules is provided later: see Table 4.1). The chemolithotrophs are discussed in Chapter 12.

Respiration coupled to sodium ion efflux

Although respiratory chains coupled to proton translocation appears to be the rule in most bacteria, a respiration-linked Na^+ pump (a Na^+-dependent NADH–quinone reductase) has been reported in several halophilic marine bacteria that require high concentrations of Na^+ (0.5 M) for optimal growth.[34,35] The situation has been well studied with *Vibrio alginolyticus*, an alkalotolerant marine bacterium that uses a $\Delta\mu_{Na^+}$ for solute transport, flagella rotation, and ATP synthesis (at alkaline pH).

V. alginolyticus creates the $\Delta\mu_{Na^+}$ in two ways. At pH 6.5, a respiration-driven H^+ pump generates a $\Delta\mu_{H^+}$, which drives a Na^+–H^+ antiporter that creates the $\Delta\mu_{Na^+}$. The antiporter creates the sodium ion gradient by coupling the influx of protons (down the proton electrochemical gradient) with the efflux of sodium ions. However, at pH 8.5, the $\Delta\mu_{Na^+}$ is created directly by a respiration-driven Na^+ pump, the Na^+-dependent NADH–quinone reductase. In agreement with this conclusion, the generation of the membrane potential at alkaline pH (pH 8.5) is resistant to the proton ionophore, *m*-carbonylcyanide phenylhydrazone (CCCP), which short-circuits the proton current (Section 3.4) but is sensitive to CCCP at pH 6.5.[36]

It has been suggested that switching to a Na^+–dependent respiration at alkaline pH may be energetically economical.[25] The reasoning is that when the external pH is more alkaline than the cytoplasmic pH, the only part of the Δp that contributes energy to the antiporter is the $\Delta\Psi$, and therefore the antiporter must be electrogenic. That is, the H^+/Na^+ must be greater than one. If one assumes that the antiporter is electroneutral when the external pH is acidic, then the continued use of the antiporter at alkaline pH would necessitate increased pumping of protons out of the cell by the primary proton-linked respiration pumps. Rather than do this, the cells simply switch to a Na^+-dependent respiration pump to generate the $\Delta\mu_{Na^+}$. This argument assumes that the ratios Na^+/e^- and H^+/e^- are identical so that the energy economies of the respiration-linked cation pumps are the same. (See note 37 for additional information regarding proton and sodium ion pumping in *V. alginolyticus*.) Several other bacteria have recently been found to generate sodium ion potentials by a primary process. For example, primary Na^+ pumping is also catalyzed by sodium ion translocating decarboxylases in certain anaerobic bacteria described shortly (Section 3.8.1). It must be pointed out, however, that although the $\Delta\mu_{Na^+}$ is relied upon for solute transport and motility in many other Na^+-dependent bacteria, it is usually created by Na^+/H^+ antiport rather than by a primary Na^+ pump. For example, the marine sulfate reducer *Desulfovibrio salexigens*, which uses a $\Delta\mu_{Na^+}$ for sulfate accumulation, generates the $\Delta\mu_{Na^+}$ by electrogenic Na^+/H^+ antiport driven by the $\Delta\mu_{H^+}$, which is created by a respiration-linked proton pump.[38] Similarly, nonmarine aerobic alkaliphiles belonging to the genus *Bacillus*, which rely on the $\Delta\mu_{Na^+}$ for most solute transport and for flagella rotation, create the $\Delta\mu_{Na^+}$ by using a Na^+/H^+ antiporter driven by a $\Delta\mu_{H^+}$ that is created by a respiration-linked proton pump (Section 3.10). Furthermore, in *D. salexigens* as well as the alkaliphilic *Bacillus* species, the membrane-bound ATP synthase is H^+ linked rather than Na^+ linked.

3.7.2 ATP hydrolysis as a driving reaction for creating a Δp

Electron transport reactions are the major energy source for creating a Δp in respiring organisms, but not in fermenting bacteria. (However, see note 39.) A major energy source for the creation of the Δp in fermenting bacteria is ATP hydrolysis catalyzed by the membrane-bound, proton-translocating ATP synthase, which yields considerable energy (Fig. 3.2, 7). Consider the following coupled reactions:

1. $ATP + H_2O \rightarrow ADP + P_i \qquad \Delta G_p$

2. $yH^+_{in} \rightarrow yH^+_{out} \qquad yF\Delta p$

To refer to the energy of ATP hydrolysis or synthesis by means of physiological concentrations of ADP, P_i, and ATP, the term ΔG_p (phosphorylation potential) is used instead of ΔG:

$$\Delta G_p = \Delta G^{\circ\prime} + 2.303RT \log[ATP]/[ADP][P_i]$$

$$(3.20)$$

The ATP synthase reaction is close to equilibrium and can operate in either direction. Therefore, the total force available from the proton potential equals the force available from ATP hydrolysis:

$$\Delta G_p/F = y\Delta p \qquad V \qquad (3.21)$$

The value of ΔG_p is about $-50,000$ J (518 mV), and the consensus value for y is 3. Under these circumstances, the hydrolysis of one ATP would generate a maximum Δp of -173 mV.

3.8 Other Mechanisms for Creating a $\Delta\Psi$ or a Δp

Redox reactions and ATP hydrolysis are the most common driving reactions for creating a proton potential. However, other mechanisms exist for generating proton potentials, and even sodium ion potentials. These driving reactions are not as widespread among the prokaryotes as the others. Nevertheless, they are very important for certain groups of prokaryotes, especially anaerobic bacteria and halophilic archaea. In some instances, they may be the only source of ATP. Some of these driving reactions are considered next.

3.8.1 Sodium transport decarboxylases can create a sodium potential

Although chemical reactions directly linked to Na^+ translocation (primary transport of Na^+) are not widespread among the bacteria, since most primary transport is of the proton, they can be very important for certain bacteria.[40–45] An example of primary Na^+ transport is the Na^+ pump coupled to respiration in *Vibrio alginolyticus*, discussed in Section 3.7.1. There is also the decarboxylation of organic acids coupled to sodium ion efflux. This occurs in anaerobic bacteria that generate a sodium gradient by coupling the decarboxylation of a carboxylic acid to the electrogenic efflux of sodium ions. The decarboxylases include methylmalonyl–CoA decarboxylase from *Veillonella alcalescens* and *Propionigenium modestum*, glutaconyl–SCoA decarboxylase from *Acidaminococcus fermentans*, and oxaloacetate decarboxylase from *Klebsiella pneumoniae* and *Salmonella typhimurium*. A description of these bacteria and their metabolism will explain the importance of the sodium-translocating decarboxylases as a source of energy.

Propionigenium modestum is an anaerobe isolated from marine and freshwater mud, and from human saliva.[46] It grows only on carboxylic acids (i.e., succinate, fumarate, L-aspartate, L-malate, oxaloacetate, and pyruvate). The carboxylic acids are fermented to propionate and acetate, forming methylmalonyl–CoA as an intermediate. (The description of the propionic acid fermentation in Section 14.7.1 explaining these reactions.) The methylmalonyl–CoA is decarboxylated to propionyl–CoA and CO_2 coupled to the electrogenic translocation of two moles of sodium ions per mole of methylmalonyl–CoA decarboxylated[47]:

$$Methylmalonyl–CoA + 2Na^+_{in}$$
$$\rightarrow propionyl–CoA + CO_2 + 2Na^+_{out}$$

P. modestum differs from most known bacteria in that it has a Na^+-dependent ATP synthase rather than a H^+-dependent ATP synthase, and thus it relies on a Na^+ current to make ATP. The consequence of this is that, when *P. modestum* grows on succinate, the decarboxylation of methylmalonyl–CoA is its only source of ATP.

Another example is *V. alcalescens*. This is an anaerobic gram-negative bacterium that grows in the alimentary canal and mouth of humans and other animals. Like *P. modestum*, it is unable to ferment carbohydrates, but it does ferment the carboxylic acids lactate, malate, and fumarate to propionate, acetate, H_2, and CO_2. During the fermentation, pyruvate and methylmalonyl–CoA are formed as intermediates, as in *P. modestum*.[48] Some of the pyruvate is oxidized to acetate, CO_2, and H_2 with the formation of one ATP via substrate-level phosphorylation (Sections 7.3). Pyruvate will then be converted to methylmalonyl–CoA that is decarboxylated to propionyl–CoA coupled to the extrusion of sodium ions. The sodium ion gradient that is created might be used as a source of energy for solute uptake via a process called Na^+-coupled symport, described in Chapter 16.

Acidaminococcus fermentans is an anaerobe that ferments the amino acid glutamate to acetate and butyrate. During the fermentation, glutaconyl–CoA is decarboxylated to crotonyl–CoA by a decarboxylase that is coupled to the translocation of sodium ions, as just shown for methylmalonyl–CoA[49]:

$$glutaconyl–CoA + yNa^+_{in}$$
$$\rightarrow crotonyl–CoA + CO_2 + yNa^+_{out}$$

The oxaloacetate decarboxylase in Klebsiella pneumoniae

We will examine the Na^+-dependent decarboxylation of oxaloacetate by oxaloacetate decarboxylase from the facultative anaerobe *K. pneumoniae* because this process has been well studied. A substantial sodium potential develops because the decarboxylation of oxaloacetate is coupled to the electrogenic efflux of sodium ion. (The standard free energy for the decarboxylation reaction is approximately –29,000 J.)

The enzyme from *K. pneumoniae* translocates two Na^+ ions out of the cell per oxaloacetate decarboxylated according to the following reaction:

$$2Na_{in}^+ + \text{oxaloacetate}$$
$$\rightarrow 2Na_{out}^+ + \text{pyruvate} + CO_2$$

1. Evidence that oxaloacetate decarboxylase is a sodium pump

Inverted membrane (inside-out) vesicles were prepared from *Klebsiella* and incubated with $^{22}Na^+$ and oxaloacetate. (See note 50 for a description of how to prepare the vesicles.) As seen in Fig. 3.14, oxaloacetate-dependent sodium ion influx took place. The uptake of sodium ion was prevented by avidin, an inhibitor of the oxaloacetate decarboxylase.

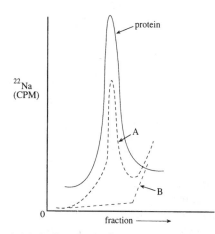

Fig. 3.14 Sodium ion influx into vesicles driven by oxaloacetate decarboxylation. Inside-out vesicles were prepared from *Klebsiella pneumoniae* and incubated with $^{22}Na^+$ in the presence (curve A) and absence (curve B) of oxaloacetate. After one minute of incubation, the vesicles were isolated by Sephadex chromatography, which separates the vesicles from the $^{22}Na^+$ in the medium. The fractions from the Sephadex columns that contained vesicles were detected by absorbance at 280 nm (protein). Only when oxaloacetate was present did the vesicle fraction contain $^{22}Na^+$ (curve A). This demonstrates the dependency of sodium uptake upon oxaloacetate. In separate experiments it was demonstrated that oxaloacetate decarboxylase was present in the vesicles, that the oxaloacetate was decarboxylated, and that inhibition of the oxaloacetate decarboxylase by avidin prevented uptake of $^{22}Na^+$. *Source*: Adapted from Dimroth, P. 1980. A new sodium-transport system energized by the decarboxylation of oxaloacetate. *FEBS Lett.* **122**:234–236.

2. The structure of the oxaloacetate decarboxylase

Oxaloacetate decarboxylase consists of two parts, a peripheral catalytic portion attached to the inner surface of the membrane and an integral membrane portion that serves as a Na^+ channel.[37] The integral membrane protein also takes part in the decarboxylation step, along with the peripheral membrane protein. How the decarboxylase translocates Na^+ to the cell surface is not understood.

3. Use of the sodium current

What does *Klebsiella* do with the energy conserved in the sodium potential? It uses the energy to actively transport the growth substrate, citrate, into the cell, as well as to drive the reduction of NAD^+ by ubiquinol (i.e., via a Na^+-dependent NADH:ubiquinone oxidoreductase). The latter is an example of reversed electron transport driven by the influx of Na^+. This all occurs during the fermentation of citrate and is depicted in Fig. 3.15.

3.8.2 Oxalate:formate exchange can create a Δp

Oxalobacter formigenes, an anaerobic bacterium that is part of the normal flora in mammalian intestines, uses dietary oxalic acid as its sole source of energy for growth. The organism has evolved a method for creating a proton potential at the expense of the free energy released from the decarboxylation of oxalic acid to formic acid and carbon dioxide.[51–53] What is especially interesting is that the enzyme is not in the membrane and therefore cannot act as an ion pump, yet a $\Delta\Psi$ is created. The reaction catalyzed by oxalate decarboxylase is

$$-OOC-COO^- + H^+ \rightarrow CO_2 + HC,OO^-$$

$$\text{oxalate} \qquad\qquad \text{formate}$$

For every mole of oxalate that enters the cell, one mole of formate leaves (Fig. 3.16). Since a dicarboxylic acid crosses the cell membrane in exchange for a monocarboxylic acid, there is net movement of negative charge toward the inside, thus creating a $\Delta\Psi$, inside negative. Also, a proton is consumed in the cytoplasm during the decarboxylation, which can contribute to a ΔpH, inside alkaline. Recently the antiporter has been purified and shown to be a 38 kDa hydrophobic polypeptide that catalyzes the

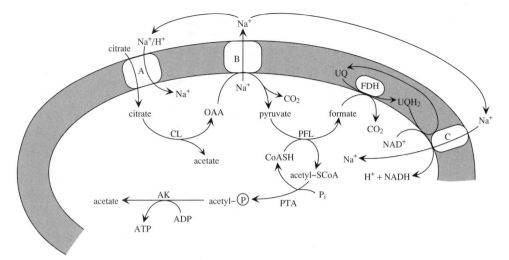

Fig. 3.15 Citrate fermentation and proposed Na$^+$ ion currents in *Klebsiella pneumoniae*. (A) Na$^+$–citrate symporter. Citrate enters the cell via symport with Na$^+$. Uptake of citrate is not dependent upon the $\Delta\Psi$ and is therefore electroneutral and dependent upon the ΔpNa$^+$. (It has been suggested that only two of the carboxyl groups of citrate are neutralized with sodium ions, and the third is protonated.) The citrate is then cleaved by citrate lyase (CL) to oxaloacetate and acetate. (B) Na$^+$-pumping oxaloacetate decarboxylase. Na$^+$ is pumped out of the cell during the decarboxylation of oxaloacetate to pyruvate and CO_2, creating a sodium ion electrochemical gradient that is used for citrate uptake (Na$^+$–citrate symporter) and NADH production (Na$^+$-pumping NADH:ubiquinone oxidoreductase). The pyruvate is used to make ATP via substrate level phosphorylation. First the pyruvate is cleaved by pyruvate–formate lyase (PFL) to acetyl–CoA and formate. Then the acetyl–CoA is converted to acetyl–phosphate by phosphotransacetylase (PTA). Finally, the acetyl–phosphate donates the phosphoryl group to ADP to form ATP and acetate in a reaction catalyzed by acetate kinase (AK). NADH is produced as follows. Some of the formate is oxidized to CO_2 by a membrane-bound formate dehydrogenase (FDH). The electron acceptor is ubiquinone. Ubiquinol is reoxidized by NAD$^+$ to form NADH + H$^+$ in a reaction catalyzed by a Na$^+$-pumping NADH:ubiquinone oxidoreductase (C). This is a case of reversed electron transport driven by the electrochemical Na$^+$ gradient. There is therefore a Na$^+$ circuit. Sodium ion is pumped out of the cell via the decarboxylase and enters via the citrate transporter and the NADH:ubiquinone oxidoreductase. *Source*: Adapted from Dimroth, P. 1997. Primary sodium ion translocating enzymes. *Biochim. Biophys. Acta* **1318**:11–51.

exchange of oxalate and formate in reconstituted proteoliposomes.[54] (For a discussion of proteoliposomes, see note 55 and Section 16.1.)

ATP synthesis

Assuming that the stoichiometry of the ATP synthase reaction is 3H$^+$/ATP and for oxalate decarboxylation it is 1H$^+$/oxalate, then a steady proton current requires the decarboxylation of three moles of oxalate per mole of ATP synthesized. In other words, one-third mole of ATP can be synthesized per mole of oxalate decarboxylated. This is apparently the only means of ATP synthesis in *Oxalobacter formigenes*.

The decarboxylation of other acids may also create a Δp

In principle, the influx and decarboxylation of any dicarboxylic acid coupled to the efflux of the monocarboxylic acid can create a proton potential (Fig. 3.17). For example, a decarboxylation and electrogenic antiport was reported for the decarboxylation of malate to lactate acid during malolactate fermentation in *Lactobacillus lactis*.[56]

3.8.3 End-product efflux as the driving reaction

Theoretically, it is possible to couple the excretion of fermentation end products down a concentration gradient to the translocation of protons out of the cell, thereby creating a Δp. This scheme, the reverse of solute uptake by proton-coupled transport systems, is called the energy recycling model.[57,58] In other words, the direction of solute transport depends upon which is greater: the energy from the proton

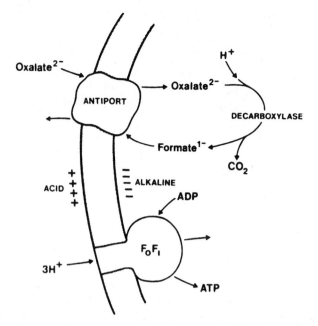

Fig. 3.16 The electrogenic oxalate:formate exchange and the synthesis by means of a proton current of ATP in *Oxalobacter formigenes*. Oxalate^{2-} enters via an oxalate^{2-}/formate$^-$ antiporter. The oxalate^{2-} is decarboxylated to formate$^-$ and CO_2, while a cytoplasmic proton is consumed. (The oxalate is first converted to oxalyl–coenzyme A, which is decarboxylated to formyl–coenzyme A. The formyl–Co A transfers the coenzyme A to incoming oxalate, thus forming formate and oxalyl–Co A.) An electrogenic exchange of oxalate^{2-} for formate$^-$ creates a membrane potential, outside positive. It is suggested that the stoichiometry of the ATP synthase is 3H$^+$/ATP. Since the decarboxylation of one mole of oxalate results in the consumption of one mole of protons, the incoming current of protons during the synthesis of one mole of ATP is balanced by the decarboxylation of three moles of oxalate. In other words, a steady state current of protons requires that $\frac{1}{3}$ of a mole of ATP be made per mole of oxalate decarboxylated. *Source*: Anantharam, V., M. J. Allison, and P. C. Maloney. 1989. Oxalate:formate exchange. *J. Biol. Chem.* **264**:7244–725. Also, see refs. 46 and 47.

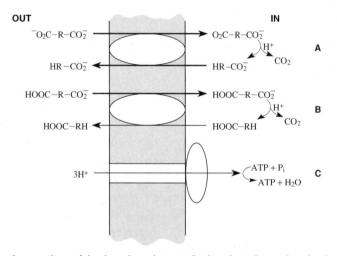

Fig. 3.17 Scheme for the coupling of the decarboxylation of a dicarboxylic acid to the development of a proton potential: dicarboxylic acid enters as a divalent anion (A) or a monovalent anion (B). In the cytoplasm, a decarboxylase cleaves off CO_2, consuming a proton in the process and producing a monocarboxylic acid with one less negative charge than the original dicarboxylic acid. Exchange of the dicarboxylic acid and the monocarboxylic acid via the antiporter is electrogenic and produces a membrane potential, inside negative. The consumption of the proton during the decarboxylation creates a ΔpH. (C) Influx of protons via the ATP synthase completes the proton circuit and results in ATP synthesis. The energetics can be understood in terms of the decarboxylase removing the dicarboxylic acid from the inside, thus maintaining a concentration gradient that stimulates influx.

potential, Δp, which drives solute uptake creating an electrochemical solute gradient $\Delta\mu/F$, or the electrochemical solute gradient, which drives solute efflux, creating a Δp. Solute efflux coupled to proton translocation can spare ATP because in fermenting bacteria, the hydrolysis of ATP (catalyzed by the ATP synthase) is used to pump protons to the outside to create the Δp. As the Δp rises, ATP hydrolysis is diminished.

Energetics

Consider solute/proton symport to be reversible. Under these circumstances, the Δp can drive the uptake of a solute against a concentration gradient; or the efflux of a solute down a concentration gradient can create a Δp. The driving force for solute transport in symport with protons is the sum of the proton potential and the electrochemical potential of the solute, that is,

$$\text{Driving force (mV)} = y\Delta p + \Delta\mu_S/F \quad (3.22)$$

where y is the number of protons in symport with the solute, S. The term $\Delta\mu_S/F$ represents the electrochemical potential of S:

$$\Delta\mu_S/F = m\Delta\Psi + 60 \log[S]_{in}/[S]_{out} \quad \text{mV} \quad (3.23)$$

where m is the charge of the solute. Therefore, the overall driving force (substituting for $\Delta\mu_S/F$ and Δp) is

Driving force (mV)

$$= 60 \log[S]_{in}/[S]_{out} + (y+m)\Delta\Psi - y60 \Delta pH$$
$$(3.24)$$

Substituting $m = -1$ for lactate, eq. 3.24 becomes

Driving force

$$= (y-1)\Delta\Psi - y60 \Delta pH + 60 \log[S]_{in}/[S]_{out}$$
$$(3.25)$$

During growth, lactate transport is near equilibrium and the net driving force is close to zero. Thus by setting eq. 3.25 equal to zero, one can solve for y:

$$y = \frac{\Delta\Psi - 60 \log[S]_{in}/[S]_{out}}{\Delta p} \quad (3.26)$$

Note that eq. 3.25 states that when y, which is the number of moles of H^+ translocated per mole of lactate, is one, then the translocation is electroneutral and a $\Delta\Psi$ does not develop; only a small ΔpH develops, owing to the acidification of the external medium by the lactic acid. We will return to these points later when we discuss the physiological significance of energy conservation via end-product efflux. Let us now consider some data that support the hypothesis that lactate/H^+ symport and succinate/Na^+ symport can create a membrane potential in membrane vesicles derived from cells.

Symport of protons and sodium ions with fermentation end products

Coupled translocation of protons and lactate can be demonstrated in membrane vesicles prepared from lactic acid bacteria belonging to the genus *Streptococcus*.[59–62] Similarly, coupled translocation of sodium ions and succinate is catalyzed by vesicles prepared from a rumen bacterium belonging to the genus *Selenemonas*.[63] In both cases a transporter exists in the membrane that simultaneously translocates the organic acid (R^-) with either protons or sodium ions out of the cell (symport) (Fig. 3.18). (See note 64.)

Lactate efflux

L-Lactate-loaded membrane vesicles were prepared from *Streptococcus cremoris* in the following way. A concentrated suspension of cells

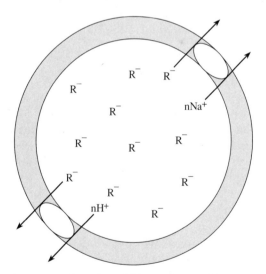

Fig. 3.18 End-product efflux in symport with protons or sodium ions. The high intracellular concentrations of R^- may drive the efflux of Na^+ or H^+ via symporters. If the ratio of protons or sodium ions to carboxyl (i.e., n/carboxyl) exceeds 1, then the symport is electrogenic and a membrane potential develops.

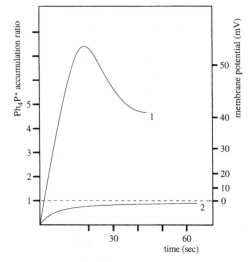

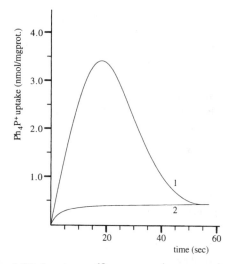

Fig. 3.19 Lactate efflux can produce a membrane potential. Lactate-loaded membrane vesicles from *Streptococcus cremoris* were incubated with the lipophilic cation Ph$_4$P$^+$ without lactate (curve 1) and with 50 mM lactate in the external medium (curve 2). In the absence of external lactate, the lipophilic probe accumulated inside the vesicles, suggesting that a membrane potential developed as lactate left the cells. The addition of lactate to the external medium prevented the formation of the membrane potential because the lactate concentration (In/out) was lowered. Additional experiments showed that the electrogenic ion was the proton. *Source*: Adapted from Otto, R., R. G. Lageveen, H. Veldkamp, and W. N. Konings. 1982. Lactate efflux-induced electrical potential in membrane vesicles of *Streptococcus cremoris*. *J. Bacteriol.* **149**:733–738.

Fig. 3.20 Succinate efflux can produce a membrane potential. Succinate-loaded membrane vesicles from *Selenomonas ruminantium* were incubated with the lipophilic cation Ph$_4$P$^+$ without succinate (curve 1) or with succinate (curve 2) in the external medium. The uptake of Ph$_4$P$^+$ indicates that a membrane potential developed when the ratio of succinate$_{in}$ to succinate$_{out}$ was high. *Source*: Adapted from Michel, T. A., and J. M. Macy. 1990. Generation of a membrane potential by sodium-dependent succinate efflux in *Selenomonas ruminantium*. *J. Bacteriol.* **172**:1430–1435.

was treated with lysozyme in buffer to degrade the cell walls. The resulting cell suspension was gently lysed (broken) by adding potassium sulfate. The cell membranes spontaneously resealed into empty vesicles. The vesicles were purified and then incubated for 1 h with 50 mM L-lactate, which equilibrated across the membrane, thus loading the vesicles with L-lactate. Then the membrane vesicles loaded with L-lactate were incubated with the lipophilic cation tetraphenylphosphonium (Ph$_4$P$^+$) in solutions containing buffer (Fig. 3.19, curve 1) and buffer + 50 mM L-lactate (Fig. 3.19, curve 2). Samples were filtered to separate the vesicles from the external medium, and the amount of Ph$_4$P$^+$ that accumulated was measured. The Ph$_4$P$^+$ accumulated inside the cells when the (the ratio of lactate)$_{in}$ to lactate$_{out}$ was high.

The accumulation of Ph$_4$P$^+$ implies that the efflux of lactate along its concentration gradient imposed a membrane potential on the vesicles.

The membrane potential was calculated by using the Nernst equation and the Ph$_4$P$^+$ accumulation ratio as explained in Section 3.5.1.

Succinate efflux

An experiment similar to the one just described for lactate efflux was done with vesicles from the rumen bacterium *Selenomonas ruminantium* loaded with succinate (Fig. 3.20). The Ph$_4$P$^+$ was taken up by the vesicles when there was a concentration gradient of succinate, indicating that a membrane potential, inside negative, developed during succinate efflux (Fig. 3.20, curve 1).

When the external buffer contained a high concentration of succinate, the membrane potential did not develop, indicating that the energy to establish the potential was derived from the efflux of succinate along its concentration gradient, [succ]$_{in}$/[succ]$_{out}$ (Fig. 3.20, curve 2).

The molvar growth yields of S. ruminantium are higher when succinate production is at a maximum, implying that more ATP is available for biosynthesis as a result of the succinate efflux.

Physiological significance of end-product efflux as a source of cellular energy

To demonstrate the generation of a membrane potential due to end-product efflux, cells or membrane vesicles must be "loaded" with high concentrations of the end product by incubating de-energized cells or membrane vesicles for several hours before dilution into end-product-free media. These are not physiological conditions, and the physiological relevance of the conditions under which the electrogenic extrusion of lactate coupled to protons has been questioned.

Assuming a given value of y, one can use eq. 3.26 to calculate the necessary lactate concentration gradient from experimentally determined values of $\Delta\Psi$ and ΔpH. When one substitutes $y = 2$ (for electrogenic lactate extrusion), a major problem appears. To substitute a measured external lactate concentration of 30 mM, a $\Delta\Psi$ of -100 mV, a $-60 \Delta pH$ of -30 mV (at pH 7.0), and $y = 2$, the calculated internal lactate concentration would have to be 14 M for lactate secretion to occur.[65] This is a nonphysiological concentration. (Internal lactate concentrations reach concentrations of about 0.2 M.) In fact, the values of y were calculated by Brink et al., who used experimentally measured values of lactate concentrations, membrane potential, and ΔpH.[65] Only when the external lactate concentrations were very low was lactate/proton efflux electrogenic (i.e., $y > 1$). For example, when the cells were grown at pH 6.34 and the lactate accumulated in the medium over time, the value of y decreased from about 1.44 to 0.9 while the external lactate concentrations increased from 8 mM to 38 mM. This means that only under certain growth conditions (low external lactate and external pH high so that the ΔpH is 0 or inverted) would one expect lactate efflux effectively to generate a Δp. This may occur in the natural habitat, where growth of the producer may be stimulated by a population of bacteria that utilize lactate, thus keeping the external concentrations low.

3.8.4 Light absorbed by bacteriorhodopsin can drive the creation of a Δp

Certain archaea, namely, the extremely halophilic archaea, have evolved a way to produce a Δp by using light energy directly (i.e., without the intervention of oxidation–reduction reactions and without chlorophyll).[66-68] See note 69 for a more complete discussion of the extreme halophiles. (It has been reported that under the proper conditions, these microorganisms can be grown photoheterotrophically, i.e., on organic carbon with light as a source of energy, but such conclusions have been questioned.[70,71])

Halophilic archaea are heterotrophic organisms that carry out an ordinary aerobic respiration, creating a Δp driven by oxidation–reduction reactions during electron transport. (See note 72.) The Δp is used to drive ATP synthesis via a membrane ATP synthase. (See Sections 3.7.1 and 3.6.2.) However, conditions for respiration are not always optimal and, in the presence of light and low oxygen levels, the halophiles adapt by making photopigments (rhodopsins), one of which (bacteriorhodopsin) functions as a proton pump that is energized directly by light energy. (See note 73.) Whereas photosynthetic electron flow is an example of an indirect transformation of light energy into an electrochemical potential (via redox reactions), bacteriorhodopsin illustrates the direct transformation of light energy into an electrochemical potential. We will first consider data that demonstrate the light-dependent pumping of protons. We will then describe the proton pump, the photocycle, and a model for the mechanism of pumping protons. Bacteriorhodopsin is examined in detail because it is the best characterized ion pump.

Evidence that halophiles can use light energy to drive a proton pump

Figure 3.21 models the results of subjecting a suspension of Halobacterium halobium to light of different intensities and for different periods of time, during which the pH of the external medium was monitored. As illustrated in Fig. 3.21A, the light produced an efflux of protons from the cell. Higher light intensities produced

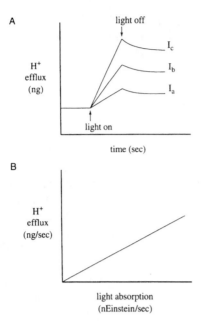

for the photocycle (i.e., the fraction of bacteriorhodopsin molecules absorbing light that undergoes the photocycle described below) is 0.64 ± 0.04.[75,76] These values suggest that one proton is pumped per photocycle.

Bacteriorhodopsin is the proton pump

Built into the cell membrane of the halophilic archaebacteria is a pigment protein called bacteriorhodopsin, which is a pump responsible for the light-driven electrogenic efflux of protons. It consists of one large polypeptide (248 amino acids, 26,486 Da) folded into seven α helices that form a transmembrane channel (Fig. 3.22). (See Ref. 77 for a review.) Located in the middle of the channel, and attached to the bacteriorhodopsin, is a pigment called retinal (a C_{20} carotenoid), which is attached via a Schiff base to a lysine residue on the protein. (Note 78 tells what a Schiff base is.) When the

Fig. 3.21 Proton pumping by bacteriorhodopsin. Expected results if *Halobacterium* cells were illuminated by light. To measure proton outflow accurately, proton inflow through the ATP synthase must be blocked either with uncouplers or nigericin, which collapse the proton electrochemical potential, or with an ATP synthase inhibitor such as DCCD. The extruded protons can be quantitated with a pH meter. (A) Proton efflux measured as a function of time at increasing light intensities $I_a < I_b < I_c$, expressed as nanoeinsteins per second. The slope of each line is the rate of proton efflux. (B) The rate of proton efflux is plotted as a function of the light intensity. From these data one can calculate the quantum yield (i.e., the number of protons extruded per photon absorbed). *Source*: Adapted from data by Bogomolni, R. A., R. A. Baker, R. H. Lozier, and W. Stoeckenius. 1980. Action spectrum and quantum efficiency for proton pumping in *Halobacterium halobium*. *Biochemistry* 19:2152–2159.

a greater rate of proton efflux (compare I_c to I_a.) In Fig. 3.21B, the rate of proton efflux, in nanograms per second, as determined from the slopes in Fig. 3.21A is plotted as a function of the light intensity in nanoeinsteins per second (an einstein is equal to a "mole" of photons). From data such as these, a quantum yield (i.e., protons ejected per photon absorbed) was calculated to be 0.52 proton/photon absorbed.[74] The reported values for the *maximum* quantum yield for proton efflux is a little higher (0.6–0.7 proton/photon absorbed). The quantum yield

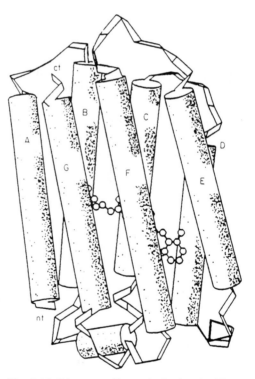

Fig. 3.22 Diagram of bacteriorhodopsin. The seven helices, shown as rods, form a central channel; the retinal is attached to a lysine residue on helix G. *Source*: Henderson, R., J. M. Baldwin, T. A. Ceska, F. Zemlin, E. Beckmann, and K. H. Downing. 1990. Model for the structure of bacteriorhodopsin based on high-resolution electron cryomicroscopy. *J. Mol. Biol.* **213**:899–929.

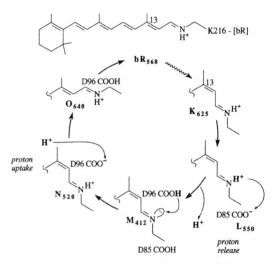

Fig. 3.23 The photochemical cycle of bacteriorhodopsin. Upon absorption of a photon of light, bR_{568} undergoes a trans-to-cis isomerization and is converted to a series of intermediates with different absorbance maxima. The Schiff base becomes deprotonated during the L-to-M transition and reprotonated during the M-to-N transition. A recent photocycle postulates two M states: L to M_1 to M_2 to N. Although not indicated, some of the steps are reversible. (See Lanyi, J. K. 1992. Proton transfer and energy coupling in the bacteriorhodopsin photocycle. *J. Bioenerg. Biomemb.* **24**:169–179). *Source*: Krebs, M. P., and H. Gobind Khorana. 1993. Mechanism of light-dependent proton translocation by bacteriorhodopsin. *J. Bacteriol.* **175**:1555–1560.

retinal absorbs light, the bacteriorhodopsin remarkably translocates protons out of the cell, and a Δp is created.

The photocycle and a model for proton pumping

The photoevents occurring in halophiles can be followed spectroscopically because when bacteriorhodopsin absorbs light, it loses its absorption peak at 568 nm (bleaches) and is converted in the dark to a series of pigments that have absorption peaks at different wavelengths. This sequence of events is called the photocycle. Also, site-specific mutagenesis of bacteriorhodopsin is being used to identify the amino acid side chains that transfer protons across the membrane.[79–82] The photocycle is shown in Fig. 3.23. The retinal is attached via a Schiff base to the ε amino group of lysine-216 (K_{216}). Before absorbing light, the retinal is protonated at the Schiff base and exists in the

all-trans (13-trans) configuration (bR_{568}). The subscript refers to the absorption maximum, in nanometers. Upon absorbing a photon of light, the retinal isomerizes to the 13-cis form (K_{625}). All subsequent steps do not require light and represent the de-energization of the bacteriorhodopsin via a series of intermediates that have absorbance maxima different from those of the original unexcited molecule. The transitions are very fast and occur in the nanosecond and millisecond ranges. These intermediates are, in the order of their appearance, K, L, M, N, and O.

In converting from L to M, the Schiff base loses its proton to aspartate-85 but regains a proton from aspartate-96 in going from M to N. Aspartate-96 acquires a proton from the cytoplasm in going from N to O. Thus, a proton has moved from the cytoplasmic side through aspartate-96 to the Schiff base to aspartate-85. From aspartate-85, the proton moves to the outside membrane surface. (See note 83.)

Precisely how the proton travels through the bacteriorhodopsin channel from the cytoplasmic side to the Schiff base in the center of the channel, and from there to the external surface of the membrane, is not known. Probably the proton is passed from one amino acid side group that can be reversibly protonated to another. Site-specific mutagenesis experiments have suggested that two of these amino acids are aspartate-85 and aspartate-96. The protonatable residues would extend along the protein from the cytoplasmic surface to the outside. In the case of bacteriorhodopsin, where the three-dimensional structure is known, the protonatable residues are oriented toward the center of the channel. Any bound water might also participate in proton translocation. For example, there might be a chain of water and hydrogen-bonded protons connecting protonatable groups. Since the Schiff base gives up its proton to the extracellular side of the channel (transition L to M) and becomes protonated with a proton from the cytoplasm (transition M to N), it has been assumed that there is a switch that reorients the Schiff base, causing it to face the cytoplasmic and extracellular sides of the channel alternately. Logically, the switch would be at M. The mechanism for the switch is unknown, although a conformational change in the bacteriorhodopsin has been suggested.[81]

It is not obviously correlated with the retinal isomerizations because the retinal is in the cis configuration throughout most of the photo-cycle, including the protonation and deprotona-tion of the Schiff base. The transfer of protons along protonatable groups has been suggested for several proton pumps besides bacteri-orhodopsin, including cytochrome oxidase and the proton-translocating ATP synthase.[84]

3.9 Halorhodopsin, a Light-Driven Chloride Pump

The halophiles have a second light-driven electrogenic ion pump, but one that does not energize the membrane. The second pump, called halorhodopsin, is structurally similar to bacteriorhodopsin and is used to accumulate Cl^- intracellularly, to maintain osmotic stab-ility.[60,74,85] Recall that the halophiles live in salt water, where the extracellular concentrations of NaCl can be 3 to 5 M. The osmotic balance is preserved by intracellular concentrations of KCl that match the extracellular Cl^- concentra-tions. (As discussed in Chapter 15, K^+ is import-ant for osmotic homeostasis in eubacteria as well.) Since the membrane is negatively charged on the inside, energy must be used to bring the Cl^- into the cell, and halorhodopsin accom-plishes this purpose. In the dark the halophiles use another energy source, probably ATP, to accumulate chloride ions actively.[86]

3.10 The Δp and ATP Synthesis in Alkaliphiles

Alkaliphilic bacteria grow in habitats having a very basic pH, usually around pH 10. These habitats include soda lakes, dilute alkaline springs, and desert soils, where the alkalinity is usually due to sodium carbonate. *Obligate* alkaliphiles are organisms that cannot grow at pH values of 8.5 or less and usually have an optimum around 9. (See note 87.) These include *Bacillus pasteurii*, *B. firmus*, and *B. alcalophilus*. The Δp has been measured in obligate aerobic alkaliphiles that grow opti-mally at pH 10 to 12, and it appears to be too low to drive the synthesis of ATP. The problem is that because the external pH is so basic, the internal pH is generally at least 2 pH units more acid. This gives the ΔpH a sign opposite to that found in the other bacteria, that is, negative. A ΔpH of 2 is equivalent to 60×2 or 120 mV. Thus, the Δp can be lowered by about 120 mV in these organisms. Typical membrane poten-tials for aerobic alkaliphiles are approximately −170 mV (positive out). Therefore, the Δp values can be as low as −170 + 120 or −50 mV.[88]

The ATPase in the alkaliphiles is a proton-translocating enzyme, as in most bacteria. Even if the ATPase translocated as many as four pro-tons, this would generate only 0.2 eV, far short of the approximately 0.4 to 0.5 eV required to synthesize an ATP under physiological condi-tions. The energy to synthesize an ATP (i.e., ΔG_p) is 40,000 to 50,000 J. Dividing this num-ber by the Faraday constant gives 0.4 to 0.5 eV.

How can this dilemma be resolved? It is poss-ible that the protons in the bulk extracellular phase are not as important for the Δp as protons on the membrane or a few angstrom units away from the membrane. One suggestion is that random collisons within the membrane, may put proton pumps in frequent contact with the ATPases, and as soon as a proton is pumped out of the cell, it may reenter via an adjoining ATPase without entering the pool of bulk pro-tons.[89] This suggestion emphasizes the activity of protons at the face of the membrane rather than the ΔpH, which is due to the concentration of protons in the bulk phase, and raises ques-tions about the details of the proton circuit.

One important tenet of the chemiosmotic theory may not apply in this situation, and that is that a delocalized Δp is used. That is, the chemiosmotic theory postulates that proton currents couple any exergonic reaction with *any* endergonic reaction (i.e., proton circuits are delocalized over the entire membrane). In that way, proton extrusion during respiration can provide the energy for several different reactions, not simply the ATP synthase (Fig. 3.2). However, it may be that in some alkaliphiles there is direct transfer of protons extruded during respiration to the ATP synthase (i.e., *localized* proton circuits).

3.11 Summary

The energetics of bacterial cell membranes can be understood for most bacteria in terms of an electrochemical proton potential established by exergonic chemical reactions or light. The protons are raised from a low electrochemical potential inside the cell to a high electrochemical potential outside the cell. When the protons circulate back into the cell through appropriate carriers, work can be done (e.g., the synthesis of ATP via the membrane ATP synthase, solute transport, flagellar rotation).

The proton potential is due to a combination of a membrane potential ($\Delta\Psi$), outside positive, and a ΔpH, outside acid. Because of the low capacitance of the membrane, not very many protons will have been extruded before a large membrane potential develops. The membrane potential seems to be the dominant component in the Δp for most bacteria, except for acidophiles, which can have a reversed membrane potential. Other cations, especially sodium ions, can use the established membrane potential for doing work, principally solute accumulation. The sodium ions must be returned to the outside of the cell, and Na^+/H^+ antiporters serve this purpose in most bacteria. Thus, although the major ion circuit is a proton circuit, the sodium circuit is also important.

A Δp is created when an exergonic chemical reaction is coupled to the electrogenic flow of charge across the cell membrane and the liberation of protons on the outer membrane surface. Energy input of at least $yF\Delta p$ joules is necessary to raise the electrochemical potential of y moles of protons to Δp volts. The three most widespread reactions that provide the energy to create the Δp are oxidation–reduction reactions during electron transport in membranes (respiration), oxidation–reduction reactions during electron transport stimulated by light absorption (photosynthesis), and ATP hydrolysis via the membrane ATP synthase. Respiration and ATP hydrolysis are reversible, and the Δp can drive reversed electron transport as well as the synthesis of ATP. During reversed electron transport, protons enter the cells rather than leave the cells. The ATP synthase is an enzyme complex that reversibly hydrolyzes ATP and pumps protons out of the cell. When protons enter via the ATP synthase, ATP is made.

Light energy can also be used directly to create a Δp without the establishment of a redox potential. This occurs in the halophilic archaea, which use a light-driven proton pump called bacteriorhodopsin to create a Δp.

Bacteriorhodopsin, which forms a proton channel through the membrane, is being studied as a model system to investigate the mechanism of ion pumping across membranes.

Fermenting bacteria have evolved additional ways to generate a Δp. Lactic acid bacteria can create a Δp via coupled efflux of protons and lactate (in addition to ATP hydrolysis) under certain growth conditions. Another anaerobic bacterium, *Oxalobacter*, creates a Δp by the oxidation of oxalic acid to formic acid coupled with the electrogenic exchange of oxalate for formate and the consumption of protons during the decarboxylation. Other examples similar to these will no doubt be discovered in the future.

Sodium potentials are also important in the prokaryotes, especially for solute transport. Although most sodium potentials are created secondarily from the proton potential via antiporters, there are some prokaryotes that couple a chemical reaction to the creation of a sodium potential. For example, some marine bacteria, as exemplified by *Vibrio alginolyticus*, couple respiration to the electrogenic translocation of sodium ions out of the cell at alkaline pH. Because these bacteria can couple the sodium potential to the membrane ATP synthase, they rely on a sodium current rather than a proton current when growing in basic solutions. When growing in slightly acidic conditions, they use a proton potential.

Several different fermenting bacteria can create a sodium potential by coupling the decarboxylation of organic acids with the translocation of sodium ions out of the cell, or by coupling the efflux of end products of fermentation with the translocation of sodium ions to the outside. In some bacteria (e.g., *Klebsiella pneumoniae*), the sodium potential is used to drive the influx of the growth substrate into the cell but not for the generation of ATP. Rather, ATP is synthesized by a substrate-level phosphorylation during the conversion of pyruvate to formate and acetate. These bacteria may also generate a membrane potential by coupled efflux of the end products of citrate degradation (formate and acetate) with protons.

It is emphasized that in fermenting bacteria, ATP is hydrolyzed via the ATP synthase to create the Δp that is necessary for membrane activities (e.g., solute transport, flagellar rotation). A decrease in the Δp should result in even more ATP hydrolysis. Thus, reactions are expected to conserve ATP if they create a membrane potential (e.g., electrogenic efflux of sodium ions or protons in symport with end products of fermentation, or during decarboxylation reactions).

Measurements of the $\Delta\Psi$ and the ΔpH are necessarily indirect because of the small size of the bacteria. The $\Delta\Psi$ is measured by using cationic or anionic fluorescent dyes that equilibrate across the membrane in response to the potential. The distribution of the dyes is monitored by fluorescence quenching. A second way to measure the membrane potential is by the equilibration of a permeant ion that achieves electrochemical equilibrium with the membrane potential. The membrane potential is computed by using the Nernst equation and the intracellular and extracellular ion concentrations. The ΔpH is measured by using a weak acid or weak base whose log ratio of concentrations inside the cell to outside the cell is a function of the ΔpH.

Study Questions

1. The E_0' for ubiquinone (ox)/ubiquinone (red) is +100 mV, and for NAD$^+$/NADH it is −320 mV. What is the ΔE_0?

   ans. 420 mV

2. In the electron transport chain, oxidation–reduction reactions with a ΔE_h of about 200 mV appear to be coupled to the extrusion of protons. Assume that 100% of the oxidation–reduction energy is converted to the Δp. For a two-electron transfer and the extrusion of two protons, what is the expected Δp?

   ans. −200 mV

3. What is the maximum Δp (i.e., 100% energy conversion) when the hydrolysis of one mole of ATP is coupled to the extrusion of four moles of protons? Three moles of protons? Assume that the free energy of hydrolysis of ATP is −50,000 J.

   ans. −130 mV, −173 mV

4. Design an experimental approach that can show that the efflux of an organic acid along its concentration gradient is coupled to proton translocation and can generate a membrane potential. (You must not only be able to demonstrate the membrane potential but also show that the proton is the conducting charge.)

5. Explain how the decarboxylation of oxalic acid by *Oxalobacter* creates a ΔpH and a $\Delta\Psi$.

6. What is the reason for stating that light creates a Δp indirectly in photosynthesis but directly in the extreme halophiles?

7. Lactate efflux was in symport with protons, whereas succinate efflux was in symport with sodium ions. Which ionophores might you use to distinguish the cations involved?

8. A reasonable figure for the actual free energy of hydrolysis of ATP inside cells is −50,000 J/mol. It is believed that the hydrolysis of ATP is coupled to the extrusion of three protons in many systems. If a Δp of −150 mV were generated, what would be the efficiency of utilization of ATP energy to create the Δp?

   ans. 87%

9. Assume a reduction potential of +400 mV for an oxidant and a potential of −100 mV for a reductant. How many joules of energy will be released when two moles of electrons flow from the reductant to the oxidant?

   ans. 96,500 J

10. Assume that 45 kJ is required to synthesize one mole of ATP. What would be the required Δp, assuming that three moles of H$^+$ entered via the ATPase per mole of ATP made?

    ans. −155 mV

11. Assume that the Δp is −225 mV. If the ΔpH at 30 °C is 1.0, what is the membrane potential?

    ans. −165 mV

12. How much energy in joules is required to move a mole of uncharged solute into the cell against a concentration gradient of 1000 at 30 °C? If transport were driven

by the Δp, what would be the minimal value of the Δp required if one H^+ were cotransported?

ans. 17,370 J, 180 mV

13. Assume membrane vesicles loaded with K^+ so that $K_{in}^+/K_{out}^+ = 1000$. The temperature is 30 °C, and valinomycin is added. What is the predicted initial membrane potential in millivolts?

ans. −180 mV

14. What is the rationale for adding valinomycin and K^+ to an experimental system in which the number of protons being pumped out of the cell is measured?

15. Briefly, what does the chemiosmotic theory state?

16. What is meant by the ΔE_h? What is the relationship between the ΔE_h and the Δp that is generated at coupling sites?

17. Assume that a concentration gradient of an uncharged solute, S, exists across the cell membrane. What is the formula that expresses the driving force in the concentration gradient, in millivolts at 30 °C? What is the common expression used for the force due to the concentration gradient when S is a proton?

18. Assume that an ion traverses a charged membrane with a potential of $\Delta\Psi$ volts. Further assume that there is no concentration gradient. What is the expression that denotes the driving force? What will the sign be if the ion moves toward the side of opposite charge?

19. What is the expression, in millivolts at 30 °C, for the Δp? What is the expression in joules?

REFERENCES AND NOTES

1. Harold, F. M. 1986. *The Vital Force: A Study of Bioenergetics.* W. H. Freeman, New York.

2. Nichols, D. G., and S. J. Ferguson. 2002. *Bioenergetics* 3. Academic Press, San Diego, CA.

3. Mitchell, P. 1961. Coupling of phosphorylation to electron and hydrogen transfer by a chemiosmotic type of mechanism. *Nature* 191:144–148.

4. Mitchell, P. 1966. Chemiosmotic coupling in oxidative and photosynthetic phosphorylation. *Biol. Rev. Cambridge Philos. Soc.* 41:445–502.

5. Mitchell, P. 1977. Vectorial chemiosmotic processes. *Annu. Rev. Biochem.* 46:996–1005.

6. Mitchell, P. 1979. Compartmentation and communication in living systems. Ligand conduction: a general catalytic principle in chemical, osmotic and chemiosmotic reaction systems. *Eur. J. Biochem.* 95:1–20.

7. Kashket, E. R. 1985. The proton motive force in bacteria: a critical assessment of methods. *Annu. Rev. Miecrobiol.* 39:219–242.

8. A single proton (or any monovalent ion, or electron) carries 1.6×10^{-19} C of charge. If we multiply this by Avogadro's number, we arrive at the charge carried by a mole, which is approximately 96,500 C, or the Faraday constant (F).

9. The actual equation is $G = G_0 + RT \ln a$, where a is activity. The activity is a product of the molal concentration (c) and the activity coefficient (γ) for the particular compound, $a = \gamma c$. In practice, concentrations are usually used instead of activities, and the concentrations are in molar units instead of molal. The symbol T is the absolute temperature in degrees Kelvin (273 + °C). The symbol R is the ideal gas constant, and G_0 is the standard free energy (i.e., when the concentration of all reactants is 1 M). When one uses 8.3144 J K^{-1} mol^{-1} as the units of R, then the free energy (G) is given in J/mol.

10. Cecchini, G., and A. L. Koch. 1975. Effect of uncouplers on "downhill" β-galactoside transport in energy-depleted cells of *Escherichia coli. J. Bacteriol.* 123:187–195.

11. Gould, J. M., and W. A. Cramer. 1977. Relationship between oxygen-induced proton efflux and membrane energization in cells of *Escherichia coli. J. Biol. Chem.* 252:5875–5882.

12. E. Padan, D. Zilberstein, and Schuldner. 1981. pH homeostasis in bacteria. *Biochim. Biophys. Acta* 650:151–166.

13. Reviewed in: Cobley, J. G., and J. C. Cox. 1983. Energy conservation in acidophilic bacteria. *Microbiol. Rev.* 47:579–595.

14. The pumping of protons out of the cell or the electrogenic influx of electrons will create a membrane potential, positive outside. However, in the aerobic acidophilic bacteria [i.e., bacteria that live in environments of extremely low pH (pH 1–4)], other events act to reverse the membrane potential. These bacteria have positive membrane potentials (i.e., inside positive with respect to outside, at low pH). It is not clear why the aerobic acidophiles have a positive $\Delta\Psi$. One possibility is that they have an energy-dependent K^+ pump that brings K^+ into the cells at a rate sufficient to establish a net influx of positive charge, creating an inside positive membrane

potential. This point is discussed further in Section 15.1.3.

15. Krulwich, T. A., and A. A. Guffanti. 1986. Regulation of internal pH in acidophilic and alkalophilic bacteria, pp. 352–365. In: *Methods in Enzymology*, Vol. 125. S. Fleischer and B. Fleischer (Eds.). Academic Press, New York.

16. Actually, what happens in the presence of nigericin is that an equalization of the K^+ and H^+ gradients occurs.

17. Padan, E., D. Zilberstein, and S. Schuldiner. 1981. pH homeostasis in bacteria. *Biochim. Biophys. Acta* 650:151–166.

18. Rottenberg, H. 1979. The measurement of membrane potential and ΔpH in cells, organelles, and vesicles. *Methods Enzymol* 55:547–569.

19. Bakker, E. P. The role of alkali-cation transport in energy coupling of neutrophilic and acidophilic bacteria: an assessment of methods and concepts. 1990. *FEMS Microbiol. Rev.* 75:319–334.

20. This is because the resonance frequency of inorganic phosphate or of the γ-phosphate of ATP in a high magnetic field is a function of the degree to which the phosphate is protonated. (Ferguson, S. J., and M. C. Sorgato. 1982. Proton electrochemical gradients and energy-transduction processes. *Annu. Rev. Biochem.* 51:185–217.)

21. Nichols, D. G., and S. J. Ferguson. 2002. *Bioenergetics 3*, pp. 195–217. Elsevier Science Ltd., London.

22. Cross, R. L. 1992. The reaction mechanism of F_0F_1–ATP synthases, pp. 317–330. In: *Molecular Mechanisms in Bioenergetics*. L. Ernster (Ed.). Elsevier Science Publishers, Amsterdam.

23. Another way of stating this is in terms of the total force. The total force is equal to $y\Delta p + \Delta G_p/F$, where $\Delta G_p/F = 518$ mV. At equilibrium the total force is 0; therefore $y\Delta p = -518$ mV, and when $y = 3$, $\Delta p = -173$ mV.

24. Maloney, P. C. 1977. Obligatory coupling between proton entry and the synthesis of adenosine 5′-triphosphate in *Streptococcus lactis*. *J. Bacteriol.* 132:564–575.

25. Maloney, P. C., and F. C. Hansen III. 1982. Stoichiometry of proton movements coupled to ATP synthesis driven by a pH gradient in *Streptococcus lactis*. *J. Membrane. Biol.* 66:63–75.

26. The actual force was about 239 mV because the temperature was 20 to 21 °C rather than 30 °C.

27. Kandpal, R. P., K. E. Stempel, and P. B. Boyer. 1987. Characteristics of the formation of enzyme-bound ATP from medium inorganic phosphate by mitochondrial F_1 adenosine triphosphatase in the presence of dimethyl sulfoxide. *Biochemistry* 26:1512–1517.

28. Grubmeyer, C., R. L. Cross, and H. S. Penefsky. 1982. Mechanism of ATP hydrolysis by beef heart mitochondrial ATPase: rate constants for elementary steps in catalysis at a single site. *J. Biol. Chem.* 257:12092–12100.

29. Vonck, J., T. K. von Nidda, T. Meier, U. Matthey, D. J. Mills, W. Kuhlbrandt, and P. Dimroth. 2002. Molecular architecture of the undecameric rotor of a bacterial Na^+-ATP synthase. *J. Mol. Biol.* 321:307–316.

30. Kaim, G., M. Prummer, B. Sick, G. Zumofen, A. Renn, U. P. Wild, and P. Dimroth. 2002. Coupled rotation within single F_0F_1 enzyme complexes during ATP synthesis or hydrolysis. *FEBS Lett.* 525:156–163.

31. Capaldi, R. A., and R. Aggeler. 2002. Mechanism of the F_1F_0 ATP synthase, a biological rotary motor. *Trends Biochem. Sci.* 27:154–160.

32. Fillingame, R. H., and O. Y. Dmitriev. 2002. Structural model of the transmembrane F_0 rotary sector of H^+-transporting ATP synthase derived by solution NMR and intersubunit cross-linking in situ. *Biochem. Biophys. Acta* 1565:232–245.

33. The ΔE_0 can be a function of the pH. This is the case for the following reaction, where m is not zero: $A_{ox} + ne^- + mH^+ \rightleftarrows A_{red}$. In some reactions $n = 1$ and $m = 0$ (cytochrome redox reactions), $n = 2$ and $m = 1$ (NAD^+ or $NADP^+$ redox reactions), $n = 2$ and $m = 2$ (fumarate–succinate redox reactions).

In reactions involving protons, if the pH is not zero, then the ΔE_0 is more negative. When $m = n$, the ΔE_0 is -60 mV/pH. When $m = 1$ and $n = 2$, the ΔE_0 is -30 mV/pH. For more discussion of this subject, see Nichols, D. G., and S. J. Ferguson. 2002. *Bioenergetics 3*. Academic Press, San Diego, CA.

34. Reviewed in Unemoto, T., H. Tokuda, and M. Hayashi. 1990. Primary sodium pumps and their signficance in bacterial energetics, pp. 33–54. In: T. A. Krulwich (Ed.). *The Bacteria*, Vol. XII. Academic Press, New York.

35. Reviewed in: Skulachev, V. P. 1992. Chemiosmotic systems and the basic principles of cell energetics, pp. 37–73. In: *Molecular Mechanisms in Bioenergetics*. Ernster, L. (Ed.). *New Comprehensive Biochemistry: Molecular Mechanisms in Bioenergetics*, Vol. 23. Elsevier, Amsterdam.

36. Tokuda, H., and T. Unemoto. 1982. Characterization of the respiration-dependent Na^+ pump in the marine bacterium *Vibrio alginolyticus*. *J. Biol Chem.* 257:10007–10014.

37. In his review of primary sodium ion translocating enzymes, Dimroth points out that *V. alginolyticus* has two different NADH:ubiquinone oxidoreductases: NQR1, which is Na^+ dependent and functions at pH 8.5 but not at pH 6.5, and NQR2, which is Na^+ independent and is not a coupling site. There is apparently no H^+-dependent NADH:ubiquinone

oxidoreductase. However, these bacteria do have a cytochrome bo oxidase that oxidizes the quinol, is not Na$^+$ dependent, and is believed to be a proton pump as in other bacteria. The presence of both pumps can explain how *V. alginolyticus* operates a Na$^+$-dependent respiratory pump at pH 8.5 and a H$^+$-dependent respiratory pump at pH 6.5. The cytochrome bo proton pump must function at both acidic and basic pH values because mutants lacking the Na$^+$-dependent NADH:ubiquinone oxidoreductase extrude Na$^+$ at pH 8.5, using a Na$^+$/H$^+$ antiporter in combination with a primary proton pump, and the wild type is known to extrude Na$^+$ at pH 6.5, using the Na$^+$/H$^+$ antiporter in combination with the primary proton pump. (Dimroth, P. 1997. Primary sodium ion translocating enzymes. *Biochim. Biophys. Acta* **1318**:11–51.)

38. Kreke, B., and H. Cypionka. 1994. Role of sodium ions for sulfate transport and energy metabolism in *Desulfovibrio salexigens*. *Arch. Microbiol.* **161**:55–61.

39. Many nonfermenting anaerobic bacteria carry out electron transport by using as electron acceptors either organic compounds such as fumarate or inorganic compounds such as nitrate. Thus, electron flow in these bacteria can be coupled to proton efflux and the establishment of a Δp. Furthermore, even fermenting bacteria can carry out some fumarate respiration generating a Δp. However, the major source of energy for the Δp in most fermenting bacteria is ATP hydrolysis.

40. Dimroth, P. 1997. Primary sodium ion translocating enzymes. *Biochim. Biophys. Acta* **1318**:11–51.

41. Dimroth, P. 1990. Energy transductions by an electrochemical gradient of sodium ions, pp. 114–127. In: *The Molecular Basis of Bacterial Metabolism.* G. Hauska and R. Thauer (Eds.). Springer-Verlag, Berlin.

42. Dimroth, P. 1980. A new sodium-transport system energized by the decarboxylation of oxaloacetate. *FEBS Lett.* **122**:234–236.

43. Dimroth, P., and A. Thomer. 1988. Dissociation of the sodium-ion-translocating oxaloacetate decarboxylase of *Klebsiella pneumoniae* and reconstitution of the active complex from the isolated subunits. *Eur. J. Biochem.* **175**:175–180.

44. Hilpert, W., and P. Dimroth. 1983. Purification and characterization of a new sodium transport decarboxylase. Methylmalonyl–CoA decarboxylase from *Veillonella alcalescens*. *Eur. J. Biochem.* **132**:579–587.

45. Buckel, W., and R. Semmler. 1983. Purification, characterization and reconstitution of glutaconyl–CoA decarboxylase. *Eur. J. Biochem.* **136**:427–434.

46. Schink, B., and N. Pfennig. 1982. *Propionigenium modestum* gen. nov. sp. nov. A new strictly anaerobic nonsporing bacterium growing on succinate. *Arch. Microbiol.* **133**:209–216.

47. Hilpert, W., B. Schink, and P. Dimroth. 1984. Life by a new decarboxylation-dependent energy conservation mechanism with Na$^+$ as coupling ion. *EMBO J.* **3**:1665–1670.

48. De Vries, W., R. Theresia, M. Rietveld-Struijk, and A. H. Stouthamer. 1977. ATP formation associated with fumarate and nitrate reduction in growing cultures of *Veillonella alcalescens*. *Antonie van Leeuwenhoek* **43**:153–167.

49. Buckel, W., and R. Semmler. 1982. A biotin-dependent sodium pump: glutaconyl–CoA decarboxylase from *Acidaminococcus fermentans*. *FEBS Lett.* **148**:35–38.

50. Inside-out vesicles are prepared by sonicating whole cells, or shearing them with a French pressure cell. Right-side-out vesicles are prepared by first removing the cell wall with lysozyme in a hypertonic medium, and then osmotically lysing the protoplasts or spheroplasts in hypotonic medium.

51. Anantharam, V., M. J. Allison, and P. C. Maloney. 1989. Oxalate:formate exchange. *J. Biol. Chem.* **264**:7244–7250.

52. Baetz, A. L., and M. J. Allison. 1990. Purification and characterization of oxalyl–coenzyme A decarboxylase from *Oxalobacter formigenes*. *J. Bacteriol.* **171**:2605–2608.

53. Baetz, A. L., and M. J. Allison. 1990. Purification and characterization of formyl–coenzyme A transferase from *Oxalobacter formigenes*. *J. Bacteriol.* **171**:3537–3540.

54. Ruan, Z., V. Anantharam, I. T. Crawford, S. V. Ambudkar, S. Y. Rhee, M. J. Allison, and P. C. Maloney. 1992. Identification, purification, and reconstitution of OxlT, the oxalate:formate antiport protein of *Oxalobacter formigenes*. *J. Biol. Chem.* **267**:10537–10543.

55. To prepare proteoliposomes, one disperses phospholipids (e.g., those isolated from *E. coli*) in water, where they spontaneously aggregate to form spherical vesicles consisting of concentric layers of phospholipid. These vesicles, called liposomes, are then subjected to high-frequency sound waves (sonic oscillation), which breaks them into smaller vesicles surrounded by a single phospholipid bilayer resembling the lipid bilayer found in natural membranes. Then purified protein (e.g., the OxlT antiporter) is mixed with the sonicated phospholipids in the presence of detergent, and the suspension is diluted into buffer. The protein becomes incorporated into the phospholipid bilayer, and membrane vesicles called proteoliposomes are formed. When the proteoliposomes are incubated with solute, they catalyze uptake of the solute into the vesicles provided the appropriate carrier protein has been incorporated. In addition, one can "load" the proteoliposomes

with solutes (e.g., oxalate) by including these in the dilution buffer.

56. Poolman, B., D. Molenaar, E. J. Smid, T. Ubbink, T. Abee, P. P. Renault, and W. N. Konings. 1991. Malolactic fermentation: Electrogenic malate uptake and malate/lactate antiport generate metabolic energy. *J. Bacteriol.* **173**:6030–6037.

57. Konings, W. N. 1985. Generation of metabolic energy by end-product efflux. *Trends Biochem. Sci* **10**:317–319.

58. Konings, W. N., J. S. Lolkema, and B. Poolman. 1995. The generation of metabolic energy by solute transport. *Arch. Microbiol.* **164**:235–242.

59. Michels, J. P., J. Michel, J. Boonstra, and W. N. Konings. 1979. Generation of an electrochemical proton gradient in bacteria by the excretion of metabolic end-products. *FEMS Microbiol. Letts.* **5**:357–364.

60. Otto, R., et al. 1982. Lactate efflux-induced electrical potential in membrane vesicles of *Streptococcus cremoris*. *J. Bacteriol.* **149**:733–738.

61. Brink, B. T., and W. N. Konings. 1982. Electrochemical proton gradient and lactate concentration gradient in *Streptococcus cremoris* cells grown in batch culture. *J. Bacteriol.* **152**:682–686.

62. Driessen, A. J. M., and W. N. Konings. 1990. Energetic problems of bacterial fermentations: extrusion of metabolic end products, pp. 449–478. In: *Bacterial Energetics.* T. A. Krulwich (Ed.). Academic Press, New York.

63. Michel, T. A., and J. M. Macy. 1990. Generation of a membrane potential by sodium-dependent succinate efflux in *Selenomonas ruminantium*. *J. Bacteriol.* **172**:1430–1435.

64. The lactic and succinic acids are presumed to be in the ionized form because the intracellular pH is much larger than the pK_a values.

65. Brink, B. T., R. Otto, U. Hansen, and W. N. Konings. 1985. Energy recycling by lactate efflux in growing and nongrowing cells of *Streptococcus cremoris*. *J. Bacteriol.* **162**:383–390.

66. Oesterhelt, D., and J. Tittor. 1989. Two pumps, one principle: Light-driven ion transport in halobacteria. *Trends Biochem. Sci* **14**:57–61.

67. Bogomoleni, R. A., R. A. Baker, R. H. Lozier, and W. Stoeckenius. 1980. Action spectrum and quantum efficiency for proton pumping in *Halobacterium halobium*. *Biochemistry* **19**:2152–2159.

68. Henderson, R., J. M. Baldwin, and T. A. Ceska. 1990. Model for the structure of bacteriorhodopsin based on high-resolution electron cryo-microscopy, *J. Mol. Biol.* **213**:899–929.

69. The extreme halophiles require unusually high external NaCl concentrations [at least 3–5 M (i.e., 17–28%)] grow. They inhabit hypersaline environ-
ments such as the solar salt evaporation ponds near San Francisco and salt lakes (e.g., the Great Salt Lake in Utah and the Dead Sea). There are now six recognized genera, two of them being the well-known *Halobacterium* and *Halococcus*. The best studied is *Hb. salinarium (halobiuml)*. The other four genera are *Haloarcula*, *Haloferax*, *Natronobacterium*, and *Natronococcus*. The majority of the known halophilic archaea are aerobic chemo-organotrophs and can grow on simple carbohydrates as well as long-chain saturated hydrocarbons. They generally grow best at pH values between 8 and 9. However, *Natronobacterium* and *Natronococcus* are also alkaliphilic and grow well at pH values up to 11. When oxygen is not present, the halobacteria will grow anaerobically by using several electron acceptors in place of oxygen. These include fumarate, dimethyl sulfoxide (DMSO), and trimethyamine N-oxide (TMAO). Members of the genera *Haloarcula* and *Haloferax* can grow on nitrate as the terminal electron acceptor. Some of the halobacteria reduce the nitrate to nitrite and some reduce it completely to nitrogen gas. Some halobacteria can also grow fermentatively in the absence of oxygen. These include *Hb. salinarium (halobium)*, which can ferment arginine to citrulline.

70. Oesterhelt, D., and G. Krippahl. 1983. Phototrophic growth of halobacteria and its use for isolation of photosynthetically deficient mutants. *Ann. Microbiol. (Inst. Pasteur).* **134B**:137–150.

71. Gest, H. 1993. Photosynthetic and quasi-photosynthetic bacteria. *FEMS Microbiol. Lett.* **112**:1–6.

72. Some halophilic archaea can use nitrate as an electron acceptor to carry out anaerobic respiration.

73. Respiration can be severely limited under certain growth conditions because the oxygen content of hypersaline waters, the normal habitat of these organisms, is usually 20% or less than is found in normal seawater, and in unstirred ponds oxygen becomes even more scarce. The halobacteria can derive energy from the fermentation of amino acids; but in the absence of a fermentable carbon source and respiration, light is the only source of energy.

74. Bogomoleni, R. A., R. A. Baker, R. H. Lozier, and W. Stoeckenius. 1980. Action spectrum and quantum efficiency for proton pumping in *Halobacterium halobium*. *Biochemistry* **19**:2152–2159.

75. Tittor, J., and D. Oesterhelt. 1990. The quantum yield of bacteriorhodopsin. *FEBS Lett.* **263**:269–273.

76. Govindjee, R., S. P. Balashov, and T. G. Ebrey. 1990. Quantum efficiency of the photochemical cycle of bacteriorhodopsin. *Biophys. J.* **58**:597–608.

77. Lanyi, J. K. 1997. Mechanism of ion transport across membranes. *J. Biol. Chem.* **272**:31209–31212.

78. A Schiff base is an imine and has the following structure: R–CH=N–R'. It is formed between a

carbonyl group and a primary amine, as follows: R–CHO + H_2N–R′ → R—CH=N—R′ + H_2O. In bacteriorhodopsin, the amino group is donated by lysine in the protein and the carbonyl is donated by the retinal.

79. Lanyi, J. K. 1992. Proton transfer and energy coupling in the bacteriorhodopsin photocycle. *J. Bioenerg. Biomemb.* **24**:169–179.

80. Oesterhelt, D., J. Tittor, and E. Bamberg. 1992. A unifying concept for ion translocation by retinal proteins. *J. Bioenerg. Biomemb.* **24**:181–191.

81. Fodor, S. P. A., J. B. Ames, R. Gebhard, E. M. M. van den Berg, W. Stoeckenius, J. Lugtenburg, and R. A. Mathies. 1988. Chromophore structure in bacteriorhodopsin's N intermediate: Implications for the proton-pumping mechanism. Biochemistry **27**:7097–7101.

82. Krebs, M. P., and H. G. Khorana. 1993. Mechanism of light-dependent proton translocation by bacteriorhodopsin. *J. Bacteriol.* **175**:1555–1560.

83. Since aspartate-85 remains protonated while a proton is released into the aqueous phase, we must conclude that the immediate source of the released proton is a different amino acid residue.

84. Senior, A. E. ATPase of *Esche* *Chem.* **19**:7–41.

85. Lanyi, J. K. electrogenic chl **70**:319–330.

86. Duschl, A secondary c *halobium. J.*

87. There are ... whose optimal growth ... never the less can grow at pH 7 or ... widely distributed among the bacterial genera, include both bacteria and archaea. There are also *alkaliphilic-tolerant* bacteria that can grow at pH values of 9 or more, but whose optimal growth pH is around neutrality.

88. Reviewed in: Krulwich, T. A., and D. M. Ivey. 1990. Bioenergetics in extreme environments, pp. 417–447. In: *The Bacteria*, Vol. XII. T. A. Krulwich (Ed.). Academic Press, New York.

89. Krulwich, T. A., and A. A. Guffanti. 1989. Alkalophilic bacteria. *Annu. Rev. Microbiol.* **43**:435–463.

4

Electron Transport

The main method by which energy is generated for growth-related physiological processes such as biosynthesis and solute transport in respiring prokaryotes is by coupling the flow of electrons in membranes to the creation of an electrochemical proton gradient (Fig. 4.1). (See note 1.) As discussed in Section 3.7.1, electrons flow spontaneously down a potential energy gradient toward acceptors that have a more positive electrode potential. (The student should review Section 3.2.2 for a discussion of the generation of membrane potentials and Section 3.7.1 for a discussion of oxidation–reduction reactions coupled to the generation of a Δp.) The electrons flow from primary electron donors to terminal electron acceptors through a series of *electron carrier proteins* and a class of lipids called *quinones*. One refers to electron flow via electron carriers in membranes as *respiration*. If the terminal electron acceptor is oxygen, then electron flow is called *aerobic respiration*. If it is not oxgyen, it is called *anaerobic respiration*. Proton translocation takes place during respiration, and a Δp is created at coupling sites described in Sections 3.7.1 and 4.5.

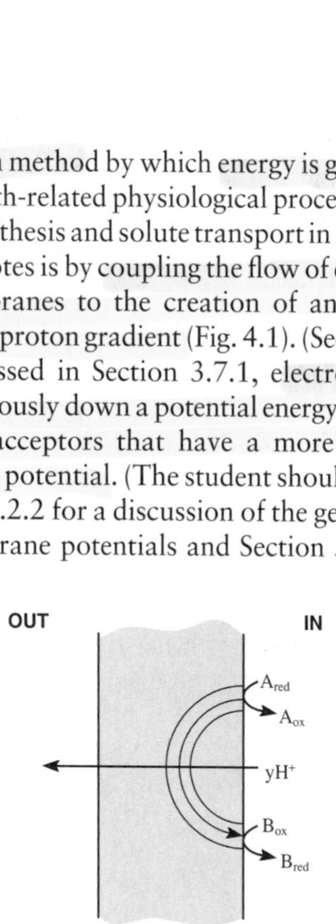

Mitochondria have three coupling sites, whereas bacteria can have one to three coupling sites depending upon the bacterium and the growth conditions. In other words, prokaryotic cell membranes (and mitochondria and chloroplasts) convert an electrode potential difference ΔE_h into a proton electrochemical potential difference (Δp). (As discussion in Section 3.7.1, E_h is the actual electrode potential of a compound at the specified concentrations of its oxidized and reduced forms.) The proton potential is then used to drive solute transport, ATP synthesis, flagella rotation, and other membrane activities. In mitochondria, all electron transport pathways are much the same. However, prokaryotes are diverse creatures and their electron transport pathways differ depending upon the primary donor and terminal acceptor. This chapter describes electron transport pathways in mitochondria and bacteria and how they are coupled to the

Fig. 4.1 Oxidation–reduction reactions in the cell membrane result in a proton potential. Electrons flow from A to B, through a series of electron carriers in the membrane, from a low potential toward a higher potential. The intermediate redox reactions between A and B are not shown. Certain of the redox reactions in the series are coupled to the translocation of protons across the cell membrane. These are called coupling sites. In this way a redox potential (ΔE_h) is converted into a proton potential (Δp): that is, $n\Delta E_h = y\Delta p$, where n is the number of electrons transferred and y is the number of protons extruded.

120

formation of a Δp. It appears that archaeal electron transport chains are similar to (eu)bacterial electron transport chains.[2]

4.1 Aerobic and Anaerobic Respiration

There is a steady current of electrons through electron carriers in prokaryotic cell membranes from low-potential electron donors (the primary donors or *reductants*) to high-potential electron acceptors (the terminal acceptor or *oxidant*). Electron acceptors can be inorganic, such as oxygen, nitrate, or sulfate. They can also be organic (e.g., fumarate). Thus, there is oxygen respiration, nitrate respiration, sulfate respiration, fumarate respiration, and so on.

4.2 The Electron Carriers

The electrons flow through a series of electron carriers. Some of these carry hydrogen as well as electrons. The electron carriers are as follows:

1. Flavoproteins (hydrogen and electron carrie) ⎫ *hydrogen carriers*
2. Quinones (hydrogen and electron carrie) ⎭
3. Iron–sulfur proteins (electron carrie)
4. Cytochromes (electron carrie) *electron carriers*

The quinones are lipids, whereas the other electron carriers are proteins, which exist in multiprotein enzyme complexes called *oxidoreductases*. (See note 3 for examples of oxidoreductases.) The electrons are not carried in the protein per se, but in a nonprotein molecule bound to the protein. The nonprotein portion that carries the electron is called a *prosthetic group*. (See note 4 for useful definitions.) The prosthetic group in iron–sulfur proteins is a cluster of iron–sulfide, which is abbreviated as FeS. The prosthetic group in flavoproteins (Fp) is a flavin, which can be either flavin adenine dinucleotide (FAD) or flavin mononucleotide (FMN). The prosthetic group in cytochromes is heme. The chemistry of the prosthetic groups is described in Section 4.2.1. Some of the prosthetic groups (flavins) carry hydrogen as well as electrons, and they are referred to as hydrogen carriers. The quinones are also hydrogen carriers. Some of the prosthetic groups (FeS and heme) carry only electrons, and they are referred to as electron carriers.

Each of the electron carriers has a different electrode potential, and the electrons are transferred sequentially to a carrier of a higher potential.

The standard potentials at pH 7 of the electron carriers and some electron donors and acceptors will be shown later (see Table 4.1).

4.2.1 Flavoproteins

A flavoprotein (Fp) is an electron carrier that has as its prosthetic group an organic molecule called a flavin. [The term is derived from the Latin word *flavius*, which means yellow, in reference to the color of flavins. They flavins FAD and FMN are synthesized by cells from the vitamin riboflavin (vitamin B_2).] See Fig. 4.2.

Phosphorylation of riboflavin at the ribityl 5'-OH yields FMN, and adenylylation of FMN yields FAD. As Fig. 4.2 illustrates, when flavins are reduced they carry 2H (two electrons and two hydrogens), one on each of two ring

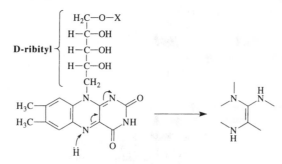

Fig. 4.2 Structures of riboflavin (X = H), FMN (X = PO_3H_2), and FAD (X = ADP). For the sake of convenience, the reduction reaction is drawn as proceeding via a hydride ion even though this need not be the actual mechanism in all flavin reductions.

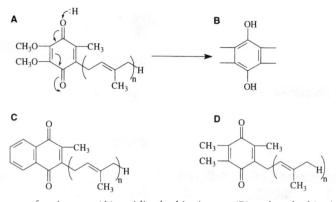

Fig. 4.3 The structure of quinones: (A) oxidized ubiquinone, (B) reduced ubiquinone, (C) oxidized menaquinone, and (D) oxidized plastoquinone. The value of n can be 4 to 10 and is 8 for both quinones in *E. coli*. In *E. coli* ubiquinone plays a major role in aerobic and nitrate respiration, whereas menaquinone is dominant during fumarate respiration. One reason for this is that ubiquinone has a potential (E_0') of +100 mV, versus +30 mV for fumarate. It is therefore at too high a potential to deliver electrons to fumarate. Menaquinone has a low potential, −74 mV, and is thus able to deliver electrons to fumarate. Plastoquinone is used in chloroplast and cyanobacterial photosynthetic electron transport.

nitrogens. There are many different flavoproteins, and they catalyze diverse oxidation–reduction reactions in the cytoplasm, not merely those of the electron transport chain in the membranes. Although all the flavoproteins have FMN or FAD as their prosthetic group, they catalyze different oxidations and have different redox potentials. These differences are due to differences in the protein component of the enzyme, not in the flavin itself.

4.2.2 Quinones

Quinones are lipid electron carriers. Owing to their hydrophobic lipid nature, some are believed to be highly mobile in the lipid phase of the membrane, carrying hydrogen and electrons to and from the complexes of protein electron carriers that are not mobile. Quinone structures and oxidation–reduction reactions are shown in Fig. 4.3. All quinones have hydrophobic isoprenoid side chains that contribute to their lipid solubility. The number of isoprene units varies but is typically 6 to 10. Bacteria make two types of quinone that function during respiration: ubiquinone (UQ), a quinone also found in mitochondria, and menaquinone (MQ, or sometimes MK). Menaquinones (Fig. 4.3C), which are derivatives of vitamin K, differ from ubiquinones in being naphthoquinones in which the addi-

tional benzene ring replaces the two methoxy groups present in ubiquinones (Fig. 4.3A,B). They also have a much lower electrode potential than ubiquinones and are used predominantly during anaerobic respiration, where the electron acceptor has a low potential (e.g., during fumarate respiration). A third type of quinone, plastoquinone (Fig. 4.3D), occurs in chloroplasts and cyanobacteria, and functions in photosynthetic electron transport. In plastoquinones, the two methoxy groups are replaced by methyl groups.

4.2.3 Iron–sulfur proteins

Iron–sulfur proteins contain nonheme iron and usually acid-labile sulfur (Fig. 4.4). The term

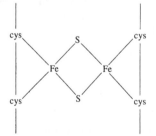

Fig. 4.4 Scheme for FeS clusters: this is an Fe_2S_2 cluster. More than one cluster may be present per protein. The sulfur atoms held only by the iron are acid labile. The iron is bonded to the protein via sulfur in cysteine residues.

"acid-labile sulfur" means that when the pH is lowered to approximately 1, hydrogen sulfide is released from the protein. This is because there is sulfide attached to iron by bonds that are ruptured in acid. Generally, the proteins contain clusters in which iron and acid–labile sulfur are present in a ratio of 1:1. However, there may be more than one iron–sulfur cluster per protein. For example, in mitochondria the enzyme complex that oxidizes NADH has at least four FeS clusters (see later, Fig. 4.9). The FeS clusters have different E_h values, and the electron travels from one FeS cluster to the next toward the higher E_h. It appears that the electron may not be localized on any particular iron atom, and the entire FeS cluster should be thought of as carrying one electron, regardless of the number of Fe atoms.

Iron–sulfurs proteins also contain cysteine sulfur, which is not acid labile, and bonds the iron to the protein. There are several different types of iron–sulfur protein, and these catalyze numerous oxidation–reduction reactions in the cytoplasm as well as in the membranes. (See note 5 for more information on iron–sulfur proteins.) The iron–sulfur proteins have characteristic electron spin resonance (ESR) spectra because of an unpaired electron in either the oxidized or reduced form of the FeS cluster in different FeS proteins. (See note 6 for a description of electron spin resonance.) The iron–sulfur proteins cover a very wide range of potentials, from approximately −400 mV to +350 mV. They therefore can carry out oxidation–reduction reactions at both the low-potential end and the high-potential end of the electron transport chain, and indeed are found in several locations. In the FeS cluster shown in Fig. 4.4, note that each Fe is bound to two acid-labile sulfurs and two cysteine sulfurs. This would be called an Fe_2S_2 cluster.

4.2.4 Cytochromes

Cytochromes are electron carriers that have heme as the prosthetic group. Heme consists of four pyrrole rings attached to each other by methene bridges (Fig. 4.5). Because hemes have four pyrroles, they are called *tetrapyrroles*. Each of the pyrrole rings is substituted by a side chain. Substituted tetrapyrroles are called *porphyrins*.

Therefore, hemes are also called porphyrins. (An unsubstituted tetrapyrrole is called a porphin.) Hemes are placed in different classes, described shortly, on the basis of the side chains attached to the pyrrole rings. In the center of each heme there is an iron atom that is bound to the nitrogen of the pyrrole rings. The iron is the electron carrier and is oxidized to ferric or reduced to ferrous ion during electron transport. Cytochromes are therefore one-electron carriers. The E_h values of the different cytochromes vary depending on the protein and the molecular interactions with surrounding molecules.

Classes of cytochromes

Figure 4.5 shows five classes of heme that distinguish the cytochromes: hemes a, b, c, d, and o. Hemes d and o have been found only in the prokaryotic cytochrome oxidases. Bacterial cytochromes include cytochromes bd (also called cytochrome d) and bo (also called cytochrome bo_3 or cytochrome o), which are quinol oxidases that reduce oxygen. (In naming cytochromes, sometimes the O_2-binding heme is given the subscript 3.)[7] As the names imply, each cytochrome has two types of heme, one being heme b and the other being heme d or o.[7] As mentioned previously, the hemes can be distinguished according to the side groups that they possess, as summarized in Fig. 4.5. For example, heme o differs from heme b in having an hydroxyethylfarnesyl group substituted for a vinyl group. However, the only difference between heme b and heme c is that the latter is covalently bound to protein by thioether linkages between the two vinyl groups and cysteine residues in the protein.

Hemes can usually be distinguished spectrophotometrically. When cytochromes are in the reduced state, absorption by the heme produces characteristic light absorption bands in the visible range: the α, β, and γ bands. The α bands absorb light between 500 and 600 nm, the β absorb at a lower wavelength, and the γ bands are in the blue region of the spectrum. The spectrum for a cytochrome c is shown in Fig. 4.6. Cytochromes are distinguished, in part, by the position of the maximum in the α band. For example, cyt b_{556} has a peak at 556 nm and cyt b_{558} a peak at 558 nm.

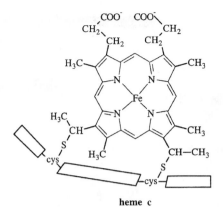

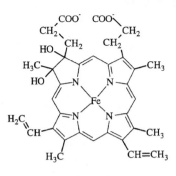

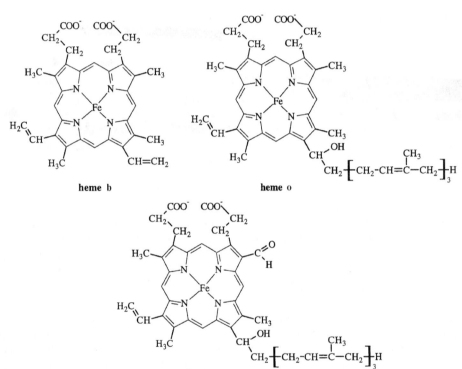

Fig. 4.5 The prosthetic groups of the different classes of cytochromes. The hemes vary according to their side groups. Heme c is covalently bound to the protein via a sulfur bridge to a cysteine residue on the protein. The structures of R_1, R_2, and R_3 are not known. *Source*: Adapted from Gottschalk, G. *Bacterial Metabolism*, Springer-Verlag, New York. 1986.

Reduced minus oxidized spectra

Because of light scattering and nonspecific absorption, it is very difficult to resolve the different peaks of individual cytochromes in whole cells unless one employs difference spectroscopy. For difference spectroscopy, the cells are placed into two cuvettes in a split-beam spectrophotometer, and monochromatic light from a single monochromator scan is split to

pass through both cuvettes. In one cuvette the cytochromes are oxidized by adding an oxidant, and in the second cuvette they are reduced by adding a reductant. The spectrophotometer subtracts the output of one cuvette from the other to give a reduced minus oxidized difference spectrum. In this way nonspecific absorption and light scattering are eliminated from the spectrum, and the cytochromes in the preparation are identified.

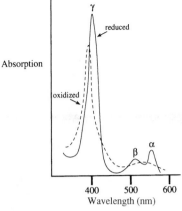

Fig. 4.6 Absorption spectra of oxidized (dashed curve) and reduced (solid curve) cytochrome c. The α band in the reduced form is used to identify cytochromes.

Dual-beam spectroscopy

To follow the kinetics of oxidation or reduction of a particular cytochrome, a dual-beam spectrophotometer is used. In a dual-beam spectrophotometer there are two monochromators. Light from one monochromator is set at a wavelength at which absorbance will change during oxidation or reduction, and the second beam of light is at a nearby wavelength for which absorbance will not change. The light is sent alternatively from both monochromators through the sample cuvette, and the difference in absorbance between the two wavelengths is automatically plotted as a function of time.

4.2.5 Standard electrode potentials of the electron carriers

Table 4.1 shows standard electrode potentials at pH 7 (E_0') of some electron donors, acceptors, and electron carriers. Notice that redox couples are generally written in the form "oxidized/reduced." Many of the oxidation–reduction reactions in the electron transport chain can be reversed by the Δp as discussed in Section 3.7.1. This means that the ox/red ratio for several of the electron carriers (flavoproteins, cytochromes, quinones, FeS proteins) must be close to 1. Thus, for these reactions the E_h (actual potential at pH 7) values of the redox couples are close to their midpoint potentials, E_m', which is the potential at pH 7 when the couple is 50% reduced: that is, [ox] = [red].

Table 4.1 Standard electrode potentials at pH 7

Couple	E_0' (mV)
Fd_{ox}/Fd_{red} (spinach)	−432
CO_2/formate	−432
H^+/H_2	−410
Fd_{ox}/Fd_{red} (*Clostridium*)	−410
NAD^+/NAD	−320
FeS(ox/red) in mitochondria	−305
Lipoic/dihydrolipophilic	−290
S^o/H_2S	−270
FAD/ $FADH_2$	−220
Acetaldehyde/ethanol	−197
FMN/$FMNH_2$	−190
Pyruvate/lactate	−185
Oxaloacetate/malate	−170
Menaquinone(ox/red)	−74
cyt b_{558}(ox/red)	−75 to −43
Fumarate/succinate	+33
Ubiquinone(ox/red)	+100
cyt b_{556} (ox/red)	+46 to +129
cyt b_{562} (ox/red)	+125 to +260
cyt d (ox/red)	+260 to +280
cyt c (ox/red)	+250
FeS (ox/red) in mitochondrin	+280
cyt a (ox/red)	+290
cyt c_{555}(ox/red)	+355
cyt a_3 (ox/red) in mitochondrin	+385
NO_3^-/NO_2^-	+421
Fe^{3+}/Fe^{2+}	+771
O_2(1 atm)/H_2O	+815

Sources: Thauer, R. K., K. Jungermann, and K. Decker. 1977. Energy conservation in chemotrophic anaerobic bacteria. *Bacteriol Rev.* 41: 100–180. Metzler, D. E. 1977. *Biochemistry: The Chemical Reactions of Living Cells.* Academic Press, New York.

4.3 Organization of the Electron Carriers in Mitochondria

For a historical perspective, see Box 4.1.

The electron carriers are organized as an electron transport chain that transfers electrons from electron donors at a low electrode potential to electron acceptors at a higher electrode potential (Fig. 4.7). Electrons can enter at the level of flavoprotein, quinone, or cytochrome, depending upon the potential of the donor. The carriers are organized in the membrane as individual complexes. The complexes can be isolated from each other by appropriate separation techniques after mild detergent extraction, which removes the lipids but does not destroy the protein–protein interactions. The separated complexes can be analyzed for their components and also can be incorporated into proteoliposomes to facilitate study of the

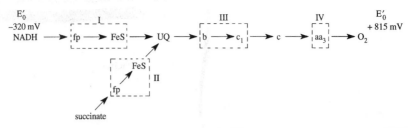

Fig. 4.7 Electron transport scheme in mitochondria. Electrons travel in the electron transport chain from a low to a high electrode potential. Complexes I to IV are enclosed in dashed lines. Complex I is NADH dehydrogenase, also called NADH–ubiquinone oxidoreductase. Complex II is succinate dehydrogenase, also called succinate–ubiquinone oxidoreductase. Complex III is the bc_1 complex, also called ubiquinol–cytochrome c oxidoreductase. Complex IV is the cytochrome aa_3 oxidase, also called cytochrome c oxidase. There are several FeS clusters in complexes I and II, and an FeS protein in complex III. Complex II has both peripheral and integral membrane protein subunits. (See note 12 for a description of complex II.) The flavin (FAD) and FeS centers are in the peripheral membrane subunits, and hemes (not shown) are in the integral membrane subunits. Electrons flow from FAD through the FeS centers to quinone, probably via heme. Abbreviations: fp, flavoprotein; FeS, iron–sulfur protein; UQ, ubiquinone; b, cytochrome b; c_1, cytochrome c_1; c, cytochrome c; aa_3, cytochrome aa_3.

BOX 4.1 HISTORICAL PERSPECTIVE: CELLULAR RESPIRATION

The elucidation of the pathway by which electrons flow from organic compounds to oxygen was the result of many years of research by different investigators. The realization that iron compounds were electron carriers came from early work on the effect of cyanide on respiration. Otto Heinrich Warburg (1883–1970), a German biochemist who studied respiration, reported the inhibitory effect of cyanide on respiration in sea urchin eggs, yeast, and bacteria before and after World War I. Because he knew that cyanide inhibits autoxidation reactions (e.g., the oxidation of cysteine to cystine, catalyzed by iron compounds), Warburg concluded that cyanide also inhibits respiration, further reasoning that an iron-containing enzyme that he called *Atmungsferment* ("respiratory ferment") catalyzes the oxidation of the organic substrates. Warburg also showed, in 1926, that carbon monoxide inhibits the uptake of oxygen by yeast cells. He knew from earlier studies by others that carbon monoxide combines with hemoglobin and that it can be dissociated

from the ferro–heme complex with visible light. He was able to show that visible light also reverses the inhibition of respiration of yeast by carbon monoxide. By measuring the effect of light of different wavelengths on reversing the inhibition, Warburg was able to determine the absorption spectrum of the pigment. The absorption spectrum, which was that of heme, supported his conclusion that *Atmungsferment* was a heme pigment. (Warburg's *Atmungsferment* turned out to be cytochrome oxidase, later studied by David Keilin.) For his discovery of the respiratory enzyme and its mode of action, Warburg was awarded the Nobel Prize in Physiology or Medicine in 1931.

Research on respiratory enzymes had actually begun in the late 1800s. Cytochromes were first described by the Englishman Charles Alexander MacMunn in three papers published between 1884 and 1887. MacMunn reported the discovery of a pigment while spectroscopically examining the tissues of various species of vertebrates and invertebrates. He called the pigment histohaematin, or myohaematin, and it had a characteristic absorption spectrum consisting of four bands. MacMunn reported that when he added the oxidizing agent hydrogen peroxide, the bands disappeared, only to reappear upon reduction. He concluded that since the substances were capable of being oxidized and reduced, they were respiratory pigments. He also concluded that they were protein. However, MacMunn's discovery was not accepted by his contemporaries, especially the famous biochemist Hoppe-Seyler, who thought MacMunn's belief that he had discovered a cellular respiratory pigment was erroneous.

MacMunn was clearly frustrated that his discovery had not been accepted, and he wrote in a book that was not published until after he died, in 1911, that "doubtless in time this pigment will find its way into the textbooks."[1] This indeed did occur, primarily through the work of David Keilin in the 1920s.

David Keilin, who left Poland to study first in Belgium, then in France, and emigrated to England in 1915, was unaware of MacMunn's work, although later when he learned of it, he acknowledged it. In 1925 Keilin, using a spectromicroscope, rediscovered the pigments in the thoracic muscles of insects, in *Bacillus subtilis*, and in yeast, and named them cytochromes (cellular pigments). Keilin discovered cytochromes a, b, and c, and later cytochrome oxidase. Keilin published several papers in the late 1920s and early 1930s describing the respiratory chain as a chain of carriers consisting of dehydrogenases that remove hydrogen from organic substrates, as well as oxidized cytochromes that are reduced by the dehydrogenases, and cytochrome oxidase, an autoxidizable heme compound that oxidizes the cytochromes and reduces oxygen. In the early 1930s it was recognized that the respiratory chain transfers electrons, rather than hydrogen, from the organic substrate to oxygen, and in 1931 Warburg correctly attributed the oxidations and reductions to a change in the valency of iron. The realization of the role of pyridine-linked dehydrogenases and flavin-linked dehydrogenases was due to the research of Warburg in the 1930s.

For more information about the history of biochemical research, see ref. 2.

REFERENCES

1. Quoted in Fruton, J. S., and S. Simmonds. 1958. *General Biochemistry*, 2nd ed. John Wiley & Sons, New York.

2. Florkin, M. 1975. A history of biochemistry. In: *Comprehensive Biochemistry*, Vol. 31. M. Florkin and E. H. Stotz (Eds.). Elsevier Scientific Publishing, Amsterdam.

oxidation–reduction reactions that each catalyzes, in addition to proton translocation. (Proteoliposomes are artificial constructs of purified lipids and proteins. They are described in Section 16.1.)

Four complexes can be recognized in mitochondria. They are complex I (NADH–ubiquinone oxidoreductase), complex II (succinate dehydrogenase), complex III (ubiquinol–cytochrome c oxidoreductase, also called the bc_1 complex), and complex IV (cytochrome c oxidase, which is cytochrome aa_3). Complexes I, III, and IV are coupling sites (Section 4.5). Each complex can have several proteins. The most intricate is complex I from mammalian mitochondria, which has about 40 polypeptide subunits, at least four iron–sulfur centers, one flavin mononucleotide (FMN), and one or two bound ubiquinones. Analogous complexes have been isolated from bacteria, but in some cases (e.g., NADH–ubiquinone oxidoreductase and the bc_1 complex), they have fewer protein components.[8–11] (See note 12 for a description of complex II and how it varies with different bacteria.) Note the pattern in the arrangement of the electron carriers in mitochondria; a dehydrogenase complex accepts electrons from a primary donor and transfers the electrons to a quinone. The quinone then transfers the electrons to an oxidase complex via intervening cytochromes. As described next, the same general pattern exists in bacteria.

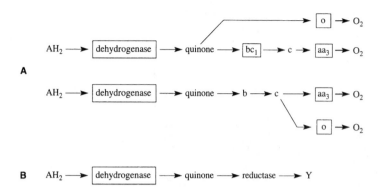

Fig. 4.8 Generalized electron transport pathways found in bacteria. The details will vary depending upon the bacterium and the growth conditions. (A) Aerobic respiration. A dehydrogenase complex removes electrons from an electron donor and transfers these to a quinone. The electrons are transferred to an oxidase complex via a branched pathway. Depending upon the bacterium, the pathway may branch at the quinone or at cytochrome. Many bacteria have bc_1, cytochrome c, and cytochrome aa_3 in one of the branches and in this way resemble mitochondria. Other bacteria do not have a bc_1 complex, and may or may not have cytochrome aa_3. (B) Anaerobic respiration. Under anaerobic conditions the electrons are transferred to reductase complexes, which are synthesized anaerobically. Several reductases exist, each one specific for the electron acceptor. Y represents either an inorganic electron acceptor other than oxygen (e.g., nitrate) or an organic electron acceptor (e.g., fumarate). More than one reductase can simultaneously exist in a bacterium.

4.4. Organization of the Electron Carriers in Bacteria

Bacterial electron transport chains vary among the different bacteria, and also according to the growth conditions. These variations will be discussed later (Section 4.7). First, we shall describe the common features in bacterial electron transport schemes and compare them with systems of mitochondrial electron transport. As with the mitochondrial electron transport chain, the bacterial chains are organized into dehydrogenase and oxidase complexes connected by quinones (Fig. 4.8). The quinones accept electrons from dehydrogenases and transfer them to oxidase complexes that reduce the terminal electron acceptor. Bacteria are capable of using electron acceptors other than oxygen (e.g., nitrate and fumarate) during anaerobic respiration. The enzyme complexes that reduce electron acceptors other than oxygen are called *reductases*, rather than oxidases.

Some of the dehydrogenase complexes are NADH and succinate dehydrogenase complexes, analogous to complex I and II in mitochondria. In addition to these dehydrogenases, several others reflect the diversity of substrates oxidized by the bacteria: H_2 dehydrogenases (called hydrogenases), formate dehydrogenase, lactate dehydrogenase, methanol dehydrogenase, methylamine dehydrogenase, and so on.

Depending upon the source of electrons and electron acceptors, bacteria can synthesize and substitute one dehydrogenase complex for another, or reductase complexes for oxidase complexes. For example, when growing anaerobically, E. coli makes the reductase complexes instead of the oxidase complexes, represses the synthesis of some dehydrogenases, and stimulates the synthesis of others. (This is discussed more fully in Chapter 18.) The electron carrier complexes in bacteria are sometimes referred to as modules, since they can be synthesized and "plugged into" the respiratory chain when needed.

4.4.1 The different terminal oxidases

A word should be said about the many different terminal oxidases found in bacteria.[13] Whereas mitochondria all have the same cytochrome c oxidase (cytochrome aa_3), bacteria have a variety of terminal oxidases, often two or three different ones in the same bacterium (i.e., two or three branches to oxygen). Some of the bacterial terminal oxidases oxidize quinols (quinol oxidases) and some oxidize cytochrome c (cytochrome c oxidases). The terminal oxidases differ in their affinities for oxygen and in whether they are proton pumps, as well as in the types of hemes and metals they contain. However, despite these differences, most belong to the heme–copper oxidase superfamily of oxidases, to which the mitochondrial cytochrome c oxidase also belongs. An exception is the cytochrome bd oxidase, discussed later, which is not a member of the heme–copper oxidase superfamily.

All members of the heme–copper oxidase superfamily share a protein subunit that is homologous to one of the subunits (subunit I) of the mitochondrial cytochrome c oxidase. This subunit has a bimetallic (binuclear) center that binds oxygen and reduces it to water, and pumps protons. (Several of the heme–copper oxidases are known to be proton pumps, whereas it is not yet known whether certain others pump protons.) The bimetallic center contains a heme iron and copper. There is a second heme in all these oxidases but it is not part of the bimetallic center and probably functions in transferring electrons to the bimetallic center. Heme–copper oxidases that are mentioned later in the context of bacterial respiratory systems include the cbb_3-type oxidases, cytochrome aa_3 oxidase, cytochrome bo_3 oxidase (cytochrome o oxidase), and cytochrome bb_3 oxidase. It should be emphasized that some of these are quinol oxidases and some are cytochrome c oxidases.

4.4.2 Bacterial electron transport chains are branched

Two major difference between mitochondrial and bacterial electron transport chains are as follows: (1) the routes to oxygen in the bacteria are branched, the branch point being at the quinone or cytochrome, and (2) many bacteria can alter their electron transport chains depending upon growth conditions (Fig. 4.8). As noted in Section 4.4.1, under aerobic conditions there are often two or three branches leading to different terminal oxidases. For example, a two-branched electron transport chain might contain a branch leading to cytochrome o oxidase (quinol oxidase) and a branch leading to cytochrome aa_3 oxidase (cytochrome c oxidase). Other bacteria (e.g., E. coli) have cytochrome o and d oxidase branches (both are quinol oxidases) but lack the cytochrome aa_3 branch. The ability to synthesize branched electron transport pathways to oxygen confers flexibility on the bacteria, since not only may the branches differ in the Δp that can be generated (because they may differ in the number of coupling sites) but their terminal oxidases also may differ with respect to affinities for oxygen. For example, in E. coli cytochrome o has a low affinity for oxygen, whereas cytochrome d has a higher affinity for oxygen. Switching to an oxidase with a higher affinity for oxygen allows the cells to continue to respire even when oxygen tensions fall to very low values. This ability is important to ensure the reoxidation of the reduced quinones and NADH so that cellular oxidations such as the oxidation of glucose to pyruvate or the oxidation of pyruvate to CO_2 may continue.

It has also been suggested that under microaerophilic conditions the use of an oxidase with a high affinity for oxygen would remove traces of oxygen that might damage oxygen-sensitive enzymes that are made under microaerophilic or anaerobic growth conditions.[14] As an example of a protective role that an oxidase might play, consider the situation with the strict aerobe Azotobacter vinelandii. This organism has a branched respiratory chain leading to cytochromes o and d as the terminal oxidases. It also fixes nitrogen aerobically. However, as with other nitrogenases, the Azotobacter nitrogenase is inactivated by oxygen. Azotobacter employs two mechanisms to protect its nitrogenase, respiratory and conformational. (See Section 12.3.2 for a discussion of this point.) It is thought that the rapid consumption by oxygen by a terminal oxidase maintains the intracellular oxygen levels low enough to prevent the inactivation of the nitrogenase. In agreement with this suggestion is the report that mutants of A. vinelandii that are deficient in the cytochrome d complex failed to fix nitrogen in air, although they did fix nitrogen when the oxygen tension was sufficiently reduced.[15] (Mutants in cytochrome o can fix nitrogen in air.)

The adaptability of the bacteria with respect to their electron transport chains can also be seen in many bacteria that can respire either aerobically or anaerobically. Under anaerobic conditions they do not make the oxidase complexes but instead synthesize reductases. For example, during anaerobic growth, E. coli synthesizes fumarate reductase, nitrate reductase, and trimethylamine-N-oxide (TMAO) reductase. (The regulation of synthesis of these reductases is discussed in Chapter 18.) The different reductases enable the bacteria to utilize alternative electron acceptors under anaerobic conditions. (Some facultative anaerobes will ferment when a terminal electron acceptor is unavailable. Fermentation is discussed in Chapter 14.)

4.5 Coupling Sites

Sites in the electron transport pathway at which redox reactions are coupled to proton extrusion creating a Δp are called *coupling sites*.[16] Each coupling site is also a site for ATP synthesis, since the protons extruded reenter via ATP synthase to make ATP. The three coupling sites in mitochondria, sites 1, 2, and 3, were shown in Fig. 4.7, where site 1 is the NADH dehydrogenase complex (complex I), site 2 is the bc_1 complex (complex III), and site 3 is the cytochrome aa_3 complex (complex IV). The succinate dehydrogenase complex (complex II) is not a coupling site.

The ratio of protons translocated for every two electrons varies, depending upon the complex. A consensus value is 10 protons extruded for every two electrons that travel from NADH to oxygen. The bc_1 complex translocates four protons for every two electrons, and depending upon the reported value, complexes I and IV translocate 2 to 4 protons for every two electrons. (The consensus value for mitochondrial complex I is $4H^+/2e^-$.) During reversed electron flow, protons enter the cell through coupling sites 1 and 2, driven by the Δp, and the electrons are driven toward the lower redox potential. This creates a positive ΔE at the expense of the Δp. (See eq. 3.19.) Coupling site 3 is not physiologically reversible. Thus, water cannot serve as a source of electrons for NAD^+ reduction by means of reversed electron flow. However, during oxygenic photosynthesis, light energy can drive electrons from water to $NADP^+$. The mechanism of photoreduction of $NADP^+$, which is different from reversed electron flow, is discussed in Chapter 5.

4.5.1 The identification of coupling sites

For an understanding of the physiology of energy metabolism during electron transport, it is necessary to study the mechanism of proton translocation, and for this the coupling sites must be identified and isolated. The coupling sites can be identified by the use of electron donors that feed electrons into the chain at different places, followed by measuring the amount of ATP made for every two electrons transferred through the respiratory chain. The number of ATPs made for every $2e^-$ transfer to oxygen is called the *P/O ratio*. It is equal to the number of ATP molecules formed per atom of oxygen taken up. When an electron acceptor other than oxygen is used, $P/2e^-$ is substituted for P/O. The P/O ratio is approximately equal to the number of coupling sites. (However, see Section 4.5.2.)

In mitochondria, the oxidation of NADH results in a P/O ratio of about 3, indicating that three coupling sites exist between NADH and O_2. The use of succinate as an electron donor results in a P/O ratio of approximately 2. Since electrons from succinate enter at the ubiquinone level, this indicates that coupling site 1 occurs between NADH and ubiquinone [i.e., the NADH–ubiquinone oxidoreductase reaction (Fig. 4.7)]. The other two coupling sites must occur between ubiquinone and oxygen. When electrons enter the respiratory chain after the bc_1 complex, the P/O ratio is reduced to 1, indicating that the bc_1 complex is the second coupling site and that site 3 is cytochrome aa_3 oxidase. Site 3 can be demonstrated by bypassing the bc_1 complex. The bc_1 complex can be bypassed via an artificial electron donor that will reduce cytochrome c [e.g., ascorbate and tetramethylphenylenediamine (TMPD)], thus channeling electrons from ascorbate to TMPD to cytochrome c to cytochrome aa_3. Alternatively, one can simply use reduced cytochrome c as an electron donor to directly reduce cytochrome aa_3.

Each coupling site is characterized by a drop in midpoint potential of about 200 mV, which is sufficient for generating the Δp (Fig. 4.9). The size of the Δp that can be generated with respect to the ΔE was discussed in Section 3.7.

4.5.2 The actual number of ATPs that can be made for every two electrons travelling through the coupling sites

According to the chemiosomotic theory, the number of ATPs made at a coupling site need

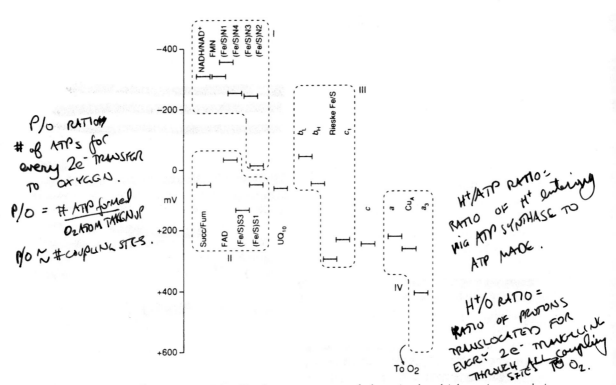

Handwritten margin notes:

P/O RATIO
# of ATP's for every 2e⁻ TRANSFER TO OXYGEN.

P/O = # ATP formed / O₂ ATOM TAKEN UP

P/O ≈ # COUPLING SITES.

H⁺/ATP RATIO = RATIO OF H⁺ entering via ATP SYNTHASE TO ATP MADE.

H⁺/O RATIO = RATIO OF PROTONS TRANSLOCATED FOR EVERY 2e⁻ TRAVELING THROUGH ALL COUPLING SITES TO O₂.

Fig. 4.9 Average midpoint potentials, E'_m, for components of the mitochondrial respiratory chain. The complexes are in dashed boxes. The actual potentials (E_h) for most of the components is not very different from the midpoint potentials. An exception is cytochrome aa_3, whose E_h is much more positive than its midpoint potential. There are changes in potential of 200 mV or more at three sites, which drive proton translocation. One site is within complex I, a second within complex III, and the third between complex IV and oxygen. *Source*: Adapted from Nicholls, D. G., and S. J. Ferguson. 1992. *Bioenergetics 2*. Academic Press, London.

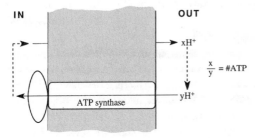

Fig. 4.10 The ratio of protons extruded to protons translocated through the ATPase determines the amount of ATP made.

not be a whole number. This is because the amount of ATP that can be made per coupling site is equal to the ratio of protons extruded at the coupling site to protons that reenter via the ATP synthase (Fig. 4.10). For example, if two protons are translocated at a coupling site and three protons enter through the ATP synthase, then two-thirds of an ATP can be made when two electrons pass through the coupling site. The ratio of protons translocated for every two electrons traveling through all the coupling sites to O_2 is called the H^+/O ratio. It can be measured by administering a pulse of a known amount of oxygen to an anaerobic suspension of mitochondria or bacteria and measuring the initial efflux of protons with a pH electrode as the small amount of oxygen is used up. The experiment requires that valinomycin plus K^+ or a permeant anion such as thiocyanate, SCN^-, be in the medium to prevent a $\Delta\Psi$ from developing. (See Section 3.2.2 for a discussion of this point.) The reported values for H^+/O for NADH oxidation vary. However, there is consensus that the true value is probably around 10. The ratio of protons entering via the ATP synthase to ATP made is called the H^+/ATP ratio. It can be measured by using inverted submitochondrial particles prepared by sonic oscillation. These particles have the ATP synthase on the outside and will pump protons into the interior upon addition of ATP. Similar inverted vesicles can be made from bacteria by first enzymatically weakening or removing the cell wall and breaking the spheroplasts or protoplasts by passage through a French press at high pressures. (See note 17 for how to make right-side-out vesicles.) Values of H+/ATP from 2 to 4 have been reported, and a consensus value of three can be used for calculations. For intact mitochondria, an additional H+ is required to bring P_i electroneutrally from the cytosol into the mitochondrial matrix in symport with H+, so H+/ATP would be 4. A value of 10 for H+/O predicts a maximum P/O ratio of 2.5 (10/4) for mitochondria. This means that the often stated value of 3 for a P/O ratio for NADH oxidation by mitochondria may be too high.

The number of protons ejected for every $2e^-$ traveling between succinate and oxygen in mitochondria is 6. Therefore, the maximal P/O ratio for this segment of the electron transport chain may be 6/4, or 1.5. The P/O ratios in bacteria can be higher, since a proton is not required to bring P_i into the cell. One significant aspect of branched aerobic respiratory chains in bacteria is that the number of coupling sites, and therefore H+/O, can differ in the branched chains. Thus, the different branches are not equally efficient in generating a Δp or making ATP. We will return to this point later.

Of course, an ATP can be made when three protons enter via the ATP synthase only if the Δp is sufficiently large. As an exercise, we can ask how large the Δp must be. Recall that $y\Delta p$ is the work that can be done in units of electron volts when y protons traverse a proton potential of Δp volts. If H+/ATP is 3, then the number of electron volts made available by proton influx through the ATP synthase is $3\Delta p$. How many electron volts is needed to synthesize an ATP? The free energy of formation of ATP at physiological concentrations of ATP, ADP, and P_i is the phosphorylation potential, ΔG_p, which is approximately 45,000 to 50,000 J. (See note 18 for an explanation of the phosphorylation potential.) Dividing by the Faraday constant (96,500 C) expresses the energy required to synthesize an ATP in electron volts. For 45,000 J this is 0.466 eV. Therefore, $3\Delta p$ must be greater than or equal to 0.466 eV. Thus, the minimum Δp is 0.466/3 or −0.155 V. Values of Δp that approximate this are easily generated during electron transport. (However,

see the discussion in Section 3.10 regarding the low Δp in alkaliphiles.)

4.6 How a Proton Potential Might Be Created at the Coupling Sites: Q Loops, Q Cycles, and Proton Pumps

The preceding discussion points out that proton translocation takes place at coupling sites when electrons travel "downhill" over a potential gradient of at least 200 mV. (See Fig. 4.9.) However, the mechanism by which the redox reaction is actually coupled to proton translocation was not explained. This is thought to occur (1) in a Q loop or Q cycle, and (2) by means of a proton pump.

In the Q loop or Q cycle, reduced quinone carries hydrogen across the membrane and becomes oxidized, releasing protons on the external face of the membrane as the electrons return electrogenically via electron transport carriers to the inner membrane surface. On the cytoplasmic side, the number of protons taken up is the same as the number that were released on the outside as the terminal oxidant (e.g., oxygen) is reduced. The result is the net translocation of protons from the inside to the outside, although protons per se do not actually traverse the membrane. Proton translocation in the Q loop or Q cycle is referred to as *scalar* translocation.

Proton pumps are electron carrier proteins that couple electron transfer to the electrogenic translocation of protons through the membrane. Such translocation of protons through the membrane is referred to as *vectorial*.

Although the two mechanisms, Q loop or Q cycle and proton pump, are fundamentally different, the result is the same: the net translocation of protons across the membrane with the establishment of a Δp. The Δp is the same regardless of whether the moving charges are electrons or protons, since both carry the same charge. The relationship between the Δp and the ΔE_h was given by eq. 3.19.

4.6.1 The Q loop

The essential feature of the Q-loop model is that the electron carriers alternate between

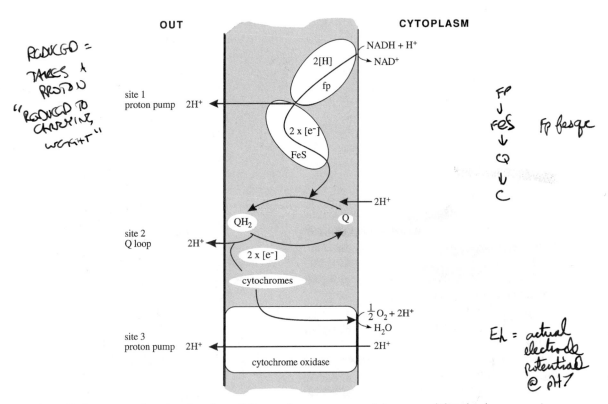

Fig. 4.11 Proton translocation showing a Q loop and a proton pump. It is proposed that the electron carriers exist in an alternating sequence of the following hydrogen [H] and electron [e⁻] carriers: flavoprotein (fp), iron–sulfur protein (FeS), quinone (Q), and cytochromes. Oxidation of the flavoprotein deposits two protons on the outer membrane surface. The electrons return to the inner membrane surface, where a quinone is reduced, taking up two protons from the cytoplasm. The reduced quinone diffuses to the outer surface of the membrane, where it is oxidized, depositing two more protons on the surface. The electrons return to the cytoplasmic surface via cytochromes, where they reduce oxygen in a reaction that consumes protons. Some cytochrome oxidases function as proton pumps. During anaerobic respiration the cytochrome oxidase is replaced by a reductase, and the electrons reduce some other electron acceptor (e.g., nitrate or fumarate). It should be noted that electrons can also enter at the level of quinone (e.g., from succinate dehydrogenase).

those that carry both hydrogen and electrons (flavoproteins and quinones) and those that carry only electrons (iron–sulfur proteins and cytochromes). This is illustrated in Fig. 4.11. The electron carriers and their sequence are flavoprotein (H carrier), FeS protein (e⁻ carrier), quinone (H carrier), and cytochromes (e⁻ carriers). The flavoprotein and FeS protein comprise the NADH dehydrogenase, which can be a coupling site (Section 4.6.3), although the mechanism of proton translocation by the NADH dehydrogenase is not understood. When electrons are transferred from the FeS protein to quinone (Q) on the inner side of the membrane, two protons are acquired from the cytoplasm. According to the model, the reduced quinone (QH_2), called quinol, then

diffuses to the outer membrane surface and becomes oxidized, releasing the two protons. The electrons then return via cytochromes to the inner membrane surface, where they reduce oxygen. This would create a $\Delta\Psi$ because the protons are left on the outer surface of the membrane as the electrons move electrogenically to the inner surface. Thus, quinol oxidation is a second coupling site.

As mentioned earlier, the energy to create the membrane potential is derived from the ΔE_h between the oxidant and the reductant. One way to view this is that the energy from the ΔE_h "pushes" the electron to the negative membrane potential on the inside surface. Note that the role of the quinone is to ferry the hydrogens across the membrane, presumably by diffusing from a reduction site on the cytoplasmic side of the membrane to an oxidation site on the outer side. For the purpose of calculating the expected Δp generated at a coupling site, it makes no difference whether one postulates the transmembrane movement of protons in one direction or electrons in the opposite direction, since both carry the same charge (eq. 3.19).

4.6.2 The Q cycle

Although the linear Q loop as just described for the oxidation of quinol may accurately describe quinol oxidation in *E. coli* and some other bacteria, it is inconsistent with experimental observations of electron transport in mitochondria, chloroplasts, and many bacteria. For example, the Q loop predicts that the ratio of H^+ released per QH_2 oxidized is 2, whereas the measured ratio in mitochondria and many bacteria is actually 4. To account for the extra two protons, Peter Mitchell suggested a new pathway for the oxidation of quinol called the *Q cycle*.[19] The Q cycle operates in an enzyme complex called the bc_1 complex (complex III). The bc_1 complex from bacteria contains three polypeptides. These are cytochrome b with two b-type hemes, an iron–sulfur protein containing a single 2Fe–2S cluster (the Rieske protein), and cytochrome c_1 with one heme. The complex spans the membrane and has a site for binding reduced ubiquinone, UQH_2, on the outer surface of the membrane called site P (for positive), and a second site on the inner surface for binding UQ, site N (for negative) (Fig. 4.12). At site P, UQH_2 is oxidized to the semiquinone anion, UQ^- (Fig. 4.12A), and the two protons are released to the membrane surface. The electron travels to the FeS protein, and from there to cytochrome c_1 on its way to the terminal electron acceptor. The UQ^- is then oxidized to UQ by the removal of the second electron, which is transferred to b_{556}, also called b_L because of its relatively low E'_m. The electron is transferred across the membrane to b_{560}, also called b_H because of its relatively high E'_h. Heme b_H transfers the electron to UQ bound at site N, reducing it to the semiquinone anion, UQ^-. *Electron flow from the Q_P site to the Q_N site is transmembrane and creates a membrane potential.* A second UQH_2 is oxidized at the P

site and the UQ_N^- is reduced to UQH_2, picking up two protons from the cytoplasm (Fig. 4.12B). The UQH_2 enters the quinone pool. Thus, for every two UQH_2 molecules that are oxidized, releasing four protons, one UQH_2 is regenerated. The net result is the oxidation of one UQH_2 to UQ with the release of four protons. The situation can be summarized as follows:

P site: $2UQH_2 + 2cyt\ c_{ox}$

$$\rightarrow 2UQ + 2cyt\ c_{red} + 4H^+_{out} + 2e^-$$

N site: $UQ + 2e^- + 2H^+_{in} \rightarrow UQH_2$

$$\overline{UQH_2 + 2cyt\ c_{ox} + 2H^+_{in}}$$

$$\rightarrow UQ + 2cyt\ c_{red} + 4H^+_{out}$$

In the presteady state, UQH_2 can be oxidized at the N site (by reversal) as well as the P site. The P site is inhibited by myxothiazol and by stigmatellin, whereas the N site is inhibited by antimycin. Therefore, as discussed later, a combination of inhibitors is required to completely inhibit the oxidation of UQH_2 in the presteady state.

Bioenergetics of the Q cycle

Since the Q cycle translocates two protons for every electron that flows to the terminal electron acceptor, it generates a larger proton current than the Q loop that translocates only one proton per electron. This can result in more ATP synthesis. Consider the situation of an ATP synthase that requires the influx of three protons to make one ATP. If the transfer of an electron through the electron transport pathway resulted in the translocation of one proton, then one-third of an ATP could be made for each electron. On the other hand, if electron transport resulted in the translocation of two protons for every electron, then two-thirds of an ATP could be made. In other words, the size of the proton current generated by respiration determines the upper value of the amount of ATP that can be made.

Distinguishing the Q loop from the Q cycle

When examining the physiology of electron transport in particular bacteria, it is important to learn whether a Q loop or a Q cycle is operating. There are several features of the Q

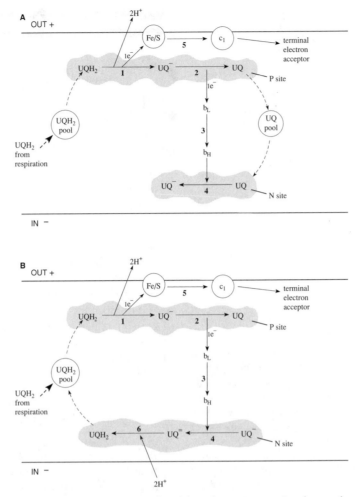

Fig. 4.12 The bc_1 complex. The bc_1 complex isolated from bacteria contains three polypeptides, which are a cytochrome b containing two heme b groups, b_H and b_L, an iron–sulfur protein, Fe/S (the Rieske protein), and cytochrome c_1. It is widely distributed among the bacteria, including the photosynthetic bacteria. (Mitochondrial bc_1 complexes are similar but contain an additional 6–8 polypeptides without prosthetic groups.) The iron–sulfur protein and cytochrome c_1 are thought to be located on the outside (positive, or P) surface of the membrane, and the cytochrome b is believed to span the membrane acting as an electron conductor. On the outer surface there is a binding site in the bc_1 complex for ubiquinol, UQH_2 (the P site). On the inner surface (negative, or N) there is a binding site in the bc_1 complex for ubiquinone, UQ (the N site). (A) Reduced ubiquinone binds to the P site and one electron is removed, forming the semiquinone anion, UQ^-, (reaction 1). At this time two protons are released on the outer membrane surface. The electron that is removed is transferred to the iron–sulfur protein and from there to cytochrome c_1 (step **5**). (Cytochrome c_1 transfers the electron to the terminal electron acceptor via a series of electron carriers in other reactions.) In reaction **2** the second electron is removed from the semiquinone anion, producing the fully oxidized quinone, UQ. The second electron is transferred transmembrane via cytochrome b to ubiquinone at site N (steps **3** and **4**). Because the electron travels transmembrane, a membrane potential is created, outside positive. (B) A second reduced ubiquinone is oxidized at the P site, releasing two more protons, and the sequence of electron transfers is repeated. The UQ^- at site N becomes reduced to UQH_2, having acquired two protons from the cytoplasm. Note that for every two UQH_2 molecules that are oxidized, four protons are released, and one UQH_2 is regenerated. Therefore, four protons are released for every UQH_2 oxidized. Another way of saying this is that the ratio H^+/e^- is 2.

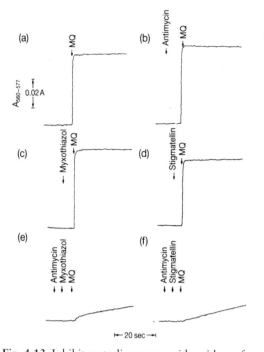

Fig. 4.13 Inhibitor studies can provide evidence for a Q cycle. One distinguishing feature of the Q cycle is that there are two sites for cytochrome b reduction, sites P and N. At site P quinol is oxidized by the iron–sulfur protein generating the semiquinone anion (Q^-). This oxidation is inhibited by stigmatellin or myxothiazol. The semiquinone anion Q^- reduces cytochrome b. Thus, stigmatellin or myxothiazol prevents the oxidation of quinol and the reduction of cytochrome b at site P. The reduced cytochrome b is reoxidized by Q or Q^- in an antimycin-sensitive step at site N. If site P is blocked (e.g., by myxothiazol or stigmatellin, or by removal of the iron–sulfur protein), cytochrome b can be reduced by quinol by a reversal of the antimycin-sensitive step at site N. There are therefore two routes for the reduction of cytochrome b as opposed to a single route in a linear pathway between quinol and cytochrome b. This experiment was done with a purified bc_1 complex isolated from the bacterium *Paracoccus denitrificans*, which was incorporated into liposomes. The reductant was menaquinol (MQ). Cytochrome b reduction is reflected by a rise in the trace that is a measure of the absorbance of reduced cytochrome b. The addition of antimycin, myxothiazol, or stigmatellin alone did not prevent the reduction of cytochrome b by MQ. However, the addition of both antimycin and stigmatellin, or antimycin and myxothiazol, blocked cytochrome b reduction. (a) Control without inhibitors. (b) Antimycin. (c) Myxothiazol. (d) stigmatellin. (e) Antimycin with myxothiazol. (f) Antimycin with stigmatellin. *Source:* Yang, X., and B. L.

cycle that help to distinguish it from the linear pathway of electron flow in the Q loop. These include the following.

1. In the Q cycle the ratio of H^+ extruded to electrons flowing is 2 rather than 1.

2. Cytochrome b can be reduced by quinol at two sites, site P or site N (by reversal), whereas in a linear respiratory chain there would be only one site, because there is only one site for UQH_2 oxidation. These sites can be distinguished by using inhibitors, as described next.

Look again at Fig 4.12. Notice that there are two sites for the oxidation of UQH: site P (steps **1** and **2**) and site N (steps **6** and **4**). (Site N will oxidize UQH_2 in a presteady state by reversing electron flow.) One can follow the oxidation of UQH_2 by measuring the reduction of cytochrome b spectrophotometrically (see Section 4.2.4). These two sites for the reduction of cytochrome b can be demonstrated with the use of inhibitors. Myxothiazol and stigmatellin block the oxidation of ubiquinol at site P, whereas antimycin blocks the oxidation of ubiquinol at site N. These inhibitions are shown in Fig. 4.13. Therefore, antimycin alone does not inhibit the reduction of cytochrome c_1 or cytochrome b in the presteady state, since the electrons are coming from site P. Also, myxothiazol and stigmatellin do not inhibit the reduction of cytochrome b, since electrons can come from site N. Thus, the presence of both myxothiazol (or stigmatellin) and antimycin is necessary to block the reduction of cytochrome b.

The fact that one requires two different inhibitors of UQH_2 oxidation to block the reduction of cytochrome b in the presteady state indicates that there are two routes for cytochrome b reduction by UQH_2 and is evidence for a Q cycle.

Trumpower. 1988. Protonmotive Q cycle pathway of electron transfer and energy transduction in the three-subunit ubiquinol–cytochrome c oxidoreductase complex of *Paracoccus denitrificans*. *J. Biol. Chem.* **263**:11962–11970.

4.6.3 Pumps

Proton pumps also exist. These catalyze the electrogenic translocation of protons across the membrane rather than electrons (Fig. 4.11). For example, proton extrusion accompanies the cytochrome aa_3 oxidase reaction when cytochrome c is oxidized by oxygen in mitochondria and some bacteria. This can be observed by feeding electrons into the respiratory chain at the level of cytochrome c, thus bypassing the quinone. The experimental procedure is to incubate the cells in lightly buffered anaerobic media with a reductant for cytochrome c and a permeant anion (e.g., SCN^- or valinomycin plus K^+). Changes in pH are measured with a pH electrode. Upon addition of a pulse of oxygen, given as an air-saturated salt solution, a sharp, transient acidification of the medium occurs, and the ratio of $H^+/$ to O can be calculated. If the electron donor itself releases protons, these scalar protons are subtracted from the total protons translocated to determine the number of vectorially translocated protons (due to proton pumping by the cytochrome c oxidase). For example, when cytochrome c is reduced by means of ascorbate plus TMPD, the ascorbate is oxidized to dehydroascorbate with the release of one scalar proton for every two electrons. Any additional protons released are due to proton pumping by the cytochrome c oxidase. When similar experiments were done with *Paracoccus denitrificans*, it was found that the *P. denitrificans* cytochrome aa_3 oxidase translocates protons with a stoichiometry of $1H^+/1e^-$.[20,21]

Cytochrome oxidase pumping activity can also be demonstrated in proteoliposomes made with purified cytochrome aa_3. Because the proton pumps move a positive charge across the membrane, leaving behind a negative charge, a membrane potential, outside positive, develops. The membrane potential should be the same as that recorded when an electron moves inward, since the proton and the electron carry the same charge (i.e., 1.6×10^{-19} C). The mechanism of pumping is not known, but probably it requires conformational changes in cytochrome oxidase resulting from its redox activity. Conformational changes probably also occur in bacteriorhodopsin during proton pumping (Section 3.8.4).

The mitochondrial NADH dehydrogenase complex (NADH:ubiquinol oxidoreductase) translocates four protons for every NADH oxidized.[22] However, the mechanism has not yet been elucidated. Two types of NADH:ubiquinol oxidoreductases exist in bacteria.[8–11] One of these, called NDH-1, is similar to the mitochondrial complex I in that it is a multisubunit enzyme complex (approximately 14 polypeptide subunits) consisting of FMN and FeS clusters, and it translocates protons across the membrane during NADH oxidation ($4H^+/2e^-$). The mechanism of proton translocation is not known, but it is referred to as a proton pump. [The marine bacterium *Vibrio alginolyticus* has a sodium-translocating NDH (Na–NADH). See Section 3.7.1.]

A second NADH:ubiquinol oxidoreductase, called NDH-2, may also be present. NDH-2 differs from NDH-1 in consisting of a single polypeptide and FAD, and in that it is not an energy-coupling site. In *E. coli*, NDH-1 and NDH-2 are simultaneously present.[23] How *E. coli* regulates the partitioning of electrons between the two NAD dehydrogenases is not known, but the mechanism clearly has important energetic consequences.

4.7 Patterns of Electron Flow in Individual Bacterial Species

Although the major principles of electron transport as outlined already apply to bacteria in general, several different patterns of electron flow exist in particular bacteria, often within the same bacterium grown under different conditions.[24] The patterns of electron flow reflect the different sources of electrons and electron acceptors that are used by the bacteria. For example, bacteria may synthesize two or three different oxidases in the presence of air and several reductases anaerobically. In addition, certain dehydrogenases are made only anaerobically because they are part of an anaerobic respiratory chain. Furthermore, whereas electron donors such as NADH and $FADH_2$ generated in the cytoplasm are oxidized inside the cell, there are many instances of oxidations in the periplasm in gram-negative bacteria. Substances that are oxidized in the periplasm include hydrogen gas, methane, methanol, methylamine, formate, perhaps ferrousion, reduced inorganic sulfur, and elemental sulfur. In most of these instances periplasmic cytochromes c accept the electrons from the electron donor and transfer them to electron carriers in the membrane. The discussion that follows is not complete, but it conveys the diversity of electron transport systems found in bacteria. The electron transport chains for the respiratory metabolism of ammonia, nitrite, inorganic sulfur, and iron are described in Chapter 12.

4.7.1 Escherichia coli

E. coli is a gram-negative heterotrophic facultative anaerobe. It can be grown aerobically by using oxygen as an electron acceptor, anaerobically (e.g., by using nitrate or fumarate as the electron acceptor), or anaerobically via fermentation (Chapter 14). The bacteria adapt to their surroundings, and the electron transport system that the cells assemble reflects the electron acceptor that is available. (See ref. 25 for a review.)

All the electron transport chains present in *E. coli* branch at the level of quinone, which connects the different dehydrogenases with the various terminal reductases and oxidases that are present under different growth conditions. *E. coli* makes three quinones, ubiquinone (UQ), menaquinone (MQ), and demethylmenaquinone (DMQ), and their relative amounts depend upon the nature of the electron acceptor. When UQ is growing aerobically, it accounts for 60% of the total quinone, DMQ is 37% of the total, and MQ is only about 3%. However, very little UQ is made anaerobically.

For anaerobic growth on nitrate, the major quinones are DMQ (70%) and MQ (30%). The major quinone grown anaerobically on fumarate or DMSO is MQ (74%), with DMQ (16%) and UQ (10%) contributing the rest.[24] There is no bc_1 complex and no cytochrome c, which may serve as additional branch points in other bacterial respiratory chains (e.g., see the discussion of *Paracoccus denitrificans* in Section 4.7.2).

Aerobic respiratory chains

When grown aerobically, *E. coli* makes two different quinol oxidase complexes, cytochrome bo complex (has heme b and heme o and is also called bo_3 or cytochrome o) and cytochrome bd complex (has heme b and heme d and is also called cytochrome d), resulting in a branched respiratory chain to oxygen (Fig. 4.14).[26–30] In these pathways electrons flow from ubiquinol to the terminal oxidase complex, which is why the oxidases are called (ubi)quinol oxidases. The cytochrome bo complex from *E. coli* is a proton pump, with a stoichiometry of $1H^+/e^-$. Cytochrome bo complex is the predominant oxidase when the oxygen levels are high. When the oxygen tensions are lowered, *E. coli* makes more cytochrome bd than cytochrome bo.

The cytochrome bd complex has a higher affinity for oxygen (lower K_m) than does the cytochrome bo complex, suggesting why it is the dominant oxidase under low oxygen tensions. Cytochrome bd is not a proton pump. In the same sense that the bc_1 complex is a coupling site because it catalyzes the oxidation of quinol resulting in the scalar extrusion of protons, both the bo and bd complexes are

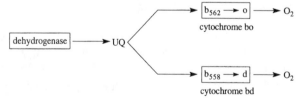

Fig. 4.14 Aerobic respiratory chain in *E. coli*. The chain branches at the level of ubiquinone (UQ) to two alternate quinol oxidases, cytochrome bo and bd. Cytochrome bd complex has a higher affinity for oxygen and is synthesized under low oxygen tensions, where it becomes the major route to oxygen. Coupling sites (i.e., sites at which protons are translocated to the outer surface) are indicated by the H^+/e^- ratio. Proton translocation occurs during NADH oxidation, catalyzed by NADH:ubiquinone oxidoreductase (also called NDH-1) and quinol oxidation, catalyzed by cytochrome bo complex or cytochrome bd complex. Additionally, cytochrome bo is a proton pump. *E. coli* has a second NADH dehydrogenase, called NADH-2, which does not translocate protons. Therefore, the number of protons translocated per NADH oxidized can vary from two (NADH-2 and bd complex) to eight (NADH-1 and bo complex).

coupling sites. The difference between the two, as mentioned, is that the bo complex is also a proton pump and catalyzes the vectorial extrusion of protons. The oxidation of quinol by the bo complex results in two protons translocated for every electron (one scalar and one vectorial), whereas the oxidation of quinol by the bd complex results in only one proton translocated per electron (scalar).

E. coli has two NADH dehydrogenases, NDH-1 and NDH-2, encoded by the *nuo* operon and *ndh*, respectively. Only NDH-I is a coupling site (proton pump). It is a complex enzyme consisting of 14 subunits and translocates two protons per electron. As discussed in Section 4.6.3, NDH-1 is similar to complex I of mitochondria. It has been shown that NDH-1 is used during fumarate respiration, whereas NDH-2 is used primarily during aerobic and nitrate respiration.[31] The regulation of transcription of the NADH dehydrogenase genes in E. coli is complex.[30,32]

Physiological significance of alternate electron routes that differ in the number of coupling sites

Since the NDH-1 dehydrogenase may translocate as many as four protons for every two electrons, whereas the NDH-2 dehydrogenase translocates none, the number of protons translocated per NADH oxidized can theoretically vary from 2 (NDH-2 and bd complex) to 8 (NDH-1 and bo complex). Assuming a H^+/ATP of 3 (i.e., the ATP synthase translocates inwardly 3 protons for every ATP made), the ATP yields per NADH oxidized can vary fourfold from 2/3, or 0.67, to 8/3, or 2.7. It also means that E. coli has great latitude in adjusting the Δp generated during respiration. Since a large Δp can drive reversed electron transport and thus slow down oxidation of NADH and quinol, it may be an advantage to be able to direct electrons along alternate routes that bypass coupling sites and translocate fewer protons. This could ensure adequate rates of reoxidation of NADH and quinol.

Anaerobic respiratory chains

In the absence of oxygen, E. coli can use either nitrate or fumarate as an electron acceptor. The nitrate (NO_3^-) is reduced to nitrite (NO_2^-), which is further reduced to ammonia (NH_4^+), and the fumarate is reduced to succinate, all of which are excreted into the medium.[33] (See note 34 for a further description of fumarate and nitrate reduction.) The anaerobic respiratory chains consists of a dehydrogenase, a reductase, and a diffusible quinone to mediate the transfer of electrons between the dehydrogenase and the reductase (Fig. 4.15).

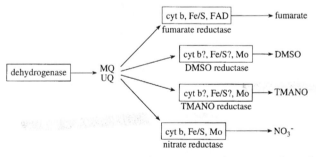

Fig. 4.15 Anaerobic respiratory chains in E. coli. When oxygen is absent, E. coli synthesizes any one of several membrane-bound reductase complexes depending upon the presence of the electron acceptors. Nitrate induces the synthesis of nitrate reductase and represses the synthesis of the other reductases. Menaquinone ($E_0' = -74$ mV) (or demethylmenaquinone, $E_0' = -40$ mV) must be used to reduce some of the reductases (e.g., fumarate reductase, because it has a sufficiently low midpoint potential). Ubiquinol ($E_0' = +100$ mV) or menaquinone can reduce nitrate reductase because the E_0' of nitrate is 421 mV. Each reductase may be a complex of several proteins and prosthetic groups through which the electrons travel to the terminal electron acceptor. The transfer of electrons from the dehydrogenases to the reductases results in the establishment of a proton potential. If the dehydrogenase has site 1 activity, there can theoretically be two coupling sites, one at the dehydrogenase step and one linked to quinol oxidation at the reductase step. Abbreviations: cyt b, cytochrome b; Fe/S, nonheme iron–sulfur protein; FAD, flavoprotein with flavin adenine dinucleotide as the prosthetic group: Mo, molybdenum; TMANO, trimethylamine N-oxide; DMSO, dimethyl sulfoxide; MQ, menaquinone; UQ, ubiquinone.

The number of coupling sites depends upon whether the electron acceptor is nitrate or fumarate.[24] When nitrate is the electron acceptor, there can be two coupling sites: that is, one at the dehydrogenase step if NDH-1 is used (site 1) and one at the quinol oxidation step [i.e., the nitrate reductase (scalar)]. However, during nitrate respiration NDH-2, which is not a coupling site, is also used.

Quinol oxidation by nitrate results in a Δp because the oxidation of one quinol and release of two protons takes place on the periplasmic side of the membrane, and the uptake of two protons during the reduction of one nitrate to nitrite takes place on the cytoplasmic side. Thus, the oxidation of quinol by nitrate catalyzed by nitrate reductase yields a proton-to-electron ratio of 1.[24] See the discussion of the reaction for nitrate reductase in Section 4.7.2.

However, the H^+/e^- ratio for quinol oxidation by fumarate in *E. coli* or *Wolinella succinogenes* is zero, indicating that both the release of protons when MQH_2 is oxidized and the uptake of protons when fumarate is reduced take place on the cytoplasmic side of the membrane. (It does not appear that either fumarate reductase or any of the nitrate reductases are proton pumps.) Thus, for *E. coli* to make ATP during the oxidation of NADH by fumarate, it must use NDH-1 rather than NDH-2, since only the former is a coupling site. In agreement with this conclusion, it has been found that the genes for NDH-1 (*nuo* genes) are essential for anaerobic respiration with fumarate as the electron acceptor. (Reviewed in ref. 35.)

4.7.2 *Paracoccus denitrificans*

P. denitrificans is a nonfermenting gram-negative facultative anaerobe that can obtain energy from either aerobic respiration or nitrate respiration.[18,36,37] It is found primarily in soil and sewage sludge. This bacterium can grow heterotrophically on a wide variety of carbon sources, or autotrophically on H_2 and CO_2 under anaerobic conditions using nitrate as the electron acceptor. It can be isolated from soil by anaerobic enrichment with media containing H_2 as the source of energy and electrons, Na_2CO_3 as the source of carbon, and nitrate as the electron acceptor. Electron transport in *P. denitrificans* receives a great deal of research

attention because certain features closely resemble electron transport in mitochondria.

Aerobic pathway

P. denitrificans differs from *E. coli* in that it has a bc_1 complex and a cytochrome aa_3 oxidase (cytochrome c oxidase), and in this way resembles mitochondria. In addition to the cytochrome aa_3, there are two other terminal oxidases in the aerobic pathway.[20,35] These are a different cytochrome c oxidase (cytochrome cbb_3) and a ubiquinol oxidase (cytochrome bb_3). (Cytochromes cbb_3 are heme–copper oxidases found in several bacteria including *Thiobacillus*, *Rhodobacter*, *Paracoccus*, and *Bradyrhizobium*.[38] Cytochrome bb_3 formerly was called cytochrome o or b_0. But no heme o was detected when the hemes were extracted from membranes and analyzed by reversed-phase high performance liquid chromatography.[20,39,40]). The cytochrome aa_3 and cytochrome bb_3 are proton pumps. (Whether cytochrome cbb_3 pumps protons apparently depends upon the assay conditions.[35]) The aerobic pathways as well as the sites of proton translocation are shown in Fig. 4.16A. Electrons traveling from NADH to oxygen can pass through as many as three coupling sites (NDH-1, bc_1 complex, cyt aa_3 or perhaps cyt cbb_3) or as few as two coupling sites (NDH-1 and cyt bb_3). Recall that the bc_1 complex and cyt bb_3 are coupling sites because they oxidize quinol, whereas the NDH-1 and cytochrome c oxidases are proton pumps. As described later, *P. denitrificans* also oxidizes methanol, and in this case the electrons enter at a cytochrome c, thus bypassing the bc_1 site.

Anaerobic pathway

P. denitrificans can also grow anaerobically by using nitrate as an electron acceptor, reducing it to nitrogen gas in a process called *denitrification* (Fig. 4.16B).[32,41] During anaerobic growth on nitrate, *P. denitrificans* has a complete citric acid cycle in which electrons are donated to the electron transport chain; but the electron transport chain is very different from that of aerobic growth. The cells contain cytochrome bb_3, nitrate reductase, nitrite reductase, nitric oxide reductase, and nitrous oxide reductase. (For a more complete

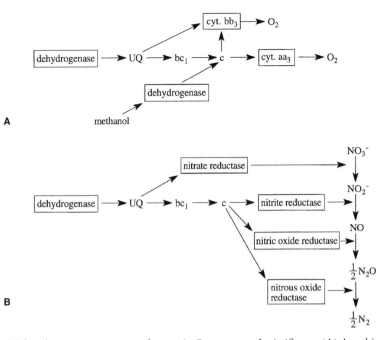

Fig. 4.16 A model for electron transport pathways in *Paracoccus denitrificans*. (A) Aerobic. The pathway has two branch points. One branch is at the level of ubiquinone leading to one of two ubiquinol oxidases (i.e., the cyt bc_1 complex or cyt bb_3). Both these quinol oxidases are coupling sites. The bc_1 complex extrudes two protons per electron via the Q cycle. Whereas cyt bb_3 is a proton pump and extrudes one proton per electron vectorially, the second proton is extruded via a Q loop. A second branch point occurs at the level of the bc_1 complex. Electrons can flow either to cyt aa_3, which is a proton pump, or to cyt cbb_3, which has been reported to pump protons under certain experimental conditions. NDH-1 is also a coupling site. (B) Anaerobic. When the bacteria are grown anaerobically, using nitrate as the electron acceptor, the cytochrome aa_3 levels are very low and the electrons travel from ubiquinone to nitrate reductase and also through the bc_1 complex to nitrite reductase, nitric oxide reductase, and nitrous oxide reductase.

description of these enzymes, see note 42.) The levels of cytochrome aa_3 are very low. The cytochrome c shown in Fig. 4.16 is periplasmic, although there is also cytochrome c in the membrane associated with some of the electron carriers. The nitrate (NO_3^-) is reduced to nitrite (NO_2^-) in a two-electron transfer via a membrane-bound nitrate reductase. The NO_2^- is reduced to nitric oxide (NO) in a one-electron transfer via a periplasmic nitrite reductase. The NO is reduced to a half-mole of nitrous acid ($\frac{1}{2}N_2O$) in a one-electron step via a membrane-bound nitric oxide reductase. And the $\frac{1}{2}N_2O$ is reduced to a half-mole of dinitrogen ($\frac{1}{2}N_2$) in a one-electron step by a periplasmic nitrous oxide reductase. Thus to reduce one mole of NO_3^- to $\frac{1}{2}N_2$, a total of five electrons flow in and out of the cell membrane through membranous electron and periplasmic electron carriers from ubiquinol to the various reductases.

As shown in Fig. 4.16, the electron transport pathway includes several branches to the individual reductases. The first branch site is at UQ, where electrons can flow either to nitrate reductase or to the bc_1 complex. Then there are three branches to the three other reductases after the bc_1 complex at the level of cytochrome c. In agreement with the model, electron flow to nitrate reductase is not sensitive to inhibitors of the bc_1 complex, whereas electron flow to the other reductases is sensitive to the inhibitors. The nitrate reductase spans the membrane, and it has been proposed that it creates a Δp via a Q loop, similar to that described in Section 4.6.1. It is proposed that the nitrate reductase accepts electrons from UQH_2 on the periplasmic side of the membrane and reduces nitrate on the cytoplasmic side. This would create a $\Delta \Psi$ as two electrons from UQH_2 flow electrogenically across the membrane, leaving two protons on the outside. In the cytoplasm, protons are taken up during nitrate reduction according to the following reaction:

$$2H^+ + 2e^- + NO_3^- \rightarrow NO_2^- + H_2O$$

The result is the net translocation of protons to the outside, although only electrons moved across the membrane, not protons. This is the same as the Q loop described in Section 4.6.1 except that the terminal electron acceptor is nitrate instead of oxygen. The electrons to the other reductases flow from UQH_2 through the bc_1 complex, and a Δp is generated via the Q cycle catalyzed by the bc_1 complex which, as described in Section 4.6.1, is a modification of the Q loop that results in the translocation of two protons per electron rather than one.

In *P. denitrificans*, *Pseudomonas aeruginosa*, and many other facultative anaerobes, both the synthesis and the activity of the denitrifying enzymes are prevented by oxygen. However, in certain other facultative anaerobes, including *Comamonas* spp., certain species of *Pseudomonas*, *Thiosphaera pantotropha*, and *Alcaligenes faecalis*, denitrifying enzymes are made and are active in the presence of oxygen.[43] In these systems, both oxygen and nitrate are used simultaneously as electron acceptors, although aeration can significantly decrease the rate of nitrate reduction. The advantage of co-respiration using both oxygen and nitrate is not obvious.

Periplasmic oxidation of methanol

Many gram-negative bacteria oxidize substances in the periplasm and transfer the electrons to membrane-bound electron carriers, often via periplasmic cytochromes c. (See Section 1.2.4 for a description of the periplasm.) An example is *P. denitrificans*, which can grow aerobically on methanol (CH_3OH) by oxidizing it to formaldehyde (HCHO) and $2H^+$ with a periplasmic dehydrogenase, methanol dehydrogenase (Fig. 4.17):

$$CH_3OH \rightarrow HCHO + 2H^+$$

(Growth on methanol is autotrophic, since the formaldehyde is eventually oxidized to CO_2, which is assimilated via the Calvin cycle.) The electrons are transferred from the dehydrogenase to c-type cytochromes in the periplasm. The cytochromes c transfer the electrons to

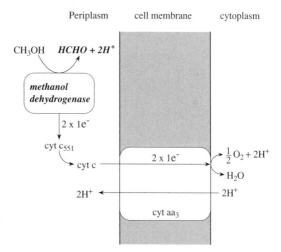

Fig. 4.17 Oxidation of methanol by *P. denitrificans*. Methanol is oxidized to formaldehyde by a periplasmic methanol dehydrogenase. The electrons are transferred to periplasmic cytochromes c and to a membrane-bound cytochrome aa_3, which is also a proton pump. A Δp is created as a result of the electrogenic influx of electrons and the electrogenic efflux of protons, accompanied by the release of protons in the periplasm (methanol oxidation) and uptake in the cytoplasm (oxygen reduction).

cytochrome aa_3 oxidase in the membrane, which reduces oxygen to water on the cytoplasmic surface. A Δp is established as a result of the inward flow of electrons and the outward pumping of protons by the cytochrome aa_3 oxidase, as well as the release of protons in the periplasm during methanol oxidation and consumption in the cytoplasm during oxygen reduction. Since the oxidation of methanol bypasses the bc_1 coupling site, the ATP yields are lower. In addition to methanol, the bacteria can grow on methylamine ($CH_3NH_3^+$), which is oxidized by methylamine dehydrogenase to formaldehyde, NH_4^+, and $2H^+$:

$$CH_3NH_3^+ + H_2O \rightarrow HCHO + NH_4^+ + 2H^+$$

Methylamine dehydrogenase is also located in the periplasm and donates electrons via cytochromes c to cytochrome aa_3, bypassing the bc_1 complex.

4.7.3 Rhodobacter sphaeroides

Rhodobacter sphaeroides is a purple photosynthetic bacterium that can be grown photoheterotrophically under anaerobic conditions or aerobically in the dark. When growing aerobically in the dark, the bacteria obtain energy from aerobic respiration. The respiratory chain resembles the mitochondrial and *P. denitrificans* respiratory chains in that it consists of NADH:ubiquinone oxidoreductase (coupling site 1), a bc_1 complex (coupling site 2), and a cytochrome aa_3 complex (coupling site 3). As in *E. coli*, there are two NADH:ubiquinone oxidoreductases: NDH-1 and NDH-2.

4.7.4 Fumarate respiration in Wolinella succinogenes

Fumarate respiration occurs in a wide range of bacteria growing anaerobically.[44] This is probably because fumarate itself is formed from carbohydrates and protein during growth. The following is a description of the electron transport pathway in *W. succinogenes*, a gram-negative anaerobe isolated from the rumen.

W. succinogens can grow at the expense of H_2 or formate, both produced in the rumen by other bacteria. The electron transport pathway is shown in Fig. 4.18A. The active sites for both the hydrogenase and the formate dehydrogenase are periplasmic, whereas the active site for the fumarate reductase is cytoplasmic. An examination of the topology of the components of the respiratory chain reveals how a Δp is generated.

Topology of the components of the electron transport pathway

The electron transport chain consists of a periplasmic enzyme that oxidizes the electron donor, a membrane-bound menaquinone (MQ) that serves as an intermediate electron carrier, and membrane-bound fumarate reductase,

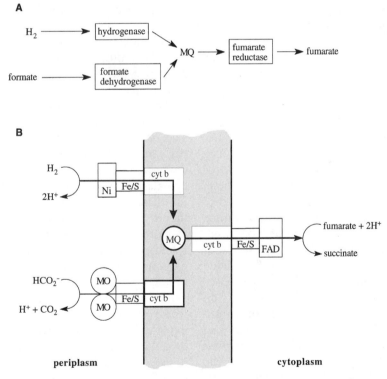

Fig. 4.18 A model for the electron transport system of *Wolinella succinogenes*. (A) Electrons flow from H_2 and formate through menaquinone (MQ) to fumarate reductase. (B) Illustration that the catalytic portions of hydrogenase and formate dehydrogenase are periplasmic, whereas fumarate reductase reduces fumarate on the cytoplasmic side. Electrons flow electrogenically to fumarate. A Δp is created because of electrogenic influx of electrons together with the release of protons in the periplasm, and their consumption in the cytoplasm. *Source*: Modified from Kroger, A., V. Geisler, E. Lemma, F. Theis, and R. Lenger. 1992. Bacterial fumarate respiration. *Microbiology* **158**:311–314.

which accepts the electrons from the mena-quinone and reduces fumarate on the cyto-plasmic side of the membrane (Fig. 4.18B). Both the hydrogenase and the formate dehy-drogenase are made of three polypeptide subunits, two facing the periplasm and one an integral membrane protein (cytochrome b). Note that cytochrome b not only serves as a conduit for electrons, but also binds the dehydrogenases into the membrane. The two periplasmic subunits of the hydrogenase are a Ni-containing protein subunit and an iron–sulfur protein. The two periplasmic subunits of the formate dehydrogenase are a Mo-containing protein subunit and an iron–sulfur protein.

The fumarate reductase is a complex con-taining three subunits. One subunit of the fumarate reductase is a flavoprotein with FAD as the prosthetic group (subunit A). A second subunit has several FeS centers (subunit B). And the third subunit has two hemes of the b type (subunit C), which binds the fumarate reduc-tase to the membrane. Fumarate reductase is similar in structure to succinate dehydrogenase isolated from several different sources, which catalyzes the oxidation of succinate to fumarate in the citric acid cycle.

Electron flow and the establishment of a Δp

Electron flow is from the dehydrogense or hydrogenase to cytochrome b to menaquinone to fumarate reductase. Two electrons are elec-trogenically transferred across the membrane to fumarate for every H_2 or formate oxidized, leaving two protons on the outside, thus estab-lishing a Δp. In a study of whole cells, a value of 1.1 was obtained for the ratio of protons to electrons during fumarate reduction, sug-gesting perhaps that a mechanism of proton translocation through the membrane may also exist.[44] If one assumes that 1.1 is a correct num-ber, and also assumes a stoichiometry for the ATP synthase of $3H^+$/ATP, then the theoretical maximum number of ATPs that can be formed from the transfer of two electrons to fumarate is 2.2/3, or 0.73. The actual number measured by experimentation was 0.56. Note that the quinone functions as an electron carrier between the cytochromes b but does not take part in hydrogen translocation across the membrane as in a Q loop or Q cycle.

4.8 Summary

All electron transport schemes can be viewed as consisting of membrane-bound dehydrogenase complexes, such as NADH dehydrogenase or succinate dehydrogenase, that remove electrons from their substrates and transfer the electrons to quinones, which in turn transfer the elec-trons to oxidase or reductase complexes. The latter complexes reduce the terminal electron acceptors. In contrast to mitochondria, which all have the same electron transport scheme, bacteria differ in the details of their electron transport pathways, although the broad out-lines of all such schemes are similar. In bacteria, the dehydrogenase, oxidase, and reductase complexes are sometimes referred to as modules because specific ones are synthesized under certain growth conditions and "plugged into" the respiratory pathway. For example, in facultative anaerobes such as E. coli, the oxidase modules are synthesized in an aerobic atmosphere and the reductase modules under anaerobic conditions.

Other dehydrogenases besides NADH dehydrogenase and succinate dehydrogenase exist. These oxidize various electron donors (e.g., methanol, hydrogen, formate, H_2, glyc-erol), and are located in the periplasm or the cytoplasm. The coenzyme or prosthetic groups for these soluble dehydrogenases vary (e.g., they may be NAD^+ or flavin). The electrons from the various dehydrogenases are trans-ferred to one of the electron carriers (e.g., quinone, cytochrome) and from there to a terminal reductase or oxidase.

An important difference between electron transport chains in bacteria and those in mitochondria is that the former are branched. Branching can occur at the level of quinone or cytochrome. The branches lead to different oxidases or reductases, depending upon whether the bacterium is growing aerobically or anaerobically. Many bacteria, including E. coli, transfer electrons from reduced quinone to cytochrome o, which is the major cytochrome used when oxygen levels are high, and to cytochrome d, which is used when oxygen becomes limiting. Other bacteria may have in addition, or instead, an electron transport pathway in which electrons travel from reduced quinone through a bc_1 complex, to cytochrome aa_3, which reduces oxygen. This is the same as the electron transport pathway found in mitochondria, although the carriers may not be identical. The alternative branches may differ in the number of coupling sites, and this could have regulatory significance regarding the rates of oxidation of reduced electron carriers, as well as ATP yields.

Another difference from mitochondria is that bacteria can have either aerobic or anaerobic electron transport chains; or, as is the case with facultative anaerobes such as *E. coli*, either can be present, depending upon the availability of oxgyen or alternative electron acceptors. A hierarchy of electron acceptors is used. For *E. coli*, oxygen is the preferred acceptor, followed by nitrate, and finally fumarate.

With respect to cytochrome c oxidase, there are two classes. Cytochrome aa_3, is the major class and has been reported in many bacteria, including *Paracoccus denitrificans*, *Nitrosomonas europaea*, *Pseudomonas AM1*, *Bacillus subtilis*, and *Rhodobacter sphaeroides*. A different cytochrome c oxidase has been reported for *Azotobacter vinelandii*, *Rhodobacter capsulata*, *R. sphaeroides*, *R. palustris*, and *Pseudomonas aeruginosa*, as well as *P. denitrificans*.[20,46] Apparently, both classes of cytochrome c oxidase coexist in the same organism and serve as alternate routes to oxygen.

The main energetic purpose of the respiratory electron transport pathways is to convert a redox potential (ΔE_h) into a proton potential (Δp). This is done at coupling sites. A membrane potential is created by electrogenic influx of electrons, leaving the positively charged proton on the outside, or during electrogenic efflux of protons during proton pumping, leaving a negative charge on the inside. Influx of electrons occurs when oxidations take place on the periplasmic membrane surface or in the periplasm, and electrons move vectorially across the membrane to the cytoplasmic surface, where reductions take place. This occurs in two situations: (1) when the substrate (e.g., H_2, methanol) is oxidized by dehydrogenases in the periplasm and the electrons move across the membrane to reduce the electron acceptor and (2) when quinones are reduced on the cytoplasmic side of the membrane, diffuse across the membrane, and are oxidized on the outside membrane surface.

Quinone oxidation can be via a Q loop or a Q cycle. A Q loop is an electron transport pathway in which a reduced quinone carries hydrogen to the outside surface of the cell membrane and releases two protons as a result of oxidation. The two electrons return to the inner surface, via cytochromes, where they reduce an electron acceptor. The inward transfer of the electrons is electrogenic (i.e., a membrane potential is created). In the Q loop, which is a linear pathway of electron flow, a proton is released for every electron.

The Q cycle, a more complicated pathway, results in the release of two protons for every electron. In the Q cycle, QH_2 gives up two protons and two electrons, but one of the electrons is recycled back to oxidized quinone. Thus, the ratio of H^+ to e^- is 2, rather than 1. The Q cycle is more efficient, since it results in the availability of more protons for influx through the ATP synthase per electron, and thus can increase the yield of ATP.

In a second method of coupling oxidation–reduction reactions to the establishment of a proton potential, some of the electron carriers act as proton pumps. A membrane potential is created when protons are pumped electrogenically out of the cell. A well-established example is cytochrome aa_3 oxidase. However, there is evidence that other oxidases are also proton pumps, including cytochrome bo in *E. coli* and cytochromes bb_3 and cbb_3 in *P. denitrificans*. Mitochondrial and certain bacterial NADH dehydrogenases are also coupling sites, and these may function as proton pumps.

Study Questions

1. What is it about the solubility properties and electrode potentials of quinones that make them suitable for their role in electron transport?

2. Design an experiment that can quantify the H^+/O for the cytochrome aa_3 reaction in proteoliposomes.

3. What are two features that distinguish the Q cycle from linear flow of electrons? How can you verify that the Q cycle is operating?

4. What is the relationship between H$^+$/O, H$^+$/ATP, and the number of ATPs that can be synthesized?

5. What is the relationship between the Δp and number of ATPs that can be synthesized?

6. Draw a schematic outline of three ways in which a cell might create a membrane potential during respiration. You can include both periplasmic and cytoplasmic oxidations.

7. Assuming a Δp of -200 mV and a H$^+$/ATP of 3, calculate what the ΔG_p must be in joules per mole if ΔG_p is in equilibrium with Δp. Assume that the temperature is 30 °C, and that $\Delta G_0'$ is 37 kcal/mol. What will be the ratio of ATP to [ADP][P$_i$]?

8. What are the similarities and differences between bacterial respiratory pathways and the mitochondrial respiratory pathway?

9. Calculate the maximum number of ATPs that can be made per NADH oxidized by *E. coli* by using the following combinations: NDH-1 plus cytochrome bo and NDH-2 plus cytochrome bd.

10. What drives reversed-electron transport? What experiment can be done to support this conclusion?

11. What is it about the sequential arrangement of electron carriers that makes proton translocation possible in a Q loop or Q cycle?

12. Consider the following data concerning proton translocation by *P. denitrificans*:

Observed H$^+$/2e$^-$ ratios for whole cells of *Paracoccus denitrificans*

Substrate	H$^+$/O	H$^+$/2 Fe(CN)$_6^{3-}$
Endogenous	7.12 ± 0.24	5.26 ± 0.34
ascorbate + TMPD	2.81 ± 0.14	1.04 ± 0.05

Source: Adapted from Van Verseveld, H. W., et al. 1981. *Biochim. Biophys. Acta* **635**:525–534.

Refer to Fig. 4.16. Ferricyanide is an oxidant that allows the terminal oxidase to be bypassed. Ascorbate feeds electrons via TMPD into cytochrome c and releases one scalar proton for every two electrons removed from it. Calculate H$^+$/2e$^-$ for the cytochrome c oxidase pump.

REFERENCES AND NOTES

1. As discussed in Section 3.7.1, some marine bacteria couple electron transport to the creation of a sodium ion gradient.

2. Lubben, M. 1995. Cytochromes of archaeal electron transfer chains. *Biochim. Biophys. Acta* **1229**:1–22.

3. For example, the enzyme complex that oxidizes NADH and reduces quinone is an NADH–quinone oxidoreductase. It consists of a flavoprotein and iron–sulfur proteins. (Reduced quinone is called quinol.) An enzyme complex that oxidizes quinol and reduces cytochrome c is called a quinol–cytochrome c oxidoreductase. It consists of cytochromes and an iron–sulfur protein.

4. Here are some definitions that are useful to know. *Cofactors* are nonprotein molecules, either metals or organic, bound with varying degrees of affinity to enzymes and required for enzyme activity. *Coenzymes* are organic cofactors that shuttle back and forth between enzymes carrying electrons, hydrogen, or organic moieties (e.g., acyl groups). *Prosthetic* groups are cofactors (organic or inorganic) that are tightly bound to the protein and do not dissociate from the protein. Thus, one difference between a coenzyme and a prosthetic group is that the former shuttles between enzymes and the latter remains tightly bound to the enzyme. Coenzyme A (carrier of acyl groups) and NAD$^+$ (electron and hydrogen carrier) are examples of coenzymes, whereas FAD (electron and hydrogen carrier) and heme (electron carrier) are examples of prosthetic groups.

5. The first iron–sulfur protein identified was a soluble protein isolated from bacteria and called ferredoxin. A protein similar to ferredoxin with a low redox potential can be isolated from chloroplasts and mediates electron transfer to NADP$^+$ during noncyclic electron flow. Iron–sulfur proteins that have no other prosthetic groups are divided into four classes based upon the number of iron atoms per molecule and whether the sulfur is acid labile. Examples from the four different classes are rubredoxin (no acid-labile sulfur), isolated from bacteria; high-potential iron protein (HiPIP), isolated from photosynthetic bacteria; chloroplast ferredoxin; and various bacterial ferredoxins.

6. Electron spin resonance spectroscopy detects unpaired electrons such as the ones that exist in the FeS cluster. Monochromatic microwave radiation is absorbed by the unpaired electron when a magnetic

field is applied. The size of the magnetic field required for radiation absorption depends upon the molecular environment of the unpaired electron. A spectrum is obtained by varying the magnetic field and keeping the frequency of the microwave radiation constant. The spectra differ for different FeS clusters. A spectroscopic constant, called the g value, is obtained which is characteristic of the FeS cluster in the protein.

7. Puustinen, A., and M. Wikstrom. 1991. The heme groups of cytochrome o from *Escherichia coli*. *Proc. Natl. Acad. Sci. USA* **88**:6122–6126.

8. Yagi, T., X. Xu, and A. Matsuno-Yagi. 1992. The energy-transducing NADH–quinone oxidoreductase (NDH-1) of *Paracoccus denitrificans*. *Biochim. Biophys. Acta* **1101**:181–183.

9. Finel, M. 1993. The proton-translocating NADH:ubiquinone oxidoreductase: a discussion of selected topics. *J. Bioenerg. Biomemb.* **25**:357–366.

10. Ohnishi, T. 1993. NADH–quinone oxidoreductase, the most complex complex. *J. Bioenerg. Biomemb.* **25**:325–329.

11. Yagi, T., T. Yano, and A. Matsuno-Yagi. 1993. Characteristics of the energy-transducing NADH–quinone oxidoreductase of *Paracoccus denitrificans* as revealed by biochemical, biophysical, and molecular biological approaches. *J. Bioenerg. Biomemb.* **25**:339–345.

12. Complex II (succinic dehydrogenase) has four polypeptides, two of which are hydrophilic peripheral membrane proteins protruding into the matrix of mitochondria or the cytoplasm of bacteria; the other two are hydrophobic integral membrane proteins. The polypeptide furthest from the membrane, called subunit A, is a flavoprotein with FAD as its prosthetic group. Subunit A is attached to the second peripheral polypeptide, called subunit B, which is closest to the membrane. Subunit B contains three FeS centers through which electrons from the flavoprotein are passed. The two integral membrane proteins (subunits C and D) are associated with one or two heme groups in many bacteria depending upon the enzyme. The hemes may transfer electrons from the FeS centers to the quinone within the membrane. There are variations of this common theme. For example, the succinic dehydrogenase from certain bacteria such as *E. coli* and some others does not contain heme. In addition, most gram-positive bacteria as well as ε proteobacteria have just one very large integral polypeptide subunit (called C), and this is believed to result from fusion of the genes for the two smaller subunits (C and D). The student is referred to a special issue of *Biochimica et Biophysica Acta*, volume 1553 (2002), which has a collection of articles on succinic dehydrogenase.

13. García-Horsman, J. A., B. Barquera, J. Rumbley, J. Ma, and R. B. Gennis. The superfamily of heme–copper respiratory oxidases. *J. Bacteriol.* **176**:5587–5600.

14. Hill, S., S. Viollet, A. T. Smith, and C. Anthony. 1990. Roles for enteric d-type cytochrome oxidase in N_2 fixation and microaerobiosis. *J. Bacteriol.* **172**:2071–2078.

15. Kelly, M. J. S., R. K. Poole, M. G. Yates, and C. Kennedy. 1990. Cloning and mutagenesis of genes encoding the cytochrome bd terminal oxidase complex in *Azotobacter vinelandii*: mutants deficient in the cytochrome d complex are unable to fix nitrogen in air. *J. Bacteriol.* **172**:6010–6019.

16. Reviewed in *J. Bioenerg. Biomembv.* **25**(4), 1993.

17. Right-side-out vesicles are made by osmotic lysis of the spheroplasts or protoplasts.

18. The phosphorylation potential, ΔG_p, is dependent upon the standard free energy of formation of ATP and the actual ratios of ATP, ADP, and P_i, according to the following equation: $\Delta G_p = \Delta G_0' + RT \ln[ATP]/[ADP][P_i]$ J/mol. The free energy of formation is proportional to the equilibrium constant (i.e., $\Delta G_0' = -RT \ln K_{eq}'$).

19. Reviewed in: Trumpower, B. L. 1990. Cytochrome bc_1 complexes of microorganisms. *Microbiol. Rev.* **54**:101–129.

20. van Verseveld, H. W., K. Krab, and A. H. Stouthamer. 1981. Proton pump coupled to cytochrome c oxidase in *Paracoccus denitrificans*. *Biochim. Biophys. Acta* **635**:525–534.

21. de Gier, J.-W. L., M. Lubben, W. N. M. Reijnders, C. A. Tipker, D-J. Slotboom, R. J. M. van Spanning, A. H. Stouthamer, and J. van der Oost, 1994. The terminal oxidases of *Paracoccus denitrificans*. *Mol. Microbiol.* **13**:183–196.

22. Hinkle, P. C., M. A. Kumar, A. Resetar, and D. L. Harrs, 1991. Mechanistic stoichiometry of mitochondrial oxidative phosphorylation. *Biochemistry* **30**:3576–3582.

23. Matsushita, K., T. Ohnishi, and H. R. Kaback. 1987. NADH–ubiquinone oxidoreductases of the *Escherichia coli* aerobic respiratory chain. *Biochemistry* **26**:7732–7737.

24. Sled, V. D., T. Freidrich, H. Leif, H. Weiss, S. W. Meinhardt, Y. Fukumori, M. W. Calhoun, R. B. Gennis, and T. Ohnishi. 1993. Bacterial NADH-quinone oxidoreductases: iron–sulfur clusters and related problems. *J. Bioenerg. Biomemb.* **25**:347–356.

25. Unden, G., and J. Bongaerts. 1997. Alternative respiratory pathways of *Escherichia coli*: energetics and transcriptional regulation in response to electron acceptors. *Biochim. Biophys. Acta* **1320**:217–234.

26. van der Oost, J., A. P. N. de Boer, J.-W. L. de Gier, W. G. Zumft, A. H. Stouthamer, and R. J. M. van Spanning. 1994. The heme–copper oxidase family consists of three distinct types of terminal oxidases and is related to nitric oxide reductase. *FEMS Microbiol. Lett.* **121**:1–10.

27. Minghetti, K. C., V. C. Goswitz, N. E. Gabriel, J. Hill, C. A. Barassi, C. D. Georgiou, S. I. Chan, and R. B. Gennis. 1992. A modified, large-scale purification of the cytochrome o complex of *Escherichia coli* yields a two heme/one copper terminal oxidase with high specific activity. *Biochemistry* **31**:6917–6924.

28. Gennis, R. B., and V. Stewart. 1996. Respiration, pp. 217–261. In: *Escherichia coli and Salmonella, Cellular and Molecular Biology*, Vol. 1. F. C. Neidhardt et al. (Eds.) ASM Press, Washington, DC.

29. Anraku, Y., and R. B. Dennis. 1987. The aerobic respiratory chain of *Escherichia coli*. *Trends Biochem. Sci.* **12**:262–266.

30. Calhoun, M. W., K. L. Oden, R. B. Gennis, M. J. Teixeira de Mattos, and O. M. Neijssel. 1993. Energetic efficiency of *Escherichia coli*: effects of mutations in components of the aerobic respiratory chain. *J. Bacteriol.* **175**:3020–3025.

31. Tran, Q. H., J. Bongaerts, D. Vlad, and G. Unden. 1997. Requirement for the proton-pumping NADH dehydrogenase I of *Escherichia coli* in respiration of NADH to fumarate and its bioenergetic implications. *Eur. J. Biochem.* **244**:155–160.

32. Green, J., M. F. Anjum, and J. R. Guest. 1997. Regulation of the *ndh* gene of *Escherichia coli* by integration host factor a novel regulator, Arr. *Microbiology* **143**:2865–2875.

33. Reviewed in: Berks, B. C., S. J. Ferguson, J. W. B. Moir, and D. J. Richardson. 1995. Enzymes and associated electron transport systems that catalyze the respiratory reduction of nitrogen oxides and oxyanions. *Biochim. Biophys. Acta* **1232**:97–173.

34. Fumarate is reduced to succinate by using the membrane-bound enzyme fumarate reductase. Nitrate is reduced to nitrite by using the membrane-bound nitrate reductase, which resembles the dissimilatory nitrate reductase in *P. denitrificans*. Nitrite is reduced to ammonia by using an NADH-linked cytoplasmic enzyme (NADH–nitrite oxidoreductase). Energy is not conserved in the reaction. If NO_3^- is the only source of nitrogen, then some of the ammonia is assimilated and the excess is excreted. Nitrite is toxic to bacteria, and probably the reduction to ammonia functions to reduce toxic levels.

35. Unden, G., and J. Schirawski. 1997. The oxygen-responsive transcriptional regulator FNR of *Escherichia coli*: the search for signals and reactions. *Mol. Microbiol.* **25**:205–210.

36. van Verseveld, H. W., and A. H. Stouthamer. 1992. The genus *Paracoccus*, pp. 2321–2334. In: *The Prokaryotes*, Vol. III, 2nd ed. A. Balows, H. G. Truper, M. Dworkin, W. Harder, and K-H Schleifer (Eds.). Springer-Verlag, Berlin.

37. van Spanning, R. J. M., A. P. N. de Boer, W. N. M. Reijnders, J.-W. L. de Gier, C. O. Delorme, A. H. Stouthamer, H. V. Westerfoff, N. Harms, and J. van der Oost. 1995. Regulation of oxidative phosphorylation: the flexible respiratory network of *Paracoccus denitrificans*. *J. Bioenerg. Biomembr.* **27**:499–512.

38. Visser, J. M., A. H. de Jong, S. de Vries, L. A. Robertson, and J. G. Kuenen. 1997. cbb$_3$-Type cytochrome oxidase in the obligately chemolithoautotrophic *Thiobacillus* sp. W5. *FEMS Microbiol. Lett.* **147**:127–132.

39. Cox, J. C., W. J. Ingledew, B. A. Haddock, and H. G. Lawford. 1978. The variable cytochrome content of *Paracoccus denitrificans* grown aerobically under different conditions. *FEBS Lett.* **93**:261–265.

40. Puustinen, A., M. Finel, M. Virkki, and M. Wikstrom. 1989. Cytochrome o (bo) is a proton pump in *Paracoccus denitrificans* and *Escherichia coli*. *FEBS Lett.* **249**:163–167.

41. Ferguson, S. J. 1987. Denitrification: a question of the control and organization of electron and ion transport. *Trends Biochem. Sci.* **12**:354–357.

42. Nitrate reductase is a membrane-bound molybdenum protein with three subunits, α, β, and γ. The γ subunit is a cytochrome b. A proposed route for electrons is from ubiquinol to the FeS centers of the β subunit, and from there to the molybdenum center at the active site in the α subunit. The α subunit then reduces the nitrate. The molybdenum is part of a cofactor (prosthetic group) called the molybdenum cofactor (Moco), which consists of molybdenum bound to a pterin. The Moco prosthetic group is in all molybdenum-containing proteins except nitrogenase. [The molybdenum cofactor from nitrogenase contains nonheme iron and is called FeMo cofactor or FeMoco (Section 11.3.2).] Nitrite reductase is a periplasmic enzyme that reduces nitrite to nitric oxide. Nitrite reductase from *P. denitrificans* has two identical subunits. These have hemes c and d$_1$. Nitric oxide reductase is a membrane-bound enzyme that reduces nitric oxide to nitrous oxide. It has heme b and heme c. Nitrous oxide reductase is a periplasmic enzyme that reduces nitrous oxide to nitrogen gas. It has two identical subunits containing copper.

43. Patureau, D., N. Bernet, and R. Moletta. 1996. Study of the denitrifying enzymatic system of *Comamonas* sp. strain SGLY2 under various aeration conditions with a particular view on nitrate and nitrite reductases. *Curr. Microbiol.* **32**:25–32.

44. Reviewed in: Kroger, A., V. Geisler, E. Lemma, F. Theis, and R. Lenger. 1992. Bacterial fumarate respiration. *Microbiology* **158**:311–314.

45. Anraku, Y. 1988. Bacterial electron transport chains, pp. 101–132. In: *Annual Review of Biochemistry*, Vol. 57. C. C. Richardson, P. D. Boyer, I. B. Dawid, and A. Meister (Eds.). Annual Reviews, Palo Alto, CA.

5

Photosynthesis

Photosynthesis is the conversion of light energy into chemical energy used for growth. Organisms that obtain most or all of their energy in this way are called phototrophic or photosynthetic. Depending upon the photosynthetic system, light energy has one or both of the following properties:

1. In all photosynthetic systems light energy can drive the phosphorylation of ADP to make ATP (the process of photophosphorylation).

2. In some photosynthetic systems light can also drive the transfer of electrons from H_2O ($\Delta E_{m,7} = +820$ mV) to $NADP^+$ ($\Delta E_{m,7} = -320$ mV). This is referred to as the photoreduction of $NADP^+$.

During photoreduction of $NADP^+$, the water becomes oxidized to oxygen gas and the $NADP^+$ becomes reduced to NADPH. The student will recognize that this is the reverse of the direction in which electrons flow spontaneously during aerobic respiration (Chapter 4). During both ATP and NADPH synthesis, electromagnetic energy is absorbed by photopigments in the photosynthetic membranes and converted into chemical energy. This chapter explains how that is done. At the heart of the process is the light-driven oxidation of chlorophyll or bacteriochlorophyll. This initiates electron transport that results in the generation of a Δp and subsequent synthesis of ATP, and in the case of chlorophyll, $NADP^+$ reduction. It is of interest that (bacterio)chlorophyll-based photosynthesis is widely distributed,

being found in bacteria, green plants, and algae, but is not present in the known archaea. (Interestingly, bacteriochlorophyll has also been found in nonphototrophic bacteria, i.e., rhizobia.[1,2]) This chapter begins with background information on the different groups of phototrophic prokaryotes and then describes photosynthetic electron transport in all the photosynthetic systems.

5.1 The Phototrophic Prokaryotes

The phototrophic prokaryotes are a diverse assemblage of organisms that share the common feature of being able to use light as a source of energy for growth. Their classification is based upon physiological differences, including whether they produce oxygen, and what they use as a source of electrons for biosynthesis. For example, there are both oxygenic phototrophs (produce oxygen) and anoxygenic phototrophs. Oxygenic phototrophs include the well-known *cyanobacteria* (formerly called blue-green algae) and members of the genera *Prochloron*, *Prochlorothrix*, and *Prochlorococcus*.[3] The latter three genera consist of organisms phylogenetically related to the cyanobacteria but having chlorophyll b as a light-harvesting pigment instead of phycobilins.

The anoxygenic phototrophs are the *purple photosynthetic bacteria*, the *green photosynthetic bacteria*, and the *heliobacteria*. (For a more complete discussion of the taxonomy of the photosynthetic bacteria, see note 4.) These microorganisms photosynthesize only under anaerobic conditions. The purple and green photosynthetic bacteria are further subdivided according to whether they use sulfur as a source of electrons. Thus there are purple sulfur and nonsulfur photosynthetic bacteria and green sulfur and nonsulfur photosynthetic bacteria. The various phototrophic prokaryotes and some of their properties are summarized in Table 5.1.

5.1.1 Oxygenic phototrophs

The oxygenic prokaryotic phototrophs use H_2O as the electron donor for the photosynthetic reduction of $NADP^+$. Most of the species belong to the cyanobacteria, which are widely distributed in nature, occurring in fresh and marine waters, and in terrestial habitats. They have only one type of chlorophyll (chlorophyll a), and light-harvesting pigments called phycobilins.

Another kind of oxygenic microbial phototroph consists of the prochlorophytes, which are a diverse group of photosynthetic pro-

karyotes that are evolutionarily related to the cyanobacteria but differ from the latter in having chlorophyll b rather than phycobilins as the light-harvesting pigment.[5] There are three genera among the prochlorophytes: *Prochloron*, *Prochlorothrix*, and *Prochlorococcus*. Members of the genus *Prochloron* are obligate symbionts of certain ascidians (sea squirts). *Prochlorothrix* is a filamentous, free-living microorganism that lives in freshwater lakes. *Prochlorococcus* is a free-living marine microorganism.

5.1.2 Anoxygenic phototrophs

The other phototrophic prokaryotes do not produce oxygen (i.e., H_2O is not a source of electrons). Instead, they use organic compounds, inorganic sulfur compounds, or hydrogen gas as a source of electrons. *These organisms will grow phototrophically only anaerobically or when oxygen tensions are low.* There are four major groups of anoxygenic phototrophs (see Table 5.1):

1. Purple photosynthetic bacteria, which includes both the purple sulfur and the purple nonsulfur bacteria

Table 5.1 Phototrophic prokaryotes

Type	Pigments[a]	Electron donor	Carbon source	Aerobic dark growth
Oxygenic				
Cyanobacteria	Chl a, phycobilins	H_2O	CO_2	No
Prochloron	Chl a, Chl b	H_2O	CO_2	No
Chlorothrix	Chl a, Chl b	H_2O	CO_2	No
Prochlorothrix	Chl a, Chl b	H_2O	CO_2	No
Anoxygenic				
Purple sulfur	Bchl a or Bchl b	H_2S^b, $S°$, $S_2O_3^{2-}$, H_2, organic	CO_2, organic	Yes[c]
Purple nonsulfur	Bchl a or Bchl b	H_2, organic, H_2S (some)	CO_2, organic	Yes
Green sulfur	Bchl a and c, d, or e	H_2S^d, $S°$, $S_2O_3^{2-}$, H_2	CO_2, organic	No
Green gliding (nonsulfur)	Bchl a and c or d	H_2S, H_2, organic	CO_2, organic	Yes
Heliobacteria	Bchl g	Organic	Organic	No

[a] Carotenoids are usually present.

[b] Members of the genus *Chromatium* accumulate $S°$ intracellularly. Members of the genus *Ectothiorhodospira* accumulate extracellular sulfur.

[c] In the natural habitat growth occurs primarily anaerobically in the light. However, several purple sulfur species (Chromatiaceae) can be grown continuously in the laboratory aerobically in the dark at low oxygen concentrations. (Overmann, J., and N. Pfennig. 1992. Continuous chemotrophic growth and respiration of Chromatiaceae species at low oxygen concentrations. *Arch. Microbiol.* **158**:59–67.)

[d] Accumulates $S°$ extracellularly.

2. Green sulfur photosynthetic bacteria

3. Green nonsulfur photosynthetic bacteria, also called green gliding bacteria

4. Heliobacteria

Purple sulfur phototrophs

The purple sulfur phototrophic bacteria grow photoautotrophically in anaerobic environments using hydrogen sulfide as the electron donor and CO_2 as the carbon source. Their natural habitats are freshwater lakes and ponds, or marine waters where the presence of sulfate-reducing bacteria produces a high sulfide content. They oxidize sulfide to elemental sulfur, which accumulates as granules intracellularly in all known genera except the *Ectothiorhodospira*, which deposits sulfur extracellularly. However, some can grow photoheterotrophically, and several have been grown chemoautotrophically under low partial pressures of oxygen with reduced inorganic sulfur as an electron donor and energy source. There are many genera, including the well-studied *Chromatium*.

Purple nonsulfur phototrophs

The purple nonsulfur phototrophic bacteria are extremely versatile with regard to sources of energy and carbon. Originally it was thought that these bacteria were not able to utilize sulfide as a source of electrons, hence the name "nonsulfur." However, it turns out that the concentrations of sulfide used by the purple sulfur bacteria are toxic to the nonsulfur purples and, provided the concentrations are sufficiently low, some purple nonsulfur bacteria (viz., *Rhodopseudomonas* and *Rhodobacter*) can use sulfide and/or thiosulfate as a source of electrons during photoautotrophic growth.

The purple nonsulfur phototrophs can grow photoautotrophically or photoheterotrophically in anaerobic environments. When growing photoautotrophically, they use H_2 as the electron donor and CO_2 as the source of carbon. Photoheterotrophic growth uses simple organic acids such as malate or succinate as the electron donor and source of carbon, and light as the source of energy. If these organisms are placed in the dark in the presence of oxygen, they will carry out ordinary aerobic respiration and grow chemoheterotrophically (e.g., on succinate or malate). A few can even grow fermentatively, very slowly, in the dark. Because of their physiological versatility, the purple nonsulfur photosynthetic bacteria have received much attention in research. They are found in lakes and ponds with low sulfide content. Representative genera are *Rhodobacter* and *Rhodospirillum*.

Green sulfur phototrophs

For a review of photosynthesis in the green sulfur and green nonsulfur photosynthetic bacteria, see Ref. 6. The green sulfur phototrophic bacteria are strict anaerobic photoautotrophs that use H_2S, $S°$, $S_2O_3^{2-}$ (thiosulfate), or H_2 as the electron donor and CO_2 as the source of carbon. The green sulfur bacteria coexist with the purple sulfur phototrophs in sulfide-rich anaerobic aquatic habitats, although they are found in a separate layer. Their light-harvesting pigments are located in special inclusion bodies called *chlorosomes*, which will be described later, and their reaction centers (the part of the photosynthetic membrane where the bacteriochlorophyll is oxidized) differ from those of the purple photosynthetic bacteria. There are several genera, including the well-studied *Chlorobium*.

Green nonsulfur phototrophs (green gliding bacteria)

The best-studied green nonsulfur photosynthetic bacterium is *Chloroflexus*, which is a thermophilic, filamentous, gliding green phototroph. *Chloroflexus* can be isolated from alkaline hot springs, whose pH can be as high as 10. Most isolates have a temperature optimum for growth from 52 to 60 °C. It can be grown as a photoheterotroph, photoautotrophically with either H_2 or H_2S as the electron donor, or chemoheterotrophically in the dark in the presence of air. However, it really grows best as a photoheterotroph.

Chloroflexus has chlorosomes, but its reaction center is a quinone type, resembling that of the purple photosynthetic bacteria. It differs from the reaction center in the green sulfur bacteria, which is an iron–sulfur type, resembling the reaction center of heliobacteria and reaction center I of cyanobacteria and chloroplasts.

Heliobacteria

The heliobacteria are recently discovered, strictly anaerobic photoheterotrophs that differ from the other phototrophs in having bacteriochlorophyll g as their reaction center chlorophyll, in containing few carotenoids, and in having no internal photosynthetic membranes or chlorosomes. The two known genera are *Heliobacterium* and *Heliobacillus*. The tetrapyrrole portion of bacteriochlorophyll g is similar to chlorophyll a, and in fact isomerizes into chlorophyll a in the presence of air. Both 16S rRNA analyses and the lack of lipopolysaccharide in the cell wall place these organisms with the gram-positive bacteria, and it has been suggested that they are phylogenetically related to the clostridia. In agreement with this, most species of heliobacteria make endospores that resemble those produced by *Bacillus* or *Clostridium*.

5.2 The Purple Photosynthetic Bacteria

In all photosynthetic systems, light is absorbed by light-harvesting pigments (also called accessory pigments), and the energy is transferred to reaction centers where bacteriochlorophyll is oxidized, creating a redox potential, ΔE. The light-harvesting pigments, the photosynthetic pigments in the reaction center, and energy transfer to the reaction center are considered in Sections 5.6 and 5.7. The structure of the photosynthetic membrane systems in the prokaryotes as revealed by electron microscopy is described in Section 5.8. The generation of the ΔE and photosynthetic electron transport are described next.[7,8] The reaction center of green nonsulfur filamentous photosynthetic bacteria is similar. See note 9 for a comparison of the two reaction centers.

5.2.1 Photosynthetic electron transport

This section discusses the oxidation of bacteriochlorophyll, the generation of the ΔE, and electron transport. Photosynthetic electron flow in the purple photosynthetic bacteria is illustrated in Fig. 5.1A. A detailed discussion of the reaction center is given in Section 5.2.2;

the photosynthetic process itself has eight steps, as follows.

Step 1. A dimer of two bacteriochlorophyll molecules (P_{870}) in the reaction center absorbs light energy, and one of its electrons becomes excited to a higher energy level (P_{870}^*). In contrast to P_{870}, P_{870}^* has a very low reduction potential (E_m').

Step 2. The electron is transferred to an acceptor molecule within the reaction center, thus oxidizing the P_{870} and reducing the acceptor molecule. The acceptor molecule is bacteriopheophytin (Bpheo), which is bacteriochlorophyll without magnesium. The light creates a $\Delta E_m'$ of about 1 V between P_{870}^+/P_{870}, which is around +0.45 V, and Bpheo$^+$/Bpheo which is around −0.6 V.

Step 3. The electron is transferred from the bacteriopheophytin to a quinone called ubiquinone A (UQ_A). The, E_m' for UQ_A/UQ_A^- is about −0.2 V.

Step 4. UQ_A transfers the electron to a ubiquinone in site B, called UQ_B, and UQ_B becomes reduced to UQ_B^-. Steps 1 to 4 are repeated so that UQ_B accepts a second electron and becomes reduced to UQ_B^{2-}.

Step 5. UQ_B^{2-} picks up two protons from the cytoplasm and is released from the reaction center as UQH_2. The UQH_2 joins the quinone pool in the membrane.

Steps 6 and 7. UQH_2 transfers the electrons to a bc_1 complex and the bc_1 complex reduces cytochrome c_2. The bc_1 complex translocates protons to the cell surface thus generating a Δp. (See Section 4.6.2 for a discussion of the bc_1 complex and the Q cycle.)

Step 8. The cytochrome c_2 returns the electron to the oxidized bacteriochlorophyll molecule in the reaction center. Thus light energy has caused a cyclic electric current in the membrane.

These eight steps describe cyclic electron flow. We can follow the path of the electron starting with cytochrome c_2 (Fig. 5.1B). Notice that an electron moves "uphill" (to a lower electrode potential) from cytochrome c_2 to ubiquinone when "boosted" or "pushed" in the reaction center by the energy from a quantum of light.

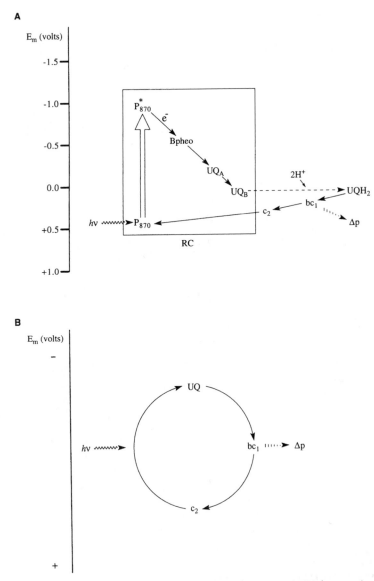

Fig. 5.1 Photosynthetic electron flow in purple photosynthetic bacteria. (A) Light energizes bacteriochlorophyll, here shown as P_{870} (bacteriochlorophyll a) in the reaction center (RC). (The number in the subscript gives the major long-wavelength absorption peak in nanometers.) Some purple photosynthetic bacteria have bacteriochlorophyll b instead, which absorbs at 1020 to 1035 nm. The energized bacteriochlorophyll reduces bacteriopheophytin (Bpheo), which is bacteriochlorophyll without Mg^{2+}. The electron then travels through two ubiquinones, UQ_A and UQ_B, in the reaction center. (Some species use menaquinone.) After a second light reaction, UQ_B becomes reduced to UQH_2 and leaves the reaction center. The electron returns to the reaction center via a proton-translocating bc_1 complex and cytochrome c_2. ATP is synthesized via a membrane ATP synthase driven by the Δp (not shown). Midpoint potentials at pH 7 (approximate): $Bchl_{870}$ (P_{870}), +450 mV; Bpheo, −600 mV; UQ_A, −200 mV; UQ_B, +80 mV; c cytochromes, +380 mV. (B) A simplified version of (A) that emphasizes the cyclic flow of electrons. *Source*: Data from Mathis, P. 1990. *Biochem. Biophys. Acta* **1018**:163–167.

As pointed out earlier, for the reaction center to reduce UQ to UQH_2, two electrons and therefore two light reactions are required. This can be written as follows,

$$2 \text{ cyt } c_{2,red} + 2H^+ + 2\,UQ$$

$$\xrightarrow{\text{2 quanta of light}} 2 \text{ cyt } c_{2,ox} + UQH_2 \quad (5.1)$$

Because two electrons are required before reduced quinone leaves the reaction center, UQ_B is referred to as a *two-electron gate*.

As illustrated in Fig. 5.1, the reduced quinone transfers the electrons to cytochrome c_2 via a bc_1 complex outside the reaction center. The bc_1 complex translocates protons across the membrane via a Q cycle:

$$UQH_2 + 2H_{in}^+ + 2 \text{ cyt } c_{2,ox}$$

$$\xrightarrow{bc_1} UQ + 4H_{out}^+ + 2 \text{ cyt } c_{2,red} \quad (5.2)$$

The Δp that is generated is used to drive the synthesis of ATP via a membrane-bound ATP synthase:

$$ADP + P_1 + 3H_{out}^+$$

$$\xrightarrow{\text{ATP synthase}} ATP + H_2O + 3H_{in}^+ \quad (5.3)$$

Thus, the net result of photosynthesis by purple photosynthetic bacteria is the synthesis of ATP. This is called *cyclic photophosphorylation*.

The actual amount of ATP made depends upon the number of protons translocated by the bc_1 complex and the number of protons that enter via the ATP synthase. For example, if four protons for every two electron transfers are translocated out of the cell by the bc_1 complex and three protons reenter via the ATP synthase, then the maximum number of ATP molecules made per two electrons is 4/3 (1.333).

Figure 5.2 illustrates the topographical relationships between the reaction center, the bc_1 complex, and the ATP synthase in photosynthetic membranes and can be compared to Fig. 5.1, which illustrates the sequential steps in electron transfer. In the reaction center, electrons travel across the membrane from a periplasmic cytochrome c_2 to the quinone at site A. This creates a membane potential, which is negative inside. After a second electron is transferred, the quinone at site B accepts two protons from the cytoplasm and leaves the reaction center as UQH_2. The UQH_2 diffuses through the lipid matrix to the bc_1 complex,

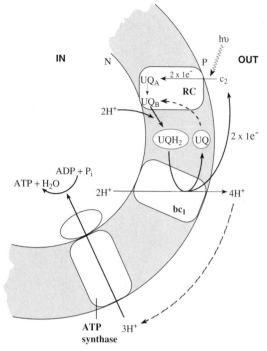

Fig. 5.2 Relationship between reaction center, bc_1 complex, and ATP synthase in photosynthetic membranes. The reaction center (RC) takes two protons from the cytoplasm and two electrons from a periplasmic cytochrome c to reduce ubiquinone (UQ) to UQH_2. The UQH_2 leaves the reaction center and diffuses to the bc_1 complex. The bc_1 complex oxidizes UQH_2 and translocates four protons to the periplasmic surface in a Q cycle, which is described in Section 4.6.2. The electrons travel back to cytochrome c_2 (cyclic flow). The translocated protons reenter via the ATP synthase, which makes ATP.

where it is oxidized. The bc_1 complex translocates four protons to the outside for every UQH_2 oxidized via a Q cycle, creating a Δp. The electrons are transferred from the bc_1 complex to a periplasmic cytochrome c_2, which returns them to the reaction center. The proton circuit is completed when the protons are returned to the cytoplasmic side via the ATP synthase accompanied by the synthesis of ATP. *Thus, light is used to drive an electron circuit, which drives a proton circuit, which in turn drives ATP synthesis.* This successful method for harnessing light energy to do chemical work is used in all phototrophic organisms from bacteria to plants.

5.2.2 A more detailed examination of the reaction center and what happens there

The description just given for electron flow in the reaction center omitted some of the components and electron carriers, and did not cover all aspects of the structure of the reaction center. A more complete description follows.

Structure and composition of the reaction center

The reaction centers from *Rhodopseudomonas viridis* and *Rhodobacter sphaeroides* have been crystallized, and the structures determined from high-resolution X-ray diffraction studies.[10,11] The reaction centers are similar. The following proteins and pigments are found in the reaction center from *R. sphaeroides*:

1. *Reaction center protein.* The reaction center protein has 11 membrane-spanning α helices, and a globular portion on the cytoplasmic side of the membrane. *It consists of three polypeptides: H, L, and M.* The reaction center protein serves as a scaffolding to which the bacteriochlorophyll and bacteriopheophytin are attached, and to which the quinones and nonheme iron are bound.

2. *Four bacteriochlorophyll molecules (Bchl).* There are four bacteriochlorophyll molecules in the reaction center. Two of the bacteriochlorophyll molecules exist as a dimer $(Bchl)_2$, and two are monomers $(Bchl_A$ and $Bchl_B)$.

3. *Two molecules of bacteriopheophytin ($Bpheo_A$ and $Bpheo_B$), which is Bchl without Mg^{2+}.*

4. *Two molecules of ubiquinone (UQ_A and UQ_B).*

5. *One molecule of nonheme ferrous iron (Fe).*

6. *One carotenoid molecule.* The function of the carotenoid is to protect the reaction center pigments from photodestruction.[12] (For a further explanation, see note 13.)

Electron transfer in the reaction center

The sequence of redox reactions in the reaction center summarized in Fig. 5.1A (boxed area) is shown in more detail in Fig. 5.3, along with the proposed time scale for the electron transfer events. Figure 5.1 shows that the terminal electron acceptors in the reaction center are quinones. These are called type II reaction centers and are found in purple bacteria (Proteobacteria) and green nonsulfur bacteria (*Chloroflexus* group). (Type I reaction centers use iron–sulfide clusters as the terminal electron acceptor, and these are present in green sulfur bacteria and heliobacteria, as discussed later.) As shown in Fig. 5.3, the arrangement of the pigment molecules and the quinones reveals twofold symmetry with right and left halves that are very similar. On the periplasmic side of the membrane there sits a pair of bacteriochlorophyll molecules $(Bchl)_2$ (Fig. 5.3). These bacteriochlorophyll molecules are P_{870} in Fig. 5.1. When the energy from a quantum of light is absorbed by $(Bchl)_2$, an electron becomes excited [i.e., $(Bchl)_2$ becomes $(Bchl)_2^*$] (Fig. 5.3, step 1). The redox potential of $(Bchl)_2^*$ is low enough to reduce bacteriochlorophyll A $(Bchl_A)$, forming $(Bchl)_2^+$ and $Bchl_A^-$ (Fig. 5.3, step 2). The electron then moves to bacteriopheophytin$_A$ ($Bpheo_A$), forming $Bpheo_A^-$ (Fig. 5.3, step 3). [As reflected by the question mark in step 2 of Fig. 5.3, the exact route of the electron is not known. It has not been unequivocally demonstrated that the electron moves through the $Bchl_A$ monomer. Some researchers suggest that the electron travels from $Bchl_2^*$ to $Bchl_A$ and very quickly moves on to $Bpheo_A$, whereas other investigators postulate that the electron moves directly from $(Bchl^*)_2$ to $Bpheo_A$.] Then the electron moves to ubiquinone bound to site A on the cytoplasmic side of the reaction center (UQ_A) forming the semiquinone, UQ_A^- (Fig. 5.3, step 4). At this time an electron is returned to $(Bchl)_2^+$ from reduced cytochrome c_2, which is periplasmic (step 5). Then, the electron is transferred from UQ_A^- to UQ_B to form UQ_B^- (Fig. 5.3, step 6). Electron flow is therefore across the membrane from periplasmic c_2 to UQ_B located on the cytoplasmic side. This generates a membrane potential, outside positive. A second light reaction occurs, and steps 1 through 6 in Fig. 5.3 are repeated so that UQ_B has two electrons (UQ_B^{2-}). Two protons are picked up from the cytoplasm to form UQH_2, which leaves the reaction center to join the quinone pool in the membrane. (Although Fig. 5.3 indicates

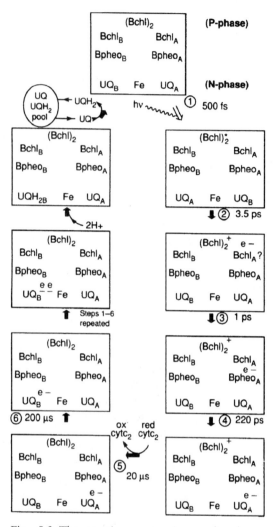

Fig. 5.3 The reaction center in purple photosynthetic bacteria. The reaction center, which spans the membrane, is represented by the boxed areas. The absorption of light creates a transient membrane potential, positive (P-phase) on the periplasmic side and negative (N-phase) on the cytoplasmic side. **Step 1:** absorption of a photon of light energizes a bacteriochlorophyll dimer $(Bchl)_2$. **Step 2:** the energized $(Bchl)_2^*$ reduces $Bchl_A$. The question mark indicates that this has not been unequivocally demonstrated. **Step 3:** bacteriopheophytin A $(Bpheo_A)$ is reduced. **Step 4:** the electron is transferred to a quinone (UQ_A). **Step 5:** the oxidized $(Bchl)_2$ is reduced via cytochrome c_2. **Step 6:** The electron moves from UQ_A to UQ_B. Steps 1 to 6 are repeated, leading to the formation of UQ_B^{2-}. Two protons are acquired from the cytoplasm to produce UQH_2, which enters the reduced quinone pool in the membrane and returns the electrons to oxidized cytochrome c_2 via the bc_1 complex. It may be that one proton is acquired by UQ_B^- and the second proton is acquired when the second electron arrives. *Source:* Nicholls, D. G., and S. J. Ferguson. 1992. *Bioenergetics* 2. Academic Press, London.

that two protons enter after both electrons have arrived at UQ_B, it has been suggested that one proton enters after UQ_B has received the first electron.) Eventually, the protons carried by UQH_2 are released on the periplasmic side during oxidation of UQH_2 by the bc_1 complex.

Thus, the reaction center and the bc_1 complex cooperate to translocate protons to the outside. As explained earlier, because UQ_B cannot leave the reaction center until it has accepted two electrons, it is called a *two-electron gate*. The length of time that it takes the electron to travel from $(Bchl)_2$ to UQ_A is a little more than 200 ps. [One picosecond (ps) is one trillionth of a second, or 10^{-12} s.] The rates of subsequent steps are slower (but still very fast) and measured in microseconds. [One microsecond (μs) is a millionth of a second, 10^{-6} s.]

Determining the pattern and timing of electron transfer in the reaction center requires the use of picosecond laser pulses and very rapid recording of the absorption spectra of the electron carriers. Interestingly, only one side of the reaction center (the A side) appears to be involved in electron transport. For example, there is photoreduction of only one of the bacteriopheophytin molecules. Why only one branch appears to function in electron transport is not known.

The contribution to the Δp by the reaction center

When reaction centers absorb light energy, a $\Delta\Psi$ and a proton gradient are created. The reason for the $\Delta\Psi$ is that the electrons move electrogenically from cytochrome c_2 to UQ_B across the membrane from outside to inside (Fig. 5.2). (This will produce a $\Delta\Psi$, outside positive, because the inward movement of a negative charge is equivalent to the outward movement of a positive charge.) The creation of the membrane potential is detected by a shift in the absorption spectra of membrane carotenoids. (For additional explanation, see note 14.) The value of this technique is that very rapid changes in membrane potential can be monitored. The $\Delta\Psi$ produced by the reaction center is delocalized, and increases the $\Delta\Psi$ are made in the bc_1 complex (Section 5.2). Additionally, two protons are taken from the cytoplasm to reduce UQ_B to UQH_2. Eventually, the two protons will be released on the outside during the oxidation of UQH_2 by the bc_1 complex. Thus, the reaction center contributes to both the $\Delta\Psi$ and the ΔpH components of the Δp, as well as creating a ΔE between UQ/UQH_2 and $c_{2,ox}/c_{2,red}$.

Summary of photosynthesis by purple photosynthetic bacteria

$$ADP + P_i \xrightarrow{\text{light and Bchl}} ATP + H_2O$$

5.2.3 Source of electrons for growth

To grow, all organisms must reduce $NAD(P)^+$ to $NAD(P)H$. This is because NADH and NADPH are the electron donors for almost all the biosynthetic reactions in the cell, including the reduction of carbon dioxide to carbohydrate. When the electron donor is of a higher potential than the $NAD^+/NADH$ couple, energy is required to reduce NAD^+. For example, this is the case for the purple photosynthetic bacteria that use certain inorganic sulfur compounds or succinate as a source of electrons. The purple photosynthetic bacteria use the Δp created by light energy to drive electron transport in reverse (Fig. 5.4). During reversed electron flow, ubiquinol reduces NAD^+ via the NADH:ubiquinone oxidoreductase. (See Section 3.7.1 for a discussion of reversed electron transport.)

Figure 5.4 illustrates the situation in which electrons from the electron donor do not pass through the reaction center but simply enter the electron transport chain and travel directly to NAD^+ via ubiquinone. That is, there is no non-cyclic electron transport. Many investigators hold this view. However, another scenario has been suggested, which is useful to consider because it illustrates the role that an increase in the Δp can play in slowing the rate of electron transfer. It should be recalled that ubiquinone can also accept electrons from bacteriopheophytin during cyclic electron transport, and reduce the bc_1 complex which generates a Δp (Fig. 5.1). It has been suggested that as the Δp grows larger, it might exert "back-pressure" on the oxidation of ubiquinol by the bc_1 complex, thus slowing down the oxidation of ubiquinol via this route and making it available for NAD^+ reduction via reversed electron transport.[7] To the extent that this might occur, the electron donor (succinate or inorganic sulfur compounds) would replenish electrons to the bacteriochlorophyll via either ubiquinone or cytochrome c.

5.3 The Green Sulfur Bacteria (Chlorbiaceae)

5.3.1 Photosynthetic electron transport

The green sulfur photosynthetic bacteria and the heliobacteria have reaction centers that are distinguished from the reaction centers of the purple photosynthetic bacteria in two respects, as detailed in the subsections that follow.

Possessing iron–sulfur centers as terminal electron acceptors

Reactions centers whose terminal electron acceptors are iron–sulfur centers are called type I to distinguish them from the reactions centers in the purple photosynthetic bacteria, which use quinones as the terminal electron acceptors and are called type II.

Reducing $NAD(P)^+$ instead of quinone

The reaction centers of green sulfur bacteria are similar to the type I reaction centers of cyanobacteria and chloroplasts.

However, the principles underlying the transformation of electrochemical energy into

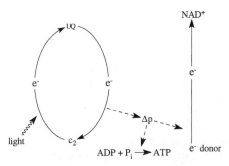

Fig. 5.4 Relationship between cyclic electron flow and reversed electron transport in the purple photosynthetic bacteria. The electron is driven by light from a cytochrome c_2 (c_2) to a ubiquinone (UQ) and then returns via electron carriers to the cytochrome c_2 with production of a Δp. The Δp is used to drive ATP synthesis as well as reversed electron transport.

a ΔE and a membrane potential are the same as for the purple photosynthetic bacteria.[15,16] Light energizes a bacteriochlorophyll a molecule (P_{840}), which reduces a primary acceptor (A_0), establishing a redox potential difference greater than 1 V (Fig. 5.5). The primary electron acceptor (A_0) in the green sulfur bacteria has recently been reported to be an isomer of chlorophyll a called bacteriochlorophyll 663.[18] It might be added that the primary acceptor of heliobacteria is also a chlorophyll a derivative, hydroxychlorophyll a, reflecting the similarities known to exist between the reaction centers of the green sulfur bacteria, heliobacteria, and photosystem I of chloroplasts and cyanobacteria. (See note 18.) The electron flows from A_0 to a quinonelike acceptor called A_1. The electron is then transferred from A_1 through three iron–sulfur centers to ferredoxin. There can be both cyclic and noncyclic electron flow. In cyclic flow the electron returns to the reaction center via menaquinone (MQ) and a bc_1 complex, creating a Δp. In noncyclic electron flow inorganic sulfur donates electrons that travel through the reaction center to NAD^+.

According to the scheme in Fig. 5.5, electrons from inorganic sulfur enter at the level of cytochrome c and the bc_1 complex is bypassed. This is a widely held view based upon available data using isolated oxidoreductases. From an energetic point of view, however, it is wasteful because a coupling site is bypassed, even though the E'_m values of some of the sulfur couples are low enough to reduce menaquinone. For example, the E'_m for sulfur/sulfide ($n = 2$) is -0.27; for sulfite/sulfide ($n = 6$) it is -0.11 V; and for sulfate/sulfite ($n = 2$) it is -0.54 V. All these couples are at a potential low enough to reduce menaquinone, which has an E'_m of -0.074 V. However, the point of entry of the electron is still an unresolved issue.[19]

5.3.2 Summary of photosynthesis by green sulfur bacteria

$$H_2S + NAD^+ + ADP + P_i \xrightarrow{\text{light and Bchl}}$$
$$S^\circ + NADH + H^+ + ATP$$

(These organisms can also use elemental sulfur, oxidizing it to sulfate.)

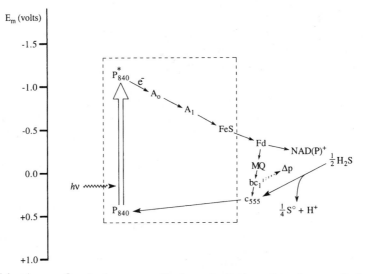

Fig. 5.5 A model for electron flow in the green sulfur bacteria. Both cyclic and noncyclic flows of electrons are possible. Reaction center bacteriochlorophyll a (P_{840}) becomes energized and reduces A_0, which is bacteriochlorophyll 663. The electron then travels to A_1, a quinonelike molecule, and then through two or three iron–sulfur centers (FeS). The first reduced product outside the reaction center is the iron–sulfur protein ferredoxin (Fd). The ferredoxin reduces $NAD(P)^+$ in the noncyclic pathway. Cyclic flow occurs when the electron reduces menaquinone (MQ) instead of $NAD(P)^+$ and returns to the reaction center via a bc_1 complex and cytochrome c_{555}. A Δp is created in the bc_1 complex. In the noncyclic pathway the electron donor is a reduced inorganic sulfur compound, here shown as hydrogen sulfide. Elemental sulfur or thiosulfate can also be used. The inorganic sulfur is oxidized by cytochrome c_{555}, which feeds electrons into the reaction center.

5.4 Cyanobacteria and Chloroplasts

Photosynthesis in cyanobacteria and chloroplasts differs from photosynthesis discussed thus far in three important respects:

1. H_2O is the electron donor and oxygen is evolved.

2. There are two light reactions in series, hence two different reaction centers.

3. Electron flow is primarily noncyclic, producing both ATP and NADPH. (Noncyclic flow also occurs in the green sulfur bacteria, as described in Section 5.3.1)

5.4.1 Two light reactions

As we shall see, chloroplasts and cyanobacteria have essentially combined the light reactions of purple photosynthetic bacteria and green sulfur photosynthetic bacteria in series, so that two light reactions energize a single electron that energizes ATP synthesis and reduces NADP+. The initial evidence for two light reactions came from early studies of photosynthesis performed with algae by Emerson and his colleagues.[20,21] They observed that the efficiency of photosynthesis, measured as the moles of oxygen evolved per einstein absorbed (i.e., the *quantum yield*) is high over all the wavelengths absorbed by chlorophyll and the light-harvesting pigments, but drops off sharply at 685 nm even though chlorophyll continues to absorb light between 680 and 700 nm (Fig. 5.6). This became known as the "red drop" effect because 700 nm light is red. One can restore the efficiency of photosynthesis of 700 nm light by supplementing it with light at shorter wavelengths (e.g., 600 nm). This is explained by pointing out that there are two reaction centers, one energized by light at a wavelength of around 700 nm (reaction center I, or RC I) and one energized by lower wavelengths of light (reaction center II, RC II). These are also called photosystems I and II (PS I and PS II). Photosystems I and II operate in series and therefore both must be energized to maintain the electron flow from water to NADP+. The lower wavelengths of light can energize both reaction centers, but 700 nm light can energize only reaction center I. It is for this reason that 700 nm light is effective only when given in combination with supplemental doses of shorter wavelengths.

5.4.2 Photosynthetic electron transport

A schematic drawing of the overall pattern of electron flow in cyanobacteria and chloroplasts

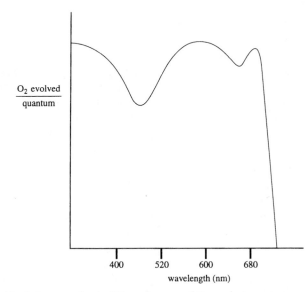

O$_2$ evolved / quantum

400   520   600   680

wavelength (nm)

Fig. 5.6 Quantum yield of photosynthesis. When the rate of O_2 evolution per quantum absorbed is plotted against wavelength, the rate drops off sharply above 680 nm. *Source*: After Stryer, L, 1988. *Biochemistry*. W. H. Freeman, New York.

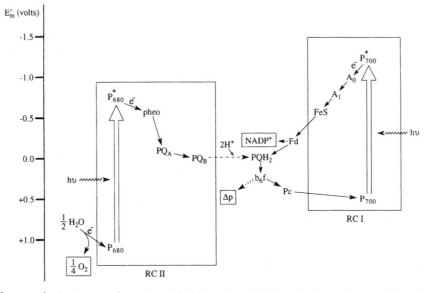

Fig. 5.7 Photosynthesis in cyanobacteria and chloroplasts. Light stimulates electron flow from water through reaction center II (RC II) to reaction center I (RC I) to NADP$^+$. Cyclic flow is possible by using RC I from ferredoxin through the b_6f complex. Abbreviations: P_{680}, chlorophyll a with a major absorption peak at 680 nm; Pheo, chlorophyll pheophytin; PQ, plastoquinone; b_6f, cytochrome b_6f complex; Pc, plastocyanin; P_{700}, chlorophyll a with a major absorption peak at 700 nm; A_0, chlorophyll a; A_1, phylloquinone; FeS, one of several iron–sulfur centers; Fd, ferredoxin.

is shown in Fig. 5.7. These systems have two reaction centers, PS I and PS II, connected by a short electron transport chain that includes the analogue to the bc_1 complex (i.e., the b_6f complex). (For a description of the b_6f complex, see note 22.) Let us begin with PS II, in which the pathway of electron flow is the same as in the type II reaction center in the purple photosynthetic bacteria (see Fig. 5.1A). Chlorophyll a, with a major absorption peak at 680 nm (i.e., P_{680}—probably a dimer), becomes energized and reduces an acceptor molecule, which is pheophytin (pheo). Having lost an electron, the P_{680}^+/P_{680} couple has a redox potential (estimated at about +1.1 V) high enough to replace the lost electron with one from water, thus oxidizing $\frac{1}{2}H_2O$ to $\frac{1}{4}O_2$. The energized electron travels to pheophytin, which has an E_m' of about –0.6 V. The pheophytin reduces plastoquinone (PQ), which is structurally similar to ubiquinone (Fig. 4.3).

A two-electron gate, similar to the two-electron gate in the reaction center of the purple photosynthetic bacteria, operates during quinone reduction. Electrons leave the reaction center in PQH$_2$ and are transferred to a copper-containing protein called plastocyanin (Pc) through a b_6f complex (structurally and

functionally similar to the bc_1 complex, except that cytochrome f, which has a c-type heme, replaces cytochrome c_1 and there are some differences in the cytochrome b). A Δp is created by the b_6f complex, but there is some controversy over whether the b_6f complex catalyzes a Q cycle.[23] The plastocyanin reduces P_{700}^+ in reaction center I, previously oxidized by the light reaction we shall describe next.

Photosystem I is similar to the type I reaction center in the green photosynthetic bacteria (see Fig. 5.5). A photon of light energizes P_{700}, which is chlorophyll a with a major absorption peak at 700 nm. The P_{700}^* reduces a chlorophyll a molecule, called A_0. For this initial redox reaction, the energized electron travels from an E_m' of about +0.5 V (P_{700}^+/P_{700}) to an E_m' of about –1.0 V (A_0/A_0^-). Note that this is a far more negative potential than that generated in reaction center II. From A_0 the electrons travel to A_1, which is a phylloquinone. Phylloquinones have a structure similar to menaquinone (Fig. 4.3), but with only one double bond in the isoprenoid chain. The electron is transferred from the quinone through several iron–sulfur centers (FeS), which reduce the iron–sulfur protein ferredoxin (Fd) that is outside the reaction center. Ferredoxin in turn reduces NADP$^+$.

Thus, the two light reactions in series energize electron flow from H_2O to $NADP^+$, which is over a net potential difference of about 1.1 V. This is called noncyclic electron flow because the electron never returns to the reaction center. However, cyclic electron flow is possible.

There is a branch point at the ferredoxin step, and it is possible for the electron to cycle back to reaction center I via the b_6f complex, augmenting the Δp, rather than reducing $NADP^+$. This may be a way to increase the amounts of ATP made relative to NADPH. The Calvin cycle, which is the pathway for reducing CO_2 to carbohydrate in oxygenic phototrophs, requires three ATPs for every two NADPHs to reduce one CO_2 to the level of carbohydrate.

5.4.3 Summary of photosynthesis by green plants, algae, cyanobacteria

$$H_2O + NADP^+ + ADP + P_i \xrightarrow{\text{light and Chl}}$$
$$\tfrac{1}{2}O_2 + NADPH + H^+ + ATP$$

5.5 Efficiency of Photosynthesis

In this section, as an exercise, we calculate the efficiencies of photosynthesis based upon input light energy and products of photosynthesis. The calculated efficiencies are only approximations based upon assumptions regarding ATP yields, actual redox potentials, standard free energies, and so on.

5.5.1 ATP synthesis

Basically, photosynthesis is work done by energized electrons. The work is the phosphorylation of ADP and the reduction of $NAD(P)^+$. Each electron is energized by a photon (quantum) of light, and each mole of electrons by an einstein (6.023×10^{23} quanta) of light. It is instructive to ask how much of this energy is conserved in ATP and NADPH.

Let us consider the synthesis of ATP. Assume that each energized electron results in the translocation of two protons (e.g., during the Q cycle in the bc_1 complex) but that three protons must reenter through the ATP synthase to make one ATP. As mentioned before, a value of $3H^+/ATP$ is reasonable in light of experimental

data. Thus, each energized electron (or each photon) results in the synthesis of two-thirds of an ATP. How much energy is required to synthesize two-thirds of an ATP? The ΔG_p for the phosphorylation of one mole of ADP to make ATP is about 45 kJ. Therefore, $\tfrac{2}{3}$ mol of ATP should require approximately $+45\,(\tfrac{2}{3})$ kJ = 30 kJ. An einstein of 870 nm light, which corresponds to the absorption maximum of bacteriochlorophyll a (found in purple phototrophs), has 138 kJ of energy. Therefore, the efficiency is $(30/138)(100)$, or about 22%. One can also calculate the efficiency by using electron volts instead of joules. The energy in a photon of light at 870 nm is 1.43 eV. The synthesis of $\tfrac{2}{3}$ mol of ATP requires 30,000 J. Dividing this number by the Faraday constant gives the energy in electron volts, (i.e., 0.31 eV).

5.5.2 ATP and NADPH synthesis

What about photosystems that reduce $NADP^+$ as well as make ATP (i.e., photosystems I and II)? The E'_m for O_2/H_2O is +0.82 V, and for $NADP^+/NADPH$ it is –0.32 V. Therefore, the energized electron must have $0.82 - (-0.32)$, or 1.14 eV, to move from water to $NADP^+$. (The answer would not be very different if E_h values were used instead of E'_m.) If two-thirds of an ATP is made, then a total of $0.31 + 1.14 = 1.45$ eV would be required to make two-thirds of an ATP and half an NADPH. A photon of light at a wavelength of 680 nm (the major long-wavelength absorption peak of chlorophyll a in reaction center II) has 1.82 eV. Two photons, or the equivalent of about 3.6 eV, are used. Therefore, approximately 40% of the light energy is conserved as ATP and NADPH.

5.5.3 Carbohydrate synthesis and oxygen production

One can also estimate the approximate efficiency by considering the number of light quanta required to produce oxygen:

$$6CO_2 + 12H_2O \rightarrow C_6H_{12}O_6 + 6H_2O$$
$$+ 6O_2 \qquad \Delta G'_0 = 2{,}870 \text{ kJ}$$

Therefore, for every mole of O_2 produced, the standard free energy requirement at pH 7 is 2870/6, or 478 kJ. The number of einsteins of light required to produce one mole of O_2 is 8. An einstein of 680 nm light carries 176 kJ of energy. Therefore, 176×8, or 1,408 kJ of light energy, is used to produce one mole of O_2. The efficiency is thus $(478/1,408)(100)$ or 34%.

5.6 Photosynthetic Pigments

5.6.1 Light-harvesting pigments

The photosynthetic pigments are divided into two categories: *reaction center pigments* (primarily chlorophylls) and *light-harvesting pigments* (carotenoids, phycobilins, chlorophylls). The light-harvesting pigments are sometimes called accessory pigments or antennae pigments. By far, most of the photosynthetic pigments are light-harvesting pigments. The light-harvesting pigments are critical to photosynthesis because they absorb light of different wavelengths and funnel the energy to the reaction center. Figure 5.8 shows whole-cell absorption spectra of a purple photoysynthetic bacterium and a cyanobacterium, including the absorption wavelengths of the pigments; the various pigments and their absorption peaks are listed in Table 5.2. The

Table 5.2 In vivo long-wavelength absorption maxima of photosynthetic pigments of phototrophic prokaryotes

Pigment	Peak wave length (nm)
Chlorophyll	
a	680
b	675
Bacteriochlorophyll	
a	800–810, 850–910
b	835–850, 1,020–1,035
c	745–760
d	725–745
e	715–725
g	670, 788
Carotenoid	
Chlorobactene	458
Isorenieratene	517
Lycopene, rhodopin	463, 490, 524
Okenone	521
Rhodopinal	497, 529
Spheroidene	450, 482, 514
Spirilloxanthin	486, 515, 552
β-carotene	433, 483
γ-carotene	433, 483
Phycobilins	
Phycocyanin	620–650
Phycoerythrin	560–566

Source: Stolz, J. F. 1991. Structure of Phototrophic Prokaryotes. CRC Press, Boca Ratoy, FL.

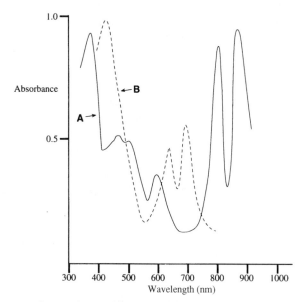

Fig. 5.8 Absorption spectra of a purple photosynthetic bacterium, *Rhodopseudomonas palustris* (A) and a cyanobacterium (B). In spectrum A, bacteriochlorophyll a peaks at 360, 600, 805, and 870 nm. Carotenoid peaks at 450–525 nm. In spectrum B, chlorophyll a peaks at 440 and 680 nm. The phycocyanin peak is around 620 nm. *Source*: Adapted from Brock, T. D., and M. T. Madigan. 1988. *Biology of Microorganisms*. Reprinted by permission of Prentice-Hall, Upper Saddle River, NJ.

nature of the light-harvesting pigments will vary with the type of organism, but some important generalizations about them can be made:

1. They absorb light at wavelengths different from reaction center chlorophyll and therefore extend the range of wavelengths over which photosynthesis is possible because they transfer energy to the reaction center (Table 5.2).

2. In the purple photosynthetic bacteria, they are embedded in the membrane as pigment–protein complexes that are in close physical association with the reaction center.

3. In the green sulfur bacteria and *Chloroflexus*, they exist in chlorosomes, which are separate "organelles," also called inclusion bodies, attached to the inner surface of the cell membrane.

4. In the cyanobacteria (and eukaryotic red algae), the light-harvesting pigments (phycobilins) that transfer energy to PS II are localized in granules called phycobilisomes that are attached to the photosynthetic membranes (also called thylakoids). It should be pointed out that as in other photosynthetic organisms, cyanobacteria also use carotenoids as light-harvesting pigments. Carotenoids have been found in isolated complexes of photosystems I and II

from cyanobacteria, and fluorescence data suggest that carotenoids absorb light and transfer energy to reaction center chlorophyll in PS I.[24] (See Section 5.7.2.) Carotenoids in the reaction centers also serve to protect the chlorophyll from photooxidation. (See note 13.)

Light-harvesting pigments of the purple photosynthetic bacteria

The light-harvesting complexes of the purple photosynthetic bacteria are *bacteriochlorophyll–protein–carotenoid complexes* localized in the cell membrane in close association with the reaction centers. The pigment–protein complexes absorb light in the range of 800 to 1,000 nm (i.e., in the near-infrared range). Upon treatment of the photosynthetic membranes with a mild detergent [e.g., lauryl dimethylamine-*N*-oxide (LDAO)], the light-harvesting complexes can be separated from the reaction centers and analyzed. In *Rhodobacter sphaeroides* there are two light-harvesting complexes (LHC I and LHC II), each containing two polypeptides (α and β) to which the pigments are attached (Fig. 5.9). Analysis of the hydrophobic domains of the polypeptides suggests that they span the membrane. It should be emphasized that for effective energy transfer to take place between

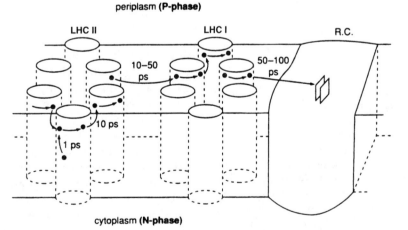

Fig. 5.9 Light-harvesting complexes LHC I and LHC II in the purple nonsulfur photosynthetic bacterium *R. sphaeroides*. An arrow shows transfer of energy to reaction center (RC). Each cylinder represents a polypeptide unit. The small solid circles symbolize bound pigments. Excitation energy begins in the LHC II units and then travels to the LHC I units. The LHC I units transfer the excitation energy to the bacterio-chlorophyll dimer in the reaction center. *Source*: Nicholls, D. G., and S. J. Ferguson. 1992. *Bioenergetics 2*. Academic Press, London.

pigment molecules, they must be positioned correctly with respect to one another (Section 5.7). Organization on protein molecules probably accomplishes this. When light is absorbed by the light-harvesting pigments, the energy is quickly transferred to the reaction center (Fig. 5.9).

Light-harvesting pigments in the green photosynthetic bacteria

The major light-harvesting pigment of the green photosynthetic bacteria is bacteriochlorophyll c, d, or e, depending upon the species (see later: Fig. 5.14B). (The green sulfur bacteria may contain Bchl c and/or d or e, but the green nonsulfur photosynthetic bacteria have only bacteriochlorophyll c.) In addition, there is bacteriochlorophyll a. As with the purple photosynthetic bacteria, the bacteriochlorophylls are associated in some way with carotenoids and protein.

In contrast to the purple bacteria, most of the light-harvesting pigments are not in the cell membrane but rather in *chlorosomes*. The chlorosomes are interesting inclusion bodies, which are about $150 \times 70 \times 30$ nm³, attached to the inner membrane surface by a baseplate, which is attached to the reaction centers (Fig. 5.10). The chlorosomes contain all the bacteriochlorophylls c, d, or e, much of the carotenoid, and some of the bacteriochlorophyll a.

Within the chlorosomes are rod-shaped elements that run lengthwise through the chlorosome; in *Chlorobium*, they are about 10 nm in diameter. The rods are composed of a polypeptide to which the bacteriochlorophyll c, d, or e is thought to be attached. However, some investigators have suggested that the bacteriochlorophyll exists as self-aggregated clusters in the chlorosomes rather than attached to protein. Bacteriochlorophyll a exists as a protein–pigment complex in the baseplate. In addition, there is bacteriochlorophyll a in the membrane which is bound to the reaction centers. Most of the light is absorbed primarily by the bacteriochlorophyll c, d, or e, in the chlorosome, and the energy is transferred to bacteriochlorophyll a in the baseplate. The bacteriochlorophyll in the baseplate then transfers the energy to membrane bacteriochlorophyll a, which transfers the energy to reaction center bacteriochlorophyll a. The green sulfur photosynthetic bacteria have a reaction center based on an iron–sulfur protein and resembling reaction center I of cyanobacteria and chloroplasts; the green nonsulfur (filamentous) photosynthetic bacteria have a quinone-based reaction center resembling that of the purple photosynthetic bacteria.

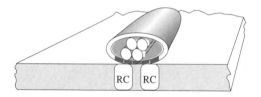

Fig. 5.10 Structure of the chlorosome. The chlorosome is surrounded by a galactolipid layer that contains some protein and is attached to the reaction centers in the cell membrane via a baseplate (shaded area). There is bacteriochlorophyll a in the baseplate and in the membrane. One model for the inside of the chlorosomes postulates rod-shaped proteins to which is attached the major light-harvesting pigment, bacteriochlorophyll c, d, or e. In an alternative model, the bacteriochlorophyll exists as an aggregate not attached to protein. It is suggested that light is absorbed by bacteriochlorophyll in the interior of the chlorosome and the energy is transmitted to bacteriochlorophyll a in the baseplate, then to bacteriochlorophyll a attached to the reaction centers, and from there to reaction center bacteriochlorophyll a.

Light-harvesting pigments in the cyanobacteria

The photosynthetic membranes of cyanobacteria are actually intracellular membrane-bound sacs called thylakoids that have both reaction centers and light-harvesting complexes called *phycobilisomes*.[25,26] The phycobilisomes are numerous granules covering the thylakoids and attached to PS II (Fig. 5.11). They can be removed from the photosynthetic membranes with nonionic detergent (Triton X-100) for purification and analysis. The phycobilisomes consist of proteins called *phycobiliproteins* that absorb light in the range of 450 to 660 nm owing to open-chain tetrapyrroles (phycobilins) covalently bound to the protein via thioether linkages to a cysteine residue in the phycobiliprotein (Fig. 5.12). Examination of the absorption spectra of chlorophylls and carotenoids reveals that the absorbance is relatively poor,

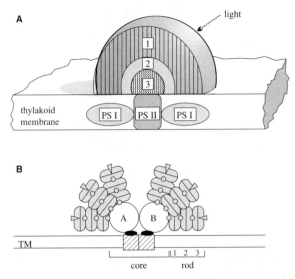

Fig. 5.11 Structure of phycobilisome. (A) Hemispherical phycobilisome common in red algae. These light-harvesting complexes are approximately 48×32 nm^2 $\times 32$ nm high. The major pigment is phycoerythrin. Light energy absorbed by the pigments in the outer region is transferred to the reaction center through an inner core: 1, phycoerythrin; 2, phycocyanin; 3, allophycocyanin. (B) Detailed model of phycobilisome from the cyanobacterium *Synechococcus*. Phycobilisomes from cyanobacteria are generally fan shaped, about 60 nm wide, 30 nm high, and 10 nm thick. The major pigment is phycocyanin. The outermost region of the phycobilisome consists of six cylindrical rods of phycocyanin. Each rod is made from three hexamers of phycocyanin: 1, 2, and 3. (*Synechococcus* does not have phycoerythrin.) The innermost region of the phycobilisome (the core) consists of two closely associated cylinders (A and B) of allophycocyanin. Each cylinder of allophycocyanin consists of two hexamers (not shown). The cross-hatched rectangle within the thylakoid membrane (TM) represents the reaction center and associated chlorophylls. The solid ovals between the core and the thylakoid membrane represent high molecular weight polypeptides thought to anchor the phycobilisome to the thylakoid membrane. There is a linker polypeptide associated with each phycocyanin hexamer that is drawn as an open circle between the hexamers. The open triangle at the terminus of each rod represents a polypeptide that may terminate the rod substructure. *Sources*: Adapted from Gantt, E. 1986. Phycobilisomes, pp. 260–268. In *Photosynthesis III: Photosynthetic Membranes and Light Harvesting Systems*. L. A. Staehelin and C. J. Arntzen (Eds.). Springer-Verlag, Berlin. Grossman, A. R., M. R. Schaefer, G. G. Chiang, and J. L. Collier. 1993. The phycobilisome, a light-harvesting complex responsive to environmental conditions. *Microbiol. Rev.* 57:725–749.

lying in the range (500–600 nm) in which the phycobiliproteins absorb.

There are three classes of phycobiliproteins: *phycoerythrin, phycocyanin*, and *allophycocyanin*. Allophycocyanin is usually at a much lower concentration than phycoerythrin and phycocyanin and, as described shortly, is considered to be an energy funnel to the reaction center. Although most species of cyanobacteria contain both phycoerythrin (red) and phycocyanin (blue), the phycocyanins generally predominate. The phycobiliproteins are arranged in layers around each other with phycoerythrin on the outside, phycocyanin in the middle, and allophycocyanin in the inside, closest to the reaction center.

A simplified drawing of a phycobilisome is shown in Fig. 5.11A. Figure 5.11B represents a detailed model at the molecular level of the phycobilisomes from the cyanobacterium *Synechococcus*, which does not contain phycoerythrin. These are fan-shaped (hemidiscoidal) phycobilisomes that consist of a dicylindrical core (A and B) containing allophycocyanin and six cylindrical phycocyanin rods radiating from the core. (In other cyanobacteria the core may consist of three cylinders.) If phycoerythrin cylinders were present, they would be an extension of the phycocyanin rods. The shorter wavelengths of light are absorbed by the phycoerythrin, the longer wavelengths by phycocyanin, and the still longer wavelengths

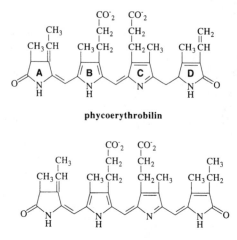

phycoerythrobilin

phycocyanobilin

Fig. 5.12 Structure of phycoerythrobilin and phycocyanobilin. In ring A, the protein is covalently bonded via a thioether linkage between one of its cysteine residues and C2 of $CH_3-CH=$.

(650 nm) by the allophycocyanin. These pigments are thus well positioned to transfer energy via inductive resonance to the reaction center. (See Section 5.7 for a discussion of energy transfer.)

A second kind of light-harvesting complex is located in the thylakoids and transmits energy to the closely associated PS I. It consists of chlorophyll a, carotenoids, and protein.

Light-harvesting pigments in algae

All the algae contain chlorophyll a and the carotenoid β-carotene. In addition, there are more than 60 other carotenoids found among algae, some of which are shown in Fig. 5.13.

1. The *green algae* (Chlorophyta) contain chlorophyll b and xanthophylls (modified carotenes) called lutein.

2. The *brown algae* (Chromophyta) have chlorophyll c_1 and c_2, as well as xanthophylls called fucoxanthin and peridinin.

3. The *red algae* (Rhodophyta) have phycobilins (e.g., phycoerythrobilin) (Fig. 5.12), chlorophylls, carotenes, and xanthophylls.

The pigment–protein complexes can be divided into two classes. One class, the inner antenna β–carotene-Chl a–protein complex, exists as part of photosystems I and II. The second class,

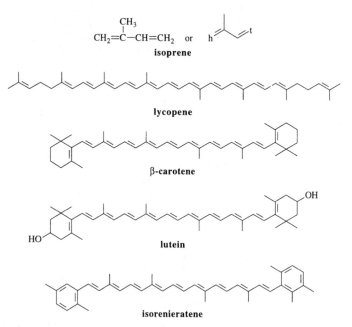

isoprene

lycopene

β-carotene

lutein

isorenieratene

Fig. 5.13 Structure of carotenoids. An isoprene molecule is shown at the top. One end can be called the head (h) and the other the tail (t). All carotenoids are made of isoprene subunits with a "tail-to-tail" connection in the middle. Carotenoids in which the ends have circularized into rings are called carotenes. Carotenes that have been modified (e.g., by hydroxylation) are called xanthophylls. Lycopene is the carotenoid responsible for the red color in tomatoes. β-Carotene is part of the inner antenna complex in all plants and algae. Lutein is a xanthophyll found in algae. Isorenieratene is a carotenoid found in green sulfur bacteria.

the outer antenna pigment–protein class, forms light-harvesting complex I (LHC I) associated with PS I, and light-harvesting complex II (LHC II) associated with PS II. The inner antenna complex that is part of the reaction centers is the same in all algae and higher plants, but LHC I and LHC II vary in composition and structure. The main functions of LHC I and II are to harvest light and to transmit its energy to the pigment complex in the reaction center.

5.6.2 Structures of the chlorophylls, bacteriochlorophylls, and carotenoids

Chlorophylls and bacteriochlorophylls

The chlorophylls are substituted tetrapyrroles related to heme (Fig. 4.5 and Fig. 5.14A). The differences are that chlorophylls have Mg^{2+} in the center of the tetrapyrrole rather than $Fe^{2+(3+)}$, and a fifth ring (ring V) is present. The different substitutions on the pyrrole rings and the number of double bonds in ring II distinguish the chlorophylls from each other. A Zn-bacteriochlorophy a was been discovered

in 1996.[27] It is present in an acidophilic photosynthetic bacterium called *Acidiphilium rubrum*.

Carotenoids

Carotenoids are long *isoprenoids*, usually C_{40} tetraterpenoids made of eight isoprene units (Fig. 5.13). (For other isoprenoids, see note 28.) Carotenoids have a system of conjugated double bonds, which accounts for their absorption spectrum in the visible range at wavelengths poorly absorbed by chlorophylls and makes them valuable as light-harvesting pigments. Light energy absorbed by carotenoids is transferred to the reaction center chlorophyll (Section 5.7). Carotenoid pigments with six-membered rings at both ends of the molecule are called carotenes (e.g., β-carotene, Fig. 5.13). Modification of carotenes (e.g., by hydroxylation) produces xanthophylls, which comprise a major portion of the light-harvesting pigments in algae. Carotenoids perform a function in addition to serving as a light-harvesting pigment. As described in Section 5.2.2 and note 13, they protect against photooxidation.

A

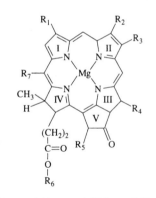

B R	bchl a	bchl b	bchl c	bchl d	bchl e	bchl g	chl a
1	COCH$_3$	COCH$_3$	CHOHCH$_3$	CHOHCH$_3$	CHOHCH$_3$	CH=CH$_2$	CH=CH$_2$
2	CH$_3$	CH$_3$	CH$_3$	CH$_3$	CHO	CH$_3$	CH$_3$
3	CH$_2$CH$_3$	=CHCH$_3$	CH$_2$CH$_3$	CH$_2$CH$_3$	CH$_2$CH$_3$	=CHCH$_3$	CH$_2$CH$_3$
4	CH$_3$	CH$_3$	CH$_2$CH$_3$	CH$_2$CH$_3$	CH$_2$CH$_3$	CH$_3$	CH$_3$
5	COOCH$_3$	COOCH$_3$	H	H	H	COOCH$_3$	COOCH$_3$
6	phytyl	phytyl	farnesyl	farnesyl	farnesyl	farnesyl	phytyl
7	H	H	CH$_3$	H	CH$_3$	H	H

Fig. 5.14 (A) Structure of bacteriochlorophyll. There is no double bond in ring II of bacteriochlorophylls a, b, and g between C3 and C4. (Nitrogen is position 1.) (B) Table identifying the R groups distinguishing the compounds schematiced in (A).

5.7 The Transfer of Energy from the Light-Harvesting Pigments to the Reaction Center

5.7.1 Mechanism of energy transfer

When light is absorbed in one region of the pigment system, excitation energy is transferred to other regions and eventually to the reaction center. The mechanism that probably accounts for the transfer of excitation energy from one pigment complex to another throughout the pigment system is called *inductive resonance transfer*.

Inductive resonance transfer

In an unexcited molecule, all the electrons occupy molecular orbitals having the lowest available energy (the ground state). When a photon of light is absorbed, an electron becomes energized and occupies a higher energy orbital that had been empty. This usually occurs without a change in the spin quantum number (Fig. 5.15). The electron is now in the excited singlet state. The excitation can be transferred from one pigment complex to another via *inductive resonance*. During resonance transfer, electrons do not travel between complexes, but energy is transferred. Essentially, when the electron in the excited molecule drops back to its ground state, the energy is transferred to a molecule close by, resulting an electron in that molecule being raised to the excited singlet state. This can occur if there is an overlap between the fluorescence spectrum of the donor molecule and the absorption spectrum of the acceptor, and if the molecules are in the proper orientation to each other. That is the reason for the critical importance of the steric relationships of the light-harvesting pigments to each other and to the reaction center. It must be emphasized that transfer by this mechanism does not occur as a result of emission and reabsorption of light.

Delocalized excitons

A second method of excitation transfer occurs over very short distances (i.e., <2 nm). When the excited electron is raised to the singlet state it is said to leave a positive "hole" in the ground state (Fig. 5.15). The combination of the excited electron and the positive hole is called an *exciton*. If two or more similar molecules are very close to each other (<2 nm), the exciton migrates over the molecular orbitals belonging to both molecules; that is, the exciton becomes *delocalized*. Thus, the excitation energy is actually *shared* by the group of interacting molecules. However, delocalized excitons in the pigment complexes extend only over very short distances and cannot connect the outer regions of the light-harvesting complex with the reaction center. On the other hand, the sharing of excitation energy by delocalized excitons is more rapid than inductive resonance transfer and is therefore expected to be an important factor for molecules at small intermolecular distances.

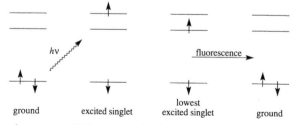

Fig. 5.15 Light raises an electron to a higher energy level in an unoccupied orbital. The electron is then said to be in the excited singlet state. As the diagram shows, the electron may drop down to a lower excited singlet state. The electron may return to the ground state, releasing energy as light (fluorescence) or heat. Under appropriate conditions, the energy made available when the electron returns to the ground state is not released but instead is used to raise an electron in a nearby pigment to an excited singlet state, thereby transferring the energy in the process called inductive resonance. In the reaction center, electrons raised to a higher energy orbital are transferred to a nearby acceptor molecule, creating a redox gradient.

5.7.2 Evidence that energy absorbed by the light-harvesting pigments is transferred to the reaction center

When isolated chloroplasts are irradiated, fluorescence from reaction center chlorophyll (chlorophyll a) can be measured, but fluorescence from the light-harvesting pigments (chlorophyll b, carotenoids) cannot. This means that the energy absorbed by the light-harvesting pigments is efficiently transferred to the reaction center chlorophyll, whereupon an electron in the reaction center chlorophyll becomes excited to a higher energy level. In the absence of fluorescence, the energized electron can reduce the primary electron acceptor. It has been suggested that the initial redox reaction is initiated in the reaction center chlorophyll by an electron in the lowest excited singlet state.[29]

5.8 The Structure of Photosynthetic Membranes in Bacteria

The anoxygenic phototrophic bacteria have photosynthetic membrane structures of a variety of types, depending upon the organism (Fig. 5.16). For example, the purple photosynthetic bacteria have numerous intracellular photosynthetic membranes in which the light-harvesting pigments and the reaction centers are closely associated. The green bacteria separate most of the light-harvesting pigments into chlorosomes.

5.9 Summary

When the energy from a photon of light reaches reaction center chlorophyll, an electron in the chlorophyll becomes energized and is transferred to a primary acceptor molecule within the reaction center. This leaves the chlorophyll in an oxidized form. The fate of the electron lost by the primary donor differentiates two different types of reaction center. One type of reaction center reduces quinone. Quinone-reducing reaction centers are the reaction centers in purple photosynthetic bacteria, and reaction center II in chloroplasts and cyanobacteria. In the photosynthetic bacteria, the quinone is ubiquinone. In chloroplasts and cyanobacteria,

it is plastoquinone. A second type of reaction center reduces an FeS protein. The latter reaction centers are reaction center I in chloroplasts and cyanobacteria, and the reaction center in green sulfur bacteria and heliobacteria.

In the purple photosynthetic bacteria the reduced quinone transfers the electrons to a bc_1 complex. The bc_1 complex reduces cytochrome c_2, which returns the electron to the reaction center. This is called cyclic electron flow. A Δp is generated by the bc_1 complex.

In cyanobacteria and chloroplasts, the reduced quinone leaves reaction center II and reduces the b_6f complex, which is similar to the bc_1 complex. The b_6f complex reduces plastocyanin, which transfers the electron to reaction center I. A Δp is established by the b_6f complex. Having lost an electron, reaction center II accepts an electron from water, thus producing oxygen as a by-product of the light reactions in the cyanobacteria and in chloroplasts. Reaction center I is also energized by light and transfers the electron to ferredoxin. The ferredoxin reduces $NADP^+$. The combination of reaction centers I and II is called noncyclic electron flow. In noncyclic flow, both ATP and NADPH are synthesized. Cyclic flow is also possible as the electron returns to reaction center I from reduced ferredoxin.

The green sulfur photosynthetic bacteria have a reaction center similar to reaction center I which reduces ferredoxin and NAD^+. The electron donors are inorganic sulfur compounds (e.g., sulfide, elemental sulfur). The electron can return to the reaction center via a quinone and a bc_1 complex in cyclic flow rather than reducing NAD^+.

Most of the light energy used in photosynthesis is not absorbed by reaction centers directly but is instead absorbed by light-harvesting pigments organized either within the membrane and closely positioned to the reaction centers, in special inclusions called chlorosomes attached to the membrane, or in granules called phycobilisomes attached to the membrane. There is considerable variability in the composition and structure of the light-harvesting complexes. However, they all consist of pigment–protein complexes that transfer energy to the reaction center. The mechanism of energy transfer between the pigment molecules demands that they be positioned very close to each other in a

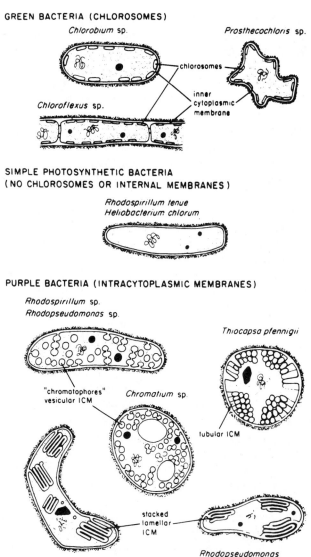

Fig. 5.16 Photosynthetic membranes in anoxygenic phototrophic bacteria. The green bacteria (*Chlorobium Chloroflexus*, *Prosthecochloris*) contain chlorosomes attached to the cytoplasmic surface of the cell membrane. Most of the purple photosynthetic bacteria have extensive invaginations of the cell membrane which can take the form of stacked lamellae (*Ectothiorhodospira* sp., *Rhodopseudomonas palustris*, *R. viridis*), vesicles (*Rhodospirillum* sp., *Rhodopseudomonas* sp.), or tubules (*Thiocapsa pfennigii*). A few photosynthetic bacteria do not have internal membranes or chlorosomes (*Rhodospirillum tenue*, *Heliobacterium chlorum*). *Source*: Sprague, S. G., and A. R. Varga. 1986. Membrane architecture of anoxygenic photosynthetic bacteria, pp. 603–619. In: *Photosynthesis III: Photosynthetic Membranes and Light Harvesting Systems*. L. A. Staehelin, and C. J. Arntzen (Eds.). Springer-Verlag, Berlin.

precise orientation. The proteins in the light-harvesting complexes and in the reaction centers fulfill this function.

Study Questions

1. Measuring oxygen production per quantum of absorbed light reveals that the efficiency falls when monochromatic light of 700 nm is used. The efficiency can be restored if the 700 nm light is supplemented with low-intensity light of shorter wavelength (e.g., 600 nm). Explain these results in terms of the light reactions of oxygenic photosynthesis.

2. Describe the similarities and differences between reaction centers I and II.

3. When bacteriochlorophyll absorbs a photon of 870 nm light (about 1.4 eV of energy), an electron travels out of the reaction center, through a bc_1 complex, and back to the reaction center (cyclic flow) with the extrusion of two protons to the outer membrane surface at the site of the bc_1 complex. Protons return to the cytoplasm through the ATP synthase with 3 as the H^+/ATP ratio. Assuming that the ΔG_p is 45 kJ, what is the efficiency of photophosphorylation?

4. For all anoxygenic photosynthetic bacteria except the green sulfur bacteria, electrons from the reductant do not pass through the reaction center. This means that light does not stimulate electron flow from the reductant to NAD^+. But many of the reductants have reduction potentials more positive than that of the $NAD^+/NADH$ couple. What energizes electron flow to NAD^+? Use ionophores to devise an experiment that would support your conclusion.

5. Energy from light absorbed by the accessory pigments is transferred to the reaction center. What is the evidence for that?

6. Explain how the absorption of light by photosynthetic membranes creates a ΔE. Explain how the ΔE is converted into a Δp. Explain how the Δp is used to drive ATP synthesis.

7. Why are two light reactions required for oxygenic photosynthesis but only one light reaction for anoxygenic photosynthesis?

REFERENCES AND NOTES

1. Fleischman, D. E., W. R. Evans, and I. M. Miller. 1995. In: *Anoxygenic Photosynthetic Bacteria*, pp. 123–136. R. E. Blankenship, M. T. Madigan, and C. E. Bauer (Eds.). Kluwer Academic Publishers, Dordrecht.

2. Kramer, D. M., A. Kanazawa, and D. Fleischman. 1997. Oxygen dependence of photosynthetic electron transport in a bacteriochlorophyll-containing rhizobium. *FEBS Lett.* **417**:275–278.

3. Reviewed in: *The Molecular Biology of Cyanobacteria*. 1995. D. A. Bryant (Ed.). Kluwer Academic Publishers, Boston.

4. These groupings do not reflect the complex taxonomy of the photosynthetic bacteria as revealed by ribosomal RNA sequencing. There are five distinct evolutionary lines of photosynthetic prokaryotes (see Fig. 1.1 and Table 1.1). The purple photosyn-

thetic bacteria are a heterogeneous assemblage that are part of the division of prokaryotes known as the *Proteobacteria* (also called the *purple bacteria*). The Proteobacteria are subdivided into four subdivisions, the $\alpha, \beta, \gamma,$ and δ groups. Photosynthetic bacteria, along with nonphotosynehetic bacteria, are in the $\alpha, \beta,$ and γ subdivisions, whereas the δ subdivision contains only nonphotosynthetic bacteria. The purple nonsulfur photosynthetic bacteria are in the α and β subdivisions, whereas the purple sulfur bacteria are in the γ subdivision. In the α group are *Rhodospirillum, Rhodopseudomonas, Rhodobacter, Rhodomicrobium,* and *Rhodopila,* all of which are nonsulfur purple bacteria. Another nonsulfur purple bacterium, *Rhodocyclus,* is in the β group. The γ group contains purple sulfur bacteria, including *Chromatium* and *Thiospirillum.* The *green sulfur bacteria,* including *Chlorobium* and *Chloroherpeton,* are only distantly related to the purple bacteria and are in a distinctly separate evolutionary line. The *green nonsulfur bacteria* (the *Chloroflexus* group) are in a separate evolutionary line, phylogenetically distinct from the purple photosynthetic bacteria and the green sulfur bacteria. The *cyanobacteria* occupy a fourth evolutionary line. The photosynthetic *heliobacteria,* which consist of *Heliobacterium, Heliospirillum,* and *Heliobacillus,* are grouped in the evolutionary line that consists of the gram-positive bacteria.

5. Bullerjahn, G. S., and A. F. Post. 1993. The prochlorophytes: are they more than just chlorophyll a/b-containing cyanobacteria? *Crit. Rev. Microbiol.* **19**:43–59.

6. Olson, J. M. 1998. Chlorophyll organization and function in green photosynthetic bacteria. *Photochem. Photobiol.* **67**:61–75.

7. Dutton, P. L. 1986. Energy transduction in anoxygenic photosynthesis, pp. 197–237. In: *Photosynthesis III: Photosynthetic Membranes and Light Harvesting Systems.* L. A. Staehelin, and C. J. Arntzen (Eds.). Springer-Verlag, Berlin.

8. Mathis, P. 1990. Compared structure of plant and bacterial photosynthetic reaction centers. Evolutionary implications. *Biochim. Biophys. Acta* **1018**:163–167.

9. Green nonsulfur photosynthetic bacteria have a reaction center resembling that of the purple photosynthetic bacteria. The primary donor of the electron, called P_{865}, is a dimer of Bchl a. The initial electron acceptor is bacteriopheophytin (Bpheo). The two quinones (Q_A and Q_B) are menaquinones rather than ubiquinones.

10. Deisenhofer, J., H. Michel, and R. Huber. 1985. The structural basis of photosynthetic light reactions in bacteria. *Trends Biochem. Sci.* **10**:243–248.

11. Komiya, H., T. O. Yeates, D. C. Rees, J. P. Allen, and G. Feher. 1988. Structure of the reaction center from *Rhodobacter sphaeroides* R-26 and 2.4.1: symmetry relations and sequence comparisons between different species. *Proc. Natl. Acad. Sci.* USA **85**:9012–9016.

12. Cogdell, R. J. 1978. Carotenoids in photosynthesis. *Philos. trans. R. Soc. Lond. B* **284**:569–579.

13. During photosynthesis under high light intensities, an electron in bacteriochlorophyll can go from the excited singlet state to the lower energy triplet state (Section 5.7). In the triplet state, the spin of the excited electron has changed, so that there are now two unpaired electrons as opposed to paired electrons in the singlet state. (Phosphorescence occurs from the triplet state.) When triplet state bacteriochlorophyll reacts with oxgyen, the oxygen becomes energized to its first excited state, singlet oxygen, $^1O_2$. Singlet oxygen is very reactive and combines with cell molecules such as unsaturated fatty acids, amino acids, and purines, thus causing oxidative damage to various cell components, including lipids, enzymes, and nucleic acids. The carotenoid in the reaction center quenches the triplet state in bacteriochlorophyll. Thus, carotenoids prevent photooxidation under high light intensities. When the carotenoid absorbs the energy from the triplet bacteriochlorophyll, the carotenoid itself becomes energized to the triplet state (triplet–triplet energy transfer). However, the energy in triplet state carotenoid is dissipated harmlessly as heat to the medium. Because the triplet state energy of carotenoids is below the singlet state energy of oxygen, carotenoids will also quench singlet oxygen.

14. Carotenoids are isoprenoid membrane pigments. They are useful in detecting changes in membrane potential. The energy levels of electrons in the conjugated-double-bond system of the carotenoids is altered by the electric field of the membrane potential. When the membrane potential changes, the carotenoid undergoes a rapid shift in its absorption spectrum peaks. This can be followed by using split-beam spectroscopy similar to that discussed for examining spectral changes of cytochromes during oxidation–reduction reactions (Section 4.2.4).

15. Nitschke, W., U. Feiler, and A. W. Rutherford. 1990. Photosynthetic reaction center of green sulfur bacteria studied by EPR. *Biochemistry* **29**:3834–3842.

16. Miller, M., X. Liu, S. W. Snyder, M. C. Thurnauer, and J. Biggins. 1992. Photosynthetic electron-transfer reactions in the green sulfur bacterium *Chlorobium vibrioforme*: evidence for the functional involvement of iron–sulfur redox centers on the acceptor side of the reaction center. *Biochemistry* **31**:4354–4363.

17. Meent, van de E. J., M. Kobayashi, C. Erkelens, P. A. van Veelen, S. C. M. Otte, K. Inoue, T. Watanabe, and J. Amesz. 1992. The nature of the primary electron acceptor in green sulfur bacteria. *Biochim. Biophys. Acta* **1102**:371–378.

18. The reaction centers of the green sulfur bacteria, heliobacteria, and photosystem I of chloroplasts differ from reaction center II and the reaction center in purple photosynthetic bacteria in producing low-potential reduced FeS proteins, rather than reducing quinone.

19. Reviewed in: ref. 7.

20. Emerson, R., and C. M. Lewis. 1943. The dependence of the quantum yield of *Chlorella* photosynthesis on wave length of light. *Am. J. Bot.* **30**:165–178.

21. Emerson, R., Chalmers, R., and Cederstrand, C. 1957. Some factors influencing the long-wave limit of photosynthesis. *Proc. Natl. Acad. Sci. USA* **43**:133–143.

22. The b_6f complex is similar to the bc_1 complex. It contains four proteins: cyt b_6, subunit IV, the Rieske Fe–S protein, and cyt f. There are four prosthetic groups: two b-type hemes, one c-type heme (cytochrome f), and a 2Fe–2S center. The b_6f complex functions in respiration as well as in photosynthetic electron transport in the cyanobacteria. See the review by Kallas, T. 1995. The cytochrome b_6f complex, pp. 259–317. In: *The Molecular Biology of Cyanobacteria*. D. A. Bryant (Ed.). Kluwer Academic Publishers. Boston.

23. Knaff, D. B. 1993. The cytochrome bc_1 complexes of photosynthetic purple bacteria. *Photosynth. Res.* **35**:117–133.

24. Hirschberg, J., and D. Chamovitz. 1995. Carotenoids in cyanobacteria, pp. 559–579. In: *The Molecular Biology of Cyanobacteria*. D. A. Bryant (Ed.). Kluwer Academic Publishers, Boston.

25. Grossman, A. R., M. R. Schaefer, G. G. Chiang, and J. L. Collier. 1993. The phycobilisome, a light-harvesting complex responsive to environmental conditions. *Microbiol. Rev.* **57**:725–749.

26. Gantt, E. 1986. Phycobilisomes, pp. 260–268. In: *Photosynthesis III: Photosynthetic Membranes and Light-Harvesting Systems*. L. A. Staehelin, and C. J. Arntzen (Eds.). Springer-Verlag, Berlin.

27. Wakao, N., N. Yokoi, N. Isoyama, A. Hiraishi, K. Shimada, M. Kobayashi, H. Kise, M. Iwaki, S. Itoh, S. Takaichi, and Y. Sakurai. 1996. Discovery of natural photosynthesis using Zn-containing bacteriochlorophyll in an aerobic bacterium *Acidiphilium rubrum*. *Plant Cell Physiol.* **37**:889–893.

28. Isoprenoids are an important class of molecule. Other isoprenoids are the side chains of quinones, the phytol chain in chlorophyll, and the retinal chromophore in bacteriorhodopsin, halorhodopsin, and rhodopsin.

29. Diner, B. A. 1986. Photosystems I and II: structure, proteins, and cofactors, pp. 422–436. In: *Photosynthesis III: Photosynthetic Membranes and Light-Harvesting System*. L. A. Staehelin, and C. J. Arntzen (Eds.). Springer-Verlag, Berlin.

6

The Regulation of Metabolic Pathways

Bacteria catalyze a very large number of chemical reactions (approximately 1000–2000) that are organized into interconnecting metabolic pathways. To avoid inefficiency, if not chaos, it is of the utmost importance that there be coordination between the pathways. *Pathways are regulated by adjusting the rate of one or more regulatory enzymes that govern the overall rate of the pathway.*

The rates at which regulatory enzymes operate are modified in two ways. One method is by the noncovalent binding to the enzyme of certain biochemical intermediates of the pathways (called *allosteric effectors*) that either stimulate or inhibit the regulatory enzyme, thus signaling to the regulatory enzyme whether the pathway is producing optimal amounts of intermediates (Sections 6.1.1 and 6.1.2). Regulatory enzymes are also called *allosteric* enzymes because in addition to having binding sites for their substrates, they also have separate binding sites for the effector molecules. (The Greek prefix *allo*, meaning "other," indicates that allosteric enzymes have a second, "other," site besides the substrate site.) A second method of altering enzyme activity is by the *covalent modification* of the enzyme (Section 6.4). Covalent modifications include the attachment and removal of chemical groups such as phosphate and nucleotides.

6.1 Patterns of Regulation of Metabolic Pathways

In addition to learning the mechanisms of enzyme regulation (allosteric or covalent modification), the student should be aware of the *patterns* of regulation. What is meant by "patterns of regulation" will be made clear shortly. Patterns of regulation will vary, *even within the same pathway in different organisms.* Nevertheless, some common patterns of metabolic regulation have evolved. These are described next.

6.1.1 Feedback inhibition by an end product of the pathway

For biosynthetic pathways, the end product is usually a negative allosteric effector for a branch point enzyme. For example, in Fig. 6.1, F and J are negative effectors for the regulatory enzymes 1 and 2, respectively. Such control is called *end-product inhibition, or feedback inhibition*, by an end product. Its role is to maintain a steady state in which the end product is utilized as rapidly as it is synthesized. When the rates of utilization of the end product increase, the concentration of the end product decreases. The decrease in concentration relieves

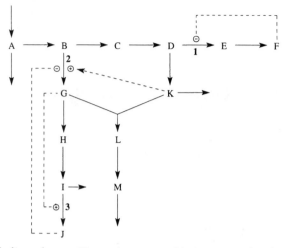

Fig. 6.1 Branched metabolic pathways. The pathways provide precursors to other pathways at the branch points. Both positive and negative regulation occurs. Here F and J are negative effectors that prevent their own overproduction by inhibiting the regulatory enzymes, 1 and 2. The positive effector K stimulates the production of G (in a different pathway), which is needed to react with K to form L. The precursor G activates a later reaction (enzyme 3) in its own pathway.

the inhibition, and the regulatory enzyme speeds up, resulting in more end-product synthesis to meet the demands dictated by more rapid utilization (e.g., during rapid growth).

There are three recognized patterns of feedback inhibition in biosynthetic pathways: *simple*, *cumulative*, and *concerted* (Fig. 6.2). Simple feedback inhibition describes what happens when the regulatory enzyme is inhibited by a single end product (Fig. 6.2A). It is encountered in linear biosynthetic pathways. In the cumulative and concerted types of inhibition, seen in branched pathways, more than one end product inhibits the enzyme (Fig. 6.2B). In concerted inhibition both end products must bind to the regulatory enzyme simultaneously to achieve any inhibition (Fig. 6.2B). In cumulative inhibition the enzyme is not completely inhibited by any single end product. For example, one end product might inhibit the enzyme by 25% and a second might inhibit the enzyme by 45%. Both end products together might inhibit the enzyme by 60% (not 70%) (Fig. 6.2B). Thus the inhibition is cumulative, but not necessarily additive. Cumulative or concerted inhibition is necessary because branched pathways may share a common regulatory enzyme prior to the point that leads to the separate branches of the pathway. Under these circumstances it is important that a single end product not be able to shut down the activity of the common enzyme, thus preventing the synthesis of all the end products. However, sometimes branched biosynthetic pathways are regulated by simple feedback inhibition. This may occur when a reaction shared by the various branchs uses enzymes that have the same catalytic sites but different effector sites

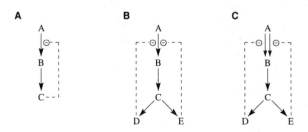

Fig. 6.2 Patterns of feedback inhibition. (A) Simple feedback inhibition in an unbranched pathway. (B) Concerted or cumulative inhibition. In concerted inhibition, there is no inhibition unless both D and E bind. In cumulative inhibition D and E independently exert partial inhibition, but the combination is less than additive. (C) Simple inhibition in a branched pathway by means of isoenzymes.

(isoenzymes) (Fig. 6.2C). (Thus, isoenzymes catalyze the same reaction but are responsive to different end products.) It can be seen that different regulatory patterns (cumulative, concerted, isoenzymes) can produce similar results (i.e., the control of the rate of the pathway by the levels of one or more of the end products).

It should not be concluded that feedback inhibition applies only to biosynthetic pathways. Catabolic pathways, including those for the breakdown of sugars and carboxylic acids, may also be subject to feedback inhibition by an end product.

6.1.2 Positive regulation

A metabolic pathway can also be positively regulated, sometimes by an intermediate in a second pathway, as shown in Fig. 6.1, where one pathway produces an intermediate, G, which combines with K in a second pathway to produce L. If insufficient G is formed, K accumulates and stimulates the enzyme that produces G. An example of this is the activation of PEP carboxylase by acetyl–CoA described in Section 8.9.1.

In another pattern of positive regulation, sometimes called "precursor activation," a precursor intermediate stimulates a regulatory enzyme "downstream" in the same pathway. This ensures that the rate of the downstream reactions matches that of the upstream reactions. In Fig. 6.1, G is a precursor that activates enzyme 3, which converts I to J. An example of precursor activation is the activation of pyurvate kinase by fructose-1,6-bisphosphate described in Section 8.1.2. We will encounter other examples of both positive and negative regulation by multiple effectors when we discuss the regulation of the enzymes of central metabolism in Chapter 8.

6.1.3 Regulatory enzymes catalyze irreversible reactions at branch points

There are certain generalizations that one can make about regulatory enzymes and the reactions that they catalyze. *The reactions catalyzed by regulatory enzymes are usually at a metabolic branch point.* For example, in Fig. 6.1, the intermediates B, D, and I are at branch points. Also, *regulatory enzymes often*

catalyze reactions that are physiologically irreversible (i.e., far from equilibrium). Thus they are poised to accelerate in one direction when stimulated.

6.2 Kinetics of Regulatory and Nonregulatory Enzymes

To appreciate how effector molecules alter the activities of regulatory enzymes, one must understand enzyme kinetics and the enzyme kinetic constants, K_m and V_{max}. This is best done by beginning with the kinetics of nonregulatory enzymes.

6.2.1 Nonregulatory enzymes

A plot of the rate of formation of product (or disappearance of substrate, S) as a function of substrate concentration for most enzyme-catalyzed reactions generates a hyperbolic curve similar to the one shown in Fig. 6.3. As the substrate concentration [S] is increased, the rate of the reaction approaches a maximum, V_{max}, because the enzyme becomes saturated with substrate. For each enzyme there is a substate concentration that gives $\frac{1}{2}V_{max}$. This is called the *Michaelis–Menten constant*, or K_m. The equation that describes the kinetics of enzyme activity is called the Michaelis–Menten equation (eq. 6.1):

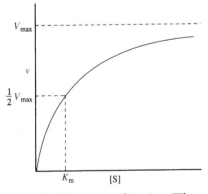

Fig. 6.3 Michaelis–Menten kinetics. When substrate concentrations [S] are plotted against initial velocity (*v*), a hyperbolic curve is obtained that approaches a maximum rate (V_{max}). The substrate concentration that yields $\frac{1}{2}V_{max}$ is a constant for each enzyme and is called the Michaelis–Menten constant or K_m.

$$v = (V_{max}S)/(K_m + S) \qquad (6.1)$$

When the substrate concentration, S, is very small compared with K_m, then the initial velocity, v, is proportional to S (actually to $V_{max}S/K_m$). However, when S becomes much larger than K_m, then S cancels out and v approaches V_{max}. There are two units for V_{max} (i.e., specific activity and turnover number). The units of specific activity are micromoles of substrate converted per minute per milligram of protein, and the units for turnover number are micromoles of substrate converted per minute per micromole of enzyme.

Derivation of the Michaelis–Menten equation

In 1913 L. Michaelis and M. L. Menten formulated a theory of enzyme action that explained the kinetics shown in Fig. 6.3. The following derivation of the Michaelis–Menten equation was developed later by Briggs and Haldane. Consider a substrate (S) being converted to a product (P) in an enzyme-catalyzed reaction. During the reaction, S combines with the enzyme (E) at the substrate site on the enzyme to form the enzyme–substrate complex (ES), which then breaks down to enzyme and product. For the derivation of Briggs and Haldane it is necessary to assume a steady state in which the enzyme is present in catalytic amounts: $S \gg$ total enzyme. Under these conditions, the rate of formation and breakdown of ES are equal, which results in a steady state level of ES. The situation can be summarized as follows:

$$S + E \text{ (free)} \underset{k_2}{\overset{k_1}{\rightleftharpoons}} (ES) \underset{k_4}{\overset{k_3}{\rightleftharpoons}} E + P$$

Notice that there are four rate constants, k_1, k_2, k_3, and k_4. If one were to measure the initial velocity (v), then the rate of the reaction (e.g., the rate of formation of P or disappearance of S) would be proportional to $k_3(ES)$. In this treatment we are ignoring the formation of (ES) from E and P, and k_4. This would be small, in any event, if the initial rates were measured before P had accumulated. If all of the enzyme were bound to S, then all of E would be in the form of (ES), and the rate would be the maximum rate. These relationships are shown in eqs. 6.2 and 6.3

$$v = dP/dt = k_3(ES) \qquad (6.2)$$

and, substituting total enzyme, E_t, for E, when all of the enzyme is saturated with substrate (eq. 6.3),

$$V_{max} = k_3(E_t S) \qquad (6.3)$$

(For enzymes that may have complex reaction pathways that depend upon more rate constants than simply k_3, the symbol k_{cat} is used for the maximal catalytic rate: i.e., $V_{max} = k_{cat}(E_t S)$. See note 1.

The instantaneous rate of formation of (ES) is given by eq. 6.4:

$$d(ES)/dt = k_1(E_t - ES)(S) \qquad (6.4)$$

The concentration of free enzyme (E) is $E_t - (ES)$, where S represents the substrate concentration.

Equation 6.4 describes the instantaneous rate of a bimolecular reaction between the substrate whose concentration is S, and the free enzyme whose concentration is $E_t - ES$. Now we must consider the rate of breakdown of ES. This is given by eq. 6.5:

$$-d(ES)/dt = k_2(ES) + k_3(ES) \qquad (6.5)$$

Let us assume the reaction reaches a steady state at the time of measurement. This is ensured by using substrate concentrations that are in large excess over the total amount of enzyme. In the steady state, the rate of formation of ES is equal to its rate of breakdown, therefore,

$$k_1(E_t - ES)(S) = k_2(ES) + k_3(ES) \qquad (6.6)$$

which on rearrangement gives

$$(S)(E_t - ES)/(ES) = (k_2 + k_3)/k_1 = K_m \qquad (6.7)$$

The constant K_m is called the Michaelis–Menten constant. Note that if k_3 is small compared with k_2 (i.e., if the rate of product formation is small with respect to the rate that the substrate dissociates from the enzyme), then the K_m is approximately equal to k_2/k_1, that is, the dissociation constant for the enzyme (K_D). This is true for some enzyme–substrate combinations. It is not always true, however, and it is a mistake to assume (as is often done) that $1/K_m$ is a direct measure of the affinity of a substrate for an enzyme. It is best to refer to the K_m as being equal to the substrate concentration that gives half maximal velocity, as described in Fig. 6.3.

Now we can solve for (ES):

$$(ES) = (E_t)(S)/(K_m + S), \qquad (6.8)$$

where (E_t) is total enzyme. Since the initial rate (v) is proportional to $k_3(ES)$, we can write eq. 6.9:

$$v = k_3(E_t)(S)/(K_m + S) \qquad (6.9)$$

Since $V_{max} = k_3(ES)$ when $(ES) = E_t$, it is also true that $V_{max} = k_3(E_t)$. Equation 6.9 then becomes eq. 6.10:

$$v = (V_{max}S)/(K_m + S) \qquad (6.10)$$

Equation 6.10 is the Michaelis–Menten equation that describes the kinetics in Fig. 6.3. Typical K_m values range from 10^{-4} M $(100 \, \mu\text{M})$ and to lower molarities.

The K_m is the substrate concentration that gives $\frac{1}{2}V_{max}$

If one substitutes $\frac{1}{2}V_{max}$ for v in the Michaelis–Menten equation and solves for S, then S is equal to K_m. That is, the K_m is equal to the substrate concentration that gives $\frac{1}{2}V_{max}$. This is shown in eq. 6.11:

$$\tfrac{1}{2}V_{max} = V_{max}(S)/(K_m + S) \qquad (6.11)$$

Divide both sides by V_{max}:

$$\tfrac{1}{2}V = S/(K_m + S) \qquad (6.12)$$

Rearrange:

$$K_m + S = 2S \qquad (6.13)$$

and,

$$K_m = S \qquad (6.14)$$

Because the V_m and therefore the K_m can only be approximated from the plot shown in Fig. 6.3, eq. 6.10 is frequently written as the reciprocal:

$$1/v = 1/V_{max} + (K_m/V_{max})(1/S) \qquad (6.15)$$

and $1/v$ is plotted against $1/[S]$ to give a *double reciprocal*, or *Lineweaver–Burk*, plot (Fig. 6.4). One can find K_m/V_{max} from the slope, or $-1/K_m$ from the intercept on the y axis, and $1/V_{max}$ from the intercept on the x axis.

6.2.2 Regulatory enzymes

Regulatory enzymes seldom follow simple Michaelis–Menten kinetics. Instead, they typi-

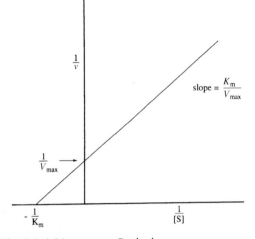

Fig. 6.4 A Lineweaver–Burk plot.

cally show sigmoidal kinetics (Fig. 6.5). An explanation for substrate-dependent sigmoidal kinetics is that the binding of one substrate molecule increases the affinity of the enzyme for a second substrate molecule or increases the rate of formation of product from sites already occupied. This is called *positive cooperativity*. Positive cooperativity makes sense for a regulatory enzyme because it makes enzyme catalysis very sensitive to small changes in substrate level concentrations when the substrate concentrations are very small in comparison to the K_m. (Notice in Fig. 6.5 how the enzyme rate rapidly changes with small changes of substrate at critical concentrations.)

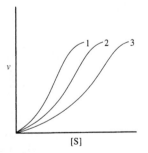

Fig. 6.5 Sigmoidal kinetics of regulatory enzymes. Initial velocities (v) are plotted against substrate concentrations [S] in the presence of a positive effector (curve 1), no effector (curve 2), or a negative effector (curve 3). The positive effector decreases the K_m as well as decreasing the sigmoidicity of the curve. The negative effector increases the K_m and the sigmoidicity of the curve. In other cases, the effectors may change the V_{max}.

Allosteric effectors (positive and negative) can also bind cooperatively to enzymes, making the enzyme more sensitive to small changes in effector concentration. Even though sigmoidal kinetics is not explained by the Michaelis–Menten equation, one can still measure the K_m and V_{max} because these are defined operationally. When one measures these "constants" for regulatory enzymes, one finds that they are subject to change, depending upon the presence or absence of effector molecules. Some of the effectors raise the K_m and some lower the K_m. An effector that raises the K_m will decrease the velocity of the reaction, and one that lowers the K_m will increase the velocity of a reaction, provided the enzyme is not saturated with substrate. This is shown in Fig. 6.5. Notice that the negative effector (curve 3) increases the sigmoidicity of the curve and also increases the K_m. The positive effector (curve 1) decreases the sigmoidicity of the curve and lowers the K_m. Effectors can also change the V_{max}. However, in Fig. 6.5 the V_{max} is not changed.

6.3 Conformational Changes in Regulatory Enzymes

When the effector binds to the allosteric site (the effector site), the protein undergoes a conformational change, and this changes its kinetic constants. Many of the regulatory enzymes are *oligomeric*, or *multimeric* (i.e., they have multiple subunits). However, for illustrative purposes we will first consider a monomeric polypeptide (Fig. 6.6). Assume that the polypeptide has three binding sites: one for the substrate (substrate site), a second for a positive effector, and a third for a negative effector. Further assume that the enzyme exists in two states, A and B. When the enzyme is in conformation A, it binds the substrate and also binds the positive effector, which locks the enzyme in state A. You might say that state A has a low K_m. State B has a high K_m and also binds the positive effector poorly. But state B does bind the negative effector, which locks it into state B. Thus, the positive effector lowers the K_m and the negative effector raises the K_m. In the cell, there is an interplay of positive and negative effectors that adjust the ratio of active and less active enzyme.

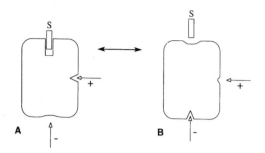

Fig. 6.6 Conformational changes in a monomeric regulatory enzyme with three binding sites: one for the substrate (S), one for the positive effector (+), and one for the negative effector (–). (A) When the enzyme binds the positive effector, a conformational change occurs that lowers the K_m. (B) When the enzyme binds to the negative effector, there is an increase in the K_m and a loss of affinity for the positive effector. In the presence of a sufficient concentration of positive effector, essentially all of the enzyme will be in conformation A and the enzyme will be turned "on." On the other hand, if sufficient negative effector is present, the enzyme will be mostly in conformation B and turned "off." The fraction of enzyme in the "on" or "off" conformation will depend upon the relative concentrations of substrate, positive effector, and negative effector. Multimeric enzymes are regulated in a similar fashion except that the effector-binding sites can be on a subunit (regulatory subunit) separate from the substrate binding site (the catalytic subunit). Conformational changes in one subunit induce conformational changes in the attached subunit.

Regulatory subunits

Many regulatory enzymes are multimeric. Some consist of a regulatory subunit and one or more catalytic subunits. The catalytic subunits bear the active sites and bind the substrate. The regulatory subunits bind the effectors. When the regulatory subunits bind the effector, the polypeptide undergoes a conformational change induced by the binding. The regulatory subunit then either inhibits or activates the catalytic subunit, resulting in changes in the kinetic constants.

6.4 Regulation by Covalent Modification

Although most regulatory proteins appear to be regulated by conformational changes induced

Table 6.1 Covalent modifications of bacterial enzymes and other proteins

Enzyme	Organism	Modification
Glutamine synthetase	*E. coli* and others	Adenylylation
Isocitrate lyase	*E. coli* and others	Phosphorylation
Isocitrate dehydrogenase	*E. coli* and others	Phosphorylation
Chemotaxis proteins	*E. coli* and others	Methylation
P_{II}	*E. coli* and others	Uridylylation
Ribosomal protein L7	*E. coli* and others	Acetylation
Citrate lyase	*Rhodopseudomonas gelatinosa*	Acetylation
Histidine protein kinase	Many bacteria	Phosphorylation
Phosphorylated response regulators	Many bacteria	Phosphorylation

Source: Neidhardt, F. C., J. L. Ingraham, and M. Schaechter. 1990. *Physiology of the Bacterial Cell*. Sinauer Associates, Sunderland, MA.

by the binding of allosteric effectors as just described, there are many important instances in both prokaryotes and eukaryotes of regulation by covalent modification of the protein. The enzyme may also be regulated by allosteric interactions. Covalent modification occurs by the reversible attachment of chemical groups such as acetyl groups, phosphate groups, methyl groups, adenyl groups, and uridyl groups. The covalent attachment of a chemical group can activate or inhibit the protein. Table 6.1 summarizes some examples of covalent modification of proteins, several of which are discussed in later chapters.

6.5 Summary

Metabolism is regulated by key regulatory enzymes that help to keep metabolism in a steady state in which concentrations of intermediates do not change. Generally, the regulatory enzymes catalyze physiologically irreversible reactions at metabolic branch points. The enzymes can be regulated by negative effectors, by positive effectors, by negative and positive effectors together, and by covalent modification. In biosynthetic pathways, the negative effectors are the end products. This is called feedback or end-product inhibition. Positive regulation by a precursor intermediate can speed up a subsequent reaction in the same metabolic pathway to avoid buildup of the precursor. Also, an intermediate in one pathway may exert positive regulation on a reaction in a second pathway if the second pathway provides a precursor metabolite for the first

pathway. Enzymes that are regulated by effector molecules that bind to effector sites are called allosteric enzymes, and the effector molecules are called allosteric effectors.

Nonregulatory enzymes display simple saturation kinetics called Michaelis–Menten kinetics. Two kinetic constants are the V_{max} and the K_m. The V_{max} is the velocity of the enzyme when it is saturated with substrate. The K_m is the substrate concentration yielding $\frac{1}{2}V_{max}$.

Regulatory enzymes usually show sigmoidal kinetics. Nevertheless a V_{max} and a K_m can be measured. A positive effector can raise the V_{max} and lower the K_m, but not necessarily both. A negative effector does just the opposite. Regulatory enzymes are frequently multimeric, consisting of catalytic subunits and regulatory subunits. The allosteric effector binds to the regulatory subunit and effects a conformational change in the catalytic subunit that alters its kinetic constants.

Study Questions

1. Plot the following data and derive a K_m and a V_{max}:

Time (min)	Product (arbitrary units) for form substrates ($M \times 10^{-6}$)			
	16	24	48	144
0.4	0.55	0.70	0.70	1.00
0.8	1.10	1.30	1.35	1.70
1.2	1.65	1.85	2.00	3.00
1.6	2.05	2.35	2.55	3.15
2.0	2.50	2.85	3.10	3.80

2. What is a rationale for having regulatory enzymes catalyze physiologically irreversible reactions at branch points?

3. What are three patterns of feedback inhibition for the regulation of a branched biosynthetic pathway?

4. Under what circumstances might you expect to see positive regulation of enzyme activity?

5. What is meant by positive cooperativity? What is the physiological advantage to it?

NOTE

1. The turnover number is equal to k_{cat} (k_3) and is related to the concentration of active sites E_t [i.e., $V_{max} = k_{cat}(E_t)$]. The turnover number is measured with pure enzyme whose molecular weight is known.

7

Bioenergetics in the Cytosol

Two different kinds of energy drive all cellular reactions. One of these is electrochemical energy, which refers to ion gradients (in bacteria primarily the proton gradient) across the cell membrane (Chapter 3). Electrochemical energy energizes solute transport, flagella rotation, ATP synthesis, and other membrane activities. The second type of energy is chemical energy in the form of high-energy molecules (e.g., ATP) in the soluble part of the cell. High-energy molecules are important because they drive biosynthesis in the cytoplasm, including the synthesis of nucleic acids, proteins, lipids, and polysaccharides. Additionally, the uptake into the cell of certain solutes is driven by high-energy molecules rather than by electrochemical energy.

This chapter introduces the various high-energy molecules used by the cell and explains why the adjective "high-energy" is applied to them, how they are made, and how they are used. The major high-energy molecules include ATP as well as other nucleotide derivatives, phosphoenolpyruvate, acyl phosphates, and acyl–CoA derivatives. The chapter begins with a discussion of the chemistry of the major high-energy molecules (Section 7.1), followed by explanations of how high-energy molecules are used to drive biosynthetic reactions (Section 7.2) and how the high-energy molecules are synthesized (Section 7.3). The information in this chapter will be referred to in later chapters, which consider in more detail the metabolic roles of the high-energy molecules. A historical perspective in given in Box 7.1.

7.1 High-Energy Molecules and Group Transfer Potential

High-energy molecules such as ATP have bonds that have a high free energy of hydrolysis that are sometimes depicted by a "squiggle" (~). These bonds are sometimes called "high-energy" bonds, but as will be discussed shortly, this is a misnomer because the bonds have normal bond energy. The important point is that the chemical group attached to the "squiggle" is readily transferred to acceptor molecules. Therefore, the high-energy molecules are said to have a *high group transfer potential*. When the chemical groups are transferred, new linkages between molecules (e.g., ester linkages, amide linkages, glycosidic linkages, ether linkages) are made. This results in the synthesis of the different small molecules in the cell (e.g., complex lipids, nucleotides, etc.), as well as polymers such as nucleic acids, proteins, polysaccharides, and fatty acids. We will first consider group transfer reactions in general, and then explain why certain molecules have a high group transfer potential.[1] Finally, we will present some examples of the use of group transfer reactions in biosynthesis.

7.1.1 Group transfer potential

A common chemical group that is transferred between molecules is the phosphoryl group, and phosphoryl group transfer will serve as an example of a group transfer reaction. Let us

BOX 7.1 HISTORICAL PERSPECTIVE: ENERGY TRANSFER IN THE CYTOSOL

In 1927 Eggleton and Eggleton reported that the contraction of vertebrate muscle was associated with an organic phosphorus compound that they named "phosphagen." When muscle contracted, the levels of phosphagen decreased, and the process was accompanied by a rise in inorganic phosphate. When oxygen was present, this sequence was reversed. In 1927 Fiske and Subbarow identified phosphagen as creatine phosphate. In 1929, Lohmann, as well as Fiske and Subbarow, isolated from striated muscle a compound that could be hydrolyzed to adenylic acid and pyrophosphate. This was ATP, and its structure was elucidated subsequently. The heat released when creatine phosphate or ATP was hydrolyzed was measured, and it was clear that either of these compounds could provide the energy for muscle contraction. However, it was not known at that time which of the two did so.

It is now known that creatine phosphate is an energy-storage compound in muscle and is synthesized via ATP, which is the molecule that drives muscle contraction. This realization was due to in large part to a paper published in 1939 by Engelhardt and Lyubimova. They showed that myosin, which was known to form the basis of muscle contraction, was an ATPase.

Between 1939 and 1941 Fritz Lipmann made a long-lasting contribution to our understanding of how energy transfer takes place in the cytosol. It was Lipmann who introduced the term "energy-rich phosphate bond" to describe the phosphate linkage in creatine phosphate, ATP, phosphoenolpyruvate, 1,3-diphosphoglyceric acid, and acetylphosphate, whose hydrolysis is associated with a large negative free-energy change (kilocalories or kilojoules per mole). Lipmann also introduced the term "group potential" to describe the ability of these compounds to donate the phosphoryl group, or other groups such as the acetyl group, to other molecules.

For additional background on these developments, see refs. 1 and 2.

REFERENCES

1. Hoffmann-Ostenhof, O. (Ed.). 1987. *Intermediary Metabolism*. Van Nostrand Reinhold, New York.

2. Fruton, J. S. 1972. *Molecules and Life: Historical Essays on the Interplay of Chemistry and Biology*. John Wiley & Sons, New York.

consider a generic phosphoryl group transfer reaction (shown in Fig. 7.1). We will examine both the chemical mechanism of transfer and the thermodynamics. Notice that the phosphorus in all phosphate groups carries a positive charge, (i.e., the P=O bond is drawn as the semipolar P^+–O^- bond). This is because phosphorus forms double bonds poorly, and the electrons in the bond are shifted toward the electron-attracting oxygen. During the phosphoryl group transfer reaction, the phosphorus atom is attacked by a nucleophile (an attacking atom with a pair of electrons seeking a positive center) shown in Fig. 7.1. The chemical group Y is displaced with its bonding electrons, and the phosphoryl group is transferred to the hydroxyl, forming ROP. This is a general scheme for group transfer reactions, not simply phosphoryl group transfers. That is, a nucleophile bonds to an electropositive center and displaces a leaving group with its bonding electrons. The reactions are called *nucleophilic displacements* or S_N2 reactions (substitution nucleophilic bimolecular). As we shall see later, various molecules such as ATP, acyl phosphates, and phosphoenolpyruvate have a high phosphoryl group transfer potential and undergo similar nucleophilic displacement reactions.

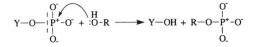

also written

where

$$(P) = \begin{matrix} O^- \\ | \\ P^+-O^- \\ | \\ O_- \end{matrix} \quad \text{(a phosphoryl group)}$$

Fig. 7.1 Phosphoryl group transfer reaction. The phosphate group is shown as ionized. Because phosphorus is a poor double-bond former, the phosphorus–oxygen bond exists as a semipolar bond. The positively charged phosphorus is attacked by the electronegative oxygen in the hydroxyl. The leaving group is YOH. If ROH is water, then the reaction is a hydrolysis and the product is inorganic phosphate and YOH. One can compare the tendency of different molecules to donate phosphoryl groups by comparing the free energies released when the acceptor is water (i.e., the free energy of hydrolysis). The group transfer potential is the negative of the free energy of hydrolysis.

But, how can one compare the phosphoryl group transfer potential of all these molecules, since the acceptors (the attacking nucleophiles) differ? A scale is used in which the standard nucleophile is the hydroxyl group of water and the phosphoryl donors are all compared with respect to the tendency to donate the phosphoryl group to water. *The group transfer potential is thus defined as the negative of the standard free energy of hydrolysis at pH 7.* It is a quantitative assessment of the tendency of a molecule to donate the chemical group to a nucleophile.

For example, suppose the standard free energy of hydrolysis of the phosphate ester bond in YOP is –29,000 J/mol. Then its phosphoryl group transfer potential is the negative of this number, or +29,000 J/mol. *Bonds that have a standard free energy of hydrolysis at pH 7 equal to or greater than –29,000 J/mol are usually called "high-energy" bonds, although as discussed later, they have normal bond energy.* The group transfer potential is not really a potential in an electrical sense, but a free energy change per mole of substrate hydrolyzed. However, the word "potential" is widely used in this context, and the convention will be followed here. The molecules with high phosphoryl group transfer potentials that we will revisit in the ensuing chapters are listed in Table 7.1. Also listed is glucose-6-phosphate, which has a low phosphoryl group transfer potential.

Group transfer potentials are a convenient way of estimating the direction in which a reaction will proceed. For example, the phosphoryl group transfer potential of ATP at pH 7 is 35 kJ/mol and for glucose-6-phosphate it is only 14 kJ/mol. This means that ATP is a more energetic donor of the phosphoryl group than is glucose-6-phosphate. It also means that ATP will transfer the phosphoryl group to glucose to form glucose-6-phosphate with the release of 35 to 14 or 21 kJ/mol under standard conditions, pH 7.

$$\text{ATP} + \text{glucose} \rightarrow \text{glucose-6-phosphate}$$

$$+ \text{ADP} \qquad \Delta G_0' = -21 \text{ kJ/mol}$$

This can also be seen by summing the hydrolysis reactions, since their sum equals the transfer of the phosphoryl group from ATP to glucose. This can be done for thermodynamic calculations even though, because the overall energy change is independent of the path of

Table 7.1 Group transfer potentials

Compound	$\Delta G_{0,\text{hyd}}'$ (kJ/mol)	Phosphoryl group transfer potential (kJ/mol)
PEP + H_2O → pyruvate + P_i	−62	+62
1,3-BPGA + H_2O → 3-PGA + P_i	−49	+49
Acetyl-P + H_2O → acetate + P_i	−47.7	+47.7
ATP + H_2O → ADP + P_i	−35	+35
Glucose-6-P + H_2O → glucose + P_i	−14	+14

the reaction, glucose-6-phosphate is not synthesized by hydrolysis reactions as written:

$$ATP + H_2O \rightarrow ADP + P_i \qquad \Delta G_0' = -35 \text{ kJ/mol}$$

$$glucose + P_i \rightarrow glucose\text{-}6\text{-}phosphate + H_2O$$

$$\Delta G_0' = +14 \text{ kJ/mol}$$

$$ATP + glucose \rightarrow glucose\text{-}6\text{-}phosphate + ADP$$

$$\Delta G_0' = -21 \text{ kJ/mol}$$

The release of 21 kJ/mol means that the equilibrium lies far in the direction of glucose-6-phosphate. Because $\Delta G_0' = -RT \ln K_{eq}' = -5.80$ $\log_{10} K_{eq}'$ kJ (at 30 °C), the equilibrium constant is 4.2×10^3 in favor of glucose-6-phosphate and ADP. However, whether a particular reaction will proceed in the direction written depends upon the actual free energy change and not the equilibrium constant. The actual free energy change at pH 7 is $\Delta G'$, which is a function of the physiological concentrations of products and reactants and in this case is

$$\Delta G' = \Delta G_0'$$

$$+ RT \ln(glucose\text{-}6\text{-}phosphate)(ADP)/(ATP)$$
$$(glucose)$$

Reactions proceed only in the direction of a negative $\Delta G'$. In the cell, the preceding reaction proceeds only in the direction of glucose-6-phosphate and ADP. This is because the ratio of glucose-6-phosphate)(ADP) to (ATP)(glucose) would have to be greater than 4.2×10^3 to change the sign of the $\Delta G'$ to cause the reaction to proceed in the direction of ATP and glucose. This does not occur, and therefore under physiological conditions the direction of phosphoryl flow is always from ATP to glucose. We will now consider why ATP and the other molecules in Table 7.1 have such high group transfer potentials (high free energies of hydrolysis).

7.1.2 Adenosine triphosphate (ATP)

As seen in Table 7.1, the standard free energy of hydrolysis of the phosphate ester bond in ATP at pH 7 is −35 kJ/mol. To understand why so much energy is released during the hydrolysis reaction, we must consider the structure of the ATP molecule (Fig. 7.2). Notice that at pH 7 the phosphate groups are ionized. (Actually, at pH 7 most of the ATP is a mixture of ATP^{3-} and

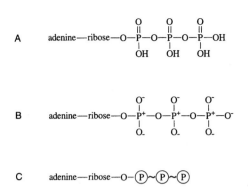

Fig. 7.2 Structure of ATP. ATP has three phosphate groups. (A) Un-ionized. (B) Although all the phosphate groups are shown ionized, giving them a net negative charge, the actual number at pH 7 is 3 or 4 for most of the ATP. Note that the phosphate–oxygen double bond is drawn as a semipolar bond, which takes into account the electronegativity of the oxygen and the low propensity of phosphorus to form double bonds. The structure predicts strong electrostatic repulsion between the phosphate groups. The electrostatic repulsion favors the transfer of phosphate to a nucleophile (e.g., water). (C) ATP drawn with squiggles to show the bonds with high free energy of hydrolysis. The phosphates are drawn as phosphoryl groups as illustrated in Fig. 7.1.

ATP^{4-}.) This produces electrostatic repulsion between the negatively charged phosphates, which accounts for much of the free energy of hydrolysis. Reactions during which phosphate is removed from ATP will be favored because the electrostatic repulsion is decreased as a result of the hydrolysis.

Transfer of a phosphoryl group to acceptors other than water

Any group that is electronegative (e.g., the hydroxyl groups in sugars) can attack the electropositive phosphorus shown in Fig. 7.2 and result in phosphoryl group transfer, provided the appropriate enzyme is present to catalyze the reaction. In this way, ATP can phosphorylate many different compounds. Enzymes that catalyze phosphoryl group transfer reactions are called *kinases*. In summary, then, the high free energy of hydrolysis of ATP is a good predictor for the tendency of ATP to donate a phosphoryl group to nucleophiles such as the hydroxyl groups in sugars. Much of the biochemistry of ATP is directly related to this

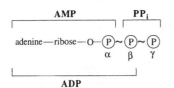

Fig. 7.3 ATP drawn with squiggles to denote a high free energy of hydrolysis (i.e., a high group transfer potential). In ATP, two of the phosphate ester bonds have a high free energy of hydrolysis owing to electrostatic repulsion. The phosphates are labeled α, β, and γ, starting with the one nearest to the ribose. If the α phosphate is attacked, then AMP is transferred and pyrophosphate (PP$_i$) is displaced. If the γ phosphate is attacked, then the phosphoryl group is transferred and ADP is displaced. Less frequently, the β phosphate is attacked and the pyrophosphoryl group is transferred, displacing the AMP. The reaction that takes place depends upon the specificity of the enzyme.

tendency. We will return to this point later, but first we must further discuss the "squiggle" and the "high-energy bond."

The "squiggle"

As mentioned, the "squiggle" symbol (~) denotes a high negative free energy of hydrolysis, and thus a greater group transfer potential. ATP is usually drawn with two squiggles because there are two phosphate ester bonds with a high free energy of hydrolysis (Fig. 7.3).

Note that the squiggle does not refer to the energy in the phosphate bond but, rather, to the free energy of hydrolysis. To make the distinction clear, consider the definition of bond energy. Bond energy is the energy required to break a bond. It is not the energy released when a bond is broken. In fact, the P–O bond energy is about +413 kJ (100 kcal). Compare this with the −35 kJ of hydrolysis energy.

7.1.3 Phosphoenolpyruvic acid

Another high-energy phosphoryl donor is phosphoenolpyruvate (PEP). In fact, PEP is a more energetic phosphoryl donor than ATP and will donate the phosphoryl group to ADP to make ATP, with release of 62 − 35 or 27 kJ/mol (under standard conditions, pH 7) (Table 7.1). As discussed in Section 7.3.4, this is an important source of ATP. PEP also donates the phosphoryl

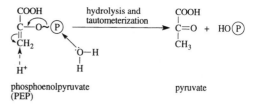

Fig. 7.4 Conversion of PEP to pyruvate. This is an enol–keto tautomerization. The removal of the phosphoryl group allows the electrons to shift into the keto form. Energy is released because the keto form has less free energy and is more stable than the enol form. The keto form of pyruvic acid is more stable than the enol form by 42 to 50 kJ/mol (10–12 kcal/mol). The following tabulation of the difference in bond energies between pyruvic acid and enolpyruvic acid shows that pyruvic acid has 76 kJ more bond energy than enolpyruvic acid:

Bond energies (kJ)

Enolpyruvic acid		Pyruvic acid	
C–O	293	C–C	247
O–H	460	C–H	364
C=C	418	C=O	636
	1,171 kJ		1,247 kJ

group to sugars during sugar transport in the phosphotransferase (PTS) system (Chapter 16).

Why does PEP have a high phosphoryl group transfer potential?

The reason for PEP's high phosphoryl group transfer potential is different from that for ATP. Consider the following reaction, which is illustrated in Fig. 7.4:

$$PEP + H_2O \rightarrow \text{pyruvic acid} + P_i$$

$$\Delta G_0' \approx -62 \text{ kJ/mol}$$

The hydrolysis removes the phosphate and allows the enol form of pyruvic acid to tautomerize into the keto form. Energy is released because the keto form is more stable than the enol form. One can account for the energy released during hydrolysis by the difference in bond energies between the keto and enol forms of pyruvic acid. A summation of bond energies reveals that the keto form has 76 kJ/mol (18 kcal/mol) more bond energy than the enol form. (Recall that bond energy is the energy required to break a bond and therefore is equal to the energy released when

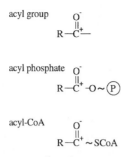

acyl group

acyl phosphate

acyl-CoA

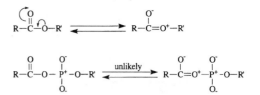

Fig. 7.6 (A) Resonance of an oxygen ester. Note that electrons shift from the oxygen in the C–OR′ to form a double bond. This is unlikely in phosphate esters (B) because the phosphorus atom bears a positive charge and prevents electrons from shifting in from the oxygen atom to form the double bond. Resonance is made less likely in thioesters because the sulfur atom does not form double bonds.

Fig. 7.5 Structures of acyl group, acyl–CoA, and acyl phosphates. Notice that the carbonyl group is polarized, and the carbon is subject to nucleophilic attack during acyl group transfer reactions. The phosphate ester and thioester bonds in the acyl phosphates and acyl–CoAs have high free energy of hydrolysis.

the bond is formed. Hence the formation of a molecule with higher bond energy will result in the *release* of energy.) Thus the hydrolysis of the phosphate ester bond results in the release of energy because it promotes the enol–keto tautomerization.

7.1.4 Acyl derivatives of phosphate and coenzyme A

Acyl derivatives of phosphate and coenzyme A also have a high free energy of hydrolysis. An acyl group is a derivative of a carboxylic acid and has the structure shown in Fig. 7.5, where R may be an alkyl or aryl group. The acyl derivatives of coenzyme A (CoA) and of phosphate are also shown in Fig. 7.5.

Why do acyl derivatives have high group transfer potentials?

The acyl derivatives of phosphate esters and thioesters have high group transfer potential because they do not resonate well. Consider the resonance forms of a normal ester (Fig. 7.6A) and a phosphate ester (Fig. 7.6B). In a normal ester, when two electrons shift from the oxygen to form a double bond during resonance, oxygen acquires a positive charge. In acyl–CoA and acyl phosphate derivatives, the oxygen attached to R′ (Fig. 7.6B) is replaced by a phosphorus or sulfur atom. Resonance is hindered in phosphate esters because the positive charge on the phosphorus atom prevents electrons from shifting in from the oxygen to form a double bond, which would leave two adjacent positive

centers. Thioesters also do not resonate as well as normal esters. The reason for this is that sulfur forms double bonds poorly. The poor resonance of the phosphate and thioesters is in sharp contrast to the high resonance of the free carboxylate group that forms when the phosphoryl group or CoA is transferred to an acceptor molecule. Thus, the hydrolysis of the acyl derivatives of phosphate and coenzyme A is energetically favored because it leads to products stabilized by resonance with respect to the reactants.

The importance of acyl derivatives in group transfer reactions

Acyl derivatives are very versatile. Depending on the specificity of the enzyme catalyzing the reaction, they donate *acyl groups, CoA groups,* or *phosphoryl groups*. For example, fatty acids and proteins are synthesized via acyl group transfer reactions, CoA transfer takes place during fermentative reactions in bacteria, and acyl phosphates donate their phosphoryl groups to ADP to form ATP. Enzymes that catalyze the transfer of acyl groups are called *transacylases*. Those that transfer CoA or phosphoryl groups are called *CoA transferases* and *kinases*, respectively.

7.2 The Central Role of Group Transfer Reactions in Biosynthesis

Group transfer reactions are central to all of metabolism because biological molecules such as proteins, lipids, carbohydrates, and nucleic acids, are synthesized as a result of group transfer reactions by means of high-energy donors such as ATP or other nucleotide derivatives, acyl–coenzyme As, and acyl–phosphates.

7.2.1 How ATP can be used to form amide linkages, thioester bonds, and ester bonds

ATP as a donor of AMP, phosphoryl groups, or pyrophosphoryl groups

ATP can donate other parts of the molecule (e.g., AMP, PP$_i$) besides the phosphoryl group. This is very important for the synthesis of many polymers (e.g., proteins, polysaccharides, nucleic acids) as well as other biochemical reactions. That is because when these groups are transferred to an acceptor molecule, the acceptor molecule itself becomes a high-energy donor for subsequent biosynthetic reactions. In other words, the energy from ATP can be transferred to other molecules, which can then drive biosynthetic reactions (i.e., the formation of new covalent bonds).

Let us look again at the structure of ATP and examine some of the group transfer reactions that it can undergo (Fig. 7.3). The phosphates are labeled α, β, and γ, when counting from the ribose moiety, with α being the phosphate closest to the ribose. If the α phosphate is attacked, then AMP is the group transferred and PP$_i$ is the leaving group. For example, this occurs during protein synthesis, discussed in Section 7.2.2. Some enzymes catalyze the transfer of the pyrophosphoryl group when the β phosphate is attacked. For example, enzymes that synthesize phosphoenolpyruvate from pyruvate make an enzyme–pyrophosphate derivative by transferring the pyrophosphoryl group from ATP to the enzyme (Section 8.13.2).

Another example is the synthesis of phosphoribosylpyrophosphate (PRPP), during which the pyrophosphoryl group is transferred from ATP to the C1 of ribose-5-phosphate (Section 9.2.2). In a subsequent reaction, the ribose-5-phosphate is transferred from PRPP to an appropriate acceptor molecule in the synthesis of purines and pyrimidines. If the γ phosphate is attacked, the phosphoryl group is transferred, forming phosphorylated derivatives and ADP as the leaving group. This is a very common reaction in metabolism (e.g., the synthesis of glucose-6-phosphate).

As stated, all the foregoing reactions can occur. The reaction that takes place depends upon which enzyme is the catalyst, since it is the enzyme that determines the specificity of the attack. As examples, we will consider the cases in which ATP is used to form amide linkages, thioesters, and esters. In all these reactions, a carboxyl group accepts either AMP or phosphate from ATP in a group transfer reaction to form a *high-energy intermediate*. The high-energy intermediate is an acyl derivative with a high group transfer potential. The acyl–AMP or acyl phosphate can donate the acyl group in a subsequent reaction, forming an amide, ester, or thioester bond, depending upon whether the attacking nucleophile is N:, O:, or S:, respectively.

Consider the reactions in Fig. 7.7. Notice the sequence of reactions. First there is a displacement on either the α or γ phosphate of ATP to form the acyl derivative. Then there is a displacement on the carbonyl carbon to transfer the acyl moiety to the nucleophile to form the ester or amide bond. Notice that a pyrophosphatase is associated with reactions in which pyrophosphate is displaced. Because of the high free energy of hydrolysis of the pyrophosphate bond, the group transfer reaction is driven to completion.

7.2.2 How ATP is used to form peptide bonds during protein synthesis

As an example of the principles described in Sections 7.2 and 7.2.1, we will consider the formation of peptide bonds during protein synthesis. The acyl donor is made by derivatizing the α-carboxyl group on the amino acid with AMP, using ATP as the AMP donor (Fig. 7.8A). The acyl–AMP is the high-energy intermediate. In the next reaction (Fig. 7.8B) the acyl group is transferred to transfer RNA (tRNA) so that the carboxyl group becomes derivatized with tRNA. The acyl–tRNA also has a high group transfer potential. Thus, energy has flowed from ATP to aminoacyl–AMP to aminoacyl–tRNA. The aminoacyl–tRNA is attacked by the nucleophilic nitrogen of an amino group from another amino acid, and the acyl portion is transferred to the amino group, forming the peptide bond. This last reaction (Fig. 7.8C) takes place on the

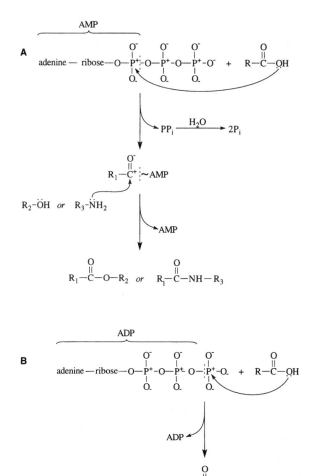

Fig. 7.7 ATP provides the energy to make ester and amide linkages. (A) A carboxyl attacks the α phosphate of ATP, displacing pyrophosphate, and forming the AMP derivative. The hydrolysis of pyrophosphate catalyzed by pyrophosphatase drives the reaction to completion. The AMP is then displaced by an attack on the carbonyl carbon by a hydroxyl or amino group, resulting in the transfer of the acyl group to form the ester or the substituted amide. (B) Similar to (A) except that the attack is on the γ phosphate of ATP, displacing ADP, and forming the acyl phosphate.

ribosome. The reaction goes to completion because of the large difference in group transfer potential between the aminoacyl–tRNA and the peptide that is formed.

The reactions shown in Fig. 7.8 exemplify the principle that the energy to make covalent bonds (in this case a peptide bond) derives from a series of group transfer reactions starting with ATP, in which ATP provides the energy to make high-energy intermediates that serve as group donors. In this way, all of the large complex molecules (proteins, nucleic acids, polysaccharides, lipids, etc.) are synthesized.

7.3 ATP Synthesis by Substrate Level Phosphorylation

We have seen how ATP can drive the synthesis of biological molecules via a coupled series of group transfer reactions. But how is ATP itself made? The answer depends upon whether the ATP is synthesized in the membranes or in the cytosol.

In the membranes, the phosphorylation of ADP is coupled to oxidation–reduction reactions via the generation of an electrochemical gradient of protons (Δp), which is then used to

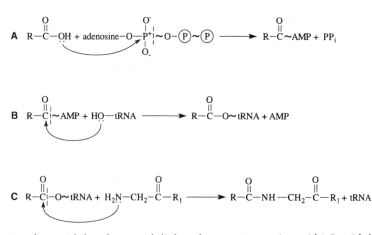

Fig. 7.8 Formation of a peptide bond (an amide linkage between two amino acids). Peptide bonds are formed as a result of a series of group transfer reactions. (A) Transfer of AMP from ATP to the carboxyl group of the amino acid. In this reaction the α phosphorus of ATP is attacked by the OH in the carboxyl group. Recall that the P=O bonds in ATP are semipolar and that the phosphorus is an electropositive center. Most of the group transfer potential of ATP is trapped in the product and the reaction is freely reversible. However, the reaction is driven to completion by the hydrolysis of the pyrophosphate (not shown). (B) The displacement of the AMP by tRNA. This is not done for energetic reasons but rather because the tRNA is an adaptor molecule that aids in placing the amino acid in the correct position with respect to the mRNA on the ribosome. The synthesis of the aminoacyl–tRNA is reversible, indicating that the group transfer potential of the aminoacyl–tRNA is similar to that of the aminoacyl–AMP. (C) The displacement of the tRNA by the amino group of a second amino acid, resulting in the synthesis of a peptide bond, takes place on the ribosome. This reaction proceeds with the release of a relatively large amount of free energy and is irreversible. On the ribosome, it is the amino group of the incoming aminoacyl–tRNA at the "a" site that attacks the carbonyl of the resident aminoacyl–tRNA at the "p" site. The polypeptide is thus transferred from the "p" site to the "a" site as the tRNA at the "p" site is released. The tRNA of the incoming amino acid at the "a" site is represented by R_1, and the incoming amino acid is glycine.

drive the phosphorylation of ADP via the membrane ATP synthase. That process, called oxidative phosphorylation or electron transport phosphorylation, was discussed in Sections 3.6.2 and 3.7.1.

In electron transport phosphorylation, electrons traveling over a potential difference of ΔE_h volts provide the energy to establish the Δp ($n\Delta E_h = y\Delta p$, where n is the number of electrons transferred and y is the number of protons extruded). ATP in the soluble part of the cell is made by phosphorylating ADP in a process called *substrate-level phosphorylation*. We can define a substrate-level phosphorylation as the phosphorylation of ADP in the soluble part of the cell by means of a high-energy phosphoryl donor. Substrate-level phosphorylations are catalyzed by enzymes called *kinases*:

$$Y \sim P + ADP \xrightarrow{\text{kinase}} Y + ATP$$

During a substrate-level phosphorylation, an oxygen in the β phosphate of ADP acts as

a nucleophile and bonds to the phosphate phosphorus in the high-energy donor. The phosphoryl group is transferred to ADP, making ATP. (This is a phosphoryl group transfer reaction similar to that shown in Fig. 7.1.) Consider the phosphoryl group transfer from an acyl phosphate to ADP (Fig. 7.9). Phosphoryl donors for ATP synthesis during substrate-level phosphorylations include 1,3-bisphosphoglycerate (BPGA), phosphoenolpyruvate (PEP), acetyl phosphate, and succinyl–CoA plus inorganic phosphate. These high-energy phosphoryl donors are listed in Table 7.1. The

$$\underset{R}{\overset{O}{\underset{||}{C}}}-O\sim\text{(P)} + ADP \longrightarrow \underset{R}{\overset{O}{\underset{||}{C}}}-OH + ATP$$

Fig. 7.9 A substrate-level phosphorylation. This is an example of an acyl phosphate donating a phosphoryl group to ADP, specifically, the phosphorylation of ADP by 1,3-bisphosphoglycerate. The carboxylic acid is displaced.

Table 7.2 Four substrate-level phosphorylations

1,3-BPGA + ADP	→ 3-PGA + ATP
PEP + ADP	→ pyruvic acid + ATP
Acetyl-P + ADP	→ acetic acid + ATP
Succinyl–CoA + P_i + ADP	→ succinic acid + ATP + CoASH

four major substrate-level phosphorylations, listed in Table 7.2, occur in the following metabolic pathways:

1. The substrate-level phosphorylations that use BPGA and PEP take place during glycolysis.

2. The succinyl–CoA reaction is part of the citric acid cycle.

3. Acetyl phosphate is formed from acetyl–CoA, itself formed from pyruvate. This is an important source of ATP in anaerobically growing bacteria.

The synthesis of ATP via substrate-level phosphorylation first requires the synthesis of one of the high-energy phosphoryl donors listed in Table 7.2. All the reactions that synthesize a high-energy molecule are oxidations, with one exception. The single exception is the synthesis of phosphoenopyruvate, which results from a dehydration. During the oxidation–reduction reaction, $-nF\Delta E$ J is used to create a molecule with a high phosphoryl group transfer potential. The synthesis of the phosphoryl donors and the substrate-level phosphorylations of ADP are described next.

7.3.1 1,3-Bisphosphoglycerate

During the degradation of sugars in the metabolic pathway called glycolysis, the 6-carbon sugar glucose is cleaved into two 3-carbon fragments called phosphoglyceraldehyde (PGALD)

(Chapter 8). The phosphoglyceraldehyde is then oxidized to 1,3-bisphosphoglycerate, using inorganic phosphate as the source of phosphate and NAD^+ as the electron acceptor (Fig. 7.10). The reaction is catalyzed by *phosphoglyceraldehyde dehydrogenase*. The energy that would normally be released as heat from the oxidation ($-2F\Delta E'_0 = -44,000$ J) is used to drive the synthesis of 1,3-bisphosphoglycerate, which has a phosphoryl group transfer potential of 49,000 J ($-\Delta C'_{0,hyd}$). (These are standard free energy changes. The actual free energy changes depend upon the ratios of concentrations of products to reactants.) The 1,3-bisphosphoglycerate then donates a phosphoryl group to ADP, in a substrate-level phosphorylation, to form ATP and 3-phosphoglycerate in a reaction catalyzed by *phosphoglycerate kinase* (Fig. 7.11). Thus, the oxidation of phosphoglyceraldehyde by NAD^+ drives ATP synthesis.

7.3.2 Acetyl phosphate

Acetyl phosphate, another high-energy phosphoryl donor, can be made from pyruvate via acetyl–CoA. The sequence is

Pyruvate → acetyl–CoA → acetyl phosphate

The acetyl phosphate donates the phosphoryl group to ADP (to form ATP), and the acetate that is produced from the acetyl phosphate is excreted into the medium. These reactions are extremely important for fermenting bacteria, and they account for the acetic acid that is produced during certain fermentations. The oxidation of pyruvate to acetyl–CoA will be discussed first (the *pyruvate dehydrogenase*, *pyruvate–ferredoxin reductase*, and *pyruvate–formate lyase* reactions). This will be followed by a description of the conversion of acetyl–CoA

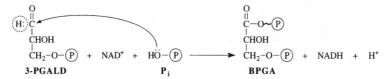

Fig. 7.10 Oxidation of phosphoglyceraldehyde (PGALD). The incorporation of inorganic phosphate (P_i) into a high-energy phosphoryl donor occurs during the oxidation of PGALD. The product is an acyl phosphate, 1,3-bisphosphoglycerate (BPGA). Energy that would normally be released as heat is trapped in the BPGA because inorganic phosphate, rather than water, is the nucleophile. As a consequence, an acyl phosphate, rather than a free carboxylic acid, is formed.

1,3-BPGA   **3-PGA**

Fig. 7.11 The phosphoglycerate kinase reaction. ADP carries out a nucleophilic attack on the phosphoryl group of 1,3-bisphosphoglycerate (BPGA), displacing the free carboxylic acid (3-phosphoglycerate, 3-PGA).

to acetyl phosphate (the *phosphotransacetylase* reaction). Then the synthesis of ATP and acetate from acetyl phosphate and ADP will be described (the *acetate kinase* reaction).

Formation of acetyl–CoA from pyruvate

Acetyl–CoA is usually made by the oxidative decarboxylation of pyruvate, which is a key intermediate in the breakdown of sugars. There are three well-characterized enzyme systems in the bacteria that decarboxylate pyruvate to acetyl–CoA. One is found in aerobic bacteria (and mitochondria) and is called *pyruvate dehydrogenase*. It is usually not present in anaerobically growing bacteria. The pyruvate dehydrogenase reaction is an oxidative decarboxylation of pyruvate to acetyl–CoA in which the electron acceptor is NAD+. Acetyl–CoA formed aerobically via pyruvate dehydrogenase is not a source of acetyl phosphate but usually enters the citric acid cycle, where it is oxidized to CO_2 (Chapter 8).

The other two enzyme systems that oxidize pyruvate to acetyl–CoA are found only in bacteria growing anaerobically. These are *pyruvate–ferredoxin oxidoreductase* and *pyruvate–formate lyase* (Fig. 7.12). (Pyruvate-ferredoxin oxidoreductase has been found in several archaea, but pyruvate–formate lyase and pyruvate dehydrogenase have not been found in any of the archaae thus far examined.[2]) The pyruvate–ferredoxin oxidoreductase reaction is an oxidative decarboxylation of pyruvate to acetyl–CoA in which the electron acceptor is ferredoxin. The pyruvate–formate lyase is an oxidative decarboxylation of pyruvate to acetyl–CoA in which the electrons remain with the carboxyl group (rather than being transferred to an acceptor such as NAD+ or ferredoxin), which is released as formate.

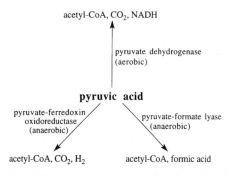

Fig. 7.12 Three enzyme systems that decarboxylate pyruvic acid to acetyl–CoA. Aerobically growing bacteria and mitochondria use the pyruvate dehydrogenase complex. The acetyl–CoA that is produced is oxidized to carbon dioxide in the citric acid cycle. Some anaerobic bacteria may also have a pyruvate dehydrogenase. Anaerobically growing bacteria generally use pyruvate–ferredoxin oxidoreductase or pyruvate–formate lyase instead of pyruvate dehydrogenase. The acetyl–CoA that is produced anaerobically (during fermentations) is converted to acetyl phosphate via the phosphotransacetylase reaction. The acetyl phosphate serves as a phosphoryl donor for ATP synthesis (the acetate kinase reaction), and the product acetate is excreted.

An important difference between pyruvate dehydrogenase and the other two enzymes is that the pyruvate dehydrogenase reaction produces NADH. This result can be disadvantageous to fermenting bacteria because there is often no externally provided electron acceptor to reoxidize the NADH. This may be why pyruvate dehydrogenase is usually not found in fermenting bacteria.

1. The pyruvate dehydrogenase reaction
Bacteria that are respiring aerobically use pyruvate dehydrogenase to decarboxylate pyruvic acid to acetyl–CoA. This enzyme reaction (Fig. 7.13) is also found in mitochondria. A more detailed description of the pyruvate dehydrogenase reaction can be found in Section 8.7.

2. Pyruvate–ferredoxin oxidoreductase
Most anaerobically growing bacteria do not use pyruvate dehydrogenase to oxidize pyruvate to acetyl–CoA and CO_2. Instead, they use *pyruvate–ferredoxin oxidoreductase* or *pyruvate–formate lyase*. Pyruvate–ferredoxin oxidoreductase is found in the clostridia, sulfate-reducing bacteria, and some other anaerobes. The enzyme catalyzes a reaction similar to that

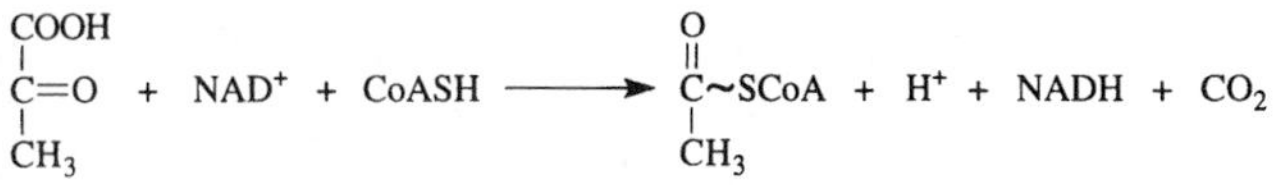

$$\underset{\underset{CH_3}{|}}{\overset{\overset{COOH}{|}}{C=O}} + NAD^+ + CoASH \longrightarrow \underset{\underset{CH_3}{|}}{\overset{\overset{O}{\parallel}}{C}}{\sim}SCoA + H^+ + NADH + CO_2$$

Fig. 7.13 The pyruvate dehydrogenase reaction. In this reaction, the two electrons that bond the carboxyl group to the rest of the molecule are transferred by the enzyme to NAD^+. At the same time, coenzyme A attaches to the carbonyl group to form the acylated coenzyme A derivative. If the oxidation were to take place by means of the :OH from water instead of the :SH from CoASH to supply the fourth bond to the carbonyl carbon, the product would be acetic acid and a great deal of energy would be lost as heat. But the thioester of the carboxyl group cannot resonate as well as the free carboxyl group and, for this reason, the energy that normally would have been released during the oxidation is "trapped" in the acetyl–SCoA, a molecule with a high group transfer potential.

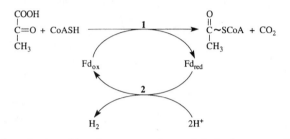

Fig. 7.14 The pyruvate–ferredoxin oxidoreductase reaction and the hydrogenase. The electrons travel from the ferredoxin (Fd) to protons via the enzyme hydrogenase. Enzymes: **1**, pyruvate–ferredoxin oxidoreductase; **2**, hydrogenase.

of pyruvate dehydrogenase except that the electron acceptor is not NAD^+. Instead, it is an iron–sulfur protein called ferredoxin (Fig. 7.14).

An important feature of the pyruvate-ferredoxin oxidoreductase in fermenting bacteria is that the enzyme is coupled to a second enzyme called *hydrogenase*. The hydrogenase catalyzes the transfer of electrons from reduced ferredoxin to H^+ to form hydrogen gas, accounting for much of the hydrogen gas produced during fermentations. The importance of the hydrogenase reaction is that it reoxidizes the reduced ferredoxin, thus allowing the continued oxidation of pyruvate. The ferredoxin-linked decarboxylation of pyruvate to acetyl–CoA is reversible and is used for autotrophic CO_2 fixation in certain anaerobic bacteria (Sections 13.1.3 and 13.1.4).

3. Pyruvate–formate lyase

Pyruvate–formate lyase is an enzyme found in some fermenting bacteria (e.g., the enteric bacteria and certain lactic acid bacteria.). In the reaction, the electrons stay with the carboxyl that is removed, and therefore formate is formed instead of carbon dioxide. Among the advantages of using pyruvate–formate lyase are that neither reduced ferredoxin nor NADH

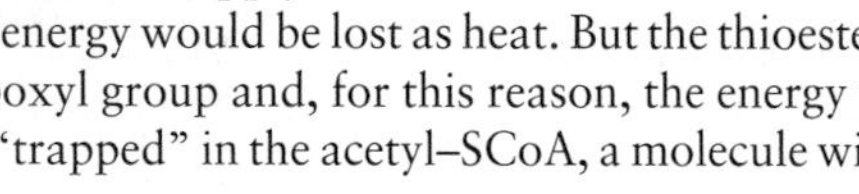

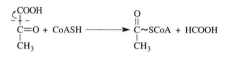

Fig. 7.15 The pyruvate–formate lyase reaction. Part of the molecule becomes oxidized and part becomes reduced. The part that becomes reduced is the carboxyl group that leaves as formic acid.

is produced and the electrons are disposed of as part of the formate. The reaction is illustrated in Fig. 7.15.

Formation of acetyl phosphate from acetyl–CoA

The acetyl–CoA that is made by using either pyruvate–ferredoxin oxidoreductase or pyruvate–formate lyase is converted to acetyl phosphate by the displacement of the CoASH by inorganic phosphate in a reaction catalyzed by *phosphotransacetylase*. Sometimes referred to as the PTA enzyme, phosphotransacetylase:

$$\text{Acetyl–CoA} + P_i \leftrightarrow \text{acetyl-P} + \text{CoASH}$$

(Acetyl-P can also be made directly from pyruvate and inorganic phosphate by using *pyruvate oxidase*. This is a flavoprotein enzyme found in certain *Lactobacillus* species. See Section 14.9.)

Formation of ATP from acetyl phosphate

The acetyl phosphate then donates the phosphoryl group to ADP in a substrate-level phosphorylation catalyzed by *acetate kinase*, sometimes referred to as the ACK enzyme. The route to the formation of acetate from acetyl–CoA via ACK and PTA is referred to as the ACK–PTA pathway. Fermenting bacteria oxidize pyruvate to acetate by using both the phosphotransacetylase and the acetate kinase, and they derive an ATP from the process while excreting acetate into the medium. We will return to this subject in Chapter 14, where fermentations are discussed in more detail.

$$\text{Acetyl-P} + \text{ADP} \xleftrightarrow{\text{acetate kinase}} \text{acetate} + \text{ATP}$$

Making acetyl–CoA during growth on acetate

Note that acetate kinase and phosphotransacetylase catalyze reversible reactions and can be used during growth on acetate to make acetyl–CoA, which is then incorporated into cell material or oxidized to CO_2. Another means of making acetyl–CoA when bacteria are growing on acetate is by use of the enzyme acetyl–CoA synthetase (the ACS enzyme).

$$\text{Acetate} + \text{ATP} + \text{CoA}$$

$$\xleftrightarrow{\text{AMP-dependent acetyl–CoA synthetase}}$$

$$\text{acetyl–CoA} + \text{AMP} + \text{PP}_i$$

An ADP-dependent acetyl–CoA synthetase

Although essentially all acetyl–CoA synthetases that have been studied generate AMP and PP_i (the AMP-dependent acetyl–CoA synthetases) and have been isolated from bacteria, methanogenic archaea, and eukaryotes, there also exist the so-called ADP-dependent acetyl–CoA synthetases, which generate ADP and P_i. The latter type of enzyme has thus far not been found to be widely distributed and has been reported in *Entamoeba histolytica* and *Giardia lamblia*, two eukaryotic microorganisms, and in certain archaea, including the hypothermophilic archaeon *Pyrococcus furiosus*, which ferments carbohydrates and peptides and grows at temperatures as high as 105 °C.[3] The latter three organisms excrete acetate into the medium and appear to use the ADP-dependent acetyl–CoA synthetase as a means of generating ATP. Note that the AMP-dependent acetyl–CoA synthetase is generally used for acetate utilization, whereas the ADP-dependent acetyl–CoA synthetase may be involved primarily with acetate production (and ATP synthesis), and can be viewed as an alternative to the ACK-PTA pathway.

$$\text{Acetate} + \text{ATP} + \text{CoA}$$

$$\xleftrightarrow{\text{ADP-dependent acetyl–CoA synthetase}}$$

$$\text{acetyl–CoA} + \text{ADP} + \text{P}_i$$

7.3.3 Succinyl–Coenzyme A

Succinyl–CoA is made by the oxidative decarboxylation of α-ketoglutarate, a reaction that occurs in the citric acid cycle (Fig. 7.16). It is strictly analogous in its mechanism and cofactor requirements to the oxidative decarboxylation of pyruvate by pyruvate dehydrogenase. The enzyme that carries out the oxidation of α-ketoglutarate is called *α-ketoglutarate dehydrogenase*. Succinyl–CoA then drives the synthesis of ATP from inorganic phosphate and ADP in a reaction catalyzed by the citric acid cycle enzyme *succinate thiokinase*. The reaction is as follows:

$$\text{Succinyl–CoA} + \text{ADP} + \text{P}_i$$

$$\rightarrow \text{succinate} + \text{CoASH} + \text{ATP}$$

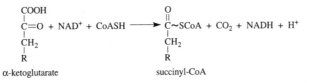

α-ketoglutarate   succinyl-CoA

Fig. 7.16 The α-ketoglutarate dehydrogenase reaction. Notice that the substrate molecule resembles pyruvic acid. The difference is that in pyruvic acid the R group is H, whereas in α-ketoglutaric acid, it is CH_2–COOH.

Succinyl phosphate is not a free intermediate. Perhaps the CoASH is transferred from succinyl–CoA to the enzyme, where it is displaced by phosphate. The phosphorylated enzyme would then be the phosphoryl donor for ADP in ATP synthesis. It should be pointed out that whereas bacteria and plants produce ATP from succinyl–CoA, animals produce GTP instead. Succinyl–CoA is important not only for ATP synthesis but also as a precursor for heme synthesis.

7.3.4 Phosphoenolpyruvate

Cells that are using the glycolytic pathway to grow on sugars make phosphoenolpyruvate from 2-phosphoglycerate (2-PGA), an intermediate in the breakdown of the sugars. The reaction is a dehydration and is catalyzed by the enzyme *enolase* (Fig. 7.17). Phosphoenolpyruvate then phosphorylates ADP in a reaction catalyzed by *pyruvate kinase*:

$$PEP + ADP \rightarrow pyruvate + ATP$$

The enolase and pyruvate kinase reactions are important in energy metabolism because they serve to regenerate the ATP that is used to phosphorylate the sugars during the initial stages of sugar catabolism. (With rare exceptions, sugars must be phosphorylated before they can be metabolized.) However, these reactions cannot account for the synthesis of net ATP from ADP and inorganic phosphate, since the phosphate in the PEP originated from ATP (or PEP), rather than from inorganic phosphate. Phosphoenolpyruvate is also necessary for the synthesis of muramic acid and certain amino acids (Chapters 9 and 11).

Fig. 7.17 The enolase reaction, in which 2-phosphoglycerate (2-PEA) is dehydrated to phosphoenolpyruvate. Isotope exchange studies suggest that the first step is the removal of a proton from C2 to form a carbanion intermediate, which loses the hydroxyl and becomes phosphoenolpyruvate (PEP).

7.4 Summary

Biochemical reactions in the cytosol are driven by high-energy molecules, so called because they have a bond with a high free energy of hydrolysis. The reasons for this depend upon the structure of the whole molecule, not on any particular bond. High free energies of hydrolysis can be due to electrostatic repulsion between adjacent phosphate groups and/or diminished resonance. There are several high-energy molecules (e.g., ATP, BPGA, PEP, acetyl–P, acetyl–CoA, succinyl–CoA).

The term "high-energy bond" is a misnomer because the free energy of hydrolysis is not bond energy. The term "group transfer potential" refers to the negative of the free energy of hydrolysis and is a useful concept when one is comparing the tendency of chemical groups to be transferred to attacking nucleophiles. Thus, ATP has a high phosphoryl group transfer potential (around 35 kJ understandard conditions, pH 7) whereas glucose-6-phosphate has a low phosphoryl group transfer potential, (around 14 kJ). Hence, ATP will transfer the phosphoryl group to glucose with the release of $35 - 14 = 21$ kJ of energy.

Group transfer reactions can occur with conservation of energy to form high-energy intermediates, which themselves can be group donors in coupled reactions. This explains how ATP can provide the energy to drive a series of coupled chemical reactions that result in the synthesis of nucleic acids, proteins, polysaccharides, lipids, and so on.

The formation of a high-energy molecule usually involves an oxidation of an aldehyde (3-phosphoglyceraldehyde) or the oxidative decarboxylation of a β-keto carboxylic acid (pyruvate or succinate). In substrate-level phosphorylation, the energy of the redox reaction is trapped in an acyl phosphate or an acyl–CoA derivative. This is in contrast to respiratory phosphorylation, where the energy from the redox reaction is trapped in a Δp, which drives ATP synthesis.

Phosphoenolpyruvate is not formed as a result of an oxidation–reduction reaction but from a dehydration. However, the phosphate in phosphoenolpyruvate was already present in 2-phosphoglycerate, having been donated by ATP or phosphenolpyruvate during the sugar phosphorylations. Therefore, synthesis of ATP from phosphenolpyruvate does not represent the formation of ATP from ADP and inorganic phosphate but, rather, the regeneration of ATP that had been used to phosphorylate the sugars.

Study Questions

1. What is the definition of group transfer potential? Why do ATP, acyl phosphates, and PEP have high phosphoryl group transfer potentials?

2. Write a series of hypothetical reactions in which PEP drives the synthesis of A–B from A, B, and ADP.

3. What are two features that distinguish substrate-level phosphorylations from electron transport phosphorylation?

4. What features do the synthesis of BPGA, acetyl–CoA, and succinyl–SCoA have in common with each other but not with PEP?

5. What do the syntheses of acetyl–CoA and succinyl–CoA have in common?

6. Write a series of reactions that result in the synthesis of a substituted amide or an ester from a carboxylic acid. Use ATP as the source of energy. How is protein synthesis a modification of this reaction?

REFERENCES AND NOTE

1. For a discussion of high-energy molecules, see: Ingraham, L. L. 1962. *Biochemical Mechanisms*. John Wiley & Sons, New York.

2. Selig, M., and P. Schonheit. 1994. Oxidation of organic compounds to CO_2 with sulfur or thiosulfate as electron acceptor in the anaerobic hyperthermophilic archaea *Thermoproteus tenax* and *Pyrobaculum islandicum* proceeds via the citric acid cycle. *Arch. Microbiol.* **162**:286–294.

3. Mail, X., and M. W. W. Adams. 1996. Purification and characterization of two reversible and ADP-dependent acetyl–coenzyme A synthetases from the hyperthermophilic archaeon *Pyrococcus furiosus*. *J. Bacteriol.* **178**:5897–5903.

8

Central Metabolic Pathways

The central metabolic pathways are the pathways that provide the precursor metabolites to all the other pathways. They are the pathways for the metabolism of carbohydrates and carboxylic acids, such as C_4 dicarboxylic acids and acetic acid. The major carbohydrate pathways are the *Embden–Meyerhof–Parnas pathway* (also called the EMP pathway or glycolysis), the *pentose phosphate pathway* (PPP), and the *Entner–Doudoroff pathway* (ED). The Entner–Doudoroff pathway has been found to be restricted almost entirely to the prokaryotes, being present in gram-negative and gram-positive bacteria, as well as archaea. (It has been reported in the amoeba *Entamoeba histolytica* and in two fungi, *Aspergillus niger* and *Penicillium notatum*.[1]) The three pathways differ in many ways, but two generalizations can be made:

1. All three pathways convert glucose to phosphoglyceraldehyde, albeit by different routes.

2. The phosphoglyceraldehyde is oxidized to pyruvate via reactions that are the same in all three pathways.

From an energetic point of view, the reactions that convert phosphoglyceraldehyde to pyruvate are extremely important because they generate ATP from inorganic phosphate and ADP. This is because there is an oxidation in which inorganic phosphate is incorporated into an acyl phosphate (i.e., the oxidation of phosphoglyceraldehyde to 1,3-bisphosphoglyc-

erate). The 1,3-bisphosphoglycerate then donates the phosphoryl group to ADP in a substrate-level phosphorylation. (See Section 7.3.1.)

The fate of the pyruvate that is formed during the catabolism of carbohydrates depends on whether the cells are respiring. If the organisms are respiring, then the pyruvate that is formed by the carbohydrate catabolic pathways is oxidized to acetyl–CoA, which is subsequently oxidized to carbon dioxide in the *citric acid cycle*. The latter generally operates only during aerobic respiration. (However, see Section 8.8.4.) If fermentation rather than respiration is taking place, then the pyruvate is converted to fermentation end products such as alcohols, organic acids, and solvents, rather than oxidized in the citric acid cycle. Fermentations are discussed in Chapter 14.

An overview of the carbohydrate catabolic pathways and their relation to one another and to the citric acid cycle is shown in Fig. 8.1. Several points can be made about this figure. Notice that there are three substrate-level phosphorylations, two during carbohydrate catabolism and one in the citric acid cycle. Furthermore, there are six oxidation reactions, one in glycolysis, one in the pyruvate dehydrogenase reaction, and four in the citric acid cycle. These oxidations produce NADH (primarily) and $FADH_2$. The NADH and $FADH_2$ must be reoxidized to regenerate the NAD^+ and FAD that are required for the oxidations.

The route of reoxidation and the energy yield depend upon whether the organism is respiring

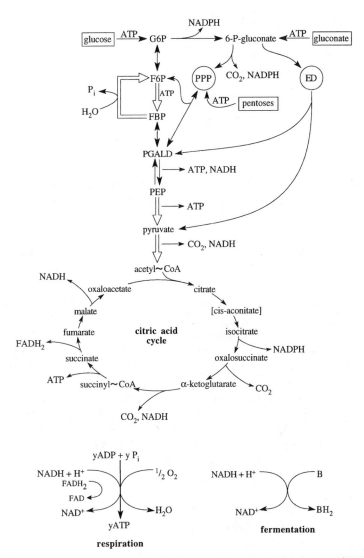

Fig. 8.1 Relationships between the major carbohydrate pathways and the citric acid cycle. The pathway from glucose-6-phosphate to pyruvate is the Embden–Meyerhof–Parnas pathway (glycolysis). The pentose phosphate pathway (PPP) and the Entner–Doudoroff pathway (ED) branch from 6-phosphogluconate. Both these pathways intersect with the glycolytic pathway at phosphoglyceraldehyde. All the carbohydrate pathways produce pyruvate, which is oxidized to acetyl–CoA. In aerobically growing organisms, the acetyl–CoA is oxidized to CO_2 in the citric acid cycle. The electrons from NAD(P)H and $FADH_2$ are transferred to the electron transport chain in respiring organisms, with the formation of ATP. In fermenting cells, the NADH is reoxidized by an organic acceptor (B) that is generated during catabolism. The citric acid cycle does not operate as an oxidative pathway during fermentative growth. Abbreviations: G6P, glucose-6-phosphate; F6P, fructose-6-phosphate; FDP, fructose-1,6-bisphosphate; PGALD, 3-phosphoglyceraldehyde; PEP, phosphoenolpyruvate; PPP, pentose phosphate pathway; ED, Entner–Doudoroff pathway; FAD, flavin adenine dinncleotide; $FADH_2$, reduced FAD.

or fermenting. During respiration, the NADH and $FADH_2$ are reoxidized via electron transport with the formation of a Δp. (The Δp is used for ATP synthesis via respiratory phosphorylation as explained in Chapter 3.) In fermenting cells, most of the NADH is reoxidized in the

cytosol by an organic acceptor, but ATP is not made. (However, see note 2.) The different pathways for the reoxidation of NADH in fermenting bacteria are discussed in Chapter 14.

The student will notice that the citric acid cycle generates a great deal of NADH and

FADH$_2$. The reoxidation of the NADH and FADH$_2$ requires adequate amounts of electron acceptor, such as is provided to respiring organisms. In fact, the oxidative citric acid cycle as illustrated in Fig. 8.1 is coupled to respiration, and during fermentative growth it becomes modified into a reductive pathway (Section 8.10).

8.1 Glycolysis

The history of biochemical research had its origins in the elucidation of the glycolytic pathway, which began with the demonstration of cell-free fermentation in yeast extracts. Some of the highlights of this period are summarized in Box 8.1.

BOX 8.1 HISTORICAL PERSPECTIVE: CELL-FREE YEAST FERMENTATION AND THE BEGINNINGS OF BIOCHEMISTRY[1–3]

The discovery of cell-free yeast fermentation

During the 1890s two German brothers, Hans Büchner, a bacteriologist, and Eduard Büchner, a chemist, were engaged in developing methods to extract proteins from microbial cells. Hans Büchner was the director of the Munich Institute of Hygiene, and Eduard Büchner was a professor of analytical and pharmaceutical chemistry at Tübingen University.

Hans Büchner wanted to break open bacterial cells to test the idea that pathogenic bacteria contained in their protoplasm both protein toxins and antitoxins and that these antitoxins in the sera of immunized animals protected them from disease. At the time it was known that animals could be made immune to diphtheria or tetanus by injecting them with sera from animals that had survived the disease.

Hans Büchner was wrong in his supposition that there were protein antitoxins in bacterial protoplasm that accounted for immunization. (It was later discovered that polysaccharides in the bacterial cell envelope act as antigens and elicit an immune response). However, his research did result in the accidental discovery of cell-free fermentation by yeast extracts.

Earlier, in 1893, Hans's brother Eduard had been working in the laboratory of Adolf von Baeyer at Munich University on a method to break yeast cells open by grinding yeast in a mortar with fine sand.

Lack of research funds, however, had put a stop to the research. Eduard later became a professor of analytical and pharmaceutical chemistry at Tübingen University.

Meanwhile, in 1894 Hans Büchner became chair of the Department of Hygiene in Munich, and in 1896 resumed his attempts to break open bacterial cells to isolate the protoplasmic proteins. Knowing of his brother's work with sand grinding of yeast, Hans suggested that his assistant Martin Hahn try the method. Martin Hahn ground yeast cells with quartz sand and added water to produce a paste, which was wrapped in cloth and pressed in a hydraulic press. A yellowish juice was produced. The plan was to study the immunological characteristics of the yellow yeast juice. To preserve the yeast juice while Hahn went on vacation, Hans Büchner suggested the addition of 40% glucose, since sugar was a known preservative, added to fruit preserves.

Meanwhile, Eduard Büchner was visiting from Tübingen, using his vacation time to work on pressed yeast juice in his brother's laboratory. He prepared yeast juice with 40% glucose, and on returning later to the laboratory, noticed bubbles emanating from the juice. He immediately realized that a cell-free fermentation was taking place. Eduard Büchner continued the research, published his conclusions in 1897, and in 1907 won the Nobel Prize in Chemistry for his work.

Eduard Büchner thought that the fermentation was carried out by a single soluble intracellular enzyme that he called "zymase." We now know that sugar fermentation requires a mixture of enzymes that take part in the glycolytic pathway.

The elucidation of the glycolytic pathway

Between Büchner's paper in 1897 and the 1940s, the chemical reactions of the entire glycolytic pathway were elucidated, as well as the enzymes catalyzing these reactions. Chemical analysis identified the metabolic intermediates, which were made to accumulate in various ways, including the use of specific enzyme inhibitors, or by the removal of specific coenzymes (e.g., by dialysis).

In addition to cell-free yeast preparations, much research was done with cell-free muscle preparations that converted glucose to lactic acid. An important discovery was made in 1905 by the English scientists Arthur Harden and William John Young. While studying fermentation by yeast extracts, they learned that the rate of fermentation could be stimulated by adding inorganic phosphate to the medium. The phosphate they added rapidly disappeared and was replaced by a form of phosphate that was not precipitable by magnesium citrate mixture. This led the investigators to search for phosphorylated intermediates and resulted in the discovery of fructose-1,6-bisphosphate (the "Harden–Young ester"). When the fructose-1,6-bisphosphate was added back to yeast extracts, the mixture fermented, proving that the ester was indeed an intermediate in the fermentation pathway.

Harden and Young also discovered that dialysis of the yeast extracts destroys the fermentative activity but that the addition of boiled yeast extract (not simply inorganic phosphate) restores fermentation.

This led to the discovery of NAD^+. Arthur Harden was knighted in 1926 and in 1929 shared the Nobel Prize in Chemistry with the Swedish Chemist Hans von Enler-Chelpin.

REFERENCES

1. Fruton, J. S. 1972. *Molecules and Life: Historical Essays on the Interplay of Chemistry and Biology*. John Wiley & Sons, New York.

2. Florkin, M. (Ed.). *Comprehensive Biochemistry*, Vol. 31, *A History of Biochemistry*, Part III. 1975. Elsevier Scientific Publishers, Amsterdam.

3. Hoffmann-Ostenhof, O. (Ed.). 1987. *Intermediary Metabolism*. Van Nostrand Reinhold, New York.

It is best to think of glycolysis as occurring in two stages:

Stage 1. This stage catalyzes the splitting of the glucose molecule (C_6) into two phosphoglyceraldehyde (C_3) molecules. It consists of four consecutive reactions. Two ATPs are used per glucose metabolized, and these donate the phosphoryl groups that become the phosphates in phosphoglyceraldehyde. (The phosphoglyceraldehyde becomes phosphoenolpyruvate in stage 2, and the phosphate that originated from ATP is returned to ATP in the pyruvate kinase step, thus regenerating the ATP.)

Stage 2. This stage catalyzes the oxidation of phosphoglyceraldehyde to pyruvate. It consists of five consecutive reactions. Stage 2 generates four ATPs per glucose metabolized, hence the net yield of ATP is 2. Stage 2 reactions are not unique to glycolysis and also occur when pyruvate is formed from phosphoglyceraldehyde in the pentose phosphate pathway and the Entner–Doudoroff pathway, accounting for ATP synthesis in these pathways.

Stage 1: Glucose + 2 ATP → 2 PGALD + 2 ADP

Stage 2: 2 PGALD + 2 P_i + 4 ADP + 2 NAD^+

→ 2 pyruvate + 4 ATP + 2 NADH + $2H^+$

Sum: Glucose + 2 ADP + 2 P_i + 2 NAD^+

→ 2 pyruvate + 2 ATP + 2 NADH + $2H^+$

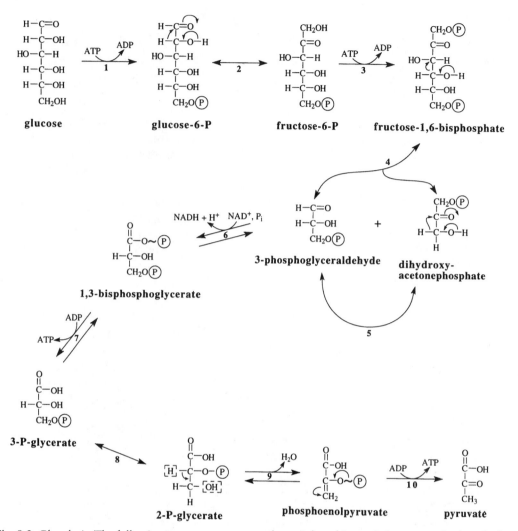

Fig. 8.2 Glycolysis. The following enzymes serve as catalysts: **1**, hexokinase; **2**, isomerase; **3**, phosphofruc-tokinase; **4**, fructose-1,6-bisphosphate aldolase; **5**, triosephosphate isomerase; **6**, triosephosphate dehydrogenase; **7**, phosphoglycerate kinase; **8**, mutase; **9**, enolase; **10**, pyruvate kinase.

The reactions are summarized in Fig. 8.2. The pathway begins with the phosphorylation of glucose to form glucose-6-phosphate (reaction 1). The phosphoryl donor is ATP in a reaction catalyzed by hexokinase. The ATP is regenerated from phosphoenolpyruvate in stage 2. Some bacteria phosphorylate glucose during transport into the cell via the phosphotransferase (PTS) system, in which case the phosphoryl donor is phosphoenolpyruvate. (See Section 16.3.4 for a discussion of the phosphotransferase system.) The glucose-6-phosphate (G6P) isomerizes to fructose-6-phosphate (F6P) in a reaction catalyzed by the enzyme isomerase (reaction 2). The isomerization is an electron

shift in which two electrons from the C2 carbon reduce the C1 aldehyde of the glucose-6-phosphate molecule to an alcohol (Section 8.1.3). The fructose-6-phosphate is phosphorylated at the expense of ATP to fructose-1, 6-bisphosphate (FBP) by the enzyme fructose-6-phosphate kinase (reaction 3). The ATP used to phosphorylate fructose-6-phosphate is also regenerated from phosphoenolpyruvate in stage 2. The fructose-1,6-bisphosphate is split into phosphoglyceraldehyde (PGALD) and dihydroxyacetone phosphate (DHAP) by fructose-1,6-bisphosphate aldolase (reaction 4). The splitting of fructose-1,6-bisphosphate is facilitated by the electron-attracting keto group at C2, thus rationalizing the isomerization of glucose-6-phosphate to fructose-6-phosphate (Section 8.1.3). The dihydroxyacetone phosphate is isomerized to phosphoglyceraldehyde (reaction 5), in a reaction similar to the earlier isomerase reaction (reaction 2). Thus, stage 1 produces two moles of phosphoglyceraldehyde per mole of glucose.

In stage 2, both moles of phosphoglyceraldehyde are oxidized to 1,3-bisphosphoglycerate (also called disphosphoglycerate, DPGA) (reaction 6). The bisphosphoglycerate serves as the phosphoryl donor for a substrate-level phosphorylation catalyzed by the enzyme phosphoglycerate kinase (reaction 7). At this point, two ATPs are made, one from each of the two bisphosphoglycerates. The product of the phosphoglycerate kinase reaction is 3-phosphoglycerate (3-PGA). The two moles of 3-phosphoglycerate are converted to two moles of 2-phosphoglycerate (2-PGA) (reaction 8), which are dehydrated to two moles of phosphoenolpyruvate (PEP) (reaction 9). The phosphoenolpyruvate serves as the phosphoryl

donor in a second site substrate-level phosphorylation to form two more moles of ATP and two moles of pyruvate (reaction 10). Notice that the phosphate in the 2-phosphoglycerate originated from ATP during the phosphorylations in stage 1 (reactions 1 and 3). Thus, the net synthesis of ATP from ADP and inorganic phosphate in glycolysis is coupled to the oxidation of phosphoglyceraldehyde to 3-phosphoglycerate (reactions 6 and 7). For a more detailed discussion of substrate-level phosphorylations see Sections 7.3.1 to 7.3.4.

8.1.1 Glycolysis as an anabolic pathway

The glycolytic pathway serves not only to oxidize carbohydrate to pyruvate and to phosphorylate ADP, but it also provides precursor metabolites for many other pathways. Figure 8.3 summarizes the glycolytic reactions and points out only a few of the branch points to other pathways. For example, glucose-6-phosphate is a precursor to polysaccharides, pentose

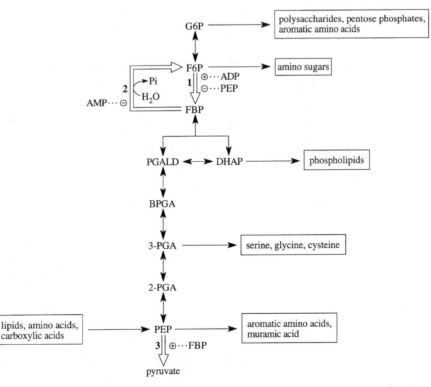

Fig. 8.3 Glycolysis as an anabolic pathway and its regulation in *E. coli*. The rationale for the pattern of regulation is that when the ADP and AMP levels are high, the ATP levels are low and therefore glycolysis is stimulated. Steady state levels of intermediates are maintained by positive and negative feedback inhibition.

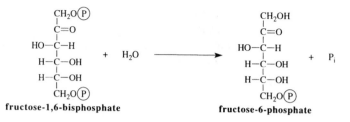

Fig. 8.4 The conversion of fructose-1,6-bisphosphate, via fructose-1,6-bisphosphatase, to fructose-6-phosphate.

phosphates, and aromatic amino acids; fructose-6-phosphate is a precursor to amino sugars (e.g., muramic acid and glucosamine found in the cell wall); dihydroxyacetone phosphate is a precursor to phospholipids; 3-phosphoglycerate is a precursor to the amino acids glycine, serine, and cysteine; and phosphoenolpyruvate is a precursor to aromatic amino acids and to the lactyl portion of muramic acid.

When the organisms are not growing on carbohydrate, they must synthesize these glycolytic intermediates from other carbon sources. Figure 8.3 shows that some of the carbon from amino acids, carboxylic acids (organic acids), and lipids are converted to phosphoenolpyruvate, from which the glycolytic intermediates can be synthesized. Also, pyruvate can serve as a carbon source and therefore must be converted to glycolytic intermediates. However, it can be seen that the glycolytic pathway can be reversed from phosphoenolpyruvate only to fructose-1,6-bisphosphate (FBP), and not at all from pyruvate. This is because the high free energy in the phosphoryl donors with respect to the phosphorylated products renders the pyruvate kinase and phosphofructokinase reactions physiologically irreversible. *Therefore, to reverse glycolysis, the kinase reactions are bypassed.* The conversion of fructose-1-1,6-bisphosphate to fructose-6-phosphate requires fructose-1,6-bisphosphate phosphatase (Fig. 8.4). Alternative ways to convert pyruvate directly to phosphoenolpyruvate without using pyruvate kinase are discussed in Section 8.13.

8.1.2 Regulation of glycolysis

Figure 8.3, in addition to presenting glycolysis as an anabolic pathway, illustrates the regulation of the process in *E. coli.* Two key enzymes

in regulating the *directionality* of carbon flow are phosphofructokinase (reaction **1**) and fructose-1,6-bisphosphate phosphatase (reaction **2**), which catalyze physiologically irreversible steps. The kinase catalyzes the phosphorylation of fructose-6-phosphate to fructose-1,6-bisphosphate, whereas the phosphatase catalyzes the dephosphorylation of fructose-1,6-bisphosphate to fructose-6-phosphate. Models for the regulation of glycolysis are based primarily on in vitro studies of the allosteric properties of the enzymes. Important effector molecules are AMP and ADP. When these are high, ATP is low, since both are derived from ATP. That is,

$$ATP \rightarrow ADP + P_i$$

$$ATP \rightarrow AMP + PP_i$$

Thus, high ADP and AMP concentrations are a signal that the ATP levels are low. (Allosteric activation and inhibition are discussed in Chapter 6.) Since glycolysis produces ATP, it makes sense to stimulate glycolysis when the ATP levels are low. *E. coli* accomplishes this by allosterically activating the phosphofructokinase with ADP, which, as mentioned, is at a higher concentration when the ATP levels are low. At the same time that glycolysis is stimulated by ADP, the reversal of glycolysis is slowed by AMP, which is also at a higher concentration when the ATP levels are low. These trends occur simultaneously because AMP inhibits the fructose-1,6-bisphosphate phosphatase reaction. The student may notice that the sum of the phosphofructokinase and fructose-1,6-bisphosphatase reaction is the hydrolysis of ATP (i.e., ATPase activity). The stimulation of the phosphofructokinase by ADP and the inhibition of the phosphatase by AMP prevents the unnecessary hydrolysis of ATP when ATP levels are low.

In *E. coli*, glycolysis is regulated not only by AMP and ADP but also by phosphoenolpyruvate and fructose-6-phosphate. As indicated in Fig. 8.3, the phosphofructokinase is feedback inhibited by phosphoenolpyruvate. This can be considered to be an example of end-product inhibition. The pyruvate kinase, another physiologically irreversible reaction, is positively regulated by fructose-1,6-bisphosphate, which is an example of a precursor metabolite activating a later step in the pathway. (Feedback inhibition and precursor activation are discussed in Sections 6.1.1 and 6.1.2.)

8.1.3 The chemical bases for the isomerization and aldol cleavage reactions in glycolysis

It is important to understand the chemistry of metabolic reactions as well as to learn the pathways and their physiological roles. To this end, the isomerization and aldol cleavage reactions will be explained because they are common reactions that we will see later in other pathways. Consider the isomerization of glucose-6-

phosphate to fructose-6-phosphate. The rationale for this isomerization is that it creates an electron-attracting keto group at C2 of the sugar, and the electron-attracting keto group is necessary to break the bond between C3 and C4 in the aldolase reaction. These reactions are shown in Fig. 8.5.

The isomerization can be viewed as the oxidation of C2 by C1, because two electrons shift from C2 to C1. This happens in two steps. A hydrogen dissociates from C2, and two electrons shift in to form the *cis*-enediol. Then, the hydrogen in the C2 hydroxyl dissociates and two electrons shift in, forcing the two electrons in the double bond to go to the C1. The result is fructose-6-phosphate. The fructose-6-phosphate becomes phosphorylated to fructose-1,6-bisphosphate. The fructose-1, 6-bisphosphate is split by the aldolase when the keto group on C2 pulls electrons away from the C–C bond between C3 and C4, as two electrons shift in from the hydroxyl on C4 (Fig. 8.5). The products of the split are phosphoglyceraldehyde and dihydroxyacetone phosphate. A second isomerization converts the dihydroxyacetone phosphate to phosphoglyceraldehyde via the

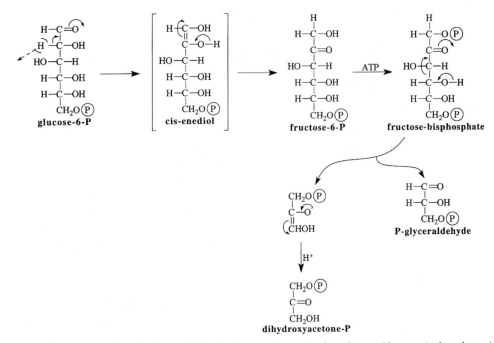

Fig. 8.5 Making two phosphoglyceraldehydes from glucose-6-phosphate. Glucose-6-phosphate itself cannot be split because there is no electron-attracting group to withdraw the electrons from the C–C bond between C3 and C4. An electron-withdrawing keto group is created on C2 when glucose-6-phosphate is isomerized to fructose-6-phosphate.

same mechanism as the isomerization between glucose-6-phosphate and fructose-6-phosphate. In this way, two phosphoglyceraldehydes can be formed from glucose-6-phosphate.

8.1.4 Why are the glycolytic intermediates phosphorylated?

In glycolysis, the phosphorylation of ADP by inorganic phosphate is due to two reactions that take place in stage 2: the oxidation of the C1 aldehyde of 3-phosphoglyceraldehyde to the acyl phosphate and the subsequent transfer of the phosphoryl group to ADP. The first reaction is catalyzed by triosephosphate dehydrogenase, and the second reaction is catalyzed by phosphoglycerate kinase. Given that these are the steps in which net ATP is made from ADP and inorganic phosphate, one can ask why the other intermediates are phosphorylated.

　　Phosphorylation of the intermediates requires the use in stage 1 of two ATPs, which are simply regenerated in the pyruvate kinase step. There is probably more than one reason for the phosphorylation of all the intermediates. One such reason may be that the kinase reactions in stage 1 are irreversible and therefore drive the reactions rapidly in the direction of pyruvate. Another possible reason has to do with the physiological role of the pathway. Glycolysis is not simply a pathway for the oxidation of glucose and the provision of ATP. Very importantly, the glycolytic pathway also provides phosphorylated precursors to many other pathways. In subsequent chapters, we will study these interconnections with other pathways.

8.2　The Fate of NADH

If the NADH were not reoxidized to NAD^+, then all pathways (including glycolysis) that require NAD^+ would stop. Clearly, glycolysis must be coupled to pathways that reoxidize NADH back to NAD^+. Bacteria have three ways of reoxidizing NADH: respiration, fermentation, and the hydrogenase reaction.

Respiration (aerobic or anaerobic)

$$NADH + H^+ + B + yADP + yP_i$$
$$\rightarrow NAD^+ + BH_2 + yATP$$

where y is approximately equal to the number of coupling sites and B is the terminal electron acceptor. (See Section 4.5.2 for a discussion of the number of ATP molecules formed per coupling site according to the chemiosmotic theory.)

Fermentation (anaerobic)

$$NADH + H^+ + B \text{ (organic)} \rightarrow NAD^+ + BH_2$$

Hydrogenase (anaerobic)

$$NADH + H^+ \rightarrow H_2 + NAD^+$$

Aerobic and anaerobic respiration are discussed in Chapter 4. In the absence of respiration, NADH can be reoxidized in the cytosol via fermentation (discussed in Chapter 14). A third way to reoxidize NADH is via the enzyme hydrogenase in the cytosol. Hydrogenases that use NADH as the electron donor are found in fermenting bacteria. However, the oxidation of NADH with the production of hydrogen gas generally proceeds only when the hydrogen gas concentration is kept low (e.g., during growth with hydrogen gas utilizers). This is because the equilibrium favors the reduction of NAD^+. Interspecies hydrogen transfer is discussed in Section 14.4.1.

8.3　Why Write NAD^+ Instead of NAD, and NADH Instead of $NADH_2$?

Oxidized nicotinamide adenine dinucleotide is written as NAD^+, and the reduced form is written NADH, not $NADH_2$. To understand why, we must examine the structures (Fig. 8.6). Notice that the molecule can accept two electrons but only one hydrogen. That is why it is written as $NADH + H^+$. The oxidized molecule is written NAD^+ because the nitrogen carries a formal positive charge.

8.4　A Modified EMP Pathway in the Hyperthermophilic Archaeon Pyrococcus furiosus

In comparison to the long history of research on bacteria, the study of archaeal metabolism is

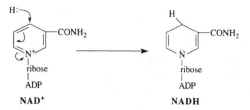

Fig. 8.6 The structures of NAD$^+$, NADH, and nicotinamide. NAD$^+$ is a derivative of nicotinamide, to which ADP–ribose is attached by the nitrogen of nicotinamide. In the oxidized form, the nitrogen has four bonds, hence carries a positive charge. NAD$^+$ accepts two electrons but only one hydrogen (hydride ion) to become NADH. The second hydrogen removed from the electron donor (the reductant) is released into the medium as a proton.

a relatively recently focus of attention, and there have been many rewarding findings that suggest the presence of several metabolic features distinct from those of the bacteria.[3,4] Examples include the novel ether-linked lipids described in Chapter 1 and their biosynthesis, discussed in Chapter 9. Additional features of archaeal metabolism that appear to be unique to these microorganisms include the synthesis of methane, and the presence of novel coenzymes for acetate and methane metabolism discussed in Chapter 13. A modified Embden–Meyerhof–Parnas pathway has been proposed for *Pyrococcus furiosus*.[5] This organism has a growth temperature optimum of 100 °C and ferments carbohydrates and peptides to acetate, H$_2$, and CO$_2$. Monosaccharides such as glucose and fructose do not support growth, but other carbohydrates such as maltose (a disaccharide of glucose) are transported into the cell and converted to glucose. If S° is present in the medium, it is used as an electron sink and reduced to H$_2$S.

A pathway proposed for the catabolism of glucose to acetate is shown in Fig. 8.7.[6] The postulated pathway resembles the classic EMP pathway but differs in several respects. Very interestingly, the phosphoryl donor in the hexokinase and fructokinase reactions appears to be not ATP but, rather, ADP. Another difference is that the enzyme that oxidizes glyceraldehyde-3-phosphate to 3-phosphoglycerate is suggested to be a ferredoxin-linked enzyme rather than an NAD$^+$-linked enzyme, and it appears that 1, 3-bisphosphoglycerate is not an intermediate.

Pyruvate is oxidized to acetyl–CoA and CO$_2$ by pyruvate:ferredoxin oxidoreductase. (See Section 7.3.2 for a description of this reaction.) Finally, an ADP-dependent acetyl–CoA synthetase catalyzes a reaction in which ADP is phosphorylated and acetate is formed. See Section 7.3.2 for a description of the ADP-dependent acetyl–CoA synthetase and its distribution. All the enzymes proposed for the modified EMP pathway have been detected in extracts of *P. furiosus*, and *in vivo* NMR studies of the products formed from [^{13}C]glucose are in agreement with the pathway drawn in Fig. 8.7.[3]

8.5 The Pentose Phosphate Pathway

Another important pathway for carbohydrate metabolism is the pentose phosphate pathway. The pentose phosphate pathway is important first because it produces the pentose phosphates, which are the precursors to the ribose and deoxyribose in the nucleic acids, and second because it provides erythrose phosphate, which is the precursor to the aromatic amino acids phenylalanine, tyrosine, and tryptophan. Also, the NADPH produced in the pentose phosphate pathway is a major source of electrons for biosynthesis in most of the pathways in which reductions occur. (See note 7 for an alternative means of generating NADPH.)

The pentose phosphate pathway is important to learn for yet one more reason. Several of its reactions are the same as the reactions in the Calvin cycle, which is used by many autotrophic organisms to incorporate CO$_2$ into organic carbon (Chapter 13).

The overall reaction of the pentose phosphate pathway is

$$G6P + 6NADP^+ \rightarrow 3CO_2 + 1PGALD$$
$$+ 6NADPH + 6H^+ \qquad (8.1)$$

8.5.1 The reactions of the pentose phosphate pathway

The pentose phosphate pathway is complex and can be best learned by dividing the reactions into three stages. Stage, which consists of oxidation–decarboxylation reactions, produces CO$_2$ and NADPH. Stage 2 consists of

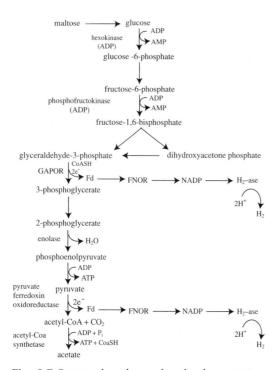

Fig. 8.7 Proposed pathway for the fermentation of maltose, from glucose, to acetate, CO_2, and H_2 in *Pyrococcus furiosus*: GAPOR, glyceraldehyde-3-phosphate ferredoxin oxidoreductase; Fd, ferredoxin; FNOR, ferredoxin NADP oxidoreductase; H_2ase, hydrogenase. *Source*: Adapted from Mukund, S., and M. W. W. Adams. 1995. Glyceraldehyde-3-phosphate ferredoxin oxidoreductase, a novel tungsten-containing enzyme with a potential glycolytic role in the hyperthermophilic archaeon *Pyrococcus furiosus*. *J. Biol. Chem.* 270:8389–8392.

isomerization reactions that make the precursors for stage 3. The stage 3 reactions are sugar rearrangements. The phosphoglyceraldehyde is produced in stage 3.

Stage 1: Oxidation–decarboxylation reactions

The oxidation–decarboxylation reactions are shown in Fig. 8.8. These reactions oxidize the C1 in glucose-6-phosphate to a carboxyl group and remove it as carbon dioxide. Glucose actually exists as a ring structure, which forms because the aldehyde group at C1 reacts with the C5 hydroxyl group, forming a hemiacetal. The C1 is therefore not a typical aldehyde

in that it does not react in the Schiff test and does not form a bisulfite addition product. Nevertheless, it is easily oxidized. The glucose-6-phosphate is oxidized by $NADP^+$ to 6-P-gluconolactone by glucose-6-phosphate dehydrogenase (reaction 1). The lactone is then hydrolyzed to 6-P-gluconate by gluconolactonase (reaction 2). In the oxidation of glucose-6-phosphate to 6-phosphogluconate, water contributes the second oxygen in the carboxyl group and an acyl phosphate intermediate is not formed. This means that the energy from the oxidation is lost as heat, and the reaction is physiologically irreversible, as is often the case for the first reaction in a metabolic pathway. (Recall that during the oxidation of phosphoglyceraldehyde, inorganic phosphate is added and 1,3-bisphosphoglycerate is formed. The energy of oxidation is trapped in the acyl phosphate rather than being lost as heat. A subsequent substrate-level phosphorylation recovers the energy of oxidation in the form of ATP.) The product of the oxidation, 6-P-gluconate, is then oxidized on the C3 to generate a keto group β to the carboxyl (reaction 3). A β-decarboxylation then occurs, generating ribulose-5-phosphate (reaction 4). (The mechanism of β-decarboxylations is described in Section 8.11.2.) Therefore, the products of the three stages are as follows: 1, carbon dioxide; 2, NADPH, and 3, the five-carbon sugar phosphate, ribulose-5-phosphate (RuMP). The rest of the pathway continues with ribulose-5-phosphate.

Stage 2: The isomerization reactions

During the second stage, some of the ribulose-5-phosphate is isomerized to ribose-5-phosphate and to xylulose-5-phosphate. Isomers are molecules having the same chemical formula but different structural formulas; that is, their parts have been switched around. For example, the chemical formula for ribulose-5-phosphate is $C_5H_{11}O_8P$. Ribose-5-phosphate and xylulose-5-phosphate have the same chemical formula. However, their structures are different (Fig. 8.9).

One of the isomerases is an epimerase, RuMP epimerase, which catalyzes a movement of the hydroxyl group from one side of the C3 in ribulose-5-phosphate to the other. The

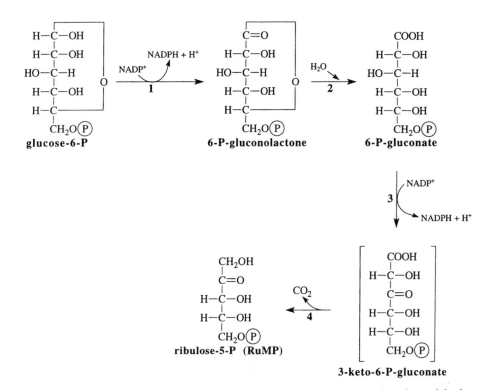

Fig. 8.8 The oxidation–decarboxylation reactions. Enzymes: **1**, glucose-6-phosphate dehydrogenase; **2**, gluconolactonase; **3** and **4**, 6-phosphogluconate dehydrogenase.

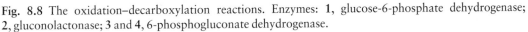

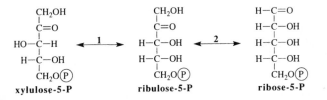

Fig. 8.9 The isomerization reactions of the pentose phosphate pathway. Enzymes: **1**, RuMP epimerase; **2**, ribose-5-phosphate isomerase.

product is xylulose-5-phosphate (the epimer[8] of ribulose-5-phosphate). The other isomerase converts ribulose-5-phosphate to ribose-5-phosphate.

Stage 3: The sugar rearrangement reactions

Stage 3 of the pentose phosphate pathway involves sugar rearrangement reactions. There are two basic types of reaction. One kind transfers a *two*-carbon fragment from a ketose to an aldose. The enzyme that catalyzes the transfer of the two-carbon fragment is called a *transke-*

tolase (TK). A second kind of reaction transfers a *three*-carbon fragment from a ketose to an aldose. The enzyme that catalyzes the transfer of a three-carbon fragment is called a *transaldolase* (TA). The rule is that the donor is always a ketose (with the OH group of the third carbon "on the left," as in xyulose-5-phosphate) and the acceptor is always an aldose. This rule is important to learn because we shall see other transketolase and transaldolase reactions later. Knowing the requirements will make it easier to remember the reactions. The transketolase and transaldolase reactions are summarized in Fig. 8.10.

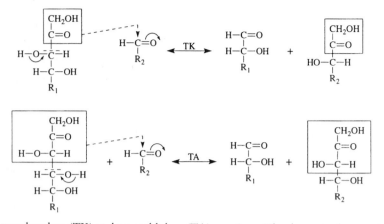

Fig. 8.10 The transketolase (TK) and transaldolase (TA) reactions. The donor is always a ketose with the keto group on C2, and the hydroxyl on C3 on the "left." In the transketolase reaction a C_2 unit is transferred with its bonding electrons to the carbonyl group on an aldehyde acceptor. The transaldolase transfers a three-carbon fragment. In the transketolase reaction, the newly formed alcohol group is on the "left," which means that the products of both the transketolase and transaldolase reactions can act as donors in a subsequent transfer.

Summarizing the pentose phosphate pathway

Figure 8.11 summarizes the pentose phosphate pathway. Reactions 1 through 3 comprise the oxidative decarboxylation reactions of stage 1. Three moles of glucose-6-phosphate must be oxidized to produce three moles of CO_2 and one mole of phosphoglyceraldehyde. Therefore, stage 1 produces three moles of ribulose-5-phosphate. Reactions 4 and 5 are the isomerization reactions of stage 2, in which the three moles of ribulose-5-phosphate are converted to one mole of ribose-5-phosphate and two moles of xylulose-5-phosphate. Reactions 6, 7, and 8 comprise stage 3. Reaction 6 is a transketolase reaction in which a xylulose-5-phosphate (C_5) transfers a two-carbon moiety to ribose-5-phosphate (C_5) with the formation of sedoheptulose-7-phosphate (C_7) and phosphoglyceraldehyde (C_3). The two-carbon moiety is highlighted as a boxed area. Reaction 7 is a transaldolase reaction in which the sedoheptulose-7-phosphate transfers a three-carbon moiety to the phosphoglyceraldehyde to form erythrose-4-phosphate (C_4) and fructose-6-phosphate (C_6). The three-carbon moiety is highlighted as a dashed box. Reaction 8 is a transketolase reaction in which xylulose-5-phosphate transfers a two-carbon moiety to the erythrose-4-phosphate forming phosphoglyc-

eraldehyde and fructose-6-phosphate. You will notice that the sequence of reactions is

Transketolase → transaldolase → transketolase

The result is that three moles of glucose-6-phosphate are converted to two moles of fructose-6-phosphate and one mole of phosphoglyceraldehyde. The two moles of fructose-6-phosphate become glucose-6-phosphate by isomerization, and the net result is the conversion of one mole of glucose-6-phosphate to one mole of phosphoglyceraldehyde, three moles of carbon dioxide, and six moles of NADPH. This is shown in the following carbon balance.

Carbon balance for the pentose phosphate pathway

| Oxidative decarboxylation | 3 glucose-6-P $\rightarrow$ 3 ribulose-5-P + 3CO$_2$ |
| | $3C_6$ $\qquad$ $3C_5$ $\qquad$ $3C_1$ |

Isomerizations
3 ribulose-5-P $\rightarrow$ 2 xylulose-5-P + ribose-5-P
$3C_5$ $\qquad$ $2C_5$ $\qquad$ C_5

Transketolase
xylulose-5-P + ribose-5-P $\rightarrow$ sedoheptulose-7-P + phosphoglyceraldehyde
C_5 $\qquad$ C_5 $\qquad$ C_7 $\qquad$ C_3

Transaldolase
sedoheptulose-7-P + phosphoglyceraldehyde $\rightarrow$ fructose-6-P + erythrose-4-P
C_7 $\qquad$ C_3 $\qquad$ C_6 $\qquad$ C_4

Transketolase
xylulose-5-P + erythrose-4-P $\rightarrow$ fructose-6-P + phosphoglyceraldehyde
C_5 $\qquad$ C_4 $\qquad$ C_6 $\qquad$ C_3

Sum:
glucose-6-P $\rightarrow$ phosphoglyceraldehyde + 3CO$_2$
C_6 $\qquad$ C_3 $\qquad$ $3C_1$

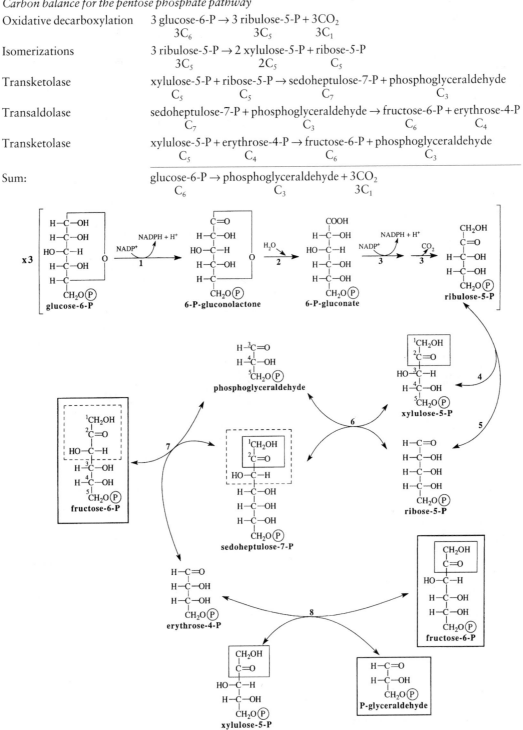

Fig. 8.11 The pentose phosphate pathway. Enzymes: **1**, glucose-6-phosphate dehydrogenase; **2**, lactonase; **3**, 6-phosphogluconate dehydrogenase; **4**, ribulose-5-phosphate epimerase; **5**, ribose-5-phosphate isomerase; **6** and **8**, transketolase; **7**, transaldolase. The two-carbon moiety transferred by the transketolase is shown in the boxed area. The three-carbon fragment transferred by the transaldolase is shown in the dashed box. The product of the glucose-6-phosphate dehydrogenase reaction is the lactone, which is unstable and hydrolyzes spontaneously to the free acid. However, there exists a specific lactonase that catalyzes the reaction.

Later we will study the Calvin cycle, a pathway by which many organisms can grow on CO_2 as the sole source of carbon. In the Calvin cycle, CO_2 is first reduced to phosphoglyceraldehyde. The phosphoglyceraldehyde is then converted via sedoheptulose-7-phosphate to pentose phosphates, essentially by means of a reversal of the pentose phosphate pathway. However, the Calvin cycle has no transaldolase and synthesizes sedoheptulose-7-phosphate by an alternate route, which runs irreversibly in the direction of pentose phosphates (Section 13.1.1).

The pentose phosphate pathway serves important biosynthetic functions. Notice that stage 1 (the oxidative decarboxylation reactions) and stage 2 (the isomerization reactions) generate the pentose phosphates required for nucleic acid synthesis. Stage 1 also produces NADPH, which is used in several biosynthetic pathways. Stage 3 generates the erythrose-4-phosphate necessary for aromatic amino acid biosynthesis.

Some bacteria rely completely on the pentose phosphate pathway for sugar catabolism

Thiobacillus novellus and *Brucella abortus* lack both stage 1 of the Embden–Meyerhof–Parnas pathway and the enzymes of the Entner–Doudoroff pathway. These organisms use only an oxidative pentose phosphate pathway to grow on glucose. They oxidize the glucose to phosphoglyceraldehyde via the pentose phosphate pathway. The phosphoglyceraldehyde is then oxidized to pyruvate via reactions that are the same as stage 2 of the EMP pathway; then the pyruvate is oxidized to CO_2 via the citric acid cycle.

Relationship of the pentose phosphate pathway to glycolysis

The pentose phosphate pathway and glycolysis interconnect at phosphoglyceraldehyde and fructose-6-phosphate (Fig. 8.1). Thus organisms growing on pentoses can make hexose phosphates. Furthermore, because stages 2 and 3 of the pentose phosphate pathway are reversible, it is possible to synthesize pentose phosphates from phosphoglyceraldehyde and avoid the oxidative decarboxylation reactions of stage 1. This would uncouple pentose phosphate synthesis from NADPH production,

possibly conferring an advantageous metabolic flexibility on the cells.

8.6 The Entner–Doudoroff Pathway

Many prokaryotes have another pathway for the degradation of carbohydrates called the Entner–Doudoroff (ED) pathway. The other pathways that we have been studying are common to all cells, whether they be prokaryotes or eukaryotes. But the Entner–Doudoroff pathway has been found almost entirely among the prokaryotes, including bacteria and archaea. (Reviewed in ref. 1.) The pathway is widespread, particularly among the aerobic gram-negative bacteria. It is usually not found among anaerobic bacteria, perhaps because of the low ATP yields, as discussed shortly.

Most bacteria degrade sugars via the Embden–Meyerhof–Parnas pathway, but when grown on certain compounds (e.g., gluconic acid), they use the Entner–Doudoroff pathway. However, some strictly aerobic bacteria cannot carry out stage 1 of the Embden–Meyerhof–Parnas pathway and rely entirely on the Entner–Doudoroff pathway for sugar degradation (Table 8.1). The overall reaction for the Entner–Doudoroff pathway is

$$Glucose + NADP^+ + NAD^+ + ADP + P_i$$

$$\rightarrow 2 \text{ pyruvic acid} + NADPH + 2H^+$$

$$+ NADH + ATP$$

Table 8.1 Distribution of the EMP and ED pathways in certain bacteria[a]

Bacterium	EMP	ED
Arthrobacter species	+	−
Azotobacter chromococcus	+	−
Alcaligenes eutrophus	−	+
Bacillus species	+	−
Escherichia coli and other enteric bacteria[b]	+	−
Pseudomonas species	−	+
Rhizobium species	−	+
Thiobacillus species	−	+
Xanthomonas species	−	+

[a] The absence of the EMP pathway means that the bacteria rely exclusively on the ED pathway to grow on glucose.
[b] Organisms such as *E. coli* synthesize the enzymes of the ED pathway when growing on gluconate.

Source: Gottschalk, G. 1986. *Bacterial Metabolism*. Springer-Verlag, Berlin.

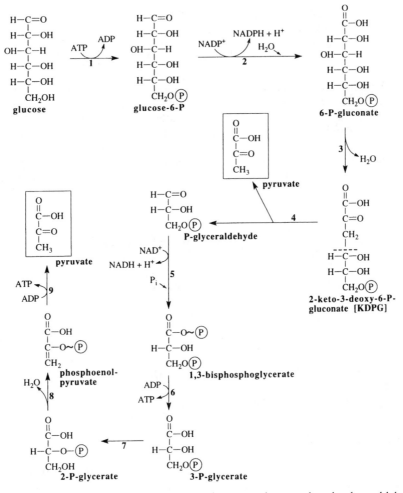

Fig. 8.12 The Entner–Doudoroff pathway. Because there is only one phosphoglyceraldehyde formed, there is only one ATP made. The enzymes unique to this pathway are the 6-phosphogluconate dehydratase (reaction 3) and the KDPG aldolase (reaction 4). The other enzymes are present in the pentose phosphate pathway and the glycolytic pathway. Enzymes: 1, hexokinase; 2, glucose-6-phosphate dehydrogenase; 3, 6-phosphogluconate dehydratase; 4, KDPG aldolase; 5, triosephosphate dehydrogenase; 6, PGA kinase; 7, mutase; 8, enolase; 9, pyruvate kinase.

It can be seen that the pathway catalyzes the same overall reaction as the Embden–Meyerhof–Parnas pathway (i.e., the oxidation of one mole of glucose to two moles of pyruvic acid), except that only one ATP is made, and one NADPH and one NADH are made instead of two NADHs. Only one ATP is made because only one phosphoglyceraldehyde is made from glucose (Fig. 8.12).

8.6.1 The reactions of the Entner–Doudoroff pathway

The first oxidation is the oxidation of the C1 in glucose-6-phosphate to the carboxyl in 6-P-gluconate (Fig. 8.12, reaction 2). These are the same enzymatic reactions that oxidize glucose-6-phosphate in the pentose phosphate pathway, and they proceed through the gluconolactone. The pathway diverges from the pentose phosphate pathway at this point because some of the 6-P-gluconate is dehydrated to 2-keto-3-deoxy-6-P-gluconate (KDPG), rather than being oxidized to ribulose-5-phosphate (reaction 3). The KDPG is split by KDPG aldolase to pyruvate and phosphoglyceraldehyde (reaction 4). The phosphoglyceraldehyde is oxidized to pyruvate in a sequence of reactions (5–9) identical to those in stage 2 of the EMP pathway.

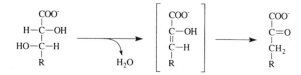

Fig. 8.13 Dehydration of a carboxylic acid with hydroxyl groups in the α and β positions. The dehydration of a carboxylic acid with hydroxyl groups in both the α and β position leads to the formation of an enol, which tautomerizes to the keto compound. That is because the hydroxyl on the C3 leaves with its bonding electrons, and the electrons bonded to the hydrogen on the C2 shift in to form the double bond. This happens when 6-phosphogluconate is dehydrated to 2-keto-3-deoxy-6-phosphogluconate in the Entner–Doudoroff pathway, and when 2-phosphoglycerate is dehydrated to phosphoenolpyruvate during glycolysis. The phosphoenolpyruvate tautomerizes to pyruvate when the phosphate is removed during the kinase reaction.

Reaction **3**, which is the dehydration of 6-phosphogluconate to KDPG, takes place via an enol intermediate that tautomerizes to KDPG. This is illustrated in Fig. 8.13. In this way, it is similar to the dehydration of 2-phosphoglycerate to phosphoenolpyruvate by enolase described in Section 7.3.4. However, in phosphoenolpyruvate, the enol derivative is stabilized by the phosphate group and the tautomerization does not take place until the phosphoryl group has been transferred.

8.6.2 Physiological role for the Entner–Doudoroff pathway

Since the Entner–Doudoroff pathway produces only one ATP, one can ask why the pathway is so common in the bacteria. Whereas hexoses are readily degraded by the Embden–Meyerhof–Parnas pathway, aldonic acids (aldoses oxidized at the aldehydic carbon) such as gluconate are not; they can, however, be degraded via the Entner–Doudoroff pathway. (Aldonic acids occur in nature and are sometimes important nutrients).

An example of degradation of aldonic acids occurs when *E. coli* is transferred from a medium containing glucose as the carbon source to one in which gluconate is the source of carbon. Growth on gluconate results in the induction of three enzymes: a gluconokinase that makes 6-P-gluconate from the gluconate (at the expense of ATP), 6-P-gluconate dehydratase, and KDPG aldolase. Thus, *E. coli* uses the Entner–Doudoroff pathway to grow on gluconate and the Embden–Meyerhof–Parnas pathway to grow on glucose.

Some bacteria do not have a complete Embden–Meyerhof–Parnas pathway and rely on the Entner-Doudoroff pathway for hexose degradation

Several prokaryotes (e.g., pseudomonads) do not make phosphofructokinase or fructose bisphosphate aldolase. Hence, they cannot carry out glucose oxidation via the Embden–Meyerhof–Parnas pathway. Instead, they use the Entner–Doudoroff pathway. (See Table 8.1 for the distribution of the EMP and ED pathways.) Some of the pseudomonads even oxidize glucose to gluconate before degrading it via the Entner–Doudoroff pathway, instead of making glucose-6-phosphate. The oxidation of glucose to gluconate may confer a competitive advantage, since it removes glucose, which is more readily utilizable by other microorganisms.

8.6.3 A partly nonphosphorylated Entner–Doudoroff pathway

A modified ED pathway has been found in the archaeon *Halobacterium saccharovorum* and in several bacteria, including members of the genera *Clostridium, Alcaligenes, and Achromobacter,* and in *Rhodopseudomonas sphaeroides.* The pathway is characterized as having nonphosphorylated intermediates prior to 2-keto-3-deoxygluconate. The first reaction is the oxidation of glucose to gluconate via an NAD$^+$-dependent dehydrogenase. The gluconate is then dehydrated to 2-keto-3-deoxygluconate by a gluconate dehydratase. The 2-keto-3-deoxygluconate is phosphorylated by a special kinase to form KDPG, which is metabolized to pyruvate by means of the ordinary ED reactions.

Bacteria that have a modified ED pathway can use it for the catabolism of gluconate, or glucose if the glucose dehydrogenase is present.

8.7 The Oxidation of Pyruvate to Acetyl–CoA: The Pyruvate Dehydrogenase Reaction

Pyruvate is the common product of sugar catabolism in all the major carbohydrate catabolic pathways (Fig. 8.1). We now examine the metabolic fate of pyruvate. One of the fates of pyruvate is to be oxidized to acetyl–CoA. As described in Section 7.3.2, this is catalyzed in anaerobically growing bacteria by either pyruvate–ferredoxin oxidoreductase or pyruvate–formate lyase. The acetyl–CoA thus formed is converted to acetyl phosphate via the enzyme phosphotransacetylase, and the acetyl phosphate donates the phosphoryl group to ADP in a substrate-level phosphorylation catalyzed by acetate kinase. The acetate thus formed is excreted. (However, certain anaerobic archaea oxidize pyruvate to acetyl–CoA by using pyruvate–ferredoxin oxidoreductase and then oxidize the acetyl–CoA to CO_2 via the citric acid cycle.[9]) The oxidation of pyruvate to acetyl–CoA during aerobic growth is carried out by the enzyme complex *pyruvate dehydrogenase*, which is widespread in both prokaryotes and eukaryotes. (The acetyl–CoA formed during aerobic growth is oxidized to CO_2 in the citric acid cycle.) The overall reaction for pyruvate dehydrogenase is:

$$CH_3COCOOH + NAD^+ + CoASH$$
$$\rightarrow CH_3COSCoA + CO_2 + NADH + H^+$$

The pyruvate dehydrogenase complex is a very large enzyme complex (in *E. coli*, about 1.7 times the size of the ribosome) located in the mitochondria of eukaryotic cells and in the cytosol of prokaryotes. The pyruvate dehydrogenase from *E. coli* consists of 24 molecules of enzyme E_1 (pyruvate dehydrogenase), 24 molecules of enzyme E_2 (dihydrolipoate transacetylase), and 12 molecules of enzyme E_3 (dihydrolipoate dehydrogenase). Several very important cofactors are involved. The cofactors are thiamine pyrophosphate (TPP), derived from the vitamin thiamine; flavin adenine dinucleotide (FAD), derived from the vitamin riboflavin; lipoic acid (RS_2); nicotinamide adenine dinucleotide (NAD^+), derived from the vitamin nicotinamide; and coenzyme A, derived from the vitamin pantothenic acid. (See Box 8.2 for a more complete discussion of vitamins.) The large size of the complex is presumably designed to process the heavy stream of pyruvate that is generated during the catabolism of sugars and other compounds. As described next, the pyruvate dehydrogenase complex catalyzes a short metabolic pathway rather than simply a single reaction. The individual reactions carried out by the pyruvate dehydrogenase complex are as follows (Fig. 8.14).

Step 1. Pyruvate is decarboxylated to form "active acetaldehyde" bound to TPP (Fig. 8.14). The reaction is catalyzed by pyruvate dehydrogenase, (E_1).

Step 2. The "active acetaldehyde" is oxidized to the level of carboxyl by the disulfide in lipoic acid. The disulfide of the lipoic acid is reduced to a sulfhydryl. During the reaction, TPP is displaced and the acetyl group is transferred to the lipoic acid. The reaction is also catalyzed by pyruvate dehydrogenase.

Step 3. A transacetylation occurs in which lipoic acid is displaced by CoASH, forming acetyl–CoA and reduced lipoic acid. The reaction is catalyzed by dihydrolipoate transacetylase, E_2.

Step 4. The lipoic acid is oxidized by dihydrolipoate dehydrogenase, E_3–FAD.

Step 5. The E_3–$FADH_2$ transfers the electrons to NAD^+.

All the intermediates remain bound to the complex and are passed from one active site to another. Presumably, this offers the advantage inherent in all multienzyme complexes (i.e., there is no dilution of intermediates in the cytosol, and side reactions are minimized). The student should refer to Section 1.2.6 for a discussion of multienzyme complexes in the cytoplasm.

8.7.1 Physiological control

The pyruvate dehydrogenase reaction, which is physiologically irreversible, is under metabolic

BOX 8.2 VITAMINS

Vitamins are growth factors that cannot be made by animals. During the early years of vitamin research, vitamins were divided into two groups: the fat-soluble vitamins and the water-soluble vitamins (B complex and vitamin C.) Vitamins are generally assayed according to specific diseases that they cure in animals fed on a vitamin-deficient diet. We now know that the water-soluble vitamin most responsible for stimulating growth in rats is riboflavin (B_2). Vitamin B_6 (pyridoxine) prevents rat facial dermatitis. Pantothenic acid cures chick dermatitis. Nicotinic acid (a precursor to nicotinamide) cures human pellagra. (The symptoms of pellagra are weakness, dermatitis, diarrhea, mental disorder, and death.) Vitamin C (ascorbic acid) prevents scurvy. Folic acid and vitamin B_{12} (cobalamin) prevent anemia. Vitamin D, or calciferol (fat soluble), prevents rickets. Vitamin E, or tocopherol (fat soluble), is required for rats for full-term pregnancy and prevents sterility in male rats. Vitamin K (fat soluble) is necessary for normal blood clotting. A deficiency in vitamin A (fat soluble) leads to dry skin, conjunctivitis of the eyes, night blindness, and retardation of growth. Male rats fed a diet deficient in vitamin A do not form sperm and become blind. Vitamin A (retinol) is an important component of the light-sensitive pigment in the rod cells of the eye.

In addition to vitamins, animals cannot make polyunsaturated fatty acids ("essential fatty acids") because they lack the enzyme to desaturate monounsaturated fatty acids. A diet deficient in polyunsaturated fatty acids leads to poor growth, skin lesions, impaired fertility, and kidney damage. Lipoic acid is synthesized by animals and is therefore not a vitamin. However, certain bacteria and other microorganisms require lipoic acid for growth. It was soon recognized that lipoic acid was required for the oxidation of pyruvic acid. The chemical identification of lipoic acid followed the purification of 30 mg from 10 tons (!) of water-soluble residue of liver. It was accomplished by Lester Reed at the University of Texas in 1949, with collaboration from scientists at Eli Lilly and Company. Lipoic acid is a growth factor for some bacteria. It is an eight-carbon saturated fatty acid in which C6 and C8 are joined by a disulfide bond to form a ring when the molecule is oxidized. The reduced molecule has two sulfhydryl groups. The coenzyme functions in the oxidative decarboxylation of α-keto acids.

control by several allosteric effectors (Fig. 8.15). The *E. coli* pyruvate dehydrogenase is feedback inhibited by the products it forms, acetyl–CoA and NADH. This can be rationalized as ensuring that the enzyme produces only as much acetyl–CoA and NADH as can be used immediately. The enzyme is also stimulated by phosphoenolpyruvate (the precursor to pyruvate), presumably signaling the dehydrogenase that more pyruvate is on the way. It is also stimulated by AMP, which signals low ATP. The stimulation by AMP probably reflects the fact that the oxidation of the product acetyl–CoA in the citric acid cycle is a major source of ATP (via respiratory phosphorylation).

8.8 The Citric Acid Cycle

See Box 8.3 for a historical perspective on the research that led to the elucidation of the citric acid cycle.

The acetyl–CoA that is formed by pyruvate dehydrogenase is oxidized to CO_2 in the citric acid cycle (Fig. 8.1). The overall reaction is:

Acetyl–CoA + $2H_2O$ + ADP + P_i + FAD +

$NADP^+$ + 2 NAD^+ → $2CO_2$ + ATP + $FADH_2$ +

NADPH + 2 NADH + $3H^+$ + CoASH

Notice that there are four oxidations per acetyl–CoA, producing two NADHs, one

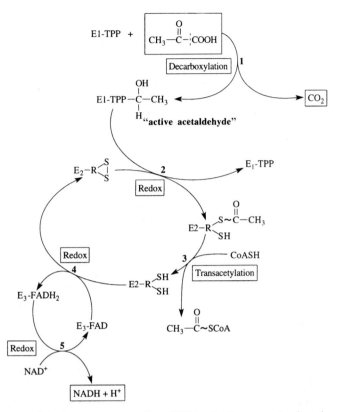

Fig. 8.14 The pyruvate dehydrogenase complex (PDH). **1**, Pyruvate is decarboxylated to "active acetaldehyde." The decarboxylation requires thiamine pyrophosphate (TPP). **2**, The "active acetaldehyde" is oxidized to an acylthioester with a high acyl group transfer potential. The oxidant is reduced lipoic acid (R–S$_2$). **3**, The lipoylacylthioester transfers the acetyl group to coenzyme A (CoASH) to form acetyl–CoA. **4**, The reduced lipoic acid is reoxidized by FAD, which in turn is reoxidized by NAD$^+$ (**5**). The products of the reaction are acetyl–CoA, CO$_2$, and NADH. Enzymes: E$_1$, pyruvate dehydrogenase; E$_2$, dihydrolipoate transacetyase; E$_3$, dihydrolipoate dehydrogenase.

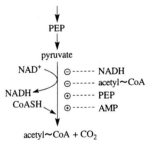

Fig. 8.15 Regulation of pyruvate dehydrogenase in *E. coli*. The activity of the enzyme in vitro is modified by several effector molecules. NADH and acetyl–CoA are negative effectors, and PEP and AMP are positive effectors.

NADPH, and one FADH$_2$, and one substrate-level phosphorylation producing ATP. The cycle usually operates in conjunction with respiration that reoxidizes the NAD(P)H and FADH$_2$. Other names for this pathway are the tricarboxylic acid (TCA) cycle and the Krebs cycle. The latter name honors Sir Hans Krebs, who did much of the pioneering work and proposed the cycle in 1937.

8.8.1 The reactions of the citric acid cycle

The citric acid cycle is outlined in Fig. 8.16. *Reaction 1* is the addition of the acetyl group from acetyl–CoA to oxaloacetate to form citrate. In this reaction the methyl group of acetyl–CoA acts as a nucleophile and bonds to the carbon in the keto group of oxaloacetate (OAA). The reaction is driven to completion by the hydrolysis of the thioester bond of acetyl–CoA, which has a high free energy of hydrolysis. Reaction 1 is catalyzed by citrate synthase. This enzyme operates irreversibly in

BOX 8.3 HISTORICAL PERSPECTIVE: THE CITRIC ACID CYCLE

During the first decades of the twentieth century, there was considerable effort by biochemists to understand how animal tissues oxidized carbohydrates to carbon dioxide and water. Between 1910 and 1920, it was learned that anaerobic suspensions of minced animal tissue transfer hydrogen from the dicarboxylic acids succinate, fumarate, malate, and the tricarboxylic acid citrate to methylene blue, reducing it to a colorless form. The enzymes catalyzing these reactions were named "dehydrogenases."

Szent-Györgyi

What people wanted to understand was the link between carbohydrate degradation and the electron transport chain, called the cytochrome–cytochrome oxidase system. It was suspected that the carboxylic acids provided the link. In 1935 the Hungarian-born scientist Albert Szent-Györgyi discovered that small amounts of fumarate, malate, or succinate, caused oxygen uptake by minced pigeon breast muscle far in excess of that which would be required to completely oxidize the carboxylic acids to carbon dioxide and water. (He determined the ratio of CO_2 produced to O_2 consumed, called the respiratory quotient.) He concluded that the carboxylic acids were acting catalytically and stimulating the oxidation of an endogenous carbohydrate, presumably glycogen or a product derived from glycogen, in the muscle tissue. Szent-Györgyi also showed that malonate, previously shown to be an inhibitor of succinate dehydrogenase, inhibits the respiration of muscle suspensions.

Krebs

In 1937, H. A. Krebs and W. A. Johnson published their classic paper on the citric acid cycle. They postulated the cycle on the bases of results published earlier by others, including Szent-Györgyi, and their own experimental results with minced pigeon breast muscle, especially the finding that citrate catalytically stimulates respiration, that citrate is converted first to α-ketoglutarate and then to succinate, and that oxaloacetate is converted to citrate by the addition of two carbons from an unidentified donor termed "triose": (Refer to Fig. 8.1.) They proposed "citric acid cycle":

$$CO_2 \uparrow$$

Citrate → isocitrate → oxalosuccinate →

$$CO_2 \uparrow$$

α-ketoglutarate → succinate → fumarate → malate → oxaloacetate → citrate

$$\uparrow$$

2C from "triose"

In their summary they conclude:

> Citric acid catalytically promotes oxidations in muscle, especially in the presence of carbohydrate. The rate of the oxidative removal of citric acid from muscle was measured. The maximum figure for $Q_{citrate}$ observed was −16.9. α-Ketoglutaric acid and succinic acid were found as products of the oxidation of citric acid. These experiments confirm Martius and Knoop's results obtained with liver citric dehydrogenase. Oxaloacetic acid, if added to muscle, condenses with an unknown substance to form citric acid. The unknown substance is in all probability a derivative of carbohydrate. The catalytic effect of citrate as well as the similar effects of succinate, fumarate, malate and oxaloacetate described by Szent-Györgyi and by Stare and Baumann are explained by the series of reactions summarized [elsewhere]. The quantitative data suggest that the "citric acid cycle" is the preferential pathway through which carbohydrate is oxidized in animal tissues.[1]

Hans Krebs was born in Germany in 1900 and worked there until 1933, when he emigrated to England. He shared the 1953 Nobel Prize in Physiology or Medicine with Fritz Lipmann for contributions toward the understanding of cellular metabolism. Hans Krebs was knighted in 1958 and died in 1981.

Lipmann

Between 1947 and 1950, Fritz Lipmann and coworkers discovered and characterized coenzyme A, the so-called coenzyme of acetylation. They learned that acetate combines with CoA to form acetyl–CoA, which is an active acetyl donor. In a paper published in 1947, Lipmann and his colleagues determined that coenzyme A is a derivative of pantothenic acid.[2]

Fritz Lipmann was born in 1899 in Germany. He earned an M.D. degree in 1924 at Berlin. He studied and did research in Germany, the Netherlands, and Denmark. In 1939 he came to the United States, where he held teaching and research positions at prestigious institutions, sharing the 1953 Nobel Prize with Hans Krebs. Fritz Lipmann died in 1986.

Ochoa

Severs Ochoa and some colleagues studied the enzyme that converted oxaloacetate to citrate, and in 1952 they reported that acetyl–CoA condenses with oxaloacetate to form citrate.[3]

Severo Ochoa was born in Spain in 1905. He did research and studied in Spain, Germany, England, and the United States, becoming an American citizen in 1956. Ochoa shared the Nobel Prize in Physiology or Medicine with Arthur Kornberg in 1959, He died in 1993.

REFERENCES

1. Krebs, H. A., and W. A. Johnson. 1937. The role of citric acid in intermediate metabolism in animal tissue. *Enzymologia* 4:148–156.

2. Lipmann, F., N. O. Kaplan, G. D. Novelli, L. C. Tuttle, and B. M. Guirard. 1947. Coenzyme for acetylation, a pantothenic acid derivative. *J. Biol. Chem.* 167:869–870.

3. Stern, J. R., S. Ochoa, and F. Lynen. 1952. Enzymatic synthesis of citric acid. V. Reaction of acetyl coenzyme A. *J. Biol. Chem.* 198:313–321.

the direction of citrate. The oxaloacetate acts catalytically in the cycle and, if it is not regenerated or replenished, the pathway stops.

Examine the structure of citrate shown in Fig. 8.16. In the subsequent reactions of the citric acid cycle, the C3 and C5 carbons will be removed as CO_2, and carbons C1 to C5 will regenerate the oxaloacetate. In *reaction 2*, catalyzed by aconitase, the citrate is dehydrated to *cis*-aconitate, which remains bound to the enzyme. *Reaction 3* (also catalyzed by aconitase) is the rehydration of *cis*-aconitate to form isocitrate, an isomer of citrate. In *reaction 4* (isocitrate dehydrogenase) the isocitrate is oxidized to oxalosuccinate. This oxidation creates a keto group β to the carboxyl group. The creation of the keto group is necessary for the decarboxylation that takes place in the next reaction. (β-Keto decarboxylations are explained in Section 8.1.2.)

Reaction 5 (isocitrate dehydrogenase) is the decarboxylation of oxalosuccinate to α-ketoglutarate. *Reaction 6* (α-ketoglutarate dehydrogenase) is the oxidative decarboxylation of α-ketoglutarate to succinyl–CoA. This is an α-decarboxylation, in contrast to a β-decarboxylation. It is a complex reaction and requires the same cofactors as does the decarboxylation of pyruvate to acetyl–CoA.

Reaction 7 (succinate thiokinase) is a substrate-level phosphorylation resulting in the formation of ATP from ADP and inorganic phosphate. This reaction was described in Section 7.3.3. *Reaction 8* (succinate dehydrogenase) is the oxidation of succinate to fumarate, catalyzed by a flavin enzyme. Succinate dehydrogenase is the only citric acid cycle enzyme that is membrane bound. In bacteria it is part of the cell membrane and transfers electrons directly to quinone in the

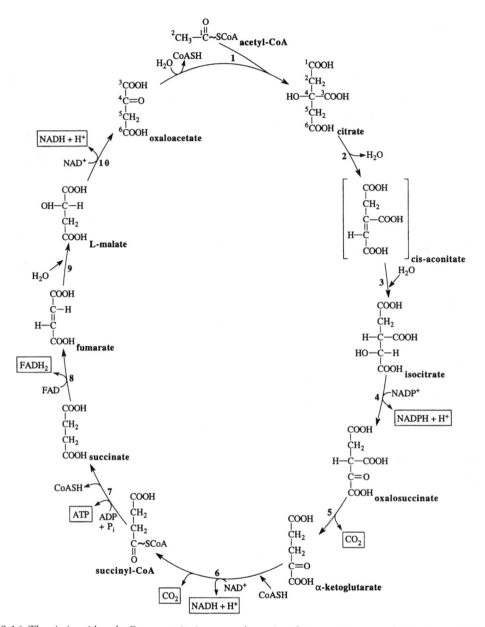

Fig. 8.16 The citric acid cycle. Enzymes: 1, citrate synthase; 2 and 3, aconitase; 4 and 5, isocitrate dehydrogenase; 6, α-ketoglutarate dehydrogenase; 7, succinate thiokinase; 8, succinate dehydrogenase; 9, fumarase; 10, malate dehydrogenase. *cis*-Aconitate is bracketed because it is an enzyme-bound intermediate.

respiratory chain. (See the description of the electron transport chain in Section 4.3.) *Reaction 9* (fumarase) is the hydration of fumarate to malate. Finally, the oxaloacetate is regenerated by oxidizing the malate to oxaloacetate in *reaction 10* (malate dehydrogenase).

The citric acid cycle proceeds in the direction of acetyl–CoA oxidation because of two irreversible steps: the citrate synthase and the α-ketoglutarate dehydrogenase reactions.

Summing up the citric acid cycle

When one examines the reactions in the citric acid cycle, it can be seen that there is no net synthesis. In other words, all the carbon that enters the cycle exits as CO_2. This is made clear by writing a carbon balance. In the carbon balance that follows, C_2 represents the two-carbon molecule acetyl-CoA, C_6 represents citrate or isocitrate, C_5 represents α-ketoglutarate, and C_4

represents either succinate fumarate, malate, or oxaloacetate. Of course, C_1 represents carbon dioxide.

Carbon balance for citric acid cycle

$$C_2 + C_4 \rightarrow C_6$$

$$C_6 \quad \rightarrow C_5 + C_1$$

$$C_5 \quad \rightarrow C_4 + C_1$$

$$\text{Sum:} \quad C_2 \quad \rightarrow 2C_1$$

8.8.2 Regulation of the citric acid cycle

The citric acid cycle is feedback inhibited by several intermediates that can be viewed as end products of the pathway. In gram-negative bacteria, the citrate synthase is allosterically inhibited by NADH, and in facultative anaerobes such as *E. coli*, also by α-ketoglutarate. The inhibition of the citrate synthase by NADH may be a way to prevent oversynthesis of NADH. The inhibition by α-ketoglutarate can also be viewed as an example of end-product inhibition, in this case to prevent overproduction of the amino acid glutamate, which is derived from α-ketoglutarate.[10] The citrate synthase from gram-positive bacteria and eukaryotes is not sensitive to NADH and α-ketoglutarate but is inhibited by ATP, another end product of the citric acid pathway. Recall the discussion in Chapter 6 emphasizing that the pattern of regulation of a particular pathway need not be the same from bacterium to bacterium.

8.8.3 The citric acid cycle as an anabolic pathway

The citric acid cycle reactions provide precursors to 10 of the 20 amino acids found in proteins. It is therefore a multifunctional pathway and is not used simply for the oxidation of acetyl–CoA. Succinyl–CoA is necessary for the synthesis of the amino acids L-lysine and L-methionine. Succinyl–CoA is also a precursor to tetrapyrroles, which are the prosthetic group in several proteins, including cytochromes and chlorophylls. Oxaloacetate is a precursor to the

amino acid aspartate, which itself is the precursor to five other amino acids. In some bacteria, fumarate is also a precursor to aspartate. α-Ketoglutarate is the precursor to the amino acid glutamate, which itself is the precursor to three other amino acids. The biosynthesis of amino acids is described in Chapter 9. As noted earlier in Section 8.8, however, since the citric acid cycle requires a constant level of oxaloacetate to function, net synthesis of these molecules requires replenishment of the oxaloacetate.

8.8.4 Distribution of the citric acid cycle

The citric acid cycle is present in most heterotrophic bacteria growing aerobically. However, not all aerobic bacteria have a complete citric acid cycle. For example, organisms that grow on C_1 compounds (methane, methanol, etc., described in Chapter 13) lack α-ketoglutarate dehydrogenase and carry out a reductive pathway as described shortly (Section 8.9). An oxidative citric acid cycle is not necessary for these organisms, since acetyl–CoA is not an intermediate in the oxidation of the C_1 compounds.

Although the oxidative citric acid cycle is a pathway usually associated with aerobic bacteria, it is also present in certain anaerobes that oxidize organic compounds completely to CO_2. These include the group II sulfate reducers (discussed in Section 12.2.2) and certain archaea. The latter are anaerobic hyperthermophilic archaea that use sulfur or thiosulfate as the terminal electron acceptor.[9]

8.9 Carboxylations that Replenish Oxaloacetate: The Pyruvate and Phosphoenolpyruvate Carboxylases

Because the citric acid cycle intermediates are constantly being removed to provide precursors for biosynthesis, they must be replaced (Section 8.8.3). Failure to do this would decrease the level of oxaloacetate that is contributed to the citrate synthase reaction, and thus for the continuation of the cycle. If the organism is growing on amino acids or organic acids (e.g., malate), replenishment of oxaloacetate is not a problem, since these molecules

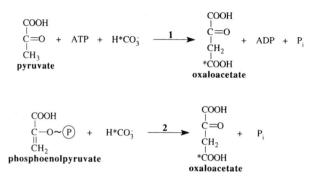

Fig. 8.17 Carboxylation reactions that replenish the supply of oxaloacetate. Enzymes: **1**, pyruvate carboxylase; **2**, PEP carboxylase. Bacteria may have one or the other.

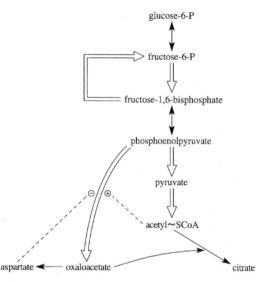

are easily converted to oxaloacetate (see later: Fig. 8.27). If the carbon source is a sugar (e.g., glucose), then the carboxylation of pyruvate or phosphoenolpyruvate replenishes the oxaloacetate (Fig. 8.17). Two enzymes that carry out the carboxylation of phosphoenolpyruvate and pyruvate are *PEP carboxylase* and *pyruvate carboxylase*, which are widespread among the bacteria. A bacterium will have one or the other. PEP carboxylase is not found in animal tissues or in fungi.

8.9.1 Regulation of PEP carboxylase

In *E. coli*, PEP carboxylase is an allosteric enzyme that is positively regulated by acetyl–CoA and negatively regulated by aspartate (Fig. 8.18). Presumably, if the oxaloacetate levels drop, acetyl–CoA will accumulate and result in the activation of the PEP carboxylase. This should produce more oxaloacetate. Aspartate can slow down its own synthesis by negatively regulating PEP carboxylase.

Fig. 8.18 The regulation of PEP carboxylase in *E. coli*. The carboxylase is positively regulated by acetyl–CoA and negatively regulated by aspartate.

8.10 Modification of the Citric Acid Cycle into a Reductive (Incomplete) Cycle During Fermentative Growth

In the presence of air, the citric acid cycle operates as an oxidative pathway coupled to aerobic respiration in respiratory organisms. Since fermenting organisms are not carrying out aerobic respiration, it seems best not to have an oxidative pathway that produces so much NADH and FADH$_2$. On the other hand, the reactions that make oxaloacetate,

succinyl–CoA, and α-ketoglutarate are necessary because these molecules are required for the biosynthesis of amino acids and tetrapyrroles. (See Sections 8.7.3 and 8.15, especially Fig. 8.27.)

The solution to the problem is to convert the citric acid cycle from an oxidative into a reductive pathway. The reductive pathway is also referred to as an incomplete citric acid cycle. Fermenting bacteria have little or no activity for the enzyme α-ketoglutarate dehydrogenase. Thus, the pathway is blocked between α-ketoglutarate and succinyl–CoA, and cannot operate in the oxidative direction (Fig. 8.19).

Succinyl–CoA is made by reversing the reactions between oxaloacetate and succinyl–CoA;

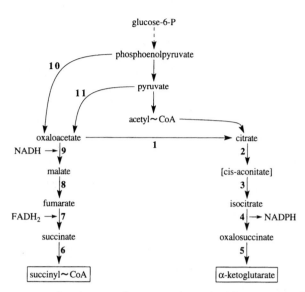

Fig. 8.19 The reductive citric acid pathway in fermenting bacteria. There are two "arms." One route oxidizes citrate to α-ketoglutarate. A second route reduces oxaloacetate to succinyl–SCoA. The enzyme α-ketoglutarate dehydrogenase is missing. The enzyme fumarase replaces succinate dehydrogenase. Enzymes: **1**, citrate synthase; **2** and **3**, aconitase; **4** and **5**, isocitrate dehydrogenase; **6**, succinate thiokinase; **7**, fumarate reductase; **8**, fumarase; **9**, malate dehydrogenase; **10**, PEP carboxylase; **11**, pyruvate carboxylase.

the enzyme fumarate reductase is used instead of succinate dehydrogenase. The latter enzyme is replaced by fumarate reductase under anaerobic conditions. These reactions consume 4H. If one includes the reactions from citrate to α-ketoglutarate that produce 2H, the net result is that the reductive pathway consumes 2H.

The reductive citric acid pathway is found not only in fermenting bacteria, but also in some other bacteria (including the enteric bacteria) that are carrying out anaerobic respiration with nitrate as the electron acceptor.[11] The reason for this is that oxygen induces the synthesis of α-ketoglutarate dehydrogenase in certain facultative anaerobes, and under anaerobic conditions, the enzyme levels are very low. (However, some nitrate respirers, e.g., *Pseudomonas stutzeri* grown under denitrifying conditions, do have an oxidative citric acid cycle.[12])

One of the consequences of an incomplete citric acid cycle is that acetate is excreted as a by-product of sugar metabolism during anaerobic growth. That is because some of the acetyl–CoA is converted to acetate concomitant with the formation of an ATP. These reactions, which are an important source of ATP, are discussed in Chapter 14. It should be pointed out that some strict anaerobes (e.g.,

the green photosynthetic sulfur bacteria) have a reductive citric acid pathway that is "complete" in that it reduces oxaloacetate to citrate. The pathway, called the reductive carboxylic acid pathway, is a CO_2 fixation pathway used for autotrophic growth and differs in some key enzymological reactions from the pathways discussed here. The reductive carboxylic acid pathway is described in Section 13.1.9.

8.11 Chemistry of Some of the Reactions in the Citric Acid Cycle

This section presents a rational basis for understanding the chemistry of some of the key reactions in the citric acid cycle. Similar reactions are seen in other pathways.

8.11.1 Acetyl–CoA condensation reactions

Acetyl–CoA is a precursor for the biosynthesis of many different molecules besides citrate. These include lipids (Chapter 9) and various fermentation end products (Chapter 14). Acetyl–CoA is so versatile because it undergoes condensations at both the methyl and carboxyl

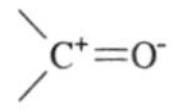

$$\text{\textbackslash}\,\text{C}^+\!=\!\text{O}^-$$

Fig. 8.20 The carbonyl group in acetyl–CoA condensation reactions is polarized. Electrons are attracted by the oxygen, leaving a partial positive charge on the carbon.

ends of the molecule. Condensations at both ends of acetyl–CoA can be understood in terms of the chemistry of the polarized carbonyl group. The electrons in the carbonyl group are not shared equally by the carbon and oxygen; rather, the oxygen is much more electronegative than the carbon and pulls the electrons in the double bond closer to itself. That is, the C=O group is polarized, making the carbon slightly positive (Fig. 8.20).

Because the oxygen in the carbonyl group is electron attracting, there is a tendency to pull electrons away from the C–H bond in the carbon adjacent to the carbonyl. This results in the formation of an enolate ion, which acts as a nucleophile (Fig. 8.21). The enolate anion seeks electrophilic centers (e.g., the carbon atoms in carbonyl groups). So acetyl–CoA can be a nucleophile at its methyl end, attacking other carbonyl groups, even other acetyl–CoA molecules. In the formation of citric acid, the methyl group of acetyl–CoA attacks a carbonyl group in oxaloacetate to form citric acid. At the same time, the thioester linkage to coenzyme A is hydrolyzed, driving the reaction to completion. Later, we will examine other pathways in

which the methyl carbon of acetyl–CoA attacks a carbonyl.

Another result of the polarization of the carbonyl group is that the carbon in the carbonyl is electron deficient and subject to attack by nucleophiles that seek a positive center (i.e., it is electrophilic). Thus, acetyl–CoA undergoes condensations at the carboxyl end in reactions in which the CoASH is displaced during a nucleophilic displacement and the acetyl portion is transferred to the nucleophile (Fig. 8.22). For example, this occurs during fatty acid synthesis and during butanol fermentations (Chapters 9 and 14), as well as in several other pathways. Because of the reactivity of acetyl–CoA at both the methyl and carboxyl ends, the molecule is widely used in building larger molecules.

8.11.2 Decarboxylation reactions

Oxalosuccinate is a β-ketocarboxylic acid (i.e., the carboxyl group is β to a keto group). The decarboxylation of β-ketocarboxylic acids occurs throughout metabolism. Another β-decarboxylation occurs in the pentose phosphate pathway. When 6-phosphogluconate is oxidized by $NADP^+$, a 3-keto intermediate is formed (Fig. 8.8).

The decarboxylation of β-ketocarboxylic acids is relatively straightforward. A single enzyme is required, and the cofactor requirements are met by Mn^{2+} or Mg^{2+}. A β-decarboxylation

$$H-\overset{\overset{\displaystyle H}{|}}{\underset{\underset{\displaystyle H}{|}}{C}}-\overset{\overset{\displaystyle O}{\|}}{C}\!\sim\!SCoA \longrightarrow \left[\; H-\overset{\overset{\displaystyle H}{|}}{C}\!=\!\overset{\overset{\displaystyle O}{|}}{C}\!\sim\!SCoA \longleftrightarrow H-\overset{\overset{\displaystyle H}{|}}{\underset{|}{C}}-\overset{\overset{\displaystyle O}{\|}}{C}\!\sim\!SCoA \;\right]$$
enolate anion

Fig. 8.21 The methylene carbon of acetyl–CoA can act as a nucleophile. Because of the electron-attracting ability of the carbonyl group, a hydrogen dissociates from the methylene carbon, forming an enolate anion that resonates. Electrons can shift to the methylene carbon, which then seeks a positive center (e.g., a carbonyl group). Because of this, acetyl–CoA will form covalent bonds to the carbon of carbonyl groups in condensation reactions.

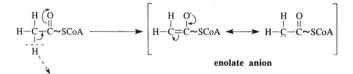

$$CH_3-\overset{\overset{\displaystyle O^-}{\|}}{C^+}\!\sim\!SCoA + :R-H \longrightarrow CH_3-\overset{\overset{\displaystyle O}{\|}}{C}-R + HSCoA$$

Fig. 8.22 Acetyl group transfer. Because the carbonyl group in acetyl–CoA is polarized, it is subject to nucleophilic attack. As a result, the acetyl group is transferred to the nucleophile and CoASH is displaced.

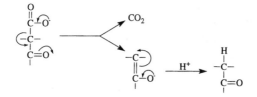

Fig. 8.23 The decarboxylation of a β-ketocarboxylic acid. The keto group attracts electrons, causing an electron shift and the breakage of the C–C bond holding the carboxyl group to the molecule. The decarboxylation of oxalosuccinate is physiologically reversible. Notice the resemblance to the aldol cleavage shown in Fig. 8.5.

is shown in Fig. 8.23. The β-keto group attracts electrons, facilitating the breakage of the bond holding the carboxyl group to the molecule. Note the similarity to the aldolase reaction, where the keto group facilitates the breakage of a C–C bond that is β to the keto group (Fig. 8.5).

Decarboxylation of α-ketoglutarate

α-Ketoglutarate is an α-ketocarboxylic acid, and its decarboxylation is more complex than

that of a β-ketocarboxylic acid. The mechanism is the same as for the decarboxylation of pyruvate, another α-ketocarboxylic acid, and is described in more detail in Section 14.10.2. (See subheading "How thiamine pyrophosphate catalyzes the decarboxylation of 2-ketocarboxylic acids.") Pyruvate and α-ketoglutarate dehydrogenases are similar in that they can be separated into three components: the TPP-containing decarboxylase, the FAD-containing dihydrolipoyl dehydrogenase, and the dihydrolipoyl transacetylase. Other α-ketocarboxylic acid dehydrogenases exist (e.g., for the catabolism of α-keto acids derived from the degradation of the branched-chain amino acids, leucine, isoleucine, and valine).

8.12 The Glyoxylate Cycle

We now come to a second pathway central to the metabolism of acetyl–CoA: the glyoxylate cycle, also called the glyoxylate bypass (Fig. 8.24). The glyoxylate cycle is required by aerobic bacteria to grow on fatty acids and

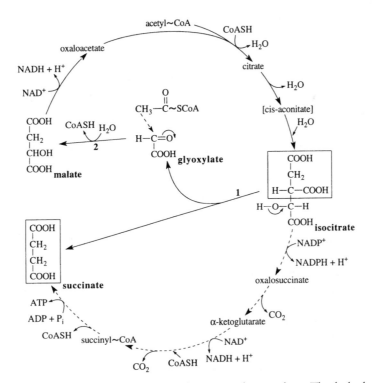

Fig. 8.24 The glyoxylate cycle. Enzymes: **1**, isocitrate lyase; **2**, malate synthase. The dashed arrows represent reactions of the citric acid cycle that are bypassed.

acetate. (Plants and protozoa also have the glyoxylate cycle. However, it is absent in animals.) The glyoxylate cycle resembles the citric acid cycle except that it bypasses the two decarboxylations in the citric acid cycle. For this reason the acetyl–CoA is not oxidized to CO_2.

Examine the summary of the glyoxylate cycle shown in Fig. 8.24. The glyoxylate cycle shares with the citric acid cycle the reactions that synthesize isocitrate from acetyl–CoA. The two pathways diverge at isocitrate. In the glyoxylate cycle the isocitrate is cleaved to succinate and glyoxylate by the enzyme *isocitrate lyase* (reaction 1). The glyoxylate condenses with acetyl–CoA to form malate, in a reaction catalyzed by *malate synthase* (reaction 2). The malate synthase reaction is of the same type as the citrate synthase (i.e., the methylene carbon of acetyl–CoA attacks the carbonyl group in glyoxylate). The reaction is driven to completion by the hydrolysis of the coenzyme A–thioester bond just as during citrate synthesis. The malate replenishes the oxaloacetate, leaving one succinate and NADH as the products. The cells incorporate the succinate into cell material by first oxidizing it to oxaloacetate, from which phosphoenolpyruvate is synthesized.

The net result of the glyoxylate cycle is the condensation of two molecules of acetyl–CoA to form succinate. The carbon balance is as follows:

Carbon balance for the glyoxylate pathway

$$C_2 + C_4 \rightarrow C_6$$

$$C_6 \rightarrow C_4 + C_2$$

$$C_2 + C_2 \rightarrow C_4$$

$$\overline{C_2 + C_2 \rightarrow C_4}$$

8.12.1 Regulation of the glyoxylate cycle

As Fig. 8.24 illustrates, isocitrate is at a branch point for both the citric acid cycle and the glyoxylate cycle. That is, it is a substrate for both the isocitrate lyase and the isocitrate dehydrogenase. What regulates the fate of the isocitrate? In *E. coli*, the isocitrate dehydrogenase activity is partially inactivated by phosphorylation when cells are grown on acetate.[13] (The regulation of enzyme activity by covalent modification is discussed in Chapter 6, and other examples are listed in Table 6.1.) Acetate also induces the enzymes of the glyoxylate cycle. Therefore, in the presence of acetate, isocitrate lyase activity increases while the isocitrate dehydrogenase is partially inactivated. However, the K_m for the isocitrate lyase is high relative to the concentrations of isocitrate. This means that the isocitrate lyase requires a high intracellular concentration of isocitrate. It is presumed that the partial inactivation of isocitrate dehydrogenase increases the concentration of isocitrate resulting in an increase in flux through the glyoxylate cycle.

8.13 Formation of Phosphoenolpyruvate

Organic acids such as lactate, pyruvate, acetate, succinate, malate, and the amino acids can be used for growth because pathways exist to convert them to phosphoenolpyruvate, which is a precursor to the glycolytic intermediates. The phosphoenolpyruvate is generally made in two ways. It can be made via the decarboxylation of oxaloacetate or via the phosphorylation of pyruvate. These two reactions are described next.

8.13.1 Formation of phosphoenopyruvate from oxaloacetate

The ATP-dependent decarboxylation catalyzed by PEP carboxykinase, shown in Fig. 8.25, results in the formation of phosphoenopyruvate from oxaloacetate. The enzyme is widespread and accounts for phosphoenolpyruvate synthesis in both eukaryotes and prokaryotes. This enzyme, along with malic enzyme, is important for growth on succinate and malate (Section 8.14).

Although PEP carboxykinase is an important enzyme for the synthesis of phosphoenolpyruvate from oxaloacetate, it catalyzes the synthesis of oxaloacetate from phosphoenolpyruvate in some anaerobic bacteria. For example, anaerobic bacteria that ferment glucose to succinate may use this enzyme to carboxylate phosphoenolpyruvate to oxaloacetate and then in other reactions reduce the oxaloacetate to succinate.[14]

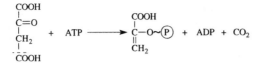

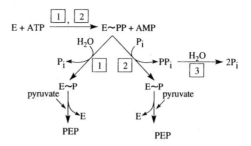

Fig. 8.25 The PEP carboxykinase reaction. The PEP carboxykinase generally operates in the direction of PEP synthesis, although during fermentation in anaerobes it can work in the direction of oxaloacetate, which is subsequently reduced to the fermentation end product, succinate.

Fig. 8.26 The PEP synthetase (1), pyruvate–phosphate dikinase (2), and pyrophosphatase (3) reactions.

8.13.2 Formation of phosphoenolpyruvate from pyruvate

Prokaryotes are able to synthesize phosphoenolpyruvate by phosphorylating pyruvate, instead of converting the pyruvate to oxaloacetate and then decarboxylating the oxaloacetate. This is necessary for growth on pyruvate for bacteria that cannot synthesize oxaloacetate from pyruvate. Prokaryotes that fall into the latter category do not have a glyoxylate pathway and also lack pyruvate carboxylase. For example, E. coli does not have a glyoxylate cycle unless growing on acetate, and it has PEP carboxylase instead of pyruvate carboxylase. Furthermore, there are strict anaerobes (e.g., methanogens and green sulfur photosynthetic bacteria) that grow autotrophically, converting CO_2 to acetyl–CoA, which is carboxylated to pyruvate (Sections 13.1.2 and 13.1.3). These microorganisms do not have a glyoxylate cycle and must phosphorylate the pyruvate to form phosphoenolpyruvate. The phosphorylation of pyruvate to phosphoenolpyruvate is described next.

PEP synthetase and pyruvate–phosphate dikinase reactions

Prokaryotes that convert pyruvate directly to phosphoenolpyruvate use one of two enzymes: PEP synthetase or pyruvate–phosphate dikinase. These enzymes are found in prokaryotes and plants, but not in animals. The following reactions are illustrated in Fig. 8.26:

PEP synthetase (reaction 1)

$$\text{Pyruvate} + H_2O + \text{ATP} \rightarrow \text{PEP} + \text{AMP} + P_i$$

Pyruvate–phosphate dikinase (reaction 2)

$$\text{Pyruvate} + \text{ATP} + P_i \rightarrow \text{PEP} + \text{AMP} + PP_i$$

Pyrophosphatase (reaction 3)

$$PP_i + H_2O \rightarrow 2P_i$$

In the PEP synthetase reaction, a pyrophosphoryl group is transferred from ATP to the enzyme. (See Fig. 7.3 for a description of pyrophosphoryl group transfer reactions.) One phosphate is removed by hydrolysis, leaving a phosphorylated enzyme. The phosphorylated enzyme then donates the phosphoryl group to pyruvate to form PEP. The reaction catalyzed by the pyruvate–phosphate dikinase is similar, except that instead of being hydrolytically removed from the pyrophosphorylated enzyme, the phosphate is transferred to inorganic phosphate to form pyrophosphate. The pyrophosphate is hydrolyzed by a pyrophosphatase, pulling the reaction to completion. The net result from both reactions is the same (i.e., the sum of the pyruvate–phosphate dikinase and pyrophosphatase reactions is the same as the PEP synthetase reaction). In either case, the synthesis of phosphoenolpyruvate from pyruvate requires the hydrolysis of two phosphodiester bonds with a high free energy of hydrolysis.

8.14 Formation of Pyruvate from Malate

Bacteria and mitochondria can convert malate directly to pyruvate by using the malic enzyme:

Malic enzyme

$$\text{L-Malate} + \text{NAD}^+ \rightleftharpoons \text{pyruvate} + \text{NADH} + CO_2$$

Bacteria and mitochondria actually possess two malic enzymes: one specific for NAD^+ and the other for $NADP^+$. The latter provides

NADPH for reductions that occur during biosynthesis (e.g., the biosynthesis of fatty acids, as discussed in Chapter 9). Malic enzyme, along with PEP carboxykinase (Section 8.13.1), is an important enzyme for growth on citric acid cycle intermediates such as succinate or malate. *E. coli* mutants that lack both PEP carboxykinase and malic enzyme will not grow on succinate or malate, although strains lacking only one of these enzymes will grow.

8.15 Summary of the Relationships Between the Pathways

Figure 8.27 illustrates the relationship between the different pathways discussed in this chapter. Reactions 1 and 2 are key enzymes in the glyoxylate cycle. Reaction 1 is isocitrate lyase. Reaction 2 is malate synthase. Fatty acids and acetate are converted to acetyl–CoA, which enters the citric acid cycle and the glyoxylate cycle. Dicarboxylic acids and amino acids are eventually degraded to citric acid cycle intermediates. Sugars can be catabolized via the Embden–Meyerhof–Parnas pathway, the pentose phosphate pathway, or the Entner–Doudoroff pathway. All the sugar pathways intersect at phosphoglyceraldehyde.

8.16 Summary

The nearly ubiquitous pathway for glucose degradation is the Embden–Meyerhof–Parnas pathway, also called the glycolytic pathway. The pathway oxidizes one mole of glucose to two moles of pyruvate, and produces two moles of NADH and two moles of ATP. There is only one oxidation, and that is the oxidation

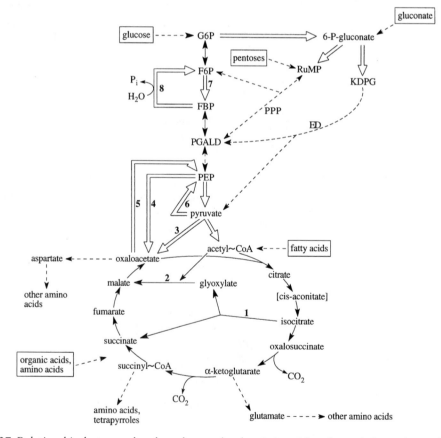

Fig. 8.27 Relationship between the glyoxylate cycle, the citric acid cycle, and the major carbohydrate pathways. Enzymes: 1, isocitrate lyase; 2, malate synthase; 3, pyruvate carboxylase; 4, PEP carboxylase; 5, PEP carboxykinase; 6, PEP synthetase or pyruvate phosphodikinase; 7, phosphofructokinase; 8, fructose-1,6-bisphosphatase.

of phosphoglyceraldehyde to phosphoglycerate. An intermediate, 1,3-bisphosphoglycerate, is formed, which donates a phosphoryl group to ADP in a substrate-level phosphorylation. Since two phosphoglyceraldehydes are formed from one glucose, two ATPs are made.

The pathway can be considered as two stages. Stage 1 generates two phosphoglyceraldehydes. Stage 2 is the oxidation of phosphoglyceraldehyde to pyruvate. Not only does glycolysis provide ATP, NADH, and pyruvate, but its intermediates are used for biosynthesis in other pathways; this will become more evident in subsequent chapters when we examine other metabolic pathways. The pathway cannot be reversed from pyruvate to glucose-6-phosphate because of two irreversible reactions, the pyruvate kinase reaction (phosphoenolpyruvate to pyruvate) and the phosphofructokinase reaction (fructose-6-phosphate to fructose-1, 6-bisphosphate). Both these reactions proceed so far in the direction of products that they are not physiologically reversible. However, the pathway can be reversed from phosphoenolpyruvate, provided fructose-1,6-bisphosphate phosphatase is present to convert fructose-1,6-bisphosphate to fructose-6-phosphate. A modification of the EMP pathway exists in the archaea, where ADP serves as the phosphoryl donor in the kinase reactions and ferredoxin is the electron acceptor in the oxidation reactions.

A second sugar-catabolizing pathway is the Entner–Doudoroff (ED) pathway. The ED pathway is widespread among prokaryotes, where it can be used for growth on gluconate. An intermediate of this pathway is phosphoglyceraldehyde, which is oxidized to pyruvate via the same reactions that occur in stage 2 of glycolysis. Because only one phosphoglyceraldehyde is produced from glucose-6-phosphate in the ED pathway, only one ATP is made. Several bacteria lack phosphofructokinase and fructose-1,6-bisphosphate aldolase, key enzymes in the first stage of the EMP pathway and so rely on the ED pathway for growth on glucose. Interestingly, modifications of the ED pathway exist in a few bacteria and archaea in which some of the intermediates are not phosphorylated.

Among the many pathways that intersect with the second stage of glycolysis is the pentose phosphate pathway. The pentose phosphate pathway oxidizes glucose-6-phosphate to pentose phosphates. The pentose phosphates can be used for the synthesis of nucleic acids, or they can be converted via the transaldolase and transketolase reactions to phosphoglyceraldehyde. The pathway also produces NADPH for biosynthetic reductions and erythrose-4-phosphate for aromatic amino acid biosynthesis. The pentose phosphate pathway is also used when pentoses are the source of carbon and energy.

In aerobic prokaryotes, pyruvate is oxidatively decarboxylated via pyruvate dehydrogenase to acetyl–CoA, CO_2, and NADH. The acetyl–CoA enters the citric acid cycle, where it is oxidized completely to CO_2. The electrons generated during the oxidations are transferred to the electron transport chain, where a Δp is created.

The citric acid cycle is also an anabolic pathway. Several biosynthetic pathways draw citric acid cycle intermediates out of the pools. For example, oxaloacetate and α-ketoglutarate are used for amino acid biosynthesis, and succinyl–CoA is used for the synthesis of tetrapyrrole, lysine, and methionine. Oxaloacetate is also used to synthesize phosphoenolpyruvate for glucogenesis. Since oxaloacetate is used catalytically in the cycle, it must be replenished during growth. When organisms are growing on carbohydrates, the oxaloacetate is replenished by carboxylating pyruvate or phosphoenolpyruvate. Growth on proteins or amino acids poses no problem, since these substances are degraded to citric acid cycle intermediates.

Pyruvate must be converted to phosphoenolpyruvate before it can be used as a source of cell carbon. The reason for this is that phosphoenopyruvate is a precursor to several metabolites, including the intermediates of glycolysis. One route is the ATP-dependent carboxylation of pyruvate to oxaloacetate (pyruvate carboxylase) combined with the ATP-dependent decarboxylation to phosphoenolpyruvate via the enzyme PEP carboxykinase. Two ATPs are therefore used to convert one pyruvate to phosphoenolpyruvate. A second route is the direct phosphorylation of pyruvate to phosphoenolpyruvate via either PEP synthetase or pyruvate–phosphate dikinase combined with a pyrophosphatase. The products are phosphoenolpyruvate, AMP, and P_i. Thus, regardless of the pathway used, two ATPs are required to synthesize phosphoenolpyruvate from pyruvate.

Bacteria growing aerobically on fatty acids or acetate use the glyoxylate cycle for net incorporation of C_2 units into cell material. This pathway is a modification of the citric acid cycle, in which the two decarboxylation reactions are bypassed. The decarboxylation reactions are bypassed by using isocitrate lyase and malate synthetase two enzymes unique to the glyoxylate cycle. The net result is that acetyl–CoA is converted to succinate rather than to carbon dioxide.

The citric acid cycle is usually present only during aerobic growth. Under anaerobic conditions, prokaryotes have a modified pathway called the reductive citric acid pathway. However, some sulfate reducers, which are anaerobic bacteria that use sulfate as an electron acceptor, have an oxidative citric acid cycle. In the reductive citric acid pathway there are two major changes. The α-ketoglutarate dehydrogenase activity is low or missing, and fumarate reductase replaces succinate dehydrogenase. The result is that oxaloacetate is reduced to succinyl–CoA, and citrate is oxidized to α-ketoglutarate. Thus, the intermediates necessary for biosynthesis of amino acids and tetrapyrroles are made, but the number of reductions exceeds the number of oxidations.

Study Questions

1. Suppose you isolated a mutant that did not grow on glucose as the sole source of carbon. However, the mutant did grow on the glucose when it was supplemented with succinate, fumarate, or malate. An examination of broken cell extracts revealed that all the citric acid cycle enzymes were present. What might be the metabolic defect in the mutant?

2. There are two physiologically irreversible reactions in the EMP pathway starting with glucose-6-phosphate and ending with pyruvate. Which ones are they? How are they regulated in *E. coli*?

3. Do you think that the glyoxylate cycle and the citric acid cycle can operate at the same time in an organism growing on acetic acid? How is this regulated in *E. coli*?

4. Fermenting bacteria have little or no α-ketoglutarate dehydrogenase activity and therefore have a block in the citric acid cycle. Under these conditions, the pathway operates reductively instead of oxidatively. Draw a reductive citric acid pathway showing how these organisms make oxaloacetate, succinyl–CoA, and α-ketoglutarate. Why is it important to be able to synthesize these compounds under all growth conditions?

5. Which nonglycolytic enzyme is necessary to reverse glycolysis from PEP to G6P?

6. Some bacteria (e.g., *Brucella abortus*) lack key enzymes in the EMP and ED pathways but can still grow on glucose. *Brucella abortus* is an animal pathogen that causes spontaneous abortion in cattle and can also infect humans, causing fever, headache, and joint pains. It has a citric acid cycle. Write a series of reactions showing how the pentose phosphate pathway can result in glucose oxidation completely to CO_2 without the involvement of stage 1 of either the EMP or the ED pathway.

7. The ED pathway is only 50% as efficient as the EMP pathway in making ATP from glucose. Why is that?

8. *E. coli* mutants lacking glucose-6-P dehydrogenase can be grown on glucose as the sole source of carbon. Since this is the enzyme that catalyzes the entrance of G6P into the pentose phosphate pathway, how can these mutants make NADPH or pentose phosphates?

9. Show how the pentose phosphate pathway might oxidize glucose completely to CO_2 without using the citric acid cycle.

10. The combination of phosphofructokinase (fructose-6-phosphate kinase) and fructose-1,6-bisphosphatase has ATPase activity. Write these reactions and sum them to satisfy yourself that this is so. How does *E. coli* ensure that these reactions do not use up all the ATP in the cell?

11. When *E. coli* is grown on pyruvate as its sole source of carbon, it does not synthesize the glyoxylate cycle enzymes. How is the pyruvate converted to PEP?

[*Hint*: For Question 11–14, start by examining Fig. 8.27.]

12. Suppose you are growing a bacterial culture on ribose as the only source of carbon. Write a series of reactions that will convert ribose to glucose-6-phosphate. Why is it important to be able to synthesize glucose-6-phosphate?

13. Suppose you are growing a culture on gluconate. How would the cells convert the gluconate to phosphoglyceraldehyde if they lacked 6-phosphogluconate dehydrogenase. How might they make ribose-5-phosphate and erythrose-4-phosphate? Why is it necessary to have a supply of the latter two compounds?

14. Suppose you are growing a bacterial culture on succinate as the source of carbon. Write the reactions by which succinate is converted to PEP. Why is it important to be able to synthesize PEP?

REFERENCES AND NOTES

1. Reviewed in: Conway, T. 1992. The Entner–Doudoroff pathway: history, physiology, and molecular biology. *FEMS Microbiol. Rev.* **103**:1–28.

2. During some fermentations, a portion of the NADH is reoxidized by fumarate via membrane-bound electron carriers and a Δp is created. This is called fumarate respiration. Fumarate respiration occurs during mixed acid fermentation discussed in Section 14.10; propionate fermentation is discussed in Section 14.7.

3. Danson, M. J. 1988. Archaebacteria: the comparative enzymology of their central metabolic pathways, pp. 166–231. In: *Advances in Microbial Physiology*, Vol. 29. A. H. Rose and D. W. Tempest (Eds.). Academic Press, New York.

4. Verhees, C. H., S. W. M. Kengen, J. E. Tuininga, G. J. Schut, M. W. W. Adams, W. M. de Vos, and V. der Oost. 2003. The unique features of glycolytic pathways in Archaea. *Biochem. J.* **375**:231–246.

5. Kengen, S. W. M., F. A. M., de Bok, N-D. van Loo, C. Dijkema, A. J. M. Stams, and W. M. de Vos. 1994. Evidence for the operation of a novel Embden–Meyerhof pathway that involves ADP-dependent kinases during sugar fermentation by *Pyrococcus furiosus*. *J. Biol. Chem.* **269**:17537–17541.

6. Mukund, S., and M. W. W. Adams. 1995. Glyceraldehyde-3-phosphate ferredoxin oxidoreductase, a novel tungsten-containing enzyme with a potential glycolytic role in the hyperthermophilic archaeon *Pyrococcus furiosus*. *J. Biol. Chem.* **270**: 8389–8392.

7. However, some bacteria (e.g., *E. coli*) also contain an enzyme called transhydrogenase, which catalyzes the reduction of $NADP^+$ by NADH. Therefore, for those bacteria the pentose phosphate pathway is not essential for NADPH synthesis.

8. Sugars that differ in the configuration at a single asymmetric carbon are called *epimers*. For example, xylulose-5-phosphate and ribulose-5-phosphate are epimers of each other at C3.

9. Selig, M., and P. Schonhet. 1994. Oxidation of organic compounds to CO_2 with sulfur or thiosulfate as electron acceptor in the anaerobic hyperthermophilic archaea *Thermoproteus tenax* and *Pyrobaculum islandicum* proceeds via the citric acid cycle. *Arch. Microbiol.* **162**:286–294.

10. Danson, M. J., S. Harford, and P. D. J. Weitzman. 1979. Studies on a mutant form of *Escherichia coli* citrate synthase desensitized to allosteric effectors. *Eur. J. Biochem.* **101**:515–521.

11. Stewart, V. 1988. Nitrate respiration in relation to facultative metabolism in Enterobacteria. *Microbiol. Rev.* **52**:190–232.

12. Spangler, W. J., and C. M. Gilmour. 1966. Biochemistry of nitrate respiration in *Pseudomonas stutzeri*. Aerobic and nitrate respiration routes of carbohydrate catabolism. *J. Bacteriol.* **91**:245–250.

13. LaPorte, D. C. 1993. The isocitrate dehydrogenase phosphorylation cycle: regulation and enzymology. *J. Cell. Biochem.* **51**:14–18.

14. Podkovyrov, S. M., and J. G. Zeikus. 1993. Purification and characterization of phosphoenolpyruvate carboxykinase, a catabolic CO_2-fixing enzyme, from *Anaerobiospirillum succiniciproducens*. *J. Gen. Microbiol.* **139**:223–228.

9

Metabolism of Lipids, Nucleotides, Amino Acids, and Hydrocarbons

The central pathways described in Chapter 8 (i.e., glycolysis, the pentose phosphate pathway, the Entner–Doudoroff pathway, and the citric acid cycle) provide precursors to all of the other metabolic pathways. This chapter focuses on some of those metabolic pathways that feed off the central pathways and are necessary for lipid, protein, and nucleic acid metabolism.

9.1 Lipids

Lipids are a structurally heterogeneous group of substances that share the common property of being highly soluble in nonpolar solvents (e.g., methanol, chloroform) and relatively insoluble in water. Essentially all of the lipids in prokaryotes are in the membranes. The major membrane lipids in bacteria and eukaryotes are phospholipids consisting of fatty acids

esterified to glycerol phosphate derivatives, and are called phosphoglycerides. Archaea can also have phosphoglycerides, but their structure is different, as is their mode of synthesis. The structures of archaeal lipids are summarized in Section 1.2.5 and their synthesis in Section 9.1.3. The metabolism of fatty acids will be discussed first.

9.1.1 Fatty acids

Types of fatty acid found in bacteria

Fatty acids are chains of methylene carbons with a carboxyl group at one end. They can be branched (methyl), saturated, unsaturated, or hydroxylated. Some examples are shown in Table 9.1. Fatty acids differ in the number of carbon atoms, in the number of double bonds they contain, in where the double bonds are placed in the molecule, and in whether the

Table 9.1 Some fatty acids

Fatty acid		Number of carbon atoms
Palmitic	$CH_3(CH_2)_{14}COOH$	16
Stearic	$CH_3(CH_2)_{16}COOH$	18
Oleic	$CH_3(CH_2)_7CH=CH(CH_2)_7COOH$	18
Linoleic	$CH_3(CH_2)_4CH=CHCH_2CH=CH(CH_2)_7COOH$	18
Lactobacillic	$CH_3(CH_2)_5CH-CH(CH_2)_9COOH$	9
	$\backslash\ /$	
	CH_2	
Tuberculostearic	$CH_3(CH_2)_7CH(CH_2)_8COOH$	19
	$\mid$	
	CH_3	

molecule is branched (e.g., tuberculostearic acid), or whether it contains cyclopropane (e.g., lactobacillic acid). However, some generalizations can be made about the fatty acids. They are usually 16 or 18 carbons long and are either saturated or have one double bond. Generally, gram-positive bacteria are richer in branched-chain fatty acids than are gram-negative bacteria.

The role of fatty acids

Fatty acids do not occur free in bacteria but are covalently attached to other molecules. Most of the fatty acids are esterified to glycerolphosphate derivatives to make *phosphoglycerides*, which are an important structural component of membranes. Fatty acids may also be esterified to carbohydrate. For example, the *lipid A* portion of lipopolysaccharide consists of fatty acids esterified to glucosamine (see later: Section 11.2.1). Fatty acids can also be esterified to protein. For example, gram-negative bacteria have a *lipoprotein* that is covalently attached to the peptidoglycan and protrudes into the outer membrane. The lipoprotein apparently is important for the stability of the outer membrane (Section 1.2.3). Because fatty acids are nonpolar, *they anchor the molecules to which they are attached to the membrane.* Furthermore, whether a fatty acid is unsaturated, saturated, or branched has important significance for the physical properties of the membrane. Unsaturated fatty acids and branched-chain fatty acids make the membrane more fluid, which is a necessary condition for membrane function.

β-Oxidation of fatty acids

Many bacteria can grow on long-chain fatty acids (e.g., pseudomonads, various bacilli, *E. coli*). The fatty acids are oxidized to acetyl–CoA via a pathway called *β-oxidation* (Fig. 9.1). Before such oxidation can take place, the fatty acid must be converted to the acyl–CoA derivative in a reaction catalyzed by *acyl–CoA synthetase*. This is a two-step process. First, pyrophosphate is displaced from ATP to make the AMP derivative of the carboxyl group (acyl adenylate), which remains tightly bound to the enzyme (Fig. 9.1, reaction 1). Then the AMP is displaced by CoASH to make the fatty

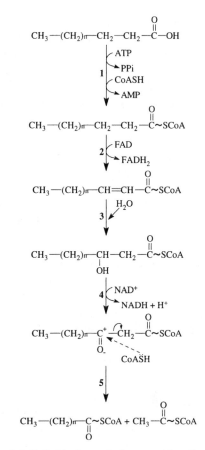

Fig. 9.1 β-Oxidation of fatty acids. Enzymes: 1, acyl–CoA synthetase; 2, fatty acyl–CoA dehydrogenase; 3, 3-hydroxyacyl–CoA hydrolyase; 4, L-3-hydroxyacyl–CoA dehydrogenase; 5, β-ketothiolase.

acyl–CoA. The activation of a carboxyl group by formation of an acyl adenylate is common in metabolism and is discussed in Section 7.2.1. (See Fig. 7.3.) For example, amino acids are activated for protein synthesis in this way. As with other reactions in which pyrophosphate is displaced from ATP, the reaction is driven to completion by hydrolysis of the pyrophosphate. A keto group is then generated β to the carboxyl. The following sequence of reactions leads to the keto group: (1) an oxidation to form a double bond (reaction 2), catalyzed by acyl–CoA dehydrogenase; (2) hydration of the double bond to form the hydroxyl (reaction 3), catalyzed by 3-hydroxyacyl–CoA hydrolyase; and (3) oxidation of the hydroxyl to form the keto group (reaction 4), catalyzed by L-3-hydroxyacyl–CoA dehydrogenase. The chemistry is the same as the conversion of succinate

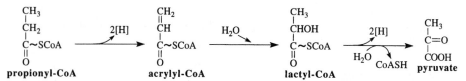

Fig. 9.2 Oxidation of propionyl–CoA to pyruvate via acrylyl–CoA.

to oxaloacetate in the citric acid cycle (Fig. 8.16). The carbonyl of the β-keto acyl–CoA is then attacked by CoASH, displacing an acetyl–CoA (reaction 5) catalyzed by β-ketothiolase. Usually the bacterium is growing aerobically and the acetyl–CoA is oxidized to carbon dioxide in the citric acid cycle. The acetyl–CoA can also be a source of cell carbon when it is assimilated via the glyoxylate cycle (Section 8.12). The displacement of the acetyl–CoA results in the generation of a fatty acid acyl–CoA that is recycled through the β-oxidation pathway.

If the fatty acid has an even number of carbon atoms, the entire chain is degraded to acetyl–CoA. However, if the fatty acid is an odd-chain

fatty acid, then the last fragment is propionyl–CoA rather than acetyl–CoA. Two of the various possible pathways for the oxidation of propionyl–CoA to acetyl–CoA are shown in Figs 9.2 and 9.3. *E. coli* forms pyruvate from propionate via methylcitrate (Fig. 9.3).[1,2] Review the glyoxylate cycle in Section 8.12 and Fig. 8.24. Notice the similarity to the methylcitrate pathway.

E. coli has two β-oxidation pathways, one that operates only under aerobic conditions and a second that operates anaerobically.[3] Thus, the bacteria can be grown anaerobically on fatty acids if a terminal electron acceptor, such as nitrate, is available.

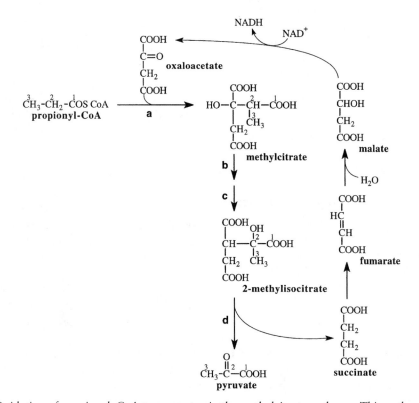

Fig. 9.3 Oxidation of propionyl–CoA to pyruvate via the methylcitrate pathway. This pathway seems to operate in *E. coli*. Enzymes: **a**, 2-methylcitrate synthase; **b** and **c**, 2-methylaconitase; **d**, 2-methylisocitrate lyase. *Source*: Adapted from Textor, S., V. F. Wendisch, A. A. De Graff, U. Müller, M. I. Linder, D. Linder, and W. Buckel. 1997. Propionate oxidation in *Escherichia coli*: evidence for operation of a methylcitrate cycle in bacteria. *Arch. Microbiol.* **168**:428–436.

The synthesis of fatty acids

An examination of metabolic pathways reveals that catabolic pathways are usually different from synthetic pathways. In other words, biosynthesis is not due simply to reversing the reactions of catabolism. This is true because there is usually at least one irreversible step in the catabolic pathway. For example, glycolysis has two irreversible reactions, the phosphofructokinase and pyruvate kinase reactions, and these must be bypassed to reverse glycolysis. Often, an entirely different set of reactions is used for biosynthesis. This is seen in fatty acid metabolism, where the biosynthetic pathway consists of reactions completely different from the β-oxidation pathway.[4] *The biosynthetic pathway differs from the β-oxidation pathway in the following ways*:

1. The reductant is NADPH, rather than NADH.

2. Biosynthesis requires CO_2 and proceeds via malonyl–CoA, a carboxylated derivative of acetyl–CoA.

3. The acyl carrier is not CoA, as it is in the degradative pathway, but a protein called the acyl carrier protein (ACP). The acyl carrier protein is a small protein (MW 10,000 Da in *E. coli*), having a residue similar to CoASH attached to one end. (See note 5 for comparison of the chemical structures of CoASH and ACP.)

In eukaryotic cells, fatty acid biosynthesis takes place in the cytosol and the degradation takes place in the matrix of the mitochondria. Since prokaryotes do not have similar organelles, both synthesis and degradation take place in the same compartment, the cytosol. The fatty acid biosynthetic enzymes exist as multienzyme complexes in eukaryotes, including yeast. However, in *E. coli* they are present in the cytosol as separate enzymes.

The sequence of reactions begins with the carboxylation of acetyl~CoA to make malonyl~CoA (Fig. 9.4A, reaction 1). The carboxylation, which requires ATP and a vitamin, biotin, is catalyzed by acetyl–CoA carboxylase. (See note 6 for more information on acetyl–CoA carboxylase.) Then the CoA is displaced by the acyl carrier protein, ACP, to form malonyl–

ACP in a reaction catalyzed by malonyl–CoA:ACP transacetylase (reaction 2). A similar reaction, catalyzed by acetyl–CoA:ACP transacetylase, displaces CoA from another molecule of acetyl–CoA to form the acetyl–ACP derivative (reaction 3). (See note 7.) The methylene carbon in malonyl–ACP acts as a nucleophile and displaces the ACP from acetyl–ACP to form the β-ketoacyl–ACP derivative in a reaction catalyzed by β-ketoacyl–ACP synthase (reaction 4). (See Section 8.1.1 for a discussion of condensation reactions of these types.) At the same time, the newly added carboxyl is removed, driving the reaction to completion. (In *E. coli*, since there is more than one condensing enzyme, either acetyl–CoA or acetyl–ACP can be used for the condensation with malonyl–ACP. See note 8 for more information on fatty acid synthesis in *E. coli*.) The β-ketoacyl–ACP is then reduced to the hydroxy derivative by an NADPH-dependent β-ketoacyl–ACP reductase (reaction 5), and dehydrated to the α-β derivative by the β-hydroxylacyl–ACP dehydrase to form the unsaturated acyl–ACP derivative (reaction 6). (The double bond is between C2, the carbon α to the carboxyl carbon, and C3, the carbon β to the carboxyl carbon.) The unsaturated acyl–ACP derivative is then reduced by enoyl–ACP reductase to the saturated acyl–ACP (reaction 7). The acyl–ACP chain is elongated by a series of identical reactions initiated by the attack of malonyl–ACP on the carboxyl end of the growing acyl–ACP chain, displacing the ACP. Fatty acid synthesis must be regulated because long-chain acyl–ACPs do not accumulate. The mechanism of regulation has not been elucidated, but it is reasonable to suggest that it involves feedback inhibition of a regulatory fatty acid biosynthetic enzyme by long-chain acyl–ACPs. (See Section 6.1.1 for a discussion of feedback inhibition.)

When the acyl–ACP chain is completed, the acyl portion is immediately transferred to membrane phospholipids by the glycerol phosphate acyltransferase reactions described in Section 9.1.2. Of course, not all the fatty acids are incorporated into phospholipids. For example, some acyl–ACPs are used for lipid A biosynthesis (Section 11.2.2). These include β-hydroxymyristoyl–ACP, lauroyl–ACP, and myristoyl–ACP.

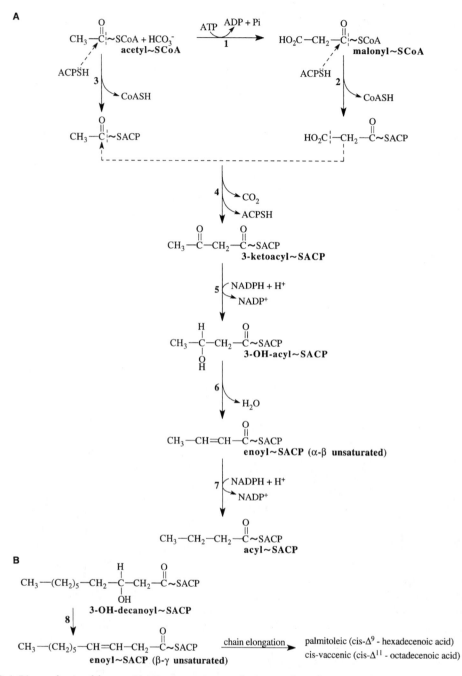

Fig. 9.4 Biosynthesis of fatty acids. Enyzmes: **1**, acetyl–CoA carboxylase; **2**, malonyl transacetylase; **3**, an acetyl transacetylase; **4**, β-ketoacyl–ACP synthase; **5**, β-ketoacyl–ACP reductase; **6**, β-hydroxyacyl–ACP dehydrase; **7**, enoyl–ACP reductase; **8**, β-hydroxydecenoyl–ACP dehydrase.

Synthesis of unsaturated fatty acids

If the fatty acid is unsaturated, it is almost always monounsaturated (one double bond). Polyunsaturated fatty acids are more typical of eukaryotic organisms. (As explained in note 9, polyunsaturated acids are required in the diets of animals.) Depending upon the organism, unsaturated fatty acids are formed in two different ways: the *anaerobic* pathway and the *aerobic* pathway. The anaerobic pathway is restricted to prokaryotes. It is widespread in bacteria, being found in *Clostridium*, *Lactobacillus*, *Escherichia*, *Pseudomonas*, the cyanobacteria, and the photosynthetic bacteria. The aerobic pathway is found in *Bacillus*, *Mycobacterium*, *Corynebacterium*, and *Micrococcus*, and in eukaryotes.

In the anaerobic pathway, a special dehydrase desaturates the C_{10} hydroxyacyl–ACP intermediate to the *trans*-α,β-decenoyl–ACP which is isomerized while bound to the enzyme to *cis*-β,γ-decenoyl–ACP (Fig. 9.4B), reaction 8). In the β,γ derivative, the double bond is between C3 and C4, whereas in the α,β derivative the double bond is between C2 and C3. *Very importantly, the cis-β,γ does not serve as a substrate for the enoyl reductase.* As a consequence, the *cis*-β,γ derivative is not reduced. Instead, it is elongated, leading to a fatty acid with a double bond. Two common monounsaturated fatty acids in bacteria synthesized in this way are palmitoleic acid, which has 16 carbons (*cis*-Δ⁹-hexadecenoic acid), and *cis*-vaccenic acid, which has 18 carbons (*cis*-Δ¹¹-octadecenoic acid). (See note 10 for a description of cis and trans fatty acids.) In palmitoleic acid, the double bond is between C9 and C10 in the final product. This is because six carbons are added to the C1 of the C10 β,γ derivative. Since vaccenic acid is two carbons longer, the double bond is between C11 and C12.

The aerobic pathway makes use of special desaturases (oxidases) that introduce a double bond into the completed fatty acyl–CoA or fatty acyl–ACP derivative. The enzyme system requires molecular oxygen and NADPH. In a multienzyme reaction sequence, two electrons are removed from the fatty acyl derivative forming the double bond, and two electrons are removed from NADH. The four electrons are transferred to O_2, forming two H_2O molecules. Bacteria that use the aerobic pathway make oleic acid from its C_{18} saturated progenitor, stearic acid, rather than *cis*-vaccenic. Oleic acid has a double bond between C9 and C10 and is called *cis*-Δ⁹-octadecenoic acid.

9.1.2 Phospholipid synthesis in bacteria

Phospholipids are lipids with covalently attached phosphate groups. They are an important constituent of cell membranes.

Because the phosphate groups are ionized, phospholipids always have negatively charged groups. There may also be positively charged groups (e.g., the protonated amino group in phosphatidylethanolamine).

There are several different types of phospholipid. However, the ones that concern us here are the major phospholipids in bacteria. These are the *phosphoglycerides*, phospholipids that contain fatty acids and phosphate esterified to glycerol, and usually some other molecule (e.g., an amino acid, an amine, or a sugar covalently bound to the phosphate). The structure of a typical phosphoglyceride is shown in Fig. 9.5. The fatty acids in positions C1 and C2 within the same molecule need not be the same.

Phospholipids and the structure of the cell membrane

A major structural feature of phospholipids is that they are amphibolic. That is, one part of the molecule is hydrophobic (apolar) and another part is hydrophilic (polar). The hydrophobic area (tail) is the part where the fatty acids are located, and the hydrophilic area (head) is the end containing the phosphate and its attached group.

The amphibolic nature of the phospholipids explains their orientation in membranes. Phospholipids are oriented in membranes as a bilayer with the charged head groups pointing out into the aqueous phase and the hydrophobic fatty acid tails interacting with each other in a hydrophobic interior. The cell membrane is completed by proteins (Fig. 1.15). Some properties of phospholipid membranes that should be remembered are as follows:

1. They are permeable only to water, gases, and small hydrophobic molecules.

2. They have a low ionic conductance and are capable of doing work when ions (especially protons and sodium ions) are transported through them via special carriers along an electrochemical gradient.

3. For a phospholipid membrane to function, its lipid portion must remain fluid.

4. The fluidity of the membrane is due to the presence of unsaturated fatty acids or

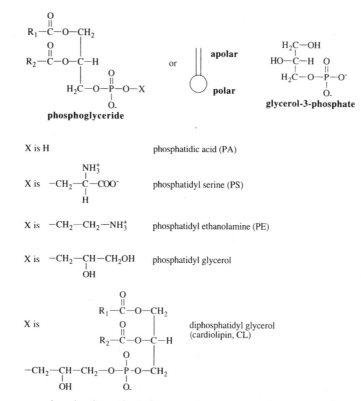

Fig. 9.5 Some common phosphoglycerides in bacteria; R represents a fatty acyl moiety esterified to the glycerol phosphate. Note that phospholipids have an apolar and a polar end. The configuration consisting of the glycerol-3-phosphate and the glycerol-3-phosphate moiety of phosphoglycerides belongs to the L-stereochemical series and is called *sn*-glycerol-3-phosphate, where *sn* indicates stereo specific numbering. Horizontal bonds are projected to the front and vertical bonds behind the plane of the page. Thus, the H and OH are in front of the plane of the page, and C1 and C3 are behind.

branched-chain (methyl) fatty acids in the phospholipids.

The synthesis of the phosphoglycerides

Phosphoglyceride synthesis starts with dihydroxyacetone phosphate (DHAP), an intermediate in glycolysis (Fig. 9.6). *Step 1* is the reduction of dihydroxyacetone phosphate to glycerol phosphate by the enzyme *glycerol phosphate dehydrogenase*. *Step 2* is the transfer to glycerol phosphate of fatty acids from fatty acyl–S–ACP (newly synthesized as described in Section 9.1.1). The reactions are catalyzed by membrane-bound enzymes called *G3P acyl transferases*. (Some bacteria are also capable of using acyl–SCoA derivatives as the acyl donor.) The first phospholipid made is phosphatidic acid, which is glycerol phosphate esterified to two fatty acids. The other phospholipids are made from phosphatidic acid.

Step 3 is a reaction in which the phosphate on the phosphatidic acid reacts with cytidine triphosphate (CTP) and displaces PP_i to form CDP–diacylglycerol. The reaction is catalyzed by *CDP–diglyceride synthase*. The formation of CDP–diacylglycerol is driven to completion by the hydrolysis of the PP_i catalyzed by a pyrophosphatase. *Step 4* is the displacement of CMP by serine, catalyzed by *phosphatidylserine synthase* (PS synthase). *Step 5* is the decarboxylation of phosphatidylserine to yield phosphatidylethanolamine, which is the major phospholipid in several bacteria. Bacteria make other phospholipids from CDP–diglyceride. The displacement of CDP by α-glycerolphosphate yields phosphatidylglycerol phosphate (*step 6*). Phosphatidylglycerol phosphate is dephosphorylated to yield phosphatidylglycerol (*step 7*). The latter displaces glycerol from a second molecule of phosphatidylglycerol to form diphosphatidylglycerol (cardiolipin) (*step 8*).

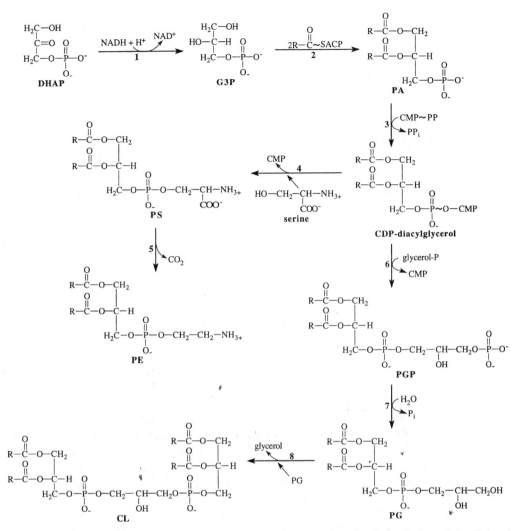

Fig. 9.6 Phosphoglyceride synthesis. The biosynthetic pathway for phosphoglycerides branches off glycolysis at DHAP. The G3P dehydrogenase is a soluble enzyme, but the G3P acyl transacylase and all subsequent reactions take place in the cell membrane. The fatty acyl–ACP derivatives are made in the cytosol, diffuse to the membrane, and transfer the acyl portion to the phospholipid on the inner surface of the membrane. Abbreviations: DHAP, dihydroxyacetone phosphate; G3P, glycerol phosphate; PA, phosphatidic acid; PS, phosphatidylserine; PGP, phosphatidylglycerol phosphate; PG, phosphatidylglycerol; CL, cardiolipin (diphosphatidylglycerol); CDL, cardiodieipid. Enzymes: 1, G3P dehydrogenase; 2, G3P acyltransferase and 1-acyl-G3P acyltransferase; 3, CDP–diglyceride synthase; 4, phosphatidylserine synthase; 5, phosphatidylserine decarboxylase; 6, PGP synthase; 7, PGP phosphatase; 8, CL synthase.

Note that modification of the head groups in all cases occurs via a nucleophilic attack by a hydroxyl on the electropositive phosphorus in the phosphate group.

The inhibition of phospholipid biosynthesis during the stringent response

The student may recall that in Section 2.2.2 it was pointed out that the stringent response, which is a response to starvation for a required amino acid or a carbon and energy source, is correlated with an accumulation of intracellular (p)ppGpp and is accompanied by an inhibition of phospholipid synthesis as well as stable RNA synthesis. During the stringent response imposed by amino acid starvation in *E. coli*, long-chain acyl–ACPs accumulate, suggesting that the acyltransferase is inhibited, which can account for the inhibition of phospholipid biosynthesis.[11]

How (p)ppGpp might be involved in the inhibition of the acyltransferase is not known.

9.1.3 Synthesis of archaeal lipids

The student should review the structure of archaeal lipids and membranes, as described in Section 1.2.5. The lipids of archaea differ in the following two ways from those of bacteria:

1. Instead of fatty acids, archaeal lipids have long-chain alcohols called isopranyl alcohols.

2. The linkage to glycerol is via an ether bond rather than an ester bond.

The metabolic pathway for the synthesis of the glycerol ethers in the archaea is not well understood at all. Figure 9.7 illustrates a proposal for *Halobacterium halobium* published in 1993. The glycerol backbone is synthesized from either glycerol-3-phosphate or dihydroxyacetone phosphate. The alcohol is believed to be derived from geranylgeranyl pyrophosphate, which is synthesized from acetyl–CoA via mevalonic acid in a well-known pathway that is widespread among the bacteria and higher organisms. (The alcohol group is esterified to the pyrophosphate.) The C1 hydroxyl on glycerol-3-phosphate or the dihydroxyacetone phosphate makes a nucleophilic attack on the geranylgeranyl pyrophosphate, displacing pyrophosphate and forming the ether linkage. If dihydroxyacetone phosphate is used, then 3-monoalkenyl-DHAP is formed. If glycerol-3-phosphate is used, then the monoalkenylglycerol-1-phosphate is formed. The monoalkenylglycerol-1-phosphate, which is formed either directly or in two steps, reacts with another geranylgeranyl pyrophosphate to form 2,3-dialkenyl-glycerol-1-phosphate.

It is proposed that in *Methanobacterium* and *Sulfolobus*, glycerol-1-phosphate, rather than dihydroxacetone phosphate or glycerol-3-phosphate, attacks the geranylgeranyl pyrophosphate. The rest of the pathway is unknown but involves a reduction of the double bonds in the hydrocarbons, the attachment of the polar head groups, and the formation of the tetraether lipids. It is not known whether the condensation between the alkyl groups to form the tetraethers occurs before or after substitution with polar head groups.

9.2 Nucleotides

9.2.1 Nomenclature and structures

A nucleotide is a molecule containing a pyrimidine or purine, a sugar (ribose or deoxyribose), and phosphate. If there were no phosphate, the molecule would be called a nucleoside. Thus, nucleotides are nucleoside phosphates. If one phosphate is present, the molecule is called a nucleoside monophosphate. If two phosphates are present, then it is called a nucleoside diphosphate, and so on.

The three major pyrimidines are cytosine, thymine, and uracil. The corresponding nucleosides are called cytidine, thymidine, and uridine, respectively. The cytidine nucleotides are called cytidine monophosphate (CMP), cytidine diphosphate (CDP), and cytidine triphosphate (CTP). A similar naming system is used for the uridine nucleotides (UMP, UDP, UTP) and the thymidine nucleotides (TMP, TDP, TTP). If the sugar is deoxyribose rather than ribose, then the nucleotide is written dCTP, and so on.

The two major purines are adenine and guanine. The nucleosides are adenosine and guanosine. The nucleotides are adenosine mono,- di-, and triphosphate and guanosine mono-, di-, and triphosphate. There are even separate names for the nucleoside monophosphates. For example, adenosine monophosphate is also called adenylic acid. Guanosine monophosphate is guanylic acid. We also have cytidylic acid, uridylic acid, and thymidylic acid. The structures of the purines, pyrimidines, and sugars are shown in Fig. 9.8. Notice that deoxyribose differs from ribose in having a hydrogen substituted for the hydroxyl group on C2 of the sugar.

9.2.2 The synthesis of the pyrimidine nucleotides

The pyrimidine part of the nucleotide is made from aspartic acid, ammonia, and carbon dioxide (Fig. 9.9). *Step 1* is the biotin-dependent synthesis of carbamoyl phosphate from HCO_3^-, glutamine, and ATP (Fig. 9.10). The enzyme that catalyzes the reaction is called *carbamoyl phosphate synthetase*. (See note 12.) It requires two ATPs. One ATP is used to form a carboxylated biotin intermediate, and the second ATP

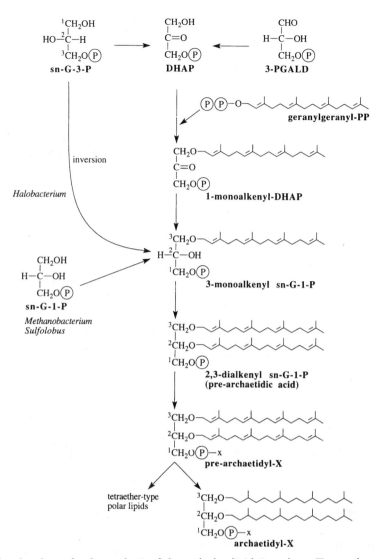

Fig. 9.7 Postulated pathway for the synthesis of glycerol ether lipids in archaea. Two pathways are depicted, one for *Halobacterium* and one for *Methanobacterium* and *Sulfolobus*. It is proposed that a nucleophilic displacement takes place on geranylgeranyl-PP displacing the pyrophosphate and forming the ether linkage. The molecule that condenses with the geranylgeranyl-PP might be glycerol-3-phosphate, glycerol-1-phosphate, or dihydroxyacetone phosphate. The rest of the pathway is unknown. *Source*: Koga et al. 1993. Ether polar lipids of methanogenic bacteria: structures, comparative aspects, and biosynthesis. *Microbiol. Rev.* 57:164–182.

is used to phosphorylate the carboxyl group. Notice that carbamoyl phosphate is an acyl phosphate and therefore has high group transfer potential.

In *step 2*, the carbamoyl group is transferred to aspartate as the phosphate is displaced by the nitrogen atom of aspartate. The product is *N*-carbamoylaspartate. The enzyme that catalyzes step 2 is *aspartate transcarbamylase* (ATCase), which is feedback inhibited by the end product, CTP. *Step 3* is the cyclization of *N*-carbamoylaspartate with loss of water to form the first pyrimidine, dihydroorotate. *Step 4* is the oxidation of dihydroorotate to orotate.

In *step 5* the orotate displaces PP_i from phosphoribosylpyrophosphate (PPRP) to form the first nucleotide, orotidine-5′-phosphate, also called orotidylate. The reaction is driven to completion by the hydrolysis of PP_i catalyzed by a pyrophosphatase. Note that the PRPP

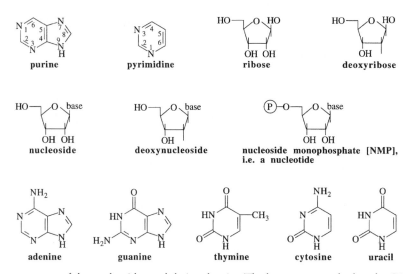

Fig. 9.8 The structures of the nucleotides and their subunits. The bases are attached to the C1 of the sugars and the phosphates to the C5. Not shown are the nucleoside di- and triphosphates.

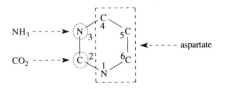

Fig. 9.9 Origins of atoms in the pyrimidine ring: C2 is derived from CO_2, and N3 comes from ammonia, via carbamoyl phosphate. The rest of the atoms are derived from aspartate.

itself is synthesized from ribose-5-phosphate and ATP via a phosphoryl group transfer reaction from ATP to the C1 carbon of ribose-5-phosphate.

Orotodine-5′-phosphate is decarboxlated in *step 6* to uridine monophosphate (UMP). Phosphorylation of UMP, using ATP as the donor, yields UDP and UTP (*steps 7 and 8*). CTP is synthesized from UTP by an ATP-dependent amination via NH_4^+ (*step 9*). (See note 13.)

The CTP can be dephosphorylated to CDP, which is the precursor to the deoxyribonucleotides dCTP and dTTP. The sequence of reactions is

CDP → dCDP → dCTP → dUTP → dUMP

→ dTMP → dTDP → dTTP

Also, UDP can be reduced to dUDP, which can be converted to dUTP. Thus, there are two routes to dUTP, one from CDP and the other (a minor route) from UDP.

9.2.3 The synthesis of the purine nucleotides

A purine is drawn in Fig. 9.11, showing the origin of the atoms. The synthesis of the purine ring requires as precursors glutamine and aspartic acid (to donate amino groups), glycine, carbon dioxide, and a C1 unit at the oxidation state of formic acid. The latter three donate all of the carbon atoms in the purine. The C1 unit at the oxidation level of formic acid can come from formic acid itself, or from serine.

Enzymatic reactions

The biosynthetic pathway for purines is shown in Fig. 9.12. The purine molecule is synthesized in stages while attached to ribose phosphate, which in turn comes from PRPP. *Step 1* is the attachment of an amino group to PRPP, donated by glutamine. The product is 5-phosphoribosylamine. This is a reaction in which the amide nitrogen of glutamine displaces the pyrophosphate in a nucleophilic displacement. The pyrophosphate that is released has a high free energy of hydrolysis, and the amination is driven to completion by the hydrolysis of the pyrophosphate by pyrophosphatase.

In *step 2*, glycine is added to the amino group of phosphoribosylamine to form 5-phosphoribosyl-glycineamide in an ATP-dependent step. In this reaction the carboxyl group of glycine is added to the amino group of phosphoribosy-

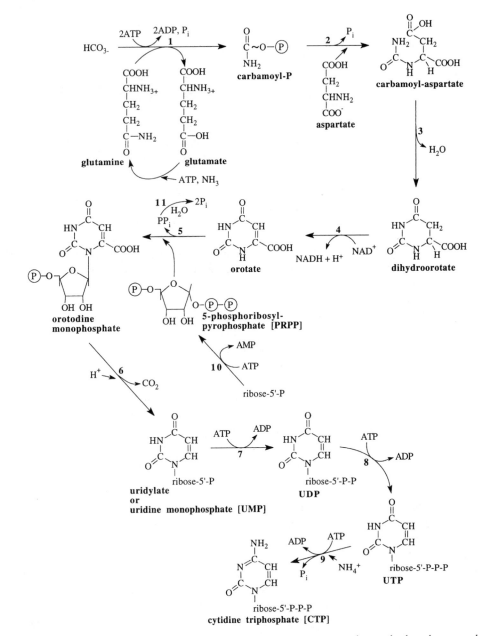

Fig. 9.10 The biosynthesis of pyrimidine nucleotides. Enzymes: **1**, carbamoyl phosphate synthetase; **2**, aspartate transcarbamoylase; **3**, dihydroorotase; **4**, dihydroorotate dehydrogenase; **5**, orotate phosphoribosyltransferase; **6**, orotidine-5-phosphate decarboxylase; **7**, nucleoside monophosphate kinase; **8**, nucleoside diphosphate kinase; **9**, CTP synthetase; **10**, PRPP synthetase; **11**, pyrophosphatase.

lamine to make the amide bond. The ATP first phosphorylates the carboxyl group, forming an acyl phosphate. Then the nitrogen in the amino group displaces the phosphate forming the amide bond. The student should review these kinds of reaction in Section 7.2.1.

Step 3 is the formylation of the amino group on the glycine moiety. This requires a

molecule to donate the formyl group (i.e., a molecule with high formyl group transfer potential). The donor of the formyl group is formyl–tetrahydrofolic acid (formyl–THF). The product is 5-phosphoribosyl-*N*-formyl-glycineamide.

In *step 4* the amide group is changed into an amidine, the nitrogen being donated by

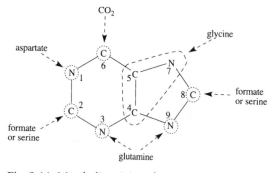

Fig. 9.11 Metabolic origins of atoms in purines.

glutamine in an ATP-dependent reaction. The product is 5-phosphoribosyl-N-formylglycineamidine. In *step 5* the 5-phosphoribosyl-N-formylglycineamidine cyclizes to 5-phosphoribosyl-5-aminoimidazole in a reaction that requires ATP. Probably the carbonyl is phosphorylated and the phosphate is displaced by the nitrogen to form the C–N bond. The cyclized product tautomerizes to form the amino group on aminoimidazole ribonucleotide.

Reaction 6 is the carboxylation of 5-phosphoribosyl-5-aminoimidazole to form 5-phosphoribosyl-5-aminoimidazole-4-carboxylic acid. In *step 7* aspartic acid combines with 5-phosphoribosyl-5-aminoimidazole-4-carboxylic acid to form 5-phosphoribosyl-4-(N-succinocarboxyamide)-5-amionoimidazole in an ATP-dependent step. In *step 8* fumarate is removed, leaving the nitrogen, to form 5-phosphoribosyl-4-carboxamide-5-aminoimidazole, which is formylated in *step 9* to form 5-phosphoribosyl-4-carboxamide-5-formaminoimidazole. The latter is cyclized in *step 10* to inosinic acid (IMP). The purine itself is called hypoxanthine. Inosinic acid is the precursor to all the purine nucleotides.

Biosynthesis of AMP and GMP from IMP

IMP is converted to AMP by substitution of the carbonyl oxygen at C6 with an amino group (Fig. 9.13). The amino donor is aspartate, and the reaction requires GTP. AMP can then be phosphorylated to ADP by using ATP as the phosphoryl donor. The ADP can be reduced to dADP or phosphorylated to ATP via substrate-level phosphorylation or respiratory phosphorylation. GMP is synthesized by an oxidation of IMP at C2 followed by an amination at that carbon. The oxidation produces a carbonyl group, which becomes aminated at the expense of ATP.

9.2.4 The role of tetrahydrofolic acid

Derivatives of tetrahydrofolic acid (THF; synthesized from the B vitamin folic acid) are coenzymes that carry single-carbon groups and are very important in metabolism. For example, the C_1 units used in purine biosynthesis are carried by derivatives of tetrahydrofolic acid. For an explanation of how THF is used, examine Fig. 9.14. The C_1 units can be donated to THF from serine, which donates its β carbon at the oxidation level of formaldehyde to THF acid to form methylene–THF and glycine. (This is also the pathway for glycine biosynthesis, as described in Section 9.3.1.) Methylene–THF is then oxidized to *formyl–THF*, which donates the formyl group in purine biosynthesis.

Bacteria can also use formate for purine biosynthesis. The formate is attached to THF acid to make formyl–THF in an ATP-dependent reaction. THF is also an important methyl carrier. Instead of methylene–THF being oxidized to formyl–THF, it is reduced to *methyl–THF*. Methyl–THF donates the methyl group in various biosynthetic reactions, including the biosynthesis of methionine.

Sulfanilamide and its derivatives, called sulfonamides (the "sulfa drugs"), inhibit the formation of folic acid and therefore inhibit purine biosynthesis in bacteria. Sulfonamides resemble p-aminobenzoic acid, an intermediate in folic acid biosynthesis, and the sulfonamides inhibit the enzyme that utilizes p-aminobenzoic acid for folic acid synthesis. Because animals obtain their folic acid from their diet, whereas most bacteria synthesize folic acid, the sulfonamides are toxic to bacteria but not to animals.

9.2.5 Synthesis of deoxyribonucleotides

The deoxyribonucleotides are synthesized from the ribonucleoside diphosphates by a reductive dehydration catalyzed by *ribonucleoside diphosphate reductase* (Fig. 9.15). The electron donor is a sulfhydryl protein called thioredoxin that obtains its electrons from NADPH.

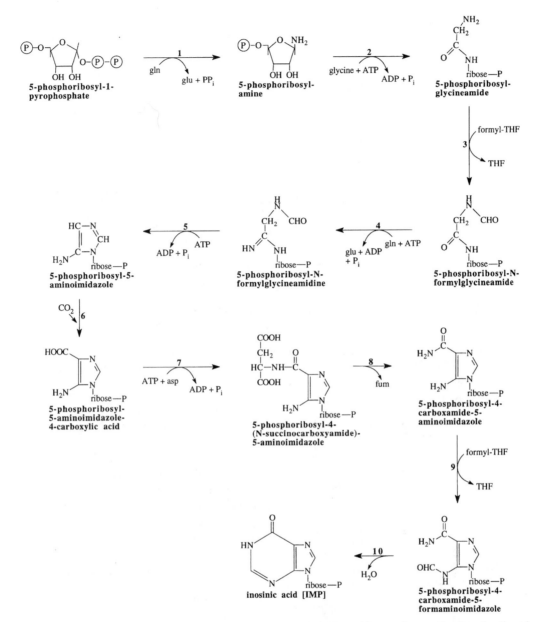

Fig. 9.12 Biosynthesis of purine nucleotides. Enzymes: **1**, PRPP amidotransferase; **2**, phosphoribosyl-glycineamide synthetase; **3**, phosphoribosylglycineamide formyltransferase; **4**, phosphoribosyl–formylglycineamidine synthetase; **5**, phosphoribosyl–aminoimidazole synthetase; **6**, phosphoribosy-laminoimidazole carboxylase; **7**, phosphoribosylaminoimidazole succinocarboxamide synthetase; **8**, adenylosuccinate lyase; **9**, phosphoribosylaminoimidazolecarboxamide formyltransferase; **10**, IMP cyclohydrolase. Abbreviations: gln, glutamine; glu, glutamate; asp, aspartate; fum, fumarate; THF, tetrahydrofolate.

9.3 Amino Acids

There are 20 different amino acids in proteins. Inspection of Table 9.2 reveals that six amino acids are synthesized from oxaloacetate and four from α-ketoglutarate, two intermediates

of the citric acid cycle. In addition, succinyl–CoA donates a succinyl group in the formation of intermediates in the biosynthesis of lysine, methionine, and diaminopimelic acid. (The succinate is removed at a later step and does not appear in the product.) This again emphasizes

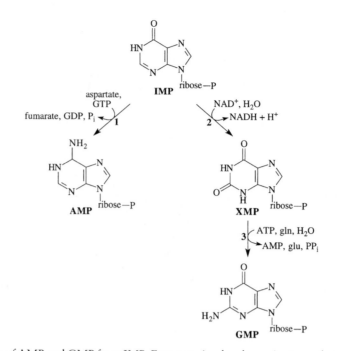

Fig. 9.13 Synthesis of AMP and GMP from IMP. Enzymes: **1**, adenylosuccinate synthetase and adenylosuccinate lyase; **2**, IMP dehydrogenase; **3**, GMP synthetase. Abbreviations: IMP, inosinic acid; AMP, adenylic acid; XMP, xanthylic acid; GMP, guanylic acid.

Table 9.2 Precursors for amino acid biosynthesis

Precursor	Amino acid
Pyruvic acid	L-Alanine, L-valine, L-leucine
Oxaloacetic acid	L-Aspartate, L-asparagine, L-methionine, L-lysine, L-threonine, L-isoleucine
α-Ketoglutaric acid	L-Glutamate, L-glutamine, L-arginine, L-proline
3-PGA	L-Serine, L-glycine, L-cysteine
PEP and erythrose-4-P	L-Phenylalanine, L-tyrosine, L-Tryptophan
PRPP and ATP	L-Histidine

that the citric acid cycle is important not only for energy generation, but also for biosynthesis.

Table 9.2 also points out that three glycolytic intermediates (pyruvate, 3-PGA, and PEP) are precursors to nine more amino acids. The pentose phosphate pathway provides erythrose-4-P, which is a precursor to aromatic amino acids. (It is actually used with PEP in the synthesis of the aromatic amino acids.) Finally, 5-phosphoribosyl-1-pyrophosphate (PRPP), which donates the sugar for nucleotide synthesis, is also the precursor to the amino acid histidine. Therefore, the three central pathways,

glycolysis, the pentose phosphate pathway, and the citric acid cycle, provide the precursors to all the amino acids. To illuminate some principles of amino acid metabolism, we will now examine the biosynthesis and catabolism of a select group of amino acids.

9.3.1 Synthesis

Glutamate and glutamine synthesis

The ability to synthesize glutamate and glutamine is of extreme importance because this is the only route for incorporation of inorganic nitrogen into cell material. All inorganic nitrogen must first be converted to ammonia, which is then incorporated as an amino group into glutamate and glutamine. (The synthesis of ammonia from nitrate and from nitrogen gas is described in Chapter 12). The amino group is then donated from these amino acids to all the other nitrogen-containing compounds in the cell. Glutamate is the amino donor for most of the amino acids, and glutamine is the amino donor for the synthesis of purines, pyrimidines, amino sugars, histidine, tryptophan, asparagine, NAD^+, and p-aminobenzoate.

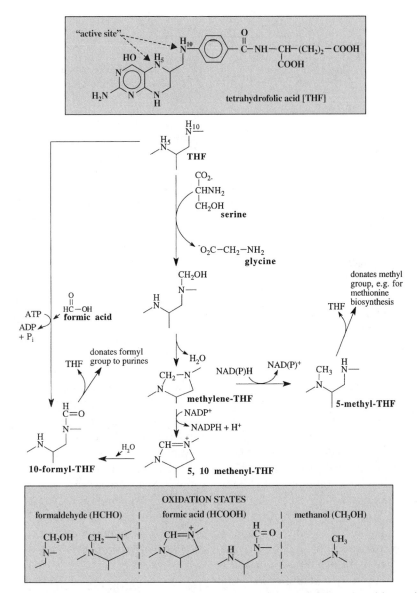

Fig. 9.14 Tetrahydrofolic acid (THF) as a C_1 carrier. Tetrahydrofolic acid is the reduced form of the vitamin folic acid. Tetrahydrofolic acid and its derivatives are important coenzymes that function in many pathways to carry one-carbon units. The one-carbon units can be at the oxidation level of formic acid (HCOOH), formaldehyde (HCHO), or methyl (CH_3-). This arrangement is essential for purine biosynthesis because it transfers groups at the oxidation level of formic acid in two separate steps. The most common precursor of the C_1 unit is the amino acid serine, which donates the C_1 unit as formaldehyde to THF. The product can then be oxidized to the level of formate to form 10-formyl–THF. Some bacteria can use formic acid itself as the source of the C_1 unit. Methylene–THF can also be reduced to methyl–THF, which serves as a methyl donor.

Glutamate is synthesized by two alternate routes. One requires the enzyme *glutamate dehydrogenase*, and the second requires two enzymes: *glutamine synthetase* (GS) and *glutamate synthase*, which is also called *glutamine oxoglutarate aminotransferase*, or the *GOGAT* enzyme.

Glutamate dehydrogenase catalyzes the reductive amination of α-ketoglutarate (Fig. 9.16). However, it should be pointed out that because of its high K_m for ammonia, the glutamate dehydrogenase reaction can be used for glutamate synthesis only when ammonia concentrations are high (>1 mM), otherwise ammonia is

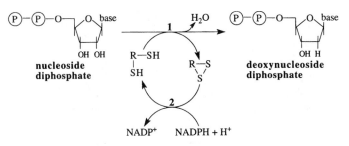

Fig. 9.15 Formation of deoxynucleotides by ribonucleoside diphosphate reductase. The deoxynucleotides are made from the nucleoside diphosphates. The OH on C2 of the ribose leaves with its bonding electrons to form water. The OH is replaced by a hydride ion (H:$^-$) donated by reduced thioredoxin, a dithioprotein. Thioredoxin, R–(SH)$_2$, is re-reduced by thioredoxin reductase, a flavoprotein. Thioredoxin reductase, in turn, accepts electrons from NADPH. Enzymes: **1**, ribonucleotide diphosphate reductase; **2**, thioredoxin reductase.

$$
\begin{array}{c}
\text{COOH} \\
|\\
\text{C}=\text{O} \\
|\\
\text{CH}_2 \\
|\\
\text{CH}_2 \\
|\\
\text{COOH}
\end{array}
\;+\; \text{NADPH} + \text{H}^+ + \text{NH}_3
\;\rightleftharpoons\;
\begin{array}{c}
\text{COOH} \\
|\\
\text{H}_2\text{N}-\text{C}-\text{H} \\
|\\
\text{CH}_2 \\
|\\
\text{CH}_2 \\
|\\
\text{COOH}
\end{array}
\;+\; \text{NADP}^+ + \text{H}_2\text{O}
$$

Fig. 9.16 The reductive amination of α-ketoglutarate, catalyzed by glutamate dehydrogenase.

incorporated into glutamate via glutamine. This is the situation in many natural environments.

Under conditions of low ammonia concentration, bacteria use two enzymes in combination for ammonia incorporation. One is L-glutamine synthetase (Fig. 9.17, reaction **1**). This enzyme incorporates ammonia into glutamate to form glutamine. The glutamine synthetase reaction uses ATP, which drives the reaction to completion. The other enzyme is glutamate synthase (reaction **2**). This enzyme transfers the newly incorporated ammonia from glutamine to α-ketoglutarate to form glutamate. Notice in Fig. 9.17 that glutamine is used catalytically. This is also seen in the following reactions:

The glutamine synthetase and glutamate synthase reactions: (The use of ATP drives the reaction to completion)

Glutamate + ATP + NH$_3$

   → glutamine + ADP + P$_i$

Glutamine + α-ketoglutarate + NADPH + H$^+$

   → 2 glutamate + NADP$^+$

α-Ketoglutarate + ATP + NH$_3$ + NADPH + H$^+$

   → glutamate + ADP + P$_i$ + NADP$^+$

As mentioned, glutamine is an amino donor for many compounds, including pyrimidines and purines. Therefore, glutamine synthetase is an indispensable enzyme unless the medium is supplemented with glutamine. Glutamate dehydrogenase, however, is not necessary for growth, since the organism can synthesize glutamate from glutamine and α-ketoglutarate by using glutamate synthase. The complex regulation of glutamine synthetase activity is discussed in Section 18.4.

Transamination reactions from glutamate are required for the synthesis of the other amino acids

The synthesis of the other amino acids requires that glutamate donate an amino group to an α-ketocarboxylic acid. The enzyme that catalyzes the amino group transfer is called a *transaminase*. A generalized transamination reaction is shown in Fig. 9.18A. Note that after the glutamate has donated its amino group, it becomes α-ketoglutarate again, which then is aminated once more to form another molecule of glutamate. Thus, glutamate can be thought of as a conduit through which ammonia passes into other amino acids. Transaminases use a coenzyme called pyridoxal phosphate (derived from

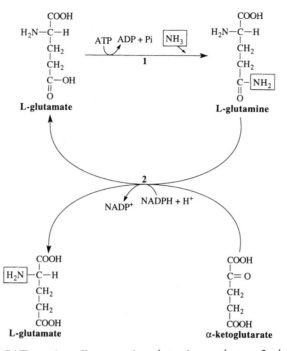

Fig. 9.17 The GS and GOGAT reactions. Enzymes: **1**, L-glutamine synthetase; **2**, glutamine: α-oxoglutarate aminotransferase, also called the GOGAT enzyme, or glutamate synthase.

vitamin B_6) to carry the amino group from glutamate to the α-ketocarboxylic acids.

Synthesis of aspartate and alanine

Aspartate is synthesized via a transamination from glutamate to oxaloacetate (Fig. 9.18B), and alanine is made via a transamination from glutamate to pyruvate (Fig. 9.18C).

Synthesis of serine, glycine, and cysteine

Serine is synthesized from 3-phosphoglycerate (Fig. 9.19). The phosphoglycerate is first oxidized to the α-keto acid, phosphohydroxypyruvate. A transamination, using glutamate as the donor, converts the phosphohydroxypyruvate to phosphoserine. Then the phosphate is hydrolytically removed to form L-serine. The hydrolysis of the phosphate ester bond drives the reaction to completion.

Serine is a precursor to both glycine and cysteine. Glycine is formed by transfer of the β carbon of serine to tetrahydrofolic acid (Fig. 9.14). The route to cysteine is initiated when serine accepts an acetyl group from acetyl–CoA to become O-acetylated. Sulfide then displaces the acetyl group from O-acetylserine, forming cysteine (see later: Fig. 12.2).

9.3.2 Catabolism

A scheme showing the overall pattern of carbon flow during amino acid catabolism is shown in Fig. 9.20. The first reaction is always the removal of the amino group to generate the α-keto acid, which eventually enters the citric acid cycle. All 20 amino acids are degraded to seven intermediates that enter the citric acid cycle. These seven intermediates are pyruvate, acetyl–CoA, acetoacetyl–CoA, α-ketoglutarate, succinyl–CoA, fumarate, and oxaloacetate. Acetoacetyl–CoA itself is a precursor to acetyl–CoA.

Removal of the amino group

There are several ways of removing amino groups from amino acids during their catabolism (Fig. 9.21). Usually amino acids are oxidatively deaminated to their corresponding keto acid. The oxidation may be catalyzed by a nonspecific *flavoprotein oxidase* that oxidizes any one of a number of amino acids and feeds the electrons directly into the electron transport chain (Fig. 9.21A). These oxidases can be D-amino oxidases as well as L-amino oxidases. The D-amino acid oxidases are useful because

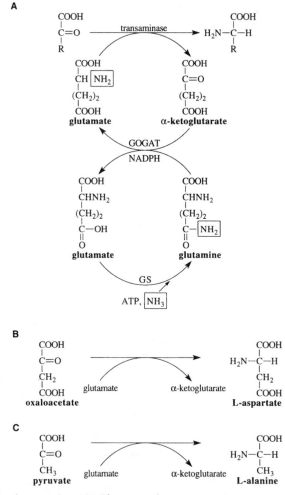

Fig. 9.18 The transamination reaction. (A) Glutamate donates an amino group to an α-ketocarboxylic acid to form an α-amino acid. The glutamate becomes α-ketoglutarate, which is aminated via either the glutamate dehydrogenase or the GS and GOGAT reactions. (B) Formation of L-aspartate from oxaloacetate. (C) Formation of L-alanine from pyruvate.

several biological molecules (e.g., peptidoglycan, certain antibiotics) have D-amino acids.

There are also *NAD(P)⁺-linked dehydrogenases* (Fig. 9.21B), which are more specific than the flavoprotein oxidases. The dehydrogenases catalyze the reversible reductive amination of the keto acid. A transaminase coupled to a dehydrogenase can lead to the deamination of amino acids. In the example shown in Fig. 9.21C, a nonspecific transaminase transfers an amino group to pyruvate-forming L-alanine. The amino group is released as ammonia, and pyruvate is regenerated by L-alanine dehydrogenase. α-Ketoglutarate and its dehydrogenase serve the same function in amino acid catabolism.

Certain amino acids are deaminated by specific *deaminases* (Fig. 9.21D). In these cases,

a redox reaction is not involved. These amino acids include serine, threonine, aspartate, and histidine. The deamination of serine and of threonine proceeds by the elimination of water.

9.4 Aliphatic Hydrocarbons

Many bacteria can grow on long-chain hydrocarbons, for example, alkanes (C_{10}–C_{18}). A few bacteria (mycobacteria, flavobacteria, *Nocardia*) grow on short-chain hydrocarbons (C_2–C_8). The hydrocarbons exist as droplets of oil outside the cell, and a major problem in using these water-insoluble molecules entails transferring them from the oil layer across the cell wall to the cell membrane, where they can

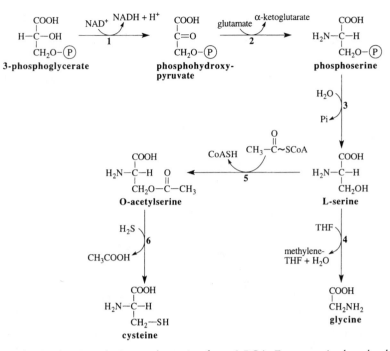

Fig. 9.19 The synthesis of serine, glycine, and cysteine from 3-PGA. Enzymes: **1**, phosphoglycerate dehydrogenase; **2**, phosphoserine aminotransferase; **3**, phosphoserine phosphatase; **4**, serine hydroxymethyltransferase; **5**, serine transacetylase; **6**, O-acetylserine sulfhydrylase.

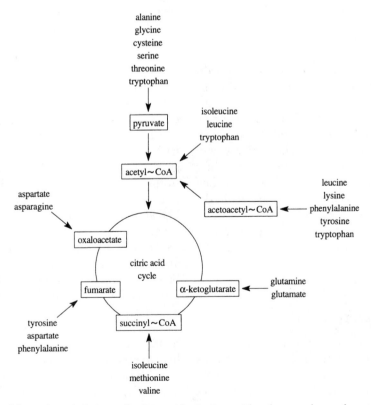

Fig. 9.20 Fates of the carbon skeletons of amino acids. Amino acid carbon can be used to synthesize the cell components, or it can be oxidized to CO_2. In the absence of the glyoxylate cycle, carbon entering at acetyl–CoA cannot be used for net glucogenesis except in some strict anaerobic bacteria that can carboxylate acetyl–CoA to pyruvate (Chapter 13).

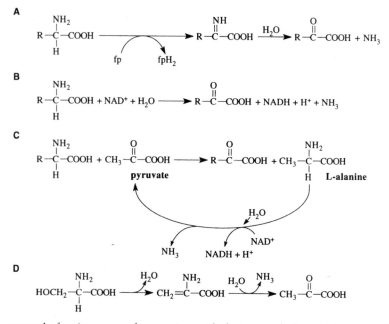

Fig. 9.21 The removal of amino groups from amino acids during catabolism. (A) Amino acid oxidases are flavoproteins that are specific for L or D amino acids, but generally not for the particular amino acid. Electrons are transferred from the flavoprotein to the electron transport chain. (B) An amino acid dehydrogenase. These molecules are NAD(P) linked and more specific for the amino acid than are the flavoprotein oxidases. (C) Transamination catalyzed by a nonspecific transaminase that uses pyruvate as the amino acid acceptor. The amino group is removed from the alanine by alanine dehydrogenase. α-Ketoglutarate can also accept amino groups and be deaminated by α-ketoglutarate dehydrogenase. (D) The deamination of serine by serine dehydratase.

be metabolized. Some bacteria have cell walls containing glycolipids in which the hydrocarbons can dissolve and be transported to the membrane. *Acinetobacter* strains secrete particles resembling the outer membrane in which the hydrocarbons dissolve. The particles with the dissolved hydrocarbon fuse with the outer membrane, and the hydrocarbons are then transferred to the cell membrane. Once in the cell membrane, the degradation of the alkanes requires their hydroxylation, via molecular oxygen and an enzyme called *monooxygenase*.

9.4.1 Degradative pathways

Hydroxylation at one end

After the hydrocarbon dissolves in the membrane, it is hydroxylated at one end by means of molecular oxygen, in a reaction catalyzed by a monooxygenase. Membrane-bound enzymes then oxidize the long-chain alcohol to the carboxylic acid (Fig. 9.22A). The carboxylic acid is derivatized with CoASH in an ATP-dependent reaction and oxidized via β-oxidation in the cytoplasm.

Since aliphatic hydrocarbons can be oxidized to CO_2 by certain sulfate-reducing bacteria in the absence of oxygen, there must exist an alternative mechanism that does not require oxygen.[14] However, the mechanism is not known. In the sequence shown in Fig. 9.22A, a short electron transport chain carries electrons from NADH via flavoprotein and an FeS protein to a *b*-type cytochrome called P450.[15] The P450 is in a complex with the hydrocarbon substrate and is called a cosubstrate. The reduced P450 binds O_2 and activates it, presumably converting it to bound O_2^- (superoxide) or O_2^{2-} (peroxide), which attacks the C1 carbon on the hydrocarbon. One of the oxygen atoms replaces a hydrogen on the hydrocarbon and becomes a hydroxyl group, while the other oxygen atom is reduced to H_2O. Some bacteria use an *n*-alkane monooxygenase instead of P450 to hydroxylate the hydrocarbon. The iron protein rubridoxin (Fe^{2+} or Fe^{3+}) is reduced by NADH and is a cosubstrate in the monooxygenase reaction.

Hydroxylation on the penultimate carbon

An alternative degradative pathway is found in *Nocardia* species (Fig. 9.22B). The hydroxylation takes place on the second carbon, forming a secondary alcohol, which is then oxidized to form a ketone. Then, in an unusual reaction, a second monooxygenase reaction creates an acetyl ester, which is subsequently hydrolyzed to acetate and the long-chain alcohol. The alcohol is oxidized to the carboxylic acid, which is degraded via the β-oxidation pathway.

9.5 Summary

Many aerobic bacteria can grow on long-chain fatty acids. The major pathway for degradation is the β-oxidative pathway. First, the fatty acid is converted to the acyl–CoA derivative in an ATP-dependent reaction catalyzed by acyl–CoA synthetase. The first reaction is a displacement of pyrophosphate from ATP to form the acyl–AMP, which remains bound to the enzyme. Then the AMP is displaced by CoASH to form the acyl–CoA. The reaction is driven to completion by hydrolysis of the pyrophosphate, catalyzed by pyrophosphatase. Thus, the activation of fatty acids requires the equivalent of two ATP molecules. A keto group is then introduced β to the carboxyl via an oxidation, hydration, and oxidation to form the β-ketoacyl–CoA derivative. Finally, the β-ketoacyl–CoA derivative is cleaved with CoASH to yield acetyl–CoA.

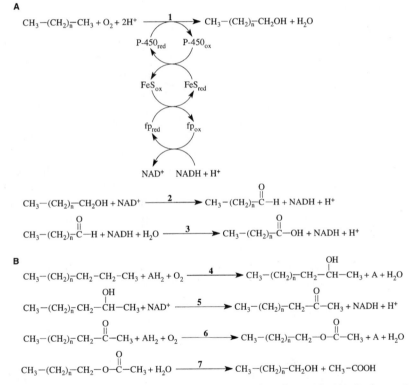

Fig. 9.22 Oxidation of an aliphatic hydrocarbon to a carboxylic acid. (A) Pathway in yeast and *Corynebacterium* species. The hydrocarbon is hydroxylated by oxygen, which is activated by cytochrome P450. The second oxygen atom is reduced to water. Other bacteria (e.g., *Pseudomonas* spp.) use a monooxygenase instead of P450. In the latter case the electron transport is from NADH to rubredoxin:NADH oxidoreductase, to rubredoxin, which is a cosubstrate in the monooxygenase reaction, rather than P450. The primary alcohol is oxidized to an aldehyde, which is then oxidized to a carboxylic acid. The carboxylic acid is oxidized via the β-oxidation pathway. (B) Pathway in *Nocardia* species. The C2 carbon is hydroxylated, forming a secondary alcohol, which is then oxidized to a ketone. A second monooxygenase reaction converts the ketone into an acetyl ester. The acetyl ester is hydrolyzed to acetate and a long-chain alcohol, which is oxidized as in (A). Enzymes: **1, 4, 6,** monooxygenase; **2, 5,** alcohol dehydrogenase; **3, 6,** aldehyde dehydrogenase; **7,** acetyl esterase.

The process is repeated while two-carbon fragments are sequentially removed as acetyl–CoA from the fatty acid. If the fatty acid is an odd-chain fatty acid, then the last fragment is propionyl–CoA, which can be oxidized to pyruvate. Because fatty acids are degraded to acetyl–CoA, growth on fatty acids uses the same metabolic pathways as growth on acetyl–CoA.

Fatty acid synthesis occurs via an entirely different set of reactions. Acetyl–CoA is carboxylated to malonyl–CoA. A transacetylase then converts malonyl–CoA to malonyl–ACP. Acetyl–CoA is also transacylated to acetyl–ACP. Malonyl–ACP then condenses with acetyl–ACP (or acetyl–CoA, depending upon which synthase is used) to form a β-ketoacyl–ACP. The condensation with malonyl–ACP is accompanied by a decarboxylation reaction that drives the condensation. The β-ketoacyl–ACP is reduced via NADPH, dehydrated, and reduced again to form the saturated acyl–ACP. The acyl–ACP is elongated by condensation with malonyl–ACP.

There are two systems found among the bacteria for synthesizing unsaturated fatty acids: an anaerobic and an aerobic pathway. In the anaerobic pathway the C_{10} hydroxyacyl–ACP is desaturated and elongated rather than being desaturated and reduced. In the aerobic pathway, which is found in eukaryotes and certain aerobic bacteria, unsaturated fatty acids are synthesized in a pathway requiring oxygen and NADPH. The double bond is introduced into the completed fatty acyl–ACP or –CoA derivative by special desaturases (oxidases) that require O_2 and NAD(P)H as substrates. During the introduction of the double bond, four electrons are transferred to O_2 to form two H_2O molecules. Two electrons are contributed by NAD(P)H, and two electrons are derived from the fatty acid when the double bond is formed. Bacteria that use this pathway make oleic acid rather than cis-vaccenic acid, which is made in the anaerobic pathway. They are both C_{18} carboxylic acids, but in oleic acid the double bond is between C9 and C10, whereas in cis-vaccenic acid it is between C11 and C12.

The phosphoglycerides are synthesized from the fatty acyl–ACP derivatives and glycerol phosphate, the latter being the reduced product of dihydroxyacetone phosphate. The phosphoglycerides differ with respect to the group that is substituted on the phosphate. The major phospholipid in cell membranes from several bacteria is phosphatidylethanolamine. Substitution on the phosphate begins with the displacement of pyrophosphate from CTP by the phosphate on phosphatidic acid, the product being CDP–diglyceride. The pyrophosphate is hydrolyzed by a pyrophosphatase, which makes the pathway irreversible. CMP is then displaced by serine to form phosphatidylserine. A decarboxylation leads to phosphatidylethanolamine. Phosphatidylglycerol phosphate is formed by displacing CMP from CDP–diacylglyceride with glycerol phosphate. A subsequent dephosphorylation produces phosphatidylglycerol. Diphosphatidylglycerol is made from two phosphatidylglycerol molecules. The transfer of the fatty acyl groups to the glycerol and all subsequent steps in phospholipid synthesis take place in the cell membrane. Archaeal phospholipids are synthesized from either dihydroxyacetone phosphate or glycerol phosphate. The precursor to the alcohol portion is geranylgeranyl pyrophosphate, which forms an ether linkage to the glycerol backbone.

The ribose and deoxyribose moieties of the nucleotides are derived from 5-phosphoribosyl-1-pyrophosphate (PPRP), which is synthesized from ribose-5-phosphate and ATP. In this this reaction, the OH on the C1 of ribose-5-phosphate displaces AMP from ATP to form the pyrophosphate derivative. It is therefore a pyrophosphoryl group transfer rather than the usual phosphoryl or AMP group transfer. The pyrophosphate itself is displaced from PPRP by orotic acid during pyrimidine synthesis or by an amino group from glutamine during purine biosynthesis. A pyrophosphatase hydrolyzes the pyrophosphate to inorganic phosphate, thus driving nucleotide synthesis.

Whenever pyrophosphate is hydrolyzed to inorganic phosphate, the equivalent of two ATPs is necessary to restore both the phosphates to ATP. Pyrimidine and purine biosynthesis differ in that the pyrimidines are made separately and then attached to the ribose phosphate, whereas the purine ring is built piece by piece while attached to the ribose phosphate. The deoxyribonucleotides are formed by reduction of the ribonucleotide diphosphates.

All the phosphates in the nucleotides are donated by ATP. The α-phosphate is derived from ribose-5-phosphate, which in turn is synthesized from glucose-6-phosphate. In some bacteria (e.g., *E. coli*), the phosphate in glucose-6-phosphate may come from PEP during transport into the cell via the phosphotransferase system, whereas in other bacteria it is transferred to glucose from ATP via hexokinase or glucokinase. The β and γ phosphates are derived from kinase reactions in which ATP is the donor. The inorganic phosphate is incorporated into ATP via substrate-level phosphorylation or electron transport phosphorylation. Since the various nucleotide triphosphates provide the energy for the synthesis of nucleic acids, protein, lipids, and polysaccharides, as well as other reactions, it is clear that ATP fuels all the biochemical reactions in the cytosol.

To incorporate ammonia into cell material, cells must synthesize both glutamate and glutamine. The glutamate donates amino groups to the amino acids, and the glutamine donates amino groups to purines, pyrimidines, amino sugars, and some amino acids. Glutamine is synthesized by the ATP-dependent amination of glutamate to form glutamine. This is catalyzed by glutamine synthetase. The glutamine then donates the amino group to α-ketoglutarate to form glutamate. The latter reaction is catalyzed by glutamate synthase, also known as the GOGAT enzyme. Glutamate can also be synthesized by the reductive amination of α-ketoglutarate to glutamate by glutamate dehydrogenase, a reaction that requires high concentrations of ammonia.

The degradation of amino acids occurs by pathways different from the biosynthetic ones. The first step is the removal of the amino group, by an amino acid oxidase, an amino acid dehydrogenase, or a deaminase. The carbon skeleton eventually enters the citric acid cycle.

Hydrocarbon catabolism begins with a hydroxylation to form the alcohol, which is oxidized to the carboxylic acid. The carboxylic acid is degraded via the β-oxidative pathway. Some bacteria use an alternative pathway to initiate hydrocarbon degradation, in which the hydrocarbon is oxidized to a ketone, which is eventually hydrolyzed to acetate and a long-chain alcohol.

Study Questions

1. Fatty acid synthesis is not simply the reverse of oxidation. What features distinguish the two pathways from each other?

2. What is it about the structure of phospholipids that causes them to form bilayers spontaneously?

3. Glycerol can be incorporated into phospholipids. Write a pathway showing the synthesis of phosphatidic acid from glycerol.

4. The C3 of glycerol becomes the C2 and C8 of purines. Write a sequence of reactions showing how this can occur.

5. Show how three carbons of succinic acid can become C4, C5, and C6 of pyrimidines. What happens to the fourth carbon from succinate?

6. What drives the condensation reaction in fatty acid synthesis?

7. Write a reaction sequence by which bacteria incorporate the nitrogen from ammonia into glutamate and glutamine when ammonia concentrations are low. How might this occur when ammonia concentrations are high? What is the fate of the nitrogen incorporated into glutamate, and that incorporated into glutamine?

8. ATP drives the synthesis of phospholipids. Write the reactions showing the incorporation of glycerol into phosphatidylserine. Focus on the steps that require a high-energy donor. How is ATP involved? (You must account for the synthesis of CTP.)

9. Bacteria that utilize aliphatic hydrocarbons as a carbon and energy source are usually aerobes. What is the explanation for the requirement for oxygen?

REFERENCES AND NOTES

1. Textor, S., V. F. Wendisch, A. A. De Graff, U. Müller, M. I. Linder, D. Linder, and W. Buckel. 1997. Propionate oxidation in *Escherichia coli*: evidence for operation of a methylcitrate cycle in bacteria. *Arch. Microbiol.* 168:428–436.

2. Gerike, U., D. W. Hough, N. J. Russell, M. L. Dyall-Smith, and M. J. Danson. 1998. Citrate synthase and 2-methylcitrate synthase: structural, functional and evolutionary relationships. *Microbiology* 144:929–935.

3. Cambell, J. W., R. M. Morgan-Kiss, and J. E. Cronan Jr. 2003. A new *Escherichia coli* metabolic competency: Growth on fatty acids by a novel anaerobic β-oxidation pathway. *Mol. Microbiol.* 47:793–805.

4. Magnunson, K., S. Jackowski, C. O. Rock, and J. E. Cronan Jr. 1993. Regulation of fatty acid biosynthesis in *Escherichia coli*. *Microbiol. Rev.* 57:522–542.

5. CoASH = P–AMP–pantothenic acid–β-mercaptoethylamine–SH. Acyl carrier protein = protein–pantothenic acid–β-mercaptoethylamine–SH. Coenzyme A has a phosphorylated derivative of AMP (AMP-3′-phosphate) attached via a pyrophosphate linkage to the vitamin pantothenic acid (a B_2 vitamin), which is covalently bound to β-mercaptoethylamine via an amide linkage. The β-mercaptoethylamine provides the SH group at the end of the molecule. In the acyl carrier protein, the AMP is missing and the pantothenic acid is bound directly to the protein. Therefore, the functional end of the acyl carrier protein is identical to CoASH, but the end that binds to the enzymes is different.

6. Acetyl–CoA carboxylase consists of four subunit proteins. These are biotin carboxylase, biotin carboxyl carrier protein, and two proteins that carry out the transcarboxylation of the carboxy group from biotin to acetyl–CoA.

7. Whether there is a separate acetyl–CoA:ACP transacetylase is not certain. Reaction 3 can be catalyzed by β-ketoacyl–ACP synthase III (acetoacetyl–ACP synthase), which has transacetylase activity.

8. There are actually three synthases and three possible routes to acetoacetyl–ACP in *E. coli*. In one pathway, β-ketoacyl–ACP synthase III catalyzes the condensation of acetyl–CoA with malonyl–ACP. A defect in synthase III leads to overproduction of 18C fatty acids, whereas overproduction of synthase III leads to a decrease in the average chain lengths of the fatty acids synthesized. The decrease in the fatty acid chain lengths in strains overproducing synthase III has been rationalized by assuming that synthase III is primarily active in the initial condensation of acetyl–CoA and malonyl–ACP and that the increased levels of synthase III stimulate the initial condensation reaction and divert malonyl–ACP from the terminal elongation reactions. In a second initiation pathway, acetyl–ACP is a substrate for β-ketoacetyl–ACP synthase I or II. In a third pathway, which is not thought to be physiologically significant under most growth conditions, malonyl–ACP is decarboxylated by synthase I and the resultant acetyl–ACP condenses with malonyl–ACP. The different synthases appear to be involved in determining the types of fatty acid made. In particular, synthases I and II, which catalyze condensations in both saturated and unsaturated fatty acid synthesis, appear to have specific roles in the synthesis of unsaturated fatty acids. For example, mutants of *E. coli* that lack synthase I do not make any unsaturated fatty acids, suggesting that only synthase I is capable of catalyzing the elongation of *cis*-3-decenoyl–ACP.

Mutations in synthase II lead to an inability to synthesize *cis*-vaccenate, in agreement with the finding that synthase II can elongate palmitoleoyl–ACP but synthase I cannot.

9. Mammals require linoleate (18:2 *cis*-Δ^9, Δ^{12}) and linolenate (18:3 *cis*-Δ^9, Δ^{12}, Δ^{15}). Hence these are essential fatty acids and must be supplied in the diet of mammals. Mammals cannot synthesize these fatty acids because they do not have the enzymes to introduce double bonds in fatty acids longer than C9. Arachidonate, a C20 fatty acid with four double bonds, can be synthesized by mammals from linolenate. Arachidonate, in turn, is a precursor to various signaling molecules such as prostaglandins.

10. Naturally occurring fatty acids are mostly cis with respect to the configuration of the double bond:

This cis configuration produces a bend of about 30° in the chain, whereas the trans configuration is a straight chain, as is the saturated.

Unsaturated (*cis*)   Saturated

11. Heath, R. J., S. Jackowski, and C. O. Rock. 1994. Guanosine tetraphosphate inhibition of fatty acid and phospholipid synthesis in *Escherichia coli* is relieved by overexpression of glycerol-3-phosphate acyltransferase (*plsB*). *J. Biol. Chem.* 266:26584–26590.

12. In addition to the carbamoyl synthetase described here, vertebrates have a carbamoyl synthetase that combines NH_4^+, CO_2, 2 ATP, and H_2O to form carbamoyl phosphate that is used to convert NH_4^+ to urea in the urea cycle.

13. In mammals the amino group is donated from the amide group of glutamine. In *E. coli* the enzyme can use either NH_4^+ or glutamine. In both cases the requirement for ATP can be explained by the formation of a phosphate ester with the carbonyl oxygen. The amino group then displaces the phosphate.

14. Aeckersberg, F., F. Bak, and F. Widdel. 1991. Anaerobic oxidation of saturated hydrocarbons to CO_2 by a new type of sulfate-reducing bacterium. *Arch. Microbiol.* 156:5–14.

15. Munro, A. W., and J. G. Lindsay. 1996. Bacterial cytochromes P-450. *Mol. Microbiol.* 20:1115–1125.

10

Macromolecular Synthesis

From a biochemical point of view, nucleic acids and proteins are polymerized by donating the subunit (i.e., a nucleotide or amino acid), from a donor with a high group transfer potential to the growing chain via a nucleophilic displacement reaction. (See Section 7.2 for a description of group transfer reactions.) This condensation reaction is generally referred to as chain elongation. For nucleic acids, the donors of the nucleotide monophosphates, (d)NMP, are the respective nucleotide triphosphates, (d)NTP. During the condensation reaction in nucleic acid biosynthesis, the α phosphate in (d)NTP is attacked by the 3′-hydroxyl group on ribose or deoxyribose (at the 3′ end of the growing polynucleotide) and the released pyrophosphate (PP_i) is subsequently hydrolyzed by a pyrophosphatase. (See Figs. 7.3 and 7.7.) This pulls the reaction to completion. These condensation reactions are discussed in Sections 7.1.2 and 7.2.2.

For protein synthesis, the donors of the amino acids are the aminoacylated tRNAs (Fig. 7.8). However, because of specific requirements that have to do with the structure of DNA as well as the need to use nucleic acid as a template during transcription (RNA synthesis) and translation (protein synthesis), the biosynthesis of DNA, RNA, and protein involves much more than simply the condensation reactions. This chapter discusses important features of the biosynthesis of DNA, RNA, and protein, including metabolic regulation of the pathways.

10.1 DNA Replication, Chromosome Separation, and Chromosome Partitioning

10.1.1 Semiconservative replication

DNA consists of two strands wound around each other in a right-handed double helix (Fig. 10.1). For a review of the history of DNA research that led to the realization that it was the genetic material; see Box 10.1; for a summary of the contributions of Francis Crick, Rosalind Franklin, James Watson, and Maurice

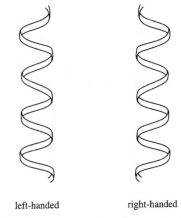

left-handed   right-handed

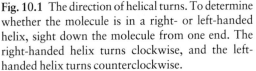

Fig. 10.1 The direction of helical turns. To determine whether the molecule is in a right- or left-handed helix, sight down the molecule from one end. The right-handed helix turns clockwise, and the left-handed helix turns counterclockwise.

BOX 10.1 HISTORICAL PERSPECTIVE:
THE DISCOVERY OF DNA AND ITS ROLE

The discovery of DNA

The discovery of DNA, and the realization that it is the genetic material, has a long history. After its identification in 1869 in the nuclei of human white blood cells, DNA was later found in cells of many types. Its chemical composition, as well as that of RNA, was learned in 1910. By the 1920s, microscopic studies of stained preparations demonstrated that DNA, along with protein, is contained in chromosomes. However, since DNA is made of only four different nucleotides, and its enormous diversity was not suspected, DNA was not then considered to be the genetic material. In fact, it was protein that was thought to be the genetic material because there was such a diversity in proteins. It was thought that the DNA played a structural role in the chromosomes.

DNA is the genetic material

It was not until 1944 that experiments were published indicating that the genetic material was indeed DNA. These experiments were done at the Rockefeller Institute in New York by Oswald Avery, Colin MacLeod, and Maclyn McCarty. They were investigating a report published in 1928 by Fred Griffith, a British scientist. Griffith had studied pneumonia caused by the bacterium *Streptococcus pneumoniae* (at that time called *Pneumococcus*). He had two strains of *S. pneumoniae*, called R and S. One difference between the two strains was that when the S strain was injected into mice, the mice died of pneumonia. However, when the R strain was injected, the mice lived. Another difference between the R and S strains was that the S strain had a capsule and the R strain did not. (The S stands for smooth, which is how the colonies looked when grown on an agar plate, and the R stands for rough, which describes the colonies formed by unencapsulated strains.)

Griffith found that if he heat-killed the S bacteria, they did not cause pneumonia unless they were injected with live R bacteria. Furthermore, he saw that the bacteria recovered from the dead mice all had capsules. Thus, the R bacteria were transformed into S bacteria by dead S bacteria. This is because genes to make capsule were transferred from the dead S bacteria to the live R bacteria, a phenomenon now called transformation (since the R bacteria are transformed into S bacteria).

Avery, MacLeod, and McCarty were able to transform R bacteria into S bacteria in a test tube by adding DNA purified from S bacteria. They thus provided the first evidence that genes were made of DNA. The DNA preparations had small amounts of protein, and to prove that it was the DNA and not the protein that was the genetic material, Avery and his coworkers showed that the transforming activity was not destroyed by proteases, which are enzymes that degrade protein, nor was the activity destroyed by an enzyme that degrades RNA; it was, however, destroyed by deoxyribonuclease, an enzyme that destroys DNA.

A second experiment, which also showed that the genetic material was made of DNA, was published in 1952 by Alfred Hershey and Martha Chase. Hershey and Chase studied the infection of the bacterium *Escherichia coli* by the bacterial virus T2. When the virus infects the bacterium, the viral genes are injected into the bacterial cell, and those genes direct the synthesis of new viruses. Hershey and Chase demonstrated, via radioisotopes that labeled the phosphorus in DNA and the sulfur in proteins, that viral DNA and not viral protein was injected into the cell, thus providing evidence that its genes were made of DNA and not protein.

Together, the experiments of Avery, MacLeod, and McCarty, and of Hershey and Chase, convinced everyone that genes are made of DNA. (We now know that genes in some viruses are made of RNA.)

BOX 10.2 HISTORICAL PERSPECTIVE:
THE STRUCTURE OF DNA

In 1953 James Watson and Francis Crick at Cambridge University in England published their model of DNA as a double helix held together by hydrogen bonding between the bases. Based upon their model of the structure of DNA, Watson and Crick suggested that one of the two strands serves as a template for a new complementary strand during DNA replication, and that each daughter cell receives a duplex consisting of a parental strand and a new strand. This process is known as *semiconservative replication*, and its occurrence was later demonstrated experimentally.

The Watson–Crick model of the structure of DNA was based In part upon the examination of X-ray diffraction photographs of DNA taken by Rosalind Franklin, a research fellow in the laboratory of Maurice Wilkins at King's College in London, as well as her presentation at a colloquium in 1952 that showed that the phosphate groups were on the outside of the DNA molecule and the bases faced inward toward each other. According to the account by Watson, in his book *The Double Helix*,[1] Wilkins showed Watson one of Franklin's recent X-ray photographs of DNA molecules surrounded by a large amount of water. The DNA was in a new form called the "B" form, which is far simpler than the pattern obtained earlier, called the "A" form, which exists in solutions that have relatively little water to hydrate the DNA. It was immediately clear to Watson from Franklin's photograph that the B form was due to a helical structure.

In 2004, in an article on the history of the discovery of the DNA double helix,[2] Aaron Klug revealed that one of Rosalind Franklin's laboratory notebooks, written in 1952, showed that she also thought that the B form was helical. Franklin's photograph helped Watson and Crick to construct the double helix model of DNA that was published in *Nature* in 1953. In 1962 Watson, Crick, and Wilkins received the Nobel Prize for their contributions in deciphering the structure of DNA. Rosalind Franklin had died of cancer in 1958, at the age of 37. For more information about the fascinating history of the elucidation of the structure of DNA, read refs. 1 through 3.

REFERENCES

1. Watson, J. D. 1968. *The Double Helix*. Penquin Books Ltd., London.

2. Klug, A. 2004. The discovery of the DNA double helix. *J. Mol. Biol.* 335:3–26.

3. Kass-Simon, G., and Farnes, P. (Eds.). 1990. *Women of Science*. Indiana University Press, Bloomington.

Wilkins toward the elucidation of the three-dimensional structure of DNA, see Box 10.2.

The DNA strands are held together by base pairing between A–T and G–C residues (Fig. 10.2). DNA is replicated via semiconservative replication. This means that each strand acts as a template for the synthesis of a daughter strand. Semiconservative replication was suggested by Watson and Crick in their paper on the structure of DNA published in *Nature* in 1953, and was later demonstrated in experiment by Messelson and Stahl, described in Fig. 10.3.

10.1.2 *The topological problem*

DNA replication is a complex topological problem because the DNA in a typical bacterium exists as a covalently closed circle of a right-handed double helix that may be 500 to 600 times longer than the cell and is tightly folded into supercoiled loops. (However, not all bacteria have circular chromosomes. See note 1.)

Supercoiling refers to the twisting of the DNA double helix around its central axis (Fig. 10.4). To visualize supercoiling, think of a telephone

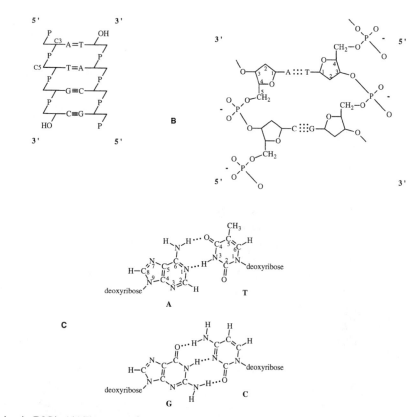

Fig. 10.2 Base pairing in DNA. (A) Two complementary strands of the double helix are in opposite polarity and held together by hydrogen bonds between A–T and G–C pairs: A, adenine; T, thymine; G, guanine; C, cytosine; P, phosphate. The deoxyribose moieties are attached via phosphodiester linkages between the C3 hydroxyl of one sugar and the C5 hydroxyl of another. (B) A more detailed examination showing the structure of the deoxyribose connected by phosphodiester linkages. The phosphate–oxygen double bond is drawn as a semipolar bond because of the high electronegativity of oxygen and the low propensity of phosphorus to form double bonds. (C) The structures of the bases showing the hydrogen bonds. Note that two hydrogen bonds hold the A–T base pairs together, whereas three hydrogen bonds hold the G–C base pairs together.

cord wound into secondary coils. (See note 2 for a further explanation of supercoiling.)

Pulling apart of the duplex strands at the replication fork in the closed circle makes the unreplicated portion ahead of the replication fork twist so that the helix becomes overwound and more tightly coiled into a positive supercoil (Fig. 10.5). Unless something were done about this, the unreplicated portion of the DNA helix would become bunched up in positive supercoils and further unwinding would stop. As described shortly, DNA gyrase solves the problem.

Positive versus negative supercoiling

Positively supercoiled DNA has more than 10.5 base pairs per helical turn (i.e., it is overwound); negatively supercoiled DNA has fewer than 10.5 base pairs per helical turn (i.e., it is underwound).

The DNA duplex itself is a right-handed helix, and therefore positive supercoils are twisted in the same direction as the helix (overwound). If the twist of the coil is counterclockwise (left-handed), the coil is negatively supercoiled (opposite to the helix) and underwound.

DNA gyrase solves the problem of overwinding

The problem of overwinding the double helix during DNA replication is solved by using topoisomerase II. This enzyme, called DNA gyrase, continuously removes the positive supercoils (overwound DNA) that form in the unreplicated DNA ahead of the replication fork and converts them into negative supercoils (underwound DNA). This activity is probably advantageous, since the duplex must unwind if

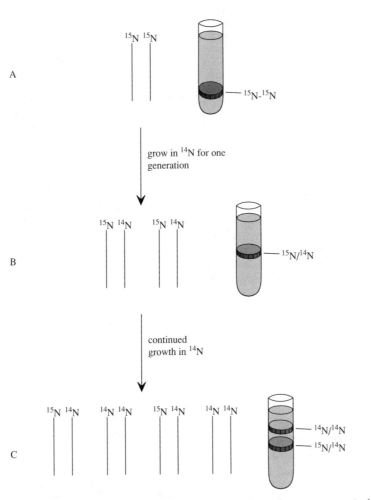

Fig. 10.3 The Messelson–Stahl experiment showing that DNA is replicated semiconservatively. (A) Cells are grown in ^{15}N for many generations so that all the nitrogen in the DNA is heavy (^{15}N). The DNA is isolated and centrifuged to equilibrium in a cesium chloride (CsCl) density gradient that separates the molecules according to their density. If both strands have ^{15}N, the duplex DNA sediments to a position near the bottom of the tube. (B) The ^{15}N cells are then grown in ^{14}N (light) media, and after one generation the DNA is isolated and centrifuged in the CsCl gradient. Semiconservative replication predicts that "hybrid" DNA would be formed, one strand being labeled with ^{15}N and the other strand with ^{14}N. The $^{15}N/^{14}N$ DNA occupies a position in the gradient higher than the $^{15}N/^{15}N$ DNA. (C) Further growth in ^{14}N results in the formation of "light" DNA (i.e., $^{14}N/^{14}N$), which occupies the highest position in the gradient.

DNA replication and RNA transcription are to take place. As shown in Fig. 10.6, DNA gyrase works by making a double-stranded break in the DNA, passing an unbroken portion through the gap, and resealing the break. Note 3 provides more information about topoisomerases and how they work.

More problems to be solved

Not only must the DNA be unwound and copied rapidly (in *E. coli*, about 1,000 nucleotides per second must be polymerized, to result in replication of a chromosome in about 40 min), but the strands must be separated from each other without getting entangled. Finally, the strands must be partitioned into the daughter cells, a process that is not completely understood.

10.1.3 Creating the replication fork

At least 30 different proteins are required to initiate replication and to replicate the DNA

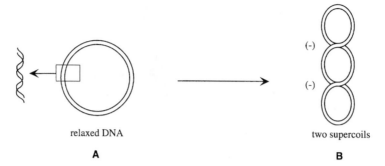

relaxed DNA

A

two supercoils

B

Fig. 10.4 Supercoiled DNA. (A) When there are 10.5 base pairs per helical turn, the DNA is relaxed. Native DNA is generally underwound (i.e., 10.5 base pairs per helical turn). Underwinding introduces a strain in the molecule, and to reduce the strain, the molecule twists upon itself to form supercoils (B). Supercoiling resulting from underwound DNA is referred to as negative supercoiling. Supercoils will also form if DNA is overwound (i.e., >10.5 base pairs per turn). Supercoiling due to overwound DNA is called positive supercoiling. In negative supercoiling the DNA is twisted in a direction opposite to that of the right-handed double helix, and in positive supercoiling the DNA is twisted in the same direction as the right-handed double helix.

in *E. coli.* For more information, the student should consult refs. 4 through 9.

DNA replication takes place in a DNA-synthesizing "factory" called a *replisome*, which consists of an assemblage of enzymes and proteins that will soon be described. Within the replisome there are replication forks created on the DNA where replication takes place. (The student should review Section 2.2.3 for a discussion of the relationship between the timing of initiation of DNA replication and the growth rate.) To make the replication fork, the strands must first unwind so that each strand can act as a template. The unwinding does not begin at just any place but rather at a particular site in the DNA duplex termed the *origin*, often referred to as the *oriC* locus. When the duplex is unwound, a Y is created. The arms of the Y are single stranded because the duplex has become unwound there; but the region downstream of the juncture, where the two arms come together, is still double stranded (Fig. 10.7). The juncture is called the *replication fork*. DNA is usually replicated bidirectionally; that is, there are two replication forks (Fig. 10.8). (Note 10 explains how directionality can be determined.) Bidirectional replication, which halves the time needed to replicate the DNA molecule, generally takes place with phages, plasmids, bacteria, and eukaryotic cells. The replication forks are thought to be stationary, and the unreplicated DNA appears to thread through them. (This is discussed in

Section 10.1.4.) The detailed steps in the creation of a replication fork are described next.

Unwinding the duplex: Creation of the prepriming complex

The prepriming complex is formed first and it is created in two stages.[9] In stage 1 the *open complex* is formed, and in stage 2 the open complex develops into the *prepriming complex*. DNA synthesis, which will be described later, actually begins with the prepriming complex.

1. Formation of the open complex

Creation of the open complex is initiated at the origin of replication (*oriC*) with ATP and two DNA-binding proteins: DnaA and a histonelike protein called HU (Fig. 10.9). See the subsection entitled *DNA-binding proteins, nucleoid structure, and gene expression* in Section 1.2.6. Within the origin (*oriC*) there are multiple sites to which DnaA binds. Approximately 30 DnaA molecules bind to the sites as the DNA wraps around a core of DnaA molecules. Then, in an ATP-dependent reaction that is aided by HU, the adjacent A–T-rich region at the 5′ end of the origin sequence unwinds to form the 45 bp *open complex*. However, something must be done to prevent the single strands from coming together again to re-form the duplex. This is the task of single-stranded binding (SSB) proteins that coat the strands.

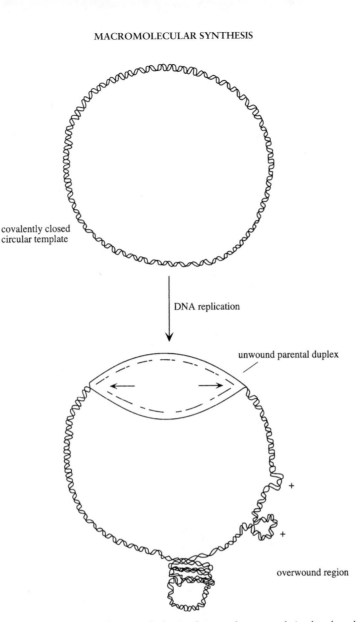

covalently closed
circular template

DNA replication

unwound parental duplex

+

+

overwound region

Fig. 10.5 Supercoiling ahead of the replication fork. As the template strands in the closed circle are pulled apart, the duplex ahead of the replication fork overwinds as positive supercoils are formed. The twists in the overwound regions are removed by DNA gyrase, which produces negative supercoils and underwinds the duplex.

2. Formation of the prepriming complex

The open complex unwinds into the *prepriming complex*. The unwinding is performed by a protein called helicase (DnaB), which must first bind to the DNA. However, DnaB does not bind to the DNA on its own but must be transferred to the open complex from a DnaB:DnaC:ATP complex. After binding to the DNA, DnaB further unwinds the strands bidirectionally to form the prepriming complex, with two replication forks ready for the initiation of DNA replication.

Timing of initiation

Precisely what determines the timing of initiation of bacterial DNA replication is still a matter of speculation. As discussed in Section 2.2.3, the cell mass per *chromosomal* replication origin (as opposed to plasmid replication origins) at the time of initiation is constant, and all *oriC* origins (even plasmid origins), begin replication at the same time. This mass is called the *initiation mass* or the initiation volume, and some have suggested that it corresponds to

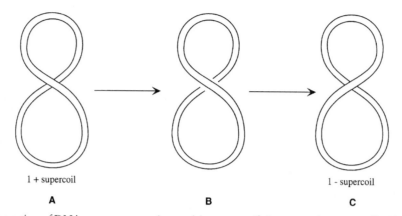

Fig. 10.6 The action of DNA gyrase: converting positive supercoils into negative supercoils. (A) A positively supercoiled node; that is, the duplex is twisted around its central axis. For example, this happens during DNA replication as the duplex is being unwound by helicase at the replication fork and the duplex ahead of the replication fork spontaneously becomes overwound by being twisted in a clockwise direction. (B) Both strands are cut by DNA gyrase, and then the gyrase passes the uncut portion through the gap and the gap is sealed. (C) The duplex is now twisted in the opposite direction; that is, it is negative supercoiled and underwound. DNA gyrase, an ATP-dependent enzyme, is sometimes referred to as providing a "swivel" that allows the replication fork to continue.

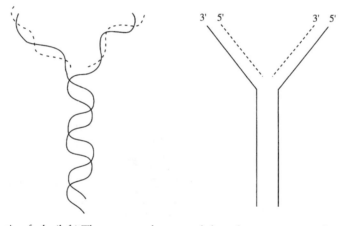

Fig. 10.7 A replication fork. (left) The two template strands have become unwound and are being copied. The unwinding is caused by DnaB protein, which is the replication fork helicase (not shown) and requires ATP. Note that there are single-stranded regions near the fork. A DNA-binding protein, called SSB (not shown), binds to the single-stranded regions, preventing them from coming together. (right) Now the duplex has both unwound and double-stranded components.

the activity of DnaA. Another question is, what prevents the newly replicated origins from reinitiating in the same cell cycle? For a discussion of what regulates the initiation of DNA replication at *oriC*, see note 11.

10.1.4 Replicating the DNA

DNA polymerases

For a review, see ref. 12. The enzymes that synthesize DNA by using a strand of DNA as a template are called DNA polymerases. Two important characteristics of DNA polymerases are (1) they can only extend nucleic acid chains (i.e., they cannot initiate new ones), and (2) they add mononucleotides to the 3' hydroxyl of deoxyribose and therefore elongate nucleic acid only at the 3' end. *E. coli* has five DNA polymerases: DNA polymerase I (Pol I) encoded by *polA*; DNA polymerase II (Pol II), encoded by *polB*; DNA polymerase III (Pol III), encoded by *polC* (*dnaE*); DNA polymerase IV (Pol IV), encoded by *dinB*; and DNA polymerase

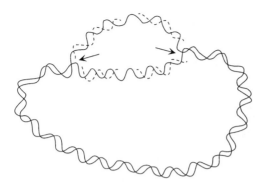

Fig. 10.8 Bidirectional replication of DNA: solid lines, template strands; broken lines, daughter strands. In most organisms and viruses, the DNA is replicated bidirectionally, as indicated by the arrows. Bidirectionality reduces the time required to replicate the DNA. Available data indicate that the replication forks are stationary in a replication factory called a replisome, and the unreplicated DNA threads through them.

V (Pol V), encoded by *umuC*. Here is what they do.

1. DNA polymerase I has a role in DNA replication, which will be discussed later. (See subsection entitled *Attaching the Okazaki fragments to each other*.) DNA polymerase I is also important for DNA repair (e.g., see Sections 19.2.3 and 19.2.5).

2. DNA polymerase II functions in the repair of damaged DNA. After UV damage, DNA polymerase II catalyzes a fast reinitiation of synthesis called "replication restart," and then DNA polymerase III takes over. (See note 30 in Chapter 19.)

3. DNA polymerase III replicates DNA at the fork. See note 13 for a description of the subunits in DNA polymerase III.

4. DNA polymerase IV functions in repairing damaged DNA. As discussed in note 30 in Chapter 19, DNA polymerase IV replicates templates that have a bulged-out nucleotide, and as a consequence, −1 frameshift mutations occur.

5. DNA polymerase V functions during DNA repair. As pointed out in Section 19.3.2, DNA polymerase V inserts nucleotides nonspecifically opposite lesions, such as abasic sites and UV-induced thymine dimers, in

the template strand. The process is called translesion synthesis, and it is associated with increased mutations because the nucleotides are inserted nonspecifically. Because of the increase in mutation rate, translesion synthesis is referred to as SOS mutagenesis or error-prone repair.

The problem in synthesizing strands of opposite polarity at the replication fork

At the replication fork the DNA templates are antiparallel; that is, one strand is 5′ to 3′ and the other strand is 3′ to 5′. In other words, the 5′-phosphate end of one strand is paired with the 3′-hydroxyl end of the other strand (Fig. 10.7). Since the DNA copy strand is antiparallel to the template strand, this means that the new strands of DNA must be of opposite polarity too; that is, one is 3′–5′ whereas the other is 5′–3′. That is, at the replication fork one of the copy strands has its 3′ end pointed toward the fork, whereas the other copy strand has its 3′ end pointed away from the fork. This presents a problem because the DNA polymerase remains at the fork, whereas the growing end (the 3′ end) of one of the strands keeps moving further and further away from the fork.

Okazaki fragments

As just mentioned, the DNA polymerase manages to remain at the replication fork and yet synthesize a strand of DNA whose 3′ end keeps moving further and further away from the replication fork. How is this accomplished? As will be explained next, the answer lies in synthesizing the strand whose 3′ end faces away from the fork. The synthesis proceeds in short fragments (about 1,000 nucleotides long), called Okazaki fragments after the investigators who discovered them. As implied in Fig. 10.10, the 3′ end then returns to the fork to initiate replication of another short fragment. (See note 14 for a description of the Okazaki experiments.) The strand copied in short fragments is called the *lagging strand* template. The other strand, which is copied in one piece, is called the *leading strand* template. As discussed later, in the subsection entitled *Model of how the DNA III polymerase synthesizing the lagging strand can stay with the replication fork*, the DNA polymerases that synthesize the leading and

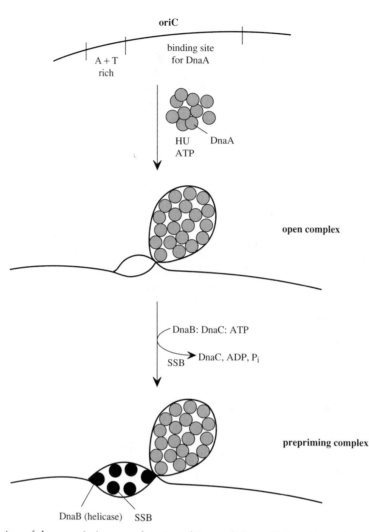

Fig. 10.9 Creation of the prepriming complex. A multimer of about 20 DnaA proteins recognizes specific sequences and binds to the duplex DNA at the origin adjacent to an A–T-rich region. The A–T region adjacent to the DnaA proteins then unwinds to form the *open complex*. This step is aided by the HU protein and requires ATP. DnaB protein (helicase) then binds in a DnaC-dependent reaction and further unwinds the duplex. Single-stranded DNA-binding proteins (SSBs) bind to the single strands behind the helicase, preventing the single strands from coming together again.

lagging strand are actually associated with each other as a dimer and do not leave the replication fork.

Synthesis of leading strand

Recall that DNA polymerase cannot initiate new polynucleotide strands but can only elongate the 3' end of an existing strand. This means that the initiation of replication of the leading strand at *oriC* (as well as the lagging strand) must begin with the synthesis of a a relatively short RNA primer (5–60 nucleotides) by an RNA polymerase. An RNA polymerase called primase (or DnaG) synthesizes the RNA primer for the Okazaki fragments, and perhaps also the RNA primer for the leading strand. DNA polymerase III then extends the RNA primer at its 3' end and continues the elongation of the DNA (Fig. 10.10). The DNA polymerase that synthesizes the leading strand stays with the replication fork as it adds nucleotides and dissociates rarely, if at all. At one time it was thought that the DNA polymerase moved

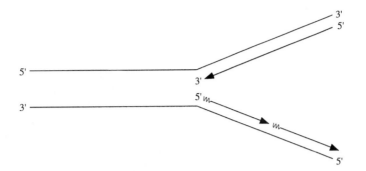

Fig. 10.10 Synthesis of leading and lagging strands. Both template strands must be copied in the 3′-to-5′ direction so that the copy strands, which grow at the 3′ ends, are of opposite polarity to their respective templates. Since the template strands are of opposite polarity, the copy strands must be synthesized in opposite directions: one strand in the direction of replication fork movement, and the other strand in the opposite direction. However, it is known that the DNA polymerase, which is at the 3′ end of the growing strand, remains at the replication fork. The strand whose 3′ end faces the replication fork is synthesized by a polymerase that stays attached to the template and moves with the replication fork. This template is called the leading-strand template, and the strand being synthesized is called the leading strand. The strand whose 3′ end faces away from the replication fork is synthesized in short fragments called Okazaki fragments, which are about 1,000 nucleotides long. Upon completion of an Okazaki fragment, the DNA polymerase leaves the template and returns to the replication fork to synthesize another fragment. The template that is copied in short fragments is called the lagging-strand template, and the strand being synthesized is called the lagging strand. Each Okazaki fragment begins with the synthesis of an RNA primer that is subsequently elongated by the DNA polymerase. The RNA primers are drawn as wavy lines. Eventually, the RNA primers are removed and the fragments are elongated by DNA polymerase and sealed. See text for details.

along the DNA with the replication fork to the terminus, much like a train moves along tracks. However, as described in Section 10.1.5, it now appears that the replication fork and the replicative DNA polymerase reside in a replication "factory" that does not move along the DNA. The replication factory is called a replisome, and the unreplicated DNA threads through it, with the terminus the last part to enter. Later, in connection with topics covered in Section 10.1.6, we describe the experiments that support this model (see note 37).

Synthesis of the lagging strand

The opposite template strand is called the *lagging* template strand. It cannot be copied in the same way as the leading template strand is copied because it is of opposite polarity, and the 3′ end of the RNA primer (i.e., the growing end) is facing away from the replication fork. Instead, the polymerase periodically disengages from the template strand to return to a newly synthesized primer at the replication fork, thus synthesizing short polynucleotide fragments, the Okazaki fragments (Fig. 10.10).

The DnaB and DnaG complex unwinds the DNA at the replication fork and synthesizes the RNA primer for the lagging strand

As replication proceeds, RNA primers for the Okazaki fragments are synthesized by the special RNA polymerase called the primase or DnaG (Fig. 10.11). The primase is associated with the helicase (DnaB) to form a complex that stays with the replication fork as the DNA is replicated. The primase must synthesize the Okazaki primers in the direction away from the replication fork. The DnaB helicase unwinds the duplex. Then the DNA polymerase III elongates the RNA primer at its 3′ end to synthesize the Okazaki fragment.

Model explaining how the DNA polymerase III synthesizing the lagging strand can stay with the replication fork

The two DNA polymerases III (Pol III) that synthesize the leading and lagging stands are physically associated with each other as a dimer that does not leave the fork. This ensures that both strands are replicated simultaneously and

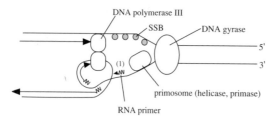

Fig. 10.11 Replication fork. The model explains how DNA polymerase might synthesize the lagging strand and stay with the replication fork. It is suggested that the lagging-strand template loops around a dimeric polymerase (two DNA polymerases physically associated with each other) so that the polymerase bound to the lagging strand moves along it in the direction of the replication fork. Thus, one of the polymerases of the dimer extends an RNA primer (wavy line) on the lagging strand, and the other polymerase elongates the leading strand. Another RNA primer, labeled (1), is made by primase ahead of the polymerase at the replication fork. The polymerase keeps moving on the template until it encounters the 5' end of a previously made Okazaki fragment (wavy line). Upon encountering the Okazaki fragment, the polymerase that is synthesizing the lagging strand disengages from the template and reinitiates at the new RNA primer with a new loop. The arrows point in the direction of strand elongation at a stationary replication fork. See Section 10.1.3 for a discussion of the stationary replication fork.

at the same rate. However, there is a problem. The template strands are of opposite polarity, and this implies that the two DNA polymerases III must move in opposite directions. Yet, they stay together as a dimer at the replication fork. It may be that the dimer stays with the replication fork and still synthesizes the lagging strand whose growing 3' end extends away from the fork because the lagging-strand template loops around one of the polymerases at the replication fork so that its 5' end faces the replication fork rather than away from it. This suggested mechanism is illustrated in Fig. 10.11. If the postulated scheme is accurate, then the Okazaki fragments will be elongated at their 3' ends at the replication fork and the dimeric DNA polymerase III can elongate both the leading and lagging strands at the same time. According to the model, after polymerizing approximately 1,000 nucleotides, or one Okazaki fragment, the polymerase synthesizing the lagging strand would disengage from the template and a new loop would form.

Attaching the Okazaki fragments to each other

When DNA polymerase III reaches the RNA at the 5' end of the previously synthesized segment of DNA in the lagging strand, it stops and leaves the 3' end of the newly synthesized DNA. The result is a nick in the DNA strand between the DNA and RNA. The very short RNA primer fragments (10–12 nucleotides long) in the Okazaki fragments are removed by the 5'-to-3' exonuclease activity of DNA polymerase I, and each ribonucleotide is simultaneously replaced with a deoxyribonucleotide polymerized by DNA polymerase I. When this happens, the nick moves in the direction of DNA synthesis and the process is called nick translation. The formation of the phosphodiester bond between the 3' hydroxyl at the end of one strand of DNA and the 5' phosphate of the other strand has been is catalyzed by *DNA ligase*. The overall scheme is shown in Fig. 10.12. The ligase makes the covalent bond between the 3' phosphate of the newly synthesized DNA and the 3'

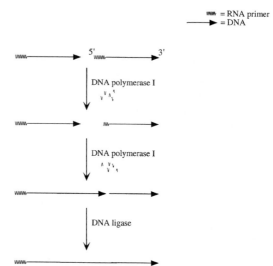

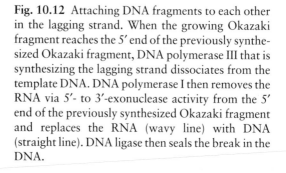

Fig. 10.12 Attaching DNA fragments to each other in the lagging strand. When the growing Okazaki fragment reaches the 5' end of the previously synthesized Okazaki fragment, DNA polymerase III that is synthesizing the lagging strand dissociates from the template DNA. DNA polymerase I then removes the RNA via 5'- to 3'-exonuclease activity from the 5' end of the previously synthesized Okazaki fragment and replaces the RNA (wavy line) with DNA (straight line). DNA ligase then seals the break in the DNA.

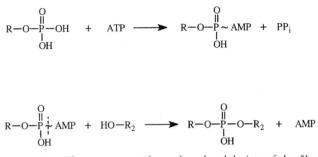

Fig. 10.13 DNA ligase reaction. The enzyme catalyzes the adenylylation of the 5'-phosphate. Depending upon the ligase, the AMP can be derived either from ATP, in which case pyrophosphate is displaced, or NAD⁺, in which case nicotinamide monophosphate (NMN) is displaced. *E. coli* DNA ligase uses NAD⁺ as the AMP donor. Then the 3'-hydroxyl of the ribose attacks the AMP derivative and displaces the AMP. The result is a phosphodiester bond.

hydroxyl of the previously synthesized DNA by transfering AMP from either ATP or NAD⁺ (depending upon the organism) to the phosphate (Fig. 10.13). This produces an AMP derivative with a high group transfer potential. The ligase then catalyzes the displacement of the AMP by the 3' hydroxyl, and the phosphodiester bond is formed.

Sliding clamps

DNA polymerase III, which is the replicative polymerase in *E. coli*, consists of three subassemblies. The total complex is referred to as DNA polymerase III holoenzyme. The subassemblies are the polymerase itself, a sliding β clamp, and a clamp loader. The sliding β clamp has a large central pore through which the DNA template is threaded. Very importantly, the sliding clamp complex binds to the DNA polymerase III core, which consists of the α, ε, and θ subunits, and keeps it tethered to the DNA, allowing the processing capacity of the polymerase to increase from approximately 12 nucleotides to thousands of nucleotides before the DNA polymerase dissociates from the template. Thus, the sliding clamp is very important for the efficiency of DNA replication. Since DNA is synthesized continuously on the leading strand, the clamp is loaded only once. However, on the lagging strand, where DNA is synthesized in Okazaki fragments of approximately 1,000 nucleotides, the clamp is unloaded and loaded frequently. For a discussion of the clamps and clamp loaders, read ref. 15 and note 16.

10.1.5 Termination

Ter sites and Tus protein

In *E. coli* the two replication forks that begin at *oriC* and polymerize bidirectionally stop in a region of the chromosome opposite *oriC*, called *Ter*, the termination region (Fig. 10.14). After termination, the sister chromosomes separate and partition (segregate) to opposite halves of the cell. The termination region contains specific sequences of nucleotides called *Ter* sequences.

The *Ter* sites are quite unusual because they allow replication to proceed in only one direction! (Note that in Fig. 10.14 the arrows refer to the direction of replication, not the movement of the replication forks, which are stationary within the replisome.) This can account for termination at the *Ter* sites. For example, assume that there are two *Ter* sites, which are located next to each other, with perhaps a short segment of DNA between them. Suppose *Ter* site 1 allows replication to proceed clockwise but not counterclockwise, and *Ter* site 2 allows counterclockwise replication but not clockwise replication. Thus, if clockwise replication arrives at site 2, it will stall; and if counter-clockwise replication arrives at site 1, it will stall. Consequently, bidirectional replication will meet at site 1 or site 2 or somewhere between them.

As shown in Fig. 10.14, the situation in *E. coli* is a little bit more complicated because it has six *Ter* sites, sites are divided into two groups of three: one group (*TerA*, *TerD*, and *TerE*) prevents counterclockwise replication,

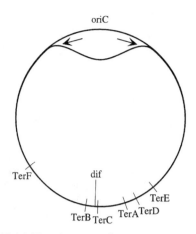

Fig. 10.14 Termination of DNA replication in
E. coli. The arrows refer to the direction of replication,
rather than to movement of the replication forks. A
terminator region exists 180° opposite the origin. In
the terminator region there are nucleotide sequences
that allow replication in only one direction. Thus,
the terminator sequences are said to be polar. *E. coli*
has six terminator (Ter) sites. These are, in the
sequence in which they exist in the DNA, TerE,
TerD, TerA, TerC, TerB, and TerF. TerE, TerD, and
TerA occupy one side of the terminator region and
TerC, TerB, and TerF occupy the opposite side. The
figure shows two replication forks. Data discussed in
Section 10.1.5 support a model in which the replication forks are stationary within a DNA synthesizing
"factory" called the replisome. Unreplicated DNA
enters the replisome and passes through the stationary replication forks. TerC, TerB, and TerF inhibit
clockwise replication, and TerA, TerD, and TerE
inhibit counterclockwise replication. Therefore,
there are three chances to stop replication in the
termination region. For example, if counterclockwise replication moves past TerA, it will stop at TerD
or TerE. Consequently, replication from both directions will stop at one of the termination sites or a site
between them. A protein, called the Tus protein in
E. coli, binds to Ter and prevents the helicase from
proceeding in one direction, thus accounting for the
polarity of inhibition of movement of the replication
fork. Notice the site marked *dif*. It is here that recombination takes place to separate chromosomes that
have undergone an unequal number of recombinations during replication.

and the second group (*TerC*, *TerB*, and *TerF*),
prevents clockwise replication. As explained in
the legend to Fig. 10.14, the presence of three
Ter sites per replication fork direction provides
a backup in case the replication fork gets by
one of these sites. A protein, which in *E. coli* is
called the Tus (terminator utilization substance)

protein binds to the *Ter* site and imposes the
one-way travel. Tus does this by inhibiting the
replicative helicase (DnaB).[17–19] Both *Ter* sites
and the Ter-binding protein have been well
studied in *B. subtilis*. See note 20 for similarities
and differences with respect to *E. coli*.

10.1.6 Chromosome separation and partitioning; some general principles

For reviews, see refs. 21 through 24. First, some
definitions: chromosome *separation* refers to
the detachment of the newly replicated chromosomes from one another, and chromosome
partitioning refers to the segregation of sister
chromosomes to opposite halves of the cell
prior to cell division.

Chromosome separation can take place in
one of two ways. If the linkage between the
daughter chromosomes is noncovalent, then
topoisomerases such as topoisomerase IV separate the sister chromosomes, as discussed
shortly. If an unequal number of recombinations have occurred between the sister
chromosomes, then the sister chromosomes are
covalently linked and a site-specific recombination must occur. (See later subsection
entitled *Site-specific recombination at dif*.)

Chromosome partitioning is not a well-
understood process in prokaryotes because
a spindle apparatus similar to that which
separates sister chromosomes, in eukaryotes
is not present. This section discusses current
ideas about how separation and partitioning of
daughter chromosomes occur in prokaryotes.
Section 10.1.7 describes chromosome partitioning in *B. subtilis*, and Section 10.1.8 describes
chromosome partitioning in *C. crescentus*.

Phenotypes of partition mutants

The genes involved in chromosome partitioning have been discovered by analyzing mutants.
Mutants defective in chromosome partitioning
can be defective in any one of several genes:
some are involved in detaching the chromosomes from each other, some are involved in
the partition of the detached chromosomes,
and some are even involved in the replication
of DNA, because only fully replicated DNA
molecules can detach from one another.
Depending upon whether the defect is in

detachment or partition of chromosomes, the phenotype differs. If the defect is in detaching the sister chromosomes from each other, then the phenotype includes elongated cells with an unusually large nucleoid mass positioned near the cell center. Anucleate cells may also form, as discussed next.

On the other hand, if the sister chromosomes can detach but cannot partition to opposite poles, the phenotype is an increased production of *anucleate cells* of the size of a newborn cell, reflecting the fact that because the sister chromosomes do not partition to opposite cell poles, they can both be in the same daughter cell upon cell division. Under these conditions, the nucleoids are single nucleoids and are not larger than normal.

Chromosome separation

Chromosome separation requires the separation of the two sister chromosome dimers, said to be catenated, by topoisomerase IV. The process is called decatenation. If there has been an unequal number of crossover events during DNA replication, then the sister chromosomes are covalently linked and must be separated by recombination by XerC and XerD.

Recombination occurs at *dif* sites. A key protein here is the cell division protein FtsK, which is localized to the septum and recruits other cell division proteins before cell division. FtsK ensures that XerCD and topoisomerase IV act on the DNA at midcell during the final stages of cell division. A discussion of how this might occur can be found later (see subsection 3).

1. Topoisomerase IV

When replication is complete, which is about 180° from the start site, the two completed DNA molecules are linked as two monomeric circles in a chain (i.e., they are catenated) and must be separated from each other (Fig. 10.15). The process is called decatenation. In *E. coli* and *Salmonella typhimurium*, the decatenation of two monomeric interlinked chromosomes requires the action of a DNA gyrase–like enzyme called topoisomerase IV.[25] Topoisomerase IV mutants are temperature sensitive for growth; that is, they grow at 30 °C but die at 42 °C. At the restrictive temperature, cell division is inhibited and the mutants form elongated cells,

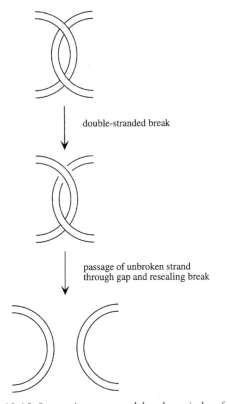

double-stranded break

passage of unbroken strand through gap and resealing break

Fig. 10.15 Separating catenated daughter circles after DNA replication by using type IV topoisomerase. The reaction is catalyzed by type IV topoisomerase (similar to DNA gyrase). The enzyme catalyzes a double-stranded cut in one of the duplexes. Then the unbroken duplex passes through the gap. The gap is sealed, and the DNA molecules have become separated. Mutants defective in topoisomerase IV are temperature sensitive for growth and show abnormal nucleoid segration. The phenotype is called Par⁻. See ref. 25.

often with unusually large nucleoids (revealed by DNA staining) in midcell or as several nucleoid masses unequally distributed in the elongated cells.[26] The nucleoids are large because the two daughter nucleoids are not able to separate. The filamentous cells may divide in regions where there is no DNA to produce anucleate cells. Inhibition of cell division resulting in elongated or filamentous cells is typical of mutants that have a block in DNA replication or separation of daughter chromosomes. This reflects the coupling between cell division and DNA metabolism as described in note 27. (Topoisomerase IV mutants have been isolated in *C. crescentus*, but the phenotype differs from that of *E. coli* and *S. typhimurium* mutants

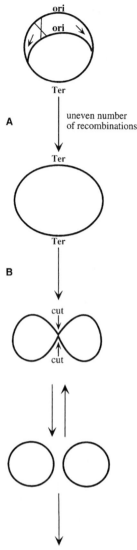

A

uneven number
of recombinations

B

cut

cut

segregation

Fig. 10.16 Site-specific recombination at termination to separate chromosomes that have undergone unequal numbers of recombinations. Simplified drawing of recombination at *dif* to resolve a DNA dimer into monomers. As discussed in Section 10.1.5, the replication forks are stationary and remain close to each other in the replisome, a DNA-synthesizing "factory." The unreplicated DNA is fed through the forks. The arrows depict the direction of polymerization of DNA. (A) Unequal numbers of recombinations have taken place covalently, linking the two circles to form a dimer. (B) Recombination is catalyzed by recombinases acting at the *dif* sites in the termination (Ter) regions. The chromosomes are then segregated to opposite cell poles. A *dif⁻* mutant produces filaments and anucleate cells as well as normal cells. As the figure indicates, it is believed that the recombination event at *dif* can proceed in

because checkpoints exist in the *Caulobacter* cell cycle.[28]) Topoisomerase IV makes a double-stranded break in one of the DNA molecules, and the unbroken molecule is passed through the break (Fig. 10.15). This is followed by sealing the break.

2. Site-specific recombination at dif

If an unequal number of recombinations have occurred between sister *circular* chromosomes, they are covalently linked as a circular dimer to each other and decatenation cannot separate them. Under these circumstances, the DNA molecules must be separated by *site-specific recombination* by using enzymes called *recombinases*. *E. coli* has a locus called *dif* (deletion induced filamentation), which lies in the replication terminus (Ter) region.[29] The recombinational event that separates the two sister chromosomes that have undergone an unequal number of recombinations during replication takes place in this Ter region (Fig. 10.16).

The *site-specific* recombination is catalyzed by two recombinase proteins, XerC and XerD, both of which are required.[30] The recombinases are activated by an *E. coli* cell division protein called FtsK, which is localized at midcell with other cell division proteins. (See Section 2.3.3 for a discussion of FtsK and the other cell division proteins in *E. coli*, and ref. 24 for a discussion of the multiple roles of FtsK in cell division and chromosome segregation.) Homologues of XerC and XerD have been found in a wide variety of bacteria. Mutations in *dif* are viable, and most of the cells are normal. This is to be expected, since *dif* is required only when there has been an unequal number of recombinations. However, approximately 10% of the cells are filamentous with unusually large nucleoids, indicative of a failure of the newly replicated chromosomes to separate. As expected, *xerC* and *xerD* mutants show the same phenotype as *dif⁻* mutants.

both directions. However, monomer formation is favored. Perhaps the segregation of the monomers ensures that the newly replicated *dif* sites cannot recombine. For recombination at *dif* to resolve the dimers, *dif* must be at its original site in the terminus region. If a copy of *dif* is reinserted in a different chromosomal site in a *dif⁻* strain, the cells still show the *dif⁻* phenotype.

3. Topoisomerase IV catalyzes decatenation primarily at dif in the presence of XerC, XerD, and FtsK

The activity of topoisomerase IV can take place at several chromosomal sites, but in the presence of FtsK, XerC, and XerD, decatenation takes place preferentially at midcell at the dif site.[31] However, unlike site-specific recombination, which requires XerC and XerD, decatenation at dif does not require the cell division protein FtsK, but is simply favored at dif because of FtsK. Why should topoisomerase IV preferentially cleave at dif? There have been two suggestions.[31] One suggestion is that the dif/XerC/XerD complex might increase the affinity of topoisomerase IV for the dif site. A second suggestion is that topoisomerase IV is part of a decatenation/resolution complex that forms at dif.

Chromosome partitioning in eukaryotes

It is worthwhile to summarize how chromosome partitioning takes place in eukaryotes before discussing how the process occurs in prokaryotes. In eukaryotes, chromosome replication is completed to form two sister chromatids bound along their length in the S phase of the cell cycle. The sister chromatids remain together during the G_2 phase, which follows the S phase. Following the G_2 phase is the M phase (mitosis), and this is when the sister chromatids partition. The M phase consists of nuclear division and cytokinesis. Partitioning occurs because the sister chromatids are attached to microtubule spindle fibers that pull them apart. This occurs during the anaphase portion of mitosis. Specifically, the sister chromatids are separated (and are now called chromosomes) and pulled to opposite cell poles by the shortening of the microtubular spindle fibers in the mitotic spindle. The attachment of each chromatid to the mcrotubule spindle fibers is via a protein complex called a kinetochore, which binds to a subset of microtubules in the mitotic spindle. The kinetochore is located at a highly condensed, constricted region called a centromere.

Chromosome partitioning in prokaryotes

In prokaryotes, "chromosome partitioning" refers to the segregation of sister chromosomes toward opposite cell poles in preparation for cell division. (Bacterial chromosomes are often referred to as nucleoids.) There is no evidence in bacteria for a device similar to the eukaryotic mitotic spindle to partition sister chromosomes. So how do bacteria partition sister chromosomes faithfully? This is a very active area of research, and the student is referred to the literature for detailed information about the similarities and differences with respect to what has been learned about B. subtilis, E.coli, and C. crescentus.[32–36] Certain generalizations can be made. First, it is noted that partitioning (segregation) takes place concurrent with replication of the DNA. Additional generalizations that emerge from a study of include the following.

The DNA is replicated in a stationary "factory," called the replisome, which in E. coli or B. subtilis is assembled at or near midcell when DNA replication is initiated, and is thought to be anchored there. When DNA synthesis is finished, the replisome is disassembled. The unreplicated DNA threads through the replisome, and the newly duplicated portions move away from the anchored replication forks within the replisome and partition toward opposite poles of the cell. Since the origins (ori sites) are replicated first, they leave the replisome first and very quickly are moved to opposite halves of the cell near the cell poles.

As will be discussed later, the "extrusion–capture" model proposes a mechanism(s) that provides the force that "extrudes" the newly replicated origins from the replisome and "captures" the origins at or near opposite cell poles. As will also be discussed, for this to occur successfully, certain proteins must compact the chromosomes and guide the origin regions toward the cell poles. Eventually, the terminus is drawn into the replisome, where it becomes replicated; then the sister chromosomes separate, and segregation is completed as the duplicated termini remain near the cell center. As a result, newborn cells in slowly growing cultures of E. coli, B. subtilis, and C. crescentus have a terminus near the new cell pole and an origin near the old cell pole (Fig. 10.17). Finally, the terminus in newborn cells is drawn toward the center of the cell by DNA replication in the centrally located replisome factory. There can be small differences among bacteria in the cellular location of chromosome regions. For

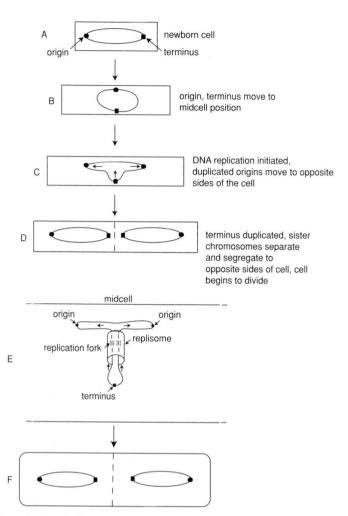

Fig. 10.17 General model for duplication, position, and partitioning of bacterial chromosomes in slowly growing cells. (A) Newborn cell showing chromosome with single origin (circle) and single terminus (square) at opposite cell poles. (B) Origin and terminus move and locate to the midcell position, where the replisome is assembled and DNA replication is initiated. (C) DNA replication is initiated, and copies of origin segregate to opposite halves of cell. In *E. coli* and *C. crescentus*, the origins are near opposite cell poles for most of the cell cycle and move to the midcell position in the newborn cell. In *B. subtilis*, origin copies segregate to cell quarter-positions that will become midcell following cell division. (D) Cell division begins. (E) Stationary replisome at midcell within which are two replication forks. DNA is pulled into the replisome, origin first and terminus last. The sister chromosomes are pushed toward opposite cell halves. (F) Cell division completed.

example, in *E. coli* the origin regions localize at quarter-cell positions instead of at the cell poles, whereas in *B. subtilis* growing vegetatively or sporulating, they localize at the cell poles.

For a description of how the replisome as well as specific regions of the chromosome such as origins, termini, and regions between them, can be microscopically visualized, see note 37, as well as the references cited earlier concerning chromosome segregation in *B. subtilis*, *E. coli*, and *C. crescentus*. For more information about

the location of the origins and termini, specifically for fast-growing cells that initiate DNA replication in the previous cell cycle, the student is referred to ref. 23.

There are important differences between the positioning of the replisome in *E. coli* and *B. subtilis* on the one hand, and *Caulobacter crescentus* on the other hand. These differences, which are related to differences in the cell cycle, are described later. (See Sections 10.1.7 and 10.1.8.)

What moves the sister chromosomes apart, and what directs their movement?

Clearly, prokaryotic cells do not have a spindle apparatus. What then is responsible for the movement of sister chromosomes to opposite poles of the cell prior to cell division? Recently, there has been a great deal of research and much speculation. Several proteins have been implicated. Some of the proteins are thought to comprise the "motor" that moves the chromosomes, some of the proteins may attach the *oriC* regions to the opposite cell pole regions, and some of the proteins may compact the chromosomes and "guide" the chromosomes toward the opposite poles. Which proteins are these? We present an overview of current ideas, followed by a more detailed examination of the proteins thought to be involved in chromosome partitioning.

Sister chromosomes might be pushed out by the replisome to opposite halves of the cell

Because of the absence of a spindle apparatus similar to that found in eukaryotes, the mechanism for the partitioning of sister chromosomes toward opposite poles of the cell in prokaryotes is not well understood. Various models have been proposed. One model postulates the replisome as a DNA-synthesizing "factory" that has motor proteins generating force that pushes the newly replicated sister chromosomes apart toward opposite cell poles. According to this model, referred to as the "extrusion–capture" model (Fig. 10.17), in *E. coli* and *B. subtilis* the replisome is anchored to the membrane at the cell center; that unreplicated DNA enters the replisome and threads through it, whereupon the newly replicated sister chromosomes exit the replisome and partition toward opposite poles of the cell. As discussed later, the situation with *C. crescentus* is slightly different. Here the replisome is initially at the stalked pole of the cell, where DNA synthesis is initiated, and it moves toward midcell during DNA replication. As mentioned earlier, the replisome is assembled when DNA replication is initiated and disassembled when DNA replication is finished.

Candidate proteins that might be involved in providing the force to move the DNA

What are the postulated motor proteins that segregate the sister chromosomes? A cooperative mechanism has been proposed whereby energy released by the combined action of the DNA polymerase and the DNA helicase, both of which can be considered to be force-generating motors, coupled "pulls" the unreplicated DNA into the stationary replisome and "pushes" the replicated sister chromosomes, *origin* regions first, out of the replisome into opposite halves of the cell. It has been suggested that because the inhibition of transcription also inhibits sister chromosome partitioning, the force-generating motor may also include stationary RNA polymerases.

What confers directionality to chromosome movements during partitioning?

In addition to the "motor" that moves sister chromosomes apart, one may consider what confers directionality to chromosome movement. One may ask, in other words, why sister chromosomes move to opposite poles. One region of the chromosome that is directed toward opposite cell poles is the *oriC* region, located at the origin of replication. In addition to *oriC*, the chromosomes of *B. subtilis* have another region near the origin, the *polar localization region* (PLR), which directs the sister chromosomes to cell poles. It may work independently of *oriC*, or it may be responsible for the migration of the nearby *oriC* to the cell poles when chromosomes are partitioned during sporulation.[38]

The question now is: what might "drag" the localization regions to the cell poles? One possibility is that proteins bind to the DNA at the localization regions and that these proteins tether the chromosome to receptor proteins in the membrane near or at the cell poles. Two proteins mentioned in this regard are the RacA protein in *B. subtilis*, which binds to the PLR as well as to regions near *oriC*, and the Par proteins (partitioning proteins) in many bacteria, including *B. subtilis* and *C. crescentus* (but not *E. coli* or *Haemophilus influenzae*), which bind to *oriC*.

In addition to proteins that have been suggested to tether the localization regions to the pole, other proteins function in chromosome partitioning. These include SMC (structural maintenance of chromosomes) proteins in *B. subtilis* and *C. crescentus*, and MukB proteins in *E. coli*. The SMC and Muk proteins condense the chromosome. It appears that condensation of the chromosome might be important in moving it toward the cell pole. In addition, there are the SetB and MreB proteins. SetB is an integral membrane protein that physically interacts with MreB, a cytoplasmic actinlike filament. It has been suggested that MreB forms dynamic filament cables that might be a mitotic-like DNA segregation apparatus.[39] Exactly how this works is not known. However, mutations in *mreB* in *E. coli* result in severe effects on nucleoid segregation.

A more detailed description of proteins that play a role in chromosome partitioning

As mentioned, several different proteins appear to have a role in chromosome partitioning. This is clearly not a well understood process. However, the following discussion and the references cited should spark the student's curiosity to explore this very important area of research.

1. RacA
RacA is a DNA-binding protein in *B. subtilis* that functions in chromosome partitioning only during sporulation.[40] It binds preferentially at the PLR region, near *oriC*, and less specifically throughout the DNA. RacA is required for two reasons: (1) for the extension of the nucleoid into an axial filament during sporulation and (2) for the anchoring of the chromosome near its origin region to the cell pole during sporulation (Fig. 10.18A). RacA works in conjunction with SoJ in moving the SpoOJ–*oriC* complex to the cell pole via interaction with DivIVA at the pole. (See Section 10.1.7.) During vegetative growth of *B. subtilis*, the two chromosomal origins become anchored at cell quarter positions rather than at the extreme poles, and RacA is not involved.

2. The Par proteins
In most bacteria, and in low copy number plasmids, an early stage of the partitioning process involves the Par (partitioning) proteins, which bind to the chromosome near the origin of replication and may contribute toward the directionality of movement of the chromosome origins to opposite cell poles. (For a review, see ref. 41, and for a discussion of the role of Par proteins in plasmid partitioning, see note 42.)

The Par proteins discussed here that are encoded by the bacterial chromosome are called ParA and ParB. ParB is a DNA-binding protein that binds to specific sequences, called *parS* sequences, near the origin of replication. The binding is cooperative with ParA, an ATPase that interacts with ParB and results in a large nucleoprotein complex that is important for the positioning of sister chromosome and plasmid copies in opposite halves of the cell. As reviewed in ref. 21, ParA is membrane associated and may help tether the ParB–*parS* complex to the membrane. Proteins homologous to the chromosomally encoded ParA and ParB proteins have been detected in many bacterial species, including *B. subtilis*, where they are called Soj and Spo0J, respectively, as well as in *Caulobacter crescentus* and *Pseudomonas putida*. However, *E. coli* and *H. influenzae* do not have chromosomal *parABS* genes. Other proteins are presumably present in these bacteria that serve a function similar to that of the Par proteins.

3. Bacterial SMC and Muk proteins
Regardless of the mechanism of partitioning, the chromosomes partitioning cannot take place unless the chromosomes are compacted and maintained in a higher order structure. One speculation is that condensation of the DNA contributes towards a "pulling force" that complements the "pushing force" generated in the replisome.[24] Proteins important for compaction and higher order structure are called structural maintenance of chromosome (SMC) proteins, which are analogues of similar proteins in eukaryotes. Not every bacterium has an SMC protein. Bacteria such as *E. coli*, which lack an SMC protein, usually have a protein called MukB, which is similar in structure and function to SMC and sometimes is referred to as a "distant relative." MukB (see Fig. 10.19) has thus far been detected in *E. coli* and *H. influenzae*.[43–45] MukB forms a complex with two other Muk proteins, MukE and MukF, and the three

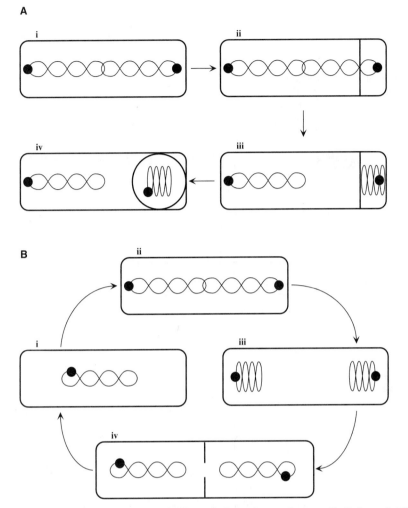

Fig. 10.18 Partitioning of chromosomes in *Bacillus subtilis*. (A) Sporulating cell. (i) Stage 0. The two chromosomes are a single axial filament attached by their origins to opposite poles of the cell. (ii) A septum forms, trapping the region proximal to the origin in the forespore. (iii) DNA translocation moves the rest of the chromosome into the forespore. (iv) After engulfment, the origins detach from the poles. (B) Vegetative cell. (i) Prior to DNA replication there is one origin. (ii) The DNA replicates, and both origins move toward the opposite cell halves. (iii) The chromosomes condense, moving away from each other toward the poles. (iv) A septum forms in the middle, dividing the cell into two daughter cells. *Source*: Adapted from Webb, C. D., A. Teleman, S. Gordon, A. Straight, A. Belmont, D. C.-H. Lin, A. D. Grossman, A. Wright, and R. Losick. 1997. Bipolar localization of the replication origin regions of chromosomes in vegetative and sporulating cells of *B. subtilis*. *Cell* **88**:667–674.

act together. For more information about the phenotype of mutations in the *muk* and *smc* genes, see note 46.

4. SetB and MreB
Another protein that appears to be involved in chromosome partitioning is SetB, an integral membrane protein in *E. coli*.[47] The deletion of *setB* results in a delay in sister chromosome separation from the cell center, and over-

production causes stretching and breaking up of the nucleoid. Interestingly, SetB physically interacts with MreB, a member of the family of actinlike proteins that form cytoplasmic helical filaments along the inner cytoplasmic membrane of cylindrical cells and are important for maintaining cell shape.[48] (One reason for believing that MreB is important for cell shape is that deletion of the gene *mreB* causes the rod-shaped *E. coli* to become spherical.) The

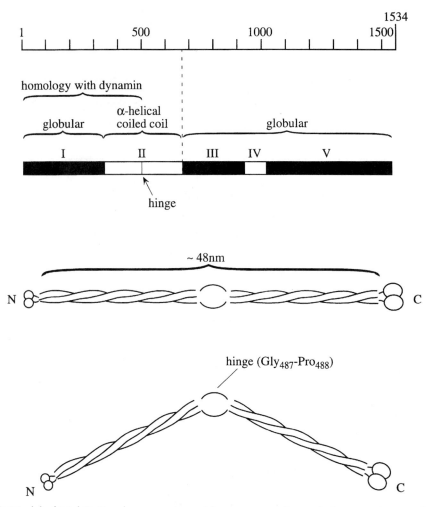

Fig. 10.19 Model of MukB. Based upon amino acid sequence studies and electron microscopic data, the model postulates that MukB is a homodimer with a rod-and-hinge structure. At the C-terminal end lies a pair of large globular domains. The N-terminal end has a pair of smaller globular domains. Electron micrographs suggest that MukB can bend at a hinge site in the middle of the rod section. MukB binds to DNA, ATP, and GTP. The amino acid sequence of MukB indicates homology with dynamin, a microtubule-associated protein from rats. It has been speculated that MukB attaches to the DNA and to some cellular structure (cytoskeletal filaments?) and moves the DNA to opposite poles. *Source*: Niki, H., R. Imamura, M. Kitaoka, K. Yamanaka, T. Ogura, and S. Hiraga. 1992. *E. coli* MukB protein involved in chromosome partition forms a homodimer with a rod-and-hinge structure having DNA binding and ATP/GTP binding activities. *EMBO J.* **11**:5101–5109.

MreB protein, which is discussed in Section 1.2.7, has been found in several different bacteria, including *E. coli*, *B. subtilis*, *Thermotoga maritima*, and *Caulobacter crescentus*.[49]

10.1.7 Chromosome partitioning in Bacillus subtilis

As specific examples of chromosome partitioning in bacteria, we will first consider *B. subtilis*, and then *C. crescentus* (Section 10.1.8). With

respect to sporulation and vegetative growth in *B. subtilis*. Section 10.1.8 focuses on *C. crescentus*.

Sister chromosome partitioning in Bacillus subtilis during sporulation (Fig. 10.18A)

The student should skip to the discussion of the partitioning of daughter chromosomes during *B. subtilis* sporulation in Section 18.18.6 and read refs. 38 and 50. It is proposed that at stage 0 (Spo0), prior to the initiation of sporulation, the two chromosomes form a single axial filament with their origin regions attached to opposite poles of the cell (Fig. 10.18Ai). Septation occurs close to one pole of the pre-divisional cell trapping approximately a third of the forespore chromosome in the forespore (Fig. 10.18Aii). It is important for the origin of the chromosome to be very close to the pole of the cell so that it is captured in the prespore when the septum forms. DNA translocation across the forespore septum takes place, resulting in the forespore chromosome being moved entirely into the forespore (Fig. 10.18Aiii). Translocation of the DNA requires a trans-porter called SpoIIIE DNA translocase, which forms a pore in the septum. Engulfment occurs and the origins are detached from the poles (Fig. 10.18Aiv).

Sister chromosome partitioning during vegetative growth of Bacillus subtilis

Prior to replication, a single chromosome with one origin exists (Fig. 10.18Bi). After repli-cation, the origins (called *oriC* as in *E. coli*) are moved to opposite halves of the cell, where the origin regions become attached at the cell quarter-positions (Fig. 10.18Bii).[51] The chromo-somes condense (Fig. 10.18Biii), and a septum forms (Fig. 10.18Biv). The stationary replicative fork has been visualized by means of replicative DNA polymerase fused in frame to the green fluorescent protein (GFP).

Some of the factors important for chromosome structure and partitioning in B. subtilis

Some of the proteins previously discussed are important for chromosome orientation and partitioning during sporulation and vegetative cell division.

1. ParA (Soj) and ParB (Spo0J)

The ParA homologue in *B. subtilis* is Soj, and the ParB homologue is Spo0J. Spo0J binds to at least eight *parS* sites located on either side of *oriC*, and in the presence of Soj the Spo0J–*parS* complexes are organized into a condensed, higher order *oriC* supercomplex, which persists throughout the cell cycle (growth, sporulation, germination) and is one of the factors that is required for chromosome partitioning during vegetative growth as well as sporulation. Mutations in *soj–spo0J* result in cells that are deficient in prespore chromosome partitioning (segregation), as well as a small fraction of anucleate cells during vegetative growth. The Soj–Spo0J system and RacA (described next) appear to have partially redundant roles in partitioning the prespore chromosome in that they both function to move *oriC* to the prespore cell pole via an interaction with DivIVA at the pole.

2. RacA

Another important protein is RacA (remodel-ing and anchoring of the chromosome A) that is synthesized for a period during the beginning of sporulation but not during vegetative growth.[40,50] RacA binds non specifically throughout the chromosome and is required for formation of the axial filament. Mutants of *racA* fail to form an extended axial filament and instead form a "stubby" nucleoid. RacA also binds preferen-tially near *oriC* and is important for efficient trapping of the *oriC* region at the cell pole in the prespore. Mutations in *racA* result in approxi-mately 50% of prespores without DNA. RacA also appears to be important for the formation of the septum at the cell pole during sporulation. Formation of the polar septum is significantly delayed in *racA* mutants. It has been suggested that the binding of RacA to DivIVA at the cell pole displaces the division inhibitor, MinCD, thus allowing septum formation at the pole.[52] The suggestion is based upon an earlier publi-cation showing that DivIVA sequesters MinCD at the poles so that cell division occurs at midcell during vegetative growth.[53] (See Section 2.3.6.)

3. DivIVA

DivIVA is an anchor protein at the cell poles. It probably binds directly or indirectly to RacA and to Spo0J which themselves are bound to the chromosome in the area of the *oriC* region. Thus, DivIVA, RacA, and Spo0J recruit the *oriC* region to the cell poles during sporulation. When RacA is absent, anucleate prespore com-partments form, resulting in a 50% drop in the frequency of sporulation. In *racA* mutant cells that do sporulate, the DivIVA–Spo0J system recruits the chromosome origins to the cell poles. Thus, the functions of RacA and the Spo0J system are partially redundant.

10.1.8 Chromosome partitioning in Caulobacter crescentus

Chromosome partitioning in *C. crescentus* is similar in some ways to the same procedure in *B. subtilis* and *E. coli*, and different in other ways. In all three species, when DNA replication is complete, the origins have moved toward opposite cell poles and the termini copies are segregated from each other closer to the cell center. In *C. crescentus*, however, replication begins at a cell pole (the stalked cell pole) rather than at midcell, and the replisome is not stationary but moves from the pole to the cell center during DNA replication. Before this is aspect of partitioning described further, a brief account of the cell cycle in *C. crescentus* will be presented.

The Caulobacter cell cycle

Normal cell division in *C. crescentus* produces two cell types, a motile swarmer cell with pili and a polar flagellum, and a stalked cell. (See Section 18.17.1 for a detailed description.) The stalked cell synthesizes DNA and undergoes repeated divisions to give rise to swarmer cells, which do not synthesize DNA and do not divide. After a while the swarmer cell sheds the flagellum, grows a stalk, begins synthesizing DNA, and becomes a dividing cell.

ParA and ParB

As with *B. subtilis*, ParA and ParB form a complex attached to the *Caulobacter* origin region (called *Cori* or *ori*) and can be seen to localize to both cell poles when chromosomal replication is complete in the stalked cell.[54] This can be demonstrated by using immunofluorescence microscopy with anti-ParA or anti-ParB antibody. (See note 55 for an explanation of how these experiments were done.)

A moving replisome

Rather that being initiated in a centrally located, stationary replisome as in *B. subtilis* and *E. coli*, DNA synthesis in *C. crescentus* is initiated in a replisome assembled at the stalked-cell pole. Soon after replication has been initiated, one of the origin copies rapidly moves to the opposite cell pole. During DNA replication, the replisome slowly moves toward the midcell position,

where DNA replication, including the replication of the terminus, continues.[56] It has been suggested that perhaps the replisome is moved passively from the pole to midcell by the newly replicated DNA accumulating at the cell pole.[56] When DNA replication is complete, the replisome disassembles. The result is that prior to cell division there is a copy of each origin near the opposite cell poles.

10.1.9 Inhibitors of DNA replication

A variety of antibiotics made by microorganisms, and chemically synthesized antimicrobial agents, inhibit DNA replication itself or inhibit the synthesis of substrates for DNA replication. Two antibiotics produced by *Streptomyces* include novobiocin, which inhibits DNA gyrase, and mitomycin C, which cross-links the guanine bases in DNA, sometimes in the same strand and sometimes in opposite strands. Cross-linking guanine in opposite strands prevents strand separation. Another antibacterial agent is nalidixic acid, a synthetic quinolone compound that inhibits DNA gyrase. Its more potent fluoroquinolone derivatives (norfloxacin, ciprofloxacin) are used to treat urinary tract infections, for example, those caused by uropathogenic *E. coli*.

There are several other chemically produced drugs that are widely used to inhibit DNA synthesis. These include acridine dyes (e.g., ethidium, proflavin, chloroquine), which insert between the bases in the same strand of DNA and distort the double helix. At low concentrations the dyes inhibit plasmid DNA replication but not chromosomal replication, although they can cause frameshift mutations. (See note 57 for an explanation.) If the concentration is sufficiently high, chromosomal replication is inhibited. Chemicals that inhibit the synthesis of precursors to DNA include the monophosphate of 5-fluorodeoxyuridine, aminopterin, and methotrexate.

5-Fluorodeoxyuridine-5′-phosphate [FdUMP, made by cells from fluorouracil (FU)] inhibits thymidylate synthase, which is the enzyme that converts dUMP to dTMP (Section 9.2.2). (See Fig. 10.20.) In this reaction a methylene group ($-CH_2$) and hydride (H:) are transferred from methylenetetrahydrofolate (methylene–THF) to dUMP to form a methyl ($-CH_3$) and thus

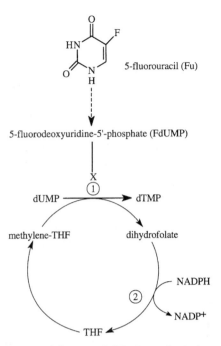

Fig. 10.20 Inhibition of dTMP synthesis by anti-cancer drugs. 5-Fluorouracil is converted to 5-fluorodeoxyuridine-5′-phosphate (fluorodeoxyuridylate, or FdUMP), which inhibits thymidylate synthase (**1**), the enzyme that converts dUMP to dTMP. A methylene group (–CH$_2$) and a hydride (H:) are transferred from methylenetetrahydrofolate (methylene–THF) to dUMP to form a methyl (–CH$_3$), which converts dUMP to dTMP. As a consequence of losing the hydride, the THF is oxidized to dihydrofolate. The dihydrofolate is reduced back to THF by NADPH in a reaction catalyzed by dihydrofolate reductase (**2**). The dihydrofolate reductase is inhibited by dihydrofolate analogues such as methotrexate and aminopterin.

converts dUMP to dTMP. As a consequence of losing the hydride ion, the tetrahydrofolate is oxidized to dihydrofolate. The THF is regenerated by reducing the dihydrofolate with NADPH in a reaction catalyzed by dihydrofolate reductase. The dihydrofolate analogues aminopterin and methotrexate inhibit dihydrofolate reductase. (For other reactions of tetrahydrofolate, see Fig. 9.14.)

10.1.10 Repairing errors during replication

All cells have repair mechanisms that fix DNA that has been damaged or has suffered errors during replication. Repair systems in bacteria

that operate during DNA replication and correct erroneous insertions of nucleotides are described next. Such errors cause relatively minor topological changes in the DNA. Damage to DNA due to ultraviolet light or mutagens, both of which can cause severe distortion of the double helix, is repaired by mechanisms described in Section 19.2.1.

Editing repair

Occasionally the wrong base is inserted in the copy strand (e.g., a T rather than a C opposite a G). This can occur if a base tautomerizes to a form that allows hydrogen bonding to the wrong partner. See Fig. 10.21 for an explanation. If the wrong nucleotide is added to the growing chain (e.g., a T opposite a G), the T must be removed; otherwise, when the strand

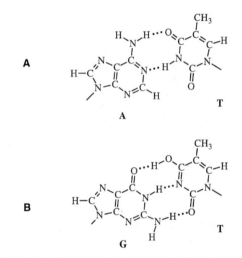

Fig. 10.21 Tautomerization can lead to incorrect base pairing. Two isomers in equilibrium that differ in the arrangements of their atoms are called tautomers. Commonly, tautomers differ in the placement of hydrogen. For example, one isomer might exist in the enol form (–OH attached to a carbon–carbon double bond), whereas the other isomer might exist in the keto form (contains a C=O group). Keto–enol tautomerizations greatly favor the keto form. When thymine tautomerizes from the keto to the enol form, a proton dissociates from the nitrogen in the ring and moves to the oxygen in the keto group to form the hydroxyl. When this happens, two electrons shift in from the nitrogen to form the C=N. (A) Adenine (A) correctly base pairs with the keto form of thymine (T). (B) Thymine has tautomerized to the enol form and base pairs with guanine (G) to form a mismatch.

containing the G–T pair is replicated, one of its progeny duplexes will have an A–T pair in place of the G–C pair: that is, a mutation will result. When the wrong base is inserted in the growing chain so that a mismatched pair results (e.g., G–T), DNA replication stops: that is, the DNA polymerase does not continue to the next position. Presumably this occurs because of the resultant topological change in the DNA (distortion in the double-stranded helix). The incorrect nucleotide is then removed via a 3′-exonuclease. DNA polymerases have 3′-exonuclease activity that functions in *proofreading* and removes mismatched nucleotides when they are added. Mutations in the *dnaQ* gene (*mutD*), which codes for a 3′-exonuclease subunit in the DNA polymerase III holoenzyme, result in greatly increased rates of spontaneous mutations.

Methyl-directed mismatch repair

The proofreading system is not perfect, and a certain number of wrong bases do get inserted into the newly replicated DNA. These can be removed, and the fidelity of DNA replication improved 10^2- to 10^3-fold (Fig. 10.22), by what is called the methyl-directed mismatch repair (MMR) system (reviewed in refs. 58 and 59). The nucleotide that is removed is the incorrect nucleotide in the copy strand rather than the template strand; thus the template strand is not changed, and a mutation does not occur.

The repair system can distinguish the copy strand from the template strand because the template strand is marked by methylation. *E. coli* has an enzyme called deoxyadenosine methylase (*Dam methylase*) that methylates all adenines at the N_6 position within 5′-GATC-3′ sequences. (The sequence is a palindrome and therefore is present in both strands but in opposite orientation i.e., 3′-CTAG-5′.) However, the enzyme does not begin to methylate the DNA until a short period after replication of the region of DNA containing the GATC sequence. Thus, for a few seconds or minutes after replication, the template strand is fully methylated but the copy strand is undermethylated. During this brief period, the copy strand can be repaired. What happens is that the copy strand with the incorrect base is cut at the 5′ side of the G in the unmethylated GATC and

the newly synthesized DNA is removed by an exonuclease to a point just beyond the mismatch (Fig. 10.22). The gap is then filled with DNA polymerase III and sealed with DNA ligase.

Several proteins are involved in mismatch repair. In *E. coli*, these include the products of *mutH*, *mutL*, and *mutS*. MutH, MutL, and MutS are thought to form a complex with the DNA. In this complex, MutH is the endonuclease. According to the model, MutS binds to the single base pair mismatch. Then MutL binds to MutS. MutH binds to a nearby GATC sequence, and its endonuclease activity is stimulated by the MutS/MutL complex. (Another *mut* gene, *mutD*, encodes the DNA polymerase III 3′-exonuclease, which is important for editing, as described earlier in the subsection entitled *Editing repair*.) DNA helicase II (the product of the *mutU* gene) is also required, and it is thought that its role is to unwind the cut strand so that it can be degraded by exonuclease.

More than one exonuclease can be used. If the endonucleolytic cut by MutH is made 3′ to the mismatch, then exonuclease I (ExoI) or ExoX, which is a 3′- to 5′-exonuclease is used. If the cut is made 5′ to the mismatch, then either exonuclease VII (ExoVII) or RecJ, which are 5′- to 3′-exonucleases, are used. DNA polymerase III fills in the gap, and DNA ligase seals the gap.

Mismatch repair can also be used to repair DNA damaged by the incorporation of base analogues or by certain types of alkylating agent as long as the distortion of the double helix is not severe. Otherwise, repair mechanisms described in Section 19.2.1 are used. Null mutations in any of the *mut* genes involved in mismatch repair results in a 10^2- to 10^3-fold increase in mutation frequency in *E. coli*. Homologues of the *mut* genes can be found in eukaryotes such as yeast, mice, and humans, where loss of function also results in increased mutation rates and, in the case of humans, is correlated with cancer in tumor cell lines from certain tissues (e.g., colorectal carcinomas). (For reviews, see refs. 60–62.)

10.2 RNA Synthesis

RNA synthesis is called *transcription* and involves the use of one of the strands of DNA as

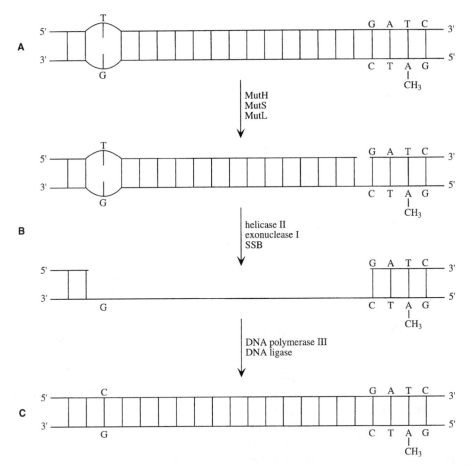

Fig. 10.22 Mismatch repair can occur when the wrong nucleotide has been inserted. For example, a T instead of a C might be inserted opposite a G. The template strand is methylated at an adenine in a CTAG sequence. The newly synthesized strand is not yet methylated, and this aids the repair enzymes in distinguishing between the newly synthesized strand and the template strand. (There is a slight delay in the methylation of newly synthesized DNA. However, the A in GATC eventually becomes methylated.) (A) An endonucleolytic cut is made either on the 5′ side or the 3′ side of the mismatch in the GATC sequence in the nonmethylated strand. (B) An exonuclease then removes the newly synthesized DNA past the point of the mismatch. This requires helicase II (MutU). (Helicase II is not identical to DnaB, which unwinds the strands during DNA replication.) If the mismatch is on the 5′ side of the cut, then exonuclease I degrades the DNA 3′ to 5′ through the mismatch. If the mismatch is on the 3′ side of the cut, then exonuclease VII or RecJ protein degrades the DNA 5′ to 3′ through the mismatch. (C) DNA polymerase III then fills in the missing DNA, and the gap is sealed with DNA ligase. Single-stranded DNA-binding protein is also required. The proteins involved in recognizing the mismatch and making the cuts are MutH, MutS, and MutL. The MutS protein recognizes the mismatch. The MutH protein is the endonuclease that makes the cut. It recognizes the GATC sequence and cleaves the unmethylated DNA on the 5′ side of the G in the GATC. MutS and MutH form a complex in which they are linked by MutL.

a template to make an RNA copy. The template strand is sometimes called the coding strand or the minus (−) strand. The nontemplate strand is sometimes referred to as the plus (+) strand. The RNA is synthesized in opposite polarity to the template DNA strand; that is, the 3′ end of the RNA (the growing end) faces in the same direction as the 5′ end of the DNA. For transcription to occur, the DNA must unwind at the site of transcription so that DNA-dependent RNA polymerase, the enzyme that synthesizes RNA, has access to the single-stranded DNA. (See note 63 for more information on the different types of RNA polymerase.) There are three stages in transcription: initiation, chain elongation, and termination.

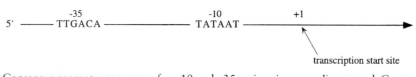

Fig. 10.23 Consensus promoter sequences for −10 and −35 regions in noncoding strand. Centered about 10 base pairs upstream of the mRNA start site (at position +1) is a 6-base region that is usually 5′-TATAAT-3′ in the noncoding strand or some minor variation thereof. It is called the −10 sequence or Pribnow box. A second consensus sequence is centered approximately 35 base pairs upstream of the transcription start site. It is generally referred to as the −35 sequence and is 5′-TTGACA-3′. The RNA polymerase binds to the entire promoter, not simply to the −35 and −10 regions. Unwinding of the DNA to initiate transcription occurs at the −10 region and extends for about 20 base pairs past the mRNA transcription start site. It is believed that the RNA polymerase itself unwinds the DNA.

10.2.1 Initiation, chain elongation, and termination

Intiation begins at the promoter region

Initiation takes place at the promoter site, which is a region of DNA at the beginning of the gene, where the RNA polymerase binds (Fig. 10.23).[64] Within the promoter region of most genes there exist short similar sequences of nucleotides centered approximately −10 and −35 base pairs upstream from the site at which transcription is actually initiated, that is, where the first nucleotide is incorporated (the +1 position, or the *start site*). The polymerase is a very large enzyme and covers the entire promoter, not simply the −10 and −35 regions. However, the −10 and −35 regions are critical because mutations that affect promoter function usually occur in one of these sequences. Furthermore, as discussed later, the frequency of initiation of transcription is specific for promoters and reflects variations in the sequences in the −35 and −10 regions.

In *E. coli*, the most common sequence in most of the genes at −10, sometimes called the Pribnow box, is 5′-TATAAT-3′ and the sequence at −35 is 5′-TTGACA-3′. These are called the consensus sequences, and although they are not identical in all promoters, there is sufficient similarity to permit the investigator (and the major RNA polymerase, called the sigma 70 polymerase) to recognize these sequences.

Promoter sequences other than the consensus −35 and −10 sequences exist, but these are not recognized by the major RNA polymerase holoenzyme (core plus sigma factor). Instead, they are recognized by RNA polymerases complexed with other sigma factors. (See the discussion of the sigma subunit later.) The binding of the RNA polymerase to these sequences and to the rest of the promoter region leads to the localized unwinding of the DNA strands, which allows the RNA polymerase to move down the template strand, making an RNA copy starting at the start site. In contrast to DNA replication, a helicase is not required.

Steps in initiation and chain elongation

1. Formation of the closed complex
RNA polymerase holoenzyme binds to the promoter region. At this stage the DNA exists as a double helix, and the DNA–polymerase complex is referred to as a "closed complex (Fig. 10.24).

2. Formation of open complex
The DNA duplex unwinds at the promoter region, including the start site, to form the *open complex*. There is no requirement for a helicase or a nucleotide triphosphate such as ATP.

3. Binding of initiating ribonucleotides
Transcription is usualy initiated at the start site (+1) with ATP or GTP, whose ribose provides the free 3′-hydroxyl that attacks the α-phosphate in the incoming nucleotide, displacing the pyrophosphate. The newly synthesized RNA therefore has a triphosphate at its 5′ end.

4. Elongation
RNA synthesis is said to be in its elongation stage after a few nucleotides (about 12) have been added to the growing chain of RNA, the sigma subunit has dissociated from the polymerase, and the polymerase is moving forward from the promoter region as it elongates the RNA. In the DNA duplex, there is an opening about 18 bases long, called a *transcription bubble*, in which the elongating RNA forms

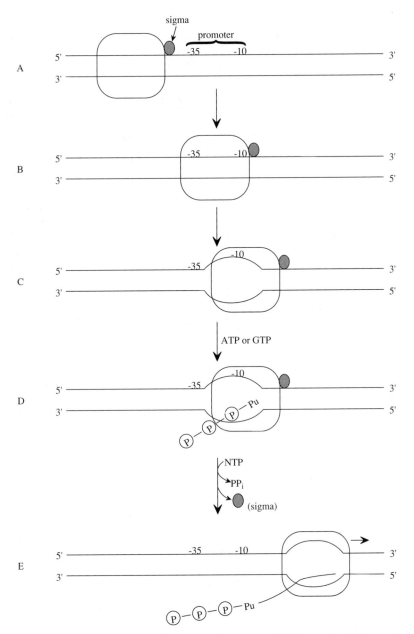

Fig. 10.24 Initiation of transcription. (A) The RNA polymerase, with its attached sigma subunit, binds to the DNA upstream of the −35 region. (B) The RNA polymerase moves to cover the promoter. This is called the *closed complex*. (C) The polymerase moves past the −10 region as it unwinds the duplex beyond the transcription start site to form the *open complex*. (D) RNA synthesis is initiated at the start site with either ATP or GTP. (E) As the polymerase migrates past the promoter, the sigma subunit is released. There is an opening in the DNA duplex of about 18 bases called a transcription bubble in which the elongating RNA forms an RNA/DNA hybrid.

an RNA/DNA hybrid with the template DNA strand. The RNA/DNA hybrid helps to keep the RNA polymerase attached to the DNA. The average rate of elongation of messenger RNA for *E. coli* is between 40 and 50 nucleotides per second, versus approximately 1,000 nucleotides per second for elongation of DNA. The rate of elongation of mRNA matches the rate of protein synthesis because approximately 16 amino acids per second are added to a growing polypeptide chain. For this to occur, the messenger RNA must move through the ribosome at a rate of 48 nucleotides (i.e., 16 codons), per second, or approximately the same rate as its synthesis. See the discussion of coupled transcription/translation in Section 10.3.6. Ribosomal RNA in fast-growing cells is synthesized at a rate of about 90 nucleotides per second.

Termination

At termination the RNA and RNA polymerase are released from the DNA, and the DNA duplex re-forms at that site.[65,66] Transcription is terminated by a DNA sequence, called a terminator. There are two patterns of termination. One requires an additional protein factor and is called factor-dependent termination. The best studied termination factor is a protein called Rho. We begin, however, with a type of termination that does not require a protein factor: factor-independent termination.

1. Factor-independent type

Factor-independent termination occurs when the RNA polymerase transcribes a self-complementary sequence of bases (Fig. 10.25A) that promotes the localized hybridization of the RNA to itself (formation of a hairpin loop) rather than to the DNA template (Fig. 10.25B). The loop is followed by a short RNA–DNA

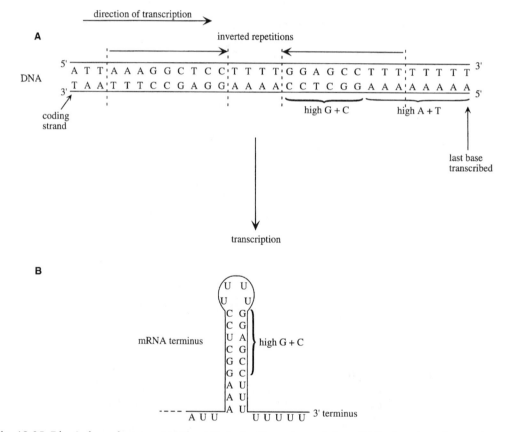

Fig. 10.25 Rho-independent termination. (A) At the 3′ terminus of the mRNA there is an inverted repeat separating two nonrepeating sequences. (B) For example, the sequence in the *E. coli trp* mRNA (encoding enzymes to synthesize tryptophan) is AAAGGCUCC-UUUU-GGAGCCUUU, which causes the mRNA to form a stem-and-loop (hairpin) structure near the 3′ end of the mRNA. Near the loop or within the stem there is a sequence rich in G + C base pairs. There are also 6 to 8 uracils at the 3′ end of the mRNA (corresponding to a string of adenines in the template DNA). The uracils are necessary for termination, as shown by mutants in which some of the A–T base pairs in this region of the DNA have been removed. According to the model, when the RNA hairpin forms, the RNA hybridizes to itself rather than to the DNA and thus the RNA–DNA duplex is disrupted in the transcription complex. In addition, the U–A duplex after the hairpin is unstable. This leads to the dissociation of the RNA from the DNA.

duplex consisting of a string of A–U base pairs that is unstable. (The interactions between A and U are weaker than the other base pair interactions.) The consequence is that the RNA spontaneously dissociates from the template along the A–U region and from the polymerase. The RNA polymerase also dissociates from the DNA and the DNA duplex reforms.

2. Rho-dependent type

Other terminator regions require protein factors for termination and do not rely on RNA–RNA hairpin loops to disrupt the RNA/DNA hybrid in the transcription bubble. The nucleotide sequences in the terminator regions are not closely related and are not always easy to recognize. There are three termination factors in *E. coli*. They are Rho, Tau, and NusA. The most well studied is Rho, which is an RNA-dependent ATPase and an ATP-dependent RNA/DNA helicase. As we shall see, these two activities of Rho are necessary for termination.

According to one model, Rho works by binding to particular sequences, called *rut* (<u>r</u>ho <u>ut</u>ilization sites) in the RNA directly behind the RNA polymerase. If a translating ribosome is present at that sequence, then Rho cannot bind. If, however, translation is terminated and Rho binds, then it moves along the mRNA behind the RNA polymerase in the 5′-to-3′ direction. The energy to move along the mRNA is provided by the hydrolysis of ATP. When the RNA polymerase reaches a termination site, it stops moving. Rho reaches the polymerase, and the Rho helicase activity unwinds the RNA–DNA duplex in the transcription bubble. This results in the disengagement of the RNA polymerase from the DNA and the re-formation of the DNA–DNA duplex. Thus, when the ribosome stops translating at the translational stop site in the mRNA, Rho can bind at the next accessible *rut* site and transcription will stop at the next Rho-dependent termination site. Rho-dependent termination can also explain polarity, as discussed in Section 10.3.8.

10.2.2 Frequency of initiation

As stated, promoter regions are not identical, and they do differ in the extent to which they influence the frequency of initiation of transcription ("promoter strength"). Strong promoters cause initiation every few seconds and weak promoters every few minutes. This discrepancy is related to the consensus sequences at the −10 and −35 regions because strong promoters have sequences that are more similar to the consensus sequences, whereas weak promoters have base substitutions in these regions.

Various transcription factors, both positive and negative, also influence the frequency of transcription. The transcription factors bind to regions near the promoter and either promote RNA polymerase activity (e.g., by stimulating its binding or the formation of the open complex, or inhibit the activity of the polymerase). These transcription factors are extremely important for the regulation of the timing of gene expression. See Section 10.2.5 for a discussion of the regulation of RNA transcription.

10.2.3 Role of topoisomerases

Just as during DNA replication, the unwinding of the duplex produces positive supercoils downstream of the unwinding region. If nothing were done with these supercoils, DNA would stop unwinding and transcription would stop. Topoisomerases change the positive supercoils to negative supercoils within which the DNA strands are more readily unwound.

10.2.4 The sigma subunit

For a review, see ref. 67. The core RNA polymerase (RNAP) in the Bacteria usually has five subunits ($\alpha_2\beta\beta'\omega$). For transcription to be initiated, a sixth subunit, called sigma (σ) factor, must attach to the core enzyme to form the RNAP holoenzyme. Sigma factor is responsible for recognition of the promoter. Different sigma factors recognize different consensus sequences in the promoter region. Thus, sigma factors determine whether or not a gene will be transcribed at any point in time. This is aided by a large number of transcription factors that activate and repress the transcription of specific genes. Clearly this is not a simple task: a bacterium such as *Escherichia coli* has approximately 4.6 million base pairs and 2,000 genes. After initiation, sigma factor dissociates from the polymerase; that is, it is not needed for chain elongation.

A bacterium can have more than one sigma factor, and the core polymerase can exchange one for another, thus contributing toward the timing and specificity of gene transcription. For example, the main sigma factor in *E. coli* is σ^{70} (having a molecular weight of 70 kDa), which recognizes the consensus sequence in most of the promoters and is termed the "housekeeping" sigma. However, several alternative sigma factors may be present, adding specificity to which genes are transcribed.

Sequence analyses has revealed that all the sigma factors fall into two families: σ^{70} and σ^{54}. Most bacteria will have several different sigma factors of the σ^{70} family but only one representative of the σ^{54} family. Bacteria vary in the number of sigma factors. Two spore-forming bacteria, *Bacillus subtilis* and *Streptomyces coelicolor*, have 18 and 63 members, respectively, of the σ^{70} family. Not all bacteria have a σ^{54} gene. For example, high-GC, gram-positive bacteria, and cyanobacteria, are not known to possess one.

Sigma factors belonging to the σ^{70} family can initiate transcription without any additional energy source or other proteins. Members of the σ^{54} family are different. They require an activator protein that binds about 100 to 150 base pairs upstream of the promoter region at "enhancer sites" to ensure that an open complex forms and transcription is initiated at the promoter region. The activator proteins are ATP dependent. As an example, read the discussion in Section 18.4.1 of the use of σ^{54} by *E. coli* to transcribe genes in the Ntr regulon when the organism is growing under limiting ammonia conditions, and examine Fig. 18.8.

E. coli has six sigma factors in the σ^{70} family: σ^{70} (RpoD), σ^{s} (RpoS), σ^{32} (RpoH), σ^{F} (FliA), σ^{E} (RpoE), and σ^{FecI}, and they differ in the specificity of the genes that they transcribe. As mentioned previously, σ^{70} is the "housekeeping" sigma, and the alternative sigma factors are used for the transcription of genes activated by stress conditions, growth transitions, and morphological changes such as flagella biogenesis. For example, when *E. coli* enters stationary phase as a result of starvation, it increases the amount of σ^{s} (also called RpoS or σ^{38}), which recognizes promoters for genes that are expressed during stationary phase (Section 2.2.2). The amounts of σ^{s} also increase during

growth in high osmolarity, and this reflects, at least in part, the role of σ^{s} in transcribing the genes encoding mechanosensitive channel proteins discussed in Section 15.2.5. The mechanosensitive channels protect the cells from hypo-osmotic shock when they are diluted into low osmolarity media. The promoter sequence recognized by σ^{s} is actually the same as the core consensus promoter sequence recognized by σ^{70}. However, there exist regulatory factors that enhance selectivity of σ^{s}. These are reviewed in ref. 67.

The factor σ^{s} was discovered because of its role in transcribing genes during stationary phase. However, activation of σ^{s} is also triggered by high osmolarity and by stress conditions such as oxidative stress, UV radiation, and heat shock. Indeed, σ^{s} has been called a master regulator of the general stress response. The regulation of σ^{s} is complex and occurs at the transcriptional, translational, and posttranslational levels. (See Section 2.2.2.)

When subjected to heat stress, *E. coli* increases the amount of another sigma factor in the σ^{70} family, σ^{32}, which stimulates the transcription of genes encoding heat-shock proteins that fold or degrade cytoplasmic proteins damaged by stresses such as heat shock (Section 19.1). Another sigma factor in *E. coli* that responds to stress is σ^{E}. It is induced by unfolded proteins in the cell envelope and stimulates the expression of genes that repair the envelope.[67] The genes stimulated by σ^{E} include those that encode chaperones and proteases that refold or degrade misfolded proteins in the envelope.

Still another sigma factor in *E. coli* that responds to stress, σ^{FecI}, stimulates the transcription of the *fecABCDE* operon that encodes a transport system for bringing iron citrate into the cell. The stress signal for this sigma factor is iron starvation. The sigma factor σ^{F} (FliA), also called σ^{28}, initiates transcription of genes necessary for flagellum biosynthesis and chemotaxis. These include genes that encode the flagellar filament protein (HAPS protein) and components of the chemotaxis system.

Anti-sigma factors

How do bacteria regulate the activities of the sigma factors? One way that bacteria control the activities of its sigma factors is by using anti-sigma factors that bind to them and inhibit their activity. For a description of an anti-sigma factor that is part of the regulatory system for sporulation genes in *Bacillus subtilis*, see the discussion of SpoIIAB in Section 18.18.1.

An anti-sigma factor in *Escherichia coli* is anti-σ^E, called RseA. (For a review of this factor, whose short form stands for regulators of sigma E, read ref. 68.) RseA is a transmembrane cell membrane protein with a cytoplasmic domain that binds to σ^E and keeps it inactive. Recall that σ^E is used to transcribe genes that are important for repairing the cell envelope when its proteins become damaged. Under these circumstances, two cell membrane proteases, DegS and Yael, sequentially degrade RseA, releasing σ^E into the cytoplasm, where it binds to core RNA polymerase and stimulates transcription of the genes in the σ^E regulon.

As mentioned earlier, the genes stimulated by σ^E include those that refold or degrade misfolded envelope and periplasmic proteins. A potent signal for the proteolysis of RseA is the presence of unfolded porin proteins in the outer membrane. It has been suggested that when the porin, which is a trimer, unfolds as a consequence of membrane stress, part of the protein that is normally buried in the folded porin trimer is exposed and binds to the membrane protease DegS. This binding activates the proteolytic activity of DegS, leading to partial proteolysis of RseA. The partially degraded RseA, in turn is a substrate for the membrane protease Yael, which completes the degradation of RseA, in turn resulting in the release of σ^E. For a more complete description of this model, see ref. 68.

10.2.5 Regulation of transcription

Bacteria generally regulate protein synthesis at the transcriptional level. (See ref. 69 for a review.) Since mRNA in bacteria is generally unstable (i.e., is rapidly enzymatically degraded), the inhibition of transcription of a particular gene generally means that the synthesis of the protein encoded by that gene ceases quickly. Conversely, stimulation of transcription of a particular gene usually results in a rapid synthesis of the protein encoded by that gene. In this way, bacteria are able to modulate the mix of enzymes and other proteins in response to environmental challenges such as the presence or absence of specific carbon and energy sources, or inorganic compounds such as inorganic phosphate or ammonium ion, and environmental stress such as starvation, changes in pH, changes in osmolarity, and changes in temperature. The responses to environmental challenges and the signal transduction pathways important for these responses are described in Chapters 18 and 19.

Operons

Frequently bacterial genes are arranged in *operons*. An operon consists of two or more genes that are cotranscribed into a polycistronic mRNA from a single promoter at the beginning of the first gene. All the genes in the operon are coordinately regulated from a single promoter.

Transcription factors

Transcription is commonly regulated at the initiation step by proteins that affect in some way the binding of RNA polymerase to the promoter, the "melting" of the DNA to form the transcription bubble, or the movement of the polymerase along the DNA. The transcription factors bind to specific sequences in or near the promoter region or, if the transcription factor stimulates gene expression, in so-called *enhancer regions*, upstream from the promoter.

Although most of the known transcription factors freely diffuse in the cytoplasm, two transcription factors that activate the ToxT promoter in *Vibrio cholerae*, ToxR and TcpP, are bound to the cell membrane. This is described in Section 18.11.1. The transcription factors generally have a helix–turn–helix motif, as described later. The majority of transcription factors that activate transcription bind to the RNA polymerase and recruit the polymerase to the promoter. An example of this type of transcription factor is the cyclic AMP receptor protein (CRP) discussed in Section 16.3.4. Some transcription factors work by binding to the DNA, altering its structure, and in this way positively influencing the initiation of transcription. An example is IHF (integration host factor), discussed in Section 18.4.1. IHF binds to DNA and bends it so that a second bound transcription activator is brought to the promoter and activates transcription. IHF can also repress transcription by binding at or near the promoter region of certain genes, as described in Section 18.7.2.

E. coli has about 350 transcription factors. (See ref. 67.) The regulation of transcription by transcription factors is especially important when the levels of the sigma factor for the particular gene are not changed. Under these circumstances it is only the transcription factors that determine the rates of transcription of specific genes. Global transcription factors regulate the transcription of genes in several different operons. All the operons regulated by a single transcription factor are collectively called a *regulon*.

Helix–turn–helix motif

Many prokaryotic DNA-binding proteins, including activators and repressors of transcription, have a *helix–turn–helix motif* in part of the protein (Fig. 10.26A).[70] In one part of the protein approximately 7 to 9 amino acids form an α-helix, which is called helix 1. Helix 1 is connected by about 4 amino acids to a second α-helix of about the same size. The helices are approximately at right angles to each other. Helix 1 lies across the DNA and helix 2 (the recognition helix) lies in the major groove of the DNA (Fig. 10.26B).

Because helix 2 lies in the major groove, the amino acid side chains that protrude from the surface of the protein can contact and form noncovalent bonds (e.g., hydrogen bonds) with base pairs in the DNA, and therefore can bind to specific nucleotide sequences (Fig. 10.26C). Hydrogen bonds can form between the carbonyl and amino groups in the protruding side chains of certain amino acids in the DNA-binding region of the protein, and amino groups, ring

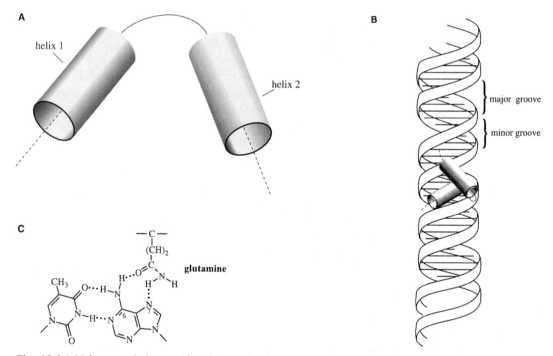

Fig. 10.26 Helix–turn–helix motif and DNA binding. (A) A DNA-binding protein typically has in one portion two α-helices connected by a short segment of about four amino acids. (B) Helix 2 lies in the major groove of the DNA duplex, and helix 1 lies at approximately right angles to helix 2. (C) Because helix 2 lies in the major groove, it makes contact with the bases. Certain amino acids in helix 2 with amino or carbonyl groups in their side chains can hydrogen-bond to specific bases. Here glutamine ($-CH_2-CH_2-CONH_2$) is hydrogen-bonded to adenine in an A:T base pair. Other amino acids, for example, asparagine ($-CH_2-CONH_2$), glutamate ($-CH_2-CH_2-COOH$), lysine ($-CH_2-CH_2-CH_2-CH_2-NH_2$), and arginine ($-CH_2-CH_2-CH_2-NH-C(NH_2)=N^+H_2$), can also hydrogen-bond to specific base pairs, and in this way the DNA-binding protein can bind to specific nucleotide sequences. DNA-binding proteins are frequently dimers and bind to inverted repeats in the DNA.

nitrogen, and carbonyl groups in specific bases. The amino acids that are involved in the bonding to the bases are generally asparagine, glutamine, glutamate, lysine, and arginine. For example, glutamine can bind to A–T base pairs because the carbonyl group and amino group in glutamine can hydrogen-bond with the amino group on C6 and the N7 in adenine, respectively (Fig. 10.26C). This does not interfere with the hydrogen bonding of the bases to each other.

How transcription factors work

Activators bind to specific sites in the DNA upstream of the RNA polymerase and make contact with the RNA polymerase while enhancing its activity. Repressors bind downstream of the RNA polymerase (in the operator region) and block progression of the RNA polymerase. Figure 10.27 illustrates general models of how transcription activators and repressors are believed to work. Some specific examples of positive and negative transcription factors are described next.

1. Positive regulators

Positive transcription regulators bind to the promoter region or to *enhancer sites*, which are sequences upstream from the promoter region. They may make contact with RNA polymerase

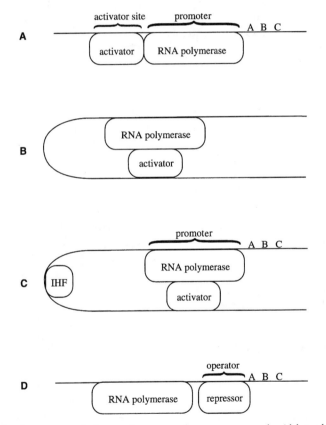

Fig. 10.27 Models for how transcription activators and repressors work. Although not shown, several transcription factors are known to function as dimers. (A) Activator proteins can bind at specific sites, called activator sites, upstream of the promoter, where they make contact with the RNA polymerase, facilitating its binding to the promoter. Some activators must bind to a small effector molecule (inducer) before they can bind to the DNA. (B) If the activator binds a distance away, upstream of the promoter, then the DNA must bend to bring the activator to the polymerase. (C) Sometimes a second protein (e.g., integration host factor, IHF), is required to bend the DNA to bring the activator site to the RNA polymerase. Here IHF binds to a site between the promoter region and the activator site and brings the activator to the RNA polymerase. (D) Repressor proteins bind to the operator region, which may or may not be in the promoter region. The repressors block progression of the RNA polymerase. Some repressor proteins (called aporepressors) must bind to an effector molecule (corepressor) before they can bind to the DNA.

and promote its binding to the promoter region, and some may facilitate the "melting" of DNA to form the transcription bubble. Consider the cyclic AMP receptor protein. When it binds to cAMP, the CRP can then bind to specific sites in the DNA and stimulate transcription.

For example, consider the stimulation by cAMP–CRP of the *lac* operon, which encodes proteins required for the uptake and metabolism of lactose. cAMP–CRP binds to specific nucleotide sequences upstream of the *lac* promoter as well as to an incoming RNA polymerase. The binding of cAMP–CRP to the RNA polymerase enhances its binding to the promoter and thus stimulates transcription.

Many unrelated genes are positively regulated by cAMP–CRP and are said to be part of the CRP regulon. They are called cAMP-dependent genes. cAMP-dependent genes may encode proteins required for the catabolism of several different carbon and energy sources, such as lactose, galactose, arabinose, and maltose. In such bacteria, mutants unable to synthesize cAMP or CRP are unable to grow on these carbon sources.

Because cAMP–CRP stimulates the transcription of many unrelated genes, it is called a *global regulator*. The levels of cAMP are lowered by glucose uptake in many bacteria. This lowers the transcription of cAMP-dependent genes, explaining glucose repression of genes required for the catabolism of carbon sources other than glucose. Glucose repression mediated by cAMP is discussed in Section 16.3.4. (Refer to Sections 18.10.1 and 18.10.2, respectively, for glucose repression that does *not* involve cAMP and for a general discussion of catabolite repression.)

Another example of a global transcription activator is NR_I–P, which is part of a two-component regulatory system that activates the *ntr regulon* as discussed in Section 18.4.1. NR_I–P binds to enhancer sites 100 to 130 base pairs upstream from the promoter region of the target genes and interacts with the RNA polymerase to form the open complex, thus stimulating transcription. For this to happen, the DNA must bend enough to bring NR_I–P to the polymerase (Fig. 10.27). As shown in Fig. 10.27C, integration host factor (IHF) promotes DNA bending. NR_I–P helps the RNA polymerase (σ^{54} polymerase) to separate

the strands of DNA around the promoter so that the polymerase can gain access to the transcription start site. Several other positive transcriptional activators are described in Chapter 18.

Another example of a positive transcription factor that requires the IHF DNA binding protein for activity is NarL–P, which stimulates the transcription of the nitrate reductase gene (Section 18.3). According to the model, IHF bends the DNA to bring the bound NarL–P to the promoter, where it stimulates transcription (Fig. 10.27). IHF can also inhibit transcription if it binds to the promoter region. (See Section 18.7.2.) Also, see the discussion of the nucleoid in Section 1.2.6 for a description of IHF and other DNA binding proteins that bind to bent DNA.

Many operons are specifically activated by a protein (activator) only when the protein binds an inducer. Examples include the activation of the L-*ara* operon, which encodes proteins required for the conversion of L-arabinose into xylulose-5′-phosphate, an intermediate in the pentose phosphate pathway. The activator protein is AraC, which must bind to L-arabinose before it can bind to the DNA and activate transcription of the operon. This requirement is believed to be due to a conformational change in the activator increasing its affinity for the DNA sequence upon binding of the positive effector. As described later, some repressor molecules must also bind to small effector molecules before they bind to DNA.

2. Negative regulators

Negative regulators are called *repressors*. They bind to nucleotide sequences in the DNA called *operator* regions and inhibit transcription (Fig. 10.27D). For some genes, the operator is within the promoter region; but for others it is outside the promoter and may even be upstream of the start site. An important negative repressor is LexA, which is inactivated by proteolytic cleavage during the SOS response, as discussed in Chapter 19. The SOS response is activated when DNA is damaged (e.g., by UV radiation). Single-stranded DNA that results from such damage activates RecA, which then activates the cleavage of LexA. As a consequence of the cleavage of LexA, many genes important for DNA repair are induced.

Another well-known repressor is the *lac* repressor, which binds to the operator region that overlaps the start site of the first gene in the *lac* operon. Transcription is inhibited when the repressor is bound because the repressor blocks RNA polymerase from proceeding into the first gene of the operon. The repressor is inactivated when it binds an isomeric product of lactose metabolism (allolactose), accounting for the induction of the lactose operon by lactose. (The allolactose binds to the repressor, causing a conformational change in the repressor protein, which lowers its affinity for the DNA. As a consequence, the repressor comes off the DNA and the *lac* operon is induced.) Often investigators use isopropyl-β-D-thiogalactoside (IPTG) as an inducer of the *lac* operon because IPTG is not metabolized.

Sometimes a repressor must bind to a small molecule before it can bind to the operator. This is the case for biosynthetic operons, operons that encode enzymes for the biosynthesis of amino acids and other small molecules. As an example, consider the Trp repressor in *E. coli*, which represses the *trp* operon. (The *trp* operon encodes the enzymes required to synthesize tryptophan from chorismic acid.) The repressor can bind to the operator only if it first binds to tryptophan, which is a *corepressor*. (In the absence of the corepressor, the inactive repressor is called an *aporepressor*.) In this way, tryptophan regulates the synthesis of its biosynthetic enzymes by feedback repression of transcription.

Some transcription regulators can be activators or repressors

Some transcription regulators activate certain genes and repress others. Several examples are discussed in Chapter 18. For example, see Section 18.10.1, which discusses the catabolite repressor/activator (Cra) system.

Attenuation of the trp operon

Whereas positive and negative regulators of transcription act at the promoter region to influence the initiation of transcription, attenuation refers to regulation of transcription *after* initiation. In attenuation, transcription is terminated before the first gene in the operon has been transcribed. Operons can be regulated by attenuation as well as by positive and negative regulation of initiation of transcription. Several operons that encode enzymes required for the biosynthesis of amino acids are known to be regulated by attenuation, including the *trp* operon in *E. coli*. (As suggested earlier, regulation of the *trp* operon occurs via feedback repression by tryptophan.) Downstream of the promoter region in the *trp* operon there exists a region called the leader region (*trpL*), which lies between the promoter and the first gene of the *trp* operon (*trpE*). When the tryptophan concentrations rise, this results in an increased amount of aminoacylated tRNA^{Trp}, which terminates transcription in *trpL*. As a consequence, the operon is not transcribed. How this occurs is described next.

The model for attenuation of the *trp* operon is diagrammed in Fig. 10.28. The four regions in the *trpL* RNA, numbered 1, 2, 3, and 4, can form hairpin loops with one another (i.e., 1:2, 2:3, and 3:4). There are two adjacent tryptophan codons in *trpL*, and these are in region 1. When the ribosome reaches these codons, it will stall if there is insufficient aminoacylated tRNA^{Trp}. If the ribosome stalls at the *trp* codons in region 1, then a hairpin loop forms between the newly synthesized regions 2 and 3. (Region 4 has not yet been synthesized.) If, on the other hand, there is sufficient tryptophan to prevent the ribosome from staling in region 1, it moves to the stop codon in region 2. This frees region 3 to pair with region 4 of the RNA when it is synthesized. If regions 3 and 4 form a hairpin loop, transcription stops because this hairpin is a factor-independent termination signal. (See Section 10.2.1 for a discussion of factor-independent termination.)

10.2.6 *Processing of ribosomal and transfer RNAs*

Ribosomal RNA

To synthesize rRNA, bacteria first make a 30S preribosomal transcript from a single gene. The 30S transcript contains the 16S, 23S, and 5S rRNAs, as well as one or more tRNAs, and is processed according to the diagram in Fig. 10.29. There are actually seven sets of rRNA genes in *E. coli* that yield the 30S preribosomal transcript. The genes all code for the same rRNAs

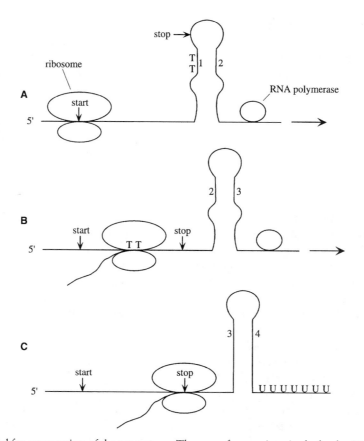

Fig. 10.28 Model for attenuation of the *trp* operon. There are four regions in the leader RNA that precedes the transcript for the first gene in the *trp* operon. Region 1 has two *trp* codons (T). (A) The RNA polymerase transcribes *trpL*, just past region 2 and pauses briefly as regions 1 and 2 form a hairpin loop. The translational start and stop sites are marked, as are the tryptophan codons. (B) While the polymerase is paused, a ribosome begins translating the mRNA at the 5′ end. As the ribosome moves, region 1 of the mRNA is drawn into the ribosome. As translation proceeds, the polymerase begins to move again and transcribes region 3. (B) Region 1 has been drawn into the ribosome and is being translated. This frees region 2 to pair with the newly transcribed region 3. If the ribosome stalls at region 1 because of an insufficiency of tryptophan, region 2 remains paired with region 3. This prevents a termination loop between regions 3 and 4 from forming, and transcription of the *trp* operon continues. If the ribosome does not stall at region 1, it will move to the end of *trpL*, where it stops at the UGA codon in region 2, which is now in the ribosome. While the ribosome occupies the UGA codon in region 2, it prevents region 2 from pairing with region 3. (C) Meanwhile, the polymerase continues to move and transcribes region 4. A termination loop forms between regions 3 and 4. Thus, in the presence of excess tryptophan transcription of the *trp* operon is terminated. See Fig. 10.25.

but differ with respect to the number and type of tRNAs encoded. Eukaryotes also process rRNA transcripts, as described in note 71.

Transfer RNA

Transfer RNAs are extensively processed after transcription in both bacteria and eukaryotes (Fig. 10.30).[72] The initial transcript has segments at both the 3′ and 5′ ends that are enzymatically removed. The removal of the 5′ end segment is catalyzed by RNase P, an endonuclease, which

is present in both prokaryotes and eukaryotes. RNAse P is a ribonucleoprotein in which the RNA has catalytic activity. It is therefore an example of a *ribozyme*. (The peptidyl transferase is another example of a ribozyme, Section 10.3.5.) The 3′ end is processed by an endonuclease that removes a terminal piece, followed by RNase D, an exonuclease, that trims the remaining nucleotides to generate the mature 3′ end. In addition to exo- and endonucleolytic cleavage, processing may occur at the 3′ end. All eukaryotic tRNAs and some

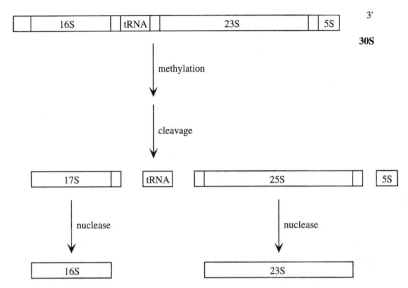

Fig. 10.29 Processing of preribosomal RNA transcripts. *E. coli* has seven genes that code for a 30S transcript containing the 16S, 23S, and 5S rRNAs and one or more tRNAs. For example, one of the *E. coli* 30S transcripts is 5′-16S-tRNA^{Ile}-tRNA^{Ala}-23S-5S-tRNA^{Asp}-tRNA^{Trp}-3′. After methylation of specific bases, the 30S transcript is cleaved into 17S, 25S, and 5S rRNA transcripts and the tRNAs. Specific nucleases then generate the 16S rRNA from the 17S transcript and the 23S rRNA from the 25S transcript. Recall that each ribosome has one 23S, one 16S, and one 5S rRNA. Cleaving the three rRNA transcripts from a single transcript rather than having the three genes transcribed separately ensures that the rRNA molecules are produced in a 1:1:1 ratio. Not all tRNA genes are present within the rRNA transcript; several of the tRNA genes are encoded outside the rRNA genes.

bacterial tRNAs lack the CCA-3′ terminus, and this must be added to the molecule after transcription by the enzyme tRNA nucleotidyl-transferase. In addition, all tRNAs have bases at specific sites that are modified after transcription (Fig. 10.31).

10.2.7 Some antibiotic and other chemical inhibitors of transcription

Transcription in both prokaryotes and eukaryotes is inhibited by the antibiotic *actinomycin D*, which inserts into the DNA between base pairs and deforms the DNA. The result is that the RNA polymerase cannot continue to move down the template, and elongation is inhibited. Another inhibitor of elongation is *acridine*, which behaves in a similar way. A commonly used antibiotic to inhibit transcription in prokaryotes is *rifamycin*, or the chemically modified derivative *rifampicin*, which is more effective. Rifamycin (and rifampicin) binds to a subunit in bacterial RNA polymerases (the β subunit) and inhibits the initiation of transcription. It is not an inhibitor of RNA synthesis

in eukaryotic nuclei. A related antibiotic, which also binds to the β subunit of bacterial RNA polymerase, is *streptolydigin*. *Amanitin* inhibits eukaryotic RNA polymerase II (synthesizes mRNA) but not bacterial RNA polymerase.

10.3 Protein Synthesis

10.3.1 Overview

Protein synthesis is called *translation* because the sequence of bases in the mRNA is translated into a sequence of amino acids in the protein. It occurs in three stages on the ribosome: (1) initiation, (2) elongation, and (3) termination.[73] A summary of each of the stages is given first, followed by a more detailed description.

Initiation

During initiation, an initiator transfer RNA (tRNA$_f$) carrying formylmethionine binds to the start site at the 5′ end of the mRNA and at a tRNA-binding site called the P (peptidyl) site on the ribosome. Amino acids do not recognize

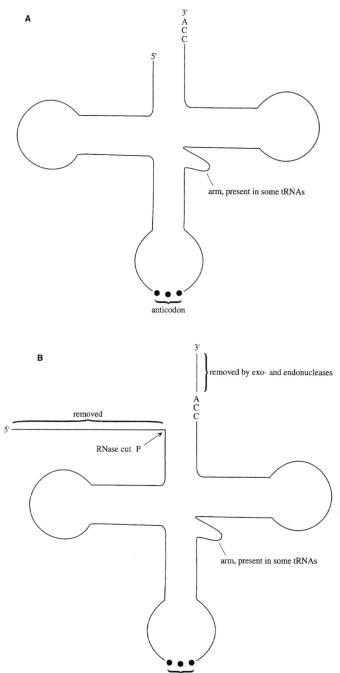

Fig. 10.30 Processing of transfer RNA. (A) Processed transfer RNA molecules have between 73 and 93 nucleotides. They consist of a 3′ end that ends in CCA. The amino acid is esterified to a hydroxyl in the ribose portion of the terminal AMP. There are four arms, three of which end in loops. Often a short fifth arm exists between two of the arms. The anticodon is present in the loop of the anticodon arm. (B) Transfer RNA is made as a larger precursor and processed at both the 3′ and 5′ ends. An exonuclease called RNase D removes the nucleotides at the 3′ end up to the CCA trinucleotide. Some tRNAs do not have CCA in the original transcript. Thus, after the 3′ end has been trimmed, tRNA nucleotidyltransferase is used to add CCA-3′-OH. The 5′ end is processed by a ribozyme called RNase P, which trims the 5′ end to produce the 5′ terminus in mature tRNAs. After the 3′ and 5′ ends have been trimmed, certain of the bases are enzymatically modified. (See text.)

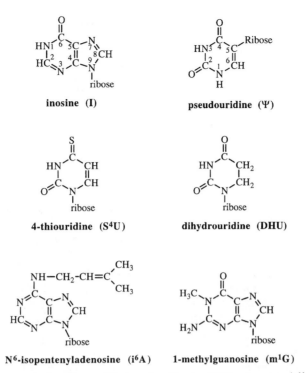

Fig. 10.31 Some post-transcriptionally modified nucleosides in tRNAs. Inosine differs from adenosine in having hypoxanthine as its base rather than adenine. The difference is that adenine has an amino group attached to C6, whereas hypoxanthine has an oxygen. Pseudouridine differs from uridine in having the ribose attached to C5 rather than N1. Thiouridine differs from uridine in having a sulfur attached to C4 rather than an oxygen. Dihydroxyuridine differs from uridine in having the ring reduced between C5 and C6. N^6-Isopentenyladenosine differs from adenosine in having an isopentenyl group attached to the amino group at the C6 position. Methylguanosine differs from guanosine in having a methyl group attached to N1.

or bind to mRNA, and it is the tRNA that is responsible for placing the correct amino acid into the polypeptide sequence. For this reason, rRNAs are called adaptor molecules. Each one has an anticodon that base pairs with the codon in the mRNA, thus ensuring that the amino acid sequence in the completed protein reflects the nucleotide sequence in the mRNA (Fig. 10.30).

Elongation

Elongation begins with the binding of an aminoacylated tRNA to a second tRNA binding site on the ribosome called the aminoacyl (A) A site. Then the α-amino group of the amino acid bound to the tRNA at the A site displaces the tRNA from the formylmethionyl–tRNA so that a dipeptidyl–tRNA is formed at the A site. (See Fig. 7.8.) This is followed by the transfer of the dipeptidyl–tRNA from the A site to the P site and the release of the initiator tRNA from the ribosome. In addition, the

ribosome moves down the mRNA in the 3′ direction so that the next codon is available at the A site for the next incoming aminoacylated tRNA. The part of chain elongation that includes the movement of the peptidyl–tRNA from the A site to the P site and the movement of the ribosome one codon down the mRNA is called *translocation*.

Termination

The process of chain elongation is repeated until a stop signal on the mRNA arrives at the A site. At this time the elongation is terminated and the polypeptide is released from the ribosome. Termination requires specific protein factors called termination factors.

Removal of the formyl group and of methionine

Most completed proteins do not have formyl methionine, or even methionine, at the N terminus. Two enzymes remove these moieties. *Peptide deformylase* removes the formyl group, and *methionine aminopeptidase* removes methionine.

Elongation rate

For *E. coli* growing at 37 °C, the polypeptide elongation rate is about 16 amino acids per second. The average protein (MW 40,000) has around 364 amino acids. This means that it takes only about 20 to make an average size protein.

Comparison to cytosolic eukaryotic protein synthesis

Overall, protein synthesis by cytosolic eukaryotic ribosomes is quite similar to protein synthesis by bacterial ribosomes. The two processes differ in certain details, however, as described in note 74.

10.3.2 Ribosomes

Ribosomes are ribonucleoprotein particles that catalyze protein synthesis (Section 1.2.6). Rapidly growing bacteria contain about 70,000 ribosomes per cell, all of which are engaged in the synthesis of protein. Each ribosome has a diameter of approximately 20 nm (200 Å) and a sedimentation coefficient of 70S. The bacterial ribosome has two subunits named after their sedimentation coefficients. These are the 50S subunit and the 30S subunit. (See Section 1.1.1 and Table 1.2 for a comparison with archaeal and eukaryotic ribosomes.) Together, the 50S and 30S subunits form the 70S ribosome, which makes the proteins.

The 30S subunit has 21 different proteins and one 16S rRNA molecule; its proteins are named S1, S2, and so on, with S standing for "small subunit." The 16S rRNA helps to align the mRNA so that it is positioned correctly on the ribosome for the initiation of protein synthesis. The 50S subunit has 31 different proteins, one 23S rRNA molecule, and one 5S rRNA molecule. Its proteins (L1, L2, etc.) are named for the large subunit. The 23S rRNA catalyzes the formation of the peptide bonds. The ribosome should not be viewed as a static structure but rather as a machine that synthesizes peptide bonds and undergoes conforma-tional changes that result in the threading of the mRNA through the ribosome (translocation) as the protein is being made.

10.3.3 Charging of the tRNA (making the aminoacyl–tRNA)

Overview

Prior to the initiation of protein synthesis on the ribosome, each amino acid must be covalently bound to a tRNA. Each of the 20 different amino acids must be activated in this way, and the process is catalyzed by 20 different enzymes called aminoacyl–tRNA synthetases.[75] The number of tRNAs can be more than 20 because the code is degenerate; that is, there exists more than one codon for certain amino acids. For example, *E. coli* can have from 1 to 6 codons for each of its amino acids, resulting in 61 codons. (However, there are fewer than 61 different tRNAs in *E. coli*. The reason for this is explained in note 76.)

The aminoacyl–tRNA synthetases recognize both the amino acid and certain structural features of the tRNA. Some synthetases recognize the anticodon, whereas others recognize nucleotide sequences in other parts of the tRNA. The process of attaching the amino acid to the tRNA is called charging the tRNA; it takes place in the cytosol, and the product is called aminoacyl–tRNA. Of particular importance is the requirement that the amino acid be attached to the right tRNA: it is the tRNA that recognizes the codon, and if the wrong amino acid is attached, then a protein with the wrong sequence of amino acids will be made.

Transfer RNA

Transfer RNAs are small single-stranded RNAs that are folded with three hairpin loops resembling a cloverleaf (Fig. 10.30). The amino acid is esterified via its carboxyl group to the 3′-OH in the ribose part of the adenylyic acid residue at the 3′ end, which is 5′-CCA-OH-3′. One of the loops holds the anticodon triplet, which hybridizes to the codon in the RNA template and thus brings the correct amino acid to the growing polypeptide chain. All tRNAs have bases that are modified after transcription during the processing of the tRNA, and these bases appear in or near the hairpin loops. The

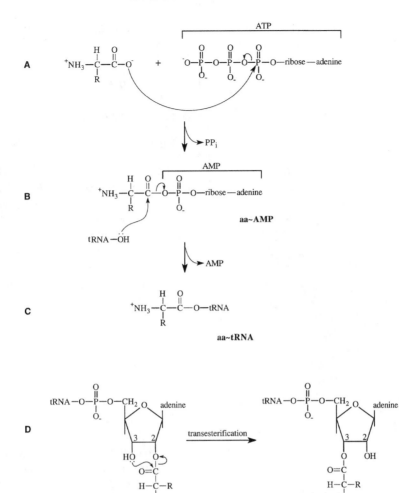

Fig. 10.32 The aminoacyl–synthetase reaction (A) The carboxyl group of the amino acid attacks the α-phosphate in ATP and displaces pyrophosphate to form the AMP derivative of the amino acid. (B) A hydroxyl on the ribose in the terminal adenylate residue attacks the carbonyl group of the amino acid and displaces the AMP. (C) The aminoacyl~tRNA. The amino acid is attached to the 3'-hydroxyl of the terminal adenylate residue in the tRNA prior to formation of the peptide bond on the ribosome. (D) Transesterification reaction. Some synthetases catalyze an attack by the 2'-hydroxyl of the terminal adenylate residue in the tRNA. If this occurs, the synthetase transfers the amino acid by a transesterification reaction to the 3'-hydroxyl. If the synthetase catalyzes an attack by the 3'-hydroxyl, a transesterification step is not necessary.

structures of some of these modified bases are shown in Fig. 10.31.

Attaching the amino acid to the tRNA (charging the tRNA)

The charging of the tRNA takes place in two steps, both of which are catalyzed by the same enzyme (Fig. 10.32). The enzyme is called aminoacyl–tRNA synthetase, and each enzyme recognizes a specific amino acid and its corresponding tRNA. (There are some examples of the same aminoacyl synthetase charging two different tRNAs, followed by the appropriate modification of the amino acid in one of the tRNAs.[77] There are also examples of two different aminoacyl–tRNA synthetases activating the same amino acid and charging the same tRNA.[78] All these cases are summarized in note 79.)

Some aminoacyl–tRNA synthetases recognize simply the anticodon region of the tRNA, which determines the specificity of the synthetase for the tRNA, whereas for other synthetases the specificity has little to do with the anticodon but is determined by other parts of the tRNA molecule. The synthesis of the aminoacyl–tRNA takes place via a route similar to the synthesis of acyl–CoA derivatives, that is, via an acyl–AMP intermediate. The first reaction is an attack of an oxygen atom of the carboxyl group of the amino acid on the α-phosphate of ATP displacing pyrophosphate from the ATP and forming an aminoacyl–AMP derivative. Then an oxygen atom of a hydroxyl group on the ribose of the tRNA attacks the aminoacyl–AMP, displacing the AMP and forming the aminoacyl–tRNA derivative. The bond in the final product is an ester linkage between the carboxyl group of the amino acid and the 3′-hydroxyl group on the ribose portion of the tRNA.[80] (See note 81 for a further description of this reaction.) As explained in Chapter 7, the reaction is driven to completion by a pyrophosphatase which hydrolyzes the pyrophosphate. (See Fig. 7.8.) Therefore, the equivalent of 2 ATPs is required to make one aminoacyl–tRNA.

Making fMet–tRNA^fMet

Making fMet–tRNA^fMet

The first amino acid in the growing polypeptide chain in bacteria is formylmethionine (fMet), and protein synthesis begins with the synthesis in the cytosol of fMet–tRNA^fMet. (The formyl group is removed after the protein has been synthesized. Sometimes the methionine is also removed.) The fMet–tRNA^fMet is made in the cytosol when methionine is attached to a special initiator tRNA (tRNA^fMet). The reaction is catalyzed by the same aminoacyl–tRNA synthetase that attaches methionine to the tRNA (tRNA^Met) responsible for inserting methionine in internal positions in the protein. Then a *transformylase* (tRNA methionyl transformylase) binds to the methionine on the Met–tRNA^fMet and catalyzes the transfer of a formyl group from N^{10}-formyltetrahydrofolate (10-formyl-THF) to the Met–tRNA^fMet to form fMet–tRNA^fMet. (See Fig. 9.14 for a description of THF as a C_1 carrier.) The transformylase recognizes not only the methionine but the tRNA^fMet as well, and is able to distinguish between tRNA^fMet and tRNA^Met. The stage is

now set for the initiation of protein synthesis on the ribosome.

10.3.4 Initiation

Making the preinitiation complex

A preinitiation complex is formed with the 30S ribosomal subunit, messenger RNA, formylmethionyl–tRNA^fMet (fMet–tRNA^fMet), and initiation factors IF-1, IF-2, and IF-3 (Fig. 10.33). First IF-3 binds to the 30S ribosomal subunit, preventing premature association between the 30S and 50S subunits. Then IF-2-GTP helps the fMet–tRNA^fMet bind to the start codon, via the anticodon on the tRNA_f, at what will become the P site when the 50S subunit binds. IF-1 is somehow involved in promoting optimal activity of the other initiation factors.

Making the 70S initiation complex

After the mRNA, fMet–tRNA^fMet, and 30S subunit have formed a complex, the 50S subunit joins to form the *70S initiation complex* as IF-1 IF-2, IF-3, GDP, and P_i are released. It is very important that the mRNA bind to the 30S subunit in the correct position to ensure that the start codon is at the P site. As described next, for many mRNAs, it is the Shine–Dalgarno sequence that positions the mRNA on the ribosome.

Shine–Dalgarno sequence

The Shine–Dalgarno sequence is a sequence of nucleotides in the initiation region of mRNA that specifies a ribosome-binding site. It is a string of about 5 to 10 bases, approximately 6 to 10 nucleotides downstream of the start codon in many mRNAs (Fig. 10.34). The Shine–Dalgarno sequence in the mRNA base pairs to a complementary sequence at the 3′ region of the 16S rRNA. This places the start codon in the mRNA at the P site in the ribosome, which is where fMet–tRNA^fMet will bind. Not all mRNAs have a Shine–Dalgarno sequence, and in fact the start codon may be at the extreme 5′ end of the mRNA. Presumably, under the latter circumstances, the sequence in the mRNA that binds to the 16S rRNA to position the start codon in the mRNA at the P site is downstream of the start codon.

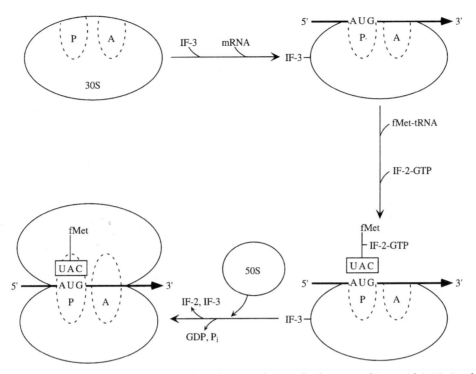

Fig. 10.33 Formation of the initiation complex. The 30S ribosomal subunit combines with initiating factor IF-3 and mRNA to form the 30S initiation complex. There are two sites in the 30S subunit, the A site and the P site, which are completed upon binding of the 50S subunit. The mRNA is positioned such that its start codon (AUG for formylmethionine) is at the P site. The mRNA binds correctly at the P site because near the 5′ end of the mRNA there is a consensus sequence called the Shine–Dalgarno sequence, and this sequence base pairs to a complementary sequence near the 3′ end of the 16S RNA in the 30S subunit. Then the initiating aminoacylated tRNA (i.e., fMet–tRNA^{fMet}) binds to initiating factor, IF-2 which itself is complexed to GTP (IF2–GTP), and the ternary complex binds to the 30S subunit such that the anticodon in the fMet–tRNA hydrogen-bonds to the AUG at the P site. The 30S initiation complex then binds to the 50S ribosomal subunit to form the 70S initiation complex. When this happens, the GTP is hydrolyzed to GDP and P_i, and the GDP, P_i, and initation factors leave the ribosome. The role of IF-3 is to prevent premature binding of the 50S subunit to the 30S subunit. IF-2 is required for correct binding of the fMet-tRNA^{fMet} to the 30S subunit. A third initiation factor, IF-1, is also present. It stimulates the activities of IF-3 and IF-2. As mentioned, the A and P sites involve both the 30S and the 50S ribosomal subunits.

Fig. 10.34 Binding of initiator region of mRNA to 16S rRNA. Upstream from the initiation codon in the mRNA there is a short sequence of nucleotides called the Shine–Dalgarno sequence. The Shine–Dalgarno sequence hybridizes to a complementary sequence of nucleotides at the 3′ end of the 16S rRNA, and this places the initiation codon at the P site, where the fMet–tRNA^{fMet} binds. Shine–Dalgarno sequences vary and can be complementary to different regions of the 16S rRNA. Some bacterial genes do not have a Shine–Dalgarno sequence; that is, the initiation codon is at the very end of the mRNA, as revealed by sequencing the N-terminal region of the protein and comparing it with the nucleotide sequence at the 5′ end of the mRNA. Presumably, mRNAs that do not have a Shine–Dalgarno sequence upstream of the initiation codon have a 16S rRNA binding site downstream of the initiation codon.

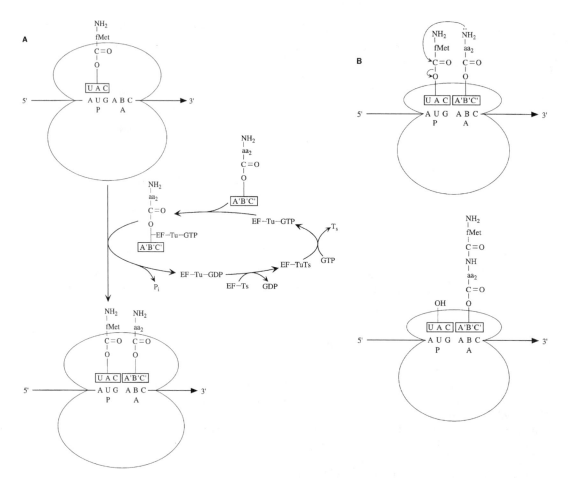

Fig. 10.35 Chain elongation and termination. (A) Binding of aminoacylated tRNA to the A site. Elongation factor Tu bound to GTP forms a ternary complex with the aminoacylated tRNA, and this complex binds to the ribosome such that the anticodon of the tRNA base pairs with the codon at the A site. The GTP is hydrolyzed to GDP and P_i and the Tu–GDP and P_i are released from the ribosome. The GDP is then displaced from the Tu–GDP complex by elongation factor Ts, and the Ts is subsequently displaced from the complex by GTP to regenerate Tu–GTP which can then bind to the next aminoacylated tRNA. (B) Formation of the peptide bond. The enzyme peptidyltransferase, which is actually the 23S RNA in the 50S ribosomal subunit, catalyzes the displacement of the tRNA at the P site by the free amino group of the incoming amino acid. The result is that the growing peptide is transferred from the P site to the A site. (RNA molecules such as the 23S RNA that act as enzymes are called ribozymes.) (C) Translocation. The ribosome moves one codon towards the 3' end of the mRNA as the tRNA in the P site is released. This positions the growing peptide at the P site and leaves an empty A site for the incoming aminoacylated tRNA. Translocation requires elongation factor G (EF-G) as well as energy from the hydrolysis of GTP to GDP and P_i. (D) Termination. When a stop codon reaches the A site, a protein release factor (RF) binds to the termination codon and stimulates peptidyltransferase to hydrolyze the linkage between the peptide and the tRNA at the P site. The ribosome dissociates into its subunits, and the newly synthesized protein is released. IF-3 binds to the 30S subunit to prevent its premature reassociation with the 50S subunit.

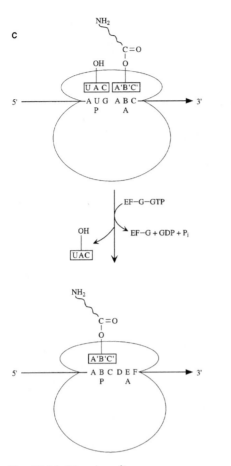

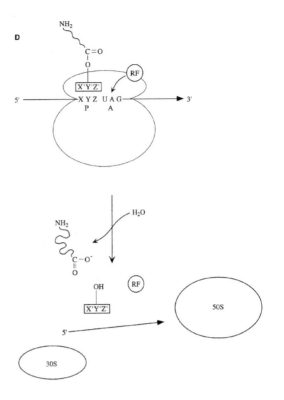

Fig. 10.35 (*Continued*)

10.3.5 Chain elongation

Start codon

Translation usually begins at an AUG codon (the start codon), which codes for methionine in the *initiator region of the mRNA*. [In a relatively small number of cases the fMet is encoded by the valine codon (GUG) or by the leucine codons (CUG, UUG) if they are at the start site. These codons encode valine or leucine, respectively, if they are downstream from the start site.] There may be internal (downstream) start codons, but these do not serve as translation initiation sites.

Leader sequence

RNA transcription may begin some distance upstream of the translational start codon. Under these circumstances an untranslated region at the 5′ end of the mRNA exists. It is called the *leader sequence*.

Binding of aminoacylated tRNA to the A site

Chain elongation begins when an incoming aminoacylated tRNA binds to the A site on the ribosome as the anticodon on the tRNA base pairs with the codon in the mRNA. The correct positioning of the incoming aminoacyl–tRNA is facilitated by elongation factor Tu bound to GTP (i.e., EF-Tu–GTP), which forms a complex with the incoming aminoacyl–tRNA and guides the aminoacyl–tRNA to the A site (Fig. 10.35A).

Recycling of EF-Tu–GTP

After the aminoacyl–tRNA has arrived at the A site, GTP is hydrolyzed and EF-Tu–GDP is released from the ribosome. The EF-Tu–GDP then reacts with EF-Ts, which stimulates the exchange of GDP for GTP so that the EF-Tu–GTP is regenerated.

Formation of the peptide bond

A peptide bond forms when the free α-amino group on the amino acid bound to the tRNA at the A site displaces the tRNA$_f$ at the P site (Fig. 10.35B). The reaction is catalyzed by peptidyltransferase, which is part of the 50S ribosomal subunit. Very interestingly, the peptidyltransferase activity appears to reside in the 23S rRNA rather than in a protein.[82] (For other examples of RNA molecules acting as enzymes, see note 83.) The result is a peptide attached to the tRNA at the A site.

The formation of the peptide bond as the ester linkage is broken between the amino acid and the tRNA does not require any additional energy. Recall that ATP was used to form the ester linkage in the first place. Thus, the energy to make the peptide bond is derived from ATP. In fact, the formation of the peptide bond uses the equivalent of two ATPs, since the formation of the aminoacyl–tRNA is driven to completion by the hydrolysis of pyrophosphate.

Translocation

The formation of the peptide bond is followed by translocation (Fig. 10.35C). During translocation, the uncharged tRNA leaves the P site, the peptidyl–tRNA moves from the A site to the P site, and the mRNA moves one codon with respect to the ribosome so that a new codon is positioned at the A site. The elongation factor EF-G (translocase) is required for translocation. During translocation, GTP is hydrolyzed to GDP and P$_i$. The empty A site is now ready to receive a new aminoacyl–tRNA. Notice that the ribosome moves down the message from the 5′ end to the 3′ end and that the protein grows at the carboxy end.

Termination

Termination occurs when a stop codon reaches the A site (Fig. 10.35D). The stop codons are UAA, UGA, and UAG. There are no tRNAs with anticodons for these codons, and because of this no aminoacyl–tRNA enters the A site when a stop codon is present. Release factors (proteins) bind to the stop codons and cause the release of the completed polypeptide from the ribosome. There are three release factors in E. coli: RF1, RF2, and RF3. The first release factor binds to UAA and UAG; RF2 binds to UAA and UGA, and RF3 facilitates the activity of RF1 and RF2. It has been suggested that release factors cause the peptidyltransferase to catalyze an attack by water on the ester linkage between the peptide and the tRNA at the P site, releasing the protein. After the release of the protein, the tRNA and mRNA are released, and the ribosome dissociates into its subunits. The initiation factor IF3 is important at this stage because it prevents the 30S and 50S subunits from reassociating before the next initiation complex has formed.

The requirement for GTP

Guanosine triphosphate provides the energy for translocation. It is also necessary for the binding of certain proteins to either tRNA or to the ribosome. How does GTP act? GTP does not act in an analogous manner to ATP. Rather, in contrast to ATP, which drives the formation of covalent bonds, GTP is thought to activate proteins by causing a conformational change in the protein. One result of this conformation change is that the protein may then be in an active shape, enabling it to bind to certain targets. Thus, the hydrolysis of GTP is associated with the release of GDP and P$_i$ from protein and its return to the inactive state, rather than with the activation of protein.

10.3.6 Polysomes: Coupled transcription and translation

Messenger RNAs are translated by several ribosomes (typically around 50) in series, ensuring that several copies of the protein can be made from a single mRNA. The complex of multiple ribosomes attached to an mRNA is called a *polysome*. This arrangement increases the rate of protein synthesis, since the amount of protein made per unit time from a particular mRNA is proportional to the number of ribosomes translating the mRNA.

Furthermore, translation and transcription are coupled in bacteria. What happens is that ribosomes attach to the 5′ end of the mRNA even before the message is completely transcribed; then they move along the mRNA toward the 3′ end as transcription continues (Fig. 10.36). Because translation is coupled to

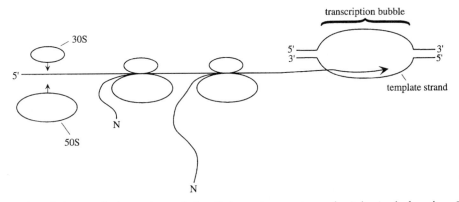

Fig. 10.36 Coupled transcription and translation. In bacteria, protein synthesis begins before the mRNA is completed. The result is a string of ribosomes on the same mRNA. There may be up to 100 ribosomes on the same mRNA, comprising what is called a polysome.

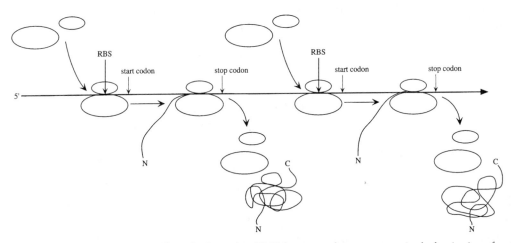

Fig. 10.37 Polycistronic mRNA. The polycistronic mRNA has more than one gene. At the beginning of each gene there is a ribosome-binding site (RBS) at which initiation begins. The ribosome then begins translating at the initiation codon and stops at the stop codon at the end of the gene. Upon termination, the ribosome and the polypeptide are released, and initiation begins anew at the beginning of the next gene. Between the genes there is an intercistronic spacer region, which may have as few as one or as many as 30 to 40 bases. The amino and carboxy terminals of the polypeptide are labeled N and C, respectively.

transcription, proteins can begin to be made even before the mRNA is completed, thus shortening the delay between the onset of transcription and the appearance of the protein. Coupled transcription and translation cannot take place in eukaryotes, since transcription and translation occur in separate cell compartments (i.e., the nucleus and cytoplasm, respectively).

10.3.7 Polycistronic messages

Bacteria frequently make polycistronic messages. In this kind of messaging, which eukaryotes do not engage in, several genes are transcribed in series into a single mRNA molecule. Initiation of transcription of the cotranscribed genes is from a single promoter upstream of the gene cluster, and the cluster of genes is called an *operon*. Usually genes that encode proteins that function in a common pathway are in the same operon. Operons in bacteria are the rule, and they may have as few as 2 genes or as many as 10 or more.

For individual proteins to be made from a polycistronic message, there must be translational start sites at the begining of each gene transcript and translational stop sites at the end (Fig. 10.37). Furthermore, there must be initiating regions (i.e., ribosome-binding sites) at the begining of each gene transcript. Thus, the translation initiation complex forms at the beginning of each gene transcript in the polycistronic message, and the ribosome, mRNA, tRNA, and polypeptide dissociate at the termination site of each gene transcript in the polycistronic message.

10.3.8 Polarity

Polarity refers to the phenomenon in polycistronic mRNAs in which a block in the translation of an upstream gene can prevent *transcription* of a downstream gene. This can happen if, for example, a mutation results in a nonsense codon (i.e., a codon that does not bind to a tRNA) in the upstream gene and there is a Rho-dependent transcription termination site in the upstream gene after the translational stop site. Under these circumstances the RNA polymerase will disengage from the DNA and transcription will stop before the polymerase has reached the downstream gene. For a discussion of Rho-dependent termination of transcription see Section 10.2.1.

10.3.9 Coupled translation

In some instances translation can be coupled. This means that in a polycistronic mRNA, translation of the immediate upstream gene is required for translation of the downstream gene. For example, the translation of the genes in the operons coding for ribosomal proteins in *E. coli* is coupled. This affords the opportunity to regulate the translation of the entire operon by regulating the translation of one of the genes. (See note 22 in Chapter 2 for a further discussion of the regulation of translation of the ribosomal protein operons.) A model for translational coupling postulates that the translational start region, including the start codon, for the downstream gene is inside a hairpin loop in the mRNA so that it cannot bind to a ribosome. When the ribosome translating the upstream gene comes to the stop codon of the upstream gene, the ribosome may disrupt the secondary structure of the mRNA, making the translational start region for the downstream gene accessible to a second ribosome (Fig. 10.38).

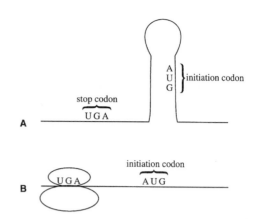

Fig. 10.38 Model for translational coupling. (A) It is postulated that in a polycistronic mRNA, the start codon for the downstream gene is in a hairpin loop, which thus it is not accessible to the ribosome. (B) When the ribosome translating the upstream gene arrives at the stop codon, it disrupts the secondary structure of the mRNA, giving a second ribosome access to the initiation site in the downstream gene.

10.3.10 Folding of newly synthesized proteins: The role of chaperone proteins

Chaperone proteins are proteins that temporarily bind to other proteins and, depending upon the chaperone protein, aid in proper folding of newly made proteins, transport of proteins to and across membranes (Sections 17.1.2 and 19.1.6), and refolding of proteins that have become unfolded as a consequence of stress such as heat shock. For reviews, the student should consult refs. 84 and 85.

The role of chaperone proteins in folding newly synthesized proteins

The information to fold into the active tertiary structure is contained in the sequence of amino acids, that is, the primary structure of the protein. (See note 86 for definitions of primary, secondary, tertiary, and quaternary structures of proteins.) However, the proper folding of many newly synthesized proteins cannot proceed without chaperone proteins, which transiently associate with the newly synthesized proteins and assist in the folding.

One role of chaperone proteins is to prevent misfolding and aggregation. Aggregation will occur if hydrophobic regions of the newly synthesized, partially unfolded proteins remain at the aqueous interface, where protein–protein interactions can occur, rather than moving into a hydrophobic core during folding. (Electrostatic repulsion between proteins with a net charge decreases the propensity for aggregation.) It has been estimated that at least 50% of newly synthesized proteins are aided in their folding by chaperone proteins that ensure rapid folding. Some proteins, however, fold independently of chaperone proteins. These are probably rapidly folding small proteins.

In addition to helping newly synthesized proteins to rapidly fold, there are chaperone proteins that prevent the N-terminal portions of nascent polypeptides, emerging from the ribosomes, from misfolding and from aggregating with identical nascent chains being synthesized on neighboring ribosomes in a polysome.

Measuring the flux of proteins through the chaperonins

One way to quantitate the flux of newly synthesized proteins through the chaperone proteins is to incubate the bacteria for a few seconds with a radioactive amino acid and then add a large excess of the same amino acid is nonradioactive form. The addition of the nonradioactive amino acid lowers the specific activity of the radioactive amino acid sufficiently to halt further incorporation of radioactivity. This is called "pulse-chase" measurement. A specific chaperone protein can be immunoprecipitated from cell extracts at various times during the "chase," and the different radioactive proteins transiently associated with the chaperone protein, as well as the transit time, can be visualized by autoradiography after gel electrophoresis. The flux of proteins through the chaperone protein can be estimated by measuring the radioactivity in individual protein bands separated by gel electrophoresis, as well as the total amount of radioactivity that was immunoprecipitated during the "chase."

GroEL/GroES, the GroE system

In experiments such as the ones just described, E. coli and anti-GroEL antibody were used to monitor the flux of proteins through the GroEL chaperonin, which cooperates with GroES as a complex, called the GroE complex, to produce proper folding.[87] (Chaperone proteins such as GroEL and GroES exist as multisubunit complexes called *chaperonins*. See note 88 for a more complete description of GroEL and GroES.) The major conclusions from the pulse-chase experiment were as follows.

1. A variety of newly synthesized proteins (as judged by the number of bands that migrated during gel electrophoresis) transiently associate with GroEL, and this can account for 10 to 15% of the protein during normal growth temperatures.

2. Upon exposure to heat stress, during which the relative amounts of GroEL and other chaperonins increase in the cell, at least 30% of the newly synthesized proteins associate with GroEL. (See the discussion of chaperonins and heat-shock proteins in Section 19.1.6.)

It appears that GroEL binds to the protein *after* synthesis is complete. This conclusion is justified by noting that when the polypeptides bound to GroEL were examined, to determine their size distribution, they were found to be of distinct sizes, rather than a continuum, as might be expected if nascent polypeptide chains bound to the chaperonin. However, although GroEL becomes associated with polypeptides after their release from ribosomes, it appears that three other chaperone proteins, DnaK, DnaJ, and GrpE, associate with nascent polypeptides while on the ribosome.

A model: First DnaK–DnaJ–GrpE, then GroE

There is evidence that a chaperone complex of DnaK, DnaJ, and GrpE can interact with a least some nascent polypeptides on ribosomes and that GroE (GroEL and GroES) binds to the polypepeptides after their release from ribosomes.[89] (See note 90 for a description of the experiments that support this conclusion and Section 19.1.6 for the role of these proteins, as well as the GroE system, in the heat-shock response.) Perhaps what happens is that the DnaK chaperone complex prevents the polypeptides from premature folding and aggregation while they are being synthesized, and releases them in an unfolded form. Then, the actual folding of the completed polypeptide takes place when it is transferred to a cavity in GroEL. GroES is a protein that caps the cavity. (See ref. 91 for a review.)

As described next, another chaperone protein, called trigger factor (encoded by the *tig* gene), is also ribosome bound and attaches to nascent polypeptides as they emerge from the ribosome, perhaps prior to DnaK. Trigger factor can replace DnaK in *dnaK* Δ mutants, and DnaK can replace trigger factor in *tig* Δ mutants. Thus they have overlapping functions.

Trigger factor

A ribosome-associated protein called *trigger factor* (TF) has been implicated in the folding of certain nascent proteins as they emerge from the ribosome, and it may also take part in transferring nascent proteins to the GroEL complex.[92] Trigger factor is believed to be functionally redundant with DnaK because mutants of *E. coli* that are heat sensitive and do not make trigger factors have no defects in growth at intermediate temperatures or in protein folding as long as the bacteria make DnaK.[93] Likewise, mutants that do not make DnaK are fine as long as they make trigger factor. In the absence of trigger factor, there is a two- to threefold increase in the amount of nascent polypeptide chains that bind to DnaK.

In addition to its chaperone activity, trigger factor is a peptidyl–prolyl *cis/trans* isomerase. It is attached to the ribosomal 50S subunit of bacterial ribosomes. It binds to nascent proteins on the ribosomes and promotes proline-limited folding of nascent proteins, probably cotranslationally. Trigger factor catalyzes the cis–trans isomerization of the peptide bond that is on the N-terminal side of proline residues.

Normally the hydrogen attached to the substituted amino group in peptide bonds is opposite (trans) to the oxygen of the carbonyl group. This is the most stable configuration. However, when the nitrogen atom is contributed by proline, the peptide linkage is occasionally cis (in about 5% of cases) in native proteins. Probably a certain proportion of the prolyl bonds must isomerize from trans to cis for proper folding to be initiated.

Actually, there are two distinct sites on trigger factor that bind to substrate proteins. One site, the proline residue site, catalyzes prolyl isomerizations. The second site lies elsewhere on trigger factor and binds in a nonspecific fashion to the unfolded portions of newly synthesized proteins. In fact, model experiments have demonstrated that unfolded proteins will bind to trigger factor regardless of the presence of proline residues; that is, the nonproline sites are sufficient. The binding to the nonproline residue sites is probably important in aiding the proline-limited folding of the proteins. One can imagine that binding to the unfolded protein chains might block folding, and in this way allow trigger factor to isomerize the proline residues. Binding to the unfolded protein chains may also be part of a trigger factor chaperone function that is important to prevent aggregation or degradation of unfolded nascent protein chains.[94,95] For a review of prolyl isomerases, see ref. 96.

10.3.11 Inhibitors

Some antibiotic inhibitors of protein synthesis and their sites of action are listed in Table 10.1. Note that chloramphenicol, erythromycin, and streptomycin inhibit bacterial but not archaeal or eukaryotic (cytosolic) protein synthesis, and that diphtheria toxin does not inhibit bacterial but does inhibit archaeal and eukaryotic protein synthesis. These properties reflect differences in the ribosomal and accessory proteins required for translation in these systems.

10.4 Summary

To replicate DNA, several problems must be attended to. These include unwinding the double helix without causing it to overwind in a positive supercoil in the unreplicated portion, creating a primer that can be extended by DNA polymerase III, and keeping the polymerase that synthesizes the lagging strand at the replication fork. These and other factors necessitate using about 20 different proteins to replicate the DNA. The replication fork is created at a specific sequence in the DNA (called the origin) when DNaA, HU, and ATP begin to unwind the helix. Further unwinding of the DNA during replication requires replication fork helicase (DnaB). Single-stranded binding protein binds to the single strands to prevent them from coming together again, thus ensuring that they can serve as templates for new DNA. Meanwhile, DNA gyrase binds to the DNA downstream of the replication fork and

Table 10.1 Antibiotics that inhibit protein synthesis

Antibiotic	Site of action	Bacteria	Archaea	Eukaryotes
Erythromycin	Inhibits translocation	+	–	–
Streptomycin[a]	Inhibits initiation and causes mistakes in reading mRNA at low concentrations	+	–	–
Tetracycline[b]	Inhibits binding of aminoacyl–tRNA	+	+	+
Chloramphenicol	Inhibits peptidyltransferase	+	–	–
Puromycin[c]	Resembles charged tRNA and causes premature chain termination	+	+	+
Cycloheximide	Inhibits peptidyltransferase	–	–	+
Diphtheria toxin	Inhibits elongation factor 2 (EF-2), which is analogous to bacterial EF-G, by ADP ribosylation	–	+	+

[a] Streptomycin binds to the S12 protein in the 30S ribosomal subunit and inhibits the binding of fMet–tRNA^{fMet} to the P site. Neomycin and kanamycin act in a manner similar to streptomycin.
[b] Tetracyclines inhibit bacterial protein synthesis by binding to the 30S ribosomal subunit and preventing attachment of aminoacylated tRNA. They also inhibit eukaryotic ribosomes, but not at the concentrations used to treat bacterial infections. This is apparently because eukaryotic membranes are not very permeable to tetracycline.
[c] Puromycin resembles the aminoacyl portion of aminoacylated tRNA. It has a free amino group that attacks the ester linkage in the growing peptide in the P site, with the result that a peptidyl puromycin forms at the A site. However, the peptidyl puromycin cannot be transferred to the P site (i.e., translocation does not occur), and therefore chain elongation is terminated as the peptidyl puromycin dissociates from the ribosome.

continuously converts the positive supercoils into negative supercoils so that the DNA does not become overwound.

Replication of the DNA is initiated when the enzyme primase synthesizes a short RNA oligonucleotide at the replication fork on both the single strands. One copy strand, called the leading strand, is synthesized continuously by elongating the 3′ end of the RNA primer, the end that faces the replication fork. The other copy strand, called the lagging strand, is synthesized in short fragments of about 1,000 nucleotides called Okazaki fragments, growing in the direction opposite to the movement of the replication fork. Each fragment begins with a short oligonucleotide RNA primer. The 3′ end of the lagging strand faces away from the replication fork. The same DNA polymerase would be able to synthesize both copy strands if the polymerase were dimeric and the template for the lagging strand looped around the polymerase so that its 5′ end faced the replication fork. The RNA primers at the 5′ end of the copy strands are removed by DNA polymerase I, which also fills the gap. The break is sealed by DNA ligase. Replication is usually bidirectional.

Termination occurs at *ter* sites and requires Tus protein. The enzyme topoisomerase IV separates the newly synthesized duplexes. In the event that an unequal number of recombinations have taken place between sister chromosomes during replication, the daughter chromosomes are covalently linked and a site-specific recombination must take place at the *dif* site, which is in the Ter region.

The process of partitioning nucleoids to opposite poles of the cells is not understood. One model posits a stationary, centrally located replicon that provides the force to partition sister chromosomes to opposite poles. Additionally, several different proteins outside the replisome appear to be involved in chromosome positioning and partitioning. These include the SMC, Muk, Par, SetB, and MreB proteins. The MreB proteins are related to actin.

Errors made during replication are repaired in two ways. Editing repair involves the immediate removal of the incorrect nucleotide by the 3′-exonuclease activity of DNA polymerase III. The second method of repair, called mismatch repair, can occur if the polymerase fails to remove the incorrect nucleotide and proceeds beyond the error site. During mismatch repair an exonuclease removes the segment of copy DNA that includes the mismatched nucleotide pair, and the gap is filled with DNA polymerase III and sealed with DNA ligase. The copy strand is recognized because it is undermethylated for a short period after synthesis.

RNA synthesis also requires localized unwinding of the DNA duplex, but the unwinding is done by RNA polymerase rather than helicase.

Initiation takes place at promoter sites, which are nucleotide sequences in which RNA polymerase binds upstream of the transcription start site. The promoter is recognized by a subunit in the polymerase called sigma factor. The sigma 70 polymerase recognizes consensus sequences at the −10 and −35 regions where the transcription start site is +1. Other sigma factors exist but do not recognize the same sequence recognized by sigma 70.

DNA gyrase is required for transcription, to prevent the DNA from being overwound downstream from the transcribed region. Special termination codons exist that signal the end of the gene and the release of the RNA polymerase. Transcription is regulated by a variety of positive and negative transcription factors that bind to the DNA in the promoter region or upstream of the promoter (enhancer regions). The transcription factors can influence the binding of the RNA polymerase to the promoter or the rate at which the open complex forms. In many operons, RNA synthesis is also regulated by attenuation.

Protein synthesis begins with the aminoacylation of tRNA^{fMet} in the cytosol. The enzymes that aminoacylate the tRNAs are called aminoacyl–tRNA synthetases. The tRNA is an adaptor molecule, and its anticodon base pairs with the codon in the mRNA, ensuring that the amino acid is placed in the correct sequence. The aminoacyl–tRNA synthetases that attach the amino acids to the tRNA distinguish between the different tRNAs and attach the amino acid only to the tRNA with the correct anticodon for that amino acid. Then, in the presence of initiation factors IF-1, IF-2, and IF-3, as well as mRNA and the 30S and 50S ribosomal subunits, the 70S initiation complex is formed. In this complex the fMet–tRNA^{fMet} is positioned at the P site in the ribosome. This positioning is directed by a sequence of nucleotides called the Shine–Dalgarno sequence, 6 to 10 nucleotides downstream of the transcription start codon.

The Shine–Dalgarno sequence hybridizes with the 3′ end of the 16S rRNA in the 30S ribosomal subunit and, because of this, the start codon is positioned at the P site on the ribosome. The next aminoacylated tRNA enters and is positioned at the A site. This requires EF-Tu and GTP. Once the P site and the A site

have been occupied by aminoacylated tRNAs, the α-amino group of the amino acid in the A site displaces the tRNA at the P site and an aminoacylated peptide is made at the A site.

After formation of the aminoacylated peptide, translocation takes place. During translocation, the ribosome moves one codon with respect to the mRNA and the aminoacylated peptide is transferred from the A site to the P site, thus enabling the A site to receive the next aminoacylated tRNA. Translocation requires elongation factor EF-G and GTP.

The process is repeated until the ribosome reaches the end of the gene. At the end of the gene there is a termination codon for which there is no tRNA. When the A site of the ribosome reaches the termination codon, protein termination factors cause peptidyltransferase to hydrolyze the linkage between the protein and the tRNA in the P site, thus releasing the protein. The ribosome also dissociates into its subunits, which can then initiate a new round of translation at the beginning of a different gene.

Once synthesized, the protein folds into its tertiary structure. This process is frequently aided by a class of proteins called chaperone proteins.

Study Questions

1. Describe the roles of the following proteins in DNA replication: helicase, DNA gyrase, DNA ligase, DNA polymerase III, DNA polymerase I, primase, DnaA, SSB.

2. Compare the biochemical reactions catalyzed by DNA polymerase and DNA ligase. How do they differ?

3. Describe the Messelson–Stahl experiment, and explain how it proves that DNA is replicated semiconservatively.

4. DNA exists in a negative supercoil. How might this be advantageous to DNA replication and RNA synthesis? During DNA replication the unreplicated portion of the DNA winds tighter. Why is this the case?

5. What are Okazaki fragments, and why are they necessary?

6. Outline an experimental approach to determine whether DNA synthesis takes

place in a centrally located "factory," and how the origins and termini move during partitioning.

7. Why is the 3′-exonuclease activity of DNA polymerase III important?

8. What is the role of the sigma subunit in RNA synthesis? Under what circumstances might a bacterium use different sigma factors?

9. During RNA synthesis, what unwinds the DNA? Compare this process with the unwinding during DNA replication.

10. How might transcription factors increase or decrease the rates of transcription?

11. How is RNA synthesis terminated?

12. The addition of tryptophan to the growth medium of an *E. coli trpR* mutant lowers the expression of the *trp* operon. What is the explanation for this?

13. What is a termination codon? How is it involved in the termination of protein synthesis?

14. What is meant by ribozyme? Give an example of a ribozyme in protein synthesis.

15. What ensures that the correct amino acid is added to the correct tRNA? If an incorrect amino acid is added, can the mistake be corrected?

16. Describe the roles of IF-1, IF-2, IF-3, and EF-G in protein synthesis.

17. What is the direction in which mRNA is translated (i.e., 5′ to 3′ or 3′ to 5′)? How is that relevant to the coupling between transcription and translation?

18. Explain the role of chaperone proteins in protein folding. What is the evidence that the binding to chaperone proteins is transient?

19. What is the experimental evidence for a stationary replisome in *B. subtilis* and *E. coli*? What is the evidence that the origins migrate to the cell poles?

20. Describe an experiment in which antibodies can be used to visualize ParA and ParB.

REFERENCES AND NOTES

1. Most bacteria have circular chromosomes. Exceptions include *Borrelia burgdorferi*, which causes Lyme disease, *Rhodococcus fasciens*, and *Streptomyces lividans*. See: Hinnebursch, J., and K. Tilly. 1993. Linear plasmids and chromosomes in bacteria. *Mol. Microbiol.* **10**:917–922.

2. Supercoiling of DNA is an important aspect of its structure in part because it makes the DNA more compact. Consider DNA as consisting of two strands coiled around a central axis. Supercoiling refers to the twisting or coiling of the central axis. This is analogous to a coiled telephone cord that twists on itself to form additional coils. Suppose the DNA exists as a closed double-helical circle. If one strand is cut and unwound while the other strand is prevented from turning, and then the cut is sealed, the circular DNA will have fewer helical turns. Moreover, the DNA will then be under structural strain and, as a consequence, will supercoil to relieve the strain. If the DNA is underwound (i.e., fewer helical turns), it will twist on itself into *negative* (left-handed) supercoils to relieve the stress. If the DNA is overwound (i.e., more helical turns are introduced before the cut is sealed), it will twist upon itself to produce *positive* (right-handed) supercoils to relieve the strain. Most cellular DNA is underwound and negatively supercoiled. It is believed that the strands of the underwound DNA are more readily separated for replication and transcription. The production of negative supercoils is accomplished by a topoisomerase II enzyme called DNA gyrase and does not occur by unwinding the cut strands before rejoining them. (See text for how DNA gyrase introduces negative supercoils.)

3. Topoisomerases change the linking number of a covalently closed duplex of DNA. The linking number is the number of times that the chains in the duplex of a covalently closed circle cross one another; that is, it is the number of helical turns. As noted in the text, DNA is said to be overwound when it has more than 10.5 base pairs per helical turn and underwound when it has fewer than 10.5 base pairs per helical turn. The original model proposed that the topoisomerase cut one strand of the duplex and, while holding both ends of the nicked strands, rotated a free end around the intact strand, thus changing the number of helical turns. This was followed by sealing the break. A more recent model suggests that rotation is not necessary, and the number of helical turns can be changed simply by passing the unbroken strand through the break in the complementary strand. An example is type I topoisomerase. The type I topoisomerases catalyze a break in one of the strands, which allows the passage through the break of the unbroken partner strand. The break is then sealed. This results in changing the linking number (number of helical turns) by multiples of one. Type II topoisomerases do it somewhat differently and change the linking number by multiples

of 2. Type II topoisomerases catalyze breakage of *both* strands of the duplex and allow the passage of *two* unbroken complementary strands from a nearby region of the molecule through the breaks before sealing. (The breaks are staggered breaks, i.e., not directly opposite each other.) Because type II topoisomerases change the number of times two strands are crossed per catalytic event, they alter the linking number in multiples of 2. An example of a type II topoisomerase is DNA gyrase from *E. coli*. It produces negative supercoils, unwinding the double helix in the following way: (1) It binds to the DNA, wrapping approximately 120 base pairs of duplex around itself. (2) It cleaves both strands with a four-base-pair stagger. When it does this, the enzyme covalently bonds to the 5′ ends of both strands. (3) It moves the uncut complementary strands through the break. The region that is moved is either within the region that is wrapped around the enzyme or close by. (4) It reseals the breaks. Passing the unbroken portion through the broken portion to produce negative supercoils unwinds the DNA. ATP is required for DNA gyrase activity.

To see how negative supercoiling can unwind a helix, take two pieces of rubber tubing and wrap them around each other in a right-hand helix. This would be clockwise, sighting down the helix away from you. Clamp both ends. Now twist the linear helix so that the central axis twists counterclockwise. This produces negative supercoils. The helix will unwind. Twisting the helix so that the central axis turns clockwise (positive supercoils) makes the helix wind tighter.

Topoisomerase I and DNA gyrase adjust the supercoiling of DNA. Gyrase adds negative supercoils and topoisomerase I removes negative supercoils. For a review of topoisomerases, see: Luttinger, A. 1995. The twisted "life" of DNA in the cell: bacterial topoisomerases. *Mol. Microbiol.* **15**:601–608.

4. Marians, K. J. 1992. Prokaryotic DNA replication. *Annu. Rev. Biochem.* **61**:673–719.

5. Baker, T. A., and S. H. Wickner. 1992. Genetics and enzymology of DNA replication in *Escherichia coli*. *Annu. Rev. Gen.* **26**:447–477.

6. Messer, W., and C. Weigel. 1996. Initiation of chromosome replication, pp. 1579–1601. In: *Escherichia coli and Salmonella: Cellular and Molecular Biology*, Vol. 1. F. C. Neidhardt et al. (Eds.). ASM Press, Washington, DC.

7. Marians, K. J. 1996. Replication fork propagation, pp. 749–763. In: *Escherichia coli and Salmonella: Cellular and Molecular Biology*, Vol. 1. F. G. Neidhardt et al. (Eds.). ASM Press, Washington, DC.

8. Ryan, V. T., J. E. Grimwade, J. E. Camara, E. Crooke, and A. C. Leonard. 2004. *Escherichia coli* pre-replication complex assembly is regulated by dynamic interplay among Fis, IHF, and DnaA. *Mol. Microbiol.* **51**:1347–1359.

9. Leonard, A. C., and J. E. Grimwade. 2005. Building a bacterial orisome: emergence of new regulator features for replication origin unwinding. *Mol. Microbiol.* **55**:1365–2958.

10. Bidirectional replication can be demonstrated in two ways. One way is to insert into the chromosome prophage I at the *att* site and prophage Mu within various mapped genes around the chromosome. The Mu can be located because when it inserts into a gene, it causes a mutation in that gene, and the location of the gene is known. If such mutants are also auxotrophic for certain amino acids, the experimenter can stop the initiation of new rounds of replication by removing the required amino acid. By adding the amino acid back again, the experimenter can initiate new rounds of replication at the replication origin. If new rounds of replication are begun with bromouracil in the medium, then bromouracil is incorporated into the DNA in place of thymine. The presence of the bromouracil makes the newly synthesized DNA denser than the parental strand. The newly synthesized strands can then be separated at various times after the initiation of replication by density gradient centrifugation and hybridized with both prophages, I and Mu, whereupon the ratio of Mu DNA to I DNA can be measured. Knowing the map position of the various Mu inserts, it is possible to demonstrate that replication proceeds bidirectionally from the origin of replication. Bidirectional replication forks have also been observed by using radioautography and electron microscopy after a pulse of tritiated thymidine. Both replication forks can be seen to be labeled when replication is examined in several bacteria. If DNA replication were unidirectional, only one replication fork would be labeled.

11. This note summarized what is known about the regulation of the initiation of DNA replication in *E. coli*.

Several DNA-binding proteins are involved. One critical protein is DnaA, which binds to the origin first, begins the process of strand separation, and then recruits helicase. The active form of DnaA that initiates unwinding is DnaA–ATP. The binding of ATP to DnaA promotes an allosteric modification in DnaA, allowing it to function. Nonhydrolyzable analogues of ATP also work. After the initiation of DNA replication, DnaA hydrolyzes the bound ATP to bound ADP. DnaA–ADP is inactive in initiating DNA replication. When the activity of DnaA in the cell is sufficiently high, DNA replication is initiated. One line of evidence for the role of DnaA in initiation is derived from studying a temperature-sensitive *dnaA* mutant harboring a plasmid containing wild-type *dnaA* under the control of the inducible *lac* promoter. When the cells are grown at 42 °C, replication is dependent upon the expression of the plasmid *dnaA* gene. The timing of initiation varies with the levels of expression of the gene. When DnaA levels are increased, initiation occurs earlier in the cell cycle.

What does DnaA do? DnaA binds to eight sites in *oriC*, five of which are called R boxes. The five R boxes are nucleotide repeats containing 9 base pains; accordingly, they are called 9-mers. The three non-R boxes are called I boxes (because, as we shall see, the binding of DnaA to these sites is stimulated by integration host factors), and these are 11-mers. These sites differ in their affinity for DnaA. In vitro studies indicate that binding initiates at the tighter binding sites and progresses to the weaker binding sites in the following order: R4,R1 > R2 > R5(M) > I3 > I2 > I1 > R3. Studies indicate that DnaA is bound to R4, R1, and R2 throughout most of the cell cycle, and binds to the weaker sites just before DNA replication is initiated. The binding of DnaA at the weaker sites (R5, R3, and the integration sites) results in the unwinding of DNA to form an open complex within three contiguous 13-mer at the left boundary of *oriC*; all these nucleotide repeats are rich is A–T. Is there is a mechanism that ensures that DnaA–ATP binds to the weaker sites at the same time in all the *oriC* copies? The answer is yes. Two histonelike DNA-bending proteins, Fis (factor for inversion stimulation) and IHF (integration host factor) are involved. These have binding sites in the right and left half of *oriC*, respectively.

Fis and IHF bind at *oriC* and bend the DNA. Presumably this is important for the mechanism of their action at *oriC*. Mutants that do not make either of these proteins are viable, but they cannot initiate replication synchronously at multiple origins in fast growing *E. coli*. In vitro studies have shed light on the mechanism of action of IHF and Fis. It was reported that Fis binds to *oriC* in vitro and forms a complex that blocks the binding of DnaA to low-affinity sites in *oriC*, as well as the binding of IHF to *oriC*, possibly inhibiting *oriC* unwinding. It has been suggested that increased levels of active DnaA displace Fis and allow IHF to bind to DNA, resulting in the redistribution of DnaA from high-affinity sites to low-affinity sites. Whereas Fis binds to *oriC* during most of the cell cycle, IHF binds to *oriC* only at the beginning of DNA replication. The regulation of the binding of IHF to DNA is not understood. As mentioned, the binding of DnaA to the low-affinity sites is necessary for unwinding, and as a consequence IHF promotes unwinding at *oriC*. Thus, Fis and IHF have opposite effects and help to regulate the initiation and synchrony of DNA replication.

Because Fis is growth rate regulated (lower levels at slow growth rates), it has been suggested that it functions only in rapidly growing *E. coli* with more than one copy of *oriC*. The levels of IHF are not growth rate related, and it has been suggested that IHF functions under all growth conditions to promote DnaA binding to the weaker sites in *oriC*. For more discussion, see ref. 8.

Interestingly, the levels of DnaA do not change very much during the cell cycle, and part of the regulation of the initiation of DNA replication is due to the regulation of the activity of existing DnaA. One model proposes that the phospholipids in the membrane keep sequestered DnaA in an inactive form until it is bound to *oriC*. In support of this model is the observation that a fraction of the cellular DnaA can be isolated with the membrane fractions. Furthermore, DnaA forms complexes with phospholipids and the phospholipid can inhibit DnaA activity.

Another factor that blocks the reinitiation of new rounds of DNA replication is sequestration of *oriC*, which refers to any mechanism that prevents reinitiation at *oriC*. To discuss sequestration, it must be pointed out that the origin region is rich in GATC/CTAG sequences and that these sequences are methylated at the adenine residues by Dam methylase shortly after a strand is copied. For reinitiation to occur, both strands at *oriC* must be methylated.

However, for a period of time after initiation (about one-third of the cell cycle), the origin remains hemimethylated; that is, the template strand is methylated but the copy strand is not. This is because newly replicated origins are sequestered from Dam methylase. The sequestration is due to a protein called SeqA, which binds to *oriC* and prevents methylation of the new strand. One of the lines of evidence for this is that purified SeqA binds to *oriC* and prevents DNA replication in vitro. There is also some evidence that in *E. coli*, sequestration of *oriC* may also involve binding of the SeqA/DNA complex to outer membrane components. Cell fractionation studies have revealed that certain subcellular fractions containing outer membrane components are enriched for hemimethylated *oriC* sequences and that hemimethylated *oriC* regions bind more readily to membrane fractions than do fully methylated *oriC* regions.

Eventually however, newly replicated DNA becomes fully methylated by Dam methylase. This takes place well before the next round of initiation, and therefore other controls besides sequestration by SeqA prevent reinitiation during the cell cycle. The nature of these other controls in *E. coli* is a matter of speculation. However, there is evidence for a repressor (CtrA) of the initiation of replication in *Caulobacter crescentus*. See Section 18.17.2.

For reviews, see: Bravo, A., G. Serrano-Heras, and M. Salas. 2005. Compartmentalization of prokaryotic DNA replication. *FEMS Microbiol. Rev.* 29:25–47, and Funnel, B. E. 1996. *The Role of the Bacterial Membrane in Chromosome Replication and Partition.* Chapman & Hall, London. Sequestration of the origin also plays a role in regulating transcription in the *oriC* region, including the transcription of *dnaA*. For a discussion of this point, see: Bogan, J. A., and C. E. Helmstetter. 1997. DNA sequestration and transcription in the *oriC* region of *Escherichia coli. Mol. Microbiol.* 26:889–896. For information on the role of SeqA in sequestering *oriC*, read: Lu, M., J. L. Campbell, E. Boye, and N. Kleckner. 1994. SeqA: a negative modulator of replication initiation in *E. coli. Cell.* 77:413–426, and Bach, T., M. A. Krekling, and K. Skarstad. 2003. Excess SeqA prolongs sequestration of *oriC* and delays nucleoid segregation and cell division. *EMBO J.* 22:315–323.

12. Benkovic, S. J., A. M. Valentine, and F. Salinas. 2001. Replisome-mediated DNA replication. *Annu. Rev. Biochem.* **70**:181–208.

13. DNA polymerase III has 10 different protein subunits coded for by separate genes. One of these, the α subunit, is encoded by the *polC* (*dnaE*) gene and is responsible for polymerization. The subunit responsible for the 3′-exonuclease activity is the ε subunit encoded by the *dnaQ* (*mutD*) gene. These two subunits, along with the φ subunit, comprise the core of the polymerase. The τ subunit dimerizes the core. The other subunits (i.e., γ, δ, δ′, χ, ψ) form a clamp that keeps the polymerase on the template. The subunits that are not part of the core are sometimes called accessory proteins. The core plus the accessory proteins is referred to as the holoenzyme.

14. The realization that DNA copied from one template is made in short fragments came from experiments in which bacteriophage T4 that was infecting *E. coli* was labeled with tritiated thymidine for various lengths of time (2–60s). The DNA was denatured with base, and the intermediates were separated from the template DNA by centrifugation in a sucrose gradient. It was found that when the DNA was labeled for a short period of time, substantial radioactivity was recovered in small pieces of DNA (about 1,500 nucleotides long). As the labeling time was increased, the radioactivity in the smaller pieces remained constant but accumulated in high molecular weight DNA, as would be expected if the smaller pieces were precursors to the high molecular weight DNA. Okazaki fragments are made during the synthesis of all DNA molecules in viruses, prokaryotes, and eukaryotes.

15. Williams, C. R., A. K. Snyder, P. Kusmi, M. O'Donnell, and L. B. Bloom. 2004. Mechanism of loading the *Escherichia coli* DNA polymerase III sliding clamp. *J. Biol. Chem.* **279**:4376–4385.

16. The *E. coli* sliding β clamp, which is part of the DNA polymerase III holoenzyme, is actually a ring of protein with a hollow core through which the DNA template is threaded. The β clamp tethers the DNA polymerase to the template, thus increasing the processivity of replication. It does this by binding to the catalytic subunit (α subunit) of the core polymerase (αεθ subunits), tethering it to the template DNA. The clamp itself is assembled by a clamp loader, also a part of the holoenzyme, which consists of seven different protein subunits: three copies of the *dnaX* gene product, and a single copy each of the δ, δ′, χ, and φ subunits. The *dnaX* gene encodes two proteins, a full-length protein (τ) and a truncated version (γ) formed by a translational frameshift. The τ, subunit of the clamp loader binds to the DNA polymerase, tethering it to the clamp loader, and also interacts with DnaB helicase to coordinate helicase activity with fork progression. In addition, the τ subunit in the clamp loader stimulates the release of the polymerase from the Okazaki fragment, allowing the lagging-strand polymerase to be used for the extension of the next primer.

17. Kuempel, P. L., A. J. Pelletier, and T. M. Hill. 1989. Tus and the terminators: the arrest of replication in prokaryotes. *Cell* **59**:581–583.

18. Khatri, G. S., T. MacAllister, P. R. Sista, and D. Bastia. 1989. The replication terminator protein of *E. coli* is a DNA sequence–specific contrahelicase. *Cell* **59**:667–674.

19. Manna, A. C., K. S. Pai, D. E. Bussiere, C. Davies, S. W. White, and D. Bastia. 1996. Helicase–contrahelicase interaction and the mechanism of termination of DNA replication. *Cell* **87**:881–891.

20. *B. subtilis* also has 6 Ter sites (TerI–TerVI), and they are polar as in *E. coli*. The protein in *B. subtilis* that binds to the Ter sequences is called RTP (replication terminator protein). The *B. subtilis* Ter sites and RTP do not seem to be related in sequences to the *E. coli* terminators or Tus.

21. Lemon, K. P., and A. D. Grossman. 2001. The extrusion–capture model for chromosome partitioning in bacteria. *Genes Dev.* **15**:2031–2041.

22. Gordon, G. S., and A. Wright. 2000. DNA segregation in bacteria. *Annu. Rev. Microbiol.* **54**:681–708.

23. Draper, G. C., and J. W. Gober. 2002. Bacterial chromosome segregation. 2002. *Annu. Rev. Microbiol.* **56**:567–597.

24. Pogliano, K., J. Pogliano, and E. Becker. 2003. Chromosome segregation in eubacteria. *Curr. Opin. Microbiol.* **6**:586–593.

25. Kato, J.-I., Y. Nishimura, R. Imamura, H. Niki, S. Hiraga, and H. Suzuki. 1990. New topoisomerase essential for chromosome segregation in *E. coli*. *Cell* **63**:393–404.

26. Kato, J.-I., Y. Nishimura, M. Yamada, H. Suzuki, and Y. Hirota. 1988. Gene organization in the region containing a new gene involved in chromosome partition in *Escherichia coli*. *J. Bacteriol.* **170**:3967–3977.

27. Cell division and DNA metabolism are coupled. That is, cell division is inhibited whenever DNA replication or partitioning is blocked. This is advantageous in that it lowers the frequency of formation of anucleate cells. When cell division is inhibited, long cells or filaments form. If chromosome separation or segregation is impaired, then the nucleoids are abnormally placed in the elongated cells or filaments. There are two mechanisms responsible for inhibiting cell division. One of these requires SulA. Blocking DNA replication results in the synthesis of SulA as part of the SOS response. (See Section 19.3.2 for a discussion of the SOS response and SulA.) SulA inhibits the formation of the FtsZ septal ring; hence cell division is inhibited. There is also a SulA-independent block in cell division whenever DNA replication or segregation are inhibited.

28. Ward, D., and A. Newton. 1997. Requirement of topoisomerase IV *parC* and *parE* genes for cell cycle progression and developmental regulation in *Caulobacter crescentus. Mol. Microbiol.* 26:897–910.

29. Kuempel, P. L., J. M. Henson, L. Dirks, M. Tecklenburg, and D. F. Lim. 1991. *dif*, a recA-independent recombination site in the terminus region of the chromosome of *Escherichia coli. New Biol.* 3:799–811.

30. Blakely, G., G. May, R. McCulloch, L. K. Arciszewska, M. Burke, S. T. Lovett, and D. J. Sherratt. 1993. Two related recombinases are required for site-specific recombination at *dif* and *cer* in *E. coli* K12. *Cell* 75:351–361.

31. Hojgaard, A., H. Szerlong, C. Tabor, and P. Kuempel. 1999. Norfloxacin-induced cleavage occurs at the *dif* resolvase locus in *Escherichia coli* and is the result of interaction with topoisomerase IV. *Mol. Microbiol.* 33:1365–2958.

32. Errington, J. 1998. Dramatic new view of bacterial chromosome segregation. *ASM News* 64:210–217.

33. Webb, C. D., A. Teleman, S. Gordon, A. Straight, and A. Belmont. 1997. Bipolar localization of the replication origins of chromosomes in vegetative and sporulating cells of *B. subtilis. Cell* 88:667–674.

34. Yongfang, L., B. Youngren, K. Serqueev, and S. Austin. 2003. Segregation of the *Escherichia coli* chromosome terminus. *Mol. Microbiol.* 50:825–834.

35. Mohl, D. A., and J. W. Gober. 1997. Cell cycle–dependent polar localization of chromosome partitioning proteins in *Caulobacter crescentus. Cell* 88:675–684.

36. Jensen, R. B., S. C. Wang, and L. Shapiro. 2001. A moving DNA replication factory in *Caulobacter crescentus. EMBO J.* 20:4952–4963.

37. The replisome can be visualized via fluorescence microscopy by fusing the DNA polymerase to the green fluorescent protein (GFP) from the jellyfish *Aequorea victoria.* These experiments support the idea of a stationary replisome in *B. subtilis* and *E. coli.* For more information on how the fusions are made and introduced into the cell, the student should consult the discussion of ZipA in Section 2.3.3.

DNA can be visualized by using the fluorescent protein GFP. The DNA can be labeled in specific regions by inserting into the DNA a cassette of many copies of *lac* operators (*lacO*) and labeling the *lacO* copies with a fusion of the Lac repressor (LacI) to GFP (LacI–GFP). This has been done to label and follow unreplicated DNA into the replisome and replicated DNA out of the replisome. This technique has allowed investigators to demonstrate that the origin is replicated first and the copies move away from the replisome toward opposite cell poles, that chromosomal regions located midway between *oriC* and *terC* move toward the centrally located replisome, and that the termini follow last into the

replisome and, after replication, become positioned in opposite halves of the cell, closer to the cell center than the origins. The terminus has also been visualized by inserting a single *parS* site near it and labeling the *parS* site with GFP–ParB. ParA and ParB have also been visualized via immunofluorescence microscopy bound as a complex to the *parS* site at the origin in *C. crescentus.* The bacteria were fixed with glutaraldehyde and formaldehyde and then washed and treated with lysozyme. After further treatment with methanol and acetone, they were incubated with antibody. The antibody was visualized with secondary antibody, which was goat anti-rabbit IgG conjugated to either fluorescein or biotin. The biotin-conjugated antibody was visualized with streptavidin-conjugated Texas red dye, which binds to avidin.

38. Wu, J. L., and J. Errington. 2002. A large dispersed chromosomal region required for chromosome segregation in sporulating cells of *Bacillus subtilis. EMBO J.* 21:4001–4011.

39. Gerdes, K., J. Møller-Jensen, G. Ebersbach, T. Kruse, and K. Nordström. 2004. Bacterial mitotic machineries. *Cell* 116:359–366.

40. Ben-Yehuda, S., D. Z. Rudner, and R. Losick. 2003. RacA, a bacterial protein that anchors chromosomes to the cell poles. *Science* 299:532–536.

41. Wheeler, R. T., and L. Shapiro. 1997. Bacterial chromosome segregation: is there a mitotic apparatus? *Cell* 88:577–579.

42. Research with low-copy plasmids has revealed plasmid segregation systems based upon *par* systems encoded by the plasmid. Generally, the *par* system consist of three components: (1) a Par protein that has ATPase activity, which will be referred to here as ParA, (2) a Par DNA-binding protein, referred to here as ParB, and (3) a centromere-like region in the DNA to which ParB attaches. As will be explained later, plasmids can differ in their Par proteins, and the reference to ParA and Par B here is for the sake of convenience. Plasmid pairs move to the midregion of the cell prior to septation and then are moved bidirectionally (segregated) to opposite cell poles. The Par proteins are required for the segregation of the plasmids. ParA and ParB assemble at the centromere-like region, and the resulting complex of DNA and Par proteins interacts with a bacterial scaffoldlike structure that is involved in moving plasmids bidirectionally toward the cell poles. Some specific examples of plasmid segregation will now be described.

The system for the *E. coli* R1 plasmid consists of the ParM ATPase, the ParR DNA-binding protein, and *parC*, the putative centromere-like region. The separated paired plasmids at midcell move to opposite cell poles. The ParM protein is not related to ParA. Rather, it is a member of the MreB family and forms actinlike helical filaments at the nucleation point of ParR/*parC*. (In addition to the ParR/*parC* complex, polymerization requires ATP and Mg^{2+}, and depolymerization requires ATP hydrolysis.) It has been suggested that the polymerization of ParM provides the driving force that moves the plasmid bound to the ParM protein toward a cell pole so that each daughter cell receives at least one copy of the plasmid. This could occur if the plasmids were located at the tips of the filaments (which they are), and if polymerization occurred between plasmid copies. Another possibility is that the ParM filaments are a scaffolding along which the plasmids move toward opposite cell halves. For a discussion of how ParM might be involved in plasmid segregation, see: Møller-Jensen, J., R. B. Jensen, J. Löwe, and K. Gerdes. 2002. Prokaryotic DNA segregation by an actin-like filament. *EMBO J.* **21**:3119–3127.

The *par* system in the *E. coli* multidrug-resistant plasmid TP228 consists of (1) an ATPase called ParF, which is part of a subgroup of the ParA superfamily of proteins (not related to ParM), (2) a DNA-binding protein called ParG, which is not related to ParB, and (3) a region upstream of the *parFG* genes to which ParG attaches. References that can be consulted include the following: Golovanov, A. P., D. Barillà, M. Golovanova, F. Hayes, and L.-Y. Lian. 2003. ParG, a protein required for active partition of bacteria plasmids, has a dimeric ribbon–helix–helix structure. *Mol. Microbiol.* **50**:1141, and Barillà, D., and F. Hayes. 2003. Architecture of the ParF/ParG protein complex involved in prokaryotic DNA segregation. *Mol. Microbiol.* **49**:487–499.

43. Niki, H., A. Jaffé, R. Imamura, T. Ogura, and S. Hiraga. 1991. The new gene *mukB* codes for a 177 kD protein with coiled-coil domains involved in chromosome partitioning of *E. coli*. *EMBO J.* **10**:183–193.

44. Niki, H., R. Imamura, M. Kitaoka, K. Yamanaka, T. Ogura, and S. Hiraga. 1992. *E. coli* MukB protein involved in chromosome partition forms a homodimer with a rod-and-hinge structure having DNA binding and ATP/GTP binding activities. *EMBO J.* **11**:5101–5109.

45. Yamanaka, K., T. Ogura, H. Niki, and S. Hiraga. 1996. Identification of two new genes, *mukE* and *mukF*, involved in chromosome partitioning in *Escherichia coli*. *Mol. Gen. Genet.* **250**:241–251.

46. Muk is derived from the Japanese word *mukaku*, which means anucleate. When grown at 22 °C, *mukB* mutants form normal nucleated rods with a small percentage (5%) of anucleate cells. At higher temperatures (e.g., 42 °C) the cells die, and anucleate cells and multinucleate filaments with abnormal patterns of nucleoids accumulate (clumps and isolated nucleoids). Examination of two paired daughter cells revealed that some of the anucleate cells were paired with cells that appeared to have more than one nucleoid, reflecting the fact that both daughter chromosomes remained with one of the daughter cells upon division.

Mutations in two other genes also lead to the production of anucleate cells. These are the *mukF* and *mukE* genes. There probably encode alternatives to the Muk proteins. This conclusion is based upon the finding that null mutants in *mukB*, *mukF*, or *mukE* show a decreased plating efficiency and a very large number of anucleate cells only at temperatures higher than 22 °C, indicating that the proteins are not absolutely essential at the lower temperatures. Similarly, strains of *B. subtilis* and *C. crescentus* bearing null mutations in *smc* have a severe plating deficiency in rich media at higher temperatures. The *B. subtilis* mutants form a significant number of anucleate cells, and the majority of *C. crescentus* mutants are "stalled" at a predivisional stage, do not elongate, and are defective in nucleoid distribution.

47. Espeli, O., P. Nurse, C. Levine, C. Lee, and K. J. Marians. 2003. SetB: an integral membrane protein that affects chromosome ion in *Escherichia coli*. *Mol. Microbiol.* **50**:495–509.

48. Van den Ent, F., L. A. Amos, and J. Löwe. 2001. Prokaryotic origin of the actin cytoskeleton. *Nature* **413**:39–44.

49. Figge, R. M., and J. W. Gober. 2003. Cell shape, division and development: the 2002 American Society for Microbiology (ASM) Conference on Prokaryotic Development. **47**:1475–1483.

50. Wu, L. J., and J. Errington. 2003. RacA and the Soj–SpoOJ system combine to effect polar chromosome segregation in sporulating *Bacillus subtilis*. *Mol. Microbiol.* **49**:1463.

51. Real, G., S. Autret, J. E. Harry, J. Errington, and A. O. Henriques. 2005. Cell division protein DivlB influences the SpoOJ/Soj system of chromosome segregation in *Bacillus subtilis*. *Mol. Microbiol.* **55**:349–367.

52. Ben-Yehuda, S., D. Z. Rudner, and R. Losick. 2003. RacA, a bacterial protein that anchors chromosomes to the cell poles. *Science* **299**:532–536.

53. Marston, A. L., H. B. Thomaides, D. H. Edwards, M. E. Sharpe, and J. Errington. 1998. Polar localization of the MinD protein of *Bacillus subtilis* and its role in selection of the mid-cell division site. *Genes Dev.* **12**:3419–3430.

54. Mohl, D. A., and J. W. Gober. 1997. Cell cycle–dependent polar localization of chromosome partitioning proteins in *Caulobacter crescentus*. *Cell* **88**:675–684.

55. Immunofluorescence microscopy to visualize ParA and ParB was performed with synchronized cultures and the appropriate antibodies. The procedure used is detailed in the last paragraphy of note 37.

56. Jensen, R. B., S. C. Wang, and L. Shapiro. 2001. A moving DNA replication factory in *Caulobacter crescentus*. *EMBO J.* **20**:4952–4963.

57. Frameshift mutations are mutations due to the insertion or removal of one or more (but not three) bases from DNA. Acridine dyes such as ethidium bromide and proflavine cause frameshift mutations.

They insert between the bases, thus increasing the distance between bases and preventing them from aligning correctly. As a consequence, the two strands of DNA slip with respect to each other. This leads to the insertion or deletion of a base.

58. Modrich, P. 1991. Mechanisms and biological effects of mismatch repair. *Annu. Rev. Genet.* 25:229–253.

59. Schofield, M. J., and P. Hsieh. 2003. DNA mismatch repair: molecular mechanisms and biological function. *Annu. Rev. Microbiol.* 57:579–608.

60. Kolodner, R. 1996. Biochemistry and genetics of eukaryotic mismatch repair. *Genes Dev.* 10:1433–1442.

61. Modrich, P. 1995. Mismatch repair, genetic stability and tumour avoidance. *Philos. Trans. R. Soc. Lond. B* 347:89–95.

62. Modrich, P. 1994. Mismatch repair, genetic stability and cancer. *Science* 266:1959–1960.

63. Bacteria have one RNA polymerase that synthesizes ribosomal, messenger, and transfer RNA. The enzyme that synthesizes the RNA in Okazaki fragments is called primase. Eukaryotes have three nuclear RNA polymerases and a mitochondrial RNA polymerase.

64. McClure, W. R. 1985. Mechanism and control of transcription initiation in prokaryotes. *Annu. Rev. Biochem.* 54:171–204.

65. Platt, T. 1986. Transcription termination and the regulation of gene expression. *Annu. Rev. Biochem.* 55:339–372.

66. Henkin, T. M. 1996. Control of transcription termination in prokaryotes. *Annu. Rev. Genet.* 30:35–37.

67. Gruber, T. M., and C. A. Gross. 2003. Multiple sigma subunits and the partitioning of bacterial transcription space. *Annu. Rev. Microbiol.* 57:441–466.

68. Ades, S. E. 2004. Control of the alternative sigma factor σ^E in *Escherichia coli*. *Curr. Opin. Microbiol.* 7:157–162.

69. Ishihama, A. 1997. Promoter selectivity control of RNA polymerase, pp. 53–70. In: *Mechanisms of Transcription.* F. Eckstein and D. M. J. Lilley (Eds.). Springer-Verlag. Berlin.

70. Steitz, T. A., D. H. Ohlendorf, D. B. McKay, W. F. Anderson, and B. W. Matthews. 1982. Structural similarity in the DNA-binding domains of catabolite gene activator and *CRO* repressor proteins. *Proc. Natl. Acad. Sci. USA* 79:3097–3100.

71. The eukaryotic cytosolic ribosome is 80S and has two subunits, 40S and 60S. The 40S subunit contains an 18S rRNA. The 60S subunit contains a 28S, a 5S, and a 5.8S rRNA. The rRNA transcript in eukaryotes is 45S. It is processed in the nucleus to produce the 18S, 28S, and 5.8S rRNAs, which are present in the ribosomes. The 5S rRNA is on a separate transcript.

72. Björk, G. R., J. U. Ericson, C. E. D. Gustafsson, T. G. Hagervall, Y. H. Jöhnsson, and P. M. Wilkström. 1987. Transfer RNA modification. *Annu. Rev. Biochem.* 56:263–287.

73. Hershey, J. W. B. 1987. Protein synthesis, pp. 613–647. In: *Escherichia coli and Salmonella typhimurium: Cellular and Molecular Biology*, Vol. 1. F. C. Neidhardt et al. (Eds.). ASM Press, Washington, DC.

74. Protein synthesis by cytosolic eukaryotic ribosomes differs in some respects from protein synthesis by bacterial ribosomes. (1) The ribosomal subunits are 40S and 60S. (2) The first amino acid is methionine rather than formyl methionine. (But still, a special tRNA^{Met} is used for initiation.) (3) There are at least nine initiation factors, rather than three. These are elF3 and el4C, which bind to the 40S subunit and stimulate further steps in initiation; the cap protein (or CBPI), which binds to the 5′ end of the mRNA and aids in binding the mRNA to the 40S subunit; elF2, which aids in binding of Met-tRNA^{Met} to the 40S subunit; elF4a, elF4B, and elF4F, which help to locate the start AUG; elF5, which facilitates the dissociation of the other initiation factors from the 40S subunit so that it can combine with the 60S subunit to form the 80S ribosome; and elF6, which stimulates the dissociation of the 80S ribosome not engaged in protein synthesis to 40S and 60S subunits. The elongation and translocation phases are analogous to what occurs in bacterial ribosomes. The cytosolic eukaryotic ribosomes use three elongation factors, eEF1α, eEF1βγ, and eEF2, which are analogous to EF-Tu, EF-Ts, and EF-G. Termination is also similar. Whereas in bacteria there are three release factors that are specific for the three termination codons, protein synthesis by cytosolic eukaryotic ribosomes uses a single release factor, eRF, which recognizes three different termination codons. Eukaryotic mRNA differs from prokaryotic mRNA in not being polycistronic and in not having a special ribosome binding analogous to the Shine–Dalgarno sequence upstream of the initiation codon. Eukaryotes use the first AUG at the 5′ end of the mRNA.

75. Arnez, J. G., and D. Moras. 1997. Stuctural and functional considerations of the aminoacylation reaction. *Trends Biochem. Sci.* 22:211–216.

76. Although there are 61 codons for the 20 amino acids, there are fewer than 61 tRNAs. This is because some tRNAs can read more than one codon for a particular amino acid. To understand the following discussion, bear in mind that the anticodon and the codon have opposite polarity, and thus the third nucleotide in the codon always pairs with the first nucleotide in the anticodon. Crick realized that the first two nucleotides in the codon form strong pairs with their partners in the anticodon and in fact are responsible for most of the codon specificity. That is, the third nucleotide in the codon frequently does not change the specificity of the codon. Therefore, codons with the same nucleotide in the first two

positions but differing in the third nucleotide can often code for the same amino acid. This actually depends upon the base at the first position in the anticodon. If the first base in the anticodon is either C or A, then the base in the third position of the codon is always G or U, respectively, and there is only one codon per amino acid and therefore only one tRNA per codon. However, if the base in the first position in the anticodon is I (inosinate), G, or U, then the third base in the codon forms a loose pair or "wobble" with its partner in the anticodon. This allows different codons for a single amino acid to be read by the same tRNA. For example, inosinate, which has hypoxanthine as its base, can be found in the first position of the anticodon for some amino acids. Inosinate can hydrogen bond weakly to U, C, or A. This means that the third base in the codon can be either U, C, or A, and all three codons will bind to the same anticodon (i.e., to the same tRNA). Likewise, U in the first position in the anticodon can pair with either A or G in the third position of the codon, so that in this case two codons can be read by the same tRNA. Similarly, G in the first position of the anticodon can pair with C or U. Thus it is possible to have fewer tRNAs than codons. At least 32 different tRNAs are required to read the 61 codons. It has been suggested that the advantages of introducing a "wobble" in the third position are that the bonding between the codon and anticodon becomes less tight and the rate of dissociation of the tRNAs from the ribosome during protein synthesis is increased.

77. Ibba, M., A. W. Curnow, and D. Söll. 1997. Aminoacyl–tRNA synthesis: divergent routes to a common goal. *Trends Biochem. Sci.* 22:39–42.

78. Becker, H. D., J. Reinbolt, R. Kreutzer, R. Giegé, and D. Kern. 1997. Existence of two distinct aspartyl–tRNA synthetases in *Thermus thermophilus*. Structural and biochemical properties of the two enzymes. *Biochemistry* 36:8785–8797.

79. Many organisms possess fewer than 20 different aminoacyl–tRNA synthetases to activate the 20 standard amino acids. That is, in a few instances the same aminoacyl–tRNA synthetase charges two different tRNAs with the same amino acid. For example, it has been found that *Bacillus subtilis*, other gram-positive bacteria, the gram-negative *Rhizobium meliloti*, some archaebacteria, chloroplasts, and mitochondria, the same enzyme i.e., GluRS, to charge tRNA[Gln] (has anticodon for Glu codon) and tRNA[Glu] (has anticodon for Gln codon) with glutamate. Then an amidotransferase recognizes the Glu–tRNA[Gln], and ω amidates the glutamate to form Gln–tRNA[Gln], which places the Gln at the Gln codon. A similar situation occurs in the halophilic archaebacterium *Haloferax volcanii* with Asn–tRNA[Asn], where both tRNA[Asp] and tRNA[Asn] are charged by the same enzyme with aspartate. The aspartate is then converted to asparagine on tRNA[Asn], which places Asn at the Asn codon. Note the similarity to the fate of methionine that is charged onto tRNA[fMet]. A special transformylase recognizes the Met–tRNA[fMet] and formylates the Met to fMet, which is used to initiation translation. In all these cases the tRNA is charged with the wrong amino acid, which is subsequently modified to the correct amino acid. There is also an example of two different aminoacyl synthetases activating the same amino acid and charging the same tRNA. The thermophilic bacterium *Thermus thermophilus* has two aspartyl–tRNA synthetases (AspRS1 and AspRS2) that charge tRNA[Asp]. They are encoded by separate genes.

80. Arnez, J. G., and D. Moras. 1997. Structural and functional considerations of the aminoacylation reaction. *Trends Biochem. Sci.* 22:211–216.

81. The amino acids are attached to the 3′-terminal CCA of tRNAs. Some synthetases attach the amino acid directly to the 3′-hydroxyl, whereas other synthetases attach the amino acid to the 2′-hydroxyl, and then a transesterification reaction moves the amino acid to the 3′-hydroxyl. A description of the structural and functional features of the aminoacyl–tRNA synthetases is reviewed in ref. 80.

82. Noller, H. F., V. Hoffarth, and L. Zimniak. 1992. Unusual resistance of peptidyl transferase to protein extraction procedures. *Science* 256:1416–1419.

83. Certain RNA molecules are known to have enzymatic activity. They are called *ribozymes*. Two well-studied examples are RNase P and self-splicing group I introns. The self-splicing intron is an endonuclease that cuts itself out of RNA precursor molecules and joins the exons together to form the mature RNA. Another ribozyme is RNase P, which is actually a nucleoprotein. However, the catalytic activity resides in the RNA. RNase P acts on tRNA precursors. Transfer RNA is made from longer precursors that have had nucleotides removed from the ends. RNase P is an RNA molecule that acts as an endonuclease and removes nucleotides from the 5′ end of tRNA molecules. Self-splicing introns and RNase P are widespread in prokaryotes and eukaryotes.

84. Frydman, J. 2001. Folding of newly translated proteins in vivo: the role of molecular chaperones. *Annu. Rev. Biochem.* 70:603–647.

85. Lund, P. A. 2001. Microbial molecular chaperones, pp. 95–140. In: *Advances in Microbial Physiology*, Vol. 44. R. K. Poole (Ed.). Academic Press, San Diego, CA.

86. The primary structure of proteins refers to the sequence of amino acids. The secondary structure refers to regular arrangements of the polypeptide chain (e.g., an α-helix in which the polypeptide backbone is wound around the long axis of the molecule in a helix and the R groups of the amino acids point out). The tertiary structure refers to the folding of the polypeptide chain. The quaternary structure refers to the spatial arrangement of folded polypeptide subunits in a multimeric protein.

87. Ewalt, K. L., J. P. Hendrick, W. A. Houry, and F. U. Hartle. 1997. In vivo observation of polypeptide flux through the bacterial chaperonin system. *Cell* 90:491–500.

88. The *groE* operon contains two genes, *groES* and *groEL*. *E. coli* requires both GroEL and GroES (sometimes referred to as the GroE system) for growth at all temperatures, although as discussed in Chapter 19, their synthesis is greatly stimulated by heat shock. Possibly, the excess GroEL at higher temperatures protects unstable proteins or aids in refolding partially denatured proteins when normal temperatures are restored. GroEL (also called Hsp60) is a double-ring structure ("double doughnut") with 14 identical subunits; there are 7 subunits per ring. GroEL binds to the protein that is to be folded, and the proteins are believed to be folded in the cavity of the "double doughnut." ATP is also necessary. ATP binds to GroEL, producing a conformational change, and increases the affinity of GroEL for the protein substrate. The hydrolysis of ATP lowers the affinity and aids in the release of the folded protein. The role of GroES (also called Hsp10), which is a single ring of seven subunits, is to aid in the release of the folded protein from GroEL One might view the GroEL cavity within which the protein folds as a "safe haven," where a protein may fold spontaneously without interacting with another protein to form aggregates.

89. Gaitanaris, G. A., A. Vysokanov, S.-C. Hung, M. E. Gottesman, and A. Gragerov. 1994. Successive action of *Escherichia coli* chaperones *in vivo*. *Mol. Microbiol.* 14:861–869.

90. Following are the methods used by Gaitanaris et al. in ref. 89 to show that DnaK and DnaJ associate with nascent polypeptides on ribosomes and that GroEL transiently associates with completed polypeptides after their release from ribosomes. Cell lysates of *E. coli* were centrifuged through a sucrose gradient. The ribosome fractions were isolated and the proteins were separated by gel electrophoresis and then transferred (blotted) onto nitrocellulose filters. The filters were treated with anti-DnaK, anti-DnaJ, and anti-GroEL antibodies, and bound antibody was detected by means of alkaline phosphatase–coupled secondary antibodies. Both DnaK and DnaJ were associated with the ribosomal fractions isolated from the sucrose density gradients, but GroEL was not. When cells were briefly treated with puromycin to release nascent polypeptides from ribosomes, DnaK and DnaJ were released from the ribosomes. Newly synthesized polypeptides were associated with GroEL. This was determined by pulse-labeling cells with [^{35}S]methionine, chasing with unlabeled methionine for various times, and separating cell fractions on a sucrose gradient. The proteins were separated by gel electrophoresis and blotted onto nitrocellulose, whereupon GroEL was detected by using anti-GroEL antibody as just described. Radioactive proteins were transiently associated with fractions containing GroEL. However, none of the GroEL-associated proteins were in the ribosomal fractions, indicating that they became associated with GroEL after their release from the ribosomes.

91. Fedorov, A. N., and T. O. Baldwin. 1997. Cotranslational protein folding. *J. Biol. Chem.* 272:32715–32718.

92. Stoller, G., K. P. Rücknagel, K. H. Nierhaus, F. X. Schmid, G. Fischer, and J.-U. Rahfeld. 1995. A ribosome-associated peptidyl-prolyl *cis/trans* isomerase identified as the trigger factor. *EMBO J.* 14:4939–4948.

93. Deuerling, E., A. Schulze-Specking, T. Tomoyasu, A. Mogk, and B. Bukau. 1999. Trigger factor and DnaK cooperate in folding of newly synthesized proteins. *Nature* 400:693–696.

94. Scholz, C., G. Stoller, T. Zarnt, G. Fischer, and F. X. Schmid. 1997. Cooperation of enzymatic and chaperone functions of trigger factor in the catalysis of protein folding. *EMBO J.* 16:54–58.

95. Scholz, C., M. Mücke, M. Rape, A. Pecht, A. Pahl, H. Bang, and F. X. Schmid. 1997. Recognition of protein substrates by the prolyl isomerase trigger factor is independent of proline residues. *J. Mol. Biol.* 277:723–732.

96. Hunter, T. 1998. Prolyl isomerases and nuclear function. *Cell* 92:141–143.

11

Cell Wall and Capsule Biosynthesis

Studying cell wall and capsule synthesis in bacteria is instructive in showing how logistic problems of extracellular biosynthesis are solved. For example, the subunits of the cell wall and capsule polymers are synthesized as water-soluble precursors in the cytosol. How do the subunits traverse the lipid barrier in the cell membrane to the sites of polymer assembly? A second problem concerns the final stages of peptidoglycan synthesis. During peptidoglycan synthesis, the newly polymerized glycan chains become cross-linked by peptide bonds on the outside cell surface. What is the source of energy for making the peptide cross-links at a site that lacks ATP? This chapter considers these and other aspects of the biosynthesis of peptidoglycan and lipopolysaccharide, as well as capsular and other extracellular polysaccharides. The biosynthesis of cytosolic polysaccharides is also considered.

11.1 Peptidoglycan

11.1.1 Structure

Peptidoglycan is a heteropolymer of glycan chains cross-linked by amino acids (Section 1.2.3). The peptidoglycan is a huge molecule, since it surrounds the entire cell and appears to be covalently bonded throughout. A schematic drawing of how this might look in gram-negative bacteria is shown in Fig. 11.1 (see also Fig. 1.8). Peptidoglycan confers strength to the

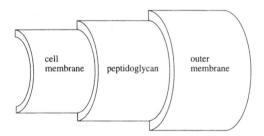

Fig. 11.1 The topological relationship of the peptidoglycan to the cell membrane and rest of the cell wall. In gram-negative bacteria such as *E. coli*, the peptidoglycan is a thin layer sandwiched between the inner and outer membranes. In gram-positive bacteria there is no outer membrane, and the peptidoglycan is a thick layer usually covalently bonded to other molecules (e.g., teichoic acids).

cell wall, and if one were to enzymatically destroy the integrity of the peptidoglycan (with lysozyme) or prevent its synthesis (with antibiotics), the cell would likely swell through the weak areas and lyse as a result of the internal turgor pressures.

The chemical composition of peptidoglycan

Peptidoglycan is made of glycan strands of alternating residues of N-acetylmuramic acid and N-acetylglucosamine linked by β-1,4 glycosidic bonds between the C_1 of N-acetylmuramic acid and the C_4 of N-acetylglucosamine (Fig. 11.2). N-Acetylmuramic acid is a modified form of N-acetylglucosamine in which a lactyl group

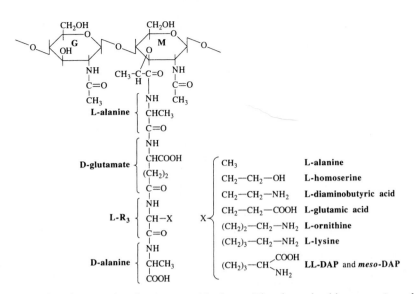

Fig. 11.2 The disaccharide–peptide subunit in peptidoglycan. The glycan backbone consists of alternating residues of N-acetylglucosamine (**G**) and N-acetylmuramic acid (**M**) linked β-1,4. The carboxyl group of the lactyl moiety in N-acetylmuramic acid is substituted with a tetrapeptide. The amino acids in the tetrapeptide are usually L-alanine, D-glutamic, L-R$_3$ (residue 3), which is an amino acid that varies with the species, and D-alanine. The peptide linkages are all α except for that between D-glutamic and the amino acid in position 3, which is γ linked. The X can be any one of a large number of side chains, some examples of which are shown. The α carboxyl of glutamic acid can be free, an amide, or substituted (e.g., by glycine).

has been attached to the C3 carbon. Attached to each N-acetylmuramic acid is a tetrapeptide. The tetrapeptide is L-alanyl-D-glutamyl-γ-L-R$_3$-D-alanine. The amino acid in position 3 varies with the species of bacterium. Gram-negative bacteria generally have *meso*-diaminopimelic acid. (However, some spirochaetes contain ornithine instead of *meso*-diaminopimelic acid.) In contrast, there is much more variability in the amino acids in position 3 in gram-positive bacteria (Fig. 11.2).

Cross-linking

The tetrapeptide chains are cross-linked by peptide bonds (Fig. 11.3A,B). There is a great deal of variability in the composition of the cross-links among the different groups of bacteria. In fact, amino acid composition and cross-link location have been used for taxonomic purposes. In most instances the peptide bridge is from the carboxyl in the terminal D-alanine in one tetrapeptide to an amino group in the amino acid in the L-R$_3$ position in another tetrapeptide. In some bacteria the cross-linking is direct—for example, between D-alanine and diaminopimelic acid in gram-negative bacteria

and many *Bacillus* species. However, in most gram-positive bacteria there is a bridge of one or more amino acids. Some examples are a bridge of five glycine residues in *Staphylococcus aureus*, three L-alanines and one L-threonine in *Micrococcus roseus*, and three glycines and two L-serines in *Staphylococcus epidermidis*. Sometimes the bridge is from the terminal D-alanine to the α-carboxyl of D-glutamic acid of another tetrapeptide. Since this is a connection between two carboxyl groups, a bridge of amino acids containing a diamino acid is necessary.

11.1.2 Synthesis

Peptidoglycan is made in several stages: (1) the precursors to the peptidoglycan are UDP derivatives of the amino sugars made in the cytosol; (2) the amino sugars are then transferred to a lipid carrier in the membrane, which carries the amino sugars across the membrane; (3) the peptidoglycan is polymerized on the outer surface of the membrane; and (4) a transpeptidation reaction cross-links the peptidoglycan. (For reviews, see refs. 1 and 2.)

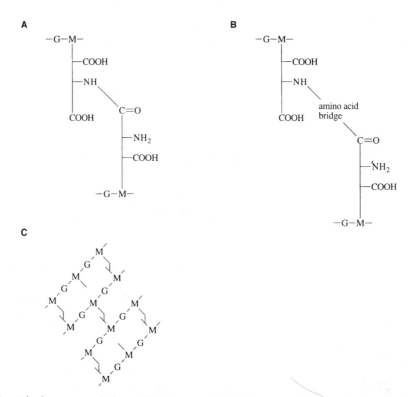

Fig. 11.3 Peptidoglycan cross-linking. (A) Direct cross-link between diaminopimelate and D-alanine as it occurs in gram-negative bacteria. (B) An amino acid bridge between two tetrapeptides, as in many gram-positive bacteria. Sometimes the bridge is between the C-terminal D-alanine and the α-carboxyl of D-glutamic acid. When this occurs, there must be a diamino acid in the bridge. (C) Schematic drawing of cross-linked peptidoglycan. Tetrapeptides are indicated by straight lines. The glycan chains in some bacteria have been found to contain about 20 to 100 sugar residues, apparently varying in length within the same bacterium. The degree of cross-linking depends upon the bacterium. In *E. coli*, which is considered to be a relatively un-cross-linked peptidoglycan, about 50% of the peptide chains are cross-linked. In some bacteria, about 90% of the chains are cross-linked, Abbreviations: G, *N*-acetylglucosamine; M, *N*-acetylmuramic acid.

Synthesis of the UDP derivatives: UDP-N-acetylglucosamine and UDP-N-acetylmuramyl-pentapeptide

The two amino sugars that are precursors to the peptidoglycan are *N*-acetylglucosamine and *N*-acetylmuramic acid. Both amino sugars are made from fructose-6-phosphate (Fig. 11.4). In step **1**, glutamine donates an amino group to fructose-6-phosphate, converting it to glucosamine-6-phosphate. Then in step **2**, a transacylase transfers an acetyl group from acetyl–CoA to the amino group on glucosamine-6-phosphate to make *N*-acetylglucosamine-6-phosphate, which is isomerized in step **3** to *N*-acetylglucosamine-1-phosphate. The monophosphate attacks UTP, displacing pyrophosphate in step **4** to form UDP-*N*-

acetylglucosamine. The reaction is driven to completion by a pyrophosphatase.

Some of the UDP-*N*-acetylglucosamine (UDP-GlcNAc) is used as the precursor to the *N*-acetylglucosamine in peptidoglycan, and some is converted to UDP-*N*-acetylmuramic acid (UDP-MurNAc). The UDP-GlcNAc is converted to UDP-MurNAc by the addition of a lactyl group to the sugar in step **5**. In this reaction, the C3 hydroxyl of the sugar displaces the phosphate from the α carbon of phosphoenolpyruvate, forming the enol pyruvate ether derivative of the UDP-*N*-acetylmuramic acid. Then, in step **6**, the enol derivative is reduced to the lactyl moiety by NADPH. The UDP-MurNAc is converted into UDP-MurNAc-pentapeptide by the sequential addition of five amino acids, L-alanine, D-glutamate, L-R₃

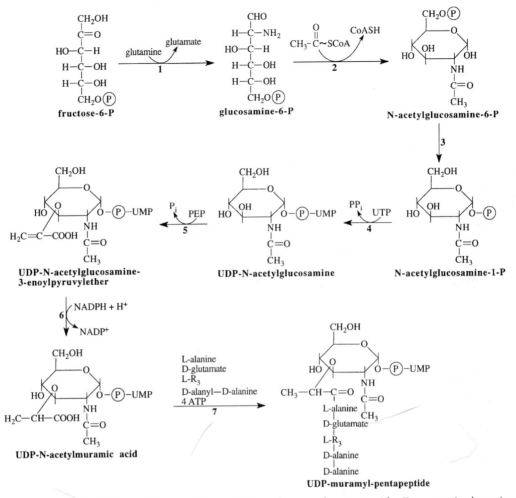

Fig. 11.4 Synthesis of *N*-acetylglucosamine and *N*-acetylmuramyl-pentapeptide. Enzymes: **1**, glutamine: fructose-6-phosphate aminotransferase; **2**, glucosamine phosphate transacetylase; **3**, *N*-acetylglucosamine phosphomutase; **4**, UDP-*N*-acetylglucosamine pyrophosphorylase; **5**, enoylpyruvate transferase; **6**, UDP-*N*-acetylglucosamincenol-pyruvate reductase. In **7**, the UDP-*N*-acetylmuramic acid is converted to the pentapeptide derivative by the sequential additions of L-alanine, D-glutamate, L-R₃, and D-alanine–D-alanine by separate enzymes.

(residue 3), and the dipeptide D-alanyl-D-alanine in step 7.

Each reaction is catalyzed by a separate enzyme and requires ATP to activate the carboxyl group of the amino acid. The activated carboxyl is probably an acyl phosphate, which is attacked by the incoming amino group displacing the phosphate. (See Fig. 7.7.) The products are ADP and inorganic phosphate. (The fifth D-alanine is removed during a cross-linking reaction described later.) [The synthesis of the peptide subunit in the archaeal pseudomurein (Fig. 1.14) takes place via UDP-peptide intermediates which also appear to form peptide bonds via acyl phosphates.[3] The D-alanine–

D-alanine is made by separate enzymes. The first of these is a racemase that converts L-alanine to D-alanine. Then an ATP-dependent D-alanyl–D-alanyl synthetase makes D-alanyl–D-alanine from two D-alanines. The racemase and synthetase are inhibited by the antibiotic D-cycloserine. The MurNAc-pentapeptide is transferred to the lipid carrier in the membrane, as described next.

Reactions in the membrane

The lipid carrier is called *undecaprenyl phosphate* or *bactoprenol*. Undecaprenyl phosphate is a C₅₅ isoprenoid phosphate whose structure

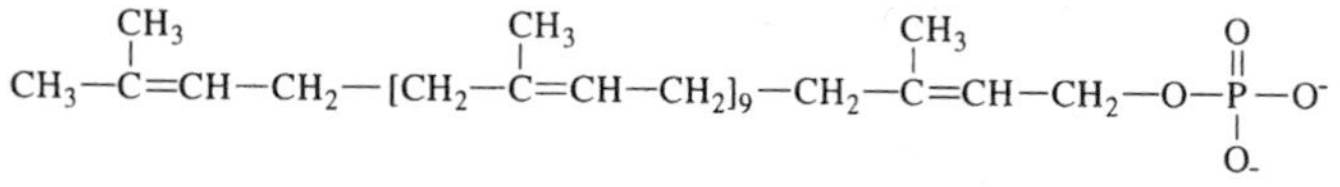

$$CH_3-\underset{\underset{CH_3}{|}}{C}=CH-CH_2-[CH_2-\underset{\underset{CH_3}{|}}{C}=CH-CH_2]_9-CH_2-\underset{\underset{CH_3}{|}}{C}=CH-CH_2-O-\underset{\underset{O_-}{|}}{\overset{\overset{O}{\|}}{P}}-O^-$$

Fig. 11.5 The structure of undecaprenyl phosphate, a C_{55} isoprenoid phosphate that carries precursors to peptidoglycan, lipopolysaccharide, and teichoic acids through the cell membrane.

is shown in Fig. 11.5. Undecaprenyl phosphate serves not only as a carrier for peptidoglycan precursors but also serves as a carrier for the precursors of other cell wall polymers (e.g., lipopolysaccharide and teichoic acids). Interestingly, eukaryotes also use an isoprenoid phosphate, dolichol phosphate, to carry oligosaccharide subunits across the endoplasmic reticulum (ER) membrane to be attached to glycoproteins in the ER lumen. Dolichol phosphate is larger than undecaprenyl phosphate but has the same structure. The nucleotide sugars diffuse to the membrane, where undecaprenyl phosphate (lipid-P) attacks the UDP-MurNAc-pentapeptide displacing UMP (Fig. 11.6, 1). The product is lipid-PP-MurNAc-pentapeptide. The GlcNAc is then transferred from UDP-GlcNAc to the MurNAc on the lipid carrier (Fig. 11.6, 2). This occurs when the C4 hydroxyl in the MurNAc attacks the C1 carbon in UDP-GlcNAc, displacing the UDP. The

product is the disaccharide precursor to the peptidoglycan, lipid-PP-MurNAc(pentapeptide)-GlcNAc. The disaccharide–lipid moves to the other side of the membrane (Fig. 11.6, 3). Exactly how the disaccharide–lipid moves through the membrane is not understood, but it has been speculated that it is aided by unidentified proteins. On the outside surface of the membrane, the lipid–disaccharide is transferred to the growing end of the acceptor glycan chain (Fig. 11.6, 4). Reaction 4 is a transglycosylation in which the C4 hydroxyl of the incoming GlcNAc attacks the C1 of the MurNAc in the glycan, displacing the lipid-PP from the growing glycan chain. This reaction is catalyzed by a membrane-bound enzyme called transglycosylase.

Notice that the growing glycan chain remains anchored via the lipid carrier to the membrane at the site of the transglycosylase. The lipid-PP released from the growing glycan

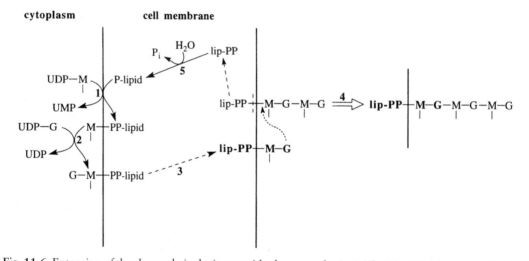

Fig. 11.6 Extension of the glycan chain during peptidoglycan synthesis. **1**, The MurNAc-pentapeptide (–M) is transferred to the phospholipid carrier (undecaprenyl phosphate) on the cytoplasmic side of the cell membrane. **2**, The GlcNAc (G) is transferred to the MurNAc-pentapeptide to form the disaccharide–PP–lipid precursor. **3**, The disaccharide–PP–lipid precursor moves to the external face of the membrane. **4**, A transglycosylase transfers the incoming disaccharide to the growing glycan, displacing the lipid-PP from the growing chain. Thus, the growing chain remains anchored to the membrane by the lipid carrier at the site of the transglycosylase. **5**, The lipid-PP released from the growing chain is hydrolyzed to lipid-P by a membrane-bound pyrophosphatase. Note that the glycan chain grows at the reducing end; that is, displacements occur on the C1 of muramic acid.

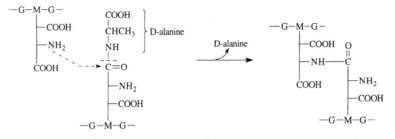

Fig. 11.7 The transpeptidation reaction. In this nucleophilic displacement, the nucleophilic nitrogen attacks the carbonyl, displacing the terminal D-alanine. The reaction is inhibited by penicillin. Terminal D-alanine residues in chains not participating in cross-linking are removed by a D-alanine carboxypeptidase.

chain is hydrolyzed by a membrane-bound pyrophosphatase to lipid-P and P_i (Fig. 11.6, 5). This reaction is very important because it helps drive the transglycosylation reaction to completion, since the hydrolysis of the phosphodiester results in the release of substantial energy. This hydrolysis also regenerates the lipid-P, which is necessary for continued growth of the peptidoglycan as well as other cell wall polymers (e.g., lipopolysaccharide and teichoic acids). The hydrolysis is inhibited by the antibiotic bacitracin.

Making the peptide cross-link

The problem of providing the energy to make the peptide cross-link outside the cell membrane is solved by using a reaction called transpeptidation. In the transpeptidation reaction, an :NH_2 group from the diamino acid in the 3 position (e.g., DAP) attacks the carbonyl carbon in the peptide bond holding the two D-alanine residues together in the pentapeptide, and displaces the terminal D-alanine. The result is a new peptide bond (Fig. 11.7). The transpeptidation reaction is inhibited by penicillin.

Penicillin-binding proteins

Bacteria that have peptidoglycan have in their membranes proteins that bind penicillin and are called penicillin-binding proteins, or PBPs. For example, E. coli has at least seven such proteins, which are involved in catalyzing the late steps of peptidoglycan synthesis. The high molecular weight PBPs from E. coli (1a, 1b, 2, 3) have been shown to catalyze the transglycosylation and transpeptidation steps; that is, they are bifunctional enzymes. (The low molecular weight PBPs from E. coli, 4, 5, and 6, appear to be dispensable.) One of the proteins, PBP3, is specifically required for peptidoglycan synthesis in the septum. (See the discussion of cell division in Section 2.3.)

11.2 Lipopolysaccharide

11.2.1 Structure

Lipopolysaccharide (LPS) is a complex polymer of polysaccharide and lipid in the outer membrane of gram-negative bacteria. The outer membrane is described in more detail in Section 1.2.3. The best characterized lipopolysaccharides are those of E. coli and Salmonella typhimurium (Fig. 11.8).[4] The lipolysaccharides of other bacteria are very similar.

As seen in Fig. 11.8, lipopolysaccharide is composed of three parts: (1) a hydrophobic region embedded in the outer membrane called

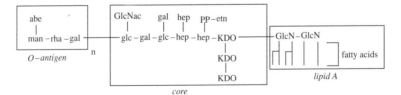

Fig. 11.8 The LPS of Salmonella typhimurium. Abbreviations: abe, abequose; man, mannose; rha, rhamnose; gal, galactose; GlcNAc, N-acetylglucosamine; glc, glucose; hep, heptose (L-glycero-D-mannoheptose); KDO, 3-deoxy-D-manno-octulosonic acid; etn, ethanolamine; P, phosphate; GlcN, glucosamine.

lipid A, which is composed of a backbone of two glucosamine residues linked β-1,6 and esterified via the hydroxyl groups to fatty acids; (2) a core polysaccharide region projecting from the outer membrane surface, whose composition is similar in all the Enterobacteriaceae and which is connected to lipid A via 3-deoxy-D-manno-octulosonate (KDO: see note 5); and (3) an outermost polysaccharide region connected to the core. The outermost polysaccharide region, sometimes called the O-antigen or the repeat oligosaccharide, is made up of repeating units of four to six sugars that vary considerably in composition among different strains of bacteria. The repeating oligosaccharide may have as many as 30 units.

There are 167 different serotypes of O-antigen in *Escherichia coli*. The structure of KDO is shown in Fig. 11.9. The arrangement of the LPS in the outer envelope is shown in Fig. 11.10 which shows that the LPS is anchored in the outer envelope by the hydrophobic lipid

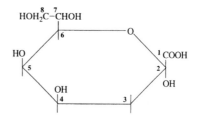

Fig. 11.9 The structure of KDO. The linkage between KDO residues has been reported to be a glycosidic linkage between a C2 hydroxyl in one KDO with either the C4 or C5 hydroxyl in a second KDO. The linkage to lipid A has been suggested to be from the C2 hydroxyl of KDO to the C6 hydroxyl of lipid A.

A region, while the repeating oligosaccharide protrudes into the medium.

The fatty acids in lipid A

Four identical fatty acids are attached directly to the glucosamine in lipid A from *E. coli* and

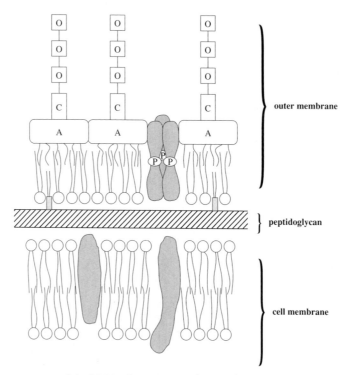

Fig. 11.10 The arrangement of the LPS in the outer membrane of some gram-negative bacteria: A, lipid A, C, core, O, oligosaccharide, P, porin. The outer envelope in the enteric gram-negative bacteria is asymmetric with the lipopolysaccharide confined to the outer leaflet of the lipid bilayer. However, some bacteria (e.g., penicillin-sensitive strains of *Neisseria* and *Treponema*) also have phospholipid in the outer leaflet. Phospholipids in the outer membrane make the bacterium more sensitive to hydrophobic antibiotics. Refer to Section 1.2.3 for a more complete discussion of the outer membrane.

Salmonella typhimurium. These are a C_{14} hydroxy fatty acid, β-hydroxymyristic acid (3-hydroxytetradecanoic acid), linked via ester bonds to the 3′-hydroxyls of the glucosamine, and via amide bonds to the nitrogen of the glucosamine. These fatty acids appear to be found uniquely in lipid A, and their presence in a bacterium implies the presence of a lipopolysaccharide-containing lipid A. Esterified to the hydroxyls of two of the β-hydroxymyristic acids are long-chain saturated fatty acids. In *E. coli*, these are lauric acid (C_{12}) and myristic acid (C_{14}). The fatty acids attached to the glucosamine anchor the lipopolysaccharide into the outer membrane.

11.2.2 Synthesis

Lipid A

The lipid A portion of the lipopolysaccharide is synthesized from UDP-GlcNAc, which as described in Fig. 11.4 is made from fructose-6-phosphate.[6] A model for lipid A synthesis in *E. coli* proposed by Raetz is shown in Fig. 11.11.[7] Lipid A is synthesized in the cytoplasmic membrane, but the early steps are catalyzed by three cytoplasmic enzymes. The first enzyme, UDP-GlcNAc acyltransferase, transfers β-hydroxymyristic acid from an ACP derivative to C3 of UDP-GlcNAc to form the monoacyl derivative. (Recall from Chapter 9 that fatty acids are synthesized as ACP derivatives.) We note parenthetically that the hydroxy acid transfer is actually a branch point in the metabolism of UDP-GlcNAc because some of the UDP-GlcNAc is incorporated into peptidoglycan. It reacts with PEP to form UDP-MurNAc, and it also condenses with UDP-MurNAc to form UDP-MurNAc-GlcNAc. Upon formation of the monoacyl derivative, a deacetylase removes the acetate from the nitrogen on C2. Interestingly, the gene for the deacetylase is *envA*, which had previously been known to be required for cell separation after septum formation. (See Section 2.3.)

After deacetylation a second molecule of β-hydroxymyristic acid is transferred from the ACP derivative by a third enzyme (an N-acyltransferase) to the nitrogen to form the 2,3-diacyl derivative. Some of the latter loses UMP and is converted to 2,3-diacylglucosamine-1-P (also called lipid X), which condenses with UDP-2,3-diacylglucosamine to form the disaccharide linked in β-1,6 linkage. A phosphate is added from ATP to form the 1,4-diphosphate derivative, and this is then modified by the addition of KDO from CMP derivatives and the esterification of fatty acids (lauryl and myristoyl) to the hydroxyl of the β-hydroxymyristic moieties. The fatty acids are donated by ACP derivatives.

Core

The core in the Enterobacteriaceae contains an inner region consisting of KDO, heptose, ethanolamine, and phosphate, and an outer region that consists of hexoses. The biosynthesis of the inner region is not fully understood. The outer core region grows as hexose units that are donated one at a time from nucleoside diphosphate derivatives to the nonreducing end of the growing glycan chain attached to the KDO. The addition of each sugar is catalyzed by a specific, is membrane-bound glycosyltransferase. The core–lipid A portion of the LPS is translocated across the cell membrane to the periplasmic surface, where the LPS is completed by attachment of the O-antigen.[8] (This is discussed later. See subsection entitled *How the lipopolysaccharide might be assembled* and Fig. 11.13.)

O-Antigen

The O-antigen region is synthesized by a mechanism that is different from that of core synthesis (Fig. 11.12). Whereas the core is synthesized via the addition of sugars one at a time to the growing end of the glycan chain, the O-antigen is synthesized as a separate polymer on a lipid carrier and then transferred as a unit to the core. The lipid carrier is undecaprenyl phosphate, the same molecule that carries the peptidoglycan precursors across the membrane. First, the repeat unit of the O-antigen is synthesized on the lipid carrier. This is done by a series of consecutive reactions in which a sugar moiety is transferred from a nucleoside diphosphate carrier to the nonreducing end of the growing repeat unit (Fig. 11.12, 1–4). Each of these reactions is catalyzed by a different enzyme. Then the repeat unit is transferred as a block to the growing oligosaccharide chain (Fig. 11.12, 5). The lipid-PP of the acceptor oligosaccharide chain is displaced and enters

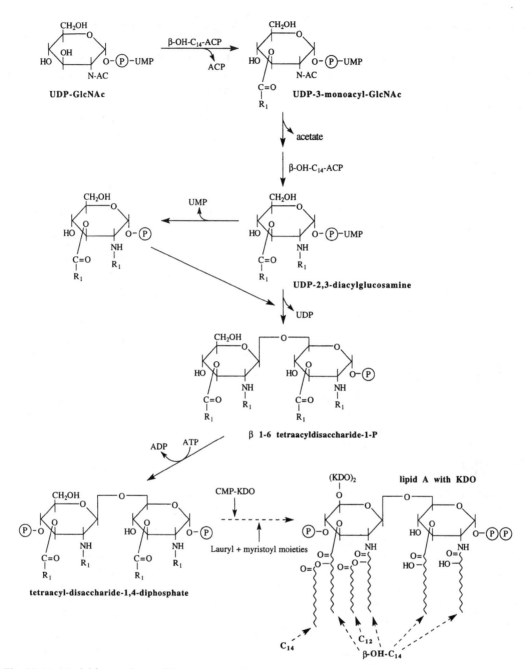

Fig. 11.11 Model for synthesis of lipid A in *E. coli*. See text for details. *Source*: Adapted from Raetz, C. R. H. 1987. Structure and biosynthesis of lipid A in *Escherichia coli*, pp. 498–503. In: *Escherichia coli and Salmonella typhimurium: Cellular and Molecular Biology*, Vol. 1. F. C. Neidhardt et al. (Eds.). ASM, Press, Washington, DC.

the lipid pyrophosphate pool, where it is hydrolyzed to lipid-P by a bacitracin-sensitive enzyme. Finally, the completed oligosaccharide chain is transferred to the lipid A–core region (Fig. 11.12, **6**), displacing the lipid pyrophosphate.

How the lipopolysaccharide might be assembled

In the model for how the LPS might be assembled depicted in Fig. 11.13 the O-antigen tetrasaccharide subunit is probably synthesized

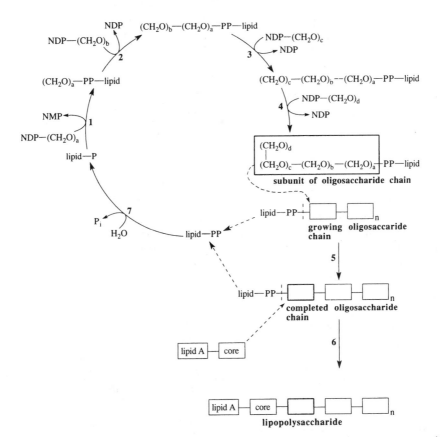

Fig. 11.12 Synthesis of O-antigen and attachment to core in *S. typhimurium*. The O-antigen is polymerized on the same phospholipid carrier that is used for peptidoglycan synthesis. Each sugar is added from a nucleotide diphosphate derivative by means of a specific sugar transferase (reactions **1–4**). The nucleotide that is used depends upon the sugar and the specific glycosyltransferase. For example, in *S. typhimurium* the nucleotide precursors are UDP-galactose, TDP-rhamnose, GDP-mannose, and CDP-abequose, in that order. The first sugar that is added (e.g., galactose in *S. typhimurium*) is transferred as a phosphorylated derivative to the lipid carrier. Hence the nucleoside monophosphate (NMP, or UMP in the case of *E. coli* and *S. typhimurium*) is released. When the O-antigen tetrasaccharide subunit is finished, it is transferred to the growing O-antigen chain by an enzyme called O-antigen polymerase (reaction **5**). The completed O-antigen is transferred from its lipid carrier to core–lipid A region by an enzyme called O-antigen: lipopolysaccharide ligase (reaction **6**). The lipid-PP that is displaced is hydrolyzed by a bacitracin-sensitive phosphatase (reaction **7**). All the reactions take place in the cell membrane. There exist immunoelectron microscopy data to suggest that the ligase reaction and perhaps the polymerase reaction take place on the periplasmic surface of the cell membrane. Presumably, the lipid carrier ferries the subunits across the cell membrane. It is not known how the LPS crosses the periplasm to enter the outer envelope. (See: Mulford, C. A., and M. J. Osborn. 1983. An intermediate step in translocation of lipopolysaccharide to the outer membrane of *Salmonella typhimurium*. *Proc. Natl. Acad. Sci. USA* **80**:1159–1163.)

on the lipid carrier on the cytoplasmic side of the membrane, then moves to the periplasmic side, where it is added to the growing O-antigen anchored to the membrane by its lipid carrier. The lipid carrier on the growing oligosaccharide is displaced. The core–lipid A region may also be assembled on the cytoplasmic surface and translocated to the periplasmic surface. The final completion of the lipo-polysaccharide would take place on the periplasmic surface, where the O-antigen is transferred to the core–lipid A, displacing the lipid-PP. The lipid-P is regenerated via a phosphatase and enters a common pool of lipid-P also used in the biosynthesis of peptidoglycan. It is not known how the lipopolysaccharide moves through the periplasm into the outer envelope.

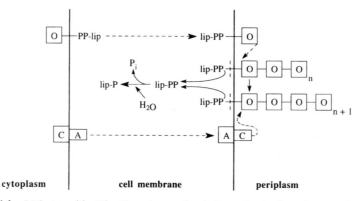

Fig. 11.13 Model for LPS assembly. The O-antigen subunit is synthesized on the cytoplasmic surface and then moves on the lipid carrier to the periplasmic surface of the membrane. A polymerase then transfers the O-antigen subunit to the growing O-antigen chain. The core–lipid A region is also synthesized on the cytoplasmic surface and is translocated to the periplasmic surface. The O-antigen is transferred to the core to complete the LPS. The entire LPS is translocated to the outer membrane. Abbreviations: O, O-antigen; C, core; lip, lipid carrier; A, lipid A.

11.3 Extracellular Polysaccharide Synthesis and Export in Gram-Negative Bacteria

11.3.1 Overview

The student should refer to Section 1.2.2 for a description of extracellular polysaccharides synthesized by bacteria and the biological roles that they play, and to recent reviews.[9–13] The extracellular polysaccharides vary in molecular weight from approximately 10^4 Da to about 10^6 Da. These extracellular polysaccharides are critical for cell survival in the natural habitat, although mutants lacking them can survive in the laboratory. They either are closely associated with the cell wall (capsular polysaccharides) or exist as an amorphous layer loosely attached to the cell (slime polysaccharides). Capsules offer protection from desiccation because they can be hydrated, and in some cases they protect pathogens from the host immune system by preventing phagocytosis.

The means of attachment of the extracellular polysaccharides to the bacterial surface varies according to the polysaccharide. Slime polysaccharides are not attached at all and are readily released from the cell surface. Capsular polysaccharides in gram-negative bacteria are generally attached via a hydrophobic molecule that anchors them to the lipid portion of the outer membrane. The hydrophobic anchor

might be lipid A or phosphatidic acid. Lipopolysaccharides (discussed in Section 11.2) are attached to the outer membrane via lipid A as shown earlier (Fig. 11.10). Figure 11.14 summarizes the modes of attachments.

Escherichia coli capsules

E. coli is an example of a gram-negative bacterium that may possesses an extracellular capsule surrounding its outer membrane. The capsule masks the antigenic determinants of the O-polysaccharides in the LPS. An easy way to detect the presence of the capsule is to heat the cells. Heating removes the capsule and makes the cells agglutinable with anti-O antigen antibody. Most capsular antigens are called K antigens (from the German *Kapselantigene*), and there are at least 80 types that differ in their antigenicity and composition.

The K antigens are acidic polysaccharides. Some contain either glucuronic or galacturonic acids, and others consist of other constituents in addition to, or instead of, glucuronic acid, including *N*-acetylneuraminic acid (NeuNAc), also called sialic acid, 3-deoxy-D-manno-octulosonic acid (KDO), *N*-acetylmannosamine, and phosphate. The repeating unit varies from two to six monosaccharides, depending upon the polysaccharide. Many appear to be substituted at their reducing end with lipid A or KDO–phosphatidic acid, presumably to anchor them in the outer membrane (Fig. 11.14B,C). (Another extracellular polysaccharide that is

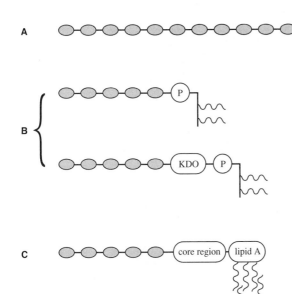

Fig. 11.14 Cell surface association of polysaccharides in gram-negative bacteria. (A) No membrane anchor. (B) Attachment through diacylglycerolphosphate. (C) Attachment through lipid A. KDO, 3-deoxy-D-manno-octulosonic acid; P, phosphate. *Source*: Whitfield, C., and M. A. Valvano. 1993. Biosynthesis and expression of cell-surface polysaccharides in gram-negative bacteria. *Adv. Microb. Physiol.* 35:135–246.

anchored to the outer membrane via lipid A is LPS. See Section 11.2.)

Interestingly, other bacteria may synthesize similar or identical capsular polysaccharides. For example, *Neisseria meningitidis* makes a capsular polysaccharide that is identical to the K1 capsule of *E. coli*. *Haemophilus influenzae* also makes a capsule quite similar to some of the K polysaccharides of *E. coli*. However, not all *E. coli* capsules are made from K-antigens. Some are made from molecules identical to the lipopolysaccharide O-antigen but without the lipid A or core regions. A classification of *E. coli* capsules that places them in four groups is based upon criteria such as the organization of their biosynthetic genes, the regulation of the expression of these genes, and the assembly of the polysaccharides and their translocation to the outer surface. These groups are as follows: group 1 (K30 serotype), group 2 (K1, K5 serotypes), group 3 (K10, K54 serotypes), and group 4 (K40, O111 serotypes).[13] The O-antigen in the O111 and other O-antigen capsules is not linked to LPS.

Summary of synthesis of extracellular polysaccharides

Most extracellular polysaccharides are synthesized from intracellular nucleoside diphos-phate–sugar precursors and must be transported through the cell membrane to the outside of the cell. The nucleoside diphosphate sugars are synthesized from sugar-1-P and nucleoside triphosphates as described in Section 11.1.2 for the synthesis of *N*-acetylglucosamine. If the extracellular polysaccharides are synthesized by gram-negative bacteria, they must be transported through the cell membrane, periplasm, and outer membrane. As we shall see, not all polysaccharides are synthesized and exported in the same way.

Steps in synthesis and assembly of extracellular polysaccharides in gram-negative bacteria

The synthesis of some extracellular polysaccharides takes place via undecaprenol intermediates in a pathway similar to the synthesis of the oligosaccharide of lipopolysaccharide, and the molecules are translocated through the membrane in an undecaprenol-dependent manner. However, other polysaccharides are not synthesized via lipid intermediates and are translocated through the membrane in an entirely different way, using specific transporters belonging to the family of ATP-binding cassette (ABC) transporters.

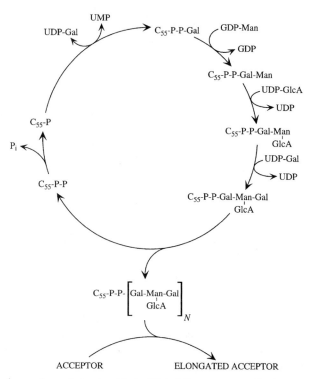

Fig. 11.15 Synthesis of capsular polysaccharide in *Klebsiella aeroqenes*. Each monosaccharide is donated to the tetrasaccharide repeat unit from a UDP derivative. This takes place on the inner surface of the cell membrane. Polymerization takes place while the sugars are attached to undecaprenol diphosphate. The tetrasaccharide repeat units are transferred from the lipid carrier to the growing chain, presumably on the outer surface of the cell membrane. See Figures 11.12 and 11.13 for a model suggesting how this might occur. Abbreviations: UDP, uridine diphosphate; Gal, galactose; Man, mannose; GlcUA, glucuronic acid, C_{55}, undecaprenol; P, Phosphate.

11.3.2 Polysaccharide synthesis via undecaprenol diphosphate intermediates

Several bacteria synthesize certain exopolysaccharides via undecaprenol intermediates similar to the pathway for the biosynthesis of the oligosaccharide repeat unit in lipopolysaccharide (Fig. 11.13) and the disaccharide repeat unit in peptidoglycan (Fig. 11.6). Polysaccharides that are synthesized via this pathway include the group 1 K-antigens of *E. coli*, the extracellular polysaccharide xanthan made by *Xanthomonas campestris*, and the capsular polysaccharide of *Klebsiella (Aerobacter) aerogenes*. As an example, we will consider the biosynthesis of the *K. aerogenes* capsular polysaccharide.

K. aerogenes capsular polysaccharide

The capsule made by *K. aerogenes* is a polymer of repeating tetrasaccharides composed of galactose, mannose, and glucuronic acid in a molar ratio of 2:1:1. The glucuronic is attached as a branch at each mannose residue. Its synthesis resembles the synthesis of the oligosaccharide in lipopolysaccharide and is depicted in Fig. 11.15.[14] The tetrasaccharide repeating unit is synthesized on undecaprenol diphosphate from nucleotide diphosphate sugar intermediates. Each repeating unit is then added as a block to the growing oligosaccharide attached to undecaprenol diphosphate.

11.3.3 Synthesis of E. coli group 2 K1 antigen: A pathway that may use an undecaprenol monophosphate intermediate

Chemical nature and biological role of the K1 capsule

The *E. coli* K1 capsule (a group 2 antigen), *Neisseria meningitidis* group B capsular polysaccharide, and capsular polymers of *Pasteurella haemolytica* and *Moraxella non-liquefaciens*, are homopolymers of sialic acid approximately 200 residues long. (For a review, see ref. 15.) The polysialic acid capsule of *E. coli* is a virulence factor for strains of *E. coli* that cause neonatal meningitis, septicemia, and urinary tract infections in children. For this and other reasons, it has been of interest for a long time. Polysialic acid capsules are virulence factors in part because they resemble poly-saccharides on host tissue and consequently are poorly immunogenic. Thus they make the bacteria more resistant to the immune response. Additionally, capsules containing sialic acid make the cells more resistant to complement-mediated killing, apparently because they inhibit the activation of complement by the alternative system.

Biosynthesis of the K1 capsule

The mechanism of polysialic acid chain elonga-tion in *E. coli* K1 has been well studied both biochemically and genetically. (See refs. 16 and 17 and references therein.) The polymer grows by the stepwise addition of single sialic acid residues, apparently from cytidine monophos-phate–sialic acid. This can be demonstrated by adding CMP [^{14}C] sialic acid to incubation mixtures containing washed membranes. The enzyme that catalyzes the reaction is called *sialytransferase*, which is loosely attached to the cell membrane. The cell extracts make undecaprenol monophosphate linked sialic acid residues by transferring sialic acid from CMP–sialic acid to undecaprenol phosphate, whereupon the sialic acid is transferred to endogenous acceptors in the membrane (reviewed in ref. 9.)

However, doubts have been raised about whether the lipid-linked sialyl residues are an obligate intermediate in the elongation reac-tion.[16] In one model, which does make this assumption, the sialic acid is transferred from CMP to undecaprenol phosphate to form sialyl monophosphorylundecaprenol, which trans-fers the sialic acid to the nonreducing end of the growing polysialic chain. It is not known whether the growing polymer is attached to undecaprenol. Neither is the identity of the initial receptor for the first sialic acid known,

being referred to simply as the endogenous receptor. The endogenous receptor does not contain sialic acid.[18] The completed poly-saccharide has phospholipid attached to its reducing end, which presumably anchors the polysaccharide to the outer membrane. It is not known at what stage of the biosynthesis of the polysialyl polymer the phospholipid is attached.

11.3.4 Pathways not involving undecaprenol derivatives

The synthesis of some extracellular polysac-charides, including some of the group 2 K-antigen capsules such as K5 in *E. coli*, alginate synthesized by *Azotobacter vinelandii* and *Pseudomonas aeruginosa*, and cellulose syn-thesized by *Acetobacter xylinum*, does not involve undecaprenol intermediates. Where such synthesis has been characterized, it is found that the polysaccharides are synthesized from nucleotide diphosphate precursors added to the growing oligosaccharide chain. For example, alginate, which is a linear copolymer of D-mannuronic and D-guluronic acid, is synthesized by brown algae from GDP–mannuronic and GDP–guluronic acid. Another example is cellulose, which is polymerized from UDP–glucose by a membrane-bound cellulose synthetase in *Acetobacter xylinum*.

11.3.5 Export of polysaccharides

Undecaprenol phosphate based translocation

Because polysaccharides are polymerized from nucleotide derivatives of the sugars that are made in the cytoplasm, one can assume that at least the first steps in polymerization take place on the cytoplasmic surface of the cell membrane. Because of this, there must be a mechanism to transport newly synthesized polysaccharide across the cytoplasmic mem-brane. One model postulates that the role of undecaprenol phosphate is to serve as part of a transmembrane assembly process that synthe-sizes and moves completed polysaccharides to the periplasmic surface of the cell membrane. This is depicted in Fig. 11.16A.

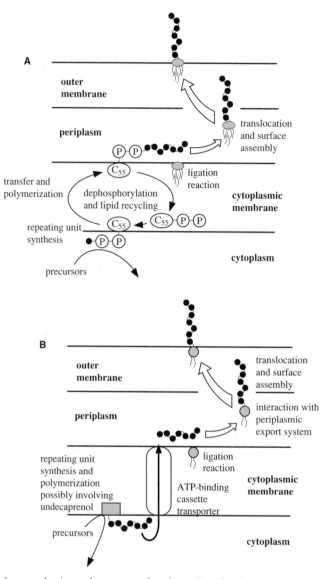

Fig. 11.16 Models for synthesis and export of polysaccharides in gram-negative bacteria. (A) *rfe*-Independent O-polysaccharide biosynthesis in *Salmonella enterica*. (B) Group 2 capsular polysaccharide biosynthesis in *Escherichia coli*. Abbreviations: C_{55}, undecaprenol; P, phosphate. *Source*: Whitfield C., and M. A. Valvano. 1993. Biosynthesis and expression of cell-surface polysaccharides in gram-negative bacteria. *Adv. Microb. Physiol.* **35**:135–246.

Translocation via ABC transporters

There is evidence for a specific transport mechanism that does not require undecaprenol. This evidence has been derived from examining the genes (*kps* cluster) for biosynthesis and export of group 2 capsular polysaccharides (including K1 and K5) in *E. coli* and is depicted in Fig. 11.16B. (See ref. 15 for a review.) Mutations in a particular region of the *kps* gene cluster (region 3) required for biosynthesis of the

polysaccharide result in the accumulation of the polysaccharide in the cytoplasm. The deduced amino acid sequences of the protein products of these genes (KpsM and KpsT) indicate that they belong to the family of ABC transporters. Accordingly, they are referred to as the KpsMT transporter and probably form a protein channel through which the polysaccharides are translocated either during or after biosynthesis. (See Section 16.3.3 for a discussion of ABC transporters.)

Similar genes have been found in the biosynthetic gene cluster for extracellular polysaccharide synthesis in other bacteria, including *Haemophilus*, *Neisseria*, *Rhizobium*, *Agrobacterium*, and may be widespread. Thus, it can be concluded that many extracellular polysaccharides are exported through the cell membrane in protein channels by means of energy derived at least in part from the hydrolysis of ATP. (A Δp is also required.) How these polysaccharides reach the outer membrane surface is not clear. Various possible mechanisms are reviewed in refs. 9 and 15.

11.4 Levan and Dextran Synthesis

Certain bacteria (e.g., the lactic acid bacteria) synthesize from sucrose extracellular polymers of fructose, called *levans*, or glucose, called *dextrans*. The synthesis is catalyzed by extracellular enzymes called *levansucrase* or *dextransucrase*. The reactions are as follows:

$$y \text{ sucrose} + (\text{fructose})_n$$
$$\rightarrow (\text{fructose})_{n+y} + y \text{ glucose} \qquad (11.1)$$

$$y \text{ sucrose} + (\text{glucose})_n$$
$$\rightarrow (\text{glucose})_{n+y} + y \text{ fructose} \qquad (11.2)$$

In both reactions a monosaccharide is transferred from the disaccharide to the reducing end of the growing oligosaccharide chain. The polysaccharide chain remains attached to the enzyme during its elongation. The dextran is a glucose polymer (glucan) in which the glucose residues are attached via $\alpha,1\text{-}6$ linkages. A different enzyme makes a glucan called mutan, which has $\alpha,1\text{-}3$ linkages. *Streptococcus mutans*, the bacterium that is a principal cause of tooth decay, forms dextrans, mutans, and fructans. The glucans enable the bacteria to adhere to the surface of the teeth.

11.5 Glycogen Synthesis

Certain bacteria synthesize glycogen, a polymer of glucose, as an intracellular carbon and energy reserve. Glycogen is a branched polysaccharide. The linear portion consists of glucose residues connected by $\alpha,1\text{-}4$ linkages and the branches are attached via $\alpha,1\text{-}6$ linkages. The oligosaccharides are extended by the addition of glucose units donated by ADP–glucose in a reaction catalyzed by *glycogen synthetase*. A *branching enzyme* transfers six to eight glucose segments from the linear portion of the oligosaccharide to form an $\alpha,1\text{-}6$ linked branch. The synthesis of nucleoside diphosphate derivatives of sugars from sugar-1-P and nucleoside triphosphates was described in Section 11.1.2.

11.6 Summary

There are two major problems that must be solved in connection with the synthesis of the cell wall: how to move the precursors through the cell membrane and how to make peptide bonds outside the cell membrane, far from the cellular ATP pools. Bacteria employ a lipid carrier, undecaprenyl phosphate, to carry the cell wall precursors through the membrane. Interestingly, eukaryotes employ a similar compound, dolichol phosphate, to move oligosaccharide precursors through the ER membrane to synthesize glycoproteins.

The peptidoglycan peptide cross-link forms as a result of a transpeptidation reaction. During the transpeptidation reaction, an amino group from the diamino acid (e.g., DAP) displaces a terminal D-alanine. This is an exchange of a peptide bond for one that was made in the cytosol at the expense of ATP. ATP also provides the energy to make the glycosidic linkages in the cell wall polymers because ATP drives the synthesis of the sugar-PP–lipid intermediates.

The formation of the glycosidic linkage is a straightforward displacement of the lipid pyrophosphate from the C1 carbon of the *N*-acetylmuramic acid by the hydroxyl on the incoming C4 carbon of *N*-acetylglucosamine. The reaction is driven to completion by the subsequent hydrolysis of the lipid pyrophosphate to the lipid phosphate and inorganic phosphate.

Lipopolysaccharide synthesis can be thought of as occurring in four stages: (1) synthesis of the lipid A portion, (2) synthesis of the core region by adding one sugar at a time to lipid A from nucleoside diphosphate precursors,

(3) synthesis of the complete repeat oligosaccharide, and (4) attachment of the repeat oligosaccharide to the core. All these events are associated with the membrane. It appears that the lipid A and core portions are synthesized on the cytoplasmic side of the membrane and then translocated in an unknown manner to the periplasmic surface. The oligosaccharide (O-antigen) subunits are assembled on undecaprenyl pyrophosphate and transferred to a growing oligosaccharide chain. When the oligosaccharide is complete, it is attached to the core. How the lipopolysaccharide enters the outer envelope is not known.

Capsular and extracellular polysaccharides are synthesized and exported in at least two ways. One pathway is similar to the pathway for the synthesis of the oligosaccharide in lipopolysaccharide and involves undecaprenol intermediates. A second pathway appears to be more complex, perhaps involving ABC-type transporters to move the polysaccharide through the cell membrane.

Study Questions

1. O-Antigen synthesis requires a lipid carrier, but core synthesis does not. Offer a plausible explanation for the difference.

2. During the synthesis of the pentapeptide in the peptidoglycan precursor, an ATP is expended to make the peptide bond as each amino acid is added to the growing pentapeptide. The products are ADP and P_i. Write a plausible mechanism by which ATP is used to provide the energy to make a peptide bond between two amino acids. (*Note*: mRNA and ribosomes are not involved. The addition of each amino acid is catalyzed by a separate enzyme, specific for that amino acid.)

3. Peptidoglycan and lipid A share a common pathway early in their syntheses. Outline the early stages of both pathways up to the branch point.

4. Carriers of subunit moieties play important roles in biosynthesis. Usually, the carriers are involved in more than one pathway. What carrier is common to both peptidoglycan and lipopolysaccharide synthesis?

5. Important enzymes in cell wall peptidoglycan synthesis are membrane-bound transglycosidases that transfer carbohydrate subunits to the growing polymer. What ensures that the growing end of the polymer remains at the site of the transglycosidase?

6. Show how ATP drives the synthesis of UDP–GlcNAc from glucose. Focus on reactions in which phosphoryl and nucleotide groups are transferred. You must show how ATP drives the synthesis of UTP. There are two phosphate groups in MurNAc(pentapeptide)-PP-lipid. Show how one of them is derived from ATP. What eventually happens to this phosphate?

7. Covalent bond formation in the cytoplasm is driven by high-energy molecules such as ATP or other nucleoside triphosphates. Peptidoglycan synthesis offers some examples of how covalent bonds can be formed outside the cytoplasm without access to a source of high-energy molecules such as nucleoside triphosphates, which are cytoplasmic. Two examples are the transpeptidation reaction, which results in the synthesis of a peptide bond, and the transglycosidase reaction, which results in the synthesis of a glycosidic bond. These reactions are displacement reactions in which a diamino acid (e.g., diaminopimelic acid) displaces D-alanine on the pentapeptide (transpeptidation) and the N-acetylglucosamine moiety of the incoming lipid–disaccharide pentapeptide displaces the lipid pyrophosphate in the growing glycan chain (transglycosidase). If one carefully examines the pathway for peptidoglycan synthesis, one can see that nucleoside triphosphates are indeed used in the cytoplasm in reactions that ultimately provide the energy for the transpeptidation and transglycosidase reactions. Write these cytoplasmic, reactions, and show how they drive the transpeptidation and transglycosidase reactions.

REFERENCES AND NOTES

1. van Heijenoort, J. 1996. Murein synthesis, pp. 1025–1034. In: *Escherichia coli and Salmonella: Cellular and Molecular Biology*, Vol. 1. F. C. Neidhardt et al. (Eds.). ASM Press, Washington, DC.

2. Höltje, J.-V. 1998. Growth of the stress-bearing and shape-maintaining murein sacculus of *Escherichia coli*. *Microbiol. Mol. Biol. Rev.* **62**:181–203.

3. Hartmann, E., and H. Konig. 1994. A novel pathway of peptide biosynthesis found in methanogenic archaea. *Arch. Microbiol.* **162**:430–432.

4. Raetz, C. R. 1996. Bacterial lipopolysaccharide: a remarkable family of bioactive macroamphiphiles, pp. 1035–1063. In: *Escherichia coli and Salmonella: Cellular and Molecular Biology*, Vol. 1. F. C. Neidhardt et al. (Eds.). ASM Press, Washington, DC.

5. This was formerly called 2-keto-3-deoxyoctonate.

6. Rick, P. D. 1987. Lipopolysaccharide biosynthesis, pp. 648–662. In: *Escherichia coli and Salmonella typhimurium: Cellular and Molecular Biology*, Vol. 1. F. C. Neidhardt et al. (Eds.). ASM Press, Washington, DC.

7. Raetz, C. R. H. 1987. Structure and biosynthesis of lipid A, pp. 498–503. In: *Escherichia coli and Salmonella typhimurium: Cellular and Molecular Biology*, Vol. 1. F. C. Neidhardt et al. (Eds.). ASM Press, Washington, DC.

8. Mulford, C. A. and M. J. Osborn. 1983. An intermediate step in translocation of lipopolysaccharide to the outer membrane of *Salmonella typhimurium*. *Proc. Natl. Acad. Sci. USA* **80**:1159–1163.

9. Whitfield, C., and M. A. Valvano. 1993. Biosynthesis and expression of cell-surface polysaccharides in gram-negative bacteria, pp. 135–246. In: *Advances in Microbial Physics*, Vol. 35. A. H. Rose (Ed.) Academic Press, New York.

10. Rick, P. D., and R. P. Silver. 1996. Enterobacterial common antigen and capsular polysaccharides, pp. 104–122. In: *Escherichia coli and Salmonella: Cellular and Molecular Biology*, Vol. 1. F. C. Neidhardt et al. (Ed.). ASM Press, Washington, DC.

11. Roberts, I. S. 1995. Bacterial polysaccharides in sickness and in health. *Microbiology* **141**:2023–2031.

12. Roberts, I. S. 1996. The biochemistry and genetics of capsular polysaccharide production in bacteria. *Annu. Rev. Microbiol.* **50**:285–315.

13. Whitfield, C., and I. S. Roberts. 1999. Structure, assembly and regulation of expression of capsules in *Escherichia coli*. *Mol. Microbiol.* **31**:1307–1319.

14. Troy, F. A., F. E., Frerman, and E. C. Heath. 1971. The biosynthesis of capsular polysaccharide in *Aerobacter aerogenes*. *J. Biol. Chem.* **246**:118–133.

15. Bliss, J. M., and R. P. Silver. 1996. Coating the surface: a model for expression of capsular polysialic acid in *Escherichia coli* K1. *Mol. Microbiol.* **21**:221–231.

16. Steenbergen, S. M., and E. R. Vimr. 1990. Mechanism of polysialic acid chain elongation in *Escherichia coli* K1. *Mol. Microbiol.* **4**:603–611.

17. Cieslewicz, M., and E. Vimr. 1997. Reduced polysialic acid capsule expression in *Escherichia coli* K1 mutants with chromosomal defects in *kpsF*. *Mol. Microbiol.* **26**:237–249.

18. Rohr, T. E., and F. A. Troy. 1980. Structure and biosynthesis of surface polymers containing polysialic acid in *Escherichia coli*. *J. Biol. Chem.* **255**:2332–2342.

12

Inorganic Metabolism

Inorganic molecules such as derivatives of sulfur, nitrogen, and iron are used by prokaryotes in a variety of metabolic ways related to energy metabolism and biosynthesis.

1. There are *assimilatory pathways* in which inorganic nitrogen and sulfur are incorporated into organic materials (e.g., amino acids and nucleotides) that contain sulfur and/or nitrogen. Most prokaryotes can do this. In addition, many prokaryotes can also utilize nitrogen gas as a source of nitrogen, a process called *nitrogen fixation*.

2. There are *dissimilatory pathways* in which inorganic compounds are used instead of oxygen as electron acceptors, a process called *anaerobic respiration*. The reduced products are excreted into the environment. During anaerobic respiration a Δp is created in the same way as during aerobic respiration (see Section 4.6). For example, many facultative anaerobes can use nitrate as an electron acceptor, reducing it to ammonia or nitrogen gas. Several obligate anaerobes use sulfate as an electron acceptor, reducing it to hydrogen sulfide. There also exist bacteria that can use Fe^{3+} or Mn^{4+} as an electron acceptor during anaerobic growth.[1-3] The latter organisms are responsible for most of the reduction of iron and manganese that takes place in sedimentary organic matter (e.g., lake sediments). However, only a few of the iron and manganese reducers have

been isolated, and little is known about their physiology.

3. There are *oxidative pathways* in which inorganic compounds such as H_2, NH_3, NO_2^-, S°, H_2S, and Fe^{2+}, rather than organic compounds, are oxidized as a source of electrons and energy. Organisms that derive their energy and electrons for biosynthesis in this way are called *chemolithotrophs*. (If they use CO_2 as the sole or major source of carbon, they are called *chemolithoautotrophs*.)

12.1 Assimilation of Nitrate and Sulfate

Many bacteria can grow in media in which the only sources of nitrogen and sulfur are inorganic nitrate salts and sulfate salts. The nitrate is reduced to ammonia, and the ammonia is incorporated into the amino acids glutamine and glutamate using the GS/GOGAT system (see Section 9.3.1). Glutamate and glutamine are the sources of amino groups for the other nitrogen-containing organic compounds. The sulfate is reduced to H_2S, which is immediately incorporated into the amino acid cysteine via the O-acetylserine pathway shown earlier (Fig. 9.19). Cysteine, in turn, is the source of sulfur for other organic molecules (e.g., such as methionine, coenzyme A, as CoASH; and acyl carrier protein, as ACPSH).

12.1.1 Nitrate assimilation

Nitrate can serve as the source of cellular nitrogen for plants, fungi, and many bacteria. Although some bacteria (e.g., *Klebsiella pneumoniae*) assimilate nitrate during aerobic growth, certain other bacteria (e.g., *E. coli*, *Salmonella typhimurium*) assimilate nitrate only during anaerobic growth. (See ref. 4 for a review of nitrate transport and reduction.) The nitrate is first reduced to ammonia. Since the oxidation state for the nitrogen in nitrate (NO_3^-) is +5 and for the nitrogen in ammonia (NH_3^+) it is −3, eight electrons must be transferred to nitrate to reduce it to ammonia. The enzymes involved in the assimilatory reduction of nitrate to ammonia are *cytoplasmic nitrate reductase*, which reduces nitrate to nitrite, and *cytoplasmic nitrite reductase*, which reduces nitrite to ammonia. The electron donors for the nitrate reductase can be NADH (via a separate NADH oxidoreductase), ferredoxin, or flavodoxin, depending upon the bacterium, as explained in note 5. The electron donor for the major nitrite reductase in *E. coli* is NADH. The electron transport pathway is shown in Fig. 12.1 as proceeding in two-electron transfer steps. The ammonia that is formed is incorporated into glutamine via *glutamine synthase* (GS). The glutamine then serves as the amino donor for pyrimidine, purine, and amino sugar biosynthesis (Sections 9.2.2, 9.2.3, and 11.1.2),

as well as for the synthesis of glutamate (glutamate synthetase). Glutamate is the amino donor for amino acid biosynthesis via the transamination reactions described in Section 9.3.1.

12.1.2 Sulfate assimilation

Many bacteria can use sulfate (SO_4^{2-}) as their principal source of sulfur. The sulfate is first reduced to sulfide (H_2S and HS^-) and then incorporated into cysteine. (See note 6 for an explanation why H_2S and HS^- are the major species of sulfide in the cytoplasm.) Since the oxidation level of sulfur in SO_4^{2-} is +6 and in S^{2-} is −2, a total of eight electrons is required to reduce sulfate to sulfide.

The first step in the reduction is the formation of adenosine-5′-phosphosulfate (APS) catalyzed by the enzyme ATP sulfurylase (Fig. 12.2, reaction 1). Here sulfate acts as a nucleophile and displaces pyrophosphate (PP_i). (See Section 7.1.1 for a discussion of nucleophilic displacements.) The pyrophosphate is subsequently hydrolyzed to inorganic phosphate, thus driving the synthesis of APS to completion. There is a sound thermodynamic reason for making the AMP derivative of sulfate prior to its reduction. Attaching the sulfate to AMP raises the reduction potential, making APS a better electron acceptor than free sulfate. (The E_0' for the reduction potential of sulfate to sulfite is very low (−520 mV) such that its

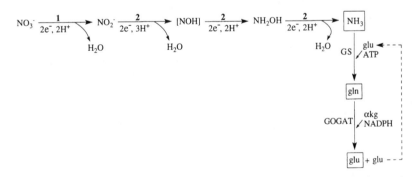

Fig. 12.1 Assimilatory nitrate reduction. This pathway is present in all bacteria that reduce nitrate to ammonia, which is then incorporated into cell material. The enzymes are found in the cytosol and are not coupled to ATP formation. Nitrate is reduced via two-electron steps to nitrite, nitroxyl, hydoxylamine, and ammonia. The ammonia is incorporated into organic carbon via glutamine synthetase (GS) and the GOGAT enzyme, or via glutamate dehydrogenase. (See Section 9.3.1.) Enzymes: **1**, nitrate reductase; **2**, nitrite reductase. Abbreviations: glu, glutamate; α-kg, α-ketoglutarate.

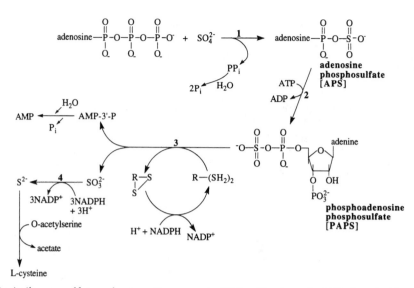

Fig. 12.2 Assimilatory sulfate reduction. Enzymes: **1**, ATP sulfurylase; **2**, APS phosphokinase; **3**, PAPS reductase; **4**, sulfite reductase; **5**, O-acetylserine sulfhydrylase. R(SH)$_2$ is reduced thioredoxin.

reduction, even by H$_2$ (E'_0 = −420 mV) is endergonic.

The next reaction, **2**, is the phosphorylation of APS to form adenosine-3′-phosphate-5′ phosphosulfate (PAPS), catalyzed by APS kinase. The reaction is an attack by the 3′-hydroxyl of the ribose of APS on the terminal phosphate of ATP, displacing ADP. The PAPS is then reduced to sulfite with the release of AMP-3′-phosphate (reaction 3). The reductant is a sulfhydryl protein called thioredoxin, which in turn accepts electrons from NADPH. (Thioredoxin is used in other metabolic pathways as a reductant. For example, recall from Section 9.2.5 that thioredoxin also reduces the nucleoside diphosphates to the deoxynucleotides. The sulfite is then reduced by NADPH to hydrogen sulfide (H$_2$S) (reaction 4). Hydrogen sulfide is very toxic and does not accumulate. The sulfide enzymatically displaces acetate from O-acetylserine to form cysteine (see Fig. 9.19). The AMP-3′-phosphate is hydrolyzed to AMP and P$_i$, thus helping to drive the overall reaction to completion. Note that three ATPs are used to reduce sulfate to sulfide, two to make the PAPS derivative, and a third to phosphorylate the AMP released from AMP-3′-phosphate to ADP (reaction 3). (The ADP is then converted to ATP via respiratory phosphorylation or substrate-level phosphorylation.)

12.2 Dissimilation of Nitrate and Sulfate

In the *dissimilatory* pathways, the nitrate and sulfate are used as electron acceptors during anaerobic respiration. The reduced products are excreted rather than being incorporated into cell material. Whereas many facultative anaerobes are capable of dissimilatory nitrate reduction and employ this pathway when oxygen is not available, the use of sulfate as an electron acceptor is restricted to obligate anaerobes, bacteria called *sulfate reducers*.

12.2.1 Dissimilatory nitrate reduction

Dissimilatory nitrate reduction (nitrate respiration) is usually facultative and occurs as a substitute for aerobic respiration when the oxygen levels become very low. It takes place in membranes, and a Δp is usually made. The products of nitrate respiration can be nitrite, ammonia, or nitrogen gas.

Denitrification

When the nitrate (or nitrite) is reduced to nitrogen gas (or nitric oxide gas, NO, or nitrous oxide gas, N$_2$O), the process is called *denitrification*. Denitrification, which can be an ecologically important drain of nitrogen from the soil, occurs when conditions become anaerobic (e.g., in water-logged soil) and also during composting and sludge digestion. The denitrifiers reduce the available nitrate in the soil, but they are also important in anaerobic niches for the breakdown of undesirable biodegradable materials, plant materials, complex organic compounds, and so forth.[7] The breakdown of organic materials by denitrifiers is usually more complete than decomposition via a fermentative process. Many bacteria denitrify, but the most commonly isolated denitrifiers are *Alcaligenes* and *Pseudomonas* species, and to a lesser extent *Paracoccus denitrificans*. The electron transport pathway for denitrification by *P. denitrificans* is summarized in Section 4.7.2.

12.2.2 Dissimilatory sulfate reduction

General description of the sulfate reducers

The sulfate reducers are heterotrophic anaerobes that grow in anaerobic muds, mostly in anaerobic parts of fresh water, and in seawater.[8] They carry out anaerobic respiration during which sulfate is reduced to H_2S. Some can be grown autotrophically with H_2 as the source of electrons and SO_4^{2-} as the electron acceptor. Formerly the sulfate reducers were believed to use as carbon sources only a limited variety of compounds (e.g., formate, lactate, pyruvate, malate, fumarate, ethanol, and a few other simple compounds). Recently, however, it has been realized that, depending upon the species, many other carbon sources can be used, including straight-chain alkanes and a variety of aromatic compounds. Sulfate reducers comprise a very diverse group of organisms that include both gram-positive and gram-negative bacteria, as well as archaea. An example of the latter is *Archaeoglobus*, a hyperthermophile isolated from sediments near hydrothermal vents. Gram-positive, spore-forming sulfate reducers belong to the genus *Desulfotomaculum*, which is very diverse. The most prominent of the gram-negative sulfate reducers belong to the genus *Desulfovibrio*, which is also phylogenetically diverse.

Traditionally, the sulfate reducers are divided into two physiological groups, I and II. Those in group I cannot oxidize acetyl–CoA to CO_2 and therefore excrete acetate when growing on certain carbon sources (e.g., lactate, ethanol). The group I genera include *Desulfovibrio* and most *Desulfotomaculum* species. Group II organisms can oxidize acetyl–CoA to CO_2. Group II sulfate reducers are found in several genera, including *Desulfotomaculum* and *Desulfobacter*.

There exist two pathways for oxidizing acetyl–CoA anaerobically to CO_2: a modified citric acid cycle and the acetyl–CoA pathway. *Desulfobacter* has a modified citric acid cycle that resembles that found in aerobes except that (1) instead of a citrate synthase there is an ATP-citrate lyase, and (2) the NAD^+-linked α-ketoglutarate dehydrogenase is replaced by a ferredoxin-dependent enzyme. The pathway is called the reductive tricarboxylic acid pathway, although it can be used in the oxidative direction (Section 13.1.9). Other sulfate reducers (e.g., *Desulfobacterium autotrophicum*, *Desulfotomaculum acetooxidans*, and the archaeon *Archaeoglobus fulgidus*) oxidize acetyl–CoA to CO_2 by means of the acetyl–CoA pathway described in Section 13.1.3. The use of these pathways in the reductive direction in autotrophic CO_2 fixation by facultatively autotrophic sulfate reducers is also described in Sections 13.1.3 and 13.1.4. Many sulfate reducers are known to be able to ferment pyruvate to acetate, or to acetate and propionate in the absence of sulfate. This is described in Section 14.8.

The path of electrons to sulfate in Desulfovibrio

Desulfovibrio carries out an anaerobic respiration during which electrons flow in the cell membrane to sulfate as the terminal electron acceptor, reducing it to H_2S. Electron flow is coupled to the generation of a Δp, which is used for ATP synthesis via respiratory phosphorylation. As stated earlier, these electrons may come from the oxidation of organic compounds (e.g., lactate). Since dissimilatory sulfate reduction takes place in membranes, involves cytochromes, and generates a Δp, it is very different from the assimilatory pathway, which is a soluble pathway and does not generate a Δp or ATP.

A pathway of electron transport has been proposed for the genus *Desulfovibrio* (Fig. 12.3). It is called the hydrogen cycling model. Lactate is oxidized to pyruvate In the cytoplasm yielding

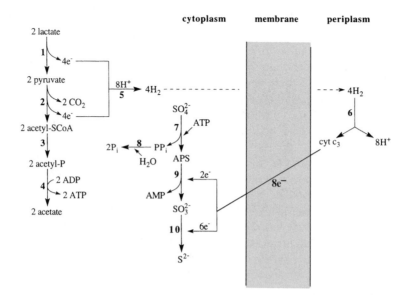

Fig. 12.3 Pathway for dissimilatory sulfate reduction in *Desulfovibrio*. Enzymes: **1**, lactate dehydrogenase; **2**, pyruvate–ferredoxin oxidoreductase; **3**, phosphotransacetylase; **4**, acetate kinase; **5**, cytoplasmic hydrogenase; **6**, periplasmic hydrogenase; **7**, ATP sulfurylase; **8**, pyrophosphatase; **9**, APS reductase; **10** sulfite reductase.

two electrons (Fig. 12.3, reaction **1**). The oxidation of lactate to pyruvate is catalyzed by a membrane-bound lactate dehydrogenase that probably is a flavoprotein. The pyruvate is then oxidized to acetyl–CoA and CO_2 by pyruvate–ferredoxin oxidoreductase, an enzyme found in other anaerobes (reaction **2**). (See Section 7.3.2 for a description of the pyruvate–ferredoxin oxidoreductase reaction.) The acetyl–CoA is used to generate ATP via a substrate-level phosphorylation, using two enzymes common in bacteria, phosphotransacetylase and acetate kinase (reactions **3** and **4**). (See Section 7.3.2 for a description of the phosphotransacetylase and acetate kinase reactions.)

Since each lactate that is oxidized to acetyl–CoA yields four electrons, two lactates must be oxidized to provide the eight electrons to reduce one sulfate to sulfide. The model proposes that the electrons are transferred from lactate dehydrogenase and pyruvate–ferredoxin oxidoreductase to a cytoplasmic hydrogenase and then to H^+, producing H_2 (reaction **5**), and the H_2 diffuses out of the cell into the periplasm. In the periplasm, the H_2 is oxidized by a periplasmic hydrogenase and the electrons are transferred to cytochrome c_3 (reaction **6**). From cyt c_3 the electrons travel through a series of membrane-bound electron carriers to APS

reductase and sulfite reductase in the cytosol (reactions **7**, **9**, **10**). A pyrophosphatase pulls the sulfurylation of ATP to completion (reaction **8**).

Note that according to the scheme proposed in Fig. 12.3, the inward flow of electrons across the membrane leaves the protons from the hydrogen on the outside, thus generating a Δp. An examination of the scheme reveals that the Δp is necessary for growth. The two ATPs made via substrate-level phosphorylation from the two moles of acetyl–CoA are used up in reducing the SO_4^{2-}. This follows because, after reduction to sulfite, AMP is produced, and the energy equivalent of two ATPs is required to make one ATP from one AMP. Thus, without using the Δp, there would be no ATP left over for growth. Some strains of *Desulfovibrio* can grow on CO_2 and acetate as the sole sources of carbon, and the Δp produced by the redox reaction between H_2 and SO_4^{2-} is the sole source of energy.

There are also facultatively autotrophic strains of sulfate reducers that grow on CO_2, H_2, and SO_4^{2-}, (e.g., *Desulfobacterium autotrophicum*). All these strains derive their ATP from the Δp created during sulfate reduction. Serious reservations, however, have been expressed over whether free H_2 is actually an electron carrier during lactate oxidation by sulfate in *Desulfovibrio* (i.e., whether the hydrogen-cycling model is generally valid). (See note 9.) Clearly, there is still much to be learned regarding how various species of *Desulfovibrio* couple electron transport to the generation of a Δp.

12.3 Nitrogen Fixation

For reviews, see refs. 10 through 14. From an ecological point of view, one of the most important metabolic processes carried out by prokaryotes is nitrogen fixation (i.e., the reduction of N_2 to NH_3). As far as is known, eukaryotes have not evolved this capability. Since fixed nitrogen is usually limiting for plant growth, the ability of prokaryotes to fix nitrogen is necessary to maintain the food chain. The ammonia that is produced via nitrogen fixation is incorporated into cell material by means of glutamine synthetase and glutamate synthase (Section 9.3.1).

It used to be thought that nitrogen fixation was restricted to a few bacteria such as *Azotobacter*, *Rhizobium*, *Clostridium*, and the cyanobacteria. It is now realized that nitrogen fixation is a capability widespread among many different families of bacteria and also occurs in archaea. The enzyme responsible for nitrogen fixation, nitrogenase, is very similar in the different bacteria, and the nitrogen fixation genes have homologous regions. This has led to the suggestion that the nitrogenase gene may have been transferred laterally between different groups of bacteria. Organisms that fix nitrogen encompass a wide range of physiological types and include aerobes, anaerobes, facultative anaerobes, autotrophs, heterotrophs, and phototrophs.

Nitrogen fixation takes place when N_2 is the only or the major source of nitrogen because the genes for nitrogen fixation are repressed by exogenously supplied sources of fixed nitrogen (e.g., ammonia.) The signal transduction pathway responsible for repression is discussed in Chapter 18. It is remarkable that biological nitrogen reduction takes place at all. The nitrogen molecule is so stable that very high pressures and temperatures in the presence of inorganic catalysts are necessary to make it reactive in nonbiological systems. Industrially, nitrogen gas is reduced to ammonia by means of the Haber process, which requires 200 atm and 800 °C. Yet prokaryotes carry out the reduction at atmospheric pressures and ordinary temperatures.

12.3.1 The nitrogen-fixing systems

Biological nitrogen-fixing systems are found in the following organisms.

1. *Rhizobium, Sinorhizobium, Bradyrhizobium, and Azorhizobium in symbiotic relationships with leguminous plants* (soybeans, clover, alfalfa, string beans, peas, i.e., plants that bear seeds in pods). The bacteria infect the roots of the plants and stimulate the production of *root nodules*, within which the bacteria fix nitrogen. The plant responds by feeding the bacteria organic nutrients made during photosynthesis.

2. *Nonleguminous plants in symbiotic relationships with nitrogen-fixing bacteria.* For example, the water fern *Azolla* makes small pores in its fronds within which a nitrogen-fixing cyanobacterium, *Anaebaena azollae*, lives. The *Azolla–Anabaena* symbiotic system is used to enrich rice paddies with fixed nitrogen. Another example is the alder tree, which has nitrogen-fixing nodules containing *Frankia*, a bacterium resembling the streptomycetes. A third example is *Azospirillum lipoferum*, which is a nitrogen-fixing rhizosphere bacterium that is found around the roots of tropical grasses.

3. *Many free-living soil and aquatic prokaryotes.* As indicated in the introduction, many different prokaryotes fix nitrogen. They include *Azotobacter*, *Clostridium*, certain species of *Desulfovibrio*, the photosynthetic bacteria, and various cyanobacteria. Nitrogen fixation is not confined to the (eu)bacteria. Some methanogens have been reported to be nitrogen fixers.[15]

Nitrogen fixation is sensitive to oxygen. The enzyme that fixes nitrogen, *nitrogenase*, is inhibited by oxygen. Thus for many prokaryotes, nitrogen fixation takes place only under anaerobic or microaerophilic conditions. However, some prokaryotes can fix nitrogen while growing in air. As described later, they have evolved systems to protect the nitrogenase from oxygen.

12.3.2 The nitrogen fixation pathway

Nitrogenase

As mentioned, the enzyme that reduces nitrogen gas is called *nitrogenase*.[16,17] The major nitrogenase in nitrogen-fixing organisms is a molybdenum-containing enzyme that consists of two multimeric proteins. One of these, usually called the molybdenum–iron protein (MoFe protein), is also known as dinitrogenase, or component I. The second protein is called the iron protein (Fe protein), and is also known as dinitrogenase reductase, or component II. Both proteins contain FeS centers.

The MoFe protein is a tetramer $(\alpha_2\beta_2)$ of four polypeptides. When it is extracted with certain solvents, a cofactor called the iron–molybdenum cluster (FeMoco) is removed. The cofactor contains approximately half of the iron and labile sulfide of the protein. Thus, the MoFe protein contains FeMoco plus additional FeS centers. (See note 18 for more information about the MoFe protein.)

The Fe protein is a dimer (γ_2) of two identical polypeptides. The dimer contains a single Fe_4S_4 cluster, which is responsible for reducing FeMoco during nitrogen fixation.

Twenty-one genes, called the *nif* genes, have been identified as necessary for the expression and regulation of the nitrogenase enzyme system in *Klebsiella pneumoniae*. The presence of sufficient NH_4^+ represses the synthesis of nitrogenase. Other nitrogen sources (nitrates, amino acids, urea) also suppress the synthesis

of nitrogenase, probably by producing ammonia. The regulation of expression of the *nif* genes is described in Section 18.4.

The nitrogenase reaction

The nitrogenase reaction is a series of reductions during which half a mole of N_2 and one mole of H^+ are reduced to one mole of NH_3 and half a mole of H_2 (Fig. 12.4).

Nitrogenase reaction

$$4e^- + 0.5N_2 + 4H^+ + 8\ ATP$$
$$\rightarrow NH_3 + 0.5H_2 + 8\ ADP + 8\ P_i$$

Since the oxidation state of N_2 is 0 and the oxidation state of the nitrogen in NH_3 is -3, this reaction requires three electrons for every nitrogen atom. A fourth electron is transferred to a proton to reduce it to hydrogen gas. The electrons are transferred one at a time in an ATP-dependent reaction from the Fe_4S_4 cluster in the Fe protein to the MoFe cluster in the MoFe protein and from there to N_2.

The details of the electron transport pathway through the proteins and the role of ATP are not well understood. (See note 19 for a more complete discussion of electron flow in the MoFe protein and the role of ATP.) However, it is clear that the hydrolysis of at least two ATP molecules is required for every electron transferred between the two proteins. Therefore, about 16 moles of ATP are necessary to convert a mole of nitrogen gas to 2 moles of ammonia. This is a great deal of energy (about 800 kJ). Recall that only 2 moles of ATP are generated during the fermentation of one mole of glucose to lactic acid, and 38 moles of ATP are produced during the complete oxidation of one

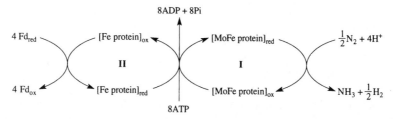

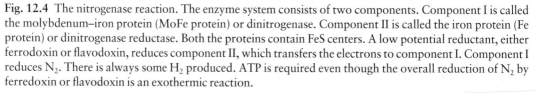

Fig. 12.4 The nitrogenase reaction. The enzyme system consists of two components. Component I is called the molybdenum–iron protein (MoFe protein) or dinitrogenase. Component II is called the iron protein (Fe protein) or dinitrogenase reductase. Both the proteins contain FeS centers. A low potential reductant, either ferrodoxin or flavodoxin, reduces component II, which transfers the electrons to component I. Component I reduces N_2. There is always some H_2 produced. ATP is required even though the overall reduction of N_2 by ferredoxin or flavodoxin is an exothermic reaction.

mole of glucose to carbon dioxide and water. Therefore, an organism growing on nitrogen gas must consume a large fraction of the ATP it produces to reduce the nitrogen to ammonia. Not surprisingly, bacteria do not fix nitrogen gas if an alternative source of nitrogen is present. As stated previously, this is because the nitrogen fixation genes are repressed when nitrogen sources other than N_2 are available.

During nitrogen reduction, protons are also reduced to hydrogen gas. The production of hydrogen gas appears to be wasteful of electrons and ATP. Indeed, some bacteria (e.g., *Azotobacter*) are very good at scavenging the hydrogen gas with a hydrogenase. The hydrogen gas is used to generate electrons for the nitrogenase.

Other nitrogenases

Although the nitrogenase just described is certainly the major one, other nitrogenases have been discovered. For example, *Azotobacter vinelandii* can synthesize three nitrogenases, encoded by three different genes. The three nitrogenases have different metal requirements. The molybdenum-containing nitrogenase (nitrogenase I) is made when the organism is grown in media containing molybdenum. Nitrogenase II is a vanadium-containing nitrogenase that is synthesized when the cells are grown in media lacking molybdenum but containing vanadium. Instead of FeMoco, the nitrogenase contains a vanadium cofactor called FeVaco. When both molybdenum and vanadium are lacking in the media, *Azotobacter* makes a third nitrogenase, called nitrogenase III, which requires only iron.

The source of electrons for nitrogen reduction

In most known systems the nitrogenases are reduced by ferredoxins (FeS proteins) or flavodoxins (flavoproteins). These electron carriers have midpoint potentials sufficiently low to reduce nitrogenase (−400 to −500 mV). The source of electrons for the ferredoxins and flavodoxins varies with the metabolism of the organism. For example, during heterotrophic anaerobic growth, the oxidation of pyruvate to acetyl–CoA and carbon dioxide generates reduced ferredoxin (pyruvate:ferredoxin oxidoreductase, Section 7.3.2) or flavodoxin (pyruvate:flavodoxin oxidoreductase), which donates electrons to nitrogenase. (*Clostridium pasteurianum* uses the ferredoxin enzyme, whereas *Klebsiella pneumoniae* uses the flavodoxin enzyme.)

The path of electrons to nitrogenase in aerobic and phototrophic bacteria and in cyanobacteria is not as well understood but is thought to involve ferredoxin or flavodoxin as the immediate electron donor. The ferredoxin or flavodoxin might be reduced by NAD(P)H (or some other electron carrier) generated during metabolism, for example, during the oxidation of carbohydrate. However, a source of energy (i.e., the proton motive force) would be necessary to drive the reduction of ferredoxin and flavodoxin by NAD(P)H (reversed electron transport, Section 4.5) because the midpoint potential of the NAD(P)$^+$/NAD(P)H couple is −320 mV, versus −400 to −500 mV for the ferredoxins and flavodoxins.

Alternatively, the cyanobacteria and green sulfur bacteria could use light energy to reduce the electron donor for nitrogenase during photosynthetic noncyclic electron flow. For example, it has been suggested that photosystem I and an electron donor other than water might reduce nitrogenase in the heterocyst, and similarly that a light-generated reductant might drive the reduction of nitrogenase in the green sulfur bacteria. Recall that these two photosystems have reaction centers that produce a reductant at a sufficiently low potential to reduce ferredoxin (Sections 5.3 and 5.4).

Protecting the nitrogenase from oxygen

All nitrogenases are rapidly inactivated by oxygen in vitro. Some nitrogen-fixing microorganisms (e.g., the clostridia or nitrogen-fixing sulfate reducers) are strict anaerobes. Others (e.g., the purple photosynthetic bacteria, *Klebsiella* spp.) are facultative anaerobes; that is, they can grow aerobically or anaerobically, although they fix nitrogen only when they are living anaerobically. Some microorganisms are microaerophilic; that is, they can grow only in low levels of oxygen. Some of these are nitrogen fixers and can grow on nitrogen gas under microaerophilic conditions. But, what about strict aerobes or the cyanobacteria that produce oxygen in the light? Various strategies have evolved to protect the nitrogenase in microorganisms that fix nitrogen in air.[20,21]

The *Azotobacter* species have a very active respiratory system that is suggested to utilize oxygen rapidly enough to lower the intracellular concentrations in the vicinity of the nitrogenase. These organisms are also able to protect their nitrogenase from inactivation by associating it with protective proteins. For example, the nitrogenase of *Azotobacter* can be isolated as an air-tolerant complex with a redox protein called the Shethna, FeS II, or protective protein.

The rhizobia in root nodules exist in plant vesicles in the inner cortex of the nodule as modified cells called *bacteroids*. Oxygen-sensitive microelectrodes can be used to show that the oxygen concentrations in the inner cortex are much lower than those in the surrounding tissue. The oxygen levels in the vicinity of the bacteroid-containing cells are controlled by a boundary of densely packed plant cells between the inner and outer cortex. The control of oxygen access to the inner cortex is achieved by regulating the intercellular spaces, which either are filled with air or contain variable amounts of water, within the boundary layer. However, the bacteroids are dependent upon oxygen for respiration, and within the nodule a plant protein called leghemoglobin binds oxygen and delivers it to the bacteroids.

It is not understood how the unicellular cyanobacteria or the nonheterocystous filamentous cyanobacteria protect their nitrogenase from photosynthetically produced or atmospheric oxygen. In fact, most of these strains fix nitrogen only when grown under anoxic or micro-oxic conditions. Anoxic conditions are maintained in the light only when photosynthetic oxygen production is experimentally inhibited. Under micro-oxic (also called microaerobic) growth conditions, atmospheric oxygen is absent, but photosynthetically produced oxygen is present. It is presumed that intracellular oxygen levels under these conditions are significantly lower than atmospheric levels, hence the term "micro-oxic." Whether this is indeed the case, especially under conditions of high illumination, has not been shown. However, some unicelullular and nonheterocystous filamentous cyanobacteria do fix nitrogen under aerobic conditions (i.e., at oxygen concentrations that are approximately equal to

atmospheric). How they protect their nitrogenase from oxygen is not known, but the topic has generated much interest. Such cyanobacteria include the unicellular *Gloethece* (formerly called *Gloeocapsa*), certain strains of the unicellular *Synechococcus*, and the marine filamentous organism *Trichodesmum*. Nitrogen fixation by unicellular and filamentous non-heterocystous cyanobacteria, including protection of the nitrogenase, have been reviewed by Bergman et al.[22] Several filamentous cyanobacteria protect their nitrogenase from oxygen by means of special cells called heterocysts, in which the organisms are able to fix nitrogen under aerobic growth conditions.

Heterocysts

Filamentous cyanobacteria such as *Anabaena* and *Nostoc* protect their nitrogenase by differentiating approximately 5 to 10% of their vegetative cells into special nitrogen-fixing cells called heterocysts in the absence of combined nitrogen[23] (Fig. 12.5). Heterocysts differ from vegetative cells in the following respects.

1. Heterocysts possess the nitrogenase enzymes.

2. They have only photosystem I (PS I), and therefore do not produce oxygen.

3. They do not fix CO_2.

4. They are surrounded by a thick cell wall consisting of glycolipid and polysaccharide that is believed to serve as a permeability barrier to atmospheric oxygen.

5. They are not dividing.

The ATP made during cyclic photophosphorylation by photosystem I is used to fix the nitrogen. But where do the heterocysts get the reducing power? What happens is that the heterocysts fix N_2 and feed reduced nitrogen to the rest of the filament (Fig. 12.5). In turn, the heterocyst is fed carbohydrate made from CO_2 by the vegetative cells. The heterocyst oxidizes the carbohydate, reducing ferredoxin, which, in turn, reduces the nitrogenase. Therefore, this is a complex situation in which two different cell types in the filament are feeding each other.

Since the heterocysts do not have photosystem II, they do not produce oxygen. However,

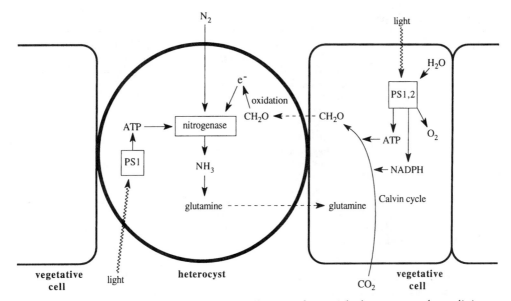

Fig. 12.5 Heterocyst interactions with vegetative cells in *Anabaena*. The heterocyst reduces dinitrogen to ammonia, which is then incorporated into glutamine via glutamine synthetase. The glutamine then enters the vegetative cells, where it serves as a source of fixed nitrogen for growth. The vegetative cells fix carbon dioxide into carbohydrate by means of the Calvin cycle. Some of the carbohydrate enters the heterocyst, where it serves as a source of carbon and NADPH. The NADPH reduces the nitrogenase via ferredoxin:NADP oxidoreductase and ferredoxin. ATP is made via cyclic photophosphorylation in the heterocyst using PS I. Since PS II is lacking in the heterocyst, oxygen is not produced there.

there is still the problem of protecting the nitrogenase from atmospheric oxygen. It has been suggested that the crystalline glycolipid and polysaccharide cell wall may present a diffusion barrier to oxygen. Heterocysts also have a high rate of respiration, which presumably also contributes to a low internal oxygen environment.

12.4 Lithotrophy

While most organisms derive energy from oxidizing organic nutrients (chemo-organotrophs) or from the absorption of light (phototrophs), there exist many prokaryotes that derive energy from the oxidation of inorganic compounds; examples are H_2, CO, NH_3, NO_2^-, H_2S, S°, $S_2O_3^{2-}$, and Fe^{2+}. This type of metabolism is called *lithotrophy*, and the organisms are called *lithotrophs*.[24]

12.4.1 The lithotrophs

The lithotrophs are physiologically diverse and exist among several different groups of bacteria and archaea. Many of the lithotrophs are aerobes; that is, they carry out electron transport with oxygen as the terminal electron acceptor. However, some are facultative anaerobes, using nitrate or nitrite as the electron acceptor when oxygen is unavailable, and a few are obligate anaerobes that use sulfate or CO_2 as the electron acceptor. For most of the lithotrophs, the sole or major source of carbon is CO_2. Such organisms are called *chemoautotrophs* or *chemolithoautotrophs*. The lithotrophs vary with regard to the autotrophic CO_2 fixation pathway that they use (Chapter 13). Other lithotrophs are facultatively heterotrophic. The facultative heterotrophs include all the bacterial hydrogen oxidizers, some sulfur-oxidizing thiobacilli, and some thermophilic iron-oxidizing bacteria. Some representative species are listed in Table 12.1.

Table 12.2 lists midpoint potentials of the inorganic substrates at pH 7. Theoretically, all these organisms, with the possible exception of the hydrogen oxidizers and the CO oxidizers, must carry out reversed electron transport to generate NAD(P)H for biosynthesis because the electron donor is more electropositive

Table 12.1 Chemoautotrophs

Bacterial group	Typical species	Electron donor	Electron acceptor	Carbon source	Product
Hydrogen oxidizing	*Alcaligenes eutrophus*	H_2	O_2	CO_2	H_2O
Carbon monoxide oxidizing (carboxydobacteria)	*Pseudomonas carboxydovorans*	CO	O_2	CO_2	CO_2
Ammonium oxidizing	*Nitrosomonas europaea*	NH_4^+	O_2	CO_2	NO_2^-
Nitrite oxidizing	*Nitrobacter winogradskyi*	NO_2^-	O_2	CO_2	NO_3^-
Sulfur oxidizing	*Thiobacillus thiooxidans*	$S, S_2O_3^{2-}$	O_2	CO_2	SO_4^{2-}
Iron–oxidizing	*Thiobacillus ferrooxidans*	Fe^{2+}	O_2	CO_2	Fe^{3+}
Methanogenic	*Methanobacterium thermoautotrophicum*	H_2	CO_2	CO_2	CH_4
Acetogenic	*Acetobacterium woodii*	H_2	CO_2	CO_2	CH_3COOH

Source: Schlegel, H. G., and H. W. Jannasch. 1992. Prokaryotes and their habitats, pp. 75–125. In: *The Prokaryotes*, Vol. I. Balows et al. (Eds.). Springer-Verlag, Berlin.

Table 12.2 Redox potentials of inorganic compounds[a]

Compound	E_0'(mV)
CO_2/CO	−540
SO_4^{2-}/HSO_3^-	−516[b]
H^+/H_2	−414[b]
S^o/HS^-	−270[b]
HSO_3^-/HS^-	−116[b]
NO_3^-/NO_2^-	+420
NO_2^-/NH_3	+440
Fe^{3+}/Fe^{2+}	+772[b]
O_2/H_2O	+818[b]

[a] For comparison, the standard potential at pH 7 for $NAD^+/NADH$ is −320 mV.
[b] Data from Thauer, R. K., K., Jungermann, and K. Decker. 1977. Energy conservation in chemotrophic anaerobic bacteria. *Bacteriol. Rev.* 41:100–180.

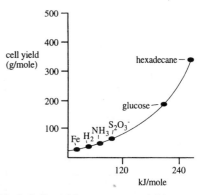

Fig. 12.6 Cell yields versus available energy in inorganic and organic electron sources. *Source*: Adapted from Brock, T. D., and M. T. Madigan. 1991. *Biology of Microorganisms*. Reprinted by permission of Prentice Hall, Upper Saddle River, NJ.)

than the $NAD^+/NADH$ couple ($E_0' = -0.32$ V). Reversed electron flow is driven by the Δp (Section 4.5). Because of the relatively small ΔE_h between the inorganic electron donor and oxgyen, and the need to reverse electron transport, the energy yields, and therefore the cell yields, are relatively small in comparison to growth on organic substrates (Fig. 12.6).

Aerobic hydrogen-oxidizing bacteria and carboxydobacteria

The hydrogen-oxidizing bacteria are usually facultative and can live either autotrophically or heterotrophically. However, some always require organic carbon for growth, and these are called chemolithoheterotrophs. The hydrogen

oxidizers can be found in aerobic or anaerobic environments where H_2 is available. The hydrogen gas itself is produced as a by-product of nitrogen fixation (e.g., in the rhizosphere of nitrogen-fixing plants and in cyanobacterial blooms). Hydrogen gas is also produced in anaerobic environments via fermentations, where some of it escapes into the aerobic atmosphere. (However, most of the H_2 produced anaerobically is utilized by the sulfate reducers and methanogens.) Among the hydrogen-oxidizing bacteria are some, called *carboxydobacteria*, that can also grow on carbon monoxide (CO) as the sole source of energy and carbon, using oxygen, or in some cases nitrates (denitrifiers), as the electron acceptor.

The hydrogen-oxidizing bacteria and the carboxydobacteria are represented by several genera, including representatives from *Pseudomonas*, *Arthrobacter*, *Bacillus*, and *Rhizobium*. Anaerobic hydrogen oxidizers include some sulfate-reducing bacteria, as well as some archaea, including methanogens growing autotrophically on CO_2 and certain sulfur-dependent archaea that use elemental sulfur as the electron acceptor, reducing it to hydrogen sulfide.

Ammonia-oxidizing bacteria

Bacteria that oxidize ammonia as a source of energy are called nitrifiers.[25] There are at least five genera of nitrifiers: *Nitrosomonas*, *Nitrosococcus*, *Nitrosospira*, *Nitrosolobus*, and *Nitrosovibrio*. All these nitrifiers are aerobic obligate chemolithoautotrophs that assimilate CO_2 via the Calvin cycle. Ammonia that is produced in the anaerobic niches by deamination of amino acids, urea, or uric acid, or via dissimilatory nitrate reduction diffuses into the aerobic environment, where it is oxidized by the aerobic nitrifiers. The aerobic nitrifying bacteria are often found at the aerobic–anaerobic interfaces, where they capture the ammonia as it diffuses from the anaerobic environments, as well as in the more highly aerobic parts of the soil and water. (There are also anaerobic ammonia-oxidizing bacteria that use nitrite as the terminal electron acceptor and produce nitrogen gas. This reaction is discussed in note 26, and reviewed in ref. 27. The bacteria grow on carbon dioxide, ammonium, and nitrite.)

Nitrosomonas oxidizes ammonia to nitrite. Along with *Nitrobacter*, which oxidizes nitrite to nitrate, it is responsible for a major portion of the conversion of ammonia to nitrate, a process called *nitrification*. The first oxidation is the oxidation of ammonia to hydroxylamine, catalyzed by *ammonia monooxygenase* (AMO). The reaction is:

$$2H^+ + NH_3 + 2e^- + O_2 \rightarrow NH_2OH + H_2O$$

In this reaction, one oxygen atom is incorporated into hydroxylamine and the other is reduced to water. The oxidation of hydroxylamine to nitrite is catalyzed by the enzyme *hydroxylamine oxidoreductase* (HAO):

$$NH_2OH + H_2O \rightarrow HONO + 4\,e^- + 4H^+$$

Of the four moles of electrons removed from one mole of hydroxylamine, two are used in the ammonia monooxygenase reaction, approximately 1.7 are transferred to oxygen via cytochrome oxidase, and about 0.3 is used to generate NAD(P)H via reversed electron transport:

$$2H^+ + 0.5O_2 + 2e^- \rightarrow H_2O$$

A proposed electron transport scheme for the topological arrangement of the electron carriers is shown in Fig. 12.7. The electron transport scheme shown in Fig. 12.7 is speculation based upon the known location of the enzymes and their redox potentials. The scheme proposes that ammonia oxidation to hydroxylamine takes place in the cytoplasm, although it may occur either there or in the periplasm. (Ammonia monooxygenase is in the cell membrane, but it is unknown whether the substrate-binding site is exposed to the cytoplasm or the periplasm.) Assuming cytoplasmic oxidation, the hydroxylamine diffuses across the cell membrane to the periplasm, where it is oxidized to nitrite by hydroxylamine oxidoreductase, a periplasmic enzyme.

During this oxidation four electrons are transferred to periplasmic cytochrome c_{554}. The four electrons are then transferred from cytochrome c_{554} to ubiquinone in the membrane. It is suggested that two electrons travel from ubiquinone to the ammonia monooxygenase enzyme and two electrons to cytochrome aa_3 via membrane cytochrome c_{553} and periplasmic cytochrome c_{552}. Cytochrome aa_3 is presumed to act as a proton pump. The Δp that is created as a result of the oxidation of ammonia and hydroxyamine and the reduction of oxygen is the sole source of energy for these bacteria.

Nitrite-oxidizing bacteria

The nitrite oxidizers are *Nitrobacter*, *Nitrococcus*, *Nitrospina*, and *Nitrospira*. They are aerobic obligate chemolithoautotrophs, with the exception of *Nitrobacter*, which is a facultative autotroph (i.e., it can also be grown heterotrophically). The details of the electron transport scheme have not been fully elucidated. However, a proposed model is shown in Fig. 12.8. Electrons travel from nitrite to

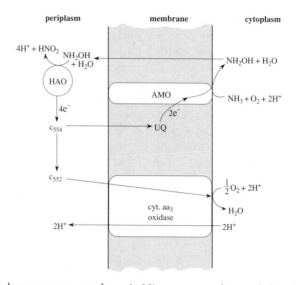

Fig. 12.7 Model for the electron transport scheme in *Nitrosomonas*. Ammonia is oxidized to hydroxylamine by the enzyme ammonia monooxygenase (AMO). Although the oxidation of ammonia is depicted as occurring in the cytoplasm, the possibility that the reaction takes place in the periplasm has not been ruled out. Two electrons are required to reduce one of the oxygen atoms to water. The hydroxylamine diffuses across the membrane to the periplasm, where it is oxidized to nitrite by a complex cytochrome called hydroxylamine oxidoreductase (HAO). The electrons are passed to a periplasmic cytochrome c and from there to ubiquinone in the membrane. Electrons travel from ubiquinone in two branches. One branch passes two electrons to the ammonia monooxygenase, and the second branch leads to oxygen via cytochromes c and aa_3. All four electrons end up in water. *Source*: adapted from Hooper, A. B. 1989. Biochemistry of the nitrifying lithoautotrophic bacteria, pp. 239–265. In: *Autotrophic Bacteria*. H. G. Schlegel and B. Bowien (Eds.). Springer-Verlag, Berlin.

oxygen via a periplasmic cytochrome c. There is a thermodynamic problem here. Cytochrome c exists at a midpoint potential (E'_m) of +270 mV, whereas the midpoint potential of the nitrate/nitrite couple is more electropositive, at +420 mV. Because electrons do not spontaneously flow toward the lower redox potential, energy must be provided to drive the electrons over the 150 mV difference for nitrite to reduce cytochrome c.

Nicholls and Ferguson suggested that the membrane potential drives the electrons from nitrite to cytochrome c.[28] They proposed that nitrite oxidation takes place on the cytoplasmic surface of the membrane and that the electrons flow across the membrane to cytochrome c located on the periplasmic side, which is typically 170 mV more positive than the cytoplasmic side (Fig. 12.8). In this way, the membrane potential lowers the potential difference between the nitrate/nitrite couple and the c_{ox}/c_{red} couple by approximately 170 mV. The student will recognize this model as *reversed*

electron transport driven by the membrane potential (which is consumed in the process).

The role of the membrane potential in driving electron transport from nitrite to cytochrome c is consistent with the observation that experimental procedures that lower the membrane potential (e.g., incubation with proton ionophores) decrease electron transfer from nitrite to oxygen in inverted membrane vesicles from *Nitrobacter*. The electrons then flow from cytochrome c back across the membrane to oxygen through cytochrome aa_3 oxidase, driven by a favorable midpoint potential difference of about +440 mV. A Δp is created by the outward pumping of protons by the cytochrome aa_3 oxidase. Electron flow is also reversed from cytochrome c to NAD^+ through two coupling sites: a ubiquinone–cytochrome c oxidoreductase (probably a bc_1 complex) and a reversible NADH–ubiquinone oxidoreductase. The model proposes that reversed electron flow to NAD^+ is coupled to the *influx* of protons (i.e., it is driven by the Δp).

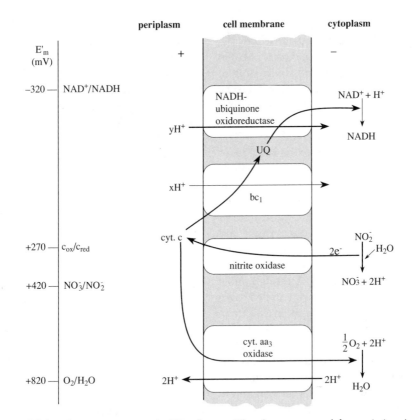

Fig. 12.8 A model for electron transport in *Nitrobacter*. The electrons travel from nitrite via membrane-bound nitrite oxidase to cytochrome c, which is at a more negative potential. It is proposed that nitrite oxidation takes place on the cytoplasmic side of the membrane and that the membrane potential drives the electrons transmembrane to cytochrome c. From cytochrome c, the electrons diverge. Most travel to oxygen at a more positive potential and a Δp is created. The coupling site is the cytochrome aa$_3$ oxidase, which is a proton pump. Other electrons travel to NAD$^+$, which is at a more negative potential. The Δp drives the electrons in reverse flow to NAD$^+$. This is accomplished by coupling electron transport with the return of protons down the proton potential to the cytoplasmic side. The scheme presumes the presence of a bc$_1$ complex as well as a reversible NADH dehydrogenase complex. *Source*: Adapted from Nicholls, D. G., and S. J. Ferguson. 1992. *Bioenergetics 2*. Academic Press, London.

Sulfur-oxidizing prokaryotes

The sulfur-oxidizing prokaryotes include the photosynthetic sulfur oxidizers (Chapter 5) and the nonphotosynthetic sulfur oxidizers. (See refs. 29–31 for reviews.) It is the nonphotosynthetic type that concerns us here, and almost all the known examples are gram-negative bacteria. Sulfur oxidizers comprise a physiologically heterogeneous group. They may be obligate autotrophs, which grow only on CO_2 as the carbon source, or facultative heterotrophs, which can also use organic carbon as the source of carbon. Some are neutrophiles, growing best around pH 7, whereas others are acidophiles, growing best between pH 1 and 5).

Sulfur-oxidizing acidophiles can be isolated from sulfur mines and coal mines, which produce sulfuric acid. An example is *Thiobacillus thiooxidans*, which can grow at a pH of 1, although the optimum is between 2 and 3. These bacteria can be mesophilic (growth temperature optimum 25–40 °C) or thermophilic (growth temperature optimum >55 °C). Some sulfur bacteria (e.g., most *Beggiatoa* strains) can be grown only mixotrophically, that is, using H_2S as the energy source and organic carbon as the source of carbon.

The nonphotosynthetic sulfur prokaryotes include the bacteria *Beggiatoa* and *Thiothrix*, bacteria belonging to the genus *Thiobacillus* and *Paracoccus*, and an archaeon belonging to the genus *Sulfolobus*, a thermophilic acidophile that grows in sulfur acid springs, where the temperature can be 90 °C and the pH can be between 1 and 5. (See the discussion of *Sulfolobus* and other archaea in Section 1.1.1). The thiobacilli are the most prominent of the sulfur oxidizers, and their sulfur oxidation pathways are the most well studied.

Sulfur compounds commonly used as sources of energy and electrons include hydrogen sulfide (H_2S), elemental sulfur ($S°$), and thiosulfate ($S_2O_3^{2-}$), all of which can be oxidized to sulfate. (Laboratory studies often utilize thiosulfate or tetrathionate because these compounds are stable in air. Tetrathionate has an additional advantage because unlike thiosulfate, it is stable at acidic pH values.) Although most sulfur bacteria are aerobes, a few can be grown anaerobically by using nitrate as the electron acceptor. The sulfur bacteria can be found in nature growing near such sources as sulfur deposits, sulfide ores, hot sulfur springs, sulfur mines, and coal mines that are sites of iron pyrite (FeS_2) deposits. Sulfide oxidizers can sometimes be found in large accumulations in thin layers between the aerobic and anaerobic environ-ments. They are sometimes confined to the aerobic–anaerobic interface because H_2S, which is produced anaerobically by sulfate reducers, is rapidly oxidized by oxygen. Thus, to be able to compete with the rapid chemical oxidation of sulfide, the sulfide-oxidizing bacteria grow where oxygen levels are relatively low.

There has been much confusion in the litera-ture regarding the intermediate stages of sulfur oxidation. Probably, more than one pathway exists. Two pathways that are receiving much research attention are described next.

Paracoccus versutus is a neutrophilic facul-tative lithoautotroph. In addition to growth on thiosulfate (but not polythionates), it can be grown on various organic carbon sources. (As reviewed in ref. 29, this organism was for-merly called *Thiobacillus versutus*.) Oxidation of thiosulfate takes place in the periplasm on a multienzyme complex (Fig. 12.9). No free intermediates are released, and both atoms of thiosulfate are oxidized to sulfate. Protons are released in the periplasm during the oxidations, and the electrons are transferred *electrogenically* from the periplasmic side of the membrane to the cytoplasmic side through cytochrome c_{552}

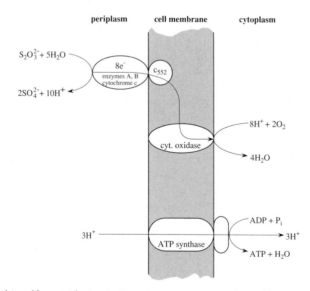

Fig. 12.9 Model for thiosulfate oxidation in *Parococcus versutus*. Thiosulfate is oxidized to sulfate in the periplasm by a multienzyme complex consisting of enzyme A, enzyme B, and cytochromes c. The electrons are electrogenically transferred to oxygen via a membrane-bound cytochrome c_{552} and a cytochrome oxidase (cytochrome aa_3). A Δp is created by the release of protons in the periplasm via the oxidations, the con-sumption of protons in the cytoplasm during oxygen reduction, and electrogenic flow of electrons across the membrane to oxygen. A proton-translocating ATP synthase makes ATP. *Source*: Adapted from Kelly, D. P. 1989. Physiology and biochemistry of unicellular sulfur bacteria, pp. 193–217. In: *Autotrophic Bacteria*. H. G. Schlegel and B. Bowien (Eds.). Springer-Verlag, Berlin.

and then cytochrome aa$_3$ oxidase to oxygen. (See Section 3.2.1 for a discussion of electrogenic movement of electrons.) A proton current is maintained by the release of protons in the periplasm during the oxidations and their consumption in the cytoplasm during the reduction of oxygen.

The expected H$^+$/O ratio (protons produced in the periplasm or translocated from the cytoplasm to the periplasm for every oxygen atom reduced) can be obtained from Fig. 12.9 (and confirmed experimentally). The ratio is 2.5 for thiosulfate oxidation. ATP is synthesized by a proton-translocating ATP synthase driven by the Δp. If one assumes that H$^+$/ATP for the ATP synthase is 3 (a consensus value), then the maximum P/O ratio would be 2.5/3, or 0.83, for thiosulfate oxidation. (See Section 4.5.2 for a discussion of the relationship between the size of the proton current and the upper limit of ATP that can be made.)

Thiobacillus tepidarius is a neutrophilic thermophilic lithoautotroph that has been isolated from hot springs. It can be grown on hydrogen sufide, thiosulfate, trithionate, and tetrathionate, as electron donors. The pathway of oxidation of thiosulfate to sulfate (called the polythionate pathway) begins in the periplasm, but the bulk of the oxidations take place in the cytoplasm. Two molecules of thiosulfate are oxidized to tetrathionate in the periplasm (Fig. 12.10). A periplasmic cytochrome c accepts the two electrons generated from the thiosulfate oxidation and transfers these electrons to a membrane-bound cytochrome c. The tetrathionate is believed to be transported into the cell, where it is oxidizing to four sulfates via the intermediate sulfite (generating 14 electrons). The 14 electrons generated during the tetrathionate oxidations in the cytoplasm are transferred to a proton-translocating ubiquinone–cytochrome b complex in the cell

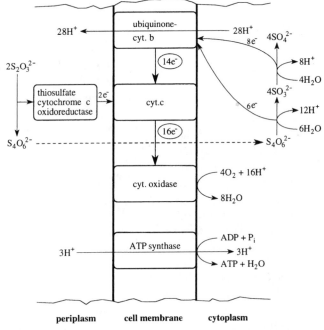

Fig. 12.10 Model for thiosulfate oxidation in *Thiobacillus tepidarius*. Thiosulfate is oxidized to tetrathionate in the periplasm by a thiosulfate–cytochrome c oxidoreductase. It is hypothesized that the tetrathionate is transported into the cell in symport with protons, where it is oxidized to sulfate. The electrons travel to oxygen via a proton-translocating ubiquinone–cytochrome b system (probably a bc$_1$ complex) and cytochrome oxidase, which appears to be a cytochrome o. *Source*: Adapted from Kelly, D. P. 1989. Physiology and biochemistry of unicellular sulfur bacteria, pp. 193–217. In: *Autotrophic Bacteria*. H. G. Schlegel and B. Bowien (Eds.). Springer-Verlag, Berlin, and Smith, D. W., and W. R. Strohl, 1991. Sulfur-oxidizing bacteria, pp. 121–146. In: *Variations in Autotrophic Life*. J. M. Shively and L. L. Barton (Eds.). Academic Press, New York.

membrane, which reduces membrane-bound cytochrome c. Cytochrome c transfers all of the electrons to cytochrome oxidase (probably cytochrome o), which reduces oxygen. Because the oxidations of tetrathionate and sulfite are cytoplasmic rather than periplasmic, the proton current is due to proton translocation by the ubiquinone–cytochrome b complex (coupled to the cytoplasmic oxidations), rather than to periplasmic oxidations, as is the case for *P. versutus*.

Some thiobacilli can incorporate a substrate-level phosphorylation when oxidizing sulfur. The substrate-level phosphorylation occurs at the level of sulfite. The sulfite reacts with adenosine monophosphate (AMP) to form adenosine phosphosulfate (APS), in a reaction catalyzed by APS reductase. The APS is oxidized to sulfate, producing ADP or ATP, using either ADP sulfurylase or ATP sulfurylase.

APS reductase

$$SO_3^{2-} + AMP \rightarrow APS + 2e^-$$

ADP sulfurylase

$$APS + P_i \rightarrow ADP + SO_4^{2-}$$

ATP sulfurylase

$$APS + PP_i \rightarrow ATP + SO_4^{2-}$$

Recall that ATP sulfurylase and APS reductase are used by dissimilatory sulfate reducers to reduce sulfate to sulfite in reactions that consume ATP (Section 12.2.2). By running these reactions in the oxidative direction, sulfur oxidizers can use the energy in pyrophosphate to make ATP.

Figure 12.11 summarizes the inorganic sulfur oxidation pathways. Elemental sulfur exists as an octet ring of insoluble sulfur (S_8^0). It is first activated by reduced glutathione (GSH) to form a linear polysulfide (G-S-S_8-H). Sulfide (S^{2-}) also reacts with GSH and is oxidized to linear polysulfide. The sulfur atoms are removed from the polysulfide one at a time during the oxidation to sulfite (SO_3^{2-}).

In another elemental sulfur oxidation pathway, reported for some thiobacilli, S° is oxygenated by a sulfur oxygenase:

Sulfur oxygenase

$$S° + O_2 + H_2O \rightarrow H_2SO_3$$

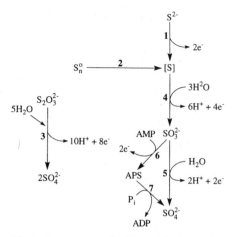

Fig. 12.11 A summary of sulfur oxidation pathways: 1, oxidation of sulfide to linear polysulfide [S]; 2, conversion of elemental sulfur to linear polysulfide; 3, thiosulfate multienzyme complex; 4, sulfur oxidase; 5, sulfite oxidase; 6, APS reductase; 7, ADP–sulfurylase. *Source*: Adapted from Gottschalk, G. 1985. *Bacterial Metabolism*, Springer-Verlag, Berlin.

However, the oxygenase reaction cannot account for sulfur oxidation under anaerobic conditions when the oxidant provided is nitrate (*T. denitrificans*) or Fe^{3+} (*T. ferrooxidans*). Furthermore, the oxygenase reaction cannot conserve energy for the cell, since the electrons do not enter the respiratory chain. (Even though the periplasm might be acidified with H_2SO_3 and thus generate a ΔpH, this by itself cannot generate net ATP in a growing cell because protons entering via the ATP synthase must not accumulate in the cytoplasm. The extrusion of protons from the cells or the utilization of protons to form water requires electron transport.)

Iron-oxidizing bacteria

A few bacteria derive energy from the aerobic oxidation of ferrous ion to ferric ion.[32] Most of these are also acidophilic sulfur oxidizers (i.e., they oxidize sulfide to sulfuric acid). The acidophilic iron oxidizers can be found growing at the sites of geological deposits of iron sulfide minerals [e.g., pyrite (FeS_2) and chalcopyrite ($CuFeS_2$)], where water and oxygen are also present. The iron–sulfide minerals are uncovered during mining operations, and the presence of acid mine water at these sites is due to the growth of the iron–sulfide oxidizers. Since Fe^{2+} is rapidly oxidized chemically by oxygen to Fe^{3+} at neutral pH but only slowly at acid pH, the acidic environment is conducive to growth of the iron oxidizers. An example is *Thiobacillus ferrooxidans*, which is an autotroph able to use either ferrous ion or inorganic sulfur compounds as a source of energy and electrons. *Thiobacillus ferrooxidans* can be grown on ferrous ion (e.g., $FeSO_4$) at optimal external pH values between 1.8 and 2.4. It maintains an internal pH of around 6.5 and therefore a ΔpH of about 4.5. The membrane potential ($\Delta\Psi$) at low pH is reversed, and energy in the Δp is due entirely to the ΔpH component (Section 3.3). (As discussed in Section 15.1.3, the inversion of the $\Delta\Psi$ is necessary for maintenance of the ΔpH and may be due to electrogenic influx of K^+.)

1. Growth on ferrous sulfate

Figure 12.12 illustrates a model for the electron transport pathway in *T. ferrooxidans* growing on the aerobic oxidation of ferrous sulfate. In this proposed pathway, extracellular Fe^{2+} is oxidized to Fe^{3+} by an Fe^{3+} complex in the outer membrane. The electrons travel to a periplasmic cytochrome c. The periplasm also contains a copper protein called rusticyanin, which is thought to be part of the electron transport scheme. From cytochrome c, the electrons flow across the membrane via cytochrome oxidase (cytochrome a_1) to oxygen, generating a $\Delta\Psi$, inside negative. The oxidation of ferrous ion drives the consumption of protons in the cytoplasm during oxygen reduction. The ΔpH is

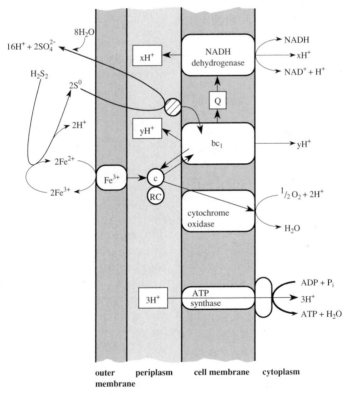

Fig. 12.12 Electron transfer in *Thiobacillus ferrooxidans*. Ferrous ion is oxidized to Fe^{3+} by an Fe^{3+} complex in the outer membrane, and the electrons flow through periplasmic cytochrome c and a copper protein, rusticyanin (RC) to a membrane cytochrome oxidase of the a_1 type. The Δp is maintained by the inward flow of electrons (contributing toward a negative $\Delta\Psi$) and by the consumption of cytoplasmic protons during oxygen reduction, maintaining a ΔpH. The consumption of two protons for every two electrons in the cytoplasm and the uptake of three protons through the ATP synthase indicate that a maximum of two-thirds of an ATP can be made for every oxidation of two ferrous ions. When the bacteria are oxidizing iron pyrite (FeS_2), the Fe^{3+} that is produced outside the cell envelope is chemically reduced to Fe^{2+} by S^{2-}. The resultant S^0 is oxidized to SO_4^{2-}, and the electrons pass through a proton-translocating bc_1 complex to periplasmic cytochrome c and thence to cytochrome oxidase. The figure also illustrates Δp-driven reverse electron transport from S^0 and from Fe^{2+} to NAD^+ through coupling sites that bring protons into the cell.

maintained because the rate of respiration and proton consumption matches the rate at which protons enter the cell through the ATPase or through leakage.

From an energetic point of view, one would not expect any proton pumping by the cytochrome oxidase. The reason for this is that the difference in midpoint potentials between the Fe^{3+}/Fe^{2+} couple (pH 2) and the O_2/H_2O (presumed to be at pH 6.5) is very small, perhaps only 0.08 V or less.[33] (See note 34 for a more complete discussion.) Under these circumstances, a two-electron transfer would generate at the most only 0.16 eV. A simple calculation reveals that this would not be enough energy to pump protons against the pH gradient. At an external pH of 2 and an internal pH of 6.5, the ΔpH is 4.5. This is equivalent to 0.06×4.5 or 0.27 V at 30 °C (eq. 3.10). Thus, each proton would have to be energized by approximately 0.27 eV to be pumped out of the cell, even in the presence of a small $\Delta\Psi$, inside positive, which may be on the order of +0.01 to +0.02 V.

ATP is synthesized by a membrane H^+-translocating ATP synthase driven by a Δp of approximately −250 mV. Electrons from Fe^{2+} must also move toward a lower redox potential to generate NAD(P)H. Figure 12.12 also illustrates how reversed electron flow from Fe^{2+} to NAD^+ might occur through a bc_1 complex and quinone. Reversed electron flow is probably driven by the Δp as protons enter the cell down the Δp gradient through the bc_1 complex and the NADH dehydrogenase complex.

2. Growth on pyrite

Pyrite is a crystalline ore of iron that is usually written as ferrous disulfide (FeS_2). It is actually a stable crystal of discrete $^-$S–S$^-$ disulfide ions (S_2^{2-}) and ferrous ions (Fe^{2+}). The ferrous and disulfide ions are oxidized to ferric ion and sulfuric acid, respectively, according to the following overall reaction:

$$4FeS_2 + 2H_2O + 15O_2 \rightarrow 4Fe^{3+} + 8SO_4^{2-} + 4H^+$$

However, the oxidation is not straightforward.[35] The oxidation of FeS_2 is the result of several redox reactions (Fig. 12.12.). When the bacteria are growing on iron pyrite, the Fe^{2+} is oxidized extracellularly by a complex of Fe^{3+} in the outer membrane (Fig. 12.12). The outer membrane iron complex then transfers the electrons to the periplasmic cytochrome c, which transfers the electron to cytochrome oxidase. The Fe^{3+} that is formed extracellularly is recycled to Fe^{2+} by disulfide from the iron pyrite, according to the following reaction, which occurs spontaneously:

$$H_2S_2 + 2Fe^{3+} \rightarrow 2S° + 2Fe^{2+} + 2H^+$$

The elemental sulfur that is produced is oxidized by the bacteria to sulfate, with the electrons traveling through a proton-translocating bc_1 complex to the periplasmic cytochrome c and from there to cytochrome oxidase.

12.4.2 Review of the energetic considerations for lithotrophic growth

ATP synthesis

As discussed in Section 3.7.1, the energy from respiratory oxidation–reduction reactions is first converted into a Δp, which is then used to drive ATP synthesis via the ATP synthase. For illustrative purposes, we will consider growth on nitrite in an aerated culture as a source of energy and electrons. The E'_m for NO_3^-/NO_2^- is +0.42 V. The E'_m for O_2/H_2O is +0.82 V. The difference in potential is therefore +0.82 − 0.42 or +0.40 V. Recall that in the respiratory chain, each coupling site is associated with pairs of electrons traveling over a midpoint potential difference of approximately 0.2 V or more. The conclusion is that there is sufficient energy for a coupling site between nitrite and oxygen.

Another way of examining this question is to consider the amount of energy required to synthesize an ATP. The energy required to synthesize ATP will vary with the concentrations of ATP, ADP, and P_i. However, we will assume a value of 0.4 to 0.5 eV, which is a reasonable estimate. Therefore, two electrons traveling over a redox gradient of 0.40 V should give 0.80 eV of energy, which is sufficient for the synthesis of an ATP. (See note 36.)

NADH reduction requires reversed electron transport

To grow, all cells must be able to make NAD(P)H because this is a major source of electrons for reductions that occur during biosynthesis. For example, the Calvin cycle, which reduces CO_2 to the level of carbohydrate, requires two moles of NADH for every CO_2. What can the cells use as a source of electrons to reduce NAD^+ when they are growing on NO_2^- and CO_2? Clearly not CO_2, since that is already the highest oxidized form of carbon. So we are left with NO_2^-. However, a comparison of the electrode potentials points to a problem. The E'_m for NO_3^-/NO_2^- is +0.42 V and the E'_m for $NAD^+/NADH$ is −0.32 V. Electrons flow spontaneously only to the more electropositive acceptor. Thus, to make the electrons flow from NO_2^- to NAD^+, the flow must be reversed, and this requires energy. How much energy is required? The potential difference ($\Delta E'_0$) is −0.32 − 0.42 or −0.7 V. Thus each electron would have to be energized by 0.7 eV. The source of energy is the Δp. The inward flow of protons through the coupling sites down the Δp gradient drives electron transport in reverse (see Section 3.7.1).

12.5 Summary

When it comes to inorganic metabolism, prokaryotes are versatile creatures. They can reduce oxidized forms of sulfur and nitrogen, as well as nitrogen gas, for incorporation into cell material, and they can oxidize a variety of inorganic substrates, trapping the energy released as a Δp. They can also carry out anaerobic respiration, using oxidized forms of iron, manganese, nitrogen, and sulfur as electron acceptors. One can add CO_2, which is used by the methanogens, to this list of electron acceptors.

There is both an assimilatory and a dissimilatory route for nitrate reduction. The assimilatory route is catalyzed by cytosolic enzymes that reduce nitrate to ammonia. The ammonia is then incorporated into amino acids via the GOGAT enzyme system. All bacteria that grow on nitrate as a source of cell nitrogen must have the assimilatory pathway.

The dissimilatory route differs in being membraneous and in producing a Δp that can be used for ATP synthesis. A common dissimilatory route reduces nitrate to N_2. Nitrate dissimilation to N_2 is also called denitrification. Organisms that carry out denitrification are widespread. However, the enzymes for denitrification are oxygen sensitive and furthermore are formed only under anaerobic conditions or when oxygen tensions are very low. These bacteria are generally facultative anaerobes and will carry out an aerobic respiration when oxygen is available. Several bacteria that are not denitrifiers can carry out an anaerobic respiration in which nitrate is reduced to nitrite or to ammonia by a membrane-bound nitrate reductase, creating a Δp.

Most bacteria can use sulfate as a source of cell sulfur. They employ a cytosolic assimilatory pathway in which the sulfate is activated by ATP to form adenosine-3′-phosphate-5′-phosphosulfate (PAPS). The formation of PAPS raises the E_h of sulfate so that it is a better electron acceptor. The reduction of PAPS yields sulfite, which is reduced to H_2S via a soluble sulfite reductase. The H_2S does not accumulate but is immediately incorporated into O-acetylserine to form L-cysteine. The L-cysteine donates the sulfur to the other sulfur-containing compounds.

Certain strict anaerobes can use sulfate as an electron acceptor for anaerobic respiration and reduce it to H_2S. These are the sulfate reducers. Organic carbon is oxidized, and the electrons are transferred to protons via an hydrogenase to form H_2. It is suggested that the H_2 diffuses into the periplasm, where it is oxidized by a cytochrome c that transfers the electrons to membrane carriers. Electrons travel across the membrane to the inner surface, creating a membrane potential. The protons resulting from the periplasmic oxidation of H_2 remain on the outer surface or in the periplasm, contributing toward the proton gradient. Dissimilatory sulfate reduction differs from assimilatory reduction in that APS is reduced rather than PAPS, the electron carriers are membranous, and a Δp is created.

Nitrogen fixation is carried out by a diversity of prokaryotes, including both bacteria and archeae. These include cyanobacteria, photosynthetic bacteria, strict heterotrophic anaerobes (*Clostridium pasteuranium, Desulfovibrio vulgaris*), and several obligate and facultative aerobes (e.g., rhizobia, *Azotobacter, Klebsiella*, methanogens). The reduction of N_2 is an ATP-dependent process catalyzed by the enzyme nitrogenase. Hydrogen gas is always a by-product of nitrogen fixation. The nitrogenase is oxygen sensitive and therefore must be protected from oxygen, either in specialized cells (heterocysts) in the case of cyanobacteria, or in leguminous nodules, or perhaps by an unusually high respiratory rate (*Azotobacter*), by binding to protective proteins, or by growth in an anaerobic environment (e.g., photosynthetic bacteria).

The oxidation of inorganic substances by oxygen is the source of energy for lithotrophs. Many lithotrophs are aerobic autotrophs. Electron transport takes place in the membrane, and a Δp is created. The Δp is used to drive ATP synthesis and reversed electron flow so that NADH can be generated for biosynthesis.

Iron-oxidizing bacteria such as *Thiobacillus ferrooxidans* live in acid environments (around pH 2) where iron–sulfide minerals and oxygen are available. By taking up cytoplasmic protons during the reduction of oxygen to water, the bacteria maintain a ΔpH of around 4.5 units. The ΔpH drives the synthesis of ATP via a membrane ATP synthase.

Lithotrophic activities are of immense ecological significance because they are necessary for the recycling of inorganic nutrients through the biosphere. For example, consider the nitrogen cycle. All organisms use ammonia and nitrate, or amino acids, as the source of nitrogen. These are called "fixed" forms nitrogen. Approximately 97% of all the nitrogen incorporated in living tissue comes from fixed nitrogen. However, the majority of the nitrogen on this planet is in the form of N_2 and unavailable to most organisms. Furthermore, the denitrification activities of bacteria living anaerobically cause a constant drain of fixed nitrogen from the biosphere. That is, living systems eventually oxidize reduced forms of nitrogen to nitrate, which is then reduced to nitrogen gas by the denitrifying bacteria. Therefore, we all depend upon the prokaryotes that reduce N_2 for our supply of fixed nitrogen.

The sulfate-reducing bacteria, which are responsible for much of the H_2S produced, feed sulfide to the photosynthetic sulfur bacteria and aerobic sulfur-oxidizing bacteria. The combined activities of the sulfate reducers and the sulfide oxidizers account for much of the elemental sulfur found in deposits worldwide. Hydrogen sulfide can also have deleterious effects. Since H_2S is toxic, it can occasionally produce harmful effects on fish, waterfowl, and even plants when the soil becomes anaerobic, as may occur in rice paddies. The H_2S can also cause corrosion of metal pipes in anaerobic soils and waters.

Study Questions

1. What are the major features distinguishing assimilatory and dissimilatory nitrate reduction and assimilatory and dissimilatory sulfate reduction?

2. During assimilatory sulfate reduction, O-acetylserine reacts with H_2S to form cysteine. The O-acetylserine is formed from serine and acetyl–CoA. Write a reaction showing the chemical structures to suggest how O-acetylserine might be formed from serine and acetyl–CoA.

3. Clostridia can reduce nitrogen gas to ammonia by means of electrons derived from pyruvate (generated from glucose via the Embden–Meyerhof–Parnas pathway). The immediate reductant for the nitrogenase is reduced ferredoxin. How is reduced ferredoxin generated from pyruvate?

4. What is the minimum number of Fe^{2+} ions that must be oxidized by oxygen to Fe^{3+} to reduce one NAD^+? Solve this problem by focusing on electron volts.

5. The ΔpH drives the synthesis of ATP in the acidophilic iron-oxidizing bacteria. How do they maintain the ΔpH?

6. What are some of the mechanisms used by nitrogen fixers to protect the nitrogenase from oxygen?

REFERENCES AND NOTES

1. Lovley, D. R. 1991. Dissimilatory Fe(III) and Mn(IV) reduction. *Microbiol. Rev.* 55:259–287.

2. Lovley, D. R. 1993. Dissimilatory metal reduction. *Annu. Rev. Microbiol.* 47:263–290.

3. Nealson, K. H., and D. Saffarini. 1994. Iron and manganese in anaerobic respiration: environmental significance, physiology, and regulation. *Annu. Rev. Microbiol.* 48:311–343.

4. Lin, J. T., and V. Stewart. 1998. Nitrate assimilation by bacteria. *Adv. Microbial Physiol.* 39:1–30.

5. In *Klebsiella pneumoniae* the electron donor for nitrate reductase appears to be an NADH-dependent oxidoreductase. Thus, the electrons travel from NADH to the oxidoreductase to the nitrate reductase. Other bacterial nitrate reductases use reduced ferredoxin or flavodoxin. The major cytoplasmic nitrite reductase in *E. coli* uses NADH as the electron donor. For example, see the following references: Lin, J. T., B. S. Goldman, and V. Stewart. 1994. The *nas* FEDCBA operon for nitrate and nitrite assimilation in *Klebsiella pneumoniae* M5a1. *J. Bacteriol.* 176:2551–2559; Gangeswaran, R., D. J. Lowe, and R. R. Eady. 1993. Purification and characterization of the assimilatory nitrate reductase of *Azotobacter vinelandii. Biochem. J.* 289:335–342.

6. Most of the reduced intracellular sulfur consists of H_2S and HS^-, rather than S^{2-}. This is because the pK_1 of H_2S is 7.04, the pK_2 is 11.96, and the cytoplasmic pH is usually close to 7. For example, neutrophilic bacteria have a cytoplasmic pH of approximately 7.5.

7. Casella, S., and W. J. Payne. 1996. Potential of denitrifiers for soil environment protection. *FEMS Microbiol. Lett.* 140:1–8.

8. Hansen, T. 1994. Metabolism of sulfate-reducing prokaryotes. *Antonie van Leeuwenhoek* 66:165–185.

9. Thauer has argued that free H_2 may not be the electron carrier for lactate oxidation by sulfate. Even at very low H_2 partial pressures, the reduction of H^+ by lactate is thermodyamically unfavorable. Furthermore, it has been demonstrated that H_2 does not affect lactate oxidation by sulfate by *Desulfovibrio vulgaris*, and *Desulfovibrio sapovorans* growing on lactate plus sulfate does not contain hydrogenase. Reviewed in: Thauer, R. K. 1989. Energy metabolism of sulfate-reducing bacteria, pp. 397–413. In: *Autotrophic Bacteria*. H. G. Schlegel and B. Bowien (Eds.). Science Tech Publishers, Madison, WI.

10. Reviewed by: Postgate, J. 1987. *Nitrogen Fixation*. Edward Arnold, London.

11. Reviewed in: Stacey, G., R. H. Burris, and H. J. Evans (Eds.). 1992 *Biological Nitrogen Fixation*. Chapman & Hall, New York.

12. Reviewed in: Dilworth, J. J., and A. R. Glenn (Eds.). 1991. *Biology and Biochemistry of Nitrogen Fixation*. Elsevier, Amsterdam.

13. Peuppke, S. G. 1996. The genetic and biochemical basis for nodulation of legumes by rhizobia. *Crit. Rev. Biotechnol.* 16:1–51.

14. Dixon, R., and D. Kahn. 2004. Genetic regulation of biological nitrogen fixation. *Natu. Rev. Microbiol.* 2:621–634.

15. Lobo, A. L. and S. H. Zinder. 1992. Nitrogen fixation by methanogenic bacteria, pp. 191–211. In: *Biological Nitrogen Fixation*. G. Stacey, H. R. Burris, and H. J. Evans (Eds.). Chapman & Hall, New York.

16. Dean, Dennis R., J. T., Bolin, and L. Zheng. 1993. Nitrogenase metalloclusters: structures, organization, and synthesis. *J. Bacteriol.* 175:6737–6744.

17. Peters, J. W., K. Fisher, and D. R. Dean. 1995. Nitrogenase structure and function: a biochemical–genetic perspective. *Annu. Rev. Microbiol.* 49:335–366.

18. The MoFe protein has pairs of metalloclusters: a pair of iron–sulfur clusters (Fe_8S_{7-8}) known as P clusters, and a pair of iron–sulfur–molybdate clusters (Fe_7S_9Mo-homocitrate), known as FeMo cofactors.

19. There are two types of metallocluster in the MoFe protein, a FeMo cofactor and a P cluster. The P cluster consists of two Fe_4S_4 clusters. Each $\alpha\beta$ dimer has a FeMo cofactor and a P cluster, making two copies of each type of cluster for every tetramer. One model of electron flow proposes that the electron travels in the Fe protein in an ATP-dependent reaction from the Fe_4S_4 cluster to a P cluster in the MoFe protein. From the P cluster the electron moves to the FeMo cofactor, and from there to N_2. The precise role of ATP in electron transfer is not known, although it has been shown to bind as the Mg^{2+} chelate to the Fe protein. It is presumed that binding and/or hydrolysis of MgATP causes conformational changes in the nitrogenase that facilitate electron transfer. For example, conformational changes in the proteins might alter the environment and redox potential of the metal clusters, or bring the Fe_4S_4 cluster in the Fe protein and the P cluster in the FeMo protein into closer proximity. Specifically, it has been proposed that binding of ATP to the Fe protein lowers the redox potential of the Fe_4S_4 cluster and also leads to association (complex formation) between the Fe protein and the MoFe protein. Association of the two proteins causes ATP hydrolysis and electron transfer. One model proposes that complex formation and hydrolysis of ATP is accompanied by a configurational change in the MoFe protein that moves a P cluster close to a Fe_4S_4 cluster in the Fe protein, thus facilitating electron transfer. The two proteins dissociate after electron transfer. Proposals about how the MoFe protein might reduce N_2 are derived from model systems. The chemical reactivity of N_2 is increased when it binds to a transition metal, especially in a complex with other ligands, such as sulfur. This has been studied in model systems, where chemically synthesized metal complexes containing molybdenum bound to N_2 were reduced with artificial reductants. It is thought that the N_2 becomes activated when it binds to the molybdenum in the FeMo cofactor in nitrogenase and, while bound, is reduced to ammonia.

20. Eady, R. R. 1992. The dinitrogen-fixing bacteria, pp. 535–553. In: *The Prokaryotes*, Vol. I. A. Balows, H. G. Truper, M. Dworkin, W. Harder, and K.-H. Schleifer (Eds.). Springer-Verlag, Berlin.

21. Yates, M. G., and F. O. Cambell. 1989. The role of oxygen and hydrogen in nitrogen fixation, pp. 383–416. In: *SGM Symposium*, Vol. 42, *The*

Nitrogen and Sulphur Cycles. J. A. Cole, and S. Ferguson (Eds.). Cambridge University Press, Cambridge.

22. Bergman, B., J. R. Gallon, A. N. Rai, and L. J. Stal. 1997. N_2 fixation by non-heterocystous cyanobacteria. *FEMS Microbiol. Rev.* **19**:139–185.

23. Haselkorn, R., and W. J. Bulikema. 1992. Nitrogen fixation in cyanobacteria, pp. 166–190. In: *Biological Nitrogen Fixation.* G. Stacey, R. H. Burris, and H. J. Evans (Eds.). Chapman & Hall, New York.

24. Kelly, D. P. 1990. Energetics of chemolithotrophs, pp. 479–503. In: *The Bacteria,* Vol. XII. T. A. Krulwich (Ed.). Academic Press, New York.

25. Hooper, A. B. 1989. Biochemistry of the nitrifiying lithoautotrophic bacteria, pp. 239–265. In: *Autotrophic Bacteria.* H. G. Schlegel, and B. Bowien (Eds.). Springer-Verlag, Berlin.

26. Bacteria that oxidize ammonia anaerobically belong to the genus *Candidatus* in the order *Planctomycetales.* They carry out the following reaction:

$$NH_4^+ + NO_2^- \rightarrow N_2 + 2H_2O$$

They have been detected in anaerobic parts of marine waters and sediments, and they make a significant contribution to the production of N_2 in these environments. Probably, the nitrite is produced from nitrate in the anaerobic zone by heterotrophic denitrifiers. The source of the nitrate is ammonia oxidized by the aerobic nitrifiers living in the oxic water zones above the sediments.

27. Strous, M., and M. S. M. Jetten. 2004. Anaerobic oxidation of methane and ammonium. *Annu. Rev. Microbiol.* **58**:99–117.

28. Nicholls, D. G., and S. J. Ferguson. 1992. *Bioenergetics 2.* Academic Press, London.

29. Kelly, D. P. 1989. Physiology and biochemistry of unicellular sulfur bacteria, pp. 193–217. In: *Autotrophic Bacteria.* H. G. Schlegel, and B. Bowien (Eds.). Springer-Verlag, Berlin.

30. Kelly, D. P., W.-P. Lu, and R. K. Poole. 1993. Cytochromes in *Thiobacillus tepidarius* and the respiratory chain involved in the oxidation of thiosulphate and tetrathionate. *Arch. Microbiol.* **160**:87–95.

31. Friedrich, C. G., 1998. Physiology and genetics of sulfur-oxidizing bacteria, pp. 235–289. In: Advances in *Microbial Physiology.* R. K. Poole (Ed.). Academic Press, New York.

32. Ingledew, W. J. 1990. Acidophiles, pp. 33–54. In: *Microbiology of Extreme Environments.* C. Edwards (Ed.). McGraw-Hill, New York.

33. Ingledew, W. J. 1982. *Thiobacillus ferrooxidans*: the bioenergetics of an acidophilic chemolithotroph. *Biochim. Biophys. Acta* **683**:89–117.

34. The amount of energy available from the oxidation of Fe^{2+} by O_2 depends upon whether the oxygen is reduced on the periplasmic side of the membrane or on the cytoplasmic side. At pH 2, the E_m of Fe^{3+}/Fe^{2+} is usually stated to be about 0.77 to 0.78 V. At pH 7, the E_m of O_2/H_2O is 0.82 V, but at pH 2, it is 1.12 V. If the oxygen were reduced in the cytoplasm, then $0.82 - 0.77$, or 0.05 eV, would be available per electron. If the oxygen were reduced in the periplasm, then $1.12 - 0.77$, or 0.35 eV would be available per electron. However, as pointed out by Nicholls and Ferguson,[28] periplasmic reduction of oxygen with consumption of periplasmic protons would not be helpful because a Δp would not develop. (There would be no electrogenic electron flow and no differential consumption of protons between periplasm and cytoplasm.) Attempting to reduce oxygen by bringing protons to the periplasm from the cytoplasm would require energy to overcome the proton concentration gradient between periplasm and cytoplasm.

35. Ehrlich, H. L., J. W., Ingledew, and J. C. Salerno. 1991. Iron- and manganese-oxidizing bacteria, pp. 147–170. In: *Variations in Autotrophic Life.* J. M. Shively, and L. L. Barton (Eds.). Academic Press, New York.

36. According to the chemiosmotic theory, the actual number of ATPs made will depend upon the ratio of protons translocated per electron to the number translocated by the ATP synthase, that is, $(H^+/e^-)/H^+/ATP)$. (See Section 4.5.2.)

13

C₁ Metabolism

Many prokaryotes can grow on C_1 compounds as their sole source of carbon. The following are some common C_1 compounds that support growth:

1. Carbon dioxide (CO_2)

2. Methane (CH_4)

3. Methanol (CH_3OH)

4. Methylamine (CH_3NH_2)

A few strictly anaerobic prokaryotes that use carbon dioxide as a source of cell carbon also use it as an electron acceptor, reducing it to methane and deriving ATP from the process. These are the methanogens, which are placed in the Archaea, one of the two prokaryotic domains (Section 1.1.1).

13.1 Carbon Dioxide Fixation Systems

Organisms that use CO_2 as the sole or major carbon source are called *autotrophs*, and the pathway of CO_2 assimilation is called autotrophic CO_2 fixation. The autotrophs include the photoautotrophs and the chemoautotrophs. The prefix indicates the source of energy and the term concludes with the name of the source of carbon. The *photo*autotrophs, which use light as their source of energy, include the plants, algae, and photosynthetic bacteria. The *chemo*autotrophs use inorganic chemicals as their source

of energy (e.g., hydrogen gas, ammonia, nitrite, ferrous ion, inorganic sulfur). Thus far, the only known chemoautotrophs are bacteria.

There are three major autotrophic CO_2 fixation pathways in prokaryotes. These are the *Calvin cycle (also known as the Calvin–Benson–Bassham, or CBB, cycle)*, the *acetyl–CoA pathway*, and the *reductive tricarboxylic acid pathway*. The Calvin cycle is the most prominent of the autotrophic CO_2 fixation systems and is found in photosynthetic eukaryotes, most photosynthetic bacteria, cyanobacteria, and chemoautotrophs. It does not occur in the archaea or in certain obligately anaerobic or microaerophilic bacteria. Instead, the acetyl–CoA pathway and the reductive carboxylic acid pathway are found in these organisms. The reductive carboxylic acid pathway occurs in the green photosynthetic bacteria *Chlorobium* (anaerobes), in *Hydrogenobacter* (microaerophilic), in *Desulfobacter* (anaerobes), and in the archaeons *Sulfolobus* and *Thermoproteus*. The acetyl–CoA pathway is more widespread and is found in methanogenic archaea, some sulfate-reducing bacteria, and in facultative heterotrophs that synthesize acetic acid from CO_2 during fermentation. The latter bacteria are called *acetogens*.

13.1.1 The Calvin cycle

The Calvin cycle is a pathway for making phosphoglyceraldehyde (PGALD) completely from

CO_2.[1] The pathway operates in the chloroplasts of plants and algae. However, in bacteria, where there are no similar organelles, the Calvin cycle takes place in the cytosol.

Summing the reactions of the Calvin cycle

During the Calvin cycle, three CO_2 molecules are reduced to PGALD, which is at the oxidation level of glyceraldehyde ($C_3H_6O_3$). Twelve electrons are required, and these are provided by 6 moles of NAD(P)H. [Recall that each NAD(P)H carries a hydride ion (one proton and two electrons) (Section 8.3)]. However, NAD(P)H is not a sufficiently strong reductant to reduce CO_2 without an additional source of energy. Each reduction thus requires an ATP. Therefore, six ATPs are required for the six reductions. However, the overall reaction, which follows, indicates that nine ATPs are required:

$$3CO_2 + 9\,ATP + 6\,NAD(P)H + 6H^+ + 5H_2O$$

$$\rightarrow PGALD + 9\,ADP + 8\,P_i + 6\,NAD(P)^+$$

The three extra ATPs are used to recycle an intermediate, ribulose-1,5-bisphosphate (RuBP), which is catalytically required for the pathway.

The Calvin cycle can be divided into two stages

The Calvin cycle is a complicated pathway, but it will be familiar because it resembles the pentose phosphate pathway in certain key reactions (Section 8.5). It is convenient to think of the Calvin cycle as occurring in two stages:

1. Stage 1 is a reductive carboxylation of ribulose-1,5-bisphosphate (RuBP) to form phosphoglyceraldehyde (PGALD).

2. Stage 2 consists of sugar rearrangements regenerating RuBP by using some of the PGALD produced in stage 1.

Stage 1 The first reaction is the carboxylation of RuBP, forming two moles of 3-phosphoglycerate (PGA). The enzyme is called RuBP carboxylase or ribulose-1,5-bisphosphate carboxylase ("rubisco"). The reaction is shown in Fig. 13.1, along with its probable mechanism. An enolate anion probably forms that is carboxylated on the C2. Hydrolysis of the intermediate yields two moles of 3-phosphoglycerate (PGA).

The PGA is then reduced to PGALD via 1,3-bisphosphoglycerate (BPGA). These reactions also take place in glycolysis. The reactions of stage 1 can be summarized as follows:

$$3CO_2 + 3\,RuBP + 3H_2O \rightarrow 6\,PGA$$

$$6\,PGA + 6\,ATP \rightarrow 6\,BPGA + 6\,ADP$$

$$6\,BPGA + 6\,NAD(P)H + 6H^+$$

$$\rightarrow 6\,PGALD + 6\,NAD(P)^+ + 6\,P_i$$

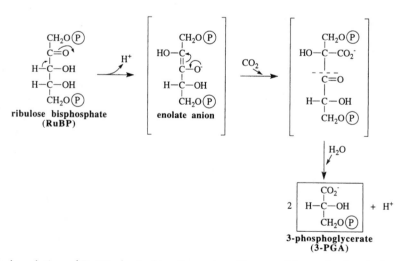

Fig. 13.1 Carboxylation of RuBP: the "rubisco" reaction. One possible mechanism is that the carbonyl group in RuBP attracts electrons, resulting in the dissociation of a hydrogen from the carbon on the *cis*-HCOH group. Electrons then shift to form the enolate anion, which becomes carboxylated at the C2. Hydrolysis of the carboxylated intermediate yields two 3-PGA molecules.

$$3CO_2 + 3\,RuBP + 3H_2O + 6\,ATP$$
$$+ 6NAD(P)H + 6H^+$$
$$\rightarrow 6\,PGALD + 6\,ADP + 6\,P_i + 6\,NAD(P)^+$$

Stage 1 is summarized as reactions **1** and **2** in Fig. 13.2.

Stage 2 The whole point of the sugar rearrangements of stage 2 is to regenerate three RuBPs from five of the six PGALDs made in stage 1. For reference, see Fig. 13.2. (The Calvin cycle is summarized without chemical structures later. See the subsection entitled. *The relationship of the Calvin cycle to glycolysis and the pentose phosphate pathway.*) Since RuBP has five carbons and PGALD has three carbons, a two-carbon fragment must be transferred to PGALD. That is, a transketolase reaction is required. The C_2 donor is fructose-6-phosphate, which is made from PGALD (reactions **3, 4,** and **5,** Fig. 13.2). The synthesis of fructose-6-phosphate from PGALD uses glycolytic enzymes and fructose-1,6-bisphosphate phosphatase. One molecule of PGALD first isomerizes to dihydroxyacetone phosphate (DHAP), which condenses with a second molecule of PGALD to form fructose-1,6-bisphosphate. The fructose-1,6-bisphosphate is then dephosphorylated to fructose-6-phosphate by the phosphatase (reaction **5**). The fructose-6-phosphate donates a two-carbon fragment in a transketolase reaction to PGALD, forming erythrose-4-phosphate and xylulose-5-phosphate (reaction **6**).

The xylulose-5-phosphate is isomerized to ribulose-5-phosphate (reaction **10**). The erythrose-4-phosphate is condensed with the second dihydroxyacetone phosphate to form a seven-carbon ketose diphosphate, sedoheptulose-1,7-bisphosphate (reaction **7**). The formation of sedoheptulose-1,7-bisphosphate is analogous to the condensation of dihydroxyacetone phosphate with PGALD to form fructose-1,6-bisphosphate, but it is catalyzed by a different aldolase, sedoheptulose-1,7-bisphosphate aldolase. (However, as discussed later, the bacterial sedoheptulose-1,7-bisphosphate aldolase and the fructose-1,6-bisphosphate aldolase may actually be a single bifunctional enzyme.) A phosphatase hydrolytically removes the phosphate from the C_1 to form sedoheptulose-7-phosphate (reaction **8**). This is an irreversible reaction and prevents the pentose phosphates that are formed from sedoheptulose-7-phosphate in the next step from being converted back to phosphoglyceraldehyde. The sedoheptulose-7-phosphate is a two-carbon donor in a second transketolase reaction in which the last of the phosphoglyceraldehyde is used as an acceptor (reaction **9**). The products are the pentose phosphates ribose-5-phosphate and xylulose-5-phosphate, both of which are isomerized to ribulose-5-phosphate (reactions **10** and **11**). The ribulose-5-phosphate is then phosphorylated to ribulose-1,5-bisphosphate by a ribulose-5-phosphate kinase that is unique to the Calvin cycle (reaction **12**). Thus, the pentose isomerase reactions and the transketolase reactions are the same in both the Calvin cycle and the pentose phosphate pathway. The reactions of stage 2 are summarized as follows:

$$2\,PGALD \rightarrow 2\,DHAP$$

$$PGALD + DHAP \rightarrow FBP$$

$$FBP + H_2O \rightarrow F6P + P_i$$

$$F6P + PGALD \rightarrow erythrose\text{-}4\text{-}P + xylulose\text{-}5\text{-}P$$

$$Erythrose\text{-}4\text{-}P + DHAP$$
$$\rightarrow sedoheptulose\text{-}1,7\text{-}bisphosphate$$

$$Sedoheptulose\text{-}1,7\text{-}bisphosphate + H_2O$$
$$\rightarrow sedoheptulose\text{-}7\text{-}P + P_i$$

$$Sedoheptulose\text{-}7\text{-}P + PGALD$$
$$\rightarrow ribose\text{-}5\text{-}P + xylulose\text{-}5\text{-}P$$

$$2\,Xylulose\text{-}5\text{-}P \rightarrow 2\,ribulose\text{-}5\text{-}P$$

$$Ribose\text{-}5\text{-}P \rightarrow ribulose\text{-}5\text{-}P$$

$$3\,Ribulose\text{-}5\text{-}P + 3\,ATP$$
$$\rightarrow 3\,ribulose\text{-}1,5\text{-}bisphosphate + 3\,ADP$$

$$5\,PGALD + 2H_2O + 3\,ATP$$
$$\rightarrow 3\,ribulose\text{-}1,5\text{-}bisphosphate + 2\,P_i + 3\,ADP$$

Summing stages 1 and 2 yields

$$3CO_2 + 9\,ATP + 6\,NAD(P)H + 6H^+ + 5H_2O$$
$$\rightarrow PGALD + 9\,ADP + 8\,P_i + 6\,NAD(P)^+$$

The carbon balance

Complex pathways can be seen in simpler perspective by examining the carbon balance. The carbon balance for the Calvin cycle is as follows:

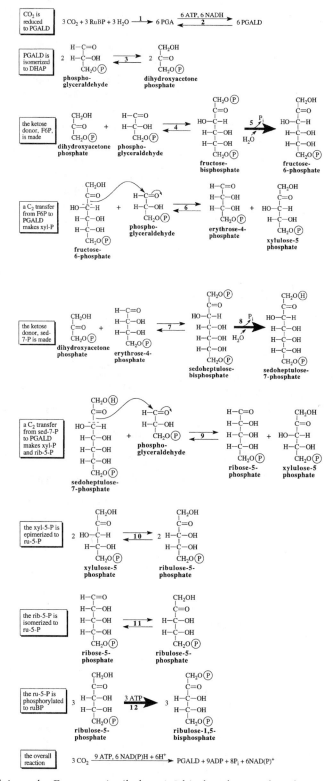

Fig. 13.2 The Calvin cycle. Enzymes: **1**, ribulose-1,5-bisphosphate carboxylase; **2**, 3-phosphoglycerate kinase and triosephosphate dehydrogenase; **3**, triosephosphate isomerase; **4**, fructose-1,6-bisphosphate aldolase; **5**, fructose-1,6-bisphosphate phosphatase; **6 and 9**, transketolase; **7**, sedoheptulose-1,7-bisphosphate aldolase; **8**, sedoheptulose-1,7-bisphosphatase; **10**, phosphopentose epimerase; **11**, ribose phosphate isomerase; **12**, phosphoribulokinase.

$$3\,C_1 + 3\,C_5 \rightarrow 6\,C_3$$

$$2\,C_3 \rightarrow C_6$$

$$C_6 + C_3 \rightarrow C_4 + C_5$$

$$C_4 + C_3 \rightarrow C_7$$

$$\underline{C_7 + C_3 \rightarrow 2\,C_5}$$

$$3\,C_1 \rightarrow C_3$$

The carbon balance for the Calvin cycle illustrates that the pathway produces one C_3 from three C_1 molecules.

The relationship of the Calvin cycle to glycolysis and the pentose phosphate pathway

Most of the reactions of the Calvin cycle also take place in glycolysis and the pentose phosphate pathway. The 12 reactions in Fig. 13.3 comprise the Calvin cycle. Reaction **1** is the carboxylation of ribulose-1,5-bisphosphate to form 3-phosphoglycerate. This reaction is unique to the Calvin cycle. The 3-phosphoglycerate enters the glycolytic pathway. Reactions **2** through **6** take place during the reversal of glycolysis, whereby 3-phosphoglycerate is transformed into fructose-6-phosphate (see Section 8.1). The fructose-6-phosphate enters the pentose phosphate pathway at the level of the sugar rearrangement reactions (reaction **7**) (Section 8.5). Reaction **7** is the transketolase reaction that forms erythrose-4-phosphate and xylulose-5-phosphate.

Now the Calvin cycle diverges from the pentose phosphate pathway. Whereas the pentose phosphate pathway synthesizes sedoheptulose-7-phosphate from erythrose-4-phosphate via the reversible transaldolase reaction (TA), the

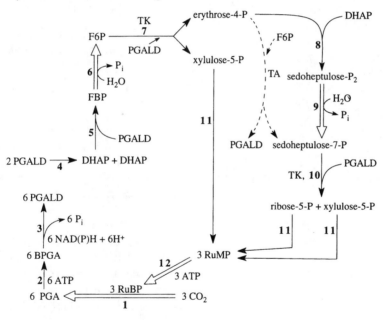

sum: $3\,CO_2 + 9\,ATP + 6\,NAD(P)H + 6\,H^+ \longrightarrow PGALD + 9\,ADP + 6\,NAD(P)^+ + 8\,P_i$

Fig. 13.3 Relationships between the Calvin cycle, glycolysis, and the pentose phosphate pathway. Reactions **1** through **12** are the Calvin cycle. Note that the only reactions unique to the Calvin cycle are reactions **1**, **8**, **9**, and **12**. Reaction **1** is catalyzed by ribulose-1,5-bisphosphate carboxylase. Reactions **2** through **5** are glycolytic reactions. Reaction **6** is catalyzed by fructose-1,6-bisphosphatase. Reaction **7** is the transketolase reaction also found in the pentose phosphate pathway. Reactions **8** and **9** are the sedoheptulose–bisphosphate aldolase and the sedoheptulose-1,7-bisphosphatase reactions. Reaction **10** is the transketolase reaction, also found in the pentose phosphate pathway. Reactions **11** are the pentose epimerase and isomerase reactions also present in the pentose phosphate pathway. Reaction **12** is the RuMP kinase reaction. The pentose phosphate pathway differs from the Calvin cycle only in that the pentose phosphate pathway synthesizes sedoheptulose-7-phosphate via a reversible transaldolase reaction (dashed lines), whereas the Calvin cycle synthesizes sedoheptulose-7-phosphate via aldolase and phosphatase reactions (reactions **8** and **9**). Since the phosphatase is irreversible, the Calvin cycle irreversibly converts phosphoglyceraldehyde to pentose phosphates.

Calvin cycle uses the aldolase (reaction 8) and the irreversible phosphatase (reaction 9) to synthesize sedoheptulose-7-phosphate. This has important consequences regarding the directionality of the Calvin cycle. Because the phosphatase (reaction 9) is an irreversible reaction, the Calvin cycle proceeds only in the direction of pentose phosphates, whereas the pentose phosphate pathway is reversible from sedoheptulose-7-phosphate. (An irreversible Calvin cycle makes physiological sense, since the sole purpose of the Calvin cycle is to regenerate ribulose-1,5-bisphosphate for the initial carboxylation reaction.)

Once the sedoheptulose-7-phosphate has formed, the synthesis of ribulose-5-phosphate (reactions 10 and 11) takes place via the same reactions as in the pentose phosphate pathway. Reaction 12 of the Calvin cycle is the phosphorylation of ribulose-5-phosphate to form ribulose-1,5-bisphosphate catalyzed by RuMP kinase, a second reaction that is unique to the Calvin cycle. The sedoheptulose-1,7-bisphosphate aldolase and the sedoheptulose-1,7-bisphosphate phosphatase also occur in the ribulose monophosphate pathway, which is a formaldehyde-fixing pathway that uses Calvin cycle reactions (Section 13.2.1).[2]

13.1.2 The acetyl–CoA pathway

Bacteria that use the acetyl–CoA pathway include methanogens, acetogenic bacteria, and most autotrophic sulfate-reducing bacteria.[3] (See note 4 for a definition of the acetogenic bacteria.) This section presents a general summary of the acetyl–CoA pathway without describing the individual reactions. The first part explains how acetyl–CoA is made from CO_2 and H_2, the second part explains how the acetyl–CoA is incorporated into cell material, and the third part summarizes how the acetyl–CoA pathway is used by methanogens for methanogenesis. Sections 13.1.3 and 13.1.4 describe the individual reactions of the acetyl–CoA pathway in acetogens and in methanogens.

Making acetyl–CoA from CO_2 via the acetyl–CoA pathway

The acetyl–CoA pathway can be viewed as occurring in four steps. The first step is a series of reactions that result in the reduction of CO_2 to a methyl group $[CH_3]$. The methyl group is bound to tetrahydrofolic acid (THF) in bacteria and to tetrahydromethanopterin (THMP) in archaea. The methyl group is then transferred to the enzyme *carbon monoxide dehydrogenase*. Carbon monoxide dehydrogenase also catalyzes the reduction of a second molecule of carbon dioxide to a bound carbonyl group $[CO]$, which will become the carboxyl group of acetate. During autotrophic growth, the electrons are provided by hydrogen gas via a hydrogenase.

$$CO_2 + 3H_2 + H^+ \rightarrow [CH_3] + 2H_2O \quad (13.1)$$

$$CO_2 + H_2 \rightarrow [CO] + H_2O \quad (13.2)$$

The bound $[CH_3]$ and $[CO]$ are then condensed by carbon monoxide dehydrogenase to form bound acetyl, $[CH_3CO]$.

$$[CO] + [CH_3] \rightarrow [CH_3CO] \quad (13.3)$$

The bound acetyl reacts with bound CoASH, [CoAS], to form acetyl–CoA, which is released from the enzyme:

$$[CH_3CO] + [CoAS] \rightarrow CH_3COSCoA \quad (13.4)$$

The acetyl–CoA either is converted to acetate via phosphotransacetylase and acetate kinase (generating an ATP) and excreted or is assimilated into cell material. As mentioned, many anaerobic bacteria can grow autotrophically with the acetyl–CoA pathway, using H_2 as the source of electrons. These bacteria must generate ATP from the Δp generated during the reduction of the CO_2 by H_2. The Δp is required for net ATP synthesis because the ATP generated during the conversion of acetyl–CoA to acetate is balanced by the ATP used for the formation of formyl–THF (see Section 13.1.3, Fig. 13.4).

Incorporating acetyl–CoA into cell material

For acetogens growing autotrophically, part of the acetyl–CoA that is produced must be incorporated into cell material. The glyoxylate cycle, which is generally associated with aerobic metabolism, is not present in these organisms, and therefore a different pathway must operate. As shown in eqs. 13.5 and 13.6, the acetyl–CoA is reductively carboxylated to pyruvate via *pyruvate synthase*, a ferredoxin-linked enzyme. The pyruvate can be phosphorylated to phosphoenolpyruvate by using PEP synthetase. The phosphoenolpyruvate is used for biosynthesis in the usual way. A more detailed discussion of acetyl–CoA assimilation in methanogens is given in Section 13.1.4.

Pyruvate synthase

$$Acetyl–CoA + Fd_{red} + CO_2$$

$$\rightarrow pyruvate + Fd_{ox} + CoASH \qquad (13.5)$$

PEP synthetase

$$Pyruvate + ATP + H_2O$$

$$\rightarrow PEP + AMP + P_i \qquad (13.6)$$

13.1.3 The acetyl–CoA pathway in Clostridium thermoaceticum

The acetyl–CoA pathway was first investigated in *C. thermoaceticum*, an acetogenic bacterium that converts one mole of glucose to three moles of acetate.[5] Under these circumstances, the production of acetate is used as an electron sink during fermentation rather than for autotrophic growth. The pathway is illustrated in Fig. 13.4. Glucose is converted to pyruvate via the Embden–Meyerhof–Parnas pathway (reactions 1). Two of the acetates are synthesized from the decarboxylation of pyruvate by means of pyruvate–ferredoxin oxidoreductase, phosphotransacetylase, and acetate kinase (reactions 2–4). The third acetate is synthesized from CO$_2$ via the acetyl–CoA pathway.

In the acetyl–CoA pathway, one of the carbon dioxides that is removed from pyruvate is not set free but instead becomes bound to the enzyme carbon monoxide dehydrogenase (CODH), where it will be used for acetate synthesis (reaction 2a). This bound CO$_2$ will eventually become the carbonyl group in acetyl–CoA. The second pyuvate is decarboxylated to release free CO$_2$ (reaction 2b). In reactions 5 through 9, the free CO$_2$ is reduced to bound methyl. The free CO$_2$ first becomes reduced to formate (HCOOH) by formate dehydrogenase (reaction 5). The formate is then attached to tetrahydrofolic acid (THF) in an ATP-dependent reaction to make formyl–THF (HCO–THF) (reaction 6). (See Section 9.2.4 for a discussion of tetrahydrofolate reactions.)

The formyl–THF is then dehydrated to form methenyl–THF (CH–THF) (reaction 7). The methenyl–THF is reduced by NADPH to methylene–THF (CH$_2$–THF) (reaction 8). The methylene–THF is reduced by ferredoxin to methyl–THF (CH$_3$–THF) (reaction 9), a reaction thought to be coupled to ATP formation by a mechanism involving Δp. The methyl group is then transferred to a corrinoid enzyme, [Co]-E, which transfers the methyl group to the CO–CODH (reactions 10 and 11).[6] The CODH makes the acetyl moiety, [CH$_3$CO] (reaction 12). The CODH also has a binding site for CoASH and synthesizes acetyl–CoA from the bound intermediates (reactions 13 and 14). The CODH can also bind free CO$_2$ and reduce it to the level of carbon monoxide (reaction 15).

Autotrophic growth

Following the entry of CO$_2$ at reactions 5 and 15 (Fig. 13.4), the acetyl–CoA pathway allows autotrophic growth on CO$_2$ and H$_2$. Carbon dioxide is reduced to [CH$_3$] and [CO] (reactions 5–10 and 15). The [CH$_3$] and [CO] combine with CoASH to form acetyl–CoA (reactions 11–14). During acetogenesis, an ATP can be made via substrate-level phosphorylation from acetyl–CoA (reactions 3 and 4). However, because an ATP is required to incorporate formate (reaction 6), there can be no net ATP synthesis via substrate-level phosphorylation alone.

How do organisms using the acetyl–CoA pathway during autotrophic growth make net ATP? It is necessary to postulate that ATP is made by a chemiosmotic mechanism coupled to the reduction of CO$_2$ to acetate. For example, there may be electrogenic electron flow following periplasmic oxidation of H$_2$; or perhaps a quinone loop or a proton pump may be operating. (Mechanisms for generating a Δp were reviewed in Sections 3.7.2 and 4.6.)

It appears that homoacetogenic clostridia do indeed create a Δp during the reduction of CO$_2$ to acetate and that the Δp drives the synthesis of ATP via a H$^+$–ATP synthase.[7]

Not all acetogenic bacteria use a proton circuit, however. *Acetobacterium woodii* couples the reduction of CO$_2$ to acetyl–CoA to a primary sodium ion pump.[8] The sodium ion translocating step has not been identified

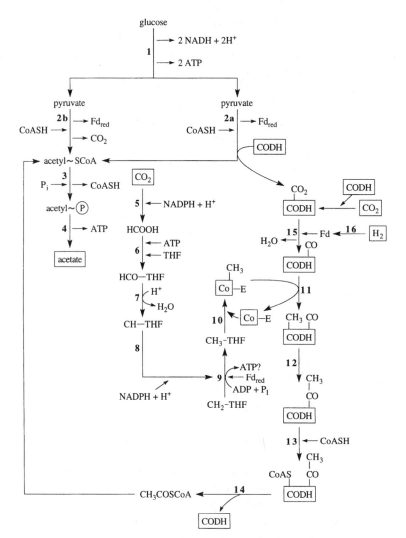

Fig. 13.4 The acetyl–CoA pathway in *Clostridium thermoaceticum*. During heterotrophic growth, the carboxyl group in some pyruvate molecules is directly transferred to CODH without being released as free CO_2. The evidence for this is that even in the presence of CO_2, pyruvate is required in cell-free extracts for the synthesis of acetic acid from CH_3–THF. Furthermore, radioisotope experiments confirm that the carboxyl in acetate is derived from the carboxyl in pyruvate without going through free CO_2. However, the cells can be grown on CO_2 and H_2. Under these circumstances, the CODH reduces CO_2 to CO–CODH via hydrogenase and electrons from H_2 (reactions 15 and 16). Enzymes: 1, glycolytic enzymes; 2, pyruvate:ferredoxin oxidoreductase; 3, phosphotransacetylase; 4, acetate kinase; 5, formate dehydrogenase; 6, formyltetrahydrofolate (HCO–THF) synthetase; 7, methenyltetrahydrofolate (CH–THF) cyclohydrolase; 8, methylenetetrahydrofolate (CH_2–THF) dehydrogenase; 9, methylenetetrahydrofolate reductase; 10, methyl-transferase; 11, corrinoid enzyme; 12–15, carbon monoxide dehydrogenase (CODH); 16, hydrogenase.

but is thought to be either the reduction of methylene–THF to methyl–THF (reaction 9) or some later step in the synthesis of acetyl–CoA because these are highly exergonic reactions. *A. woodii* also has an ATP synthase that is sodium ion dependent. Thus when growing on CO_2 and H_2, these bacteria rely on a sodium ion current to make ATP. A similar situation exists

in *Propionigenium modestum*, where a Na⁺-translocating methylmalonyl–CoA decarboxylase generates a Na⁺ electrochemical potential that drives the synthesis of ATP (Section 3.8.1).

Carbon monoxide dehydrogenase

It is clear that carbon monoxide dehydrogenase (CODH), also called carbon monoxide dehydrogenase/acetyl–CoA synthase (CODH/ACS), is a crucial enzyme for acetogenesis, for autotrophic growth in anaerobes, and for methanogenesis from acetate. Because of the importance of this enzyme, some of the evidence for its participation in synthesizing and degrading acetyl–CoA is described next.

The CODH catalyzes the breakage and formation of the carbon–carbon bond between the methyl group [CH₃] and the carbonyl group [CO] in the acetyl moiety. Evidence for this is the following exchange reaction carried out by CODH[9]:

$$^{14}CH_3\text{–}CODH + CH_3COSCoA$$
$$\rightleftharpoons CH_3\text{–}CODH + {}^{14}CH_3COSCoA$$

The CODH catalyzes the breakage and formation of the thioester bond between CoASH and the carbonyl group of the acetyl moiety. Evidence for this is the following exchange reaction carried out by CODH in conjunction with another enzyme called disulfide reductase,[10] in which an asterisk desionates radioactive CoASH:

$$CH_3COSCoA + {}^*CoASH$$
$$\rightleftharpoons CH_3COS^*CoA + CoASH$$

The following exchange has also been demonstrated to be catalyzed by CO dehydrogenase[11]:

$$[1\text{-}^{14}C]Acetyl\text{-}CoA + {}^{12}CO$$
$$\rightleftharpoons {}^{14}CO + [1\text{-}^{12}C]acetyl\text{-}CoA$$

The three exchange reactions just described show that the CODH is capable of disassembling acetyl–CoA into its components, [CH₃], [CO], and [CoASH], and resynthesizing the molecule. Therefore the presence of CODH is sufficient for making acetyl–CoA from the bound components. A recent article describes the crystal structure of the enzyme and a Ni-Fe-Cu center at the active site.[12]

13.1.4 The acetyl–CoA pathway in methanogens

Acetyl–CoA synthesis from CO_2 and H_2

Among the archaea are a group called methanogens that can grow autotrophically on CO_2 and H_2 (Section 1.1.1). During autotrophic growth, the CO_2 is first incorporated into acetyl–CoA in a pathway very similar to the acetyl–CoA pathway that exists in the bacteria (e.g., in C. thermoaceticum).[13–16] In other words, one molecule of CO_2 is reduced to [CO] bound to CODH. A second molecule of CO_2 is reduced to bound methyl, [CH₃], which is transferred to the CODH. Then the CODH makes acetyl–CoA from the bound CO, CH₃, and CoASH. There are, however, differences between the acetyl–CoA pathway found in the methanogenic archaea and that in the bacteria. The differences have to do with the cofactors and the utilization of formate. For reference, the structures of the cofactors are drawn in Fig. 13.5. The differences between the two pathways are as follows:

1. The carriers for the formyl and more reduced C_1 moieties in the methanogens is not tetrahydrofolic acid (THF) but rather tetrahydromethanopterin (THMP), a molecule resembling THF (For a comparison of the structures, see Fig. 13.5, where the TAMP is labeled "tetrahydromethanopterin").

2. The methanogens do not reduce free CO_2 to formate. Nor do they incorporate formate into THMP to form formyl–THMP. What the methanogens do instead is fix CO_2 onto a C_1 carrier called methanofuran (MFR) and reduce it to formyl–MFR. The formyl group is then transferred to THMP. (Recall that the acetogenic bacteria reduce free CO_2 to formate and then incorporate the formate into formyl–THF.)

Figure 13.6 is a diagram showing how the acetyl–CoA pathway is thought to operate in methanogens. (The same pathway is shown in Fig. 13.7, but with the chemical structures drawn.) The CO_2 is first condensed with methanofuran (MFR) and reduced to formyl–MFR (reaction 1). The formyl group is then transferred to tetrahydromethanopterin (THMP) to form formyl–THMP (reaction 2). Then a dehydration produces methenyl–THMP (reaction 3). The methenyl–THMP is reduced to methylene–THMP (reaction 4), which is reduced to methyl–THMP (reaction 5). The pathway diverges at the methyl level. One branch synthesizes acetyl–CoA in reactions quite similar to the synthesis of acetyl–CoA in

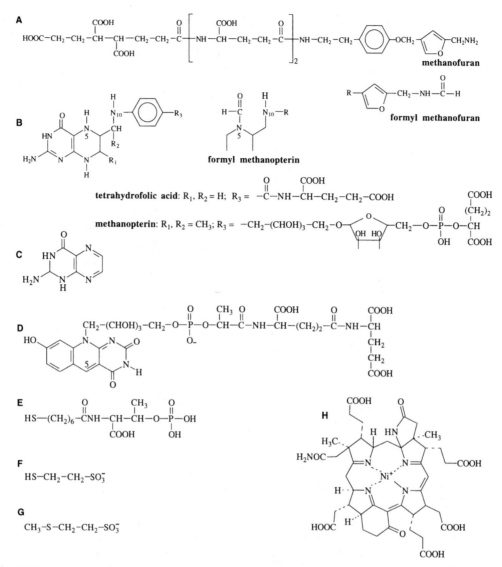

Fig. 13.5 Structures of coenzymes in the acetyl–CoA pathway and in methanogenesis. (A) Methano-furan (4-[N-(4,5,7-tricarboxyheptanoyl-n-L-glutamyl-n-L-glutamyl)-p-(β-aminoethyl) phenoxymethyl]-2-(aminomethyl)furan). The CO_2 becomes attached to the amino group on the furan and is reduced to the oxidation level of formic acid (formyl–methanofuran). (B) Tetrahydrofolic acid and methanopterin; both are derivatives of pterin, shown in (C). In formyl–methanopterin the formyl group is bound to N5, whereas in tetrahydrofolic acid it is bound to N10. (C) Pterin. (D) Oxidized coenzyme F_{420}. This is a 5-deazaflavin and carries two electrons but only one hydrogen (on N1). Compare this structure to flavins that have a nitrogen at position 5 and therefore carry two hydrogens (Section 4.2.1). (E) Factor B or HS-HTP (7-mercaptoheptanoylthreonine phosphate). (F) Coenzyme M (2-mercaptoethanesulfonic acid. (G) Methyl–coenzyme M. (H) Coenzyme F_{430}. This prosthetic group is a nickel–tetrapyrrole.

the bacteria (Fig. 13.4), and the other branch synthesizes methane (Section 13.1.5).

For the synthesis of acetyl–CoA, the methyl group is transferred to a corrinoid enzyme (Figs. 13.6 and 13.7, reaction **9**). A second molecule of CO_2 is reduced by carbon monoxide dehydrogenase (CODH) to bound carbon monoxide [CO] (reaction **14**). The CODH then

catalyzes the synthesis of acetyl–CoA from [CH₃], [CO], and CoASH (reactions **11–13**).

13.1.5 Methanogenesis from CO_2 and H_2

In addition to incorporating the [CH₃] into acetyl–CoA as just described, the methanogens can reduce it to methane (CH₄) in a series

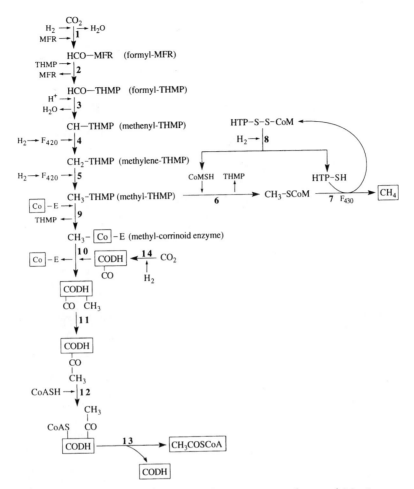

Fig. 13.6 The acetyl–CoA pathway in methanogens. There are routes of entry of CO$_2$. In one sequence, CO$_2$ is bound to carbon monoxide dehydrogenase and is reduced to the oxidation level of carbon monoxide [CO] (reaction **14**). The electron flow is from H$_2$ via a hydrogenase to ferredoxin to CODH. In a second sequence, CO$_2$ is attached to methanofuran (MF) and becomes reduced to the level of formic acid (formyl–MF) (reaction **1**). The acceptor of electrons from H$_2$ in this reaction is not known. The formyl group is then transferred to a second carrier, tetrahydromethanopterin (THMP) to form formyl–THMP (reaction **2**). The formyl–THMP is reduced to methyl–THMP (CH$_3$–THMP) (reactions **3–5**). The CH$_3$–THMP donates the methyl group to a corrinoid enzyme for the synthesis of acetyl–CoA (reactions **9–13**), or to CoMSH for the synthesis of methane (reactions **6–8**). Enzymes: 1, formyl–MFR dehydrogenase (an iron–sulfur protein); 2, formyl–MFR:H$_4$MPT formyltransferase; 3, N^5,N^{10}–methenyl–H$_4$MPT cyclohydrolase; 4, N^5,N^{10}–methylene–H$_4$MPT dehydrogenase; 5, N^5,N^{10}–methylene–H$_4$MPT reductase; 6, $N5$–methyl–H$_4$MPT: CoMSH methyltransferase; 7, methyl–S–CoM reductase; 8, heterodisulfide reductase; 9, methyltransferase; 10–13, carbon monoxide dehydrogenase. THMP is also written as H$_4$MPT.

of reactions not found among the bacteria. For methane synthesis, the methyl group is transferred from CH$_3$–THMP to CoMSH to form CH$_3$–SCoM (Figs. 13.6 and 13.7, reaction 6). The terminal reduction is catalyzed by the *methylreductase system*, (reactions 7 and 8).[17] The methylreductase system has two components. One component is a methylreductase that reduces CH$_3$–SCoM to CH$_4$ and CoM–S–S–HTP. The electron donor for the methyl-reductase is HTP–SH.[18] A nickel-containing tetrapyrrole, F$_{430}$, is an electron carrier that is part of the methyl reductase. The second component is an FAD-containing heterodisulfide reductase that reduces CoM–S–S–HTP to CoMSH and HTP–SH. The crystal structure of methylreductase has been published, and a model for how it works has been proposed.[19]

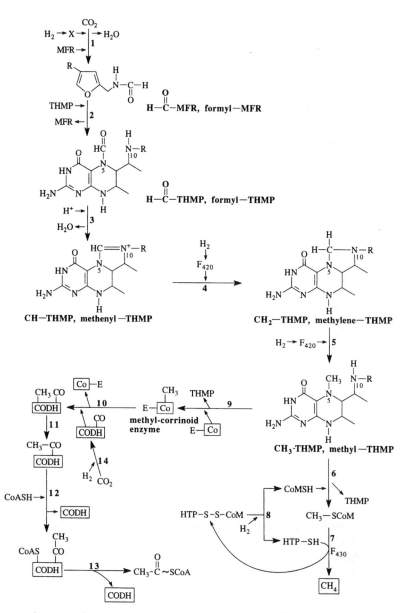

Fig. 13.7 The acetyl–CoA pathway in methanogens. See the legend to Fig. 13.6.

Energy conservation during methanogenesis

The production of methane yields ATP and is the only means of ATP formation for the methanogens. Probably, energy is conserved for ATP synthesis at two sites in the pathway of methanogenesis. It appears that electron transfer to CoM–S–S–HTP (Fig. 13.7, reaction 8) occurs in the membranes and is accompanied by the generation of a Δp that drives ATP synthesis via a H^+-translocating ATP synthase. (See note 20 for experimental evidence.) The mechanism of the generation of the Δp is not known, but could include H_2 oxidation on the outer membrane surface depositing protons on the outside, followed by electrogenic movement of electrons across the membrane to CoM–S–S–HTP on the inside surface. This is a tactic often used by bacteria to generate a Δp coupled to periplasmic oxidations. For example, recall the periplasmic oxidation of H_2 by *Wolinella succinogenes* coupled to the cytoplasmic reduction of fumarate described in Section 4.7.4, or the reduction of sulfate by *Desulfovibrio* described in Section 12.2.2.

There may also be a second energy-coupling site. The transfer of the methyl group from CH$_3$–THMP to CoMSH (Fig. 13.7, reaction **6**) is an exergonic reaction accompanied by the uptake of Na$^+$ into inverted vesicles, creating a primary sodium motive force (inside positive). How might the sodium motive force be used? The sodium motive force probably drives ATP synthesis in whole cells, since it is known to energize the synthesis of ATP in washed inverted vesicles made from *Methanosarcina mazei* Gö1. There appear to be two different ATP synthases present, one coupled to protons and the other coupled to sodium ions.[21] It has been speculated that the sodium motive force created during methyl transfer to CoMSH may also serve to drive the formation of formyl–MFR from free CO$_2$, using H$_2$ as the reductant (Fig. 13.7, reaction **1**). The latter reaction has a standard free energy change of approximately +8 kJ/mol.

Unique coenzymes in the Archaea

As discussed in Section 1.1.1, there are several unique aspects to the biochemistry of the Archaea. For example, there are several coenzymes in archaea that are not found in bacteria or eucarya. These are represented by five coenzymes used in methanogenesis: methanofuran (formerly called CDR, for carbon dioxide reduction enzyme), methanopterin, coenzyme M (CoSM), F$_{430}$, and HS–HTP, also known as factor B or coenzyme B (Fig. 13.5). Coenzyme M, F$_{430}$, and HS–HTP are found only in methanogens. The other coenzymes occur in other archaea. The electron carrier, F$_{430}$, also exists in bacteria and eucarya.

13.1.6 Methanogenesis from acetate

Species of methanogens within the genera *Methanosarcina* and *Methanotrhix* can use acetate as a source of carbon and energy.[22] In fact, acetate accounts for approximately two-thirds of the biologically produced methane. (See note 23.) The methanogens that convert acetate to methane and CO$_2$ carry out a dismutation of acetate in which the carbonyl group is oxidized to CO$_2$, and the electrons are used to reduce the methyl group to methane according to the following overall reaction:

$$CH_3CO_2^- + H^+ \rightarrow CH_4 + CO_2$$

The acetate is first converted to acetyl–CoA. *Methanosarcina* uses acetate kinase and phosphotransacetylase to make acetyl–CoA from acetate:

Acetate kinase

Acetate + ATP → acetyl phosphate + ADP

Phosphotransacetylase

Acetyl phosphate + CoA → acetyl–CoA

    + inorganic phosphate

Methanotrhix uses acetyl–CoA synthetase to make acetyl–CoA:

Acetyl-CoA synthetase

Acetate + ATP → acetyl–AMP +

    pyrophosphate

Acetyl–AMP + CoA → acetyl–CoA + AMP

Reactions similar to those of the acetyl–CoA pathway in methanogens convert acetyl–CoA to methane and CO$_2$, deriving ATP from the process. Examine Fig. 13.6 again, beginning with acetyl–CoA: in reactions **13** and **12**, acetyl–CoA combines with carbon monoxide dehydrogenase (CODH), which catalyzes the breakage of the C–S bond and the release of CoA. In reaction **11**, the CODH catalyzes the breakage of the C–C bond forming the carbonyl [CO] and methyl [CH$_3$] groups. Then the carbonyl group is oxidized to CO$_2$ (reaction **14**) and the methyl group is transferred to a corrinoid enzyme (reaction **10**). (Note that the electrons removed from the carbonyl group are not released in hydrogen gas as implied in reaction **14** but are used to reduce the methyl group to methane.) The methyl group is transferred to tetrahydromethanopterin (THMP) (reaction **9**)[24]. (Acetate-grown cells of *Methanosarcina* species also contain large amounts of a derivative of THMP, tetrahydrosarcinapterin (H$_4$SPT), which has also been reported to serve as a methyl group carrier during methanogenesis from acetate.[25]) The methyl group is then transferred to coenzyme M (CoMSH) (reaction **6**) and reduced to methane by means of the electrons derived from the oxidation of the carbonyl group (reaction **7**). The electron donors for the reduction of the methyl group are HTP–SH and CH$_3$–S–CoM. Each of these contributes one electron from its respective sulfur atom, and the oxidized product is the heterodisulfide HTP–S–S–CoM. The reduction of HTP–S–S–CoM to CoMSH and HTP–SH (reaction 8) is coupled to the oxidation of the carbonyl group (reaction 14). The electron transport pathway that links the oxidation of the carbonyl group with the reduction of HTP–S–S–CoM is thought to generate ATP, presumably via a Δp.[19]

Methanogenesis from acetate is summarized as follows:

$$CH_3CO\text{–}CoA \rightarrow [CH_3] + [CO] + CoA \quad (13.7)$$

$$[CO] + H_2O \rightarrow CO_2 + 2[H] \quad (13.8)$$

$$[CH_3] + 2[H] + ADP + P_i \rightarrow CH_4 + ATP \quad (13.9)$$

It was reported that ATP can also be made during the oxidation of [CO] to CO_2 and H_2.[26]

13.1.7 Incorporation of acetyl–CoA into cell carbon by methanogens

The glyoxylate cycle, one of the means of incorporating net acetyl–CoA into cell material, is not present in prokaryotes that use the acetyl–CoA pathway. How do they grow on the acetyl–CoA? The acetyl–CoA must be converted to phosphoenolpyruvate, which feeds into an incomplete citric acid pathway and into gluconeogenesis. The enzymes to make phosphoenolpyruvate from acetyl–CoA are widespread in anaerobes. A suggested pathway for acetyl–CoA incorporation by methanogens is discussed next.

Figure 13.8 shows a proposed pathway for acetyl–CoA incorporation by methanogens.[27] The first reaction is the carboxylation of acetyl–CoA to form pyruvate, a reaction catalyzed by pyruvate synthase. The pyruvate is then phosphorylated via PEP synthase to form phosphoenolpyruvate. The phosphoenolpyruvate has two fates. Some of it enters the gluconeogenic pathway, and some is carboxylated to oxaloacetate via PEP carboxylase. In M. thermoautotrophicum, the oxaloacetate is reduced to α-ketoglutarate via an incomplete reductive citric acid pathway. That is, the reactions between citrate and α-ketoglutarate

do not take place. M. thermoautotrophicum lacks citrate synthase and so must make its α-ketoglutarate this way. However, another methanogen, M. barkeri, does have citrate synthase and synthesizes α-ketoglutarate in an incomplete oxidative pathway via citrate and isocitrate. Thus, no methanogen seems to have a complete citric acid pathway.[28]

13.1.8 Using the acetyl–CoA pathway to oxidize acetate to CO_2 anaerobically

Several anaerobes can oxidize acetate to CO_2 without the involvement of the citric acid cycle. They do this by operating the acetyl–CoA pathway in reverse. These include some of the sulfate reducers that grow heterotrophically on acetate and oxidize it to CO_2.[29] The anaerobes begin the oxidation with reactions 4 and 3 to make acetyl–CoA (Fig. 13.4) and then reverse reactions 14 through 11. This produces CH$_3$–[Co-E] and CO–[CODH]. In reaction 15 of Fig. 13.4, CO–[CODH] is oxidized to CO_2. The CH$_3$ is oxidized to CO_2 via reactions 10 through 5. In this way, both carbons in acetate are converted to CO_2. A Δp is made during electron transport. (However, it may be that the enzymes involved in the oxidative acetyl–CoA pathway are not identical to those of the reductive acetyl–CoA pathway that operates in autotrophs.) This pathway is used by Desulfotomaculum acetoxidans and some other group II sulfate reducers (complete oxidizers). Still other group II sulfate reducers use the reductive tricarboxylic acid pathway to oxidize acetate (Section 13.1.9). Archaeas sulfate reducers (e.g., Archaeoglobus fulgidus) may also use the acetyl–CoA pathway to oxidize acetate to CO_2 anaerobically.

13.1.9 The reductive tricarboxylic acid pathway (reductive citric acid cycle)

Bacteria that use the reductive TCA pathway are strict anaerobes belonging to the genera Desulfobacter and Chlorobium, and the aerobic Hydrogenobacter.[30] (See note 31.) The pathway is also present in the Archaea and is thought to have evolved earlier than the Calvin cycle as an autotrophic carbon dioxide fixation

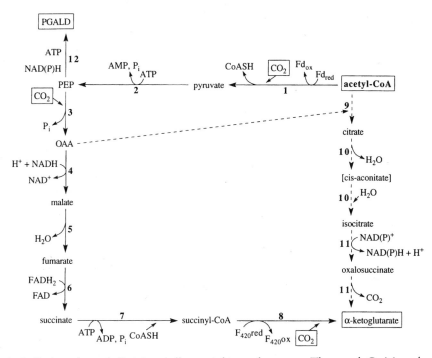

Fig. 13.8 Assimilation of acetyl–CoA into cell material in methanogens. The acetyl–CoA is carboxylated to form pyruvate, which is then phosphorylated to make phosphoenolpyruvate (PEP). In *M. thermoautotrophicum*, the PEP is carboxylated to oxaloacetate (OAA), which is reduced to α-ketoglutarate or to phosphoglyceraldehyde. *M. barkeri* synthesizes α-ketoglutarate by a different pathway: the oxaloacetate condenses with another acetyl–CoA to form citrate, and the citrate is then oxidized to α-ketoglutarate via aconitate, isocitrate, and oxalosuccinate (dashed lines). Enzymes: 1, pyruvate synthase; 2, PEP synthetase; 3, PEP carboxylase; 4, malate dehydrogenase; 5, fumarase; 6, fumarate reductase; 7, succinyl–CoA synthetase; 8, α-ketoglutarate synthase; 9, citrate synthase; 10, aconitase; 11, isocitrate dehydrogenase; 12, enolase, mutase, phosphoglycerate kinase, and in triosephosphate dehydrogenase.

pathway.[32] The overall reaction is the synthesis of one mole of oxaloacetate from four moles of carbon dioxide. In the reductive tricarboxylic acid pathway, phosphoenolpyruvate (PEP) is carboxylated to form oxaloacetate (OAA) via the enzyme PEP carboxylase (Fig. 13.9, reaction 1). The oxaloacetate is reduced to succinate, which is derivatized to succinyl–CoA. The succinyl–CoA is carboxylated to α-ketoglutarate (the precursor to glutamate). This is a reductive carboxylation requiring reduced ferredoxin and is carried out by α-ketoglutarate synthase. The α-ketoglutarate is carboxylated to isocitrate, which is converted to citrate. Therefore, this is really a reversal of the citric acid pathway, substituting fumarate reductase for succinate dehydrogenase and α-ketoglutarate synthase for α-ketoglutarate dehydrogenase. Thus far, one mole of phosphoenolpyruvate (C₃) has been converted to one mole of citrate (C₆).

The citrate is split by a special enzyme, called *ATP-dependent citrate lyase*, to acetyl–CoA (C₂) and oxaloacetate (C₄). [The ATP-dependent citrate lyase is found in eukaryotic cells but is rare in prokaryotes. (See note 33.) Most prokaryotes use the citrate synthase reaction, which proceeds only in the direction of citrate.] The oxaloacetate can be used for growth.

The acetyl–CoA is used to regenerate the phosphoenolpyruvate as follows. The acetyl–CoA (C₂) is reductively carboxylated to pyruvate (C₃) using reduced ferredoxin and the enzyme pyruvate synthase. There is a thermodynamic reason to explain why reduced ferredoxin and not NADH is the electron donor. NADH is not a sufficiently strong reductant to reduce acetyl–CoA and CO₂ to pyruvate. Reduced ferredoxin has a potential more electronegative than that of NADH and is

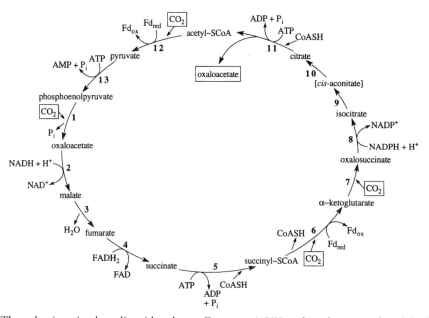

Fig. 13.9 The reductive tricarboxylic acid pathway. Enzymes: **1**, PEP carboxylase; **2**, malate dehydrogenase; **3**, fumarase; **4**, fumarate reductase; **5**, succinyl–CoA synthetase (succinate thiokinase); **6**, α-ketoglutarate synthase; **7** and **8** isocitrate dehydrogenase; **9**, **10**, aconitase; **11**, ATP–citrate lyase; **12**, pyruvate synthase; **13**, PEP synthetase.

therefore a stronger reductant. The pyruvate is then converted to phosphoenolpyruvate via PEP synthetase, thus regenerating the phosphoenolpyruvate. This discussion and Section 13.1.2 emphasize the importance of pyruvate synthase and PEP synthetase for anaerobic growth on acetate.

Note that to reverse the citric acid cycle, three new enzymes are required:

1. *Fumarate reductase* replaces succinate dehydrogenase. This commonly occurs in other prokaryotes that synthesize succinyl–

CoA from oxaloacetate under anaerobic conditions.

2. *α-Ketoglutarate synthase* replaces the NAD$^+$-linked α-ketoglutarate dehydrogenase.

3. *ATP-dependent citrate lyase* replaces citrate synthase.

In addition, pyruvate synthase replaces pyruvate dehydrogenase, and PEP synthetase replaces pyruvate kinase.

Summary of the reductive tricarboxylic acid pathway

$$CO_2 + PEP \rightarrow OAA + P_i$$

$$OAA + NADH + H^+ + FADH_2 + Fd_{red} + ATP + CO_2 \rightarrow \alpha\text{-Ketoglutarate} + NAD^+ + FAD + Fd_{ox} + ADP + P_i$$

$$\alpha\text{-Ketoglutarate} + NADPH + H^+ + CO_2 \rightarrow citrate$$

$$Citrate + ATP + CoASH \rightarrow OAA + acetyl\text{–}CoA + ADP + P_i$$

$$Acetyl\text{–}CoA + Fd_{red} + CO_2 \rightarrow pyruvate + Fd_{ox}$$

$$Pyruvate + ATP \rightarrow PEP + AMP + P_i$$

$$4CO_2 + 2\,NAD(P)H + 2H^+ + 2\,Fd_{red} + FADH_2 + 3\,ATP \rightarrow OAA + 2\,NAD(P)^+ + FAD + 2\,Fd_{ox} + 2\,ADP + AMP + 4\,P_i$$

Anaerobic acetate oxidation by reversal of the reductive tricarboxylic acid pathway

As mentioned previously, some group II sulfate reducers reverse the reductive tricarboxylic acid pathway to oxidize acetate to CO$_2$.[28] An example is *Desulfobacter postgatei*. As shown in Fig. 13.9, by reversing reactions **11** through **6**, acetyl–CoA is oxidized to CO$_2$. The oxaloacetate used in reaction **11** is regenerated in reactions **5** through **2**. Acetyl–CoA is made from acetate by transferring the CoA from succinyl–CoA. Hence, reaction **5** is replaced by a CoA transferase.

13.2 Growth on C$_1$ Compounds Other than CO$_2$: The Methylotrophs

Many aerobic bacteria can grow on compounds other than CO$_2$ that do not have carbon–carbon bonds. These bacteria are called *methylotrophs*.[34,35] Compounds used for methylotrophic growth include single-carbon compounds such as methane (CH$_4$), methanol (CH$_3$OH), formaldehyde (HCHO), formate (HCOOH), and methylamine (CH$_3$NH$_2$), as well as multicarbon compounds without C–C bonds such as trimethylamine [(CH$_3$)$_3$N], dimethyl ether [(CH$_3$)$_2$O], and dimethyl carbonate (CH$_3$OCOOCH$_3$). These compounds appear in the natural habitat as a result of fermentations and breakdown of plant and animal products, and pesticides. (Note 36 tells where these compounds are found.)

The methylotrophs are divided according to whether they can also grow on multicarbon compounds. Those that *cannot* are called *obligate* methylotrophs. The obligate methylotrophs that grow on methanol or methylamine but not on methane are aerobic gram-negative bacteria of the genera *Methylophilus* and *Methylobacillus*. (Some strains of *Methylophilus* can use glucose as a sole carbon and energy source.[37]) The obligate methylotrophs that grow on methane or methanol are called *methanotrophs*. (See note 38.) They are gram-negative and fall into five genera, *Methylomonas*, *Methylococcus*, *Methylobacter*, *Methylosinus*, and *Methylocystis*. All the methanotrophs form extensive intracellular membranes and resting cells, either cysts or exospores. The intracellular membranes, which are not present in methylotrophs that grow on methanol but not on methane, are believed to be involved in methane oxidation.

The methyltrophs that grow on either C$_1$ compounds (methanol or methylamine) or multicarbon compounds are called *facultative* methylotrophs. The facultative methylotrophs are found in many genera and consist of both gram-positive and gram-negative bacteria. They include species belonging to the genera *Bacillus*, *Acetobacter*, *Mycobacterium*, *Arthrobacter*, *Hyphomicrobium*, *Methylobacterium*, and *Nocardia*. Some species of *Mycobacterium* can grow on methane as well as on methanol or multicarbon compounds.

Methylotrophs assimilate the C$_1$ carbon source via either the ribulose–monophosphate (RuMP) pathway or the serine pathway. The RuMP pathway assimilates formaldehyde into cell material, whereas the serine pathway assimilates carbon dioxide *and* formaldehyde. (See note 39.) There also exist bacteria that grow on methanol and oxidize it to CO$_2$, which is assimilated via the ribulose bisphosphate (RuBP) pathway (Calvin cycle). These microorganisms have been called "pseudomethylotrophs" or autotrophic methylotrophs.

13.2.1 Growth on methane

Methane is oxidized aerobically to CO$_2$ in a series of four reactions. (Methane can also be oxidized anaerobically. See note 40 for an explanation and a reference to a review article.) The first oxidation is to methanol, and if it uses a mixed-function oxidase called methane monooxygenase. There are two different methane monooxygenases: one in the membrane and one that is soluble. All methanotrophs have the membrane-bound enzyme. It is not yet known whether the soluble monooxygenase, which is the better characterized enzyme, is present in all methanotrophs.

In the first oxidation reaction, one atom of oxygen is incorporated into methanol and the other atom is reduced to water, using NADH as the reductant:

$$CH_4 + NADH + H^+ + O_2$$
$$\rightarrow CH_3OH + NAD^+ + H_2O$$

The second reaction is the oxidation of methanol to formaldehyde, catalyzed by

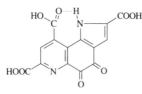

Fig. 13.10 Structure of pyrroloquinoline–quinone (PQQ).

methanol dehydrogenase, which has as its prosthetic group a quinone called pyrroloquinoline quinone (PQQ) (Fig. 13.10):

$$CH_3OH + PQQ \rightarrow HCHO + PQQH_2$$

In the third reaction the formaldehyde is oxidized to formate by formaldehyde dehydrogenase:

$$HCHO + NAD^+ + H_2O$$

$$\rightarrow HCOOH + NADH + H^+$$

And the formate is oxidized to CO_2 by formate dehydrogenase:

$$HCOOH + NAD^+ \rightarrow CO_2 + NADH + H^+$$

Because the formaldehyde and formate dehydrogenases are soluble enzymes that use NAD^+, they are probably located in the cytoplasm. Of the two NADHs produced, one is reutilized in the monooxygenase reaction and the second is fed into the respiratory chain in the usual way. However, methanol dehydrogenase is a periplasmic enzyme in gram-negative bacteria. It is thought that the methanol diffuses into the periplasm, where it is oxidized by the dehydrogenase. The electrons are transferred to periplasmic cytochromes c, which in turn transfer the electrons to a membrane cytochrome oxidase, which probably pumps protons out of the cell during inward flow of electrons to oxygen.

A Δp is established by the inward flow of electrons, the outward pumping of protons by the cytochrome aa_3 oxidase, and the release of protons in the periplasm and consumption in the cytoplasm. (See Section 4.7.2 for a description of electron transport and the generation of a Δp during methanol oxidation in *Paracoccus denitrificans*.) A small number of gram-positive bacteria are also methylotrophic. However, the biochemistry of methanol oxidation may not be the same as in gram-negative bacteria.

Some of the formaldehyde produced during methane oxidation is incorporated into cell material rather than being oxidized to CO_2. Depending upon the particular bacterium, one of two formaldehyde fixation pathways: the serine pathway or the ribulose–monophosphate cycle.

The serine pathway

The serine pathway produces acetyl–CoA from formaldehyde and carbon dioxide (Fig. 13.11). The formaldehyde is incorporated into glycine to form serine in a reaction catalyzed by serine hydroxymethylase (reaction 1). In this reaction, the formaldehyde is first attached to tetrahydrofolic acid to form methylene–THF. (Tetrahydrofolic acid reactions are described in Section 9.2.4). Methylene–THF then donates the C_1 unit to glycine to form serine, regenerating THF. The serine is then converted to hydroxypyruvate via a transaminase, which aminates glyoxylate, regenerating glycine (reaction 2). Hydroxypyruvate is reduced to glycerate (reaction 3), which is phosphorylated to 3-phosphoglycerate (reaction 4). The 3-phosphoglycerate is then converted to 2-phosphoglycerate (reaction 5), which is dehydrated to phosphoenolpyruvate (reaction 6). The phosphoenolpyruvate is carboxylated to oxaloacetate (reaction 7). The oxaloacetate is reduced to malate (reaction 8). Malyl–CoA synthetase then converts malate to malyl–CoA (reaction 9), which is split by malyl–CoA lyase to acetyl–CoA and glyoxylate (reaction 10). Thus, the glyoxylate is regenerated and the product is acetyl–CoA. So far, the following has happened:

$$HCHO + CO_2 + 2\,NADH + 2H^+ + 2\,ATP$$

$$+ CoASH \rightarrow acetyl–CoA + 2\,NAD^+$$

$$+ 2\,ADP + 2\,P_i$$

But how is acetyl–CoA incorporated into cell material? In *some* methylotrophs, the serine pathway goes around a second time to generate a second oxaloacetate. Then the second oxaloacetate condenses with the acetyl–CoA to form citrate (reaction 11). The citrate isomerizes to isocitrate (reaction 12). The isocitrate is cleaved by isocitrate lyase to form succinate and glyoxylate (reaction 13). The succinate can be assimilated into cell material via oxaloacetate and PEP (Section 8.13). The second glyoxylate can be used for the second

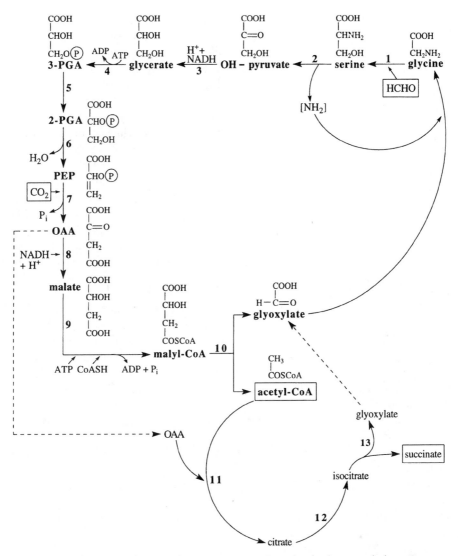

Fig. 13.11 The serine–isocitrate lyase pathway. Enzymes: **1**, serine hydroxymethylase; **2**, transaminase; **3**, hydroxypyruvate reductase; **4**, glycerate kinase; **5**, phosphoglycerate mutase; **6**, enolase; **7**, PEP carboxylase; **8**, malate dehydrogenase; **9**, malyl–CoA synthetase; **10**, malyl–CoA lyase; **11**, citrate synthase; **12**, *cis*-aconitase; **13**, isocitrate lyase.

round of the serine pathway, which produces the second oxaloacetate. Therefore, the serine pathway and the isocitrate lyase pathway (called the serine–isocitrate lyase pathway) can be described by the following overall reaction:

$$2HCHO + 2CO_2 + 3\,NADH + 3H^+ + 3\,ATP$$

$$\rightarrow succinate + 3\,NAD^+ + 3\,ADP + 3\,P_i + 2H_2O$$

Alternatively, the succinate can be converted to the second oxaloacetate via fumarate and malate, and the second glyoxylate can be converted via the serine pathway to 3-phosphoglycerate, which is assimilated into cell material.

However, only a few methylotrophs have isocitrate lyase, and it has not been established how the acetyl–CoA is converted to glyoxylate or otherwise assimilated in the strains lacking isocitrate lyase.

The reactions of Fig. 13.11 are summarized as follows:

1. $2CH_2O + 2$ glycine $\rightarrow 2$ serine

2. 2 Serine + 2 glyoxylate

    $\rightarrow 2$ glycine + 2 hydroxypyruvate

3. 2 Hydroxypyruvate + 2 NADH + 2H⁺

    $\rightarrow 2$ glycerate + 2 NAD⁺

4. 2 Glycerate + 2 ATP

$\quad\quad$ → 2 3-PGA + 2 ADP + 2 P_i

5. 2 3-PGA → 2 2-PGA

6. 2 2-PGA → 2 PEP + 2 H_2O

7. 2 PEP + 2CO_2 → 2 OAA + 2 P_i

8. OAA + NADH + H^+ → malate + NAD^+

9. Malate + ATP + CoASH

$\quad\quad$ → malyl–CoA + ADP + P_i

10. Malyl–CoA → glyoxylate + acetyl–CoA

11. Acetyl–CoA + OAA → citrate

12. Citrate → isocitrate

13. Isocitrate → glyoxylate + succinate

2CH_2O + 2CO_2 + 3 NADH + 3H^+ + 3 ATP →

succinate + 3 NAD^+ + 3 ADP + 3 P_i + 2H_2O

The ribulose–monophosphate cycle

There are several methanotrophs that use the *ribulose–monophosphate* (RuMP) pathway instead of the serine pathway for formaldehyde assimilation.[41] The pathway is shown in Fig. 13.12. It is convenient to divide the RuMP pathway into three stages. Stage 1 begins with condensation of formaldehyde with ribulose-5-phosphate to form hexulose-6-phosphate (reaction **1**). The reaction is catalyzed by hexulose phosphate synthase. The hexulose phosphate is then isomerized to fructose-6-phosphate by hexulose phosphate isomerase (reaction **2**). To synthesize a three-carbon compound (e.g., dihydroxyacetone phosphate or pyruvate) from formaldehyde (a one-carbon compound), stage 1 must be repeated three times.

In stage 2, one of the three fructose-6-phosphates is split into two C_3 compounds. There are two pathways for the cleavage, depending upon the organism. In one pathway, fructose-6-phosphate is phosphorylated to fructose-1,6-bisphosphate, which is cleaved via the fructose-1,6-bisphosphate aldolase to phosphoglyceraldehyde and dihydroxyacetone phosphate. This is called the FBPA pathway, for fructose bisphosphate aldolase. If this route is taken, the net product of the pathway is dihydroxyacetone phosphate. In the second

pathway, fructose-6-phosphate is isomerized to glucose-6-phosphate, which is oxidatively cleaved to phosphoglyceraldehyde and pyruvate by means of the enzymes of the Entner–Doudoroff pathway (Section 8.6). This is called the KDPGA pathway, for the 2-keto-3-deoxy-6-phosphogluconate aldolase. If this route is taken, the net product of the pathway is pyruvate. Both pathways also produce phosphoglyceraldehyde.

Stage 3 is a sugar rearrangement stage during which the phosphoglyceraldehyde produced in stage 2 and two fructose-6-phosphates produced in stage 1 are used to regenerate the three ribulose-5-phosphates. In some bacteria the rearrangements take place via the pentose phosphate cycle enzymes (called the TA pathway, for transaldolase), whereas in other bacteria the rearrangements use the enzymes of the closely related Calvin cycle (called the SBPase pathway, for sedoheptulose-1,7-bisphosphate aldolase).

Taking into consideration the alternative pathways in stages 2 and 3, there are four variations of the RuMP cycle that might occur in the methylotrophs. The obligate methanotrophs (*Methylococcus* and *Methylomonas*) and obligate methylotrophs (*Methylophilus*, *Methylobacillus*) that have been examined all use the KDPGA mode of cleavage and the TA pathway of sugar arrangement. The facultative methylotrophs thus far examined use the FBPA mode of cleavage. Some use the SBPase sugar rearrangment pathway, and some use the TA pathway. Use of the KDPG aldolase and the SBPase pathways in combination has not yet been found.

13.3 Summary

There are three characterized CO_2 fixation pathways used for autotrophic growth: the Calvin cycle, the acetyl–CoA pathway, and the reductive tricarboxylic acid pathway. The Calvin cycle is the only one in the aerobic biosphere. It is present in green plants, algae, cyanobacteria, chemoautotrophs, and most photosynthetic bacteria. The Calvin cycle uses the transketolase and isomerization reactions of the pentose phosphate pathway and several glycolytic reactions to reduce CO_2 to phosphoglyceraldehdye. It bypasses the transaldolase

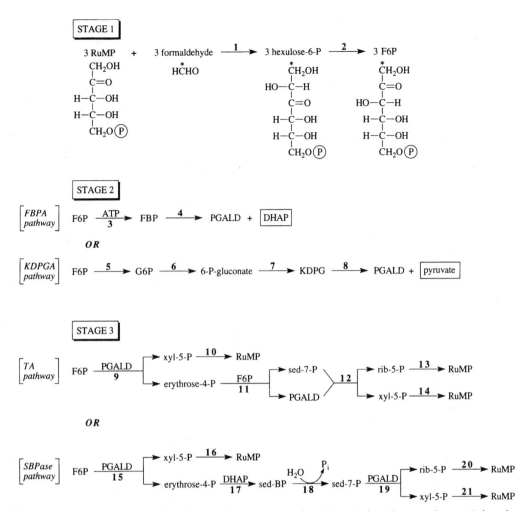

Fig. 13.12 The ribulose–monophosphate cycle. Enzymes: **1**, hexulose-6-phosphate synthetase; **2**, hexulose-6-phosphate isomerase; **3**, phosphofructokinase; **4**, fructose-1,6-bisphosphate (FBP) aldolase; **5**, glucose phosphate isomerase; **6**, glucose-6-phosphate dehydrogenase; **7**, 6-phosphogluconate dehydratase; **8**, 2-keto-3-deoxy-6-phosphogluconate (KDPG) aldolase; **9, 12, 15,** and **19**, transketolase; **10, 14, 16,** and **21**, ribulose-5-phosphate epimerase; **11**, transaldolase; **13** and **20**, ribose-5-phosphate isomerase; **17**, sedoheptulose-1,7-bisphosphate aldolase; **18**, sedoheptulose-1,7-bisphosphatase. Abbreviations: RuMP, ribulose-5-phosphate; F6P, fructose-6-phosphate; FBP, fructose-1,6-bisphosphate; PGALD, phosphogly-ceraldehyde; DHAP, dihydroxyacetone phosphate; KDPG, 2-keto-3-deoxy-6-phosphogluconate; xyl-5-P, xylulose-5-phosphate; rib-5-P, ribose-5-phosphate; sed-BP, sedoheptulose-1,7-bisphosphate.

reaction and synthesizes sedoheptulose-7-P via an aldolase and an irreversible phosphatase and proceeds only in the direction of pentose phosphate. The pathway has two unique enzymes, RuMP kinase and RuBP carboxylase.

The acetyl–CoA pathway is widespread among anaerobes and occurs in both the archaea and the bacteria. However, the archaea use different coenzymes to carry the C₁ units. The pathway is used not only reductively for autotrophic growth but can also be used for anaerobic oxidation of acetate to CO_2 (e.g., by certain group II sulfate reducers). An analogous pathway is used by methanogens to form CH_4 and CO_2 from acetate. A key enzyme in the acetyl–CoA pathway is carbon monoxide dehydrogenase (CODH), which is capable of reducing CO_2 to the level of carbon monoxide and catalyzing its condensation with bound methyl and CoA to form acetyl–CoA. The bound methyl is made by reducing a second molecule of CO_2 via a separate series of enzymatic reactions and transferring the methyl group to the CODH.

The reduction of CO_2 to bound methyl also takes place during methanogenesis. In methanogenesis, the bound methyl is further reduced to methane, an energy-yielding reaction, rather than being transferred to CODH. Growth on acetyl–CoA requires the synthesis of phosphoenolpyruvate because the glyoxylate pathway is not present in these organisms. It involves a ferredoxin-linked carboxylation of acetyl–CoA to pyruvate catalyzed by pyruvate synthase, followed by the phosphorylation of pyruvate to phosphoenolpyruvate via the PEP synthetase. The incorporation of phosphoenolpyruvate into cell material takes place by means of reactions common to heterotrophs.

The third CO_2 fixation pathway has been found among the photosynthetic green sulfur bacteria, *Hydrogenobacter*, and *Desulfobacter* species. The pathway is a reductive citric acid pathway and synthesizes oxaloacetate from four moles of CO_2. There are two carboxylation reactions. One carboxylation is common, even among heterotrophs. It is the carboxylation of phosphoenolpyruvate to form oxaloacetate catalyzed by PEP carboxylase. The oxaloacetate is reduced to succinyl–CoA. The second carboxylation is found only in some strict anaerobes. It is the ferredoxin-linked carboxylation of succinyl–CoA to form α-ketoglutarate catalyzed by α-ketoglutarate synthase. The α-ketoglutarate is converted to citrate via a reversal of the reactions of the citric acid cycle, and the citrate is cleaved to oxaloacetate and acetyl–CoA via an ATP-dependent citrate lyase. The bacteria can incorporate the oxaloacetate into cell carbon. The phosphoenolpyruvate is regenerated from acetyl–CoA via its carboxylation to pyruvate by means of pyruvate synthase and phosphorylation of the latter to phosphoenolpyruvate using PEP synthetase. Therefore, the bacteria are synthesizing oxaloacetate by carboxylating phosphoenolpyruvate and regenerating the phosphoenolpyruvate via a reductive citric acid pathway. In some group II sulfate reducers, the pathway can operate in the oxidative direction and oxidize acetate to CO_2.

There are many bacteria that grow aerobically on C_1 compounds such as methane or methanol. They oxidize the C_1 compounds to CO_2, deriving ATP from the respiratory pathway in the usual way, and incorporate the rest of the C_1 at the level of formaldehyde, into cell carbon. Bacteria that do this are called methylotrophs. A subclass of methylotrophs are the methanotrophs, bacteria that are able to grow on methane. Two pathways for formaldehyde incorporation have been found in these bacteria. They are the serine–isocitrate lyase pathway, which incorporates both formaldehyde and carbon dioxide, and the ribulosemonophosphate cycle, which incorporates only formaldehyde.

The use of C_1 compounds is an important part of the carbon cycle. The methane produced by methanogens escapes into the aerobic atmosphere and is transformed back into carbon dioxide by aerobic methane oxidizers. The carbon dioxide is reduced to organic carbon in the aerobic environments by both photosynthetic eukaryotes and chemoautotrophic prokaryotes. It is also reduced to organic carbon anaerobically by photosynthetic prokaryotes, acetogens, and methanogens.

Methanogenesis occurs in a variety of anaerobic environments, including swamps and marshes, the mammalian rumen, the intestine of termites, and anaerobic microenvironments in the soil, lake muds, and rice paddies. Indeed, close to 70% of the atmospheric methane is produced by methanogens. This is approximately 10^8 tons of methane per year. (The other 30% or so originates from abiogenic sources such as biomass burning and coal mines.) Vast amounts of methane are produced by the methanogens, which are responsible for cycling much of the earth's carbon into a gasous form for reutilization by the methane oxidizers and eventually the CO_2-fixing organisms. Clearly these microorganisms play a critical life-supporting role in the biosphere.

Study Questions

1. Determine the number of ATPs required to fix three moles of CO_2 into phosphoglyceraldehyde by means of the Calvin cycle, the acetyl–CoA pathway, and the reductive carboxylic acid pathway.

2. Describe the role of carbon monoxide dehydrogenase in methane and carbon dioxide production from acetate, and in acetyl–CoA synthesis from CO_2.

3. Why can anaerobes but not aerobes carboxylate acetyl–CoA and succinyl–CoA?

REFERENCES AND NOTES

1. The properties of the Calvin cycle enzymes and their regulation are reviewed in: Bowien, B. 1989. Molecular biology of carbon dioxide assimilation in aerobic chemolithotrophs, pp. 437–460. In: *Autotrophic Bacteria.* H. G. Schlegel, and B. Bowien (Eds.). Science Tech Publishers, Madison, WI, and Springer-Verlag, Berlin.

2. It appears that the sedoheptulose-1,7-bisphosphatase and the fructose-1,6-bisphosphatase may be a single bifunctional enzyme, capable of using either fructose-1,6-bisphosphate or sedoheptulose-1,7-bisphosphate as the substrate. In addition, the fructose-1,6-bisphosphate aldolase and the sedoheptulose-1,7-bisphosphate aldolase may also be a single bifunctional enzyme.

3. Reviewed in: Wood, H. G., and L. G. Ljungdahl. 1991. Autotrophic character of the acetogenic bacteria, pp. 201–250. In: *Variations in Autotrophic Life.* J. M. Shively, and L. L. Barton (Eds.). Academic Press, New York.

4. The "acetogenic bacteria" are defined as anaerobic bacteria that synthesize acetic acid solely from CO_2 and secrete the acetic acid into the media. Many of the acetogenic bacteria are facultative heterotrophs that use the acetyl–CoA pathway to reduce CO_2 to acetic acid as an electron sink, but these can also be grown autotrophically on CO_2 and H_2.

5. The early history is reviewed in: Wood, H. G. 1985. Then and now. *Annu. Rev. Biochem.* **54**:1–41.

6. Corrinoid enzymes are proteins containing vitamin B_{12} derivatives as the prosthetic group. Vitamin B_{12} is a cobalt-containing coenzyme.

7. Reviewed in ref. 3.

8. Heise, R., V. Muller, and G. Gottschalk. 1993. Acetogenesis and ATP synthesis in *Acetobacterium woodii* are coupled via a transmembrane primary sodium ion gradient. *FEMS Microbiol. Lett.* **112**:261–268.

9. Lu, W.-P., S. R. Harder, and S. W. Ragsdale. 1990. Controlled potential enzymology of methyl transfer reactions involved in acetyl–CoA synthesis by CO dehydrogenase and the corrinoid/iron–sulfur protein from *Clostridium thermoaceticum. J. Biol. Chem.* **265**:3124–3133.

10. Pezacka, E., and H. G. Wood. 1986. The autotrophic pathway of acetogenic bacteria: role of CO dehydrogenase disulfide reductase. *J. Biol. Chem.* **261**:1609–1615.

11. Ragsdale, S. W., and H. G. Wood. 1985. Acetate biosynthesis by acetogenic bacteria: evidence that carbon monoxide dehydrogenase is the condensing enzyme that catalyzes the final steps of the synthesis. *J. Biol. Chem.* **260**:3970–3977.

12. Doukov, T. I., T. M. Iverson, J. Seravalli, S. W. Ragsdale, and C. L. Drennan. 2002. A Ni-Fe-Cu center in a bifunctional carbon monoxide dehydrogenase/acetyl–CoA synthase. *Science* **298**:567–572.

13. Fuchs, G. 1990. Alternatives to the Calvin cycle and the Krebs cycle in anaerobic bacteria: pathways with carbonylation chemistry, pp. 13–20. In: *The Molecular Basis of Bacterial Metabolism.* G. Hauska, and R. Thauer (Eds.). Springer-Verlag, Berlin.

14. Reviewed in: Jetten, M. S. M., A. J. M. Stams, and A. J. B. Zehnder. 1992. Methanogenesis from acetate: a comparison of the acetate metabolism in *Methanothrix soehngenii* and *Methanosarcina* spp. *FEMS Microbiol. Rev.* **88**:181–198.

15. Wolfe, R. S. 1990. Novel coenzymes of archaebacteria, pp. 1–12. In: *The Molecular Basis of Bacterial Metabolism.* G. Hauska, and R. Thauer (Eds.). Springer-Verlag, Berlin.

16. Weiss, D. S., and R. K. Thauer. 1993. Methanogenesis and the unity of biochemistry. *Cell* **72**:819–822.

17. Olson, K. D., L. Chmurkowska-Cichowlas, C. W. McMahon, and R. S. Wolfe. 1992. Structural modifications and kinetic studies of the substrates involved in the final step of methane formation in *Methanobacterium thermoautotrophicum. J. Bacteriol.* **174**:1007–1012.

18. HTP–SH is also known as factor B or coenzyme B. It is 7-mercaptoheptanolythreonine phosphate.

19. Ermler, U., W. Grabarse, S. Shima, M. Goubeaud, and R. K. Thauer. 1997. Crystal structure of methyl–coenzyme M reductase: the key enzyme of biological methane formation. *Science* **278**:1457–1462.

20. There are several reasons for suggesting that a Δp is generated during electron transfer to the mixed disulfide: (1) *M. barkeri* creates a Δp during catabolism of acetate; (2) the membranes of *M. thermophila* and *M. barkeri* contain electron carriers, including cytochrome b and FeS proteins; and (3) over 50% of the heterodisulfide reductase is in the membranes.

21. Becher, B., and V. Müller. 1994. $\Delta\mu_{Na^+}$ drives the synthesis of ATP via an $\Delta\mu_{Na^+}$-translocating F_1F_0–ATP synthase in membrane vesicles of the archaeon *Methanosarcina mazei* Göl. *J. Bacteriol.* **176**:2543–2550.

22. Ferry, J. G. 1992. Methane from acetate. *J. Bacteriol.* **174**:5489–5495.

23. The methanogens are paramount in recycling organic carbon into gaseous forms of carbon in anaerobic habitats (Section 13.4.) Most of the methane that enters the aerobic atmosphere is oxidized to CO_2 by the aerobic methanotrophs (Section 12.2).

24. Fischer, R., and R. K. Thauer. 1989. Methyltetrahydromethanopterin as an intermediate

in methogenesis from acetate in *Methanosarcina barkeri*. *Arch. Microbiol.* **151**:459–465.

25. Grahame, D. A. 1991. Catalysis of acetyl–CoA cleavage and tetrahydrosarcinapterin methylation by a carbon monoxide dehydrogenase–corrinoid enzyme complex. *J. Biol. Chem.* **266**:22227–22233.

26. Bott, M., B. Eikmanns, and R. K. Thauer. 1986. Coupling of carbon monoxide oxidation to CO_2 and H_2 with the phosphorylation of ADP in acetate-grown *Methanosarcina barkeri*. *Eur. J. Biochem.* **159**:393–398.

27. Stupperich, E., and G. Fuchs. 1984. Autotrophic synthesis of activated acetic acid from two CO_2 in *Methanobacterium thermoautotrophicum*. I. Properties of the in vitro system. *Arch. Microbiol.* **139**:8–13.

28. Reviewed in: Danson, M. J. 1988. Archaebacteria: the comparative enzymology of their central metabolic pathways, pp. 165–231. In: *Advances in Microbial Physiology*, Vol. 29. A. H. Rose, and D. W. Tempest (Eds.). Academic Press, New York.

29. Hansen, T. A. 1993. Carbon metabolism of sulfate-reducing bacteria, pp. 21–40. In: *The Sulfate-Reducing Bacteria: Contemporary Perspectives*. J. M. Odom, and R. Singleton Jr. (Eds.). Springer-Verlag, New York.

30. Reviewed in: Amesz, J. 1991. Green photosynthetic bacteria and heliobacteria, pp. 99–119. In: *Variations in Autotrophic Life*. J. M. Shively, and L. L. Barton (Eds.). Academic Press, New York.

31. *Desulfobacter* is a group II sulfate-reducing bacterium that grows on fatty acids, especially acetate; it uses sulfate as an electron acceptor and produces sulfide (Section 11.2.2). *Chlorobium* is a green sulfur photoautotroph. *Hydrogenobacter* is an obligate aerobic, chemolithotrophic hydrogen oxidizer.

32. Reviewed in: Schonheit, P., and T. Schafer. 1995. Metabolism of hyperthermophiles. *World J. Microbiol. Biotechnol.* **11**:26–57.

33. In eukaryotic cells the ATP-dependent citrate lyase is used to generate acetyl–CoA from citrate. The acetyl–CoA is then used as a precursor to fatty acids and steroids in the cytosol.

34. Reviewed in: Colin Murrell, J., and H. Dalton (Eds.). 1992. *Methane and Methanol Utilizers*. Plenum Press, New York and London.

35. Reviewed in: Lidstrom, M. E. 1992. The aerobic methylotrophic bacteria, pp. 431–445. In: *The Prokaryotes*, Vol. 1, 2nd ed. A. Balows, H. G. Truper, M. Dworkin, W. Harder, and K.-H. Schleifer (Eds.). Springer-Verlag, Berlin.

36. Methanol appears in the environment as a result of the microbial breakdown of plant products with methoxy groups (e.g., pectins and lignins). Formate is a fermentation end product excreted by fermenting bacteria. Methylamines can result from the breakdown products of some pesticides and certain other compounds, including carnitine and lecithin derivatives. Carnitine is a trimethylamine derivative that is present in many organisms and in all animal tissues, especially muscle. It functions as a carrier for fatty acids across the mitochondrial membrane into the mitochondria, where the fatty acids are oxidized. Lecithin is phosphatidylcholine, which is a phospholipid containing a trimethylamine group. (Section 9.1.2.)

37. Green, P. N. 1992. Taxonomy of methylotrophic bacteria, pp. 23–84. In: *Methane and Methanol Utilizers*. J. Colin Murrell, and H. Dalton (Eds.). Plenum Press, New York and London.

38. The methanotrophs are responsible for oxidizing about half the methane produced by methanogens. The rest of the methane escapes to the atmosphere.

39. Most of the nonmethanotrophs that are obligate methylotrophs use the RuMP pathway. Within the methanotrophs, some use the RuMP pathway and some the serine pathway.

40. For many years it was thought that methane was oxidized only aerobically using a monooxygenase via the pathway described in Section 13.2.1. However, recent observations indicate that anaerobic oxidation of methane exists in anaerobic sediments where methane production (methanogenesis) is taking place.

$$CH_4 + SO_4^{2-} + H^+ \rightarrow CO_2 + HS^- + 2H_2O$$

This has been reported in sediments at the bottom of the Black Sea and in deep-sea sediments that overlie natural gas deposits where methane is produced. At this time, the anaerobic oxidation of methane is believed to require two different organisms that live in mutual dependence (syntrophism) but have not yet been grown in pure culture. The methane oxidizer is a methanogenic archaean that is thought to oxidize methane via a pathway very similar to the reverse of the methanogenic pathway shown in Fig. 13.6. The electrons removed from the methane are transferred via an unknown carrier to sulfate-reducing bacteria living in the consortium and are used to reduce sulfate to sulfide. The "electron shuttles" suggested include formate and acetate. How the methanogen derives energy from this reversal of methanogenesis is not known. For a review of anaerobic oxidation of methane, as well as ammonium, read: Strous, M., and M. S. M. Jetten. 2004. Anaerobic oxidation of methane and ammonium. *Annu. Rev. Microbiol.* **58**:99–117.

41. Dijkhuizen, L., P. R. Levering, and G. E. De Vries. 1990. The physiology and biochemistry of aerobic methanol-utilizing gram-negative and gram-positive bacteria, pp. 149–181. In: *Methane and Methanol Utilizers. Biotechnology Handbooks*, Vol. 5. J. Colin Murrell, and H. Dalton (Eds.). Plenum Press, New York and London.

14

Fermentations

In the numerous anaerobic niches in the biosphere (muds, sewage, swamps, etc.) are prokaryotes that can grow indefinitely in the complete absence of oxygen, a capability that is rare among eukaryotes. (Eukaryotes require oxygen to synthesize unsaturated fatty acids and sterols.) Anaerobically growing prokaryotes reoxidize NADH and other reduced electron carriers either by anaerobic respiration [e.g., using nitrate, sulfate, or fumarate as the electron acceptor (see Chapters 4 and 11)] or by carrying out a fermentation.

A *fermentation* is defined as a pathway in which NADH (or some other reduced electron acceptor that is generated by oxidation reactions in the pathway) is reoxidized by metabolites produced by the pathway. The redox reactions occur in the cytosol rather than in the membranes, and ATP is produced via substrate-level phosphorylation.

It should not be concluded that fermentation occurs only among prokaryotes. Eukaryotic microorganisms (e.g., yeast) can live fermentatively, although as mentioned, oxygen is usually necessary unless the medium is supplemented with sterols and unsaturated fatty acids. Furthermore, certain animal cells (e.g., muscle cells, human red blood cells) are capable of fermentation.

Fermentations are named after the major end products they generate. For example, yeast carry out an ethanol fermentation, and muscle cells and red blood cells carry out a lactic acid fermentation. Microorganisms perform many different types of fermentation. However, the carbohydrate fermentations can be grouped into six main classes: *lactic, ethanol, butyric, mixed acid, propionic,* and *homoacetic.* A homoacetic fermentation of glucose is described in Section 13.1.3 and Fig. 13.4.

14.1 Oxygen Toxicity

Many anaerobic prokaryotes (strict anaerobes) are killed by even small traces of oxygen. Strict anaerobes are killed by oxygen because toxic products of oxygen reduction accumulate in the cell. The toxic products of oxygen are produced when single electrons are added to oxygen sequentially.[1] These toxic products are hydroxyl radical ($OH^{\cdot}$), superoxide radical (O_2^-), and hydrogen peroxide (H_2O_2). A more complete discussion of oxygen toxicity can be found in Section 19.4.

The superoxide radical forms because oxygen is reduced by single-electron steps:

$$O_2 + e^- \rightarrow O_2^-$$

A small amount of superoxide radical is always released from the enzyme when oxygen is reduced by electron carriers such as flavoproteins or cytochromes. This is because the electrons are transferred to oxygen one at a time. The hydroxyl radical and hydrogen peroxide are derived from the superoxide radical.

Aerobic and aerotolerant organisms do not accumulate superoxide radicals because they

383

Table 14.1 The distribution of catalase and superoxide dismutase

Bacterium	Superoxide dismutase	Catalase
Aerobes or facultative anaerobes		
Escherichia coli	+	+
Pseudomonas species	+	+
Deinococcus radiodurans	+	+
Aerotolerant bacteria		
Butyribacterium rettgeri	+	−
Streptococcus faecalis	+	−
Streptococcus lactis	+	−
Strict anaerobes		
Clostridium pasteurianum	−	−
Clostridium acetobutylicum	−	−

Source: Stanier, R. Y., J. L. Ingraham, M. L. Wheelis, and P. R. Painter. 1986. *The Microbial World*. Reprinted by permission of Prentice-Hall, Upper Saddle River, NJ.

have an enzyme called *superoxide dismutase* that is missing in strict anaerobes. The superoxide dismutase catalyzes the following reaction:

$$O_2^- + O_2^- + 2H^+ \rightarrow H_2O_2 + O_2$$

Notice that one superoxide radical transfers its extra electron to the second radical, which is reduced to hydrogen peroxide. Strict anaerobes also lack catalase, the enzyme that converts hydrogen peroxide to water and oxygen:

$$H_2O_2 + H_2O_2 \rightarrow 2H_2O + O_2$$

Catalase catalyzes the transfer of two electrons from one hydrogen peroxide molecule to the second, oxidizing the first to oxygen and reducing the second to two molecules of water. Table 14.1 shows the distribution of catalase and superoxide dismutase in aerobes and anaerobes.

If the H_2O_2 is not disposed of, it can oxidize transition metals such as free iron (II) in the *Fenton reaction* and form the free hydroxyl radical, OH˙:

$$Fe^{2+} + H^+ + H_2O_2 \rightarrow Fe^{3+} + OH˙ + H_2O$$

(See note 2 for a definition of transition metals.)

14.2 Energy Conservation by Anaerobic Bacteria

An examination of mechanisms of energy conservation and ATP production in anaerobic bacteria reveals a variety of methods, which were described in earlier chapters. Some of these are reviewed in Fig. 14.1. Most fermenting bacteria make most or all of their ATP via substrate-level phosphorylations, and they create a Δp (needed for solute transport, motility, etc.) by reversing the membrane-bound ATPase. Many anaerobic bacteria also generate a Δp by reducing fumarate. This is an anaerobic respiration that can occur during fermentative metabolism when fumarate is produced. During fumarate reduction, NADH dehydrogenases donate electrons to fumarate via menaquinone and a Δp is created, perhaps via a quinone loop. (See Section 4.6.1 for a discussion of quinone loops.) The production of a Δp by fumarate respiration in fermenting bacteria probably spares ATP that would normally be hydrolyzed to maintain the Δp.

Some anaerobes (e.g., *Wolinella succinogenes*) carry out a periplasmic oxidation of electron donors, such as H_2 or formate in the case of *W. succinogenes*, to create a Δp. (See Section 4.7.4.) There are several other means available to anaerobic bacteria for generating a proton motive force or a sodium motive force. Some anaerobes and facultative anaerobes are capable of creating an electrochemical ion gradient by symport of organic acids out of the cell with protons or sodium ions. The organic acids are produced during fermentation, and the energy to create the electrochemical gradient is due to the concentration gradient of the excreted organic acid (high inside). This has been demonstrated for lactate excretion (proton symport) by the lactate bacteria and for succinate excretion (sodium ion symport) by a

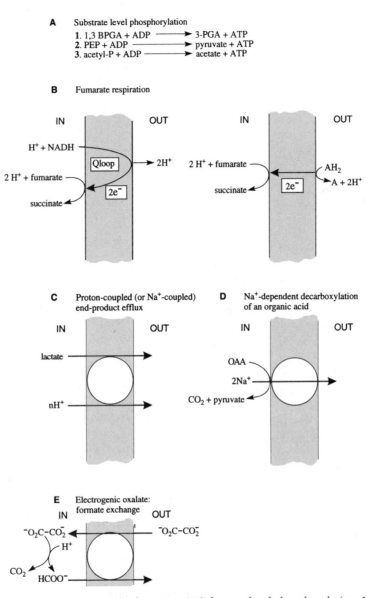

Fig. 14.1 Energy conservation in anaerobic bacteria. (A) Substrate-level phosphorylation: **1**, the PGA kinase reaction; **2**, the pyruvate kinase reaction; **3**, the acetate kinase reaction. (B) Fumarate respiration. When the electron donor is NADH, a Q loop probably operates to translocate protons out of the cell. When the electron donor is periplasmic, proton translocation is not necessary. (C) Efflux of an organic acid coupled to protons or sodium ions (e.g., the coupled efflux of protons and lactate by the lactate bacteria). (D) Decarboxylation of an organic acid coupled to Na^+ efflux (e.g., the decarboxylation of oxaloacetate by *Klebsiella*). (E) Electrogenic oxalate:formate exchange in *Oxalobacter*.

rumen bacterium, *Selenomonas ruminantium*. (See Section 3.8.3 for a discussion of lactate and succinate efflux in symport with protons and sodium ions.)

Klebsiella pneumoniae is capable of generating an electrochemical sodium ion gradient by using the energy released from the decarboxylation of oxaloacetate to pyruvate (Sec-

tion 3.8.1). The decarboxylase pumps Na^+ out of the cell. The sodium potential that is created is used to drive the uptake of oxaloacetate, which is used as a carbon and energy source. *Oxalobacter formigenes* creates a proton potential by catalyzing an electrogenic anion exchange coupled to the decarboxylation of oxalate to formate (Section 3.8.2).

14.3 Electron Sinks

What to do with the electrons removed in the course of oxidations poses a major problem that must be addressed during fermentation. For example, how is the NADH reoxidized? Respiring organisms do not have this problem, since the electrons travel to an exogenous electron acceptor (e.g., oxygen or nitrate). Since fermentations usually occur in the absence of an exogenously supplied electron acceptor, the fermentation pathways themselves must produce the electron acceptors for the electrons produced during oxidations. The electron acceptors are called "electron sinks" because they dispose of the electrons removed during the oxidations, and the reduced products are excreted into the medium. Consequently, fermentations are characterized by the excretion of large quantities of reduced organic compounds such as alcohols, organic acids, and solvents. Frequently, hydrogen gas is also produced, since hydrogenases use protons as electron acceptors. Generally, one can view fermentations in the following way, where AH_2 is the compound being fermented, and B is the electron sink generated from AH_2:

$$AH_2 + NAD^+ P_i + ADP +$$
$$\rightarrow B + NADH + H^+ + ATP$$
$$B + NADH + H^+$$
$$\rightarrow BH_2 (excreted) + NAD^+$$

and

Hydrogenase

$$NADH + H^+ \rightarrow H_2 + NAD^+$$

Thus there is the lactate fermentation

$$Glucose + 2\ ADP + 2\ P_i \rightarrow 2\ lactate + 2\ ATP$$

or, to choose another example, an ethanol fermentation:

$$Glucose + 2\ NAD^+ + 2\ ADP + 2\ P_i$$
$$\rightarrow 2\ pyruvate + 2\ NADH + 2H^+ + 2\ ATP$$
$$2\ Pyruvate \rightarrow 2\ acetaldehyde + 2CO_2$$
$$2\ Acetaldehyde + 2\ NADH + 2H^+$$
$$\rightarrow 2\ ethanol + 2\ NAD^+$$

$$Glucose + 2\ ADP + 2\ P_i$$
$$\rightarrow 2\ ethanol + 2CO_2 + 2\ ATP$$

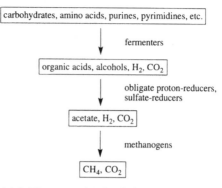

Fig. 14.2 The anaerobic food chain.

14.4 The Anaerobic Food Chain

The fermentation of amino acids, carbohydrates, purines, and pyrimidines to organic acids and alcohols (acetate, ethanol, butanol, propionate, succinate, butyrate, etc.) by prokaryotes, and the conversion of these fermentation end products to CO_2, CH_4, and H_2 by the combined action of bacteria of several different types, is called the anaerobic food chain (Fig. 14.2). Anaerobic environments include muds, lake bottoms, and sewage treatment plants.

The anaerobic food chain is an important part of the carbon cycle and serves to regenerate gaseous carbon (i.e., carbon dioxide and methane), which is reutilized by other microorganisms and plants throughout the biosphere. As illustrated in Fig. 14.2, the process can be viewed as occurring in three stages. Fermenters produce organic acids, alcohols, hydrogen gas, and carbon dioxide. These fermentation end products are oxidized to CO_2, H_2, and acetate by organisms that have been only partially identified. Finally, the methanogens grow on the acetate, H_2, and CO_2, converting these to CH_4 and CO_2.

14.4.1 Interspecies hydrogen transfer

Some anaerobic bacteria use protons as the major or sole electron sink. They include the *obligate proton-reducing acetogens* that oxidize butyrate, propionate, ethanol, and other compounds to acetate, H_2, and CO_2. The physiology of these organisms is not well understood. Probably the electrons travel from the organic substrate to an intermediate electron carrier (e.g., from NAD^+ to hydrogenase to H^+). Some of these oxidations are as follows:

$$CH_3CH_2CH_2COO^- + 2H_2O \rightarrow$$

$$2CH_3COO^- + H^+ + 2H_2 \quad \Delta G_0' = +48.1 \text{ kJ}$$

$$CH_3CH_2COO^- + 3H_2O \rightarrow$$

$$CH_3COO^- + HCO_3^- + H^+ + 3H_2$$

$$\Delta G_0' = +76.1 \text{ kJ}$$

$$CH_3CH_2OH + H_2O \rightarrow$$

$$CH_3COO^- + H^+ + 2H_2 \quad \Delta G_0' = +9.6 \text{ kJ}$$

Notice that these reactions are thermodynamically unfavorable under standard conditions at pH 7. In fact, the proton reducers live symbiotically with H_2 utilizers that keep the H_2 levels low and pull the reaction toward H_2 production. The H_2 utilizers are methanogens and sulfate reducers. (Under conditions of sufficient sulfate, the sulfate reducers predominate over methanogens. The sulfate reducers also oxidize ethanol, lactate, and other organic acids to acetate.)

The symbiotic relationship between the obligate proton reducers and the hydrogen utilizers is called a *syntrophic* association.[3] It is also called *interspecies hydrogen transfer*. In fact, many other fermenting bacteria besides the obligate proton reducers have dehydrogenases that transfer electrons from NADH to protons and can use protons as an electron sink when grown in the presence of hydrogen gas utilizers. An example is *Ruminococcus albus*, which is discussed in Section 14.12.

Hydrogenases are also coupled to the thermodynamically favored oxidation of reduced ferredoxin, as in pyruvate:ferredoxin oxidoreductase found in the clostridia, sulfate reducers (Section 12.2.2), and *Ruminococcus albus* (see later: Fig. 14.12). Ferredoxins often have two iron–sulfur clusters, each one capable of carrying an electron (also indicated in Fig. 14.12).

14.5 How to Balance a Fermentation

A written fermentation is said to be balanced when the hydrogens produced during the oxidations equal the hydrogens transferred to the fermentation end products. Only under these conditions can all of the NADH and reduced ferredoxin be recycled to the oxidized

forms. It is important to know whether a fermentation is balanced because if it is not, the overall written reaction is incorrect. There are two methods used to balance fermentations, the oxidation/reduction method and the available hydrogen method; both are bookkeeping methods to keep track of the hydrogens.

14.5.1 The O/R method

The O/R method entails the computation of an oxidation/reduction (O/R) balance is computed as described here. One arbitrarily designates as zero the O/R value for formaldehyde (CH_2O) and multiples thereof [($CH_2O)_n$] and uses that formula as a standard with which to compare the reduction level of other molecules. Following are the steps in determining the O/R value of any molecule:

1. *Add or subtract water to the molecule in question to make the C/O ratio 1. This will allow a comparison to* $(CH_2O)_n$. For example, the formula for ethanol is C_2H_6O. The C/O ratio is 2. One water must be added so that the C/O ratio is 1. This changes C_2H_6O to $C_2H_8O_2$. Acetic acid is $C_2H_4O_2$. Since the C/O ratio is already 1, nothing further need be done. The formula for carbon dioxide is CO_2. To make the C/O ratio 1, it is necessary to subtract H_2O, that is: $[CO_2 - H_2O]$. The result is $C(-2H)O$.

2. *Now compare the number of hydrogens in the modified formula with* $(CH_2O)_n$, *which has the same number of carbons as in the modified formula.* For ethanol, $C_2H_8O_2$ is compared with $(CH_2O)_2$. There are 8 hydrogens in $C_2H_8O_2$ but only 4 in $(CH_2O)_2$. Thus, ethanol has an additional 4 hydrogens. Carbon dioxide, $C(-2H)O$, however, has 4 fewer hydrogens than CH_2O. Acetic acid ($C_2H_4O_2$) and $(CH_2O)_2$ have the same number of hydrogens.

3. *Add −1 for each additional 2H and +1 for a decrease in 2H.* Thus, the O/R for ethanol is −2, for CO_2 it is +2, and for acetic acid it is 0.

Since both the oxidized and reduced fermentation end products originate from the substrate, in a balanced fermentation, the sum of the O/R of the products equals the O/R of the substrate. For example, the O/R for glucose is 0. When one mole of glucose is fermented to two moles of ethanol and two moles of carbon dioxide, the O/R of the products is $(-2 \times 2) + (+2 \times 2) = 0$. Often one simply takes the ratio (+/−) of the O/R of the products when a carbohydrate is fermented. For a balanced fermentation, +/− should be 1.

14.5.2 The available hydrogen method

Like the O/R method, the available hydrogen procedure is merely one of bookkeeping. It has nothing to do with the chemistry of the reactions. According to this method, one "oxidizes" the molecule to CO_2 by using water to obtain the "available hydrogen." For example:

$$C_6H_{12}O_6 + 6H_2O \rightarrow 24H + 6CO_2$$

Thus glucose has 24 available hydrogens. The available H in all the products must add up to the available H in the starting material. Table 14.2 gives the concentration of products per 100 moles of glucose used. The available H in the glucose is $24 \times 100 = 2400$. The available H in the products adds up to 2,242. Thus, the balance is $2,400/2,242 = 1.07$.

14.6 Propionate Fermentation via the Acrylate Pathway

The genus *Clostridium* comprises a heterogeneous group of bacteria consisting of gram-positive, anaerobic, spore-forming bacteria that cannot use sulfate as a terminal electron acceptor. The clostridia can be isolated from anaerobic regions (or areas of low oxygen levels) in soil. They ferment organic nutrients to products that can include alcohols, organic acids, hydrogen gas, and carbon dioxide. *C. propionicum* oxidizes three moles of lactate to two moles of propionate, one mole of acetate, and one mole of carbon dioxide, to produce one mole of ATP. The pathway is called the *acrylate pathway* because one of the intermediates is acrylyl–CoA. The bacteria derive ATP via a substrate-level phosphorylation during the conversion of acetyl–P to acetate catalyzed by acetate kinase. Since only one acetate is made for every three lactates used, the pathway yields one-third of an ATP per lactate. Growth yields are proportional to the amount of ATP produced, and it is to be expected that the growth yields for these organisms are very low. (The molar growth yield for ATP is 10.5 g of cells per mole of ATP synthesized, Section 2.4.)

14.6.1 The fermentation pathway of Clostridium propionicum

A molecule of lactate is oxidized to pyruvate, yielding 2[H] (Fig. 14.3, reaction 1). The pyruvate is then oxidized to acetyl–CoA and CO_2, yielding 2[H] again (reaction 2). The acetyl–CoA is converted to acetate and ATP via acetyl–P (reactions 3 and 4). During the oxidations, 4[H] are produced, which must be

Table 14.2 Balancing an acetone–butanol fermentation

Substrate and products	Moles per 100 moles of substrate	Carbon (mol)	O/R value	O/R value (mol/100 mol)	Available H	Available H (mol/100 mol)
Substrate						
Glucose	100	600	0	—	24	2,400
Products						
Butyrate	4	16	−2	−8	20	80
Acetate	14	28	0	—	8	112
CO_2	221	221	+2	+442	0	—
H_2	135	—	−1	−135	2	270
Ethanol	7	14	−2	−14	12	84
Butanol	56	224	−4	−224	24	1,344
Acetone	22	66	−2	−44	16	352
Total		569		−425, +442		2,242

Source: Gottschalk, G. 1986. *Bacterial Metabolism*. Springer-Verlag, Berlin.

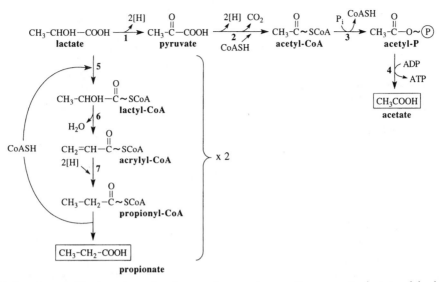

Fig. 14.3 Propionate fermentation via the acrylate pathway. Enzymes: **1**, lactate dehydrogenase; **2**, pyruvate–ferredoxin oxidoreductase; **3**, phosphotransacetylase; **4**, acetate kinase; **5**, CoA transferase; **7**, a dehydrogenase. Reaction 6 has not been sufficiently characterized.

reutilized. The electron acceptor is created from a second and third molecule of lactate (actually lactyl–CoA). The lactate acquires a CoA from propionyl–CoA (reaction 5). The lactyl–CoA is dehydrated to yield the unsaturated molecule, acrylyl–CoA (reaction 6). Each acrylyl–CoA is then reduced to propionyl–CoA, using up the 4[H] (reaction 7). The fermentation is thus balanced. The propionate, which is produced during the CoA transfer step (reaction 5), is catalyzed by CoA transferase, an enzyme that occurs in many anaerobes.

What can we learn from this pathway? The bacteria use a standard method for making ATP under anaerobic conditions. They decarboxylate pyruvate to acetyl–CoA and then, using a phosphotransacetylase (to make acetyl–P) and an acetate kinase, they make ATP and acetate. These reactions are widespread among fermenting bacteria. The production of acetate is presumed to be associated always with the synthesis of two moles of ATP per mole of acetate if the bacteria are growing on glucose and using the Embden–Meyerhof–Parnas pathway. One ATP is produced from acetyl–CoA, and the second ATP is produced during the production of the pyruvate in the EMP pathway.

Another common reaction among fermenting bacteria is the transfer of coenzyme A from one organic molecule to another, a reaction catalyzed by CoA transferase. The other way of attaching a coenzyme A molecule to a carboxyl group is to transfer an AMP or a phosphate to the carboxyl group from ATP, making an acyl phosphate or an acyl–AMP, and then displacing the AMP or phosphate with CoASH. (Recall the activation of fatty acids prior to their degradation, Section 9.1.1.) However, fermenting organisms must conserve ATP. The CoA transferase reaction is one way this can be done.

14.7 Propionate Fermentation via the Succinate–Propionate Pathway

Many bacteria that yield propionic acid as a product of fermentation use a pathway different from the acrylate pathway. The other pathway, called the *succinate–propionate pathway*, yields more ATP than the acrylate pathway per mole of propionate formed. One of the organisms that utilizes this pathway, *Propionibacterium*, ferments lactate as well as hexoses to a mixture of propionate, acetate, and CO_2. *Propionibacterium* is a gram-positive anaerobic, nonsporulating, nonmotile, pleomorphic rod that is part of the normal flora in the rumen of herbivores; it is also found on human skin and in dairy products (e.g., cheese). *Propionibacterium* is used in the fermentation

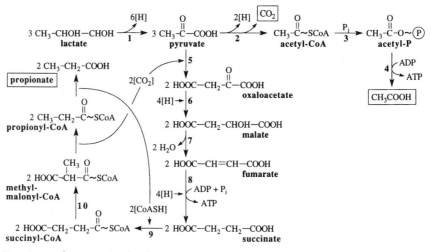

Fig. 14.4 Propionate fermentation by the succinate–propionate pathway. Enzymes: 1, lactate dehydroge-nase (a flavoprotein); 2, pyruvate dehydrogenase (an NAD$^+$ enzyme); 3, phosphotransacetylase; 4, acetate kinase; 5, methylmalonyl–CoA–pyruvate transcarboxylase; 6, malate dehydrogenase; 7, fumarase; 8, fumarate reductase; 9, CoA transferase; 10, methylmalonyl–CoA racemase.

process that produces Swiss cheese. The char-acteristic flavor of this cheese is due to the propionate, and the holes in the cheese are due to the carbon dioxide produced.

The pathway illustrated in Fig. 14.4 shows that three molecules of lactate are oxidized to pyruvate (reaction 1). This yields six electrons. Then, one pyruvate is oxidized to acetate, CO_2, and ATP, yielding two more electrons (reaction 2). We now have eight electrons to use up. The other two pyruvates are carboxylated to yield two molecules of oxaloacetate (reaction 5). The reaction is catalyzed by methyl malonyl–CoA transcarboxylase. The two oxaloacetates are reduced to two malates, consuming a total of four electrons (reaction 6). The two malates are dehydrated to two fumarates (reaction 7). The two fumarates are reduced via fumarate reductase to two succinates, consuming four electrons (reaction 8). The latter reduction is coupled to the generation of a Δp. The fer-mentation is now balanced.

The two succinates are converted to two molecules of succinyl–CoA via a CoA trans-ferase (reaction 9). The two molecules of succinyl–CoA are isomerized to two molecules of methylmalonyl–CoA in an unusual reaction in which COSCoA moves from the α-carbon to the β-carbon in succinyl–CoA to form methylmalonyl–CoA (reaction 10). The reac-tion can be viewed as an exchange between adjacent carbons of a hydrogen for a COSCoA.

The enzyme that carries out this reaction is methylmalonyl–CoA racemase, and it requires vitamin B_{12} as a cofactor. The two molecules of methylmalonyl–CoA donate the carboxyl groups to pyruvate via the transcarboxylase and in turn become propionyl–CoA (reaction 5). The propionyl–CoA donates the CoA to succinate via the CoA transferase and becomes propionate (reaction 9). Notice the important role of transcarboxylases and CoA trans-ferases. These enzymes allow the attachment of CO_2 and CoA to molecules without the need for ATP.

When comparing Figs. 14.3 and 14.4, we learn that in metabolism there is sometimes more than one route to take from point A to point B. This is demonstrated in the following remarks about two important enzyme reactions.

1. *The fumarate reductase as a coupling site Propionibacterium* and other bacteria that use the succinate–propionate pathway use a circuitous route, but one that sends elec-trons through an energy-coupling site via the membrane-bound fumarate reductase. Electron flow to fumarate requires a quinone, and presumably the Δp is generated via a redox loop involving the quinone (Sec-tion 4.6). The use of fumarate as an elec-tron acceptor during anaerobic growth is widespread among bacteria. (See the later discussion of the mixed-acid fermentation

in Section 14.10.) The electron donors besides lactate include NADH, H_2, formate, and glycerol-3-phosphate. The Δp that is established can be used for ATP synthesis, for solute uptake, or to spare ATP that might be hydrolyzed to maintain the Δp.

2. *The transcarboxylase reaction spares an ATP* One way to carboxylate pyruvate is to use pyruvate carboxylase and CO_2 (Section 8.9). However, this requires an ATP. *Propionibacterium* has a transcarboxylase called methylmalonyl–CoA–pyruvate transcarboxylase that transfers a carboxyl group from methylmalonyl–CoA to pyruvate, hence an ATP is not used (Fig. 14.4, reaction 5). By using the transcarboxylase, the bacteria save energy by substituting one covalent bond for another. However, not all fermenting bacteria that produce propionate from pyruvate use the methylmalonyl–CoA–pyruvate transcarboxylase. For example, *Veillonella alcalescens* and *Propionigenum modestum* use a sodium-dependent decarboxylase to remove the carboxyl group from methylmalonyl–CoA while generating an electrochemical potential (Section 3.8.1). The sodium-dependent decarboxylase pumps sodium ions out of the cell, generating a sodium ion potential, which can be used as a source of electrochemical energy (e.g., for solute uptake or ATP synthesis).

14.7.1 The PEP carboxytransphosphorylase of propionibacteria and its physiological significance

The reaction

Propionibacteria growing on carbon sources such as glucose that enter the glycolytic pathway can produce succinate as well as propionate as an end product of fermentation. This means that they must have an enzyme to carboxylate a C_3 intermediate to form the C_4 product. The C_3 intermediate that is carboxylated is phosphoenolpyruvate (an intermediate in glycolysis). The phosphoenolpyruvate is carboxylated to oxaloacetate, which is then reduced to succinate via reactions 6, 7, and 8 shown in Fig. 14.4. The enzyme that catalyzes the carboxylation of phosphoenolpyruvate is

called PEP carboxytransphosphorylase, and it catalyzes the following reaction: (See note 4 for a description of other C_3 carboxylases.)

$$PEP + CO_2 + P_i \rightarrow oxaloacetate + PP_i$$

During the carboxylation, a phosphoryl group is transferred from PEP to inorganic phosphate to form pyrophosphate. Pyrophosphate is a high-energy compound, and propionibacteria have enzymes that phosphorylate fructose-6-phosphate to fructose-1,6-bisphosphate and serine to phosphoserine, using pyrophosphate as the phosphoryl donor.

Physiological significance

The carboxylation of phosphoenolpyruvate or pyruvate to oxaloacetate and the reduction of the oxaloacetate to succinate is a widespread pathway among fermenting bacteria (see, e.g., the later discussion of mixed-acid fermentation in Section 14.10). These reactions were also discussed in the context of the reductive citric acid pathway (Section 8.8).

The pathway from PEP or pyruvate to succinate is extremely important for anaerobes and serves three purposes: (1) the fumarate is an electron sink, enabling NADH to be reoxidized; (2) the fumarate reductase is a coupling site (i.e., a Δp is generated); and (3) the succinate can be converted to succinyl–CoA, which is required for the biosynthesis of tetrapyrroles, lysine, diaminopimellc acid, and methionine. With respect to fumarate acting as an electron sink, Gest has suggested that the carboxylation of the C_3 glycolytic intermediate and the reductive pathway from oxaloacetate to succinate may have evolved when the earth's atmosphere was still anaerobic, serving the purpose of balancing fermentations and thus sparing one of the two pyruvates derived from glucose for biosynthesis.[5] The reactions between oxaloacetate and succinate may have become part of the oxidative citric acid cycle later, during the evolution of aerobic metabolism.

14.8 Acetate Fermentation (Acetogenesis)

As discussed in Section 13.1.3 in connection with the acetogenic bacterium *Clostridium thermoaceticum*, some bacteria use CO_2 as

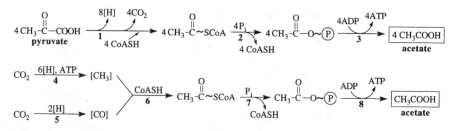

Fig. 14.5 Acetogenesis from pyruvate by *Desulfotomaculum thermobenzoicum*. Enzymes: **1**, pyruvate dehydrogenase; **2** and **7**, phosphotransacetylase; **3** and **8**, acetate kinase; **4**, enzymes of the acetyl–CoA pathway; **5** and **6**, carbon monoxide dehydrogenase.

an electron sink and reduce it to acetate as a fermentation end product via the acetyl–CoA pathway. This is called acetogenesis. Another acetogenic bacterium is the sulfate reducer *Desulfotomaculum thermobenzoicum*, when it is growing on pyruvate in the absence of sulfate.[6] (Another sulfate reducer, *Desulfobulbus propionicus*, uses the succinate–propionate pathway to ferment pyruvate to a mixture of acetate and propionate.) The acetogenic pathway is shown in Fig. 14.5. Four pyruvate molecules are oxidatively decarboxylated to four acetyl–CoA molecules, producing eight electrons (reaction **1**). The acetyl–CoA is converted to acetyl phosphate via phosphotransacetylase (reaction **2**). The acetyl phosphate is converted to acetate and ATP via the acetate kinase (reaction **3**). Six of the electrons are used to reduce CO_2 to bound methyl, $[CH_3]$ (reaction **4**). This requires an ATP in the acetyl–CoA pathway to attach the formic acid to the THF (see Section 13.1.3). The remaining two electrons are used to reduce a second molecule of CO_2 to bound carbon monoxide, $[CO]$ (reaction **5**). The $[CH_3]$, $[CO]$, and CoASH combine to yield acetyl–CoA (reaction **6**). The fourth acetyl–CoA is converted to acetate with the formation of an ATP (reactions **7** and **8**).

14.9 Lactate Fermentation

The lactic acid bacteria are a heterogeneous group of aerotolerant anaerobes that ferment glucose to lactate as the sole or major product of fermentation. They include the genera *Lactobacillus*, *Sporolactobacillus*, *Streptococcus*, *Leuconostoc*, *Pediococcus*, and *Bifidobacterium*. Lactic acid bacteria are found living on the skin of animals, in the gastrointestinal tract, and in other places (e.g., mouth and throat).

Some genera live in vegetation and in dairy products. Several lactic acid bacteria are medically and commercially important organisms. These include the genus *Streptococcus*, several species of which are pathogenic.

The lactic acid bacteria are also important in various food fermentations (e.g., the manufacture of butter, cheese, yogurt, pickles, and sauerkraut). Although they can live in the presence of air, they metabolize glucose only fermentatively and derive most or all of their ATP from substrate-level phosphorylation. Under certain growth conditions they may transport lactate out of the cell in electrogenic symport with H^+, creating a $\Delta\Psi$ (Section 3.8.3). There are two major types of lactate fermentation: *homofermentative* and *heterofermentative*. The former uses the Embden–Meyerhof–Parnas pathway (glycolysis), and the latter uses the pentose phosphate pathway. A third pathway, called the *bifidum pathway*, is found in *Bifidobacterium bifidum*.

14.9.1 Homofermentative lactate fermentation

The homofermentative pathway produces primarily lactate. The bacteria use the glycolytic pathway to oxidize glucose to pyruvate. This nets them two ATPs per mole of glucose via the oxidation of phosphoglyceraldehyde. The NADHs produced during this oxidative step are used to reduce the pyruvate, forming lactate. The overall reaction is as follows:

$$\text{Glucose} + 2\,\text{ADP} + 2\,\text{P}_i \rightarrow 2\,\text{lactate} + 2\,\text{ATP}$$

Whenever lactate is produced during fermentations, it is always the result of the reduction of pyruvate. Since the ATP yield per pyruvate is one, it can be assumed that one ATP is made per mole of lactate produced.

14.9.2 *Heterofermentative lactate fermentation*

The heterofermentative lactate fermentation produces lactate via the decarboxylation and isomerase reactions of the pentose phosphate pathway (Fig. 14.6). The glucose is oxidized to ribulose-5-phosphate by means of glucose-6-phosphate dehydrogenase and 6-phosphogluconate dehydrogenase (reactions 1–3). Four electrons are produced in the form of two NADHs. The ribulose-5-phosphate is

isomerized to xylulose-5-pho̶
the epimerase (reaction 4). Th̶
reaction occurs during whic̶
phosphate is cleaved with the ̶
phosphate to form phosphoglyce̶
acetyl phosphate (reaction 5). The enzy̶
catalyzes this reaction is called *phosphoketo-lase*, and it requires thiamine pyrophosphate (TPP) as a cofactor. The phosphoglyceralde-hyde is oxidized to pyruvate via reactions also found in the glycolytic pathway yielding an ATP and a third NADH, and the pyruvate

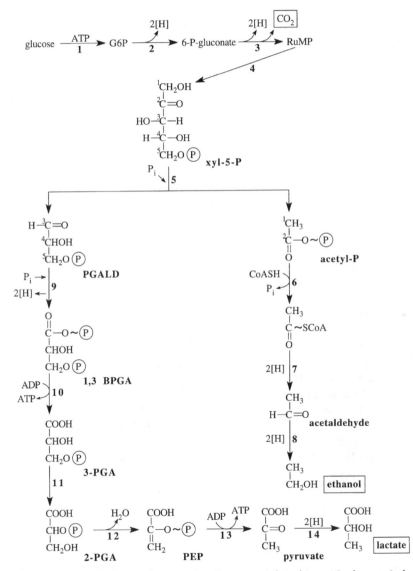

Fig. 14.6 Heterofermentative lactate fermentation. Enzymes: **1**, hexokinase; **2**, glucose-6-phosphate dehydrogenase; **3**, 6-phosphogluconate dehydrogenase; **4**, ribulose-5-phophate epimerase; **5**, phosphoketolase; **6**, phosphotransacetylase; **7**, acetaldehyde dehydrogenase; **8**, alcohol dehydrogenase; **9**, PGALD dehydrogenase; **10**, PGA kinase; **11**, phosphoglycerate mutase; **12**, enolase; **13**, pyruvate kinase; **14**, lactate dehydrogenase.

THE PHYSIOLOGY AND BIOCHEMISTRY OF PROKARYOTES

educed to lactate by one of the three ADHs (reactions **9–14**). The acetyl phosphate produced in the phosphoketolase reaction is reduced to ethanol via the second and third NADH (reactions **6–8**). Thus, the fermentation is balanced. The overall reaction is as follows:

$$Glucose + ADP + P_1$$
$$\rightarrow ethanol + lactate + CO_2 + ATP$$

Note that the heterofermentative pathway produces only one ATP per glucose, in contrast to the homofermentative pathway, which produces two ATPs for every glucose.

14.9.3 Bifidum pathway

The bifidum pathway ferments two glucoses to two lactates and three acetates with the production of 2.5 ATPs per glucose. The overall reaction is:

$$2\ Glucose + 5\ ADP + P_i$$
$$\rightarrow 3\ acetate + 2\ lactate + 5\ ATP$$

The ATP yields are therefore greater than for the homo- or heterofermentative pathways. The pathway uses reactions of the pentose phosphate pathway (Section 8.5) and the homofermentative pathway. Two glucose molecules are converted to two molecules of fructose-6-P, requiring two ATPs. One of the fructose-6-P molecules is cleaved by fructose-6-phosphate phosphoketolase to erythrose-4-P and acetyl–P. The acetyl–P is converted to acetate via acetate kinase, with the formation of an ATP. The erythrose-4-P reacts with the second fructose-6-P in a transaldolase reaction to form glyceraldehyde-3-P and sedoheptulose-7-P. These then react in a transketolase reaction to form xylulose-5-P and ribose-5-P. The latter isomerizes to form a second xylulose-5-P. The two xylulose-5-P molecules are cleaved by xylulose-5-phosphate phosphoketolase to two molecules of glyceraldehyde-3-P and two of acetyl–P. The two glyceraldehyde-3-P molecules are converted to two lactates with the production of four ATPs, using reactions of the homofermentative pathway, and the two acetyl–P molecules are converted to two acetates with the production of two more ATPs. Thus, seven ATPs are produced for every two glucose molecules fermented, but since two ATPs were

used to make the two fructose-6-P molecules, the net gain in ATP per glucose is 5/2, or 2.5. Note that since glucose-6-P is not oxidized to 6-phosphogluconate, the acetyl–P can serve as a phosphoryl donor for ATP synthesis rather than being reduced to ethanol.

14.9.4 Synthesis of acetyl–CoA or acetyl–P from pyruvate by lactic acid bacteria

Not all the pyruvate produced by lactic acid bacteria needs be converted to lactate. Depending upon the species, the lactic acid bacteria may have one of three enzymes or enzyme complexes for decarboxylating pyruvate (Section 7.3.2). Streptococci have pyruvate dehydrogenase, usually found in aerobically respiring bacteria; several lactic acid bacteria are known to have pyruvate formate lyase, an enzyme also found in the Enterobacteriaceae; *Lactobacillus plantarum* and *L. delbruckii* use the flavoproteins pyruvate oxidase and lactate oxidase in coupled reactions that convert two pyruvates to one acetyl–P and one lactate, as follows:

Pyruvate oxidase

$$Pyruvate + P_i + FAD$$
$$\rightarrow acetyl–P + CO_2 + FADH_2$$

Lactate oxidase

$$Pyruvate + FADH_2 \rightarrow lactate + FAD$$

14.10 Mixed-Acid and Butanediol Fermentation

The enteric bacteria are facultative anaerobes. In the absence of oxygen, as part of the adaptation to anaerobic growth, the following physiological changes take place:

1. Terminal reductases replace the oxidases in the electron transport chain.

2. The citric acid cycle is modified to become a reductive pathway. α-Ketoglutarate dehydrogenase and succinate dehydrogenase are missing or occur at low levels, the latter being replaced by fumarate reductase.

3. Pyruvate–formate lyase is substituted for pyruvate dehydrogenase. This means that the cells oxidize pyruvate to acetyl–CoA

and formate, rather than to acetyl–CoA, CO_2, and NADH.

4. The bacteria then carry out a mixed-acid or butanediol fermentation.

The mixed-acid and butanediol fermentations are similar in that both produce a mixture of organic acids, CO_2, H_2, and ethanol. The butanediol fermentation is distinguished by producing large amounts of 2,3-butanediol, acetoin, more CO_2 and ethanol, and less acid. The mixed-acid fermenters belong to the genera *Escherichia*, *Salmonella*, and *Shigella*. All three can be pathogenic, causing intestinal infections such as dysentery, typhoid fever (*Salmonella typhimurium*), or food poisoning. Butanediol fermenters are *Serratia*, *Erwinia*, and *Enterobacter*.

14.10.1 *Mixed-acid fermentation*

The products of the mixed-acid fermentation are succinate, lactate, acetate, ethanol, formate,

carbon dioxide, and hydrogen gas (Fig. 14.7). Each of these are made from one phos-phoenolpyruvate and CO_2 or one pyruvate. For example, the formation of succinate is due to a carboxylation of phosphoenolpyruvate to oxaloacetate followed by two reductions to form succinate (reactions 10–13). All the other products are formed from pyruvate. The formation of lactate from pyruvate is simply a reduction (reaction 4). Pyruvate is also decarboxylated to acetyl–CoA and formate by means of the pyruvate–formate lyase (reaction 3). The acetyl–CoA can be reduced to ethanol (reactions 6 and 7) or converted to acetate and ATP via acetyl–P (reactions 8 and 9). The formate can be oxidized to CO_2 and H_2 by the enzyme system formate–hydrogen lyase (reaction 5). This system actually consists of two enzymes; formate dehydrogenase and an associated hydrogenase. The formate dehydro-genase oxidizes the formate to CO_2 and reduces the hydrogenase, which transfers the electrons to two protons to form hydrogen gas. *Shigella*

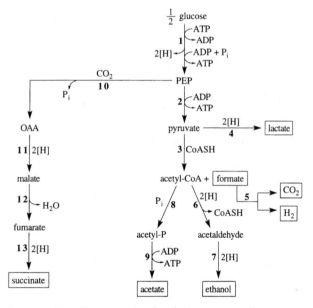

Fig. 14.7 Mixed-acid fermentation. Enzymes: **1**, glycolytic enzymes; **2**, pyruvate kinase; **3**, pyruvate–formate lyase; **4**, lactate dehydrogenase; **5**, formate–hydrogen lyase; **6**, acetaldehyde dehydrogenase; **7**, alcohol dehydrogenase; **8**, phosphotransacetylase; **9**, acetate kinase; **10**, PEP carboxylase; **11**, malate dehydrogenase; **12**, fumarase; **13**, fumarate reductase. Note the ATP yields: per succinate, approximately 1; per ethanol, 1; per acetate, 2; per formate, 1; per CO_2 and H_2, 1; per lactate, 1. Energy equivalent to approximately 1 ATP is conserved per succinate formed because the fumarate reductase reaction takes place in the cell membrane and generates a Δp. Note also the reducing equivalents used in the production of the end products: per succinate, 4; per ethanol, 4; per acetate, 0; per lactate, 2; per formate, 0. The number of reducing equivalents used must equal the number produced during glycolysis. Therefore, only certain ratios of end products are compatible with a balanced fermentation.

and *Erwinia* do not contain formate–hydrogen lyase, and therefore do not produce gas. (See note 7.)

Each of the pathways following phospho-enolpyruvate or pyruvate can be viewed as a metabolic branch that accepts different amounts of reducing equivalent, that is, 0, 2[H], or 4[H], depending upon the pathway. The reducing equivalents in the different branches are succinate (4), ethanol (4), lactate (2), acetate (0), and formate (0). During glycolysis, two electrons are produced for each phosphoenolpyruvate or pyruvate formed. Therefore, to balance the fermentation, two electrons must be used for each phosphoenolpyruvate or pyruvate formed. The pathways that utilize four electrons per phosphoenolpyruvate or pyruvate formed spare the second phosphoenolpyruvate or pyruvate for biosynthesis. However, they may do this at the expense of an ATP. For example, the reduction of acetyl–CoA to ethanol uses 4H, but this is at the expense of an ATP that can be formed when acetyl–CoA is converted to acetate. In this context, the production of succinate is particularly valuable. The pathway utilizes 4H and also includes a coupling site (fumarate reductase).

14.10.2 Butanediol fermentation

The butanediol fermentation is characterized by the production of 2,3-butanediol and acetoin (Fig. 14.8). Glucose is oxidized via the glycolytic pathway to pyruvate (reactions 1). There are three fates for the pyruvate. Some of it is reduced to lactate (reaction 10), some is converted to acetyl–CoA and formate (reaction 2), and some is used for the synthesis of 2,3-butanediol (reactions 6–9). The formate is converted to CO_2 and H_2 (reaction 3), and the acetyl–CoA is reduced to ethanol (reactions 4 and 5). The first free intermediate in the butanediol pathway is α-acetolactate, formed by the enzyme α-acetolactate synthase, which decarboxylates pyruvate to enzyme-bound "active acetaldehyde" (reaction 6); this is a reaction that depends upon thiamine pyrophosphate (TPP). The active acetaldehyde is transferred by

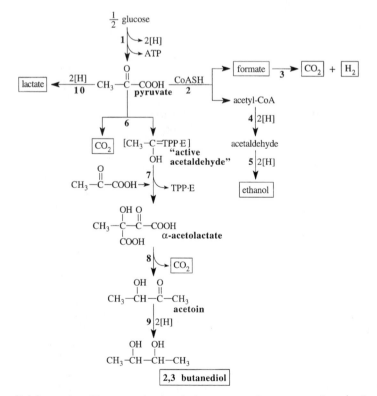

Fig. 14.8 Butanediol formation. Enzymes: 1, glycolytic enzymes; 2, pyruvate–formate lyase; 3, formate–hydrogen lyase; 4, acetaldehyde dehydrogenase; 5, alcohol dehydrogenase; 6 and 7, α-acetolactate synthase; 8, α-acetolactate decarboxylase; 9, 2,3-butanediol dehydrogenase; 10, lactate dehydrogenase.

the α-acetolactate synthase to pyruvate to form α-acetolactate (reaction 7). The α-acetolactate, a β-ketocarboxylic acid, is decarboxylated to acetoin (reaction 8). The acetoin is reduced by NADH to 2,3-butanediol (reaction 9). The production of butanediol is favored under slightly acidic conditions and is a way for the bacteria to limit the decrease in external pH caused by the synthesis of organic acids from pyruvate.

How thiamine pyrophosphate catalyzes the decarboxylation of α-ketocarboxylic acids

The decarboxylation of α-ketocarboxylic acids presents a problem because there is no electron-attracting carbonyl group β to the C–C bond to withdraw electrons, as there is in β-ketocarboxylic acids. (See Section 8.11.2 for a description of the decarboxylation of β-ketocarboxylic acids.) The problem is solved by using thiamine pyrophosphate (TPP). A proposed mechanism is illustrated in Fig. 14.9. A proton dissociates from the thiamine pyrophosphate to form a dipolar ion, which is stabilized by the positive charge on the nitrogen atom. The negative center of the dipolar ion adds to the keto group of the α-ketocarboxylic acid. Then, the electron-attracting $=N^+-$ group pulls electrons away from the C–C bond, facilitating the decarboxylation. The product is an α-ketol called "active aldehyde." The flow of electrons is reversed, and the "active aldehyde" then attacks a positive center on another molecule. The product varies, depending upon the enzyme.

As shown in Fig. 14.9, α-acetolactate synthase catalyzes the condensation of "active acetaldehyde" with pyruvate to form α-acetolactate, and pyruvate decarboxylase catalyzes the addition of a proton to form acetaldehyde. Pyruvate dehydrogenase and α-ketoglutarate dehydrogenase catalyze the condensation with lipoic acid to form the acyl–lipoate derivative.

14.11 Butyrate Fermentation

Butyrate fermentations are carried out by the butyric acid clostridia. The clostridia are a heterogeneous group of anaerobic spore-forming bacteria that can be isolated from anaerobic muds, sewage, feces, or other anaerobic environments. They all are classified in the genus *Clostridium*. Some are saccharolytic (i.e., they ferment carbohydrates) and/or proteolytic. The proteolytic clostridia are important in the anaerobic decomposition of proteins, called "putrefaction." Other clostridia are more specialized and will ferment only a few substrates (e.g., ethanol, acetate, certain purines, certain amino acids).

The butyric acid clostridia ferment carbohydrates to butyric acid. The fermentation products also include hydrogen gas, carbon dioxide, and small amounts of acetate. The bacteria first oxidize the glucose to two moles of pyruvate via the Embden–Meyerhof–Parnas pathway (Fig. 14.10). This produces two molecules of NADH, plus two of ATP. The pyruvate is then decarboxylated to acetyl–CoA, CO_2 and H_2 by means of pyruvate–ferredoxin oxidoreductase and hydrogenase (reaction 2). The acetyl–CoA is condensed to form acetoacetyl–CoA (reaction 3), which is reduced to β-hydroxybutyryl–CoA by means of one of the two NADHs (reaction 4). The β-hydroxybutyryl–CoA is reduced to butyryl–CoA by using the second NADH (reactions 5 and 6). The CoASH is displaced by inorganic phosphate (reaction 7) and the butyryl-phosphate donates the phosphoryl group to ADP to form ATP and butyrate (reaction 8). This pathway therefore utilizes three substrate-level phosphorylations, the phosphoryl donors being 1,3-bisphosphoglycerate, phosphoenolpyruvate, and butyryl–P. Note the role of the hydrogenase (reaction 9) in the hydrogen sink. (For a discussion of acetyl–CoA condensations, see Section 8.11.1.)

14.11.1 Butyrate and butanol–acetone fermentation in Clostridium acetobutylicum

Some clostridia initially make butyrate during fermentation and, when the butyrate accumulates and the pH drops (owing to the butyric acid), the fermentation switches to a butanol–acetone fermentation. As we shall see, the butyrate is actually taken up by the cells and converted to butanol and acetone. The accumulation of butyric acid in media of low pH can be toxic because the undissociated form

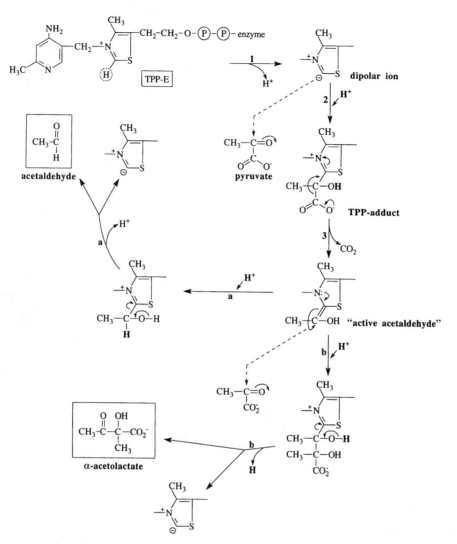

Fig. 14.9 Proposed mechanism for reactions catalyzed by thiamine pyrophosphate (TPP). Step **1**. The TPP enzyme loses a proton to form a dipolar ion. The anion is stabilized by the positive charge on the nitrogen. The anionic center is nucleophilic and can attack positive centers such as carbonyl carbons. Step **2**. The dipolar ion condenses with pyruvate to form a TPP adduct. Step **3**. The electron-attracting N in the TPP facilitates the decarboxylation to form "active acetaldehyde." The "active acetaldehyde" can form acetaldehyde (step **a**) or α-acetolactate (step **b**).

of the acid is lipophilic and can enter the cell, acting as an uncoupler and also resulting in a decrease in the ΔpH. (The pK of butyric acid is 4.82.) Butanol and acetone production by the clostridia was at one time the second largest industrial fermentation process, after ethanol. The solvents are now synthesized chemically.

As an example of the butyrate fermentation and the shift to the butanol–acetone fermentation, we will consider the fermentation of carbohydrates carried out by *C. acetobutylicum*.[8] During exponential growth, in what is called

the acidogenic phase, the bacteria produce butyrate, acetate, H_2, and CO_2. When the culture enters stationary phase, the acids are taken up by the cells, concomitant with the fermentation of the carbohydrate, and are converted to butanol, acetone, and ethanol. This is called the solventogenic phase.

Pentoses are also fermented, and these are converted to fructose-6-phosphate and phosphoglyceraldehyde via the pentose phosphate pathway (Fig. 14.11). The pyruvate formed from the sugars is oxidatively decarboxylated

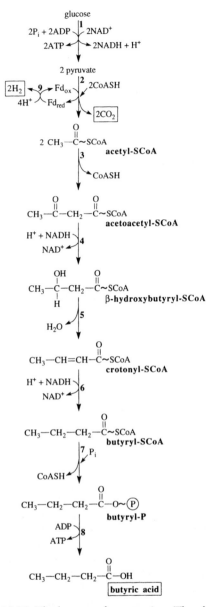

to acetyl–CoA and CO_2 via the pyruvate–ferredoxin oxidoreductase (reaction 1). The acetyl–CoA has two fates. Some of it is converted to butyrate and ATP as described in Fig. 14.10 (reactions 2–7). Some acetyl–CoA is also converted to acetate via acetyl–P in a reaction that yields an additional ATP (Fig. 14.11, reactions 8 and 9). The ability to send electrons to hydrogen via the NADH:ferredoxin oxidoreductase (reaction 16) allows the bacteria to produce more acetate, hence more ATP, rather than reducing acetyl–CoA to butyrate. Notice that the reaction generates twice as much ATP as butyrate per acetate. However, reduction of protons by NADH is not favored energetically and is limited by increasing concentrations of hydrogen gas.

The NADH oxidoreductase can be pulled in the direction of hydrogen gas by other bacteria that utilize hydrogen in a process called interspecies hydrogen transfer, explained in Section 14.4.1. In the solventogenic phase, butyrate and acetate are taken up by the cells and reduced to butanol and ethanol (dashed lines in Fig. 14.11). The acids are converted to their CoA derivatives by accepting a CoA from acetoacetyl–CoA (reaction 10). The reaction is catalyzed by CoA transferase. (Recall that CoA transferase is also used in the acrylate pathway for propionate fermentation, Section 14.6.) The acetoacetate that is formed is decarboxylated to acetone and CO_2 (reaction 11). The acetyl–CoA is reduced to ethanol (reactions 12 and 13), and the butyryl–CoA is reduced to butanol (reactions 14 and 15).

The molar ratios of the fermentation end products in clostridial fermentations will vary according to the strain.[9] For example, C. acetobutylicum is an important solvent-producing strain, and when grown at pH_0 below 5 produces butanol and acetone in the molar ratio of 2:1 with small amounts of isopropanol, whereas C. sporogenes and C. pasteurianum produce very little solvent. Clostridium butyricum forms butyrate and acetate in a ratio of about 2:1, whereas C. perfringens produces these acids in a ratio of 1:2, with significant amounts of ethanol and lactate.

Fig. 14.10 The butyrate fermentation. The glucose is degraded via glycolysis to pyruvate, which is then oxidatively decarboxylated to acetyl–CoA. Two molecules of acetyl–CoA condense to form acetoacetyl–CoA, which is reduced to butyryl–CoA. A phosphotransacetylase makes butyryl–P, and a kinase produces butyrate and ATP. Two ATPs are produced in the glycolytic pathway for every glucose and one ATP from butyryl–P. Note the production of hydrogen gas as an electron sink. This actually allows the production of the third ATP from butyryl–P, rather than reducing the butyryl–SCoA to butanol to balance the fermentation. Enzymes: 1, glycolytic enzymes; 2, pyruvate–ferredoxin oxidoreductase; 3, acetyl–CoA acetyltransferase (thiolase); 4, β-hydroxybutyryl–CoA dehydrogenase; 5, crotonase; 6, butyryl–CoA dehydrogenase; 7, phosphotransbutyrylase; 8, butyrate kinase; 9, hydrogenase.

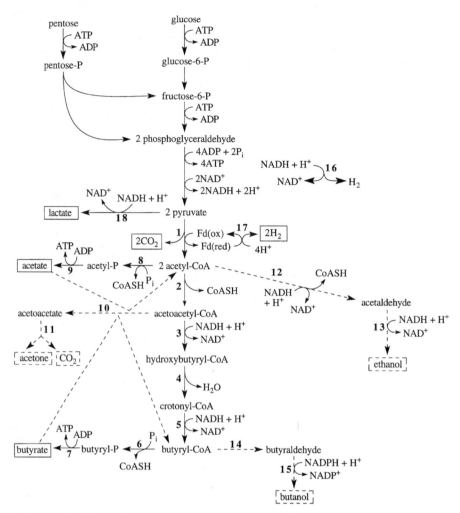

Fig. 14.11 Butyrate and butanol–acetone fermentation in *C. acetobutylicum*. Carbohydrates are oxidized to acetyl–SCoA. Pentose phosphates are converted to fructose-6-phosphate and phosphoglyceraldehyde via the pentose phosphate reactions. Glucose is oxidized to pyruvate via the Embden–Meyerhof–Parnas pathway. The pyruvate is oxidized to acetyl–SCoA. In the butyric acid fermentation, the acetyl–SCoA is converted to acetate and butyrate (solid lines). When the acetate and butyrate levels rise, they are taken up by the cells and converted to butanol, and ethanol, while carbohydrates continue to be fermented (dashed lines). During the butanol–acetone fermentation, the acetoacetyl–SCoA donates the CoASH to butyrate and acetate and becomes acetoacetate, which is decarboxylated to acetone. Enzymes: **1**, pyruvate–ferredoxin oxidoreductase; **2**, acetyl–CoA acetyltransferase (thiolase); **3**, hydroxybutyryl–CoA dehydrogenase; **4**, crotonase; **5**, butyryl–CoA dehydrogenase; **6**, phosphotransbutyrylase; **7**, butyrate kinase; **8**, phosphotransacetylase; **9**, acetate kinase; **10**, acetoacetyl–SCoA:acetate/butyrate:CoA transferase; **11**, acetoacetate decarboxylase; **12**, acetaldehyde dehydrogenase; **13**, ethanol dehydrogenase; **14**, butyraldehyde dehydrogenase; **15**, butanol dehydrogenase; **16**, NADH–ferredoxin oxidoreductase and hydrogenase; **17**, hydrogenase. *Source*: Adapted from Jones, D. T., and D. R. Woods. 1986. Acetone–butanol fermentation revisited. *Microbiol. Rev.* 50:484–524.

14.12 *Ruminococcus albus*

Ruminococcus albus is a rumen bacterium that can use the glycolytic pathway to ferment glucose to ethanol, acetate, H_2, and CO_2 (Fig. 14.12). It is a good fermentation to exam-

ine because it illustrates how growth of bacteria in mixed populations can influence fermentation end products. The fermentation is easy to understand. The glucose is first oxidized to two moles of pyruvate, yielding two moles of NADH. Then each mole of pyruvate is oxidized

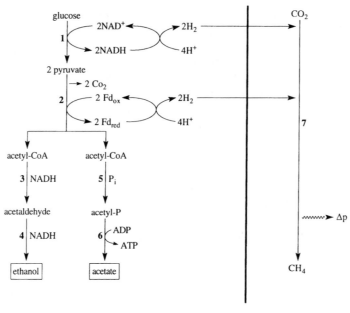

Fig. 14.12 Fermentation of glucose by *Ruminococcus albus*. *R. albus* ferments glucose to a mixture of ethanol, acetate, CO_2, and H_2. Methanogens draw off the H_2, thus stimulating electron flow to H_2. The result is a shift in the fermentation end products toward acetate, accompanied by a greater ATP yield. The production of H_2 by one species and its utilization by another is called interspecies hydrogen transfer. The methanogens can also utilize the acetate produced by *R. albus*.

to acetyl–CoA, CO_2, and H_2 by means of the pyruvate–ferredoxin oxidoreductase and hydrogenase. One of the acetyl–CoA molecules is converted to acetate, allowing an ATP to be made. The second acetyl–CoA is reduced to ethanol via the two NADHs. Thus the fermentation is balanced. The overall reaction is the conversion of one mole of glucose to one mole of ethanol, one mole of acetate, two moles of hydrogen, and two moles of carbon dioxide yielding, three ATPs.

Something different happens when the bacterium is grown in a coculture with a methanogen. Growth with a methanogen shifts the fermentation in the direction of acetate with the concomitant production of more ATP. This is explained in the following way. *R. albus* has a hydrogenase that transfers electrons from NADH to H$^+$ to produce hydrogen gas. When hydrogen accumulates in the medium, the hydrogenase does not oxidize NADH because the equilibrium favors NAD$^+$ reduction. The NADH therefore reduces acetyl–CoA to ethanol. However, the methanogen utilizes the hydrogen gas for methane production and keeps the hydrogen levels very low. In the presence of the methanogen, the NADH in *R. albus* reduces protons to hydrogen gas instead of reducing acetyl–CoA. The result is that *R. albus* converts the acetyl–CoA to acetate. This is also an advantage to the methanogen, since methanogens can also use acetate as a carbon and energy source. The transfer of hydrogen gas from one species to another is called *interspecies hydrogen transfer* and is an example of nutritional synergism found among mixed populations of bacteria. (See the discussion of interspecies hydrogen transfer in Section 14.4.1.)

14.13 Summary

Fermentations are cytosolic oxidation–reduction pathways in which the electron acceptor is an organic compound, usually generated in the pathway.

The source of ATP in fermentative pathways is substrate-level phosphorylation. For sugar fermentations, these are the phosphoglycerate kinase, pyruvate kinase, acetate kinase, and butyrate kinase reactions. In other words, ATP is made from bisphosphoglycerate, phosphoenolpyruvate, acetyl–P, and butyryl–P. Butyryl–P is a phosphoryl donor during butyrate fermentations.

However, other means of conserving energy are available for fermenting bacteria. For example, a Δp can be created during electron flow to fumarate, the fumarate being generated during fermentation. Other means of creating a proton potential or sodium potential exist in certain groups of bacteria. These include efflux of carboxylic acids in symport with protons or sodium ions, decarboxylases that function as sodium ion pumps, and oxalate:formate exchange.

Besides making ATP, fermenting bacteria must have a place to unload the electrons removed during oxidation reactions. Of course, the reason for this is that they must regenerate the NAD$^+$, oxidized ferredoxin, and FAD to continue the fermentation. The electron acceptors are sometimes referred to as "electron sinks." During a fermentation the electron sinks are created from the carbon source. In fact, all the electron sinks for the major carbohydrate fermentations either are pyruvate itself or are synthesized from pyruvate or phosphoenol pyruvate plus CO_2. Protons can also be used as electron sinks, and many fermenting bacteria have hydrogenases that reduce protons to hydrogen gas.

The excreted end products of fermentations, including hydrogen gas, are used by other anaerobic bacteria so that an anaerobic food chain develops. At the bottom of the food chain are the methanogens, which convert hydrogen gas, carbon dioxide, and acetate to CO_2 and CH_4. These are recycled to organic carbon in the biosphere for use by autotrophic and methanotrophic organisms as a source of carbon.

Study Questions

1. Write a fermentation balance using both the O/R and the available hydrogen method for the following:

$C_6H_{12}O_6$ (glucose) + H_2O

$\rightarrow C_2H_4O_2$ (acetate) + C_2H_6O (ethanol)

+ $2H_2 + 2CO_2$

If the EMP pathway is used, what is the yield of ATP?

2. *C. propionicum* and *Propionibacterium* both carry out the following reaction:

3 Lactate $\rightarrow$ acetate + 2 propionate + CO_2

C. propionicum nets one ATP, but *Propionibacterium* derives more. How might *Propionibacterium* make more ATP from the same overall reaction?

3. Use the Entner–Doudoroff pathway to write an ethanol fermentation. Contrast the ATP yields with an ethanol fermentation by means of the EMP pathway.

4. Consider the following fermentation data for *Selenomonas ruminantium* products formed per millimole of glucose:

Product	Amount formed (nmol)
Lactate	0.31
Acetate	0.70
Propionate	0.36
Succinate	0.61

Source: Michel, T. A., and J. M. Macy. 1990. Generation of a membrane potential by sodium-dependent succinate efflux in *Selenomonas ruminantium*. *J. Bacteriol.* **172**:1430–1435.

Find the fermentation balance according to both the O/R method and the available hydrogen method. What percentage of the glucose carbon is recovered in end products?

5. Consider the following data for fermentation products made by *Selenomonas rumanantium* in pure culture and in coculture with *Methanobacterium rumanantium* (in moles per 100 moles of glucose).

Product	Proportion (mol/100 mol glucose)	
	Selenomonas	*Selenomonas* + *Methanobacterium*
Lactate	156	68
Acetate	46	99
Propionate	27	20
Formate	4	0
Methane	0	51
CO$_2$	42	48

Source: Chen, M., and M. J. Wolin. 1977. Influence of CH_4 production by *Methanobacterium ruminantium* on the fermentation of glucose and lactate by *Selenomonas ruminantium*. *Appl. Environ. Microbiol.* **34**:756–759.

Offer an explanation to account for the shift from lactate to acetate in the coculture compared to the pure culture. How might you expect this to affect the growth yields of *Selenomonas*?

REFERENCES AND NOTES

1. Henle, E. S., and S. Linn. 1997. Formation, prevention, and repair of DNA damage by iron/hydrogen peroxide. *J. Biol. Chem.* **272**:19095–19098.

2. Transition elements are defined as those elements in which electrons are added to an outer electronic shell before one of the inner shells is complete. There are nine of them. They are all metals, and they include iron, which is the one that occurs in highest amounts in living systems.

3. Shink, B. 1997. Energetics of syntrophic cooperation in methanogenic degradation. *Microbiol. Mol. Biol. Rev.* **61**:262–280.

4. Other enzymes besides PEP carboxytransphosphorylase that carboxylate C_3 glycolytic intermediates to oxaloacetate are PEP carboxylase and pyruvate carboxylase (Section 8.9), and PEP carboxykinase (Section 8.13). The latter enzyme usually operates in the direction of PEP synthesis; however, in some anaerobes (e.g., Bacteroides) it functions to synthesize oxaloacetate.

5. Gest, H. 1983. Evolutionary roots of anoxygenic photosynthetic energy conversion, pp. 215–234. In: *The Phototrophic Bacteria: Anaerobic Life in the Light. Studies in Microbiology*, Vol. 4. J. G. Ormerod (Ed.). Blackwell Scientific Publications, Oxford.

6. Tasaki, M., Y. Kamagata, K. Nakamura, K. Okamura, and K. Minami. 1993. Acetogenesis from pyruvate by *Desulfotomaculum thermobenzoicum* and differences in pyruvate metabolism among three sulfate-reducing bacteria in the absence of sulfate. *FEMS Microbiol. Lett.* **106**:259–264.

7. Gas production is generally observed as a bubble in an inverted vial placed in the fermentation tube. The bubble is due to H_2, since CO_2 is very soluble in water.

8. Jones, D. T., and D. R. Woods. 1986. Acetone–butanol fermentation revisited. *Microbiol. Rev.* **50**:484–524.

9. Hamilton, W. A. 1988. Energy transduction in anaerobic bacteria, pp. 83–149. In: *Bacterial Energy Transduction*. C. Anthony (Ed.). Academic Press, New York.

15

Homeostasis

Homeostasis refers to the ability of living organisms to maintain an approximately constant internal environment despite changes in the external milieu. As applied to the bacterial cell, this means (among other things) the ability to maintain a steady intracellular pH and a constant osmotic differential across the cell membrane, despite fluctuations in external pH and osmolarity. The regulatory mechanisms underlying homeostasis are a subject of active research.

15.1 Maintaining a ΔpH

15.1.1 Neutrophiles, acidophiles, and alkaliphiles

Bacteria can be found growing in habitats that vary in pH from approximately pH 1 to 2 in acid springs to as high as pH 11 in soda lakes and alkaline soils. However, regardless of the external pH, the internal pH is usually maintained within 1 to 2 units of neutrality, which is necessary to maintain viability (Table 15.1).[1-6] Thus, bacteria maintain a pH gradient (ΔpH) across the cell membrane. For example, bacteria that grow optimally between pH 1 and 4 (i.e., *acidophiles*) have an internal pH of about 6.5 to 7. That is, they maintain a ΔpH of greater than 2.5 units, inside alkaline. Those that grow optimally between pH 6 and 8 (i.e., *neutrophiles*) have an internal pH of around 7.5 to 8.0 and

Table 15.1 pH homeostasis in bacteria

Bacteria	pH$_{out}$	pH$_{in}$	ΔpH (pH$_{in}$ − pH$_{out}$)
Neutrophile	6–8	7.5–8	+
Acidophile	1–4	6.5–7	+
Alkaliphile	9–12	8.4–9	−

can maintain a ΔpH of 0.5 to 1.5 units, inside alkaline. Those that grow best at pH 9 or above (often in the range of pH 10–12, i.e., *alkaliphiles*), have an internal pH of 8.4 to 9. Therefore, alkaliphiles maintain a negative ΔpH of about 1.5 to 2 units, inside acid. (This lowers the Δp. For a discussion of this point, see Section 3.10.)

15.1.2 Demonstrating pH homeostasis

One method to demonstrate pH homeostasis is to perturb the cytoplasmic pH by changing the external pH, and then study the recovery process. When *E. coli* is exposed to a rapid change in the external pH (e.g., of 1.5–2.5 units), the internal pH initially changes in the direction of the external pH. However, a recovery soon occurs as the internal pH bounces back to its initial value (Fig. 15.1). In other bacteria, pH homeostasis may be so rapid that a temporary change in the internal pH is not even measurable.

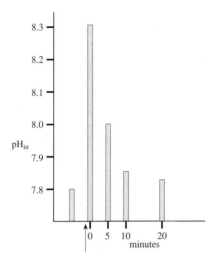

Fig. 15.1 pH homeostasis in *E. coli*. Growing *E. coli* cells were shifted from pH 7.2 to pH 8.3 (arrow). The ΔpH (pH$_{in}$-pH$_{out}$) was measured by using a weak acid or a weak base as described in Chapter 3, and the pH$_{in}$ was calculated. Immediately after the cells had shifted to pH 8.3, the cytoplasmic pH rose to 8.3. However, within a few minutes the cytoplasmic pH was restored to approximately its original value. The mechanism by which *E. coli* acidifies the cytoplasm to maintain pH homeostasis is uncertain. Data from Zilberstein, D., V. Agmon, S. Schuldiner, and E. Padan. 1984. *Escherichia coli* intracellular pH, membrane potential, and cell growth. *J. Bacteriol.* **158**:246–252.

15.1.3 The mechanism of pH homeostasis

Proton pumping

One would expect to find that many factors influence the intracellular pH. These include the buffering capacity of the cytoplasm and metabolic reactions that produce acids and bases. However, it is generally believed that the regulation of intracellular pH is, to a large extent, a consequence of controlling the flow of protons across the cell membrane.

The reason for believing this is that the ΔpH is dissipated when the proton pumps are inhibited or when proton ionophores are added to the medium. (Recall that proton ionophores equilibrate protons across the cell membrane.) The idea is that when the cytoplasm becomes too acid, protons are pumped out. For bulk proton flow to take place (e.g., by bringing K$^+$ into the cell), the pumping must be done electroneutrally. When the cytoplasm becomes too basic, protons are brought in via exchange with outgoing K$^+$ or Na$^+$ (Fig. 15.2). This implies the existence of feedback mechanisms by which the intracellular pH can signal proton pumps and antiporters. Regulation at this level is not understood. Moreover, even the influx of

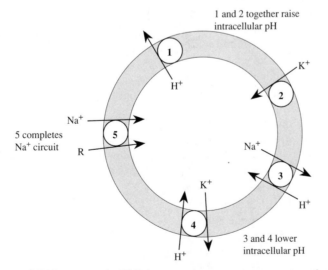

Fig. 15.2 Mechanisms of pH homeostasis. (**1**) Primary proton pumps create a membrane potential. (**2**) The uptake of K$^+$ dissipates the membrane potential, allowing extrusion of protons via the pumps and an alkalinization of the cytoplasm. The bulk solution is kept electrically neutral because of the counterions that had neutralized the K$^+$. Therefore, a $\Delta\Psi$ is changed into a ΔpH, inside alkaline. (**3**) and (**4**) Cation/proton antiporters pump Na$^+$ and K$^+$ out and bring in H$^+$. These are suggested to be the major mechanism for acidifying the cytoplasm. (**5**) Sodium ion uptake systems complete the sodium circuit.

protons depends upon the outgoing proton pumps. This is because the antiporters that bring in protons in exchange for sodium or potassium ions are driven by the Δp: by the ΔpH, outside acid, and/or by the membrane potential, $\Delta \Psi$. When the inflow of protons is electrogenic (i.e., when the ratio of H^+ to Na^+, or K^+, on the antiporter exceeds 1), $\Delta \Psi$ drives the antiporter. Therefore, inhibition of the proton pumps should lead to a situation where $pH_{in} = pH_{out}$ (i.e., a collapse of the ΔpH), and this can be tested.

The two major classes of proton pumps are those coupled to respiration and the proton-translocating ATPase. The former can be inhibited by respiratory poisons such as cyanide, and the latter by inhibitors of the ATPase [e.g., N, N'-dicyclohexylcarbodiimide (DCCD)] or by mutation. One can also short-circuit the proton flow by using proton ionophores. The ionophores will dissipate the proton potential, thus neutralizing the pumps.[7] These experiments show that in the absence of proton pumping, the ΔpH falls as the protons tend to equilibrate across the cell membrane.

The role of K^+ in maintaining a ΔpH

A problem in using the proton pumps to move protons out of the cell is that the pumps are electrogenic, and the number of protons that can be pumped out of the cell is limited by the membrane potential that develops. (See the discussion of membrane capacitance, Section 3.2.2.) Thus, for proton pumping to raise the intracellular pH, the excess membrane potential must be dissipated either by the influx of cations or by the efflux of anions. In neutrophilic bacteria, K^+ influx dissipates the membrane potential, allowing more protons to be pumped out of the cell. This means that K^+ influx is required to raise the intracellular pH.

This is seen in Fig. 15.3, where the addition of K^+ to E. coli cells suspended in media of low pH caused an immediate increase in intracellular pH, which stabilized at approximately pH 7.6. There is much more uncertainty about how neutrophiles might *decrease* their intracellular pH. Potassium ions and sodium ions may play a role in lowering the intracellular pH by bringing protons back into the cell via the H^+/K^+ and H^+/Na^+ antiporters (Fig. 15.2). However,

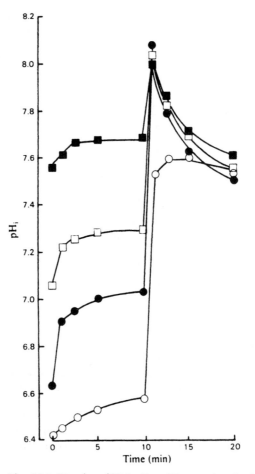

Fig. 15.3 Uptake of K^+ in E. coli is associated with pH homeostasis. Washed E. coli cells were suspended in buffer without K^+ at pH 5.3 (open circles), 6.8 (solid circles), 7.15 (open squares), or 7.6 (solid squares). The cytoplasmic pH (pH_{in}) was determined by using the distribution of a weak acid. Glucose was added shortly after 0 min, and K^+ was added at 10 min. Potassium uptake was complete within 10 min of being added. Immediately upon the addition of K^+, the intracellular pH, which had been lowered by the extracellular pH, rose and stabilized at approximately pH 7.6. The rise in the cytoplasmic pH was due to pumping protons out of the cell in response to the depolarization of the membrane by the influx of potassium ions. The mechanism for acidification of the cytoplasm (i.e., recovery from overshoot of the pH_{in}) is unknown. *Source*: Kroll, R. G., and I. R. Booth. 1983. The relationship between intracellular pH, the pH gradient and potassium transport in *Escherichia coli*. *Biochem. J.* 216:709–716.

the evidence for this suggestion is not as strong in neutrophiles as it is in alkaliphiles.

In alkaliphiles the Na+/H+ antiporter acidifies the cytoplasm

Since alkaliphiles live in a very basic medium, their main problem is keeping a cytoplasmic pH more acid than the external pH, perhaps by as much as 2 units. In other words, they must always be bringing protons into the cell. (See ref. 8 for a review of bioenergetics in alkaliphiles, including pH homeostasis.) In contrast to research with neutrophiles, a strong case (reviewed in ref. 5) can be made for the acidification of the cytoplasm of alkaliphiles by Na+/H+ antiporters (Fig. 15.2). When alkaliphiles are placed in a medium without Na+ at pH 10 or 10.5, the internal pH quickly rises to the value of the external pH. However, when Na+ is present in the external medium, the internal pH does not rise upon shifting to the more basic medium. (See note 9 for more detail.)

Furthermore, mutants of alkaliphiles that cannot grow at pH values above 9 are defective in Na+/H+ antiporter activity. The antiporter is electrogenic (H+ > Na+) and in this case is driven by the $\Delta\Psi$, the membrane potential, which is generated by the primary proton pumps of the respiratory chain. The sodium ion circuit is completed when sodium ion enters the cell via Na+/solute symporters that are also driven by the $\Delta\Psi$. The use of Na+/solute symporters is advantageous because solute transport is driven by the sodium potential rather than the proton potential, the latter being low because of the inverted ΔpH.

pH Homeostasis in acidophiles

Acidophiles differ from other bacteria in that the external pH is several units *lower* than the cytoplasmic pH.[6] The maintenance of the large ΔpH requires an *inverted $\Delta\Psi$ at low pH$_{out}$*; otherwise proton efflux would be limited by a positive $\Delta\Psi$ as well as a low pH$_{out}$, and proton influx would be promoted by a negative $\Delta\Psi$ as well as a high pH$_{in}$. Accordingly, acidophilic bacteria have small membrane potentials, which can be inside positive at acidic pH$_{out}$. For example, the membrane potential of *Thiobacillus ferrooxidans* is +10 mV at an outside of 2, and the membrane potential of *Bacillus acidocaldarius* is +20 to +30 mV at an outside pH of 2.5. However, acidophiles pump protons out of the cell during electron transport, generating a membrane potential that is outside positive as in other bacteria.

How is the membrane potential inverted? It has been suggested that the maintenance of the inverted membrane potential in acidophiles is due to an *inward flux of K+* greater than an outward flux of protons. This might be due to the electrogenic influx of K+ catalyzed by a known ATP-dependent K+ pump. Thus, the method of maintaining a ΔpH in acidophiles and neutrophiles may be similar in relying on K+ influx to depolarize the membrane. It has also been suggested that the *efflux* of K+ in acidophiles may be slowed when pH$_{in}$ falls, thus limiting the entry of protons against a positive $\Delta\Psi$.

Although the protons are not being pumped against a membrane potential when the external pH is acid, they are being pumped against a proton gradient. The ΔpH in acidophiles can be 4 to 5 units, which is equivalent to 240 to 300 mV (i.e., 60 ΔpH). This is a large concentration gradient against which to pump protons. The energy to pump the protons is derived from aerobic respiration. However, iron-oxidizing acidophilic bacteria do not generate sufficient energy during electron transport to pump protons out of the cell because the ΔE_h between Fe^{3+}/Fe^{2+} and O$_2$/H$_2$O is very small (<100 mV). These bacteria appear to regulate their ΔpH by consuming cytoplasmic protons during respiration, rather than by proton pumping. This is discussed in Section 12.4.1.

15.2 Osmotic Pressure and Osmotic Potential

15.2.1 Osmotic pressure

When two solutions are separated by a membrane that allows the passage of water but not of solute, the water will diffuse from the less concentrated to the more concentrated side, thereby equalizing the water concentration. What is happening is that the concentration of water (actually the thermodynamic activity of water) in the less concentrated solution is higher than in the concentrated solution. Thus the water is simply following its concentration gradient. The diffusion of water into the more concentrated solution is called *osmosis*. When water is diffusing from a side with pure water, the pressure that must be applied to stop this osmotic flow is called the *osmotic pressure* and is given the symbol Π. If a solution is sufficiently dilute so that one can discount the interactions between the solute molecules, then $\Pi = RTC_s$, where R is the gas constant, T is temperature in kelvins, and C_s is the molar concentration of solute particles. It is important to point out that C_s represents the concentration of independent particles that contribute to the osmotic pressure. For example, C_s is equal to the sum of the concentration of ions produced when a salt completely ionizes in solution. Thus, C_s is equal to the osmolarity of the solution (osM), and the units can be expressed as moles per liter or osmoles per liter. For example, a one-molar (1 M) solution of NaCl is 2 osmoles per liter because there is one mole each of Na^+ and Cl^- per liter. The gas constant, R, is equal to 0.0281 liter-atm K^{-1} mol^{-1}, and the units of osmotic pressure are usually expressed as millimeters of mercury (mmHg). At higher solute concentrations one must take into account the interactions between solute molecules, and Π/RT is not equal to C_s but rather to the sum of the *effective* molar concentrations of the solutes. Often concentrations are expressed as molality (moles of solute per kilogram of solvent) rather than molarity, and the units of osmotic pressure are given as *osmolality*.

15.2.2 Osmotic potential

For a review of osmotic potential and its regulation, see ref. 10. Csonka has pointed out that the term "osmotic potential" is more useful than "osmotic pressure" in that it emphasizes that water flows from solutions of a high osmotic potential to solutions of a low osmotic potential.[11] [The osmotic potential is numerically equal to the osmotic pressure but, as shown shortly, has a negative sign (eq. 15.3).] The osmotic potential, π, is a function of the activity (a) of the solvent. For water, this would be,

$$\pi = (RT/V_w) \ln a_w \qquad (15.1)$$

where R is the universal gas constant, T is the absolute temperature, V_w is the partial molal

volume of water, and a_w is the activity of water. The activity of pure water is defined as one, making the osmotic potential zero. Solutes tend to lower the activity of water, therefore making the osmotic potential of solutions negative. This is because in an ideal solution (i.e., one in which the interactions between solute and solvent molecules are independent of concentration), the activity of the solvent is equal to its mole fraction. For example, for water,

$$a_w = n_w/(n_w + n_s) \qquad (15.2)$$

where n_w is the number of water molecules and n_s is the number of solute molecules. For dilute solutions, the osmotic potential is related to the molar concentration (moles of solute per liter), C_s, as follows:

$$\pi = -RTC_s \qquad (15.3)$$

Equation 15.3 points out that solutions with higher concentrations of solutes have more negative osmotic potentials.[12] Thus, water flows into these solutions.

15.2.3 Turgor pressure and its importance for growth

Because the cytoplasm of most bacterial cells is much more concentrated in particles than is the medium in which the cells are suspended, the cytoplasm has a more negative osmotic potential (a more positive osmotic pressure) than the medium, and water flows into the cell. The incoming water expands the cell membrane, which exerts a pressure directed outward against the cell wall. The pressure exerted against the cell wall is called the turgor pressure. The turgor pressure is equal to the difference in osmotic pressure between the medium and the cytoplasm.

$$P = \Delta\Pi = \Pi_{in} - \Pi_{out} = RT(osm_{in} - osm_{out}) \qquad (15.4)$$

where P is the turgor pressure, and osm is the total concentration (in units of molarity) of osmotically active solutes in the cells and in the medium.

The turgor pressures in gram-positive bacteria are about 15 to 20 atm and in gram-negative bacteria between 0.8 and 5 atm.[13-15] This, of course, is the reason that bacterial cell walls must be so strong. In bacteria, the tensile strength of the cell wall is due to the peptido-glycan. It is important to realize that the turgor pressure provides the force that expands the cell wall and is necessary for the growth of the wall and cell division.[16] In fact, sudden decreases in turgor pressure brought on by increasing the osmolarity of the suspending medium result in a cessation of growth accompanied by the inhibition of a variety of physiological activities (e.g., nutrient uptake and DNA synthesis). Therefore, the physiological significance of osmotic homeostasis is that it maintains an internal turgor pressure necessary for growth.

Bacteria have the capability of adjusting their internal osmolarity to changes in external osmolarity for the sake of maintaining cell turgor. How bacteria detect differences in external osmolarity and transfer the appropriate signals to the adaptive cellular machinery is largely unknown. The problems associated with analyzing the signaling are discussed in Section 15.2.7. Before we address these problems, we will examine the evidence for osmotic regulation and identify the molecules primarily responsible for maintaining the osmotic differential between the cytoplasm and the external medium.

15.2.4 Adaptation to high-osmolarity media

When cells are placed in media of high osmolarity, they increase the intracellular concentrations of certain solutes called *osmolytes*, thus ensuring that the internal osmolarity is always higher than the external and that cell turgor is therefore maintained. The osmolytes used by bacteria are sometimes called *compatible solutes* to reflect their relative nontoxicity. Some compatible solutes are not synthesized by the cells but are accumulated intracellularly from the medium via transport. Organic compatible solutes that are accumulated via transport but are not synthesized are called *osmoprotectants*.

One of the most important compatible solutes in bacteria is K^+, which has an intracellular concentration high enough to make this ion a major contributor to the internal osmolarity and hence turgor. (See note 17.) Bacteria

that live in high-osmolarity media have proportionally higher intracellular concentrations of K^+ because of uptake. Therefore, K^+ is important for two major aspects of homeostasis: maintenance of cell turgor and maintenance of a ΔpH (Section 15.1.3). Bacteria employ other compatible solutes in addition to K^+. The situation with regard to compatible solutes can be summarized as follows.

Solutes that increase intracellularly in many different bacteria in response to high external osmolarity include K^+, the amino acids glutamate, glutamine, and proline, the quaternary amine betaine (also called trimethylglycine or glycinebetaine), and certain sugars (e.g., trehalose, which is a disaccharide of glucose). Very few chemoheterotrophic bacteria can synthesize betaine de novo; probably the others find it in the environment. Betaine is, however, synthesized by cyanobacteria and phototrophic bacteria and is probably abundant in saline environments, where these phototrophs grow. Betaine is also synthesized by several halophilic and halotolerant archaebacteria. Pathways for the biosynthesis of organic compatible solutes have been reviewed.[15]

Osmotic homeostasis in halobacteria

As described in Section 1.1.1, the halobacteria (archaeons) live in waters having NaCl concentrations from 3 to 5 M. To prevent water from exiting the cell via osmosis and also to maintain a positive internal turgor pressure, the cytoplasm is kept quite salty. However, the cation is K^+ (KCl), rather than Na^+ (NaCl). Potassium ion uptake systems maintain internal K^+ concentrations on the order of 3 M. The Na^+ is exported from the cytoplasm.[18] Of course for the proteins to cope with such a high ionic strength, there must be adaptations in the cytoplasm. It is well known that ionic interactions are weakened in the presence of high concentrations of salt, and this can have a profound effect on the tertiary and quaternary structure of proteins, leading to unfolding and dissociation of subunits. In fact, however, the proteins and structures in extreme halophiles *require* high concentrations of salt (at least 1 M) for stability and activity. Even the cell envelope of halobacteria disintegrates when the salt concentration is lowered. (See ref. 19

for a review of the effect of salt on halophilic macromolecules and ref. 20 for a general review of the biology of the halophilic bacteria.)

Osmotic homeostasis in E. coli

It is possible to observe osmotic homeostasis by rapidly increasing the external osmolarity of the medium and identifying the intracellular compatible solutes that maintain turgor. To illustrate osmotic homeostasis, some experiments performed with *E. coli* will be considered.[21] When *E. coli* is shifted to a medium of high osmolarity, a series of adjustments take place (Fig. 15.4). First, there is an influx of K^+ via the potassium uptake systems discussed in Chapter 16. These are thought to respond to a decrease in turgor pressure.[22]

For there to be an influx of K^+, however, two problems must be solved by the cell. Steps must be taken to (1) preserve electrical neutrality in the cytoplasm and (2) prevent depolarization of the cell membrane, which would lead to a drop in the Δp. In fact, *E. coli* handles these problems in two ways. A transient alkalinization of the cytoplasm as protons are pumped out both prevents depolarization of the membrane and aids in maintaining cytoplasmic neutrality during the early rapid uptake of K^+. However, the cytoplasm reacidifies, and the main reason for cytoplasmic neutrality is the rapid accumulation during reacidification of glutamic acid, which ionizes, providing counterions to the increased levels of K^+.[16,23] The accumulation of glutamic acid is due to either increased synthesis, decreased utilization, or a combination of the two. Thus, the initial major compatible solute is potassium glutamate. However, after several minutes the potassium glutamate is replaced by the newly synthesized sugar, trehalose, which then becomes the major compatible solute. This is due to the excretion of the K^+ and the catabolism or excretion of glutamate, as well as the synthesis of trehalose.

If proline or betaine is present in the medium, then *E. coli* transports these into the cell and replaces the trehalose or the excess K^+, indicating that *E. coli* preferentially uses proline and betaine as osmotic stabilizers rather than trehalose or K^+. (See note 24.) (The K^+ is excreted and the trehalose is catabolized.) The preferential use of some osmoprotectants over others is common among bacteria. In many bacteria betaine is a preferred osmoprotectant that is readily accumulated from media of high osmolarity and can even suppress the uptake of other osmoprotectants, as well as causing the excretion of K^+ and the catabolism of trehalose, thus replacing these osmolytes.

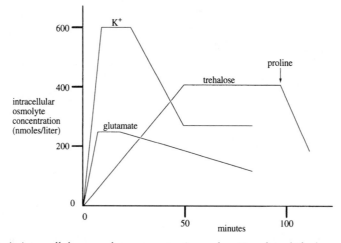

Fig. 15.4 Changes in intracellular osmolyte concentrations when *E. coli* is shifted to a medium of higher osmolarity. Growing cells were shifted to a medium containing 0.5 M NaCl at time zero. The intracellular osmolyte concentrations are the differences between the concentrations in cells subjected to osmotic upshock and control cells that were not. Upon a shift to a medium of higher osmolarity, the cells accumulated potassium ion and synthesized glutamate. The potassium glutamate was replaced by newly synthesized trehalose. The addition of proline to the medium (arrow) caused the displacement of trehalose by proline. Proline was also capable of inducing an early efflux of K^+ if added earlier. *Source:* Data adapted from ref. 21.

Effect of osmolarity on transcription and on activities of enzymes

As part of adaptation to a change in media osmolarity, bacteria synthesize new enzymes or transporters, which are responsible for the biosynthesis of compatible solutes or the transport of these into the cells. Increased transcription of some of the genes activated by an osmotic upshift is due to increased levels of the transcription factor σ^s (sigma S). The transcription factor σ^s is a subunit of RNA polymerase holoenzyme that is responsible for recognizing σ^s-dependent promoters that are activated during osmotic upshifts as well as during starvation. (See Section 2.2.2 and ref. 25 for a review.) The increase in σ^s is regulated by medium osmolarity at the level of translation and protein turnover. Following are a few examples of the effect of osmolarity on transcription and enzyme acitivities.

1. *E. coli* activates enzymes for the synthesis of periplasmic oligosaccharides when shifted to low-osmolarity media. This will be explained in Section 15.2.5.

2. *Staphylococcus aureus* activates a preexisting proline uptake system when shifted to high-osmolarity media.[26] Proline is an osmoprotectant.

3. *E. coli* and *S. typhimurium* increase the transcription of *proU*, an operon that codes for a proline (and betaine) transport system when shifted to high-osmolarity media.

4. Another set of genes whose transcription is increased in high-osmolarity media is the *kdp* operon in *Escherichia*, which codes for a high-affinity K$^+$ transport system. (See Section 18.11 for a discussion of the regulation of the *kdpABC* operon.) As mentioned, K$^+$ is a major compatible osmolyte in most bacteria.

One can ask how bacteria sense changes in the external osmolarity and transmit the appropriate signals to the genome or to certain enzymes. The answers are largely unknown. Perhaps the best-studied signaling system that responds to changes in external osmolarity includes the genes for the OmpF and OmpC porins in *Escherichia* and *Salmonella* (see Section 1.2.3). The total amounts of OmpF and OmpC are fairly constant, but the ratio changes with the osmolarity and temperature of the medium. In high-osmolarity media, the transcription of the *ompF* gene, which codes for the larger OmpF channel, is repressed and the transcription of the *ompC* gene, which codes for the smaller OmpC channel, is increased. The result is a switch to a smaller porin channel in high-osmolarity media.

Why bacteria should switch from one porin to the other is not clear, but it probably has nothing to do with osmotic homeostasis. The smaller OmpC channel is probably an advantage in the intestinal tract, where these bacteria live, because of the presence of toxic molecules (e.g., bile salts).[27] The argument is that in the intestinal tract, the osmotic pressures are higher thus favoring the smaller OmpC channels, whereas in ponds and streams, where the bacteria are also found, the lower osmotic pressures favor the larger OmpF channels, which may allow more efficient uptake of nutrients. Consistent with this hypothesis is the observation that higher temperatures, expected in the intestines of animals as opposed to habitats outside the body, also repress the transcription of *ompF*. Regardless of the physiological significance of the switch in porins, the system is of interest to us here because it is regulated in some way by osmotic pressure, and the signaling pathway from the membrane to the genome is being dissected experimentally.[28] The signaling pathway, which involves a membrane sensor protein called EnvZ, is reviewed in Section 18.7.1.

15.2.5 Adaptation to low-osmolarity media

Thus far we have been considering the adaptation of bacteria to *high*-osmolarity media, which tend to suck water out of the cytoplasm and lower the turgor pressure. As just discussed, bacteria respond to high osmolarity by raising the intracellular concentrations of compatible solutes. Many bacteria have means for adjusting to low osmolarity media, thus limiting cell turgor pressure. Few details are known.

One response of bacteria to low-osmolarity media is to decrease the concentration of cytoplasmic osmolytes. This might occur via specific excretion of osmolytes or as a result of their catabolism. For example, *E. coli* excretes K^+ via special transporters that may respond to cell turgor.[29] An important means of adjusting to a drop in external osmolarity involves mechanosensitive channels, as discussed next.

Mechanosensitive channels

Mechanosensitive (MS) channels have been found in cell membranes of eukaryotes, archaea, and bacteria, and have been most studied in *Escherichia coli*.[30,31] They offer protection against osmotic stress when cells are placed in a sufficiently dilute (hypo-osmotic) medium, that is, when they are hypo-osmotically shocked. The channels are gated, which means that when water rushes into the cell from an hypo-osmotic (dilute) environment, thus increasing the turgor pressure, the increased tension in the cell membrane causes a conformational change in the MS channel proteins such that the channels transiently open. As a consequence, numerous solutes (e.g., K^+, ATP, glutamate, and compatible solutes) exit. This lowers the internal osmotic pressure, and thus less water rushes in and the cells are not osmotically lysed. The channels actually allow transport in both directions. At the same time that there is an efflux of solutes from the cytoplasm, Na^+ and H^+ enter the cell.

Osmotic homeostasis in the periplasm

The periplasm is reportedly filled with a gel whose volume is still a matter of controversy (Section 1.6). (See note 32.) It has been suggested that gram-negative bacteria adapt to low-osmolarity media by raising the osmolarity of the periplasm so that the cytoplasm never actually "sees" the low-osmolarity external medium. In this way, swelling of the cell membrane and concomitant compression of the periplasm are minimized, as well as the turgor pressure across the cell membrane. (The option of lowering the osmolarity of the cytoplasm by excreting solutes is of limited usefulness. The cytoplasm must maintain a minimum concentration of salts and other solutes, approximately 300 mosM, to support growth.) In fact, it has been reported that the periplasm remains as a separate compartment under all conditions

of external osmolarity, with an osmolarity apparently iso-osmotic with that of the cytoplasm.[33] However, the nonspecific diffusion channels formed by the porins, render the outer membrane of *Escherichia* and *Salmonella*, and by inference other gram-negative bacteria, permeable to small molecules of molecular weight less than 600. This being the case, one can ask how bacteria maintain an osmolarity in the periplasm higher than that of the external medium. One possibility is that gram-negative bacteria synthesize and/or accumulate periplasmic osmolytes when grown in low-osmolarity media. Indeed, many gram-negative bacteria besides *E. coli*, including those belonging to the genera *Salmonella, Pseudomonas, Agrobacterium, Acetobacter, Klebsiella, Enterobacter, Bradyrhizobium, Brucella, Xanthomonas, Alcaligenes,* and *Rhizobium*, synthesize periplasmic β-glucan oligosaccharides (called membrane-derived oligosaccharides, or MDO, in *E. coli*) when grown in media of *low* osmolarity.[34–36] It has been shown that *E. coli, Agrobacterium, Rhizobium,* and *Bradyrhizobium* increase the synthesis of periplasmic β-glucans when grown in low-osmolarity media. The enzymes that synthesize the oligosaccharides are constitutive; therefore their *activities* are increased when the cells are grown in low-osmolarity media.[29]

The increase in the synthesis of the periplasmic oligosaccharides when certain bacteria are grown in media of low osmolarity has led to the suggestion that the role of the oligosaccharides is to raise the osmolarity in the periplasm. The glucans generally have anionic groups because they are substituted with phosphorylated compounds (phosphoglycerol, phosphoethanolamine, and phosphocholine, and sometimes succinic acid). The anionic oligosaccharides would be expected to be very effective in raising the osmolarity of the periplasm because cations would accumulate in the periplasm in response to the negatively charged nonpermeable oligosaccharides. (See note 37.) However, *E. coli* mutants unable to synthesize the MDOs show no growth defects in low-osmolarity media.[38] If it is necessary to maintain a high periplasmic osmolarity when the external osmolarity is decreased, then there must exist alternative mechanisms besides the synthesis of MDOs.

15.2.6 Conceptual problems

It has been suggested that the periplasm is maintained iso-osmotic with respect to the cytoplasm. If indeed this is the case, there are important consequences for our understanding of how the cell walls of gram-negative bacteria are able to resist high turgor pressures. It is generally believed that the turgor pressure is exerted across the cell membrane, not the outer membrane, and that the overlying peptidoglycan acts as a strong retainer against which the cell membrane is pressed. If the periplasm were iso-osmotic with the cytoplasm, the turgor pressure would not be across the cell membrane and against the peptidoglycan but, rather, against the outer membrane. However, the outer membrane is not built to withstand high turgor pressures.[39]

If indeed the turgor pressure were exerted against the outer membrane, one must suppose that the peptidoglycan reinforces the outer membrane by being tightly bonded to it at numerous sites. Perhaps the lipoprotein molecules that are covalently bonded to the peptidoglycan are anchored sufficiently to the outer membrane via hydrophobic bonding to provide the needed stability (Section 1.2.3). Another conceptual difficulty regarding a periplasm iso-osmotic with the cytoplasm entails the sensing of turgor pressure by cell membrane proteins: without differential pressure across the cell membrane, it is uncertain how iso-osmoticity could be maintained. Clearly, much more needs to be learned about turgor pressure and osmotic regulation in the periplasm.

15.2.7 What is the nature of the signal sensed by the osmosensors?

The signals to which the putative osmosensors respond are not understood.[5] Indeed, the different osmosensors may not even respond to the same type of signal. Several possibilities for osmosensor responses have been discussed, including membrane proteins that are sensitive to (1) pressure against the peptidoglycan sacculus, (2) membrane stretch, (3) changes in the concentrations of intracellular solutes (e.g., K^+), and (4) water activity. It is important to know whether the periplasm is truly iso-osmotic with the cytoplasm as just discussed. For under these conditions, there should be no pressure differential across the cell membrane, and mechanisms (1) and (2) just listed would not apply.

It should also be understood that after the cells adapt to an osmotic upshift and begin to grow again, the osmotic differential between the cytoplasm and medium is presumably restored, and therefore systems still activated under these circumstances cannot be responding to a change in turgor pressure or related events such as membrane stretching. They could, however, be responding to increased concentrations of specific solutes or some other parameters not dependent upon a changed turgor pressure.

For example, when E. coli is shifted to a high-osmolarity medium, it continues to repress the ompF gene and stimulate the ompC gene, even though adaptation to the high-osmolarity medium has taken place. Similarly, the proU genes that specify proline and glycinebetaine uptake systems in E. coli and S. typhimurium remain induced in high-osmolarity media, also implying that the sensor is not responding to a changed osmotic pressure differential on both sides of the cell membrane. On the other hand, the kdp genes for potassium ion uptake are quickly but only transiently induced by an upshift in osmolarity. This fact, plus the finding that only nonpermeable solutes induce transcription of the kdp operon, suggests that the osmosensor for these genes may indeed detect turgor pressure or something closely related, such as membrane stretch. (See Section 18.8 for a discussion of the kdp gene products.)

15.3 Summary

Bacteria must maintain a fairly constant internal pH and adjust their osmolarity in response to the external osmolarity. Very little is known about the details of how this is done. In broad outline, however, the internal pH is understood to be adjusted by using proton pumps to extrude protons and antiporters that bring protons into the cell in exchange for sodium or potassium ions. What is not clear is precisely how the activities of the pumps and antiporters are regulated by the external pH, although it is reasonable to suggest that they respond to the internal pH, by some sort of feedback mechanism. The acidophilic iron-oxidizing bacteria represent a special problem in that there is very little energy obtainable by Fe^{2+} oxidation to pump protons out of the cell. These bacteria maintain a ΔpH by consuming cytoplasmic protons during oxygen respiration.

Bacteria must maintain an osmotic differential across the cell membrane because the resulting turgor pressure is essential for growth of the cell wall and for cell division. When the turgor pressure is suddenly decreased, a variety of physiological activities come to a halt (e.g., nutrient uptake, DNA synthesis). When the external osmotic pressure is increased (causing decreased turgor), bacteria respond by increasing the concentration of internal osmolytes to raise the internal osmotic pressure and turgor. An important molecule in this regard is potassium ion, which increases, in some cases transiently, in response to an increase in external osmolarity. A counterion must also increase, and in E. coli this is glutamate. In E. coli, the K^+ is subsequently replaced by trehalose. If proline or betaine is in the medium, it replaces the K^+ and trehalose.

It is clear that transcriptional changes occur when bacteria are shifted into media of different osmolarities. For example, when E. coli is grown in high-osmolarity media, the transcription of several genes is increased. These include proU, the operon that codes for the uptake of two osmoprotectants, proline and betaine, and kdp, the operon that codes for the proteins required for the uptake of K^+. Also, when E. coli and S. typhimurium are grown under conditions of high external osmolarity, the gene for the OmpC porin is activated, whereas the gene for the OmpF porin is repressed. This leads to more OmpC and less OmpF. A protein in the cell membrane called EnvZ has a periplasmic domain and is thought to be an osmosensor (Chapter 17).

There is also osmoregulation of the periplasm in gram-negative bacteria. Gram-negative bacteria adapt to media of low osmolarity by synthesizing membrane-derived oligosaccharides in the periplasm. The idea is that the multiple-charged anionic MDO molecules accumulate cations in the periplasm, thus raising its osmo-larity. However, mutants of E. coli defective in MDO synthesis show no growth defects in low-osmolarity media. It must be concluded that other mechanisms of osmoregulation of the periplasm are present and that osmoregulation of the periplasm is not well understood at this time.

Study Questions

1. Na^+/H^+ and K^+/H^+ antiporters are an important way to lower the intracellular pH. What is the evidence for this in alkaliphiles?

2. No one knows for sure how acidophiles establish an inverted membrane potential. How might an inverted membrane potential affect pH homeostasis? What experiments would support the hypothesis that an electrogenic K^+ pump was responsible for the inverted membrane potential in acidophiles?

3. Upon shifting E. coli to a higher osmolarity, there is a transient uptake of K^+. What is the role of K^+ in this regard? What keeps the cytoplasm neutral? How is the membrane potential maintained? Is there a complication with the ΔpH? Describe the role of K^+ in pH homeostasis.

4. What is meant by turgor pressure? What is the evidence that turgor pressure is necessary for cell growth?

5. What is meant by periplasmic osmotic homeostasis? What might be the role of MDOs?

REFERENCES AND NOTES

1. Padan, E., D. Zilberstein, and S. Schuldner. 1981. pH homeostasis in bacteria. Biochim. Biophys. Acta 650:151–166.

2. Bakker, E. P. 1990. The role of alkali-cation transport in energy coupling of neutrophilic and acidophilic bacteria: an assessment of methods and concepts. FEMS Microbiol. Rev. 75:319–334.

3. Krulwich, T. A., A. A. Guffanti, and D. Seto-Young. 1990. pH homeostasis and bioenergetic work in alkaliphiles. FEMS Microbiol. Rev. 75:271–278.

4. Matin, A. 1990. Keeping a neutral cytoplasm; the bioenergetics of obligate acidophiles. FEMS Microbiol. Rev. 75:307–318.

5. Krulwich, T. A., and D. M. Ivey. 1990. Bioenergetics in extreme environments, pp. 417–447.

In: *The Bacteria*, Vol. XII. T. A. Krulwich (Ed.). Academic Press, San Diego, CA.

6. Booth, I. R. 1985. Regulation of cytoplasmic pH in bacteria. *Microbiol. Rev.* **49**:359–378.

7. Harold, F. M., E. Pavlasova, and J. R. Baarda. 1970. A transmembrane pH gradient in *Streptococcus faecalis*: origin, and dissipation by proton conductors and *N,N'*-dicyclohexylcarbodiimide. *Biochim. Biophys. Acta* **196**:235–244.

8. Ivey, D. M., M. Ito, R. Gilmour, J. Zemsky, A. A. Guffanti, M. G. Sturr, D. B. Hicks, and T. A. Krulwich. 1998. Alkaliphile bioenergetics, pp. 181–210. In: *Extremophiles: Microbial Life in Extreme Environments*. K. Horikoshi and W. D. Grant (Eds.). John Wiley & Sons, New York.

9. Careful measurements have been made of the ΔpH of the facultative alkaliphile *Bacillus firmus* to show that the bacteria are capable of maintaining a ΔpH of up to 2 units as the external pH is raised. When pH_{out} was 7.5, pH_{in} was also 7.5. When the external pH was increased from 7.5 to 10.5, the cytoplasmic pH increased only to 8.2. At an external pH of 11.4, the cytoplasmic pH was only 9.5. Growth rates did not begin to slow until the external pH was greater than 10.6. For a review, see ref. 8.

10. Csonka, L. N., and A. D. Hanson. 1991. Prokaryotic osmoregulation: genetics and physiology. *Annu. Rev. Microbiol.* **45**:569–606.

11. Csonka, L. N. 1989. Physiological and genetic responses of bacteria to osmotic stress. *Microbiol. Rev.* **53**:121–147.

12. A derivation of eq. 15.3 can be found in the review article by Csonka.[11]

13. Ingraham, J. L. 1986. Effect of temperature, pH, water activity and pressure on growth, pp. 1543–1554. In: *Escherichia coli and Salmonella typhimurium: Cellular and Molecular Bidogy*, Vol. 1. F. C. Neidhardt et al. (Eds.). ASM Press, Washington, DC.

14. Koch, A. L., and M. F. S. Pinette. 1987. Nephelometric determination of turgor pressure in growing gram-negative bacteria. *J. Bacteriol.* **169**:3654–3668.

15. Walsby, A. E. 1986. The pressure relationships of halophilic and non-halophilic prokaryotic cells determined by using gas vesicles as pressure probes. *FEMS Microbiol. Rev.* **39**:45–49.

16. Koch, A. L. 1991. Effective growth by the simplest means: the bacterial way. *ASM News* **57**:633–637.

17. Glutamate and K^+ are present in high concentrations in most (eu)bacteria. The former is the major organic anion, and the latter is the major inorganic cation. Glutamate is generally synthesized rather than accumulated from the medium, using either glutamate dehydrogenase or glutamate synthase discussed in Section 9.3.1. In *E. coli*, potassium ion is taken up by four different transport systems, including the Kdp and Trk systems discussed in Section 15.3.3. Potassium ion efflux is catalyzed by two systems, the KefB and KefC transporters.

18. Galinski, E. A. 1995. Osmoadaption in bacteria. *Adv. Microbial Physiol.* **37**:273–328.

19. Lanyi, J. K. 1974. Salt-dependent properties of proteins from extremely halophilic bacteria. *Bacteriol. Rev.* **38**:272–290.

20. Vreeland, R. H., and L. I. Hochstein (Eds.). 1992. *The Biology of Halophilic Bacteria.* CRC Press, Boca Raton, FL.

21. Dinnbier, U., E. Limpinsel, R. Schmid, and E. P. Bakker. 1988. Transient accumulation of potassium glutamate and its replacement by trehalose during adaptation of growing cells of *Escherichia coli* K-12 to elevated sodium chloride concentrations. *Arch. Microbiol.* **150**:348–357.

22. Epstein, W. 1986. Osmoregulation by potassium transport in *Escherichia coli*. *FEMS Microbiol. Rev.* **39**:73–78.

23. McLaggan, D., T. M. Logan, D. G. Lynn, and W. Epstein. 1990. Involvement of γ-glutamyl peptides in osmoadaptation of *Escherichia coli*. *J. Bacteriol.* **172**:3631–3636.

24. Betaine is synthesized by oxidizing the hydroxyl group in choline to a carboxyl group, that is:

$$(CH_3)_3N^+\text{–}CH_2CH_2OH \text{ (choline)} \rightarrow$$

$$(CH_3)_3N^+\text{–}CH_2COOH \text{ (betaine)}$$

Betaine and glycinebetaine (*N,N,N*-trimethylglycine) are different names for the same molecule.

25. Hengge-Aronis, R. 1996. Back to log phase: σ^s as a global regulator in the osmotic control of gene expression in *Escherichia coli*. *Mol. Microbiol.* **21**:887–893.

26. Townsend, D. E., and B. J. Wilkinson. 1992. Proline transport in *Staphylococcus aureus*: a high-affinity system and a low-affinity system involved in osmoregulation. *J. Bacteriol.* **174**:2702–2710.

27. Nikaido, H., and M. Vaara. 1985. Molecular basis of bacterial outer membrane permeability. *Microbiol. Rev.* **49**:1–32.

28. Stock, J. B., A. J. Ninfa, and A. M. Stock. 1989. Protein phosphorylation and regulation of adaptive responses in bacteria. *Microbiol. Rev.* **53**:450–490.

29. Bakker, E. P., I. R. Booth, U. Dinnbier, W. Epstein, and A. Gajewska. 1987. Evidence for multiple potassium export systems in *Escherichia coli*. *J. Bacteriol.* **169**:3743–3749.

30. Stokes, N. R., H. D. Murray, C. Subramaniam, R. L. Gourse, P. Louis, W. Bartlett, S. Miller, and I. R. Booth. 2003. A role for mechanosensitive channels in survival of stationary phase: regulation

of channel expression by RpoS. *Proc. Natl. Acad. Sci. USA* 100:15959–15964.

31. Booth, I. R., and P. Louis. 1999. Managing hypoosmotic stress: aquaporins and mechanosensitive channels in *Escherichia coli*. *Curr. Opin. Microbiol.* 2:166–169.

32. The volume of the periplasm in *Escherichia* and *Salmonella* has been reported by some investigators to be as high as 20 to 40% of the total cell volume and as low as 5% of the total cell volume by others. One must conclude that there is some uncertainty regarding the size of the periplasm and that, furthermore, the periplasm of different types of bacteria may vary significantly in physical dimensions.

33. Stock, J. B., B. Rauch, and S. Roseman. 1977. Periplasmic space in *Salmonella typhimurium* and *Escherichia coli*. *J. Biol. Chem.* 252:7850–7861.

34. Miller, K. J., E. P. Kennedy, V. N. Reinhold. 1986. Osmotic adaptation by gram-negative bacteria: possible role for periplasmic oligosaccharides. *Science* 231:48–51.

35. Weissborn, A. C., M. K. Rumley, and E. P. Kennedy. 1992. Isolation and characterization of *Escherichia coli* mutants blocked in production of membrane-derived oligosaccharides. *J. Bacteriol.* 174:4856–4859.

36. Kennedy, E. P. 1996. Membrane-derived oligosaccharides (periplasmic β-D-glucans) of *Escherichia coli*. pp. 1064–1071. In: *Escherichia coli and Salmonella: Cellular and Molecular Biology*. F. C. Neidhardt (Ed.). ASM Press, Washington, DC.

37. Whenever a solution (1) of a nonpermeant ion is separated from another solution (2) (e.g., by a membrane permeable to other ions), there will be an unequal distribution of ions at equilibrium. The compartment with the nonpermeant ion will contain a higher concentration of permeant ions and hence a higher osmotic pressure (or lower osmotic potential). For the case of the periplasm and the outer membrane, suppose the nondiffusible oligosaccharide is R^- and the diffusible ions are K^+ and Cl^-. Then K^+ will passively accumulate in the periplasm as the counterion to R^-. At equilibrium, the concentration of K^+ will be greater on the same side of the membrane as R^- (the periplasm), whereas the concentration of Cl^- will be greater on the other side (the medium). Because the K^+ can diffuse across the outer membrane and R^- cannot, a voltage potential, outside positive, develops across the membrane. This is the Donnan potential, which is really a K^+ diffusion potential. The following equations explain the relationships between the diffusible and nondiffusible ions. At equilibrium,

$$[K^+]_1 = [R^-]_1 + [Cl^-]_1 \qquad (a)$$

and

$$[K^+]_2 = [Cl^-]_2 \qquad (b)$$

Equations a and b describe electrical neutrality in the two solutions.

Because almost all the K^+ that diffuses across the membrane must be coupled with Cl^-, or else the diffusion potentials would prevent further cation (or anion) flow, the diffusion kinetics can be described by a second-order rate equation

$$k[K^+]_1[Cl^-]_1 = k[K^+]_2[Cl^-]_2$$

where k is the rate constant for diffusion through the membrane. Since, however, k cancels out, we have

$$[K^+]_1[Cl]_1 = [K^+]_2[Cl]_2 \qquad (c)$$

But, eq. b states that $[K^+]_2 = [Cl^-]_2$, therefore,

$$[K^+]_1[Cl]_1 = [K^+]_2^2 \qquad (d)$$

But, eq. a states that $[Cl^-]_1 = [K^+]_1 - [R^-]_1$, therefore,

$$[K^+]_1([K^+]_1 - [R^-]_1) = [K^+]_2^2 \qquad (e)$$

and rearranging, we have

$$[K^+]_1^2 - [K^+]_1[R^-]_1 = [K^+]_2^2$$

Thus, the concentration of K^+ on the side without R^- (side 2) is less than the concentration on the side with R^- (side 1) by $[K^+]_1[R^-]_1$. Note that if R^- is multivalent, then even more K^+ will accumulate, because each negative charge on R^- is balanced by a K^+.

38. Kennedy, E. P. 1982. Osmotic regulation and the biosynthesis of membrane-derived oligosaccharides in *Escherichia coli*. *Proc. Natl. Acad. Sci. USA* 79:1092–1095.

39. The difficulty in reconciling a periplasm iso-osmotic with the cytoplasm with the fragility of the outer membrane was suggested to me by Arthur Koch.

16

Solute Transport

Bacterial cell membranes consist in large part of a phospholipid matrix that acts as a permeability barrier blocking the diffusion of water-soluble molecules into and out of cells (see Section 1.2.5). This has the advantage of allowing the bacterium to maintain an internal environment different from the external, and one conducive to growth. For example, metabolites can be maintained at an intracellular concentration that is orders of magnitude higher than the extracellular concentration. This has two important consequences: (1) the promotion of more rapid enzymatic reactions and (2) the retention of metabolic intermediates within the cell. The lipid barrier also minimizes the passive diffusion of ions, including protons, and thus functions to maintain the electrochemical proton and sodium ion gradients that are important for driving ATP synthesis, solute transport, and other membrane activities.

Since the phospholipid presents a permeability barrier, virtually everything that is not lipid soluble enters and leaves the cell through integral membrane proteins that have various names, including transporters, carriers, porters, or permeases. The amino acid sequences of a few transporters, deduced from nucleotide sequences, suggest that these proteins form multiple transmembrane loops that fold in the membrane, perhaps making an internal channel through which the solute is passed. The solute might move along amino acid side chains that face into the channel. For example,

the lactose/proton symporter from *E. coli* probably forms 12 transmembrane loops.[1] The importance of transporters is easy to demonstrate. Mutants that lack a transporter for a required nutrient will grow very poorly in low concentrations of the nutrient. In the natural environment, such mutants would not be expected to survive. But they can be isolated during a screen for slow growers and maintained as laboratory cultures. Transport through the outer membrane of gram-negative bacteria is also an important area of research. The role of porins and of TonB in outer membrane transport is discussed in Sections 1.2.3 and 1.2.4, respectively.

16.1 The Use of Proteoliposomes to Study Solute Transport

Proteoliposomes are artificial membrane vesicles of protein and phospholipid that are of enormous value in studying solute transport, and it is instructive to describe how they are made and some of their properties. There are several methods to prepare proteoliposomes, but they are all similar and rely upon the ability of membrane proteins solubilized in detergent to integrate into phospholipid bilayers when the detergent is removed by dilution or dialysis.[2]

One way to prepare phospholipid bilayers is to first disperse the phospholipids in water, where they spontaneously aggregate to form

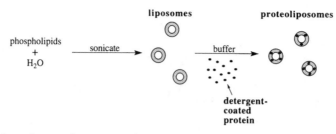

Fig. 16.1 Preparation of proteoliposomes. Phospholipids are dispersed in water and sonicated. This is followed by the formation of small vesicles, each surrounded by a lipid bilayer. The vesicles are mixed with detergent-solubilized protein and diluted into buffer. Proteoliposomes form, with the protein incorporated into the bilayer. The proteoliposomes can be loaded with substrate or ATP by including these in the dilution buffer.

spherical vesicles called liposomes, which consist of concentric layers of phospholipid (Fig. 16.1). The liposomes are then subjected to high-frequency sound waves (sonic oscillation), which breaks them into smaller vesicles surrounded by a single phospholipid bilayer resembling the lipid bilayer found in natural membranes. Then the protein, which has been solubilized in detergent, is mixed with the sonicated phospholipids in the presence of detergent and buffer, and the suspension is diluted into buffer, which lowers the concentration of detergent. The protein leaves the detergent and becomes incorporated into the phospholipid bilayer. The protein and lipid membrane vesicles that form are called proteoliposomes.

One can "load" the proteoliposomes with ATP or other substrates by including these in the dilution buffer. When the proteoliposomes are incubated with solute, they catalyze uptake of the solute into the vesicles, provided the appropriate transporter has been incorporated. The vesicles can be reisolated (e.g., by centrifugation or filtration), and the amount of solute taken into the vesicles can be quantified. Some transporters that have been reconstituted into proteoliposomes and used to demonstrate catalyzed transport are the lactose permease from *E. coli*, the oxalate/formate antiporter from *Oxalobacter formigenes*, the Na^+/H^+ antiporter from *E. coli*, and the histidine permease from *Salmonella typhimurium*.[3–6] Proteoliposomes are also used to study electron transport. For example, they have been important in experiments that establish that certain cytochrome oxidases are proton pumps.

16.2 Kinetics of Solute Uptake

16.2.1 Transporter-mediated uptake

The existence of transporters can be revealed by the kinetics of solute uptake. If one were to add a solute (e.g., an amino acid or sugar) at different concentrations to a bacterial suspension, and plot the initial rate of uptake into the cell as a function of the external concentration of solute, a curve such as that shown in Fig. 16.2 would be generated. Notice that the curve is a hyperbola that approaches a maximum rate. The kinetics for transporter-mediated solute

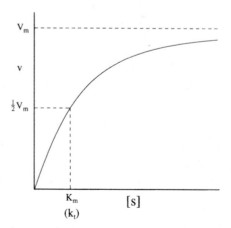

Fig. 16.2 Kinetics of transport. Bacteria are incubated with different concentrations of solute (S), and the initial rate of solute uptake is measured for each solute concentration. The rate (v) approaches a maximum (V_m) as the fraction of transporter molecules bound to solute reaches a maximum. The concentration of solute that gives $\frac{1}{2}V_m$, the K_m (sometimes called k_t), is characteristic for each transporter.

uptake can be rationalized by assuming that the only significant route of entry for the solute is on a limited number of transporters. That is, the solute does not passively leak into the cell to any significant extent. Therefore, the rate at which the solute enters is directly proportional to the fraction of transporters that are occupied with solute. As the external concentration of solute increases, a progressively larger fraction of the transporters binds solute, and the rate of transport increases to a maximum rate (V_{max}), the point at which there are no unloaded transporters.

The concentration of solute that produces one-half the maximum initial rate of transport is called the K_m or sometimes the K_t. Because K_m is frequently assumed to be a measure of the affinity of the solute for the transporter, it is called the affinity constant. However, we will refer to this concentration simply as the solute concentration that gives $\frac{1}{2}V_{max}$. The value of the K_m is characteristic of the transporter and can range from less than 1 µM to several hundred micromolar. The kinetics shown in Fig. 16.2 are described by the Michaelis–Menten equation for enzyme catalysis (see Section 6.2.1).

16.2.2 Uptake in the absence of a transporter

What if there were no transporter and the nutrient simply diffused into the cell? Slow-growing mutants have been isolated that have no functional transporter for a particular nutrient. The kinetics of uptake in these mutants (Fig. 16.3) reflects what one would expect in the absence

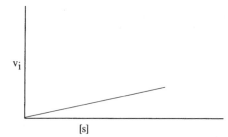

Fig. 16.3 The initial rate of uptake (V_i) of solute (S) in the absence of a transporter. In the absence of a transporter, the rate of solute entry is relatively slow and does not approach a maximum even at very high concentrations of [S]. Rather, it is proportional to the concentration gradient.

of a transporter. Note that the rates of uptake are low and are proportional to the concentration gradient, with no saturation at very high external concentrations of solute.

16.3 Energy-Dependent Transport

A transporter simply facilitates the entry and exit of the solute across the membrane. At equilibrium, it does not bias the transport in any direction and therefore cannot, on its own, cause the accumulation of solute against a concentration gradient. However, we know that many transport mechanisms catalyze the accumulation of solutes into the cell. The internal concentration when the steady state is reached can be several orders of magnitude higher than the external concentration. Of course, this requires energy. The source of energy can be either *chemical*, *light*, or *electrochemical*.

Bacterial transport systems are now divided into two categories, primary and secondary. *Primary transport systems*, driven by an energy-producing (exergonic) metabolic event, include proton translocation drive by ATP, light, or oxidation–reduction reactions (Sections 3.7.1, 3.7.2, 3.8.4, 4.5), light-driven chloride transport (Section 3.9), sodium ion transporting decarboxylases (Section 3.8.1), the uptake of inorganic or organic solutes driven by ATP hydrolysis (described later in this chapter), and the uptake of sugars driven by phosphoryl group transfer from phosphoenolpyruvate, called the phosphotransferase system (described later in this chapter). During transport by the phosphotransferase system, the sugar accumulates inside the cell as the phosphorylated derivative.

Secondary transport systems are driven by electrochemical gradients (e.g., proton and sodium ion gradients). During secondary transport the solute moves "down" an electrochemical ion gradient, usually of protons or sodium ions. (See note 7.) Both primary and secondary transport systems also exist for the efflux of drugs and other toxic substances, and this is described in Section 16.3.5.

Secondary transport is coupled to primary transport by the proton in the following way:

Primary transport

$$H^+_{in} + energy \rightarrow H^+_{out}$$

Secondary transport

$$H_{out}^+ + S_{out} \rightarrow H_{in}^+ + S_{in}$$

The distinction between primary and secondary transport is an important one because it emphasizes a central feature of the chemiosmosis theory: that the coupling between energy-yielding reactions and energy-requiring reactions in the membrane is via ion currents. In the preceding example, the coupling between the energy-dependent uptake of S and the primary energy-yielding reaction is via the proton current, which is most commonly used.

"Active transport" refers to primary transport during which the solute is not chemically modified (e.g., the ATP-dependent uptake of histidine). Active transport is therefore a subclass of primary transport. It should be pointed out that, before the chemiosmotic theory was developed, the distinction between primary and secondary transport was not made. In fact, an older definition of active transport was *any* transport that results in a concentration gradient where the solute is accumulated in a chemically unmodified form. However, the current definition restricts the use of "active transport" to primary transport in which the solute does not chemically change.

Secondary transport is catalyzed by uniporters, symporters and antiporters that use electrochemical ion gradients to accumulate solutes. Examples are illustrated later (see Fig. 16.4). *Uniport* refers to solute translocation in the absence of a coupling ion. For example, the uptake of K^+ down its electrochemical gradient is uniport. *Symport* refers to solute uptake in which two solutes are carried on the carrier in the same direction. An example of this would be the uptake of a solute coupled to the uptake of one or more protons, or sodium ions. *Antiport* refers to the coupled movement of two solutes in opposite directions. For example, an exchange of H^+ for Na^+ on the same carrier is antiport.

16.3.1 Secondary transport

A general description

The transporter functions only when it transports both the ion and the solute in one direction (symport) or the solute in one direction *and* the

ion in the other (antiport) (Fig. 16.4).[8] This is how solute transport is coupled to ion currents. However, there are also transporters that move just an ion along its electrochemical gradient (uniport). In all these cases, the transporter is part of the electrical circuit. This is analogous to saying that for an electrical current to make a motor turn, the current must flow through the motor.

Most bacteria employ both proton and sodium symporters, but mainly the former. (See note 9 for more information.) (Sodium ion symporters are common in bacteria for amino acid transporters.) However, certain bacteria (e.g., alkaliphiles, halophiles, marine bacteria) rely more heavily on sodium symporters. The reason for this becomes clear when one considers their ecological niches. The alkaliphiles live in environments with pH values in the range of 9 to 11. Because their cytoplasmic pH is below 8.5, the ΔpH ($pH_{in} - pH_{out}$) is negative rather than positive. This decreases the Δp. These organisms therefore depend upon the sodium ion and the membrane potential ($\Delta\Psi$) for solute transport. Some of these bacteria also have flagella motors powered by a sodium potential rather than a proton potential.[10,11] Halophilic bacteria require high external NaCl concentrations (3–5 M) to grow. They use predominantly Na^+/solute symporters. Marine bacteria also live in high concentrations of Na^+ (close to 0.5 M), and they rely heavily on Na^+/solute symporters.[12]

The use of ionophores can aid in identifying the ion current that is responsible for secondary transport. This is discussed in Section 3.4. It might be added that in contrast to transporters driven by ATP described in Section 16.3.3, the characterized transporters driven by the Δp (e.g., the Lac permease), are relatively simple in composition, consisting of a single protein that traverses the membrane in several loops.

Energetics of transport

The free energy in joules required to move a mole of solute from outside a cell to inside the cell is given by eq. 16.1, where $[S]_{in}$ is the concentration inside the cell and $[S]_{out}$ is the concentration outside the cell.

$$\Delta G = RT \ln[S]_{in}/[S]_{out} \qquad J/mol$$

$$= 60 \log [S]_{in}/[S]_{out} \qquad mV \text{ at } 30 \,°C \quad (16.1)$$

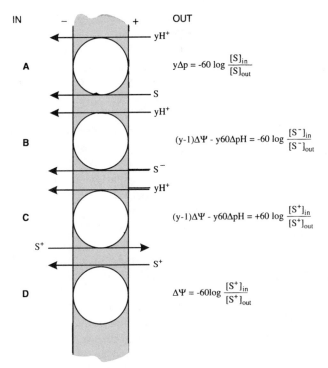

Fig. 16.4 Some examples of solute transport driven by the $\Delta\Psi$ and the ΔpH. (A) Symport of protons with an uncharged solute. (B) Symport of protons with a monovalent anion. (C) Antiport of protons with a monovalent cation. (D) Uniport of a monovalent cation. Another possibility, not shown, is symport of protons with cations. The ratio of protons to solute is given by y, which is assumed to have a value of one or greater.

For example, if $[S]_{in}/[S]_{out} = 10^3$, then $\Delta G = 17.4$ kJ/mol at 30 °C. (See note 13.) This means that at least 17.4 kJ of work must be applied against the concentration gradient (at 30 °C) to move one mole of solute into the cell when the concentration ratio, in/out, is 10^3 and remains as such. This would be the case in a steady state situation, where S_{in} is used as fast as it is brought into the cell. Note that 17.4 kJ does *not* refer to the energy required to move 10^3 moles of S to one side of the membrane to establish a ratio of $10^3/1$. Equation 16.1 can also be expressed as an electrical potential in volts, since $(RT/F)(2.303) = 0.06$ V, or 60 mV at 30 °C. We now discuss some examples of solute transport coupled to an electrochemical gradient (Fig. 16.4).

Some examples of solute transport coupled to an electrochemical gradient

1. Symport of an uncharged solute with protons (Fig. 16.4A)

Many solutes are transported by symport. Assume that the stoichiometric ratio of H$^+$/S is

y. At equilibrium, the force accompanying the diffusion of the solute down its concentration gradient (out of the cell) is $-60 \log [S]_{in}/[S]_{out}$ millivolts, and is balanced by the Δp drawing the solute in the opposite direction (into the cell), so that at 30 °C

$$y(\Delta\Psi - 60 \ \Delta pH)$$

$$= y\Delta p = -60 \log[S]_{in}/[S]_{out} \qquad (16.2)$$

[Alternatively, one can write the total driving force, which in this case would be

$$y\Delta p + 60 \log[S]_{in}/[S]_{out}$$

and set this equal to 0 (equilibrium). Rearrangement of the equation then yields eq. 16.2. (A similar procedure produces eqs. 16.3–16.5 given shortly.)

For example, if the ratio (y) of protons to solute transported is 1:1 and the concentration gradient $[S]_{in}/[S]_{out}$ is 10^3, then the minimum Δp required to maintain that concentration gradient would be -180 mV. Notice that $[S]_{in}/[S]_{out}$ is an exponential function (logarithmic function) of $y\Delta p$. For example, when y is increased to 2, the maximum concentration gradient attained at equilibrium is squared. Theoretically, very large concentration gradients can be maintained by the Δp.

2. Symport of a monovalent anion with protons (Fig. 16.4B)

The Δp can also drive the uptake of anions. However, whether the $\Delta\Psi$ is part of the driving force depends upon whether a net charge is transported. For example, for a monovalent anion, the ratio H^+/R^- must be greater than one if the $\Delta\Psi$ is to be part of the driving force. The relative contributions of the $\Delta\Psi$ and the ΔpH to the driving force are as follows:

$$(y-1)\Delta\Psi - y60\,\Delta pH$$
$$= -60\log[S^-]_{in}/[S^-]_{out} \qquad (16.3)$$

Equation 16.3 is the same as eq. 3.25, whose derivation can be found in Section 3.8.3. (See also eq. 3.23 in Section 3.8.3 for a multivalent anion.)

3. Antiport of a monovalent cation with H^+ (Fig. 16.4C)

The proton:sodium ion antiporter is widespread among the bacteria. It is used for pumping sodium ions out of the cell. The equation is similar to eq. 16.3 except that the sign is changed for the expression for the electrochemical potential of S^+ because it is moving out of the cell; that is, $[S^+]_{in}/[S^+]_{out} < 1$:

$$(y-1)\Delta\Psi - y60\,\Delta pH$$
$$= +60\log[S^+]_{in}/[S^+]_{out} \qquad (16.4)$$

4. Electrogenic uniport of a cation (Fig. 16.4D)

For some transporters, the membrane potential alone can provide the driving force for the uptake of cations (e.g., for K^+ uptake). The relationship between $\Delta\Psi$ and the uptake of cations is

$$\Delta\Psi = -60\log[S^+]_{in}/[S^+]_{out} \qquad (16.5)$$

For a divalent cation, the driving force would be $2\Delta\Psi$, since there is twice as much charge per ion. Equation 16.5 is one form of the Nernst equation.

16.3.2 Evidence for solute/proton or solute/sodium symport

One way to demonstrate coupling of transport to proton or sodium ion influx is to measure the alkalinization of the medium (decrease

in protons) or a decrease in the extracellular Na^+ concentration when bacteria or membrane vesicles are incubated with the appropriate solutes.[14] The changes in the external pH or Na^+ concentration occur as a result of symport with the solute and can be measured with H^+- or Na^+-selective electrodes. It is necessary to ensure that H^+ or Na^+ influx is electroneutral and a membrane potential, which would impede further influx of cations, does not develop. Thus the experiments are done in the presence of a permeant anion (e.g., CNS^-) that serves as a counterion to protons or sodium ions, or K^+ plus valinomycin, whose efflux can exchange for the incoming protons or sodium ions. (This is explained in note 15.)

The experiments may be done so that the ion influx is driven by the solute concentration (i.e., symport) is demonstrated rather than accumulation of the solute against a concentration gradient. An example is the experiment reported by West in 1970.[16] West suspended *E. coli* in dilute buffer and added the sugar lactose, which promoted an increase in the external pH (Fig. 16.5). This result implied that the lactose permease catalyzed a symport of lactose with protons. Other investigators have since performed similar experiments, demonstrating that many sugars and amino acids enter bacteria in symport with either protons or sodium ions.

16.3.3 Primary transport driven by ATP

The following transport systems are driven by ATP (or some phosphorylated derivative in equilibrium with ATP): (1) H^+ transport [i.e., the ATP synthase (ATPase)], which uses the proton motive force (PMF) to make ATP; (2) K^+ influx in *E. coli*; (3) some transport systems in gram-positive bacteria; and (4) transport systems in gram-negative bacteria that use periplasmic binding proteins and are also called shock-sensitive transport systems. (The student should review the discussion of the periplasm in Section 1.2.4.)

Shock-sensitive transport systems in gram-negative bacteria

The shock-sensitive systems rely on periplasmic binding proteins that combine with sugars and

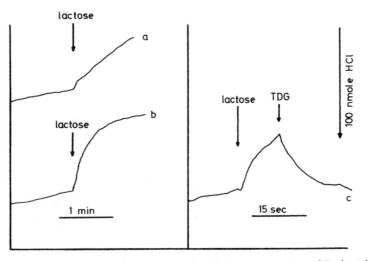

Fig. 16.5 Lactose-dependent proton uptake. Lactose was added to a suspension of *Escherichia coli* without (**a**) and with (**b**) CNS⁻, and the pH was measured with a pH meter. Lactose caused the immediate uptake of protons. Uptake was stimulated by CNS⁻, which prevented a membrane potential, inside positive, from developing. Lactose uptake was inhibited by thiodigalactoside (TDG), which is a competitive inhibitor of the *lac* permease (**c**). These data reflect symport between lactose and protons. *Source*: West, I. C. 1970. Lactose transport coupled to proton movements in *Escherichia coli. Biochem. Biophys. Res. Commun.* **41**:655–661.

amino acids in the periplasm and transfer these to the actual transporters in the cell membrane. They are distinguished from all other transport systems in that they are not functional in cells that have been osmotically shocked, a treatment that makes the outer envelope permeable and causes the release of periplasmic proteins. (The procedure for osmotic shock is described in note 17.)

Shock-sensitive transport systems, also called *periplasmic permeases*, are characteristic of gram-negative bacteria and are responsible for the uptake of a wide range of solutes, including sugars, amino acids, and ions.[18–22] The shock-sensitive transport systems are characterized by very high efficiencies. They are capable of maintaining concentration gradients of approximately 10^5, and K_m values for uptake are in the range of 0.1 to 1 μM. This means that they can scavenge very low concentrations of solute and accumulate these to relatively high internal concentrations. The shock-sensitive systems generally utilize a membrane transporter that is part of a superfamily of transporters found in both gram-negative and gram-positive bacteria, as well as in eukaryotes. We turn next to a discussion of the so-called ATP-binding cassette (ABC) transporters.

ATP-binding cassette (ABC) transporters

For reviews, see refs. 23 and 24. The ABC transporters are found in a range of organisms from bacteria to humans; the name designates their ATP-binding domains. The ABC transporter is a membrane complex consisting of four subunits (Fig. 16.6). Two subunits of the transporter are nonidentical hydrophobic, transmembrane moieties (**a** and **c** in Fig. 16.6.) The other two subunits (**b** in Fig. 16.6) are identical and hydrophilic, each has a nucleotide-binding domain (NBD) that binds ATP on the cytoplasmic side of the membrane. (For the names of these subunits in the histidine transport system, see note 25.)

In addition, there is an ABC signature sequence of amino acids that indicates the presence of the ATP-binding domains. Probably, the binding and/or hydrolysis of ATP at both sites causes a conformational change in the transporter that results in the translocation of the substrate across the membrane. (For evidence that ATP drives uptake, read note 26 and see Fig. 16.7.)

Smaller solutes, such as sugars, amino acids, and small peptides, enter the periplasm by diffusing through a porin pore in the outer

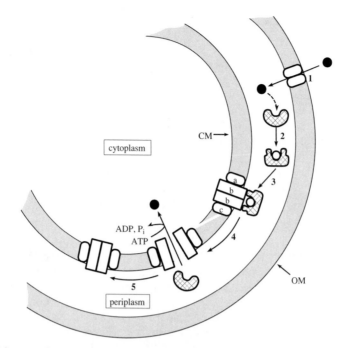

Fig. 16.6 Model for periplasmic transport. (**1**) The solute enters the periplasm through a porin pore in the outer membrane. (**2**) Inside the periplasm the solute binds to a binding protein, which undergoes a conformational change when it binds the solute. (**3**) The binding protein carrying the solute binds to the transporters (**a, b, c**) located in the inner (cell) membrane. (**4**) The transporter is thought to undergo a conformational change that may result in the opening of a pore through which the solute diffuses to the cytoplasm. (**5**) The putative pore closes again when the binding protein is released from the permease. The transporter has a binding site for ATP, which has been demonstrated to be hydrolyzed during histidine transport, and presumably is hydrolyzed during the transport of other solutes. An alternative hypothesis speculates that the binding protein triggers conformational changes in the permease that make a binding site available for the solute. The solute is then passed from one binding site to another on the permease until it is released inside the cell, rather than diffusing through a pore.

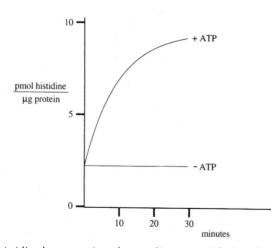

Fig. 16.7 Transport of histidine by reconstituted proteoliposomes. The histidine carrier proteins plus other membrane proteins were solubilized in detergent and incorporated into proteoliposomes in the presence or absence of ATP. Incubation mixtures contained the histidine-binding protein. Histidine was taken up by the proteoliposomes that were loaded with ATP. *Source*: Adapted from ref. 6.

membrane. Larger solutes, such as vitamin B_{12}, are actively transported across the outer membrane into the periplasm by special transporters. The energy to actively transport solute across the outer membrane is provided by the electrochemical gradient of the cell membrane. Once inside the periplasm, the solutes bind to a periplasmic solute-binding protein that delivers the solute to the transporter in the cell membrane. Thus the ABC transporters are more complex than Δp-driven transporters, which consist of a single protein that crosses the cell membrane in several loops. Solute transport via the ABC transporter in gram-negative bacteria can be visualized as occurring in several sequential steps, as illustrated in Fig. 16.6.

Step 1. The solute enters the periplasm through an outer membrane pore (e.g., through a porin). (Consult Section 1.2.3 for a description of the outer membrane and of porins.)

Step 2. The solute binds to a specific periplasmic binding protein to form a solute:binding protein complex. The binding protein undergoes a conformational change that allows it to bind productively to the transporter in the membrane.

Step 3. The liganded binding protein binds to the ABC transporter in the cell membrane, delivering the solute to the transporter. As noted earlier, the membrane-bound transporter is a complex consisting of four proteins, two of which are identical.

Step 4. The transporter complex translocates the solute across the cell membrane, perhaps through a channel that opens transiently. ATP hydrolysis, catalyzed by the transporter complex, occurs at this step. The binding protein is essential not only for delivering the solute, but also for stimulating the ATPase activity. The ATP-binding domains of the transporter are exposed to the cytoplasm.

Step 5. The transporter complex returns to its unstimulated state.

In some cases the membrane-bound transporter interacts with more than one type of binding protein and can transport more than one kind of solute. For example, the histidine permease, besides transporting histidine, also transports arginine via the arginine-ornithine binding protein. Several different binding proteins for branched-chain amino acids use the same membrane transporter.

It is important to note that ABC transporters are part of a superfamily of structurally related proteins and are not confined to gram-negative bacteria. They also occur in gram-positive bacteria, where an equivalent of the periplasmic binding protein that delivers the solute to the transporter is attached to the outside surface of the cell membrane or to the transporter, and in eukaryotic cells.[27] The ABC transporters can be used for either uptake or export, and because of this they are sometimes referred to as "traffic ATPases." ABC transporters also catalyze the export of surface components such as capsular polysaccharides and teichoic acids, as well as drugs and proteins. See Section 16.5, which discusses drug-export systems; the subsection of Section 11.3.5 entitled *Translocation via ABC transporters*; and the coverage of the type I pathway of extracellular protein secretion in Section 17.5.1.

ATP-driven K+ influx

Potassium ion is the principal cation in bacteria, and it not only plays a role in osmotic and pH homeostasis (Chapter 15) but is also a cofactor for many enzymes as well as ribosomes. Bacteria accumulate K^+ to a level several orders of magnitude higher than the external concentrations. Most of what we know about K^+ transport is derived from studies of *E. coli*.[28–30]

There are two transport systems. The major route for K^+ uptake occurs via the TrK system, which is always present (constitutive) and operates at a high rate, but relatively low affinity (high K_m: i.e., 2 mM) for K^+. This transporter requires both a Δp and ATP to function, and the reasons for the dual requirement are not clear. (It has been speculated that the energy source is actually the Δp and that ATP acts as a positive regulator.[31]) When *E. coli* is grown in media in which the K^+ concentrations are very low, the cells synthesize a second K^+ transport system called the Kdp system, which serves to scavenge K^+. However, the induction depends upon the salt-dependent osmolarity of the medium. The threshold concentration of K^+ that prevents induction is higher when the osmolarity is higher.

This means that the Kdp system can be induced by raising the osmolarity of the medium by using salts without altering the external K$^+$ concentrations, as long as the external K$^+$ concentration is not too high. This reflects the important role for K$^+$ as an intracellular osmolyte.[32] (See Section 15.2.4 for a discussion of K$^+$ as an osmolyte.) The Kdp system has a very low K_m for K$^+$ (2 µM) and uses ATP as the energy source.

There are three structural proteins in the Kdp system, KdpA, KdpB, and KdpC, all of which are located in the cell membrane. The genes are encoded in the *kdpABC* operon. KdpA is a membrane-spanning protein exposed to the periplasm (Fig. 16.7). Since mutations in this protein produce changes in the K_m for K$^+$ uptake, it is believed that KdpA binds and translocates K$^+$.[33] KdpA may form a channel through which the K$^+$ moves across the membrane. KdpB, an integral membrane protein, is an ATPase that couples ATP hydrolysis with K$^+$ uptake (Fig. 16.8). The function of KdpC, a peripheral membrane protein, is not known.

There are two proteins that are required for the transcription of the *kdpABC* operon. These are the KdpD protein, located in the cytoplasmic membrane with cytoplasmic domains, and the KdpE protein, located in the cytosol. These two proteins are part of a two-component signaling system that may sense intracellular signals such as low intracellular K$^+$ concentration or an increase in intracellular ionic strength, both of which are thought to stimulate transcription of the *kdpABC* promoter.[34–38] (See ref. 39 and Section 18.8 for a discussion of this point.)

16.3.4 The phosphotransferase system

The PTS differs from the systems driven by ATP or electrochemical gradients in that it catalyzes the accumulation of carbohydrates as the *phosphorylated derivatives* instead of as the free sugar.[40–42] The phosphoryl donor in these transport systems is phosphoenolpyruvate (PEP), an intermediate in glycolysis (Section 8.1). Because the carbohydrate is modified (by phosphorylation), this type of transport is not referred to as active transport but rather as group translocation. The phosphotransferase system is characteristic of anaerobic and facultatively anaerobic bacteria, but it seems to be lacking in many aerobes. It also does not occur in eukaryotes and has not been found in archaea. The overall reaction of the PTS for glucose transport is

$$PEP + carbohydrate_{out}$$
$$\rightarrow pyruvic\ acid + carbohydrate\text{–}P_{in}$$

During the reaction, PEP donates a phosphoryl group to the carbohydrate, which accumulates inside the cell as the phosphorylated derivative. The phosphoryl group is transferred to the carbohydrate via a consecutive series of reactions beginning with PEP as the initial donor. The series, illustrated in Fig. 16.9, consists of the following steps.

Step 1. In the soluble part of the cell, the phosphoryl group is transferred from phosphoenolpyruvate to enzyme I (EI):

$$PEP + E_I \rightarrow P\text{~}E_I + pyruvate$$

Step 2. Enzyme I transfers the phosphoryl group to a small cytoplasmic protein called HPr:

$$P\text{~}E_I + HPr \rightarrow P\text{~}HPr + E_I$$

Enzyme I and HPr are common to all the PTS carbohydrate uptake systems.

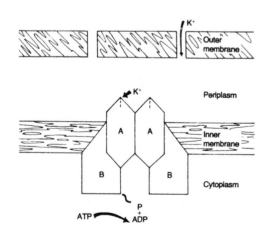

Fig. 16.8 The Kdp system for K$^+$ uptake. K$^+$ is thought to bind to KdpA in the periplasm, after which it is transported through the membrane. The energy for K$^+$ transport is derived from ATP hydrolysis mediated by KdpB. The function of KdpC (not shown) is unknown. *Source:* Epstein, W., and L. Laimins. 1980. Potassium transport in *Escherichia coli*: diverse systems with common control by osmotic forces. *Trends Biochem. Sci.* 5:21–23.

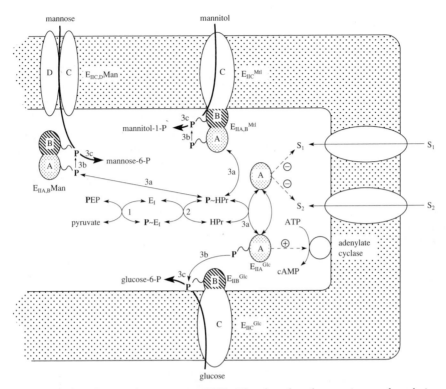

Fig. 16.9 The sugar–phosphotransferase system (PTS). The phosphoryl group is transferred via a series of proteins to the sugar. The sugar-specific carrier proteins in the membrane, II^{Man}, II^{Mtl}, and II^{Glc}, accept the phosphoryl group from P–HPr and transfer it to the sugar-forming mannose-6-P mannitol-1-P, and glucose-6-P. Enzyme II can be a single protein with three domains, A, B, and C, as in the mannitol system, or separate proteins, as in the glucose and mannose systems. The mannose carrier consists of two proteins, C and D. Note that the phosphoryl group travels from HPr to IIA to IIB to the sugar, which is translocated by IIC into the cell in an unknown manner. At some stage during translocation the sugar becomes phosphorylated. However, phosphorylation of the sugar need not take place during translocation per se. That is, the sugar may be phosphorylated on the inside surface of the cell membrane prior to its release into the cytoplasm. Also shown is the stimulation of adenylate cyclase by P–EII^{GlcA} (formerly called P–III^{Glc}) and the inhibition of non-PTS sugar carriers (S_1 and S_2 carriers) by EII^{GlcA}. *Source:* Adapted from Postma, P. W., J. W. Lengeler, and G. R. Jacobson, 1993. Phosphenolpyruvate:carbohydrate phosphotransferase systems of bacteria. *Microbiol. Rev.* **57**:543–594.

Steps 3a, 3b, 3c. The HPr then transfers the phosphoryl group to a carbohydrate specific permease complex in the membrane called enzyme II, which transfers the phosphoryl group to the carbohydrate during carbohydrate uptake into the cell:

$$P\sim HPr + E_{II} \rightarrow HPr + P\sim E_{II}$$

$$P\sim E_{II} + [CH_2O]_{out} \rightarrow E_{II} + [CH_2O–P]_{in}$$

Enzyme II has three domains, A, B, and C. The phosphoryl group is transferred from HPr to domain A, then to domain B, and finally to the carbohydrate in a reaction that requires domain C, which is always an integral membrane protein:

$$P\sim HPr + E_{IIA} \rightarrow P\sim E_{IIA} + HPr$$

$$P\sim E_{IIA} + E_{IIB} \rightarrow E_{IIA} + P\sim E_{IIB}$$

$$P\sim E_{IIB} + carbohydrate_{out} \xrightarrow{E_{IIC}}$$

$$E_{IIB} + carbohydrate–P_{in}$$

How E_{IIC} brings the carbohydrate into the cell is not understood.

Different carbohydrate uptake systems differ with respect to the number of separate proteins that constitute "enzyme II." There can be from one to four proteins, one of which (IIC), is always membrane bound and catalyzes the transport of the carbohydrate into the cell. For example, enzyme II for mannitol uptake

(E_{II}^{Mtl} in Fig. 16.9) is a single membrane-bound protein with three domains, A, B, C. Domain C is in the membrane, whereas domains A and B project into the cytoplasm. However, in some transport systems the three domains are on separate enzyme II proteins. Thus, enzyme II for glucose consists of two proteins, IIA and IIBC (Fig. 16.9). In this case, IIBC is membrane bound, whereas IIA (formerly called III^{Glc}) is cytoplasmic. In a third case ($E_{II}Man$), E_{IIAB} exist as a single cytoplasmic protein, whereas there are two membrane-bound E_{II} proteins, IIC and IID. In the cellobiose PTS in *E. coli*, the enzymes II are three proteins: IIA and IIB in the cytoplasm and IIC in the membrane.

Catabolite regulation by the PTS in enteric bacteria

It has been known for many years that when bacteria using the phosphotransferase system for the transport of glucose are presented with a choice of glucose and another carbon source, they will preferentially utilize the glucose and delay the use of the other carbon source until the glucose has been depleted.[43] This is called glucose repression, or catabolite repression, and is responsible for diauxic growth, described in Section 2.2.4. The phenomenon is not restricted to glucose, since many PTS carbohydrates are used in preference to other carbon sources. (See note 44.) (See Section 18.10.1 for a discussion of a different catabolite repression system in *E. coli* that is due to the Cra system, and Section 18.10.2 for a discussion of catabolite repression in gram-positive bacteria.)

1. The model
The widely accepted model for regulation by the phosphotransferase system in enteric bacteria illustrated in Fig. 16.8 rests on the following postulates.

1. IIA^{Glc} inhibits several enzymes required for carbohydrate metabolism, including certain non-PTS sugar transporters such as the lactose and melibiose transporter, the MalK protein, which is essential for the maltose transport system, and glycerol kinase. $P–IIA^{Glc}$ is dephosphorylated to IIA^{Glc} by $IICB^{Glc}$ during glucose transport. Therefore, according to the model, glucose transport

into the cell inhibits the above-mentioned enzymes. The inhibition by glucose of transport and metabolism of non-PTS carbohydrates is called *inducer exclusion*.

2. $P–IIA^{Glc}$ stimulates the membrane-bound enzyme adenylate cyclase, which makes cyclic AMP (cAMP), which in turn stimulates the transcription of many genes that code for catabolic enzymes. It is actually a complex of cAMP and the cAMP receptor protein (CRP) that binds to the polymerase and regulates transcription. (The CRP protein is also called the catabolite activator protein or CAP.) During glucose uptake by the PTS system, $P–IIA^{Glc}$ is dephosphorylated; hence adenylate cyclase is no longer stimulated, and transcription of cAMP-dependent genes is inhibited. However, there is always a basal level of cAMP synthesized regardless of the carbon source. This is necessary because the transcription of genes required for the metabolism of many PTS sugars also requires cAMP.

3. The model also explains how PTS carbohydrates in addition to glucose can also depress the entry of non-PTS sugars and inhibit the expression of cAMP-dependent genes. The uptake of PTS carbohydrates would be expected to draw phosphoryl groups away from P–HPr toward the sugars. This should decrease the phosphorylation state of IIA^{Glc}, which is phosphorylated by P–HPr. Because the phosphorylation of IIA^{Glc} by P–HPr is reversible, phosphate would be expected to flow from $P–IIA^{Glc}$ to the PTS carbohydrates. The subsequent increase in IIA^{Glc} would be expected to inhibit the enzymes required for uptake and metabolism of non-PTS sugars, and the decrease in $P–IIA^{Glc}$ would be expected to inhibit the adenylate cyclase.

2. Rationale for the model
The phosphotransferase system is believed to be involved in the utilization of glucose-repressed carbon sources on the basis of the original finding that mutants lacking HPr or enzyme I are unable to grow on glucose-repressed carbon sources. In *E. coli*, these carbon sources include lactose, maltose, melibiose, glycerol,

rhamnose, xylose, and citric acid cycle intermediates. The reason for this is that the carbon sources cannot phosphorylate IIA^{Glc}, which stimulates the production of cAMP by activating the adenylate cyclase. The requirement for EI and Hpr for growth on non-PTS sugars is not restricted to *E. coli* or other enterics, although it is less well studied in other bacteria.

Because adding glucose to wild-type cells has the same repressive effect on growth due to non-PTS sugars and citric acid cycle intermediates as does mutations in HPr or enzyme I, it was suggested that P–IIA^{Glc} is required for growth on these substances. The reasoning is that (1) a defect in enzyme I or HPr will result in the inability to phosphorylate IIA^{Glc} and (2) the addition of glucose to wild-type cells should result in the dephosphorylation of P–IIA^{Glc}.

Adenylate cyclase was implicated in the process when it was discovered that the addition of cAMP, the product of adenylate cyclase, could overcome the Hpr mutant phenotype with respect to growth. Knowledge of the requirement for cAMP in the transcription of several genes led to a model postulating a stimulation by P–IIA^{Glc} of the enzyme that synthesizes cAMP (adenylate cyclase) (Fig. 16.9). Mutations in enzyme I or HPr, or the addition of glucose to wild-type cells, should lower the amounts of P–IIA^{Glc} and thus cause a decrease in the levels of cAMP.

Of course, it is possible that the inhibition of the adenylate cyclase is due to an increase in amounts of IIA^{Glc} rather than to a decrease in levels of P–IIA^{Glc}. However, mutations that result in lowered IIA^{Glc} activity (*crr* mutants) do not relieve the inhibition of adenylate cylase by glucose. Also, it was reported in the mid-1970s that the addition of PTS carbohydrates to *E. coli* cells that had been made permeable with toluene inhibited adenylate cyclase activity, which according to the current model can be attributed to the lowering of the levels of P–IIA^{Glc} by the PTS carbohydrates.[45,46] However, it should be emphasized that the evidence for stimulation of adenylate cyclase by P–IIA^{Glc} is thus far indirect, and based primarily on genetic evidence. What is lacking is direct evidence that P–EII^{Glc} binds to the adenylate cyclase and/or stimulates its activity.

Mutants defective in HPr are also defective in the *uptake* of certain non-PTS sugars. The model explains this by postulating that IIA^{Glc} inhibits the carriers for these sugars. The evidence for this is abundant. Uptake in HPr mutants is restored by a mutation in the gene for IIA^{Glc}, which results in lowered activity of IIA^{Glc} (*crr* mutants). Also, IIA^{Glc} is capable of binding to the lactose carrier (permease) and of using liposomes made with the lactose permease to inhibit the transport of the lactose analogue thiomethylgalactoside (TMG).[47,48]

16.4 How to Determine the Source of Energy for Transport

Methods are available to distinguish whether bacterial transport is driven by the electrochemical proton potential (Δp) or by ATP.

1. *The ATP synthase must be inactivated* To determine whether the source of energy is ATP hydrolysis or the Δp, it is necessary to isolate these sources of energy from each other and to perturb them independently (i.e., to increase and decrease the ATP and Δp levels independently of each other). Since bacteria use ATP synthase to interconvert the proton potential and ATP, one cannot vary the ATP levels and Δp independently while the ATP synthase is functioning. To circumvent this problem, investigators either use mutants defective in the ATP synthase ("*unc*" mutants) or add inhibitors of the ATP synthase [e.g., N,N'-dicyclohexylcarbodiimide (DCCD)]. Once this has been done, the ATP *or* the Δp can be decreased or increased independently of the other.

2. *Perturbing the intracellular levels of ATP* How are intracellular ATP levels manipulated? Once the ATP synthase has been inactivated, bacteria must rely completely on substrate-level phosphorylation to synthesize ATP, and the problem becomes one of interfering with substrate-level phosphorylation. To lower the levels of substrate-level phosphorylation, one can (1) starve the cells of endogenous energy reserves whose metabolism produces ATP via substrate-level phosphorylation and (2) add an inhibitor of substrate-level phosphorylation (e.g., arsenate). (See note 49.) Conversely, one can

increase the production of ATP from substrate-level phosphorylation in starved cells by feeding them an energy source (e.g., glucose.) However, one must be careful. Glucose can also stimulate respiration, and respiration can be a source of Δp. Therefore to stimulate substrate-level phosphorylation with glucose without encouraging respiration, respiration should be prevented by using a respiratory inhibitor (e.g., cyanide) or by incubating the cells anaerobically.

3. *Perturbing the Δp* How is the Δp manipulated? One can use ionophores that collapse the proton potential. (However, see note 50.) (See Section 3.4 for a discussion of ionophores.) It is also possible to *increase* electrochemical potentials in de-energized cells (e.g., starved cells, or cells in vesicles). For example, if there is a high K^+ concentration in the cells, the addition of valinomycin will produce a potassium diffusion potential. (See Fig. 3.5.) Also, the addition of certain substrates that feed electrons directly into the electron transport chain (e.g., D-lactate or succinate) will produce a proton potential without increasing substrate-level phosphorylation. A summary of some of these methods is presented in Table 16.1. However, caution must be observed in the interpretation of data derived from the use of inhibitors on whole cells because of indirect effects of the inhibitors.[21,42] Also, ATP cannot be distinguished from other high-energy molecules in equilibrium with ATP (e.g., PEP) when one is studying whole cells. Ideally, one should incorporate the carrier into proteoliposomes and directly test potential energy sources as was described in Section 16.3.3 for the histidine permease.

16.5 Drug-Export Systems

For a review of drug-export systems, the student should consult refs. 51 through 54. Drug-export systems are an important way by which bacteria and fungi become resistant to antimicrobial agents, such as antibiotics, dyes, detergents, disinfectants, and antiseptics. Pathogenic bacteria have a variety of drug-export systems, and these are described in ref. 54. Other mechanisms of resistance are described in note 55. Some drug-export systems (e.g., the system that exports tetracycline antibiotics) are specific for the compound transported. These are called *dedicated drug-export systems*. Most are not specific for the compound transported and export a range of substances that are not necessarily structurally related. These are called *multidrug resistance (MDR) transport systems*.

MDR transport systems exist in gram-negative and gram-positive bacteria, and in fungi, such as the pathogenic yeast *Candida albicans*. An example of an MDR system is the TAP system in *Mycobacterium tuberculosis*. The TAP system pumps out both aminoglycoside and tetracycline antibiotics. Other MDR systems pump out disinfectants and antiseptic compounds as well as antibiotics. Here is an interesting question that the student might consider: how is it that a single transporter can recognize and transport such a variety of chemically unrelated compounds? This is discussed in ref. 56. As one might imagine, bacteria with these MDR systems pose a severe problem in hospital settings and are responsible for nosocomial infections (infections acquired in hospitals).

The two major classes of MDR systems, distinguished from each other by the source of energy for transport, are called primary and secondary. The *secondary transport systems* are energized by proton antiport (efflux of toxic

Table 16.1 Summary of data that can distinguish between energy sources for transport using whole cells

Effect on transport	Δp	ATP
Stimulated by		
Glucose	No[a]	Yes
D-Lactate	Yes	No
Inhibited by		
DNP[b]	Yes	No[c]
Arsenate	No	Yes[d]

[a] Respiration must be inhibited.

[b] One can use any combination of ionophores that abolishes both the ΔpH and the $\Delta \Psi$.

[c] The ATP synthase must be inhibited.

[d] Arsenate prevents ATP formation via substrate-level phosphorylation. This is the only source of ATP when the ATP synthase is not operating.

substance coupled to influx of protons), or in some cases sodium antiport, via the proton or sodium electrochemical potential. The proton antiporter drug efflux systems are the major systems for exporting drugs in bacteria. However, sodium antiporters are also important and have been discovered in several bacteria, including *Escherichia coli*, *Neisseria gonorrhoeae*, *N. meningitidis*, *Vibrio cholerae*, and *V. parahaemolyticus*. The sodium antiporter drug efflux systems are called the MATE (multidrug and toxic compound extrusion) systems. The *primary transport systems*, which are energized by ATP hydrolysis are present in prokaryotes and are the dominant drug efflux systems in eukaryotes, including pathogenic fungi and parasitic protozoa. The primary transport systems use an ATP-binding cassette transporter, as described in Sections 16.3.3 and 17.5.1. All the transporters are integral cell membrane proteins with several transmembrane loops.

Drugs and certain other substances can be exported without entering the periplasm by means of a protein called the *membrane fusion protein (MFP)*. The MFP, a periplasmic protein anchored to the cell membrane, links the cell membrane-bound transporter with an outer membrane protein that facilitates exit through the outer membrane.[57] It can be thought of as part of a channel bridging the periplasm. An example of an MFP system is the type I protein secretion system diagrammed in Chapter 17 (see Fig. 17.4).

As an example of a multidrug system that uses an MFP, and is powered by the Δp, consider the AcrAB–TolC system in *Escherichia coli*, which is the major system for multiple drug resistance in *E. coli*.[58] (The homologue in *Pseudomonas aeruginosa* is MexAB–OprM.) The AcrAB–TolC system pumps out dyes, detergents, bile salts, and antibiotics such as tetracycline, chloramphenicol, β-lactams, novobiocin, and erythromycin, all coupled to the influx of protons. AcrB is the cell membrane transporter, AcrA is the periplasmic MFP, and TolC is a transperiplasmic and outer membrane "channel" protein. Figure 16.10 presents a model for how the AcrAB–TolC complex captures drugs laterally from the outer leaflet of the inner membrane and pumps them out of the cell.

(There is evidence that certain drugs, such as aminoglycosides, can be captured from the periplasm without entering the cell membrane.[59]) TolC is suggested to be a tunnel that extends through the outer membrane and at least half the periplasmic space, where it is thought to link up with the periplasmic domain of AcrB. For more discussion about TolC, see the discussion of the type I pathway for protein secretion in Section 17.5.1.

16.6 Bacterial Transport Systems in Summary

Figure 16.11 summarizes the transport systems used by bacteria. These include primary transport systems (driven by chemical energy) and secondary transport systems (driven by electrochemical ion gradients). Note the role of the proton circuit for secondary transport. Most uptake transport is in symport with protons, although several symporters use Na^+ instead of H^+. The return of the Na^+ to the extracellular space requires a proton circuit in most bacteria.

16.7 Summary

To bring solutes into or out of the cell, bacteria employ several different types of energy-dependent transport system. A single bacterium may have representatives of all the transport systems except the shock-sensitive systems, which are present only in gram-negative bacteria. Transport systems may be classified as being either primary or secondary, depending on how they are coupled to the energy source. The difference is that primary transport systems are directly coupled to an energy-generating reaction (e.g., ATP or PEP hydrolysis), whereas secondary transport systems are energized by an existing electrochemical gradient (e.g., a proton or sodium ion gradient), itself having been produced by energy-yielding reactions such as ATP hydrolysis or redox reactions. That is, secondary transport is indirectly coupled to an energy-yielding chemical reaction by ion currents. (This is one of the predictions of the chemiosmotic theory.) Active transport is defined as primary transport

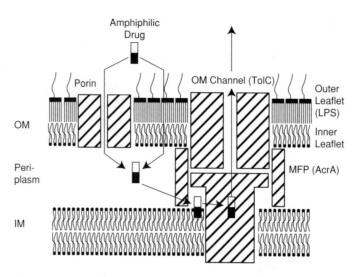

Fig. 16.10 The AcrAB–TolC drug efflux system in *Escherichia coli*, where AcrB is the cell membrane transporter, AcrA is the membrane fusion protein, and TolC is the outer membrane channel. The model shows an amphiphilic drug entering the periplasm through a porin; the lipophilic portion is solid black. The drug then partitions into the lipid phase of the inner membrane and from there travels into AcrB, from where it is pumped out of the cell through TolC. Thus, the drug can be removed from the periplasm and need not enter the cytoplasm. Abbreviations: OM, outer membrane; IM, inner membrane; MFP, membrane fusion protein; LPS, lipopolysaccharide; RND, resistance-nodulation-division. *Source*: Yu, E. W., J. R. Aires, and H. Nikaido. 2003. AcrB multidrug efflux pump of *Escherichia coli*: composite substrate-binding cavity of exceptional flexibility generates its extremely wide substrate specificity. *J. Bacteriol.* **185**:5657–5664.

in which the solute accumulates in a chemically unmodified form.

ATP hydrolysis drives some primary transport systems, including osmotic shock-sensitive (periplasmic permeases) systems, present in gram-negative bacteria, that transport a range of solutes similar to those systems driven by electrochemical gradients. Primary transport systems consist of a periplasmic binding protein plus three or four membrane proteins. Primary transport systems utilizing ATP include the ATP synthase (translocates protons), the Kdp K^+–ATPase in *E. coli* (K^+ influx), and the multidrug export systems that use an ATP-binding cassette (ABC) system.

The transport of carbohydrates via the sugar–phosphotransferase system driven by phosphoenolpyruvate is also a primary transport system. It is widespread in bacteria, being found in anaerobic and facultative anaerobes. It is absent from strict aerobes, however, and from eukaryotes. This is not an active transport mode, since the sugar is phosphorylated during transport, and therefore the solute accumulates

in altered form. Sometimes this type of transport is called group translocation. Transport by the PTS results in the inhibition of transport of non-PTS sugars and the inhibition of adenylate cyclase, leading to catabolite repression by glucose. This can be explained in terms of the regulatory roles of A^{Glc} (III^{Glc}) and $P–A^{Glc}$ ($P–III^{Glc}$).

There exist other catabolite repression systems besides the one mediated by IIA^{Glc}. Enteric bacteria have a cAMP-independent catabolite repression system mediated by FruR, which is also called Cra. FruR is also an activator of certain genes. Thus FruR represses enzymes required for sugar catabolism and activates enzymes required for gluconeogenesis. Glycolytic intermediates such as fructose-1,6-bisphosphate remove FruR from the FruR-regulated operons, resulting in repression of gluconeogenic genes and stimulation of glycolytic genes. Gram-positive bacteria have a catabolite repression system mediated by HPr(Ser-P), which combines with the protein CcpA to repress catabolite-repressible genes. Hpr(Ser-P) is a phosphorylated

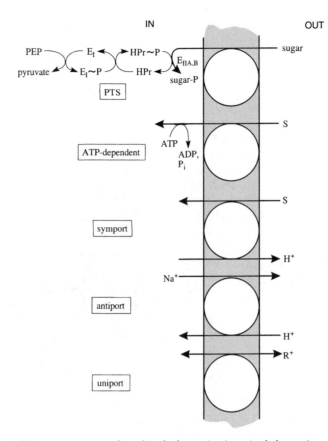

Fig. 16.11 Summary of transport systems found in the bacteria. Any single bacterium uses diverse transport systems to take up sugars, amino acids, ions, vitamins, organic acids, and so on. Many of these systems are powered by electrochemical energy (ion gradients). They catalyze symport, antiport, and uniport. There are also transport systems that use chemical energy instead of electrochemical energy. An example of the latter is the PTS, which is specific for carbohydrates, is driven by phosphoenolpyruvate, and accumulates the sugar as the phosphorylated derivative. Other chemically driven transport systems use ATP or a high-energy molecule in equilibrium with ATP. These include shock-sensitive transport systems (periplasmic permeases) that are found in gram-negative bacteria, and other chemically driven uptake systems found in both gram-negative and gram-positive bacteria.

form of HPr made by a special ATP-dependent kinase that is stimulated by fructose-1, 6-bisphosphate. All these catabolite repression systems involve the repression of genes by carbohydrates.

There also exist catabolite repression systems in which genes are repressed when cells are grown on organic acids, rather than carbohydrates. As discussed in Section 2.2.5 several obligately aerobic bacteria (e.g., *Rhizobium*) will grow first on organic acids when given a mixture of glucose and the organic acid. Clearly systems of catabolite repression mediated by intracellular signals remain to be discovered.

Secondary transport systems require ion symport, antiport, or uniport, and they make use of electrochemical ion gradients to transport a diverse range of solutes including sugars, amino acids, and ions. The coupling ions are either protons or sodium ions, depending upon the transporter. This type of transport is found in bacteria, fungi, plants, protozoa, and higher eukaryotes. It is the simplest of the transport systems and consists of a single membrane-spanning protein.

The type of transport system used by a bacterium depends upon the organism, not the solute being transported. For example, *Pseudomonas* (a strict aerobe) actively transports glucose by means of symport with H^+, whereas the facultative anaerobe *E. coli* transports the same sugar via the phosphotransferase system. Furthermore, a single bacterium can use several different types of transport for the same class of compound. For example, *E. coli* uses PTS for some sugars but uses H$^+$ symport, Na$^+$ symport, or ATP for others.

Harold has speculated regarding the diversity of transport systems in the bacteria, pointing out that whereas transport systems coupled to electrochemical gradients have the advantage of being simple in composition, they are limited by the Δp with respect to the concentration gradients that can be attained.[60] Furthermore, they operate close to equilibrium and can theoretically be reversed if the Δp suffers a transient decrease (e.g., during starvation). On the other hand, transport systems powered by ATP or PEP are structurally more complex but have the advantage of being driven unidirectionally by the relatively large free energies available in ATP and PEP.

Study Questions

1. Suppose the transport of X is driven by symport with H$^+$ in a 1:1 ratio. Assume that X is not charged, the $\Delta\Psi$ is −120 mV, and the ΔpH is 1. What is the expected X_{in}/X_{out}? What would be the answer if X were X^-? If X were X^- and the ΔpH were 2?

   ans. 1,000, 10, 100

2. What is the explanation of the curious fact that mutants in the phosphotransferase system that are defective in enzyme I or HPr are also defective in the transport of some sugars that do not use the PTS?

3. What are proteoliposomes, and how are they prepared?

4. Solute transport might be driven by the Δp or by ATP. Describe some experiments that could distinguish between the two sources of energy. (*Hint*: You will have to manipulate the ATP and Δp separately, and you might want to use proteoliposomes for some of your experiments.)

5. What is the procedure for inducing "osmotic shock"? What is the cellular location of the proteins released by osmotic shock? What are the functions of some of the proteins released by osmotic shock?

REFERENCES AND NOTES

1. Brooker, R. J. 1990. The lactose permease of *Escherichia coli. Res. Microbiol.* **141**:309–316.

2. Racker, E., B. Violand, S. O'Neal, M. Alfonzo, and J. Telford. 1979. Reconstitution, a way of biochemical research; some new approaches to membrane-bound enzymes. *Arch. Biochem. Biophys.* **198**:470–477.

3. Viitanen, P., M. Garcia, and H. Kaback. 1984. Purified reconstituted *lac* carrier protein from *Escherichia coli* is fully functional. *Proc. Natl. Acad. Sci. USA* **81**:1629–1633.

4. Maloney, P. C., V. Anantharam, and M. J. Allison. 1992. Measurement of the substrate dissociation constant of a solubilized membrane carrier. *J. Biol. Chem.* **267**:10531–10536.

5. Taglich, D., E. Padan, and S. Schuldiner. 1991. Overproduction and purification of a functional Na$^+$/H$^+$ antiporter coded by *nhaA* (ant) from *Escherichia coli. J. Biol. Chem.* **266**:11289–11294.

6. Bishop, L., R. Agbayani Jr., S. V. Ambudkar, P. C. Maloney, and G. F.-L. Ames. 1989. Reconstitution of a bacterial periplasmic permease in proteoliposomes and demonstraton of ATP hydrolysis concomitant with transport. *Proc. Natl. Acad. Sci. USA* **86**:6953–6957.

7. The proton gradient is created by a primary transport system. Some bacteria can create sodium gradients with a primary transport system, but most create a sodium gradient by converting a proton gradient into a sodium ion gradient by using a H$^+$/Na$^+$ antiporter.

8. Wright, J. K., R. Seckler, and P. Overath. 1986. Molecular aspects of sugar:ion cotransport. *Annu. Rev. Biochem.* **55**:225–248.

9. Bacteria, yeast, and plants use primarily the proton as the coupling ion for symport, whereas animals rely on the sodium ion. This may be because the cell membranes of most bacteria, fungi, and plants create a proton potential by pumping protons out of the cell via a proton-translocating ATPase, whereas the cell membranes of animal cells create a sodium potential by pumping sodium ions out of the cell via a Na$^+$/K$^+$-ATPase, which exchanges Na$^+$ for K$^+$.

10. Hirota, N., and Y. Imae. 1983. Na$^+$-driven flagellar motors of an alkalophilic *Bacillus* strain YN-1, *J. Biol. Chem.* **258**:10577–10581.

11. Imae, Y., and T. Atsumi. 1989. Na$^+$-driven bacterial flagellar motors. *J. Bioenerg. Biomemb.* **21**:705–716.

12. Maloy, S. R. 1990. Sodium-coupled cotransport, pp. 203–224. In: *The Bacteria*, Vol. XII. T. A. Krulwich (Ed.). Academic Press, New York.

13. $R = 8.3144 \, \text{J K}^{-1}\text{mol}^{-1}$; $T = 273.16\text{K} + °\text{C}$; to convert natural logarithms to log$_{10}$, multiply by 2.303.

14. Wilson, D. M., T. Tsuchiya, and T. H. Wilson. 1986. Methods for the study of the melibiose carrier of *Escherichia coli*, pp. 377–387. In: *Methods in Enzymology*. S. Fleischer, and B. Fleischer (Eds.), Vol. 125. Academic Press, New York.

15. The negative charge on SCN$^-$ is delocalized over the three atoms of the molecule, and this allows it to penetrate the lipid bilayer.

16. West, I. C. 1970. Lactose transport coupled to proton movements in *Escherichia coli. Biochem. Biophys. Res. Commun.* **41**:655–661.

17. Gram-negative cells are shocked in the following way. They are first suspended in a hypertonic solution of Tris buffer, EDTA, and sucrose. Such treatment removes much of the supply of divalent metal cations that are holding the lipopolysaccharide together, along with the lipopolysaccharide, and plasmolyzes the cells. Then the cells are rapidly diluted into water or dilute MgCl$_2$ to neutralize the EDTA. This results in the release of the periplasmic proteins. The treatment inhibits cellular functions that depend on periplasmic binding proteins (e.g., ATP-dependent transport of sugars and amino acids). Other cellular functions, including other transport systems, are retained.

18. Ames, G. F.-L. 1988. Structure and mechanism of bacterial periplasmic transport systems. *J. Bioenerg. Biomembr.* **20**:1–18.

19. Ames, G. F.-L. 1990. Energetics of periplasmic transport systems, pp. 225–245. In: *The Bacteria*, Vol. 12. T. A. Krulwich (Ed.). Academic Press, New York.

20. Ames, G. F.-L. 1986. Bacterial periplasmic transport systems: structure, mechanism, and evolution. *Annu. Rev. Biochem.* **55**:397–425.

21. Ames, G. F.-L., and A. K. Joshi. 1990. Energy coupling in bacterial periplasmic permeases. *J. Bacteriol.* **172**:4133–4137.

22. Furlong, C. E. 1987. Osmotic-shock-sensitive transport systems, pp. 768–796. In: *Escherichia coli and Salmonella typhimurium: Cellular and Molecular Biology*, Vol. 1. F. C. Neidhardt et al. (Eds.). ASM Press, Washington, DC.

23. Fath, M. J., and R. Kolter. 1993. ABC transporters: bacterial exporters. *Microbiol. Rev.* **57**:995–1017.

24. Davidson, A. L., and J. Chen. 2004. ATP-binding cassette transporters in bacteria. *Annu. Rev. Biochem.* **73**:241–268.

25. In the specific case of the histidine permease system, the transporter consists of two hydrophobic proteins, HisQ and HisM, which span the membrane, and two identical hydrophilic proteins, HisP. The HisP protein may be a peripheral membrane protein bound to the inner surface of the membrane, or it may span the membrane, being separated from the hydrophobic lipids by the HisQ and HisM proteins. The HisP protein binds ATP and is responsible for ATP hydrolysis. The periplasmic histidine binding protein is called HisJ.

26. Direct evidence that ATP is the source of energy for the transport of histidine via an ABC transporter was obtained by Bishop et al. (see ref. 6). It is instructive to learn how it was done. The investigators extracted the proteins, including the histidine carrier proteins, from the membranes of *E. coli* and incorporated these proteins into proteoliposomes. (See Section 16.1 for a description of proteoliposomes.) The proteoliposome vesicles were sometimes loaded with ATP by including the ATP in the dilution step. When the reconstituted proteoliposomes were incubated with the histidine-binding protein (HisJ) and histidine, they transported histidine, but only when ATP was present (Fig. 16.7). The transport of histidine was not affected by the ionophores valinomycin (and K$^+$), nigericin, or the proton ionophore FCCP, all of which collapse the electrochemical potentials. (Ionophores and their physiological activities are described in Section 3.4.) These experiments demonstrated that histidine uptake via its periplasmic binding protein is driven by ATP and not an electrochemical potential.

27. Boos, W., and J. M. Lucht. 1996. Periplasmic binding protein–dependent ABC transporters. pp. 1175–1209. In: *Escherichia coli and Salmonella: Cellular and Molecular Biology*, Vol. 1, 2nd ed. F. C. Neidhardt et al. (Eds.). ASM Press, Washington, DC.

28. Epstein, W., and L. Laimins. 1980. Potassium transport in *Escherichia coli*: diverse systems with common control by osmotic forces. *Trends Biochem. Sci.* **5**:21–23.

29. Rosen, B. 1987. ATP-coupled solute transport systems, pp. 760–767. In: *Escherichia coli and Salmonella typhimurium: Cellular and Molecular Biology*, Vol. 1. F. C. Neidhardt et al. (Eds.). ASM Press, Washington, DC.

30. Wood, J. M. 1999. Osmosensing by bacteria: signals and membrane-based sensors. *Microbiol. Mol. Biol. Rev.* **63**:230–262.

31. Stewart, L. M. D., E. P. Bakker, and I. R. Booth. 1985. Energy coupling to K$^+$ uptake via the Trk system in *Escherichia coli*: the role of ATP. *J. Gen. Microbiol.* **131**:77–85.

32. Epstein, W. 1986. Osmoregulation by potassium transport in *Escherichia coli. FEMS Microbiol. Rev.* **39**:73–78.

33. Buurman, E. T., K.-T., Kim, and W. Epstein. 1995. Genetic evidence for two sequentially occupied K$^+$ binding sites in the Kdp transport ATPase. *J. Biol. Chem.* **270**:6678–6685.

34. Walderhaug, M. O., J. W. Polarek, P. Voelkner, J. M. Daniel, J. E. Hesse, K. Altendorf, and W. Epstein. 1992. KdpD and KdpE, proteins that control expression of the *kdpABC* operon, are

members of the two-component sensor–effector class of regulators. *J. Bacteriol.* **174**:2152–2159.

35. Voelkner, P., W. Puppe, and K. Altendorf. 1993. Characterization of the KdpD protein, the sensor kinase of the K⁺-translocating Kdp system of *Escherichia coli. Eur. J. Biochem.* **217**:1019–1026.

36. Nakashima, K., H. Sugiura, H. Momoi, and T. Mizuno. 1992. Phosphotransfer signal transduction between two regulatory factors involved in the osmoregulated *kdp* operon in *Escherichia coli. Mol. Microbiol.* **6**:1777–1784.

37. Sugiura, A., K. Hirokawa, K. Nakashima, and T. Mizuno. 1994. Signal-sensing mechanisms of the putative osmosensor KdpD in *Escherichia coli. Mol. Microbiol.* **14**:929–938.

38. Sugiura, A., K. Nakashima, K. T. Tanaka, and T. Mizuno. 1992. Clarification of the structural and functional features of the osmoregulated *kdp* operon of *Escherichia coli. Mol. Microbiol.* **6**:1769–1776.

39. Heermann, R., A. Fohrmann, K. Altendorf, and K. Jung. 2003. The transmembrane domains of the sensor kinase KdpD of *Escherichia coli* are not essential for sensing K⁺ limitation. *Mol. Microbiol.* **47**:839–848.

40. Saier, M. J., Jr, and A. M. Chin. 1990. Energetics of the bacterial phosphotransferase system in sugar transport and the regulation of carbon metabolism, pp. 273–299. In: *The Bacteria*, Vol. XII. T. A. Krulwich (Ed.). Academic Press, New York.

41. Meadow, N. D., D. K. Fox, and S. Roseman. 1990. The bacterial phosphoenolpyruvate:glycose phosphotransferase system. *Annu. Rev. Biochem.* **59**:497–542.

42. Postma, P. W., J. W. Lengeler, and G. R. Jacobson. 1993. Phosphoenolpyruvate:carbohydrate phosphotransferase systems of bacteria. *Microbiol. Rev.* **57**:543–594.

43. Saier, M. H., Jr. 1989. Protein phosphorlation and allosteric control of inducer exclusion and catabolite repression by the bacterial phosphoenolphyruvate:sugar phosphotransferase system. *Microbiol. Rev.* **53**:109–120.

44. Inhibition of transport and metabolism of non-PTS carbohydrates such as lactose, melibiose, maltose, and glycerol (the class I PTS carbohydrates) by PTS carbohydrates is enhanced in mutants of *E. coli* that have less EI activity (leaky *ptsI* strains).

45. Harwood, J. P., C. Gazdar, C. Prasad, A. Peterkofsky, S. J. Curtis, and W. Epstein. 1976. Involvement of the glucose enzymes II of the sugar phosphotransferase system in the regulation of adenylate cyclase by glucose in *Escherichia coli. J. Biol. Chem.* **251**:2462–2468.

46. Peterkofsky, A. and C. Gazdar. 1975. Interaction of enzyme I of the phosphoenolpyruvate:sugar phosphotransferase system with adenylate cyclase of *Escherichia coli. Proc. Natl. Acad. Sci. USA* **72**:2920–2924.

47. Nelson, S. O., J. K. Wright, and P. W. Postma. 1983. The mechanism of inducer exclusion. Direct interaction between purified III^Glc of the phosphoenolpyruvate:sugar phosphotransferase system and the lactose carrier of *Escherichia coli. EMBO J.* **2**:715–720.

48. Osumi, T., and M. H. Saier Jr. 1982. Regulation of lactose permease activity by the phosphoenolpyruvate:sugar phosphotransferase system: evidence for direct binding of the glucose-specific enzyme III to the lactose permease. *Proc. Natl. Acad. Sci. USA* **79**:1457–1461.

49. Arsenate can substitute for inorganic phosphate in the synthesis of 1,3-diphosphoglycerate. The acyl arsenate is quickly chemically hydrolyzed to the carboxylic acid, and ATP is not made.

50. As long as the ATP synthase is inhibited, the ATP levels should not be affected. However, if the ATP synthase is functioning, then collapsing the Δp will shift the equilibrium of the ATP synthase in the direction of ATP hydrolysis.

51. Borges-Walmsley, M. I., and A. R. Walmsley. 2001. The structure and function of drug pumps. *Trends Microbiol.* **9**:71–79.

52. Nikaido, H. 1996. Multidrug efflux pumps of gram-negative bacteria. *J. Bacteriol.* **178**:5853–5859.

53. Putman, M., H. W. van Veen, and W. N. Konings. 2000. Molecular properties of bacterial multidrug transporters. *Microbiol. Mol. Biol. Rev.* **64**:672–693.

54. Borges-Walmsley, M. I., K. S. McKeegan, and A. R. Walmsley. 2003. Structure and function of efflux pumps that confer resistance to drugs. *Biochem. J.* **376**:313–338.

55. Bacteria become resistant to antimicrobial drugs for a variety of reasons in addition to those due to drug-export systems. These include enzymatic alteration of the drug. An example of enzymatic inactivation of drugs is the acylation or phosphorylation of aminoglycoside antibiotics that in the unaltered form bind to the 30S ribosomal subunit and inhibit protein synthesis. For example, resistance to the aminoglycoside streptomycin can be due to phosphorylation of a hydroxyl group on the antibiotic. The aminoglycoside kanamycin has a free amino group and is inactivated by N-acetylation. Similarly, resistance to the antibiotic chloramphenicol can be due to enzymatic acylation of hydroxyl groups in the chloramphenicol which can no longer bind to the 50S ribosomal subunit, with the result that protein synthesis is inhibited. Another example of an enzymatic alteration that destroys the activity of drugs is the opening of the β-lactam ring in penicillin and cephalosporins antibiotics by β-lactamase. In all these cases, the bacterium usually harbors a plasmid that encodes an enzyme that inactivates the

drug, although in some cases the encoding gene is chromosomal. Bacteria also may become resistant to a drug if the drug target is altered by chromosomal genes. For example, bacteria resistant to the antibiotic erythromycin, which ordinarily binds to the 50S ribosomal subunit and inhibits protein synthesis, may make an altered form of the 50S subunit that does not bind to erythromycin. Bacteria resistant to the semisynthetic antibiotic rifampin are known to synthesize an altered RNA polymerase that does not bind rifampin. Another example of a metabolic change that confers resistance is the mechanism of resistance to sulfonamides, drugs that inhibit the synthesis of folic acid. Resistant bacteria develop a pathway to take up preformed folic acid from the medium. A third way for bacteria to become resistant to a drug may enteric the prevention, due to chromosomal genes, of the entry of the drug into the cell. For example, increased resistance to penicillin may be due to alteration of outer membrane porin proteins, resulting in decreased entrance of penicillin.

56. Adler, J., O. Lewinson, and E. Bibi. 2004. Role of a conserved membrane-embedded acidic residue in the multidrug transporter MdfA. *Biochemistry* **43**:518–525.

57. Zgurskaya, H. I., and H. Nikaido. 1999. AcrA is a highly asymmetric protein capable of spanning the periplasm. *J. Mol. Biol.* **285**:409–420.

58. Yu, E. W., J. R. Aires, and H. Nikaido. 2003. A multidrug efflux pump of *Escherichia coli*: composite substrate-binding cavity of exceptional flexibility generates its extremely wide substrate specificity. *J. Bacteriol.* **185**:5657–5664.

59. Lomovskaya, O., and M. Totrov. 2005. Vacuuming the periplasm. *J. Bacteriol* **187**:1879–1883.

60. Harold, F. M. 1986. *The Vital Force: A Study of Bioenergetics*. W. H. Freeman, San Francisco.

17

Protein Transport

Many hundreds of different proteins synthesized by cytosolic ribosomes are destined for transport to various cellular locations, such as the cell membrane, periplasm, outer envelope, and cell wall, or secretion out of the cell. The traffic in transporting these proteins to their correct locations is considerable, and mechanisms must exist to ensure that the proteins go to their correct final destination. For example, cell membranes alone contain approximately 300 different proteins. In gram-negative bacteria there may be 100 various proteins in the periplasm. The outer envelope of gram-negative bacteria is the site of perhaps 50 different proteins. In addition, both gram-negative and gram-positive bacteria transport proteins that are part of surface layers (glycocalyx) and appendages (flagella, fibrils, pili), as well as extracellular hydrolytic enzymes (e.g., proteases, lipases, nucleases, saccharidases).

Pathogenic bacteria may also secrete protein toxins that adversely affect host cells. The toxins may be secreted into the extracellular medium or directly into target cells. In addition to toxins, pathogenic bacteria may secrete hydrolytic enzymes that degrade host connective tissue. The degradation of host connective tissue presumably facilitates the spread of the bacteria. For a recent review of protein secretion in bacteria, the student is referred to ref. 1.

We will begin with some definitions. The transport of proteins into or through a membrane is referred to as *translocation*. If the protein is translocated into the periplasm of a gram-negative bacterium, the process is referred to as *export*. If the protein is transported to the extracellular medium, into another cell, or to the bacterial cell surface, the process is called *secretion*. [The extracellular transport of nonproteinaceous compounds (e.g., end products of fermentation) is called *excretion*.]

What is it about the structure of a protein that determines whether it will be translocated at all, and if so, to what place? What mechanisms are responsible for the translocation of exportable proteins through the cell membrane, which is generally nonpermeable to proteins? What is the source of energy for protein translocation? These are important areas of research not only in prokaryotes but in eukaryotes as well, where proteins synthesized on cytosolic ribosomes are transported to different cell compartments such as mitochondria and chloroplasts, or are secreted out of the cell via the endoplasmic reticulum.

By far the majority of the research on bacterial protein translocation through the cell membrane has been with *Escherichia coli*, which has served as a model system primarily because of the ease of genetic manipulation. Indeed, virtually all the proteins involved in *E. coli* protein translocation through the cell membrane have been purified, and the genes have been cloned. This has allowed the formulation of a model that describes the general features by which most proteins are translocated through the cell membrane. It is called the *Sec system*, or general secretory pathway (GSP). (Even though the Sec system is sometimes referred to as the GSP system, we will see that it transports proteins into the cell membrane and periplasm, but not out of the cell.) Proteins to be translocated by the Sec system have an N-terminal signal peptide that directs them to the membrane-bound Sec translocation apparatus, called SecYEG, and they are often bound to a chaperone protein, for example, SecB, which prevents protein folding and brings the unfolded protein to the translocation apparatus. The proteins are translocated unfolded through the cell membrane.

A second general protein export pathway that is Sec-independent is called the twin-arginine translocation (Tat) protein export pathway because the signal peptides on the proteins have a highly conserved twin-arginine motif (Arg-Arg) near the N-terminus. One of the significant differences between the Tat system and the Sec system is that the Tat system translocates folded proteins, whereas the Sec system translocates unfolded proteins. The Tat system translocates cofactor-bound respiratory and photosynthetic electron transport enzymes (redox proteins) to the periplasm, as well as integral inner membrane proteins. The Sec system as well as the Tat system (see Section 17.4) are found in both gram-positive and gram-negative bacteria.

There are six major secretion pathways to the outside of the cell, and these pathways are described in Section 17.5. The pathways are concerned with gram-negative bacteria, because they deal with how proteins are translocated across the outer membrane. Some of these are Sec dependent, and some are Sec independent.

When proteins are exported into the periplasm, they become folded there. Section 17.6 explains how this occurs.

17.1 The Sec System

17.1.1 The components

Before describing the model for protein export by means of the Sec system (Section 17.1.2.), we will introduced the four components of the system. These are (1) a leader peptide, (2) a chaperone protein, (3) a membrane-bound complex of three proteins (called SecYEG), and (4) a cytoplasmic ATPase, called SecA, that is bound peripherally to SecYEG.[2-8] The combination of SecYEG and SecA, referred to as the *translocase*, is responsible for moving the protein through the cytoplasmic membrane. The SecYE complex forms a protein-conducting channel sometimes referred to as the Sec translocon. (The word "translocon" refers to a protein-conducting channel.) SecG stimulates the transport. Other membrane proteins that are involved in translocation (SecD, SecF, and yajC) play a regulatory role as discussed in note 9 and described in refs, 10 through 12.

The leader peptide

Proteins that are translocated by the Sec system are synthesized with a leader peptide (also called leader sequence or signal sequence) at the amino-terminal end that is removed during or after translocation. The leader peptide plays an important role in the Sec system. It is necessary for attachment and insertion of the protein into the cell membrane. The protein with the leader peptide is called a precursor protein or a *pre-protein*. If it is secreted to the outside, it is sometimes called a *presecretory* protein. There is ample evidence that the leader peptide is necessary for the initial stages of translocation through the membrane. For example, if the leader peptide is altered by amino acid substitutions, or deleted, then translocation is impaired or does not occur at all.

So, what do we know about leader peptides? Leader peptides have three regions: (1) a basic region (positively charged near neutral pH) at the extreme N-terminal end, called the N-domain (1–5 amino acids), which attaches to the cytoplasmic side of the membrane, perhaps to negatively charged phospholipids; (2) a central hydrophobic region (the core), called the H-domain (7–15 amino acids), which inserts into the membrane; and (3) an uncharged C-terminal region, called the C-domain (3–7 amino acids), which contains a recognition site for a peptidase that removes the leader peptide during or after translocation (Fig. 17.1). The primary amino acid sequences of different leader peptides can vary significantly. Although the leader peptide is important for the initial stages

a Met Lys Ala Thr Lys|Leu Val Leu Gly Ala Val Ile Leu Gly Ser Thr Leu Leu Ala Gly△Cys

b (Met) Met Ile Thr Leu Arg Lys|Leu Pro Leu Ala Val Ala Val Ala Ala Gly Val Met Ser Ala Gln Ala Met Ala△Val

c Met Lys Ile Lys Thr Gly Ala Arg|Ile Leu Ala Leu Ser Ala Leu Thr Thr Met Met Phe Ser Ala Ser Ala Leu Ala△Lys

d Met Ser Ile Gln His Phe Arg|Val Ala Leu Ile Pro Phe Phe Ala Ala Phe Cys Leu Pro Val Phe Ala△His

e Met Lys Thr Lys|Leu Val Leu Gly Ala Val Ile Leu Thr Ala Gly Leu Ser Gly Ala Ala△Glu

f Met Lys Lys Ser Leu Val Leu Lys|Ala Ser Val Ala Val Ala Thr Leu Val Pro Met Leu Ser Phe Ala△Ala

g Met Lys Lys|Leu Leu Phe Ala Ile Pro Leu Val Val Pro Phe Tyr Ser His Ser△Ala

Fig. 17.1 Leader peptides of exported proteins in *E. coli*. The leader sequence consists of a basic amino-terminal end that has positively charged lysine and arginine residues, followed by a hydrophobic region. The boundary between the hydrophilic basic region and the hydrophobic region is denoted with a vertical line. The cleavage site for the leader peptide peptidase is at the carboxy-terminal end (△). **a**, Lipoprotein, located in outer membrane. **b**, Phage lambda receptor, located in outer membrane. **c**, Maltose-binding protein located in periplasm. **d**, β-Lactamase, located in periplasm. **e**, Arabinose-binding protein, located in periplasm. **f**, fd phage major coat protein, located in cell membrane. **g**, fd phage minor coat protein, located in cell membrane. Note that although the leader peptides share similar features with respect to charged and hydrophobic regions, the amino acid sequences are not the same. Nor are there any differences that can distinguish between leader sequences of outer membrane proteins, periplasmic proteins, and cell membrane proteins. *Source*: (After Osborn, M. J., and H. C. P. Wu. 1980. Proteins of the outer membrane of gram-negative bacteria. *Annu. Rev. Microbiol.* **34**:369–422. Reproduced, with permission, from the *Annual Review of Microbiology*, vol. 34, 1980, by Annual Reviews Inc.)

of translocation through the membrane, it is clear that it does not determine the final destination of the protein. Two reasons for believing this are as follows.

1. There are no obvious differences in amino acid sequences between leader peptides of periplasmic and outer membrane proteins. One would expect to find such differences if the leader peptide specified the final location of the protein.

2. Exchange of the leader peptide between proteins destined for two different compartments using recombinant DNA technology does not influence their final destination.

Chaperone proteins

When exportable proteins are synthesized, the nascent preprotein binds to a soluble chaperone protein upon leaving the ribosome. The chaperone protein brings the preprotein to the membrane and, very importantly, also prevents the preprotein from assuming a tightly folded configuration or aggregating into a complex that cannot be translocated. The leader peptide

is very important in this regard because it retards the folding of the preprotein, giving the chaperone protein time to bind.

A major chaperone protein in *E. coli* is SecB. SecB binds to the mature domain (not the leader peptide) of the preprotein and prevents premature folding in the cytoplasm and aggregation. The SecB protein recognizes the unfolded topology of the proteins rather than specific amino acid sequences. The SecB protein has an additional role: it binds to SecA at the membrane site of translocation and delivers the preprotein to SecA, which is a peripheral component of the translocase. Although SecB is very important, many proteins translocate by means of the Sec system without requiring SecB; rather, they use other chaperone proteins described in Section 19.1.6.

The translocase: Sec A and SecYEG

The preprotein is transferred from SecB to SecA, which is attached to the membrane-bound portion of the translocase, SecYEG.[13] For a review of the translocase, see ref. 14. It is not known exactly how the translocase moves proteins through the membrane. For this discussion, we will assume that the translocase forms a hydrophilic channel for proteins. The formation of a hydrophilic channel can be rationalized because many of the proteins secreted into the external medium do not have significantly hydrophobic regions that would facilitate their movement through the lipid bilayer.

The translocase consists of an integral membrane protein complex. The purified complex is composed of three different polypeptides, SecY, SecE, and SecG. The SecY protein spans the membrane several times and along with SecE is part of the protein-conducting channel. (SecYE has been referred to as the Sec translocon because it is a protein-conducting channel.) Although SecG does not appear to be absolutely necessary for protein translocation, it is clearly important. When SecG is incorporated into proteoliposomes along with SecY and SecE it has a large stimulatory effect on protein translocation, and antibodies against SecG inhibit protein translocation in everted membrane vesicles.[15]

17.1.2 A model for protein export

Protein export by the Sec system can be divided into several stages, as illustrated in Fig. 17.2. For a review of the mechanism of the Sec transport system, see ref. 14. As mentioned, before the protein is exported, it must bind to a chaperone protein, which brings the protein in an unfolded state to the membrane-bound translocase. One of the chaperone proteins is SecB, which brings many preproteins to the membrane for export through the cytoplasmic membrane.

Step 1. SecA binds to the lipid bilayer and to the protein trimer SecYEG in the cytoplasmic membrane. The combination of SecA and SecYEG is called the *translocase*. After binding to SecYEG, SecA develops a high affinity for the preprotein–SecB complex.

Step 2. SecB delivers the preprotein to SecA. The SecB protein may be released at this step.

Step 3. After binding to the preprotein, SecA undergoes a conformational change that results in the binding of ATP. Although not

shown in Fig. 17.2, it may be that ATP exchanges for a bound ADP that had been previously generated by the ATPase activity of SecA.

Steps 4 and 5. The binding of ATP causes translocation of part of the preprotein through the translocase. It has been reported that a portion of SecA also inserts into the membrane at this time, and deinserts later upon the hydrolysis of the ATP.[16,17] (For a review of the evidence for this, see note 18.) A further discussion of whether SecA becomes inserted into the membrane can be found in ref. 14. The model for translocation is as follows. The amino terminus of the leader peptide leaves SecA and enters the membrane. When this happens, the positively charged amino terminus binds to the negatively charged phospholipid head groups or perhaps to the SecYEG complex and remains attached to the cytoplasmic side of the membrane. The hydrophobic region of the leader sequence spontaneously inserts into the membrane, forming a loop. The N terminus of the leader peptide remains on the cytoplasmic side of the membrane, possibly attached to the negatively charged phospholipid head groups, while the carboxy-terminal end "flips" into the lipid bilayer as the preprotein enters the membrane portion of the translocase (SecYEG). It is envisaged that a channel opens up within the translocase to accommodate the protein. During these initial stages, a small segment of the preprotein (perhaps 20–30 amino acids) is translocated into the membrane. The precise mechanism by which all this occurs is not known. Limited translocation occurs in vitro via nonhydrolyzable analogues of ATP, indicating that it is the binding of ATP rather than its hydrolysis that provides the energy for translocation.[19,20]

Step 6. ATP is hydrolyzed, and this results in the release of the preprotein from SecA and the release of SecA from the membrane. Figure 17.2A indicates that both ADP and P_i are released from SecA as a consequence of the hydrolysis. However, it has been suggested that only P_i is released, and that ADP remains bound and is exchanged for ATP

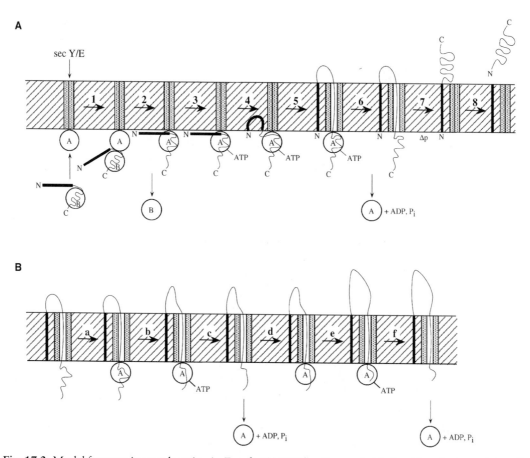

Fig. 17.2 Model for protein translocation in *E. coli*. (A) (**1**) The preprotein binds, either during or immediately after its synthesis, to SecB, forming a preprotein–SecB complex, which moves to the SecA–SecY–E membrane complex (the translocase) and binds to SecA. (**2**) The preprotein is transferred to SecA and SecB is released. (**3**) SecA binds ATP. (**4**) The leader peptide leaves SecA and is inserted into the lipid bilayer as a loop. (**5**) The carboxy terminus of the leader peptide "flips" to the periplasmic side, and the preprotein enters the translocase channel. A short segment of the preprotein is translocated. (**6**) ATP is hydrolyzed and SecA is released into the cytosol. (**7**) Translocation continues, but now driven by the Δp alone. (**8**) The leader peptide is cleaved, releasing the protein into the periplasm. (B) In the absence of a Δp in vitro there can be several rounds of SecA binding and release, promoting ATP-dependent translocation: **a**, binding SecA; **b**, binding of ATP to SecA accompanied by the translocation of a small segment; **c**, hydrolysis of ATP and release of SecA; **d**, rebinding of SecA as the cycle continues (**e**, **f**). Symbols: A, SecA; B, SecB.

during the next cycle of SecA use. It is known that SecA has two binding sites for ATP. See note 21.

Step 7. The rest of the protein is translocated, driven by the Δp. In agreement with this, translocation in respiring cells is immediately inhibited by uncouplers that dissipate the Δp, but the intracellular ATP levels do not change. Furthermore, the direction of translocation is reversed when the polarity of the Δp is reversed in proteoliposomes translocating an outer membrane protein

(proOmpA).[22] However, it may also be that there occur cycles of ATP-driven SecA binding and translocation followed by periods of Δp-driven translocation after SecA is released. (See later subsection entitled *Recycling of SecA*.)

Step 8. The leader sequence is cleaved by a leader peptidase, and the translocated protein is released into the periplasm. The leader peptide must be removed from periplasmic and outer membrane proteins for them to leave the membrane surface, but it need not be removed for translocation per se to take place. According to this model, the initiation of translocation requires the hydrolysis of ATP. The rest of translocation is driven by Δp and does not require ATP. However, see the discussion of the recycling of SecA next.

Recycling of SecA

The initial stages of translocation, which consist of the insertion of the leader peptide into the membrane and a limited amount of translocation (steps 1–5), require SecA and ATP binding. ATP hydrolysis releases SecA (step 6), and the remainder of translocation is driven by the Δp (step 7). However, experiments have demonstrated, that in the absence of a Δp, ATP can drive translocation to completion in vitro. (See note 23 for how this can be demonstrated.) This is because SecA can rebind to the portion of the protein not yet translocated and promote successive rounds of ATP-dependent translocation (Fig. 17.2B, steps a–d). For some proteins, it may be necessary to increase the SecA concentrations to demonstrate complete translocation in the absence of a Δp.[24] The extent of the in vitro requirement for the Δp apparently varies with the preprotein that is used.[25] Although SecA and ATP can be shown to drive translocation to completion in certain in vitro experiments in which there is no Δp, the evidence favors the conclusion that the major driving force for proton translocation in vivo is the Δp, not ATP. (See the preceding description of step 7.)

The foregoing does not imply that SecA binding and rebinding may not be important beyond the initiation stages of translocation. Cycles of SecA rebinding and dissociation may, in fact, occur in vivo, promoting limited translocation in conjunction with translocation driven by the Δp. One possible role for cycles of SecA binding and ATPase-dependent dissociation from the preprotein as it is being translocated might be to unfold untranslocated regions that have assumed a tight tertiary configuration that cannot be threaded through the Sec translocase. One might expect such foldings to occur, since protein secretion need not be coupled to translation, and the untranslocated regions of the protein that are still in the cytosol may fold in a way not suitable for translocation.

Cotranslational translocation

Polysomes translating presecretory proteins associate with membranes, and it seems that some proteins are translocated while they are being translated, although in E. coli most proteins are exported post-translationally. The idea is that the ribosome translating a messenger RNA for a preprotein is so closely associated with a membrane translocase that the polypeptide is threaded through the translocase during translation. If the nascent polypeptide enters the cytosol before engaging the translocase, chaperones might be necessary. Cotranslational translocation may involve the signal recognition particle (SRP), which is described later (Section 17.3).

Protein translocation independent of the Sec proteins

It would be misleading to imply that all proteins translocated by E. coli require the Sec proteins. See the discussion of the Tat system, which translocates folded proteins, in Section 17.4.

17.2 The Translocation of Membrane-Bound Proteins

Thus far the description of protein translocation has been confined to those proteins that cross through the cytoplasmic membrane (also called inner membrane) and ultimately reside in either the periplasm or the outer membrane. What about proteins whose final destination is the cytoplasmic membrane? The export of several of the cytoplasmic membrane proteins is also Sec dependent and occurs in a similar fashion. The question is, what keeps these proteins from being translocated through the membrane into the periplasm?

The answer lies in internal hydrophobic regions of the protein that stop translocation and anchor the protein into the membrane because they bind to the lipid. Sometimes it is the signal sequence itself that anchors the protein in the membrane (Fig. 17.3A,a). These signal sequences differ from those just discussed in not being recognized by the signal peptidase and are therefore not cleaved. At other times it is an internal hydrophobic region called the "signal–membrane anchor" or "stop-transfer"

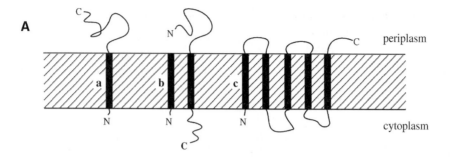

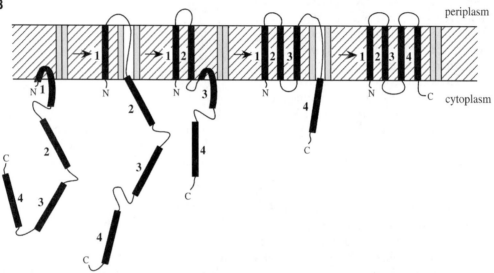

Fig. 17.3 Model for the Sec-dependent translocation of membrane proteins. (A) (a) Protein anchored to the membrane by its leader peptide. (b) Protein anchored to the membrane by a "stop-transfer" signal. The leader sequence has been proteolytically removed and is shown in the membrane. (c) Protein anchored to membrane by alternating leader peptides and stop-transfer signals. (B) Insertion of a protein with four hydrophobic domains. Domain **1** inserts into the membrane as a signal peptide. This opens the translocase channel and initiates translocation. When domain **2** (stop-transfer signal) enters the translocase, the putative channel opens laterally, allowing domain **2** to diffuse laterally into the lipid matrix. The channel then closes. Domain **3** (secondary signal sequence) reinitiates translocation by inserting into the lipid bilayer and reopening the channel. When domain **4** (stop-transfer signal) enters the translocase, it stops translocation and moves laterally out of the translocase channel into the lipid matrix. *Source*: Adapted from Pugsley, A. P. 1993. The complete general secretory pathway in gram-negative bacteria. *Microbiol. Rev.* 57:50–108.

sequence, that anchors the protein after the signal sequence has been cleaved (Fig. 17.3A,b). The "stop-transfer" regions have a stretch of 15 or more hydrophobic amino acids, which vary from one protein to another. Proteins that span the membrane several times do so via a series of alternating uncleaved signal sequences and "stop-transfer" signals (Fig. 17.3A,c). (These include the membrane-bound solute transporters discussed in Chapter 16.)

A model for the insertion of proteins that span the membrane multiple times is shown in Fig. 17.3B. A noncleavable signal sequence initiates insertion into the membrane. At some point, translocation stops as an internal hydrophobic "stop-transfer" signal enters the translocase. According to the model, the translocase opens laterally when the "stop-transfer" signal enters, allowing the diffusion of the signal into the lipid bilayer. Then the channel closes. A downstream secondary signal sequence reinitiates translocation, and the process is repeated until the entire protein has been translocated into the membrane. This is not a well-understood process.

An inner membrane protein called YidC has been suggested to play a role in the movement of transmembrane segments of inner membrane proteins from the Sec-translocase into the lipid bilayer, as well as independently of the Sec translocase. (See refs. 26 and 27.) As discussed in Section 17.4, some integral membrane proteins are translocated in a folded state by a Sec-independent system called the Tat system.

17.3 The E. coli SRP

The components required for the insertion of inner membrane proteins in *E. coli* are not very understood; more is known about the translocation of proteins into the periplasm. In *E. coli* the placement of many of the inner membrane proteins, especially those that do not have extensive periplasmic loops, involves a "signal recognition particle" (SRP) that delivers the preprotein to SecYEG.[28] Homologues of SRP and its receptor are present in all prokaryotes thus far examined, and in all eukaryotes. (See ref. 29 for a review of SRP.) The *E. coli* SRP is a ribonucleoprotein particle that consists of a 4.5S RNA and a 48 kDa protein subunit. The term SRP indicates that the particle recognizes a hydrophobic signal sequence in the protein that will be transferred into the membrane. These proteins do not have a cleavable signal peptide, but instead have hydrophobic N-terminal signal sequences recognized by the SRP. The hydrophobic regions become transmembrane helices in the membrane-integrated protein. A membrane receptor for the SRP, homely FtsY, is also present. Mutants lacking either SRP or FtsY are not viable, although there is no profound effect on the export of proteins to the periplasm or outer membrane. This is consistent with the conclusion that SRP is primarily involved with the assembly of certain inner membrane proteins. (Reviewed in ref. 30.) Both SRP and FtsY are homologous to SRP and its receptor found in eukaryotes.

It has been speculated that the *E. coli* SRP functions in a manner similar to the functioning of SRP in eukaryotes, and that protein translocation and ribosomal translation take place at the same time (cotranslational targeting). According to this model, SRP binds to the hydrophobic N-terminal signal sequences as they emerge from the ribosome, and the complex consisting of SRP and the ribosome-bound nascent chain binds to the SRP receptor, FtsY, which docks at the membrane and delivers the nascent protein–ribosome complex to the SecYEG translocase. Translation continues at the membrane. The newly synthesized protein moves into the SecYEG translocase, and from there into the membrane via the membrane protein YidC. In some cases the protein is delivered directly from SRP to YidC without entering SecYEG. The SRP system of cotranslational protein export is a means of preventing the hydrophobic regions of the newly synthesized membrane proteins from self-aggregating in the aqueous cytoplasm. A brief account of the role of SRP in eukaryotes is summarized in Box 17.1.

17.4 Protein Translocation of Folded Proteins; the Tat System

There is a general protein export pathway that is Sec-independent, post-translational, and driven by the Δp; it translocates folded proteins, as opposed to the Sec system, which translocates only unfolded proteins. It is called the Tat (twin-arginine translocation) protein export pathway because the cleavable N-terminal signal peptides on the proteins have a highly conserved twin-arginine motif, which consists of a pair of consecutive arginine residues (Ser-Arg-Arg-X-Phe-Leu-Lys: see note 31). The translocase is called the Tat translocase. It consists of integral membrane proteins TatA, TatB, TatC, and TatE, and it translocates primarily cofactor-bound respiratory and photosynthetic electron transport enzymes (redox proteins) to the periplasm, but also integral inner membrane proteins. Because these proteins often have cofactors such as $NADP^+$, FAD, molybdocofactors, iron–sulfur and iron–nickel clusters, or heme that are bound in the cytoplasm, they are folded in a tertiary structure prior to translocation by the Tat system. The evidence for this includes data demonstrating that mutants that fail to bind or synthesize a

BOX 17.1 SRP-DEPENDENT PROTEIN TRANSLOCATION ACROSS THE ENDOPLASMIC RETICULUM MEMBRANE IN EUKARYOTES

In eukaryotic cells, proteins that are destined for locations other than the cytosol, nucleus, mitochondria, or chloroplasts are synthesized on ribosomes attached to the endoplasmic reticulum (ER). The proteins are secreted into the lumen of the ER as they are being translated; that is, secretion and translation are coupled. After being secreted into the lumen of the ER, the proteins move to the Golgi complex and from there to various destinations such as the cell membrane, secretory vesicles, or lysosomes.

It has been known since the 1970s that proteins destined to be translocated across the ER membrane have a *signal sequence* at their N-terminal ends which is later removed by a *signal peptidase* on the luminal side of the ER. This is analogous to the leader peptide found on the presecretory proteins in prokaryotes. (Some secreted or membrane proteins in eukaryotes, e.g., hen ovalbumin, have a signal sequence that is internal rather than at the N-terminal end.)

Eukaryotes have a 16S ribonucleoprotein particle called the signal recognition particle (SRP), which binds to the signal sequence as the protein is being synthesized on cytosolic ribosomes and also binds to the ribosome directly. The binding of SRP stops translation. The SRP–ribosome–nascent polypeptide complex then diffuses to the ER, where the SRP binds to a receptor (docking protein) and delivers the ribosome–nascent polypeptide complex to the translocase in the ER. Once the SRP has been removed from the ribosome, translation resumes, and the newly synthesized polypeptide is translocated through the translocase as it is being translated.

The translocase is a heterotrimeric machine called the Sec61 complex, which is similar in some respects to the bacterial SecYEG. Meanwhile the SRP is released into the cytosol and can guide another ribosome–nascent polypeptide complex to the ER.

Eukaryotes also have a post-translational secretory pathway that is SRP independent. It has been speculated that in eukaryotes the post-translational pathways come into play only if the SRP fails to bind to the nascent polypeptide. This can happen because the time of affinity of SRP for the signal sequence decreases as the polypeptide gets longer, and thus there is only a short period during which SRP can bind effectively. As in bacteria, the post-translational pathway appears to involve the use of chaperones analogous to SecB, which prevent the presecretory protein from folding into a conformation that prevents translocation.

References 1 and 2 compare protein translocation in eukaryotes and prokaryotes (Bacteria and Archea). The student is encouraged to read these reviews for a more detailed account of the similarities and differences between the secretory pathways in eukaryotes and prokaryotes.

REFERENCES

1. Rapoport, T. A., B. Jungnickel, and U. Kutay. 1996. Protein transport across the eukaryotic endoplasmic reticulum and bacterial inner membranes. *Annu. Rev. Biochem.* 65:271–303.

2. Pholschröder, M., W. A. Prinz, E. Hartmann, and J. Beckwith. 1997. Protein translocation in the three domains of life: variations on a theme. *Cell* 91:563–566.

cofactor do not export the protein. For a review of the Tat pathway, see refs. 32 through 35, Certain cofactor-containing proteins, such as periplasmic c-type cytochromes, are translocated unfolded by means of the Sec system and acquire the cofactor after export. However, most cofactor-containing proteins use the Tat system.

17.5 Extracellular Protein Secretion

A variety of different proteins are secreted out of the cell, including hydrolytic enzymes such as proteases, nucleases, lipases, and carbohydrases; nonhydrolytic enzymes such as cholera toxin, diphtheria toxin, and pertussis toxin; structural proteins such as pilin and flagellar proteins; and virulence proteins secreted directly into host cells. The hydrolytic enzymes can generate nutrient for the bacterium, and/or they can facilitate the invasion of plant or animal tissues. For a description of what some of the nonhydrolytic toxic proteins do to target cells, see note 36.

There are six major secretion systems that are known for gram-negative bacteria. For a review of these systems see refs. 37 through 42. Two of the secretion systems are *Sec independent*, that is, independent of the GSP. These are the type I and type III systems, and they will be described first. Three of the secretion systems are *Sec dependent*. These are the type II system, the type V (also called autotransporter) system, and the chaperone/usher pathway. These will be discussed second. The sixth system, called the type IV system, can be Sec dependent or Sec independent, depending upon the specific system. This will be discussed last.

The Sec-dependent pathways work in two stages; stage I translocates proteins via the Sec system across the cell membrane into the periplasm, and stage II translocates the proteins from the periplasm across the outer envelope to the outside. See Fig. 17.4 for an overview of some of the secretion systems. Protein secretion has been studied in gram-positive bacteria as well. These include penicillinase and the secretion of various proteases by *Bacillus* species.

17.5.1 The type I pathway

Examples of proteins secreted by the type I system include hemolysin (the toxin that lyses red blood cells and kills other cell types), secreted by pathogenic strains of *E. coli*, proteases by *Erwinia chrysanthemi* and *Pseudomonas aeruginosa*, and leukotoxin (toxin that damages white blood cells) by *Pasteurella haemolytica*. These proteins (1) do not have leader sequences, although the carboxyl terminus of the secreted proteins is recognized by the secretion apparatus and (2) do not require any of

the *sec* genes for secretion. A third characteristic of the system is that the proteins are secreted directly from the cytoplasm to the outside of the cell without ever entering the periplasm. This is done via a secretion apparatus that spans the periplasm.

The secretion apparatus

The proteins required for the secretion apparatus are encoded by secretory genes that have been identified by mutant analysis. The secretory genes are usually contiguous with the structural gene for the protein that is secreted. Since the carboxyl terminus of the secreted protein is required for secretion, it would appear that the secretion apparatus recognizes this region of the protein.

The secretion apparatus consists of three secretory proteins. Two are inner membrane proteins and one is an outer membrane protein. One of the inner membrane proteins is an ATP-binding protein that belongs to the *ABC exporters (ATP-binding cassette)* superfamily of proteins, which also catalyze extracellular polysaccharide export, solute uptake, and drug export. (See Sections 11.3.1, 16.2.2, and 16.5). This protein is sometimes referred to as the *ABC transporter*. ABC transporters actually have two ATP-binding sites. (See the discussion of the structure of the ABC transporter in Section 16.3.3.) The fact that the ABC transporter has ATP-binding sites indicates that ATP binding and/or hydrolysis drives the secretion of the proteins, presumably by causing a conformational change in the transporter. However, it is important to point out that the mechanism of translocation is not well understood.

The other two secretory proteins are called auxiliary proteins. One of these, the membrane fusion protein (MFP), is anchored in the inner membrane and has a periplasmic domain that is believed to connect to the other auxiliary protein and to the outer membrane protein (OMP). The cellular location of the secretory proteins suggest that they may form a continuous channel through the inner and outer membrane, allowing proteins to move directly from the cytoplasm to the external medium without entering the periplasm. The three transporter proteins in the different type I systems have different names, but they are homologous proteins, and the secretion systems can substitute for one another in mutants.[43,44] There are actually families of type I systems. Within a family there is sequence similarity between the components of the secretion apparatus, and substitution is possible. When components from different families are interchanged, there is a significant drop in protein secretion. An illustrative example of a type I system, hemolysin secretion by *Escherichia coli*, is described next.

Hemolysin secretion by E. coli

Hemolysin toxin (HlyA) is an important virulence factor secreted by uropathogenic and enterohemorrhagic *E. coli*. (See note 45 for a description of these diseases.) It inserts into eukaryotic cell membranes, creating pores that result in the leaking out of cytoplasmic contents and the death of the cells. Reference to Fig. 17.4 will clarify the following discussion.

In *E. coli*, the structural gene for hemolysin is *hlyA*. It is located next to *hlyB* and *hlyD*, the genes for the ABC transporter (HlyB) and the auxiliary inner membrane fusion protein (HlyD), respectively. The outer membrane protein, TolC, is a channel that spans the outer membrane and extends deep into the periplasm; it is a narrow, tunnel-like structure, 100 Å long and 100 to 150 Å wide. (TolC is reviewed in ref. 46.) TolC homologues are widespread among gram-

negative bacteria and play an important role in exporting proteins as well as drugs and peptides. (See note 47 for a brief summary.) For a model of TolC functions in the export of drugs, see Section 16.5 and Fig. 16.11. The TolC channel is open to the outside but is ordinarily closed at its periplasmic entrance until a substrate to be transported is brought to it by the translocation machinery. The role of the inner membrane fusion protein, HlyD, which has a periplasmic portion, is to provide contact between the ABC transporter in the inner membrane (HlyB) and the opening of the TolC channel in the periplasm. The model shown in Fig. 17.4 shows that HlyD is part of the transperiplasmic channel. Thus, the ABC transporter (HlyB) is thought to transport the unfolded protein across the cell membrane and into the periplasmic TolC tunnel via HlyD. This is a very dynamic situation, and it has been suggested that the attachment of the transporter apparatus (HlyD) to TolC takes place only when the substrate to be transported binds to the inner membrane ABC transporter (HlyB); this in turn is believed to occur after the membrane fusion protein (HlyD) and the ABC transporter (HlyB) have interacted. A secretion signal in the C-terminal region of the hemolysin recognizes the ABC transporter. It has been suggested that folding of the protein begins in the TolC tunnel. According to the model, the bridge between HylD and TolC collapses after

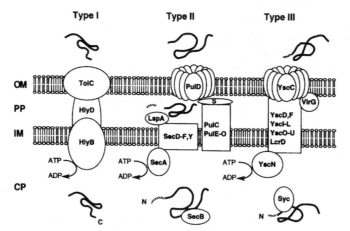

Fig. 17.4 Overview of types I, II, and III secretion systems. Type I is represented by hemolysin secretion by *E. coli*. As pointed out in the text, TolC forms a channel in the outer membrane that projects as a tunnel into at least 50% of the periplasmic space. Type II is represented by pulluninase secretion by *Klebsiella oxytoca*. Type III is represented by Yop secretion by *Yersinia*. OM, outer membrane; PP, periplasm; IM, inner membrane; CP, cytoplasm. SecB and Syc are cytoplasmic chaperone proteins. LspA is a periplasmic peptidase that cleaves off the amino-terminal signal sequence. PulD and YscC (types II and III outer membrane proteins) are homologous. *Source*: Hueck, C. J., 1998. Type III protein secretion systems in bacterial pathogens of animals and plants. *Microbiol. Mol. Biol. Rev.* **62**:379–433.

the protein has been secreted and is rebuilt for the next secretion event. Hydrolysis of ATP via the ABC transporter provides the energy to drive the secretion of the protein.

The gene (*tolC*) for the auxiliary outer membrane secretory protein is not linked to the *hly* genes. This is an exception to the rule that all three secretory genes are linked to the gene for the secreted protein, but TolC is a multifunctional protein and is also used for the transport of other molecules. Following recognition of the protein, the ABC transporter in the inner membrane translocates the protein into the transperiplasmic channel, where folding might begin.

17.5.2 The type III pathway

Type III secretion systems (TTSS) are Sec-independent systems that form an apparatus used for the injection of virulence proteins into eukaryotic host cells in some gram-negative animal pathogenic bacteria, including *Yersinia* spp., *Salmonella* spp., *Shigella* spp., enteropathogenic *Escherichia coli* (EPEC), *Bordetella*

pertussis, and *Pseudomonas aeruginosa*, and plant pathogens such as *Pseudomonas syringae*, *Ralstonia solanacearum*, and *Xanthomonas campestris*. The virulence proteins aid in the survival of the pathogen in the host. For a list of diseases caused by the animal pathogens see note 48. For reviews, see refs. 49 through 56. The archetype of the type III systems is the Ysc (Yop secretion) system in *Yersinia*. Genes equivalent to the *ysc* genes are called the *psc* genes in *Pseudomonas*, the *bsc* genes in *Bordetella*, the *esc* genes in EPECs, the *hrc* genes in plant pathogens, and the *rhc* genes in *Rhizobium*.

The secretion apparatus

The supramolecular structure of the type III secretion apparatus is similar in the different bacteria and includes a hollow cylindrical basal body sometimes referred to as a syringe. The "syringe" is anchored in both the inner and outer membranes, and spans the periplasm. (See the simplified drawing of the *Yersinia* type III system in Fig. 17.5.) Each "syringe" is attached to a needlelike structure that protrudes from

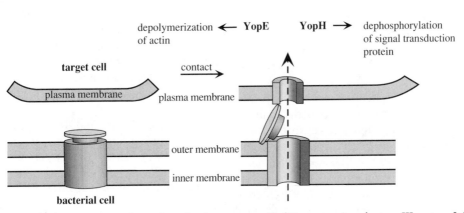

Fig. 17.5 Model for secretion and translocation into target cell of Yops proteins, the type III system. It is suggested that the protein secretion apparatus in *Yersinia*, called Ysc, or the Ysc injectisome, forms a channel that crosses the inner and outer bacterial membranes. The entire secretion apparatus is very complicated and consists of approximately 27 proteins of both the inner membrane and outer membrane varieties. When there is no contact between *Yersinia* and the target cell, a plug, called YopN, closes the opening of the Ysc channel. When the bacterium adheres to the target cell (but not necessarily directly to YopN), the channel is no longer plugged. A hollow needle grows out from Ysc and penetrates the target cell. (For evidence of needle formation, see ref. 58.) Certain Yop proteins (YopB, YopD, and LcrV) are injected through the needle into the target cell plasma membrane, where they form a channel, called a translocon. Other injected Yop proteins move via the needle through the translocon into the cytosol and cause several toxic effects. These include the depolymerization of actin and inactivation of host cell signaling systems. *Source*: Adapted from Cornelis, G. R., and H. Wolf-Watz. 1997. The *Yersinia* Yop virulon: a bacterial system for subverting eukaryotic cells. *Mol. Microbiol.* 23:861–867.

the bacterial cell surface, and several of these are found on each cell. This has been reported for *Salmonella typhimurium*, *Shigella flexneri*, and *Yersinia enterocolitica*. A somewhat larger needlelike structure has been reported for enteropathogenic *Escherichia coli* (EPEC).[50,57,58] When the bacterium attaches to a target cell, it constructs a proteinaceous channel called a *translocon* in the target cell plasma membrane. The translocon is presumed to be the channel through which bacterial toxic proteins are injected into the cytosol of the target cell. It is not known for certain whether the needle forms a continuous channel with the translocon, or whether there is a discontinuity between the tip of the needle and the translocon. (See ref. 54 and note 59 for a further discussion of translocons and needles.)

Proteins secreted

The secretion apparatus functions to secrete toxic proteins from the cytoplasm of the bacterium across the inner and outer membranes through the surface-protruding needle into the cytosol of target eukaryotic cells to which the pathogenic bacteria adhere. The toxic proteins secreted into the cytosol of the target cells are called *effector proteins*, and there are several different types. Effector proteins interfere with the defense mechanisms of the particular target cell, including phagocytosis, the production of toxic forms of oxygen, and the production of inflammatory cytokines, which are part of the host's defense response to infection. The effector proteins can also stimulate the entry into nonphagocytic cells of certain pathogens, such as *Shigella* and *Salmonella*, that can multiply intracellulary in host cells. (Not all pathogens with type III systems are intracellular pathogens. For example, enteropathogenic *E. coli*, do not enter the cells, but instead grow on the surface of the target cells.) For more information about effector proteins, see note 60 and refs. 61 through 63. The entire apparatus, including the translocon, is a complex structure made from about 25 different proteins.

Relationship to flagellin secretion apparatus

Interestingly, it appears that perhaps the type III system evolved from the flagellin secretion system. The reason for drawing this conclusion is that most of the type III proteins that are located in the bacterial cell membrane are homologues to the flagellar biosynthetic proteins that are also located in the cell membrane. In addition, electron microscopic studies of the suprastructure of the type III systems in *Salmonella typhimurium*, *Shigella flexneri*, *Yersinia enterocolitica*, and enteropathogenic *Escherichia coli* resembles that of purified flagella. This means that the TTSS have a structure embedded in the inner and outer membrane resembling that of the flagellar basal body from which protrudes a "needle" that is analogous to the hook and flagellar filament.

A model comparing flagella and the type III secretion system is shown in Fig. 17.6. An additional point about the type III secretion system is that one of the outer membrane proteins of the type III secretion system is actually homologous to PulD, an outer membrane protein that is part of the type II secretion system. (See Fig. 17.4.)

Summary of the common properties of the type III systems

There is no leader sequence typical of the Sec pathway, and furthermore, there is no processing of the N-terminal end during secretion. There are two hypotheses that attempt to explain the nature of the signal that directs the proteins to be secreted to the translocation machinery: (a) the 5′ end of the mRNA is the signal, and (b) the N-terminal 20 amino acids comprise the signal, and it functions with chaperone proteins that bind to it and bring the presecretory proteins to the TTSS.

Hypothesis (a). Because the amino acid sequences at the N-terminal ends that determines secretion of proteins through the TTSS are not very similar, some have suggested that the signal for secretion by the type III system is actually in the 5′ end of the mRNA that encodes the secreted protein, and that the 5′ mRNA targets the ribosome–RNA complex to the TTSS.[64] On the other hand, it has been pointed out that this would not apply to situations in which presynthesized proteins have been known to be secreted.[65]

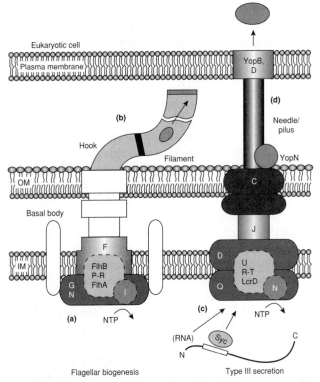

Fig. 17.6 Models for flagellar and type III secretion systems. Many similarities between these systems have been discovered by comparing the individual proteins, and also by examining the suprastructure of the type III secretion apparatus. As discussed in Section 1.2.1, the flagellar basal body is anchored in the cell membrane, and in the case of gram-negative bacteria it spans the periplasm as a hollow cylinder also anchored in the outer membrane. The flagellin monomers are secreted through the central channel of the basal body into a hollow flagellum filament that extends from the cell surface, and they are assembled into the flagellum at the distal end. The suprastructure of the type III secretion systems, has been visualized by electron microscopy. A close resemblance to the flagellum basal body has been observed. Instead of a projecting flagellum, however, the secretion apparatus has a hollow straight extension called a needle that projects from the cell surface. Similar to the secretion of flagellin proteins, proteins secreted by the type III system are thought to travel through the hollow needle extending from the cell surface. The needle is believed to insert into target cells and deliver the proteins. In some cases (e.g., in *Yersinia*), a penetrating needle may form in certain media only after the bacterium has made contact with the target cell. For a discussion of the conditions that allow needle formation in *Yersinia*, see ref. 58. *Source*: Thanassi, D. G., and S. J. Hultgren. 2000. Multiple pathways allow protein secretion across the bacterial outer membrane. *Curr. Opin. Cell Biol.* **12**:420–430.

Hypothesis (b). Research with *Yersinia* spp. has indicated that signals for secretion/translocation reside in the amino termini of secreted proteins. Gene fusion experiments have demonstrated that the N-terminal 15 to 17 amino acids are able to direct the export of heterologous reporter proteins through the TTSS into the culture supernate. However, precisely how these signals operate is not understood, since there is relatively little amino acid sequence similarity between the signals. As discussed next, there is evidence that chaperone proteins also take part

in targeting the presecretory proteins to the TTSS.

The following evidence that the chaperone targets some presecretory proteins to the secretion apparatus has been presented by Lee and Galán.[66] There is a chaperone-binding domain within approximately the first 140 amino acids of the presecretory protein, and in the absence of these domains, the *Salmonella typhimurium* SptP and SopE proteins are not secreted through their cognate TTSS. Interestingly, they are instead secreted through the flagellar export pathway, which as described earlier is thought to be an ancestral precursor to the TTSS systems.

Is there a normally unused ancestral flagellar secretion signal in at least some presecretory proteins destined for the TTSS? And does this signal operate to direct proteins to the flagellar secretion apparatus when the chaperone-binding domain is not present? The signals that target the presecretory protein to the TTSS are an important area of current research. Earlier publications using *Yersinia* have drawn different conclusions from that just described for the *Salmonella* SptP and SopE proteins, and the student should read note 67 for a further discussion of this very interesting area of research.

A second role for chaperone proteins has been proposed. It is suggested that they prevent premature folding of the presecretory protein, thus keeping the protein in a nonglobular, secretion-competent state that allows introduction of the protein into the secretion channel and movement through the needle complex, as described next.

Proteins are secreted through both the inner and outer membranes, and into the cytosol of target cells. In pathogenic *Yersinia* spp., *S. typhimurium*, and *S. flexneri*, a hollow cylindrical conduit resembling a needle is part of the secretory apparatus and projects from the bacterial surface. The needle inserts into the target cell. Enteropathogenic *E. coli* (cells) have a similar needlelike projection, except that it is a larger filament.

We conclude our summary of the common properties of type III secretory systems by noting that in the absence of binding of the bacterium to the target cell, there is no secretion of effector proteins. One of the reasons for this is that the channel (called an injectisome channel) is plugged, as described for *Yersinia* in the subsection that follows.

Yersinia and its two type III secretion systems, Ysc and Ysa

The prototype type III secretion system is the Ysc system of pathogenic *Yersinia* spp., and we turn now to some of the details of this particular system.

There are three pathogenic, or disease-causing, species of *Yersinia*. *Yersinia pestis* is the causative agent of plague. Both *Y. enterocolitica* and *Y. pseudotuberculosis* can cause severe gastroenteritis, local abscess formation, and peritonitis in humans. (See note 68 for a description of these diseases.) All three species tend to grow primarily extracellularly in lymphatic tissue in the mammalian host, and they use the type III system to secrete toxic proteins (the effector proteins) into the cytosol of macrophages to prevent phagocytosis and killing of the infectious bacteria by the macrophages. The bacteria thus evade the host defense mechanisms and multiply primarily extracellularly in lymphoid tissues. However, they can also invade and replicate in nonactivated macrophages. This is important for the spread of the bacteria within the body, as described earlier (see note 67). The effector proteins secreted by type III systems into the target cells have cytotoxic effects. As an example, we will consider proteins of *Yersinia*'s outer membrane, the Yops. (See refs. 69–71 for reviews.)

The major type III system used by the pathogenic *Yersinia* is called the Ysc (Yop secretion) system and is plasmid encoded. *Y. enterocolitica* has recently been reported to possess a second contact-dependent type III secretion system, which associated with virulence. The second system, which is chromosomally encoded, is called the Ysa system, and it secretes Ysps (*Yersinia* secreted proteins). The Ysa system is described in ref. 72.

Examine Fig. 17.6 again. It was reported that *Y. enterocolitica* forms needles that extend from the secretion apparatus in the bacterial cell membrane, similar to those formed by *S. typhimurium* and *S. flexneri*. It appears that the Yop proteins, including the translocon proteins (YopB, YopD, and LcrV), are injected through the type III apparatus into target cells through the needles, which are made of the YscF protein. It has been suggested that the needle is made upon adhesion of the bacterium to the target cell and penetrates the target cell at that time, although needles can also be made when bacteria are growing in certain media in the absence of target cells. See note 73 for experimental evidence in support of this conclusion. When the needle inserts into the plasma membrane of target cells, it encounters low Ca^{2+} levels that Lee et al. have suggested are responsible for activating the injection of Yop proteins.[74] In extracellular host fluids the Ca^{2+} levels are about 1.2 mM, but in intracellular fluids they are only about 100 nM.

Several proteins in *Yersinia* block secretion of Yops prior to attachment to the target cell, even when calcium is present. Hela cells can be used to study this action. In *Yersinia*, one of the plugs is a protein called YopN, which is also translocated into Hela cells when secretion takes place (Fig. 17.5).[75] Secretion after cell contact, including the unplugging of the injectisome channel to allow passage of proteins, is initiated upon reception of an activation signal, which occurs when the bacterium adheres to the target cell. *Yersinia* can adhere via the binding of outer membrane proteins called adhesins to integrins, which are proteins projecting from the eukaryotic cell. Alternatively, secretion can be initiated when *Yersinia* binds to phagocytic receptors, without the involvement of adhesins.

The first proteins introduced are probably the Yop proteins (YopB, YopD, LcrV) that form the channel (translocon) in the target cell membrane. After this, at least six Yops [YopE, YopH, YopJ (YopP), YopM, YopO (YpkA), and YopT] are injected into the cytosol of the host cell through the channel, producing a cytotoxic effect. They inhibit phagocytosis and the production of proinflammatory cytokines. (See note 76 for an explanation of how cytokines promote inflammation and why inflammation is important for the host defense against infections.) The student can refer to ref. 77 for a summary of the effects of the Yop proteins on host cells, as well as to the previously cited review articles. A brief summary of these proteins and their functions is given in note 78. Note 79 discusses the mechanism of action of YopE and YopT.

17.5.3 The type II pathway

The type II pathways appear to be the *major route* by which proteins are secreted by gram-negative bacteria.

Two stages

Type II pathways secrete proteins in two stages (Figs. 17.4 and 17.7). Stage 1 translocates the protein into the periplasm, and stage 2 secretes the protein from the periplasm across the outer membrane to the external medium. Stage 1 uses the general secretory pathway (the Sec-dependent pathway) to secrete the protein into the periplasm. Once the protein has crossed the inner membrane (cell membrane), the leader

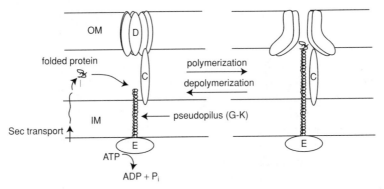

Fig. 17.7 Model of how type II systems might work. The protein to be secreted is transported in an unfolded state via the Sec system across the inner membrane (IM). The N-terminal signal sequence is removed and the protein folds in the periplasm. Protein D is an outer membrane (OM) complex called a secreton, consisting of 12 to 14 subunits that form a pore. The folded protein is pushed out through the D-protein pore by what is suggested to be a polymerizing pseudopilus. The pseudopilus is comprised of proteins G–K, with G being the dominant protein. Proteins G–K show homology to type IV pili. Protein E is suggested to be an ATPase that provides energy for the translocation and assembly of the pseudopilus. Protein C is suggested to be involved in transferring energy for transport from the inner membrane to the protein D complex. Not shown are proteins O and S. Protein O, an integral cell membrane protein, has two activities. It is a prepilin peptidase that cleaves the N-terminal end of the pseudopilin proteins, and it also methylates the newly created N-terminal amino acid. This all takes place on the cytoplasmic side of the membrane. Protein S is an outer membrane lipoprotein that stabilizes the protein D complex. In addition, there are several other proteins that are present in some type II pathways and not others. The latter proteins may be involved in specific requirements, such as substrate recognition.

sequence is proteolytically removed on the external face of the inner membrane, and the protein enters the periplasm for the second step of secretion through the outer membrane to the extracellular milieu. Although the type II pathway is sometimes referred to as a GSP pathway, the student should understand that this does not mean the GSP pathway in general, but merely one of its terminal branches.

Secretion apparatus for second stage

The translocation machinery and mechanism of the second stage are not well understood. After the proteins have been translocated in an unfolded state by the Sec system into the periplasm in the first stage, the signal peptide at the N-terminal end is removed. In the second stage, the proteins fold into a near-native conformation and are translocated through the outer membrane. The secretory apparatus for the second stage of secretion is called the *secreton*, and it consists of 12 to 16 proteins. Surprisingly, the proteins of the secreton are located in *both* the inner and outer membranes, with most being located in the inner membrane. Several of the inner membrane secreton proteins have extensive periplasmic domains.

The location of secreton proteins in the inner membrane is unexpected because the GSP system is responsible for translocating the proteins across the inner membrane into the periplasm (stage 1). One possible role for secreton proteins in the inner membrane may be related to transferring energy to the outer membrane for transport in stage 2. Each component of the type II secreton has been designated Gsp, followed by a letter corresponding to a homologous protein in the *pul* system for the secretion of pullanase. (See note 80 for a description of the pullulanases.) However, it has been argued that the notation Gsp is inappropriate, and sometimes the word "protein" is used instead. See note 81 for a discussion of this point. According to Fig. 17.7, PulC (protein C) connects the inner membrane proteins with the PulD (protein D) complex in the outer membrane.

PulD proteins are called secretins, and a popular model proposes that the proteins form a ring with a central channel in the outer membrane, in some ways similar to the TolC channel. PulS (protein S, not shown in Fig. 17.7) is an outer membrane lipoprotein that is suggested to stabilize PulD. One model postulates that energy for transport through the outer membrane channel is transmitted by PulC to the outer membrane complex. This may be similar to how TonB energizes solute uptake. (See the discussion of TonB in Section 1.2.4.) Interestingly, some of the proteins (proteins G–K) anchored in the cell membrane are believed to form a periplasmic pseudopilus that is required for the secretion of proteins (Fig. 17.7). For more information about the postulated roles of the proteins, including a role for the pseudopilus, see note 82 and the legend to Fig. 17.7.

The Out system in Erwinia chrysanthemi and E. carotovora

There are several type II pathways, and each can secrete more than one kind of protein. For example, a secretion machinery called the Out system exists in *Erwinia chrysanthemi* and *E. carotovora*, the causative agents of soft-rot disease in plants. The enzymes secreted by the Out system, which destroy plant tissue, are pectate lyase, *exo*-poly-α-D-galacturonosidase, pectin methylesterase, and cellulase. Mutants in the *out* genes have a defect in the secretion of cellulases and pectinases and accumulate these enzymes in the periplasm.

Cholera toxin and protease secretion by Vibrio cholerae

Other type II systems exist in other bacteria and are responsible for the secretion of different proteins. For example, the *eps* (extracellular protein secretion) genes are required for the secretion of cholera toxin and protease by *Vibrio cholerae*.[83] The *eps* genes are also required for correct assembly of the proteins in the outer membrane.

Specificity between each type II system and the protein secreted

There is specificity between each type II system and the proteins that it secretes. A resident type II system need not secrete extracellular proteins of related bacteria when the structural genes are introduced, even though there may be homology between the secretory proteins from the donor and recipient bacteria. For example, there is homology in the proteins between the *out* gene products from *Erwinia* and the *pul* gene products from *Klebsiella*. Yet, when the pectate lyase gene from *Erwinia* is expressed in *Klebsiella*, the enzyme is not secreted.

There can even be specificity between Out systems from two different *Erwinia* species. For example, the *E. chrysanthemi* Out system does not secrete an extracellular pectate lyase encoded by a gene from a different species of *Erwinia*, even though both species of *Erwinia* use the Out system.

What can we conclude from all of this? Each type II system not only distinguishes between different secreted proteins but also recognizes "self" proteins. The basis for this distinction is not understood. However, if the genes for a type II system as well as its cognate protein are transferred to a different genus of bacterium, then protein secretion can occur. For example, when the structural gene for pullulanase as well as the adjoining secretion genes are transferred from *Klebsiella pneumoniae* to *E. coli*, pullulanase is synthesized and secreted.[84]

17.5.4 The type V pathway: Autotransporters

The type V pathway (see Fig. 17.8) is also called the autotransporter pathway because the proteins transport themselves across the outer membrane of gram-negative bacteria without the aid of an obvious secretion system. The secreted proteins, referred to as autotransporters, include proteases, toxins, and adhesins.

Two stages

Translocation uses the Sec (GSP) pathway to move the autotransporter through the cell membrane into the periplasm; but in contrast to the type II system, there is no requirement for a separate secretion apparatus to move the protein from the periplasm through the outer membrane to the outside. So, how is this done? Autotransporters have three domains: an amino-terminal domain recognized by the Sec system, an internal domain called the "passenger domain," and a carboxyl domain called the β or "helper" domain. Once in the periplasm, the amino-terminal signal domain is cleaved by a peptidase. The helper domain inserts into the

outer membrane and forms a pore through which the internal passenger domain travels to the outside surface. The pore has been compared with porin pores formed by porins in the outer membrane.

Depending upon the protein, the passenger domain can remain bound to the membrane via the helper domain, or it can be severed by proteolytic cleavage and released into the extracellular medium. Some models hold that since the outer membrane pore is too narrow to accommodate large completely folded proteins, folding of the passenger domain is completed on the outer membrane surface. In summary, the autotransporter forms its own channel in the outer membrane.

Immunoglobulin A protease, an example of an autotransporter

As an example of an autotransporter, we will consider the secretion of immunoglobulin A protease by *Neisseria gonorrhoeae*, the causative agent of gonorrhea. The protease is synthesized as a precursor protein with an N-terminal leader sequence typical for proteins translocated by the Sec system. The leader sequence functions with the Sec machinery and is removed by the peptidase after translocation through the cell membrane. Up to this stage, the secretion is similar to the secretion of proteins via the type II system. However, the immunoglobulin A protease has information in its sequence that directs it through the outer membrane to the outside.

At the carboxy end of the protein, the helper domain aids the protein in traversing the outer membrane where it anchors the protein. The protein is released into the medium by proteolytic cleavage (autocatalytic) at the C-terminal end, leaving the helper peptide embedded in the membrane. Some other proteins are secreted in a similar fashion, although the proteolytic cleavage may be catalyzed by a separate enzyme.

How the secreted proteins cross the periplasm and insert into the outer envelope is not known. In *E. coli*, the helper domain can promote translocation across the outer membrane of hybrid proteins that would not normally release the passenger protein into the medium. It thus appears that the helper region may be sufficient for translocation through the

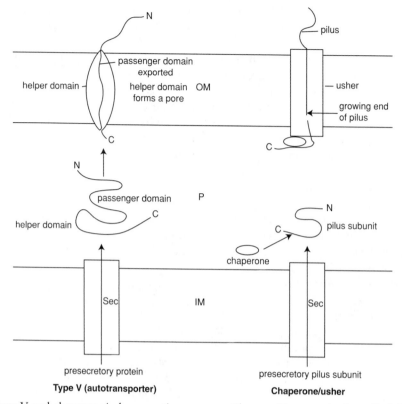

Fig. 17.8 Type V and chaperone/usher secretion systems. The type V system is also called the autotransporter system. The presecretory protein is exported via the Sec system to the periplasm (P), where the N-terminal signal sequence is cleaved by a peptidase. The protein has a C-terminal helper domain, followed by a passenger domain. The C-terminal helper domain inserts into the outer membrane (OM) to form a secretion channel through which the passenger domain is exported to the cell surface. The passenger domain can be released into the extracellular medium by proteolysis or can remain attached to the outer membrane. The chaperone/usher system is used for the formation of adhesive pili by pathogenic bacteria. The pilus subunit is translocated through the inner membrane (IM) via the Sec system. In the periplasm, the amino-terminal signal sequences of the pilus subunits are cleaved. When in the periplasm, these subunits bind at their C-terminal ends to chaperone proteins that allow the subunits to fold properly and also prevent them from binding to each other in the periplasm. The chaperone brings the subunits to the outer membrane usher. The pilus subunits are assembled into a linear fiber of folded subunits that threads through the usher. The pilus rod assumes its final helical shape at the cell surface.

outer envelope without the need for an accessory secretion system.

17.5.5 The chaperone/usher pathway

Two stages

The chaperone/usher pathway (see Fig. 17.8) uses the Sec system to secrete across the cell membrane to the periplasm such adhesive virulence structures as the P and type 1 pili in uropathogenic *E. coli*. The pilus subunits are exported individually across the cell membrane via the Sec system, and their amino-terminal signal sequences are cleaved. Once in the periplasm, each pilus subunit binds at its C-terminal end to a periplasmic chaperone protein. The chaperone–pilus complex migrates to an outer membrane "usher," which forms a channel in the outer membrane. The chaperone binds to the "usher" and transfers the pilus subunit to it. More pilus subunits are brought by chaperone proteins to the "usher," where they assemble at the periplasmic surface onto a pilus fiber that is threaded through the "usher" channel to the outside.

17.5.6 The type IV pathway

For a review of the type IV secretion systems (T4SS), see refs. 41 and 85; for a discussion of the structure of the proteins and how they interact, read ref. 86. Type IV systems translocate DNA and/or protein across the inner and outer membranes in gram-negative bacteria via a Sec-dependent (two stage) or a Sec-independent (one-stage) mechanism. The targets of delivery of substances transported by type IV systems include cells of bacteria, fungi, plants, and animals, and the extracellular medium. It is thus a ubiquitous and very important pathway. Substances secreted by type IV systems into the extracellular medium, or transferred into host cells, include virulence factors such as pertussis toxin. The gram-negative bacterium *Bordetella pertussis* secretes this toxin into the extracellular medium during infection, where it is taken up by cells of the upper respiratory tract, causing some of the symptoms of whooping cough (see earlier: note 36). Other virulence factors are T-DNA, the oncogenic portion of the Ti plasmid that is transferred from the gram-negative bacterium *Agrobacterium tumefaciens* into the plant host cells and induces crown gall tumors; the conjugative IncN plasmid pKM101, transferred via the Tra system between cells of *A. tumefaciens*, and several other proteins and plasmids by a variety of bacteria.

The type IV systems have different names. For example, the *B. pertussis* system is called the Ptl system, and the nine proteins for the secretion apparatus are encoded by the *ptl* locus. The *Agrobacterium* system that transfers T-DNA is called the VirB system, and the 11 proteins for the secretion are encoded by the *virB* locus. The system for transferring the pKM101 plasmid is called the IncN system.

Secretion apparatus of the VirB system in Agrobacterium that translocates oncogenic plasmid DNA into plant cells

Much of what is known about type IV secretion comes from studying the VirB system of *Agrobacterium tumefaciens*, the prototypical type IV system, which is responsible for secreting virulence factors that cause crown gall in infected plants. (See Fig. 17.9.) Export of oncogenic plasmid DNA (T-DNA) into plant cells by the VirB system of *Agrobacterium* is Sec independent and takes place in a single step from the cytoplasm through a translocation channel that spans the periplasm.

Several proteins make up the secretory apparatus. The functions of the individual proteins are summarized in the legend to Fig. 17.9 and discussed in detail in refs. 87 and 88. Some of the proteins form a complex attached to both the inner and outer membranes and are thought to assemble as a translocation channel through the inner membrane, periplasm, and outer membrane; some form a pilus that transfers the DNA from the translocation channel into the plant cell. As explained next, pertussis toxin is secreted into the periplasm by the Sec system and then enters the type IV translocation channel to be transported into target eukaryotic cells.

Example of a Sec-dependent system: Export of protein toxins

Figure 17.9 illustrates how pertussis toxin from *Bordetella pertussis*, is exported in two stages. In the first stage, which is Sec dependent, the protein enters the periplasm and then moves into the translocation channel. Pertussis toxin is not secreted into cells; rather, it is secreted into the extracellular medium.

17.6 Folding of Periplasmic Proteins

17.6.1 The importance of intramolecular disulfide bonds for periplasmic proteins

For a review, see ref. 89. In *E. coli*, newly synthesized proteins that are translocated across the cell membrane into the periplasm often form intramolecular disulfide bonds (S–S bonds) when they arrive in the periplasm. (Actually, disulfide bonds are very important for folding and stability of extracytosolic proteins in all organisms, not simply bacteria.) The S–S bonds form between –SH groups in the side chains of nearby cysteine residues. These intramolecular disulfide bonds are important for proper folding and three-dimensional stability of the polypeptide once it has been translocated into the periplasm. Such disulfide bonds are much less common in cytoplasmic proteins because of the existence of high concentrations in the

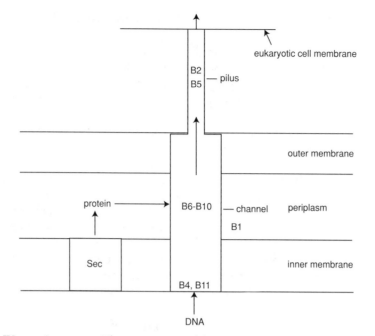

Fig. 17.9 Type IV secretion system. The components of the secretion apparatus are labeled according to the nomenclature for the VirB system (i.e., the prefix Vir is not used). The secretion of certain proteins, such as pertussis toxin, takes place in two stages. Stage 1 uses the Sec system to translocate the protein into the periplasm. From there, the protein enters the translocation apparatus that spans the inner membrane, periplasm, and outer membrane. Pertussis toxin is secreted into the extracellular medium during infections and causes the symptoms of whooping cough. DNA secretion takes place in one stage, as the DNA enters the secretion apparatus from the cytoplasm. VirB1 (B1 in the figure) is found in the periplasm. It has a motif seen in lytic transglycosylases and may be important for hydrolysis of the peptidoglycan so that the transporter can be assembled. A proteolytic product of VirB1 (not shown) is on the surface of the outer membrane and may be important for contact between donors and recipient cells. It is not part of the pilus. The pilus is composed of a major protein called VirB2 and a minor protein called VirB5. It is thought that the pilus is a tube through which proteins and DNA are moved into target eukaryotic cells. The channel subunits are VirB6 through VirB10. VirB6 is an inner membrane protein. VirB7 is a lipoprotein in the outer membrane that is complexed with VirB9. Both VirB7 and VirB9 are thought to function in the assembly of the secretion apparatus in the outer membrane. VirB10 is believed to provide a link between VirB components in the inner and outer membrane. VirB4 and VirB11 are in the inner membrane and are believed to couple ATP hydrolysis with transport. (VirB11 proteins may also be found in the cytoplasm.)

cytosol of reducing substances that break S–S bonds.

As an example of disulfide bonds in periplasmic proteins, consider *E. coli* alkaline phosphatase. This is a periplasmic enzyme that in its active form consists of two identical subunits, each of which has intrachain disulfide bonds that enable the subunits to fold properly in the periplasm.

17.6.2 The role of thiol–redox enzymes in catalyzing the formation of disulfide bonds in the periplasm

Disulfide bond formation is enzymatically catalyzed in the periplasm by *thiol–redox*

enzymes, also called *thiol–disulfide* exchange proteins. This is diagrammed in Fig. 17.10 for *E. coli*. The periplasmic thiol–redox enzyme in *E. coli* is DsbA (disulfide bond), which is required for disulfide bond formation in vivo and has been demonstrated in vitro to carry out this activity. DsbA catalyzes the formation of disulfide bonds in newly synthesized and exported periplasmic proteins, causing them to fold properly. In the process, disulfide groups in DsbA become reduced to sulfhydryl groups. Reduced DsbA is reoxidized by donating electrons to DsbB, a cytoplasmic membrane thiol–redox protein with periplasmic loops. DsbB itself is reoxidized by ubiquinone or menaquinone in the electron transport chain,

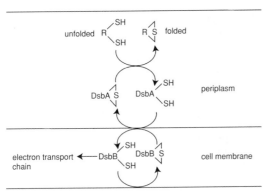

Fig. 17.10 Formation of disulfide bonds in periplasmic proteins. Periplasmic proteins (R) are secreted into the periplasm unfolded. Many of them form intramolecular disulfide bonds, which are important for their folding and stability. This activity is catalyzed by a periplasmic protein called DsbA. Electrons are transferred from the periplasmic proteins to DsbA to a membrane protein called DsbB, and from there to the electron transport chain. The overall sum of the reactions is the equivalent of transferring disulfide bonds between cysteine residues from DsbB to DsbA to R. The cysteine residues in DsbB are located in periplasmic loops. Each loop has two pairs of cysteine residues, and electrons are transferred from DsbA in sequence from one pair to the next before being transferred to the electron transport chain.

and the electrons eventually go to a terminal electron acceptor.[90] For more information about the structure and function of DsB, as well as its relationship to the electron transport chain, see note 91.

Thiol–redox enzymes such as DsbA have pairs of cysteine residues separated by two amino acids, Cys-X1-X2-Cys. This is often referred to as a thioredoxin-like motifs because it is also present in thioredoxin and is characteristic of disulfide oxidoreductases. (For a discussion of the role of thioredoxin, see Section 19.4.3.) In DsbA, the thioredoxin-like motif is Cys30-Pro31-His32-Cys33.

17.6.3 Correcting the formation of disulfide bonds between the wrong cysteine residues

Occasionally, the wrong cysteine residues will take part in the formation of a disulfide bond, causing the polypeptide to fold incorrectly. *E. coli* has a mechanism to repair this, and it involves the periplasmic thiol–redox protein called DsbC.[92] DsbC binds to improperly folded proteins, breaks (reduces) the disulfide bond, and catalyzes the formation of the correct disulfide bond. DsbC is called an *isomerase-reductase*, and it has a thioredoxin-like motif (Cys98-X1-X2-Cys101).

17.7 Summary

Bacteria translocate (export) proteins into the cell membrane, periplasm, and outer envelope, or secrete them into the extracellular medium. The translocation machinery and requirements for targeting to different destination sites are being studied in several bacteria, especially in *E. coli*.

The Sec system is necessary for the translocation of most proteins across the cell membrane. A preprotein is made with a leader sequence at the amino-terminal end. The role of the leader sequence is to initiate translocation by inserting into the lipid bilayer. The leader sequence also aids in preventing premature folding of the newly synthesized protein. The preprotein binds to a chaperone protein in the cytosol.

An important chaperone protein is SecB. Two roles of the chaperone protein are to prevent folding of the preprotein into a configuration that disallows translocation and to prevent the formation of protein aggregates. A third role for the chaperone protein is to deliver the preprotein to the translocase.

The preprotein–chaperone protein complex binds to the translocase on the inner membrane surface, and the translocation of the preprotein into the membrane is initiated. The early stages of translocation require ATP and SecA, but the remainder of translocation can be driven by the Δp.

Once the preprotein has been translocated through the membrane, the leader sequence is removed by a peptidase on the outer surface of the cell membrane. *E. coli* also has a signal recognition particle (SRP) that may guide certain proteins destined to remain in the inner membrane. There exists a Sec-independent pathway for translocating folded proteins with bound cofactors into the cell membrane and the periplasm. It is called the twin-arginine translocation (Tat) pathway. Respiratory and photosynthetic electron transport proteins are translocated via the Tat pathway.

There are six secretion pathways found among gram-negative bacteria. Some use the Sec-dependent pathway for translocating proteins through the cell membrane into the periplasm prior to secretion out of the cell, and others do not. These pathways secrete various enzymes, pilus proteins, and flagellar proteins into the extracellular space, as well as virulence proteins that aid infection into target cells. The secretion pathways are type I, type II, type III, type IV, type V (autotransporter), and the chaperone/usher pathway.

Many periplasmic proteins fold properly by forming intrapeptide disulfide bonds between cysteine residues. This activity is catalyzed in *E. coli* by a periplasmic protein called DsbA. Improperly folded proteins are repaired by DsbC, another *E. coli* periplasmic protein.

Study Questions

1. What is the experimental basis for believing that the leader peptide is necessary for translocation? What are two reasons for believing that the leader peptide is not necessary for determining the final destination of the protein?

2. What are the postulated roles for chaperone proteins in protein translocation?

3. What is the role of Sec A in protein translocation? What is the evidence that ATP binding, and not hydrolysis, drives the insertion of SecA into the membrane?

4. What are the functions of the different domains of the leader sequence (i.e., the basic N terminal, middle hydrophobic, and carboxy terminus)?

5. What is the evidence, based upon mutational analysis, that there are two secretory systems for proteins secreted into the medium and that one is *sec* dependent and one is *sec* independent.

6. How does the type III protein secretion system differ from types I and II? How is it similar?

7. What is the evidence that SecDFyajC plays a regulatory role in protein export via the Sec system?

8. What is the evidence that in the absence of a Δp, ATP can drive Sec-dependent translocation to completion?

9. What is the evidence that translocons form pores in target cell membranes?

10. How does the Tat system differ from the GSP system?

REFERENCES AND NOTES

1. Fisher, A. C., and M. P. DeLisa. 2004. A little help from my friends: quality control of presecretory proteins in bacteria. *J. Bacteriol.* **186**:7467–7473.

2. Driessen, A. J. M. 1992. Bacterial protein translocation: kinetic and thermodynamic role of ATP and the protonmotive force. *Trends in Biochem. Sci.* **17**:219–223.

3. Wickner, W., A. J. M. Driessen, and F.-U. Hartl. 1991. The enzymology of protein translocation across the *Escherichia coli* plasma membrane. *Annu. Rev. Biochem.* **60**:101–124.

4. Saier, Jr, M. H., P. K. Werner, and M. Muller. 1989. Insertion of proteins into bacterial membranes: mechanism, characteristics, and comparisons with the eucaryotic process. *Microbiol. Rev.* **53**:333–366.

5. Pugsley, A. P. 1993. The complete general secretory pathway in gram-negative bacteria. *Microbiol. Rev.* **57**:50–108.

6. Murphy, C. K., and J. Beckwith. 1996. Export of proteins to the cell envelope in *Escherichia coli*, pp. 967–978. In: *Escherichia coli and Salmonella: Cellular and Molecular Biology*. F. C. Neidhardt et al. (Eds.). ASM Press. Washington, DC.

7. Ito, K. 1992. SecY and integral membrane components of the *Escherichia coli* protein translocation system. *Mol. Microbiol.* **6**:2423–2428.

8. Rapoport, T. A., B. Jungnickel, and U. Kutay. 1996. Protein transport across the eukaryotic endoplasmic reticulum and bacterial inner membranes. *Annu. Rev. Biochem.* **65**:271–303.

9. In *E. coli* the SecYEG translocase can be isolated from the membrane in a complex with three other proteins, SecD, SecF, and yajC. SecDFyajC plays a regulatory role in protein secretion and facilitates the coordination between ATP and Δp-driven translocation, lowering the amount of ATP used for translocation. Recent in vitro studies using isolated membranes with depleted, wild-type, or enriched DF levels indicate that SecDFyajC stabilizes the inserted form of SecA in the membrane and slows the movement of the preprotein, proOmpA. As a consequence, translocation intermediates accumulate in the membrane, and these are translocated by means of the driving force of the Δp. Mutants lacking SecDF

are defective in translocation and growth. Over-expression of *secDF* suppresses mutations in the leader peptide sequence; anti-SecD antibodies have been reported to block translocation in spheroplasts. There appears to be only 10% as much SecDFyajC as SecYEG in *E. coli*, suggesting that either some SecYEG translocases operate without SecDFyajC or that SecDFyajC cycles between operating SecYEG translocases. For a discussion of the role of SecDFyajC in protein secretion, see: Duong, F., and W. Wickner. 1997. The SecDFyajC domain of preprotein translocase controls preprotein movement by regulating SecA membrane cycling. *EMBO J.* 16:4871–4879.

10. Gardel, C., K. Johnson, A. Jacq, and J. Beckwith. 1987. *secD*, a new gene involved in protein export in *Escherichia coli*. *J. Bacteriol.* 169:1286–1290.

11. Wickner, W., and M. R. Leonard. 1996. *Escherichia coli* preprotein translocase. *J. Biol. Chem.* 271:29514–29516.

12. Duong, F., and W. Wickner. 1997. The SecDFyajC domain of preprotein translocase controls preprotein movement by regulating SecA membrane cycling. *EMBO J.* 16:4871–4879.

13. Manting, E. H., C. van der Does, and A. J. M. Driessen. 1997. In vivo cross-linking of the SecA and SecY subunits of the *Escherichia coli* preprotein translocase. *J. Bacteriol.* 179:5699–5704.

14. Manting, E. H., and A. J. M. Driessen. 2000. *Escherichia coli* translocase: the unravelling of a molecular machine. *Mol. Microbiol.* 37:226–241.

15. Nishiyama, K.-I., S. Mizushima, and H. Tokuda. 1993. A novel membrane protein involved in protein translocation across the cytoplasmic membrane of *Escherichia coli*. *EMBO J.* 12:3409–3415.

16. Economou, A., and W. Wickner. 1994. SecA promotes preprotein translocation by undergoing ATP-driven cycles of membrane insertion and deinsertion. *Cell* 78:835–843.

17. Economou, A., J. A. Pogliano, J. Beckwith, D. B. Oliver, and W. Wickner. 1995. SecA membrane cycling at SecYEG is driven by distinct ATP binding and hydrolysis events and is regulated by SecD and SecF. *Cell* 83:1171–1181.

18. SecA insertion and deinsertion can be studied by means of vesicles prepared from the inner membrane. Instead of ATP, a nonhydrolyzable form of ATP, adenyl–imidodiphosphate (AMP-PNP), can be used for insertion of SecA. It is possible to determine the portion of the protein that has inserted because it becomes resistant to protease, but is not so after detergent treatment of the vesicles. For more information about these experiments, see ref. 17.

19. Schiebel, A., A. J. M. Driessen, F.-U. Hartl, and W. Wickner. 1991. $\Delta\mu_{H^+}$ and ATP function at different steps of the catalytic cycle of preprotein translocase. *Cell* 64:927–939.

20. The in vitro system consists of inverted *Escherichia coli* inner membrane vesicles translocating the outer membrane protein proOmpA.

21. SecA has two nucleotide binding sites: NBS I ($K_{d,ADP} = 0.1$–0.3 μM) and NBS II ($K_{d,ADP} = 300$–500 μM). Both sites play a role in translocation. This is further discussed in ref. 14.

22. Driessen, A. J. M. 1992. Precursor protein translocation by the *Escherichia coli* translocase is directed by the protonmotive force. *EMBO J.* 11:847–853.

23. There are several ways to separate effects of the Δp from those of SecA and ATP in vitro. For example, membrane vesicles or proteoliposomes will not have a Δp unless one is imposed. Different methods can be used to impose a Δp. For example, the stimulation of electron transport by incubating inner membrane vesicles with NADH has been used. Vesicles are prepared from *unc⁻* cells so that the Δp and ATP are not interconvertible. A Δp has also been created by light in proteoliposomes in which bacteriorhodopsin was incorporated, or by adding reduced cytochrome c to proteoliposomes in which cytochrome c oxidase has been incorporated. Proton ionophores can be used to dissipate the Δp. SecA can be inactivated with anti-SecA antibody, and ATP levels can be experimentally manipulated.

24. Yamada, H., S-ichi Matsuyama, H. Tokuda, and S. Mizushima. 1989. A high concentration of SecA allows proton motive force–independent translocation of a model secretory protein into *Escherichia coli* membrane vesicles. *J. Biol. Chem.* 264:18577–18581.

25. Yamada, H., H. Tokuda, and S. Mizushima. 1989. Protonmotive force–dependent and –independent protein translocation revealed by an efficient in vitro assay system of *Escherichia coli*. *J. Biol. Chem.* 264:1723–1728.

26. Fröderberg, L., E. Houben, J. C. Samuelson, M. Chen, S.-K. Park, G. J. Phillips, R. Dalbey, J. Luirink, and J.-W. L. De Gier. 2003. Versatility of inner membrane protein biogenesis in *Escherichia coli*. *Mol. Microbiol.* 47:1015–1027.

27. van der Laan, M., M. L. Urbanus, C. Ten Hagen-Jongman, N. Nouwen, B. Oudega, N. Harms, A. J. M. Driessen, and J. Luirink. 2003. A conserved function of YidC in the biogenesis of respiratory chain complexes. *Proc. Natl. Acad. Sci. USA* 100:5801–5812.

28. De Gier, J.-W. L., Q. A. Valent, G. V. Heijne, and J. Luirink. 1997. The *E. coli* SRP: preferences of a targeting factor. *FEBS Lett.* 408:1–4.

29. Keenan, R. J., D. M. Freymann, R. M. Stroud, and P. Walter. 2001. The signal recognition particle. *Annu. Rev. Biochem.* 70:755–775.

30. Duong, F., J. Eichler, A. Price, M. R. Leonard, and W. Wickner. 1997. Biogenesis of the gram-negative bacterial envelope. 1997. *Cell* 91:567–573.

31. This is the consensus sequence in Proteobacteria; the Ser, Phe, Leu, and Lys residues are present in a frequency higher than 50%, and X is a polar amino acid.

32. Palmer, T., F. Sargent, and B. C. Berks. 2005. Export of complex cofactor-containing proteins by the bacterial Tat pathway. *Trends Microbiol.* **13**:175–180.

33. Berks, B. C., T. Palmer, and F. Sargent. 2003. The Tat protein translocation pathway and its role in microbial physiology. *Adv. Microbial Physiol.* **47**:187–254.

34. Berks, B. C., F. Sargent, and T. Palmer. 2000. The TAT protein export pathway. *Mol. Microbiol.* **35**:260–272.

35. Palmer, T., and B. C. Berks. 2003. Moving folded proteins across the bacterial cell membrane. *Microbiology* **149**:547–556.

36. Cholera toxin (CT) is a protein produced by *Vibrio cholerae* and is responsible for the watery diarrhea associated with cholera. It is an enzyme that enters intestinal cells and transfers ADP–ribose from NAD$^+$ to a regulatory enzyme in the target cell (a process called ADP-ribosylation); this results in the stimulation of the enzyme adenyl cyclase in the intestinal cells. Adenyl cyclase converts ATP to cAMP, and the increased amounts of cAMP stimulate the secretion of Cl$^-$ and inhibit the absorption of NaCl. As a consequence, the osmolarity of the intestinal fluid is increased, and large amounts of water as well as bicarbonate and K$^+$ enter the intestinal fluid and are lost from the body. Enterotoxigenic *Escherichia coli* (ETEC) produces a similar toxin, called heat-labile toxin (LT). Pertussis toxin (PT) is produced by *Bordetella pertussis* and is responsible for some of the symptoms of pertussis (whooping cough). It is an enzyme that enters target cells in the upper respiratory tract and ADP-ribosylates a regulatory protein, causing a rise in cellular cAMP. This is probably the cause of the increased secretions and mucus production in the upper respiratory tract colonized by the bacteria. Diphtheria toxin is produced by *Corynebacterium diphtheriae*, the causative agent of diphtheria. The toxin is an enzyme that ADP-ribosylates elongation factor 2 in target cells and stops protein synthesis. Many cell types, including those in the heart and other organs, can be affected.

37. Lory, S. 1992. Determinants of extracellular protein secretion in gram-negative bacteria. *J. Bacteriol.* **174**:3423–3428.

38. Salmond, G. P. C., and P. J. Reeves. 1993. Membrane traffic wardens and protein secretion in gram-negative bacteria. *Trends Biochem. Sci.* **18**:7–12.

39. Wandersman, C. 1996. Secretion across the bacterial outer membrane. pp. 955–966. In: *Escherichia coli and Salmonella: Cellulular and Molecular Biology*. F. C. Neidhardt et al. (Eds.). ASM Press, Washington, DC.

40. Thanassi, D. G., and S. J. Hultgren. 2000. Multiple pathways allow protein secretion across the bacterial outer membrane. *Curr. Opin. Cell Biol.* **12**:420–430.

41. Henderson, I. R., F. Navarro-Garcia, M. Desvaux, R. Fernandez, and D. Ala'Aldeen. 2004. Type V protein secretion pathway: the autotransporter story. *Microbiol. Mol. Biol. Rev.* **68**:692–744.

42. Ghosh, P. 2004. Process of protein transport by the type III secretion system. *Microbiol. Mol. Biol. Rev.* **68**:771–795.

43. For example, colicin was exported through the α-hemolysin export system (HlyBD system) in strains of *E. coli* defective in the colicin export system.

44. Fath, M. J., R. C. Skvirsky, and R. Kolter. 1991. Functional complementation between bacterial MDR-like export systems: colicin V, alpha-hemolysis, and *Erwinia* protease. *J. Bacteriol.* **173**:7549–7556.

45. Enterohemorrhagic *E. coli* (EHEC) causes hemorrhagic colitis, the symptoms of which are bloody diarrhea with little or no fever. EHEC produces a toxin similar to Shiga toxin, produced by *Shigella dysenteriae*. An important EHEC serotype is 0157:H7. It is usually transmitted to humans from animals via contaminated raw milk or undercooked meat. Outbreaks have occurred in fast-food restaurants. Hemorrhagic colitis can be very dangerous in children under 5 years of age because the disease sometimes develops into hemolytic–uremic syndrome, in which the toxin causes kidney damage. *E. coli* is the most common cause of urinary tract infections. The most important virulence factor is the ability to adhere to uroepithelial cells via pilus adhesins.

46. Koronakis, V., V. K. J. Eswaran, and C. Hughes. 2004. Structure and function of TolC: the bacterial exit duct for proteins and drugs. *Annu. Rev. Biochem.* **73**:467–489.

47. Besides participating in the export of proteins via the type I system, TolC is used for the export of drugs (Section 16.5). In addition, TolC is used for the export of low molecular weight peptides such as heat-stable enterotoxin and cationic antimicrobial peptides, which are transported to the periplasm via the Sec system or other transporter, whereupon they access the TolC channel from the periplasm. Homologues of TolC have been found among many gram-negative bacteria and appear to be widespread.

48. The following are some animal pathogens that use the type III system to inject virulence proteins into eukaryotic cells, and the diseases they cause: *Yersinia* spp. (bubonic plague, pneumotnic plague, enterocolitis), *Salmonella* spp. (enteritis, septicemia, typhoid fever), *Shigella* spp. (gastroenteritis), *Escherichia coli* (enteropathogenic, enteroinvosive,

and enterohemorrhagic *E. coli* gastroenteritis), *Bordetella pertussis* (pertussis, also called whooping cough), *Pseudomonas aeruginosa* (burn wound infections; infections of the respiratory tract, urinary tract, ear, and eye; gastroenteritis; bacteremia; endocarditis).

49. Rosqvist, R., S. Håkansson, Å. Forsberg, and H. Wolf-Watz. 1995. Functional conservation of the secretion and translocation machinery for virulence proteins of yersiniae, salmonellae, and shigellae. *EMBO J.* **14**:4187–4195.

50. Kubori, T., Y. Matsushima, D. Nakamura, J. Uralil, M. Lara-Tejero, A. Sukhan, J. E. Galán, and S.-I. Aizawa. 1998. Supramolecular structure of the *Salmonella typhimurium* type III protein secretion system. *Science* **280**:602–605.

51. Wattiau, P., S. Woestyn, and G. R. Cornelis. 1996. Customized secretion chaperones in pathogenic bacteria. *Mol. Microbiol.* **20**:255–262.

52. Galán, J. E. 1996. Molecular genetic bases of *Salmonella* entry into host cells. *Mol. Microbiol.* **20**:263–271.

53. Hueck, C. J. 1998. Type III protein secretion systems in bacterial pathogens of animals and plants. *Microbiol. Mol. Biol. Rev.* **62**:379–433.

54. Büttner, D., and U. Bonas. 2002. Port of entry—the type III secretion translocon. *Trends Microbiol.* **10**:186–192.

55. Cornelis, G. R., and F. Van Gijsegem. 2000. Assembly and function of type III secretory systems. *Annu. Rev. Microbiol.* **54**:735–774.

56. Tampakaki, A. P., V. E. Fadouloglou, A. D. Gazi, N. J. Panopoulos, and M. Kokkinidis. 2004. Conserved features of type III secretion. *Cell. Microbiol.* **6**:805–816.

57. Blocker, A., P. Gounon, E. Larquet, K. Niebuhr, V. Cabiaux, C. Parsot, and P. Sansonetti. 1999. The tripartite type III secreton of *Shigella flexneri* inserts IpaB and IpaC into host membranes. *J. Cell Biol.* **147**:683–693.

58. Hoiczyk, E., and G. Blobel. 2001. Polymerization of a single protein of the pathogen *Yersinia enterocolitica* into needles punctures eukaryotic cells. *Proc. Natl. Acad. Sci. USA* **98**:4669–4674.

59. Evidence suggests that when the bacteria adhere to the target cell, some of the secreted proteins form a channel called a translocon in the target cell plasma membrane. Evidence for translocons includes the creation of pores by the insertion of putative translocon proteins into artificial lipid bilayers and the requirement of these proteins for the lysis of erythrocytes by *Yersinia* spp., *Pseudomonas aeruginosa*, *Shigella flexneri*, and enteropathogenic *Escherichia coli* (EPEC). In addition, cells infected with mutant *Y. enterocolitica* that can make translocon proteins but not Yop effector proteins have pores in their membranes. That is, if the infected cells are loaded with a fluorescent dye before infection, they release

the dye (but not cytosolic components) into the medium upon infection. This indicates that the infected cells did not lyse; rather, the translocon proteins formed narrow pores in the membrane of the infected cells. The needle may form a continuous passage with the translocon, or there may be a small space between the tip of the needle and the translocon.

60. Bacteria generally produce two to eight different effector molecules secreted by the type III system. There have been more than 20 such molecules described for various bacteria. For example, the *Yersinia* effectors delivered by the Yop apparatus are YopE, YopH, YopM, YopJ/P, YopO/YpkA, and YopT, and the *Pseudomonas aeruginosa* effectors delivered by the Psc apparatus are ExoS, ExoT, ExoU, and ExoY. The targets of the effector molecules nad the effects on the animal cell vary, but there are two major targets: the host cell's cytoskeleton system and the inflammatory response and cell signaling. The text and the referenced notes describe the Yops and how they block phagocytosis and interfere with cell signaling and the inflammatory response. This note gives more information about effector proteins produced by *Shigella* and *Salmonella*, and how they stimulate the uptake of the bacteria into nonphagocytic host cells.

As an example of the promotion of uptake, we will consider *Shigella* and *Salmonella*. *Shigella* invades epithelial cells in the colon, destroys the colonic mucosa, and causes bacillary dysentery. *Salmonella* invades nonphagocytic epithelial cells in the intestinal tract and causes gastroenteritis in humans. Both bacteria enter the nonphagocytic cells during infection by stimulating endocytosis with effector molecules secreted by their type III systems. *Endocytosis* refers to the uptake by localized regions of the plasma membrane that surround what is at the cell surface and pinch off an intracellular vesicle containing the ingested material. For certain types of endocytosis, such as the uptake of pathogens, membrane protrusions surround the pathogen and fuse to form the vesicle. All forms of endocytosis require remodeling of the cell cortex because shape changes must occur. The actin cytoskeleton is an important part of the cell cortex, and it takes part in the remodeling, including the formation of membrane protrusions required for the uptake of pathogens. The mammalian cell contains a pool of actin monomers in the cytoplasm, and these monomers can be recruited to form actin filaments that can take part in shape change. *Shigella* binds to the surface of the epithelial cells and induces the formation of microspike-like protrusions from the cell surface, containing actin filaments aligned lengthwise. The protrusions envelop the bacterium and pinch off an endocytic vesicle containing the bacterium. The process resembles phagocytosis. The formation of the protrusions and the uptake of *Shigella* are dependent upon localized actin polymerization, which itself is stimulated by the effector protein IpaC secreted by the type III apparatus called the Mxi–Spa system, encoded by genes on a resident plasmid. (Ipa stands for invasion plasmid antigen, and Mxi stands for membrane expression of invasion.) The engulfment into an endocytic vesicle by nonphagocytic cells induced by *Salmonella* is due to the production of membrane protrusions that are part of a process called membrane ruffling or macropinocytosis. *Salmonella* effector proteins delivered by the type III system after cell contact include four Sips (*Salmonella* invasion proteins), proteins that are homologous to the *Shigella* Ipa proteins. The formation of protrusions is stimulated by an effector protein SipC, which binds to actin, nucleates its polymerization into submembranous filaments (F-actin), and bundles the filaments into cables. Once inside the cells, *Shigella* and *Salmonella* replicate, although in different parts of the cell. *Shigella* escapes from the vesicle and replicates in the cytoplasm, killing the cell. *Shigella* spreads inter- and extracellularly to adjacent colonic epithelial cells and to the underlying layers of connective tissue, causing tissue destruction. *Salmonella* replicates inside the endocytic vesicle until the cell lyses.

61. Nhieu, G. T. V., E. Caron, A. Hall, and P. J. Sansonetti. 1999. IpaC induces actin polymerization and filopodia formation during *Shigella* entry into epithelial cells. *EMBO J.* **18**:3249–3262.

62. Hayward, R. D., and V. Koronakis. 1999. Direct nucleation and bundling of actin by the SipC protein of invasive *Salmonella*. *EMBO J.* **18**:4926–4934.

63. Zhou, D., M. S. Mooseker, and J. E. Galán. 1999. Role of the *S. typhimurium* actin-binding protein SipA in bacterial internalization. *Science* **283**:2092–2095.

64. Anderson, D. M., and O. Schneewind. 1997. A mRNA signal for the type III secretion of Yop proteins by *Yersinia enterocolitica*. *Science* **278**:1140–1143.

65. Page, A.-L., and C. Parsot. 2002. Chaperones of the type III secretion pathway: jacks of all trades. *Mol. Microbiol.* **46**:1–11.

66. Lee, S. H., and J. E. Galán. 2004. *Salmonella* type III secretion-associated chaperones confer secretion-pathway specificity. *Mol. Microbiol.* **51**:483–495.

67. San Ho Lee and J. E. Galán, in ref. 66, their publication showing the requirement for the chaperone-binding domain as well as the N-terminal signal for secretion of *Salmonella* proteins through the type III system, cite earlier publications on *Yersinia* indicating that about 20 amino-terminal acids will mediate the secretion of some reporter proteins via the type III system into the extracellular environment, and that some *Yersinia* effector proteins can be secreted in the absence of the chaperone-binding region or in the absence of these amino-terminal acids. They suggest that perhaps that secretion through the flagellar apparatus may have occurred in the *Yersinia* studies. See: Lee, S. H., and J. E. Galán. 2004.

68. *Yersinia* multiply outside the host cells and can also multiply within certain cell types, such as nonactivated macrophages in the host. *Yersinia enterocolitica* causes a severe enterocolitis, especially in young children. The symptoms include fever, diarrhea, and abdominal pain. The disease reservoir is other animals (e.g., farm animals, cats, dogs). This enterocolitis is transmitted to humans via contaminated foods (e.g., milk, undercooked meat). The bacterium, which can also be found in well water and lakes, produces an enterotoxin that causes diarrhea. *Y. enterocolitica* grows in the lumen of the bowel, but can also adhere to and invade specialized epithelial cells, called M cells, of the intestinal mucous membranes. The M cells overlie clusters of lymphoid tissue (T cells, B cells, macrophages) called Peyer's patches. *Y. enterocolitica* (and *Y. pseudotuberculosis*) invade the M cells but do not replicate there. Instead, they pass through the M cells to the underlying lymphoid tissue, enter macrophages, and spread via the macrophages to mesenteric lymph nodes, where they replicate. *Y. pseudotuberculosis* is primarily a pathogen of wild and domestic animals and is transmitted to humans by the eating of infected meat. The bacteria cause necrotic lesions in the liver, spleen, and lymph nodes. In humans these is swelling of the mesenteric lymph nodes, accompanied by severe abdominal pain and, usually, diarrhea and fever. *Yersinia pestis* causes plague. The bacteria live in wild rodents (rats, mice, ground squirrels, chipmunks, prairie dogs) and are transmitted to humans by fleas that have bitten these rodents and then bite humans. The wild rodents are relatively resistant to disease. The bacteria spread to the draining lymph nodes, where they grow. A high fever and swollen lymph nodes (buboes) develop, and the infection at this stage is called bubonic plague. If the infection is not treated, the bacteria spread to the blood and cause a septicemia. Once in the blood, the bacteria spread throughout the body to all the organs. Subcutaneous bleeding from ruptured blood cells produces black spots on the skin, and the disease at this stage has been called Black Death. If the lungs become infected, the infection is called pneumonic plague. Mortality from pneumonic plague approaches 100%. All three pathogenic species of *Yersinia* can prevent uptake by specialized phagocytic cells such as macrophages and polymorphonuclear leukocytes. Hence they replicate extracellularly and are found primarily outside the cells during an infection. Prevention of phagocytosis is due to the type III secretion system, as described in the text.

69. Cornelis, G. R., and H. Wolf-Watz. 1997. The *Yersinia* Yop virulon: a bacterial system for subverting eukaryotic cells. *Mol. Microbiol.* **23**:861–867.

70. Silhavy, T. J. 1997. Death by lethal injection. *Science* **278**:1085–1086.

71. Cornelis, G. R. 2002. *Yersinia* type III secretion: send in the effectors. *J. Cell Biol.* **158**:401–408.

72. Young, B. M., and G. M. Young. 2002. Evidence for targeting of Yop effectors by the chromosomally

encoded Ysa type III secretion system of *Yersinia enterocolitica. J. Bacteriol.* 184:5563–5571.

73. When *Y. enterocolitica* were grown in certain media, they formed needles composed of the YscF protein. When cells with preformed needles were incubated with Hela cells in media that did not allow the formation of needles, the Hela cells did not undergo cytotoxic damage. However, if the medium allowed needle formation, the Hela cells were damaged. Thus, contact with the Hela cells was presumed to trigger the formation of new needles, and as a consequence the Hela cells were punctured and toxins were translocated. It was suggested that the polymerization of protein subunits into needles provides the force to penetrate the target cell. See ref. 58.

74. Lee, V. T., S. K. Mazmanian, and O. Schneewind. 2001. A program of *Yersinia enterocolitica* type III secretion reactions is activated by specific signals. *J. Bacteriol.* 183:4970–4978.

75. Day, J. B., F. Ferracci, and G. V. Plano. 2003. Translocation of YopE and YopN into eukaryotic cells by *Yersinia pestis yopN, tyeA, sycN, yscB*, and *IcreG* deletion mutants measured using a phosphorylatable peptide tag and phosphospecific antibodies. *Mol. Microbiol.* 47:807–823.

76. The inflammatory response, which is important for fighting infections, includes localized vasodilation and increased capillary permeability. These effects are associated with the influx from the blood vessels of phagocytic cells into the infected tissue. Cytokines are very important in the infection-fighting process. Cytokines are low molecular weight proteins secreted primarily by activated lymphocytes and activated macrophages, as well as by other cells in the body. They are cell-to-cell signaling molecules that activate the recipient cell. Cytokines produced by white blood cells such as lymphocytes (responsible for the immune response), eosinophils, and basophils, as well as by macrophages (phagocytic cells found in the tissues), are called interleukins. Interleukin 8 (IL-8) is secreted by activated macrophages, as well as by endothelial cells that compose the inner lining of blood vessels. IL-8 targets the phagocytic white blood cells called neutrophils, inducing the adherence of circulating neutrophils to vascular endothelial cells. This action promotes the exiting of the neutrophils from the blood vessels into tissues spaces, where they phagocytize and kill infectious agents. IL-8 is also a chemoattractant for the neutrophils. Tumor necrosis factor α (TNF-α) is a cytokine secreted by activated macrophages during an infection. Macrophages infected by gram-negative bacteria can be stimulated to secrete increased amounts of TNF-α by bacterial lipopolysaccharide (LPS). (The secretion of IL-8 is LPS independent.) TNF-α stimulates the adherence of circulating neutrophils to the endothelial lining of blood vessels and their exit into the tissue spaces. TNF-α also stimulates macrophages and endothelial cells to produce IL-8, which as mentioned, also promotes the exit of neutrophils into the tissue spaces to fight infections. Since certain Yop proteins, such as YopJ, inhibit IL-8 and TNF-α secretion; they are anti-inflammatory.

77. Viboud, G. I., S. S. Kin, M. B. Ryndak, and J. B. Bliska. 2003. Proinflammatory signaling by the type III translocation factor YopB is counteracted by multiple effectors in epithelial cells infected with *Yersinia pseudotuberculosis. Mol. Microbiol.* 47:1305–1315.

78. A summary of the functions of the Yop proteins: YopE, YopH, YopO (also known as YpkA), and YopT inhibit phagocytosis by macrophages and polymorphonuclear leukocytes. YopE and YopT disrupt the actin-containing microfilaments in the target cells and in this way collapse the cytoskeleton, as well as inhibiting phagocytosis. YopH is a protein tyrosine phosphatase. It dephosphorylates macrophage proteins, and in this way has been suggested to disrupt phosphate-dependent signal transduction necessary for normal phagocytosis. YopH has also been reported to dephosphorylate scaffolding proteins that are important for the formation of focal adhesions. When these proteins are dephosphorylated, integrin-mediated phagocytosis of bacteria by host cells is inhibited. YopH also prevents the production of toxic forms of oxygen that are produced by macrophages upon phagocytosis of pathogens. YopO (YpkA) binds to actin, but the mechanism by which it inhibits phagocytosis is not clear. YopJ (also known as YopP) inhibits the production of proinflammatory cytokines such as TNF-α and IL-8 by infected cells (e.g., macrophages). Thus, inflammation, which is a way for the body to fight infections, is decreased. The intracellular action of YopM is unknown, although the protein does localize to the nucleus.

79. YopE activates the GTPase activity of certain other proteins called RhoA, Rac, and Cdc42. These proteins bind GTP and as a consequence accelerate actin polymerization and phagocytosis. When YopE accelerates the hydrolysis of GTP by these proteins, it inactivates them with respect to actin polymerization. YopT inactivates RhoA, Rac, and Cdc42 but by a different mechanism. It is a protease that detaches these proteins from the membrane, and thus inactivates them.

80. Pullulan is a starchlike polysaccharide formed by the fungus *Pullularia pullulans*. Pullulanases are enzymes that catalyze the cleavage of certain α-1,4 and/or α-1,6 linkages in starch and related oligosaccharides such as pullulan. These enzymes have been isolated from bacteria, archaea, and fungi. Some pullulanases are used industrially as starch-debranching enzymes as an early step in the conversion of starch to saccharides. Further saccharification is catalyzed by α or β-amylases. There are four types of pullulanase, and they differ in the specificities of the bonds that are cleaved. The pullulanase from *Klebsiella pneumoniae* hydrolyzes α-1,6 linkages.

81. It has been pointed out that the use of Gsp to denote secreton proteins can be misleading because

the secreton is not part of the GSP pathway. Furthermore, the Tat pathway can be used, as well as the GSP pathway for the translocation of proteins across the cell membrane prior to secretion by the secreton. Because of this, the type II pathway is sometimes referred to as the secreton-dependent pathway (SDP), and the proteins are denoted simply as "protein C," "protein D," and so on. For more discussion of this point, see: Desvaux, M., N. J. Parham, A. Scott-Tucker, and I. R. Henderson. 2004. The general secretory pathway: a general misnomer? *Trends Microbiol.* 12:306–309.

82. Protein D forms an outer membrane pore (secreton). Protein S is an outer membrane lipoprotein thought to stabilize the protein D complex. Protein E appears to be an ATPase involved in providing energy for the translocation and assembly of proteins G to K, which form a pseudopilus in the cytoplasm. The role for protein F is not known. Proteins G to K are homologous to type IV pili and are suggested to form a pseudopilus in the periplasm. Protein O is a prepilin peptidase involved in processing and methylating the N-terminal ends of proteins G to K. It is held to be unlikely that the pseudopilus provides a channel in the periplasm because the secreted proteins are folded and the pilus is too narrow for the folded proteins. One model postulates that the pseudopilus pushes the secreted protein through the outer membrane pore. The idea is that the cytoplasmic membrane-bound pseudopilus would retract (depolymerize) and extend (polymerize), and in doing so would push the folded protein into the protein D pore in the outer membrane, or open the gated channel. This might be similar to the retraction and extension of type IV pili described in twitching motility for *Pseudomonas* and *Myxococcus* in Chapter 18. For a discussion of this model, as well as other aspects of type II secretion, the student is referred to: Sandkvist, M. 2001. Biology of type II secretion. *Mol. Microbiol.* 40:271–283; Camberg, J. L., and M. Sandkvist. 2005. Molecular analysis of the *Vibrio cholerae* type II secretion ATPase EpsE. *J. Bacteriol.* 187:249–256.

83. Sandkvist, M., L. O. Michel, L. P. Hough, V. M. Morales, M. Bagdasarian, M. Koomey, V. J. Dirita, and M. Bagdasarian. 1997. General secretion pathway (*eps*) genes required for toxin secretion and outer membrane biogenesis in *Vibrio cholerae*. *J. Bacteriol.* 179:6994–7003.

84. d'Enfert, C., A. Ryter, and A. P. Pugsley. 1987. Cloning and expression in *Escherichia coli* of the *Klebsiella pneumoniae* genes for production, surface localization and secretion of the lipoprotein pullulanase. *EMBO J.* 6:3531–3538.

85. Christie, P. J., and J. P. Vogel. 2000. Bacterial type IV secretion: conjugation systems adapted to deliver effect or molecules to host cells. *Trends Microbiol.* 8:354–360.

86. Yeo, H.-J., and G. Waksman. 2004. Unveiling molecular scaffolds of the type IV secretion system. *J. Bacteriol.* 186:1919–1926.

87. Cascales, E., and P. J. Christie. 2004. Definition of a bacterial type IV secretion pathway for a DNA substrate. *Science* 304:1170–1173.

88. Jakubowski, S. J., V. Krishnamoorthy, E. Cascales, and P. J. Christie. 2004. *Agrobacterium tumefaciens* VirB6 domains direct the ordered export of a DNA substrate through a type IV secretion system. *J. Mol. Biol.* 341:96–977.

89. Ritz, D., and J. Beckwith. 2001. Roles of thiol–redox pathways in bacteria. *Annu. Rev. Microbiol.* 55:21–48.

90. Kadokura, H., M. Bader, H. Tian, J. C. A. Bardwell, and J. Beckwith. 2000. Roles of a conserved arginine residue of DsbB in linking protein disulfide-bond-formation pathway to the respiratory chain of *Escherichia coli*. *Proc. Natl. Acad. Sci. USA* 97:10884–10889.

91. During aerobic growth, reduced DsbB is oxidized by ubiquinone. Electrons are transferred from ubiquinone to oxygen via cytochrome oxidase *bd* or *bo*. During anaerobic growth reduced DsbB is oxidized by menaquinone, and sometimes ubiquinone, depending upon the terminal electron acceptor. For a description of these two electron carriers and the roles that they play in electron transport, see Section 4.7.1. As far as the structure of DsbB is concerned, the DsbB protein has four transmembrane and two periplasmic loops. Each of the periplasmic loops has four cysteine residues. These are Cys41 and Cys44 (Cys41-X1-X2-Cys44), which form a disulfide bond, and Cys104 and Cys130, which also form a disulfide bond. One way to describe what is happening is in terms of transferring disulfide bonds. Thus, the disulfide bond in DsbB between Cys41 and Cys44 is transferred to Cys104 and Cys130, and from there to Cys30 and Cys33 in DsbA. DsbA transfers its disulfide bond to the target periplasmic protein which then folds appropriately. The disulfide bond between Cys41 and 44 in DsbB is regenerated by transferring electrons to quinones in the electron transport chain. Another way of stating this is that DsbA reduces the disulfide bond at Cys104–Cys130 in DsbB, and in so doing regenerates its own disulfide bond that can accept electrons from cysteine residues in the target periplasmic protein. The electrons are then transferred from the sulfhydryl groups at Cys104 and Cys130 in DsbB to the disulfide bond at Cys41-Cys44. the resulting sulfhydryl groups at Cys41 and Cys44 are then reoxidized by the electron transport chain, regenerating the disulfide bond.

92. Sone, M., Y. Akiyama, and K. Ito. 1997. Differential *in vivo* roles played by DsbA and DsbC in the formation of protein disulfide bonds. *J. Biol. Chem.* 272:10349–10352.

18

Microbial Development and Physiological Adaptation: Varied Responses to Environmental Cues and Intercellular Signals

Bacteria may alter cell morphology, the production of flagella, cell metabolism, gene transcription, and cell behavior in response to environmental fluctuations such as the availability of respiratory electron acceptors, the supply of carbon, nitrogen, or phosphate, changes in the osmolarity and temperature of the medium, and whether the microorganisms are growing in liquid or on a solid surface. In this way bacteria survive in and adapt to ever-changing environmental conditions.

Indeed, bacteria can be found in essentially all environments where life is possible, and from this point of view they can be considered to be one of the most successful forms of life. In addition, individual bacteria that undergo physiological or developmental responses to signaling molecules produced by other bacteria of their own type in the population may then go on to engage in cooperative behavior. Such cooperative behavior includes biofilm formation, which is a common occurrence among bacteria, and fruiting body formation by myxobacteria. Underlying the adaptations and responses are sophisticated detection systems with which the bacterium continuously monitors the environment and transmits signals across its cell membrane to specific intracellular targets, which can be the transcriptional machinery, enzymes, or a cellular component such as the flagellum motor. This chapter describes some of these signaling circuits, and metabolic, behavioral, and developmental responses to the signals.

There are many signaling pathways that bacteria use to transmit environmental and intercellular signals to a cell target, be it cellular proteins (e.g., flagellar motors) or transcription factors that regulate gene expression. In most cases the signal causes the cells to activate or inactivate a cytoplasmic transcription factor that regulates the transcription of a gene. Some genes are induced by activated transcription factors, and some genes are repressed. The activation or inactivation of the transcription factor itself as a result of the signaling pathway usually occurs in one of two ways: by covalent modification [e.g., phosphorylation (activation) or dephosphorylation (inactivation)] of the transcription factor or by a conformational change of the transcription factor upon binding a coinducer.

Two-component systems, which will be described in Section 18.1, are the most common type of signaling system. They include a histidine kinase (HK) protein that autophosphorylates at a histidine residue in response to a signal, then transfers the phosphoryl group to an aspartate residue in a response regulator (RR) protein that regulates gene transcription in its phosphorylated form. There may be intermediate protein carriers that transfer the phosphoryl group from the HK to the RR. Among the examples of two-component systems discussed are the following:

1. The Arc system, which regulates gene transcription in response to oxygen supply

2. The Nar system, which regulates the transcription of genes required for nitrate and nitrite metabolism

3. The Che system, which regulates the rotation of flagellar motors (rather than gene transcription)

4. The RegB/RegA system in photosynthetic bacteria, which regulates the transcription of genes for the light-harvesting complex and the photoreaction center under anaerobic conditions

5. The Pho system, which regulates genes for phosphate assimilation when the supply of phosphate is limiting

6. The EnvZ/OmpR system, which regulates the expression of genes for porin synthesis

7. The KdpABC system, which regulates expression of genes for potassium ion transport

8. The Spo phosphorelay system for the regulation of sporulation genes in *Bacillus subtilis*

9. The CtrA system, which regulates gene expression during the *Caulobacter crescentus* cell cycle

10. The Asg system in *Myxococcus xanthus*, which regulates gene expression during multicellular development

11. The PhoQ/PhoP system, which regulates gene expression important for the virulence of *Salmonella*

12. The Bvg system, which regulates the expression of virulence genes in *Bordetella pertussis*

13. The Agr system in *Staphylococcus aureus*, which regulates virulence gene expression

14. The Vir system in *Agrobacterium tumefaciens*, which regulates the expression of genes necessary for the transfer of plasmid DNA to plant cells

Also described is the FNR (fumarate nitrate reductase) system. It regulates the expression of genes required for fermentation and anaerobic respiration but is not a two-component system.

Other signaling systems discussed in this chapter include quorum-sensing systems, which are widespread in bacteria and are important for cell-to-cell signaling at high population densities. Some of the quorum-sensing systems send signals to two-component signaling systems, and others do not. Quorum-sensing systems rely on the production of extracellular signaling molecules that accumulate in the external medium as the population density increases.

The quorum-signaling molecules were first discovered as signals that induced luminescence in bioluminescent bacteria; and because the signal induced the gene that encoded the enzyme that made the signal in a positive feedback loop, the quorum-sensing signals were originally called "autoinducers." The detection by the producing bacteria of a threshold concentration of the signaling molecule in the external medium lets the bacteria know how dense their own population is and stimulates adaptive changes in the bacteria when the cell population is dense enough for these changes to occur. In this sense, quorum sensing reflects a primitive form of multicellularity that has evolved among the bacteria, and as we shall see, is very important for the survival of bacteria in their natural habitat.

Quorum sensing regulates different processes, depending upon the species of bacteria: bioluminescence, biofilm formation, fruiting body formation in myxobacteria, the production of virulence factors, and conjugation are among the processes shown to be so regulated. The signaling molecules used by gram-positive bacteria are usually oligopeptides. The peptides generally work by binding to the externally exposed regions of specific membrane-bound two-component sensor kinases and activating the phosphorylation of the respective partner response regulator. In that way the peptides regulate gene transcription.

The signaling molecules used by gram-negative bacteria are generally acylated homoserine lactones (acyl–HSLs). Most of these function by binding to and activating a positive transcription factor in the cytoplasm, and in that way the signal serves as a coinducer. In some cases, the extracellular signals produced by bacteria derepress or repress the transcription of genes. This chapter will also give examples of other signaling molecules used by gram-positive and gram-negative bacteria. These include a furanosyl borate diester called

AI-2, used by both gram-positive and gram-negative bacteria, and a subset of amino acids used in the developmental signaling pathway by *Myxococcus xanthus*.

This chapter shows how the various signaling systems underlie adaptive metabolic changes (Sections 18.2–18.10); it also covers the production of virulence factors by pathogenic bacteria (Section 18.11), chemotaxis, aerotaxis, and photoresponses (Sections 18.12–18.14). After an introduction to bacterial development and to quorum sensing (Section 18.15), we discuss myxobacteria multicellular development (Section 18.16), cell-cycle-dependent DNA replication and transcription in *Caulobacter* (Section 18.17), *Bacillus* sporulation and competence (Sections 18.18 and 18.19), bioluminescence (Section 18.20), LuxR/LuxI-like systems in nonluminescent bacteria (Section 18.21), and the formation of biofilms (Section 18.22). Since many of the signaling systems feed into two-component signaling systems, we begin with an introduction to two-component systems (Section 18.1).

18.1 Introduction to Two-Component Signaling Systems

In several systems, a signal transduction pathway exists, called a two-component system. Each two-component system includes a *histidine kinase* (HK) protein (often called a *sensor kinase*) that receives a signal and transmits it to a partner *response regulator* (RR) protein, sometimes via other proteins that take part in a phosphorelay.[1] The response regulator protein in turn transmits the signal to the target. This is summarized in Fig. 18.1.

Histidine kinases have two domains, an *input domain* and a *transmitter domain*. Response regulators have two domains, a *receiver domain* and an *output domain*. See note 2 for a further discussion of domains in sensor kinases and response regulators. Specifically, the histidine kinase receives a signal at its input domain and autophosphorylates (using ATP as the phosphoryl donor) at a histidine residue in its transmitter domain. The transmitter domain is the carboxy-terminal region comprising approximately 240 amino acids. The histidine kinase then transfers the phosphoryl group to an aspartate residue in the receiver domain of the partner response regulator protein. The receiver domain is the amino-terminal region comprising about 120 amino acids. This activates the response regulator, which transmits the signal to its target via its output domain (Section 18.1.2 and Fig. 18.1). Most of the known

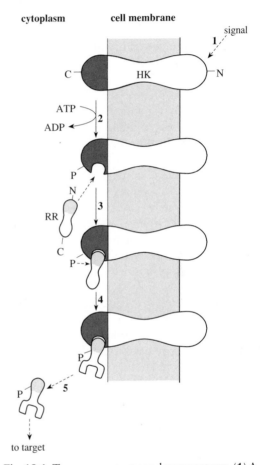

Fig. 18.1 Two-component regulatory systems. (1) A transmembrane histidine kinase (HK) is activated by a signal at its N-terminal domain. (2) The activated protein autophosphorylates in the C-terminal domain. (3) The response regulator protein (RR) binds to the C-terminal end of the histidine kinase and the phosphoryl group is transferred from the histidine kinase to the response regulator, thus activating the latter. (4) The activated response regulator leaves the histidine kinase and stimulates its target. Shaded and stippled areas of the histidine kinase and response regulator represent conserved amino acid sequences typical for the respective class of protein. The change in shape of the proteins represents a presumed conformational change. In some systems the histidine kinase is cytoplasmic and detects signals within the cytoplasm.

phosphorylated response regulators (RR-P) bind to DNA and stimulate or repress the transcription of specific genes. Exceptions include P-CheB and P-CheY, which affect the chemotaxis machinery (Section 18.12). Histidine kinases may reside in the cell membrane (usually transmembrane) or in the cytoplasm, although they are often in the membrane. The response regulators are in the cytoplasm.

The signaling pathway also includes a phosphatase that dephosphorylates the response regulator, returning it to the nonstimulated state, where it once again can respond to the signal. The phosphatase may be the histidine kinase itself, the response regulator, or a separate protein. Even though signaling systems that consist of a histidine kinase and a response regulator protein are called "two-component" systems, often there are more than two proteins in the signal transduction pathway, since additional proteins may exist that carry the phosphate from the histidine kinase to the response regulator protein. Proteins that make up the phosphorelay pathway between the histidine kinase and the partner response regulator protein are called *phosphotransferases*. Two-component signaling systems involving histidine kinases also occur in the archaea, fungi, plants, slime molds, and presumably other eukaryotes.[3,4]

It must be emphasized that the histidine kinase need not be the first protein in the signal transduction pathway to respond to the signal. In other words, it need not be the sensor. In many systems, signals first interact with protein(s) other than the histidine kinase, and the stimulus is relayed to the histidine kinase. For example, in the E. coli chemotaxis system described in Section 18.12, the transmembrane proteins called chemoreceptors, or MCP proteins, are sensor proteins that respond to chemoeffectors, and as a consequence, change the activity of a cytoplasmic histidine kinase (CheA). Section 18.5 presents another example of an initial receiver of the signal that is not the histidine kinase occurs in the PHO regulon control system, which is repressed by inorganic phosphate. (A regulon is a set of noncontiguous genes or operons controlled by the same transcription regulator.) The proteins that initially bind inorganic phosphate are in the phosphate transport system (Pts), which is believed to bind inorganic phosphate and then stimulate

the enzymatic activity of the membrane-bound PhoR histidine kinase. A third (and more complicated) example is found in the Ntr regulon, which is repressed by ammonia (Section 18.4). The ammonia levels determine the concentrations of glutamine and α-ketoglutarate via the enzymes glutamine synthetase and glutamate synthase, respectively. (As discussed in Section 9.3.1, these enzymes together function to incorporate ammonia into glutamate and glutamine, which donate amino groups to other molecules during biosynthesis.) The α-ketoglutarate and glutamine in turn influence the activity of a bifunctional enzyme, uridylyltransferase–uridylyl-removing (UT-UR) enzyme, which modifies a signal transduction protein (P_{II}), which in turn regulates the activity of a cytoplasmic histidine kinase (N_{II}). In this case the histidine kinase is indeed far removed from the initial signal, ammonia.

Two-component signaling systems have been discovered in many bacteria, both gram-negative and gram-positive, and have been implicated in a wide range of physiological responses.[5-7] These include nitrogen assimilation, outer membrane porin synthesis, chemotaxis in E. coli and S. typhimurium, nitrogen fixation in Klebsiella and Rhizobium, sporulation in Bacillus, fruiting body formation in myxobacteria, oxygen regulation of gene expression in E. coli, the uptake of carboxylic acids in Rhizobium and Salmonella, the production of virulence factors by Salmonella and Bordetella, and bioluminescence in some Vibrio species. It is clearly a widespread and important signal transduction system that enables bacteria to adapt to changes in the external milieu. As mentioned, there is also evidence to suggest that similar systems occur in eukaryotes.[8,9]

18.1.1 Components of two-component signaling systems

We noted at the outset that the "two-component" systems contain proteins of three types.

1. A *histidine kinase (HK)*, sometimes called a *sensor kinase*. The histidine kinase receives a signal at its input domain and autophosphorylates at a histidine residue in its transmitter domain:

$$HK + ATP \xrightarrow{\text{signal}} HK\text{-}P + ADP$$

2. *A partner response regulator (RR)*, also called a cognate response regulator, which is phosphorylated at an aspartate residue in its receiver domain by HK-P and sends a signal to its target (e.g., the genome or the flagella motor) via its output domain:

$$RR + HK\text{-}P \rightarrow RR\text{-}P_{active} + HK$$

However, not all signals result in increased synthesis of RR-P. See Section 18.1.2. Furthermore, as noted earlier, there may exist enzymes called phosphotransferases that carry the phosphate from the HK to the RR in what is referred to as a phosphorelay pathway.

3. *A phosphatase*, which inactivates the RR-P:

$$RR\text{-}P + H_2O \rightarrow RR + P_i$$

The phosphatase may be the histidine kinase, the response regulator, or a separate protein. As mentioned, some signals stimulate phosphatase activity and thus act as inhibitors or repressors rather than stimulators or inducers. We will see some examples of this later.

18.1.2 Signal transduction in two-component systems

Figure 18.1 illustrates a simplified model of signal transduction in two-component systems. In this model the signal stimulates phosphorylation of the response regulator protein. (The model will be modified later to accommodate differences between signaling systems, such as the inclusion of a separate sensor protein that receives the signal and transmits it to the histidine kinase, and to show systems in which the signal actually results in less phosphorylation of the regulatory protein, hence in a suppression of a particular response rather than an activation.)

In Fig. 18.1, the histidine kinase (HK) is depicted as a transmembrane protein composed of three domains: an N-terminal domain that is presumed to be at the outer surface of the cell membrane and to bind to an external signaling ligand; a hydrophobic domain that is transmembrane; and a conserved C-terminal domain that is cytoplasmic. Some histidine kinases (e.g., NR_{II} and CheA) are cytoplasmic proteins (Section 18.12.4). Signal transduction as depicted in Fig. 18.1 can be conveniently thought of as occurring in three steps.

Step 1. In response to a stimulus at the N-terminal domain, the histidine kinase auto-phosphorylates at the C-terminal domain. The phosphoryl donor is ATP.

Step 2. The phosphoryl group is transferred from the histidine kinase to its partner response regulator protein. All the response regulator proteins are related in having conserved amino acid sequences in the N-terminal domain (usually) that may bind to the conserved C-terminal region of the histidine kinases.

Step 3. After phosphorylation, the response regulator becomes activated and changes the activity of its target. The effect is usually the stimulation or repression of gene transcription. As we shall see, some response regulator proteins (e.g., NarL) do both, depending upon the target gene (Section 18.3). Other response regulators may have targets other than the genome. For example, in chemotaxis, the phosphorylated derivatives of the response regulators CheY and CheB change the rotational direction of the flagella motors, and the extent of methylation of the chemoreceptor proteins in the membrane, respectively (Section 18.12.4). The response regulator proteins differ at their C-terminal domains, and this probably confers specificity with regard to their targets and activities.

Not all two-component systems respond to the signal by increasing the phosphorylation of the response regulator protein. Sometimes the signal results in either dephosphorylation of the response regulator protein or inhibition of the phosphorylation of the response regulator protein. In either case, the level of RR-P falls rather than rises in response to the signal. The signal can result in the dephosphorylation of RR-P when the histidine kinases are bifunctional enzymes that can act either as kinases or as phosphatases when stimulated, depending upon the particular histidine kinase. For example, in the repression of the Ntr regulon by ammonia, the signal causes stimulation of the phosphatase activity of the histidine kinase, and as a consequence, represses gene transcription. In the presence of excess ammonia, the histidine kinase NR_{II} acts as a phosphatase rather than as a kinase and inactivates the response regulator, NR_I, which in its phosphorylated form is a positive transcription factor (Section 18.4.1). Another example may be the repression of the PHO regulon by inorganic phosphate. The histidine kinase, PhoR, appears to respond to excess inorganic phosphate by dephosphorylating the response regulator, PhoB (Section 18.5). In the case of chemotaxis to an attractant signal, the signal *suppresses* histidine kinase activity rather than stimulating the activity. This results in the formation of less RR-P (i.e., CheY-P, the protein that causes the cells to tumble and swim randomly: see Section 18.12.2), and consequently the cells swim smoothly toward the attractant.

18.1.3 Amino acid sequences define histidine kinases and response regulator proteins

The histidine kinases are defined by a conserved sequence of about 200 amino acids at the C-terminal end. The C-terminal domain is the site of the conserved histidine residue that becomes phosphorylated in response to a stimulus (Fig. 18.2). As indicated in Fig. 18.1, most of the known histidine kinases are transmembrane. The C-terminal end is in the cytoplasm, where it interacts with the response regulator protein. The N-terminal end may be exposed on the extracellular membrane surface (the periplasm in gram-negative bacteria). As stated earlier, the amino acid sequence at the N-terminal domain varies with the different histidine kinases, presumably because they

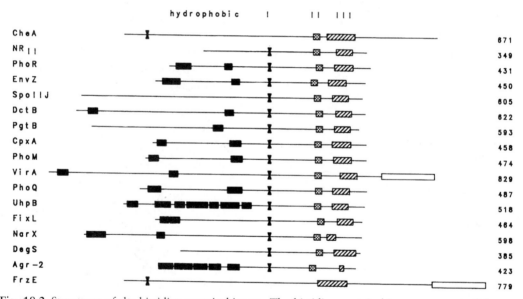

Fig. 18.2 Structures of the histidine protein kinases. The histidine protein kinases are part of the two-component regulatory systems. They autophosphorylate at a histidine residue and then transfer the phosphate to a response regulator protein. Most of the histidine kinases are believed to be transmembrane proteins with an extracytoplasmic amino terminus that responds to stimuli. Exceptions are CheA, NR_{II}, and FrzE, which are cytoplasmic. Hydrophobic domains that presumably span the membrane are indicated by the solid boxes at the amino-terminal end. Domain I, the cytoplasmic domain that includes the phosphorylated histidine residue, is indicated by the solid hours loss symbol. Regions II and III (stippled and hatched boxes) represent regions in the carboxy-terminal domain where certain amino acids appear with high frequency at specific locations in the sequence. These are called conserved regions. For example, when the amino acid sequences are lined up for comparison, position 43 may be a glycine in all the proteins (totally conserved), whereas position 44 may be arginine in, say, 60% of the proteins (partially conserved), and so on. The open boxes at the extreme carboxy ends are homologous to response regulator domains at their amino-terminal ends. Lengths of proteins in amino acid residues, as predicted from nucleotide sequences, shown in right-most column. *Source*: Stock, J. B., A. J. Ninfa, and A. M. Stock. 1989. Protein phosphorylation and regulation of adaptive responses in bacteria. *Microbiol. Rev.* 53:450–490.

respond to different stimuli. The response to the stimulus within the N-terminal region causes the enzyme to autophosphorylate the conserved histidine residue in the cytoplasmic domain (Fig. 18.1).

The *response regulator proteins* are defined by a conserved amino-terminal domain of about 100 amino acids (Fig. 18.3). The conserved amino-terminal end of the response regulator protein is thought to interact with the conserved carboxy end of the histidine kinase, becoming phosphorylated at an aspartate residue. The phosphate is eventually removed by a phosphatase, which may be the histidine

kinase, the response regulator protein, or perhaps a third protein. Since the carboxy ends of the histidine kinases and the amino ends of the different response regulator proteins are conserved, it is theoretically possible that a histidine kinase in one signaling system may activate a response regulator of a different system. This possible interaction between different signaling systems would be an example of what has been termed "cross-regulation," the regulation of a response regulator by a signal that comes from a source other than the cognate histidine kinase.[10] Cross-regulation is discussed in Section 18.5.1.

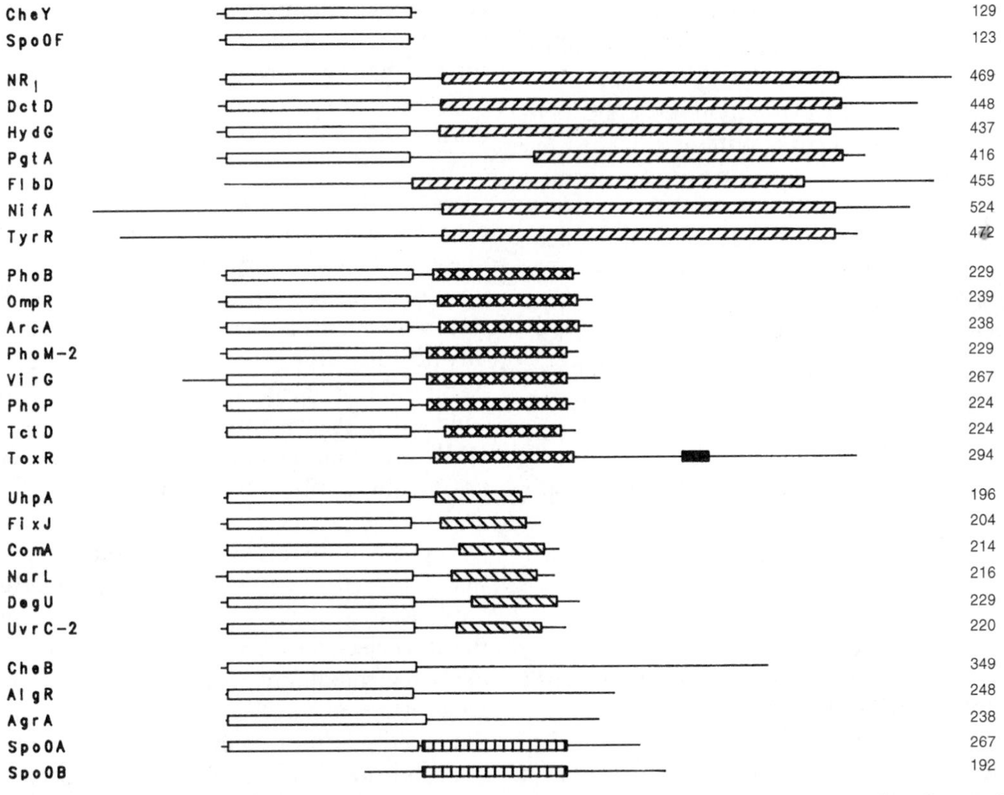

Fig. 18.3 Structures of the response regulator proteins. Response regulator proteins are cytoplasmic proteins that are phosphorylated, or presumed to be phosphorylated, by the histidine kinase proteins. The phosphorylated regulator proteins transmit the signal to the genome or to some other cellular machinery (e.g., the flagellar motor). The amino-terminal region of the response regulator protein has conserved amino acids (open boxes). Approximately 20 to 30% of the amino acids are identical at corresponding posititions when the sequences are aligned. This is the region thought to interact with the carboxy domain of the histidine kinase, and to become phosphorylated. Other conserved regions are indicated by cross-hatched boxes. These are in the carboxy-terminal domain, which is thought to interact with target molecules. For example, NR$_I$, DctD, and NifA share a homologous carboxy-terminal domain that is thought to interact with one of the *E. coli* sigma factors, σ^{54}. NR$_I$ and NifA have a common carboxy-terminal region that is thought to bind to DNA. One of the proteins, ToxR, spans the membrane, and the hydrophobic region is indicated by the solid box. Lengths of the protein in amino acid residues shown as in Fig. 18.2. *Source*: Stock, J. B., A. J. Ninfa, and A. M. Stock. 1989. Protein phosphorylation and regulation of adaptive responses in bacteria. *Microbiol. Rev.* 53:450–490.

18.2 Responses by Facultative Anaerobes to Anaerobiosis

A shift from an aerobic to an anaerobic atmosphere results in extensive changes in the metabolism of facultative anaerobes due to the repression of genes required for aerobic growth and the induction of genes necessary for anaerobic growth.[11] These adaptive responses, which have stimulated much interest, are described next. Most of the discussion focuses on what has been learned from studying E. coli and related bacteria. However, similar systems exist in other bacteria. The metabolic changes will be discussed first, followed by a description of the regulation of the relevant genes. Many genes that are responsive to anaerobiosis are globally regulated by two systems: the Arc (two-component) system, and the FNR (not two-component), system.[12,13] Then we shall describe the regulation of the formate hydrogen–lyase system, which also responds to anaerobiosis but is regulated by formate and a transcription activator, FhlA. The RegB/RegA system will also be described; this two-component system stimulates the transcription of genes for the light-harvesting complex and photoreaction center of certain photosynthetic bacteria under anaerobic incubation conditions. A fifth system, the (NarL/NarP/NarX/NarQ) system, is a two-component system that regulates the response to nitrate and nitrite as electron acceptors under anaerobic conditions, is discussed in Section 18.3.

18.2.1 Metabolic changes accompanying the shift to anaerobiosis

During aerobic respiration, the citric acid pathway is cyclic and oxidative (Fig. 18.4A). However, in the absence of oxygen several important changes take place (Fig. 18.4B). These include the replacement of fumarase A and succinate dehydrogenase by fumarase B and fumarate reductase, respectively, and the repression of the synthesis of α-ketoglutarate dehydrogenase. As described in Section 8.10, the result of these enzymatic changes is the conversion of the oxidative citric acid cycle into a reductive noncyclic pathway. (Not all bacteria have a reductive citric acid pathway

when they respire anaerobically. For example, when Pseudomonas stutzeri grows anerobically, using nitrate as an electron acceptor, it oxidizes glucose completely to CO_2 and appears to have an oxidative citric acid cycle.[14] Similarly, Paracoccus denitrificans has a complete citric acid cycle when it grows anaerobically, with nitrate as the electron acceptor.[15])

Additionally, when E. coli is growing anaerobically, acetyl–CoA is no longer made by pyruvate dehydrogenase but rather by pyruvate–formate lyase. This is an advantage under anaerobic conditions because it decreases the amount of NADH that must be reoxidized. The acetyl–CoA thus formed is converted to acetate and ethanol, and the formate is converted to hydrogen gas and carbon dioxide via formate hydrogen–lyase. In E. coli there is also a decrease in synthesis of other enzymes used during aerobic growth (e.g., those of the glyoxylate cycle and fatty acid oxidation).

Depending upon the presence of particular electron acceptors, major changes also take place in the respiratory pathway.[16] Facultative anaerobes such as E. coli carry out aerobic respiration in the presence of oxygen; but in the absence of oxygen, they carry out anaerobic respiration by means of nitrate, fumarate, or some other electron acceptor (e.g., TMAO or DMSO, compounds that occur in nature and presumably exist in the intestine, where E. coli grows).

There is a hierarchy of electron acceptors that are used: oxygen is the most preferred, with nitrate second, followed by fumarate and the other electron acceptors. The hierarchy parallels the maximum work that can be done when electrons travel over the electrode potential gradient to the terminal acceptor. The work is proportional to the difference in electrode potential between the electron acceptor and donor. For example, the maximum work that can be done is greatest when oxygen is the electron acceptor ($E'_m = +0.82$ V), less when nitrate is the electron acceptor ($E'_m = +0.42$ V), and least when fumarate is the electron acceptor ($E'_m = +0.03$ V). As described in Section 4.4, the electrons are passed to these terminal electron acceptors from reduced quinone. For example, ubiquinone ($E'_m = +0.1$ V) transfers electrons from the reductant to either the cytochrome oxidase or the nitrate reductase module.

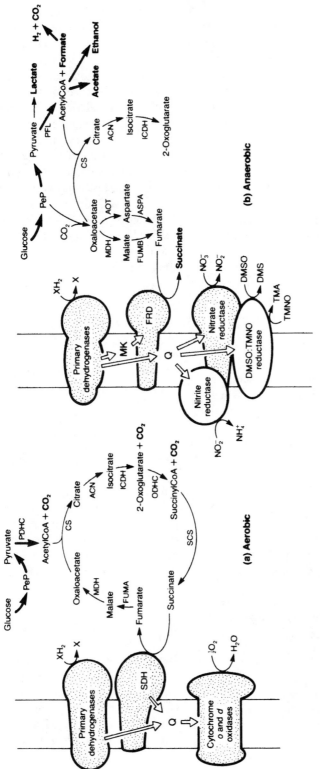

Fig. 18.4 Comparison of respiratory systems and carbon metabolism in *E. coli*. (A) An oxidative citric acid cycle feeds electrons into succinic dehydrogenase (SDH) and NADH dehydrogenase (a primary dehydrogenase). (B) The citric acid pathway is noncyclic and reductive because the α-ketoglutarate dehydrogenase activity is absent (or severely diminished). The electron transport chain has been altered so that electrons flow to acceptors other than oxygen. When nitrate is present, nitrate reductase (NR) is synthesized. Nitrate represses the synthesis of fumarate reductase (FRD) and DMSO:TMNO reductase. In the absence of an electron acceptor for anaerobic respiration, the major energy yielding pathways are fermentative. *Source:* Spiro, S., and J. R. Guest. 1991. Adaptive responses to oxygen limitation in *Escherichia coli. Trends Biochem. Sci.* 16:310–314.

Menaquinone ($E'_m = -0.074$ V), rather than ubiquinone, transfers electrons to the fumarate reductase complex. (Fumarate reductase is at too low a potential to accept electrons from ubiquinone.)

E. coli determines which electron acceptors will be used, in part by regulating the transcription of genes coding for the electron acceptors. Thus, in the presence of oxygen, the genes for nitrate reductase, fumarate reductase, and the other reductases are repressed, and therefore only aerobic respiration can take place. (However, see the discussion in Section 4.7.2 regarding denitrifying enzymes not sensitive to oxygen in some facultative anaerobes.) In the absence of oxygen but in the presence of nitrate, nitrate reductase genes are induced (by nitrate), but the genes for fumarate reductase and the other reductases are repressed by nitrate. Therefore, nitrate respiration takes place. In the absence of both oxygen and nitrate there is no longer any repression of fumarate reductase and so fumarate respiration takes place.

Changes also take place within the aerobic respiratory pathway. When oxygen levels are high, E. coli uses cytochrome o, encoded by the cyo operon, as the terminal oxidase. When oxygen becomes limiting (and during stationary phase), cytochrome o is replaced by cytochrome d, encoded by the cyd operon. One advantage to this is that cytochrome d has a higher affinity for oxygen ($K_m = 0.23$–0.38 μM) than has cytochrome o ($K_m = 1.4$–2.9 μM).

In summary, then, oxygen represses the synthesis of the anaerobic reductases, ensuring that oxygen is used as the electron acceptor in air. Nitrate induces the synthesis of nitrate reductase and represses the synthesis of the other reductases, ensuring that nitrate is used as the electron acceptor in the presence of nitrate anaerobically. In the absence of any exogenously supplied electron acceptor, E. coli relies on fermentation as its major source of ATP.

18.2.2 Regulatory systems that govern gene expression accompanying the shift to anaerobiosis

Four systems for the regulation of gene expression by oxygen or nitrate in E. coli will be described. Two of these are activated by anaerobiosis (ArcA/B and FNR), the third by nitrate

and nitrite (NarL/NarP/NarX/NarQ system); the fourth is activated by anaerobiosis and formate and repressed by nitrate (the FhlA regulon). Photosynthetic bacteria have a regulatory system, called the RegA/RegB system, which is activated under anaerobic conditions and controls the induction of photosynthetic genes, as decribed in Section 18.6.1. A common way to test the regulation of gene transcription is to construct a fusion between the promoter of the gene of interest and the gene for β-galactosidase (lacZ). Then strains harboring the fusion are tested under different conditions to see how the test conditions affect the production of β-galactosidase. (A more detailed explanation of lacZ fusions is given later, in note 25.) The four systems in E. coli for the regulation of gene expression by oxygen or nitrate are as follows:

1. *The Arc (aerobic respiratory control, or anoxic redox control) system.* This two-component, system represses under anaerobic conditions the transcription of several genes that are expressed only during aerobic growth. In addition, it stimulates the transcription of a much smaller number of genes that are expressed during microaerophilic or anaerobic conditions. The Arc system functions during aerobic and anaerobic growth.

2. *The FNR (fumarate nitrate reductase) system.* This system stimulates the transcription of many genes that are required for fermentation and anaerobic respiraton, and represses the transcription of some genes that function only during aerobic growth. More than 75 genes are regulated by FNR in E. coli. The FNR system is active only during anaerobic growth; it is not a two-component system.

3. *The NarL/NarP/NarX/NarQ system, also called the Nar system.* This two-component system stimulates the transcription of the nitrate reductase and other genes required for nitrate and nitrite metabolism, and represses the transcription of the other terminal reductase genes. The system is active only under anaerobic growth conditions in the presence of nitrate or nitrite. The Nar system is described in Section 18.3.

4. *The FhlA regulon, also called the formate regulon.* This systems consists of genes

repressed by oxygen and nitrate and induced by formate. The regulon includes several genes including those for formate hydrogenlyase, which is necessary to convert the fermentation end product formate to H_2 and CO_2.

The Arc system

The Arc system consists of a transmembrane sensor kinase (histidine kinase), ArcB, and a partner response regulator, ArcA. Mutations in the *arc* genes cause derepression of several genes and repression of a relatively short list of others, indicating that phosphorylated ArcA is both a repressor and an activator of gene transcription under anaerobic conditions. By analyzing these mutants, it has been possible to conclude that the Arc system is responsible for the following activities:

1. Anaerobic repression of genes for

Citric acid cycle enzymes

Glyoxylate cycle enzymes

Several dehydrogenases for aerobic growth (e.g., pyruvate dehydrogenase)

Fatty acid oxidation enzymes

Cytochrome o oxidase

2. Anaerobic induction of the gene for pyruvate formate–lyase[17,18]

3. Induction in low oxygen of the genes for cytochrome d oxidase and cobalamin synthesis[19] and (along with FNR), the pyruvate formate–lyase gene[17]

Interestingly, ArcA-P, but not ArcB, may also activate the transcription of the mating system genes, as explained in note 20.

When oxygen levels are sufficiently low, the ArcB protein autophosphorylates, using ATP as the phosphoryl donor, and then transfers the phosphoryl group to ArcA (Fig. 18.5).[21,22] How does ArcB detect changes in the levels of oxygen? Apparently, it is not oxygen itself that is the signaling molecule. This conclusion has been reached from two lines of evidence: (1) the

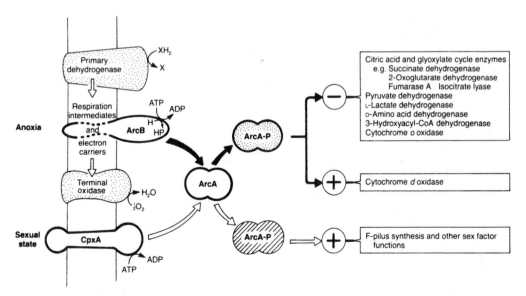

Fig. 18.5 The ArcA/ArcB regulatory system in *E. coli*. ArcB is a membrane protein activated by anoxia, perhaps by a reduced form of an electron carrier. The model postulates that the activated form of ArcB becomes phosphorylated and then phosphorylates ArcA, which then becomes a repressor of aerobically expressed enzymes and an inducer of cytochrome d oxidase. It has been found that ArcA-P also activates the genes for cobalamin synthesis (*cob* genes) in *Salmonella typhimurium*. (See: Andersson, D. I. 1992. Involvement of the Arc system in redox regulation of the *cob* operon in *Salmonella typhimurium*. *Mol. Microbiol.* 6:1491–1494.) ArcA also responds to CpxA, a membrane-bound sensor protein that is necessary for the synthesis of the F-pilus and other sex factor functions. *Source*: Spiro, S., and J. R. Guest. 1991. Adaptive responses to oxygen limitation in *Escherichia coli*. *Trends Biochem. Sci.* 16:310–314.

level of expression of the *sdh* (succinate dehydrogenase) operon in cells grown with different terminal electron acceptors and (2) the study of mutants with deletions in the cytochrome oxidase genes. These experiments are described next.

The level of expression of the *sdh* operon, which is under ArcA/ArcB control, varies with the midpoint potential of the electron acceptor. (The expression of the *sdh* operon is complex and is controlled by factors in addition to ArcA/ArcB. See note 23 for further comment.) The level of expression is highest with oxygen, lower with nitrate, and lowest with fumarate. This parallels the midpoint potential of the electron acceptor, which is most positive for oxygen, less positive for nitrate, and least positive for fumarate. Instead of responding to the terminal electron acceptors fumarate, nitrate, and oxygen per se, the ArcB protein may respond to a reduced carbon compound in the cell, such as the reduced form of an electron transport carrier (e.g., flavoprotein, quinone, cytochrome), or NADH, or some metabolic intermediate that might accumulate anaerobically.

One way to test this hypothesis is to delete the cytochrome o and d genes and measure the expression of genes under the control of the Arc system. Deletion of the cytochrome oxidase genes inhibits electron transport at the terminal step and would be expected to have extensive metabolic consequences, including, for example, an increase in the ratio of reduced to oxidized forms of electron carriers, and the accumulation of metabolites that are formed in the absence of an exogenous terminal electron acceptor. If this is the case, then deletion of the cytochrome oxidase genes would be expected to mimic the absence of oxygen. This proposition was tested, using *cyo–lacZ* and *cyd–lacZ* fusions as probes to monitor the expression of the *cyo* and *cyd* genes.[24] (Gene fusions are explained in note 25.) When both the *cyo* and *cyd* genes were deleted and the cells were grown in air, *cyo–lacZ* expression was lowered and *cyd–lacZ* expression increased, as if oxygen were absent. (See note 26 for a description of the control experiments.)

One interpretation of these experiments is that the reduced form of an electron transport carrier (e.g., a quinone) may signal ArcB, which in turn autophosphorylates and then phosphorylates ArcA (Fig. 18.5). The control of the

phosphorylating activity of ArcB also appears to involve certain cellular metabolites, such as pyruvate, acetate, and D-lactate, that increase when oxygen is lacking. These products increase the autophosphorylation activity of ArcB in vitro. It has been demonstrated in vivo that D-lactate allosterically enhances the autophosphorylation activity of ArcB that has already been activated by anaerobiosis, rather than signaling inactive Arcb.[27] Other intermediates may have similar effects. These and other aspects of the regulation of the ArcA/ArcB system are discussed by Iuchi et al.[28] and by Lynch and Lin.[29]

The FNR system

When *E. coli* is shifted from aerobic to anaerobic growth, a number of genes required to grow anaerobically are induced. At the same time, genes required for aerobic growth are repressed (Fig. 18.4). Part of the regulation is due to a protein called FNR, which is encoded by the *fnr* gene. The FNR protein is a positive regulator of transcription for many genes that are expressed only during anaerobic growth, and a repressor for certain genes that are expressed only during aerobic growth (Fig. 18.6).[30,31] There is no evidence that FNR is phosphorylated or is part of a two-component regulatory system, and it is discussed here in the context of a protein involved in aerobic/anaerobic regulation of gene expression.

Genes whose expression requires FNR include those coding for the anaerobic respiratory enzymes fumarate reductase (*frdABCD*) and nitrate reductase (*narGHJI*), as well as several other enzymes, including pyruvate formate lyase (*pfl* genes), formate dehydrogenase-N (the respiratory formate dehydrogenase, *fdnGHI*), aspartase (*aspA*), anaerobic fumarase B (*fumB*), and glycerol-3-phosphate dehydrogenase (*glpA*).

In agreement with the role of activator of gene expression, mutations in the *fnr* gene result in an inability to grow anaerobically on fumarate or nitrate as electron acceptors. FNR is also a negative regulator for several aerobically expressed genes, including those for cytochrome o oxidase (*cyoABCDE*), cytochrome d oxidase (*cydAB*), succinate dehydrogenase (*sdhCDAB*), and superoxide dismutase (*sodA*). Thus, FNR is a global regulator of gene

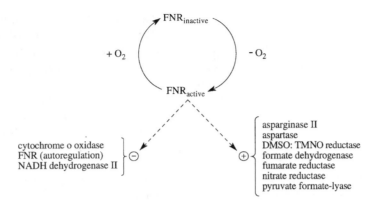

Fig. 18.6 FNR is a transcription regulator protein during anaerobic growth. In the absence of oxygen, FNR becomes activated to become an inducer for many anaerobically expressed genes and a repressor for certain aerobically expressed genes. It has been speculated that FNR might become activated by the reduction of bound ferric ion, causing a conformational change in the protein. This is not an example of a two-component regulatory system.

expression during anaerobic growth. It should be pointed out, however, that many of the genes regulated by FNR are also regulated by the ArcA/ArcB system and the Nar system.

Although FNR regulates gene activity during anaerobic growth, studies have shown that the FNR protein in *E. coli* is present in comparable amounts in both aerobically and anaerobically growing cells. However, it is believed to be largely in an inactive state during aerobic growth.[32,33] So, how does *E. coli* regulate the activity of FNR? The answer entails an iron–sulfur cluster in FNR. Active FNR is a homodimer of an iron–sulfur protein with an oxygen-labile [4Fe–4S] cluster in each monomer. Upon exposure to oxygen the [4Fe–4S] cluster is oxidized and can even be lost from the protein. (See note 34 for a further explanation.) When this happens a fraction of the FNR loses its iron clusters and becomes an apoprotein.[35–37] Both the protein bearing an oxidized cluster and the apoprotein are inactive. They bind with low affinity to DNA and do not stimulate transcription. Both the oxidation of the iron–sulfur clusters and their loss are reversible, and under anaerobic conditions the Fe–S clusters are restored.

However, FNR expression need not be regulated exactly the same in all bacteria. For example, whereas in *E. coli* the transcription of *fnr* seems to be similar under both aerobic and anaerobic growth conditions, and the amounts of active FNR increase anaerobically solely at the post-translational level, transcription of *fnr* is strongly stimulated in *B. subtilis* during oxygen limitation. The anaerobic activation of transcription of *fnr* in *B. subtilis* requires *resDE*, a two-component system (*resD* is a response regulator gene and *resE* is a histidine kinase gene).[38] The activity of FNR in *B. subtilis* may also be regulated by anaerobiosis in a fashion similar to its regulation in *E. coli*.

An interesting parallel exists between the FNR protein and another transcriptional regulator, the cAMP receptor protein, CRP (cyclic AMP receptor protein, also called the catabolite activator protein, CAP). The CRP protein is a positive transcriptional regulator for catabolite-sensitive genes (i.e., genes repressed by glucose). CRP, in response to binding to cAMP, binds to specific sites on the promoter region of the target gene to activate transcription. A comparison between the nucleotide-derived amino acid sequences of FNR and CRP reveals that FNR and CRP are very similar in structure: both have a DNA-binding domain that allows the dimer to bind and a nucleotide-binding domain (although numerous attempts to show specific binding of nucleotides to FNR have failed). This has led to the suggestion that FNR and CRP are examples of a family of proteins that regulate transcription at the promoter region of target genes. Indeed, there have been several reports of proteins resembling FNR in bacteria other than the enterics, and these FNR-like proteins take part in the regulation of a variety of metabolic activities. (See note 39 and refs. 40 and 41.)

In summary, ArcA and FNR are activated by anaerobiosis and regulate the enzymological changes that accompany the shift from aerobic to anaerobic growth. For example, activated ArcA (ArcA-P) primarily represses several aerobically expressed genes, including pyruvate dehydrogenase, succinate dehydrogenase, α-ketoglutarate dehydrogenase, and cytochrome o oxidase (Fig.18.5). It is also a positive regulator for some genes expressed when oxygen is low or lacking, including the cytochrome d oxidase gene and the pyruvate–formate lyase gene. Similarly, activated FNR represses several genes normally expressed during aerobic growth, including the genes for cytochrome o oxidase, succinate dehydrogenase, and pyruvate dehydrogenase, and is required for the induction of many genes during anaerobic growth, including pyruvate–formate lyase, fumarate reductase, and nitrate reductase (cf. Figs. 18.5 and 18.6. Also, see note 23.)

It is important to point out that in several cases, FNR and ArcA regulate the same gene. For example, they are both positive regulators for the pyruvate–formate lyase gene. Sometimes they have opposite effects on the same gene. For example, ArcA is a positive regulator for the cytochrome d oxidase gene, whereas FNR is a negative regulator for that gene. Multiple controls of the same gene are a common theme in bacterial physiology. However, a note of caution must be introduced when one is attempting to extract from physiological studies of mutants conclusions about coregulation by ArcA and FNR. This is because FNR has been reported to stimulate anaerobic *arcA* expression when this activity was monitored via *arcA–lacZ* fusions.[41]

Regulation of the formate hydrogen–lyase pathway

When enterobacteria such as *E. coli* are grown anaerobically, formic acid is converted to H_2 and CO_2 via the enzyme complex formate hydrogen–lyase (Figs. 14.7 and 18.4B). Formate hydrogen–lyase consists of two enzymes, formate dehydrogenase H and hydrogenase 3, whose genes are repressed by oxygen and nitrate, and induced by formate. Induction also requires an acidic pH in the external medium, which probably means that the cells make formate

hydrogen–lyase to prevent the medium from becoming too acidic owing to the production of formic acid from pyruvate (the pyruvate formate–lyase reaction). (*E. coli* actually makes three distinct formate dehydrogenases and three different hydrogenases. See note 42 for a discussion of this point.)

The genes for formate hydrogen–lyase are part of a regulon that has two names: the FhlA regulon and the formate regulon. The regulon is controlled by a transcription factor called FhlA and by formate, which is a coactivator of FhlA. (See ref. 43 for a review; see note 44 for a further description of the regulon.) Under aerobic conditions, formate levels are kept low, and because of this the regulon is not induced.

Two different mechanisms keep the formate levels low under aerobic conditions. The synthesis of formate from pyruvate is prevented under aerobic conditions because the *pfl* gene, which codes for pyruvate formate–lyase, which in turn synthesizes acetyl–CoA and formate from pyruvate and CoASH, requires active FNR as a positive regulator. Since oxygen prevents activation of FNR, oxygen represses the *pfl* gene, thus preventing formate synthesis. Oxygen and nitrate also lower the levels of formate in the cell because they induce the synthesis of the respiratory formate dehydrogenases FDH-O and FDH-N, which oxidize formate to CO_2.

18.3 Response to Nitrate and Nitrite: The Nar Regulatory System

E. coli can be grown anaerobically by using nitrate (NO_3^-) as the electron acceptor. The process is called nitrate respiration and is reviewed in Section 4.7.1. (See the subsection entitled *Anaerobic respiratory chains*.) In fact, when *E. coli* is given a choice of electron acceptors such as nitrate, nitrite, or fumarate under anaerobic conditions, it will utilize the nitrate. This property may be advantageous because nitrate has a more positive redox potential than the alternative electron acceptors, and therefore more energy is potentially available from electron transport when nitrate is the electron acceptor. Nitrate is preferentially used as an electron acceptor during anaerobic respiration because it induces the transcription of genes resulting in the synthesis of a membrane-bound nitrate reductase and represses the transcription of genes encoding the other reductases (e.g., fumarate reductase, DMSO/TMNO reductase, formate-dependent nitrite reductase). The formate-dependent nitrite reductase operon *nrf* (nitrite reduction by formate) encodes a periplasmic nitrite reductase that catalyzes nitrite reduction via formate as the electron donor. A Δp is produced.[45] (See note 46, which lists the various genes in these operons.) When nitrite is the electron acceptor, it induces transcription of genes encoding both the cytoplasmic and periplasmic nitrite reductases.

After a brief summary of the pathway for nitrate reduction (Section 18.3.1), we discuss the enzymes involved (Section 18.3.2). A simplified overview of the Nar system and how it regulates the transcription of genes required to use nitrate and nitrite as electron acceptors during anaerobic respiration (Section 18.3.3) is followed by a more detailed account of the Nar regulatory system (Section 18.3.4). Then we consider the role of integration host factor (IHF) in the regulation of gene transcription by the Nar system (Section 18.3.5) and the biochemical and genetic evidence for some of the conclusions regarding how the genes are regulated by the Nar system (Section 18.3.6).

18.3.1 Pathway of nitrate reduction

The nitrate (NO_3^-) is first reduced to nitrite (NO_2^-) which, in turn, is reduced to ammonia (NH_4^+). A total of eight electrons is required: two to reduce nitrate to nitrite, and six to reduce nitrite to ammonia. The enzymes involved are described in Section 18.3.2. Part of the ammonia is assimilated into cellular nitrogeneous compounds, and part is excreted into the medium. Thus, under anaerobic conditions *E. coli* can use nitrate both as an electron acceptor during anaerobic respiration and as a source of cellular nitrogen.

18.3.2 The enzymes involved

The two major enzymes for the reduction of nitrate to ammonia during nitrate respiration in *E. coli* are a membrane-bound nitrate reductase that reduces nitrate to nitrite (a two-

electron transfer) and a cytoplasmic NADH-linked nitrite reductase that reduces nitrite to ammonia (a six-electron transfer). Nitrate reductase is also called *respiratory nitrate reductase* and is encoded by the *narG* operon (*narGHJI*). It translocates one proton per electron from the cytoplasm to the periplasm during quinol oxidation and is therefore a coupling site. This topic, which was discussed in Section 4.7.1, is reviewed in ref. 47. The role of the cytoplasmic NADH-linked nitrite reductase, which is encoded by the *nirB* operon (*nirBDC*), is to remove nitrite (which is toxic) produced from nitrate, and to regenerate NAD+.

As mentioned earlier, *E. coli* also makes a periplasmic nitrite reductase involved in anaerobic respiration, cytochrome c_{552}, encoded by the *nrfA* operon (*nrfABCDEFG*). It is also called *respiratory nitrite reductase* and is a coupling site, but the mechanism is not understood. The enzyme is induced by nitrite and repressed by nitrate. For a review of nitrate and nitrite respiration, see ref. 48.

18.3.3 The Nar system and gene regulation, an overview

For the following discussion, refer to Fig. 18.7. The responses to both nitrate and nitrite are regulated by dual two-component signaling systems that consist of two membrane-bound sensor kinases (NarX and NarQ) and two cytoplasmic response regulator proteins (NarL and NarP), rather than one sensor kinase and one response regulator protein.[49–52] The systems are NarX/NarL and NarQ/NarP. (Note 53 explains how the *nar* genes were discovered.) The biochemical evidence for this is summarized later. The result of this regulation is that in the presence of excess nitrate there is preferential synthesis of enzymes that use nitrate as an electron acceptor (i.e., nitrate reductase), and in the presence of excess nitrite there is preferential synthesis of enzymes that use nitrite as an electron acceptor (i.e., respiratory nitrite reductase, cytochrome c_{522}).

Both nitrate and nitrite induce the formation of a cytoplasmic nitrite reductase, and this enzyme probably serves to prevent the accumulation of toxic nitrite in the cytoplasm. Nitrate stimulates transcription of the respiratory

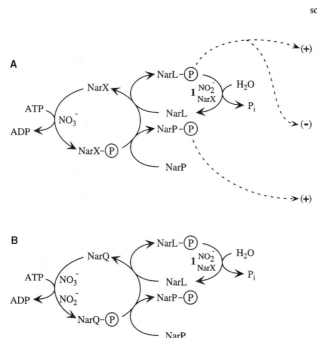

Fig. 18.7 Model for nitrate/nitrite control of anaerobic gene expression. The membrane-bound histidine kinase sensor proteins are NarQ and NarX. The cytoplasmic response regulator proteins are NarL and NarP. Nitrate stimulates the phosphorylation of NarL and NarP via NarX (A) and NarQ (B). Nitrite stimulates the phosphorylation of NarL and NarP via NarQ, but causes the dephosphorylation of NarL-P by stimulating the phosphatase activity of NarX. The result is that nitrate increases the amounts of NarL-P and nitrite lowers the amounts of NarL-P. (C) Some of the genes regulated by the response regulator proteins. In the presence of nitrate, the higher levels of NarL-P induce the *narG* operon, which encodes respiratory nitrate reductase, and repress the *nrfA* operon, which encodes respiratory nitrite reductase. The *nrfA* operon is induced, however, in the presence of nitrite. This is because (1) nitrite stimulates the phosphorylation of NarP via NarQ, and NarP-P stimulates transcription of the *nrfA* operon, and (2) the lower concentrations of NarL-P caused by nitrite actually induce the *nrfA* operon. The second effect has been explained by a second site for the *nrfA* operon (an activation site) that has a high affinity for NarL-P, which stimulates transcription when bound to this site. See: Darwin, A. J., K. L. Tyson, J. W. Busby, and V. Stewart. 1997. Differential regulation by the homologous response regulators NarL and NarP of *Escherichia coli* K12 depends on the DNA binding site arrangement. *Mol. Microbiol.* **25**:583–595.

nitrate reductase gene (and inhibits transcription of the respiratory nitrite reductase gene) because nitrate causes increased amounts of the response regulator NarL-P, and NarL-P stimulates the expression of *narG*, the operon that encodes nitrate reductase, and inhibits expression of *nrfA*, the operon that encodes respiratory nitrite reductase. On the other hand, nitrite stimulates expression of the respiratory nitrite reductase operon and lowers the expression of the nitrate reductase operon. This is because nitrite lowers the amounts of the response regulator NarL-P (hence depressing transcription of *narG*) and increases the

amounts of the response regulator NarP-P. The higher concentrations of NarP-P stimulate expression of the respiratory nitrite reductase operon (*nrfA*). The lower concentrations of NarL-P also stimulate expression of *nrfA*, and this has been explained by an activation site for the *nrfA* operon that has a high affinity for NarL-P.[50]

18.3.4 The roles of NarX and NarQ

Evidence supporting this section's descriptions of NarX and NarQ do is provided in Section 18.3.6.

NarX

In the presence of nitrate, NarX phosphorylates NarL. However, very importantly, in the presence of nitrite, NarX *dephosphorylates* NarL-P (Fig. 18.7A, reaction 1). This is why there is less NarL-P in the presence of excess nitrite. As a consequence, the nitrate reductase operon (*narG*) is stimulated by nitrate but not by nitrite, and the periplasmic nitrite reductase operon (*nrfA*) is repressed by nitrate but not by nitrite. The levels of NarP-P, on the other hand, are the same for both nitrate and nitrite, and therefore the differential effects of nitrate and nitrite on gene transcription depend upon NarL-P and not NarP-P. There are several other examples of a signaling molecule lowering the levels of a phosphorylated response regulator protein by stimulating its dephosphorylation. Another example of a signal that does this is P_{II}, which represses the *glnALG* operon by stimulating the dephosphorylation of NR_I–P by NR_{II} (Fig. 18.8).

NarQ

In contrast to the differential effect of nitrate and nitrite on the sensor kinase, NarX activity with respect to the response regulator NarL (nitrate stimulates the kinase activity, whereas nitrite stimulates the phosphatase activity), the response of NarQ to either nitrate or nitrite is the same: that is, phosphorylation of NarL and NarP (Fig. 18.7B). Therefore, it is the NarX/NarL couple that is responsible for the differential response to nitrate and nitrite.

18.3.5 The regulatory role of IHF

We know that IHF (integration host factor) activates as well as represses genes. See the discussion of the nucleoid in Section 1.2.6 for a description of IHF and its role in DNA bending and the regulation of gene activity, as well as Sections 10.2.5, 18.4.2, and 18.7.2 for more discussion of IHF. The transcription of the

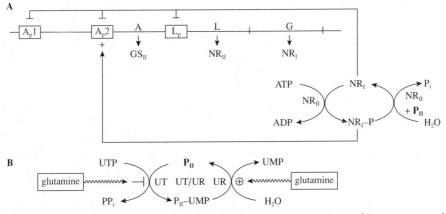

Fig. 18.8 Model for transcriptional regulation of *glnALG*. The protein products of the operon are glutamine synthetase (GS), a histidine protein kinase/phosphatase, nitrogen regulator II, (NR_{II}), and a response regulator (NR_I, also called NtrC). (A) Under ammonia-limiting conditions, NR_{II} phosphorylates NR_I, forming NR_I–P. The increased levels of NR_I–P stimulate a large increase in transcription rate from *glnAp2* (labeled *Ap2*), which uses sigma 54 RNA polymerase. As the levels of NR_I–P increase, other Ntr operons are also stimulated. Under conditions of excess ammonia, P_{II} levels rise, and NR_{II} is stimulated by P_{II} to decrease phosphorylation of NR_I (not shown), and to dephosphorylate NR_I–P. As a consequence, transcription from *glnAp2* stops and sigma 70 RNA polymerase transcribes from promoters *glnAp1* and *glnLp* (labeled *Ap1*, *Lp*) at a low rate. Transcription from these promoters is further regulated by repression by NR_I. Not shown is the effect of α-ketoglutarate, which is to inhibit the activity of P_{II} on NR_{II}, resulting in more NR_I–P. (B) The levels of P_{II} are increased by glutamine (signaling high ammonia). The bifunctional enzyme uridylyltransferase/uridyl removal (UT/UR) enzyme, which can both add and remove UMP from P_{II}, is inhibited by glutamine in adding UMP to P_{II}, and is stimulated by glutamine in removing UMP from P_{II}–UMP. The net result is that more P_{II} is generated. For a description model, see: Ninfa, A. J., and P. Jiang. 2005. P_{II} signal transduction proteins: sensors of α-ketoglutarate that regulate nitrogen metabolism. *Curr. Opin. Microbiol.* 8:168–173.

nitrate reductase gene (*narG*) and the nitrite extrusion protein gene (*narK*), which catalyzes the electrogenic excretion of nitrite, also is stimulated by integration host factor.[54] This indicates that IHF brings a loop of DNA to which NARL-P is attached close to the promoter of these genes. (See Fig. 10.27 in Chapter 10.)

18.3.6 Some biochemical and genetic evidence for roles of NarX, NarQ, NarL, and NarP

There exist biochemical data in support of the model that NarX and NarQ are kinases that autophosphorylate and transfer the phosphoryl group to the response regulator proteins. It has been demonstrated with purified proteins that NarX and NarQ can autophosphorylate, using ATP as the phosphoryl donor, and transfer the phosphoryl group to NarL.[55,56] NarX and NarQ can also negatively regulate the NarL protein by dephosphorylating NarL-P, although NarX is more active than NarQ in this respect.[57]

NarX and NarQ have two transmembrane regions near the N-terminal end and a region in between, exposed to the periplasm. Mutations in the periplasmic region of NarX or NarQ result in mutants that either have lost the ability to respond to nitrate and nitrite or behave as if nitrate or nitrite were present even in its absence.[53,58] The mutants were analyzed by means of *narG–lacZ* and *frdA–lacZ* fusions. (Note 25 gave a description of gene fusions.) From these data it has been concluded that nitrate and nitrite are detected in the periplasmic region and that the signal is transduced across the cell membrane to the response regulator proteins. Both NarL and NarP have been demonstrated to bind to specific sites in target operons.[50]

18.4 Response to Nitrogen Supply: The Ntr Regulon

Bacteria use inorganic and organic nitrogenous compounds as a source of nitrogen for growth. The inorganic nitrogen is in the form of ammonia, nitrate, or dinitrogen gas, and the organic nitrogen is in the form of amino acids, urea, and other nitrogenous compounds. Ultimately,

all these nitrogenous compounds are either reduced or catabolized by the bacteria to ammonia. The enteric bacteria have two enzymatic reactions for the assimilation of ammonia; the glutamine synthetase reaction and the glutamate dehydrogenase reaction.

Glutamine synthetase

$$\text{L-Glutamate} + \text{ATP} + \text{NH}_3$$
$$\rightarrow \text{L-glutamine} + \text{ADP} + \text{P}_i$$

Glutamate dehydrogenase

$$\alpha\text{-Ketoglutarate} + \text{NADPH} + \text{H}^+ + \text{NH}_3$$
$$\rightarrow \text{L-glutamate} + \text{NADP}^+ + \text{H}_2\text{O}$$

A third reaction which will become important for the discussion that follows, is the conversion of α-ketoglutarate to L-glutamate by the enzyme glutamate synthase.

Glutamate synthase

$$\alpha\text{-Ketoglutarate} + \text{glutamine} + \text{NADPH} + \text{H}^+$$
$$\rightarrow 2 \text{ L-glutamate} + \text{NADP}^+$$

The three foregoing reactions, which were described in Section 9.3.1, are crucial because glutamate supplies nitrogen to approximately 85% of the nitrogen-containing molecules in the cell, and glutamine supplies the remaining 15%.

When the supply of ammonia is restricted, glutamine synthetase is the main reaction for the assimilation of ammonia because the enzyme has a low K_m for ammonia. On the other hand, glutamate dehydrogenase has a high K_m for ammonia (about 1 mM) and operates only under conditions of high external ammonia concentrations. When the ammonia supply is adequate, then ammonia is the preferred source of nitrogen and it represses genes required for the assimilation of other nitrogenous compounds. However, when ammonia in the growth medium becomes limiting, genes required for ammonia transport and ammonia production from external nitrogen sources are induced. Furthermore, under limiting ammonia conditions, glutamine synthetase becomes more active and the transcription of the gene for glutamine synthetase is also stimulated, resulting in more glutamine synthetase to scavenge the small amounts of ammonia for the synthesis of glutamine and glutamate.

The genes that are regulated by the ammonia supply belong to the Ntr regulon.[59-62] As mentioned previously, a *regulon* is a set of noncontiguous operons or genes controlled by a common regulator, and for the *Ntr regulon*, the regulator is a response regulator protein, called NR_I (the nitrogen regulatory protein; also called the nitrogen regulatory protein C, or NtrC), which must be phosphorylated to activate gene transcription.

The Ntr regulon consists of many genes in several operons. (See note 63 for a further explanation.) These include the genes for glutamine synthetase, as well as NR_I (NtrC), and NR_{II} (also called NtrB), a histidine kinase that phosphorylates Ntr_I, all of which are in an operon that is called *glnALG* in *Escherichia coli* and *glnABC* in *Salmonella typhimurium*.

Other genes in the Ntr regulon are the gene that encodes another positive transcription factor, Nac (see note 63), the genes for the nitrogenase system in *Klebsiella* (*nif* genes), the genes for the degradation of urea in *Klebsiella*, the genes for periplasmic binding proteins required for amino acid transport [e.g., for glutamine (*glnH*)], arginine, lysine, ornithine (*argT*), and histidine (*hisY*) in *Salmonella*, the genes for histidine degradation (*hut*) in *Klebsiella*, and the genes for proline degradation (*put*) in *Salmonella*.[64] It should not be concluded automatically that if a gene system is under Ntr regulation in one genus of bacterium, it is also under Ntr regulation in all bacteria. This is clearly not the case. For example, the *hut* genes are not regulated by ammonia levels in *Salmonella typhimurium*, although they are so regulated in *Klebsiella pneumoniae*. Along the same lines, it should be emphasized that the following discussion regarding the Ntr system applies only to the enteric bacteria, in which the regulation of nitrogen metabolism has been studied in detail. There is much less known about the regulation of nitrogen metabolism in other prokaryotes, and one should be careful about extrapolating from the enterics to other prokaryotes.

18.4.1 Model for the regulation of the Ntr regulon

Before describing the regulation of the Ntr regulon, it will be helpful to introduce a regulatory protein called P_{II}. As will be made clear shortly, P_{II} has two important roles: (1) it represses transcription of the Ntr regulon, and (2) it inhibits the activity of glutamine synthetase. In both cases P_{II} acts indirectly, as will be explained later. This is a complex signaling pathway. See ref. 65 for a review. Refer to Fig. 18.8 in reading the following set of explanations; the version with fewer details precedes the subsection providing more complete coverage.

The less detailed explanation

Transcription of the Ntr regulon is induced when the extracellular ammonia levels fall below 1 mM, whereupon P_{II}, which inhibits expression of the Ntr regulon, is inactivated and, as a result, the Ntr regulon is activated. This inactivation/activation sequence leads to the synthesis of proteins required for nitrogen assimilation, described earlier. The reason for this is that active P_{II} stimulates the dephosphorylation (and thus inactivation) of NR_I–P, which is the positive transcription regulator for the Ntr regulon. In summary, then, high levels of ammonia cause the activation P_{II}, thus repressing the Ntr regulon, and low levels of ammonia cause the inactivation P_{II}, thus inducing the Ntr regulon. However, as described next, the effects of ammonia are not direct. Rather, they are mediated via glutamine and α-ketoglutarate, whose levels reflect the ammonia levels.

When the environmental levels of ammonia go sharply from high to low, there is a decrease in the intracellular concentration of glutamine, and an increase in the intracellular concentration of α-ketoglutarate. Glutamine drops when ammonia levels fall sharply because while the cells are growing in high ammonia, they have low amounts of glutamine synthetase, the enzyme that synthesizes glutamine from ammonia and glutamic acid (Chapter 9). Thus, when the cells experience a sharp drop in external ammonia levels, there are low amounts of glutamine synthetase, hence of glutamine. The increase in α-ketoglutarate occurs because the conversion of α-ketoglutarate to glutamate requires glutamine or ammonia to donate the amino group. The low glutamine and high α-ketoglutarate levels signal the cells that ammonia levels are low; the result is increased transcription of the Ntr regulon, which encodes enzymes required to increase ammonia assimilation. Glutamine inhibits the Ntr regulon because glutamine stimulates the production of P_{II}, which as was noted, inhibits the regulon, by causing the dephosphorylation of the positive response regulator, NR_I–P. It has been proposed that α-ketoglutarate stimulates the Ntr regulon because α-ketoglutarate inhibits the activity of P_{II} (rather than its production). This line of reasoning has been reviewed in ref. 66.

As stated earlier, P_{II} influences transcription indirectly, as shown in Fig. 18.8. An increase in the amount of P_{II} inhibits transcription of the Ntr regulon because P_{II} activates the phosphatase activity of the histidine kinase, NR_{II}, and the phosphatase activity of NR_{II} inactivates the response regulator (positive transcription factor), NR_I–P. When the levels of P_{II} decrease, NR_{II} acts as a kinase, rather than as a phosphatase, and phosphorylates NR_I, which then activates transcription.

Thus, the sequence of events is as follows: (1) low ammonia results in an increase in the concentrations of α-ketoglutarate because its conversion to glutamine and glutamate is slowed; (2) the α-ketoglutarate lowers the concentration of P_{II} by stimulating its conversion to P_{II}–UMP; (3) when the P_{II} levels fall, NR_{II} no longer acts as a phosphatase but rather as a kinase and phosphorylates NR_I; and (4) NR_I–P stimulates transcription. When ammonia levels are high, transcription is repressed because glutamine levels rise and the glutamine stimulates the conversion of P_{II}–UMP to P_{II}, which stimulates the phosphatase activity of NR_{II}, which results in a decrease in NR_I–P. These events are described in more detail next.

The more detailed explanation

The central operon involved in the regulation of the Ntr regulon, the *glnALG* operon, encodes three genes (Fig. 18.8A): *glnA*, the gene for glutamine synthetase; *glnL*, the gene for NR_{II}, which is a bifunctional histidine kinase/phosphatase whose substrate is NR_I; and *glnG*, the gene for NR_I, a response regulator. Under nitrogen excess (high ammonia), the operon is transcribed at a low level from the *glnAp1* and *glnLp* promoters by sigma 70 RNA polymerase. The small amount of transcription

from *glnAp1* is sufficient to guarantee the synthesis of enough glutamine synthetase to meet the cell's needs for glutamine when the ammonia concentrations are high. Part of the reason that only a small amount of transcription takes place from these promoters is that transcription from them is repressed by NR_I (Fig. 18.9A).[67] Under nitrogen-limiting conditions, the *glnALG* operon is transcribed at a high frequency from the *glnAp2* promoter by the sigma 54 RNA polymerase. (See note 68 for a discussion of the sigma 54 polymerase and more on transcriptional regulation.) Transcription from the *glnAp2* promoter requires the phosphorylated form of the response regulator NR_I (i.e., NR_I–P), reflecting the requirement of sigma 54 RNA polymerases for the binding of a transcription regulator upstream to the promoter to form an open complex. Since NR_I–P binds more than 100 base pairs upstream of the promotes region, the DNA must bend for NR_I–P to make contact with the RNA polymerase (Fig. 18.9A). (See note 69 for a model of how NR_I–P stimulates transcription.)

The protein kinase, NR_{II}, is itself regulated by another protein, called P_{II} (product of the *glnB* gene). When P_{II} levels are high, NR_{II} acts like a phosphatase rather than a kinase, and the levels of NR_I–P drop. Hence, P_{II} inhibits transcription of the *glnALG* operon.

α-Ketoglutarate inhibits the activity of P_{II}. Under conditions of excess ammonia, P_{II} is generated from P_{II}–UMP because glutamine stimulates the removal of UMP from P_{II}–UMP catalyzed by the bifunctional enzyme uridylyl-transferase–uridylyl removal (UT/UR) (product of the *glnD* gene) (Fig. 18.8B). When ammonia levels fall, the P_{II} is inactivated by being converted to P_{II}–UMP in a reaction catalyzed by UT/UR and inhibited by glutamine. Under these circumstances, NR_{II} acts as a kinase rather than as a phosphatase and transcription is stimulated. In summary then, we have four steps.

1. High ammonia leads to high P_{II} because glutamine stimulates the removal of UMP from P_{II}–UMP.

2. High P_{II} converts NR_{II} from a kinase to a phosphatase.

3. The phosphatase activity of NR_{II} leads to the dephosphorylation of NR_I–P.

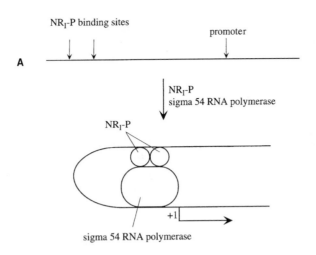

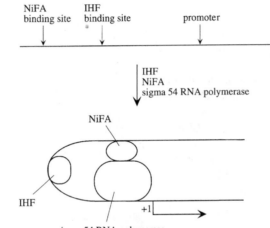

Fig. 18.9 Activation by NR_I–P and NifA plus IHF. Promoters read by sigma 54 polymerase require an activator that binds at an upstream site, sometimes over more than 100 base pairs away from the promoter. According to one model for how the activator contacts the polymerase, the DNA bends. (A) Activation of *glnA* promoter by NR_I–P. Because the binding site for NR_I–P has an inverted repeat, each one capable of binding NR_I–P, it has been proposed that two NR_I–P molecules bind. (B) Activation of promoter by NifA. It is known that IHF is necessary for the transcriptional activation by NifA of most NifA-dependent genes. The model supposes that IHF binds at a site between the NifA-binding site and the polymerase-binding site and bends the DNA, bringing the activator to the RNA polymerase. See: Magasanik, B. 1996. Regulation of nitrogen utilization, pp. 1344–1356. In: *Escherichia coli and Salmonella, Cellular and Molecular Biology*, Vol. 1. F. C. Neidhardt et al. (Eds.). ASM Press, Washington, DC.

4. Dephosphorylation of NR_I–P leads to a lowering of transcription of the *glnALG* operon, since NR_I–P is a positive transcription factor.

Thus, high ammonia leads to the dephosphorylation of the response regulator, hence to repression of the *glnALG* operon, As we shall see later, a similar situation exists for the regulation of the PHO regulon. In this case a high concentration of inorganic phosphate leads to the dephosphorylation of the response regulator and a consequent repression of the PHO regulon (Section 18.5.1).

A key enzyme in the signal transduction pathway is the bifunctional enzyme UT/UR, which either adds or removes UMP from P_{II}. It can be considered to be a sensor protein that responds to the α-ketoglutarate/glutamine ratio and modifies P_{II}, which in turn regulates the activity of the histidine kinase/phosphatase (NR_{II}). Alternatively, one might consider glutamine synthetase to be the sensor because it responds to the ammonia supply and determines the levels of glutamine and α-ketoglutarate, which signal the UT/UR.

Regulation of the other operons in the Ntr regulon

It has been estimated that the level of NR_I in ammonia-repressed cells is about five molecules per cell, which when phosphorylated under conditions of ammonia starvation will activate transcription of glnALG. When the levels of NR_I–P become sufficiently high (around 70 molecules per cell) as a result of increased transcription of glnALG, transcription of the other operons in the Ntr regulon system is activated, as well.

NR_I–P directly stimulates some of the Ntr operons [e.g., glnALG and operons encoding proteins for the uptake of glutamine (glnHPQ) in E. coli, as well as histidine (hisJQMP) and the gene for arginine uptake (argT) in S. typhimurium]. However, in other instances, NR_I–P activates the transcription of a gene encoding a second positive regulator, which activates transcription of the target promoters. For example, in Klebsiella pneumoniae, NR_I–P activates the transcription of nifAL. (See ref. 70 for a review.) The product of nifA is NifA, which is a positive regulator of the nitrogen fixation (nif) genes. NifA binds to upstream activator sequences (UAS) and makes contact with σ^{54}-RNA polymerase, which is bound downstream. The contact between NifA and the polymerase is facilitated by a bending of the intervening DNA, as described later. (See Section 18.4.2, and Fig. 18.9B.) [In K. pneumoniae, the product of nifL inactivates the nifA product in response to oxygen and ammonia, thus making certain that the nif genes (except for nifAL) are expressed only during anaerobiosis and ammonia starvation.] In Klebsiella species, but not in Salmonella typhimurium, P–NR$_I$ also activates the transcription of nac, which encodes a positive regulator of the hut operons that encode enzymes for the catabolism of histidine as a source of nitrogen. The hut genes use a sigma 70 RNA polymerase (the major polymerase), rather than the sigma 54 polymerase.

18.4.2 Regulatory Role of IHF

The student should review the discussion of the nucleoid in Section 1.2.6, where the DNA-binding protein known as integrating host factor is described. See ref. 71 for a review of IHF, a protein that is known to bend DNA. (Any bent DNA, including DNA to which IHF is bound in vitro, can be analyzed via gel electrophoresis because bent DNA fragments travel at different rates.) IHF activates certain genes and represses others. In some genes it binds at or near the promoter region, and it may interfere with the binding or activity of RNA polymerase. It also binds to specific locations on the DNA between the promoters and NifA-binding sites of genes activated by NifA (with the exception of one NifA activated gene, nifB). IHF is also required for the activation of one NR_I-dependent promoter, glnHp2, but not the others. According to the model, IHF bends the DNA, bringing bound activator such as NifA or NR_I–P to the promoter, where the activator stimulates transcription (Fig. 18.9B). (This material is reviewed in ref. 4. See Sections 18.7.2 and 18.11.5 for other examples of IHF regulatory activity.)

18.4.3 Effect of P_{II}–UMP and P_{II} on glutamine synthetase activity

When ammonia levels are low, not only is transcription of the gene for glutamine synthetase (GS) stimulated, but the activity of the existing enzyme is stimulated, ensuring that the low ammonia levels do not preclude incorporation of ammonia into glutamine and glutamic acid. A model for the regulation of GS is illustrated in Fig. 18.10 and explained in the legend. Basically, what happens is that GS becomes adenylylated and inactivated in the presence of high amounts of glutamine (gln) and P_{II}, and activated (deadenylylated) in the presence of P_{II}–UMP. Levels of P_{II} rise when ammonia levels are low. The adenylylated enzyme is further inactivated because it is susceptible to feedback inhibition by a variety of nitrogenous

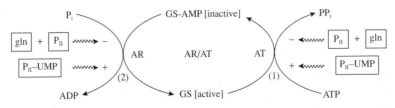

Fig. 18.10 Adenylylation of glutamine synthetase (GS) inhibits its activity, and deadenylylation increases its activity. The reactions are catalyzed by AT/AR, the bifunctional adenylyltransferasel/adenylyl removal enzyme. In this model, glutamine (gln) and P_{II} act together to inactivate GS by stimulating its adenylylation (1) and by inhibiting its deadenylation (2). P_{II}–UMP activates GS by inhibiting its adenylylation (1) and stimulating its deadenylylation (2). Furthermore, the levels of P_{II} are increased by glutamine, as described in Fig. 18.8. For a discussion of this model, see: Ninfa, A. J., and P. Jiang. 2005. P_{II} signal transduction proteins: sensors of α-ketoglutarate that regulate nitrogen metabolism. *Curr. Opin. Microbiol.* 8:168–173.

Table 18.1 Regulation of transcription of *glnALG* and glutamine synthetase

NH_3	P_{II}/P_{II}–UMP	Transcription[a]	Glutamine synthetase[b]
High	High	Low	Inactive
Low	Low	High	Active

[a] Indirectly repressed by P_{II}.
[b] Indirectly inactivated by P_{II}; indirectly activated by P_{II}–UMP.

compounds that depend upon glutamine for their synthesis (e.g., glucosamine-6-phosphate, carbamoyl phosphate, CTP, AMP, tryptophan, histidine). Thus P_{II} turns out to be an important protein that signals the ammonia levels to both the transcription machinery and glutamine synthetase.

Table 18.1 presents an overview of the regulation of transcription and glutamine synthetase activities by ammonia and P_{II}. While the post-translational control of glutamine synthetase by adenylylation occurs in *E. coli* and related bacteria, as well as in species of *Vibrio*, *Thiobacillus*, and *Streptomyces*, it does not occur in all bacteria (e.g., *Bacillus* and *Clostridium*) (Reviewed in ref. 61.) This is another reason for the need to use caution in extrapolating results from some bacteria to all bacteria, especially with regard to the regulation of metabolism.

18.5 Response to Inorganic Phosphate Supply: The Pho Regulon

Bacteria have evolved a signaling system to induce the formation of phosphate assimilation pathways when the supply of phosphate becomes limiting. The following discussion pertains specifically to *E. coli* and related bacteria, although homologues of the histidine kinase (PhoR) and response regulator (PhoB) have been found in other bacteria including *Bacillus*, *Pseudomonas*, *Rhizobium*, *Agrobacterium*, and *Synechococcus*. (For references to homologues of PhoR and PhoB, see ref. 72.) Under low inorganic phosphate conditions, *E. coli* stimulates the transcription of at least 38 genes (most of them in operons) involved in phosphate assimilation, including genes encoding a periplasmic alkaline phosphatase that can generate phosphate from organic phosphate esters (*phoA*); an outer membrane porin channel for anions, including phosphate (*phoE*); a high-affinity inner membrane phosphate uptake system called the Pst system and a protein required for phosphate repression (*pstSCAB-phoU*); a histidine kinase and response regulator (*phoBR*), 14 genes for phosphonate[73] uptake and breakdown (*phn*CDEFGHIJKLMNOP); genes for glyceraldehyde-3-phosphate uptake; and a gene encoding a phosphodiesterase that hydrolyzes glycerophosphoryl diesters (deacylated phospholipids) (*ugpBAECQ*). All these genes (except for *pit*, which functions in the presence of excess phosphate) are repressed by phosphate and are in the PHO regulon. The PHO regulon is controlled by PhoR, which appears to be a histidine kinase, and a response regulator, PhoB. The phosphorylated form of PhoB (i.e., PhoB-P) activates the transcription of the genes in the PHO regulon. The *pho* promoters use the sigma 70 RNA polymerase (encoded by *rpoD*). (See note 74 for a description of how PhoR activates transcription.)

When inorganic phosphate is in excess, it is transported into the cell by a low-affinity transporter called Pit. (See ref. 75 for a comprehensive review of phosphate transport.) Pit consists of a single transmembrane protein that is driven by the Δp. (See the discussion of secondary transport in Section 16.3.1.)

18.5.1 The signal transduction pathway

A model for the regulation of the PHO regulon in *E. coli* by P_i is shown in Fig. 18.11.[72,76] The components required for regulation are (1) PstS, a periplasmic P_i binding protein, (2) PstA, PstB, and PstC, integral membrane proteins required for P_i uptake (PstB is the permease), (3) PhoU, (4) PhoR, and (5) PhoB. PhoR detects P_i. However, it is not known whether P_i interacts with PhoR indirectly by binding to the Pst system or directly by binding to PhoR. What is known is that under conditions of P_i excess (about 4 μM), repression of the PHO regulon requires the Pst system, PhoU, and a form of PhoR called PhoR[R]. When P_i becomes limiting, the PHO regulon is induced and this requires a form of PhoR called PhoR[A], as well as PhoB, which is the response regulator.

One model proposes that P_i binds to PstS and that the P_i-bound PstS binds to PstABC, which is a member of the ABC transporter family. (See note 77 for further comment on Pst and its relationship to the ABC-type transporters.) A "repressor complex" is postulated that consists of P_i bound to PstS, PstABC, PhoU, and PhoR. PhoR is thought to be a histidine kinase/phosphatase bifunctional enzyme. (See note 78.) In the "repressor complex" PhoR is suggested to be a monomer with phosphatase activity and to maintain PhoB in its dephosphorylated (inactive) state. (This would be analogous to the Ntr regulon, where P_{II} stimulates the phosphatase activity of NR_{II}. Section 18.4.1.)

The model further proposes that when phosphate becomes limiting (<4 μM), PhoR is released from the "repressor complex" and functions as a histidine kinase, which is depicted in Fig. 18.11 as a dimer, although that has not been demonstrated. The dimer form of PhoR autophosphorylates and activates PhoB by transferring to it the phosphoryl group. In agreement with this model, Pst and PhoU are required for repression of the PHO regulon, but not for its activation. However, it must be added that the interaction of the Pst system and

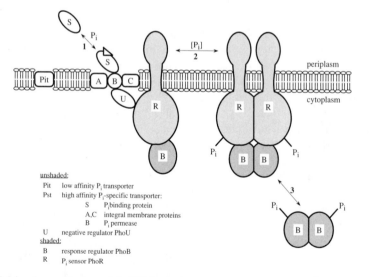

Fig. 18.11 Model for the regulation of the PHO regulon. Inorganic phosphate in the periplasm binds to the phosphate-binding protein, PstS. The P_i–PstS binds to the phosphate transporter, PstABC, in the cell membrane. A "repressor complex" forms between P_i–PstS, PstABC, PhoU, and PhoR. In the repressor complex, PhoR acts as a phosphatase and maintains PhoB in the dephosphorylated state. When P_i becomes limiting, PhoR is released from the repressor complex, autophosphorylates, and then phosphorylates PhoB. Phosphorylated PhoB is a positive transcription regulator for the PHO regulon. PhoR as a phosphatase is depicted as a monomer and PhoR as a kinase is drawn as a dimer, although this has not been demonstrated. *Source*: Wanner, B. L. 1993. Gene regulation by phosphate in enteric bacteria. *J. Cell. Biochem.* **51**:47–54.

PhoU with PhoR to cause P_i repression is not really understood. In mutants that do not have PhoR, the Pho regulon is expressed constitutively; that is, the genes are not repressed by inorganic phosphate. This is because PhoB is always phosphorylated. This may be due to the postulated phenomenon of *cross-regulation*, in which a response regulator of one two-component system might be controlled by a different regulatory system (e.g., a related histidine kinase) or by acetyl phosphate, whose levels may reflect growth conditions (Section 18.9).[72] In this case the different regulatory systems that may phosphorylate PhoB are CreC (formerly called PhoM), which is induced by growth on glucose and is not regulated by phosphate, and the phosphoryl donor acetyl phosphate produced during growth on pyruvate.[72] The evidence for acetyl phosphate as a global signal for response regulator proteins is discussed in Section 18.9.

18.6 Effect of Oxygen and Light on the Expression of Photosynthetic Genes in the Purple Photosynthetic Bacterium Rhodobacter capsulatus

Rhodobacter capsulatus is a purple nonsulfur photosynthetic bacterium that can grow heterotrophically in the dark, deriving its energy from aerobic respiration, or photoheterotrophically in the light without oxygen. The synthesis of the photosynthetic apparatus is under the control of light and oxygen with regulation at the level of transcription. There are at least three regulatory systems that control transcription of the photosynthetic genes. One of these is a repressor system (the CrtJ system) activated by oxygen. A second is an inducer system (the RegB/RegA system) activated by anaerobiosis. And the third is an inducer system (the HvrA system) activated by low light. These systems are described next.

18.6.1 Response to oxygen

The student should review the description of the light-harvesting complex in photosynthetic purple bacteria presented in Section 5.6.1 and Fig. 5.9. The photosynthetic genes in the purple photosynthetic bacteria such as *R. capsulatus* exist in several operons collectively known as the *photosynthetic gene cluster*. These genes code for proteins required for the biosynthesis of the photosynthetic pigments (bacteriochlorophyll and carotenoids) as well as proteins that make up part of the reaction center and light-harvesting complexes. Although not discussed here, some of the regulators of photosynthetic genes expression described next (RegB/RegA, Aer, CrtJ, HvrA) also regulate the expression of terminal oxidases in the electron transport chain. For more information, the student is referred to ref. 79.

Aerobic repression

Oxygen represses the synthesis of photosynthetic pigments, ensuring that photosynthesis occurs only under anoxic conditions. (The regulation of photosynthetic genes is reviewed in ref. 80.) One of the genes in the photosynthetic gene cluster encodes carotenoid, an aerobic transcription repressor of genes for bacteriochlorophyll, as well as the genes for synthesis of light-harvesting II complex (*puc* genes). The aerobic repressor is called CrtJ (or PpsR), and it is activated in the presence of oxygen. When CrtJ is exposed to oxygen, an intramolecular disulfide bond forms, causing a conformational change in CrtJ that results in a significantly increased ability to bind to its target promoters. (This is cited in ref. 81.) The student should review the discussion of another redox-sensitive transcription regulator, the FNR protein, which is inactivated by oxygen rather than being activated. See Section 18.2.2 and note 34. A second aerobic repressor of photosynthetic genes, Aer, was recently in 1996.[80] Some of the genes repressed by CrtJ are also repressed by Aer; that is, they are co-repressed. Other genes are repressed by CrtJ or Aer, but not by both.

Anaerobic induction

The photosynthetic bacterium *Rhodobacter capsulatus* uses a two-component system (RegB/RegA) to stimulate the transcription of genes for its light-harvesting and photoreaction center complexes under anaerobic incubation conditions.[82] (For a review of the RegB/RegA global regulatory system, see ref. 83.) Under anaerobic conditions, a signal is sent to RegB, causing it to autophosphorylate. RegB-P then phosphorylates RegA.[81] DNA-binding studies indicate that RegA-P is a transcriptional activator.[84] It has been concluded that the RegB/RegA system is a conserved global regulatory system involved in redox control of transcription of a variety of the genes in *R. capsulaus* and *R. sphaeroides* that are related to energy-generating and energy-utilizing processes, including expression of genes for CO_2 fixation, nitrogen fixation, hydrogen oxidation, denitrification, electron transport, and aerotaxis.[82]

Thus, RegB/RegA is a global regulator for many genes that serve a function in anaerobically growing *R. capsulatus and R. sphaeroides* that is analogous to that of the ArcB/ArcA system in *E. coli*. RegB autophosphorylates, activating the RegB/RegA system, in the absence of oxygen. It may be that the signal preventing autophosphorylation originates with the redox state of a component of the respiratory chain rather than with oxygen itself. Recall that ArcB also autophosphorylates under anaerobic conditions. (See Fig. 18.5.) See the discussion in Section 18.2.2 of how ArcB might sense anaerobiosis. Homologues of RegA and RegB have been found in a wide variety of photosynthetic and nonphotosynthetic bacteria, where they are presumed to exert redox control over gene expression (cited in ref. 82.)

18.6.2 Response to light

Photosynthetic pigment synthesis is increased anaerobically when the light intensity is decreased, allowing the cells to capture more light energy despite the lower light intensities. *R. capsulatus* has a gene, called the *hvrA* gene, whose product (HvrA) is required for the stimulation of transcription of some genes in the photosynthetic gene cluster by low light intensity. (See note 85 for more information about HvrA.) This conclusion is supported by the finding that mutations in *hvrA* cause a loss in low-intensity light stimulation of transcription of target operons. It has been shown that HvrA binds to target promoters in the photosynthetic gene cluster, and it has been demonstrated that high light reduces the intracellular levels of HvrA. However, not all the low-light-inducible genes require HvrA for transcription, and thus there is clearly another light-sensitive signaling system.

In addition to transcriptional regulation by light, there appears to be post-transcriptional regulation.[86] This was reported for the synthesis of the peptides of the light-harvesting complex I (B800-850 peptides), which increase around fourfold in parallel with the number of B800-850–bacteriochlorophyll complexes when the light intensity is diminished. It was found, however, that low-light-grown cells had only one-fourth the mRNA for the B800-850 peptides. We therefore have more protein present under conditions of less mRNA. Perhaps there is decreased turnover of the protein under low light conditions. This is discussed in ref. 85.

18.7 Response to Osmotic Pressure and Temperature: Regulation of Porin Synthesis

When *E. coli* is growing in higher osmolarity or at high temperature, the synthesis of the bacterinm's slightly smaller porin channel, OmpC, increases relative to the larger OmpF channel. For example, when *E. coli* is grown at 30 °C and in the presence of 1% NaCl (the temperature and approximate osmolarity of the intestine), only OmpC is made.[87] When *E. coli* is grown at lower temperatures and at osmolarities that approximate the conditions in lakes and streams, OmpF is preferentially made. The changes in porin composition of the outer membrane do not change the intracellular osmotic pressure, and therefore they are not part of a homeostatic response to osmotic pressure changes.

One can rationalize the change in porins by assuming that the smaller OmpC channel is advantageous in the intestinal tract because it may retard the inflow of toxic substances such as bile salts, whereas in lower osmolarity environments, such as would be experienced outside the body, the larger OmpF channel might be an advantage to increase inward diffusion of dilute nutrients.

18.7.1 Regulation of expression of ompF and ompC by a two-component system, EnvZ/OmpR

EnvZ is an inner membrane histidine protein kinase that has been postulated to function also as an osmotic sensor.[88] However, very little is known concerning the signals to which EnvZ responds when it phosphorylates or dephosphorylates OmpR. EnvZ is a transmembrane protein, part of which is exposed to the periplasm (a loop near the N-terminal end) and part to the cytoplasm (the C-terminal domain). It is thought that the domain that senses osmolarity signals resides either in the periplasm or in the membrane. The response regulator is a cytoplasmic protein called OmpR.

One model proposes that increased external osmolarity activates EnvZ so that it phosphorylates OmpR (Fig. 18.12). However, mutations in EnvZ result in a high-osmolarity phenotype (constitutive OmpF⁻OmpC⁺ phenotype), indicating that perhaps EnvZ detects a signal in low-osmolarity media rather than in high-osmolarity media. The putative signal in low-osmolarity media would prevent the phosphorylation of OmpR by EnvZ.[89] The model further proposes that high levels of P-OmpR repress *ompF* and stimulate *ompC* by binding to low-affinity sites upstream from the respective promoters, and that low levels of P-OmpR stimulate transcription of *ompF* by binding to

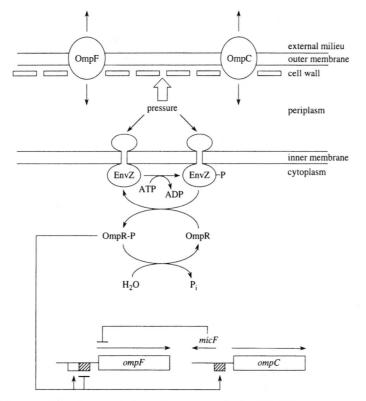

Fig. 18.12 Model to explain the regulation of porin synthesis. EnvZ is a transmembrane histidine kinase/phosphatase whose substrate is the cytoplasmic response regulator OmpR. To explain how a shift to high-osmolarity media represses *ompF* and stimulates *ompC*, it is assumed that high osmotic pressures activate (not necessarily directly) the kinase activity of EnvZ. The resulting high levels of OmpR-P repress transcription of *ompF* and stimulate transcription of *ompC*. OmpR-P can effect these changes by binding to DNA sequences upstream of the respective promoters. A third gene, *micF*, is transcribed from the *ompC* promoter but in the opposite direction. Its transcription is also activated by OmpR-P. The gene *micF* codes for an RNA transcript that is complementary to the 5′ end of the *ompF* mRNA (i.e., the region where protein translation is initiated) and blocks translation. *MicF* RNA is therefore "antisense" RNA because it binds to the *ompF* RNA transcript that is translated into protein (i.e., the "sense" RNA). It may be that an additional control of *ompF* expression is the inhibition of translation of *ompF* mRNA by *micF* RNA. When the external osmotic pressures are lower, the concentrations of OmpR-P are too low to bind to the DNA sites for repression of *ompF* and stimulation of *ompC*. However, low levels of OmpR-P can stimulate the transcription of *ompF* because of high-affinity binding sites for OmpR-P upstream of the *ompF* promoter. *Source*: Stock, J. B., A. J. Ninfa, and A. M. Stock. 1989. Protein phosphorylation and regulation of adaptive responses in bacteria. *Microbiol. Rev.* 53:450–490.

high-affinity sites upstream from the *ompF* promoter. Thus, when the external osmotic pressure is raised, the levels of OmpR-P should increase and repress *ompF* while stimulating *ompC*. Several lines of evidence support the model: (1) it has been demonstrated in vitro that EnvZ can accept a phosphoryl group from ATP and in turn transfer it to OmpR, forming OmpR-P; (2) it is also known that when *E. coli* is shifted to a medium of high osmolarity, the levels of OmpR-P increase; (3) in a mutant of *E. coli* that has elevated levels of OmpR-P, *ompF* is repressed and *ompC* is expressed all the time (constitutively); and (4) OmpR-P stimulates transcription from *ompF* and *ompC* promoters.[90–92]

18.7.2 Repression of transcription of ompF and ompC by IHF

Also repressing *ompF* and *ompC* is integration host factor. IHF binds to nucleotide sequences in the promoter regions of the *ompF* and *ompC* genes in *Escherichia coli* and represses transcription. (See Section 18.4.2 for a discussion of IHF.) This was demonstrated in vitro for *ompF*, and the investigators also showed that OmpR reverses the inhibition, suggesting that IHF might somehow interfere with the activity of OmpR at the promoter site.[93]

18.7.3 Inhibition of translation of ompF mRNA by micF RNA, an antisense RNA

There appears to be an additional mode of inhibition of *ompF* expression, particularly at levels of osmolarity that are not very high. Another regulatory gene, *micF*, is also stimulated by high concentrations of OmpR-P and produces an RNA molecule that inhibits *ompF* mRNA translation by binding to the mRNA (Fig. 18.12).[94] One example of an antisense RNA is *micF*. Other examples of antisense RNAs are known that function in regulating gene activity.[95,96] Support for the conclusion that MicF inhibits *ompF* mRNA translation derives from the observation that when multiple copies of *micF* are introduced on a high-copy-number plasmid, there is inhibition of in vivo synthesis of OmpF.

18.8 Response to Potassium Ion and External Osmolarity: Stimulation of Transcription of the kdpABC Operon by a Two-Component Regulatory System

When *E. coli* is placed in a medium that is high in osmolarity due to salts, or if the concentrations of K^+ are growth limiting, the bacterium responds by synthesizing a high-affinity K^+ uptake system to correct the situation. The uptake system, called the KdpABC transporter, is described in Section 16.3.3. (See subsection entitled *ATP-driven K^+ influx*.) For a discussion of the role of K^+ uptake in pH and osmotic homeostasis, the student is referred to Sections 15.1.3 and 15.2.2.

There are two regulatory genes for the transcription of the *kdpABC* operon, *kdpD* and *kdpE*, both of which are required for expression of the *kdpABC* genes. The KdpD and KdpE proteins are part of a two-component regulatory system, with KdpD being a membrane sensor kinase that autophosphorylates when stimulated, and KdpE being a cytoplasmic response regulator protein that is phosphorylated by KdpD. The phosphorylated form of KdpE activates the kdpABC promoter.[97–99] What signal does KdpD sense? It appears that KdpD senses both a decrease in intracellular K^+ concentration and an increase in intracellular ionic strength.[100]

18.9 Acetyl Phosphate Is a Possible Global Signal in Certain Two-Component Systems

It has been proposed that in the absence of the cognate HK proteins, that is, in mutants lacking specific histidine kinases, acetyl phosphate donates the phosphoryl group to the cognate response regulator protein, thus counteracting the effect of the mutation. (Reviewed in ref. 101.) The extent to which acetyl phosphate does this in vivo especially in wild-type cells, is not known. Nevertheless, it is an interesting phenomenon and potentially very significant. The following discussion pertains to the role that acetyl phosphate might play as a global regulator, and to the methods used to investigate this role.

18.9.1 Changing the intracellular levels of acetyl phosphate to investigate its possible role

To study the role of acetyl phosphate in regulation in vivo, the investigator must be able to change its intracellular concentrations. Mutants can be used for this purpose. Mutants lacking phosphotransacetylase (encoded by the pta gene) or acetate kinase (encoded by the ackA gene) have higher levels of acetyl phosphate, whereas mutants lacking both enzymes have lower levels of acetyl phosphate. To understand why this is so, we must first recall how E. coli makes acetyl phosphate. (See Section 7.3.2 for a review.) Phosphotransacetylase reversibly converts acetyl–CoA to acetyl phosphate:

Phosphotransacetylase

$$Acetyl–CoASH + P_i \rightarrow acetyl–P + CoASH$$

Acetate kinase makes acetyl phosphate directly from acetate and ATP. This is also a reversible reaction:

Acetate kinase

$$Acetate + ATP \rightarrow acetyl–P + ADP$$

Mutants in ackA (acetate kinase gene) have increased levels of acetyl phosphate when growing on carbon sources such as glucose or pyruvate that yield acetyl–CoA (via pyruvate dehydrogenase), since they can use the phosphotransacetylase to form acetyl phosphate from acetyl–CoA but cannot convert the acetyl phosphate to acetate without the acetate kinase. The same result might be achieved by growing pta (phosphotransacetylase) mutants on acetate, since these should be able to synthesize acetyl phosphate by using the acetate kinase but could not convert it to acetyl–CoA. On the other hand, mutants lacking both pta and ackA genes cannot make acetyl phosphate from pyruvate or acetate. Such mutants show the regulation that would be expected if acetyl phosphate were able to phosphorylate response regulator proteins in mutants lacking the cognate histidine kinases.

The choice of carbon source also influences the levels of acetyl phosphate. Acetyl phosphate pools are lowest in cells grown on glycerol (<0.04 mM), higher in cells grown on glucose (0.3 mM), and highest in cells grown on acetate (1.5 mM).[100]

Sometimes increasing the copy number of ackA, the gene that codes for acetate kinase, can indicate whether acetyl phosphate can phosphorylate the regulatory protein. For example, one of the genes of the Pho regulon, phoA (codes for alkaline phosphatase), could be expressed in the absence of the cognate HK proteins (mutants lacking both PhoR and CreC) by increasing the copy number of ackA.

In vitro phosphorylation of a response regulator protein by acetyl phosphate

Some investigators have demonstrates in vitro phosphorylation of a response regulator protein by acetyl phosphate. For example, NR_I (the response regulator protein that regulates transcription of the Ntr regulon) has been phosphorylated by acetyl phosphate in vitro.[102] Mutants lacking NR_{II}, the histidine kinase/phosphatase for the Ntr regulon, have residual regulation of expression of glnA (codes for glutamine synthetase), which is part of the Ntr regulon. This suggests the presence of an alternative donor of the phosphoryl group for the response regulator, NR_I. The alternative donor may be acetyl phosphate.

18.9.2 Regulatory role for OmpR-P and acetyl phosphate in the repression of transcription of flagellar biosynthesis genes in E. coli

In addition to regulating transcription of the porin genes (Section 18.7.1), OmpR regulates the transcription of other genes, including flagellar genes. When E. coli is grown under stressful conditions (e.g., high osmolarity, elevated temperatures), flagellar synthesis is inhibited. The repression of the flagellar genes by high osmolarity is apparently due to a repressive effect of OmpR-P, which is known to regulate other genes in addition to OmpF and OmpC. Under conditions of high osmolarity, OmpR is phosphorylated by the sensor kinase EnvZ. However, acetyl phosphate can also phosphorylate OmpR, even in wild-type cells, leading to the repression of flagellar genes.[103]

How the effect of acetyl phosphate on the transcription of flagellar genes is measured

To explain how the effect of acetyl phosphate on the transcription of flagellar genes can be measured, it is necessary to introduce the master operon for flagellar genes and to describe how the transcription of the operon can be monitored. The master operon for flagellar biosynthesis is *flhDC*, which codes for FlhD and FlhC, two DNA-binding proteins that are required for transcription of all the genes in the flagellar regulon. The use of *flhDC–lacZ* fusions can be used to monitor the activity of the *flhDC* promoter by doing β-galactosidase assays. (See note 25 for a description of *lacZ* fusions.) It was shown that the expression of *flhDC* is substantially increased in *pta ack* double mutants or *pta* mutants compared with wild-type cells.[103] These mutants can convert the carbon source to acetyl–CoA, but without phosphotransacetylase (encoded by *pta*), they cannot convert the acetyl–CoA to acetyl phosphate. Hence the conclusion that in wild-type cells acetyl phosphate represses the operon.

On the other hand, expression of the *flhDC* operon was decreased in *ack* mutants. The *ack* gene encodes acetate kinase, and in the absence of this enzyme, the cells use the phosphotransacetylase to make acetyl phosphate from acetyl–CoA but cannot metabolize the acetyl phosphate further. Hence the pool size of acetyl phosphate increases. All this is consistent with the hypothesis that acetyl phosphate represses the *flhDC* operon.[102]

How the involvement of OmpR in the repression of flagellar genes is detected

The involvement of OmpR in the repression of *flhDC–lacZ* by acetyl phosphate was shown by using *ompR⁻* strains. *OmpR⁻* mutants showed an enhanced expression of *flhDC–lacZ*, indicating that OmpR is a negative regulator of *flhDC–lacZ*, and this enhanced expression was not reduced in *ackA* mutants that had higher levels of acetyl phosphate, indicating that repression by acetyl phosphate requires OmpR.

Acetyl phosphate has been shown to phosphorylate OmpR in vitro, and presumably it also does so in vivo. Since OmpR is phosphorylated by EnvZ when cells are grown at high osmolarities (see Section 18.7.1), it might be expected that *flhDC* expression would be decreased in bacteria grown at higher osmolarities, and this in fact is the case in wild-type cells but not in *ompR⁻* cells.

Finally, it was demonstrated by a DNA mobility shift assay that phosphorylated OmpR binds to the *flhDC* promoter. The student should recall that in other instances in which it was inferred that acetyl phosphate had caused the phosphorylation of a response regulator in vivo, the effects were seen in mutants lacking the cognate histidine kinase. However, acetyl phosphate appears to phosphorylate OmpR even in cells that have the kinase, which is EnvZ.

Possible explanation of why E. coli produces fewer flagella at higher temperatures

It has been known that *E. coli* grown at elevated temperatures produces fewer flagella. This can be explained by the repression of the flagellar operon by acetyl phosphate. The acetyl phosphate levels in cells grown at higher temperatures is elevated because the acetate kinase is less active. Therefore, the inhibition of flagellar synthesis at higher temperatures might be due to the increased levels of acetyl phosphate, which may phosphorylate OmpR, which may then repress the activity of *flhDC*.

18.10 Response to Carbon Sources: Catabolite Repression, Inducer Expulsion, Permease Synthesis

Catabolite repression refers to the preferential use of one carbon source for growth over another when bacteria are grown in the presence of both carbon sources. It results in diauxic growth as described in Section 2.2.4. There is more than one mechanism responsible for catabolite repression. One mechanism was discussed in Section 16.3.4, which describes catabolite repression by glucose as a consequence of glucose transport by the phosphotransferase system (PTS) in *E. coli* and related bacteria. Here the uptake of glucose by the PTS lowers the cAMP pools, and this slows the transcription of cAMP-dependent genes required to metabolize alternative carbon sources. Section 16.3.4 also points out that glucose uptake by the PTS inhibits permeases required for the uptake of other carbohydrate carbon sources (inducer exclusion).

However, inducer exclusion by the PTS and adjustment of cAMP pools are not the only means of catabolite repression. Sections 18.10.1 and 18.10.2 describe two more catabolite repression systems. The Cra system, in *E. coli*, does not involve either PTS or cAMP. The second system, the CRE system, is found in the gram-positive bacterium *B. subtilis*. The following discussion is restricted to catabolite repression by sugars. However, there exist many bacteria that preferentially use carboxylic compounds rather than carbohydrates as a source of carbon and will show diauxic growth with the appropriate mixtures. In these instances it is the carboxylic acid that represses the expression of genes required to metabolize the carbohydrate. Some of these systems have been recently reviewed.[104]

In addition to catabolite repression, bacteria may use other means to favor growth on particular carbon sources. One of these is called *inducer expulsion* (Section 18.10.3), which is characterized by glucose inhibition of the accumulation of alternative carbon sources. A second is the induction of permeases that brings certain carbon sources into the cell (Section 18.10.4).

18.10.1 *Catabolite repression in E. coli that does not involve cAMP or CRP: The Cra system*

E. coli and *Salmonella* have an additional catabolite repressor system that does not involve cAMP. It has been called the FruR (fructose repressor) system and more recently the Cra (catabolite repressor/activator) system.[105,106]

Cra is a transcriptional regulator

As described in Section 10.2.5, transcriptional regulators exist that induce certain genes and repress others. Cra is such a regulator. It represses genes coding for enzymes required for growth on sugars, that is, enzymes of the Embden–Meyerhof–Parnas (EMP, or glycolytic) pathway and the Entner–Doudoroff (ED) pathway, while inducing genes encoding enzymes required for growth on organic acids

and amino acids, that is, enzymes of the citric acid cycle and the glyoxylate cycle, and gluconeogenic enzymes.

The reason for believing that Cra is an activator for genes required for growth on organic acids is that *cra* mutants cannot grow on lactate, pyruvate, and citric acid cycle intermediates and show lower activities of enzymes of the citric acid cycle, glyoxylate cycle, as well as gluconeogenic enzymes.[107] Recall that growth on these carbon sources requires gluconeogenesis, the citric acid cycle, and the glyoxylate pathway (acetate) (Sections 8.1, 8.8.3, and 8.12). The reason for believing that Cra is a repressor of genes required for growth on sugars is that mutations in *cra* lead to higher levels of enzymes required for the uptake and catabolism of sugars such as glucose and fructose. (For more background information on Cra, see note 108.)

Model for how Cra activates certain genes and represses other

One model proposes that Cra activates certain genes and inhibits others depending on where it binds to the DNA in relationship to the RNA polymerase binding site. If it sits upstream of the binding site, then it is postulated to be an activator; if it sits downstream of the binding site, it is suggested to be a transcription inhibitor (Fig. 18.13). (See the discussion of the regulation of transcription in Section 10.2.5.) This model is in fact supported by sequence analysis studies of Cra-binding sites that place these sites downstream of negatively controlled promoters and upstream of positively controlled promoters. (Reviewed in ref. 109.)

Glucose and fructose remove Cra from DNA, explaining catabolite repression

When glucose or fructose is in the growth medium, Cra no longer binds to the promoter regions of the target genes. Thus growth on the sugars can occur because the repressor (Cra) for the genes of the EMP and ED pathways is removed. However, the presence of glucose or fructose prevents growth on organic acids, such as acetate or succinate, or amino acids because the activator (Cra) of genes that encode enzymes required for growth on these carbon sources is removed.

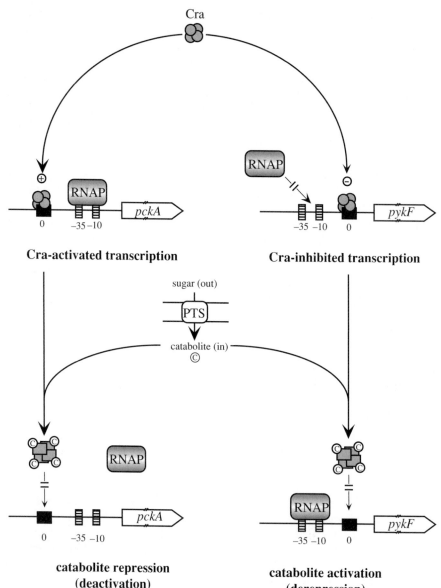

Fig. 18.13 Model for the regulation of gene transcription by Cra. It is proposed that Cra activates gene transcription if it binds upstream of the RNA polymerase-binding site and inhibits gene transcription if it binds downstream of this site. Catabolites derived from glucose or fructose are proposed to bind to Cra, preventing it from binding to DNA. Thus, these catabolites would repress genes whose transcription is stimulated by Cra (gluconeogenic genes) and activate genes whose transcription is repressed by Cra (glycolytic genes). See text for details. *Source*: Saier, Jr., M. H. 1996. Cyclic AMP-independent catabolite repression in bacteria. *FEMS Microbiol. Lett.* **138**:97–103.

Model for how glucose and fructose remove Cra from DNA

Two catabolites derived from glucose or fructose, fructose-1-phosphate (F1B) and fructose-1,6-bisphosphate (FBP), are thought to bind to Cra and remove it from the DNA, therefore diminishing the effect of Cra. Thus, these catabolites would repress genes activated by Cra (gluconeogenic genes) and activate genes inhibited by Cra (glycolytic genes). The evidence in favor of this model includes the finding that

Cra does bind to Cra-regulated operons in vitro and that fructose-1-phosphate or fructose-1,6-bisphosphate displaces Cra from DNA.

Comparison of the Cra system and the cAMP system

The Cra system is similar to the cAMP system in that the sugar prevents catabolite-repressed genes from being activated. (See Section 16.3.4 and Fig. 16.8.) In one system the activator is cAMP–CRP whose levels are lowered by glucose because the levels of the adenylyl cyclase activator, IIA^{Glc}–P are lowered when incoming glucose is phosphorylated by IIA^{Glc}–P, whereas in the other system the activator of gluconeogenic genes is Cra, which is removed from the DNA by F1P or FBP, leading to catabolite repression of these genes. The Cra system differs from the cAMP system in that Cra is also a negative transcription regulator of genes required for sugar catabolism and provides for the activation of these genes by glucose and fructose because F1P and FBP remove Cra from these genes. Hundreds of genes are regulated by cAMP and Cra in the enteric bacteria.

Repression of a nitrite reductase gene by Cra and catabolite activation

As discussed in Section 18.3, E. coli makes two different nitrite reductases under anaerobic conditions regulated by the Nar system. These are the formate-dependent periplasmic nitrite reductase encoded by the nrf operon and the NADH-linked cytoplasmic nitrite reductase encoded by the nir operon. Both nitrate and nitrite stimulate transcription of nir, whose product is required for detoxifying the nitrite formed from nitrate during nitrate respiration. Transcription of nrf is induced by nitrite but repressed by nitrate. In addition to regulation by the Nar system, these operons are induced anaerobically by FNR. (See Section 18.3.2.)

In addition to stimulation by FNR, nitrate, and nitrite, the nir operon is repressed by Cra and activated by catabolites.[110] (Transcription from the nrf operon is not regulated by Cra and is catabolite repressed in an unknown manner.) These experiments were done in E. coli strains carrying nir–lacZ and nrf–lacZ transcriptional fusions. Thus, the nir and nrf promoters could be monitored by β-galactosidase assays. Transcription of the nir operon was repressed in poor medium but increased in rich medium containing glucose. Furthermore, transcription of the nir gene was increased in a cra⁻ mutant (Fig. 18.14). Thus it was concluded that Cra inhibits transcription of the nir operon and that the inhibition is relieved by glucose. This is similar to the inhibition by Cra of the genes responsible for the uptake and catabolism of sugars such as glucose and fructose, and the release of

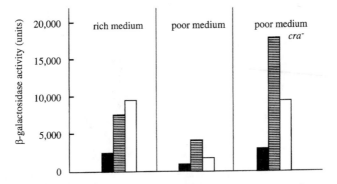

Fig. 18.14 Repression of nir (NADH-dependent nitrite reductase) by Cra. E. coli was grown anaerobically in rich [Luria broth (LB) plus glucose] or poor medium (minimal salts plus glycerol and fumarate) without nitrite (solid bars), with 2.5 mM nitrite (lined bars), or with 20 mM nitrate (open bars). The first two sets of bars show the results with the wild-type strain; the results using a cra⁻ mutant are shown on the right. The expression of the nir–lacZ fusion was monitored via β-galactosidase assays. The expression of the nir operon was enhanced in rich medium or by a mutation in the cra gene. This is consistent with the hypothesis that Cra represses the operon and that a catabolite derived from glucose reverses the repression. See text for further details. *Source*: Adapted from Tyson, K., S. Busby, and J. Cole. 1997. Catabolite regulation of two *Escherichia coli* operons encoding nitrite reductases: role of the Cra protein. *Arch. Microbiol.* **168**:240–244.

the inhibition by fructose-1-phosphate (F1B) and fructose-1,6-bisphosphate (FBP) derived from glucose. (See earlier subsection entitled *A model for how Cra activates certain genes and represses other* and Fig. 18.13.) The binding of Cra to the *nir* promoter and the release of Cra from the DNA by fructose-1-phosphate was demonstrated in vitro.[109]

18.10.2 Catabolite repression in gram-positive bacteria via the CcpA system

A third system of catabolite repression has been discovered in low-GC (30–40%) gram-positive bacteria such as *Bacillus*, *Staphylococcus*, *Streptococcus*, *Enterococcus*, *Lactococcus*, and *Lactobacillus*. (These bacteria do not synthesize detectable cAMP or CRP.[111,112]) It is called the CcpA (catabolite control protein A) system. CcpA is a transcription inhibitor of genes that are catabolite repressed by glucose.

A second transcription repressor, CcpB, has been identified in *B. subtilis* and proposed to act in parallel with CcpA via the same mechanism.[113]

The model

A proposed model is drawn in Fig. 18.15. When cells are grown on glucose, the HPr protein that is also part of the phosphotransferase system becomes phosphorylated at a serine residue, Hpr(Ser-P), and binds to the transcription regulator CcpA. The Hpr-P–CcpA complex binds to a nucleotide sequence called CRE (catabolite responsive elements) (possibly as a ternary complex with fructose-1,6-bisphosphate) that is proximal to the promoter of genes susceptible to catabolite repression, and inhibits transcription. (See note 114.) Thus the key event with respect to catabolite repression is that glycolytic intermediates stimulate the phosphorylation of Hpr and therefore the repression of CcpA-sensitive operons.

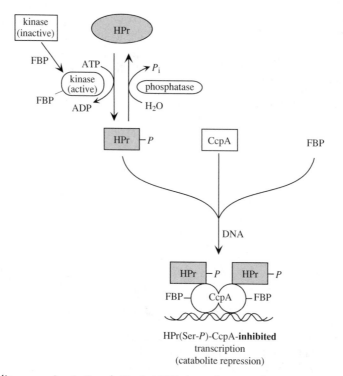

Fig. 18.15 Catabolite repression in *B. subtilis*. An ATP-dependent Hpr kinase is activated by fructose-1,6-bisphosphate (FBP). Phosphorylated Hpr (Hpr-P) binds to the transcription regulator CcpA, and possibly also to FBP, to form a ternary complex. The ternary complex binds to the CRE region near the promoters of catabolite-repressible genes and represses transcription. *Source*: Adapted from Saier, Jr., M. H., S. Chauvaux, J. Deutscher, J. Reizer, and J.-J. Ye. 1995. Protein phosphorylation and regulation of carbon metabolism in gram-negative versus gram-positive bacteria. *Trends Biochem. Sci.* 20:267–269.

*The key regulatory event, stimulated
by glycolytic intermediates, is
the phosphorylation of Hpr*

It is important to understand that the phosphorylation of Hpr is not due to enzyme I, the PEP-dependent kinase that phosphorylates a histidine residue, resulting in Hpr(His-P), which functions in the PTS. Rather, the phosphorylation is due to an ATP-dependent kinase that phosphorylates a serine residue in Hpr, forming Hpr(Ser-P). See note 115 for a further explanation.

Hpr is phosphorylated by a specific ATP-dependent kinase that is activated by the glycolytic intermediates fructose-1,6-bisphosphate and 2-phosphoglycerate, as well as by gluconate-6-P. Thus, when the cells are growing on glucose and the levels of glycolytic intermediates rise, Hpr becomes phosphorylated and represses the transcription of genes having the CRE sequences. A phosphatase removes the phosphate from Hpr-P under starvation conditions. The model also provides a role for inorganic phosphate (P_i). According to the model, inorganic phosphate would prevent catabolite repression by inhibiting the kinase and by stimulating a phosphatase that dephosphorylates HPr(Ser-P).

Genetic evidence for the model

Mutations in CRE, or the genes coding for CcpA or CcpB, or mutations that lead to a failure to phosphorylate serine 46 in Hpr, will cause failure of catabolite repression, and such failures comprise genetic evidence in support of the model. (See note 116 for a general description of how some of these mutants can be isolated.) Also, the ATP-dependent kinase that phosphorylates HPr at Ser-46 has been purified and shown to be stimulated by FBP and inhibited by inorganic phosphate. A P_i-activated phosphatase that dephosphorylates HPr-Ser-P has been isolated from several gram-positive bacteria.

Comparison to E. coli

A comparison of the catabolite repression mechanism of *E. coli* with that of *B. subtilis* shows that whereas catabolite repression in *E. coli* is due to a decrease in a *positive* regulator of catabolite-repressible genes (i.e., cAMP–CRP,

or Cra), catabolite repression in *B. subtilis* is due to an increase in a *negative* regulator factor (i.e., CcpA–Hpr-Ser-P or CcpB–Hpr-Ser-P).

*Does the CcpA-dependent catabolite
repression system exist in gram-negative
bacteria?*

A BLAST (<u>B</u>asic <u>L</u>ocal <u>A</u>lignment <u>S</u>earch <u>T</u>ool) search of the protein data-base revealed that homologues of the Hpr kinase are present in some gram-negative bacteria.[117] Future research might reveal that CcpA-dependent catabolite repression occurs in gram-negative bacteria as well as in gram-positive bacteria.

18.10.3 Two mechanisms leading to the prevention by glucose of the intracellular accumulation of other sugars in some gram-positive bacteria: Uncoupling from the proton gradient, and inducer expulsion

Glucose prevents the accumulation of other sugars in some gram-positive bacteria, but the mechanism is not via the inhibition of sugar uptake by IIA^{Glc}, as is the case with *E. coli*. Two known mechanisms lead to the equilibration of the sugar across the cell membrane, hence no intracellular accumulation. One of these is the uncoupling of sugar transport from the proton motive force, and the second is the dephosphorylation of the incoming sugar–P. Consider the situation in *Lactobrevis*, where Hpr(Ser-P) appears to bind to the lactose/proton symporter and uncouples sugar transport from proton symport.[118] The consequence of uncoupling the symporter from the proton gradient is that the symporter catalyzes facilitated diffusion rather than active transport. The result is that lactose equilibrates across the membrane and cannot be accumulated. In some gram-positive bacteria (streptococci, lactococci, and enterococci), but not others (bacilli, staphylococci), there exists a membrane-associated sugar–P phosphatase activated by HPr(Ser-P).[119] It is suggested that when the sugar–Ps are accumulated in the cell via the PTS, they are dephosphorylated and then exit along their concentration gradient. This phenomenon, called *inducer expulsion*, occurs upon the addition of glucose. The purified phosphatase has broad substrate specificity and dephosphory-lates a long list of sugar–P molecules including glucose-1-P, glucose-6-P, and fructose-6-P. Presumably only certain sugar–Ps are attacked because the enzyme's location on the membrane determines its accessibility to specific sugar–Ps being transported via sugar transporters into the cell.

18.10.4 Response to carbon source: Induction of a permease for dicarboxylic acids in Rhizobium meliloti

R. (Sinorhizobium) meliloti are nitrogen-fixing gram-negative α-proteobacteria that live in the soil or symbiotically in the root nodules of leguminous plants (e.g., the agriculturally important alfalfa plants). (See Section 18.12.8 for a discussion of chemotaxis in R. meliloti.) The bacteria grow most rapidly on C_4 dicarboxylic acids such as succinate, malate, and fumarate. R. meliloti has a well-characterized two-component system that regulates the production of a permease (transporter) that brings dicarboxylic acids into the cell. When R. meliloti is cultured with C_4 dicarboxylic acids, the gene for the dicarboxylic acid permease (dctA) is induced.

The model

A transmembrane histidine kinase (sensor kinase) called DctB autophosphorylates at a histidine residue when the cells are incubated with dicarboxylic acids; subsequently, it transfers the phosphoryl group to an aspartate residue in a cytoplasmic response regulator protein, DctD. DctD-P then activates the transcription of dctA, the dicarboxylic acid permease gene.

Support for the model

Support for this model includes demonstration in vitro of autophosphorylation of the purified DctB protein, along with the transfer of the phosphoryl group to DctD and the binding of phosphorylated DctD to DNA.[120]

The dicarboxylic acids themselves are not sufficient to activate the sensor kinase

The inclusion of dicarboxlic acids in the incubation mixture does not affect the auto-phosphorylation reaction, indicating that there may be another component (perhaps in the periplasm) that binds to the dicarboxylic acid and transmits the signal to the histidine kinase, or that the histidine kinase responds directly to the dicarboxyic acids only when integrated into the membrane.

18.11 Virulence Factors: Synthesis in Response to Temperature, pH, Nutrient, Osmolarity, and Quorum Sensors

This section discusses some of the regulatory systems that transduce environmental signals to control the transcription of genes that encode virulence factors. Virulence factors are structures or substances that aid pathogenic bacteria in the infection process and/or contribute toward the symptoms of disease (reviewed in refs. 121 and 122). Virulence factors are generally exported to the bacterial cell surface or beyond. They include pili for adsorption to host tissue, toxins, flagella to aid the bacterium in arriving at the site of colonization, and extracellular enzymes such as proteases (see, e.g., ref. 123.). As discussed in Section 17.5.2, some virulence factors produced by Yersinia spp. are actually injected into target cells.

Virulence factors are not necessarily constitutively produced, and it appears that several are made only during the course of an infection, not when the pathogen is growing outside the host. Certainly this is the case with laboratory cultures, where only under specific conditions of temperature, pH, nutrient composition, iron availability, and osmolarity are certain virulence factors made.[124] How bacteria sense environmental factors such as temperature and osmolarity is not well understood. The student should review Section 18.7, entitled Response to osmotic pressure and temperature: Regulation of porin synthesis, for information about a regulatory system influenced by osmotic pressure and temperature. The following discussion focuses on the effect of environmental factors, including extracellular ones (quorum sensors), on the expression of virulence genes.

18.11.1 ToxR and cholera

ToxR is a transcription regulatory protein produced by Vibrio cholerae, the bacterium that causes cholera. It stimulates the transcription of virulence genes that are in the ToxR regulon. (For reviews, see refs. 125 and 126.) To understand the importance of ToxR, it is necessary to briefly describe cholera pathogenesis.

Cholera

Vibrio cholerae bacteria grow in the small intestine, attached to the intestinal epthelial layer. The microorganisms swim through the mucus layer and attach to intestinal cells via type IV pili called TCP pili (toxin-coregulated pilus). The symptoms of cholera include an extensive watery diarrhea, which can result in the loss of as much as 10 to 20 liters of fluid a day, and if untreated results in dehydration, loss of electolytes, and death in approximately 60% of infected people. The diarrhea is caused by cholera toxin (CT), a potent enterotoxin that is secreted by *V. cholerae*. (Cholera toxin consists of one A subunit and five identical B subunits and is encoded by the *ctxAB* operon, which resides in the CTX lysogenic phage. See note 127 for a description of how cholera toxin works.) *V. cholerae* is spread via the oral–fecal route, and people become ill with cholera when they ingest food or water contain contaminated with feces containing the bacteria.

Response to temperature, pH, osmolarity, and amino acids

Exactly which environmental stimuli in the intestine influence the activity of ToxR within infected animals is a subject of study.[123,128] Toxin production is lowered in pure culture when the temperature is increased to 37 °C, which is opposite to what one would intuitively expect. Additionally, low pH in culture stimulates cholera toxin production, but the pH in the intestine where *V. cholerae* grows is believed to be alkaline. This suggests that factors other than temperature and pH increase the activity of ToxR during growth in the intestine. Two of these factors might be osmolarity and the presence of certain amino acids, since toxin production in pure culture is increased under osmolar conditions similar to those in mucus, as well as in the presence of amino acids likely to be present in mucosal secretions. However, the expression of the ToxR regulon in response to environmental cues can differ with the strain of *V. cholerae*. See note 129.

The role of ToxR in pathogenesis

The importance of ToxR for pathogenesis is clear, since mutants in which the *toxR* gene has been deleted do not colonize the intestine in human volunteers. ToxR is an inner membrane bifunctional protein that can sense environmental signals and subsequently activate the transcription of 30 to 40 genes in the ToxR regulon. The genes activated by ToxR are called the ToxR-activated, or *tag*, genes, and these are required for efficient colonization of the intestine, toxin production, and survival within the host.

The *tag* genes have been identified by screening Tn*phoA* fusions whose transcriptional activity under different growth conditions parallels that of cholera toxin production (See note 25 for an explanation of gene fusions and note 130 for how such fusions were used to detect ToxR-regulated genes.) Genes whose transcription is increased by ToxR include the *ctxAB* operon, which encodes the A and B subunits of cholera toxin; *tcpA*, which encodes the major subunit of the *V. cholerae* pilus; several genes required for pilus assembly and transport; and other genes, including *toxT*. As mentioned, the pilus is required for *V. cholerae* to bind to the intestinal epithelium and subsequent colonization. Thus, it can be considered to be a virulence factor. As described next, except for the transcription of *ctxAB*, the genes in the ToxR regulon are not activated directly by ToxR.

How ToxR activates gene expression

For study purposes, the *toxR* gene was cloned in *E. coli*, where it activates the transcription of a *ctx–lacZ* fusion. (See note 25.) ToxR binds to the promoter region of the *ctxAB* operon, resulting in increased transcription, but its activation of the other genes in the ToxR regulon is indirect. That is, it appears that ToxR increases the transcription of a second regulatory gene, *toxT*, whose product directly activates the other genes in the ToxR regulon, including *ctxAB* and and the toxin-coregulated pilus (TCP) operon.[131] Figure 18.16 shows one model of the regulatory pathways.

The mystery of how ToxR transmits environmental signals to the toxR regulon

It is unclear exactly how ToxR receives and transmits environmental signals to the ToxR regulon. It has been concluded, based in part

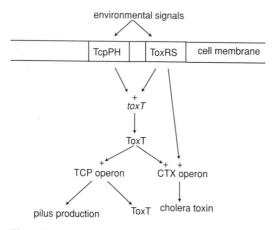

Fig. 18.16 Model showing regulatory pathways for ToxR regulon. TcpPH and ToxRS represent pairs of inner membrane proteins (TcpP/TcpH and ToxR/ToxS) that respond to environmental signals and coactivate the transcription of *toxT*. ToxT, along with ToxR, activates the transcription of the cholera toxin (CTX) operon, encoded on a lysogenic phage (CTX phage). ToxT also activates the transcription of the toxin-coregulated pilus (TCP) operon, which is responsible for the biogenesis of the pilus required to colonize the intestine. The TCP operon contains the gene for the main pilin subunit (TcpA, not shown), as well as genes required for the regulation of pilin gene expression, pilin secretion, and the assembly of the pilus. The *toxT* gene is also within the TCP operon, and thus more ToxT is produced when the operon is induced.

upon amino acid sequence data and the construction of ToxR–PhoA fusions that have alkaline phosphatase activity and also regulate the cholera toxin gene, that ToxR spans the cell membrane with a periplasmic C-terminal region and an N-terminal cytoplasmic region that has a DNA-binding domain. (See ref. 133 and references therein and note 25 for an explanation of why alkaline phosphatase activity from a *toxR–phoA* fusion can indicate a C-terminal periplasmic domain.) The C-terminal domain may be sensitive to certain environmental stimuli discussed next. The N-terminal domain is thought to bind to promoter sequences in the DNA of target genes, and it may also respond to intracellular signals.

A signal-transducing cascade

How the ToxR protein senses the environmental signals, and whether other components besides ToxR are involved in sensing these

signals, are open questions. It has been suggested that another protein, ToxS, encoded in the toxRS operon, is a membrane-bound protein with a periplasmic domain that may actually receive environmental signals and activate ToxR by interacting with its periplasmic domain. The cytoplasmic domain of ToxR would then activate the *toxT* gene, whose product in turn would activate the genes in the ToxR regulon (Fig. 18.16). However, other signaling proteins are also involved. Two of these, TcpP and TcpH, are proteins encoded by the TcpPH operon, as described next.

The *toxT* gene is also positively regulated by TcpP

The *toxT* gene is also positively regulated by another membrane-bound transcriptional activator, TcpP, which itself may be activated by a membrane-associated protein with a large periplasmic domain, TcpH.[132] Optimal expression of *toxT* requires both ToxR and TcpP. As shown in Fig. 18.16, *both* ToxR and TcpP are involved in sensing environmental and cytoplasmic signals and *cooperatively* activate the *toxT* gene.

18.11.2 The PhoP/PhoQ two-component regulatory system

A two-component system called the PhoP/PhoQ system exists in many nonpathogenic and pathogenic gram-negative bacteria, including *E. coli*, *P. aeruginosa*, and *Salmonella* spp. PhoQ is a transmembrane sensor histidine kinase, and PhoP is a cytoplasmic response regulator. For a review of the PhoP/PhoQ system, see refs. 133 and 134. The PhoP/PhoQ system is not related to the two-component PhoB/PhoR system discussed in Section 18.6.1. The gene *phoP* was originally discovered as a positive regulator for the synthesis of an acid phosphatase, the product of the *phoN* gene. See note 135 for a description of how *phoP* was discovered.

The PhoP/PhoQ system is activated by low Mg²⁺ concentrations and inactivated by high Mg²⁺ concentrations

The PhoP/PhoQ system is stimulated by low extracellular Mg^{2+} and is required to adapt to low extracellular Mg^{2+} concentrations (micromolar range).[136,137] Two of the genes activated by PhoP/PhoQ encode proteins for two of the three Mg^{2+} uptake systems. The model shown in Fig. 18.17 proposes that when the Mg^{2+} concentrations are sufficiently low, PhoQ autophosphorylates at a conserved histidine residue in its cytoplasmic domain and then acts as a kinase and phosphorylates PhoP at a conserved aspartate residue. Phosphorylated PhoP then interacts with promoters on target genes and activates transcription. However, high Mg^{2+} concentrations repress genes regulated by PhoP-P because PhoQ is a Mg^{2+} sensor, and when the periplasmic domain of PhoQ binds Mg^{2+}, PhoQ becomes a phosphatase and dephosphorylates PhoP-P. In addition to high Mg^{2+}, Ca^{2+} binds to PhoQ and represses the transcription of genes activated by PhoP, and these two cations should be viewed as being

physiological signals controlling the PhoQ/PhoP system. See ref. 138 for a review. This discussion will focus on what has been learned from studying *Salmonella* because this genus has been investigated the most.

Salmonella pathogenesis

We will begin with a brief summary of *Salmonella* pathogenesis followed by the role of PhoP/PhoQ in regulating the expression of virulence genes. *Salmonella* spp. cause gastroenteritis, enteric fever, and septicemia (blood infections). During the course of the infections, the bacteria are ingested by phagocytic cells such as macrophages and epithelial cells and must survive intracellularly. Indeed, the intestinal epithelial cells must be invaded to allow the spread of the bacteria from the intestinal mucosal surface to underlying tissue

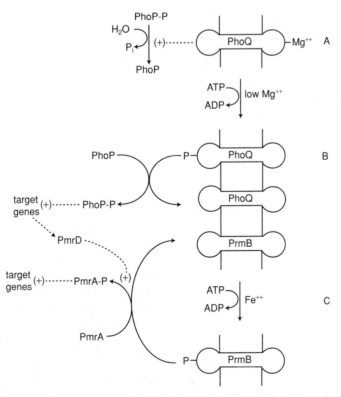

Fig. 18.17 The regulation of gene transcription by PhoP/PhoQ and PmrA/PmrB. (A) All genes activated by PhoP are repressed by high concentrations of Mg^{2+} because the sensor histidine kinase PhoQ dephosphorylates PhoP-P when the Mg^{2+} concentrations are high. (B) PhoQ autophosphorylates when the Mg^{2+} concentrations are low and transfers the phosphoryl group to PhoP. PhoP-P activates the transcription of many genes, including *pmrD*. In a way that is not understood, PmrD increases the levels of PmrA-P, which in turn stimulates transcription of target genes. (C) Fe^{2+} binds to PmrB and activates its autophosphorylation activity, thus stimulating transcription of PmrA-dependent genes independently of PhoP/PhoQ.

and the development of gastroenteritis, enteric fever, and septicemia.

The PhoP/PhoQ system regulates virulence genes

There are at least 40 genes regulated by the PhoP/PhoQ system in *Salmonella*, and they are in the PhoP/PhoQ regulon. (Recall that a regulon is a set of noncontiguous operons or genes controlled by a common regulator. The regulator in this case is PhoP.) Some of the genes regulated by the PhoP/PhoQ system govern virulence. For example, PhoQ and PhoP are responsible for the induction in *Salmonella* spp. of the five *pag* genes (PhoP-activated genes), which are important for the survival of *Salmonella* inside the acid bacteriocidal environment of phagolysosomes of macrophages (they are not induced inside epithelial cells) and for the repression of several other genes called *prg* (PhoP-repressed genes).[139] One of the *pag* genes, namely *pagC*, is responsible for the synthesis of an envelope protein that supports virulence as well as survival inside macrophages.

Mutants in *phoQ/phoP* are less virulent and have decreased survivability inside macrophages. They are also less resistant to cationic antimicrobial peptides such as defensins, which the bacteria can encounter in the intestinal lumen and within phagolysosomes of neutrophiles and macrophages. (See Section 18.11.3.) Some of the genes activated by PhoP-P modify the lipopolysaccharide, whereas others are responsible for the synthesis of extracellular proteases that cleave the antimicrobial peptides.

18.11.3 PmrA/PmrB: A two-component system that interacts with the PhoP/PhoQ system

In addition to regulating the activity of several genes directly, PhoP is part of a signaling hierarchy that indirectly regulates the expression of several in *Salmonella* genes by activating a second two-component signaling pathway called the PmrA/PmrB system, where PmrB is the histidine kinase and PmrA is the response regulator. The PmrA/PmrB system in *Salmonella* spp. and *Pseudomonas aeruginosa* activates genes conferring resistance to the antibiotic polymyxin, a cationic antimicrobial peptide (CAP).[140] See note 141 for an explanation of CAPs and related antimicrobial compounds called defensins.

Resistance to polymyxin is due at least in part to a modification of the lipopolysaccharide, to which polymyxin and other CAPs bind. In *Salmonella*, PhoP-P activates the PmrA/PmrB system by promoting the transcription of the *pmrD* gene, and PmrD stimulates the PmrA/PmrB system at the post-transcriptional level so that more PmrA becomes phosphorylated.[142] Exactly how this is done is not yet known, but either increased phosphorylation or decreased dephosphorylation may be involved. The PmrA/PmrB system can also be activated independently of PhoP/PhoQ by growth in high Fe^{2+} concentrations. This stimulates the synthesis of proteins that protect the bacterium from the toxic effects of Fe^{2+}. Thus, the PmrA/PmrB system in *Salmonella* is activated by low Mg^{2+} concentrations via PhoP/PhoQ, or directly by high Fe^{2+} concentrations sensed by the histidine kinase sensor protein PmrB. This is outlined in Fig. 18.17. The *pmrD* gene is also present in several gram-negative bacteria, indicating that control of PmrA/PmrB by PhoP/PhoQ is not restricted to *Salmonella* spp.

18.11.4 The bvg genes and pertussis

There are several other regulatory genes that respond to environmental stimuli and promote virulence. (Reviewed in ref. 123.) These include the *bvgAS* genes (Bordetella virulence genes), which are required for the expression of virulence genes (the *vir* regulon) in *Bordetella pertussis*, the causative agent of whooping cough (pertussis).[143] (The *bvgAS* locus was formerly called the *vir* locus.) Growth at 37 °C as opposed to 25 °C results in increased expression of virulence genes.

Pertussis

Whooping cough is primarily a childhood disease in which the bacteria grow attached to the ciliated epithelial cells of the bronchi and trachea. (Virulence is dependent upon bacterial surface adhesins that bind the bacteria to the cilia.) Episodes of severe, spasmodic coughing are followed by a "whoop" during inspiration. The organism is transmitted via respiratory discharges. Children are now vaccinated with the DPT (diphtheria, pertussis, tetanus) vaccine. However, prior to the vaccine pertussis was a major childhood killer.

The vir Regulon

Genes induced by the *bvg* genes (i.e., the genes in the *vir* regulon) include the pertussis toxin operon *ptxA-E*, the adenylate cyclase toxin–hemolysis gene *cyaA*, the filamentous hemagglutinin locus *fhaB*, the fimbrial subunit genes *fim*, and the *bvg* operon.

The bvg genes encode a two-component signaling system that regulates the vir regulon

According to deduced amino acid sequences, BvgS is a transmembrane sensor kinase with a periplasmic N-terminus region and a cytoplasmic C-terminus region, Similarly, BvgA is proposed to be a response regulator protein that is phosphorylated by BvgS-P and then activates transcription of the virulence genes, perhaps in some cases by activating other genes that are transcriptional activators. (In *E. coli*, the cloned *bvgAS* locus will activate the *fhaB* and *bvgA* promoters but not the *ptxA–E* or *cyaA* promoters, suggesting that perhaps a signal relay is required for the latter promoters.) Increased expression of virulence genes at 37 °C has also been studied in *Shigella*, as discussed next.

18.11.5 Virulence genes and bacillary dysentery

Bacillary dysentery

Shigella spp., including *S. dysenteriae* and *S. flexneri*, cause bacillary dysentery (shigellosis), which can be manifested simply as a watery diarrhea or, in more serious cases, as a severe diarrheal disease of the colon accompanied by passage of bloody, mucoid stools. In the severe cases, the bacteria invade and multiply within the mucosal epithelial cells, causing degeneration of the epithelium and intense inflammation of the colon (colitis). (*Shigella* are facultative intracellular pathogens.) Most *Shigella* species produce a toxin, and the toxin produced by *Shigella dysenteriae* is *shiga toxin*, which is at

least partly responsible for damaging the blood vessels, for inflammation, and probably for the watery diarrhea.

Virulence genes

Virulence of *Shigella* is due primarily to a plasmid that encodes the virulence genes. Virulence genes are required for adherence and invasion (via endocytosis) of host cells, spreading (via polymerization of actin filaments), and synthesis of toxin. (For a review of virulence genes in *Shigella* and information on how the bacteria invade host cells and spread intercellularly, see ref. 144. Also, see note 145.)

1. hns (virR): Response to temperature
Shigella has a chromosomal regulatory gene called *hns* (*virR*) that encodes a repressor for the virulence genes. The significance of *hns* for this discussion is that the regulation of virulence genes by *hns* is controlled by temperature. *Shigella* strains grown at 30 °C are avirulent (i.e., the virulence genes are repressed by *hns*), whereas those grown at 37 °C are virulent in tissue culture assays and in animal assays (i.e., they cause invasion of tissue culture cells and production of keratoconjunctivitis in guinea pigs). Consistent with the role of *hns* as a repressor gene at 30 °C, mutations in this gene allow expression of the invasive phenotype at both 30 and 37 °C. Presumably, turning off virulence genes is an advantage to the bacteria when they are growing outside a host. The product of the *hns* gene is a histonelike protein called H-NS (H1), and binding of H-NS near target genes leads to transcriptional repression, perhaps by interfering with the binding of RNA polymerase. (See the discussion of H-NS and the nucleoid in Section 1.2.6, and the discussion of the control of rRNA synthesis in Section 2.2.2.)

2. virF and virB
The virulence genes are also positively regulated. One of the regulatory genes, *virF*, encodes a positive regulator (VirF) of the virulence gene regulon. It activates some of the *vir* genes directly, and others indirectly, by activating the transcription of another positive regulatory gene, *virB*. It is thought that (H-NS), the product of the *hns* gene, represses *virF* expression at 30 °C and that this accounts for the repression

of the other virulence genes. The virulence genes are also repressed when the cells are grown in low-osmolarity media at 37 °C, and this was also shown to be due to repression by H-NS in *S. flexneri*.[146]

3. IHF; responses to temperature and osmolarity

When *Shigella* grow outside the host in nature, they are generally found in media of low osmolarity as well as at temperatures lower than 37 °C. In addition to being repressed by H-NS at 30 °C, the expression of virulence genes is positively stimulated upon a temperature upshift. The positive stimulation appears to be due to another DNA-binding protein, integration host factor (IHF).[147] (See the discussion of IHF in Section 10.2.5 and the discussion of DNA-binding proteins in the nucleoid in Section 1.2.6.) This was determined by examining the expression of a *lacZ* fusion to a virulence gene upon a temperature upshift in wild-type cells and *ihf:Tn10* mutants.

Thus, DNA binding proteins H-NS (a repressor) and IHF (an activator) appear to take part in signaling pathways that detect environmental changes (i.e., temperature and osmolarity) that result in alterations in the expression of virulence genes.

18.11.6 Virulence gene expression in Staphylococcus: Stimulation of a two-component system by a peptide pheromone

Staphyloccus aureus and other *Staphylococcus* strains secrete several extracellular protein virulence factors. For example, *S. aureus* secretes enterotoxin B and toxic shock syndrome toxin 1, as well as other toxins in the postexponential phase of growth. The transcription of the toxin genes is regulated by a two-component system consisting of a membrane histidine kinase called AgrC, a response regulator protein called AgrA, and a small peptide secreted by the cells into the medium.[148] The peptide is a quorum-sensing signal because the amount secreted increases with cell density; and when the concentration of the peptide reaches a threshold concentration, it activates AgrC, which leads to increased transcription of virulence genes.

18.11.7 Virulence gene expression in Agrobacterium tumefaciens: Stimulation of a two-component system by phenolic plant exudates

One of the best-characterized systems for the regulation of expression of virulence genes in bacterial–plant interactions is the VirA–VirG two-component system in *Agrobacterium tumefaciens*, the bacterium that causes crown gall tumors in plants. (For reviews, see refs. 123 and 149.)

Stages in the formation of crown gall tumors

The bacteria gain entrance to the plant tissue via a wound site on the plant and grow extracellularly at the site of infection. Bacteria that are directly attached to plant cells transfer into the plant cell nucleus a small piece of oncogenic DNA, called T-DNA, from a plasmid, called the Ti (tumor inducing) plasmid. The process of transfer of the T-DNA from the bacterium into the plant resembles bacterial conjugation and takes place via a type IV secretory system described in Section 17.5.6. The T-DNA integrates into the plant genome, and a tumor results. The tumor forms because T-DNA encodes plant growth hormones (i.e., auxin and cytokinin), which stimulate growth and division of the plant cells.

When the T-DNA integrates into the plant genome, the plant makes nutrients called opines, which feed the bacteria

The crown gall tumor actually feeds the bacteria. There are genes in the T-DNA that encode enzymes to make organic molecules called opines, which when released by the plant cell can be used as a source of nutrient by the bacteria. Opines are small organic molecules covalently bonded to amino acids. For example, one opine, called octopine, is N^2-(1,3-dicarboxyethyl)-L-arginine. The genes that encode enzymes to utilize the opines are in the region of the Ti plasmid that is not transferred. Thus, when the T-DNA becomes incorporated into the plant genome, the plant makes the opines and the bacteria use these opines for growth.

Interestingly, the genes from the Ti plasmid that are expressed in the plant have eukaryotic-type promoters and eukaryotic-type translation initiation regions, whereas the plasmid genes expressed in the bacterium have prokaryotic-type promoters and prokaryotic-type translation initiation regions. The transfer of T-DNA from the Ti plasmid into the plant cell does not result in loss of T-DNA from the plasmid. This is because only one strand of the T-DNA is transferred to the plant cell, and a new complementary strand is synthesized in the donor bacterium. The process is very much like the transfer of a single plasmid strand during bacterial conjugation.

The plant signals the bacteria to transfer T-DNA into the plant

For T-DNA to be transferred from the bacterium into the plant, virulence genes (*vir* genes) on the plasmid must be activated. The *vir* gene products include an endonuclease (VirD) that nicks the plasmid DNA, a DNA-binding protein (VirE) that attaches to the single strand of DNA produced by the endonuclease and brings the DNA into the plant cell, and a bacterial membrane protein (VirB) that functions as a bridge in transporting the DNA across the bacterial membrane into the plant cell. [The VirB protein(s) are actually part of a type IV secretion system. See Section 17.5.6 for a discussion of this point.]

The plant activates the expression of the *vir* genes in the bacterium. What happens is that phenolic signaling compounds (e.g., *p*-hydroxybenzoic acid) produced by the wounded plant tissue activate a two-component system called the VirA–VirG system in the bacterium. VirA is a transmembrane sensor kinase whose periplasmic domain is thought to interact with the phenolic compounds. VirA is then proposed to autophosphorylate and transfer the phosphoryl group to VirG, a cytoplasmic regulator protein. The phosphorylated form of VirG is believed to be a positive transcription factor for the *vir* genes, including *virA and virG*, which are also plasmid encoded. The result is transfer of T-DNA into the plant cell.

Role of plant opines and a bacterial quorum-sensing system in stimulating plasmid transfer between bacteria in the tumor

As described in Section 18.21.1, the Ti plasmid is transferred between *Agrobacterium* cells in a conjugation process that requires the expression of *tra* genes on the plasmid. When the density of the cell population is sufficiently high, the transfer is stimulated by a bacterial quorum-sensing system called the TraR–TraI system, which is homologous to the Lux system in luminescent bacteria. TraI is the acyl–HSL synthase, and it synthesizes the signaling molecule 3-oxo-C_8-HSL. The acyl–HSL binds to and activates TraR. Active TraR then stimulates the expression of genes required for conjugation, as well as the transcription of *traI*, the gene responsible for making the acyl–HSL. The plant opines stimulate the transcription of *traR*, and therefore the production of the acyl–HSL, since TraR stimulates the transcription of *traI*. Because of this, the plant opines stimulate plasmid transfer between bacterial cells in the tumor.

Since the genes to metabolize the opines are part of the Ti plasmid, the spread of the plasmid between the bacteria distributes the ability to catabolize the opines, which are used as nutrient by the bacteria within the tumor. TraR also stimulates the genes for plasmid replication. Thus, the plant opines stimulate plasmid replication and therefore tumorigenesis, as well as conjugation and plasmid transfer. For more information about how all this occurs, see note 150.

18.12 Chemotaxis

Chemotaxis refers to the ability of bacteria to move along a concentration gradient toward a chemical attractant (positive chemotaxis) or away from a chemical repellent (negative chemotaxis). For reviews see refs. 151 through 155. The attractants and repellents are called *chemoeffectors*. As we shall see, the chemotaxis signaling system is a two-component system that includes a histidine kinase (CheA) and a response regulator (CheY) that affects the switch mechanism in the flagellar motor. The student should refer to the description of bacterial flagella in Section 1.2.1 while reading this section.

As noted earlier, the histidine kinase is cytoplasmic, and the signal is first sensed by a transmembrane sensor protein. The transmembrane sensor protein is usually referred to as the methyl-accepting chemotaxis protein (MCP), but sometimes the term "receptor–transducer protein" is used. Adaptation plays an important role in chemotaxis, and this will also be discussed.

Most of what is known about chemotaxis is derived from the original studies with the enteric bacteria *Escherichia* and *Salmonella*, and most of what follows is derived from research with these organisms. Systems that differ from the enteric bacteria are described in Section 18.12.8.

18.12.1 Bacteria measure changes in concentration over time

One might think that bacteria swimming along a chemical concentration gradient are detecting the spatial gradient itself, but this is not the case. Calculations indicate that because of the small size of bacteria, the difference in concentration of a chemoeffector between the ends of the cell would be too small to be measured accurately.[156] What the bacterium actually measures is the absolute concentration of chemoeffector, which it then compares with the concentration previously measured. In other words, *bacteria measure changes in concentration over time*. They actually "remember" the previous concentration. If a bacterium finds that it is in a higher concentration of an attractant or a lower concentration of repellent than at a previous time, it will continue to move in that direction. If the bacterium finds that it is in a lower concentration of attractant or a higher concentration of repellent than at a previous time, it will move in a *randomly* different direction.

The propensity to swim randomly is fundamental to chemotaxis. Some bacteria swim randomly because they periodically tumble, as discussed next, whereas other bacteria swim randomly without tumbling. Chemotaxis in bacteria that tumble is understood best.

Random swimming without tumbling is discussed in Section 18.12.8.

18.12.2 Tumbling

E. coli and many other bacteria swim along a path for a few seconds or less and then tumble. Very importantly, when they recover from the tumble, they swim in a *randomly* different direction. If nothing affected the frequency of tumbles and swimming, bacteria would just swim in random directions. This is the situation when they are not responding to chemoeffectors. However, *attractants decrease the frequency of tumbles, whereas repellents increase the frequency of tumbles*.[156,157] This is illustrated in Fig. 18.18. If a bacterium moves into an area in which the concentration of chemoattractant is higher than at a previous moment, the cell detects the higher concentration (because of increased binding to chemoreceptors), tumbles less frequently, and continues to move in that direction. Likewise, if the cell swims into an area in which the concentration of repellent is higher, tumbling increases and the cell changes its direction, moving away from the repellent. But why does the cell tumble?

E. coli and *S. typhimurium* are peritrichously flagellated bacteria in which helical flagella

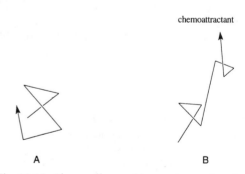

Fig. 18.18 Chemoeffectors bias random swimming. (A) The swimming pattern is smooth swimming for a short period interspersed with brief periods of tumbling. After tumbling, the bacterium resumes swimming in a randomly different direction. (B) A chemoattractant decreases the frequency of tumbling, thus prolonging the periods of smooth swimming. The result is that the bacterium swims in the direction of higher concentrations of chemoattractant. A chemorepellent increases the frequency of tumbling, causing the bacterium to swim away from the repellent.

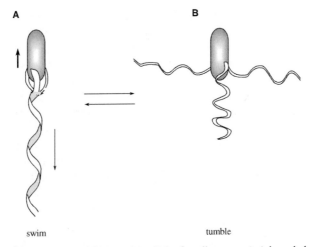

Fig. 18.19 Flagella fly apart during tumbling. In *E. coli* the flagella are peritrichously located and form a trailing bundle while the cell is swimming. (A) The filaments are in a left-handed helix. When the motors rotate counterclockwise, a helical wave travels proximal to distal (outward from the cell), pushing the cell forward. (B) When the motors rotate clockwise, the filaments undergo a transition to a right-handed waveform, causing them to fly apart and the cells to tumble. *Source*: Adapted from MacNab, R. M. 1987. Motility and chemotaxis, pp. 732–759. In: *Escherichia coli and Salmonella typhimurium*. F. C. Neidhardt et al. (Eds.). ASM Press, Washington, DC.

(an average of four in *E. coli*) protruding randomly from the cell sides are wrapped around one an other in a helical bundle that extends several cell lengths at the rear of the swimming cell (Fig. 18.19). The flagella are usually in the conformation of a left-handed helix. When the flagella rotate counterclockwise (CCW) (viewed from the tip of the flagellum toward the cell), they remain in a bundle as a helical wave moves from the proximal to the distal portion of the flagellum. This pushes the bacterium forward. The forward movement is called a run. Such "runs" by *E. coli* last about one second. However, when one or more flagella rotate clockwise (CW), they unwind and leave the filament bundle. The result is that there is no longer any coordinated flagella movement, and the cell moves erratically in place for a fraction of a second. This is called tumbling. Then the flagella move CCW again to re-form the bundle and the cell "runs" again, but in a randomly different direction.

As mentioned, repellents cause the flagellar motor to reverse and the cells to tumble, whereas attractants decrease the frequency of reversal and promote smooth swimming. The tumbling and smooth swimming responses to changes in chemoeffector concentrations can be seen by observing the swimming of individual cells under a microscope. It is also possible to monitor the change in direction of flagellar rotation by tethering the bacteria to a glass slide by a flagellum. One way to do this is to use an antibody to the flagellum that sticks to both the flagellum and the glass slide.[158] The cells are grown in minimal media with glucose as the carbon source. Under these conditions, the average number of flagella per cell is reduced and cells tethered by a single flagellum can be observed. When the flagellar motor rotates, the cell rotates because the flagellum is fixed to the slide. It is then possible to infer that the addition of attractants causes the flagella to rotate CCW and the addition of repellents causes CW rotation.[159]

18.12.3 Adaptation

An important aspect of the chemotactic response is *adaptation*. Bacteria adapt (i.e., become desensitized) to a chemoeffector, and thus after a short period of time (seconds to minutes), they no longer respond to it at the initial concentration but may respond to a higher concentration. In fact, one could say that adaptation is the way the bacterium "remembers" the previous concentration of chemoeffector.

Why is an adaptation circuit built into the signaling pathway? Adaptation makes good sense. When a bacterium adapts to a chemoattractant, it resumes tumbling at the nonstimulated frequency and will stay in the area; but if it encounters an increasing concentration gradient of attractant, tumbling will again be suppressed. Consequently, the bacterium will swim toward and remain in the area of the higher concentration of chemoattractant. Another way of saying this is that adaptation allows the bacterium to respond to any local gradients it may encounter while randomly swimming. Adaptation to repellents is also easily rationalized. If a bacterium did not adapt to repellents, it would continue to tumble at the stimulated rate even when swimming toward the lower concentration of a given repellent, hence would be trapped in the area of that repellent.

18.12.4 Proteins required for chemotaxis

Chemotaxis employs complex regulatory circuits involving both cytoplasmic and membrane proteins. In *E. coli*, these proteins (MCPs, CheA, CheW, CheR, CheZ) are all located as a cluster at the cell poles.[160–162] A description of the proteins and their functions will be given first, followed by an explanation of the regulatory circuits. As we shall see in Sections 18.13 and 18.14, homologous proteins are also part of the signaling transduction pathways that operate during phototaxis and aerotaxis.

Cytoplasmic proteins involved in chemotaxis

Mutants of *E. coli* and *Salmonella typhimurium* have been isolated that fail to show chemotaxis. These mutants have resulted in the identification of six genes required for chemotaxis. The genes are *cheA*, *cheB*, *cheR*, *cheW*, *cheY*, and *cheZ*. Deletion of any one of these genes prevents chemotaxis without affecting motility.

The Che proteins are cytoplasmic proteins that are part of a signal transduction pathway between the attractant or repellent and the flagellar motor switch.[163] (See note 164 for a further description of the flagellar switch

proteins and the Mot proteins.) CheA belongs to the class of signaling proteins called histidine kinases. It has conserved amino acid sequences at the carboxy-terminal domain similar to the other histidine kinases, is phosphorylated at a histidine residue, and it transfers the phosphoryl group to a response regulator protein (Fig. 18.1).[165] CheY and CheB are *response regulator proteins*. They have amino-terminal domains similar to the other response regulators and undergo phosphorylation and dephosphorylation.

Membrane proteins involved in chemotaxis

Chemoreceptor proteins for the chemoeffectors are built into the bacterial cell membrane. The chemoreceptor proteins are also called receptor–transducer proteins, or simply transducer proteins. The chemoreceptor proteins are thought to loop across the membrane with a periplasmic domain separating two hydrophobic membrane-spanning regions that connect to cytoplasmic domains. This is similar to the configuration illustrated earlier (Fig. 17.3A, B) for a protein anchored to the membrane by a stop-transfer signal and a leader peptide, which for the case of the chemoreceptor proteins protein would be uncleaved. The cytoplasmic domain interacts with cytoplasmic components such as the CheW and CheA proteins, whereas the periplasmic domain interacts with chemoeffectors or periplasmic proteins to which the chemoeffectors are bound. The cytoplasmic domain of the chemoreceptor proteins also contains four or five glutamate residues that become methylated and demethylated as part of the adaptation response described in Section 18.12.5. Therefore, the chemoreceptor proteins are also called *methyl-accepting chemotaxis proteins* (MCPs).

E. coli has four different MCPs. These are the Tsr protein (taxis to serine and away from repellents), Tar (taxis to aspartate and maltose and away from repellents), Trg (taxis to ribose, glucose, and galactose), and Tap (taxis to dipeptides). *Salmonella* does not have Tap, but it does have Tcp, which senses citrate. Some chemoeffectors, primarily certain sugars, first bind to periplasmic binding proteins, also called primary chemotaxis receptors, and

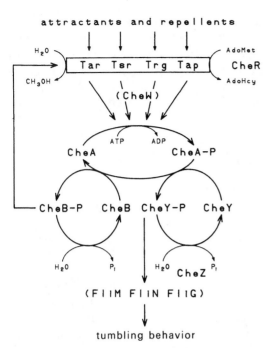

attractants and repellents

tumbling behavior

Fig. 18.20 Model for signal transduction during chemotaxis in *E. coli* and *S. typhimurium*. Attractants bind to the receptor–transducer proteins (Tar, Tsr, etc.) in the membrane. The receptors also mediate the response to certain repellents (leucine, indole, acetate, Co^{2+}, Ni^{2+}). This activates a CheW-dependent change in the rate of autophosphorylation of CheA and the proteins that are phosphorylated by CheA (CheB and CheY). Attractants reduce the rate of autophosphorylaton, and repellents increase the rate of autophosphorylation. The phosphoryl group is transferred from CheA-P to CheY and CheB. CheY-P interacts with the switch proteins (FliM, FliN, FliG) to cause clockwise rotation of the motor and tumbling. Thus repellents increase tumbling because they increase the level of CheY-P, and attractants reduce tumbling because they reduce the level of CheY-P. CheB-P is a methylesterase whose activity results in demethylation of the receptor–transducer proteins. As CheB-P goes up, methylation goes down. Thus, repellents reduce the level of methylation because they increase the level of CheB-P and attractants increase methylation because they reduce the levels of CheB-P. CheB-P autodephosphorylates. CheR is a methyltransferase that methylates the receptor–transducer proteins. The more highly methylated receptor–transducer protein does not transmit the chemoattracant signal to CheA. Hence adaptation to a chemoattractant is due to increased methylation of the receptor–transducer proteins. Adaptation to a repellent is due to undermethylation of the receptor–transducer protein. CheZ is thought to be a phosphatase that removes the phosphate from CheY-P. The relative activities

the sugar-bound periplasmic binding proteins interact with specific receptor–transducers in the membrane. Under these circumstances, the receptor–transducer (MCP) acts as a *secondary* chemoreceptor rather than a primary chemoreceptor. These are the same periplasmic binding proteins that bind sugars and interact with membrane-bound transporters in the shock-sensitive permease systems described in Section 16.3.3. They therefore function in both chemotaxis and solute transport. However, the majority of periplasmic binding proteins in *E. coli* function only in transport, not in chemotaxis. (See note 166 for a more complete discussion of the MCP proteins.)

18.12.5 A model for chemotaxis

Stimulation

The cytoplasmic domains of the receptor–transducer proteins (Tar, Tsr, Trg, Tap) control the autophosphorylation activity of CheA in an activity that requires CheW (Fig. 18.20):

$$CheA + ATP \rightarrow CheA\text{-}P + ADP + P_i$$

CheA-P is a protein kinase that phosphorylates CheY:

$$CheA\text{-}P + CheY \rightarrow CheA + CheY\text{-}P$$

CheY-P has autophosphatase activity. CheZ binds to CheY-P, and as a consequence phosphatase activity is increased. It has been suggested that CheZ is an allosteric effector that enhances CheY-P autodephosphorlation[167]:

$$CheY\text{-}P + H_2O \xrightarrow{CheZ} CheY + P_i$$

However, recent evidence indicates that CheZ itself can catalyze the dephosphorylation of CheY-P.[168,169] Thus, phosphatase activity is due to both CheY-P autodephosphorylation and CheZ-catalyzed dephosphorylation. In the nonstimulated state there is a steady state

---

of CheA-P and CheZ determine the level of CheY-P, hence the frequency of tumbling. Theoretically, changes in the activity of CheZ could also alter the frequency of tumbling. *Source*: Stock, J. B., A. J. Ninfa, and A. M. Stock. 1989. Protein phosphorylation and regulation of adaptive responses in bacteria. *Microbiol. Rev.* **53**:450–490.

level of CheY-P that is determined by the rate of phosphorylation of CheY and the rate of dephosphorylation of CheY-P.

CheY-P is a response regulator protein. It binds to the flagellar motor switch, FliM, FliG, FliN (primarily FliM), and initiates a reversal of the motor (Mot) so that it rotates in a clockwise direction, causing the cell to tumble.[170] In the nonstimulated cell, there is a certain intermediate level of CheY-P that supports normal run–tumble behavior.

When an attractant or repellent stimulates a chemoreceptor protein, a signal is sent to the cytoplasmic region of the receptor–transducer protein. The signal is presumably a conformational change in the periplasmic region of the chemoreceptor protein that is propagated to the cytoplasmic region of the chemoreceptor protein, which interacts with CheW/CheA. An attractant decreases the rate of autophosphorylation of CheA, and a repellent increases the rate of autophosphorylation. Thus, attractants promote counterclockwise rotation and smooth swimming, whereas repellents promote clockwise rotation and tumbling. (Chemotaxis in the gram-positive *subtilis* differs from that in the enterics in several ways. For example, the attractant causes an increase in CheY phosphorylation, and CheY-P causes the flagellar motor to rotate CCW resulting in smooth swimming.[171])

Signal gain

The interesting subject of signal gain is reviewed in ref. 172. Chemotactic signals are amplified. This means that a very small change in receptor occupancy to a particular chemotaxis ligand can result in a large change in the CheA kinase activity, and therefore a relatively large change in motor bias and the probability of tumbling. Signal gain is defined as the "ratio of the fractional change in motor bias to the fractional change in receptor occupancy."[171] Somehow, a small change in receptor occupancy results in a change in the activity of many CheA molecules. How might this occur?

One clue, previously noted, is that the five different chemotaxis receptors in *E. coli* (Tsr, Tar, Tap, Trg, and Aer) are clustered together at the cell poles. A model proposed on the basis of the physical interactions between the different receptor molecules in a cluster suggests that the binding of a chemoattractant to one receptor molecules induces conformational changes in other receptor molecules, not necessarily of the same type, producing a large signal gain. In other words, the receptor cluster behaves as an integrated signaling unit affecting the activity of CheA.

Acetyl phosphate can phosphorylate CheY

It has been suggested that acetyl phosphate can donate its phosphoryl group in vivo to CheY in the absence of the cognate histidine kinase (CheA). The initial observation was that mutants of *E. coli* that have none of the cytoplasmic chemotaxis proteins other than CheY swim smoothly except when incubated with acetate. Then they tumble. The response to acetate requires acetate kinase, the enzyme that synthesizes acetyl phosphate from acetate and ATP. This suggests that acetyl phosphate can phosphorylate CheY, and indeed this has been demonstrated in vitro (as discussed in ref. 173). Acetate has no effect on chemotaxis to aspartate or serine in wild-type cells, indicating that the levels of CheY-P are not significantly influenced by acetyl phosphate in wild-type cells and are regulated entirely by CheA and CheZ in response to the appropriate chemoeffectors.[173] Thus, the physiological function of acetate regarding CheY phosphorylation and chemotaxis in general is not clear at all. Perhaps it plays a subtle regulatory role in chemotaxis, somehow connecting chemotaxis to metabolism. It was reported in 1998 that acetyl–AMP can acetylate CheY with results on flagellar rotation similar to those of the phosphorylation of CheY, and therefore it appears that there are two pathways for stimulating CheY by acetate in *E. coli*.[174]

Model for adaptation

CheA-P also phosphorylates a second response regulator protein, CheB:

$$CheA\text{-}P + CheB \rightarrow CheA + CheB\text{-}P$$

CheB-P is the active form of a methylesterase that removes methyl groups from glutamate residues in the receptor–transducer proteins. There is also a methyltransferase, CheR, that adds methyl groups to these residues. As a result of the stimulation of the chemoreceptor protein by the chemoattractant, the levels of CheB-P also decrease because there is less of the phosphoryl donor, CheA-P. When autophosphorylation of CheA, hence phosphorylation of CheB, is inhibited by binding of the attractant to the chemoreceptor protein, the removal of methyl groups is slowed relative to the addition of methyl groups, and *all* the chemoreceptor proteins *become more highly methylated.* (There are four or five glutamate residues that can be methylated.)

The more highly methylated chemoreceptor protein stimulates tumbling, hence adaptation to that concentration of attractant. There is ample evidence, both in vivo and in vitro, that the methylated forms of the chemoreceptor proteins stimulate the activity of CheA, hence the phosphoryation of CheY, clockwise rotation, and tumbling.[175,176] (See note 177 for experimental evidence.) (Another way of saying this is that methylation of the chemoreceptor protein cancels the attractant-induced signal.) As a result of the methylation, the prestimulus levels of CheY-P and CheB-P are restored, and the cell resumes the prestimulus run–tumble behavior, despite the presence of bound attractant. As long as the attractant is bound to the receptor–transducer protein, it remains highly methylated, and the cell is said to be *adapted* to that particular concentration of attractant.

However, the *other* (nonoccupied) chemoreceptor proteins quickly lose their extra methyl groups so that the cell is adapted only to the specific attractant. If the chemoreceptor protein is not fully saturated with bound attractant, the cells will respond to the attractant when it swims into areas of higher concentration of attractant. However, as the chemoreceptor protein binds more chemoattractant, the average number of methyl groups per transducer is increased two- to fourfold, thus balancing the increased binding of attractant and resulting in adaptation to the higher concentration of chemoattractant. (See note 178 for an explanation of how this might occur.) When the chemoattractant is removed, the overmethylated chemoreceptor protein stimulates CheA autophosphorylation. This results in increased CheY-P and consequent tumbling.[179] However, the methylesterase CheB-P also increases and returns the chemoreceptor protein to the methylation level of the prestimulus state and normal run–tumble behavior.

Adaptation to a repellent differs from adaptation to an attractant in that during repellent adaptation the activated CheB-P *demethylates* the chemoreceptor protein that responds to the repellent. In summary, then, binding of the attractant to the chemoreceptor inhibits the kinase activity of CheA, which leads to a fall in the levels of CheY-P, hence smooth swimming. However, binding of the attractant to the chemoreceptor also promotes the methylation of the chemoreceptor. The methylated chemoreceptor stimulates the kinase activity of CheA, which leads to a rise in the levels of CheY-P and tumbling; thus adaptation takes place.

Some evidence in favor of the model

The phenotypes of chemotaxis mutants support the model. This is summarized in Table 18.2. For example, *che*W, *che*A, and *che*Y mutants do not tumble, whereas *che*B and *che*Z mutants tumble constantly. This is in agreement with the model, which stipulates that CheW and CheA promote the phosphorylation of CheY and CheY-P causes the motor to reverse and tumbling to take place. CheB competes with CheY for the phosphoryl group on CheA-P and would therefore be expected to promote smooth swimming. Hence a mutation in *che*B should make the cells tumble. CheZ dephosphorylates CheY, and therefore a mutation in *che*Z should result in high levels of CheY-P and therefore increased tumbling.

Table 18.2 Effects of chemotaxis mutants

Gene	Mutant phenotype	Rationale for mutant phenotype
*che*Y⁻	Smooth swimming	CheY-P is required for tumbling
*che*A⁻, *che*W⁻	Smooth swimming	CheA-P and CheW phosphorylate CheY
*che*B⁻	Tumbling	CheB competes with CheY for phosphoryl group from CheA-P
*che*Z⁻	Tumbling	CheZ dephosphorylates ChY-P

Biochemical data also support the model. When an attractant binds to the extracellular domain of the chemoreceptor protein, several events occur: the rate of phosphorylation of CheY decreases; the rate of demethylation is reduced; and the rate of methylation is increased. There is also a great deal of evidence indicating that in their methylated form, the chemoreceptors stimulate CheY phosphorylation and tumbling.[173,175,176]

Adaptation mechanism that does not rely on methylation

If cells are experimentally exposed to a large and abrupt change in the concentration of a stimulus (e.g., the addition of an attractant), then adaptation, which can take minutes, is certainly correlated with an increased level of receptor methylation, as described earlier. On the other hand, it has been pointed out that the methylation-dependent adaptation system may be too slow for adaptation by cells swimming in gradients, and in fact a methylation-independent system may exist.[180] During chemotaxis along an attractant concentration gradient, the cells encounter very small changes in stimulus concentrations (as opposed to the experiments in which large concentrations of attractant are added to a cell suspension and the behavior of the cells is monitored), and adaptation takes only about one second. It has been argued that the increase in methylation of the MCP proteins in a second is too small to account for adaptation to attractants in gradients. Furthermore, cheR⁻cheB⁻ double mutants that show no changes in methylation are nevertheless chemotactic in swarm plates and also show adaptation to the addition of very small concentrations of attractant when CWW rotation of tethered cells is measured.[179,181-183] (See note 184 for a description of chemotaxis assays and how to measure responses to attractants and repellents using tethered cells.)

Apparently there exists a methylation-independent system that allows cells to respond chemotactically to small concentrations of stimulus and to be chemotactic in gradients. Furthermore, since chemotaxis to sugars transported by the PTS does not involve MCP proteins, adaptation cannot be via methylation of these proteins. (See Section 18.12.7)

18.12.6 Mechanism of repellent action

Repellents for *E. coli* and *S. typhimurium* include indole, glycerol, ethylene glycol, phenol (for *S. typhimurium* only), organic acids (e.g., formic, acetic, benzoic, salicylic), alcohols, Co^{2+} and Ni^{2+} (for *E. coli* only), and hydrophobic amino acids (leucine, isoleucine, tryptophan, valine). In general, the mechanism of action of organic repellents is not well understood, except that the response requires the receptor–transducer (MCP) proteins. Yet, there is no direct evidence for specific binding between an organic repellent and a specific MCP. In some cases (e.g., glycerol, ethylene glycol, aliphatic alcohols) there is not even specificity regarding which receptor–transducer protein is required. This has been learned by using mutants lacking one or more of the receptor-transducer proteins and showing that any of the remaining receptor–transducer proteins can mediate the repellent response.

However, in other instances there is specificity. For example, the response to leucine, isoleucine, and valine requires the Tsr protein, and the response to the cations Co^{2+} and Ni^{2+} requires Tar. Because much higher concentrations of organic repellents (in the millimolar range) are usually required than for attractants (in the micromolar range), it has been suggested that the organic repellents (which all have some hydrophobic character as well as polar groups) do not actually bind to the receptor–transducer proteins but act indirectly by perturbing the membrane at the site of the receptor–transducer proteins. For example, they might make the membrane more fluid, which could result in the alteration of receptor–transducer activity.

In 1990 a study was carried out to investigate the action of repellents on the membrane fluidity of *E. coli*.[185] It was concluded that changes in membrane fluidity do *not* account for repellent activity, suggesting that repellents may indeed bind to receptor–transducer proteins, however, with low affinity, and in some cases low specificity. In summary, the response to repellents *does* involve the receptor–transducer proteins (MCP proteins), but the mechanism of how they affect the activities of the receptor–transducer proteins is not understood. However, there is more known about the repellent activity of weak organic acids. These are believed to affect the activities of the receptor–transducer proteins Tsr and Tap by lowering the intracellular pH.

18.12.7 Chemotaxis that does not use MCPs: The phosphotransferase system is involved in chemotaxis toward PTS sugars

In addition to functioning in sugar uptake, the PTS (see Section 16.3.4) is important for chemotaxis toward PTS sugars.[186] Some of the evidence for this is as follows:

1. Mutants of E. coli defective in EI and HPr are usually defective in chemotaxis toward PTS sugars.

2. EII mutants defective in chemotaxis toward the sugar specifically recognized by EII also exist.

Comparison of PTS-mediated and MCP-dependent chemotaxis

E. coli mutants lacking CheB, CheR, or the MCP proteins display normal chemotaxis to PTS sugars. This means that methylation-dependent adaptation does not occur in the system, and there must be another adaptation mechanism. However, CheA, ChW, and CheY are required for chemotaxis toward PTS sugars.[187] One model postulates that EI–P can cause the phosphorylation of CheA as well as the incoming PTS sugar. Thus, as the sugar becomes phosphorylated during its uptake, the levels of CheA-P fall, causing lower levels of CheY-P, hence smooth swimming. (Reviewed in ref. 188.)

18.12.8 Chemotaxis that is not identical with the model proposed for the enteric bacteria

Chemotaxis in many bacteria differs in several respects from chemotaxis in E. coli. For a review of this subject, see refs. 154 and 187. For example, some bacteria do not reverse the rotation of their flagella and do not tumble. Other bacteria do reverse the direction of rotation of their flagella, but this causes them to back up, not tumble. Additionally, some

bacteria must metabolize the chemoattractant to generate the chemoattractant signal; that is, the initial signal does not come from the binding of the chemoattractant to an MCP-like protein. How they adapt is not known. Of course, as discussed earlier, sometimes even E. coli does not use the MCP proteins (e.g., when it responds to PTS sugars). Several bacteria have multiple che genes, especially cheY. The following is a description of chemotaxis in bacteria other than E. coli in which the process has been studied in detail.

Bacteria that do not tumble

Not all bacteria tumble, yet some of those lacking this ability are capable of chemotaxis. For example, Rhodobacter sphaeroides and Rhizobium (Sinorhizobium) meliloti rotate their flagella only in a clockwise direction and do not tumble. (See later subsections entitled Rhodobacter sphaeroides and Rhizobium meliloti.) Yet they swim randomly and exhibit chemotaxis.

In some bacteria with polar flagella (flagella at one pole or both poles, either single or in bundles) the reversal of the flagellar motor causes the cells to reverse the direction of swimming rather than tumble. An example is Caulobacter or certain photosynthetic bacteria that change direction by reversing the direction of flagellar rotation. Because they do not back up in an absolutely straight line, and/or do not swim forward in a straight line, the direction of forward swimming is random with respect to the original forward direction. Some bacteria whose swimming behavior and chemotaxis pathway are not exactly like those of E. coli are described next.

Rhodobacter sphaeroides

R. sphaeroides is a photosynthetic bacterium that can live aerobically heterotrophically in the dark or anaerobically as a photoheterotroph in the light. (See Section 5.1.2: Purple nonsulfur phototrophs.) It has a single medially (subpolar) located flagellum that rotates in one direction only (clockwise), pushing the cell forward. The cells stop swimming intermittently, and when swimming resumes the cells swim randomly in a different direction. It appears that while the bacterium is stopped the flagellum is coiled against the cell body and rotates slowly resulting in reorientation of the cell. (Reviewed in ref. 189.) R. sphaeroides is attracted to weak organic acids, sugars and polyols, glutamate, ammonia, and certain cations (e.g., K[+] and Rb[+]). Unlike the case of the enteric bacteria, metabolism of the chemoattractant appears to be necessary for chemotaxis, because if metabolism is blocked (e.g., by a specific inhibitor), chemotaxis is also inhibited.

R. sphaeroides has homologues to all the E. coli che genes except cheZ.[190,191] (See note 192 for a description of how the che genes were detected in R. sphaeroides.) Also, although a typical enteric-like membrane-bound MCP appears to be absent, a cytoplasmic methyl-accepting chemotaxis protein called TlpA (transducer-like protein A), with some similarity to the enteric MCPs, does exist.[193] (See note 194 for a further description of how this was determined.) Recall that metabolism is required for chemotaxis in R. sphaeroides. Perhaps TlpA responds to a cytoplasmic metabolic intermediate and transfers the signal to CheA.

In contrast to E. coli, the che genes exist in multiple copies in R. sphaeroides. There are two copies of cheA and cheW and three copies of cheY genes. This is because there are two independent chemosensory pathways in R. sphaeroides as opposed to one in E. coli. (Reviewed in ref. 195.) The reason for believing that two independent chemosensory pathways exist is that in mutants that lack (owing to deletion) the originally discovered chemotaxis operon containing homologues to cheA, cheW, cheR, and homologues to two cheY genes, the cells are still capable of chemotaxis as well as the wild type.[190] When the deletion mutant was mutagenized and che[-] and pho[-] mutants were screened, a second chemotaxis operon was discovered containing homologues to cheY, cheA, chR, cheB, two homologues to cheW, and tipC, a transducer-like protein. Mutations of this second chemotaxis operon (cheA[-]) in a wild-type background produced chemotaxis mutants, suggesting that it is part of an operative chemotaxis pathway. The role of the first chemotaxis operon remains in question, since its absence or presence does not seem to markedly affect chemotaxis. [See the discussion later, in Section 18.13.2, of the involvement of chemotaxis-like proteins (Che proteins) that form part of a signal transduction pathway involved in the movement of Rhodospirillum centenum colonies toward light.] How R. sphaeroides senses carbohydrates that act as attractants is not understood. As mentioned, metabolism of the carbohydrate is necessary, suggesting that the chemotactic signal is generated during metabolism and not by binding of the sugar to an MCP-like protein.[196] (See note 197 for evidence.)

Since R. sphaeroides changes its direction as a result of stopping rather than tumbling, it is reasonable to suppose that an increased concentration of chemoattractant might decrease the frequency of stops (the equivalent of the suppression of tumbling), thus ensuring that the cells continue to swim in the direction of the chemoattractant. Likewise, a decreased concentration of chemoattractant might increase stopping frequency, hence promote changes in the direction of swimming. However, it does not happen exactly this way.

When anaerobically grown cells tethered to glass via anti-flagellin antibody were examined, an increase in attractant concentration did not cause a change in rotational behavior; but when the attractant concentration was decreased, the cells transiently stopped rotating. (Reviewed in ref. 188.) It appears that cell accumulation in response to a chemoattractant gradient by R. sphaeroides is dependent upon sensing a reduction, not an increase, in the concentration of chemoattractant. This is opposite to the situation in E. coli, which responds to an increase in attractant concentration.

Some metabolites, especially weak organic acids, cause an increase in the speed of swimming by R. sphaeroides, a response called chemokinesis.[198] Many compounds that cause chemokinesis also cause chemotaxis. However, the two phenomena are separable because (1) most amino acids and sugars that are chemoattractants do not cause chemokinesis, and (2) metabolism is required for chemotaxis but not for chemokinesis. (It appears that transport across the cell membrane may be involved in chemokinetic signaling.) The increased rate of swimming in response to certain metabolites lasts for a long time; that is, there is no rapid adaptation. It has been pointed out that in the absence of adaptation, chemokinesis should result in the spreading of the population rather than its accumulation.[199]

Rhizobium meliloti

Rhizobium meliloti (also called *Sinorhizobium meliloti*) is an aerobic bacterium that can live freely in soil, in the rhizosphere of host plants, or eventually symbiotically with leguminous plants in root nodules, where it fixes nitrogen. (See Sections 12.3.1 and Section 18.10.4.) The bacteria enter the plants via plant rootlets, to which they are attracted by root exudates that include amino acids, carbohydrates, and flavones (cyclized isoprenoid compounds made from phenylalanine and malonyl–CoA). Chemotaxis studies with *R. meliloti* employ L-amino acids and D-mannitol as attractants.

   R. meliloti has 5 to 10 peritrichously located flagella that form a bundle when the cell swims. Swimming consists of straight runs interrupted by very quick turns. Unlike, for example, *R. sphaeroides*, the cells do not seem to stop before turning. It has been suggested that the flagella never actually stop rotating. When the cell is swimming in one direction, the flagella rotate as a single bundle. The cell turns when the flagellar motors slow down and rotate at different rates, whereupon the bundle flies apart. (Reviewed in ref. 188.) Cells tethered to a microscope slide by a single flagellum (the other flagella were mechanically sheared off the cell) rotated only clockwise with very brief (<0.1 s), intermittent stops.[200]

   R. meliloti has homologues to the enteric *che* genes.[201] Present in *R. meliloti* are *cheY* (two copies), *cheA*, *cheW*, *cheR*, and *cheB*.[202] (There is no *cheZ*.) Mutations in the *che* genes make the cells defective in chemotaxis. *R. meliloti* possesses a *tlpA* gene, which as in *R. sphaeroides* encodes a cytoplasmic MCP-like protein. Other MCP-like proteins have been detected, as reviewed in ref. 188. As noted earlier, *R. meliloti* flagella turn in only one direction (clockwise), and the cell does not seem to stop momentarily before it turns in a new direction.

   R. meliloti responds to attractants by increasing its swimming speed (chemokinesis). This is equivalent to decreasing the frequency of tumbles in *E. coli*. It has been speculated that the rate of flagellar rotation is determined by CheY-P, which perhaps slows the flagellar motor. Thus the presence of the attractant might inhibit the autophosphorylation of CheA, which would decrease the levels of CheY-P and increase swimming speed.[188] This sequence would result in chemotaxis toward the attractant if the cells were also capable of adaptation.

Azospirillum brasilense

There have been reports that *Azospirillum brasilense* lacks a methylation-dependent pathway for chemotaxis toward organic compounds.[203,204] In *Azospirillum brasilense*, it appears that the strongest chemoattractants are electron donors to the aerobic redox chain. This, plus the fact that chemoattraction toward oxygen is so strong in this organism that it masks responses to other chemoattractants, has led to the suggestion that the signaling system involves the respiratory chain. It is reasonable to suggest that the signaling system in these cases (and in phototaxis and aerotaxis, as discussed next) may involve the redox level of one of the electron carriers or the proton motive force.

18.13 Photoresponses

As we shall see in Section 18.13.1, tactic responses to light as well as to electron acceptors such as oxygen and nitrate also exist among the prokaryotes. The student should consult ref. 205, which is a review article that discusses various photosensors, including a relatively recently discovered family of photosensors called photoactive yellow protein (PYP), which appears to be widespread among the bacteria. It seems that in several of these cases the signaling pathway to the flagellum motor uses the Che proteins that function in the chemotaxis signaling pathway. (See Section 18.12.4.) Thus, one might view various sensory signaling pathways as differing in the initial sensory receptors but converging at the level of the Che proteins (e.g., CheA).

   Taxis toward electron acceptors or toward light has been referred to as "energy taxis" and is reviewed in ref. 206. Photoresponses in halobacteria (Section 18.13.1) and photosynthetic bacteria (Section 18.13.2) will be discussed, followed by a description of taxis toward oxygen, that is, aerotaxis, as well as taxis toward nitrate under anaerobic conditions (Section 18.14).

18.13.1 Halobacteria

The student should review the description of the halophilic archaea in Sections 1.1.1 and 3.8.4, and the discussion of the role of light and retinal pigments in establishing a Δp and in Cl^- transport in Sections 3.8.4 and 3.9. One of the pigments is a rhodopsin called bacteriorhodopsin, which couples the absorption of light energy to pumping protons across the cell membrane from inside to outside, thus creating a Δp. A second rhodopsin that is present is halorhodopsin, a light-driven pump used to pump Cl^- into the cell to maintain osmotic stability.

Swimming behavior

Halobacteria have a single flagellum consisting of 5 to 10 filaments in a right-handed helix at one pole of the cell. (Reviewed in ref. 207.) (Actually, the cells show monopolar flagellation only during exponential growth. In the stationary phase their flagellation is mostly bipolar.) When the flagellum motor turns clockwise, the cells swim forward, and when the motor turns counterclockwise, the swimming direction is reversed. (The flagellar filaments remain together even when the direction of rotation is reversed.) During a brief period between CW and CCW rotation, the flagellum is not rotating and the bacterium changes its orientation, perhaps as a result of Brownian motion. When swimming resumes it is in a direction random to the original direction.

Photoresponse

The cells are capable of both positive and negative photoresponses depending upon the wavelength of light. The bacteria are attracted toward orange/red light (500–600 nm), which is where bacteriorhodopsin and halorhodopsin absorb, and are repelled by UV/blue light, which can damage DNA. Attracting light retards the process of switching; hence the bacterium swims in the direction of the light.

Repulsive light increases the switching frequency; hence the bacteria swim away from repulsive light. This behavior is not really phototaxis in that the bacteria do not perceive a gradient of light and are simply responding to changes in intensity. In this way, the phenomenon is similar to chemotaxis.

Signal transfer to flagellar motor

The photoreceptors are rhodopsins similar to bacteriorhodopsin.[208-210] As a consequence of absorbing light, there is a conformational change in the photoreceptor, which results in a signal (probably a conformational change) to a closely associated, membrane-bound MCP-like protein called Htr. It has been suggested that the Htr signals proteins homologous to the Che proteins, which then alter the frequency of flagellar reversal. From this perspective the signaling pathway for photosensing is similar to chemotaxis except that in photosensing, the sensor (Htr) responds to a light-activated photoreceptor, whereas the sensor (MCP) in chemotaxis responds to the binding of a chemoeffector or a periplasmic protein to which the chemoeffector is bound (Section 18.12.5). The signal in both cases would be transferred via phosphorelay to the flagellar motor via the Che proteins. Exactly how the absorption of light by the photoreceptor activates the appropriate protein, Htr I, is not known, but see note 211 for a more complete discussion of photosensing.

18.13.2 Photosynthetic bacteria

Avoidance of the dark

Photosynthetic bacteria such as *Chromatium*, *Thiospirillum*, and *Rhodospirillum* that are swimming in a cell suspension will reverse their direction of swimming when they swim out of a light area into a dark area, whereupon they accumulate in the light.[209] This is easily observed microscopically by shutting the condenser so that there is a small circle of light in an otherwise dark field. For example, when *Rhodospirillum rubrum*, a photoheterotroph with polar flagellatation, swims from an area of light into an area of darkness, the polar flagellar bundles reverse their direction of rotation and the cells back up into the light. Such behavior has been called *scotophobia* because it is an avoidance of the dark rather than a seeking of light.

Experiments with a horizontal beam of light in a microscope field indicate that free-swimming photosynthetic bacteria do not respond to the direction of propagation of the light but rather to a decrease in intensity.[212] If the cells can reverse their direction of swimming by reversing the rotation of the flagella, then they tend to back up whenever they start to swim across the light/dark boundary. They become trapped throughout the beam of light, not simply near its source, indicating that they are not responding to the direction from which the the light is coming but rather to the decrease in intensity as they cross into the dark.

Not all photosynthetic bacteria can reverse the direction of flagella rotation, and therefore some do not back up when crossing a light/dark boundary. In *R. sphaeroides*, for example, the flagellum rotates in only one direction, and therefore when the light intensity is reduced, there is an increase in the frequency of stopping. When the cells stop swimming, their orientation changes, leading to a change in swimming direction when flagellar rotation resumes. Interestingly, a narrow beam of light aimed horizontally across a microscope field, causes bacteria to accumulate in the dark, rather than in the light.[211] This is because the cells do not reverse when crossing the light/dark boundary but stop and swim off in a randomly different direction, which more often than not brings them into the dark.

Role of photosynthetic electron transport

Mutants in the photosynthetic reaction center do not respond to a decrease in light intensity, suggesting that the response involves monitoring changes in photosynthetic electron transport, the Δp, or both. (A decrease in light intensity in wild-type cells causes a decrease in Δp and in photosynthetic electron transport.) This conclusion is supported by the finding that inhibitors of electron transport prevent photoresponses even though the cells swim normally.[212]

The evidence gathered for *R. sphaeroides* suggests that it is photosynthetic electron trans-

port (probably the redox level of one of the carriers) rather than the Δp that signals a step-down in the light intensity.[213] Some of the experimental evidence for this is derived from the use of the proton ionophore FCCP, which stimulates electron transport while it collapses the Δp. (See Section 3.4.1 for an explanation of uncouplers.) The experiments were done with cells tethered to glass by anti-flagellin antibody. Under these conditions the cells rotated and the frequency of stops was measured. The tethered cells showed no response to the addition of FCCP, although the $\Delta \Psi$ decreased to the same extent caused by a step-down in light intensity. This indicates that the cells respond to a decrease in photosynthetic electron transport rather than to a decrease in th Δp upon a step-down in light intensity. Perhaps when the rate of electron transport is decreased, a signal is sent to an Aer-like protein, which results in an increase in CheY-P that stops the flagellar motor. This, however, is speculation. (Aer proteins are MCP homologues required for response to oxygen gradients. See Section 18.14. Also, see the subsection of Section 18.12.8 entitled *Rhizobium meliloti*.)

Phototaxis of colonies

Individual cells of *Rhodospirillum centenum* in liquid culture show the scotophobic response described earlier for other photosynthetic bacteria. Cells grown in liquid media have a single polar flagellum and change the direction of swimming by reversing the rotation of the flagellum. However, *R. centenum* colonies move and are capable of phototaxis in gradients of light.[214] This is a true phototaxis in that the colonies move along a light gradient toward the source of light down an intensity gradient when a converging beam of light is used. When two attractant beams of light at 90° to each other are used, the colonies move in a direction that is 45° between the beams, suggesting that they are integrating the signal from the light beams. The colonies move toward a source of infrared light and away from a source of visible light. The cells have lateral flagella when grown on agar, and they swarm, or move cooperatively, on the solid agar.

Mutants in the photoresponses have been isolated.[215] Some of the mutants that are not capable of photoresponses have defects in the synthesis of reaction center components or electron carriers that function in photosynthetic electron transport. This indicates that the photoresponses are in some way related to changes in photosynthetic electron transport as the cells move along the light gradient. Other mutants have defects in the *che* genes, indicating that the signal cascade proceeds from the photoreceptor through the Che proteins to the flagellar motors. Still other mutants, which are normal in their chemotaxis responses and in photosynthetic growth, appear to have defects in the perception of phototactic light or in the signal transduction pathway upstream of the Che proteins. (See note 216.)

18.14 Aerotaxis

Many aerobic bacteria, including *E. coli* and *Salmonella typhimurium*, swim toward a higher concentration of oxygen. The response to oxygen requires electron transport as well as a portion of the chemotaxis pathway, including CheW, CheA, and CheY.[217,218] This conclusion was based on the results of experiments with mutants lacking the terminal cytochrome oxidases o and d or the *che* genes. When wild-type *E. coli* is grown anaerobically in the presence of nitrate so that cytochrome o is absent and nitrate reductase is present, the cells are attracted to nitrate but not to oxygen. This indicates that electron transport to oxygen generates the oxygen signal and that electron transport to nitrate generates the nitrate signal. When both cytochrome o and nitrate reductase are present, the cells are attracted to both oxygen and nitrate in a competitive manner. Inhibitors of electron transport prevent taxis toward the electron acceptors. The primary signal might be a change in the Δp, or as suggested next, in the redox state of one of the electron carriers.

A protein called Aer has been referred to as an MCP homologue and seems to be required, inasmuch as mutants in *aer* show reduced responses to oxygen gradients.[219] Unlike MCPs, Aer does not have a periplasmic sensing domain. However, it does have a hydrophobic domain that anchors it in the cell membrane, as well as cytoplasmic domains that interact with CheA/CheW. One can suppose that Aer senses the change in the redox state of a component of the electron transport chain and transmits the signal to CheA.

For example, suppose that increased levels of oxidation of the electron carriers signaled Aer to lower the levels of CheA-P. This would result in lower amounts of CheY-P, hence smooth swimming and accumulation of cells in areas of higher oxygen concentration. Aer noncovalently binds cytoplasmic FAD, and perhaps the redox state of the FAD, itself altered by a component of the electron transport chain, alters the configuration and thus the activity of Aer. For a model, see ref. 218.

18.15 Introduction to Bacterial Development and to Quorum Sensing

Bacterial development in its broadest sense can refer to (1) cellular differentiation in which a cell acquires phenotypic properties that clearly differentiate it from a precursor cell, (2) cellular differentiation in which a cell divides to produce two daughter cells that can be distinguished morphologically and/or physiologically, or (3) multicellular development in which members of a population of cells interact to form specialized structures, such as fruiting bodies, or interactive communities within biofilms. Often, differential gene expression during these developmental processes calls for cell-to-cell signaling (e.g., quorum sensing, described in Section 18.15.1), as well as signaling within cells, for example, via two-component (or multiple component) phosphorelay systems. The remainder of this chapter is devoted to describing examples of bacterial development and the signaling processes that take place. We will begin with intercellular signaling, that is, quorum sensing.

18.15.1 Quorum sensing

Quorum sensing is a signaling pathway that bacteria use to become aware of how many cells of their own kind are in the population. As we shall see, it can be very important for the initiation of bacterial development. The signaling results in the expression of certain genes that are normally expressed when the cell density reaches a threshold level (for reviews, see refs. 220–223). The extracellular signaling molecules that bacteria produce in proportion to their cell density have traditionally been called autoinducers because the first ones discovered were found to stimulate the expression of genes that encoded them.

One important class of autoinducers are oligopeptides produced by gram-positive bacteria. These oligopeptides have been referred to as autoinducing polypeptides (AIPs).[222] The AIPs usually work by stimulating two-component phosphorelay systems that regulate gene transcription. A second class of autoinducers, made by gram-negative bacteria, are called acylated homoserine lactones (acyl–HSLs). These act by combining with and activating transcription factors that regulate the expression of specific genes.

The first quorum-sensing systems based upon acyl–HSLs were discovered in luminescent bacteria, and they serve to activate genes for luminescence. It is now clear that similar systems are widespread among bacteria, although the genes that are regulated differ according to the bacterium and its ecological niche. Some bacteria use molecules other than oligopeptides and acyl–HSLs as chemical intercellular signals, as reviewed in ref. 224. These include furanosyl borate diesters, also known as AI-2, synthesized by LuxS synthase. The LuxS system was originally discovered as part of the bioluminescence system in *Vibrio harveyi*, but it is now known to be present in nonluminescent bacteria, both gram positive and gram negative. (See the discussion of quorum sensing in biofilms in Section 18.22.7.) As described in later sections, quorum sensing is important for diverse functions, including myxobacteria fruiting body formation (Section 18.16.), *B. subtilis* sporulation and competence (Sections 18.18 and 18.19), bioluminescence, conjugal plasmid transfer, the activation of virulence genes, and biofilm formation (Sections 18.20–18.22).

18.16 Myxobacteria

The myxobacteria are unique among the bacteria in forming multicellular fruiting bodies. Roland Thaxter of Harvard University, recog-

nized through his extensive studies of fungi that these organisms are actually bacteria, not fungi, as reported by some in the earlier literature. Professor Thaxter published papers on myxobacteria in 1892, 1897, and 1904.

18.16.1 Life cycle

Rod-shaped, gram-negative gliding bacteria belonging to the δ-proteobacteria group, the myxobacteria can be isolated from soil, dung pellets, and decaying vegetation on the forest floor. (For reviews, see refs. 225–228.) The cells glide on solid surfaces such as soil particles and vegetation in thin, spreading populations. These populations of cells are referred to in the myxobacteria literature as *swarms*, although the mode of motility is not the same as *swarming motility* that occurs with flagellated bacteria, discussed in Section 1.2.1. Gliding motility by myxobacteria does not require flagella. Rather, it is a smooth, nonrotating forward or backward movement on a solid surface in the direction of the long axis of the cell. The mechanism is discussed in Section 18.16.2. It should be pointed out that the most common way for most bacteria, not simply myxobacteria, to live in nature populations of interacting bacteria on a solid surface. Such populations on a solid surface, including myxobacteria populations, are called biofilms. (See Section 18.22 for a discussion of biofilms.)

Myxobacteria are unique among the known prokaryotes in that the cells aggregate and construct multicellular fruiting bodies, which can be induced in the laboratory by subjecting the cells to starvation on agar. Because the shapes of the fruiting bodies are due to cell movements rather than cell growth, the movements are referred to as *morphogenetic movements*. Thus in myxobacteria, as in many microbial developmental systems, starvation is a signal for development.

During the course of multicellular development, individual cells convert to resting cells called myxospores, which are housed in the fruiting bodies. The shape of the fruiting body is species specific and ranges from simple mounds of cells containing resting cells called myxospores, as in *Myxococcus xanthus*, to elaborate fruiting bodies with stalks supporting several compartments, or sporangioles, that house the myxospores, as in *Stigmatella aurantiaca*. Figure 18.21A is a photomicrograph of an *S. aurantiaca* fruiting body. Most research is done with *M. xanthus*, and this section will focus on the results of that research.

Two stages in the population life cycle

The myxobacteria have two stages in the population life cycle, a vegetative stage and a developmental stage (Fig. 18.21B). The vegetative stage is the time of feeding and growing. During this period the cells in the swarms move in groups and feed on the source of nutrient, which in their natural habitat is usually other bacteria that they lyse by secreting lytic enzymes. In the laboratory they can be grown on mixtures of peptides or amino acids. During the developmental stage, which is triggered by nutrient depletion, the myxobacteria cells move into numerous aggregation centers consisting of approximately 100,000 cells each. As a result of cell movements, the aggregation centers develop into multicellular fruiting bodies. Within the fruiting bodies, the cells differentiate into myxospores. Eventually, the fruiting bodies are dispersed by nonspecific means, such as sticking to the bristles of roaming insects, to new locations where they germinate and produce a new population of cells that enters the vegetative stage.

Cooperative feeding

It is important to realize that myxobacteria feed cooperatively; that is, when they are together in a population, they grow best on other bacteria or proteins such as casein. This is because the digested products of the lytic enzymes secreted by the myxobacteria are shared. Thus, the dispersal of fruiting bodies can be viewed as a means of ensuring that when the myxospores germinate, a feeding population will be produced, consisting of at least 100,000 cells.

Cell-to-cell signaling

To move as coherent groups of cells in the vegetative swarms while seeking nutrients, and also to move cooperatively to construct multicellular fruiting bodies, the myxobacteria have evolved systems of intercellular signaling.

Intercellular signaling and gliding motility are discussed in Section 18.16.2, and the relationship of intercellular signaling to the construction of fruiting bodies and myxospores is discussed in Section 18.16.3.

18.16.2 Intercellular signaling and gliding motility

Two motility systems

There are two types of gliding motility in *M. xanthus*, controlled by two sets of genes: system A genes (for adventurous motility) and system S genes (for social motility).[229,230] Mutants in system A (A^-S^+) can glide only as groups; that is, system A genes are required for single-cell motility. Mutants in system S genes (A^+S^-) show a motility pattern indicating that they glide as single cells (well-separated cells, with some cell clusters); that is, system S genes are required for group motility. Mutants in the S-motility system are defective in aggregation and fruiting body formation, whereas those in the A-motility system usually show only minor developmental defects. Although, as we shall see next, the mechanisms of motility via the A- and S-systems differ, there does exist at least one gene, *nla24*, that is required for both A- and S-motility, and it has been suggested that Nla24 might be a transcriptional activator of certain genes whose products are required for either the A- or S-motility systems.[231] (See note 232 for a further explanation.) As will be explained shortly, in connection with S-motility, when both the A- and S-motility systems are functioning in the same cell, they work coordinately at opposite poles of the cell. The S-motility system works at the forward pole, pulling the cell forward, and the A-motility system works at the rear pole, pushing the cell forward. When the cell reverses its direction of motility, the A- and S-motors switch poles.

1. Social motility (S-motility)
Social motility (S-motility) requires type IV pili (also referred to as type 4 pili or TFP), which are made of PilA subunits; they are located at any one time at either one, but not both, of the cell poles.[233-235] In addition, as we shall see, extracellular cell fibrils and lipopolysaccharide are also required for S-motility.

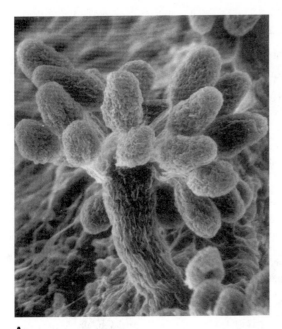

A

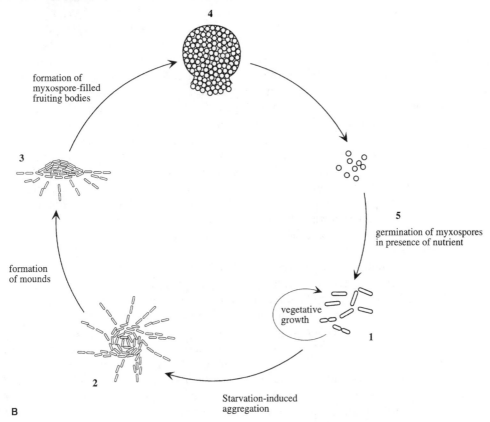

B

Fig. 18.21 Fruiting bodies and life cycles. (A) Fruiting body of *Stigmatella aurantiaca*. The fruiting body consists of a cellular stalk supporting several sporangioles. Each sporangiole houses myxospores. (B) Life cycle of *Myxococcus xanthus*. (1) Vegetative growth: cells grow on solid surfaces in dense populations called swarms. (2) Aggregation: when nutrients are depleted, cells glide into aggregation centers, each one consisting of many thousands of cells. (3) Mound formation: each aggregation center becomes a mound of cells as bacteria continue to accumulate. (4) The mound develops into a fruiting body when the cells differentiate into resting cells called myxospores. Each myxospore is surrounded by a coat (capsule). (5) When nutrient becomes available once more, the myxospores germinate into vegetative cells, returning the population to a new growth phase. Once the cells have depleted the supply of nutrients, they can aggregate on agar and form myxospore-filled fruiting bodies within 72 to 96 h.

The type IV pili constitute the motility apparatus for social motility, and mutants defective in the production and function of type IV pili cannot move via the S-system. (Read the discussion of the assembly of type IV pili in Section 17.5.6, and see note 236 for a further explanation.) According to the model for type IV pili function in S-motility, the pili at the leading

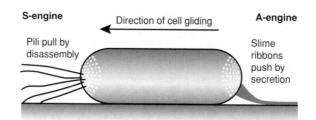

Fig. 18.22 Gliding motility via type IV pili and slime secretion. The A- and S-motor function in a cooperative manner in A⁺S⁺ cells. This is deduced from the observation that the maximum swarming rate in A⁺S⁺ cells is greater than in A⁺S⁻ or A⁻S⁺ cells. One hypothesis is that the A-motor secretes slime at one pole and this pushes the cell forward. Support for this suggestion, which has yet to appear, would include more information regarding the chemical and physical nature of the slime, and how it is attached to the substrate. Experimental support exists, however, for a model according to which the S-motor extends type IV pili at the opposite end and, when the pili attach to the cell in front (or to the substrate) and retract, the cell is pulled forward. When the cell reverses its direction of gliding, the activities of pili extension/retraction and slime secretion switch poles. *Source*: Reproduced with permission from *Nature Reviews in Microbiology*. Kaiser, D. 2003. Coupling cell movements to multicellular development in myxobacteria. *Nat. Rev. Microbiol.* 1:45–54. Macmillan Magazines Ltd.

pole adhere to the polysaccharide portion of extracellular fibrils (consisting of 50% protein and 50% polysaccharide), extending from a cell in front, and then retract, pulling the rear cell forward (Fig. 18.22).[237] For more information, including references supporting these conclusions, see note 238. When retraction is completed, the cell in front is released. Mutant studies have shown that S-motility also requires the lipopolysaccharide (LPS) O-antigen in the outer membrane.[239] It has been suggested that the LPS O-antigen mutants may not be able to retract the type IV pili that normally pass through the LPS O-antigen layer during retraction. Motility via type IV pili requires cells to be close to enough to adhere to one another, and the cells move in groups, rather than as single cells. However, on certain surfaces movement of single cells can occur. For example, on very wet agar surfaces (0.4% agar), type IV pili can propel single A⁻S⁺ cells, presumably by binding to the surface.[233] However, in general, movement by type IV pili should be thought of as group motility.

The conclusion that extracellular fibrils play a role in S-motility is based upon experimental results showing that S-motility and fruiting body formation are defective in *dif* (defective in fruiting) mutants, even if type IV pili are present.[240] The *dif* genes encode proteins that are homologous to proteins in the *E. coli* chemotaxis signaling system (Section 18.12). These are the methyl-accepting chemotaxis proteins

(MCPs), CheA, CheY, and CheW. Mutations in some of the *dif* genes have been demonstrated to lead to defects in the production of cell surface fibrils. It has been suggested that the Dif homologues of the chemotaxis-like signal proteins constitute a phosphotransfer signaling system in *Myxococcus* that regulates the biogenesis of cell surface peritrichous fibrils, and that these fibrils play a necessary role in social motility. For example, as notes earlier, polar type IV pili are thought to adhere to fibrils on the cell in front, retract, and pull the cell forward.

M. xanthus has three other systems (Frz, Che3, and Che4) that encode homologues of chemotaxis-like signal proteins. One of the systems, Frz, regulates the frequency of reversal of the direction of motility (Section 18.16.7). The Che3 system regulates developmental gene expression, and the Che4 system regulates S-motility reversal (Section 18.16.8).

Type IV pili are also found in other bacteria (e.g., *Pseudomonas aeruginosa*, *Neisseria gonorrhoeae*) in which the pili are responsible for pulling groups of bacteria across wet surfaces via a form of gliding motility called *twitching*, which is important for the colonization of these surfaces. As discussed in Section 18.22.5, twitching motility can be important for the maturation of biofilms formed by bacteria possessing type IV pili. Twitching and social gliding motility in myxobacteria are identical with respect to mechanism and requirement for type IV pili.[241,242]

2. Adventurous motility (A-motility)

The mechanism of A-motility is different from that of S-motility in that the former does not require type IV pili and is correlated with the secretion of slime (extracellular polyelectrolyte gel whose chemical nature is not well characterized) from nozzlelike structures at the rear pole of the moving cell.[243] The nozzlelike structures can be seen at both cell poles by using electron microscopy, but the slime is secreted only from the rear pole of the moving cell. (See Fig. 18.22.)

According to one model, the slime is introduced into the nozzlelike structures in a dehydrated form and then becomes hydrated, causing it to swell and be extruded.[243] The model further proposes that the extruded slime adheres to the substrate; and as a consequence, force is generated, pushing the cell forward. A similar mechanism has been proposed for gliding filamentous cyanobacteria. (See Section 1.2.1.) It should be pointed out, however, that the proposed mechanism for A-motility, although interesting, is still hypothetical. What is needed is experimental data describing the chemical and physical nature of the slime, evidence that its secretion is associated with the production of force, and proof that the slime is sufficiently rigid and anchored to the substrate to "push" the cells forward. The current model (Fig. 18.22) suggests that the motility motors at the poles function coordinately in that the pili at the front "pull" the cell forward and the secretion of slime at the rear "pushes" the cell forward. The model also predicts that the A- and S-motility systems must switch poles when the cell reverses its direction of motility. For a discussion of genes required for A-motility, and the proteins that they encode, see note 244.

Intercellular signaling stimulates swarm expansion

The rate of swarm expansion (also called colony expansion) via the gliding of cells away from the outer edge of colonies by means of either the A- or S-motility system is stimulated when the cell density is increased (Fig. 18.23). This suggests that both motility systems are stimulated by cell–cell interactions, although as we shall see later, system A drives single-cell motility and its operation does not require that

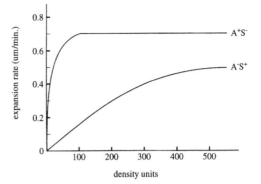

Fig. 18.23 Rate of swarm expansion as a function of initial cell density A^-S^+ and A^+S^- mutants. Microdroplets of cells at cells at various density units (100 density units = 4×10^8 cells/mL) were placed on nutrient agar and the rate of swarm expansion was measured. The swarms expanded linearly with time as the cells glided away from the center of the drop. Approximately 90% of swarm expansion was due to movement and 10% to growth, as determined by comparing swarm expansion via wild-type cells with swarm expansion via nonmotile mutants. *Source*: Data from Kaiser, D., and C. Crosby. 1983. Cell movement and its coordination in swarms of *Myxococcus xanthus*. *Cell Motil.* 3:227–245.

cells be close together, whereas system S drives group motility and requires cell–cell contact for movement to occur. For more about cell–cell stimulation and motility, see note 245. Furthermore, other experiments have shown that the maximum colony expansion rate in A^+S^+ cells is greater than in A^+S^- or A^-S^+ cells, indicating that the two motor types for motility act cooperatively, as indicated in Fig. 18.22.

18.16.3 Intercellular signaling and multicellular development

For reviews of A- and C-signaling, and the bases for the conclusions drawn in the following discussion, see refs. 246 through 249. Early evidence for cell-to-cell signaling involved in the formation of fruiting bodies and myxospores came from the isolation of *Myxococcus* mutants that were unable to form myxospores.[250] (The formation of myxospores is a convenient measure of the ability of the cells to aggregate and form fruiting bodies.) These cells had mutations that could be separated into five classes (*asg*, *bsg*, *csg*, *dsg*, and *esg*), and were able to sporulate when mixed with wild-type cells.

They would also sporulate when mixed with each other in complementary pairs (e.g., bsg^+/csg^- and bsg^-/csg^+).

The existence of five classes of mutants that are capable of extracellular complementation indicates that five extracellular signals (A, B, C, D, E) exist for development. As we shall describe later, two signals, the A- and C-signals, have been isolated, and these are the ones best characterized. Signaling via the A-, B-, D-, and E-signals is necessary for the initiation of expression of developmental genes during the first 5 h of development, whereas C-signal is essential for the initiation of expression of developmental genes after 6 h, the time at which the cells begin to aggregate into small aggregation centers that after 24 h become fruiting bodies housing the myxospores.

Overview of A- and C-signaling

For a more detailed description of the A- and C-signaling pathways, see Sections 18.16.4 and 18.16.5. The signaling pathway steps listed next correspond to the steps numbered in Fig. 18.24.

1. A-signal

The A-signal has been referred to as a cell density signal. It is a subset of extracellular amino acids that accumulates in the medium about 1 to 2 h poststarvation and functions to about 4 h poststarvation. When the combined extracellular concentration of the amino acids that comprise A-signal reaches a threshold value of about 10 μM, the expression of many developmental genes whose products are made early in development is stimulated. Mutants that do not

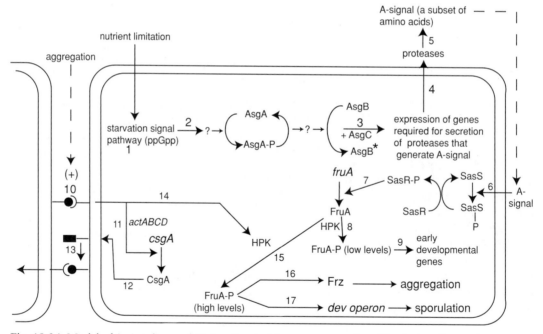

Fig. 18.24 Model of A-signaling and C-signaling in *Myxococcus xanthus*. The A-signaling pathway (steps 1–9), induced by nutrient limitation and high cell density, is responsible for the transcription of early developmental genes. AsgA is a cytoplasmic histidine protein kinase (HPK), and AsgB is a DNA-binding protein, but not a response regulator. Studies with mutants that are deficient in the production of A-signal indicate that three other proteins are involved in the A-signal generation signal pathway. The mutations are in the following genes: *asgC*, an allele of *sigA*, which encodes the major sigma factor; *asgD*, which encodes a histidine protein kinase similar to AsgA; and *asgE*, which encodes a protein whose amino acid sequence suggests that it spans the membrane. The C-signaling pathway (steps 10–17) is induced by cell–cell contact and is responsible for gene transcription after 6 h. Note that the early expressed genes (step 9) are A-signal dependent/C-signal independent, whereas the later expressed genes (steps 16 and 17) are A-signal and C-signal dependent owing to step 14. Also, steps 11 and 12 constitute an amplifying feedback loop for the expression of *csgA*, the gene encoding the C-signal. Symbols: solid circles, C-signals (p. 17); solid rectangle, precursor to C-signal (p. 25); half-circle, C-signal receptor. See text for details.

make A-signal (*asg* mutants) are arrested in development at around 2 h. Thus, A-signal can be thought of as a cell density signal that signals starving cells when the cell density is sufficient for aggregation and fruiting body formation. From this point of view, the A-signal is analogous to other quorum signals, including acylated homoserine lactones produced by a wide range of gram-negative bacteria and peptide signals used by various gram-positive bacteria. (See Section 18.15.) If the concentration of A-signal amino acids is above 10 mM, the cells continue to grow, rather than entering the developmental pathway, reflecting the fact that starvation as well as cell density stimulates development. For more about A-signaling, see Section 18.16.4.

Step 1. A-signaling begins when *Myxococcus xanthus* responds to nutrient limitation by activating a starvation signal pathway. Starvation for amino acids, carbon, energy, or phosphorus, but not for purine or pyrimidine, triggers development. The earliest stages of the signaling pathway involve the synthesis of an intracellular signal called (p)ppGpp, which refers to guanosine 3′-diphosphate, 5′-diphosphate (ppGpp), and guanosine 3′-diphosphate 5′-triphosphate (pppGpp). [For more discussion of the role of (p)ppGpp in regulating cell metabolism, read the description of the stringent response in Section 2.2.2.] There exists strong evidence that (p)ppGpp is part of the signaling pathway in *M. xanthus* that turns on the expression of developmental genes when the cells are faced with nutrient limitation. See note 251 for a summary of the evidence. The (p)ppGpp signal is a global regulator, widespread in bacteria. See note 252 for more information about (p)ppGpp.

Step 2. The starvation signal pathway in step 1, including the guanosine derivatives, activates a cytoplasmic phosphotransfer system that includes unknown elements and a protein histidine kinase called AsgA, as well as a putative DNA-binding protein called AsgB. AsgB is not a response regulatory protein.

Step 3. AsgB activates transcription of genes responsible for secretion of proteases, which as we shall see generate the A-signal. AsgB

has a helix–turn–helix motif similar to many DNA-binding proteins. (See the subsection entitled *Helix–turn–helix motif* in Section 10.2.5.) The sigma factor responsible for the AsgB-activated transcription is AsgC. For more information about the *asg* genes and the evidence in support of the model, see note 253 and refs. 254 through 256.

Step 4. Proteases are made and secreted.

Step 5. The proteases degrade cell surface proteins to generate A-signal which, as described later, is a subset of amino acids.

Step 6. A-signal accumulates in the extracellular medium. When the combined concentration of the amino acids of the A-signal exceeds 10 μM, a membrane-bound histidine kinase, SasS, is activated by autophosphorylation and in turn phosphorylates its cognate response regulator, SasR. (In these enzyme names, Sas stands for <u>s</u>uppressor of *asg*.)

Step 7. SasR-P stimulates the transcription of the gene *fruA*. This occurs early in development, that is, at approximately 6 h.

Step 8. It has been suggested that the gene product FruA is phosphorylated by a putative histidine protein kinase, HPK, to produce a low level of FruA-P.

Step 9. Low levels of FruA-P have been proposed to stimulate the transcription of developmental genes that are expressed prior to aggregation.[257]

Steps 10 through 17 are listed in the next subsection, on C-signaling.

2. C-signaling

C-signaling begins to function approximately 6 h poststarvation, after A-signaling. The C-signal is a cell surface protein at the cell poles that binds to a receptor on another cell pole. The C-signal molecule is a 17 kDa protein that is processed from a 25 kDa protein by a cell surface protease. The 25 kDa protein is encoded by the *csg*A gene. C-signaling begins at around 6 h and becomes primarily responsible either directly or indirectly for the transcription of developmental genes expressed during aggregation, fruiting body formation, and myxospore formation. Consequently, mutants in C-signaling fail to aggregate and to sporulate. C-signaling is also required for morphogenetic movements leading to the formation of aggregates. For more about C-signaling, see Section 18.16.5.

Step 10. C-signal activates a cytoplasmic signaling pathway when cells are in contact with each other, end to end. This is because the C-signal and its receptor are located on the surface of the cell poles. C-signaling is enhanced during aggregation and fruiting body formation because the cells make end-to-end contact as they move in chains and in parallel arrays called streams into aggregation centers and within the developing fruiting bodies.

Step 11. As a consequence of more end-to-end contact, there is increased C-signaling between the ends of the cells, stimulating their cognate receptors on the ends of the cells in front and back. This sets up an amplification loop resulting in increased transcription of the *act* operon, which contains the gene for C-signal, *csgA*. Transcription of the *csgA* gene is controlled by four genes in the *act* operon.[258] The four proteins encoded by these genes are ActA and ActB, whose functions are to stimulate the transcription of *csgA*, and ActC and ActD, both of which are important for the timing of *csgA* transcription. Sequence analysis indicates that *actA* encodes a response regulator and that *actB* encodes a sigma-54 activator protein.

Step 12. The product of the *csgA* gene is CsgA, a 25 kDa protein, which is inserted into the outer membranes.

Step 13. The 25 kDa protein is processed by a cell surface protease to a 17 kDa protein that is the actual C-signal.[259]

Step 14. Increased C-signaling resulting in the activation of a cognate FruA histidine protein kinase (HPK).

Step 15. The HPK increases the levels of FruA-P. (Recall from steps 6 and 7 that the *fruA* gene itself is transcribed as a result of prior activation by the A-signal.

Step 16. The C-signal transduction pathway branches at FruA-P. One branch leads to the *frz* operon and aggregation. The *frz* operon

contains genes that regulate the frequency of reversal of the direction of gliding motility. This is discussed further in the subsection of section 18.16.7 entitled *The Frz proteins regulate the frequency of reversal of the direction of gliding.*

Step 17. The second branch from FruA-P leads to the *dev* operon (*devTRS*) and sporulation. The *dev* operon consists of five genes that are expressed in the fruiting body. Mutants in these genes do not form myxospores. As the cells aggregate, more cell-to-cell end contacts occur, resulting in increased C-signaling and higher amounts of FruA-P. The aggregation pathway (step 16) is activated at a lower level of FruA-P than is the pathway of this final step, myxospore formation. This aspect of the C-signaling pathway accounts for why aggregation precedes myxospore formation.

18.16.4 A-signal identity and generation

A-signal activity has been isolated from buffer in which wild-type cells have been subjected to starvation (shaking the cells in buffer) and shown to be a specific set of amino acids.[260–262] The six most active amino acids are tyrosine, proline, phenylalanine, tryptophan, leucine, and isoleucine. It has been demonstrated that the amino acids restore development when added to *asg* mutants. When the concentration of amino acids in the A-signal exceeds 10 µM, which corresponds to what is produced at a cell density greater than 3×10^8/mL, the signal informs the population of cells that the cell density has reached the requisite number for initiating the developmental program.

Sensing A-signal

As discussed earlier, the Asg proteins appear to be involved in a starvation-induced signaling cascade that activates genes required to produce A-signal. Once A-signal has been produced and is accumulating in the external medium, how is it sensed by the cells? At least one pathway for sensing A-signal involves the product of the *sasS* gene.[263] The nucleotide sequence predicts that the product SasS will be a membrane-bound sensor histidine kinase. According to model, SasS responds to A-signal by phosphorylating a cognate response regulator protein (SasR) that regulates the transcription of genes expressed early in development. To find *sasS*, it was necessary to use a developmental reporter gene (i.e., a gene that is expressed only during development). The reporter gene that was used, gene *4521*, is fused to Tn*5 lac* such that transcription from the *4521* promoter results in the synthesis of β-galactosidase, which can be measured. (An explanation of how *lacZ* fusions are constructed and their use was given in note 25.) The expression of the reporter gene was shown to increase early in development and to require A-signal. (The addition of A-signal rescues the expression of the *4521* gene in *asg* mutants.)

However, expression of *4521* requires *both* starvation and the A-signal; that is, neither one alone is sufficient. Second site mutations in the mutant *asg* strains that expressed *4521* during growth as well as starvation were isolated, and these mapped to the *sasB* locus. Thus, mutations in the *sasB* locus expressed *4521* despite the absence of the A-signal and regardless of whether the cells were starved. These varioints are called *suppressor* mutants because they suppress the mutant *asg* phenotype.

One of the mutant genes in the *sasB* locus, was *sasS*. SasS responds to the A-signal and to starvation by phosphorylating itself (autophosphorylation) and then transferring the phosphoryl group to SasR, which then activates transcription of *4521*. (It appears that both starvation and cell density signals feed into a signaling pathway that results in the autophosphorylation of SasS during starvation, but not during growth, when the set of amino acids comprising the A signal exceeds a threshold value.) The suppressor mutation in *sasS* presumably locks it into the autophosphorylating conformation so that it does not need the A-signal or the starvation signal. Null *sasS* mutants do not express *4521* even in the presence of A-signal, in agreement with the hypothesis that SasS is part of the signal cascade that couples A-signal to the expression of *4521*. In addition, the null *sasS* mutants are not able to form fruiting bodies and sporulate.

18.16.5 C-signal identity and signaling pathway

Mutants defective in making C-signal (*csgA* mutants) do not form fruiting bodies, nor do they form myxospores. They construct abnormal aggregates, and only after a significant delay. C-signal is thus critical for the morphogenetic movements of cells during aggregate and fruiting body construction as well as the expression of developmental genes. Studies with developmentally expressed *lacZ* fusion reporter genes indicate that C-signal acts at a developmental stage after A-signal. Indeed, most of the genes that are developmentally expressed after 6 h are induced directly or indirectly by C-signal, whereas the other extracellular signals act earlier. The signaling cascade leading to developmental gene expression initiated by C-signal is being characterized. CsgA is a membrane-associated protein, and C-signal itself is a processed form of the protein that is located on the surface of cell poles. The chemical identity of C-signal is discussed in note 264 and refs. 248 and 259. The C-signal binds to a receptor at the cell pole of the recipient cell. The identity of the receptor is not known.

C-signaling regulates reversal frequency of gliding motility as well as myxospore formation

Prior to aggregate formation, when the amounts of C-signal molecules on the cell surface are relatively low, cells that run into each other and make end-to-end contact both transmit C-signal and reverse their direction of motility. This occurs when traveling waves of cells collide. As development proceeds, end-to-end contact becomes more frequent, and the production of C-signal increases as a result of the self-amplifying loop (Fig. 18.24, steps **11** and **12**). At around 6 h the amount of C-signal per cell reaches a threshold level, and at that time C-signaling decreases motility reversal. This promotes the gliding of cells within streams into aggregation centers. See note 265 for a model of how this might occur. At still higher concentrations of C-signal, sporulation genes are induced in the fruiting bodies.

C-signaling requires end-to-end cell contact

In contrast to A-signaling, C-signaling requires close cell–cell contact, in particular end-to-end contact.[266,267] Nonmotile cells do not express C-signal-dependent genes or sporulate, despite having the wild-type *csg* gene, because they cannot align themselves properly for C-signal transmission between the cells. However, when the nonmotile cells were placed on an agar surface that had been lightly scratched with emery paper and allowed to settle into the grooves, the cells in the grooves became aligned end-to-end parallel to each other and were capable of stimulating the expression of C-signal-dependent *lacZ* fusion reporter genes as well as sporulation.[268] Cells outside the grooves did not express the fusion. When exogeneous C-signal was added, the nonmotile mutants did express C-signal-dependent genes and sporulated, although they did not aggregate.[269]

As discussed later in the context of the *frz* genes, a hypothesis has been developed to explain how C-signal stimulates aggregation. It has been proposed that the exchange of C-signal between cells that are moving into aggregates decreases the frequency of motility reversals and that this is important for cells to continue moving into aggregates.[270]

18.16.6 Preventing developmental gene expression during growth: SasN

The amino acids that comprise A-signal are also present in the growth medium. Therefore, we would expect to find mechanisms for preventing A-signal-dependent genes from being expressed during growth. In fact, this is the case. The expression of the developmental gene *4521* (Ω4521 Tn*5 lac* insertion) requires both starvation and A-signal. One of the suppressors of mutant *asg* is a gene called *sasN* that encodes the SasN protein, a negative regulator of developmental gene expression during growth.[271] In the absence of SasN, owing to a null mutation in *sasN*, the gene *4521* is highly expressed during growth and development. It has been suggested that SasN, a membrane-associated protein, interacts with SasS, blocking its autophosphorylation during growth. Upon starvation, SasN is inactivated, allowing SasS to respond to the A-signal.

18.16.7 Switching of the A- and S-motility systems from one cell pole to the other

Since the A-system "pushes" the cell forward at the lagging pole, and the S-system "pulls" the cell forward at the leading pole, these two systems must not only operate at opposite cell poles but, when the cell reverses its direction of motility, the A- and S-motility motors must switch poles. This occurs on the average of 10 every min during growth, every 8 min during early development, and much less frequently later in development, when the cells are gliding into aggregation centers during fruiting body formation. For a review, see ref. 272. How is all this regulated?

Although the process is not well understood, several different proteins that are involved have been identified. These include the MglA protein, which regulates the switching of the A- and S-motors from pole to pole; the Frz proteins which regulate the activity of MglA; and C-signal, which regulates the activity of the Frz proteins. Importantly, late in development when the cells are traveling end-to-end in streams toward and within developing fruiting bodies, the increased C-signaling inhibits reversals. How this all occurs will now be described, starting with the MglA protein.

The MglA protein inhibits pole switching and the reversal of gliding

MglA decreases the frequency of motor reversals so that net movement can occur. In fact, mutants defective in *mglA* reverse motility so rapidly that there is no net movement in either direction. As a consequence, spreading of the colonies does not take place.[241,273,274] The *mglA* gene encodes a cytoplasmic GTPase. The active form of MglA is MglA-GTP, and without GTPase activity, it could not act to decrease the frequency of motility reversals. One suggestion is that MglA-GTP takes part in partially disassembling the A- and S-motility apparatus at their old poles so that at any one time, one pole (the rear pole) has only the A-motor and the other pole (the leading pole) has only the S-motor. Thus, in the absence of MglA both motility motors remain active at both poles, with the result that the cell attempts to move simultaneously in both directions.[271] (Reassembly does not require MglA because *mglA* deletion mutants are able to assemble the A- and S-systems.[271]) For more information on the inactivation of the motility machinery at the cell poles, see note 275. As described next, a

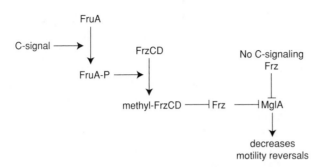

Fig. 18.25 The Frz signaling pathway for motor reversal includes the core components FrzCD and FrzE (not shown) that signal the reversal switch. There are two signaling pathways. In one pathway increased C-signaling that occurs later in development sends a signal that decreases the frequency of reversals, and in the other pathway (during growth) C-signal is not involved, and the cells reverse more frequently. Both the C-signal-dependent pathway and the C-signal-independent pathway converge at MglA, a protein that decreases the frequency of reversals by ensuring that one pole retains only a functioning A-motor, whereas the opposite pole retains only a functioning S-motor. The C-signal-dependent model results in the phosphorylation of FruA. FruA-P then stimulates the methylation of FrzCD. Increased methylation of FrzCD is correlated with a decrease in reversals. The model proposes that methyl–FrzCD inhibits reversal frequency by inhibiting the Frz system, which inhibits MglA. As a consequence, MglA remains active, and motility reversal is decreased. In the absence of C-signaling, the Frz system inhibits MglA and, as a consequence, motility reversal is increased. Some of the evidence supporting this model is that null mutations of *frzCD* and *frzE* reverse much less frequently than wild type. The model is summarized from Søgaard-Anderson, L. 2004. Cell polarity, intercellular signalling and morphogenetic cell movements in *Myxococcus xanthus*. *Curr. Opin. Microbiol.* 7:587–593.

model has been proposed suggesting that the Frz proteins inactivate MglA, and in this way cause an increase in reversal frequency.

The Frz proteins stimulate pole switching and the frequency of reversals

The Frz proteins are part of a signal transduction pathway that increases the frequency of reversals.[276,277] (For the following discussion, see Fig. 18.25.) The *frz* genes are so named because the mutations originally isolated in these genes resulted in the formation during development of abnormal aggregates that have a *frizzy* appearance rather than resembling discrete mounds. The frizzy aggregates have a tangled, swirling pattern and do not develop into fruiting bodies. The model shown in Fig. 18.25 indicates that the Frz proteins stimulate reversals by inhibiting MglA, the protein that decreases the frequency of motility reversals.

Increased C-signaling inhibits pole switching and the reversal of gliding motility by inhibiting the Frz system

Increased C-signaling that occurs later in development inhibits motility reversal. The model is shown in Fig. 18.25. C-signaling does this by increasing the methylation of FrzCD, a cytoplasmic homologue of the methyl-accepting proteins (MCPs) in the enteric bacteria; and methyl–FrzCD inhibits Frz signaling. Indeed, it has been demonstrated that the addition of purified C-signal to *csgA* mutants results in full methylation of FrzCD and that increased methylation of FrzCD is correlated with a reduced frequency of reversals. (Reviewed in ref. 278.) The inhibition of motor reversals by increased C-signaling is important for aggregation and fruiting body formation because it results in chains and streams of cells (parallel chains of cells) moving into aggregates and remaining in the aggregates as they move in concentric circles.[279] As shown in steps **14** and **15** in Fig. 18.24, and in Fig. 18.25, the signaling pathway from C-signal requires the C-signal-dependent phosphorylation of the response regulator FruA. Thus the signaling pathway is a "two-component" phosphorelay pathway.

18.16.8 Che3 and Che4 are also important for development

In addition to the Dif and Frz systems that utilize homologues of the *E. coli* chemotaxis proteins to signal extracellular matrix (ECM) formation and motility reversal, respectively, the *Myxococcus xanthus* genome encodes at least two other systems that utilize chemotaxis-like proteins for signaling. These two systems are the Che3 and Che4 systems.[280,281] The Che3 system regulates developmental gene expression, and the Che4 system regulates S-motility reversal. Mutants in the Che3 system aggregate early when starved, and show increased and premature expression of developmental genes. Deletion of the *che4* operon prevents aggregation and myxospore formation in A⁻S⁺ cells (but not A⁺S⁺ cells). Thus, *M. xanthus* has at least four sets of signaling systems utilizing homologues of the chemotaxis two-component phosphorelay signaling proteins, and these systems contribute to the developmental program. (Other bacteria, with the exception of *E. coli*, are known to have multiple homologues of Che proteins, reflecting different roles for these proteins in phosphorelay signaling pathways.[282])

18.16.9 Chemotaxis

M. xanthus moves up phosphoethanolamine (PE) gradients consisting of certain fatty acids and displays adaptation to PE, two activities that are consistent with chemotaxis.[283] Adaptation is correlated with increased methylation of FrzCD, which is part of the Frz system discussed earlier. (Reviewed in ref. 284.) The cells glide up the gradient because PE decreases the frequency of reversal of gliding motility. It has been postulated that binding of PE to a receptor on the extracellular matrix that covers the cell surface stimulates a signal transduction pathway that regulates motor reversal. Mutants that lack the ECM do not respond to PE.

In addition to the ECM, the excitatory response to PE requires a protein called DifA, which is a homologue to the MCPs and is believed to be a membrane protein. It has been proposed that PE binds to a surface receptor and that a signal that decreases reversals is sent to the motility motors via DifA. Since the A- and S-motility motors must coordinate with each other for net movement to occur, PE must affect the reversal frequency of both motors. It is not understood how that occurs, but per-

haps it is through MglA, as discussed earlier. (See Section 18.16.8, subsection entitled *The MglA protein inhibits pole switching and the reversal of gliding*.) The cells are most sensitive to PE under starvation conditions, suggesting that perhaps this represents a signaling system important for directed motility during fruiting body formation.[285]

18.17 Caulobacter Development: Control of DNA Replication and Cell Cycle Genes

Caulobacter crescentus has been a model system for the study of bacterial cell differentiation because at each cell division two different cell types are produced: a stalked cell and a swarmer cell. In the past few years much has been learned about how the predivisional cell uses intracellular signals to regulate cell cycle events. See refs. 286 through 292 for reviews. Partitioning of sister chromosomes during the cell cycle is described in Section 10.1.6.

18.17.1 The Caulobacter life cycle

Caulobacter crescentus belongs to the α subdivision of the Proteobacteria. It is an aquatic, crescent-shaped, gram-negative bacterium that undergoes an asymmetric cell division to produce two very different cell types: a motile swarmer cell and a sessile stalked cell (Fig. 18.26). The student is referred to the discussion of sporulation in Section 18.18 for another example of an asymmetric cell division yielding two different cell types.

The swarmer cell differs morphologically from the stalked cell not only in lacking a stalk but in having a flagellum and several pili at one pole. The two cells are also very different physiologically. Whereas the stalked cell synthesizes DNA and divides, the swarmer cell swims away and does not initiate DNA synthesis and cell division until it has shed its flagellum, retracted its pili, and grown a stalk at the same pole that previously harbored the flagellum and pili. Thus, the swarmer cell serves to disperse *C. crescentus*, whereas the stalked cell attaches to surfaces via an N-acetylglucosamine adhesin (called a holdfast) at the tip of the stalk, grows,

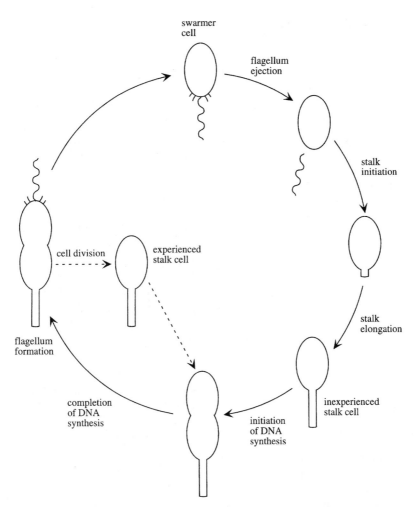

Fig. 18.26 Life cycle of *Caulobacter crescentus*. The swarmer cell is a swimming, nongrowing cell specialized for dispersal. It has a single flagellum, several pili, and chemosensory receptor proteins at the same cell pole. After a period of swimming it settles down to become a sessile stalked cell that synthesizes DNA, grows, and divides. During the transition it degrades the chemosensory receptor proteins, sheds its flagellum, retracts its pili, and produces a stalk at the same cell pole. The stalk is a cylindrical extension of the cell and has a polysaccharide adhesive cap, called a holdfast, that anchors the cell to the substratum. As the stalked cell grows, it produces a flagellum, pili, and chemosensory proteins at the pole opposite the stalk. As a consequence, cell division produces two cell types: a motile swarmer cell and a stalked cell that remains sessile. The stalked cell continues to grow and divide. *Source*: Adapted from Dworkin, M. 1985. *Developmental Biology of the Bacteria*. Benjamin/Cummings, San Francisco.

and reproduces swarmer cells. (Adhesion is a multistep process beginning with attachment of swarmer cells to surfaces. See note 293.) Organic nutrient is often concentrated at surfaces, and this is probably related to why *Caulobacter* attaches to surfaces.

Clearly the *Caulobacter* cell cycle is complex. Processes such as the replication of DNA, the synthesis of flagella and pili, the production of the chemosensory proteins, and cell septation take place at specific times and in specific

cellular locations in the stalked cell prior to production of the swarmer cell. (The chemosensory proteins are placed at the pole opposite the stalked pole in the dividing cell and are used for chemotaxis by the swarmer cell. They are degraded when the swarmer cell becomes a stalked cell.) What factors are involved in the regulation of the timing and the spatial location of these events? One important regulatory molecule is CtrA, and this is discussed next.

18.17.2 CtrA, a global regulator of gene expression

CtrA is a response regulator protein (cell cycle transcription regulator A) that is activated by phosphorylation due to a phosphorelay pathway activated by histidine kinases.[294–297] (The student should review Sections 18.1 and 18.1.1 for a description of phosphorelay pathways.) It has been estimated that CtrA-P directly or indirectly controls the expression of 26% of the *Caulobacter* cell-cycle-regulated genes. This includes the direct control of 95 genes in 55 operons. (Reviewed in ref. 291.) CtrA-P can therefore be viewed as a global regulator of gene expression during the cell cycle. We will first examine the role of CtrA in regulating gene expression in both the stalked cell and the swarmer cell. Then we will see how its levels and distribution between the stalked and swarmer cells are regulated. Finally, we will describe the pathway of phosphorylation of CtrA that results in its activation.

The role of CtrA-P, in regulating DNA replication

We will begin with the control of DNA replication by CtrA-P. Since the swarmer cell indeed contains CtrA-P, which is a repressor of the initiation of DNA replication, this process does not take place in the swarmer cell. One might ask how phosphorylated CtrA represses the initiation of DNA replication. For DNA replication to be initiated, transcription must occur from a promoter within the origin of replication. CtrA-P binds to the origin of replication in the swarmer cell and in so doing represses this transcription. Hence it prevents the inititation of DNA replication. As shown in Fig. 18.27, CtrA levels are high in the swarmer cell, drop in the young stalked cell, allowing DNA replication to be initiated, and then rise in the stalked cell, preventing the reinitiation of DNA replication.

CtrA-P regulates DNA replication in another way. When DNA is synthesized in the stalked cell, the new strand is not methylated and therefore the replicated DNA is hemimethylated. Unless DNA becomes fully methylated, replication cannot be initiated in the stalked cell in the next cell cycle. However, near the end of DNA replication in the late predivisional cell, CtrA-P activates transcription of *ccrM*, the gene for CcrM DNA methylase. As a consequence, the DNA becomes fully methylated, and thus replication can be initiated in the early stalked cell when the CtrA levels drop.

The role of CtrA-P in regulating transcription of ftsZ

CtrA-P represses the transcription of *ftsZ* in swarmer cells and predivisional cells, accounting, in part, for why there is no FtsZ in swarmer cells.[298,299] Recall that FtsZ forms the septal ring required for cytokinesis (Section 2.3.3). Since swarmer cells do not divide, they do not need FtsZ.

The repression of the *ftsZ* gene occurs only after sufficient FtsZ has accumulated for septation to occur in the stalked cell.[297] As septation progresses, the concentration of FtsZ declines owing to proteolysis. At the time of cell division, essentially all of the FtsZ has been degraded. The stalked cell resynthesizes FtsZ for the next division, but the swarmer cell does not start synthesis again until it begins to differentiate into a stalked cell.[300]

CtrA-P activates the transcription of many genes in the predivisional cell

In addition to binding to the origin of replication and preventing transcription, and thus the initiation of DNA replication in the swarmer cell, and also repressing *ftsZ* so that FtsZ is not made in the swarmer cell but only during the swarmer-to-stalked cell transition period and in the young stalked cell, CtrA-P activates the transcription of many genes. These include cell division genes (*ftsI*, *ftsQ*, *ftsA*), genes required for the synthesis of flagellar components, genes required for chemotaxis, and *pilA*, the gene for the synthesis of the pilus subunit, as well as *ccrM*, the gene for a DNA methyltransferase that is required to fully methylate newly replicated DNA so that DNA replication can be initiated in the early stalked cell.

How the levels and distribution of CtrA are regulated

Examine Fig. 18.27, in which the shaded areas denote CtrA. The swarmer cell has CtrA, but

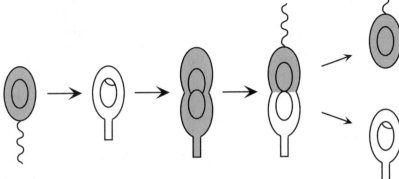

Fig. 18.27 Relative levels of CtrA (shaded areas) during the *Caulobacter* cell cycle are controlled by spatially regulated proteolysis. CtrA prevents the initiation of DNA replication and the transcription of *ftsZ*. It is present in swarmer cells and is degraded during the transition phase when swarmer cells become stalked cells. The *ctrA* gene is transcribed in the stalked cell; but proteolysis in the half of the predivisional cell destined to be the progeny stalked cell ensures that only the swarmer cell is born with CtrA. The ellipse within the cell represents DNA. *Source*: Adapted from Quon, K. C., B. Yang, I. J. Domian, L. Shapiro, and G. T. Marczynski. 1998. Negative control of bacterial DNA replication by a cell cycle regulatory protein that binds at the chromosome origin. *Proc. Natl. Acad. Sci. USA* **95**:120–125.

the early stalked cell does not. This is because the ClpXP protease complex destroys CtrA in the early stalked cell derived from a swarmer cell, allowing DNA replication to begin in the early stalked cell. Then CtrA is made and is distributed throughout the predivisional cell prior to flagellum synthesis. It is then destroyed in the stalked cell half of the predivisional cell so that once cell division has taken place, DNA replication can begin in the progeny stalked cell, but not in the progeny swarmer cell.

Thus the destruction of CtrA is due to spatially regulated proteolysis in the stalked cell half of the predivisional cell, as well as during the transition from swarmer to stalked cell. (This was shown by immunofluorescence microscopy.[296] See the discussion of FtsI in Section 2.3.3 and note 65 in Chapter 2 for an explanation of immunofluorescence microscopy. See note 301 for an explanation of how the cellular amounts of specific proteins can be measured.)

Temporal and spatially regulated proteolysis implies that there are cell cycle signals that activate proteolysis of CtrA at certain times (temporal control) and that mechanisms exist to ensure that proteolysis takes place only in certain regions of the cell (spatial control). One key player is the response regulator protein DivK, which is phosphorylated throughout

the cell cycle (Fig. 18.28). DivK activates the proteolysis of CtrA via the ClpXP protease, presumably by phosphorylating components of a phosphorelay system that activates the protease. Cold-sensitive (cs) mutants of *divK* make a mutant protein (DivK-cs) at the restrictive temperature and do not degrade CtrA (although the mutant form of DivK is phosphorylated). This results in the prevention of the initiation of DNA replication and cell cycle progression beyond the stalked stage when swarmer cells are incubated at the restrictive temperature (20 °C).[302] The phenotype is filamentous cells with abnormally long stalks.

Figure 18.28 suggests that DivK-P receives a phosphoryl group and transfers it to CtrA. However, the amount of CtrA-P in DivK-cs cells incubated at the restrictive temperature is similar to that found in wild-type cells incubated under the same conditions.[301] Because of this, it has been suggested that DivK-P transfers its phosphoryl group to a protein other than CtrA, and that this is part of a signaling pathway that activates the ClpXP protease.

The levels of CtrA are also regulated at the transcriptional level. The *ctrA* gene is transcribed in the stalked cell but not in the swarmer cell, hence the CtrA protein in the swarmer cell is inherited from the predivisional cell. However, once the swarmer cell has become a

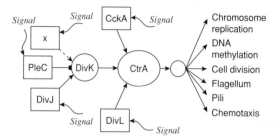

Fig. 18.28 Model of the signal transduction network controlling the activity of the master response regulator, CtrA. Response regulators and the protein kinases are shown as circles and boxes, respectively. CckA, PleC, and DivJ are histidine kinases. DivL is a tyrosine kinase. DivJ phosphorylates DivK, and PleC plays a role in dephosphorylating DivK-P or prevents the phosphorylation of DivK. CtrA is phosphorylated and stable at the restrictive temperature in cells that have a temperature-sensitive mutation in the gene encoding DivK. Because of this, it has been suggested that DivK-P participates in a phosphorelay pathway that activates the ClpXP protease that degrades CtrA, rather than phophorylating CtrA, as implied in the figure. (See text for a further discussion.) *Source*: Ausmees, N., and C. Jacobs-Wagner. 2003. Spatial and temporal control of differentiation and cell cycle progression in *Caulobacter crescentus. Annu. Rev. Microbiol.* 57:225–247. Reprinted with permission, from the *Annual Review of Microbiology*, Volume 57, 2003, by Annual Reviews, www.annualreviews.org.

stalked cell, the *ctrA* gene is transcribed and CtrA accumulates.[303] The question is: what controls the transcription of the *ctrA* gene? As described in ref. 291, there are two promoters for *ctrA*, P_1 and P_2. To be active, P_1 must be hemimethylated. Then, when P_1 is active, CtrA is made, and CtrA-P stimulates P_2. This is therefore autoinduction.

Since the DNA, including P_1, is fully methylated in the early stalked cell, CtrA is not transcribed. However, as DNA replication continues past the *ctrA* gene, P_1 is initially hemimethylated (i.e., the newly synthesized strands are not methylated), and therefore CtrA is made and phosphorylated, and the resulting CtrA-P activates P_2 so that CtrA rapidly accumulates in the predivisional cell, becomes phosphorylated, as described next, and regulates the CtrA regulon.

One of the genes activated by CtrA-P is the methyltransferase gene, *ccrM*; as a consequence, the DNA becomes fully methylated, and transcription from P_1 can no longer

take place. Before this, however, transcription from P_1 is negatively regulated by CtrA-P autorepression. Thus, there is a very complex regulation of *ctrA* transcription during the cell cycle, consisting of repression by CtrA-P at the P_1 promoter, and stimulation by CtrA-P at the P_2 promoter. The regulation of transcription, plus proteolysis by the ClpXP protease, is responsible for controlling the levels of CtrA during the cell cycle.

The activity of CtrA is regulated by phosphorylation by histidine kinases

To regulate the transcription of genes, CtrA must be phosphorylated via a phosphorelay pathway. A major histidine kinase for the phosphorylation of CtrA is CckA. Temperature-sensitive mutations in the *cckA* gene are severely defective in the phosphorylation of CtrA and are altered in gene expression similar to that of *ctrA* temperature-sensitive mutants.[304] CckA, a hybrid histidine kinase, as well as other histidine kinases and response regulator proteins, control gene expression in the *Caulobacter* cell cycle. A multipathway signal transduction network that involves several histidine kinases (CckA, PleC, DivJ) and a tyrosine kinase (DivL) that feed into the central global regulator, CtrA, is diagrammed in Fig. 18.28. PleC and DivJ have opposite effects on the phosphorylation level of DivK. The levels of DivK-P are decreased in a *divJ* mutant and are increased in a *pleC* mutant. This suggests that DivJ phosphorylates DivK, and that PleC plays a role in dephosphorylating DivK-P. See ref. 291 and note 305 for a discussion.

Current questions about the signaling systems are focused on their coordination and cellular localizations

Several aspects of the signaling systems are the subject of active research. For example, how does the cell signal the proteases and the histidine kinases, and how is all of the signaling coordinated? Additionally, most of the components of the signaling network become located in distinct subcellular locations, such as the cell poles, at certain times in the cell cycle. For example, the membrane-bound CckA histidine kinase is dispersed in the cell membrane at early stages of the cell cycle, then becoming localized at the cell pole opposite the stalk in the predivisional cell.

The timing of the localization often coincides with the period during which the protein is most active. For example, the CckA kinase is most highly phosphorylated, and therefore most active, when at the cell pole. Other proteins are dispersed in the cytoplasm and then become anchored at a particular pole during a specific stage in the cell cycle. For example, the response regulator DivK (Fig. 18.28) is dispersed in the cytoplasm of swarmer cells; but then, when the swarmer cell differentiates into a stalked cell, DivK becomes localized and attached at the pole where a stalk develops. Later, in early predivisional cells, some of the DivK migrates and becomes attached to the opposite pole, where the swarmer cell will be produced. When cell division is complete, DivK is released from the new swarmer pole and becomes dispersed in the swarmer cell cytoplasm, but remains attached at the stalked pole of the new stalked cell.

Spatial and temporal localization is not confined to the signal transduction proteins. It also occurs with the chromosomal origins of replication, proteins involved in chromosome segregation, flagellar and pili structural proteins, and proteins of the chemotaxis system. The mechanisms responsible for when and where certain proteins become localized in *Caulobacter* during its interesting cell cycle comprise an exciting area of research, and what is learned with *Caulobacter* will shed light on similar questions of what targets and/or anchors proteins at specific cellular locations, such as cell poles, in other bacterial cells. For a review, see ref. 291.

18.18 Sporulation in Bacillus subtilis

For reviews of *B. subtilis* sporulation, see refs. 306 through 308. The partitioning of the daughter chromosomes during sporulation was described in Section 10.1.7. *Bacillus subtilis* is a gram-positive bacterium that lives in the upper portions of soil, and like other bacteria has evolved ways to adjust to stressful changes in the natural environment. One of these ways is to sporulate when faced with limiting supplies of a carbon or nitrogen source. In the laboratory, this occurs when the population enters stationary phase. During a relatively brief period of the cell cycle, *B. subtilis* decides whether to form a midcell septum and continue cell division, or to form a polar septum and sporulate.[309] The decision to sporulate is regulated by a phosphorelay signal transduction system to be described later (Section 18.18.3).

B. subtilis experiences several physiological changes during adaptation to nutrient deprivation, and sporulation is only one of them. Adaptation includes synthesizing a complex motility and chemotaxis system, which in the natural habitat would increase the chances of finding nutrient, and producing antibiotics that inhibit the growth of other organisms that compete for the limiting nutrient. In addition, the motile cells secrete proteases and other degradative enzymes that might generate nutrients from carbon sources in the natural environment. If starvation continues, the cells develop competence for the uptake of exogenous DNA similar to their own DNA (and therefore in the natural environment from other *B. subtilis* cells in the population), and later the population sporulates. As explained in Section 18.18.6, sporulation is a last resort that is chosen when the population of cells has increased to a high cell density, and growth is no longer possible because there is not sufficient nutrient for the dense population of cells to continue to grow.

The spore is a dormant stage in the life cycle and is resistant to environmental stresses such as heat, ultraviolet radiation, and toxic chemicals. Spores can remain dormant for hundreds of years but will germinate into growing cells (vegetative cells) when nutrient becomes available.

It is well known that *B. subtilis* will sporulate more efficiently at high cell densities. This is correlated with the production and accumulation in the media of extracellular pentapeptides, which serve as cell density signals.[310] (Reviewed in ref. 311.) As reviewed in ref. 312, peptides are often used by gram-positive bacteria as intercellular signals. When the cell densities are low, Spo0F-P, which is part of the phosphorelay system that activates sporulation genes, is dephosphorylated by Rap phosphatases (Section 18.18.3). As the cell density increases, pentapeptide signaling molecules produced by the bacteria accumulate in the medium, enter the cells, and inhibit the Rap phosphatases so that sporulation occurs (Section 18.18.4).

18.18.1 Stages in sporulation

Morphological description

The spore forms inside the cytoplasm of the progenitor cell as a consequence of an asymmetric septation and is therefore called an *endospore*, as opposed to exospores, which are formed by some bacteria at the tips of cells by a process similar to budding. (See note 313 for an example of an exospore-producing bacterium.)

The stages of endospore formation are shown in Fig. 18.29. The growing vegetative cells are said to be in stage 0. As the cells get ready for sporulation, the two copies of the chromosomes, which may be fully or partially replicated, condense and elongate into an axial filament. This has been referred to as stage I. Then a septum is laid down asymmetrically near one pole of the cell. (The student should read Section 10.1.7 and Box 18.1, which summarize information about the RacA and DivIVA proteins and their roles in axial filament formation, the binding of the chromosomes to the cell poles, and the site of septum formation.) Upon completion of the septum, the sporulating cell is said to be in stage II.

Once the septum has been completed, the cell (now called a sporangium) is divided into two compartments: a mother cell compartment and

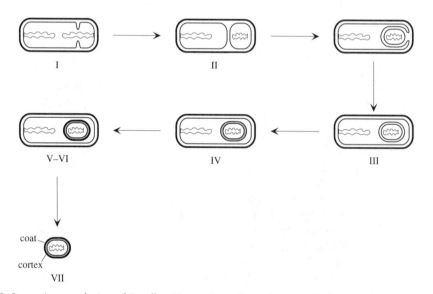

Fig. 18.29 Stages in sporulation of *Bacillus*. Vegetative cells are in stage 0. The two chromosomes from the completed round of replication become aligned along the long axis of the cell in stage I, with their origins attached to opposite poles of the cell. (See the discussion in Section 10.1.7 regarding the attachment of the nucleoid to the cell poles.) At stage II a septum has formed near one pole, dividing the cell into a mother cell compartment and a forespore, trapping approximately one-third of the forespore chromosome in the forespore. DNA translocation across the forespore septum takes place, causing the forespore chromosome to be moved entirely into the forespore. Once the septum has formed, the cell is referred to as a sporangium. The mother cell engulfs the forespore, bringing the sporangium to stage III, where the forespore is set free in the cytoplasm of the mother cell as a protoplast. There is a space between the inner and outer membranes of the spore protoplast, and in stage IV a cortex consisting of peptidoglycan is synthesized in that space. The cortex covers a peptidoglycan cell wall (the germ cell wall) that is laid down on the surface of the inner forespore membrane. A polypeptide multilayered coat is synthesized by the mother cell around the developing spore in stage V. A proteinaceous exosporium is also made. The exosporium covers the spore coat as a layer that is either loosely fitting or more closely fitting, depending upon the species of *Bacillus*. The spore (also called an endospore) matures during stage VI, developing resistance properties, and is released from the lysed sporangium in stage VII. The entire process, including the release of the spore, takes between 8 and 10 h. Each flattened circle (wavy line) represents a nucleoid.

BOX 18.1 PROTEINS INVOLVED IN FORMATION OF THE SPORE SEPTUM AND CHROMOSOME PARTITIONING

FtsZ

The student should read Section 2.3.3 for a discussion of the role of FtsZ as well as other cell division genes in septum formation and cell division in *E. coli*. Prior to the formation of the asymmetric septum in stage II, the cells form an FtsZ ring at both cell poles. The establishment of the FtsZ ring at the cell poles rather than in the cell center is an early event that distinguishes cell division from sporulation. For a discussion of how *B. subtilis* chooses the pole to form the sporulation septum, see later (Section 18.18.5).

SpoIIIE

The translocation of the chromosome into the forespore compartment after septum formation requires the product of the *spoIIIE* gene, which codes for an ATP-dependent DNA translocase located at the leading edge of the growing septum that partitions the forespore compartment from the mother cell compartment.

Septation actually bisects the prespore chromosome, trapping approximately one-third of the prespore chromosome, which includes the origin of replication, inside the forespore compartment, perhaps attached to the cell pole. The product of *spoIIIE* is required to move the remaining two-thirds through the septum into the forespore. The SpoIIIE protein forms a pore through which the remainder of the prespore DNA travels.

The mechanism of transfer of the remaining two-thirds of the chromosome into the prespore compartment through the spore septum is believed to be similar to the transfer of plasmid DNA during conjugation between cells. (The prespore compartment and mother cell compart-

ment are analogous to two separate cells.) This conclusion is supported by the finding that the carboxy-terminal domain of SpoIIIE has significant sequence similarity to DNA transfer proteins (Tra proteins) of conjugative plasmids of *Streptomyces*, and that mutations in this region block chromosome transfer into the prespore compartment.[1]

Spo0J

Spo0J is one of three proteins that are believed to recruit the *oriC* region to the cell poles. The other two proteins are DivIVA and RacA. Spo0J is homologous to ParB. (See the discussion of chromosome partitioning and the Par proteins in Section 10.1.7.)

Spo0J localizes to the cell poles along with the chromosomal origins of replication (*oriC*). It has been suggested that Spo0J aids in positioning the origin of replication of the chromosome to the cell poles, perhaps by binding to sites near *oriC* and to proteins at the pole. Indeed, it has been demonstrated that SpOJ binds to sites near *oriC* of the *B. subtilis* chromosome.[2] SpoJ also functions during chromosome partitioning during vegetative growth. Consistent with this, null mutants of *spo0J* produce a significant increase in cells without DNA during vegetative growth.

RacA

RacA (<u>r</u>emodeling and <u>a</u>nchoring of <u>c</u>hromosome <u>A</u>) is synthesized during sporulation but not during vegetative growth.[3] RacA binds nonspecifically to the DNA but preferentially near *oriC*. It may be involved, along with Spo0J, in the binding of the chromosome to the cell pole during sporulation. Mutations in *racA* frequently result in forespores without DNA. RacA interacts at the cell pole with DivIVA.

DivIVA

DivIVA is an anchor protein at the cell poles. It probably binds directly or indirectly to RacA and to Spo0J, which themselves are bound to the chromosome in the area of the *oriC* region.

Interestingly, the binding of RacA to DivIVA may promote polar septation. It has been suggested that the binding of RacA to DivIVA displaces the division inhibitor, MinCD, at the pole, thus allowing septum formation at the pole.[4] The suggestion is based upon an earlier publication showing that DivIVA sequesters MinCD at the poles so that cell division occurs at midcell during vegetative growth.[5]

REFERENCES

1. Wu, L. J., P. J. Lewis, R. Allmansberger, P. M. Hauser, and J. Ellington. 1995. A conjugation-like mechanism for prespore chromosome partitioning during sporulation in *subtilis*. *Genes Dev.* 9:1316–1326.

2. Lin, D. C.-H., and A. D. Grossman. 1998. Identification and characterization of a bacterial chromosome partitioning site. *Cell* 92:675–685.

3. Wu, L. J., and J. Errington. 2003. RacA and the Soj–Spo0J system combine to effect polar chromosome segregation in sporulating *Bacillus subtilis*. *Mol. Microbiol.* 49:1463.

4. Ben-Yehuda, S., D. Z. Rudner, and R. Losick. 2003. RacA, a bacterial protein that anchors chromosomes to the cell poles. *Science* 299:532–536.

5. Marston, A. L., H. B. Thomaides, D. H. Edwards, M. E. Sharpe, and J. Errington. 1998. Polar localization of the MinD protein of *Bacillus subtilis* and its role in selection of the mid-cell division site. *Genes Dev.* 12:3419–3430.

a smaller forespore compartment. (Sometimes the forespore compartment is referred to as a prespore compartment.) At this early stage (stage II), about one-third of the spore DNA is trapped in the forespore compartment. After this has occurred, the remaining two-thirds of the spore DNA is quickly translocated into the forespore compartment by the DNA translocase, SpoIIIE. During stage III the mother cell septum grows around the forespore, a process called engulfment, and in so doing pinches off the forespore as a protoplast floating in the cytoplasm within the mother cell compartment.

When engulfment is complete, the forespore in stage III has two sets of membranes, an inner and an outer membrane.

In stage IV, a cortex, consisting of peptidoglycan, is synthesized in the space between the inner and outer forespore membranes. The cortex covers another layer of peptidoglycan called the germ cell wall, which is made on the surface of the inner forespore membrane. During stage V a spore coat is synthesized surrounding the outer membrane. The coat is made of protein and is produced by the mother

cell. At the end of stage V, the prespore is dehydrated and phase-bright. The resistance properties of the spore develop during stage VI, and the mature spore is released as a result of lysis of the mother cell during stage VII. The entire process, including the release of the spore, takes around 8 to 10 h and requires the activities of at least 113 sporulation-specific genes.[314]

Sporulation is an example of cell division ending in two different developmental fates for the daughter cells

The polar septum that forms during sporulation is similar to the septum that forms at midcell during cell division when the cells are growing vegetatively. Both require the same genes for synthesis, and both consist of two septal membranes with a layer of peptidoglycan between them. However, there are some differences. For example, the sporulation septum is much thinner than the midcell septum, and the peptidoglycan in the sporulation septum is autolyzed with complete loss of wall material.

This does not occur during vegetative division, and the wall material in the septum remains at the poles of the new cells after vegetative cell division. Although at the time of polar septation the forespore has only one-third of its chromosome, the remaining two-thirds is rapidly moved into the forespore via the DNA translocase, SpoIIIE. Despite these differences with ordinary cell division, it is reasonable to conclude that sporulation is an example of cell division that leads to two cells with identical genes but very different developmental fates owing to differential gene expression.

A similar situation occurs in *Caulobacter*, discussed in Section 18.17, where cell division leads to two different cell types with the same genetic information. *Caulobacter* is different in that the two different cells become independent of one another, whereas during sporulation the cells do not separate, and one of the two cells, the mother cell, engulfs the cell that will become the forespore. *B. subtilis* differentially expresses genes in the mother cell and forespore, resulting in two different types of cells, by compartmentalization of sigma factors, which determine which genes are expressed. The genes that are expressed leading to the formation of spores are call sporulation genes, or *spo* genes.

Sporulation genes

B. subtilis sporulates when it expresses *spo* genes, which were discovered by examining mutants that failed to complete sporulation. The genes are named after the stage of blockage (0, I, II, etc.) and are distinguished from one another by a letter. For example, mutants in *spo0A* fail to initiate sporulation and do not proceed to stage I, and *spoIIA* mutants complete stage II (septation), but fail to proceed to stage III (engulfment).

Factors that regulate the expression of sporulation genes

1. Sigma factors

The vegetative sigma factor is σ^A, which is required for the expression of genes during vegetative growth as well as certain genes required for sporulation. In addition to σ^A there exist several sporulation-specific sigma factors that are made in a sequential order during sporula-

tion. These sigma factors, σ^H, σ^F, σ^E, σ^G, and σ^K, are located in either the mother cell compartment or the forespore compartment, where they transcribe genes. For a summary of the location and roles that these sigma factors play in their respective compartments, as well as a discussion of how their activities are regulated, see note 315 and Box 18.2. For a review of how the activities of the sporulation sigma factors are regulated, and models of how they may be compartmentalized, see refs. 307 and 318. In addition to the sigma factors, a master transcription factor called Spo0A, which is activated via a two-component phosphorelay system when the cells receive signals to sporulate, is also critical. This is described next.

2. Spo0A

Spo0A is activated by phosphorylation to Spo0A-P. Spo0A-P is a transcription factor whose amounts increase at the beginning of sporulation. Spo0A activates transcription of genes required for sporulation and represses the transcription of other genes expressed during postexponential growth. It has been aptly referred to as the master regulator of sporulation genes. Spo0A-P is responsible for axial filament formation, polar septation leading to the forespore, and compartmentalized gene expression in the mother cell and forespore. A brief summary of how Spo0A-P accomplishes all of this follows.

Spo0A-P directly regulates the expression of genes in the Spo0A regulon. (The sigma factors that transcribe the Spo0A-P regulon, are listed in note 317.) A recent estimate puts the number of genes in the Spo0A regulon at 121, of which 40 are positively regulated and 81 are negatively regulated.[318] Twenty-five of the genes in the Spo0A regulon are themselves transcription factors, setting the number of genes indirectly regulated by Spo0A at around 400. Mutations in *spoA* result in cells blocked at stage 0 of sporulation. The genes activated by Spo0A-P include genes responsible for the formation of the axial filament (including *racA*), as well as a gene (*spoIIE*) that encodes a protein that facilitates the formation of polar FtsZ rings.[319] Spo0A-P also directly activates transcription of *spoIIA* and *spoIIG*, which encode the first prespore-specific sigma factor, σ^F, and the first mother-cell-specific sigma factor, σ^E, respectively.

BOX 18.2 SPORULATION-SPECIFIC SIGMA FACTORS

σ^H

The earliest acting sporulation-specific sigma factor is σ^H, which also controls the transcription of some stationary phase genes. The main vegetative sigma factor, σ^A, transcribes the σ^H gene (*spo0H*) weakly during exponential growth but greatly increases the transcription of the gene at the initiation of sporulation. The transcription of the gene for σ^H is stimulated by the active form of the response regulator protein Spo0A. (Spo0A does this indirectly. It represses transcription of the gene for AbrB, which is a negative regulator of the gene for σ^H.) In the predivision cell, σ^H is required for the transcription of genes important for axial filament and polar septum formation. The genes transcribed by σ^H-RNA polymerase include genes encoding proteins in the phosphorelay pathway, including *spo0A*, *kinA*, *kinE*, and *spo0F*, as well as the operon that contains the gene for σ^F. (The phosphorelay pathway initiating sporulation is described later, in Section 18.18.3.) The subsequent formation of the septum at one of the two FtsZ rings also requires σ^H, which activates the transcription of the *ftsAZ* operon, whose products (FtsA and FtsZ) are required for formation of the septum. (See the discussion of FtsA and FtsZ in Section 2.3.3.) A cell septum must also form during vegetative growth. However, the promoter that initiates transcription of the *ftsAZ* operon during sporulation is different from the promoter that is used during vegetative growth and is dependent upon σ^H. In the absence of σ^H, a polar septum does not form.

Other genes regulated by σ^H include the *spoIIA* operon, which includes the gene encoding σ^F, whose product is necessary to transcribe genes in the forespore, as well as *racA*, required for axial filament formation. As described in Section 10.1.7 and Box 18.1, RacA binds preferentially near *oriC*, and less specifically throughout the DNA. RacA is required for the extension of the nucleoid into an axial filament during sporulation, and it is also required for the anchoring of the chromosome near the origin region at the cell poles during sporulation (but not during vegetative growth). The *spoIIA* operon and *racA* are also under the control of Spo0A.

σ^E

Sigma factor σ^E is important for the transcription of genes in the mother cell compartment after the forespore septum is formed. The gene for σ^E is transcribed prior to the formation of the forespore septum at about the same time as the gene for σ^F, a forespore-specific sigma factor, is transcribed. A model for how σ^E is restricted to the mother cell compartment despite being made prior to septation can be found in ref. 1. The protein is synthesized in an inactive form, called pro-σ^E, which is presumed to be degraded in the forespore compartment, accounting for the restriction of σ^E to the mother cell compartment. (An alternative suggestion is that pro-σ^E exists in both compartments and the σ^E is degraded in the forespore compartment.)

Activation of pro-σ^E to σ^E requires proteolytic cleavage of an N-terminal sequence of 27 amino acids. The putative protease, SpoIIGA, which cleaves pro-σ^E to active σ^E, is made prior to septation, is initially inactive in the septal membrane, and is activated by a protein made in the forespore as a result of transcription of by σ^F-polymerase.

The protein that activates the protease is called SpoIIR, and it migrates to the forespore side of the septum to activate the protease. This is another example of communication between the mother cell compartment and the forespore.

The functions of σ^E include triggering engulfment of the forespore and the initiation of the assembly of the spore coat, as well as stimulating the synthesis of σ^K in the mother cell compartment.

σ^F

The initiation of forespore-specific gene transcription requires σ^F. The gene for σ^F is transcribed, along with the gene for σ^E, before the forespore septum is made, and it is kept in an inactive form in both the mother cell compartment and the forespore. Transcription requires the positive transcription regulator, Spo0A, which is part of the phosphorelay signaling system, and σ^H. It is kept inactive by an anti-sigma factor (SpoIIAB) that binds to it and prevents it from binding to the RNA polymerase. The inhibition is reversed by an anti-anti-sigma factor (SpoIIAA) in the forespore, but not in the mother cell compartment.

In its dephosphorylated form, SpoIIAA interacts with SpoIIAB, resulting in the release of σ^F. This happens only in the forespore compartment because there exists a septum-bound phosphatase, called SpoIIE, produced in the predivisional cell, that dephosphorylates SpoIIAA-P in the forespore.

The functions of σ^F include transcribing *spoIIIG*, which encodes σ^G, which in turn activates late gene transcription in the forespore. As mentioned previously, σ^F participates in the expression of mother cell compartment genes by directing the expression of a forespore gene that encodes a protein that migrates to the forespore side of the septum and stimulates a septum-associated protease that converts pro-σ^E to active σ^E in the mother cell compartment.

σ^G

After engulfment, σ^G replaces σ^F as the primary sigma factor for the transcription of sporulation genes in the forespore. The transcription of the gene for σ^G is directed by σ^F, which explains why σ^G is localized to the forespore. Initially, σ^G is inactive because it binds to an inhibitor protein.

The activation of σ^G in the forespore is due to a complex of membrane proteins made in the mother cell, as directed by σ^E. The activating proteins are associated with the mother cell membrane before engulfment and become part of the outer membrane of the forespore. This is one of the ways that the mother cell ensures the proper development of the forespore.

σ^K

The later stages of sporulation depend upon the activity of σ^K in the mother cell compartment. Transcription of the gene for σ^K is directed by σ^E, which explains why σ^K is found only in the mother cell compartment. The protein product, σ^K, which is inactive, is activated by proteolytic removal of a 20 amino acid sequence from the N-terminal end. The activation of pro-σ^K in the mother cell compartment depends upon σ^G in the forespore. It has been proposed that σ^G directs the transcription of a gene encoding an inner membrane forespore protein that interacts with outer membrane forespore proteins synthesized in mother cell compartment (as a result of σ^E-polymerase), and the result is cleavage of pro-σ^K in the mother cell compartment to active σ^K. Active σ^K stimulates the transcription of genes important for cortex and coat synthesis, maturation of the spore, and release of the spore.

REFERENCE

1. Hilbert, D. W., and P. J. Piggott. 2004. Compartmentalization of gene expression during *Bacillus subtilis* spore formation. *Microbiol. Mol. Biol. Rev.* 68:234–262.

18.18.2 Proteins involved in formation of the spore septum and chromosome partitioning

There are several proteins that are important for the formation of the spore septum and for the partitioning of the chromosome between the mother cell compartment and the forespore. These proteins are FtsZ, which forms a ring that will develop into the polar septum, SpoIIIE, which moves the DNA from the mother cell into the forespore compartment, and Spo0J, RacA, and DivIVA, which bring the *oriC* regions to the cell poles during chromosome partitioning. These proteins are described in Box 18.1.

18.18.3 Phosphorelay system for initiation of sporulation

The decision to sporulate rests upon the integration of a variety of environmental and physiological signals that are transduced via various signaling pathways that result in the phosphorylation of Spo0A, which is the master response regulator protein for activating the transcription of the sporulation genes. The phosphorelay system has intermediate phosphoryl donors (phosphorelay proteins) between the histidine kinase and Spo0A.[320,321]

The phosphorelay model for the initiation of sporulation

The model shown in Fig. 18.30 stipulates that a histidine kinase receives a signal for sporulation and as a consequence autophosphorylates. There are actually five histidine kinases involved in sporulation.[322] However, two are most important for sporulation in laboratory media. One is KinA, which is cytoplasmic, and the other is KinB, which is membrane bound.[323] (See note 324 for evidence of this; for a discussion of the other three histidine kinases, KinC, KinD, and KinE, see note 325.)

It is not understood how the kinases are regulated; that is, the signals are unknown. One might imagine that KinB responds to an external signal, whereas KinA responds to a cytoplasmic signal. The histidine kinase then transfers the phosphoryl group to an aspartate residue in a cytoplasmic response regulator protein called Spo0F. Spo0F has a phosphorylated domain homologous to the phosphorylated domains of other response regulator

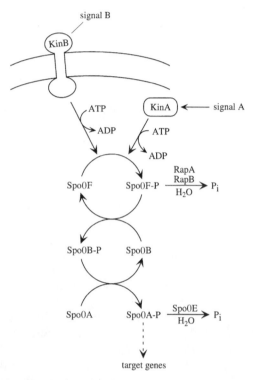

Fig. 18.30 Phosphorelay system during initiation of sporulation in *B. subtilis*. A histidine kinase autophosphorylates in response to a signal and then transfers the phosphoryl group to a phosphorelay protein, Spo0F. There are two histidine kinases shown, KinA (cytoplasmic) and KinB (membrane bound). Three other histidine kinases, Kin C, Kin D, and KinE, also exist and are discussed in the text. Spo0F-P then phosphorylates the phosphorelay protein, Spo0B, which in turn transfers the phosphoryl group to a response regulator protein, Spo0A. Spo0A-P is a transcriptional activator for sporulation genes. (Spo0A-P also represses transcription of the negative regulator gene, *abrB*, which turns off transcription of genes normally expressed during the transition phase between exponential growth and sporulation.) At least three phosphatases are important for regulating the level of Spo0A-P. They are RapA (Spo0L) and RapB (Spo0P), which dephosphorylate Spo0F-P, and Spo0E, which dephosphorylates Spo0A-P. For more information, see refs. 306 and 328.

proteins. The crystal structure of SpoOF was reported in 1997.[326]

SpoOF-P functions as a phosphorelay protein and transfers the phosphoryl group to a histidine residue in a phosphotransferase protein called SpoOB. SpoOB-P then transfers the phosphoryl group to an aspartate residue in a response regulator protein called SpoOA. As discussed earlier, SpoOA-P is the transcription factor that is the primary regulator of the expression of the sporulation genes.

Advantage to the phosphorelay

The phosphorelay system described for *B. subtilis* sporulation differs from standard two-component systems in that the phosphate is transferred between two response regulator proteins, SpoOF and SpoOA, via a phosphotransferase SpoOB. The phosphorelay system offers more sites for fine tuning or control than some of the more simple two-component systems. This may be necessary to ensure that *B. subtilis* does not stop multiplying and enter the sporulation pathway prematurely.

Regulatory role of phosphatases

In many two-component systems the level of phosphorylation of the response regulatory protein is determined by the kinase and phosphatase activities of the histidine kinase. In other words, many histidine kinases are bifunctional enzymes and have both kinase and phosphatase activities. For example, see the discussions of NarX in Section 18.3, NR_{II} in Section 18.4.1, and PhoR in Section 18.5.1. However, in the phosphorelay that controls sporulation, the levels of the phosphorylated response regulator proteins are determined by separate phosphatases. In fact, there exist several phosphatases that take part in adjusting the levels of the phosphorylated response proteins.[327] (Reviewed in refs. 307 and 328.)

The existence of multiple phosphatases allows multiple inputs into the regulation of the phosphorelay pathway. One of the phosphatases, SpoOE, dephosphorylates the master transcription regulator, SpoOA-P. It is not known how SpoOE is regulated. Two of the phosphatases, RapA (response regulator aspartyl

phosphatase A) and RapB, dephosphorylate the phosphorelay protein, SpoOF-P. A third phosphatase, RapE, which dephosphorylates SpoOF-P and whose activity is controlled in a similar fashion to that of RapA, has been reported.[329] However, it is less significant than RapA in regulating the onset of sporulation, and it appears to play an accessory role.[330] Regulation of the level of phosphorylated response regulators by separate phosphatases occurs in other phosphorelay systems. Recall that the level of CheY-P in the chemotaxis system of the enteric bacteria is regulated in part by a separate phosphatase, CheZ. See Section 18.12.5.

How the RapA phosphatase is controlled

It is clearly important to understand how *B. subtilis* regulates the amounts and activities of the phosphatases that control the phosphorelay pathway. A model for how the RapA phosphatase is regulated was published in 1996, and it involves a two-component signaling system.[331] (See Fig. 18.31.) There exists a two-component signaling system consisting of a membrane-bound histidine kinase called ComP and a response regulator protein called ComA. Upon receiving a signal, ComP phosphorylates ComA, which then activates the transcription of the *rapA* operon. There are two genes in the *rapA* operon. One is *rapA*, which encodes the RapA phosphatase protein, and the other is *phrA*, (phosphatase regulator A), which encodes the protein PhrA, which when processed inhibits the activity of RapA phosphatase.

Since the ComP/CompA two-component system produces an inhibitor of the RapA phosphatase, it stimulates sporulation. To inhibit the RapA phosphates, PhrA must be processed. PhrA is secreted from the cell and cleaved extracellularly to a pentapeptide (PhrA peptide) that is imported back into the cell through an oligopeptide transport system called Opp. See note 332 for a further description of the processing of PhrA. (The oligopeptide permease, Opp, catalyzes the transport of PhrA peptide as well as other short peptides into the cell, including another pentapeptide called CSF, discussed later.)

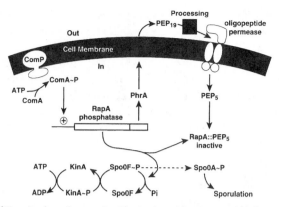

Fig. 18.31 Regulation of RapA phosphatase by PhrA. An unknown signal for competence stimulates the ComP histidine kinase and causes phosphorylation of the ComA response regulator. The phosphorylated ComA then stimulates the transcription of the *rapA* operon, which encodes RapA phosphatase and PhrA. The RapA phosphatase dephosphorylates Spo0F-P so that sporulation does not occur while the cells are in the competence process. The *phrA* gene product is secreted to the cell surface, where it is processed to a pentapeptide (PEP_5) that reenters the cell via the oligopeptide permease. PEP_5 inhibits RapA so that sporulation can take place. Certain steps in the secretion and/or processing or uptake of PhrA are probably regulated to prevent the inhibition of the RapA phosphatase while the cell is in the competence pathway. *Source*: Perego, M. 1997. A peptide export–import control circuit modulating bacterial development regulates protein phosphatases of the phosphorelay. *Proc. Natl. Acad. Sci. USA* **94**:8612–8617.

Once inside the cell, PhrA peptide acts as an inhibitor of RapA phosphatase activity and, as a consequence, sporulation is not inhibited by the RapA phosphatase. Genetic evidence for this is as follows: if the *phrA* gene is deleted, then RapA phosphatase activity is very high and sporulation is inhibited. It has also been demonstrated that the PhrA peptide inhibits RapA phosphatase activity in vitro.[333] However, there must be additional steps in the regulatory pathway. This is because the *rapA* operon encodes the RapA phosphatase, in addition to the phosphatase inhibitor. It may be that control of initiation of sporulation involves, in part, regulation of the secretion and processing of PhrA and/or its uptake into the cell. This is discussed in Section 18.16.4 In the subsection entitled *Is PhrA a quorum sensor?*

How the RapB phosphatase is controlled

There is a third phosphatase that dephosphorylates Spo0F-P, thereby preventing sporulation. It is called RapB. Cells cannot sporulate if the levels of RapB are too high. However, unlike *rapA* (and *rapE*), the ComA/ComP two-component system is not involved in the transcription of *rapB*, and a Phr peptide is not involved in inhibiting the activity of RapB. Instead, *rapB* is constitutively transcribed throughout the growth cycle, and the RapB phosphatase is inhibited by the CSF peptide, a quorum sensor that increases as the cell density increases during growth, reaches active levels at the end of exponential growth and the onset of stationary phase, and is described in Section 18.18.4. Thus, the CSF peptide should stimulate sporulation when the levels rise appropriately.

There are some similarities between CSF and the Phr peptide. Each is synthesized as a 40 amino acid protein precursor that is exported by the general secretory pathway out of the cell, where it is processed to a 5 amino acid peptide that is brought back into the cell via the same oligopeptide permease (Opp), and ultimately inhibits a phosphatase (RapB) that dephosphorylates Spo0F-P.

However, there is a potential problem here. The name of the peptide CSF designates the **c**ompetence and **s**porulation stimulating **f**actor, and as discussed later (Sections 18.19.1 and 18.19.2), CSF also stimulates competence. So, which pathway will be stimulated by CSF, sporulation or competence? As we shall see, it turns out that the threshold levels that stimulate competence are lower than the levels that stimulate sporulation.

18.18.4 Quorum sensing plays a role in the initiation of sporulation

Quorum sensing has been rationalized as a way for starving cells to become spores when the population density is so high that any subsequent entry of small amounts of nutrient would be insufficient to sustain the high density of competing bacteria.[333] One quorum sensor is CSF. There is some discussion of whether PhrA is also a quorum sensor.

CSF is a quorum sensor

As discussed in Section 18.18.3, CSF is an extracellular peptide that stimulates sporulation when it reaches a threshold level in the medium. The oligopeptide permease (Opp) mentioned earlier transports CSF (product of the *phrC* gene) as well as PhrA (product of the *phrA* gene) into the cells. Whereas PhrA stimulates sporulation by inhibiting the RapA phosphatase that inhibits sporulation by dephosphorylating Spo0F-P, CSF stimulates sporulation by inhibiting the RapB phosphatase that dephosphorylates Spo0F-P. (Recall that Spo0F-P is part of the phosphorelay pathway that produces Spo0A-P, which stimulates the expression of sporulation genes.)

Is PhrA a quorum sensor?

One suggestion is that the PhrA peptide that accumulates in the medium is a quorum sensor that serves as a signal for sporulation to take place when the cell density is sufficiently high.[334] An alternative view is that the PhrA peptide is secreted to the outside of the cell membrane, but not into the medium, and is then reimported by the same cell that produced it.[335] This has been referred to as the export–import control circuit, and may function as a timing device that regulates tells the phosphatase it must become active to inhibit sporulation and allow competence to develop. Thus, the processing, export, and import of the PhrA peptide can be viewed as a means of delaying sporulation so that competence can develop first.

18.18.5 Choosing the pole for the asymmetric division

As discussed in Box 18.1, at the initiation of sporulation, Z rings form at both poles. Thus, the cell must choose which pole to use for the asymmetric division. Let us first ask why *B. subtilis* places a Z ring at both poles. A clue comes from studying *racA* mutants that are defective in forming axial filaments. As expected, these mutants frequently form anucleate forespores. However, many of these mutants undergo a second asymmetric division at the opposite pole, and if they trap DNA in the second forespore, they form a viable spore. Thus, the second potential site of forespore formation can help the cell produce a viable spore if axial filament formation fails.

The second question seeks to determine how the cell chooses which pole to form the septum. The answer has something to do with FtsA, a protein that must join FtsZ if the septum is to form. It turns out that FtsA localizes to only one of the two FtsZ rings. What directs FtsA to only one of the two poles is not known. The pole that is chosen for septum formation, however, is usually the older of the two cell poles.

18.18.6 Sporulation is a last resort

When Spo0A-P reaches a threshold level, sporulation is induced. However, as reviewed in ref. 336, lower levels of Spo0A-P indirectly regulate many nonsporulating genes associated with the adaptation of *B. subtilis* to poor nutritional conditions as the bacterium enters stationary phase, including genes for cell motility and chemotaxis, competence, the degradation of carbohydrates, extracellular proteases, and antibiotics. Spo0A-P does this by inhibiting the transcription of *abrB*, a gene that encodes a transcriptional regulator that inhibits the transcription of many stationary phase adaptation genes. The levels of Spo0A-P required to repress the transcription of *abrB* are lower than those required to directly stimulate the expression of the *spo* genes. The levels of Spo0A-P markedly increase when starvation is imminent for a high density of *B. subtilis* cells. Part of the increase is due to increased transcription of the *spo0A* gene, as well as other genes of the phosphorelay system, via the σ^H-polymerase that increases upon starvation, and increased phosphorylation of Spo0A stimulated by the secreted peptide pheromone CSF at high cell densities. (See the discussion of CSF and the RapB phosphatase in Section 18.18.3.) Thus sporulation is a last resort that *B. subtilis* chooses when a highly dense population of cells is faced with starvation, and growth is not possible.

18.19 Competence in Bacillus subtilis

Many different bacteria exhibit competence, the state of permitting synthesis of a DNA uptake system in which the DNA being taken up is similar to the DNA of the recipient cell (e.g., when the donor cell is of the same species). It is thus a form of gene transfer between cells when cells in the population are lysing and releasing DNA. (The incorporation of extracellular DNA is referred to as transformation. See Box 10.1 for a description of how transformation was discovered and its importance for establishing that DNA is the genetic material, and see refs. 337 and 338 for reviews of material in this section.) When cultures of *B. subtilis* undergo nutritional limitation and enter the stationary phase, 10 to 20% of the cells ultimately develop competence. (If the cells are grown in defined minimal medium without a complex mixture of amino acids, competence can develop in a subpopulation of cells at high cell density during late exponential growth.[337])

Cells that have become competent have synthesized a DNA uptake system, and as a consequence become transformable and are capable of incorporating DNA from the medium. The binding and uptake of DNA into the cell relies on at least five different types of membrane-associated protein: ComA, ComE, ComC, ComF, and three copies of ComG. These proteins are postulated to provide the uptake machinery, including the cell membrane pore, that brings the DNA into the cell. The expression of the genes encoding DNA uptake as well as DNA recombination proteins is controlled by a transcription factor called ComK. ComK is induced only in late exponential or stationary phase, only at a high cell density, and only in 10 to 20% of *B. subtilis* cells in the

population. The induction of ComK requires a quorum-sensing system, the CSF/ComX system, which is described next.

18.19.1 Model for the regulation of competence

Briefly, what is described next is a signaling pathway in which high cell density signals (quorum signals) stimulate the transcription of competence genes. Specifically, two secreted peptides, CSF and ComX, enter the cells and stimulate a two-component phosphorelay pathway that results in the transcription of a gene, *comS*, that encodes a protein, ComS, that activates a transcription activator, ComK, that stimulates the transcription of competence genes.

Activation of ComK by ComS

Refer to Fig. 18.32. For *B. subtilis* to become competent to take up exogeneous DNA and incorporate it, ComK, the transcription factor for competence genes, must be activated. Under noninducing conditions, ComK is kept inactive by binding to a complex consisting of the MecA and Clp proteins, ClpC and ClpP. (ClpC was called MecB earlier, before it was realized that MecB is similar to the *E. coli* heat-shock proteins ClpC.) The activator of ComK is ComS, which causes the release of ComK from the inhibitory complex. Importantly, when ComK is in the protein complex it is targeted for proteolysis by the ClpP protease in conjunction with ClpC. Thus, ComS also stimulates the accumulation of ComK by protecting it from proteolysis.[339]

Activated ComK stimulates the transcription of competence genes, including comK

Activated ComK binds to the *comK* promoter and stimulates its own transcription. This is referred to as an autostimulatory loop. The newly synthesized ComK also activates the transcription of all the other genes that encode proteins required for the binding and uptake of exogenous DNA and its recombination into host DNA (Fig. 18.32). The autostimulation of transcription of *comK* by ComK ensures a rapid rise in the levels of ComK, and thus a

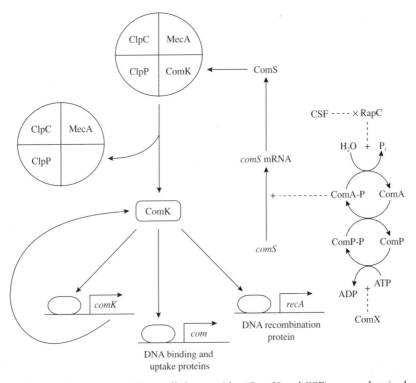

Fig. 18.32 Regulation of competence. Extracellular peptides (ComX and CSF) accumulate in the extracellular medium. The histidine kinase, ComP, is in the cell membrane, and extracellular ComX binds to it and activates it so that more of the response regulator, ComA-P, is formed inside the cell. CSF enters the cell and inhibits the activity of the RapC phosphatase, thus promoting increased amounts of ComA-P. ComA-P activates the transcription of *comS*. ComS then activates ComK by displacing it from the ComK/MecA/ClpC complex. ClpP is a protease that works in conjunction with ClpC to degrade ComK while it is in the complex. ComK stimulates the transcription of genes required for competence. *Source*: Adapted from Lazazzera, B. A., and A. D. Grossman. 1997. A regulatory switch involving a Clp ATPase. *BioEssays* **19**:455–458.

commitment to competence. (For more about the regulation of transcription of *comK*, including other regulatory factors, see note 340.)

Two secreted peptides stimulate the transcription of comS

Two secreted peptides, CSF and ComX, which are described in Section 18.19.2, accumulate at high cell densities, enter the cells, and activate the transcription of *comS* by stimulating a two-component signaling system that produces a positive transcription factor (Fig. 18.32). This process is called quorum sensing.

18.19.2 Quorum sensing and the control of competence

There is a way for *B. subtilis* to sense when the cell density is sufficiently high that trans-

cription of *comS* is appropriate. The high cell density signals that activate the transcription of *comS* are the peptides CSF (competence and sporulation factor) and ComX, which are secreted by and accumulate in the medium as the cell density increases when cells approach stationary phase. As described in Section 18.18.4, CSF also stimulates sporulation by inhibiting the RapB phosphatase. How the cell regulates whether competence or sporulation is stimulated by CSF is discussed later. However, the peptides do not directly activate the transcription of *comS*. Instead, the peptides increase the amounts of a phosphorylated response regulator protein (albeit by different mechanisms) that is part of the two-component ComP/ComA regulatory system that stimulates the transcription of *comS*. Therefore, there is a signal cascade leading to transcriptional activation of competence genes, as follows:

1. Accumulation in the medium of inducing levels of CSF and ComX at high cell densities

2. Stimulation of the ComP/ComA two-component system by CSF and ComX

3. Activation of transcription of *comS*

4. Activation of ComK by ComS

5. Activation of transcription of competence genes by ComK

The student will note the similarity to the regulation of developmental gene expression in *Myxococcus xanthus*, discussed in Section 18.16.3. *M. xanthus* aggregates and forms fruiting bodies only at high cell densities. It has been suggested that an extracellular quorum-sensing signal (the A-signal) activates a two-component signaling system (the Sas system) that activates the transcription of developmental genes.

For a review of the regulation of the initiation of competence, see refs. 312 and 341. For a more general review of cell–cell communication in gram-positive bacteria, see ref. 342. For more detail on the activation of the signaling system by CSF and ComX, see Box 18.3.

Low concentrations of CSF stimulate competence, and high concentrations stimulate sporulation

B. subtilis is capable of switching from the competence pathway to the sporulation pathway. This ability is in part regulated by the extracellular quorum sensor CSF. As we shall see, low concentrations of CSF move the cells into the competence pathway, and high concentrations of CSF move the cells into the sporulation pathway.

When CSF is present at relatively low extracellular concentrations (1–10 nM), it stimulates competence by inhibiting the activity of the RapC phosphatase that dephosphorylates ComA-P, which is a response regulator important for the transcription of competence genes (Fig. 18.32 and Box 18.3). On the other hand, higher concentrations of CSF stimulate sporulation and inhibit competence. How does this occur?

The stimulation of sporulation by higher levels of CSF is due to inhibition of the RapB phosphatase, and this results in higher levels of Spo0A-P, which is a positive transcription factor for sporulation genes. (See Section 18.18.4.) The inhibition of competence by higher concentrations (20 nM–1 μM) of CSF is due to the inhibition of transcription of *comS*. Thus, low levels of CSF stimulate the transcription of *comS*, but high levels inhibit the transcription. The decreased levels of ComS result in increased proteolysis of ComK, hence an inhibition of the competence pathway. (See the earlier discussion of how ComS increases the levels of active ComK by inhibiting its proteolysis.)

It is not known how CSF inhibits *comS* expression. One possibility is that CSF at high concentrations inhibits the histidine kinase (ComP), or perhaps it activates a phosphatase that removes the phosphate from the response regulator ComA-P. The latter would be analogous to regulation of the Ntr regulon, with P_{II} stimulating the phosphatase activity of NR_{II}. (See Section 18.4.1.)

18.20 Bioluminescence

Many bacteria employ quorum signals that stimulate the bacteria to express certain genes that are appropriately expressed only when the population has reached a threshold cell density. This was first discovered among the luminescent bacteria. A survey of bioluminescent systems is presented in Section 18.20.1, followed by an explanation of the biochemistry of bacterial luminescence in Section 18.20.2. The use of intercellular quorum signals for the expression of bioluminescent genes is described in Section 18.20.3. The use of quorum signals for nonluminescent gram-negative bacteria is described in Section 18.21 and its subsections. As we shall see, the most common quorum signals used by gram-negative bacteria are acylated homoserine lactones (acyl–HSLs) and furanosyl borate diesters.

18.20.1 Brief survey of bioluminescent systems

Bioluminescence is the emission of light by living organisms. The organisms that are bioluminescent reflect a diverse assemblage that includes representatives from the algae (dinoflagellates), fungi, shrimp, insects (fireflies), squid, fish, and bacteria.[343–345] Although some live in the soil or

BOX 18.3 ACTIVATION OF THE TWO-COMPONENT REGULATORY SYSTEM BY CSF AND COMX

The two-component signaling system activated by CSF and ComX consists of a histidine kinase called ComP, which spans the cell membrane, and a cytoplasmic response regulator protein called ComA (Fig. 18.32). When stimulated by the extracellular peptide ComX (half-maximal response occurs at approximately 5–10 nM), ComP autophosphorylates and then phosphorylates the response regulator ComA to form ComA-P. CSF, another extracellular signal, enters the cell and increases the levels of the phosphorylated response regulator, ComA-P. (The cells respond in this way to extracellular concentrations of 1–10 nM.) CSF inhibits the activity of a cytoplasmic phosphatase (RapC) that dephosphorylates ComA-P.

The result of all this is that the increased level of ComA-P results in the stimulation of transcription of *comS*. (See note 1 regarding *comS* and the *srfA* operon.) In this way, the accumulation in the medium of CSF and ComX stimulates the expression of the competence gene, *comS*. ComS activates ComK, and ComK stimulates the transcription of genes required for competence.

Notice that there are two ways in which the level of the response regulator protein ComA-P is increased. One signal (ComX) stimulates the kinase activity of ComP so that more ComA is phosphorylated. A second signal (CSF) inhibits a phosphatase (RapC), which removes the phosphate from ComA-P. It is not uncommon for the levels of aphosphorylated response regulator protein to be regulated by both phosphorylation and dephosphorylation. As explained in the text, if the levels of CSF in the extracellular medium rise sufficiently, competence is inhibited and sporulation is stimulated.

NOTE

1. The *srfA* operon encodes the three subunits of the peptide synthetase required to synthesize the cyclic lipopeptide antibiotic surfactin, which is composed of seven amino acids and a β-hydroxy fatty acid. One of the genes in the *srfA* operon is *comS*, which is the only gene in that operon required for competence. Transcription of the *srfA* operon is stimulated by the phosphorylated response regulator protein ComA-P, which binds upstream of the *srfA* promoter. ComA is phosphorylated by the histidine kinase ComP. The peptide extracellular signaling molecule ComX stimulates the kinase activity of ComP, and the extracellular polypeptide signaling molecule CSF enters the cell and inhibits the dephosphorylation of ComA-P by the RapC phosphatase. Thus, both signaling polypeptides increase the amounts of ComA-P, which stimulates the transcription of *comS*. ComS activates ComK, which activates the transcription of all the competence genes required for the putake and processing of DNA. See text for model of how ComS activates ComK. For reviews, see: Solomon, J. M., and A. D. Grossman. 1996. Who's competent and when: regulation of natural genetic competence in bacteria. *Trends Genet.* 12:150–155; Lazazzera, B. A., and A. D. Grossman. 1997. A regulatory switch involving a Clp ATPase. *BioEssays* 19:455–458.

in fresh water, most are marine organisms. In fact, the majority of animal species living at a depth of approximately 200 to 1000 m in the ocean are bioluminescent. In some cases, bioluminescence is used for intraspecies communication, and to hunt or attract prey.

With regard to luminous bacteria, it is reasonable to suggest that the ability to emit light increases the probability that a bacterium will be taken up by a host, and thereby find a beneficial environment, which includes nutrient. Thus, one can imagine that enteric bacteria that normally thrive in the intestines of fish might increase their chances of being ingested if, while colonizing particles in the dark ocean depths, they emitted light, that fish might then see. Then fish, attracted by the light, might then see the bacterial colonies and ingest them.

Luminous bacteria

Luminous bacteria live in freshwater and terrestrial environments but are most common in marine environments. Marine luminous bacteria can be isolated from the intestines of fish and invertebrates in three forms: as saprophytes growing on dead marine animals (e.g., fish, crustacea); as symbionts growing in the light organs of fish (primarily); squids, and pyrosomes (colonial tunicates), and as part of the plankton. The advantages to the bacterial host of harboring luminous bacteria are several, as the following discussion will illustrate.

Some species of luminous bacteria live symbiotically in the light organs of fish. Light organs in fish consist of tubules or canals, densely packed with extracellular bacteria, that connect to the exterior through pores from which the bacteria may exit into the intestine or seawater, depending upon the location of the light organ. The majority of bacterial light organs are inside the fish, derived as outpocketings of the intestinal tract. Probably these light organs serve the fish by pointing counterillumination, during which the fish eliminates the silhouette that would be produced by downdwelling light by emitting light from its ventral side, thus confounding predators. A similar situation occurs in the Hawaiian bobtail squid, *Euprymna scolopes*, where *Vibrio fischeri* lives in a light organ embedded in the center of the mantle cavity on the ventral surface of the squid and is used for counterillumination.[346] Some fish have light organs that are special pouches under the eyes. They are used for vision (for seeking prey) and for intraspecies communication.

Since bacterial luminescence is continuous, fish have evolved a variety of ways to control the release of the light to the outside. These include the use of pigment screens (melanophores) that block the light organ, shutters, and reflectors. For example, some fish whose light organ is a pouch under the eye can block the light by pulling tissue over the pouch (like an eyelid). Other fish can rotate the luminescent pouch downward so that the light does not exit.

Luminous bacteria that live in the oceans belong to three genera: *Vibrio*, *Photobacterium*, and *Alteromonas*. (The genus *Photobacterium* is not easily distinguished from *Vibrio*. See note 347 for a discussion of this point.) It is presumed, or has been demonstrated, that they all can live saprophytically on dead fish or meat, can be a free-living part of the plankton in the oceans, and can also live in the intestines of their respective hosts. Some (*V. fischeri*, *P. phosphoreum*, and *P. leiognathi*) can also live as symbionts in light organs of fish and (except for *P. phosphoreum*) squid. However, none of the marine luminous bacteria belonging to the *Vibrio harveyi* group, which is discussed later, are known to colonize light organs of fish, although they live in the gut of marine animals as well as in the sea.

The terrestrial and freshwater luminous bacteria are *Xenorhabdus luminescens* and certain strains of *Vibrio cholerae*, respectively. The luminescent strains of *V. cholerae* are not known to cause disease. *X. luminescens* grows as a symbiont in the intestines of a nematode that infects a variety of insects. While the nematode grows and completes its life cycle in the dying insect (e.g., a caterpillar), *X. luminescens* is expelled by the nematode into the insect hemolymph, where the bacterial growth results in a glowing insect carcass. The developing nematodes feed on the bacteria in the insect hemolymph.

18.20.2 Quorum sensing and bioluminescence

Biochemistry of bacterial luminescence

Bacterial luminescence is due to an enzyme called bacterial luciferase that is encoded in the *lux* operon. As we shall see, transcription of the *lux* operon is stimulated by quorum-signaling molecules, and thus light is produced at high cell densities. Students who would like to learn more about the very interesting biochemical reactions that produce the light are referred to Box 18.4, and to Figs. 18.33 and 18.34, which are cited in that discussion. The regulation of expression by quorum-signaling molecules of the *lux* operon is discussed next.

The lux operon

For the luciferase and other enzymes of the luminescent pathway to be made, the *lux* operon must be transcribed. The *lux* operon contains a minimum of five genes, *luxCDABE*.

BOX 18.4 THE BIOCHEMISTRY OF BACTERIAL LUMINESCENCE

The overall reaction

Luminous bacteria produce blue-green light when they simultaneously oxidize reduced riboflavin-5'-phosphate ($FMNH_2$) and a long-chain aldehyde (e.g., C_{14}) (RCHO, tetradecanal), with oxygen. The reaction is catalyzed by a flavin monooxygenase called *bacterial luciferase*, as follows:

$$FMNH_2 + RCHO + O_2 \rightarrow FMN$$
$$+ H_2O + RCOOH + light$$

Light is emitted because during the reaction the flavin becomes electronically excited and subsequently fluoresces as the electron returns to its ground state. (In certain bacterial strains there may be a second chromophore, bound to a second protein, that emits light.) This is a complicated reaction. Of the two oxygen atoms in O_2, one becomes reduced to water while the other is incorporated into the carboxyl group of the carboxylic acid.

The bioluminescent pathway is illustrated schematically in Fig. 18.33. The figure points out that for luminescence to continue, both $FMNH_2$ and RCHO must be regenerated. The aldehyde (RCHO) is regenerated from the carboxylic acid (RCOOH) via reduction by NADPH, catalyzed by a fatty acid reductase complex consisting of three proteins (a transferase, a reductase, and a synthetase). The $FMNH_2$ is regenerated by the reduction of FMN by NADH, catalyzed by an NADH–FMN oxidoreductase (flavin reductase). Notice that the luciferase pathway can be viewed as a shunt leading from the cytochrome respiratory chain, and from this point of view is energetically costly.

The biochemistry of light emission

During the luminescent reaction, oxygen reacts with $FMNH_2$ (bound to the luciferase) to form a flavin-4a-hydroperoxide (FMNH-4a-OOH). (See concluding equations and Fig. 18.34.) The flavin hydroperoxide then reacts with the aldehyde to form a postulated peroxyhemiacetal intermediate, which decomposes to the carboxylic acid and an enzyme-bound flavin-4a-hydroxide (flavin-4a-OH) with an excited electron. When the electron returns to the ground state, fluorescence occurs.[1] At some point, water is eliminated from the flavin-4a-hydroxide producing FMN. The reactions shown in Fig. 18.34 can be summarized as follows:

Formation of peroxide

$$FMNH_2 + luciferase + O_2 \rightarrow$$
$$luciferase–FMNH-4a-OOH$$

Binding of aldehyde

$$Luciferase–FMNH–OOH + RCHO$$
$$\rightarrow luciferase–FMNH–OO–RCHO$$

Oxidation of aldehyde

$$Luciferase–FMNH–OO–RCHO$$
$$\rightarrow luciferase–(FMNH–OH)^* + RCOOH$$

Emission of lgiht

$$Luciferase–(FMNH–OH)^* \rightarrow$$
$$luciferase–FMNH–OH + light$$

Elimination of water

$$Luciferase–FMNH–OH \rightarrow$$
$$luciferase + FMN + H_2O$$

Sum

$$FMNH_2 + O_2 + RCHO \rightarrow$$
$$FMN + H_2O + RCHOOH + light$$

REFERENCE

1. Kurfuerst, M., P. Macheroux, S. Ghisla, and J. W. Hastings. 1987. Isolation and characterization of the transient, luciferase-bound flavin-4a-hydroxide in the bacterial luciferase reaction. *Biochim. Biophys. Acta* **924**:104–110.

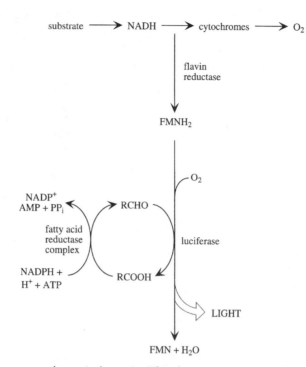

Fig. 18.33 Bioluminescent pathway in bacteria. The electron transport pathway through luciferase to oxygen is a shunt off of the cytochrome pathway. A flavin reductase reduces FMN, which then combines with luciferase and a long-chain aldehyde. The luciferase catalyzes the oxidation of $FMNH_2$ to FMN and the oxidation of the aldehyde to the carboxylic acid with the production of light. The aldehyde is regenerated from the carboxylic acid via a fatty acid reductase complex that uses ATP and NADPH. *Source*: Nealson, K. H., and J. W. Hastings. 1992. The luminous bacteria, pp. 625–639. In: *The Prokaryotes*, Vol. I, 2nd ed. A. Balows, H. G. Truper, M. Dworkin, W. Harder, and K.-H. Schleifer (Eds.). Springer-Verlag, Berlin.

(The *lux* operon is shown later, in Fig. 18.36.) The *luxAB* genes encode the α and β subunits, respectively, of the luciferase (a heterodimer), and *luxCDE* encode the three proteins of a fatty acid reductase complex, which is necessary for the synthesis of a long-chain aldehyde that must be oxidized by luciferase for light to be produced. Intercellular signaling via diffusible signaling molecules (quorum signals) regulates the expression of these genes.

Quorum signals

It has been known for some time that certain species of luminous bacteria will emit light in laboratory culture only when the cell density reaches a certain "quorum" threshold (e.g., >10^7/mL).[348] This is because the bacteria will not luminesce until they have accumulated in the extracellular medium a threshold concentration of what has historically been called an *autoinducer*, a species-specific signal for luminescence. With one known exception, the autoinducers are N-acyl-L-homoserine lactones (acyl–HSLs), that is, homoserine lactones, each one linked to an acyl side chain via an amide bond. The structure of the acyl side chain varies with the specific autoinducer (Fig. 18.35). The side chains differ in their length and according to whether there is a keto, hydroxyl, or no substitution on carbon 3. Thus the acyl–HSLs are named 3-oxo-C6-HSL, 3-hydroxy-C4 HSL, C8-HSL, and so forth. (See note 349 for more discussion of the nomenclature.) In addition to acyl–HSLs, *V. fischeri* and *V. harveyi* produce a quorum signaling molecule whose structure is unrelated to acyl–HSLs. It is a furanosyl borate diester, called AI-2 in *V. harveyi*. See note 350 for a description of the chemical structure of AI-2.

Briefly, what we will learn is that *V. fischeri* and *V. harveyi* each have two major quorum-sensing systems, but there are some differences in these systems between the two organisms. *V. fischeri* has the *lux* system that positively stimulates the *lux* genes, and the *ain* system that

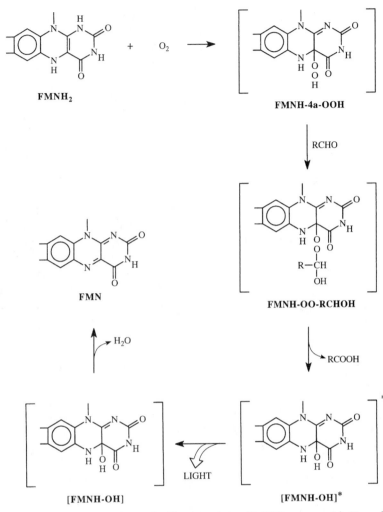

Fig. 18.34 Model for light emission during luciferase activity. $FMNH_2$ reacts with O_2 to form the hydroperoxide derivative, FMNH-4a-OOH. The FMNH-4a-OOH then combines with a long-chain aldehyde, RCHO, to form the peroxyhemiacetal, FMNH-OO-RCHO. The aldehyde becomes oxidized to the carboxylic acid and dissociates, resulting in an electronically excited flavin hydroxide, [FMNH–OH]*. The excited electron in the flavin hydroxide returns to its ground state, releasing light. The flavin hydroxide loses water to become oxidized flavin, FMN. The overall reaction is that one atom of oxygen becomes reduced to water as it accepts two electrons from $FMNH_2$, and one atom of oxygen becomes incorporated into the carboxyl group as the aldehyde is oxidized to a carboxyl. (For a discussion, see: Kurfuerst, M., P. Macheroux, S. Ghisla, and J. W. Hastings. 1987. Isolation and characterization of the transient luciferase-bound flavin-4a-hydroxide in the bacterial luciferase reaction. *Biochm. Biophys. Acta* **924**:104–110.)

derepresses the lux genes. *V. harveyi* has the *luxM* system, which is similar to the *ain* system, and the *luxS* system, both of which derepress the *lux* genes.

How *Vibrio fischeri* regulates expression of the lux operon

Vibrio fischeri (sometimes called *Photobacterium fischeri*, as discussed in note 347) occurs as a symbiont where it lives at very high cell densities (i.e., around 10^{11}/mL), in the light organs of certain marine fishes and squids. It also occurs as a free-living bacterium in seawater but at a much lower cell density (i.e., $<10^2$ cells/mL). The bacteria will luminesce if grown to a high cell density in laboratory culture. In nature, however, luminescence occurs only in the light organs of the fishes and squid because it is here that the bacterial cell density is so high that

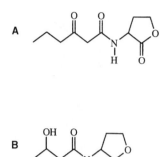

Fig. 18.35 Autoinducers produced by *Vibrio fischeri* and *V. harveyi*: (A) the *V. fischeri* autoinducer, β-ketocaproyl homoserine lactone and (B) the *V. harveyi* autoinducer, β-hydroxybutyryl homoserine lactone.

intercellular signaling molecules produced by the bacteria reach a concentration sufficient to induce luminescence. *V. fischeri* has two well-studied signaling systems based upon acyl–HSLs that regulate the *lux* operon. These are the *lux* system and the *ain* system.

1. The lux system

For the following discussion, refer to Fig. 18.35. The first acyl–HSL discovered in *V. fischeri* is part of the *lux* system, and as we shall see, it binds to and activates a positive transcription regulator called LuxR. The acyl–HSL produced by the *lux* system has an acyl chain of 6

carbons with a keto group on the third carbon. It is called β-ketocaproyl homoserine lactone, or 3-oxo-C6-HSL (Fig. 18.35A), and it is synthesized by LuxI synthase, which is encoded by the gene *luxI*. The acyl–HSL is a nonpolar substance that diffuses freely back and forth across the bacterial cell membrane and accumulates in the extracellular medium during growth. When the acyl–HSL reaches a critical concentration (about 10 nM) inside the cells (the concentrations inside and outside the cells are the same), it binds to and activates a positive transcription factor called LuxR. LuxR is encoded by the *luxR* gene, which is to the left of the *lux* operon (*luxICDABEG*) and is expressed divergently (Fig. 18.36). The function of the *luxG* gene is not known, although it is transcribed when the other *lux* genes are activated. Mutations in this gene do not appear to influence luminescence or its regulation. It has been suggested that *luxG* may encode a flavin reductase.[364]

The binding of the acyl–HSL to LuxR enables LuxR to bind to the *lux* regulatory DNA and to stimulate transcription of the *lux* operon by recruiting σ^{70} RNA polymerase.[352,353] (Although LuxR is the primary regulatory factor for the *lux* genes, several other factors besides LuxR regulate the transcription of the *lux* genes.[354,355]) Notice that the operon contains the gene for the synthesis of the acyl–HSL (i.e., *luxI*). Thus, the

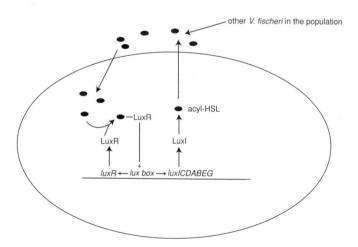

Fig. 18.36 The regulation of bioluminescence in *V. fischeri*. At high cell densities the autoinducer acyl–HSL accumulates in the medium, enters the cells, and combines with LuxR. The acyl–HSL/LuxR complex binds to the *lux* box and induces transcription of the *luxR* gene and the *luxICDABEG* genes. As a consequence, there is amplification of the acyl–HSL signal, and light is produced.

presence of acyl–HSL stimulates its own synthesis in a positive feedback loop.

The LuxI/LuxR system is necessary for bioluminescence in the squid light organ. That is, mutants in *luxI* do not express light at a detectable level in the squid light organ. However, as described next, a second important system for the regulation of bioluminescence exists, the *ain* system, which is the predominant system for bioluminescence at the lower cell densities in laboratory cell culture. As we shall see, the *ain* system acts differently from the *lux* system in that it (1) inactivates a member of a phosphorelay pathway that inhibits transcription of the lux operon and (2) mildly stimulates transcription by binding to LuxR.

Another important transcriptional activator is the protein LitR, encoded by *litR*.[356] LitR is about 60% homologous to the *V. harveyi* LuxR and *V. cholerae* HapR discussed later. When grown in culture, mutants in which *litR* is inactivated are delayed in the onset of luminescence and produce only 20% of the light produced by the wild type.

Model showing how LitR stimulates expression of the lux operon

A model based upon experimental evidence described in ref. 356 proposes that LitR binds to a portion of the *lux* operon that includes the promoter of *luxR* and stimulates transcription of *luxR*. As a consequence, more LuxR is produced. The LuxR binds to the quorum signal, AI-1, and the combination induces the expression of the *lux* operon (*luxICDABEG*), resulting in the production of light.

2. The ain system
V. fischeri produces a second acyl–HSL autoinducer via the *ain* system.[357,358] The autoinducer of the *ain* system has an 8-carbon acyl chain with no substitutions and is called C8-HSL (N-octanoyl HSL). C8-HSL is synthesized by a synthase called AinS (autoinducer synthase) encoded by *ainS*. (AinS is homologous to the *V. harveyi* Later we shall discuss AI-1 acyl–HSL synthase, called LuxM.) The *ain* system is the predominant system for the bioluminescence that occurs at the intermediate cell densities (10^8–10^9/mL) that exist in laboratory culture, whereas the *lux* system is necessary for bioluminescence at the higher cell densities ($>10^{10}$/mL) in the squid light organ. Very importantly, It has been discovered that both the *ain* and *lux* systems are needed not only for bioluminescence, but also for successful colonization of the squid light organ. The reason for believing this is that inactivation via mutation of *ainS* or *luxI* results in drastically reduced colonization of the light organ as measured by the number of colony-forming units (cfu). That both the *ain* and *lux* systems are important for successful colonization of the squid light organ indicates that they regulate the expression of genes necessary for colonization, as well as luminescence genes.

How does the *ain* system work? The *ain* system works much differently from the *lux* system. Whereas in the *lux* system the *lux* operon is stimulated when the autoinducer binds to a transcriptional activator (LuxR), this is not the case for the *ain* system. Instead, evidence suggests that the autoinducer of the *ain* system regulates gene expression by dephosphorylating a protein called LuxO-P that represses the *lux* operon.[359] LuxO is part of a phosphorelay pathway and becomes phosphorylated in the absence of the autoinducer. This is similar to the *V. harveyi* AI-1 system discussed next and summarized in Fig. 18.37.

In *V. fischeri*, the dephosphorylation of the repressor LuxO-P results in increased transcription of *litR*, which as mentioned previously encodes a positive transcription regulator of *luxR*. There is therefore a hierarchical cascade. The *ain* system acts at a lower cell density to stimulate the expression of the *lux* operon, which takes over at higher cell densities. In summary, then, the *ain* system stimulates the *lux* system principally by derepressing the *lux* operon. It has also been reported that If the levels of C8-HSL are sufficiently high, then *ain* can bind with low efficiency to LuxR, and weakly stimulate the *lux* operon in this way as well.

3. The AI-2 system
As reported in ref. 358, genome sequence and other data suggest the presence in *V. fischeri* of a third quorum-signaling system called signaling system 2 (also called the AI-2 system). The autoinducer is not an acyl–HSL but rather a

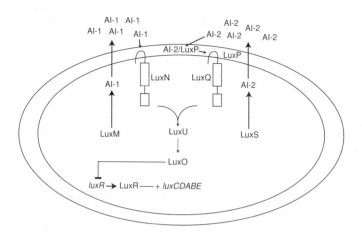

Fig. 18.37 Model for the control of bioluminescence in *Vibrio harveyi*. There are two autoinducers produced, AI-1 and AI-2. The synthases LuxM and LuxS are required for the synthesis of AI-1 and AI-2, respectively. The autoinducer AI-1 is an acylated homoserine lactone (3-OH-C4-HSL), whereas AI-2 is a furanosyl borate diester that is chemically unrelated to acyl–HSLs. The cognate hybrid sensor kinases are LuxN (responds to AI-1) and LuxQ (responds to AI-2). Each has a periplasmic loop, which is thought to bind to the autoinducer. A periplasmic protein, LuxP, is thought to bind to AI-2, forming a complex that binds to the periplasmic loop of LuxQ. The phosphotransfer pathways converge at LuxU, which phosphorylates LuxO. LuxO-P indirectly represses the *lux* operon (*luxCDABE*) in the following way: LuxO-P is proposed to stimulate the transcription of a gene ("X") that encodes a transcription repressor of *luxR* (a *V. fischeri litR* homologue), that encodes a positive transcription regulator of the *lux* operon. In the presence of the auto-inducers, LuxO-P is dephosphorylated and the *lux* operon is no longer repressed. Both AI-1 and AI-2 must be present for LuxO-P to be completely dephosphorylated.

furanosyl borate diester. (See note 350 for a description of the chemical structure of the furanosyl borate diesters.) The mechanism by which it transfers signals is a phosphorelay signaling system that appears to be the same system used by *V. harveyi* and is shown in Fig. 18.37. However, the inactivation of this system in mutants (created by introducing a mutagenized version of the gene *luxS*, which normally encodes the enzyme LuxS, required to synthesize the autoinducer, AI-2), had only a very small effect on luminescence in culture and no effect on growth or colonization of the squid light organ. Thus, its role in *V. fischeri* is not clear. However, as described in the next subsection, it does play an important role in regulating luminescence in *V. harveyi*.

Interestingly, the LuxS/AI-2 system is present in a wide range of bacteria (>55 species), both gram positive and gram negative. For some examples, see the discussion of biofilm forma-tion in Section 18.22.7. Perhaps the production of AI-2 by of bacteria different types in the same community is a way for the bacteria to engage in interspecies communication for the sake of sensing the cumulative cell density and/or changes in environmental conditions that might result in metabolic changes resulting in the production of AI-2.

How *V. harveyi* regulates expression of the *lux* operon

There are no homologues to *luxI* or *luxR* in the *V. harveyi* chromosome, reflecting their lack of a *lux* signaling system, as described earlier in *V. fischeri* for the regulation of the *lux* operon. *V. harveyi* actually has two signaling systems, called signaling system 1 and signaling system 2, that regulate the *lux* operon. Both these phosphorelay signaling systems employ histi-dine kinase and response regulator proteins. In this way they differ from the *lux* system of *V. fischeri*, which is independent of phosphory-lation. *V. harveyi* signaling system 1, which makes AI-1 (3-OH-C4-HSL), is similar to the *ain* system studied in *V. fischeri* that makes C8-HSL. The AinS homologue in *V. harveyi* is called the LuxM synthase. Signaling system 2 makes the autoinducer called autoinducer 2 (AI-2), which is a furanosyl borate diester synthesized by the LuxS synthase. (See note 350.) Interestingly, as discussed earlier, *V. fischeri* also has the LuxS/AI-2 system, but it does not play the major role in the regulation of luminescence as it does in *V. harveyi*.[358]

The control of the *luxCDABE* operon in *V. harveyi* is illustrated in Fig. 18.37. (The *luxCDABE* genes encode functions identical to the *V. fischeri* operon discussed earlier.) There are two membrane-bound sensor kinases, LuxN and LuxQ, and they respond to two separate signaling molecules, AI-1 and AI-2. signaling molecule AI-1 is an acyl–HSL ((3-OH-C4-HSL) made by LuxM synthase, but AI-2 is a chemically unrelated molecule (furanosyl borate diester), made by LuxS synthase.[360] AI-2 binds to LuxP in the periplasm, and the combination binds to LuxQ.

The sensor kinases are hybrid kinases, so called because both have a sensor domain that binds the signaling molecule and a response regulator domain that transmits the signal to the phosphotransferase. Both signaling pathways converge at the same phosphotransferase, LuxU. In the absence of bound signaling molecules, the sensor kinases autophosphorylate and phosphorylate LuxU, which phosphorylates LuxO. LuxO-P represses the expression of *luxR*, which encodes the transcriptional activator (LuxR) of the *lux* operon, and no light is produced. (The *V. harveyi* LuxR is not related to the *V. fischeri* LuxR; rather, it is homologous to the *V. fischeri* LitR.)

When the signaling molecules accumulate in the extracellular medium, they enter the periplasm, bind to periplasmic domains of their cognate histidine kinase sensors, and convert them from phosphorylating to dephosphorylating enzymes. (AI-2, the furanosyl borate diester, actually combines with a periplasmic protein, LuxP, that is similar to the ribose-binding periplasmic protein in *Escherichia coli*; the LuxP–AI-2 complex is thought to bind to the cognate sensor kinase LuxQ, in the inner membrane.) Upon conversion of the histidine kinase sensors, the phosphoryl group is transferred from LuxO-P to LuxU, and from LuxU-P to LuxN and LuxQ, where the phosphate

is hydrolyzed via the phosphatase activities of LuxN and LuxQ bound to their respective autoinducers. As a consequence, the *lux* operon is no longer repressed and light is produced. Although the presence of either AI-1 or AI-2 results in dephosphorylation of LuxO-P, the complete dephosphorylation of LuxO-P requires the presence of both AI-1 and AI-2; that is, both LuxN and LuxP must be converted from kinases to phosphatases.[361]

Model for how LuxO-P represses the lux operon

Evidence suggests that LuxO-P does not directly repress the *lux* operon, but rather activates the transcription of a repressor of the positive transcription factor, LuxR, of the *lux* operon.[348,362] Furthermore, repression appears to occur after the mRNA for LuxR has been synthesized; that is, repression is post-transcriptional.

The model for repression is as follows: (1) five genes are activated by LuxO-P, and they produce multiple small regulatory RNAs (sRNAs); (2) the sRNAs combine with an sRNA-binding protein called Hfq to form the actual repressor; (3) the repressor complex destabilizes the *luxR* mRNA.[363]

Because of the destabilization of the *luxR* mRNA, LuxR is not made, and because LuxR activates transcription of the *lux* operon, the *lux* operon is repressed, and light is not produced. A similar quorum-sensing system is described later for *V. cholerae* (Section 18.21.4), and the mode of repression is similar.

18.21 Systems Similar to LuxR/LuxI in Nonluminescent Bacteria

Quorum-signaling molecules similar to the ones produced by luminescent bacteria have been discovered in nonluminescent bacteria, where they regulate the expression of genes unrelated to the *lux* genes. (For reviews see refs. 364–366.) The components of the quorum sensing pathways in the systems described next are quite similar to those of the *V. fischeri* system. They employ proteins similar to LuxI and LuxR, and the quorum-signaling molecules are acyl–HSLs.

18.21.1 The Rhizobiaceae

The members of the family Rhizobiaceae are gram-negative bacteria that interact with plants either to cause disease or to induce symbiotic nitrogen-fixing nodules. *Agrobacterium tumefaciens* is a plant pathogen that causes crown gall disease; and various rhizobia genera, including *Rhizobium*, *Bradyrhizobium*, and *Azorhizobium*, form symbiotic nitrogen-fixing nodules on leguminous plant roots or stems, depending upon the genus. Leguminous plants are plants such as peas that form pods that split into two valves with the seeds attached to one edge of the valves. Acyl–HSL quorum sensor molecules play an important role in the plant–bacterial interactions, as well as bacterial–bacterial interactions. *Agrobacterium tumefaciens* will be discussed first, and then *Rhizobium leguminosarum*.

Agrobacterium tumefaciens

Agrobacterium tumefaciens produces crown gall tumors in plants. The bacteria cause tumors in the region where the roots and stem meet (i.e., the crown) by transferring a small piece of plasmid DNA, called T-DNA, from a resident plasmid called the Ti (tumor-inducing) virulence plasmid, directly into the plant nucleus. (For more about *Agrobacterium* and crown gall tumors, see Section 18.11.7.) In addition to transferring T-DNA to plants, the Ti plasmid itself can be transferred via conjugal transfer between *Agrobacterium* cells within and around the crown gall tumors or in culture with added opines. (Opines, described in Section 18.11.7, are derivatives of amino acids produced by the tumors used as a carbon, energy, nitrogen, and sometimes phosphorus source by the bacteria. The genes for producing the opines are on the T-DNA that is transferred into the plant from the bacteria.)

1. Quorum signaling via acyl–HSLs between the bacteria

Conjugation between the bacteria requires a high cell density. A quorum-sensing system, which is strikingly similar to the *V. fischeri* LuxR/LuxI system, is encoded in the Ti plasmid and regulates the expression of genes (the *tra* genes) required for conjugal plasmid transfer between bacteria.[367–369] The quorum-sensing genes encoded on the plasmid are *traI* and *traR*. *A. tumefaciens* produces the extracellular conjugation factor is 3-oxo-C8-HSL, which is synthesized by the TraI protein (acyl–HSL synthase). The TraI protein is homologous to LuxI, the acyl–HSL synthase that synthesizes the *V. fischeri* acyl–HSL. The *A. tumefaciens* 3-oxo-C8-HSL differs from the *V. fischeri* Lux acyl–HSL (3-oxo-C6-HSL) only in the length of the acyl chain. *A. tumefaciens* mutants that do not produce 3-oxo-C8-HSL are unable to conjugate unless 3-oxo-C8-HSL is added to the incubation medium. These results indicate that 3-oxo-C8-HSL serves as a cell-to-cell signal that prepares the population of cells for conjugation.

When 3-oxo-C8-HSL binds to TraR, a LuxR homologue, TraR becomes activated to bind to DNA (*tra* boxes) and activates transcription of the downstream *tra* genes required for conjugation. (See note 370 for the evidence that TraR is homologous to LuxR.) Furthermore, 3-oxo-C6-HSL, the *V. fischeri* acyl–HSL, will substitute for the *A. tumefaciens* acyl–HSL in activating transcription of the *tra* genes, although not as effectively. Additionally, the presumptive binding sites for TraR (*tra* boxes) in the *A. tumefaciens* genome have sequence similarity to the binding site (*lux* box) for LuxR upstream of the *luxI* promoter in the *V. fischeri* genome.[365] It is clear that there is a great deal of similarity between quorum sensing and gene activation in the LuxI–LuxR luminescence system in *V. fischeri* and the TraI–TraR conjugation system in *A. tumefaciens*.

2. Plant–bacteria signaling via opines

The plant, via the opines secreted by the crown gall tumors, determines whether conjugal transfer of the Ti plasmid occurs. The opines enter the bacteria and bind to a receptor protein (either AccR or OccR, depending upon the strain of bacterium). The complex of opines and the receptor protein initiates the transcription of *traR*, which activates the transcription of the *tra* genes upon becoming activated by the quorum-sensing system. Therefore, the plant–microbe signaling system (opines) and the microbe–microbe signaling system (acyl–HSL) combine to regulate Ti plasmid conjugation.

Rhizobium leguminosarum

Rhizobium leguminosarum forms nitrogen-fixing root nodules on leguminous plants. Initially, signals are exchanged between the plant and bacteria living next to the root and, as a consequence, the plant root tissues form nodules which the bacteria enter. This is a multistep process summarized in note 371 and reviewed in ref. 372. Within the nodules, the bacteria differentiate into cells of another type, called bacteroids. A symbiotic relationship exists between the bacteroids and the plant tissue. The bacteroids convert nitrogen gas to ammonium, which the plants use. In return, the plant converts carbon dioxide to organic carbon and feeds the bacteria, mostly dicarboxylic acids. Genes conferring the ability to stimulate nodule formation (*nod* genes) and nitrogen fixation (*nif* and *fix* genes) are encoded on a resident symbiotic bacterial plasmid. The expression of these genes, as well as genes necessary for the conjugal transfer of the plasmid between bacterial cells (*tra* genes) and genes that influence the growth rate, is regulated by a complex hierarchy of acy–HSL signaling molecules produced by *Rhizobium leguminosarum*.[373–376]

18.21.2 Pseudomonas aeruginosa

Pseudomonas aeruginosa also produces acyl–HSL signaling molecules. *P. aeruginosa* is one of the more common opportunistic pathogens and can cause infections at different body sites (e.g., urinary tract, lungs, skin), especially in immunosuppressed individuals, burn victims, or individuals who are otherwise weakened by sickness or age. The organism is virulent in part because of extracellular virulence factors (exotoxins, proteases) that it produces. The activation of the virulence genes is due to LasI–LasR, a quorum-sensing system similar to the LuxI–LuxR system.[377,378] The acyl–HSL has been identified as 3-oxo-C12-HSL, which differs from one of the *V. fischeri* HSLs in having a longer acyl chain. The quorum-sensing signal 3-oxo-C12-HSL is synthesized by LasI and interacts with LasR to activate the virulence genes. As discussed in Section 18.22.7, the Las system is also required for *P. aeruginosa* to make mature biofilms under some growth conditions.

P. aeruginosa also has a second quorum-sensing system similar to the Lux system. It is called the RhlI–RhlR, or simply the RhI system. The Rhl system is stimulated by the Las system in a hierarchical cascade. Specifically, LasR/3-oxo-C12-HSL activates the transcription of *rhlR*. This process is reviewed in ref. 364. RhlI is an acyl–synthase that makes C4-HSL (*N*-butanoyl-L-homoserine lactone), which binds to RhlR, a transcriptional regulator, to activate a subset of the virulence genes activated by the Las system as well genes not activated by the Las system.

18.21.3 Erwinia carotovora subsp. carotovora

Quorum-signaling molecules are also produced by the plant pathogen *Erwinia carotovora* subsp. *carotovora (Ecc)*, which causes soft rot disease in potatoes and other plants.[379] During infection, the bacterium produces several exoenzymes (including pectin lyase, cellulase, and a protease), which degrade the plant tissue and aid the bacteria in colonizing and propagating in the plant. Mutants that fail to synthesize any of the exoenzymes have decreased virulence. The bacteria also produce carbapenem, an antibiotic that may play a role in inhibiting the growth of competing bacteria at the site of infection when *Erwinia* begins to degrade the plant tissue.

These bacteria have a gene that encodes a LuxI homologue called CarI (also known as ExpI). (Reviewed in ref. 380.) CarI catalyzes the synthesis of a quorum-sensing signal that appears to be identical to 3-oxo-C6 HSL produced by *V. fischeri* and interacts with transcription regulators to stimulate transcription of genes that code for the exoenzymes and for the antibiotic (gene cluster *carA* to *carH*). (See note 381 for the evidence that CarI is homologous to LuxI.)

There is also a LuxR homologue called CarR that binds to 3-oxo-C6 HSL and activates the transcription of the *car* gene cluster for antibiotic synthesis. CarR is not required for the transcription of the exoenzyme genes, implying the existence of additional response regulators that interact with 3-oxo-C6 HSL. To date, no LuxR homologue has been identified in *Ecc* that regulates the transcription of the genes for the exoenzymes. Interestingly, there is a gene downstream of *carI*, called *expR* (or *eccR*), which is a homologue of *luxR*. However, inactivation of this gene does not appear to have a significant effect on exoenzyme synthesis. ExpR is probably involved in the regulation of other genes dependent upon 3-oxo-C6 HSL. See ref. 378 for a more complete discussion of this point.

18.21.4 Vibrio cholerae

As discussed in ref. 361, quorum sensing regulates biofilm formation as well as the expression of virulence genes in *V. cholerae*. Two of the three quorum-sensing systems in *V. cholerae* (systems 1 and 2) are analogous to the two systems described for *V. harveyi* and converge at LuxU (Fig. 18.37). (Less is known about system 3.)

However, the result of quorum sensing is somewhat unexpected. Instead of stimulating biofilm formation and virulence gene expression, quorum sensing in *V. cholerae* inhibits these activities. That is, biofilm formation and virulence gene expression are higher at lower cell densities. The key here is a transcriptional regulator called HapR, which is homologous to the *V. fischeri* LitR and the *V. harveyi* LuxR, two transcriptional activators of the *lux* operon discussed earlier. HapR represses the expression of virulence genes (*toxT*, *tcpPH*) and genes required for the synthesis of exopolysaccharide, an activity that is important for the maturation of biofilms. (For a description of virulence genes in *V. cholerae*, including *toxT* and *tcpPH*, see Section 18.11.1; for a description of the role that exopolysaccharide plays in biofilm formation, see Section 18.22.4.) LuxO-P represses the transcription of *hapR*. Since LuxO-P is present at low cell densities (in the absence of sufficient quorum signal), virulence gene expression and biofilm formation are stimulated at low cell densities.

One can ask why it is an advantage for *V. cholerae* to lower the expression of its virulence genes and its ability to form biofilms at high cell density. After all, the bacteria are at a high cell density when they colonize the human intestine. In other pathogens that have been studied, both virulence gene expression and biofilm formation are greater at high cell densities. Why is *V. cholerae* different in these respects?

A rationale has been proposed related to the mode of transmission of cholera.[361] The suggestion is that when *V. cholerae* first enters the intestine from contaminated water, it begins to form a biofilm on the intestinal epithelium and to secrete toxins that produce diarrhea. The inhibition of both biofilm formation and production of toxins at high cell densities in the intestine presumably aids in the exit of the bacteria from the intestine during diarrhea as part of the transmission to other hosts, and helps to prepare the bacteria for life in the marine waters where the toxins are not necessary.

18.22 Biofilms

For reviews of the subject of biofilms, see refs. 382 through 388. A very general definition of a biofilm is that it is a community of microorganisms immobilized and living on a solid surface exposed to air or liquid. It is now recognized that in their natural environment, most bacteria live this way in sessile, complex communities of interacting populations rather than as single cells suspended in liquid. Bacteria that grow suspended in liquid are referred to as planktonic bacteria, whereas those living in biofilms are called biofilm bacteria. There are many adaptive and developmental changes that occur when bacteria switch from growing in liquid to growing in a biofilm. The surfaces may be abiotic or biotic, and these types are described next.

18.22.1 Abiotic surfaces

Examples of abiotic surfaces include oil or water pipes, various containers that store liquids, and soil particles. Biofilms living on abiotic surfaces can cause problems. For example, these biofilms might corrode the inner surface of the pipes and containers. Other abiotic surfaces on which bacterial biofilms can be found are medical implants such as artificial heart valves, urinary catheters, artificial hearts, joint replacements, and contact lenses. Such biofilms are an important cause of nosocomial (hospital-acquired) infections.

There are also positive aspects of biofilms. Biofilms in sewage digestor vessels and pipes in sewage treatment systems serve a necessary role in degrading the waste material. Biofilms on rocks in lakes and streams are home to bacteria, cyanobacteria, and protists of many types.

18.22.2 Biotic surfaces

Biotic surfaces of animals naturally inhabited by biofilms include teeth (dental plaque), the epithelial lining of body tracts such as the mouth, the intestinal tract, including the rumen wall of cattle, and the distal urogenital tract, as well as the skin. There are also biofilms living on plant root surfaces and soil particles within a few millimeters of the plant roots (the rhizosphere).

The naturally occurring biofilms on biotic surfaces constitute a natural microbial flora and can be of benefit to the animal. For example, the natural flora can protect against infection by pathogens by competing for body surface and nutrients. The natural flora can also benefit the host in nutritional ways. For example, the bacterial flora in the rumen of the cow, such as species of *Bacteroides* and *Ruminococcus*, is responsible for cellulose degradation.

18.22.3 Advantages of living in a biofilm

Since many bacteria in nature live in biofilms, there must be associated ecological advantages. Indeed there are, and they include the availability of nutrients such as those resulting from being part of food webs, and protection from environmental toxins.

Nutrient availability

The supply of nutrients is often greater at the site of the biofilm. For example, bacteria in biofilms on plant roots and the soil particles surrounding the plant roots grow on the organic matter (amino acids, sugars, organic acids) that leak out of the roots (a process called exudation) as they grow in the soil. Roots also actively secrete carbohydrates. Bacteria living on animal tissues may use nutrients produced by the tissues, such as occur in sebum, saliva, and mucus, or nutrients eaten by the animal, such as those found in the mouth or intestinal tract. It has also been pointed out that organic matter in water may be adsorbed to surfaces such as rocks, where it will be more concentrated.

Food webs

A food web is a complex of interrelated food chains in a community of organisms of different types (see Section 18.23.4 for a discussion of the organization of biofilm communities). That is, the excreted end products of metabolism of one or more groups of organisms are used as nutrients by one or more other groups of organisms in the food web. This arrangement clearly benefits the organisms that receive the nutrients. However, as we shall see, often the survival of a producer organism is dependent upon the utilization of its products of metabolism by other organisms. Some biofilms have an aerobic portion near the surface and anaerobic portions at greater depths. Other biofilms are completely aerobic or anaerobic.

1. Anaerobic biofilms and sludge granules

The student should review the description of the anaerobic food chain in Section 14.4 and examine Fig. 14.2.

As a specific example of a complex anaerobic community existing in a biofilm, let us consider anaerobic *sludge granules* (also called *flocs*) that exist in the sludge bed reactors of anaerobic sewage treatment plants. These are reactors in which sewage continuously flows from bottom to top and is decomposed anaerobically by the microorganisms living in the sludge granules. The complex bacterial community within each sludge granule, which consists of fermenting bacteria, obligate proton reducers, methanogens, and sulfate-reducing bacteria (SRB), is responsible for an anaerobic food chain, as discussed in Section 14.4, that degrades organic compounds to methane and carbon dioxide. We will now consider the metabolic interdependency between these four groups of bacteria living in anaerobic biofilms.

Fermenting bacteria. These bacteria metabolize organic matter to organic acids, alcohols, hydrogen gas, and carbon dioxide that are released from the cells as end products of metabolism (waste products). Other bacteria rely on these fermentation end products as sources of carbon and energy.

Obligate proton-reducing acetogens. These bacteria, described in Section 14.4.1, use the fermentation end products as a source of carbon and energy and oxidize these compounds to acetate as a waste product. The electrons are delivered to protons, producing hydrogen gas, which serves as the electron sink.

Methanogenic bacteria. These bacteria grow on the acetate, hydrogen gas, and carbon dioxide produced by the fermenters and the obligate proton-reducing acetogens. Methane is the main waste product.

Sulfate-reducing bacteria. The SRB also grow on the acetate, hydrogen gas, and carbon dioxide, as well as some of the other fermentation end products. In fact, if it were not for the methanogens and the sulfate-reducers, the obligate proton reducers would not survive. This is because the obligate proton reducers cannot use protons as terminal electron acceptors if the concentration of hydrogen gas is allowed to rise. (See Section 14.4.1 for a discussion of interspecies hydrogen transfer.) Thus, there is an interdependent relationship between several communities of bacteria living in complex biofilms, and they depend upon one another for their survival.

2. Aerobic biofilms
Bacterial communities in aerobic biofilms can also be part of a food web. Many different types are present. Two of these are nitrifying bacteria and sulfide oxidizing bacteria.

For a review of nitrifying bacteria and sulfide-oxidizing bacteria, subject, see Section 12.4 on lithotrophy. In these biofilms ammonia-oxidizing bacteria produce nitrite, which is used by nearby communities of nitrite-oxidizing bacteria, which in turn depend upon the nitrite as a source of electrons and energy. The nitrifying bacteria utilize oxygen as an electron acceptor, and in this way can contribute toward an anaerobic environment in deeper levels of the biofilm, a contribution that is important for the survival of anaerobic bacteria. Similarly, oxygen-dependent sulfide oxidation can result in less oxygen in deeper levels of the biofilm.

Protection from chemicals and antibiotics
Existence in a biofilm supports survival because biofilm communities are much more resistant to chemicals such as chlorine, detergents, and antibiotics than are planktonic communities of similar bacteria. This topic is reviewed in ref. 389. There is no single explanation for the resistance. Possible factors that contribute to the resistance include decreased diffusion of the chemicals within the biofilm, binding of the chemicals to the polymers in the biofilm matrix, and decreased bacterial growth rate, especially in the interior of the biofilm. (Actively growing cells are more sensitive to many antibiotics.)

18.22.4 Biofilms are organized communities

The bacteria in biofilms are not randomly distributed. Rather, biofilms can consist of discrete, multilayered patches or clusters of bacteria, called microcolonies. (See Fig. 18.38.) Microcolonies can consist of a single type of bacteria, or there may be bacteria of several different types that interact with each other in a cooperative manner. (See the preceding discussion of food webs.)

Microcolonies vary in structure depending upon the conditions under which the biofilm is formed. They can be rough, round-cell aggregates, flat, elongated, mushroomlike, or pillarlike. Some of this morphology depends upon hydrodynamic shear when the biofilms form, as well as nutritional signals. The cells within the microcolonies are embedded in an extracellular matrix called extracellular polymeric substance (EPS). It consists of extracellular polysaccharide, protein, and other substances produced by the bacteria. Much of the EPS is arranged in the form of fine fibers that can be visualized by using light and electron microscopy. The extracellular polysaccharide is the best characterized of the matrix components. The cells within the matrix in the microcolonies are immobilized and attached to one another via the polymers in the matrix. The matrix material is structurally very important and can comprise between 50 and 90% of the total organic matter in the biofilm.

Interspersed between the microcolonies are open areas with much less matrix. These open areas are important because they may serve as water channels through which oxygenated water and nutrients reach the microcolonies;

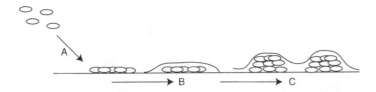

Fig. 18.38 General model for the formation of biofilms. (A) Planktonic bacteria attach to the surface. For bacteria with flagella, these appendages are important for bringing the bacteria to the surface where they may attach, and also for the spreading of the bacteria across the surface as a monolayer. For certain bacteria, such as *P. aeruginosa*, the movement of the attached bacteria from the monolayer into microcolonies (shown in C) requires a form of motility different from swimming. This type of motility, called twitching, is mediated by type IV pili. Attachment to the surface is very important at an early stage, and for some bacteria pili of certain types are used for attachment. For example, type I pili are absolutely necessary for *E. coli* to attach. Other external molecules, such as lipopolysaccharide O-antigen, teichoic acid, and exopolysaccharides can also be important for attachment to the surface. (B) Once irreversibly attached to the surface, the bacteria synthesize extracellular polymeric substances consisting of exopolysaccharide and protein. Water channels permeate the matrix. (C) As a consequence of motility, and/or cell division, the attached bacteria form three-dimensional aggregates of cells called microcolonies, whose structure varies according to the conditions under which the biofilm forms. The microcolonies can be flat, mounded, elongated, mushroomlike, or pillarlike.

in addition, they can bring fluid and organic material to deeper regions of the biofilm to other microcolonies. The open areas also serve as channels through which metabolic waste products can leave. Thus, the survival of the biofilm depends upon channels between the microcolonies. It is important to realize that the water channels are between, the microcolonies not within them. Thus water may percolate around, but not through, the microcolonies.

18.22.5 The formation of biofilms

How biofilm formation is studied

The formation of biofilms in laboratory settings has been well studied in recent years. Biofilm development can be viewed in 96-well plastic microtiter dishes to which the bacteria can attach, or by using continuous-flow cells attached to a microscope stage. The student is referred to ref. 390 for a description of the use of microtiter dishes and to ref. 391 for a description of flow cells for viewing biofilms.

The microtiter dish assay has been used to quantify the initial attachment in biofilm formation. For quantitative analysis, the planktonic cells are removed and the attached cells are stained. The continuous-flow cells can be examined by using confocal laser scanning microscopy or fluorescent reporter genes to monitor the formation of microcolonies and

the maturation of biofilms. (See note 392 for an explanation of confocal laser scanning microscopy.) The specific events that occur during biofilm formation vary with the organism as well as with the conditions under which the biofilm forms. For example, biofilms formed by *P. aeruginosa* under flow-through conditions are heterogeneous with mushroom-shaped microcolonies when glucose is the carbon source; but they are flat, uniform, and densely packed when citrate is the carbon source.

Stages

Figure 18.38 presents a general stagewise description of biofilm formation, consisting of (1) initial reversible attachment of planktonic cells to the surface, (2) formation of a monolayer, which includes irreversible attachment of the cells, spreading over the surface, and secretion of extracellular polymeric substances (EPS), and (3) maturation of the biofilm, which includes thickening and the formation of microcolonies. In addition, cells can become dispersed from mature biofilms to colonize new areas.

1. Initial reversible attachment; the role of flagella and pili

Flagella serve an important role, albeit an indirect role, in the initial reversible attachment for bacteria that swim. The reason for believing this is that *P. aeruginosa* and *E. coli* mutants unable to form functional flagella do not attach as well as the wild type. Why is this the case? Presumably, the flagella-mediated motility brings bacteria to the surface, where they can attach. However, nonflagellated bacteria also form biofilms; for example, streptococci naturally form biofilms on teeth.

Pili can also be important for initial attachment. See the discussion of pili in Section 1.2.1. *E. coli* mutants that do not form type I pili are defective in initial attachment, reflecting the importance of these pili for the attachment of *E. coli* under certain conditions of biofilm formation.

2. Formation of a monolayer

Type IV pili can play an important role for bacteria that possess them, since they allow the organisms to spread over the surface to form a monolayer. Type IV pili are required for twitching motility, a form of gliding motility on surfaces. See Section 1.2.1 and note 393 for a description of twitching motility. See ref. 394 for a discussion of the role of type IV pili in the maturation of *P. aeruginosa* biofilms.

According to the proposed model, after *P. aeruginosa* attaches to the surface, a motile subpopulation of cells spreads via type IV pili twitching motility to cover the surface; eventually, during maturation, a subpopulation of nonmotile cells that has down-regulated twitching motility forms stalklike microcolonies by clonal growth. When the motile cells reach microcolony stalks and migrate up them, mushroomlike caps form on top of the stalks. See note 395 for a description of how the experiment that supports the model was performed.

3. Maturation

The maturation of biofilms is often characterized by the cessation of motility, the thickening of the biofilm, the formation of cell clusters (microcolonies) in which the cells are held together by the EPS they produce, and the formation of water channels between the microcolonies. Some microcolonies are mounds, but they can be flat, elongated, mushroomlike, or pillarlike, depending upon the conditions under which the biofilm forms. Cell division can be important for the growth of microcolonies, especially in the early stages. In addition, for some bacteria, such as *Vibrio cholerae*

and *E. coli*, flagella are important because they promote swimming of the bacteria into microcolonies after attachment to the surface. (To designate the larger mushroomlike or pillarlike colonies that develop from the microcolonies, and are separated by water-filled channels, some researchers use the form "macrocolonies".)

4. Dispersal

Dispersal from mature biofilms involves detachment of the cells from each other and from the EPS. For example, it has been observed with *P. aeruginosa* that some cells in the microcolonies become motile, detach from the biofilm, swim out into the water channels, and colonize new surfaces. The cells left behind in the microcolonies remain nonmotile. For other examples of dispersal, see ref. 396. Environmental factors, such as depletion of carbon and nitrogen sources for several hours, can enhance detachment from biofilms and dispersal. In fact, long periods of starvation can result in the dissolution of biofilms, as pointed out in ref. 394.

18.22.6 Adaptive changes in biofilms: Gene expression

When bacteria form biofilms, the expression of many genes changes in response to the switch from planktonic existence to a sessile life. For example, *E. coli* and *P. aeruginosa* decrease the production of flagellin genes and increase the expression of genes required for the synthesis of extracellular exopolysaccharides that are part of the matrix in the biofilm. These are colanic acid (also called the M antigen) in the case of *E. coli*, and alginate in the case of *P. aeruginosa*. Genes required for the attachment to specific substrates may be induced. This has been reported for marine bacteria (*Vibrio harveyi*) that attach to chitin, a polymer of N-acetylglucosamine that is present in the exoskeleton of crustaceans.[397] Gene expression in response to low oxygen levels, as well as low inorganic phosphate levels, has been shown to influence the maturation of *Agrobacterium tumefaciens* biofilms. (See note 398.) When the proteins of *P. aeruginosa* were separated by using two-dimensional gel electrophoresis and then analyzed, it was found that the bacteria increased the production of several hundred proteins grown as a biofilm rather than in suspension.[399] Presumably, these proteins play important roles in adapting to the biofilm way of life. Similarly, when grown as a biofilm, *E. coli* induces the expression of many genes.[400] Gene expression is regulated, at least in part, by intercellular signals called quorum signals.

18.22.7 Quorum sensing in biofilms

Bacteria in biofilms signal each other in order to regulate gene expression. The process is called quorum sensing, and it is an important area of current research. (The student should review quorum sensing by bacteria in Section 18.15.1.) It was reported that the production of normal biofilms by *Pseudomonas aeruginosa* on the glass surface of a reaction chamber requires the production of extracellular N-acyl–homoserine lactones (acyl–HSLs), which are quorum signals that allow the cells to communicate with one another for the purpose of coordinating gene expression in dense populations of cells.[401,402] Under certain growth conditions, *lasI* mutants (defective in the production of the quorum signal 3-oxo-C12-HSL) formed structurally abnormal biofilms only 20% the thickness of wild type. These mutant biofilms consisted of continuous sheets of cells much more densely packed than the characteristic well-spaced microcolonies separated by water channels seen in wild-type biofilms. Furthermore, the bacteria in the mutant biofilms were dislodged and dispersed from the glass surface by 0.2% sodium dodecyl sulfate (SDS), whereas wild-type biofilms were resistant to this detergent. Addition of 3-oxo-C12-HSL restored the wild-type phenotype, including resistance to SDS.

It has been reported that the Las quorum-sensing mutants formed amounts of extracellular polysaccharide similar to those noted in the wild type; however, staining with alcian blue, which stains the extracellular polysaccharide (it binds to acidic polysaccharides), revealed that the mutant and wild type differed with respect to how the extracellular polysaccharide was situated in the biofilm.[397] Specifically, whereas the wild type formed extracellular polysaccharide located in the interstices be-

tween the bacteria, so that the spaces between the bacteria were filled with extracellular polysaccharide, the mutants made extracellular polysaccharide that was associated only with the bacterial cell surface, and the spaces between the closely packed bacteria were not filled with extracellular polysaccharide.

Another example of quorum sensing that has been shown to be important for biofilm formation is found in the *Salmonella enterica* serovas Typhimurium.[403] Biofilm formation on gallstones was studied, and examination of mutants revealed that LuxS, the enzyme that forms the quorum signal AI-2, is required for biofilm formation. Recall that the LuxS/AI-2 system was originally discovered in *Vibrio harveyi* as part of the system that regulates bioluminescence and that in this system, AI-2 is a furanosyl borate diester. It is now known that the LuxS system is present in a wide range of nonluminescent bacteria, both gram positive and gram negative. However, evidence exists that bacteria living in nonmarine environments, where borate is much less plentiful, may make and respond to AI-2 that does not contain boron.[404] However, AI-2 without boron is active in a bioassay with *V. harveyi* in borate-containing medium because the borate adds spontaneously to the molecule. (AI-2 can be detected via bioassay. See note 405.)

Quorum sensing has also been reported to be important for the formation of biofilms by the gram-positive bacteria *Streptococcus gordonii* and *S. mutans*, two of the bacteria found in dental plaques. *S. mutans* produces the quorum sensor called AI-2 by using the LuxS synthase, which as stated earlier was originally discovered in the luminescent bacterium *Vibrio harveyi*. Mutants that fail to produce AI-2 form less biofilm, and the biofilm that forms differs morphologically from the wild type in being less smooth and less confluent.[406]

Oligopeptide signals are also important for biofilm formation by *Streptococcus*.[407,408] Mutants of *S. mutans* unable to produce or secrete the quorum-signaling peptide called competence-stimulating peptide (CSP) not only were defective in competence but also formed abnormal biofilms with a three-dimensional structure different from that of wild type. (For a review of how CSP acts in competence development, see Section 18.19 and note 409.) The mutant biofilms had a spongelike architecture (weblike microcolonies), and the cells were in very long chains compared with wild type. It was suggested that the inability of the cells to separate affected the architecture of the biofilm. Thus, CSP is involved in cell density phenotypes in addition to competence, including biofilm formation.

Thus, there exists significant evidence that an important aspect of biofilm formation is the cell-to-cell signaling that takes place within bacterial populations. Such signaling can promote the successful maturation of the biofilm by influencing gene expression; it also emphasizes that biofilms are truly multicellular communities in which individual microorganisms are dependent upon one another for then survival.

18.23 Summary

Bacteria can sense environmental and cytoplasmic signals and transmit information to the genome or to other parts of the cell to elicit a response. Such signaling systems detect and transmit signals when there are changes in conditions such as pH, osmolarity, temperature, starvation, and nutrient source, as well as the absence or presence of specific inorganic ions, such as phosphate. The signaling systems also detect signaling molecules released by other cells and aid in cooperative cell behavior. Often global transcription regulators are involved. These are transcription regulators that control transcription in noncontiguous operons. A set of noncontiguous operons or genes regulated by the same transcription regulator is called a regulon.

Many of the signaling systems are called "two-component" signaling systems because they have a histidine kinase (component 1) that causes the phosphorylation (transphosphorylation) of a response regulator (component 2). However, other proteins may also be involved in signal transduction. (The histidine kinase is also called a histidine/sensor kinase.)

All the "two-component" systems have four functional components: (1) a sensor that receives the signal, (2) an autophosphorylating histidine protein kinase, (3) a partner response regulator that is the substrate for the histidine

kinase, and (4) a phosphatase that removes the phosphate from the regulator. As discussed shortly, some of the proteins can be bifunctional and carry out two of the activities. During signaling, phosphate (actually the phosphoryl group) travels from ATP to a histidine residue in the histidine kinase to an aspartate residue in the response regulator to water (the phosphatase reaction). The phosphorylated form of the response regulator is the active state and transmits the signal to the genome (stimulation or inhibition of transcription), to the flagellar motor (reversal of turning direction), or to an enzyme. Although many signals cause an increase in phosphorylation of the response regulator, some (e.g., chemoattractants, excess inorganic phosphate, excess ammonia) cause a decrease in phosphorylation (which in some cases is due to dephosphorylation of the response regulator) and therefore an inactivation of the response regulator.

All the histidine protein kinases have conserved domains (transmitter domains) that bear a startling similarity to one another and are the basis for the classification of these proteins. These conserved domains, lying in the carboxy terminus of the protein, are the site of the conserved histidine residue that is phosphorylated. The response regulators have conserved amino-terminal (receiver) domains that are believed to interact with the carboxy terminus of the kinase. For several of the response regulators, it has been shown that the phosphoryl group is transferred from the histidine to the conserved aspartate residue in the amino terminus of the response regulator.

Although the two-component systems are regulated primarily by the activities of their respective histidine kinases, it has been suggested that acetyl phosphate may be a phosphoryl donor to different response regulator proteins under certain physiological conditions. This has usually been observed in mutants that lack the cognate histidine kinase. In these mutants acetyl phosphate is able to donate the phosphoryl group to the response regulator. The extent to which this might occur in wild-type cells that have the cognate histidine kinase is generally not yet known. However, analysis of mutants in acetyl phosphate synthesis or utilization in which the levels of acetyl phosphate would be expected to be higher or lower than wild-type cells suggests that increased levels of acetyl phosphate can inhibit flagellar biosynthesis in *E. coli* wild-type cells by phosphorylating OmpR, which in its phosphorylated form is a negative transcription regulator of regulatory genes required for flagellar biosynthesis. If this sequence in fact occurs, it could account for why *E. coli* grown at higher temperatures have fewer flagella. The possibility is reasonable, since in cells grown at higher temperatures, the acetyl phosphate levels are elevated because the enzyme that converts acetyl phosphate to acetate (acetate kinase) is less active at the higher temperatures. Presumably this set of conditions leads to a phosphorylation of OmpR and a consequent repression of the flagellar regulatory genes. All these data suggest that acetyl phosphate may serve as a global metabolic signal that affects the activity of several different genes by phosphorylating response regulator proteins.

Responses that use a two-component regulatory system include the Che system in chemotaxis, the ArcA/ArcB system for oxygen regulation of gene expression, the NarL/X/Q system for nitrate regulation, the PHO system for phosphate assimilation, the Ntr system for nitrogen assimilation, the EnvZ system for porin gene expression, the KdpABC system for K⁺ uptake, the RegB/RegA system for the regulation of expression of the genes for the light-harvesting and photoreaction center proteins in *Rhodobacter capsulatus*, some bioluminesence signaling systems in *Vibrio*, and the dicarboxylic acid permease system in *R. meliloti*.

Other two-component signaling systems operate during sporulation, flagellar synthesis in *E. coli*, and cell-to-cell signaling in myxobacteria. There is also evidence that a two-component signaling system operates to regulate the initiation of DNA replication in *Caulobacter*. Two-component systems also detect environmental signals and regulate the transcription of virulence genes in certain bacteria (e.g., *Agrobacterium*, *Salmonella*, *Bordetella*).

There are several other important regulatory systems in bacterial pathogens that control the expression of virulence genes in response to environmental signals. These may involve a transmembrane sensor, which need not be part of a two-component signaling system, and which responds to environmental signals

such as temperature, osmolarity, and pH. Such systems include the ToxR/S system in *Vibrio cholerae*. *Shigella* spp. possess a regulatory gene called the *virR* gene that encodes a protein that is part of a system that regulates the temperature-dependent expression of virulence genes.

The FNR system, which regulates anaerobic gene expression, is not a two-component regulatory system. Neither is there evidence that the formate regulon, which is induced by formate under anaerobic conditions, is controlled by a two-component system. It is clear, moreover, that the Arc and the FNR systems, represent two systems of gene regulation with which *E. coli* senses anaerobiosis and activates the transcription regulators ArcA and FNR.

The situation is often complex, with multiple regulators controlling the same operon. In the control of the *cydAB* operon in *E. coli*, for example, the operon encodes cytochrome d oxidase and is regulated by FNR and ArcA, and interestingly by Cra as well. This was demonstrated by using *cyd–lacZ* fusion strains that have null mutations in either *cra*, *arcA*, or *fnr* genes.[410] (See note 411.)

Catabolite repression refers to the repression by particular carbon sources of expression of genes required for growth on other carbon sources. The result is diauxic growth. For example, in many bacteria glucose represses genes required for the uptake and/or catabolism of other carbon sources. Some bacteria utilize carboxylic acids in preference to glucose, and it is the carboxylic acids that repress the genes required for glucose catabolism.

The mechanisms that underlie catabolite repression vary. These include the cAMP–CRP-dependent system for glucose repression that has been well studied in *E. coli*, This system involves the PTS sugar uptake system. *E. coli* also has a cAMP–CRP-independent system, the Cra system. Cra is a global regulator that influences the direction of carbon flow in either the glycolytic or gluconeogenic direction by controlling the expression of genes encoding enzymes required for the relevant pathways. Cra is an activator for genes required for gluconeogenesis and a repressor for genes required for glucose catabolism. When the cells are growing on glucose or fructose, it is proposed that the levels of fructose-1-phosphate and fructose-1,6-bisphosphate rise and bind to Cra, removing it from the DNA. Thus, gluconeogenic genes required for growth on pyruvate, lactate, acetate, and citric acid cycle intermediates would be repressed and glycolytic genes required for growth on carbohydrates would be activated.

Catabolite repression in low-GC gram-positive bacteria involves yet a different system, the CcpA system. When these bacteria are growing on glucose, the levels of the glycolytic intermediates fructose-1,6-bisphosphate and 2-phosphoglycerate rise and activate a special kinase that phosphorylates Hpr, forming Hpr(Ser-P). Phosphorylated Hpr may form a ternary complex with the CcpA DNA-binding protein and fructose-1,6-bisphosphate, and this complex can bind to the CRE sequence in target genes, repressing transcription.

Many gram-negative bacteria have been demonstrated to possess a signaling system based upon acylated homoserine lactones (acyl–HSLs) that serve as cell density signals. The acylated homoserine lactone systems were first studied in luminous bacteria that luminesce only when the population of cells reaches a sufficient cell density in late logarithmic growth, and in the light organs of fish and squid hosts. They are thus called quorum-sensing systems. The system in V. fischeri includes an acyl–HSL, a positive transcription regulatory factor (LuxR) activated by the acyl–HSL, and an operon containing luxI, the gene for synthesis of the acyl–HSL.

V. fischeri has a second acyl–HSL system that is also important, called the ain system. The ain system works differently in that it derepresses the transcription of the lux genes by inactivating a component of a phosphorelay signaling system. V. harveyi has two signaling systems that regulate luminescence, one of which is similar to the ain system in V. fischeri and relies on an acyl–HSL. The second system relies on a furanosyl borate diester as a signaling molecule, and it, along with the ain-like system, derepresses the lux genes. Neither system in V. harveyi has components that are homologous to luxI or luxR in V. fischeri.

Other quorum-sensing systems based upon acyl–HSLs include signaling systems that stimulate conjugation in Agrobacterium, nodulation and conjugation in Rhizobium, and systems that regulate the transcription of virulence genes in Pseudomonas and Erwinia.

Bacterial development is a large and important area of research in which cell-to-cell signaling as well as intracellular signaling regulate changes in cell physiology, cell morphogenesis, cell division, and multicellular development. Examples discussed in this chapter are myxobacteria fruiting body formation, the Caulobacter cell cycle, Bacillus sporulation, and Bacillus competence.

The formation of biofilms by bacteria is another example of bacterial development. Thus widespread phenomenon is recognized as an important example of adaptive and developmental response to environmental challenges faced by bacteria in their natural habitat. Cooperative cell behavior and cell-to-cell signaling, including signaling via quorum sensors, are important characteristics of biofilm formation.

Study Questions

1. Describe the components of two-component regulatory systems and how they work. In your answer, explain what is meant by "cross talk" or "cross-regulation."

2. What criteria must be established to characterize a protein as a kinase or response regulator protein in a two-component system?

3. What is FNR, and what is its role? Why is it not believed to be part of a two-component regulatory system?

4. Describe the signaling pathway for chemotaxis in E. coli.

5. What causes E. coli to swim randomly? Is this the mechanism for all bacteria? Explain the differences. How is random swimming related to chemotaxis?

6. What is the relationship between adaptation and methylation of MCPs?

7. In the Arc and Nar systems, which proteins are thought to be the sensor/kinase proteins and which ones the regulator proteins? What is the evidence for this?

8. What is the phenotype of an Fnr⁻ mutant in E. coli? An arc⁻ mutant?

9. Describe the role that P_{II} plays in transcriptional regulation of the *glnALG* operon and in the regulation of activity of glutamine synthetase. How are the levels of P_{II} and P_{II}–UMP regulated?

10. In what way is catabolite repression by the Cra system similar to the same mechanism in the cAMP system? How does it differ from catabolite repression by the CcpA system in *B. subtilis*?

11. How does *B. subtilis* partition Hpr-P between the phosphotransferase system and the CcpA catabolite repression system?

12. What is the evidence that acetyl phosphate may be a global signal?

13. Assume that acetyl phosphate phosphorylates OmpR, which then represses the flagellar genes. Explain why the addition of acetate to either wild-type cells or to *pta⁻* cells represses flagellar gene expression, but no such repression occurs when acetate is added to *ackA⁻* cells.

14. Discuss the role that a two-component phosphorelay system might play in the production of virulence factors by pathogenic bacteria. Give an example of such a two-component system. Describe some experiments to test which environmental factors might provide the necessary signals.

15. The chemoattractant phosphatidylethanolamine (PE) is a lipid that is part of the cell envelope of *M. xanthus* and diffuses very poorly in aqueous environments. What are some possible ways by which recipient cells might encounter this molecule? Hint: read ref. 283.

16. What are the differences between the bioluminescent quorum-sensing systems in *V. fischeri* and *V. harveyi*? Are they regulated in the same way? Is a phosphorelay involved? Are non-HSLs involved?

17. Discuss the role that acyl–HSLs play in the luminescent bacteria as well as in specific nonluminescent bacteria. Discuss an experimental physiological/biochemical and genetic approach to investigate whether the systems in two different bacteria have interchangeable components.

REFERENCES AND NOTES

1. Hoch, J. A., and T. J. Silhavy (Eds.). 1995. *Two-Component Signal Transduction*. ASM Press, Washington, DC.

2. The transmitter domain of the sensor kinase, which is at the carboxyl terminus of the protein, is generally attached to an *input domain* (e.g., a region that might bind a signaling molecule, such as a (ligand-binding site). Thus, the sensor kinase would be N–input domain–transmitter domain–C. The input domain would be expected to be exposed to the signal; for example, it might be in the periplasm if the sensor kinase were a transmembrane protein in a gram-negative bacterium. Presumably, after binding the signal, the sensor kinase undergoes a conformational change that results in its autophosphorylation in the transmitter domain and subsequent phosphorylation of the response regulator protein in its receiver domain. As expected, input domains vary, depending upon the sensor kinase. The receiver domain of the response regulator protein is usually attached to an *output domain* at the amino terminus such as a DNA-binding region. Thus, the response regulator would be N–receiver domain–output domain–C. The sequences of the DNA-binding domains are divided into classes that place the response regulator proteins into families (e.g., the OmpR family). Not all response regulator proteins have output domains. For example, CheY and Spo0F do not.

3. Hughes, D. A. 1994. Histidine kinases hog the limelight. *Nature* 369:187–188.

4. Wurgler-Murphy, S. M., and H. Saito. 1997. Two-component signal transducers and MAPK cascades. *Trends Biochem. Sci.* 22:172–176.

5. Saier, M. H. Jr. 1993. Introduction: protein phosphorylation and signal transduction in bacteria. *J. Cell. Biochem.* 51:1–6.

6. Stock, J. B., A. J. Ninfa, and A. M. Stock. 1989. Protein phosphorylation and regulation of adaptive responses in bacteria. *Microbiol. Rev.* 53:450–490.

7. Parkinson, J. S., and E. C. Kofoid. 1992. Communication modules in bacterial signaling proteins. *Annu. Rev. Genet.* 26:71–112.

8. Ota, I. M., and A. Varshavsky. 1993. A yeast protein similar to bacterial two-component regulators. *Science* 262:566–569.

9. Chang, C., S. F. Kwok, A. B. Bleecker, and E. M. Meyerwitz. 1993. *Arabidopsis* ethylene-response gene *etr1*: similarity of product to two-component regulators. *Science* 262:539–544.

10. Wanner, B. L. 1992. Is cross regulation by phosphorylation of two-component response regulator proteins important in bacteria? *J. Bacteriol.* 174:2053–2058.

11. Reviewed in *Antonie van Leeuwenhoek*. 1994. Vol. 66, Nos. 1–3.

12. Spiro, S., and J. R. Guest. 1991. Adaptive responses to oxygen limitation in *Escherichia coli*. *Trends Biochem. Sci.* **16**:310–314.

13. Gunsalus, R. P., and S.-J. Park. 1994. Aerobic–anaerobic gene regulation in *Escherichia coli*: control by the ArcAB and FNR regulons. *Res. Microbiol.* **145**:437–449.

14. Spangler, W. J., and C. M. Gilmour. 1966. Biochemistry of nitrate respiration in *Pseudomonas stutzeri*. Aerobic and nitrate respiration routes of carbohydrate catabolism. *J. Bacteriol.* **91**:245–250.

15. Reviewed in: Berks, B. C., S. J. Ferguson, J. W. B. Moir, and D. J. Richardson. 1995. Enzymes and associated electron transport systems that catalyze the respiratory reduction of nitrogen oxides and oxyanions. *Biochim. Biophys. Acta* **1232**:97–173.

16. Gunsalus, R. P. 1992. Control of electron flow in *Escherichia coli*: coordinated transcription of respiratory pathway genes. *J. Bacteriol.* **174**:7069–7074.

17. Sawers, G., and B. Suppman. 1992. Anaerobic induction of pyruvate formate–lyase gene expression is mediated by the ArcA and FNR proteins. *J. Bacteriol.* **174**:3474–3478.

18. Sawers, G. 1993. Specific transcriptional requirements for positive regulation of the anaerobically inducible *pfl* operon by ArcA and FNR. *Mol. Microbiol.* **10**:737–747.

19. Anderson, D. I. 1992. Involvement of the Arc system in redox regulation of the *cob* operon in *Salmonella typhimurium*. *Mol. Microbiol.* **6**:1491–1494.

20. Mutations in *arcA* produce a defect in the synthesis of F pili, which results in resistance to male-specific bacteriophages that attach to the F pilus. (See Fig. 18.5.) ArcA also responds to a second sensor kinase protein (CpxA), which is necessary for the production of the F pilus in donor strains of *E. coli*. The *cpxA* gene was originally discovered as a mutation that reduced the efficiency of DNA transfer as a consequence of reduced F-plasmid *tra* gene expression. It is now known that the CpxA protein is an inner membrane protein whose amino acid sequence places it in the class of sensor kinase proteins. There also exists a cognate response regulator (CpxR) for CpxA, shown in Fig. 18.5. The complex relationship between these two regulatory systems is discussed in the following papers: Iuchi, S., D. Furlong, and E. C. C. Lin. 1989. Differentiation of *arcA*, *arcB*, and *cpxA* mutant phenotypes of *Escherichia coli* by sex pilus formation and enzyme regulation. *J. Bacteriol.* **171**:2889–2893. Dong, J.-M., S. Iuchi, H.-S. Kwan, Z. Lu, and E. C. C. Lin. 1993. The deduced amino acid sequence of the cloned *cpxR* gene suggests the protein is the cognate regulator for the membrane sensor, CpxA, in a two-component signal transductional system of *Escherichia coli*. *Gene* **136**:227–230.

21. Iuchi, S., and E. C. C. Lin. 1992. Purification and phosphorylation of the Arc regulatory components of *Escherichia coli*. *J. Bacteriol.* **174**:5617–5623.

22. Georgellis, D., A. S. Lynch, and E. C. C. Lin. 1997. In vitro phosphorylation study of the Arc two-component signal transduction system of *Escherichia coli*. *J. Bacteriol.* **179**:5429–5435.

23. With regard to the genes for succinic dehydrogenase (*sdh*CDAB), it should be pointed out that the aerobic/anaerobic regulation of transcription of this operon involves more than simply the ArcA/ArcB system. Not only are the levels of succinic dehydrogenase suppressed during anaerobic growth, but they are also lowered in *E. coli* by growth on glucose, even when the cells are grown aerobically and raised by growth on acetate. Furthermore, mutations in *FNR* increase the levels of transcription of an *sdhC–lacZ* fusion, suggesting that FNR represses the succinate dehydrogenase operon under anaerobic conditions. The mechanism by which glucose lowers the expression of the *sdh* operon is not understood. It appears to be independent of the cAMP–CAP system. See: Park, S.-J., C.-P. Tseng, and R. P. Gunsalus. 1995. Regulation of succinate dehydrogenase (*sdhCDAB*) operon expression in *Escherichia coli* in response to carbon supply and anaerobiosis: role of ArcA and FNR. *Mol. Microbiol.* **15**:473–482.

24. Iuchi, S., V. Chepuri, H.-A. Fu, R. B. Gennis, and E. C. C. Lin. 1990. Requirement for terminal cytochromes in generation of the aerobic signal for the *arc* regulatory system in *Escherichia coli*: study utilizing deletions and *lac* fusions of *cyo* and *cyd*. *J. Bacteriol.* **172**:6020–6025.

25. Gene fusions are valuable probes to monitor the expression of genes of interest. Genes are fused to reporter genes whose products are easy to identify. For example, a reporter gene might be *lacZ* (β-galactosidase), *lux* (luciferase), *cat* (chloramphenicol acetyltransferase), or *phoA* (alkaline phosphatase). Consider a *lacZ* fusion. The fused gene has the promoter region of the target gene but not the promoter for the *lacZ* gene. Expression of the fused gene is therefore under control of the promoter region of the target gene. If the translational initiating region (ribosome-binding site) is not present in the reporter gene, a translational fusion results. The fusions produce a hybrid protein whose amino-terminal end is derived from the target gene and its carboxy-terminal end from β-galactosidase. The hybrid protein has β-galactosidase activity. Therefore, an assay for β-galactosidase is a measure of the expression of the target gene. Thus, one can measure the expression of virtually any gene simply by constructing the proper gene fusion and performing an assay for β-galactosidase. One can construct gene fusions in vitro or in vivo. In vitro construction involves using restriction endonucleases to cut from a plasmid containing the cloned DNA a portion of the gene with

its promoter region and ligating it to a *lacZ* gene, without its promoter or ribosome-binding site, in a second plasmid. The plasmid containing the fused gene is then introduced into the bacterium and transformants are selected on the basis of their resistance to an antibiotic-resistant marker on the plasmid, and their production of β-galactosidase. In vivo construction of gene fusions can also be performed. In this case, one uses Tn5 transposons fused to a promoterless *lacZ* gene. The transposon is introduced into the bacterium, where it can recombine with the bacterial chromosome. Cells harboring the transposon are selected by means of the antibiotic-resistance marker on the transposon. Many of the strains have the transposon inserted into the host bacterial genes in the proper orientation and frame, so that β-galactosidase production is under the control of the promoter of the interrupted gene. Insertion of the transposon into a gene interrupts the gene so that the normal gene product is not made. The gene is identified by mutant analysis.

Cloned genes fused to the *E. coli* periplasmic alkaline phosphatase gene, *phoA*, are also used. These produce a hybrid protein consisting of the N-terminal region of the gene whose transcription is being analyzed and the C-terminal end of alkaline phosphatase. The hybrid protein is missing the N-terminal alkaline phosphatase signal sequence for protein export and relies on the protein export signal sequence of the target gene for export. When the C-terminal region of the hybrid protein containing the alkaline phosphatase region reaches the periplasm, the alkaline phosphatase activity can be measured. Such fusions can be used to examine the transcription of genes that encode secreted proteins such as virulence factors.

26. The control experiments were done to show that, in an *arc⁺ cyo⁺ cyd⁺* background, anaerobiosis repressed *cyo–lacZ* and induced *cyd–lacZ* expression, and that a mutation in the *arc* genes prevented the repression of *cyo–lacZ* and lowered the expression of *cyd–lacZ*. These experiments showed that the fusion genes are regulated by the availability of oxygen and that the Arc system is responsible for the regulation.

27. Rodriguez, C., O. Kwon, and D. Georgellis. 2004. Effect of D-lactate on the physiological activity of the ArcB sensor kinase in *Escherichia coli*. *J. Bacteriol.* **186**:2085–2090.

28. Discussed in: Iuchi, S., A. Artistarkhov, J. M. Dong, J. S. Taylor, and E. C. C. Lin. 1994. Effects of nitrate respiration on expression of the Arc-controlled operons encoding succinate dehydrogenase and flavin-linked L-lactate dehydrogenase. *J. Bacteriol.* **176**:1695–1701.

29. Lynch, A. S., and E. C. C. Lin. 1996. Responses to molecular oxygen, pp. 1526–1538. In: *Escherichia coli and Salmonella: Cellular and Molecular Biology*, Vol. 1. F. C. Neidhardt et al. (Eds.). ASM Press, Washington, DC.

30. Spiro, S., and J. R. Guest. 1990. FNR and its role in oxygen-regulated gene expression in *Escherichia coli*. *FEMS Microbiol. Rev.* **75**:399–428.

31. Unden, G., S. Becker, J. Bongaerts, G. Holighaus, J. Schirawski, and S. Six. 1995. O_2-sensing and O_2-dependent gene regulation in facultatively anaerobic bacteria. *Arch. Microbiol.* **164**:81–90.

32. Jones, H. M., and R. P. Gunsalus. 1987. Regulation of *Escherichia coli* fumarate reductase (*frd*ABCD) operon expression by respiratory electron acceptors and the *FNR* gene product. *J. Bacteriol.* **169**:3340–3349.

33. Unden, G., and J. R. Guest. 1985. Isolation and characterization of the FNR protein, the transcriptional regulator of anaerobic electron transport in *Escherichia coli*. *Eur. J. Biochem.* **146**:193–199.

34. It appears that inactivation of FNR is via the reversible oxidation of the [4Fe–4S]²⁺ clusters. The oxidation of the cluster results in the inactivation of the protein under conditions of prolonged incubation in oxygen by the disassembly and loss of the clusters from the protein. This leads to the release of sulfide (S^{2-}) and ferric ion (Fe^{3+}). Reassembly of the clusters from sulfide and Fe ions (Fe^{2+} and Fe^{3+}) has been demonstrated in vitro under anaerobic conditions and requires a special protein, the NifS protein from *Azotobacter*, which is required for the incorporation of Fe–S clusters into the *Azotobacter* nitrogenase. *E. coli* has a similar protein. See ref. 35.

35. Unden, G., and J. Schirawski. 1997. The oxygen-responsive transcriptional regulator FNR of *Escherichia coli*: the search for signals and reactions. *Mol. Microbiol.* **25**:205–210.

36. Roualt, T. A., and R. D. Klausner. 1996. Iron–sulfur clusters as biosensors of oxidants and iron. *Trends Biochem. Sci.* **21**:174–177.

37. Jordan, P. A., A. J. Thomson, E. T. Ralph, J. R. Guest, and J. Green. 1997. FNR is a direct oxygen sensor having a biphasic response curve. *FEBS Lett.* **416**:349–352.

38. Nakano, M. M., P. Zuber, P. Glaser, A. Danchin, and F. M. Hulett. 1996. Two-component regulatory proteins ResD–ResE are required for transcriptional activation of *fnr* upon oxygen limitation in *subtilis*. *J. Bacteriol.* **178**:3796–3802.

39. These proteins more closely resemble FNR than CRP on the basis of one or more of the following characteristics: DNA-binding specificity, sequence identity, failure to bind cAMP, complementation of an *E. coli fnr* mutant, or being activated under conditions of oxygen limitation. They are involved in the regulation of various physiological reactions such as luminescence (*Vibrio fischeri*), denitrification (*Pseudomonas aeruginosa*), nitrogen fixation (*Rhizobium* species), and hemolysin biosynthesis (*Bordetella pertussis*).

40. Spiro, S. 1994. The FNR family of transcriptional regulators. *Antonie van Leeuwenhoek* **66**:23–36.

41. Compan, I., and D. Touati. 1994. Anaerobic activation of *arcA* transcription in *Escherichia coli*: roles of Fnr and ArcA. *Mol. Microbiol.* **11**:955–964.

42. *E. coli* actually synthesizes three different formate dehydrogenases (FDH) and three distinct hydrogenases (Hyd). FDH-O is synthesized during aerobic growth and also anaerobically when nitrate is the electron acceptor. FDH-N is made anaerobically when nitrate is present, and FDH-H is made only under fermentation conditions. The FDH-O and FDH-N enzymes oxidize formate and transfer the electrons via quinones to oxygen and nitrate, respectively, during respiration. Accordingly, these are sometimes called respiratory formate dehydrogenases to distinguish them from FDH-H, which functions only during fermentation. (FDH-O is also called formate oxidase.) Since FDH-O is also present in cells grown with nitrate as the electron acceptor, it may also be able to reduce nitrate. FDH-H is part of the formate hydrogen–lyase enzyme complex that oxidizes formate to CO_2 and H_2 during fermentation. *E. coli* also synthesizes three hydrogenases: Hyd-1, Hyd-2, and Hyd-3. Hyd-1 and Hyd-2 (uptake hydrogenases) oxidize H_2 (e.g., during the reduction of fumarate or nitrate). Hyd-3 is part of the formate hydrogen–lyase and is responsible for the reduction of protons generating hydrogen gas.

43. Reviewed in: Unden, G., S. Becker, J. Bongaerts, J. Schirawski, and S. Six. 1994. Oxygen-regulated gene expression in facultatively anaerobic bacteria. *Antonie van Leeuwenhoek* **66**:3–23.

44. The regulon consists of the following genes: the *hyc* operon (hydrogenase 3), the *hyp* operon (uptake hydrogenases 1 and 2), and *fdhF* (formate dehydrogenase). The *fhlA* gene is encoded in the *hyp* operon.

45. Motteram, P. A. S., J. E. G. McCarthy, S. J. Fergukson, J. B. Jackson, and J. A. Cole. 1981. Energy conservation during the formate-dependent reduction of nitrite by *Escherichia coli*. *FEMS Microbiol. Lett.* **12**:317–320.

46. In *E. coli*, the membrane-bound nitrate reductase is encoded by the *narGHJI* operon. The fumarate reductase is encoded by the *frdABCD* operon. The DMSO/TMAO reductase is encoded by the *dmsABC* operon. The periplasmic formate-dependent nitrite reductase is encoded by the *nrfABCDEFG* operon. The cytoplasmic nitrate reductase is encoded by the *nirBDC* operon.

47. Unden, G., and J. Bongaerts. 1997. Alternative respiratory pathways of *Escherichia coli*: energetics and transcriptional regulation in response to electron acceptors. *Biochim. Biophys. Acta* **1320**:217–234.

48. Gennis, R. B., and V. Stewart. 1996. Respiration, pp. 217–261. In: *Escherichia coli and Salmonella: Cellular and Molecular Biology*, Vol. 1. F. C. Neidhardt et al. (Eds.). ASM Press, Washington, DC.

49. Rabin, R. S., and V. Stewart. 1993. Dual response regulators (NarL and NarP) interact with dual sensors (NarX and NarQ) to control nitrate- and nitrite-regulated gene expression in *Escherichia coli* K-12. *J. Bacteriol.* **175**:3259–3268.

50. Darwin, A. J., K. L. Tyson, S. J. W. Busby, and V. Stewart. 1997. Differential regulation by the homologous response regulators NarL and NarP of *Escherichia coli* K-12 depends on DNA binding site arrangement. *Mol. Microbiol.* **25**:583–595.

51. Williams, S. B., and V. Stewart. 1997. Discrimination between structurally related ligands nitrate and nitrite controls autokinase activity of the NarX transmembrane signal transducer of *Escherichia coli* K-12. *Mol. Microbiol.* **26**:911–925.

52. Chiang, R. C., R. Cavicchioli, and R. P. Gunsalus. 1997. "Locked-on" and "locked-off" signal transduction mutations in the periplasmic domain of the *Escherichia coli* NarQ and NarX sensors affect nitrate- and nitrite-dependent regulation by NarL and NarP. *Mol. Microbiol.* **24**:1049–1060.

53. The discovery of the *nar* genes unfolded as follows. Originally mutants that were defective in nitrate repression of the fumarate reductase gene, *frdA*, were isolated. (The actual screen was to look for mutant colonies in which nitrate did not repress the expression of the fusion gene *frdA–lacZ*.) The mutations were found to be in two genes, called *narL* and *narX*. Whereas *narL* was definitely required for nitrate repression or induction of the nitrate-regulated genes, mutations in *narX* caused only a partial loss of nitrate regulation. It therefore appeared that *narX* was dispensable. Another gene was later discovered, *narQ*, that was also a sensor for nitrate reductase. *E. coli* can use either NarX or NarQ as the sensor when one or the other is inactivated. To observe the loss of nitrate-dependent repression of fumarate reductase and loss of induction of nitrate reductase, cells must be mutated in both *narX* and *narQ*.

54. Stewart, V. 1993. Nitrate regulation of anaerobic respiratory gene expression in *Escherichia coli*. *Mol. Microbiol.* **9**:425–434.

55. Schroder, I., C. D. Wolin, R. Cavicchioli, and R. P. Gunsalus. 1994. Phosphorylation and dephosphorylation of the NarQ, NarX, and NarL proteins of the nitrate-dependent two-component regulatory system of *Escherichia coli*. *J. Bacteriol.* **176**:4985–4992.

56. Cavicchioli, R., I. Schroder, M. Constanti, and R. P. Gunsalus. 1995. The NarX and NarQ sensor–transmitter proteins of *Escherichia coli* each require two conserved histidines for nitrate-dependent signal transduction to NarL. *J. Bacteriol.* **177**:2416–2424.

57. Cavicchioli, R., I. Schröder, M. Constanti, and R. P. Gunsalus. 1995. The *Escherichia coli* NarQ and NarX regulatory proteins contain two conserved

histidines that are required for nitrate dependent signal transduction to NarL. *J. Bacteriol.* **177:**2416–2424.

58. Cavicchioli, R., R. C. Chiang, L. V. Kalman, and R. P. Gunsalus. 1996. Role of the periplasmic domain of the *Escherichia coli* NarX sensor–transmitter protein in nitrate-dependent signal transduction and gene regulation. *Mol. Microbiol.* **21:**901–911.

59. Magasanik, B. 1993. The regulation of nitrogen utilization in enteric bacteria. *J. Cell. Biochem.* **51:**34–40.

60. Magasanik, B. 1988. Reversible phosphorylation of an enhancer binding protein regulates the transcription of bacterial nitrogen utilization genes. *Trends Biochem. Sci.* **13:**475–479.

61. Merrick M. J., and R. A. Edwards. 1995. Nitrogen control in bacteria. *Microbiol. Rev.* **59:**604–622.

62. Zimmer, D. P., E. Soupene, H. L. Lee, V. F. Wendisch, A. Khodursky, B. J. Peter, R. A. Bender, and S. Kustu. 2000. Nitrogen regulatory protein C–controlled genes of *Escherichia coli*: scavenging as a defense against nitrogen limitation. *Proc. Natl. Acad. Sci. USA* **97:**14674–14679.

63. It has been estimated that Ntr_I (NitrC) activates approximately 75 genes in 25 operons in *E. coli*. Most of these are transcribed by σ^{54}-holoenzyme and activated directly. One of the activated genes is *nac*, which encodes the nitrogen assimilation control (Nac) protein that binds to DNA and activates the transcription by σ^{70}-holoenzyme of about 25 of the 75 genes (about 9 operons). For a list of the genes regulated by NitrC and Nac, see ref. 62.

64. Magasanik, B. 1982. Genetic control of nitrogen assimilation in bacteria. *Annu. Rev. Gen.* **16:**135–168.

65. Reviewed in: Magasanik, B. 1996. Regulation of nitrogen utilization, pp. 1344–1356. In: *Escherichia coli and Salmonella, Cellular and Molecular Biology*, Vol. 1. F. C. Neidhardt et al. (Eds.). ASM Press, Washington, DC.

66. Ninfa, A. J. and P. Jiang. 2005. PII signal transduction proteins: sensors of α-ketoglutarate that regulate nitrogen metabolism. *Curr. Opin. Microbiol.* **8:**168–173.

67. MacNeil, T., G. P. Roberts, D. MacNeil, and B. Tyler. 1982. The products of *gln* L and *gln* G are bifunctional regulatory proteins. *Mol. Gen. Genet.* **188:**325–333.

68. Sigma 54 (encoded by the *rpoN* gene in *E. coli*) polymerases are widely distributed among bacteria, not simply the enterics, and they bind to promoters that are unrelated to sigma 70 promoters. They transcribe many different genes (i.e., not merely genes in the Ntr regulon). Sigma 54 polymerases can bind to promoter regions of target genes but, unlike sigma 70 polymerases, are unable to produce an open complex without the aid of an activator protein which usually binds 100 to 150 base pairs upstream of the promoter. (See Fig. 18.9.) Binding of the activator protein at such long distances from the start site of transcription is required for the looping mechanism of activation. In the looping mechanism the intervening DNA loops out, allowing the activator to interact with the polymerase. For some promoters, the looping out of the DNA is facilitated by DNA-bending proteins such as integration host factor (IHF). (For sigma 70 polymerases, the transcription activators bind adjacent to the polymerase site and interact with the polymerase without looping of the DNA.) The activator proteins are ATP-dependent and, in a way that is not understood, the hydrolysis of ATP provides the energy to convert the closed complex to the open complex, thus initiating transcription. For a review of sigma 54, see Buck, M., M.-T. Gallegos, D. J. Studholme, Y. Guo, and J. D. Gralla. 2000. The bacterial enhancer-dependent σ^{54} (σ^N) transcription factor. *J. Bacteriol.* **182:**4129–4136.

69. The reason for the requirement of NR_I–P for transcription is that the sigma 54 polymerase binds to the $glnAp_2$ promoter but cannot by itself initiate transcription. It is unable to do this because it cannot form an open complex (i.e., it cannot "melt" the DNA double helix around the promoter site to gain access to the transcription start site). However, NR_I–P binds to sites on the DNA base pairs upstream from the promoter region (called "enhancer sites, " containing 100–130 bp) and interacts with the RNA polymerase to form the open complex and thus initiate transcription.

70. Dixon, R. 1998. The oxygen-responsive NIFL–NIFA complex: a novel two-component regulatory system controlling nitrogenase synthesis in γ-proteobacteria. *Arch. Microbiol.* **169:**371–380.

71. Friedman, D. I. 1988. Integration host factor: a protein for all reasons. *Cell* **55:**545–554.

72. Wanner, B. L. 1993. Gene regulation by phosphate in enteric bacteria. *J. Cell. Biochem.* **51:**47–54.

73. Phosphonates are organophosphates in which there is a direct carbon–phosphorus bond rather than a phosphate ester linkage.

74. The promoters in the PHO regulon are unusual in that they have −10 sequences but no consensus −35 sequences for σ^{70}-RNA polymerase. Thus, in the absence of PhoB, RNA polymerase does not bind to the promoter. The promoters, however, have one to three (tandemly repeated) 18 base pairs sequences called *pho* boxes, which are 10 base pairs upstream of the −10 consensus sequence. The phosphorylated PhoB protein binds to the *pho* boxes, and this enables the RNA polymerase to bind to the promoter. It has been postulated that PhoB interacts with the sigma subunit of the RNA polymerase, enabling the polymerase to bind to the promoter. See Makino, K., M. Amemura, S.-K. Kim, A. Nakata,

and H. Shinagawa. 1993. Role of the σ^{70} subunit of RNA polymerase in transcriptional activation by activator protein PhoB in *Escherichia coli*. *Genes Dev.* 7:149–160.

75. van Veen, H. W. 1997. Phosphate transport in prokaryotes: molecules, mediators and mechanisms. *Antonie van Leeuwenhoek.* 72:299–315.

76. Wanner, B. L. 1996. Phosphorus assimilation and control of the phosphate regulon. pp. 1357–1381 in: *Escherichia coli and Salmonella, Cellular and Molecular Biology*. F. C. Neidhardt et al. (Eds.). ASM Press, Washington, DC.

77. The Pst system is similar to the histidine uptake system described in Section 16.3.3 and belongs to the superfamily of ABC (ATP-binding cassette) transporters. These have four membrane-bound proteins, two of which are transmembrane and probably form a channel, whereas the other two have ATP-binding sites. In addition, there is a periplasmic binding protein. Pst A and Pst C are thought to form a transmembrane channel, PstB is the ATPase, which probably exists as a dimer, and PtsS is the periplasmic binding protein.

78. PhoR autophosphorylates in vitro using ATP as the phosphoryl donor and transfers the phosphoryl group to PhoB, but it has not been demonstrated that it dephosphorylates PhoB-P (see ref. 72).

79. Swem, D. L., and C. E. Bauer. 2002. Coordination of ubiquinol oxidase and cytochrome cbb_3 oxidase expression by multiple regulators in *Rhodobacter capsulatus*. *J. Bacteriol.* **184**:2815–2820.

80. Bauer, C. E., and T. H. Bird. 1996. Regulatory circuits controlling photosynthesis gene expression. *Cell* 85:5–8.

81. Dong, C., S. Elsen, L. R. Swem, and C. E. Bauer. 2002. Aer, a second aerobic repressor of photosynthetic gene expression in *Rhodobacter capsulatus*. *J. Bacteriol.* **184**:2805–2814.

82. Mosely, C., J. Y. Suzuki, and C. E. Bauer. 1994. Identification and molecular genetic characterization of a sensor kinase responsible for coordinately regulating light harvesting and reaction center gene expression in response to anaerobiosis. *J. Bacteriol.* **176**:7566–7573.

83. Elsen, S., L. R. Swem, D. L. Swem, and C. E. Bauer. 2004. RegB/RegA, a highly conserved redox-responding global two-component regulatory system. *Microbiol. Mol. Biol. Rev.* **68**:263–279.

84. Du, S., T. H. Bird, and C. E. Bauer. 1998. DNA binding characteristics of RegA. *J. Biol. Chem.* **273**:18509–18513.

85. HvrA is an H-NS-like protein. Such proteins are widespread in gram-negative bacteria. They bind to DNA and modulate gene expression. See the subsection of Section 2.2.2 entitled *Other Factors That Control rRNA Synthesis: Fis, H-NS*. HvrA has been reported

to be necessary for the ammonium-dependent inhibition of the transcription of certain of the nitrogenase genes (*nif* genes) in the nonsulfur purple photosynthetic bacterium *Rhodobacter capsulatus*, as well as being a low-light activator of genes for the photosynthetic apparatus. Raabe, K., T. Drepper, K.-U. Riedel, B. Masepohl, and W. Klipp. 2002. The H-NS-like HvrA modulates expression of nitrogen fixation genes in the phototrophic purple bacterium *Rhodobacter capsulatus* by binding to selected *nif* promoters. *FEMS Microbiol. Lett.* **216**:151–158.

86. Zucconi, A. P., and J. T. Beatty. 1988. Posttranscriptional regulation by light of the steady-state levels of mature B800-850 light-harvesting complexes in *Rhodobacter capsulatus*. *J. Bacteriol.* 170:877–882.

87. Nikaido, H., and M. Vaara. 1987. Outer membrane, pp. 7–22. In: *Escherichia coli and Salmonella typhimurium, Cellular and Molecular Biology*, Vol. 1. F. C. Neidhardt et al. (Eds.). ASM Press, Washington, DC.

88. Forst, S. A., and D. L. Roberts. 1994. Signal transduction by the EnvZ–OmpR phosphotransfer system in bacteria. *Res. Microbiol.* **145**:363–373.

89. Reviewed in: Pratt, L. A., and T. J. Silhavy. 1995. Porin regulon of *Escherichia coli*, pp. 105–127. In: *Two-Component Signal* Transduction. J. A. Hoch, and T. J. Silhavy (Eds.). ASM Press, Washington, DC.

90. J. Waukau and S. Forst. 1992. Molecular analysis of the signaling pathway between EnvZ and OmpR in *Escherichia coli*. *J. Bacteriol.* **174**:1522–1527.

91. H. Aiba, T. Mizuno, and S. Mizushima. 1989. Transfer of phosphoryl group between two regulatory proteins involved in osmoregulatory expression of the *ompF* and *ompC* genes in *Escherichia coli*. *J. Biol. Chem.* 264:8563–8567.

92. H. Aiba, F. Nakasai, S. Mizushima, and T. Mizuno. 1989. Evidence for the physiological importance of the phosphotransfer between two regulatory components, EnvZ and OmpR, in osmoregulation in *Escherichia coli*. *J. Boil. Chem.* **264**:14090–14094.

93. N. Ramani, L. Huang, and M. Freundlich. 1992. In vitro interactions of integration host factor with the *ompF* promoter–regulatory region of *Escherichia coli*. *Mol. Gen. Genet.* **231**:248–255.

94. Mizuno, T., M.-Y. Chou, and M. Inouye. 1984. A uniqe mechanism regulating gene expression: translational inhibition by a complementary RNA transcript (micRNA). *Proc. Natl. Acad. Sci. USA* 81:1966–1970.

95. Reviewed in: Delihas, N. 1995. Regulation of gene expression by trans-encoded antisense RNAs. *Mol. Microbiol.* 15:411–414.

96. Ramani, N., H. Hedeshian, and M. Freundlich. 1994. *micF* Antisense RNA has a major role in

osmoregulation of OmpF in *Escherichi coli.* *J. Bacteriol.* **176**:5005–5010.

97. Sugiura, A., K. Nakashima, K. T. Tanaka, and T. Mizuno. 1992. Clarification of the structural and functional features of the osmoregulated *kdp* operon in *Escherichia coli.* *Mol. Microbiol.* **6**:1769–1776.

98. Nakashima, K., H. Sugiura, H. Momoi, and T. Mizuno. 1992. Phosphotransfer signal transduction between two regulatory factors involved in the osmoregulated *kdp* operon in *Escherichia coli.* *Mol. Microbiol.* **6**:1777–1784.

99. Sugiura, A., K. Hirokawa, K. Nakashima, and T. Mizuno. 1994. Signal-sensing mechanisms of the putative osmosensor KdpD in *Escherichia coli.* *Mol. Microbiol.* **14**:929–938.

100. Heermann, R., A. Fohrmann, K. Altendorf, and K. Jung. 2003. The transmembrane domains of the sensor kinase KdpD of *Escherichia coli* are not essential for sensing K⁺ limitation. *Mol. Microbiol.* **47**:839–848.

101. McCleary, W. R., J. B. Stock, and A. J. Ninfa. 1993. Is acetyl phosphate a global signal in *Escherichia coli? J. Bacteriol.* **175**:2793–2798.

102. Feng, J., M. R. Atkinson, W. McCleary, J. B. Stock, B. L. Wanner, and A. J. Ninfa. 1992. Role of phosphorylated metabolic intermediates in the regulation of glutamine synthetase synthesis in *Escherichia coli. J. Bacteriol.* **174**:6061–6070.

103. Shin, S., and C. Park. 1995. Modulation of flagellar expression in *Escherichia coli* by acetyl phosphate and the osmoregulator OmpR. *J. Bacteriol.* **177**:4696–4702.

104. For a review of catabolite repression by carboxylic acids, see: *Research in Microbiology*, 1996, Vol. 147, No. 6–7.

105. Saier, Jr., M. H. 1996. Cyclic AMP-independent catabolite repression in bacteria. *FEMS Microbiol. Lett.* **138**:97–103.

106. Saier, Jr., M. H., T. M. Ramseier, and J. Reizer. 1996. Regulation of carbon utilization, pp. 1325–1343. In: *Escherichia coli and Salmonella typhimurium: Molecular and Cellular Biology*, Vol. 1. F. C. Neidhardt et al. (Eds.). ASM Press, Washington, DC.

107. Chin, M. A., B. U. Feucht, and M. H. Saier Jr. 1987. Evidence for regulation of gluconeogenesis by the fructose phosphotransferase system in *Salmonella typhimurium. J. Bacteriol.* **169**:897–899.

108. Mutants in *cra* show increased levels of enzymes such as the following, which are involved in the uptake and catabolism of glucose: enzyme I and HPr of the phosphotransferase system, phosphofructokinase, pyruvate kinase, 6-phosphogluconate dehydratase, and KDPG aldolase. Cra is also a repressor of the *fruBKA* operon, which consists of genes required for the fructose-specific PTS. The *fruBKA* operon codes for three proteins. FruB is an HPr-like protein called Fpr, which differs structurally from HPr used in the regular PTS but can substitute for HPr in mutants unable to synthesize HPr. In fact, *cra* mutants were originally found as suppressor mutants that allowed mutants defective in HPr synthesis (*ptsH* mutants) to grow on PTS carbohydrates. In addition to a C-terminal HPr domain, Fpr has an N-terminal IIA^Fru domain. FruK is a fructose-1-phosphate kinase that is necessary to convert the incoming fructose-1-phosphate to fructose-1,6-bisphosphate. FruA is the permease. These fructose-specific genes are required for fructose transport via the fructose–PTS system, and the conversion of the incoming fructose-1-phosphate to fructose-1,6-bisphosphate.

109. Ramseier, T. M. 1996. Cra and the control of carbon flux via metabolic pathways. *Res. Microbiol.* **147**:489–493.

110. Tyson, K., S. Busby, and J. Cole. 1997. Catabolite regulation of two *Escherichia coli* operons encoding nitrite reductases: role of the Cra protein. *Arch. Microbiol.* **168**:240–244.

111. Saier, Jr., M. H., S. Chauvaux, J. Deutscher, J. Reizer, and J.-J. Ye. 1995. Protein phosphorylation and regulation of carbon metabolism in gram-negative versus gram-positive bacteria. *Trends Biochem. Sci.* **20**:267–271.

112. Hueck, C. J., and W. Hillen. 1995. Catabolite repression in *subtilis*: a global regulatory mechanism for the gram-positive bacteria? *Mol. Microbiol.* **15**:395–401.

113. Chauvaux, S., I. T. Paulsen, and M. H. Saier Jr. 1997. CcpB, a novel transcription factor implicated in catabolite repression in *subtilis. J. Bacteriol.* **180**:491–497.

114. Depending upon the gene, CREs may overlap promoters or be located within the reading frame. One suspects that if the CRE overlapped the promoter, binding of a regulatory protein would prevent initiation of transcription; and if the CRE were in the reading frame, then binding of a regulatory protein would inhibit transcription elongation. The protein that binds to the CREs has not been unambiguously identified.

115. Also present in *B. subtilis* is the PEP-dependent kinase, which can also phosphorylate Hpr and operates during phosphotransferase-mediated glucose uptake. (Although the phosphotransferase system is characteristic of anaerobes and facultative anaerobes, it does occur in some aerobes, e.g., *B. subtilis.*) However, the PEP-dependent kinase is not part of the catabolite-repressing system. This is because the phosphorylation of Hpr by the ATP-dependent kinase is at a serine residue (Ser-46), whereas the phosphorylation of Hpr by the PEP-dependent kinase (EI) is at a histidine residue (His-15), and Hpr(His-P) is not active in catabolite repression.

116. One method of isolating mutations in genes required for catabolite repression by glucose is to mutagenize a strain carrying an operon translational fusion of the gene of interest to *lacZ*. One can then screen for blue colonies on media containing glucose plus the alternative carbon source and the indicator X-Gal (5-bromo-4-chloro-3-indoyl-β-D-galactoside). X-Gal is an analogue of lactose. It is colorless, but when it is cleaved by β-galactosidase a blue color is produced. Mutagenesis can be performed by using transposons. A transposon (Tn) is a DNA sequence that moves from one DNA molecule to another or from one part of a DNA molecule to another part. When the transposon inserts in a gene of interest, it can produce a mutation in that gene because of the disruption in the gene's DNA sequence. The transposon can be introduced into the bacterial cell on a plasmid via transformation or conjugation. Transposons can carry antibiotic resistance genes that may be used to select for transformants during the isolation of mutants as well as for selection of transformants during cloning of the gene.

117. Warner, J. B., and J. S. Lolkema. 2003. CcpA-dependent carbon catabolite repression in bacteria. *Microbiol. Mol. Biol. Rev.* 67:475–490.

118. Ye, J.-J., and M. H. Saier Jr. 1995. Cooperative binding of lactose and HPr(Ser-P) to the lactose:H+ permease of *Lactobrevis*. *Proc. Natl. Acad. Sci. USA* 92:417–421.

119. Ye, J.-J., and M. H. Saier Jr. 1995. Purification and characterization of a small membrane-associated sugar–phosphate phosphatase that is allosterically activated by HPr(Ser-P) of the phosphotransferase system in *Lactococcus lactis*. *J. Biol. Chem.* 270:16740–16744.

120. Giblin, L., B. Boesten, S. Turk, P. Hooykaas, and F. O'Gara. 1995. Signal transduction in the *Rhizobium meliloti* dicarboxylic acid transport system. *FEMS Microbiol. Lett.* 126:25–30.

121. Finlay, B. B., and S. Falkow. 1997. Common themes in microbial pathogenicity revisited. *Microbiol. Mol. Biol. Rev.* 61:136–169.

122. Falkow, S. 1997. What is a pathogen? *ASM News* 63:359–365.

123. Beier, D., G. Spohn, R. Rappuoli, and V. Scarlato. 1997. Identification and characterization of an operon of *Helicobacter pylori* that is involved in motility and stress adaptation. *J. Bacteriol.* 179:4676–4683.

124. Miller, J. F., J. J. Mekalanos, and S. Falkow. 1989. Coordinate regulation and sensory transduction in the control of bacterial virulence. *Science* 243:916–922.

125. Skorupski, K., and R. K. Taylor. 1997. Control of the ToxR virulence regulon in *Vibrio cholerae* by environmental stimuli. *Mol. Microbiol.* 25:1003–1009.

126. Raskin, D., J. Bina, and J. Mekalanos. 2004. Genomic and genetic analysis of *Vibrio cholerae*. *ASM news* 70:57–62.

127. Cholera toxin consists of one A subunit and five identical B subunits. The pentamer of B subunits binds to GM$_1$ ganglioside (a glycolipid) on the surface of host intestinal epithelial cells. The A subunit is separated from the B pentamer and enters the cytosol, where it is dissociated into two fragments, A$_1$ and A$_2$, by the reduction of a disulfide bond that links these subunits together. The A$_1$ subunit activates adenylate cyclase. The activation of adenylate cyclase leads to an increase in cAMP, which leads to secretion of chloride and water. Activation of adenylate cyclase is a multistep process. The A$_1$ subunit catalyzes the transfer of ADP-ribose from NAD$^+$ to a GTP-binding protein (G$_s$) that is associated with adenylate cyclase. ADP-ribosylation of the GTP-binding protein inhibits its GTPase activity, locking the adenylate cyclase into the "on" mode. To understand this, it is necessary to know how adenylate cyclase is regulated. Adenylate cyclase is a membrane-bound protein that exists in a complex with G$_s$. When G$_s$ binds ATP, it becomes activated and activates adneylate cyclase. But, G$_s$ is a GTPase, and ordinarily the GTP is hydrolyzed so that G$_s$ has bound GDP and is not active. Hormones such as glucagon or epinephrine bind to a hormone receptor, which then binds to the G$_s$ protein and triggers the exchange of GTP for bound GDP on the G$_s$ protein. This activates the G$_s$ protein (hence adenylate cyclase) until it hydrolyzes the bound GTP. ADP-ribosylation prevents GTP hydrolysis; hence G$_s$ continually activates adenyl cyclase in the presence of the A$_1$ subunit of cholera toxin.

128. Miller, V. L., and J. J. Mekalanos. 1988. A novel suicide vector and its use in construction of insertion mutations: osmoregulation of outer membrane proteins and virulence determinants in *Vibrio cholerae* requires *toxR*. *J. Bacteriol.* 170:2575–2583.

129. There are two major disease-causing strains of *V. cholerae*: the classical strain (*V. cholerae* O1) and the EI Tor strain. Higher expression of the ToxR regulon in the classical strain occurs at pH 6.5 and 30 °C and lower expression at pH 8.5 and 37 °C. The conditions that favor optimal expression in the classical strain are not the same conditions that favor optimal expression in the EI Tor strain. This is reviewed in Skorupski, K., and R. K. Taylor. 1997. Control of the ToxR virulence regulon in *Vibrio cholerae* by environmental stimuli. *Mol. Microbiol.* 25:1003–1009.

130. To understand how to find and monitor expression of virulence genes by means of Tn*phoA* fusions, it is necessary to know something about transposons. Transposons are DNA elements that move from one part of a DNA molecule to another, or from one DNA molecule to another. Most transposons move almost randomly. The movement is

called transposition, and it is catalyzed by enzymes called transposases, which are encoded in the transposon. Bacteria have transposons of many different kinds, some of which carry antibiotic resistance genes in addition to the transposase genes. The presence of the genes for antibiotic resistance can be used to select for cells carrying the transposon. TnphoA is a derivative of the transposon Tn5 carrying a kanamycin-resistance gene that can be used to isolate random gene fusions. The transposon contains an inserted reporter gene, phoA, which encodes the C-terminal end of periplasmic alkaline phosphatase but does not encode the promoter, translational start site, or signal sequence (for export of the protein) of the alkaline phosphatase gene. When TnphoA inserts into a gene in the bacterial chromosome in the right reading frame, a fusion protein will be synthesized. The fusion protein has the C-terminal end of alkaline phosphatase and the N-terminal end of the protein encoded by the gene into which TnphoA has inserted. The fusion protein will have alkaline phosphatase activity only if TnphoA inserts into a gene that encodes a protein to be exported, so that the N-terminal end of the fusion protein has a signal sequence for export. This ensures that the C-terminal end of the fusion protein enters the periplasm, where it can exhibit alkaline phosphatase activity. It is possible to screen the colonies for active alkaline phosphatase. An indicator (XP) is included in the agar. (XP is 5-bromo-4-chloro-3-indolylphosphate. It turns blue when the phosphate is removed.) If active PhoA is present, it cleaves XP, producing a blue color. Genes detected in this way can be mapped and cloned. To learn how TnphoA is constructed, see: Manoil, C., and J. Beckwith. 1985. TnphoA: a transponson probe for protein export signals. Proc. Natl. Acad. Sci. USA 82:8129–8133.

To screen for virulence genes that encode exported virulence proteins, Peterson and Mekalanos introduced transposon TnphoA on a plasmid (pRT291) into V. cholerae. The plasmid was introduced into a streptomycin-resistant strain of C. cholerae via conjugation, using a donor E. coli strain carrying the plasmid containing TnphoA. Transconjugates were selected on the basis of resistance to streptomycin and kanamycin. The antibiotic-resistant transconjugates were then mated with E. coli carrying a plasmid (pPH1JI) that was not compatible with pRT291. The plasmid pPH1JI carries a gentamicin resistance gene. V. cholerae colonies that were resistant to gentamicin, kanamycin, and streptomycin, and produced alkaline phosphatase, were selected on XP agar containing gentamicin, kanamycin, and streptomycin. These carried pPH1JI (conferring resistance to gentamicin) and TnphoA (conferring resistance to kanamycin). Because pRT291 and pPH1JI are incompatible, the kanamycin-resistant colonies carried random insertions of the transposon in the chromosome. In other words, TnphoA transposed (moved) from the plasmid to the bacterial chromosome. The PhoA-positive cells (detected with the XP) were grown in the absence of gentamicin to promote the loss of pPH1JI. To find the genes carrying TnphoA insertions, which are regulated by ToxR, the production of PhoA in the different isolates was monitored under conditions known to produce high or low levels of cholera toxin and TCPA. Proof that these genes were indeed positively regulated by ToxR was obtained by testing the production of PhoA in toxR⁻ strains. See: Peterson, K. M., and J. J. Mekalanos. 1988. Characterization of the Vibrio cholerae ToxR regulon: identification of novel genes involved in intestinal colonization. Infect. Immun. 56:2822–2829.

131. DiRita, V. J., C. Parsot, G. Jander, and J. J. Mekalanos. 1991. Regulatory cascade controls virulence in Vibrio cholerae. Proc. Natl. Acad. Sci. USA 88:5403–5407.

132. Häse, C. C., and J. J. Mekalanos. 1998. TcpP protein is a positive regulator of virulence gene expression in Vibrio cholerae. Proc. Natl. Acad. Sci. USA 95:730–734.

133. Groisman, E. A. 2001. The pleiotropic two-component regulatory system PhoP–PhoQ. J. Bacteriol. 183:1835–1842.

134. Groisman, E. A., and F. Heffron. 1995. Regulation of Salmonella virulence by two-component regulatory systems, pp. 319–332. In: Two-Component Signal Transduction. J. A. Hoch and T. J. Silhavy (Eds.). ASM Press, Washington, DC.

135. An acid phosphatase, the product of the phoN gene, is induced under certain growth conditions such as phosphate limitation. A mutant hunt revealed phoP, which when expressed constitutively results in high levels of phosphatase synthesis. The reason for this is that PhoP is a positive regulator of the phoN gene. It was later discovered that Tn10-generated mutants that were very sensitive to killing by macrophages mapped to the phoP gene identified earlier as being required for alkaline phosphatase synthesis. See the review by Groisman and Heffron in ref. 134.

136. Véscovi, E. G., F. C. Soncinin, and E. A. Groisman. 1996. Mg²⁺ as an extracellular signal: environmental regulation of Salmonella virulence. Cell 84:166–174.

140. Moskowitz, S. M., R. K. Ernst, and S. I. Miller. 2004. PmrAB, a two-component regulatory system of *Pseudomonas aeruginosa* that modulates resistance to cationic antimicrobial peptides and addition of aminoarabinose to lipid A. *J. Bacteriol.* **186**:575–579.

141. Cationic antimicrobial peptides (CAPs) are detergent-like peptides that kill gram-positive and gram-negative bacteria by collapsing the membrane potential of the cytoplasmic membrane. In gram-negative bacteria, CAPs bind to the lipopolysaccharide and traverse the periplasm. Pathogens can be exposed to CAPs at epithelial cell surfaces, such as in the lungs. Similar peptides, called defensins, are present in human neutrophils and macrophages and are part of the body's defense against infection. The antibiotic polymyxin, produced by *Polymyxa*, is an acylated cyclic CAP. As noted in the text, polymyxin-resistant mutants of *Pseudomonas aeruginosa* have been mapped to the PmrA/PmrB two-component signal transduction system.

142. Kox, L., F. F. Wösten, and E. A. Groisman. 2000. A small protein that mediates the activation of a two-component system by another two-component system. *EMBO J.* **19**:1861–1872.

143. Miller, J. F., S. A. Johnson, W. J. Black, D. T. Beattie, J. J. Mekalanos, and S. Falkow. 1992. Constitutive sensory transduction mutations in the *Bordetella pertussis bvgS* gene. *J. Bacteriol.* **174**:970–979.

144. Hale, T. L. 1991. Genetic basis of virulence in *Shigella* species. *Microbiol. Rev.* **55**:206–224.

145. *Shigella* spp. as well as *Listeria monocytogenes* invade eukaryotic cells and make a cell surface protein that elicits the formation of a tail of actin at one cell pole when the bacteria are growing intracellularly. As a consequence, the bacteria are propelled through the cytoplasm of the host cell. The actin tail also aids in intercellular spread of the bacteria. See: Bernardini, M. L., J. Mounier, H. D'Hauteville, M. Coquis-Rondon, and P. J. Sansonetti. 1989. Identification of *icsA*, a plasmid locus of *Shigella flexneri* that governs bacterial intra- and intercellular spread through interaction with F-actin. *Proc. Natl. Acad. Sci. USA* **86**:3867–3871.

146. Porter, M. E., and C. J. Dorman. 1994. A role for H-NS in the thermo-osmotic regulation of virulence gene expression in *Shigella flexneri*. *J. Bacteriol.* **176**:4187–4191.

147. Porter, M. E., and C. J. Dorman. 1997. Positive regulation of *Shigella flexneri* virulence genes by integration host factor. *J. Bacteriol.* **179**:6537–6550.

148. Otto, M., R. Süssmuth, G. Jung, and F. Götz. 1998. Structure of the pheromone peptide of the *Staphylococcus epidermidis agr* system. *FEBS Lett.* **424**:89–94.

149. Zhu, J., P. M. Oger, B. Schrammeijer, P. J. J. Hooykaas, S. K. Farrand, and S. C. Winans. 2000. The bases of crown gall tumorigenesis. *J. Bacteriol.* **182**:3885–3895.

150. TraR is a LuxR-type transcription factor that binds to the *A. tumefaciens* acyl–HSL and stimulates the transcription of genes required for plasmid transfer (*tra* genes), the gene whose product synthesizes the acyl–HSL (*traI*), and genes responsible for plasmid replication (*rep* genes). The increased replication of the plasmid results in increased tumorigenesis. The transcription of *traR* is stimulated by plant opines. The opines bind a receptor protein (either AccR or OccR, depending upon the strain) in the bacterium, and the complex initiates the transcription of *traR*. This ensures that conjugal transfer of the Ti plasmid occurs only when the opines are present. (AccR is a repressor. The AccR/opine combination derepresses *traR*, and the OccR/opine combination induces *traR*.) Additional regulation involves the TraM protein. TraM binds to TraR and inhibits its activity. For more information about TraM, see: Chen, G., J. W. Malenkos, M.-R. Cha, C. Fuqua, and L. Chen. 2004. Quorum-sensing antiactivator TraM forms a dimer that dissociates to inhibit TraR. *Mol. Microbiol.* **52**:1641–1651.

151. Macnab, R. M. 1987. Motility and chemotaxis, pp. 723–759. In: *Escherichia coli and Salmonella typhimurium: Cellular and Molecular Biology*, Vol. 1. F. C. Neidhardt et al. (Eds.). ASM Press, Washington, DC.

152. Stock, J. B., A. J. Ninfa, and A. M. Stock. 1989. Protein phosphorylation and regulation of adaptive responses in bacteria. *Microbiol. Rev.* **53**:450–490.

153. Manson, M. D. 1992. Bacterial motility and chemotaxis, pp. 277–346. In: *Advances in Microbial Physiology*, Vol. 33. A. H. Rose (Ed.). Academic Press, New York.

154. Blair, D. F. 1995. How bacteria sense and swim. *Annu. Rev. Microbiol.* **49**:489–522.

155. Szurmant, H., and G. W. Ordal. 2004. Diversity in chemotaxis mechanisms among Bacteria and Archaea. *Microbiol. Mol. Biol. Rev.* **68**:301–319.

156. Macnab, R. M., and D. E. Koshland Jr. 1972. The gradient-sensing mechanism in bacterial chemotaxis. *Proc. Natl. Acad. Sci. USA* **69**:2509–2512.

157. Berg, H. C., and D. A. Brown. 1972. Chemotaxis in *Escherichia coli* analysed by three-dimensional tracking. *Nature* **239**:500–504.

158. Silverman, M., and M. Simon. 1974. Flagellar rotation and the mechanism of bacterial motility. *Nature* **249**:73–74.

159. Larsen, S. H., R. W. Reader, E. N. Kort, W.-W. Tso, and J. Adler. 1974. Changes in direction of flagellar rotation is the basis of the chemotactic response in *Escherichia coli*. *Nature* **249**:74–77.

160. Maddock, J. R., and L. Shapiro. 1993. Polar location of the chemoreceptor complex in the *Escherichia coli* cell. *Science* 259:1717–1723.

161. Sourjik, V., and H. C. Berg. 2000. Localization of components of the chemotaxis machinery of *Escherichia coli* using fluorescent protein fusions. *Mol. Microbiol.* 37:740–751.

162. Shiomi, D., I. B. Zhulin, M. Homma, and I. Kawagishi. 2002. Dual recognition of the bacterial chemoreceptor by chemotaxis-specific domains of the CheR methyltransferase. *J. Biol. Chem.* 277:42325–42333.

163. Iino, T., Y. Komeda, K. Kutsukake, R. M. Macnab, P. Matsumura, J. S. Parkinson, M. I. Simon, and S. Yamaguchi. 1988. New unified nomenclature for the flagellar genes of *Escherichia coli* and *Salmonella typhimurium*. *Microbiol. Rev.* 52:533–535.

164. Three proteins called FliG, FliM, and FliN, coded for by *fliG*, *fliM*, and *fliN*, are the switch proteins. These proteins are not in the basal body and seem to be peripheral (rather than integral) cell membrane proteins closely associated with the basal body. A description of the flagellar motor is given in Section 1.2.1. Mutants that completely lack these proteins do not make flagella (Fla⁻ phenotype) even though these proteins are not part of the basal body, hook, or filament. Missense mutations in these genes cause a Fla⁻ phenotype, a Che⁻ phenotype (CW or CCW biased), or paralyzed flagella (Mot phenotype), depending upon the mutation. It therefore appears that these proteins are necessary for basal body synthesis and switching, and perhaps are involved in energy coupling. Two other proteins should be mentioned, even though they are not chemotaxis proteins. The MotA and MotB proteins are necessary for motor function. Mutations in the *motA* and *motB* genes cause paralyzed flagella. The proteins are integral membrane proteins that are thought to form a ring around the S and P rings of the basal body, coupling the proton potential to flagellar rotation. The MotA protein is a proton channel and the MotB protein may mediate the interaction of MotA with the basal body.

165. The histidine kinase, CheA is a dimer. One of the monomers acts as a kinase and phosphorylates a histidine residue on the second monomer to form CheA-P.

166. The MCP proteins are truly remarkable with respect to the different classes of signals that they transduce. Tsr responds to the chemoattractant L-serine. It also responds to the repellents L-leucine and indole, and to low external pH. Tsr also responds to weak organic acids such as acetate that act as repellents because they lower the internal pH. In addition, the Tsr protein is a thermoreceptor mediating taxis toward warmer temperatures up to 37 °C. The Tar protein in *E. coli* responds to the chemoattractant L-aspartate as well as to maltose bound to the periplasmic maltose-binding protein. Tar also detects the repellents Co^{2+} and Ni^{2+}. In mutants lacking Tsr, the Tar protein mediates taxis toward higher temperatures. The Trg protein responds to the chemoattractants ribose, glucose, and galactose when these are bound to their respective periplasmic binding proteins. The Tap protein (present in *E. coli* but not *S. typhimurium*) mediates taxis toward a variety of dipeptides when they are bound to DPP, a periplasmic dipeptide binding protein. Trg and Tap from *E. coli* serve as repellent receptors for phenol. Trg is also an attractant receptor, whereas Tap is a repellent receptor for weak organic acids.

167. Lukat, G. S., and J. B. Stock. 1993. Response regulation in bacterial chemotaxis. *J. Cell. Biochem.* 51:41–46.

168. Parkinson, J. S. 2003. Bacterial chemotaxis: a new player in response regulator dephosphorylation. *J. Bacteriol.* 185:1492–1494.

169. Silversmith, R. E., G. P. Guanga, L. Betts, C. Chu, R. Zhao, and R. B. Bourret. 2003. CheZ-mediated dephosphorylation of the *Escherichia coli* chemotaxis response regulator CheY: role for CheY glutamate 89. *J. Bacteriol.* 185:1495–1502.

170. Welch, M., K. Oosawa, S.-I. Aizawa, and M. Eisenbach. 1993. Phosphorylation-dependent binding of a signal molecule to the flagellar switch of bacteria. *Proc. Natl. Acad. Sci. USA* 90:8787–8791.

171. Ordal, G. W., L. Màrquez-Magana, and M. J. Chamberlin. 1993. Motility and chemotaxis, pp. 765–784. In: *B. subitilis and Other Gram-Positive bacteria: Biochemistry, Physiology, and Molecular Genetics.* A. L. Sonenshein, J. A. Hoch, and R. Losick (Eds.). ASM Press, Washington, DC.

172. Parkinson, J. S. 2004. Signal amplification in bacterial chemotaxis through receptor teamwork. *ASM News* 70:575–582.

173. Dailey, F. E., and H. C. Berg. 1993. Change in direction of flagellar rotation in *Escherichia coli* mediated by acetate kinase. *J. Bacteriol.* 175:3236–3239.

174. Barak, R., W. N. Abouhamad, and M. Eisenbach. 1998. Both acetate kinase and acetyl coenzyme A synthetase are involved in acetate-stimulated change in the direction of flagellar rotation in *Escherichia coli*. *J. Bacteriol.* 180:985–988.

175. Dunten, P., and D. E. Koshland Jr. 1991. Tuning the responsiveness of a sensory receptor via covalent modification. *J. Biol. Chem.* 266:1491–1496.

176. Borkovich, K. A., L. A. Alex, and M. I. Simon. 1992. Attenuation of sensory receptor signaling by covalent modification. *Proc. Natl. Acad. Sci. USA* 89:6756–6760.

177. It is possible to construct mutants in the aspartate chemoreceptor protein (Tar) so that glutamine (the amidated form of glutamate) is substituted for

the glutamate residues that are normally methylated. The chemotaxis signaling system cannot tell the difference between a chemoreceptor that has glutamine and one that has methylated glutamate. Such mutants show that substituting glutamine for glutamate in a CheR$^-$ and CheB$^-$ mutant produces a tumbling behavior. The tumbling behavior can be reversed by adding the attractant aspartate to the medium, reflecting the fact inhibition of the kinase activity of CheA by the binding of attractant to the chemoreceptor. Furthermore, CheR$^-$ mutants swim smoothly and CheB$^-$ mutants tumble more frequently than the wild type, again indicating that methylation stimulates tumbling. Because the affinity of the methylated chemoreceptor for the attractant reported by not much different from that for the nonmethylated chemoreceptor, the effect of methylation on CheA kinase activity appears to be the dominant reason for adaptation. (However, the extent to which methylation changes the affinity of the chemoreceptor to the attractant is a subject of controversy. See: Blair, D. F. 1995. How bacteria sense and swim. *Annu. Rev. Microbiol.* **49**:489–522.) Additional evidence that methylation of Tar stimulates CheA kinase activity is that membranes prepared from cells in which glutamine is substituted for glutamate in Tar, or in which the glutamate residues are methylated because the cells were incubated with aspartate, are more active in stimulating the phosphorylation of CheY than membranes containing unmodified chemoreceptor.

178. Presumably the binding of chemoattractant induces a conformational change in the receptor–transducer protein, exposing additional methylating sites on the receptor–transducer protein, thus increasing the level of methylation.

179. Park, C., D. P. Dutton, and G. L. Hazelbauer. 1990. Effects of glutamines and glutamates at sites of covalent modification of a methyl-accepting transducer. *J. Bacteriol.* **172**:7179–7187.

180. Stock, J. B., and M. G. Surette. 1996. Chemotaxis, pp. 1103–1129. In: *Escherichia coli and Salmonella: Cellular and Molecular Biology*. Vol. I. F. C. Neidhardt et al. (Eds.). ASM Press, Washington, DC.

181. Stock, J., A. Borczuk, F. Chiou, and J. E. B. Burchenal. 1985. Compensatory mutations in receptor function: a reevaluation of the role of methylation in bacterial chemotaxis. *Proc. Natl. Acad. Sci. USA* **82**:8364–8368.

182. Stock, J., G. Kersulis, and D. E. Koshland Jr. 1985. Neither methylating nor demethylating enzymes are required for bacterial chemotaxis. *Cell* **42**:683–690.

183. Stock, J., and A. Stock. 1987. What is the role of receptor methylation in bacterial chemotaxis? *Trends Biochem. Sci.* **12**:371–375.

184. There are several ways to measure chemotaxis. One way is to use swarm plates. If a chemotactic bacterium is inoculated in the center of semisolid agar fortified by attractants it can metabolize, then as the cells grow, the colony swarms outward from the center. This is because the growing cells create a gradient of attractant as they utilize it. In another assay, a capillary is filled with a solution of an attractant and placed in a suspension of cells without attractant. The cells swim into the capillary following the gradient. One then plates out the cells in the capillary to quantitate the chemotaxis. A third method is to observe tethered cells with a microscope. The cells can be tethered to a coverslip coated with antiflagellin antibody. The coverslip is then placed over a chamber through which various solutions can be added and the clockwise or counterclockwise rotation of the bacteria can be monitored. For *E. coli*, CCW rotation is equivalent to smooth swimming and CW rotation is equivalent to tumbling. One might measure the percentage of cells that rotated CCW over a specific time period. For example, suppose the measurement was for 15 s. Upon addition of attractant, the percentage of cells that rotate CCW without reversing over a 15 s period might be 100%, then falling to the prestimulus level, which might be 80%, in 1 or 2 min as adaptation occurred. (As explained in Section 18.12.2, not all bacteria reverse the direction of flagellar rotation.)

185. Eisenbach, M., C. Constantinou, H. Aloni, and M. Shinitzky. 1990. Repellents for *Escherichia coli* operate neither by changing membrane fluidity nor by being sensed by periplasmic receptors during chemotaxis. *J. Bacteriol.* **172**:5218–5224.

186. Reviewed in: Titgemeyer, F. 1993. Signal transduction in chemotaxis mediated by the bacterial phosphotransferase system. *J. Cell. Biochem.* **51**:69–74.

187. Eisenbach, M. 1996. Control of bacterial chemotaxis. *Mol. Microbiol.* **20**:903–910.

188. Alexandre, G., and I. B. Zhulin. 2001. More than one way to sense chemicals. *J. Bacteriol.* **183**:4681–4686.

189. Armitage, J. P., and R. Schmidtt. 1997. Bacterial chemotaxis: *Rhodobacter sphaeroides* and *Sinorhizobium melloti*—variations on a theme? *Microbiology* **143**:3671–3682.

190. Ward, M. J., A. W. Bell, P. A. Hamblin, H. L. Packer, and J. P. Armitage. 1995. Identification of a chemotaxis operon with two *cheY* genes in *Rhodobacter sphaeroides*. *Mol. Microbiol.* **17**:357–366.

191. Hamblin, P. A., Maguire, B. A., Grishanin, R. N., and J. P. Armitage. 1997. Evidence for two chemosensory pathways in *Rhodobacter sphaeroides*. *Mol. Microbiol.* **26**:1083–1096.

192. The chemotaxis operon from *R. sphaeroides* was detected by using a cloned *cheA* homologue from *Rhizobium meliloti*. The chemotaxis operon was sequenced, revealing the presence of the *che* genes.

193. Ward, M. J., D. M. Harrison, M. J. Ebner, and J. P. Armitage. 1995. Identification of a methyl-accepting chemotaxis protein in *Rhodobacter sphaeroides*. *Mol. Microbiol.* **18**:115–121.

194. The *R. sphaeroides* MCP protein was discovered by sequencing a fragment of DNA upstream of the chemotaxis operon. The deduced amino acid sequence of the protein product shows significant similarity to an enteric MCP (MCP Tsr). The protein cross-reacts with antisera to the McpA protein of *Canlobacter crescentus* and was located primarily in the cytoplasmic fraction, as determined by Western blotting. For Western blotting, the proteins of the membrane an cytoplasmic cell fractions are first separated by gel electrophoresis (SDS-PAGE). Then the protein bands are electrophoretically transferred to nitrocellulose. The proteins on the nitrocellulose are incubated with antisera (e.g., to anti-McpA antibody), and the bands that bind the antibody are stained by means of an anti-rabbit secondary antibody (e.g., goat anti-rabbit antibody) conjugated to horseradish peroxidase. An assay for the peroxidase stains the bands.

195. Armitage, J. P. 1997. Behavioral responses of bacteria to light and oxygen. *Arch. Microbiol.* **168**:249–261.

196. Y. Jeziore-Sassoon, P. A. Hamblin, C. A. Bootle-Wilbraham, P. S. Poole, and J. P. Armitage. 1998. Metabolism is required for chemotaxis to sugars in *Rhodobacter sphaeroides*. *Microbiology* **144**:229–239.

197. Part of the evidence that metabolism is required for the chemotactic response to carbohydrate attractants is that mutants of *R. sphaeroides* lacking glucose-6-phosphate dehydrogenase, a key enzyme in the Entner–Doudoroff (and pentose phosphate) pathway, are defective in chemotaxis towards sugars metabolized by this pathway. See ref. 196.

198. Packer, H. L., and J. P. Armitage. 1994. The chemokinetic and chemotactic behavior of *Rhodobacter sphaeroides*: two independent responses. *J. Bacteriol.* **176**:206–212.

199. Schnitzer, M. J., S. M. Block, H. C. Berg, and E. M. Purcell. 1990. Strategies for chemotaxis. *Symp. Soc. Gen. Microbiol.* **46**:15–34.

200. Gotz, R., and R. Schmitt. 1987. *Rhizobium meliloti* swims by unidirectional, intermittent rotation of right-handed flagellar helices. *J. Bacteriol.* **169**:3146–3150.

201. Schmitt, R. 2002. Sinorhizobial chemotaxis: a departure from the enterobacterial paradigm. *Microbiology* **148**:627–631.

202. Greck, M., J. Platzer, V. Sourjik, and R. Schmitt. 1955. Analysis of a chemotaxis operon in *Rhizobium meliloti*. *Mol. Microbiol.* **15**:989–1000.

203. Sockett, R. E., J. P. Armitage, and M. C. W. Evans. 1987. Methylation-independent and methylation-dependent chemotaxis in *Rhodobacter sphaeroides* and *Rhodospirillum rubrum*. *J. Bacteriol.* **169**:5808–5814.

204. Zhulin, I. G., and J. P. Armitage. 1993. Motility, chemokinesis, and methylation-independent chemotaxis in *Azospirillum brasilense*. *J. Bacteriol.* **175**:952–958.

205. Hellingwerf, K. J., W. D. Hoff, and W. Crielaard. 1996. Photobiology of microorganisms: how photosensors catch a photon to initialize signalling. *Mol. Microbiol.* **21**:683–693.

206. Alexandre, G., S. Greer-Phillips, and I. B. Zhulin. 2004. Ecological role of energy taxis in microorganisms. *FEMS Microbiol. Rev.* **28**:113–126.

207. Alam, M., and D. Oeserhelt. 1984. Morphology, function and isolation of halobacterial flagella. *J. Mol. Biol.* **176**:459–475.

208. Spudich, J. L. 1993. Color sensing in the archaea: a eukaryotic-like receptor coupled to a prokaryotic transducer. *J. Bacteriol.* **175**:7755–7761.

209. Bickel-Sandkötter, S. 1996. Conversion of energy in halobacteria: ATP synthesis and phototaxis. *Arch. Microbiol.* **166**:1–11.

210. Armitage, J. P. 1997. Behavioral responses of bacteria to light and oxygen. *Arch. Microbiol.* **168**:249–261.

211. By way of introduction, the student should know that all photosensing involves the absorption of a photon of light by a chromophore bound to a protein. When the photon is absorbed, one or more of the following four events are induced: (1) excitation transfer, (2) electron transfer, (3) H^+ transfer, (4) photoisomerization. As we shall see, photoisomerization and H^+ transfer occur when the rhodopsins absorb light energy. Two photosensory rhodopsin pigments, called SR I and SR II, are responsible for the photoresponses. SR II is made constitutively at relatively low levels, but SR I is induced to high levels under growth conditions that also induce bacteriorhodopsin (i.e., low oxygen tensions and intense illumination). It is clear that the rhodopsins that function in photoresponses are similar to bacteriorhodopsin and halorhodopsin. (See Sections 3.8.4 and 3.9 for a discussion of bacteriorhodopsin and halorhodopsin, respectively.) These pigment proteins are related structurally and presumably have a common origin. Their primary photochemistry is the same: a light-induced photoisomerization of the retinal from the all-trans isomer to the 13-*cis* form. When the rhodopsins absorb light, a signal is transmitted to the flagellum motor, probably via a membrane-bound signal-transducing protein, Htr I (halobacterial transducer for sensory rhodopsin I) to which SR I is tightly bound. (A second protein, Htr II, is complexed to SR II.) The Htr proteins are similar to the MCP proteins in that they

have two membrane-spanning domains and a cytoplasmic domain that is homologous to the MCPs. Presumably a conformational change takes place in Htr when light is absorbed by the chromophore, and a signal is thus generated to the Che proteins that signal the flagellum motor. The model proposes that SR I exists in either one of two conformations: SR_{587} or S_{373}, the where the subscripts denote maximum absorption peaks, in nanometers. The two conformations are interconvertible. When SR_{587} absorbs light (orange/red) it is converted to S_{373}. When S_{373} absorbs light (UV/blue) it is converted back to SR_{587}. There are approximately 5,000 molecules per cell of SR I, and in the presence of steady background light, an equilibrium exists between the two forms. SR_{587} is the receptor for attractant light. The absorption of light by SR_{587} suppresses flagellar reversals, causing the cells to swim toward higher intensities of orange/red light. S_{373} is the receptor for repellent light. The absorption of light by S_{373} increases flagellar reversals, causing the cells to swim away from high intensities of UV/blue light. This is an interesting case of the same photoreceptor serving as either the attractant or the repellent light receptor, depending upon its conformation. It has been presumed that Htr I undergoes a conformational change upon the absorption of light by SR I and transmits the signal to the flagellar motor. It has been demonstrated that SR I is capable of light-induced proton pumping similar to bacteriorhodopsin proton pumping, but in the presence of tightly bound Htr I the protons circulate, rather than being released into the bulk phase. Either the protons circulate through Htr I residues during SR I photocycle (a direct involvement of Htr I in proton circulation) or proton circulation is dependent upon conformational interactions between SR I and Htr I (an indirect role for Htr I in proton circulation).

212. Grishanin, R. N., D. E. Gauden, and J. P. Armitage. Photoresponses in *Rhodobacter sphaeroides*; role of photosynthetic electron transport. *J. Bacteriol.* **179**:24–30.

213. Sackett, M. J., J. P. Armitage, E. E. Sherwood, and T. P. Pitta. 1997. Photoresponses of the purple nonsulfur bacteria *Rhodospirillum centenum* and *Rhodobacter sphaeroides*. *J. Bacteriol.* **179**:6764–6768.

214. Ragatz, L., Z.-Y. Jiang, C. Bauer, and H. Gest. 1995. Macroscopic phototactic behavior of the purple photosynthetic bacterium *Rhodospirillum centenum*. *Arch. Microbiol.* **163**:1–6.

215. Jiang, Z.-Y., B. G. Rushing, Y. Bai, H. Gest, and C. E. Bauer. 1998. Isolation of *Rhodospirillum centenum* mutants defective in phototactic colony motility by transposon mutagenesis. *J. Bacteriol.* **180**:1248–1255.

216. Sequence analysis of one of the mutants indicates the presence of a photoreceptor that has extensive sequence similarity with MCPs. (Carl Bauer, personal communication.)

217. Shioi, J., R. C. Tribhuwan, S. T. Berg, and B. L. Taylor. 1988. Signal transduction in chemotaxis to oxygen in *Escherichia coli* and *Salmonella typhimurium*. *J. Bacteriol.* **170**:5507–5511.

218. Rowsell, E. H., J. M. Smith, A. Wolfe, and B. L. Taylor. 1995. CheA, CheW, and CheY are required for chemotaxis to oxygen and sugars of the phosphotransferase system in *Escherichia coli*. *J. Bacteriol.* **177**:6011–6014.

219. Bibikov, S. I., A. C. Miller, K. K. Gosink, and J. S. Parkinson. 2004. Methylation-independent aerotaxis mediated by the *Escherichia coli* Aer protein. *J. Bacteriol.* **186**:3730–3737.

220. Winans, S. C., and B. L. Bassler. 2002. Mob psychology. *J. Bacteriol.* **184**:873–883.

221. Miller, M. B., and B. L. Bassler. 2001. Quorum sensing in bacteria. *Annu. Rev. Microbiol.* **55**:165–199.

222. Fuqua, C., and E. P. Greenberg. 2002. Listening in on bacteria: acyl–homoserine lactone signalling. *Nat. Rev. Mol. Cell Biol.* **3**:685–688.

223. Taga, M. E., and B. L. Bassler. 2003. Chemical communication among bacteria. *Proc. Natl. Acad. Sci. USA* **100** (suppl. 2): 14549–14554.

224. Pappas, K. M., C. L. Weingart, and S. C. Winans. 2004. Chemical communication in proteobacteria: biochemical and structural studies of signal synthases and receptors required for intercellular signalling. *Mol. Microbiol.* **53**:755–770.

225. Dworkin, M., and D. Kaiser (Eds.). 1993. *Myxobacteria II.* ASM Press, Washington, DC.

226. Brun, Y. V., and L. J. Shimkets (Eds.). 2000. *Prokaryotic Development.* ASM Press, Washington, DC.

227. Kaplan, H. B. 2003. Multicellular development and gliding motility in *Myxococcus xanthus*. *Curr. Opin. Microbiol.* **6**:572–577.

228. Kaiser, D. 2003. Coupling cell movement to multicellular development in myxobacteria. *Nat. Rev. Microbiol.* **1**:45–54.

229. Hodgkin, J., and D. Kaiser. 1979. Genetics of gliding motility in *Myxococcus xanthus* (Myxobacterales): genes controlling movement of single cells. *Mol. Gen. Genet.* **171**:167–176.

230. Hodgkin, J., and D. Kaiser. 1979. Genetics of gliding motility in *Myxococcus xanthus* (Myxobacterales): two gene systems control movement. *Mol. Gen. Genet.* **171**:177–191.

231. Lancero, H., N. B. Caberoy, S. Castaneda, Y. Li, A. Lu, D. Dutton, X.-Y. Duan, H. B. Kaplan, W. Shi, and A. G. Garza. 2004. Characterization of a *Myxococcus xanthus* mutant that is defective for adventurous motility and social motility. *Microbiology* **150**:4085–4093.

232. Mutants in *nla24* are defective in the production of (extracellular polymeric substance (EPS), which is part of the polysaccharide/protein fibril matrix that plays a role in S-motility, as well as the expression of the *aglIU* and *calB* genes that are required for A-motility. (The EPS and associated protein are sometimes collectively referred to as the extracellular matrix, or ECM.) See ref. 231.

233. Wu, S. S., and D. Kaiser. 1997. Regulation of expression of the *pilA* gene in *Myxococcus xanthus*. *J. Bacteriol.* **179**:7748–7758.

234. Sun, H., D. R. Zusman, and W. Shi. 2000. Type IV pilus of *Myxococcus xanthus* is a motility apparatus controlled by the *frz* chemosensory system. *Curr. Biol.* **10**:1143–1146.

235. Kaiser, D. 2000. Bacterial motility: how do pili pull? *Curr. Biol.* **10**:R777–R780.

236. The type IV pilus is constructed of subunits, which in *Pseudomonas aeruginosa* and *Myxococcus xanthus* are called PilA (or pilin). One can think of the production of type IV pili as occurring in three post-transcriptional stages. Stage 1 involves the removal of the leader peptide from prepilin, the precursor to the pilin subunit. Stage 2 is the assembly of PilA into the pilus in the periplasmic space. And stage 3, which is coupled to stage two, is extrusion of the pilus through the outer envelope. As described in Chapter 17, several proteins required for pilus assembly are homologous to proteins involved with type II protein secretion. When the pilus retracts, it disassembles at its base and the PilA subunits remain in the inner membrane (cell membrane) and are recycled when pili are reassembled. *M. xanthus* and *P. aeruginosa* have a putative two-component system that regulates the expression of the *pilA* gene. In both organisms, PilR is suggested to be the positive transcription regulator that is part of the two-component signaling system. Wu and Kaiser reported that the putative sensor histidine kinase, PilS, is a negative regulator of *pilA* expression in *M. xanthus*.[233] The energy required for disassembly and retraction of the pilus is thought to be provided by ATP. One of the proteins required for retraction is PilT, which is an inner membrane protein that is thought to mediate the disassembly of the pili. PilT is a nucleotide-binding protein, possibly an ATPase. Mutations in the *pilT* gene result in hyperpiliated cells because the pili do not retract. Another protein, PilQ, forms the export pore in the outer membrane and is necessary for the extrusion of pili. Homologous proteins involved in the biogenesis and function of type IV pili in different bacteria may have different names.

As discussed in Section 18.16.7, *M. xanthus* switches the type IV pili to the opposite cell pole when the cell reverses its direction of gliding motility. This is due to inactivation of the the pilus machinery at one pole and activation at the opposite pole, rather than switching the entire machinery from pole to pole. The evidence in favor of this includes immunochemical labeling experiments that show that fluorescent PilQ and perhaps other Pil proteins remain as a protein complex in the absence of pili at the pole that is not being used for pilus extension and retraction (reviewed in Kaiser, D., and R. Yu. 2005. Reversing cell polarity: evidence and hypothesis. *Curr. Opin. Microbiol.* **8**:216–221). Furthermore, the labeling experiments indicated that little PilQ was at the sides of the cell, indicating that it did not diffuse in the outer membrane from pole to pole.

A gene required for social motility called the *tgl* gene (transient gliding) is necessary for the assembly of the polar type IV pili in *M. xanthus*. A pure culture of mutants that are *tgl⁻* cannot move via the S-system because the cell variants cannot assemble pili. Thus, there is no movement in A⁻ cells that are also *tgl⁻*. However, *tgl⁻* mutants that are A⁻ can be stimulated to move via the S-system by cells that are *tgl⁺*. The basis for the stimulation is that Tgl is an outer membrane protein that is thought to be transferred to other cells via pole-to-pole contact. Thus, *tgl⁺* cells can stimulate *tgl⁻pilA⁺* cells, which have the pilus subunits in their cell membrane, to assemble the subunits into pili upon contact at the pole, and then to move. The recipient cells remain *tgl⁻*, and therefore the movement is transient and lasts as long as the recipient cells retain sufficient amounts of Tgl that was transferred to them. One speculation is that the mechanism of transfer between cells is via some process of membrane–vesicle fusion between the poles of interacting cells. Tgl transfer is thought to also take place between wild-type cells and to stimulate pili assembly in recipient cells. In this way, the assembly of type IV pili is stimulated when cells come into end-to-end contact. This stimulation speeds up the frequency of motility reversal and appears to be important for developmental aggregation. For more information, read: Rodriguez-Soto, J. P., and D. Kaiser. 1997. The *tgl* gene: social motility and stimulation in *Myxococcus xanthus*. *J. Bacteriol.* **179**:4361–4371; Identification and localization of the Tgl protein, required for *Myxococcus xanthus* social motility. *J. Bacteriol.* **179**:4372–4381.

237. Li, Y., H. Sun, X. Ma, A. Lu, R. Lux, D. Zusman, and W. Shi. 2003. Extracellular polysaccharides mediate pilus retraction during social motility of *Myxococcus xanthus*. *Proc. Natl. Acad. Sci.* **100**:5443–5448.

238. *M. xanthus* is covered by proteoglycan fibrils (10–30 nm in diameter) consisting of 50% protein and 50% polysaccharide. The fibrils interconnect neighboring cells. It appears that the type IV pili at the forward-moving pole of *M. xanthus* cells adhere to the polysaccharide portion of the fibrils attached to adjacent cells in front, and that binding of the pili to the polysaccharide triggers pilus retraction and movement of the cell in the forward direction via a "pulling" force. This is supported, in part, by the study of mutants defective in the synthesis of the extracellular fibrils. For more information read ref. 237, as well as: Bellenger, K., M. Xiaoyuan,

W. Shi, and Z. Yang. 2002. A *cheW* homologue is required for *Myxococcus xanthus* fruiting body development, social gliding motility, and fibril biogenesis. *J. Bacteriol.* **184**:5654–5660; Lu, A., K. Cho, W. P. Black, X.-Y. Duan, R. Lux, Z. Yang, H. B. Kaplan, D. R. Zusman, and W. Shi. 2005. Exopolysaccharide biosynthesis genes required for social motility in *Myxococcus xanthus*. *Mol. Microbiol.* **55**:206–216.

239. Bowden, M. G., and H. B. Kaplan. 1998. The *Myxococcus xanthus* lipopolysaccharide O-antigen is required for social motility and multicellular development. *Mol. Microbiol.* **30**:275–284.

240. Bellenger, K., M. Xiaoyuan, W. Shi, and Z. Yang. 2002. A *cheW* homologue is required for *Myxococcus xanthus* fruiting body development, social gliding motility, and fibril biogenesis. *J. Bacteriol.* **184**:5654–5660.

241. Semmler, A. B. T., C. B. Whitchurch, and J. S. Mattick. 1999. A re-examination of twitching motility in *Pseudomonas aeruginosa*. *Microbiology* **145**:2863–2873.

242. Spormann, A. M., and D. Kaiser. 2003. Gliding mutants of *Myxococcus xanthus* with high reversal frequencies and small displacements. *J. Bacteriol.* **181**:2593–2601.

243. Wolgemuth, C., E. Hoiczyk, D. Kaiser, and G. Oster. 2002. How myxobacteria glide. *Curr. Biol.* **12**:369–377.

244. There is much less known about the mechanism of A-motility than to S-motility. To learn more about A-motility, researchers one identifying the genes and proteins the process requires. Two genes in the A-motility system that have been studied are the *CgjB* gene and the *aglU* gene. Both the encoded proteins have been suggested to be outer membrane lipoproteins. The *aglU* gene is important not only for A-motility but also for the expression of genes required to complete myxospore formation. Other genes important for A-motility include genes that encode protein homologues of the *E. coli* Tol proteins, which are part of an outer membrane channel that functions in biopolymer transport out of the cell. (See Section 17.5.1.) The Tol homologues in *M. xanthus* may function as a complex in the transport of slime through the nozzles. Recall that one model for A-locomotion is that the cells are "pushed" by slime excretion through nozzles at the rear of the cell. For a discussion of these genes, and how they are identified, read: Rodriguez, A. M., and A. M. Spormann. 1999. Genetic and molecular analysis of *clgB*, a gene essential for single-cell gliding in *Myxococcus xanthus*. *J. Bacteriol.* **181**:4381–4390; White, D. J., and P. L. Hartzell. 2000. AglU, a protein required for gliding motility and spore maturation of *Myxococcus xanthus*, is related to WD-repeat proteins. *Mol. Microbiol.* **36**:662–678; Youderian, P., N. Burke, D. J. White, and P. L. Hartzell. 2003. Identification of genes required for

adventurous gliding motility in *Myxococcus xanthus* with the transposable element *mariner*. *Mol. Microbiol.* **49**:555–570.

245. Wild-type cells in either the A- or S-motility system can stimulate mutant cells in the corresponding system to move transiently if the wild-type and mutant cells are not separated from each other by a space on the agar or by a filter membrane. For example, a cell mutated in one of five different genes, *cgl*B, *cgl*C, *cgl*D, *cgl*E, and *cgl*F (called *cgl*, for contact or conditional gliding) required for A-motility can be stimulated to move via A-motility if it makes contact with a wild-type cell that is not mutated in that particular gene. Thus, a Cgl protein is transferred to the mutant cell via cell–cell contact. The Cgl proteins have been suggested to be involved in the structure and function of the A-motor, and/or in signaling. For example, some of the proteins may be part of the motor, and some of the proteins may be outer membrane proteins. The CglB protein has been reported to be a membrane lipoprotein. In addition, a cell mutated for the *tgl* gene, which is required for S-motility, can be stimulated to move via S-motility if it makes contact with a *tgl*⁺ cell. As we shall see later, the Tgl protein is a membrane lipoprotein protein required for the synthesis of type IV pili that are required for S-motility. For more information, consult: Hodgkin, J., and D. Kaiser. 1977. Cell-to-cell stimulation of movement in nonmotile mutants of *Myxococcus*. *Proc. Natl. Acad. Sci. USA* **74**:2938–2942; Kaiser, D., and L. Kroos. 1993. Intercellular signaling. pp. 257–283. In: *Myxobacteria II*. M. Dworkin and D. Kaiser (Eds.). ASM Press, Washington, DC; Rodriguez, A. M., and A. M. Spormann. 1999. Genetic and molecular analysis of *clgB*, a gene essential for single-cell gliding in *Myxococcus xanthus*. *J. Bacteriol.* **181**:4381–4390.

246. Shimkets, L. J., 1999. Intercellular signaling during fruiting body development of *Myxococcus xanthus*. *Annu. Rev. Microbiol.* **53**:525–549.

247. Søgaard-Andersen, L., M. Overgaard, S. Lobedanz, E. Ellehauge, L. Jelsbak, and A. A. Rasmussen. 2003. Coupling gene expression and multicellular morphogenesis during fruiting body formation in *Myxococcus xanthus*. *Mol. Microbiol.* **48**:1–8.

248. Søgaard-Andersen, L. 2004. Cell polarity, intercellular signalling and morphogenetic cell movements in *Myxococcus xanthus*. *Curr. Opin. Microbiol.* **7**:587–593.

249. Kaiser, D. 2004. Signaling in myxobacteria. *Annu. Rev. Microbiol.* **58**:75–98.

250. Hagen, D. C., A. P. Bretscher, and D. Kaiser. 1978. Synergism between morphogenetic mutants of *Myxococcus xanthus*. *J. Bacteriol.* **172**:15–23.

251. The pentas and tetraphosphate nucleotide-(p)ppGpp increase in intracellular concentration in *M. xanthus* during starvation-induced development. Mutations in *relA*, the gene that encodes the enzyme

(p)ppGpp synthetase I required to synthesize ppGpp from ATP and GTP, are defective in development and do not produce A-signal. It has also been shown that genes expressed during development are stimulated by an increase in intracellular concentration of ppGpp. This was determined by introducing into *M. xanthus* a copy of the *E. coli relA* gene. The *relA* gene was placed in a plasmid under the control of a light-inducible promoter from *M. xanthus* (*carQRS*). The plasmid was introduced into *M. xanthus* by electroporation, and it integrated into the *M. xanthus* chromosome at the Mx8 prophage attachment site. Light caused an increase in the intracellular amounts of ppGpp and pppGpp either under starvation or non-starvation conditions in strains carrying the *relA* gene under the control of the light-inducible *carQRS* promoter. Importantly, the expression of genes normally expressed during development (monitored by the expression of developmentally specific transcriptional *lacZ* fusions that are normally expressed early during starvation) was increased by light in the strains carrying the light-inducible *relA* gene. The light-inducible, developmentally expressed genes included a *lacZ* fusion gene whose expression normally requires A-signal. Mutants unable to synthesize A-signal did not show the light-dependent expression of the A-signal-dependent fusion gene. This indicates that A-signaling and (p)ppGpp are in the same signaling pathway. One possibility that has been mentioned is that (p)ppGpp increases intracellularly during starvation and that the increased levels of (p)ppGpp signal the cells to make A-factor, which serves as an intercellular quorum-sensing signal resulting in the activation of expression of developmental genes. For a further discussion of the role of (p)ppGpp in *M. xanthus* development, see: Singer, M., and D. Kaiser. 1995. Ectopic production of guanosine penta- and tetraphosphate can initiate early developmental gene expression in *Myxococcus xanthus*. *Genes Dev.* 9:1633–1644.

252. Research with other bacteria has determined that the intracellular signal (p)ppGpp accumulates under conditions of nutritional stress, such as amino acid starvation, nitrogen limitation, or a shift to a poorer carbon or energy source. It has been referred to as an "alarmone." It is responsible for the stringent response, which is the inhibition of rRNA and tRNA synthesis by increased levels of (p)ppGpp triggered by amino acid starvation. See Section 2.2.2 for a discussion of the stringent response and (p)ppGpp. As discussed in Section 2.2.2, particularly note 22, it is known from studying other bacteria that (p)ppGpp does not simply inhibit the synthesis of rRNA and tRNA but is a global regulator and influences the expression of many genes. For example, it stimulates the expression of virulence genes in *Salmonella* in the intestinal lumen allowing the invasion of intestinal epithelial cells. See: Pizzaro-Cerdá, J., and K. Tedin. 2004. The bacterial signal molecule, ppGpp, regulates *Salmonella* virulence gene expression. *Mol. Microbiol.* 52:1827–1844.

There exists a model for how (p)ppGpp stimulates the expression of many genes. The (p)ppGpp binds to the core RNA polymerase (RNAP), and this results in the inhibition of synthesis of rRNA and tRNA. One suggestion is that the binding of (p)ppGpp to the core polymerase (E) destabilizes σ^{70}-E promoter complexes at the stringent promoters, increasing the pool of free RNA polymerase, as well as reducing the ability of σ^{70} to compete with alternative sigma factors for the core enzyme. [Think of this as (p)ppGpp "loosening" the binding of σ^{70} to the core.] This results in (p)ppGpp-dependent stimulation of the expression of genes dependent upon alternative sigma factors, as well as genes dependent upon σ^{70}.

253. The expression of genes responsible for generating the A-signal requires the *asg* genes. Two of the genes encode proteins that are thought to be part of a phosphorelay system containing at least one hybrid histidine kinase–response regulator protein (AsgA) and one putative DNA-binding protein (AsgB), which is not a response regulator protein. The third gene, *asgC*, encodes a sigma factor. The evidence for these conclusions is based upon nucleotide sequence analysis. The nucleotide sequence of *asgA* suggests that its protein product, AsgA, has both histidine kinase and response regulator domains; that is, it might be a bifunctional protein. Interestingly, there is no membrane-spanning domain, indicating that AsgA is a cytoplasmic protein and probably senses the accumulation of ppGpp, a small cytoplasmic molecule whose levels increase upon starvation. Cytoplasmic sensor kinases are known in other two-component-based systems. For example, as discussed in Section 18.18.3, KinA is a cytoplasmic histidine kinase in the *Bacillus subtilis* sporulation phosphorelay pathway, and we know from Section 18.4.1 that NR_{II} is a cytoplasmic histidine kinase for the *Escherichia coli* Ntr regulon. It has also been deduced from nucleotide analysis that AsgB is a DNA-binding protein that may be a transcription factor that alters gene expression, resulting in the production of A-signal, and that the *asgC* gene encodes the major sigma factor RpoD (σ^{70}). One would expect that a mutation in the major sigma factor would result in a defect in growth. However, the *asgC* mutants do not show a general growth defect, indicating that the mutation does not affect transcriptional initiation in genes required for growth. One suggestion is that perhaps the mutation prevents the RpoD sigma factor from exchanging with a developmentally specific sigma factor required for transcription of a gene(s) required for production of A-signal. Clearly, there is much to be learned about the role of the *asg* genes in coupling the starvation signal to the synthesis of A-signal.

254. Plamann, L., L. Yonghui, B. Cantwell, and J. Mayor. 1995. The *Myxococcus xanthus asgA* gene encodes a novel signal transduction protein required for multicellular development. *J. Bacteriol.* 177:2014–2022.

255. Plamann, L., M. Davis, B. Cantwell, and J. Mayor. 1994. Evidence that *asgB* encodes a DNA-binding protein essential for growth and development of *Myxococcus xanthus*. *J. Bacteriol.* **176**:2013–2020.

256. Davis, J., J. Mayor, and L. Plamann. 1995. A missense mutation in *rpoD* results in an A-signaling defect in *Myxococcus xanthus*. *Mol. Microbiol.* **18**:943–952.

257. Horiuchi, T., M. Taoka, T. Isobe, T. Komano, and S. Inouye. 2002. Role of *fruA* and *csgA* genes in gene expression during development of *Myxococcus xanthus*. *J. Biol. Chem.* **30**:26753–26760.

258. Gronewold, T. M. A., and D. Kaiser. 2002. *act* Operon control of developmental gene expression in *Myxococcus xanthus*. *J. Bacteriol.* **184**:1172–1179.

259. Lobedanz, S., and L. Søgaard-Andersen. 2003. Identification of the C-signal, a contact-dependent morphogen coordinating multiple developmental responses in *Myxococcus xanthus*. *Genes Dev.* **7**:2151–2161.

260. Plamann, L., A. Kuspa, and D. Kaiser. 1992. Proteins that rescue A-signal-defective mutants of *Myxococcus xanthus*. *J. Bacteriol.* **174**:3311–3318.

261. Kuspa, A., L. Plamann, and D. Kaiser. 1992. Identification of heat-stable A-factor from *Myxococcus xanthus*. *J. Bacteriol.* **174**:3319–3326.

262. Kuspa, A., L. Plamann, and D. Kaiser. 1992. A-signaling and the cell density requirement for *Myxococcus xanthus* development. *J. Bacteriol.* **174**:7360–7369.

263. Yang, C., and H. B. Kaplan. 1997. *Myxococcus xanthus sasS* encodes a sensor histidine kinase required for early developmental gene expression. *J. Bacteriol.* **179**:7759–7767.

264. Mutants that do not make the C-signal have mutations in the *csgA* gene. The gene encodes two proteins, one with a molecular weight of 25 kDa, and another with a molecular weight of 17 kDa. The 25 kDa protein is homologous to NAD(P)-dependent short-chain alcohol dehydrogenases and, when added to *csgA* cells, rescues their development. The 17 kDa protein corresponds to the C-terminal end of the 25 kDa protein and does not bind NAD+. However, the 17 kDa protein can also rescue *csgA* mutants, even though it is not an alcohol dehydrogenase. It appears that the actual signal is the 17 kDa protein and that it forms as a result of proteolysis of the 25 kDa protein (step **13**, Fig. 18.24).[248,259]

265. A model for the role of C-signaling in aggregate formation has been proposed. (See: Jelsbak, L., and L. Søgaard-Andersen. 2002. Pattern formation by a cell surface–associated morphogen in *Myxococcus xanthus*. *Proc. Natl. Acad. Sci. USA* **99**:2032–2037.) At high cell densities, more cells transmit C-signal because there is more end-to-end contact. This results in chains of cells that move in the same direction and at the same speed, and reverse less frequently. For each chain of cells, the direction of movement is determined by the cell at the front of the chain, and all the cells in the chain follow the lead cell. Since cells at high cell density tend to stay in parallel contact with one another, several parallel chains of cells, referred to as streams, can develop. Aggregation centers may form when chains and streams of cells entering the aggregate begin moving in large concentric circles and become trapped there. (See: White, D. 1993. Myxospore and fruiting body morphogenesis, pp. 307–346. In: *Myxobacteria II*. M. Dworkin and D. Kaiser (Eds.). ASM Press, Washington, DC.)

Actually, there are two responses to the C-signal, depending upon how many C-signal molecules are on the surface of the donor cell. If only a few C-signal molecules are present on the donor surface, the direction of gliding reverses. This occurs prior to aggregation during a period when the movement of groups of cells occurs in travelling waves and is called rippling. During rippling, C-signaling stimulates cell reversals. (For a description of rippling and other aspects of signaling in myxobacteria, see: Kaiser, D. 2004. Signaling in myxobacteria. *Annu. Rev. Microbiol.* **58**:75–98.) However, as more end-to-end cell contact is made, the recipient cell responds by making more C-signal molecules in a positive feedback loop (the *act* operon), adding these molecules to the cell surface. When a threshold number of C-signaling molecules have been added to the cell surface, the frequency of gliding reversal in responding cells is depressed, and the speed of gliding is increased as the cells move in streams, making end-to-end contact. Cells move toward the aggregation centers in these streams and continue to move as streams in concentric circles within the aggregates. The result is the creation of aggregates and the development of aggregate into fruiting bodies.

266. Kim, S. K., and D. Kaiser. 1990. Purification and properties of *Myxococcus xanthus* C-factor, an intercellular signaling protein. *Proc. Natl. Acad. Sci. USA* **87**:3635–3639.

267. Sager, B., and K. Kaiser. 1995. Intercellular C-signaling and the traveling waves of *Myxococcus*. *Genes Dev.* **8**:2793–2804.

268. Kim, S. K., and D. Kaiser. 1990. Cell alignment required in differentiation of *Myxococcus xanthus*. *Science* **249**:926–928.

269. Kim, S. K., and D. Kaiser. 1990. Cell motility is required for the transmission of C-factor, an intercellular signal that coordinates fruiting body morphogenesis of *Myxococcus xanthus*. *Genes Dev.* **4**:896–904.

270. Søgaard-Anderson, L., and D. Kaiser. 1996. C factor, a cell-surface-associated intercellular signaling protein stimulates the cytoplasmic Frz signal transduction system in *Myxococcus xanthus*. *Proc. Natl. Acad. Sci. USA* **93**:2675–2679.

271. Xu, D., C. Yang, and H. B. Kaplan. 1998. *Myxococcus xanthus sasN* encodes a regulator that prevents developmental gene expression during growth. *J. Bacteriol.* **180**:6215–6223.

272. Kaiser, D., and R. Yu. 2005. Reversing cell polarity: evidence and hypothesis. *Curr. Opin. Microbiol.* **8**:216–221.

273. Spormann, A. M., and D. Kaiser. 1999. Gliding mutants of *Myxococcus xanthus* with high reversal frequencies and small displacements. *J. Bacteriol.* **181**:2593–2601.

274. Thomasson, B., J. Link, A. G. Stassinopoulos, N. Burke, L. Plamann, and P. L. Hartzell. 2002. MglA, a small GTPase, interacts with a tryosine kinase to control type IV pili-mediated motility and development of *Myxococcus xanthus*. *Mol. Microbiol.* **46**:1399–1413.

275. MglA is responsible for the inactivation of the A- and S-motility systems at opposite poles to ensure that a single pole does not have both motors and that net movement can take place in either the forward or backward direction. The dismantling of the S-motility machinery at one pole is accompanied by the removal of the pilin and Tgl proteins, but not of other proteins involved in S-motility. Pilin proteins are the subunits of the type IV pili, and the Tgl protein is necessary for the assembly of the type IV pili. However, other proteins required for S-motility, such as PilQ, remain. PilQ forms the export pore in the outer membrane that is necessary for the extrusion of pili. The nozzles that extrude the slime associated with A-motility remain at both cell poles but are inactivated at the pole where the S-system is functioning. Presumably, one or more key proteins required for the extrusion of slime through the nozzles, such as the CglB protein, are inactivated or removed. CglB is an outer membrane lipoprotein and may be part of the active slime nozzles at the cell pole.

276. Blackhart, B. D., and D. R. Zusman. 1985. "Frizzy" genes of *Myxococcus xanthus* are involved in control of frequency of reversal of gliding motility. *Proc. Natl. Acad. Sci. USA* **82**:8767–8770.

277. Ward, M. J., and D. R. Zusman. 2000. Developmental aggregation and fruiting body formation in the gliding bacterium *Myxococcus xanthus*, pp. 243–262. In: *Prokaryotic Development*. Y. V. Brun, and L. J. Shimkets (Eds.). ASM Press. Washington, DC.

278. Søgaard-Anderson, L. 2004. Cell polarity, intercellular signalling and morphogenetic cell movements in *Myxococcus xanthus*. *Curr. Opin. Microbiol.* **7**:587–593.

279. Jelsbak, L., and L. Søgaard-Andersen. 2002. Pattern formation by a cell surface-associated morphogen in *Myxococcus xanthus*. *Proc. Natl. Acad. Sci. USA* **99**:2032–2037.

280. Kirby, J. R., and D. R. Zusman. 2003. Chemosensory regulation of developmental gene expression in *Myxococcus xanthus*. *Proc. Natl. Acad. Sci. USA* **100**:2008–2013.

281. Vlamakis, H. C., J. R. Kirby, and D. R. Zusman. 2004. The Che4 pathway of *Myxococcus xanthus* regulates type IV pilus–mediated motility. *Mol. Microbiol.* **52**:1799–1811.

282. Berleman, J. E., and C. E. Bauer. 2005. A Che-like signal transduction cascade involved in controlling flagella biosynthesis in *Rhodospirillum centenum*. *Mol. Microbiol.* **55**:1390–1402.

283. Kearns, D. B., and L. J. Shimkets. 1998. Chemotaxis in a gliding bacterium. *Proc. Natl. Acad. Sci. USA* **95**:11957–11962.

284. Kearns, D. B., and L. J. Shimkets. 2001. Lipid chemotaxis and signal transduction in *Myxococcus xanthus*. *Trends Microbiol.* **9**:126–129.

285. Kearns, D. B., A. Venot, P. J. Bonner, B. Stevens, G.-J. Boons, and L. J. Shimkets. 2001. Identification of a developmental chemoattractant in *Myxococcus xanthus* through metabolic engineering. *Proc. Natl. Acad. Sci. USA* **98**:13990–13994.

286. Hung, D., H. McAdams, and L. Shapiro. 2000. Regulation of the *Caulobacter* cell cycle, pp. 361–378. In: Y. V. Brun, and L. Shimkets (Eds.). *Prokaryotic Development*. ASM Press, Washington, DC.

287. Ohta, N., T. W. Grebe, and A. Newton. 2002. Signal transduction and cell cycle checkpoints in developmental regulation of *Caulobacter*, pp. 341–359. In Y. V. Brun, and L. Shimkets (Eds.). *Prokaryotic Development*. ASM Press, Washington, DC.

288. Jenal, U., and C. Stephens. 2002. The *Caulobacter* cell cycle: timing, spatial organization and checkpoints. *Curr. Opin. Microbiol.* **5**:558–563.

289. Jensen, R. B., S. C. Wang, and L. Shapiro. 2002. Dynamic localization of proteins and DNA during a bacterial cell cycle. *Nat. Rev. Mol. Cell Biol.* **3**:167–176.

290. Ausmees, N., and C. Jacobs-Wagner. 2003. Spatial and temporal control of differentiation and cell cycle progression in *Caulobacter crescentus*. *Annu. Rev. Microbiol.* **57**:225–247.

291. Quardokus, E. M., and Y. V. Brun. 2003. Cell cycle timing and developmental checkpoints in *Caulobacter crescentus*. *Curr. Opin. Microbiol.* **6**:541–549.

292. McGrath, P. T., P. Viollier, and H. H. McAdams. 2004. Setting the pace: mechanisms tying *Caulobacter* cell-cycle progression to macroscopic cellular events. *Curr. Opin. Microbiol.* **7**:192–197.

293. It has been proposed that the attachment of *Caulobacter crescentus* to surfaces is a multistep process beginning with the swarmer cell. According to this model, the swimming swarmer cell is best able to make contact with the surface, and the polar pili favor attachment by reducing the radius of the cell surface. These characteristics are important because inert and bacterial surfaces often have a repulsive net negative charge. The retraction of the pili and subsequent growth of the adhesive stalk provide a stronger attachment. For a discussion of these points, see: Bodenmiller, D., E. Toh, and Y. V. Brun. 2004. Development of surface adhesion in *Caulobacter crescentus*. *J. Bacteriol.* **186**:1438–1447.

294. Sciochetti, S. A., T. Lane, N. Ohta, and A. Newton. 2002. Protein sequences and cellular factors required for polar localization of a histidine kinase in *Caulobacter crescentus*. *J. Bacteriol.* **184**:6037–6049.

295. Amon, A. 1998. Controlling cell cycle and cell fate: common strategies in prokaryotes and eukaryotes. *Proc. Natl. Acad. Sci. USA* **95**:85–86.

296. Quon, K. C., B. Yang, I. J. Domian, L. Shapiro, and G. T. Marczynski. 1998. Negative control of bacteria DNA replication by a cell cycle regulatory protein that binds at the chromosome origin. *Proc. Natl. Acad. Sci. USA* **95**:120–125.

297. Domian, I. J., K. C. Quon, and L. Shapiro. 1997. Cell type-specific phosphorylation and proteolysis of a transcriptional regulator controls the G1-to-S transition in a bacterial cell cycle. *Cell* **90**:415–424.

298. Kelly, A. J., M. J. Sackett, N. Din, E. Quardokus, and Y. V. Brun. 1998. Cell cycle-dependent transcriptional and proteolytic regulation of FtsZ in *Caulobacter*. *Genes Dev.* **12**:1–13.

299. Quardokus, E. M., N. Din, and Y. V. Brun. 1996. Cell cycle regulation and cell type-specific localization of the FtsZ division initiation protein in *Caulobacter*. *Proc. Natl. Acad. Sci. USA* **93**:6314–6319.

300. Quardokus, E. M., N. Din, and Y. V. Brun. 2001. Cell cycle and positional constraints on FtsZ localization and the initiation of cell division in *Caulobacter crescentus*. *Mol. Microbiol.* **39**:949–959.

301. The cellular amounts of any protein can be followed by using immunoblot assays. Cell extracts are subjected to SDS–polyacrylamide gel electrophoresis, which separates proteins according to their molecular weights. After electrophoresis, the proteins are transferred as bands to a nitrocellulose membrane (Western blotting). Usually electroblotting is used to effect the transfer. A blotting sandwich is constructed in which the polyacrylamide gel is covered with a nitrocellulose membrane, and a constant current is applied for several hours. The proteins are electrophoretically transferred to the membrane, where they bind tightly. The protein in question can be detected and its relative amounts measured by using specific antibody that binds to the protein (immunoblotting). Changes in the amount of protein are reflected in changes in the amount of antibody that binds to the protein band. The antibody can be labeled in some way to permit measurement of the amount that binds to the protein. For example, a radioactive antibody might be selected. Sometimes an unlabeled antibody is used which is detected via a "sandwich" reaction with a second antibody. For example, the first antibody might be rabbit IgG raised against the protein to be detected, and the second antibody would be anti-rabbit IgG. The second antibody might be labeled with a radioactive atom or a fluorescent molecule, or perhaps conjugated to an enzyme such as phosphatase, which can be detected by using a colorimetric assay. The sandwich technique is usually more sensitive.
The amounts of a specific protein can also be measured by means of immunoprecipitation (i.e., by precipitating the protein with specific antibody). The cells can be labeled with a radioactive isotope such as [^{35}S]methionine, whereupon the precipitated protein can be analyzed after SDS–polyacrylamide gel electrophoresis and the radioactivity in the protein band measured.

302. Hung, D. Y., and L. Shapiro. 2002. A signal transduction protein cues proteolytic events critical to *Caulobacter* cell cycle progression. *Proc. Natl. Acad. Sci. USA* **99**:13160–13165.

303. Quon, K. C., G. T. Marczynski, and L. Shapiro. 1996. Cell cycle control by an essential bacterial two-component signal transduction protein. *Cell* **84**:83–93.

304. Jacobs, C., N. Ausmees, S. J. Cordwell, L. Shapiro, and M. T. Laub. 2003. Functions of the CckA histidine kinase in *Caulobacter* cell cycle control. *Mol. Microbiol.* **47**:11279–1290.

305. CckA is actually a hybrid histidine kinase containing both a transmitter domain and a receiver domain at its C terminus. That is, it autophosphorylates at a histidine residue (transmitter domain) and transfers the phosphoryl group to an aspartate residue (receiver domain). The phosphoryl group is then transferred from the receiver domain of CckA to the receiver domain of CtrA. This might occur directly, or perhaps phosphotransferase enzymes take part in a phosphorelay. There are two other histidine kinases in *Caulobacter*. These are PleC and DivJ. PleC is thought to act as a phosphatase. The CckA kinase is the main phosphoryl donor for CtrA. DivL is different. It autophosphorylates at a tyrosine residue rather than at a histidine residue. Thus, it is a tyrosine kinase. There is ample evidence that DivL is part of the regulatory network. Conditional mutants of *divL* are similar to those for temperature-sensitive mutants of *ctrA* in that they do not divide and develop into very long filaments at the restrictive

temperature. There is also evidence that a truncated version of DivL is able to phosphorylate CtrA in vitro.

306. Stragier, P., and R. Losick. 1996. Molecular genetics of sporulation in *subtilis*. *Annu. Rev. Genet.* **30**:297–341.

307. Hilbert, D. W., and P. J. Piggot. 2004. Compartmentalization of gene expression during *Bacillus subtilis* spore formation. *Microbiol. Mol. Biol. Rev.* **68**:234–262.

308. Piggot, P. J., and D. W. Hilbert. 2004. Sporulation of *Bacillus subtilis*. *Curr. Opin. Microbiol.* **7**:579–586.

309. Mandelstam, J., and S. A. Higgs. 1974. Induction of sporulation during synchronized chromosome replication in *Bacillus subtilis. J. Bacteriol.* **120**:38–42.

310. Lazazzera, B. A., J. M. Solomon, and A. D. Grossman. 1997. An exported peptide functions intracellularly to contribute to cell density signaling in *B. subtilis. Cell* **89**:917–925.

311. Sonenshein, A. L. 2000. Control of sporulation initiation in *Bacillus subtilis. Curr. Opin. Microbiol.* **3**:561–566.

312. Lazazzera, B. A., and A. D. Grossman. 1997. A regulatory switch involving a Clp ATPase. *BioEssays* **19**:455–458.

313. An example of an exospore-producing bacterium is the purple nonsulfur photosynthetic bacterium *Rhodomicrobium*. Rhodomicrobium has a life cycle that includes forms with hyphae that reproduce by budding, flagellated cells that do not divide, and heat-resistant, desiccation-resistant, resting cells called exospores. The different cell types form in response to changes in light intensity, carbon dioxide levels, and nutrient levels. Each reproductive cell grows a hypha (prostheca) from one pole and buds off a daughter cell at the tip of the hypha. When the carbon supply is low, the prosthecate bacteria bud off exospores from the hyphae tips.

314. Stragier, P. 1994. A few good genes: developmental loci in *Bacillus subtilis*, pp. 207–245. In: *Regulation of Bacterial Differentiation*. P. Piggot, C. P. Moran Jr., and P. Youngman (Eds.). ASM Press, Washington, DC.

315. The activities of the sigma factors important for sporulation after septum formation and then engulfment are carefully regulated so that they become active at the appropriate time and in the appropriate compartments. In the forespore, sporulation genes are initially transcribed by σ^F-RNA polymerase immediately after asymmetric septation and prior to engulfment. After engulfment, σ^G-RNA polymerase becomes important in the forespore. In the mother cell compartment, σ^E-RNA polymerase transcribes sporulation genes prior to engulfment, followed by σ^K-RNA polymerase after engulfment. The fore-spore sigma factors, σ^F and σ^G, are kept inactive by binding to an inhibitor protein (anti-sigma factor), and the mother cell sigma factors, σ^E and σ^K, are synthesized as inactive precursors that are activated by proteolytic cleavage. As described in Box 18.2, activation of the sigma factors relies upon communication through the septal membrane between the mother cell and the forespore, emphasizing that these two compartments do not develop independently of each other.

316. Levin, P. A., and R. Losick. 2000. Asymmetric division and cell fate during sporulation in *Bacillus subtilis*, pp. 167–189. In: *Prokaryotic Development*. Y. V. Brun and L. J. Shimkets (Eds.). ASM Press, Washington, DC.

317. The transcription of genes regulated by Spo0A-P during sporulation include genes transcribed either with σ^A, which is the major sigma factor in vegetative cells, or σ^H, a sigma factor responsible for the expression of sporulation genes early in sporulation. The gene for Spo0A itself is positively regulated by means of a promoter recognized by σ^H-polymerase.

318. Molle, V., M. Fujita, S. T. Jensen, P. Eichenberger, J. E. González-Pastor, J. S. Liu, and R. Losick. 2003. The Spo0A regulon of *subtilis*. *Mol. Microbiol.* **50**:1683–1701.

319. Khvorova, A., L. Zhang, M. L. Higgins, and P. J. Piggot. 1998. The *spoIIE* locus is involved in the Spo0A-dependent switch in the location of FtsZ rings in *subtilis. J. Bacteriol.* **180**:1256–1260.

320. Burbulys, D., K. A. Trach, and J. A. Hoch. 1991. Initiation of sporulation in *B. subtilis* is controlled by a multicomponent phosphorelay. *Cell* **64**:545–552.

321. Hoch, J. A. 1993. *spo0* Genes, the phosphorelay, and the initiation of sporulation, pp. 747–755. In: *subtilis and Other Gram-Positive Bacteria: Biochemistry, Physiology, and Molecular Genetics*. A. L. Sonenshein, J. A. Hoch, and R. Losick (Eds.). ASM Press, Washington, DC.

322. Jiang, M., W. Shalo, M. Perego, and J. A. Hoch. 2000. Multiple histidine kinases regulate entry into stationary phase and sporulation in *Bacillus subtilis*. *Mol. Microbiol.* **38**:535–542.

323. Dartois, V., T. Djavakhishvili, and J. A. Hoch. 1997. KapB is a lipoprotein required for KinB signal transduction and activation of the phosphorelay to sporulation in *Bacillus subtilis*. *Mol. Microbiol.* **26**:1097–1108.

324. KinA has been shown to phosphorylate Spo0F in vitro. KinB has not been purified from the membrane and shown to phosphorylate Spo0F. However, a hybrid protein (encoded by a fused gene) consisting of the N-terminal domain of KinA and the C-terminal domain (the kinase domain) of KinB was isolated and shown to phosphorylate Spo0F in vitro. For further details, see ref. 323.

325. The three histidine kinases that can phosphorylate Spo0F in addition to KinA and KinB are KinC, KinD, and KinE. In the absence of KinA and KinB, however, these kinases do not produce sufficient Spo0F-P for sporulation in laboratory media and are viewed as playing a more supporting role. It is believed that the five kinases respond to signals unique to each kinase, and perhaps KinC, KinD, and KinE respond to signals more dominant in the natural habitat, rather than in laboratory media, and play a more important role in sporulation under these circumstances. The presence of five different histidine kinases in the initiation of sporulation points to the complexity of the process.

326. Madhusudan, J. Z., J. A. Hoch, and J. M. Whiteley. 1997. A response regulatory protein with the site of phosophorylation blocked by an arginine interaction: crystal structure of Spo0F from *subtilis*. *Biochemistry* 36:12739–12745.

327. Perego, M., C. Hanstein, K. M. Welsh, T. Djavakhishvili, P. Glaser, and J. A. Hoch. 1994. Multiple protein–aspartate phosphatases provide a mechanism for the integration of diverse signals in the control of development in *B. subtilis*. *Cell* 79:1047–1055.

328. Perego, M., and J. A. Hoch. 1996. Protein aspartate phosphatases control the output of two-component signal transduction systems. *Trends Genet.* 12:97–102.

329. Ohlsen, K. L., J. K. Grimsley, and J. A. Hoch. 1994. Deactivation of the sporulation transcription factor Spo0A by the Spo0E protein phosphatase. *Proc. Natl. Acad. Sci. USA* 91:1756–1760.

330. Jiang, M., R. Grau, and M. Perego. 2000. Differential processing of propeptide inhibitors of Rap phosphatases in *Bacillus subtilis*. *J. Bacteriol.* 182:303–310.

331. Perego, M., and J. A. Hoch. 1996. Cell–cell communication regulates the effects of protein aspartate phosphatases on the phosphorelay controlling development in *subtilis*. *Proc. Natl. Acad. Sci. USA* 93:1549–1553.

332. The PhrA peptide (PhrA) in its preinhibitor form has 44 amino acids. It is exported across the cell membrane by the Sec secretion system (the general secretory pathway, GSP), and the signal peptide is removed by a signal peptidase. This results in a 19 amino acid peptide, which is on the outer membrane surface. The new peptide (PEP$_{19}$), called proinhibitor, is further cleaved to a 5-amino acid inhibitor, called the PhrA peptide (PEP$_5$). The PhrA peptide is brought into the cell via the oligopeptide permease (OPP). The PhrA peptide inhibits the RapA phosphatase, which results in the accumulation of Spo0F-P.

333. Perego, M. 1997. A peptide export–import control circuit modulating bacterial development regulates protein phosphatases of the phosphorelay. *Proc. Natl. Acad. Sci. USA* 94:8612–8617.

334. Winans, S. C., and J. Zhu. 2000. Roles of cell–cell communication in confronting the limitations and opportunities of high population densities, pp. 261–272. In: *Bacterial Stress Responses*. G. Storz, and R. Hengge-Aronis (Eds.). ASM Press, Washington, DC.

335. Perego, M. 1999. Self-signaling by Phr peptides modulates *subtilis* development, pp. 243–258. In: *Cell–Cell Signaling in Bacteria*. G. M. Dunny, and S. C. Winans (Eds.). ASM Press, Washington, DC.

336. Burkholder, W. F., and A. D. Grossman. 2000. Regulation of the initiation of endospore formation in *Bacillus subtilis*, pp. 151–166. In: *Prokaryotic Development*. Y. V. Brun and L. Shimkets (Eds.). ASM Press, Washington, DC.

337. Solomon, J. M., and A. D. Grossman. 1996. Who's competent and when: regulation of natural genetic competence in bacteria. *Trends Genet.* 12:150–155.

338. Lazazzera, B. A., T. Palmer, J. Quisel, and A. D. Grossman. 1999. Cell density control of gene expression and development in *Bacillus subtilis*, pp. 27–46. In: *Cell–Cell Signaling in Bacteria*. G. M. Dunny and S. C. Winans (Eds.). ASM Press, Washington, DC.

339. Turgay, K., J. Hahn, J. Burghoorn, and D. Dubnau. 1998. Competence in *subtilis* is controlled by regulated proteolysis of a transcription factor. *EMBO J.* 17:6730–6738.

340. It is important that the transcription of *comK* is tightly regulated, for only with such control can the process occur at the proper time. To this end, premature transcription is prevented by three repressors of *comK* that bind to the *comK* promoter. They are AbrB, CodY, and Rok. An additional control step is the formation of the complex between ComK and MecA, which results in the proteolysis of ComK by the ClpCP complex. The autostimulatory transcription of *comK* by ComK requires the response regulator DegU, which stabilizes the binding of ComK to the *comK* promoter. For more information about *comK* transcription, see: Smits, W. K., C. C. Eschevins, K. A. Susanna, S. Bron, O. P. Kuipers, and L. W. Hamoen. 2005. Stripping *Bacillus*: ComK auto-stimulation is responsible for the bistable response in competence development. *Mol. Microbiol.* 56:604–614.

341. Hamoen, L. W., G. Venema, and O. P. Kuipers. 2003. Controlling competence in *subtilis*: shared use of regulators. *Microbiology*. 149:9–17.

342. Dunny, G. M., and B. A. B. Leonard. 1997. Cell–cell communication in gram-positive bacteria. *Annu. Rev. Microbiol.* 51:527–564.

343. Haygood, M. G. 1993. Light organ symbioses in fishes. *Crit. Rev. Microbiol.* 19:191–216.

344. Nealson, K. H., and J. Woodland Hastings. 1997. The luminous bacteria, pp. 625–629. In: *The

Prokaryotes, Vol. 1, 2nd ed. A. Balows et al. (Eds.). Springer-Verlag, Berlin.

345. Meighen, E. A. 1991. Molecular biology of bacterial bioluminescence. *Microbiol. Rev.* **55**:123–142.

346. Ruby, E. G., and M. J. McFall-Ngai. 1992. A squid that glows in the night: development of an animal–bacterial mutualism. *J. Bacteriol.* **174**:4865–4870.

347. *Photobacterium* and *Vibrio* are not easily distinguished, and *Vibrio fischeri* is now often referred to as *Photobacterium fischeri*. The principal way in which *Photobacterium* is distinguished from *Vibrio* is that the former lacks a flagellar sheath, accumulates poly-β-hydroxybutyrate (PHB), and uses D-mannitol. However, there is room for confusion, since some *Vibrio* species accumulate PHB and utilize mannitol. In 1998 investigators published a method to distinguish these genera based upon restriction fragment length polymorphism (RFLP) patterns of 16S rDNA (see: Urakawa, H., K. Kita-Tsukamoto, and K. Ohwada. 1998. A new approach to separate the genus *Photobacterium* from *Vibrio* with RFLP patterns by *Hha*I digestion of PCR-amplified 16S rDNA. *Curr. Microbiol.* **36**:171–174.)

348. Greenberg, E. P. 1997. Quorum sensing in gram-negative bacteria. *ASM News* **63**:371–377.

349. There are alternative names for the acyl–HSLs mentioned in the literature. For example, consider β-hydroxybutyryl homoserine lactone, produced by *V. harveyi* and shown in Fig. 18.35B. This is sometimes written as *N*-3-hydroxybutanoyl-HSL, 3-OH-C4-HSL, and is referred to as AI–1.

350. Furans are five-membered ring structures. Four of the atoms in the ring are carbon atoms, and one is an oxygen atom. Borates are boron atoms attached to hydroxyl groups. The furanosyl borate diester is a furan derivative esterified to two hydroxyl groups in borate, forming a duplex ring structure. It is called (2*S*,4*S*)-2-methyl-2,3,3,4-tetrahydroxytetrahydro-furan-borate (*S*-THMF-borate). The structure is shown in: Chen, X., S. Schauder, N. Potier, A. Van Dorsselaer, I. Pelczer, B. L. Bassler, and F. M. Hughson. 2002. Structural identification of a bacterial quorum-sensing signal containing boron. *Nature* **415**:545–549.

351. Hammer, B. K., and B. L. Bassler. 2003. Quorum sensing controls biofilm formation in *Vibrio cholerae. Mol. Microbiol.* **50**:101–104.

352. Hanzelaka, B. L., and E. P. Greenberg. 1995. Evidence that the N-terminal region of the *Vibrio fischeri* LuxR protein constitutes an α-binding domain. *J. Bacteriol.* **177**:815–818.

353. Sitnikov, D. M., J. B. Schineller, and T. O. Baldwin. 1995. Transcriptional regulation of bioluminescence genes from *Vibrio fischeri. Mol. Microbiol.* **17**:801–812.

354. Ulitzur, S., and P. Dunlap. 1995. Regulatory circuitry controlling luminescence autoinduction in *Vibrio fischeri. Photobiol. Photochem.* **62**:625–632.

355. Ulitzur, S., A. Matin, C. Fraley, and E. Meighen. 1977. H-NS protein represses transcription of the *lux* systems of *Vibrio fischeri* and other luminous bacteria cloned into *Escherichia coli. Curr. Microbiol.* **35**:336–342.

356. Fidopiastis, P. M., C. M. Miyamoto, M. G. Jobling, E. A. Meighen, and E. G. Ruby. 2002. LitR, a new transcriptional activator in *Vibrio fischeri*, regulates luminescence and symbiotic light organ colonization. *Mol. Microbiol.* **45**:131–143.

357. Gilson, L., A. Kuo, and P. V. Dunlap. 1995. AinS and a new family of synthesis proteins. *J. Bacteriol.* **177**:6946–6951.

358. Lupp. C., and E. G. Ruby. 2004. *Vibrio fischeri* LuxS and AinsS: comparative study of two signal synthases. *J. Bacteriol.* **186**:3873–3881.

359. Lupp, C., M. Urbanowski, E. P. Greenberg, and E. G. Ruby. 2003. The *Vibrio fischeri* quorum-sensing systems *ain* and *lux* sequentially induce luminescence gene expression and are important for persistence in the squid host. *Mol. Microbiol.* **50**:319–331.

360. Chen., X., S. Schauder, N. Potier, A. Van Dorsselaer, I. Pelczer, B. L. Bassler, and F. M. Hughson. 2002. Structural identification of a bacterial quorum-sensing signal containing boron. *Nature* **415**:545–549.

361. Mok, K. C., N. S. Wingreen, and B. L. Bassler. 2003. *Vibrio harveyi* quorum sensing: a coincidence detector for two autoinducers controls gene expression. *EMBO J.* **22**:870–881.

362. Lilley, B. N., and B. L. Bassler. 2000. Regulation of quorum sensing in *Vibrio harveyi* by LuxO and sigma-54. *Mol. Microbiol.* **36**:940–954.

363. Lenz, D. H., K. C. Mok, B. N. Lilley, R. V. Kulkarni, N. S. Wingreen, and B. L. Bassler. 2004. The small RNA chaperone Hfq and multiple small RNAs control quorum sensing in *Vibrio harveyi* and *Vibrio cholerae. Cell* **118**:69–82.

364. Whitehead, N. A., A. M. L. Barnard, H. Slater, N. J. L. Simpson, and G. P. C. Salmond. 2001.

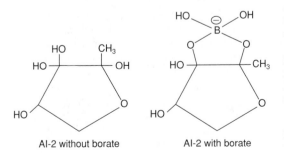

AI-2 without borate   AI-2 with borate

Quorum-sensing in gram-negative bacteria. *FEMS Microbiol. Rev.* **25**:365–404.

365. Fuqua, C., and E. P. Greenberg. 2002. Listening in on bacteria: acyl–homoserine lactone signalling. *Nat. Rev. Mol. Biol.* **3**:685–695.

366. Bassler, B. L. 2002. Small talk: cell-to-cell communication in bacteria. *Cell* **109**:421–424.

367. Fuqua, C., and S. C. Winans. 1996. Conserved *cis*-acting promoter elements are required for density-dependent transcription of *Agrobacterium tumefaciens* conjugal transfer genes. *J. Bacteriol.* **178**:435–440.

368. Zhang, L., P. J. Murphy, A. Kerr, and M. E. Tate. 1993. *Agrobacterium* conjugation and gene regulation by N-acyl-L-homoserine lactones. *Nature* **362**:446–447.

369. Piper, K. R., S. B. von Bodman, and S. K. Farrand. 1993. Conjugation factor of *Agrobacterium tumefaciens* regulates Ti plasmid transfer by autoinduction. *Nature* **362**:448–450.

370. The TraR protein shows 20% amino acid identity and 62% amino acid similarity over a 234-residue overlap with LuxR.

371. Leguminous plants include peas, soybeans, alfalfa, and vetch. *Rhizobium* species swim by chemotaxis toward amino acids and dicarboxylic acids in exudates produced by the plant roots, as well as toward flavonoids secreted by the roots. It has been reported that flavonoid production by alfalfa is increased when the soil is limited for nitrogen. The bacteria then attach to the root hairs. The plant flavonoids induce the expression of *nod* genes on a plasmid harbored by the bacteria. The *nod* genes encode approximately 25 proteins that are responsible for the synthesis and export of a lipooligosaccharide called Nod factor. Nod factor is a molecule with a chitin backbone, four to five N-acetylglucosamine molecules long, and a lipid that is attached to the nonreducing end. The Nod factor is host specific owing to modifications on the backbone. Nod factor causes the root hair to curl, and this entraps the bacteria in a pocket. At this time a lesion in the root hair cell wall is formed by hydrolysis of the cell wall, and bacteria enter the roots at the site of hydrolysis of the cell wall. The root cell plasma membrane invaginates, and the plant cell constructs a tube made of cell wall material. The tube grows through the root hair cell into the root cortex. The tube is filled with proliferating bacteria and is called an infection thread. The infectious thread penetrates and passes through root cortical cells toward cortical cells that have divided to form a nodule primordium. The bacteria leave the infection thread and enter the cytoplasm of the nodule cells. Within the infected plant cells, the bacteria multiply and differentiate into nondividing Y-shaped cells called bacteroids. Singly or in groups, the bacteroids become surrounded by portions of the plant cell membrane to form structures called symbiosomes. The bacteroids within the symbiosome fix nitrogen. The genes for nitrogen fixation are located on the plasmid. In addition to bacteroids, there are some dormant bacteria in the nodule cells; and when the nodule breaks down and releases its contents into the soil, the dormant bacteria grow and divide using nutrients released from the nodules. The released bacteria can infect new root hairs or grow in the soil.

372. van Rhijn, P., and J. Vanderleyden. 1995. The *Rhizobium*–plant symbiosis. *Microbiol. Rev.* **59**:124–142.

373. Gray, M. K., J. P. Pearson, J. A. Downie, B. E. A. Boboye, and E. P. Greenberg. 1996. Cell-to-cell signaling in the symbiotic nitrogen-fixing bacterium *Rhizobium leguminosarum*: autoinduction of a stationary phase and rhizosphere-expressed genes. *J. Bacteriol.* **178**:372–376.

374. Cubo, M. T., A. Economou, G. Murphy, A. W. B. Johnston, and J. A. Downie. 1992. Molecular characterization and regulation of the rhizosphere-expressed genes *rhiABCR* that can influence nodulation by *Rhizobium leguminosarum* biovar. *viciae*. *J. Bacteriol.* **174**:4026–4035.

375. He, X., W. Chang, D. L. Pierce, L. O. Seib, J. Wagner, and C. Fuqua. 2003. Quorum sensing in *Rhizobium* sp. strain NGR234 regulates conjugal transfer (*tra*) gene expression and influences growth rate. *J. Bacteriol.* **185**:809–822.

376. González, J. E., and M. M. Marketon. 2003. Quorum sensing in nitrogen-fixing rhizobia. *Microbiol. Mol. Biol. Rev.* **67**:574–592.

377. Pearson, J. P., K. M. Gray, L. Passador, K. D. Tucker, A. Eberhard, B. H. Iglewski, and E. P. Greenberg. 1994. Structure of the required for expression of *Pseudomonas aeruginosa* virulence genes. *Proc. Natl. Acad. Sci. USA* **91**:197–201.

378. Smith, R. S., S. G. Harris, R. Phipps, and B. Iglewski. 2002. The *Pseudomonas aeruginosa* quorum-sensing molecule N-(3-oxododecanoyl)homoserine lactone contributes to virulence and induces inflammation in vivo. *J. Bacteriol.* **184**:1132–1139.

379. Pirhonen, M., D. Flego, R. Heikinheimo, and E. T. Palva. 1993. A small diffusible signal molecule is responsible for the global control of virulence and exoenzyme production in the plant pathogen *Erwinia carotovora*. *EMBO J.* **12**:2467–2476.

380. Swift, S., J. A. Downie, N. A. Whitehead, A. M. L. Barnard, G. P. C. Salmond, and P. Williams. 2001. Quorum sensing as a population-density-dependent determinant of bacterial physiology. *Adv. Microb. Physiol.* **45**:199–270.

381. The evidence that the Car system is very similar to the LuxI–LuxR system is as follows: (1) *carI* encodes a protein whose amino acid sequence is strikingly similar to LuxI; (2) the wild-type gene of *carI* can complement a mutated *luxI*, and the

wild-type *luxI* can complement a mutated *carI*; (3) culture filtrates of *E. carotovora* contained a compound that could substitute for the *V. fischeri* acyl–HSL in stimulating bioluminescence in *E. coli* carrying a *luxI* mutation. With respect to the second point, we note that both wild-type genes were introduced into an *carI* mutant of *E. carotovora*, or a *lux* plasmid containing a mutant *luxI* gene was introduced into *E. carotovora* carrying wild-type *carI*. Then *carI*-dependent gene expression or bioluminescence was monitored. It is also important to note that the expression of the virulence genes (the genes encoding the exoenzymes) increases when the cells enter stationary phase. This suggests that the *carI* gene product is a cell density signal, as is the *V. fischeri* acyl–HSL.

382. Costerton, J. W., K. J. Cheng, G. G. Geesey, T. I. Ladd, J. C. Nickel, M. Dasgupta, and T. J. Marrie. 1987. Bacterial biofilms in nature and disease. *Annu. Rev. Microbiol.* **41**:435–464.

383. Davey, M. E., and G. A. O'Toole. 2000. Microbial biofilms: from ecology to molecular genetics. *Microbiol. Mol. Biol. Rev.* **64**:847–866.

384. Watnick, P., and R. Kolter. 2000. Biofilm, city of microbes. *J. Bacteriol.* **182**:2675–2679.

385. O'Toole, G., H. B. Kaplan, and R. Kolter. 2000. Biofilm formation as microbial development. *Annu. Rev. Microbiol.* **54**:49–79.

386. Stoodley, P., K. Sauer, D. G. Davies, and J. W. Costerton. 2002. Biofilms as complex differentiated communities. *Annu. Rev. Microbiol.* **56**:187–209.

387. Ramey, B. E., M. Koutsoudis, S. B. von Bodman, and C. Fuqua. 2004. Biofilm formation in plant–microbe associations. *Curr. Opin. Microbiol.* **7**:602–609.

388. Webb, J. S., M. Givskov, and S. Kjelleberg. 2003. Bacterial biofilms: prokaryotic adventures in multicellularity. *Curr. Opin. Microbiol.* **6**:578–585.

389. Mah, T.-F. C., and G. A. O'Toole. 2001. Mechanisms of biofilm resistance to antimicrobial agents. *Trends Microbiol.* **9**:34–39.

390. Conway, B.-A. D., V. Venu, and D. P. Speert. 2002. Biofilm formation and acyl homoserine lactone production in the *Burkholderia cepacia* complex. *J. Bacteriol.* **184**:5678–5685.

391. Sauer, K., A. K. Camper, G. D. Ehrlich, J. W. Costerton, and D. G. Davies. 2002. *Pseudomonas aeruginosa* displays multiple phenotypes during development as a biofilm *J. Bacteriol.* **184**:1140–1154.

392. Confocal scanning microscopy produces a three-dimensional image. The specimen is illuminated with the laser beam of a computerized microscope focused on one point. Mirrors scan the laser beam across the specimen along the x and y axes, and the image is reconstructed on a video monitor. By adjusting the plane of focus, cells at various depths can be seen, and a three-dimensional image can be constructed.

393. Twitching motility takes place on a solid surface and requires type IV pili. The cells move by extending the pilus, attaching the pilus to an object in front of the cell (e.g., another cell), and retracting the pilus, thereby pulling the cell toward the object to which the pilus is attached. Cells must be close to other cells for twitching motility to occur. For a description, see: Skerker, J. M., and H. C. Berg. 2001. Direct observation of extension and retraction of type IV pili. *Proc. Natl. Acad. Sci. USA* **98**:6901–6904. Also, see earlier, Section 1.2.1.

394. Klausen, M., A. Aaes-Jørgensen, S. I. Molin, and T. Tolker-Nielsen. 2003. Involvement of bacterial migration in the development of complex multicellular structures in *Pseudomonas aeruginosa* biofilms. *Mol. Microbiol.* **50**:61–68.

395. The experiment by Klausen et al. made use of two cultures of *P. aeruginosa* that could be microscopically distinguished because they had been color-coded with two different fluorescent molecules (yellow and blue). The biofilms were grown in flow chambers inoculated with 1:1 mixtures of the labeled cultures. After one day of development, the biofilms consisted of microcolonies that contained either blue or yellow fluorescent cells, indicating that they resulted from clonal growth, and the surface between the colonies had a mixture of blue and yellow cells. After 4 days the microcolonies had either blue or yellow fluorescent cells in the stalks, and the caps were a mixture of blue and yellow fluorescent cells. This indicated that the caps had formed when motile cells migrated from the general population up the stalks. Mutants that did not form type IV pili did not cover the surface, indicating that twitching motility was necessary for surface coverage. The mutants formed irregularly shaped microcolonies that had either yellow or blue fluorescent cells, but not a mixture.

396. Thormann, K. M., R. M. Saville, S. Shukla, and A. M. Spormann. 2005. Induction of rapid detachment in *Shewanella oneidensis* MR-1 biofilms. *J. Bacteriol.* **187**:1014–1021.

397. Montgomery, M. T., and D. L. Kirchman. 1994. Induction of chitin-binding proteins during the specific attachment of the marine bacterium *Vibrio harveyi* to chitin. *Appl. Environ. Microbiol.* **60**:4284–4288.

398. Normal biofilm formation by *A. tumefaciens* requires the *sinR* gene that encodes SinR, a protein that is a member of a superfamily of transcription regulators that also includes FNR. FNR is an oxygen-sensitive regulator of gene expression described in Section 18.22. It is suggested that SinR is activated in biofilms in areas that are microaerobic, including areas associated with the plant host, and as a consequence activates the expression of genes important for the maturation of biofilms. It has also been

demonstrated that *A. tumefaciens* cells increase surface attachment and surface coverage in biofilms under conditions of phosphorus limitation. The response to low phosphorus requires the PhoR–PhoB signaling system, described in Section 18.5. See: Ramey, B. E., A. G. Matthysse, and C. Fuqua. 2004. The FNR-type transcriptional regulator SinR controls maturation of *Agrobacterium tumefaciens* biofilms. *Mol. Microbiol.* **52**:1495–1511, and Danhorn, T., M. Hentzer, M. Givskov, M. R. Parsek, and C. Fuqua. 2004. Phosphorus limitation enhances biofilm formation of the plant pathogen *Agrobacterium tumefaciens* through the PhoR–PhoB regulatory system. *J. Bacteriol.* **186**:4492–4501.

399. Sauer, K., A. K. Camper, G. D. Ehrlich, J. W. Costerton, and D. G. Davies. 2002. *Pseudomonas aeruginosa* displays multiple phenotypes during development as a biofilm. *J. Bacteriol.* **184**:1140–1154.

400. Schembri, M. A., K. Kjaergaard, and P. Klemm. 2003. Global gene expression in *Escherichia coli* biofilms. *Mol. Microbiol.* **48**:253–267.

401. Davies, D. G., M. R. Parsek, J. P. Pearson, B. H. Iglewski, J. W. Costerton, and E. P. Greenberg. 1998. The involvement of cell-to-cell signals in the development of a bacterial biofilm. *Science* **280**:295–298.

402. Bollinger, N., M. Hassett, B. H. Iglewski, J. W. Costerton, and T. R. McDermott. 2001. Gene expression in *Pseudomonas aeruginosa*: evidence of iron overrieds effects on quorum sensing and biofilm-specific gene regulation. *J. Bacteriol.* **183**:1990–1996.

403. Prouty, A. M., W. H. Schwesinger, and J. S. Gunn. 2002. Biofilm formation and interaction with the surfaces of gallstones by *Salmonella* spp. *Infect. Immun.* **70**:2640–2649.

404. Miller, S. T., K. B. Xavier, S. R. Campagna, M. E. Taga, M. F. Semmelhack, B. L. Bassler, and F. M. Hughson. 2004. *Salmonella typhimurium* recognizes a chemically distinct form of the bacterial quorum-sensing signal Al-2. *Mol. Cell* **15**:677–687.

405. Rather than isolating and chemically identifying Al-2, a bioassay can be used for detection. For the bioassay, cell-free supernates are prepared and incubated with a strain of *V. harveyi* that lacks the Al–1 sensor but possesses the Al-2 sensor, and the induction of light production is measured.

406. Wen, Z. T., and R. A. Burne. 2004. LuxS-mediated signaling in *Streptococcus mutans* is involved in regulation of acid and oxidative stress tolerance and biofilm formation. *J. Bacteriol.* **186**:2682–2691.

407. Loo, C. Y., D. A. Corliss, and N. Ganeshkumar. 2000. *Streptococcus gordonii* biofilm formation: identification of genes that code for biofilm phenotypes. *J. Bacteriol.* **182**:1374–1382.

408. Li, Y.-H., N. Tang, M. B. Aspiras, P. C. Y. Lau, J. H. Lee, R. P. Ellen, and D. G. Cvitkovitch. 2002. A quorum-sensing signaling system essential for genetic competence in *Streptococcus mutans* is involved in biofilm formation. *J. Bacteriol.* **184**:2699–2708.

409. CSP is made by various streptococci, including *S. pneumoniae*, *S. gordonii*, *S. intermedius*, and *S. mutans*. When CSP accumulates in the extracellular medium to a threshold level, it triggers competence development. CSP interacts with its histidine kinase receptor (ComD), resulting in the regulation of competence gene transcription. Genes induced include those encoding the precursor to CSP, the histidine kinase (ComD), a cognate response regulator (ComE), and an alternative sigma factor (ComX) controlled by ComE. CSP induces 124 genes, and many of these are required for developmental events in biofilms in addition to competence.

410. Ramseier, T. M., S. Y. Chien, and M. H. Saier Jr. 1996. Cooperative interaction between Cra and Fnr in the regulation of the *cydAB* operon of *Escherichia coli*. *Curr. Microbiol.* **33**:270–274.

411. The absence of Cra or ArcA resulted in decreased gene expression by cells under microaerophilic or anaerobic condtions, indicating that both Cra and ArcA stimulate transcription of *cyd*. The results with the *fnr* null mutation indicated that FNR represses *cyd* transcription under anaerobic conditions (but, interestingly, activates *cyd* transcription under microaerophilic conditions). In mutants that lacked FNR, there appeared to be no effect of a deletion of *cra* on transcription of *cyd*, indicating that the regulatory activity of Cra on *cyd* expression requires FNR. The situation is not unique to *cyd*, in as much as many genes are controlled by multiple transcription factors that may function interdependently.

19

How Bacteria Respond to Environmental Stress

There are numerous physiological ways in which bacteria respond to environmental stresses such as changes in pH, osmolarity, high and low temperatures, and starvation. Their very survival in nature depends upon these abilities. The student is referred to ref. 1, which is a comprehensive review of bacterial stress responses. This chapter discusses the heat-shock response (Section 19.1), responses to damaged DNA, including the SOS response (Sections 19.2 and 19.3), and oxidative stress (Section 19.4.)

Responses to environmental stress were discussed earlier, in Section 1.2.5 (mechanosensitive channels), Section 2.2.2 (entrance into stationary phase), Section 14.1 (oxygen toxicity), and Section 15.2 (osmotic stress). E. coli uses three different sigma factors to transcribe genes in response to such stress conditions as starvation, low pH, hyperosmolarity, and high or low temperature. These are σ^{32} (RpoH), σ^E (σ^{24}), and σ^s (RpoS). (See Section 2.2.2 for a discussion of RpoS.)

19.1 Heat-Shock Response

19.1.1 Heat-shock proteins

When E. coli is shifted to a higher temperature, it responds by *transiently* increasing the rate of synthesis of a group of proteins called the heat-shock proteins (Hsps) relative to other proteins. For a review, see ref. 2. (Cold-shock proteins, which may function during adaptation to low temperature also are induced in bacteria.[3–5]) Upon a downshift from the higher temperature, the synthesis of heat-shock proteins is decreased. For example, when E. coli is shifted from 30 °C to 42 °C, the cells increase the rate of synthesis of more than 30 Hsps for 5 to 10 min. The increased rate is about 10 to 20 times greater than the rate of synthesis of the majority of proteins, and during this period over 20% of all the proteins synthesized are Hsps.

As will be explained later, the cell has a need for several of these proteins at all temperatures, not simply high temperatures. However, the need is greater at the elevated temperatures, hence the increased rate of synthesis. This type of response to a temperature shift is not unique to bacteria and occurs in cells of animals (including humans), plants, and eukaryotic microorganisms. In fact, it was first discovered in *Drosophila*, where it is manifested in the synthesis of new chromosomal puffs associated with the production of Hsps in salivary gland chromosomes.

The heat-shock proteins are classified into families according to their molecular weights. The heat-shock proteins with a molecular weight of about 70 kDa are in the Hsp70 family (e.g., DnaK), those with a molecular weight of around 60 kDa are in the Hsp60 family (e.g., GroEL), those with a molecular weight of about 40 kDa are in the Hsp40 family (e.g., DnaJ), and those with a molecular weight around 10 kDa are in the Hsp10 family (e.g., GroES). The proteins made during the heat-shock response are remarkably similar across phylogenetic lines. For example, the DnaK protein from *E. coli* is a member of the Hsp70 family and is about 50% homologous with heat-shock proteins in the Hsp70 family found in humans.

What do the Hsps do? It is clear that in *E. coli* at least some of them are absolutely required for growth to occur above 20 °C. The evidence for this is that null mutations (loss-of-function mutants) in a gene (*rpoH*) encoding a sigma factor (σ^{32}) required to transcribe the *hsp* genes prevent growth above 20 °C.[6] (Most of the *E. coli* genes are transcribed by an RNA polymerase that has σ^{70} as the sigma factor.) In addition to the absolute requirement for Hsp proteins above 20 °C, there is an enhanced effect on growth by the Hsps at *all* temperatures, as judged by the slower growth of null mutants of *rpoH* in comparison to wild-type cells.[5] Thus, as will be made clear in Section 19.1.6, the Hsps not only function to repair or eliminate proteins damaged by heat stress, but they also play important roles in growth at all temperatures. This is because several of the Hsps (e.g., DnaK, DnaJ, GrpE, GroEL, GroES) are important for proper protein folding and protein export at all temperatures, and the increased amount at higher temperatures reflects an increased demand for these proteins.

The following discussion concerns how *E. coli* uses alternative sigma factors to respond to environmental stress, one of these being heat shock. Although the heat-shock response is universal, the mechanisms that regulate the synthesis of the heat-shock proteins differ in various prokaryotes. For example, the expression of heat-shock genes in several gram-positive bacteria thus far investigated indicates that an alternative sigma factor is not involved. On the other hand, *Bacillus subtilis* does induce an alternative sigma factor during the heat-shock response.[7]

19.1.2 The σ^{32} (RpoH) regulon

Following a temperature upshift (e.g., from 30 °C to 42 °C), there is a transient increase in the amount of sigma factor σ^{32}, also called RpoH, which is responsible for the synthesis of at least 30 Hsps that act in the cytoplasm. Thus, σ^{32} recognizes the promoters of genes in a major heat-shock regulon, the σ^{32} regulon. Mutants that do not make σ^{32} cannot grow at temperatures above 20 °C. A major portion of the research discussed next has been done with *E. coli*.

Several factors contribute toward the accumulation of σ^{32} at higher temperatures. During steady state growth at any temperature, σ^{32} is an unstable protein with a half-life of only one minute. After a shift from 30 °C to 42 °C, however, the protein is stabilized for a few minutes and, as a consequence, its amounts increase. The proteases involved in the turnover of σ^{32} have been identified.[8] Apparently, the free pools of the cytoplasmic proteins DnaK and DnaJ are being sensed during heat stress. At nonstress temperatures (30 °C) the available DnaK and DnaJ bind to σ^{32}, and the sigma factor is subject to proteolysis by several proteases including FtsH, Hs1VU, ClpAP, and Lon. At high temperatures (42 °C) the DnaK and DnaJ preferentially bind to denatured proteins, and the sigma factor binds to RNAP. Binding of σ^{32} to RNAP protects the sigma factor from proteolysis and results in a holoenzyme that transcribes the Hsp σ^{32} regulon. Thus, it is not the temperature per se that increases the activity and amount of σ^{32}, but rather the amount of denatured protein. This is in agreement with the finding that increased synthesis of the Hsps is induced not only by high temperatures, but also by ethanol, starvation, and oxidative stress.

Additionally, there is an increase in the rate of translation of the mRNA for σ^{32} during the period of increased synthesis of the heat-shock proteins. Interestingly, after a temperature downshift, the *activity* of σ^{32} decreases, and this, rather than a decrease in concentration of σ^{32}, results in a lowered rate of synthesis of the heat-shock proteins. Thus, the regulation of σ^{32} expression is complex and occurs at several levels, including stability of the protein, activity of the protein, and translation of the mRNA. The student is referred to the review by Gross, which discusses possible mechanisms of regulation.[9] Also, see the discussion in Section 2.2.2 of sigma factor σ^s, which directs the transcription of genes induced by starvation and osmotic upshift, and is also regulated at several levels, as well as ref. 10 and note 11.

19.1.3 The σ^E (σ^{24}) regulon

Another regulon that in *E. coli* is activated by stress is the σ^E regulon, which is activated at very high temperatures (45–50 °C) and protects against damage to extracytoplasmic proteins.[12,13] Stress conditions such as high heat (45–50 °C) or ethanol can result in the denaturation (misfolding) of proteins in the outer membrane or periplasm of *E. coli*. As a consequence of misfolding, a signaling pathway results in the activation of sigma factor σ^E (σ^{24}) in the cytoplasm. As a consequence of the activation of σ^E, the σ^E regulon, which consists of at least 11 genes, is transcribed. The proteins synthesized include a periplasmic peptidylprolyl isomerase (FkpA) and a periplasmic protease (DegP) that are postulated to be involved in the folding, refolding, or degradation of misfolded envelope proteins. Sigma factor σ^E is required for growth at all temperatures. It is suggested that under nonstress conditions σ^E is kept at low activity by being bound to an anti-sigma factor that is localized in the inner membrane (cell membrane). Envelope stress results in release of the sigma factor so that it can bind to the RNAP.

19.1.4 The σ^s (RpoS) regulon

One of the sigma factors, σ^s (or RpoS), has been referred to as a "master regulator" of the general stress response in *E. coli*. It increases the expression of many genes in response to stress conditions, including starvation, entry into stationary phase, hyperosmolarity (0.3 M NaCl), and low pH (pH 5).[14] RpoS and some of the genes that it regulates are discussed in Section 2.2.2.

19.1.5 The Cpx system

Another system that responds to misfolded proteins is a two-component system called the Cpx system. It is induced by several conditions, including elevated pH and the production of misfolded envelope proteins. The membrane histidine kinase (HK) is CpxA, and the response regulator (RR) is CpxR. The Cpx system is also involved in controlling part of the σ^E regulon (DegP), as well as the expression of genes in the Cpx regulon. The genes in the Cpx regulon encode envelope folding factors.

19.1.6 Functions of the E. coli heat-shock proteins

Four roles for the Hsps

Let us now ask what the Hsps do in the cell. Depending upon the particular protein, an Hsp may function in the following way: (1) the *folding* of newly synthesized proteins at all temperatures, (2) the *export* of proteins at all temperatures, (3) the *refolding* of misfolded polypeptides, and (4) the *proteolysis* of improperly folded or otherwise abnormal proteins.[9,15]

Several of the Hsps are chaperone proteins

If an Hsp aids in the folding of newly synthesized proteins, the refolding of improperly folded proteins, or the export of proteins, it belongs to the class of proteins called *chaperone proteins*. These are proteins that take part in the folding (or prevent folding), assembly, or export of other proteins but are not part of the final protein or protein complex. An example of a chaperone protein is SecB, described in Section 17.1 as part of the Sec transport system that exports proteins.

The Hsps that aid in folding newly synthesized proteins

Several of the Hsps are chaperone proteins that play important roles not only at elevated temperatures, but (as discussed in Section 10.3.10) at ordinary growth temperatures as well, because they aid in the folding of newly synthesized proteins. Newly synthesized proteins usually do not spontaneously fold into their correct structures without the help of chaperone proteins that transiently bind to the proteins during the folding process. Chaperone proteins also prevent newly synthesized proteins from aggregating with each other before proper folding has occurred. This is especially important at higher temperatures, where improper folding or protein aggregation may occur more frequently.

Three important Hsp chaperone proteins that assist folding are DnaK (Hsp70 family), DnaJ (Hsp40 family), and GrpE, which act together and with two other Hsp chaperone proteins, GroEL (Hsp60 family) and GroES (Hsp10 family), which also cooperate to produce correct folding. (This is discussed in more detail in Section 10.3.10 and the referenced notes therein.) It appears that DnaK, DnaJ, and GrpE can be ribosome associated and can bind to the protein as a team while the protein is still attached to the ribosome and then, after the protein has left the ribosome, pass the protein into a cavity of the multimeric GroEL. The final stages of folding take place in the GroEL cavity, which is capped by GroES. The chaperone function of GroEl–GroES is ATP-dependent. (For a review, see ref. 9. See note 16 for more information on folding.) Chaperone proteins, such as GroEL, exist as multisubunit complexes of stacked rings called chaperonins. Apparently mutants that make no σ^{32} grow at temperatures below 20 °C (albeit much more slowly than the wild type) because the recognition of weak promoters in the σ^{32} regulon by other sigma factors results in some transcription of the *gro* and *dna* genes.[2,9]

The Hsps that aid in protein export

Some chaperone proteins seem to play a role in protein export similar to that of SecB, a non-Hsp chaperone protein discussed in Section 17.1.1. SecB is the major chaperone protein in *E. coli*. It prevents folding and brings unfolded presecretory proteins to SecA and the translocation machinery at the membrane. Some of the Hsps have been reported to function in the export of some of the SecB-independent proteins. The latter are proteins that use the Sec machinery but not SecB. For example, DnaK and DnaJ are used for the export of some (not all) SecB-*independent* proteins. and DnaK and DnaJ can substitute for SecB in strains lacking SecB.[17,18] Other unidentified Hsps can substitute for SecB in mutants that do not make SecB or have low amounts of available SecB.

The Hsps that refold denatured proteins

In *Escherichia coli*, an Hsp called ClpB is an ATPase that is thought to disentangle thermally aggregated proteins and transfer them to the DnaK–DnaJ–GrpE chaperone system. DnaK–DnaJ–GrpE chaperones and the GroEL–GroES chaperonins cooperate to refold proteins that have misfolded under stress conditions. They are aided by another chaperone, Hsp IbpB, which stabilizes partially folded proteins until they can be refolded. If the proteins are too damaged to be refolded, or if the folding chaperones are saturated, heat-shock proteases, described next, will destroy the damaged proteins.

The Hsps that are ATP-dependent proteases

At least six of the Hsps are ATP-dependent proteases.[9,19] Some of these proteases may function, in part, to degrade improperly folded or denatured proteins at all growth temperatures. Some also play a role in the degradation of specific native proteins during ordinary growth temperatures. The ATP may be required, in part, to unfold the protein prior to proteolysis. One of these, Lon, is of major importance in degrading damaged proteins.

Lon is an ATP-dependent protease that is encoded by the *capR* (*lon*) gene.[20] Like the other Hsps, Lon functions not only during the heat-shock response, but also during ordinary growth temperatures, and it is of interest to examine the different roles that this protease plays in cells regardless of the temperature. Lon is important for degrading abnormal proteins, for example, proteins that may result from nonsense or missense mutations, or denatured proteins in general. In fact, it appears that Lon is responsible for the degradation of the majority of abnormal proteins in *E. coli*. In addition, several normal cellular proteins are also degraded by Lon, two examples of which follow.

Mutations in *capR* (*lon*) give rise to strains that synthesize excess capsular polysaccharide and form mucoid colonies. This is because Lon recognizes and degrades the RcsA protein, which is a positive regulator of capsular polysaccharide synthesis. Hence, the lack of functional Lon results in increased levels of RcsA and consequently large amounts of capsular polysaccharide accumulate. Lon mutants have an additional phenotype. In the absence of Lon, cells are very sensitive to ultraviolet radiation as a consequence of the UV-induced

SOS response discussed in Section 19.3. When the SOS response is induced in *lon* mutants, cell division is inhibited, resulting in cell filamentation and death. Cell division is defective because an inhibitor of cell septation called SulA accumulates and is stable in UV-irradiated cells in the absence of Lon. (SulA inhibits the polymerization of FtsZ. See Section 2.3.2 for a description of cell division and the role of FtsZ in septum formation, and Section 19.3.1 for the role of SulA in the SOS response.) SulA is another substrate for Lon, accounting for why cells that do not make Lon accumulate SulA. In UV-irradiated wild-type cells, the inhibition of cell division is transient because Lon degrades the accumulated SulA. Thus, the Lon protease can be viewed as an enzyme that not only degrades abnormal proteins, but also helps to regulate the levels of certain regulatory proteins.

19.2 Repairing Damaged DNA

There are several ways in which bacteria repair damaged DNA. A more complete discussion of this subject can be found in refs 21 through 24. We will begin with the kinds of damage that can occur (Section 19.2.1). This brief introduction will be followed by a description of repair of the following: (1) UV-damaged DNA by photo-reactivation (Section 19.2.2), (2) damaged DNA by nucleotide excision repair (NER), (Section 19.2.3), (3) damaged DNA by recombination (Section 19.2.4), (4) damaged DNA by base excision repair (BER), (Section 19.2.5), and

(5) damaged DNA by using the GO system (Section 19.2.6).

19.2.1 Kinds of DNA damage

A particularly severe type of damage called photodimerization of pyrimidines occurs when DNA bases are excited by ultraviolet radiation (Fig. 19.1). As a consequence of the dimerization of adjacent pyrimidines such as thymine, DNA replication is blocked. Pyrimidine dimers are removed by specific excision processes or photo-reactivation, as described next. Other mal-formitations of DNA can include mismatched base pairs in the duplex that arise during DNA replication (discussed in Section 10.1.10), breaks or gaps in the DNA, and bases modified by chemicals, many of which are carcinogens.

19.2.2 Repairing UV-damaged DNA by photoreactivation

Exposure to ultraviolet radiation can cause the formation of thymine dimers (Fig. 19.2). (For a review of this subject as well as other DNA repair systems and mutagenesis, see ref. 23.) During photoreactivation a special enzyme called *DNA photolyase*, encoded by the *phr* gene in *E. coli*, reverses the dimerization reaction upon absorbing blue light (300–500 nm). The enzyme contains noncovalently bound FADH$_2$ (reduced flavin adenine dinucleotide) and, in *E. coli*, also a folic acid derivative (5,10-methenyltetrahydrofolyl)polyglutamate (MTHF), both of which absorb light. The

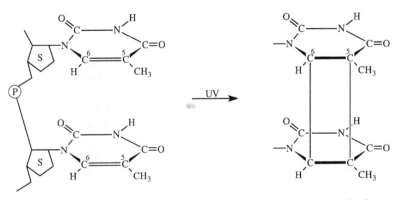

Fig. 19.1 Formation of thymine dimers by ultraviolet light. Ultraviolet light causes the formation of cyclo-butane pyrimidine dimers. Usually this occurs between adjacent thymine dimers in the same strand of DNA and involves C5 and C6 of both molecules.

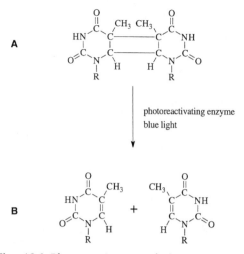

photoreactivating enzyme
blue light

Fig. 19.2 Photoreactivation of thymine dimers. Thymine dimers form in the presence of ultraviolet light. Most organisms other than placental mammals possess a flavoprotein enzyme called photolyase, which will reverse the dimerization upon absorption of blue light (350–500 nm). DNA photolyase in *E. coli* is encoded by the *phr* gene.

enzyme uses the absorbed light energy to cleave the dimer into two monomers, thus restoring the original coding properties of the DNA. The primary chromophore is MTHF, which absorbs 60 to 80% of the light absorbed by the enzyme at 385 nm. The energy is transferred from MTHF to the flavin. The excited flavin molecule cleaves the pyrimidine dimer. (For more details of the model, see note 25.)

DNA photolyase is widespread, occurring in both prokaryotic and eukaryotic microorganisms, plants, and animals. However, it is not found in placental mammals. Purified DNA photolyases from other organisms have $FADH_2$ but differ with respect to the second chromophore. Some, like *E. coli*, have MTHF, whereas others have 8-hydroxy-5-deazaflavin, instead. The folate-containing enzymes have an absorption maximum of about 380 nm, whereas the deazaflavin enzymes have a slightly higher absorption maximum (i.e., about 440 nm).

19.2.3 Repairing damaged DNA via nucleotide excision repair

In *E. coli* the products of three genes, *uvrA*, *uvrB*, and *uvrC*, encode UvrABC endonuclease, an enzyme that recognizes the distortion in the

double helix most commonly caused by the UV-induced pyrimidine dimer. This endonuclease also cuts the DNA eight nucleotides away on the 5′ side of the dimer and four nucleotides away on the 3′ side of the dimer. Removal of the segment of DNA, aided by UvrD, which is a helicase (helicase II), leaves a single-stranded gap that is filled by DNA polymerase I and sealed by DNA ligase (Fig. 19.3). The endonuclease has been shown in vitro to recognize other kinds of damage such as chemically modified bases (e.g., alkylated bases if they are bulky enough) that cause distortion and similarly remove the damaged DNA. The role of the *uvr* genes was originally discovered when it was found that mutants in the *uvr* genes are very sensitive to killing not only by ultraviolet light but also by different chemicals that damage DNA.

19.2.4 Repairing damaged DNA by recombination

The excision mechanisms thus far discussed require that one of the two so-called sister strands have no errors and be able to serve as the template. When both strands are damaged, for example, by gaps (discontinuities) in the daughter strand made during the copying of a damaged template, RecA is used in the process referred to as recombinational repair, or daughter strand gap (DSG) repair. For example, the damage can occur when pyrimidine dimers form as a consequence of UV irradiation and the region of DNA containing the pyrimidine dimer is not replicated. Replication begins again about 1,000 base pairs past the damaged DNA, resulting in a single-stranded gap. The gap is filled, but not by replication; rather a segment of the complementary strand with the correct sequence from the sister duplex moves into the gap strand exchange. (The term "sister duplex" designates a pair of DNA strands comprising a single unit and resulting from replication at the replication fork.)

In the model for the repair process outlined in Fig. 19.4, when a gap is present in the daughter strand opposite a lesion in the template strand, a RecA-dependent recombinational event takes place, resulting in the transfer of the complementary strand from the sister duplex

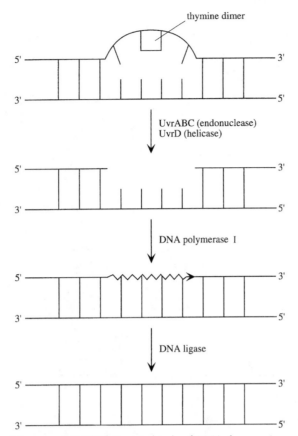

thymine dimer

UvrABC (endonuclease)
UvrD (helicase)

DNA polymerase I

DNA ligase

Fig. 19.3 Nucleotide excision repair (NER). Certain kinds of DNA damage (e.g., thymine dimers) can be repaired by excision repair. The distortion in the double helix is recognized by a complex of proteins called UvrABC, an endonuclease that cuts the damaged DNA eight nucleotides away on the 5′ side of a thymine dimer and four nucleotides away on its 3′ side. Removal of the segment of DNA, aided by the helicase UvrD, leaves a single-stranded gap, which is filled with DNA polymerase I and sealed with DNA ligase.

to the gap in the daughter strand. Then a crossover intermediate forms, which after cutting and sealing results in filling the gap with the complementary DNA from the sister strand. The resultant gap in the sister strand is filled by DNA polymerase I (Pol I). The gaps are sealed by DNA ligase. The faulty nucleotide can now be removed by using nucleotide excision repair (UvrABC) or simply diluted out during growth. Postreplicative repair is generally error free, as opposed to SOS mutagenesis described in Section 19.3.2.

19.2.5 Repair of deaminated bases (base-excision repair)

A common type of DNA damage is the deamination of bases. As a consequence of deamina-

tion, an amino group is replaced by a keto group. For example, cytosine might be deaminated to uracil, adenine to hypoxanthine, and guanine to xanthine. (See Fig. 9.7 for the structures of the bases.) The deaminated bases pair with the wrong base during replication, resulting in mutations. For example, hypoxanthine pairs with cytosine so that an AT pair is replaced by a GC pair. Also, uracil will pair with adenine, resulting in a GC pair being changed to an AT pair. Deamination can occur spontaneously, or it can be caused by various chemical reagents (e.g., nitrous acid). Deaminated bases (as well as other damaged bases such as alkylated bases) are removed by specific enzymes called *DNA glycosylases*. (See note 26 for a discussion of alkylation of bases.) The enzyme catalyzes the breakage of the *N*-glycosyl bond between the base and the

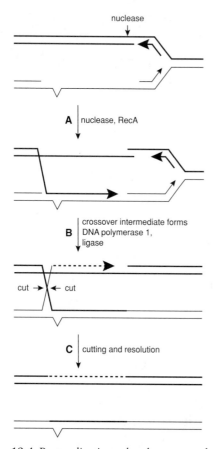

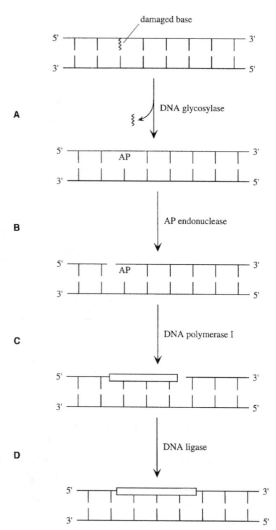

Fig. 19.4 Postreplication daughter strand gap (DSG) repair; the daughter strand is the strand that is being copied from the template strand. The presence of a structural deformation in the duplex (e.g., one imposed by the presence of a thymine dimer in the template strand) causes a gap in the daughter strand opposite the lesion. The gap results from a collapse of the replication fork and a restart of DNA replication at an upstream site, and is repaired by RecA and other enzymes. (A) The good template in the undamaged sister duplex is nicked and moves into the gap in the daughter strand of the damaged sister duplex. This is referred to as sister strand exchange and requires nuclease activity and RecA. Some DNA replication may occur in the daughter strand of the receiving duplex to fill any gap that may be present between the 3′ end of the incoming strand and the 5′ end of the strand with the gap. (B) A crossover intermediate forms. (C) The crossover intermediate is cut and the DNA is resolved into two separate duplexes. During the process, all breaks are sealed, resulting in the incorporation of the sister strand into the gap in the daughter strand, and leaving a gap in the sister template strand. As the figure indicates, DNA replication in the donor template strand probably begins before the crossover intermediate has been resolved.

Fig. 19.5 Base excision repair. (A) A damaged base is removed by DNA glycosylase, leaving an AP site (an apurinic or apyrimidinic site). (B) An AP endonuclease cleaves the phosphodiester bond on the 5′ side of the AP site. (C) DNA polymerase I extends the 3′ end while removing a portion of the damaged strand with its 5′-3′-exonuclease activity. (D) The nick is sealed by DNA ligase. Newly synthesized DNA is shown as a rectangular box.

sugar, leaving an apyrimidinic or apurinic site (Fig. 19.5). Such sites are referred to as *AP sites*. Each glycosylase is specific for each type of damaged base that is removed. Thus these repair pathways are sometimes called *specific repair pathways*. After the base has been removed, AP endonucleases cleave the phosphodiester linkage next to the AP site, usually on the 5′ side. This generates a free 3′-hydroxyl end, which is extended by DNA polymerase I (in *E. coli*) as

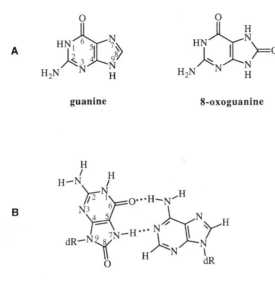

Fig. 19.6 Structures of guanine and 8-oxoguanine (8-oxoG) and 8-oxoG-A base pairs (dR, deoxyriboside). Oxygen free radicals can hydroxylate deoxyguanosine at the C8 position. (A) One of the tautomeric forms is 8-oxodeoxyguanosine, which is the major one under physiological conditions. (B) Sometimes the DNA polymerase will insert dAMP rather than CMP opposite 8-oxoG. Unless this error is repaired, when the DNA replicates there will be a change of a GC base pair to an AT base pair.

the 5′-exonuclease activity of DNA polymerase I removes a portion of the damaged strand ahead. After the damaged portion has been replaced, DNA ligase seals the nick.

19.2.6 The GO system and protection from 8-oxoguanine (7,8-dihydro-8-oxoguanine)

Reactive forms of oxygen such as superoxide radical, hydrogen peroxide, and hydroxyl radicals can oxidatively damage guanine to form 8-oxoguanine (GO). Such damage is called a lesion. (See ref. 27 for a review and Section 14.1, *Oxygen toxicity*, for a description of how reactive forms of oxygen are made in the cell.) The structures of guanine and 8-oxoguanine are shown in Fig. 19.6A. Mutations will occur if 8-oxoguanine then mispairs with adenine (Fig. 19.6B), causing GC pairs to be converted to AT pairs.[28] The GO system, however, prevents this from happening. The GO system consists of three proteins: MutM, MutT, and MutY.

MutM

MutM is an *N*-glycosylase that removes 8-oxoguanine (Fig. 19.7). (See Section 19.2.5

for a discussion of DNA glycosylases and AP site endonucleases.) After removal of the 8-oxoguanine and cutting of the strand, a portion of the damaged strand is removed by exonuclease activity and repaired by DNA polymerase I and DNA ligase.

MutY

Occasionally, as a result of replication, AMP is paired with 8-oxoguanine before the 8-oxoguanine can be removed. (See Fig. 19.6B.) When this happens, MutY, which is an *N*-glycosylase, removes the adenine to produce an AP site opposite the 8-oxoguanine (Fig. 19.7). The DNA is cut by an AP endonuclease at the AP site (apurinic site) and the DNA is repaired as diagrammed in Fig. 19.7, yielding a C–8-oxoG–C base pair, which can be repaired by MutM. This avoids the formation of an AT base pair from a GC base pair.

MutT

Reactive forms of oxygen also convert the guanine in dGTP to 8-oxoguanine to form 8-oxo-dGTP, which is a substrate for DNA polymerase. MutT is a phosphatase that

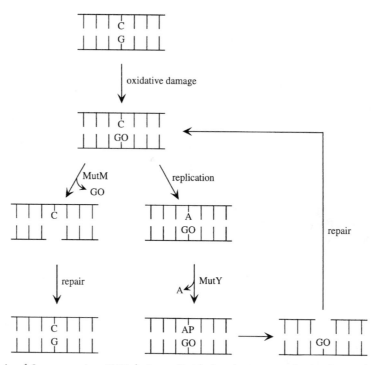

Fig. 19.7 Repair of 8-oxoguanine (GO) lesions. Oxidative damage results in the oxidation of guanine residues to 8-oxoguanine. The MutM protein removes the 8-oxoguanine and cuts the damaged DNA with its AP endonuclease activity. The damaged DNA is removed, and the gap is repaired by DNA polymerase and DNA ligase. Sometimes an AMP is inserted opposite the GO. When this happens, the MutY protein can remove the adenine, leaving an apurinic (AP) site. An AP endonuclease activity, which appears to be associated with MutY, cuts the damaged strand near the AP site. A portion of the damaged strand is removed, and the gap is filled by DNA polymerase and sealed with DNA ligase. This generates a GO–C base pair, affording the opportunity for MutM to remove the GO. The MutY protein can correct A/G mismatches as well as A/GO mismatches.

converts 8-oxo-dGTP to 8-oxo-dGMP, thus preventing the incorporation of 8-oxoguanylic acid into DNA.

Other Mut proteins

Several other Mut proteins have been described. They are MutD, which is commonly known as the ε subunit of DNA polymerase III holoenzyme, which is responsible for its 3′-exonuclease editing function, and MutS, MutL, and MutH, which function in mismatch repair. See Section 10.1.10. Mutations in any of the *mut* genes increase the spontaneous rate of mutation because they prevent the repair of damaged DNA or the removal of incorrectly paired nucleotides. They are called *mut* genes because of the <u>mut</u>ator phenotype that results from the mutations.

19.3 The SOS Response

The term "SOS," used to describe the bacterial response to stress, alludes to the Morse code signal, introduced at the beginning of the twentieth century as a call for help to be transmitted via telegraph. (For more about the interesting history of the Morse code and "SOS," see Box 19.1.) In the context of genetics, the use of "SOS" indicates that a cell is responding to a distress signal generated by DNA damage. When DNA is damaged or DNA replication is inhibited, a signal is produced that results in the induction of over 20 unlinked genes, some of which were discussed earlier in this chapter. The products of the induced genes perform various functions to cope with the DNA damage, including repairing the damaged DNA, allowing DNA replication to proceed past the

BOX 19.1 HISTORICAL PERSPECTIVE: THE MORSE CODE

The Morse code is named after Samuel Finley Breese Morse (1791–1872), an American painter and inventor. In 1838 he refined one of his inventions, the telegraph, which he patented in 1854. He also developed the telegraphic code that is named after him.

In the Morse code messages are transmitted by means of sequences of dots and dashes, or short and long sounds or flashes that represent letters of the alphabet and numbers. The code phrase SOS (in Morse code: . . . _ _ _ . . .) was first used in 1910. The letters were arbitrarily chosen as an easy way to transmit via the Morse code a call for help. They are not, as commonly thought, an acronym for "save our ship" or "save our souls."

Source: Barnhart, R. K. 1995. *The Barnhart Concise Dictionary of Etymology*. HarperCollins, New York.

---

damaged site (a process called *translesion synthesis*), and stalling cell division to give the cell time to repair the damaged DNA. We now turn to a discussion of these genes and their products.

What is the "SOS signal" that the bacteria use? The signal is generated when RecA binds to single-stranded DNA (ssDNA). This occurs when ssDNA is produced at a stalled replication fork as the cell attempts to replicate damaged DNA, or during infection with mutant ssDNA phage that are defective in synthesis of the complementary (minus) strand.[29] (RecA is also important for recombinational repair, as discussed in Section 19.2.4.) When the signal is received, the SOS regulon is induced by inactivating a repressor (LexA) of the SOS regulon. Precisely how the nucleoprotein complex of ssDNA and RecA inactivates LexA is described next.

19.3.1 Regulation of the SOS response

Role of RecA and LexA

We will begin with the transcription repressor protein LexA, which interferes with the binding of RNA polymerase by binding to LexA-binding sites in the SOS box, located near the operator of each SOS gene or operon in the SOS regulon (Fig. 19.8). In an undamaged cell LexA represses the transcription of over 20 genes in the SOS regulon, including *recA* and *lexA*.

When damage occurs to the DNA, RecA becomes activated (RecA*), and this results in the inactivation of LexA (Fig. 19.8). The sequence proceeds in the following way. RecA becomes activated to RecA* when it forms a nucleoprotein filament by polymerizing on single-stranded DNA, which can result as a consequence of an imperfect attempt to replicate a damaged DNA template. LexA, a transcription repressor, then binds to the RecA*–DNA complex, and RecA* promotes the autoproteolytic cleavage of LexA at the Ala^{84}–Gly^{85} bond, thus inactivating the repressor. The genes in the SOS regulon include genes for excision repair and recombinational repair, as well as genes that allow damaged DNA to be copied (Table 19.1). Some of these genes are described next.

Table 19.1 Some SOS responses and induced genes of *E. coli*

Induced gene	Response
umuDC, recA	UV mutagenesis of bacterial chromosome
sulA	Inhibition of cell division (filamentation)
uvrA, uvrB, uvrD	*uvr*-Dependent excision repair
recA, ruvAB	Daughter strand gap repair
recA, recN, ruvAB	Double-strand break repair

Source: Adapted from Walker, G. C. 1996. The SOS response of *Escherichia coli*, pp. 1400–1416. In: *Escherichia coli and Salmonella: Cellular and Molecular Biology*, Vol. 1. F. C. Neidhardt et al. (Eds.). ASM Press, Washington, DC.

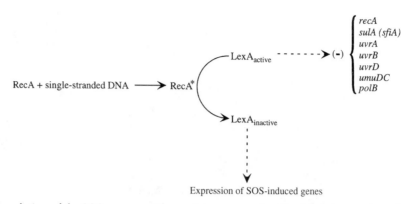

Fig. 19.8 Regulation of the SOS response. The model proposes that RecA becomes activated when it binds to single-stranded DNA resulting from replicative bypass of damaged DNA. Activated RecA (i.e., RecA*) leads to the inactivation of LexA by autoproteolytic cleavage. LexA is a repressor for all the genes induced during the SOS response. There are over 20 genes regulated by LexA. Some of these genes are shown: *recA*, for example, encodes the RecA protein; *sulA* inhibits FtsZ and thus cell division. (See the discussion of Lon in the subsection of Section 19.1.6 entitled *The Hsps that are ATP-dependent proteases.*) UvrA, UvrB, and UvrD function during excision repair. (See Section 19.2.3.) PolB is DNA polymerase II. The gene *umu*DC encodes proteins that comprise the subunits of Pol V, which copies damaged DNA, namely, UmuD' (two subunits) and UmuC (one subunit). (See Section 19.3.2.)

Genes whose transcription is stimulated during SOS response

Some of the genes activated during the SOS response as well as the physiological roles for the gene products are listed in Table 19.1; several were discussed earlier. The list does not include *uvrC* or *phr* (the gene that encodes the photoreactivating enzyme) because they are not under LexA control. The product of the *uvrD* gene (helicase II) is not actually required for endonuclease activity but is required for the release of the oligonucleotide fragment and of UvrC from the complex after the cuts in the DNA have been made. (These genes are involved in nucleotide excision repair. See Section 19.2.3.) The list also includes *sulA*, which, as discussed earlier, codes for a protein that inhibits septation and causes the cells to grow as filaments until DNA repair has been accomplished. The levels of SulA increase during the SOS response, but they are quickly reduced by the action of Lon protease when the DNA is repaired, at which time cell division resumes (Section 19.1.6). Presumably, it is an advantage to the cell not to divide while it is busy repairing its DNA.

The SOS response also includes the induction of certain prophages such as lambda. This is because RecA* promotes the cleavage of repressors of these bacteriophages in a fashion similar to the induced proteolysis of LexA. Prophage induction cannot be a benefit to the damaged host, but it benefits the virus in that after induction, the virus can find a host whose DNA is not damaged. Other genes induced include *recA*, *umuC*, and *umuD*. As we shall see, the products of these genes allow DNA replication past sites of DNA lesions, but at the same time increase the frequency of mutations.

19.3.2 SOS-induced translesion synthesis and mutagenesis

For a review, see ref. 23. Replication of the DNA past a lesion in the template strand, such as an abasic site or a UV-induced thymine dimer, is called *translesion synthesis* (TLS). TLS cannot be performed by the replicative DNA polymerase III (DNA Pol III), which stalls at such sites. A new polymerase, called DNA Pol V, is called upon for translesion synthesis. Translesion synthesis increases the survival of the damaged cell, but as we shall soon learn, also increases the rate of mutation, since DNA Pol V makes errors as nucleotides are inserted nonspecifically opposite lesions in the template strand. The process of making mutations during translesion synthesis is referred to as *SOS mutagenesis* or *error-prone repair*. As we shall see, the catalytic unit of DNA Pol V is encoded by the *umuC* gene, one of the genes in the SOS regulon.

Three genes are required for translesion synthesis. They are *recA*, *umuD*, and *umuC*. The latter two are in the *umuDC* operon. For translesion synthesis to to occur, RecA must be activated. As mentioned previously, the activation of RecA occurs when the protein binds to regions of single-stranded DNA generated when Pol III attempts to replicate the damaged DNA template. The RecA/ssDNA nucleoprotein is the signal that the DNA is damaged. UmuD interacts with the RecA nucleoprotein, which activates the autocleavage of UmuD to an active form, called UmuD'. (Recall that RecA/ssDNA also activates the autocleavage of LexA.)

The finding that UmuC could catalyze translesion synthesis in vitro in the absence of DNA Pol III has led to the following model to account for translesion synthesis and SOS mutagenesis (reviewed in ref. 23). UmuC, which when complexed with an activator, consisting of two UmuD', subunits, replicates DNA past DNA lesions in the template strand, such as abasic sites, thymine dimers, and photoproducts. In view of its catalytic activity, UmuD'$_2$C is now called Pol V. Compared with the normal replicative DNA polymerase, Pol III, Pol V is error-prone. This is because Pol V does not have a stringent proofreading capability, which is why it can insert nucleotides nonspecifically opposite DNA lesions; moreover, it does have a lower base-pairing fidelity, which is why it accepts damaged bases in its active site. Thus, mutations can occur opposite the lesion as well as downstream of the lesion as Pol V continues replication. Translesion synthesis requires, in addition to Pol V, RecA and single-stranded binding protein. It may be that Pol III also interacts with Pol V and plays a role at the replication fork during translesion synthesis, as reviewed in ref.[23].

E. coli does not rely solely on Pol V to help it replicate DNA that has been damaged. In fact, there are two other SOS-induced DNA polymerases (Pol II and Pol IV) that can be called upon to replicate DNA opposite lesions at a damaged fork. These are discussed in note 30, and reviewed in refs. 22 and 31.

19.4 Oxidative Stress

19.4.1 Toxic forms of oxygen

When cells are exposed to toxic forms of oxygen, they are said to be subject to oxidative stress.[32] As discussed in Section 14.1, aerobic organisms form several derivatives of oxygen that can be toxic to cells. (See Section 14.1 for an explanation of how toxic forms of oxygen originate inside cells.) The toxic forms include hydroxyl radical (HO$^{\cdot}$), superoxide radical (O_2^-) and hydrogen peroxide (H_2O_2), which can damage molecules such as DNA, proteins, and lipids. Since electrons are passed to oxygen one at a time via the cytochromes, the superoxide radical is an intermediate in the reduction of oxygen to water. The hydrogen peroxide is derived from superoxide and can react with transition metal ions to form hydroxyl radicals. (See the discussion of the Fenton reaction in Section 14.1.)

There does not exist a mechanism to detoxify hydroxyl radicals, and thus protection from toxic forms of oxygen must be obtained by eliminating superoxide and hydrogen peroxide. The enzyme *superoxide dismutase* (Sod) converts superoxide to hydrogen peroxide and oxygen, and the enzyme *catalase* converts hydrogen peroxide to water and oxygen. (See Section 14.1 for the reactions that these enzymes catalyze.) It is of interest that the reactive forms of oxygen as well as nitric oxide (which also induces the superoxide-induced stress response, discussed next) are also produced in animals by phagocytic cells (e.g., activated macrophages and neutrophils) as cytotoxic agents to kill bacteria that have been ingested by phagocytosis.

19.4.2 Proteins made in response to oxidative stress

To study the response to oxidative stress, one can add to cultures of bacteria H_2O_2, or redox-cycling reagents such as methyl viologen (paraquat) or phenazine methosulfate. The redox-cycling reagents accept from NADH or NADPH a single electron, which they transfer directly to O_2 to form the superoxide radical, O_2^-. This results in an increase in the synthesis of many proteins that can be detected by two-dimensional gel electrophoresis. The functions of many of the induced proteins are not known. However, some of the proteins have an obvious role in protecting the cell from oxidative damage. For example, in *E. coli* the induced enzymes include Mn^{2+}-superoxide dismutase, catalase (HPI catalase), and endonuclease IV. The latter functions in excision repair of damaged DNA, including damage caused by superoxide. (Endonuclease IV cuts the DNA at apurinic or apyrimidine sites produced by DNA glycosylases, which remove damaged bases from DNA.)

19.4.3 Transcriptional regulation of genes induced by oxidative stress

Two transcriptional regulatory systems in *E. coli* control the production of proteins in response to oxidative stress: the SoxRS system and the OxyR system, which control the *soxRS* and *oxyR* regulons, respectively.[33] These two regulatory pathways control the expression of genes that encode antioxidant enzymes. For a review of the OxyR and SoxRS systems, see refs. 34 and 35. We will begin with SoxRS, which includes SoxR, a transcriptional activator of *soxS*.

SoxR and SoxS

SoxR is an iron–sulfide protein that is activated (probably by undergoing a conformational change) in response to superoxide, which causes a redox reaction in the iron–sulfide centers, or to nitric oxide (NO), which causes

nitrosylation. For a more complete explanation, see note 36. Activated SoxR activates transcription of the *soxS* gene. The SoxS protein then activates the transcription of target genes whose products confer resistance to a diverse array of toxic elements including superoxide, nitric oxide produced by activated macrophages, multiple antibiotics, organic solvents, and heavy metals (Fig. 19.9). The activated genes include *sodA* (codes for Mn^{2+}-superoxide dismutase) and *nfo* (codes for the DNA repair enzyme, endonuclease IV).[37] (SoxS autoregulates its own synthesis by being a negative transcription regulator of the gene that encodes it, *soxS*.) Another oxygen-sensitive iron–sulfide protein transcriptional regulator is FNR, whose activity was discussed in Section 18.2.2.

OxyR

When cells are exposed to H_2O_2, OxyR becomes activated and as a consequence binds to target promoters. OxyR is also activated in vitro by O_2. OxyR is the transcriptional activator of several genes including *katG*, which codes for a catalase. (Catalases catalyze the transfer of two electrons from one molecule of hydrogen peroxide to another, resulting in the formation of one molecule of oxygen formed by the electron donor, and two molecules of water, formed by the electron acceptor. See Section 14.1.) OxyR activates transcription, apparently through contacts with one of the subunits of RNA polymerase. There is no increase in the level of OxyR protein, simply in its activity. OxyR is a homotetramer, and the activated form has an intramolecular disulfide

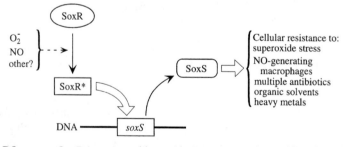

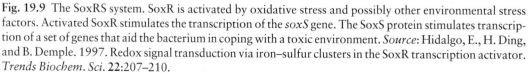

Fig. 19.9 The SoxRS system. SoxR is activated by oxidative stress and possibly other environmental stress factors. Activated SoxR stimulates the transcription of the *soxS* gene. The SoxS protein stimulates transcription of a set of genes that aid the bacterium in coping with a toxic environment. *Source*: Hidalgo, E., H. Ding, and B. Demple. 1997. Redox signal transduction via iron–sulfur clusters in the SoxR transcription activator. *Trends Biochem. Sci.* **22**:207–210.

bond between two cysteine residues, forming a cystine disulfide that results in a conformational change in the protein. The disulfide bond forms when the two cysteine residues are oxidized by H_2O_2 or another oxidant. (The student should read the discussion of CrtJ in Section 18.6.1. A repressor of photosynthetic genes in bacteria, CrtJ is activated by the formation of an intramolecular disulfide bond in the presence of oxygen.) In addition, the transcription of the *oxyS* gene is activated by OxyR. The RNA product of the *oxyS* gene is not translated, and its function is not understood at present. It has been called an antimutator because it protects against mutagenesis.

The role of thioredoxins and glutaredoxins in reducing disulfide bonds in OxyR and other proteins in the cytoplasm

The student is urged to consult the review in ref. 38 for a description of the very interesting area of research into thiol–redox enzymes in bacteria and other organisms. OxyR is only transiently activated by the formation of disulfide bonds during oxidative stress. There exist small thiol–redox enzymes in bacteria, called thioredoxins and glutaredoxins, whose function is to reduce disulfide bonds in other proteins (e.g., OxyR) to sulfhydryl groups. When this occurs to OxyR, the form of the protein that senses oxidative stress is regenerated. Other proteins whose disulfide bonds are reduced by thioredoxin and glutaredoxin are various reductases such as ribonucleotide reductase, which generates deoxynucleotides for DNA synthesis (Section 9.2.5), and phosphoadenosine phosphosulfate (PAPS) reductase, which is important for assimilatory sulfate reduction (Section 12.1).

The thioredoxins and glutaredoxins have pairs of cysteine residues separated by two amino acids, Cys-X1-X2-Cys. The cysteines donate electrons to disulfide residues in other proteins, producing sulfhydryl groups in the target proteins; in turn, their own cysteine residues are oxidized to cystine disulfides. They are sometimes called *thiol–disulfide* exchange proteins. Reduced thioredoxin is regenerated by NADPH via a reductase, and reduced glutaredoxin is regenerated by reduced glutathione

(GSH). (Actinomycetes do not contain glutathione. See note 39.) Glutathione is a tripeptide containing cysteine (γ-glutamyl-cysteine-glycine). The oxidized form of glutathione is glutathione disulfide (GSSG), which consists of two glutathione molecules linked by a cystine disulfide. Glutathione is regenerated from GSSG via an NADPH reduction catalyzed by the enzyme *glutathione reductase*.

In summary then, NADPH regenerates both reduced thioredoxin and reduced glutaredoxin, and the reduced thioredoxin and reduced glutaredoxin in turn regenerate reduced forms of other disulfide proteins (e.g., OxyR). See the discussion in Section 17.6.2 of thiol–redox enzymes in the periplasm.

19.5 Summary

Upon a temperature upshift, bacteria, as well as other organisms, transiently increase the rates of synthesis of proteins called heat-shock proteins. In *E. coli*, the increased rates of synthesis are due to increased activity of a sigma factor, σ^{32}. Although the biological roles for most of these proteins is not known, it is clear that some of them aid bacteria in coping with the increase in temperature. The heat-shock proteins include chaperone proteins that aid in the folding of newly synthesized proteins and also function in the export of certain proteins, and ATP-dependent proteases that degrade specific proteins and abnormal proteins. Several of the heat-shock proteins have been shown to play a role during growth at ordinary temperatures.

In addition to σ^{32}, σ^E is important in responding to stress, including heat stress. The σ^E regulon encodes proteins responsible for the folding, refolding, or degradation of misfolded envelope proteins.

DNA can be damaged in several different ways, and various systems exist to repair the damage. There are specific repair systems called base excision repair (BER) systems that use *N*-glycosylases to remove damaged bases and AP endonucleases, DNA polymerase I, and DNA ligase to repair the DNA. Bases can be damaged in several ways, including deamination and alkylation. Guanine can also be oxidatively damaged. The cell protects itself from oxidative guanine damage by using the MutM, MutY, and MutT proteins. MutM and MutY are *N*-glycosylases and function in base excision systems. However, MutT is not an *N*-glycosylase; rather, it works by preventing the incorporation of oxidized guanylic acid into DNA.

Pyrimidine dimers can be produced by UV irradiation, and the dimerization can be reversed by photoreactivation. Pyrimidine dimers can also be removed by a nucleotide excision repair (NER) system.

There are also general repair systems. These include nucleotide excision repair, postreplication recombination repair, and mismatch repair (Chapter 10.)

Bacteria respond to damaged DNA by inducing the genes of the SOS system. The protein RecA becomes activated and leads to the inactivation of a repressor, LexA. As a consequence, the transcription of the genes of the SOS regulon is stimulated. The protein products of some of these genes effect repairs by nucleotide excision, postreplication recombination repair, and translesion synthesis (TLS), which refers to replication of the DNA past a lesion in the template strand.

Oxidative stress is due to the production of toxic forms of oxygen that damage cell molecules. The transcription of several genes whose protein products protect bacteria from oxidative stress is stimulated by oxidative stress. In *E. coli*, these proteins include superoxide dismutase, catalase, and endonuclease IV. There are two transcriptional regulatory systems that control the expression of genes induced by oxidative stress. These are the SoxRS and the OxyR systems, which consist of transcriptional activators that are activated by superoxide, nitric oxide, and hydrogen peroxide.

Study Questions

1. What is meant by heat-shock proteins, and what is the evidence that they are required for growth?

2. Which heat-shock proteins are required for growth at all temperatures? Why is that?

3. Discuss the role of Lon in UV resistance and in polysaccharide synthesis.

4. Suppose a gap occurs in a daughter DNA strand during replication. How might this be repaired?

5. Describe two different pathways that can remove or correct thymine dimer damage in the template strand.

6. Mutations in *mutM*, *mutY*, and *mutT* increase the spontaneous rate of mutation. Explain why this is the case. The increase in the mutation rates are additive; that is, the rates are higher when two or three of the *mut* genes are mutated than when only one of the genes is mutated. Explain why this is the case.

7. What is meant by the SOS response? How are the genes regulated?

8. What is the explanation for SOS mutagenesis?

9. What is meant by oxidative stress, and how does *E. coli* respond to it?

REFERENCES AND NOTES

1. Storz, G., and R. Hengge-Aronis (Eds.). 2000. *Bacterial Stress Responses.* ASM Press, Washington, DC.

2. Yura, T., M. Kanemori, and M. T. Morita. 2000. The heat shock response: regulation and function, pp. 3–18. In: *Bacterial Stress Responses.* G. Storz, and R. Hengge-Aronis (Eds.). ASM Press, Washington, DC.

3. Phadtare, S., K. Yamanaka, and M. Inouye. 2000. The cold shock response, pp. 33–45. In: *Bacterial Stress Responses.* G. Storz, and R. Hengge-Aronis (Eds.). ASM Press, Washington, DC.

4. Graumann, P., and M. A. Marahiel. 1996. Some like it cold: response of microorganisms to cold shock. *Arch. Microbiol.* 166:293–300.

5. Graumann, P., T. M. Wendrich, M. H. W. Weber, K. Schröder, and M. A. Marahiel. 1997. A family of cold shock proteins in *Bacillus subtilis* is essential for cellular growth and for efficient protein synthesis at optimal at optimal and low temperatures. *Mol. Microbiol.* 25:741–756.

6. Zhou, Y.-N., K. Noriko, J. W. Erickson, C. A. Gross, and T. Yura. 1988. Isolation and characterization of *Escherichia coli* mutants that lack the heat-shock sigma factor σ^{32}. *J. Bacteriol.* 170:3640–3649.

7. Hecker, M., W. Schumann, and U. Völker. 1996. Heat-shock and general stress response in *Bacillus subtilis. Mol. Microbiol.* 19:417–428.

8. Kanemori, M., K. Nishihara, H. Yanagi, and T. Yura. 1997. Synergistic roles of Hs1VU and other ATP-dependent proteases in controlling in vivo turnover of σ³² and abnormal proteins in *Escherichia coli. J. Bacteriol.* 179:7219–7225.

9. Gross, C. A. 1996. Function and regulation of the heat-shock proteins, pp. 1382–1399. In: *Escherichia coli and Salmonella: Cellular and Molecular Biology.* Vol. 1. F. C. Neidhardt et al. (Eds.). ASM Press, Washington, DC.

10. Lange, R., and R. Hengge-Aronis. 1994. The cellular concentration of the σˢ subunit of RNA polymerase in *Escherichia coli* is controlled at the levels of transcription, translation, and protein stability. *Genes Dev.* 8:1600–1612.

11. The expression of the *rpoS* gene, which codes for σˢ, is regulated at the transcriptional and translational levels, as well as by altering the stability of the protein.[10] Transcription is stimulated when the cells enter stationary phase in rich media and by an increase in medium osmolarity. Translation of the *rpoS* mRNA is stimulated when cells enter stationary phase and also during an osmotic upshift. The sigma factor itself is very unstable during exponential growth and becomes stabilized when cells enter stationary phase. The molecular mechanisms responsible for the changes in the rate of translation and for the alterations in protein stability are unknown. However, it is clear that these processes, as well as transcriptional control, are responsible for the increase in σˢ in response to starvation and osmolarity signals.

12. Raivio, T. L., and T. J. Silhavy. 2000. Sensing and responding to envelope stress, pp. 19–32. In: *Bacterial Stress Responses.* G. Storz, and R. Hengge-Aronis (Eds.). ASM Press, Washington, DC.

13. Raivio, T., and T. J. Silhavy. 2001. Periplasmic stress and ECF sigma factors. *Annu. Rev. Microbiol.* 55:591–624.

14. Weber, H., T. Polen, J. Heuveling, V. F. Wendisch, and R. Hengge. 2005. Genome-wide analysis of the general stress response network in *Escherichia coli:* σˢ-dependent genes, promoters, sigma factor selectivity. *J. Bacteriol.* 187:1591–1603.

15. Mujacic, M., M. W. Bader, and F. Baneyx. 2004. *Escherichia coli* Hsp31 functions as a holding chaperone that cooperates with the DnaK–DnaJ–GrpE system in the management of protein misfolding under severe stress conditions. *Mol. Microbiol.* 51:849–859.

16. The crystal structures of GroEL and GroES have been published. This work plus electron microscopic studies have suggested that a cavity exists in GroEL within which the target protein is folded. For references, see the following review: Rassow, J., O. von Ahsen, U. Bömer, and N. Pfanner. 1997. Molecular chaperones: towards a characterization of the heat-shock protein 70 family. *Trends Cell Biol.* 7:129–132.

17. Wild, J., E. Altman, T. Yura, and C. A. Gross. 1992. DnaK and DnaJ heat-shock proteins participate in protein export in *Escherichia coli. Genes Dev.* 6:1165–1172.

18. Kusukawa, N., T. Yura, C. Ueguchi, Y. Akiyama, and K. Ito. 1989. Effects of mutations in heat-shock genes *groES* and *groEL* on protein export in *Escherichia coli. EMBO J.* 8:3517–3521.

19. Suzuki, C. K., M. Rep, J. M. van Dijl, K. Suda, L. A. Grivell, and G. Schatz. 1997. ATP-dependent proteases that also chaperone protein biogenesis. *Trends Biochem. Sci.* 22:118–123.

20. Chung, C. H., and A. L. Goldberg. 1981. The product of the *lon* (*capR*) gene in *Escherichia coli* is the ATP-dependent protease, protease La. *Proc. Natl. Acad. Sci. USA* 78:4931–4935.

21. Rupp, W. D. 1996. DNA repair mechanisms, pp. 2277–2294. In: *Escherichia coli and Salmonella: Cellular and Molecular Biology,* Vol. 2, F. C. Neidhardt et al. (Eds.). ASM Press, Washington, DC.

22. Friedberg, E. C., G. C. Walker, and W. Siede (Eds.). 1995. *DNA Repair and Mutagenesis.* ASM Press, Washington, DC.

23. Walker, G. C., B. T. Smith, and M. D. Sutton. 2000. The SOS response to DNA damage, pp. 131–144. In: *Bacterial Stress Responses.* G. Storz, and R. Hennge-Aronis (Eds.). ASM Press, Washington, DC.

24. Sutton, M. D., B. T. Smith, V. G. Godoy, and G. C. Walker. 2000. The SOS response: recent insights into *umuDC*-dependent mutagenesis and DNA damage tolerance. *Annu. Rev. Gene.* 34:479–497.

25. The model for the *E. coli* photolyase proposes that light energy is absorbed by MTHF, raising an electron to an excited state (*MTHF). When the electron returns to the ground state, energy is transferred to deprotonated reduced flavin, FADH⁻, exciting an electron to an excited state, *FADH⁻. Then *FADH⁻ is thought to transfer the excited electron to the pyrimidine dimer, resulting in the cleavage of the C–C bonds holding the dimer together. The reaction is completed when the electron is returned to FADH⁰, regenerating FADH⁻. For details of the model, see the discussion of photoreactivation of DNA in E. C. Friedberg, G. C. Walker, and W. Siede (Eds.). 1995. *DNA Repair and Mutagenesis.* ASM Press, Washington, DC.

26. Alkyl groups include methyl (CH_3–) and ethyl (CH_3CH_2–). These can be added to the bases by alkylating agents such as ethylmethanesulfonate (EMS) and nitrosoguanidine (NTG). The alkyl groups are often added to the N3 of adenine and the N7 of guanine to form *N*-methyl or *N*-ethyl derivatives of the bases. Nitrosoguanidine will also methylate the O6 of guanine and the O4 of thymine to form the *O*-methyl derivatives. The alkylated bases mispair, causing mutations.

27. Michaels, M. L., and J. H. Miller. 1992. The GO system protects organisms from the mutagenic effect of the spontaneous lesion 8-hydroxyguanine (7,8-dihydro-8-oxoguanine). *J. Bacteriol.* **174**:6321–6325.

28. Kouchakdjian, M., V. Bodepudi, S. Shibutani, M. Eisenberg, F. Johnson, A. P. Grollman, and D. J. Patel. 1991. NMR structural studies of the ionizing radiation adduct 7-hydro-8-oxodeoxyguanosine (8-oxo-7*H*-dG) opposite deoxyadenosine in a DNA duplex. 8-Oxo-7*H*-dG(*syn*)·dA(anti) alignment at lesion site. *Biochemistry* **30**:1403–1412.

29. Higashitani, N., A. Higashitani, A. Roth, and K. Horiuchi. 1992. SOS induction in *Escherichia coli* by infection with mutant filamentous phage that are defective in initiation of complementary-strand DNA synthesis. *J. Bacteriol.* **174**:1612–1618.

30. Two DNA polymerases besides Pol V are induced by the SOS response as a result of DNA damage. One of these is DinB (Pol IV), encoded by the *dinB* gene in the SOS regulon. DinB replicates templates that have a bulged-out nucleotide. The other SOS-induced DNA polymerase is Pol II, encoded by the *polB* gene. Pol II catalyzes "replication restart," a fast reinitiation of DNA replication after UV damage, after which Pol III takes over. In mutants lacking both Pol II and Pol V, replication resumes in UV-irradiated cells after a long (90 min) delay, indicating that there is yet another mechanism for restarting replication.

31. Rangarajan, S., R. Woodgate, and M. F. Goodman. 1999. A phenotype for enigmatic DNA polymerase II: a pivotal role for Pol II in replication restart in UV-irradiated *Escherichia coli*. *Proc. Natl. Acad. Sci. USA* **96**:9224–9229.

32. Storz, G., and M. Zheng. 2000. Oxidative stress, pp. 47–59. In: *Bacterial Stress Responses.* G. Storz, and R. Hengge-Aronis (Eds.). ASM Press, Washington, DC.

33. Lynch, A. S. and E. C. S. Lin. 1996. Responses to molecular oxygen, pp. 1526–1538. In: *Escherichia coli and Salmonella: Cellular and Molecular Biology*, Vol. I. F. C. Neidhardt et al. (Eds.). ASM Press, Washington, DC.

34. Pomposiello, P. J., and B. Demple. 2002. Global adjustment of microbial physiology during free radical stress, pp. 319–341. In: *Advances in Microbial Physiology*, Vol. 46. R. K. Poole (Ed.). Academic Press, San Diego, CA.

35. Hidalgo, E., H. Ding, and B. Demple. 1997. Redox signal transduction via iron–sulfur clusters in the SoxR transcription activator. *Trends Biochem. Sci.* **22**:207–210.

36. SoxR is a homodimer and has a [2Fe–2S] center in each subunit of the dimer. The activity of SoxR is regulated by redox reactions. One way that this was demonstrated is as follows. The midpoint potential for the [2Fe–2S] clusters at pH 7.6 is about −285 mV. When the protein was titrated in vitro, its transcriptional activity was inhibited at −380 mV, that is, when the FeS centers are more than 95% reduced. This was reversible; that is, when the FeS centers were reoxidized by titration, activity of the protein was restored. Candidates for the in vivo reductants and oxidants for the iron–sulfide centers in SoxR include NADPH, NADH (reductants), O_2, O_2^- (oxidants). The according to the model, SoxR works by binding to the *soxS* promoter regardless of the oxidation state of SoxR; the oxidized form of SoxR, however, undergoes an allosteric modification, enabling it to activate the *soxS* promoter, apparently by stimulating the formation of the "open" complex required to initiate transcription. Note the similarities and differences from the mechanism proposed for oxygen regulation of FNR activity discussed in Section 18.2.2. It is proposed that oxygen can inactivate FNR by oxidizing the [4Fe–4S] clusters, and that prolonged exposure to oxygen actually results in the loss of the iron–sulfide clusters. Both the protein with the oxidized iron–sulfide clusters and the apoprotein (missing the clusters) bind very poorly to DNA and do not influence transcription.

Nitric oxide also activates SoxR, but the mechanism is different from that of oxidizing the iron–sulfide centers. Nitric oxide displaces the sulfide from the iron–sulfide centers, forming a nitrosylated SoxR (dinitrosyl–iron–cysteine complex). The nitrosylated SoxR activates transcription to the same extent as SoxR with oxidized iron–sulfide centers.

37. Li, Z., and B. Demple. 1994. SoxS, an activator of superoxide stress genes in *Escherichia coli*. *J. Biol. Chem.* **269**:18371–18377.

38. Ritz. D., and J. Beckwith. 2001. Roles of thiol–redox pathways in bacteria. *Annu. Rev. Microbiol.* **55**:21–48.

39. Most bacteria contain glutathione as a component of their redox buffer. The actinomycetes, including *Mycobacteria*, which have a cysteine-containing sugar-based compound called mycothiol, are an exception.

INDEX